Statistics for Spatial Data

Statistics for Spatial Data

Revised Edition

NOEL A. C. CRESSIE
Iowa State University

A Wiley-Interscience Publication
JOHN WILEY & SONS, INC.
New York • Chichester • Toronto • Brisbane • Singapore

Library of Congress Cataloging-in-Publication Data:

Cressie, Noel A. C.

Statistics for special data revised edition / Noel A. C. Cressie

p. cm.—(Wiley series in probability and mathematical statistics. Applied probability and statistics section)

"A Wiley-Interscience publication."

Includes bibliographical references and index.

ISBN 0-471-51094-7 (cloth)

ISBN 978-1-119-11461-1 (pbk)

1. Spatial analysis (Statistics) I. Title. II. Series.

QA278.2.C75 1993

519.5—dc20

93-775

CIP

10 9 8 7 6 5 4 3 2 1

To Yoko, Amie, and Sean

Contents

*Theory
†Applications

PART II LATTICE DATA

Preface

The purpose of this book is to present Statistics for spatial data to scientists and engineers. (Notice that Statistics is capitalized to distinguish it from its other meaning: a collection of numbers that summarize a complex phenomenon—such as baseball or cricket.) In the last 10 years, much interest has been generated in the area, but its exposure in the literature has been uneven. This book attempts to take that literature and extend it, correct it, and unify it. What appears to be a gathering of unconnected subject areas can be annealed into a cohesive approach to the analysis of spatial data. Chapter 1 provides an overview of the approach and of the enormous diversity of problems involving spatial data, from the microscopic to the astronomic.

The book attempts to give a somewhat complete coverage of each of three parts, dealing with *geostatistical data*, *lattice data*, and *point patterns*. Thus, the subject areas are classified according to the type of observations encountered, reflecting my belief that the roots of statistical science are in data. Statistical models, then, try to make sense out of the data, albeit imperfectly. Design, inference, and diagnostics are natural consequences of the data–model symbiosis, and all play an important role in Statistics for spatial data.

This book grew from lecture notes for a one-semester, 3-credit Statistics graduate course that I conduct at Iowa State University. In 45 lectures, each of 50-minutes duration, I cover the following topics:

Part I (Geostatistical Data): All of Chapter 2 except Section 2.5. All of Chapter 3 except Sections 3.3 and 3.6.

Part II (Lattice Data): Chapter 6. Chapter 7, Sections 7.2, 7.3, and 7.6.

Part III (Point Patterns): Chapter 8, Sections 8.1, 8.2, 8.4, 8.5.1, 8.5.2, 8.5.3. Chapter 9, Section 9.1.

Prerequisites for the Statistics graduate course are one semester of Masters-level statistical inference and one semester of Masters-level linear models. While giving the course and preparing the book, I have benefitted from useful reference books in the area. Reading lists for the course have included Matern (1960), Bartlett (1975), Journel and Huijbregts (1978), Cliff and Ord (1981), Ripley (1981), and Upton and Fingleton (1985).

It is my hope that this book can be read in varying depths by people with varying mathematical and statistical backgrounds. However, there are sections in the book, on random processes, point processes, and random sets, that are beyond a Masters-level student and closer to the frontiers of theoretical research (these sections are denoted by an asterisk). Equally, there are sections that concentrate purely on an application and do not add to the theoretical development of the subject (these sections are denoted by a dagger). These applications-oriented sections of the book should appeal to a large number of scientists and engineers, with at least the background of a service course in Statistics (or equivalent). For this reason, most chapters begin with an application, which is meant to be an invitation to read on. The emphasis on applications has led to considerable use of graphs and illustrations.

There are some features of the book that I believe will enhance its value as a textbook and as a reference book. An attempt has been made at completeness, in terms of the topics covered, and uniformity, in terms of the depth to which they are covered, except for Chapters 5 and 9. These two chapters contain material that is either of personal interest or is speculative in nature. The reader will notice frequent referencing to a diverse literature; one of the interesting features of Statistics for spatial data is that a large proportion of it is appearing *outside* Statistics journals. The referencing allows me to pick up apparently different streams of thought and tie them together. Equally importantly, I have tried to give credit where it is due. The linear reader will also notice a certain amount of repetition between chapters (and, to a lesser extent, between sections). This is deliberate and is meant to help the sporadic reader who wants to understand the essence of a topic but who has not read all the previous pages.

We should not forget our roots. All data sets for the various spatial statistical analyses are given in the book, as well as some background to the problems being studied. Further, each of the three parts has a section devoted to sources of spatial data, both real and simulated.

No exercises are given at the end of sections; the depth of coverage within a section should allow practice exercises to suggest themselves to an instructor teaching from the book. Software is not given. (A geostatistics package, *Toolkit*, by Geostokos, London, was used for the kriging presented in Chapter 3.) Currently, those of us who work in the area tend to custom-build our own software, which is usually not very portable. Statistics for spatial data will truly realize its enormous potential when a comprehensive software package is developed.

This is a big book. I had thought of splitting it into smaller volumes; however, the present format emphasizes the subject's unity. This may be the last time spatial Statistics will be squeezed between two covers. A healthy exponential growth of the literature is apparent from the bibliography.

The future of the subject is in solving problems for spatiotemporal data; some sections are devoted to it, but a proper treatment needs another book. Statistics for spatial and temporal data would provide dynamic models for phenomena distributed through space and evolving in time. Onward into the next decade!

NOEL CRESSIE

Ames, Iowa
December 1, 1990

Acknowledgments

As a boy, knowing "why," "how," and "when" was not enough for me. With encouragement from my parents, a keen interest in the "where" question eventually led me to summer jobs with mining companies in Western Australia (while an undergraduate) and three years of doctoral work at Princeton University. There, I had the good fortune to be taught by Geof Watson, John Tukey, and, in my last year, Julian Besag. Geof gave me an appreciation for all things spatial and geometric, John showed me how to make my data speak (sometimes, even sing) to me, and Julian introduced me to the mysteries of Markov random fields (and English hockey). More recently, in his role as a series editor for Wiley and as an aficionado of Statistics for spatial data, Geof Watson has been of immense help with his comments and his support.

The influence of Georges Matheron of the École Nationale Supérieure des Mines de Paris in Part I and Chapter 9 is obvious; his work has been truly pioneering. I was fortunate to spend a post-doctoral period of five months at his center in Fontainebleau, France, in 1975.

My coauthors on spatial articles provided valuable impetus to my research in the subject and my colleagues throughout the world have (through their letters, their telephone calls, their questions at seminars, their anonymous referees' reports, and their comments in hallways) helped shape my current opinions on Statistics in general and on Statistics for spatial data in particular. Along with those mentioned in the paragraphs above, I would like to thank Peter Diggle for early suggestions on topics that a course on spatial statistics might cover, Dale Zimmerman for comments on Chapter 5, Subhash Lele for comments on parts of Chapters 6 and 7, and Daryl Daley for comments on Chapter 8.

The task of taking a diverse and uneven literature on spatial Statistics and extending it, correcting it, and unifying it has not been an easy one. The editors at Wiley have been very understanding of my desire for a complete and uniformly comprehensive product. Bea Shube provided ideas and encouragement in the early stages of this project. In the last three years, helpful advice has come from Kate Roach, and, for a brief period before her, from Maggie Irwin. My department head at Iowa State University, Dean Isaacson, has been equally understanding and in various ways helped to make an impossible task possible.

The writing of this book started in the second half of 1985 while I was an ASA/NSF/Census Fellow at the U.S. Bureau of the Census. Based on an experimental course called Spatial Statistics that I conducted at Iowa State University in 1984, I gave a series of seven seminars at the Census Bureau and distributed material that later became Chapter 6. The rest of the book (over 90%) was written at Iowa State, partially supported by the Department of Statistics and by the National Science Foundation. A special mention should be made of the superb journal and book collection of the Parks Library at Iowa State; it is one of the great resources of the University, and has helped me achieve a coverage that would otherwise have been impossible.

We are very fortunate, in our Department of Statistics, to be surrounded by intelligent and motivated graduate students. I have now offered my course on Spatial Statistics four times; it has helped me attract good students with a keen interest in the subject. As research assistants and as candidates for graduate degrees, they have been been involved in various aspects of this book. The contributions of three of them deserve special mention and thanks. Stephen Rathbun wrote a preliminary draft of Sections 8.3, 8.5 through 8.9, and 7.4, and commented on subsequent drafts of Chapter 8. Carol Gotway wrote a preliminary draft of Sections 3.6 and 5.1, Martín Grondona wrote a preliminary draft of Sections 5.6 and 5.7, and Gotway and Grondona gave comments on Part I. All three were involved in the production of the figures. My thanks also go to Renkuan Guo, Jeff Helterbrand, Fred Hulting, and Jay Ver Hoef, who contributed in various ways to the preliminary through penultimate and final drafts.

This manuscript was turned into immaculate type by Sharon Shepard; by comparing what I gave her and what she gave me, it is clear that Sharon was able to work near miracles. Jeanette LaGrange, Rose Ann Anderson, and Jan Franklin also provided valuable secretarial assistance.

A trace of my space–time line would show a trip to Tokyo, Japan, in 1987. There I met Yoko. She could rightly claim that Western marriage vows say nothing about writing a book. Yet somehow she understood. We are fortunate to have been able to create two new space–time lines in our household, Amie and Sean. It was difficult for them to understand, and even harder for me to persevere. My deepest gratitude goes to all of them for their love and patience.

I am glad I started exploring Statistics for spatial data, because I have learned a great deal. Some of the territory is now well charted. Other parts are only passable by those who are sure-footed, and by not looking down. Much more of it I can only glimpse and tell you what I see. It is an exciting area that deserves a place in every statistical scientist's repertoire; I hope you will agree.

N. A. C. C.

Statistics for Spatial Data

CHAPTER 1

Statistics for Spatial Data

Perhaps, it may turn out a Sang;
Perhaps, turn out a Sermon.
ROBERT BURNS

Statistics, the science of uncertainty, attempts to model order in disorder. It is not surprising that students (and their teachers) find the subject enigmatic. However, as life experiences and scientific experiences accumulate, Statistics is usually recognized as an extremely powerful research tool. Even when the disorder is discovered to have a perfectly rational explanation at one scale, there is very often a smaller scale where the data do not fit the theory *exactly*, and the need arises to investigate the new, residual uncertainty.

Scientists and engineers have attempted to measure the level of disorder through, among other things, a quantity called *entropy*; see Jaynes (1957) and his subsequent writings (Rosenkrantz, 1983). Suppose there are $i = 1, \ldots, k$ possible states of nature that occur at random according to a probability distribution $(p_1, p_2, \ldots, p_k)$, where $0 \le p_i \le 1$, $i = 1, \ldots, k$, and $p_1 + p_2 + \cdots + p_k = 1$. Shannon (1948) defined entropy as

$$E \equiv -\sum_{i=1}^{k} p_i \cdot \log p_i,$$

where $x \cdot \log x \equiv 0$, for $x = 0$. When the state of nature i has a numerical interpretation, the mean $\mu \equiv \sum_{i=1}^{k} i \cdot p_i$ could be thought of as the center of the distribution and the variance $\sigma^2 \equiv \sum_{i=1}^{k} (i - \mu)^2 p_i$ as the amount of variability in the distribution. It is σ^2 that statisticians mostly use to quantify disorder, but not surprisingly E and σ^2 are closely related (see, e.g., Mukherjee and Ratnaparkhi, 1986).

There is a very interesting entropy inequality that can be proved. Consider the initial probability distribution $\mathbf{p}' \equiv (p_1, \ldots, p_k)$ evolving into a subsequent probability distribution $\mathbf{p}'A$, where A is a $k \times k$ matrix of nonnegative elements each of whose rows sums to 1. Write $A = (a_{ij})$; then a_{ij} can be interpreted as the *conditional* probability that the subsequent state of nature is j *given* that the initial state of nature was i, and A is called a transition matrix. When states of nature evolve through repeated application of a stochastic mechanism described by the matrix A, the result is called a Markov chain (see, e.g., Isaacson and Madsen, 1976; Section 6.3). The entropy inequality is

$$E(\boldsymbol{\lambda}^{(1)} : \boldsymbol{\mu}^{(1)}) \geq E(\boldsymbol{\lambda}^{(0)} : \boldsymbol{\mu}^{(0)}),$$

where $E(\boldsymbol{\lambda}^{(l)} : \boldsymbol{\mu}^{(l)}) \equiv -\sum_{j=1}^{k} \lambda_j^{(l)} \cdot \log(\lambda_j^{(l)}/\mu_j^{(l)})$, $l = 0, 1$. Initially, $\lambda_j^{(0)} = p_j$ and $\mu_j^{(0)} = 1$, $j = 1, \ldots, k$, so that $E(\boldsymbol{\lambda}^{(0)} : \boldsymbol{\mu}^{(0)})$ is simply $E = -\sum_{j=1}^{k} p_j \cdot \log p_j$. The subsequent measures are $\lambda_j^{(1)} = \sum_{i=1}^{k} a_{ij} \lambda_i^{(0)} = \sum_{i=1}^{k} a_{ij} p_i$ and $\mu_j^{(1)} = \sum_{i=1}^{k} a_{ij} \mu_i^{(0)} = \sum_{i=1}^{k} a_{ij}$, $j = 1, \ldots, k$, depending on the transition matrix A.

A physical interpretation of this inequality might be that the universe tends to seek levels of entropy that are higher in relation to previous levels. Technology attempts to slow the *increase* in entropy by constraining evolving systems, but there are convincing arguments given that it will never decrease total entropy (e.g., Brooks and Wiley, 1988, Chapter 2). In short, Statistics is important because disorder is here to stay.

Independent-and-Identically Distributed-Data Model

Beginning classes in Statistics (and many of the more advanced ones) always assume that observations on a phenomenon are taken under identical conditions and that each observation is taken independently of any other. The data then form a random sample [i.e., are independent and identically distributed (i.i.d.)]; standard statistical techniques can be applied to build a statistical model and to estimate the model's parameters (see, e.g., Hogg and Craig, 1978). For example, Heyl and Cook (1936) describe experiments performed between May 1934 and July 1935 to determine the acceleration of gravity in a laboratory of the National Bureau of Standards, Washington, D.C. The method used was that of the reversible pendulum; various configurations of pendulum tube diameter and type of knife edge were used. One particular configuration, taken late May/early June 1934, yielded (*after* appropriate adjustments for flexure and clock rate were made)

$$76, 82, 83, 54, 35, 46, 87, 68,$$

expressed as deviations from $980{,}000 \times 10^{-3}$ cm/sec^2, in units of 10^{-3} cm/sec^2 (Heyl and Cook, 1936, Table 12). That these data can be modeled as arising from a random sample is probably a fair assumption, but a later experiment performed under a different configuration yielded deviations (again after appropriate adjustments)

$$76, 76, 78, 79, 72, 68, 75, 78,$$

in units of 10^{-3} cm/sec^2 (Heyl and Cook, 1936, Table 8). Again these could be modeled as a random sample, but because each experiment attempted to measure the same physical constant, the data should obviously be *combined* in some way. Is it fair to model the preceding 16 numbers as observations from a random sample?

Inhomogeneous-Data Model

Lack of homogeneity in data is usually accounted for in statistical models by a nonconstant-mean assumption; often the mean is assumed to be a linear combination of several explanatory variables. However, even after these large-scale variations are accounted for, there are often reasons to suspect inhomogeneous small-scale variations.

Cressie (1982) assumes the data from the experiment (to determine the acceleration of gravity) just described to be independent realizations from statistical distributions whose means are constant but whose *variances differ* (sometimes called heteroskedasticity) markedly, depending on the configurations of pendulum tube diameter and type of knife edge. Standard one-sample theory no longer applies, but it is still possible to construct a confidence interval for the common mean, based on a weighted t-like statistic.

The preceding example shows a relaxation of the identical-distribution assumption. Relaxation of the independence assumption is a further obvious way to generalize statistical models, but are these more general models of any scientific value? In the pages to follow, I hope to convince the reader that the answer to this question is very definitely "Yes."

Dependent-Data Model

Independence is a very convenient assumption that makes much of mathematical–statistical theory tractable. However, models that involve statistical dependence are often more realistic; two classes of models that have commonly been used involve intraclass-correlation structures and serial-correlation structures. These offer little scope for spatial data, where dependence is present in all directions and becomes weaker as data locations become more dispersed.

We have not yet been able to escape the three-dimensional world in which we live, nor the unidirectional flow of time through which we live. The notion that data close together, in time or space, are likely to be correlated (i.e., cannot be modeled as statistically independent) is a natural one and has been used successfully by statisticians to model physical and social phenomena. Purely temporal models, or time series models as they have come to be known (e.g., Box and Jenkins, 1970), are usually based on identically distributed observations that are dependent and occur at equally spaced time points. The unidirectional flow of time underlies the construction of these models.

Spatial models are a more recent addition to the Statistics literature. Geology, soil science, image processing, epidemiology, crop science, ecology, forestry, astronomy, atmospheric science, or simply any discipline that works with data collected from different spatial locations, need to develop (not necessarily statistical) models that indicate when there is dependence between measurements at different locations. However, the models need to be more flexible than their temporal counterparts, because past, present, and future have no analogy in space, and furthermore it is simply not reasonable to assume that spatial locations of data occur regularly, as do most time series models.

When dealing with data where (space–time) dependencies are likely, two approaches can be contrasted. Departures from the independence paradigm could be modeled, or statistical procedures could be constructed that would be robust to these departures. Throughout this book I shall consider mostly the modeling approach.

The Weather

It is useful to consider an example that has both the temporal and spatial component in it. The weather is a universal topic of conversation (some would say it is a resort of the desperate), and among its various facets that of rainfall is paramount. To the more than one million inhabitants of South Australia, the driest state in the driest continent, drought can be devastating. In a series of papers, Cornish, with others (1936, 1954, 1958, 1961, 1976), looked at temporal and spatial aspects of 26 rainfall recording stations in South Australia, shown in Figure 1.1. Data analyzed were monthly rainfall amounts at 26 locations that had been recording for a period of 30 years or more. Thus, at any one location, the data form a typically lengthy time series (e.g., for Adelaide, the capital city of South Australia, there were 1248 monthly observations available), and at any one time point, the data are spatial and not more than 26 in number; over time and space they form a collection of approximately 20,000 observations. More recently, meteorological space–time data sets have been collected for the purposes of studying the effects of atmospheric pollution, in particular *acid rain* (see e.g., Peters and Bonelli, 1982). Daily data collection over a number of years at various locations throughout say the northeast United States, yields a massive data set. But most of it is temporal so that, spatially speaking, the data are still rather sparse. Nevertheless, spatial prediction is just as important as temporal prediction, because people living in those cities and rural districts without monitoring stations have the same right to know how little or how much their water or their air is polluted. (Section 4.6 discusses some of the issues surrounding acid rain and analyzes a spatial data set.)

Scope and Purpose

The topics covered in this book are almost exclusively related to data analysis and statistical modeling of spatial data; the basic components are spatial locations $\{\mathbf{s}_1, \ldots, \mathbf{s}_n\}$ and data $\{Z(\mathbf{s}_1), \ldots, Z(\mathbf{s}_n)\}$ observed at those locations. Usually the data are assumed random and sometimes the locations are

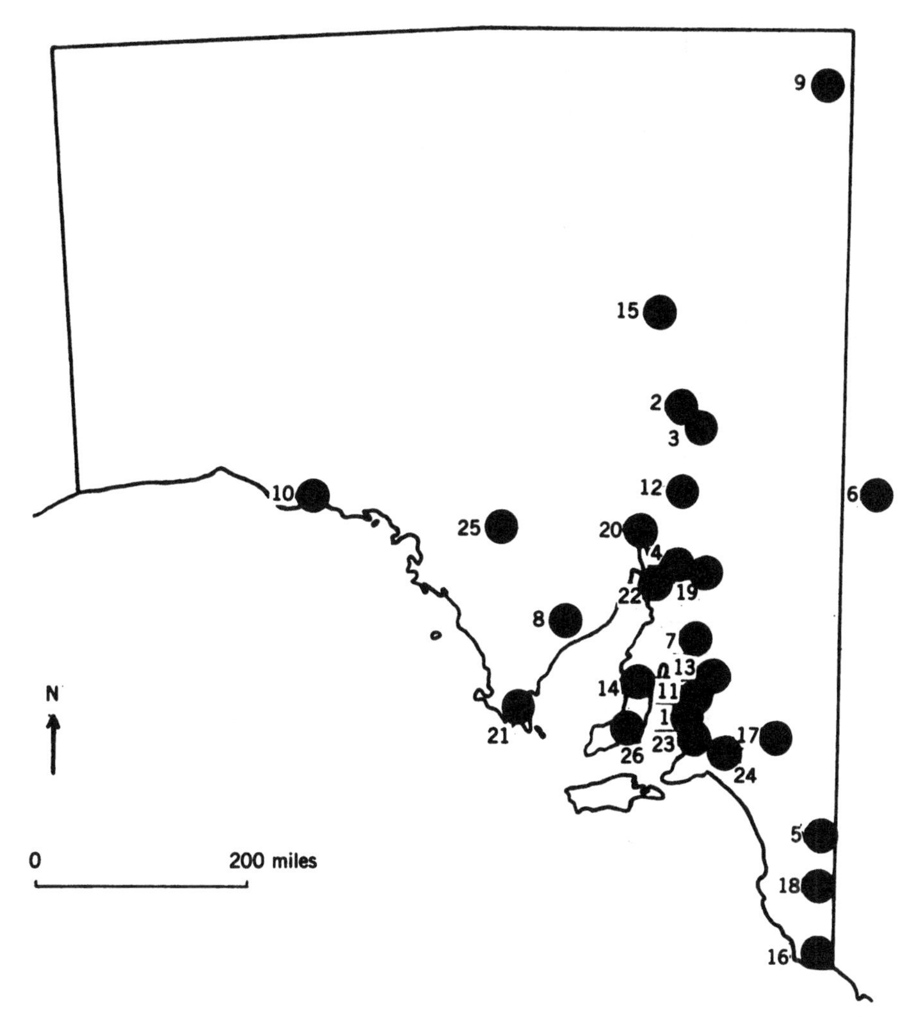

Figure 1.1 Map of South Australia showing 26 rainfall recording locations: Adelaide (1), Beltana (2), Blinman (3), Booleroo Centre (4), Bordertown (5), Broken Hill (6), Clare (7), Cleve (8), Cordillo Downs (9), Fowler's Bay (10), Gawler (11), Hawker (12), Kapunda (13), Maitland (14), Marree (15), Mount Gambier (16), Murray Bridge (17), Naracoorte (18), Peterborough (19), Port Augusta (20), Port Lincoln (21), Port Pirie (22), Stirling West (23), Strathalbyn (24), Yardea (25), Yorketown (26).

assumed random. Moreover, once the locations are given, the possibility of mistaken or imprecise positioning is generally not modeled. Some modifications to this paradigm are occasionally considered; for example, see Sections 4.4, 4.5, and 6.1.

Issues regarding the measurement, storage, and retrieval of spatial information are extremely important, although that is not the emphasis in this book. A geographic information system (GIS) is a collection of computer software tools that facilitate, through georeferencing, the integration of spatial, nonspatial, qualitative, and quantitative data into a data base that can be managed under one system environment (e.g., Burrough, 1986). Much of the research in GIS has been concerned with computational geometry, spatial discretization, spatial languages and user interfaces, systems designs and architectures for data integration, spatial data handling approaches for alternative system architectures (such as parallel processing and neural networks), and so forth. Satellites are amassing huge amounts of data, a small fraction of which is being analyzed. This embarrassment of riches must be controlled by being selective; selectivity can be guided by the types of problems solved and the models that will be needed to solve them. Geographic information systems have only recently begun to incorporate model-based spatial analyses into their information-processing subsystems; it is hoped that this book will provide ideas for further initiatives.

Another topic that receives little attention in this book is the definition of "distance" between two sites at known locations. Mostly, I represent the two sites as vectors $\mathbf{s} = (s_1, \ldots, s_d)'$ and $\mathbf{u} = (u_1, \ldots, u_d)'$ in Euclidean space $\mathbb{R}^d$ and use the Euclidean distance, $\|\mathbf{s} - \mathbf{u}\| \equiv \{\sum_{i=1}^{d}(s_i - u_i)^2\}^{1/2}$, although in Section 4.6 great–arc distances are chosen instead. In any particular problem, other distances may be more appropriate, such as road miles or a measure that combines, say, travel expense, convenience, and time. These alternative metrics lead naturally to questions of *clustering* (e.g., Mardia et al., 1979, Chapter 13). For example, which regions are "close" on a "map" drawn according to the new metric? Based only on pairwise "distances" between sites, multidimensional scaling (e.g., Kruskal and Wish, 1977; Mardia et al., 1979, Chapter 14) constructs a geometric configuration of the site "locations." A danger in the uncritical use of clustering methods for spatial data analysis is that important spatial contiguities may be violated; Legendre (1987) advocates constraining the clustering to respect these contiguities.

My principal purpose for writing this book is to make statistical theory and methods for spatial data available to scientists and engineers. I believe that the full potential of Statistics is seen in its application to substantive problems, and I have tried to reinforce this in the applications I have given throughout the book. I hope that readers will find ideas and information here that will be useful in their own areas of scientific endeavor, although I do recognize that there are large parts that may be inaccessible to all but the most theoretically inclined. Such sections have been marked with an asterisk to warn the nonstatistician to obtain help in reading them. By way of balance, those sections that are truly oriented toward applications have been marked with a dagger.

The book is divided into three parts, dealing with spatial processes indexed over continuous space (a topic sometimes referred to as geostatistics), spatial processes indexed over lattices in space (the spatial analogue of time series), and spatial point processes (including marked point processes and random set processes). These are convenient subareas that do not exhaust all of the statistical models one might use to analyze spatial data. The next section gives some background on spatial problems and defines the general model considered in this book.

1.1 SPATIAL DATA AND SPATIAL MODELS

The first manifestations of statistics for spatial data appear to have arisen in the form of data maps. For example, Halley (1686) superimposed, onto a map of land forms, directions of trade winds and monsoons between and near the tropics, and attempted to assign them a physical cause.

Spatial models appeared much later. For example, Student (1907) was concerned with the distribution of particles throughout a liquid. Instead of analyzing their spatial positions he aggregated the data into counts of particles per unit area: A hemocytometer of area 1 mm^2, divided into 400 squares, was used to count yeast cells. Student found that the distribution of the number of cells per square followed a Poisson distribution. (Daley and Vere-Jones, 1988, p. 8, can be consulted for a brief history of counting problems.)

R. A. Fisher was clearly aware of spatial dependence in agricultural field experiments, because he went to such great lengths to remove it (see Fisher, 1935, Chapter IV). In the 1920s and 1930s, at Rothamsted Experimental Station in England, he established the principles of randomization, blocking, and replication. As well as controlling for unwanted bias, randomization also neutralizes (but does not remove) the effect of spatial correlation (Yates, 1938; Section 5.6.2). However, it should be realized that randomization does *not* neutralize the spatial correlation at spatial scales larger or smaller than the plot dimensions.

Fairfield Smith (1938) was concerned with choosing plot dimensions so that any increase in plot size would yield little decrease in error variance. Although his analysis was empirical, the very formulation of the problem recognizes the presence of spatial correlation in field experiments. Models for such phenomena did not begin to appear until much later (Whittle, 1954).

Nearest-neighbor methods for analyzing agricultural field trials attempt to take spatial dependence into account, indirectly, by using residuals from neighboring plots as covariates, or by differencing (Papadakis, 1937; Bartlett, 1938, 1978; Wilkinson et al., 1983; Besag and Kempton, 1986). A review of these methods is given in Section 5.7.1.

Presently, Statistics can be found in any quantitative discipline, giving rise to rich variations on the theme that was played at Rothamsted 60 years ago. In areas such as geology, ecology, and environmental science, it is not often possible (nor always appropriate) to randomize, block, and replicate the data. There is a need for new statistical models and approaches that address new questions arising from old and new technologies. Many of the resulting problems, such as resource assessment, environmental monitoring (e.g., for global warming), and medical imaging, are spatial in nature.

The General Spatial Model

Statistics, in all its guises from exploratory data analysis to asymptotic distribution theory of parameter estimators, relies on a more or less vague stochastic model. I shall present such a model for spatial data, which has a very simple structure that is flexible enough to handle an extremely large class of problems (including the ubiquitous i.i.d. case). The data may be continuous or discrete, they may be spatial aggregations or observations at points in space, their spatial locations may be regular or irregular, and those locations may be from a spatial continuum or a discrete set. At the very least, the stochastic model is used to summarize extant data or to predict unobserved data. It may or may not explain why a particular phenomenon occurs, and so should be distinguished from the more commonly accepted use of the word "model." To many scientists, a model must have causative dynamic components, but that is not necessarily the case in this book.

Let $\mathbf{s} \in \mathbb{R}^d$ be a generic data location in d-dimensional Euclidean space and suppose that the potential datum $\mathbf{Z}(\mathbf{s})$ at spatial location $\mathbf{s}$ is a random quantity. Now let $\mathbf{s}$ vary over index set $D \subset \mathbb{R}^d$ so as to generate the multivariate random field (or random process)

$$\{\mathbf{Z}(\mathbf{s}): \mathbf{s} \in D\}; \tag{1.1.1}$$

a realization of (1.1.1) is denoted $\{\mathbf{z}(\mathbf{s}): \mathbf{s} \in D\}$. This "superpopulation" model for spatial data is used exclusively throughout the book, although brief discussion of a design-based approach (whose randomness is derived from sampling the deterministic process $\{z(\mathbf{s}): \mathbf{s} \in D\}$) is given in Section 2.3.

Usually, D is assumed to be a *fixed* (i.e., nonrandom) subset of $\mathbb{R}^d$ (see, e.g., Vanmarcke, 1983), but I shall assume more generally that D is a *random set*. Formally speaking, a random set is a measurable mapping from a probability space onto a measure space of (usually closed) subsets of $\mathbb{R}^d$ (see Section 9.3 for more details). Less formally speaking, I shall assume that D as well as $\mathbf{Z}$ may vary from realization to realization, giving another source of randomness to the problem. This simple structure allows me to talk about problems with continuous spatial index, problems with lattice index, spatial point patterns, and more, all under the same umbrella. Thus, the three seemingly distinct parts of this book alluded to earlier treat special cases of (1.1.1), where the index set D could be a random set. Specifically:

Part I (Chapters 2 through 5): Geostatistical data. D is a fixed subset of $\mathbb{R}^d$ that contains a d-dimensional rectangle of positive volume; $\mathbf{Z}(\mathbf{s})$ is a random vector at location $\mathbf{s} \in D$.

Part II (Chapters 6 and 7): Lattice data. D is a fixed (regular or irregular) collection of countably many points of $\mathbb{R}^d$ (this can be generalized to where D is a graph in $\mathbb{R}^d$; see Section 6.1); $\mathbf{Z}(\mathbf{s})$ is a random vector at location $\mathbf{s} \in D$.

Part III (Chapter 8): Point patterns. D is a point process in $\mathbb{R}^d$ or a subset of $\mathbb{R}^d$; $\mathbf{Z}(\mathbf{s})$ is a random vector at location $\mathbf{s} \in D$. In its most general form this generates a *marked spatial point process*; when no $\mathbf{Z}$ is specified the usual spatial point process is obtained [or one could think of scalar $Z(\mathbf{s}) \equiv 1$, for all $\mathbf{s} \in D$].

Part III (Chapter 9): Objects. D is a point process in $\mathbb{R}^d$; $Z(\mathbf{s})$ is *itself* a random set. This yields processes such as the Boolean model (Section 9.4) and a related cell-growth process (Section 9.7).

The flexibility of (1.1.1) is now apparent, and although it is clear that D and $\mathbf{Z}(\mathbf{s})$ could be even more general random quantities, the fact that D is a subset of $\mathbb{R}^d$ allows (1.1.1) to be called a *spatial process*. It is even possible to think of the ubiquitous i.i.d. (possibly multivariate) model as a 0-dimensional spatial process, because any spatial index is unimportant. I have not mentioned how the D process and the $\mathbf{Z}$ process might covary. For most of this book, one or the other is fixed so the question does not arise. When both are random, they are usually assumed to be independent, or one process is analyzed conditionally upon the observed values of the other. Spatial modeling occurs then within the $\mathbf{Z}$ process (Parts I and II), within the D process, or within both processes (Part III), and typically involves modeling the large- and the small-scale variations in terms of a finite number of parameters.

I prefer to keep (multivariate) *time-series* processes separate from spatial processes by using the index t and denoting them as

$$\{\mathbf{Z}(t): -\infty < t < \infty\}. \tag{1.1.2}$$

The unidirectional flow of time sometimes forces one to distinguish between (1.1.2) and (1.1.1) in $\mathbb{R}^1$ (see Section 6.3).

A *space–time process* will be denoted as

$$\{\mathbf{Z}(\mathbf{s}; t): \mathbf{s} \in D(t), t \in T\}, \tag{1.1.3}$$

where each of $\mathbf{Z}$, D, and T is possibly random. This is as general a process as will be considered in this book.

Description, Indication, and Estimation

As disciplines in Statistics mature, their statistical analyses advance through three stages of development that I describe as *description*, *indication*, and *estimation*. For the first, the goal is to summarize the data (perhaps to suggest models). For the second, estimates of model parameters are obtained from the data, but no measures of precision are available. For the third, enough distribution theory is at hand to allow (approximate) inference on the model parameters; for example, the bias and variance of an estimator can be calculated and estimated. Statistics for spatial data is still a young discipline, showing increasing signs of maturity. I hope this book will contribute to, and be an influence on, its formative years.

1.2 INTRODUCTORY EXAMPLES

In this section, I shall present examples where data are spatial in nature. Some data sets are new and some have already appeared in the literature, although they may not have been analyzed spatially. It is not my intention to present spatial statistical analyses in this section, but rather to illustrate the range of problems that can be addressed. For some of these data, more complete analyses are carried out later in the book.

As described in Section 1.1, spatial data can be thought of as resulting from observations on the stochastic process

$$\{\mathbf{Z}(\mathbf{s}): \mathbf{s} \in D\}, \tag{1.2.1}$$

where D is possibly a random set in $\mathbb{R}^d$. I shall present some special cases that are typical of spatial statistical problems one might encounter.

1.2.1 Geostatistical Data

Geostatistics (Matheron, 1962, 1963a, 1963b) emerged in the early 1980s as a hybrid discipline of mining engineering, geology, mathematics, and statistics. Its strength over more classical approaches to ore-reserve estimation is that it recognizes spatial variability at both the large scale *and* the small scale, or in statistical parlance it models both spatial trend *and* spatial correlation. Trend-surface methods (e.g., Whitten, 1970) include only large-scale variation, assuming independent errors. Watson (1972) compares the two approaches and points out that most geological problems have a small-scale variation, typically exhibiting strong positive correlation between data at nearby spatial locations. One of the most important problems in geostatistics is to predict the ore grade in a mining block from observed samples. Matheron (1963b) has called this process of prediction *kriging* (see Chapter 3).

Exploration data on coal ash (Gomez and Hazen, 1970, Tables 19 and 20) for the Robena Mine Property in Greene County, Pennsylvania, are an example of regularly spaced data in $\mathbb{R}^2$, although some locations have not been sampled or their observations are missing. Section 2.2 performs a spatial exploratory data analysis on these mining data, and kriging is carried out in Section 3.4.

The geostatistical method has found favor among soil scientists who seek to map soil properties of a field from a small number of soil samples at known locations throughout the field. For example, soil pH in water, soil electrical conductivity, exchangeable potassium in the soil, soil–water tension, and soil–water infiltration are some of the variables that could be sampled. Water erosion is of great concern to agriculturalists, because rich topsoil can be carried away in runoff water. Some forms of tillage result in greater soil–water infiltration than others, and the study described in Section 4.3 investigates this question. The greater the infiltration, the less the runoff, resulting in less soil erosion and less stream pollution by pesticides and fertilizers. Also, greater infiltration implies better soil structure that is more conducive to crop growth. In Cressie and Horton (1987) and Section 4.3, it is explained how double-ring infiltrometers were placed at regular locations in a field that had received four tillage treatments. From these data the spatial relationships can be characterized, treatments compared, and kriging maps drawn.

There are many published sources for case studies using geostatistics, although the original data are usually not given for confidentiality reasons (original data for both of the examples just introduced are available in Sections 2.2 and 4.3, respectively). The books by David (1977, 1988), Journel and Huijbregts (1978), and Clark (1979) are all devoted to applications in the mining industry. Application of geostatistics in other areas abound; for example, in rainfall precipitation (e.g., Ord and Rees, 1979), atmospheric science (e.g., Thiebaux and Pedder, 1987), soil mapping (e.g., Burgess and Webster, 1980), and predicting groundwater contaminant concentrations (e.g., Istok and Cooper, 1988). For further applications, see the various sections of Chapter 4 that contain analyses of data sets from hydrology, soil science, public health, uniformity trials, and acid rain.

Geostatistical-type problems are distinguished most clearly from lattice- and point–pattern-type problems by the ability of the spatial index $\mathbf{s}$ to vary *continuously* over a subset of $\mathbb{R}^d$. That is not to say that methods from one class of problems cannot be borrowed from methods usually associated with another class; see, for example, Sections 4.4 and 7.6, where the same public-health data set is analyzed using two approaches.

1.2.2 Lattice Data

A *lattice* of locations evokes an idea of regularly spaced points in $\mathbb{R}^d$, linked to nearest neighbors, second-nearest neighbors, and so on. In this book these will be referred to as *regular* lattices, allowing for the possibility of *irregular lattices*, whose relative displacements do not follow a predictable pattern and whose linkages are not always obvious from their geometry. Of all the possible spatial structures that (1.2.1) can generate, a data set whose spatial locations are a regular lattice in $\mathbb{R}^d$ is the closest analogue to a time series observed at equally spaced time points.

Remote sensing from satellites offers an efficient means of data gathering. For example, it allows information to be gathered rapidly on weather patterns, mineral distribution, and crop acreage without having to conduct lengthy and labor-intensive traditional surveys. Satellites orbit the earth and receive data in the form of electromagnetic reflectance waves at a number of frequencies, including those in the visible part of the spectrum. By various sampling and integration methods (see Section 7.4 for more details), the earth's surface is divided into small rectangles (e.g., 56 m × 56 m) called *pixels* (short for picture elements). An agricultural scene of interest (around, say, 34,000 km^2) has certain proportions devoted to wheat, corn, soybeans, and so forth that need to be estimated. These various crops have their own reflectance properties that, together with noise, are remotely sensed. Thus the data are received as a regular lattice in $\mathbb{R}^2$ (neglecting the earth's curvature) and are identified with the centers of their respective pixels. (This is analogous to temporal problems where economic time-series data, such as yearly export earnings, are aggregated over the whole year but are identified with single, equally spaced points on the time axis.) There is a large overlap between remote-sensing techniques and (low-level) medical-imaging techniques; although the spatial scales are vastly different, the form of the data and the questions being asked are often similar. Statistical models for such data need to express the fact that observations nearby (in time or space) tend to be alike; see Section 7.4.

Sometimes it might be expected that nearby observations tend to be dissimilar. Competition between plants for light and soil nutrients could lead to large healthy plants being surrounded by less sturdy ones. Mead (1967) analyzes data on cabbages and investigates this competition effect.

In contrast to geostatistical problems, data from lattice problems may be exhaustive of the phenomenon. Of course, sampling may also occur; for example, suppose only a small window of the full data set is available.

There have been various published studies based on lattice data (although such studies have not always exploited the spatial component) that include Mercer and Hall (1911) (see also Sections 4.5 and 7.1), Batchelor and Reed (1918), Cochran (1936), Bartlett (1974), Cliff and Ord (1981), Ripley (1981), Symons et al. (1983) (see also Sections 4.4, 6.2, and 7.6), and Cressie and Chan (1989).

1.2.3 Point Patterns

Point patterns arise when the important variable to be analyzed is the location of "events." Most often the first question to be answered is whether the pattern is exhibiting complete spatial randomness, clustering, or regularity. For example, consider the locations of longleaf pines in a natural forest in southern Georgia (given in Section 8.2). What is the biological significance of clustering of these trees? The variable, diameter at breast height, is also recorded along with the tree's location. Do large (small) trees cluster and how do large and small trees interact? The size variable is usually called a *mark variable*, and the whole process is then called a *marked spatial point process*.

But the mark variable does not have to be a real variable; it could, for example, be a set. The process then consists of $\{Z(\mathbf{s}_i): \mathbf{s}_i \in D\}$, D a spatial point process and $Z(\mathbf{s}_i)$ a *random set* (Section 9.3) located at $\mathbf{s}_i \in D$. Data from this process are often observed as a realization of

$$X = \bigcup\{Z(\mathbf{s}_i): \mathbf{s}_i \in D\}, \qquad (1.2.2)$$

which is referred to as a (generalized) Boolean model (Serra, 1980) or a mosaic process (Diggle, 1981b). The data are often images, and their assimilation to a model like (1.2.2) requires higher levels of image analysis than the methods considered in Section 7.4 of Part II. The goal is to estimate parameters of the random set *and* the point process; see Cressie and Laslett (1987) and Section 9.5 where an artificially generated Boolean model of random parallelograms is analyzed. Diggle (1981b) analyzes the incidence of heather according to a Boolean model; see also Section 9.5. A model for tumor growth that iterates the Boolean model through successive stages is developed in Section 9.7. Pictures of cell islands growing *in vitro* are presented and analyzed in a manner that takes *shape* as well as size into account.

Published studies involving analyses of spatial point patterns can be found *inter alia* in Pielou (1959, 1977), Matern (1960), Getis and Boots (1978), Marquiss et al. (1978), Ripley (1981), Diggle (1983), and Upton and Fingleton (1985). Of course, there is a large overlap of methods for patterns occurring in space and for those occurring in time; the reader interested in *temporal* methods is referred to Cox and Lewis (1966), the articles in Lewis (1972) (in particular the review by Daley and Vere-Jones, 1972), Cox and Isham (1980), and Daley and Vere-Jones (1988). A second-moment measure of spatial point patterns is adapted for the temporal animal-behavior data in Section 8.4.

1.3 STATISTICS FOR SPATIAL DATA: WHY?

Some simple spatial models will be given to show the effect of correlation on estimation, prediction, and design. The models allow closed-form expressions to be calculated, from which discussion of more general issues can be initiated.

Estimation

Consider the following simple statistical model, taught in all beginning Statistics service courses. Suppose $Z(1), \ldots, Z(n)$ are independent and identically distributed (i.i.d.) from a Gaussian (i.e., normal) distribution with unknown mean μ and known variance σ_0^2. The minimum-variance unbiased estimator of μ is

$$\bar{Z} \equiv \sum_{i=1}^{n} Z(i)/n, \tag{1.3.1}$$

and inference on μ is straightforward: The estimator $\bar{Z}$ is Gaussian with mean μ and variance σ_0^2/n. Thus, a two-sided 95% confidence interval for μ is

$$\left(\bar{Z} - (1.96)\sigma_0/\sqrt{n}\,, \bar{Z} + (1.96)\sigma_0/\sqrt{n}\right). \tag{1.3.2}$$

Instead of independent data, now suppose the data are positively correlated with a correlation that decreases as the separation between data increases:

$$\operatorname{cov}(Z(i), Z(j)) = \sigma_0^2 \cdot \rho^{|i-j|}, \qquad i, j = 1, \ldots, n, 0 < \rho < 1. \tag{1.3.3}$$

It is well known that such a correlation function results from a first-order autoregressive process $Z(i) = \rho Z(i-1) + \epsilon(i)$, $i \geq 1$, where $\epsilon(i)$ is part of an i.i.d. sequence of Gaussian random variables with zero mean and variance $\sigma_0^2(1-\rho^2)$ and is independent of $Z(i-1)$; see, e.g., Fuller (1976, Section 2.3). However, that is *not* the generating mechanism I have in mind; here, $\mathbf{Z} \equiv (Z(1), \ldots, Z(n))'$ are *spatial* data in $\mathbb{R}^1$ from which prediction of $Z(0)$ or $Z(3/2)$ is just as appropriate as that of $Z(n+1)$.

Now,

$$\begin{aligned}\operatorname{var}(\bar{Z}) &= n^{-2}\left\{\sum_{i=1}^{n}\sum_{j=1}^{n} \operatorname{cov}(Z(i), Z(j))\right\} \\ &= \{\sigma_0^2/n\}\Big[1 + 2\{\rho/(1-\rho)\}\{1-(1/n)\} \\ &\qquad -2\{\rho/(1-\rho)\}^2(1-\rho^{n-1})/n\Big]. \end{aligned} \tag{1.3.4}$$

For $n = 10$ and $\rho = 0.26$, $\operatorname{var}(\bar{Z}) = \{\sigma_0^2/10\}[1.608]$, and a two-sided 95% confidence interval for μ is

$$\left(\bar{Z} - (2.485)\sigma_0/\sqrt{10}\,, \bar{Z} + (2.485)\sigma_0/\sqrt{10}\right).$$

Thus, failure to realize the presence of positive correlation in the data leads to a confidence interval (1.3.2) that is *too narrow*; for $n = 10$ and $\rho = 0.26$, the actual coverage probability of (1.3.2) is 87.8%, not 95%. [If $-1 < \rho < 0$, (1.3.2) would be too wide, but because positive spatial dependence is seen more often, these introductory remarks concentrate on situations where ρ is positive.]

Some intuitive understanding of the effect of spatial correlation can be obtained from (1.3.4). Write it as

$$\operatorname{var}(\bar{Z}) = \sigma_0^2/n', \tag{1.3.5}$$

where

$$n' \equiv n \Big/ \Big[1 + 2\{\rho/(1-\rho)\}\{1 - (1/n)\} \\ -2\{\rho/(1-\rho)\}^2(1-\rho^{n-1})/n\Big] \tag{1.3.6}$$

can be interpreted as the *equivalent number of independent observations* (see Section 4.6.2). If $n = 10$ and $\rho = 0.26$, then $n' = 6.2$; that is, 6 independent observations achieve approximately the same precision as 10 correlated observations [correlated according to (1.3.3) with $\rho = 0.26$]. For large n, $n' \simeq n/[(1+\rho)/(1-\rho)]$, showing that correlation has an effect on inferences even in large samples.

Spatial models, more complicated than (1.3.3), show the same general behavior. Haining (1988) considers constant-mean Gaussian models in $\mathbb{R}^2$ that are conditionally specified autoregressions (Section 6.6), and simultaneously specified autoregressions and moving averages (Section 6.7.1), each with an unknown variance parameter σ^2. He compares the variance of $\bar{Z}$ assuming independence, with the variance of $\bar{Z}$ assuming positive dependence; he also compares the latter with the variance of the maximum likelihood estimator $\hat{\mu}$ of the constant mean μ. In general, classical inference based on $\bar{Z}$ and $\hat{\sigma}^2/n$ is misleading; for positive spatial dependence, $\widehat{\text{var}}(\bar{Z})$ and $\widehat{\text{var}}(\hat{\mu})$ are typically much larger than $\hat{\sigma}^2/n$.

Grenander (1954) demonstrates that, for a class of time-series models that includes (1.3.3), $\bar{Z}$ and the maximum likelihood estimator

$$\hat{\mu} \equiv \left\{Z(1) + (1-\rho)\sum_{i=2}^{n-1} Z(i) + Z(n)\right\} \Big/ \{n - (n-2)\rho\}$$

(which is also the minimum variance unbiased estimator) have the same asymptotic (100%) efficiency as $n \to \infty$. Does this contradict the preceding calculations? No, because there, inference for μ was based on $\bar{Z}$ *and* σ_0^2/n, and was found to be invalid. Grenander's result allows one to conclude that inference based on $\bar{Z}$ and σ_0^2/n' [n' is given by Eq. (1.3.6)] is not only valid but is also asymptotically *efficient*. General models for which this happens are given by Grenander and Rosenblatt (1984, Section 7.4), and they include stationary autoregressive time series.

Prediction

Suppose now that an unknown observation $Z(n+1)$ is to be *predicted* from data $\mathbf{Z} \equiv (Z(1), \ldots, Z(n))'$, where it is assumed $Z(1), \ldots, Z(n), Z(n+1)$ are jointly Gaussian, identically distributed with unknown mean μ and known variance σ_0^2, and *independent*. From Section 3.2, the predictor $p(\mathbf{Z}; n+1)$ that satisfies the unbiasedness condition $E(p(\mathbf{Z}; n+1)) = \mu$ and minimizes the mean-squared prediction error

$$E(Z(n+1) - p(\mathbf{Z}; n+1))^2 \tag{1.3.7}$$

is

$$p_0(\mathbf{Z}; n+1) = \bar{Z}, \tag{1.3.8}$$

the sample mean. The minimized mean-squared prediction error is

$$ms_0 = \sigma_0^2\{1 + (1/n)\}. \tag{1.3.9}$$

When the independence assumption is replaced with (1.3.3), the unbiased predictor that minimizes the mean-squared prediction error (1.3.7) is

$$p_\rho(\mathbf{Z}; n+1) = \rho Z(n) + (1-\rho)\frac{\{Z(1) + (1-\rho)\Sigma_{i=2}^{n-1}Z(i) + Z(n)\}}{\{n - (n-2)\rho\}}; \tag{1.3.10}$$

see Section 3.2. Abbreviate (1.3.10) to $p_\rho(\mathbf{Z})$. Notice that $\rho = 0$ in (1.3.10) yields (1.3.8). If $\bar{Z}$ is used instead of (1.3.10), the mean-squared prediction error is

$$E(Z(n+1) - \bar{Z})^2 = \sigma_0^2\Big\{1 + (1/n)\big[1 + 2\{\rho/(1-\rho)\}\{\rho^n - (1/n)\} - 2\{\rho/(1-\rho)\}^2(1-\rho^{n-1})/n\big]\Big\}. \tag{1.3.11}$$

For $n = 10$ and $\rho = 0.26$, (1.3.11) is $\sigma_0^2\{1.09051\}$. Compare this to (1.3.9), which is $\sigma_0^2\{1.1\}$: The difference is not large. Indeed, when spatial correlations decay geometrically with distance, classical prediction intervals are often approximately *valid*. However, they can be highly *inefficient*.

Assuming the dependence model given by (1.3.3), the minimized mean-squared prediction error is obtained by calculating (1.3.7) with optimal predictor (1.3.10). That is,

$$ms_\rho = \sigma_0^2\big[1 - \rho^2 + \{(1-\rho)^2(1+\rho)/(n-(n-2)\rho)\}\big]; \tag{1.3.12}$$

see Section 3.2. Notice that $\rho = 0$ in (1.3.12) yields (1.3.9).

For $n = 10$ and $\rho = 0.26$, (1.3.12) is $\sigma_0^2\{1.01952\}$. Thus a 95% prediction interval for $Z(n+1)$, based on the optimal predictor (1.3.10), is

$$\left(p_\rho(\mathbf{Z}) - (1.96)\sigma_0\{1.01952\}^{1/2},\, p_\rho(\mathbf{Z}) + (1.96)\sigma_0\{1.01952\}^{1/2}\right). \tag{1.3.13}$$

This should be compared to the *wider* 95% prediction interval for $Z(n+1)$ based on $\bar{Z}$:

$$(\bar{Z} - (1.96)\sigma_0\{1.09051\}^{1/2}, \bar{Z} + (1.96)\sigma_0\{1.09051\}^{1/2}). \quad (1.3.14)$$

For large n, the squared ratio of the prediction-interval width from using $p_\rho(\mathbf{Z})$ to that from using $\bar{Z}$ is approximately $1 - \rho^2$, which is a measure of the (asymptotic) efficiency of inference based on $\bar{Z}$ versus inference based on the optimal predictor $p_\rho(\mathbf{Z})$. Thus, for $\rho = 0.26$ the efficiency measure is 93% and for $\rho = 0.5$ it is 75%.

Experimental Design

Another example of the importance of exploiting spatial correlation is in the area of experimental design. Suppose that experimental units are laid out in a $t \times b$ array made up of b blocks (columns), each with t units in them. Units that are most alike are put in the same block so that all of the gross heterogeneity occurs between blocks. Suppose further that t treatments are to be compared as follows. A response variable is measured after applying the treatments to experimental units in such a way that each treatment appears exactly once in each block. A design (i.e., an allocation of treatments to units) of the type just described is known as a *complete block design*.

In papermaking, toward the end of the process when the paper is still a continuous sheet in a roll (that is typically 10 ft wide and 3×10^4 ft long), the sheet is *calendered*. A calender is a stack of two or more hard smooth rollers through which the paper is passed, whose purpose is to reduce the caliper of the sheet (so making it more dense) and increase the smoothness of the sheet. In calendering, high compressive stresses are used to make a more compact, thinner sheet with a smoother surface. In terms of basic smoothness and thickness, most of the variation is across the width of the paper roll. Therefore, any experiment on calender treatments (e.g., number and degree of smoothness of calender rolls, temperature of the sheet) has to take this into account, as well as the possible spatial correlation running the length of the sheet.

A well designed calendering experiment in the laboratory should split the sheet into several narrower sheets (e.g., each of width 2 ft), allowing these to form the blocks of the experiments. Within the blocks, pages closer together will tend to be more alike (in terms of gloss or caliper) than those further apart, and this can be modeled with (one-dimensional) spatial correlation. Therefore, when a calender treatment is applied to a "page," its location within the block should be useful information for performing an efficient analysis.

Let the response on the ith unit of the jth block be $Y_{ij(k)}$, $i = 1, \ldots, t$, $j = 1, \ldots, b$, $k = 1, \ldots, t$, where k denotes the treatment assigned (according to the design) to that unit. Assume an additive model

$$Y_{ij(k)} = \mu + \alpha_k + \beta_j + \delta(i, j), \quad (1.3.15)$$

where $\delta(\cdot,\cdot)$ is a Gaussian error process with zero mean. Further, assume independence of the δs between blocks but spatial dependence (1.3.3) down blocks. That is,

$$\text{cov}(\delta(i,j),\delta(i',j')) = \begin{cases} 0, & j \neq j', \\ \sigma^2\rho^{|i-i'|}, & j = j', \end{cases} \tag{1.3.16}$$

where $|\rho| < 1$.

A goal of experimental design is to find the allocation of treatments to units that will give the most precise estimates of (estimable) treatment effects, $\tau_k \equiv \alpha_k - (\sum_{l=1}^{t}\alpha_l/t)$, $k = 1,\ldots,t$. Specifically, one might minimize

$$A \equiv \sum_{1\le k<l\le t}\sum \text{var}(\hat{\tau}_k - \hat{\tau}_l)\bigg/\binom{t}{2}, \tag{1.3.17}$$

where $\{\hat{\tau}_k\colon k = 1,\ldots,t\}$ are the generalized-least-squares estimators of the treatment effects (see Section 5.6.2). Gill and Shukla (1985a) find that *first-order nearest-neighbor* (*NN*) *balanced* designs, whose end units satisfy further balance conditions, are optimal designs. First-order NN balanced means that the number of times two treatments appear adjacent in a block is constant for every possible pair of treatments. For example, consider $t = 10$ and $b = 5$; then Table 1.1 shows an assignment of treatments $0, 1,\ldots, 9$ within the five blocks that is a first-order NN balanced design.

Let A_O, A_N, and A_R denote the value of (1.3.17) for, respectively, the optimal design, the design in Table 1.1, and a randomized complete block design. Table 1.2 shows relative efficiencies (to two decimal places) for different values of ρ in (1.3.16).

Table 1.1 Example of a First-Order NN Balanced Complete Block Design

Blocks				
1	2	3	4	5
1	2	0	3	9
2	3	1	4	0
0	1	9	2	8
3	4	2	5	1
9	0	8	1	7
4	5	3	6	2
8	9	7	0	6
5	6	4	7	3
7	8	6	9	5
6	7	5	8	4

Source: Grondona and Cressie (1992).

Table 1.2 Relative Efficiencies of the First-Order NN Balanced Design Given in Table 1.1 (A_O/A_N), and a Randomized Complete Block Design (A_O/A_R)

	ρ						
	−0.75	−0.50	−0.25	0	0.25	0.50	0.75
A_O/A_N	1.00	1.00	1.00	1.00	1.00	1.00	1.00
A_O/A_R	0.33	0.67	0.92	1.00	0.87	0.63	0.36

Source: Grondona and Cressie (1992).

Optimal designs may not exist; first-order NN balanced designs are relatively easier to construct and appear to be (almost) as efficient as the optimal designs. Further, from Table 1.2, randomized complete block designs can be grossly inefficient when $|\rho| \geq 0.5$. More general spatial correlation structures are considered in Grondona and Cressie (1992) and Section 5.6.2; the same efficiency patterns emerge.

There is often a big difference between what scientists and engineers *should* do to obtain statistical optimality and what they do in practice. Randomized complete block designs may be inefficient, but they are very, very easy to construct. In Section 5.7, it is shown how to recover some of the lost efficiency after the inefficient (nonspatial) design is used. My preference is for a spatial modeling approach, such as the one described in Section 5.7.2 and in Grondona and Cressie (1991): The strategy there is to estimate and model the spatial correlation from (median-polish) residuals, and then to use generalized-least-squares estimates of the treatment effects $\{\tau_k: k = 1, \ldots, t\}$. An artificial experiment is constructed by adding treatment effects (according to a randomized complete block design) to spatially correlated uniformity data; the spatial analysis yields gains in efficiency of over 30%.

In many applications of Statistics to problems that are inherently spatial, data are observational rather than the result of a controlled experiment. Thus, the opportunity to randomize, block, and replicate is often not available. In these circumstances, the model fitted to the data may implicitly assume that some uncontrolled-for effects or interactions are zero. Hopefully, such assumptions are based on prior knowledge of the phenomenon and not on mathematical convenience. Wiens et al. (1986) report an interesting seven-year study of territorial locations, territorial sizes, and breeding densities within bird-breeding areas in central Oregon. They manipulated the habitat compostion and structure by removing 75%, 50%, 25%, and 0% of the shrubs from 625-m^2 blocks in a checkerboard design. A nearby, unaltered area served as the control. In spite of the elaborate experiment, the authors expressed concern about uncontrollable factors such as site tenacity by breeding adults and about the appropriateness of the spatial scale on which the manipulation was conducted. In fact, these same concerns are relevant to *all* designed experiments; it is simply a question of degree.

Linear Models with Spatially Dependent Error

Suppose the spatial data $Z(\mathbf{s}_1),\ldots,Z(\mathbf{s}_n)$, observed at spatial locations $\{\mathbf{s}_1,\ldots,\mathbf{s}_n\}$, are modeled as a collection of random variables, generated by the random process

$$Z(\mathbf{s}) = \sum_{l=1}^{q} \beta_l x_l(\mathbf{s}) + \delta(\mathbf{s}), \qquad \mathbf{s} \in D \subset \mathbb{R}^d, \tag{1.3.18}$$

where $\{x_l(\cdot): l = 1,\ldots,q\}$ is a collection of q nonrandom explanatory variables that may or may not depend on spatial location and $\delta(\cdot)$ is a zero-mean finite-variance error process that may or may not be spatially correlated.

For example, the model with constant mean μ and covariances given by (1.3.3) is a special case of (1.3.18) in $\mathbb{R}^1$, with just one explanatory variable $x_1(\cdot) \equiv 1$, $\beta_1 \equiv \mu$, and $\mathrm{cov}(\delta(u),\delta(v)) = \sigma_0^2 \cdot \rho^{|u-v|}$, $u, v \in \mathbb{R}^1$. Also, the complete-block-design model (1.3.15) and (1.3.16) is a special case of (1.3.18) in $\mathbb{R}^2$, where explanatory variables are $x_1(\cdot) \equiv 1$, $x_{k+1}(i,j) = 1$ if the (i,j)th plot receives the kth treatment and zero otherwise ($k = 1,\ldots,t$), and $x_{l+t+1}(i,j) = 1$ if $j = l$ and zero otherwise ($l = 1,\ldots,b$). The error process has anisotropic spatial correlation given by (1.3.16), specifying correlation within blocks but not between blocks.

It is helpful to write the observations as $\mathbf{Z} \equiv (Z(\mathbf{s}_1),\ldots,Z(\mathbf{s}_n))'$ and the model given by (1.3.18) in matrix and vector notation. That is,

$$\mathbf{Z} = X\boldsymbol{\beta} + \boldsymbol{\delta}, \tag{1.3.19}$$

where X is an $n \times q$ matrix whose (i,j)th element is $x_j(\mathbf{s}_i)$, $\boldsymbol{\beta} \equiv (\beta_1,\ldots,\beta_q)'$, and $\boldsymbol{\delta} \equiv (\delta(\mathbf{s}_1),\ldots,\delta(\mathbf{s}_n))'$.

Often (although not always), the parameters of greatest interest are the "large-scale-variation" parameters $\boldsymbol{\beta}$. For the moment, assume that any "small-scale-variation" parameters associated with the error process $\delta(\cdot)$ are known. That is, one can obtain

$$\mathrm{var}(\boldsymbol{\delta}) = \Sigma, \tag{1.3.20}$$

where Σ is a known $n \times n$, symmetric, nonnegative-definite matrix.

To keep the exposition simple, in all that is to follow, I shall assume that the regression coefficients $\boldsymbol{\beta}$ are estimable, Σ is invertible, and matrix inverses exist whenever I need them. Generalizations of this simple paradigm can be found, for example, in Searle (1971), Section 5.8.

Suppose the goal is efficient *estimation* of $\boldsymbol{\beta}$. [Efficient *prediction* of an unknown $Z(\mathbf{s}_0)$ is another possible goal; see Chapter 3.] The best linear unbiased estimator of $\boldsymbol{\beta}$ is also the *generalized-least-squares* (g.l.s.) estimator

(Aitken, 1935), that is, the value of $\boldsymbol{\beta}$ that minimizes

$$(\mathbf{Z} - X\boldsymbol{\beta})'\Sigma^{-1}(\mathbf{Z} - X\boldsymbol{\beta}). \tag{1.3.21}$$

It is

$$\hat{\boldsymbol{\beta}}_{\mathrm{gls}} = (X'\Sigma^{-1}X)^{-1}X'\Sigma^{-1}\mathbf{Z}, \tag{1.3.22}$$

which is also the maximum likelihood estimator if $\delta(\cdot)$ is a Gaussian process. The variance matrix of (1.3.22) is

$$\mathrm{var}(\hat{\boldsymbol{\beta}}_{\mathrm{gls}}) = (X'\Sigma^{-1}X)^{-1}. \tag{1.3.23}$$

When $\Sigma = \sigma^2 I$, which it is under classical (nonspatial) assumptions of i.i.d. errors, (1.3.21) reduces to $(\mathbf{Z} - X\boldsymbol{\beta})'(\mathbf{Z} - X\boldsymbol{\beta})/\sigma^2$. Then the appropriate minimizer is

$$\hat{\boldsymbol{\beta}}_{\mathrm{ols}} = (X'X)^{-1}X'\mathbf{Z}, \tag{1.3.24}$$

which is the *ordinary-least-squares* (o.l.s.) estimator of $\boldsymbol{\beta}$. However, when $\Sigma \neq \sigma^2 I$ [e.g., Σ has entries given by (1.3.16)], the o.l.s. estimator is often inefficient. Specifically,

$$\mathrm{var}(\hat{\boldsymbol{\beta}}_{\mathrm{ols}}) = (X'X)^{-1}(X'\Sigma X)(X'X)^{-1} \tag{1.3.25}$$

and the difference $\mathrm{var}(\hat{\boldsymbol{\beta}}_{\mathrm{ols}}) - \mathrm{var}(\hat{\boldsymbol{\beta}}_{\mathrm{gls}})$ is nonnegative-definite (e.g., Searle, 1971, Section 3.3). Notice that when $\Sigma = \sigma^2 I$, the difference is 0.

Because the o.l.s. estimator (1.3.24) does not require knowledge of Σ, considerable research effort has been devoted to finding conditions on X and Σ for which o.l.s. and g.l.s. are equally efficient or for which o.l.s. and g.l.s. estimators are the same. Zyskind (1967) derives eight equivalent necessary and sufficient conditions for the latter to happen; for example, one such condition is $X(X'X)^{-1}X'\Sigma = \Sigma X(X'X)^{-1}X'$, which occurs if and only if Σ is of the form $a^2 I + P_1 A P_1 + P_2 B P_2$, where $P_1 = X(X'X)^{-1}X'$ and $P_2 = I - P_1$.

However, care should be taken. Just because o.l.s. and g.l.s. yield the same estimators, it cannot be assumed that all inference can proceed as if $\Sigma = \sigma^2 I$. Indeed, under Zyskind's conditions, $\mathrm{var}(\hat{\boldsymbol{\beta}}_{\mathrm{ols}}) = \mathrm{var}(\hat{\boldsymbol{\beta}}_{\mathrm{gls}}) = (X'\Sigma^{-1}X)^{-1}$, which depends on Σ. Therefore, from a total-inference point of view, results on the equivalence of o.l.s. and g.l.s. are not all that helpful. Moreover, except for a few special cases, spatial linear models like (1.3.18) generally give rise to matrices X and Σ that do *not* satisfy Zyskind's conditions. [Kramer and Donninger, 1987, give one of these special cases when $\delta(\cdot)$ is a spatial autoregressive process of the type described in Section 6.7.2. Curiously, they show that o.l.s. can be as *efficient* as g.l.s. when the spatial-dependence parameter in $\delta(\cdot)$ tends to its largest possible value.] The generally poor performance of classical statistical-inference procedures in the presence of spatial dependence (see, e.g., the simple examples above; Cliff and Ord, 1981, Chapter 7; Dow et al., 1982; Anselin and Griffith, 1988) and the ubiquity of such spatial dependence, demonstrate why Statistics for spatial data should be part of every statistician's repertoire.

Valid Spatial Models

In reality, Σ is unknown; typically it will depend on parameters $\boldsymbol{\theta}$ obtained from a spatial parametric covariance model $C(\mathbf{u}, \mathbf{v}; \boldsymbol{\theta}) \equiv \text{cov}(\delta(\mathbf{u}), \delta(\mathbf{v}))$, $\mathbf{u}, \mathbf{v} \in D \subset \mathbb{R}^d$. For example, assuming homoskedasticity, one parameter will certainly be $\sigma^2 \equiv \text{var}(\delta(\mathbf{u}))$. Let $\Sigma(\boldsymbol{\theta})$ denote the $n \times n$ variance matrix var($\mathbf{Z}$) whose (i, j)th element is $C(\mathbf{s}_i, \mathbf{s}_j; \boldsymbol{\theta})$.

Choice of an appropriate model can be a difficult task because one must take care to use valid models; that is, use only those $\boldsymbol{\theta}$ for which $\sum_{i=1}^m \sum_{j=1}^m a_i C(\mathbf{u}_i, \mathbf{u}_j; \boldsymbol{\theta}) a_j \geq 0$, for *any* m, *any* sequence of real numbers $\{a_i: i = 1, \ldots, m\}$, and *any* sequence of spatial locations $\{\mathbf{u}_i: i = 1, \ldots, m\}$. For example, a valid covariance model in $\mathbb{R}^1$ is $C(s_i, s_j; \theta) = C_T(|s_i - s_j|; \theta)$, the "tent" covariogram

$$C_T(h; \theta) \equiv \begin{cases} \sigma^2\left(1 - \dfrac{|h|}{\theta}\right), & 0 \leq |h| \leq \theta, \\ 0, & |h| > \theta. \end{cases}$$

However, in $\mathbb{R}^2$, $C(\mathbf{s}_i, \mathbf{s}_j; \theta) \equiv C_T(\|\mathbf{s}_i - \mathbf{s}_j\|; \theta)$ is *not* a valid model; see Section 2.5.1 for a counterexample. On the other hand, if the spatial model for data $\mathbf{Z} = (Z(\mathbf{s}_1), \ldots, Z(\mathbf{s}_n))'$ has sufficient measurement error so that $\Sigma(\boldsymbol{\theta})$ is diagonally dominant, that is, if

$$C(\mathbf{s}_i, \mathbf{s}_i; \boldsymbol{\theta}) \geq \sum_{\substack{j=1 \\ j \neq i}}^{n} \left|C(\mathbf{s}_i, \mathbf{s}_j; \boldsymbol{\theta})\right|, \qquad i = 1, \ldots, n,$$

then $\Sigma(\boldsymbol{\theta})$ is always nonnegative-definite (Varga, 1962, p. 22ff) and hence a valid model is obtained. In the preceding simple examples, the models (1.3.3) and (1.3.16) are valid provided the parameter $\rho \in (-1, 1)$.

Questions of validity of the various spatial models presented in this book will be dealt with as they arise. It is important to ensure that the parameter space Θ of $\boldsymbol{\theta}$ consists only of valid models, and that an estimate $\hat{\boldsymbol{\theta}}$ of $\boldsymbol{\theta}$ belongs to Θ.

Estimated Generalized Least Squares

Assume that the spatial dependence parameter $\boldsymbol{\theta}$ is contained in a known parameter space Θ, any member of which yields a valid spatial model. Any estimator $\hat{\boldsymbol{\theta}}$ will be assumed to belong to Θ, with probability 1.

Estimated generalized-least-squares (e.g.l.s.) estimators are obtained by substituting $\Sigma(\hat{\boldsymbol{\theta}})$ for $\Sigma(\boldsymbol{\theta})$ in (1.3.22):

$$\hat{\boldsymbol{\beta}}_{\text{egls}} = \left(X'\Sigma(\hat{\boldsymbol{\theta}})^{-1}X\right)^{-1}X'\Sigma(\hat{\boldsymbol{\theta}})^{-1}\mathbf{Z}, \tag{1.3.26}$$

where $\hat{\boldsymbol{\theta}}$ is an estimator of the small-scale-variation parameters $\boldsymbol{\theta}$.

Prediction of an unknown $Z(\mathbf{s}_0)$ proceeds in a similar way, often by assuming that the estimated value $\hat{\boldsymbol{\theta}}$ is the true value (Chapter 3). The consequences of this assumption are investigated in Section 5.3.

Two different approaches are often employed to obtain $\hat{\boldsymbol{\theta}}$. If the probability density (or mass) function of $\mathbf{Z}$ is completely specified as $f(\mathbf{z}; \boldsymbol{\beta}, \boldsymbol{\theta})$, then the likelihood

$$l(\boldsymbol{\beta}, \boldsymbol{\theta}) \equiv f(\mathbf{Z}; \boldsymbol{\beta}, \boldsymbol{\theta})$$

can be maximized with respect to $\boldsymbol{\beta}$ and $\boldsymbol{\theta}$ to yield estimators $\hat{\boldsymbol{\beta}}$ and $\hat{\boldsymbol{\theta}}$. Then (1.3.26) is the maximum likelihood (m.l.) estimator of $\hat{\boldsymbol{\beta}}_{\text{gls}}$. Sections 7.2 and 7.3 deal mainly with this approach.

A second possibility is to obtain an initial estimate of $\boldsymbol{\beta}$ (e.g., $\hat{\boldsymbol{\beta}}_{\text{ols}}$), construct residuals

$$\mathbf{R} \equiv \mathbf{Z} - X\hat{\boldsymbol{\beta}}_{\text{ols}}, \tag{1.3.27}$$

and, using these residuals as a proxy for $\boldsymbol{\delta}$, estimate $\boldsymbol{\theta}$ free from the presence of $\boldsymbol{\beta}$. This method is often employed in a semiparametric setting where the joint distribution of $\mathbf{Z}$ is not specified, but the mean $X\boldsymbol{\beta}$ and the variance $\Sigma(\boldsymbol{\theta})$ are specified. Then, $\hat{\boldsymbol{\theta}}$ can be obtained from an approximate method-of-moments approach based on $\mathbf{R}$ (Section 2.6.2). From $\hat{\boldsymbol{\theta}}$ and (1.3.26), $\hat{\boldsymbol{\beta}}_{\text{egls}}$ is obtained. One then has the option of substituting this estimate back into (1.3.27) to obtain a new $\mathbf{R}$, a new $\hat{\boldsymbol{\theta}}$, and a new $\hat{\boldsymbol{\beta}}_{\text{egls}}$, and so forth. Terminating at this second step yields a two-stage estimation procedure.

The example considered throughout this section [$\mathbf{Z}$ Gaussian with covariances defined by (1.3.3)] satisfies the conditions of Mardia and Marshall's (1984) theorem given in Section 7.3.1. Hence, for $\boldsymbol{\theta} = (\sigma^2, \rho)'$ and $\hat{\boldsymbol{\theta}}$ its m.l. estimator, $\hat{\boldsymbol{\beta}}_{\text{egls}}$ given by (1.3.26) is asymptotically Gaussian with mean $\boldsymbol{\beta}$ and variance matrix $(X'\Sigma(\boldsymbol{\theta})^{-1}X)^{-1}$ and is asymptotically independent from $\hat{\boldsymbol{\theta}}$. However, for small sample sizes, the m.l. estimators of the small-scale-variation parameters are notoriously biased. A review of results for the model (1.3.3) can be found in Dielman and Pfaffenberger (1989), together with small-sample simulations and recommendations.

The exact variance of $\hat{\boldsymbol{\beta}}_{\text{egls}}$ is difficult to compute because $\mathbf{Z}$ appears in (1.3.26) nonlinearly. Preliminary tests on $\boldsymbol{\theta}$ (see Section 7.3.1) are often employed to determine whether g.l.s. (actually, e.g.l.s.) will be used in preference to o.l.s. Intuitively, as sample size increases, e.g.l.s. will be indistinguishable from g.l.s., provided $\hat{\boldsymbol{\theta}}$ is a consistent estimator of $\boldsymbol{\theta}$.

Williams (1975) can be consulted for results that formalize this intuition. In finite samples, terms of smaller orders of magnitude can improve the asymptotic distribution of $\hat{\boldsymbol{\beta}}_{\text{egls}}$ (e.g., Rothenberg, 1984; Grubb and Magee, 1988). Asymptotics in spatial problems can be of the infill type (Sections 2.6 and 5.8) or of the increasing-domain type (Section 7.3). Statisticians and probabilists are most familiar with the latter; indeed, all of the asymptotic results mentioned in this section are for increasing domains.

Some interesting finite-sample results that establish inequalities between $E(\widehat{\text{var}}(\mathbf{c}'\hat{\boldsymbol{\beta}}_{\text{gls}}))$, $\text{var}(\mathbf{c}'\hat{\boldsymbol{\beta}}_{\text{gls}})$, and $\text{var}(\mathbf{c}'\hat{\boldsymbol{\beta}}_{\text{egls}})$ are proved by Harville (1985) and Eaton (1985). For example, Harville shows that if $\hat{\boldsymbol{\theta}}$ is an even, translation-invariant estimator, and the data are Gaussian, then

$$\text{var}\left(\mathbf{c}'\hat{\boldsymbol{\beta}}_{\text{gls}}\right) \le \text{var}\left(\mathbf{c}'\hat{\boldsymbol{\beta}}_{\text{egls}}\right), \tag{1.3.28}$$

for all $\mathbf{c} \in \mathbb{R}^q$. Using a matrix form of Jensen's inequality, Eaton shows that if $E(\Sigma(\hat{\boldsymbol{\theta}})) = \Sigma(\boldsymbol{\theta})$, then

$$E\left(\widehat{\text{var}}\left(\mathbf{c}'\hat{\boldsymbol{\beta}}_{\text{gls}}\right)\right) \le \text{var}\left(\mathbf{c}'\hat{\boldsymbol{\beta}}_{\text{gls}}\right), \tag{1.3.29}$$

where $\widehat{\text{var}}(\mathbf{c}'\hat{\boldsymbol{\beta}}_{\text{gls}})$ is $\mathbf{c}'(X'\Sigma(\hat{\boldsymbol{\theta}})^{-1}X)^{-1}\mathbf{c}$, the often used estimator of $\text{var}(\mathbf{c}'\hat{\boldsymbol{\beta}}_{\text{gls}})$; see (1.3.23). Although for $\hat{\boldsymbol{\theta}}$ a consistent estimator of $\boldsymbol{\theta}$ the three quantities in (1.3.28) and (1.3.29) are asymptotically indistinguishable, the inequalities show that, as an estimator of $\text{var}(\mathbf{c}'\hat{\boldsymbol{\beta}}_{\text{egls}})$, $\widehat{\text{var}}(\mathbf{c}'\hat{\boldsymbol{\beta}}_{\text{gls}})$ has potentially two sources of negative bias. Harville and Jeske (1992) investigate estimators of $\text{var}(\mathbf{c}'\hat{\boldsymbol{\beta}}_{\text{egls}})$ that attempt to correct for this bias.

Spatial Data Analysis and Modeling: A Summary

- All data have a more-or-less precise spatial and temporal label associated with them. A purely spatial model usually has no causative component in it; such models are useful when a space–time process has reached temporal equilibrium (e.g., ore deposition), or when short-term causal effects are aggregated over a fixed time period (e.g., final presidential election returns from the states of the United States).
- Whether the spatial labels are thought to be an important part of the modeling and analysis of the data is a concern that should be addressed problem by problem.
- Data that are close together in space (and time) are often more alike than those that are far apart. A spatial model incorporates this spatial variation into the generating mechanism, in contrast to a nonspatial model.
- It is almost always true that the classical, nonspatial model is a special case of a spatial model, and so the spatial model is more general (space–time models are even more general).

- Whether one chooses to model the spatial variation through the (nonstochastic) mean structure (called large-scale variation throughout the book) or the stochastic-dependence structure (called small-scale variation) depends on the underlying scientific problem, and can sometimes be simply a trade-off between model fit and parsimony of model description. What is one person's (spatial) covariance structure may be another person's mean structure.
- Explanatory variables (suggested by the problem under investigation) should be included in the mean structure first, and great care should be taken to find all of them. A missed variable that is itself varying spatially will contribute to the spatial dependence, as can a misspecification of the functional relationship between the independent and explanatory variables. As a consequence, a model that includes a spatial-dependence component pays a low-cost premium that insures against misspecification of the mean structure (e.g., Dubin, 1988).
- Having allowed for explanatory variables, models with spatial dependence typically have a more parsimonious description than classical trend-surface models. They also have more stable spatial extrapolation properties and yield more efficient estimators of explanatory-variable effects.

Some Final Remarks

The simple examples presented in this chapter serve to illustrate a general principle in all of Statistics, namely that inferences can depend crucially on assumptions. In situations such as for spatial or temporal data, where it is often more realistic to abandon the independence assumption, classical inferences can be very misleading. Stephan (1934) saw this very clearly in problems dealing with social data; he wrote

> ...Data of geographic units are tied together, like bunches of grapes, not separate, like balls in an urn. Of course, mere contiguity in time and space does not of itself indicate lack of independence between units in a relevant variable or attribute, but in dealing with social data, we know that by virtue of their very *social* character, persons, groups and their characteristics are interrelated and not independent. Sampling error formulas may yet be developed which are applicable to these data, but until then the older formulas must be used with great caution. Likewise, other statistical measures must be carefully scrutinized when applied to these data....

In the last 40 years, considerable attention has been devoted to concerns of the type expressed by Stephan; see Section 5.6.1 for a brief review.

Among the exact sciences, it is often believed that the only source of variability is measurement error, which is usually modeled as a white-noise random process. In my opinion, this viewpoint is far too simplistic, and could lead to important small-scale (spatial or nonspatial) effects being missed.

Problems in the earth sciences, ecology, soil science, hydrology, and so on can provide the statistician with a considerable challenge. Often studies are observational rather than designed, and there is no replication because there is just one unit: the earth! Nevertheless, explanations for various phenomena can be conjectured and their consistency across time or space can be assessed. But, what is the source of random variation in the spatial models (and temporal models, for that matter) proposed? A standard tactic in science is that when there are infinitely many possible data values from which to choose, those possibilities are dealt with statistically. With no data, but with experience in the way, say, oilfields of a particular type behave, one might assume that a spatial model with trend and additive stationary error would be appropriate.

Suppose that you are confronted with a spatial data set and a scientific or engineering problem to solve. After some *descriptive* analyses of the data, the set of potential models can hopefully be narrowed to just a few. An *indication* of a model's parameters can be obtained by applying an estimation criterion, such as maximum likelihood, maximum pseudolikelihood, method-of-moments, and so forth. Whether these fitted parameters are significant is determined by statistical-inferential methods. *Estimation* of a model's parameters involves both indication and, at least, measures of the estimator's variation. In either case (indication or estimation), it is important to check the fit of the model (because no model is exactly true) using *diagnostic* statistics and plots.

The approach taken in this book is to broaden the classical models and methods to those that recognize the presence and importance of spatial information. The success of these spatial models and methods will be judged by their powers of organization, explanation, and prediction. Much has been accomplished in the last 20 years, and a vigorous future (in modeling, simulation, estimation, computation, inference, asymptotics, diagnostics, etc.) for the subject is assured.

A rereading of Section 1.1 would be helpful at this point, but for those who are ready to jump in, a brief synopsis is given as follows. Chapters 2, 3, and 4 deal with geostatistical data, and Chapter 5 is a collection of special topics of interest to me (Part I). Chapters 6 and 7 deal with lattice data (Part II). Chapters 8 and 9 deal with, respectively, point patterns and objects based on point patterns (Part III).

> We shall not cease from exploration
> And the end of all our exploring
> Will be to arrive where we started
> And know the place for the first time. (T. S. Eliot)

PART I

Geostatistical Data

CHAPTER 2

Geostatistics

In Chapter 1, a categorization of (most) spatial statistical problems into three broad types was given. The first part of this book is concerned with modeling data as a (partial) realization of a random process $\{Z(\mathbf{s}): \mathbf{s} \in D\}$, where the index set D allows $\mathbf{s}$ to vary continuously throughout a region of d-dimensional Euclidean space.

The prefix "geo" in the title of this chapter appears to refer to statistics pertaining to the earth, and indeed this was its original meaning. Hart (1954) coined the term "geostatistics" in a *geo*graphical context to denote statistical techniques that emphasize location within areal distributions. Matheron (1962, 1963a, 1963b) used the term in a *geo*logical context to denote theory and methods for inferring ore reserves from data spatially distributed throughout an ore body. In a two-volume work of 504 pages (in French), Matheron (1962, 1963a) developed a comprehensive theory of inference (mainly prediction) for geostatistical data. A very much abbreviated version (in English) appeared in Matheron (1963b).

For me, geostatistics has thrown off its earthly shackles and has taken on a more universal role, one that is concerned with statistical theory and applications for processes with continuous spatial index.

Chapter Summary

Section 2.1 describes the class of problems being considered in Part I of this book, and Section 2.2 presents a data set upon which exploratory spatial analyses are carried out. The theoretical foundations of geostatistics are laid in Section 2.3. Estimation of the variogram is discussed in Section 2.4. Intrinsically stationary processes have a spectral theory, which is presented in Section 2.5. Finally, Section 2.6 discusses inference for the variogram based on a selection of estimators.

2.1 CONTINUOUS SPATIAL INDEX

It will become clear from Chapters 2 through 5 that a wide variety of problems can be solved using geostatistical methods. The common thread that links them is that the data can be thought of as a (usually partial) realization of a random process (or stochastic process or random field)

$$\{Z(\mathbf{s}): \mathbf{s} \in D\}, \tag{2.1.1}$$

where D is a fixed subset of $\mathbb{R}^d$ with positive d-dimensional volume. In other words, the spatial index $\mathbf{s}$ varies *continuously* throughout the region D. It is possible to allow for spatiotemporal data by considering the variable $Z(\mathbf{s}, t)$, but for most of this book it will be assumed that the data are purely spatial, either having been aggregated over time or referring to a fixed instant of time. (The time dimension is present in the soil–water-tension data of Section 4.2 and the acid-rain data of Section 4.6.)

My approach throughout this chapter is to present ways to model (2.1.1) that allow inference on the model parameters from data $\{Z(\mathbf{s}_1), \ldots, Z(\mathbf{s}_n)\}$ at known spatial locations $\{\mathbf{s}_1, \ldots, \mathbf{s}_n\}$. The cornerstone is the variogram, a parameter that in the past has been either unknown or unfashionable among statisticians. In Part I, much of the emphasis will be on prediction, although estimation (Sections 4.3, 5.6.2, and 5.7) and hypothesis testing (Section 4.3) are also important.

Another approach, which will not be pursued here, is to try to characterize the random measure $\{\zeta(B): B \in \mathscr{D}\}$, where the index B varies over $\mathscr{D}$, the Borel sets of D, and $\zeta(B) \equiv \int_B Z(\mathbf{s})\, d\mathbf{s}$. In mining terms, $\zeta(B)$ represents the amount of metal in the mining block B. The more direct approach through the variogram (or covariogram) seems to be more amenable to applications.

2.2[†] SPATIAL DATA ANALYSIS OF COAL ASH IN PENNSYLVANIA

Before I give a more formal presentation of geostatistical models (in subsequent sections of this chapter) and spatial prediction (in Chapter 3), an example is introduced and its data analyzed in an exploratory manner. Much of the material presented in this section comes from Cressie (1984a). The choice of example before theory is deliberate. My intention is to motivate some of the theoretical questions that arise later, and an example in coal exploration is a fitting introduction to a subject whose origins were in mining applications.

It is my view that geostatistics can profit considerably from the philosophy and methods of modern data analysis, now to be found in much of applied statistics. A lot of time, effort and money is put into geological exploration, but only recently has the value of good *statistical exploration* been realized. Spatial data are often looked at and summarized in mostly classical ways, as if they were the result of a random sample or perhaps of a time series, rather than of a spatial collection of dependent random variables whose dependence is strongly tied to (relative) spatial locations.

Exploratory Data Analysis

By its very nature, exploratory data analysis must not trust every observation equally, but rather isolate for further perusal observations that look suspiciously atypical. Having said this, there is the implication that some underlying stochastic model is in mind, from which departures need to be clearly identified. Indeed, any of us who has drawn a histogram (or stem-and-leaf plot) of a batch of numbers must know how much simpler statistical analysis is when a histogram appears bell-shaped with no unusually large or small observations to cause concern.

In the spatial context, a common underlying stochastic model is that all the data (which are observed in a comparable way) come from a joint Gaussian (i.e., normal) distribution whose correlation structure depends on the *spatial locations* of the data. When one allows for the possibility of an initial transformation of the data, this underlying model is quite general; it is also the basic one from which linear geostatistical inference has developed.

Unfortunately, the Gaussian model may not fit all that well. Some observations may be atypical due to an anomaly or a recording error or a similar problem, which could be modeled by adding a small component of contamination to the underlying Gaussian model. Then the goal would be to estimate the parameters of the Gaussian model. (However, it may be that the atypical observations are the most informative, in which case the contamination model is inappropriate.)

Data analytic methods that graph and summarize the way the observations act and interact should, first of all, be relatively unaffected by the presence of contamination and, second, be able to highlight the contaminant from the rest. Such methods are called *resistant*.

The Statistician's Role

In any mining geostatistical study, there should ideally be inputs from many different areas of expertise. The team should at the very least include a geologist, a mining engineer, a metallurgist, a statistician, and a financial manager. This section is being written from the point of view of a statistician, whose main responsibilities are to lead the team through the following six stages.

1. Designing the sampling plan.
2. Graphing and summarizing the data.
3. Detecting and allowing for spatial nonstationarity.
4. Estimating spatial relationships, usually through the variogram (sometimes called structural analysis).
5. Estimating the *in situ* resources, usually through kriging.
6. Assessing the recoverable reserves and, if a decision is made to mine, determining the mine plan and providing current reserve assessments as the mining proceeds.

In this section, a closer look is taken at stages 2 and 3, in particular as it pertains to the central problem of variogram estimation. In sections 2.4 and 2.6, a more formal presentation of variogram estimation is made. Chapter 3 and Sections 5.1, 5.2, 5.4, and 5.9 contain methods that are relevant to the last two stages, and the (often ignored) first stage is discussed in Section 5.6.1.

Coal-Ash Data

Data obtained from Gomez and Hazen (1970, Tables 19 and 20) on coal ash for the Robena Mine Property in Greene County, Pennsylvania, serve to illustrate the spatial data analytic techniques. These data come from the Pittsburgh coal seam that is associated with a deltaic sedimentation system that includes much of southwestern Pennsylvania, northeastern Ohio, and northern West Virginia. I have used the 208 coal-ash core measurements at locations with west coordinates greater than 64,000 ft; spatially this defines an approximately square grid, with 2500-ft spacing, running roughly southwest to northeast and northwest to southeast. Figure 2.1 displays the locations. In order to talk easily and sensibly about directions, the grid was reoriented so that it appears to run in an east–west and north–south

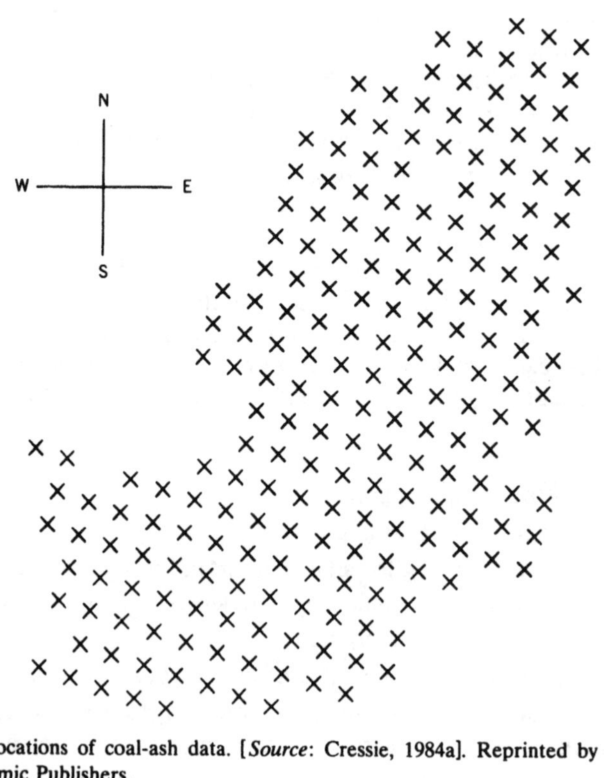

Figure 2.1 Locations of coal-ash data. [*Source*: Cressie, 1984a]. Reprinted by permission of Kluwer Academic Publishers.

direction. Figure 2.2*a* shows $\{z(x, y)\colon\ x = 1,\ldots,16;\ y = 1,\ldots,23\}$, the coal-ash measurements and their spatial locations, and Figure 2.2*b* is a three-dimensional scatter plot of the same data. Comparison of Figures 2.1 and 2.2*a* shows how the grid was reoriented. From now on I shall refer to the locations of the cores according to the reoriented grid of Figure 2.2*a*.

What follows is an exploratory analysis of the spatial data using techniques of exploratory data analysis (EDA; see Tukey, 1977), as well as spatial techniques very much in the spirit of EDA. Consider the three-dimensional scatter plot of Figure 2.2*b*; its usefulness as an exploratory tool is rather limited because interesting features in the data are easily hidden. Exploratory spatial data analysis requires some fresh approaches, suggested by univariate and multivariate EDA techniques, spatial methods of summarizing the data, and innovative graphics (such as brushing, rotating, and symbol scaling; e.g., Cleveland, 1985).

Exploratory Spatial Data Analysis

Exploratory methods need to be resistant to observations that are atypical of an underlying model. In a nonspatial setting, this is usually caused by departures from a Gaussian error model or from an assumed linear relationship between two variables. In a spatial setting, further departures might be expected from stationary-mean or stationary-dependence assumptions. (Stationarity is defined in Section 2.3; informally speaking, it refers to variational properties that do not change throughout the region of interest D.)

If the spatial nature of the observations is ignored and the data are treated simply as a batch of numbers, then the stem-and-leaf plot (see, e.g., Velleman and Hoaglin, 1981, Chapter 1) gives some idea of distribution shape. (Implicit in interpreting a stem-and-leaf plot is the belief that the data are identically distributed, something that is often not the case.) From Figure 2.3, there is clearly one unusually large observation, and several others on the high and low side that are suspect. Analyzers of *spatial* data should also be suspicious of observations when they are unusual *with respect to their neighbors*. Implicitly, this is saying that the data are governed by some sort of *local* stationarity (that does not, in general, guarantee a global stationarity across the ore body).

Local stationarity will be incorporated into the underlying Gaussian model by allowing the *expectation* of a datum to be a smooth function of its *position*; however, the mean-square differences between two data points will remain a function only of their *relative position*. (In Section 5.4, this is generalized so that mean-squared higher-order differences are functions of relative position.) Figure 2.4 is an attempt to summarize this possible nonstationarity in the mean using the sample median and sample mean across rows and down columns. Here, the spatial aspect of the data enters for the first time. (When the data are not located on a regular grid, a low-resolution grouping of observations into a two-way table, such as in Section 4.1, still

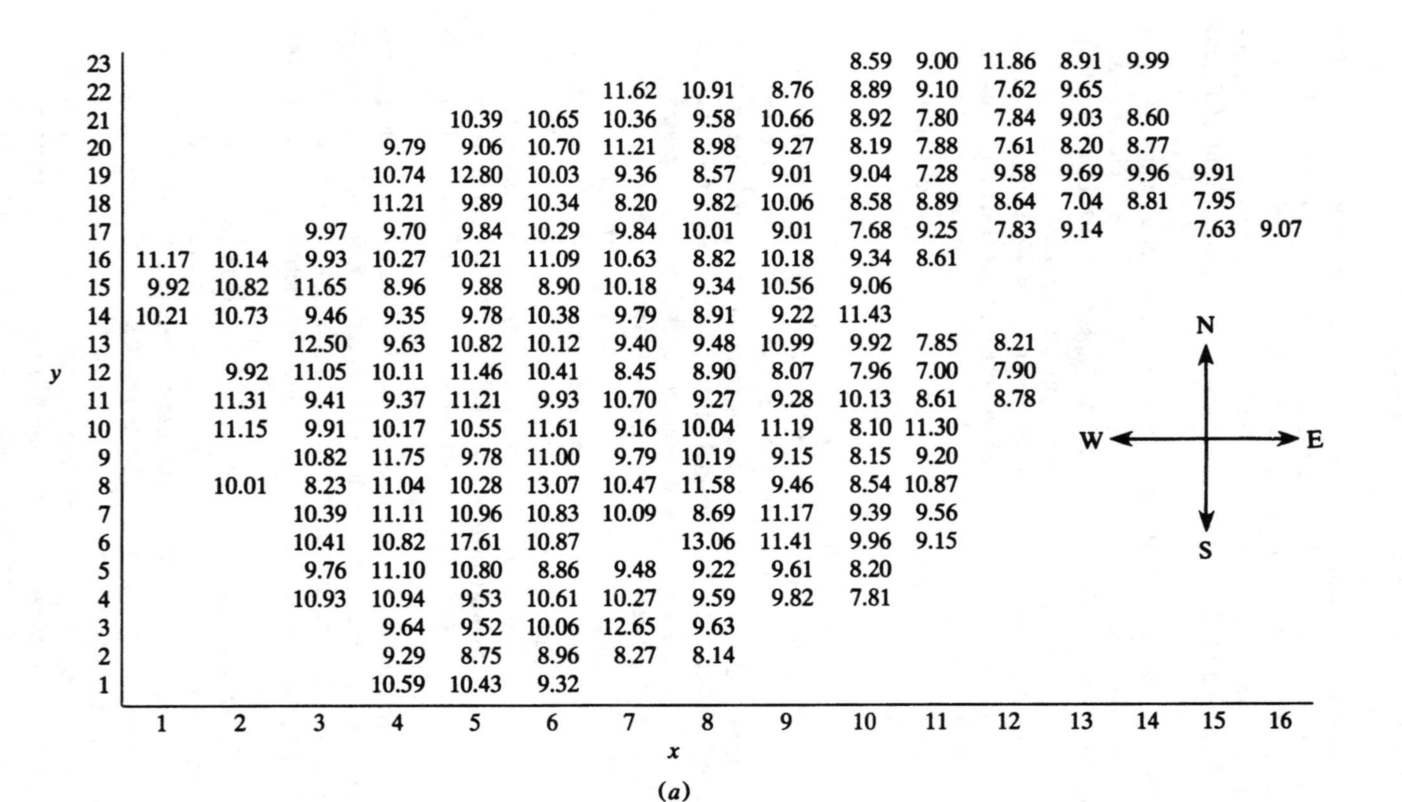

Figure 2.2 (*a*) Core measurements (in % coal ash) at reoriented locations. Units on the vertical axis are % coal ash. [*Source*: Cressie, 1984a]. Reprinted by permission of Kluwer Academic Publishers.

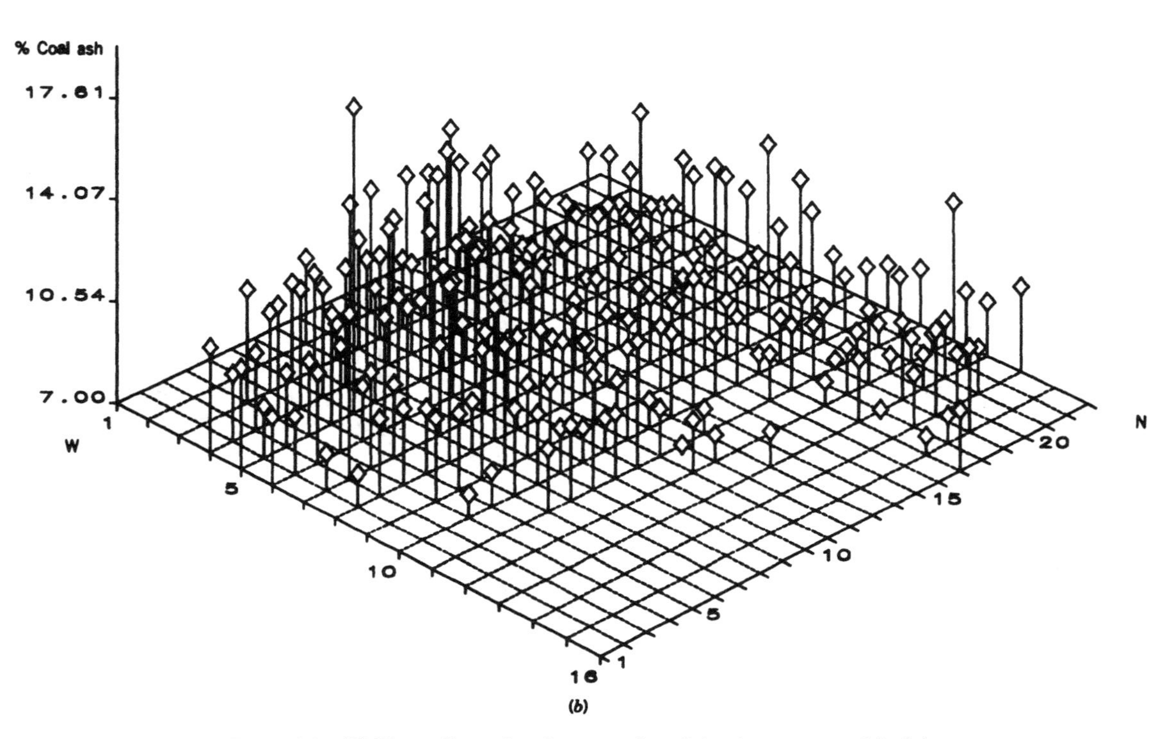

Figure 2.2 (*b*) Three-dimensional scatter plot of the data presented in (*a*).

6	
6	
7	003
7	66678888899
8	00111222222234
8	56666666788888899999999
9	000000011111222222223333333444444
9	555555666666666677888888888999999999
10	000000001111111222222333334444444
10	56666667777788888899999
11	00000111222222223344
11	5666689
12	
12	568
13	11
13	
14	
14	
15	
15	
16	
16	
17	
17	6

Figure 2.3 Stem-and-leaf plot of coal-ash data. 13|1 denotes 13.1% coal ash. [*Source*: Cressie, 1984a]. Reprinted by permission of Kluwer Academic Publishers.

allows the method to be carried out.) There appears to be a linear trend in the east–west direction but little or no trend in the north–south direction.

The use of median and mean serves two purposes. First, from an EDA point of view, the median is a resistant summary statistic, but also its comparison to the nonresistant mean summary has the additional function of highlighting rows or columns that may contain atypical observations. If (mean − median) is too big, then the row or column should be scanned for possible outliers. A small amount of formal statistics indicates what is meant by too big. Suppose for the moment that $Y_1, Y_2, \ldots, Y_n$ are independent and identically distributed (i.i.d.) from a symmetric probability density function f with mean μ and variance σ^2. Then, the median $\tilde{Y}$ can be approximated [provided $f(\mu) \neq 0$] as

$$\tilde{Y} \simeq \mu + \left(\frac{1}{n}\right) \sum_{i=1}^{n} \frac{\operatorname{sgn}(Y_i - \mu)}{2f(\mu)}, \qquad (2.2.1)$$

where $\operatorname{sgn}(x) = 1$ if $x > 0$, $= 0$ if $x = 0$, $= -1$ if $x < 0$. The degree of approximation is made more precise in Section 3.5.4.

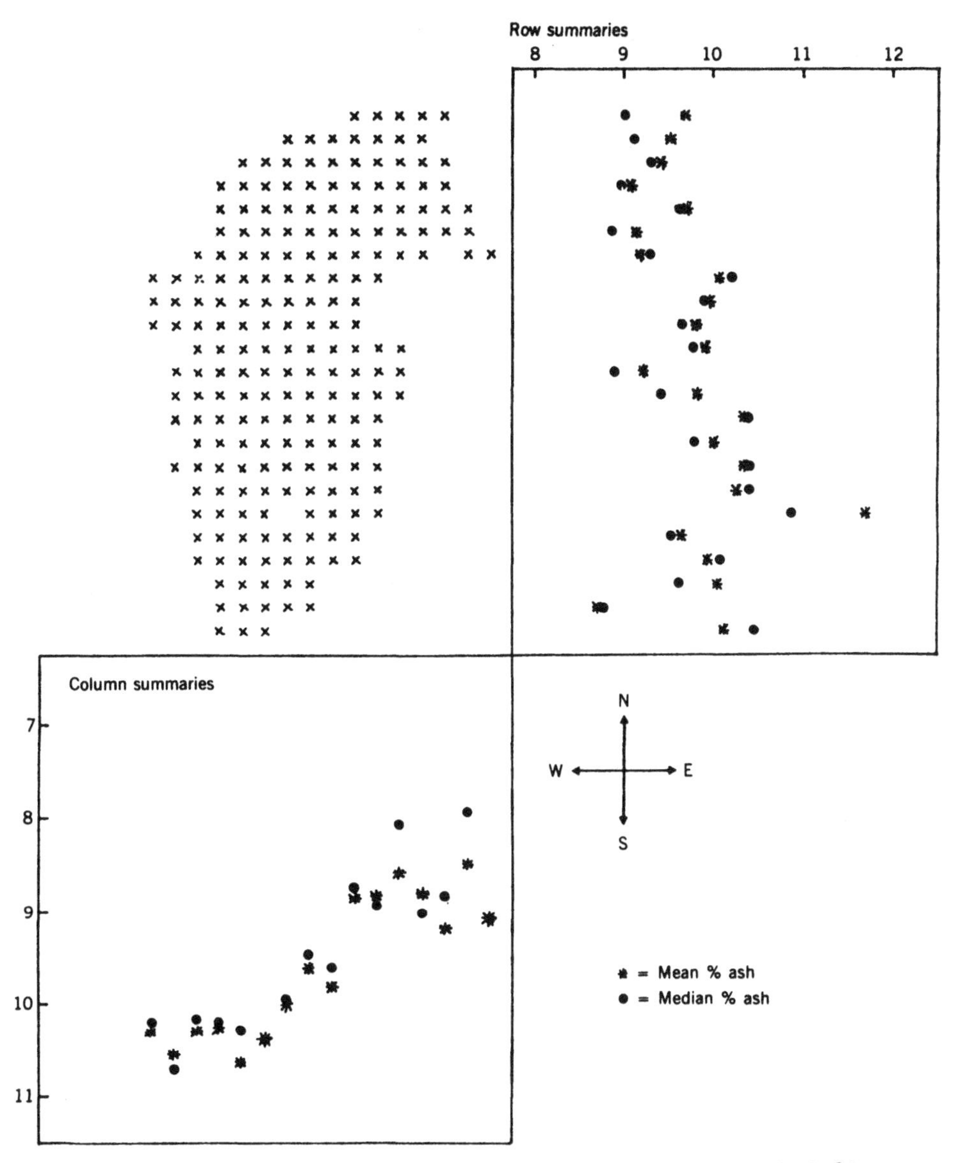

Figure 2.4 Mean and median summaries of nonstationarity. Units are % coal ash. [*Source*: Cressie, 1984a]. Reprinted by permission of Kluwer Academic Publishers.

Trivially, the sample mean has a similar representation:

$$\bar{Y} = \mu + (1/n) \sum_{i=1}^{n} (Y_i - \mu). \tag{2.2.2}$$

Then $\bar{Y} - \tilde{Y} \simeq (1/n)\Sigma_{i=1}^{n}\{\epsilon_i - (\text{sgn}(\epsilon_i))/(2f(\mu))\}$, where $\epsilon_i \equiv Y_i - \mu$, $i = 1, \ldots, n$. Now assume the ϵ_is to be Gaussian. Then it can be shown using the

Table 2.1 The Standardized (Mean − Median) Difference u

Row	1	2	3	4	5	6	7	8	9
u	−1.54	−0.40	6.12	−0.45	0.35	2.01	−0.56	−0.07	0.63
Row	10	11	12	13	14	15	16	17	18
u	−0.18	2.12	0.80	0.46	0.78	0.10	−1.05	−0.60	1.05
Row	19	20	21	22	23				
u	0.18	0.35	0.25	1.33	2.47				
Column	1	2	3	4	5	6	7	8	9
u	1.11	−0.76	0.78	0.35	2.87	0.02	0.22	1.29	1.23
Column	10	11	12	13	14	15			
u	1.03	−0.58	3.17	−1.24	1.39	1.48			

Source: Cressie (1984a). Reprinted by permission of Kluwer Academic Publishers.

methods of Section 3.5.4 that

$$\operatorname{var}(\bar{Y} - \tilde{Y}) = \frac{\sigma^2}{n}\left(\frac{\pi}{2} - 1\right). \tag{2.2.3}$$

Thus,

$$u \equiv n^{1/2}(\bar{Y} - \tilde{Y})/(0.7555\hat{\sigma}) \tag{2.2.4}$$

standardizes the (mean − median) difference; values of $|u|$ around 3 or greater warrant attention. In the spirit of EDA, use

$$\hat{\sigma} = (\text{interquartile range})/(2 \times 0.6745)$$

as a resistant measure of the standard deviation of the Y_is. (The interquartile range is the difference between the upper quartile and the lower quartile of $Y_1, \ldots, Y_n$.) Table 2.1 shows values for u and highlights columns 5 and 12 and row 3. This brings the data $z(5,6) = 17.61\%$ and $z(7,3) = 12.65\%$ to attention, and possibly $z(12,23) = 11.86\%$.

When there are positive covariances between the Ys, (2.2.3) is an underestimate of the true variance of $\bar{Y} - \tilde{Y}$ (Section 3.5.4). Hence, (2.2.4) is really a liberal diagnostic in that it highlights *more* atypical rows and columns than it should.

Another way of detecting atypical observations also relies on the belief of local stationarity. Figures 2.5 and 2.6 show bivariate plots of $Z(\mathbf{s})$ and $Z(\mathbf{s} + h\mathbf{e})$ as $\mathbf{s}$ varies over the data locations: Figure 2.5 shows the plot for $h = 1$ with $\mathbf{e}$ a unit-length vector in the east–west direction and Figure 2.6 shows the plot for $h = 1$ with $\mathbf{e}$ a unit-length vector in the north–south direction. The outlier $z(5,6) = 17.61$ is blatantly obvious, and other values such as $z(7,3) = 12.65$, $z(8,6) = 13.06$, $z(6,8) = 13.07$, $z(3,13) = 12.50$, and $z(5,19) = 12.80$ seem to be atypical, given their surrounding values.

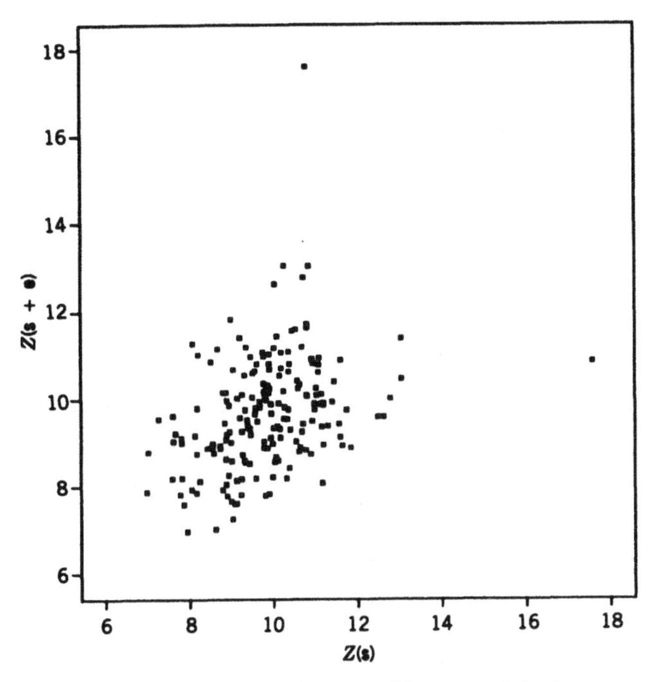

Figure 2.5 Bivariate scatter plot of $Z(\mathbf{s} + \mathbf{e})$ versus $Z(\mathbf{s})$, where $\mathbf{e}$ is in the east–west direction. Units on both axes are % coal ash. [*Source*: Cressie, 1984a]. Reprinted by permission of Kluwer Academic Publishers.

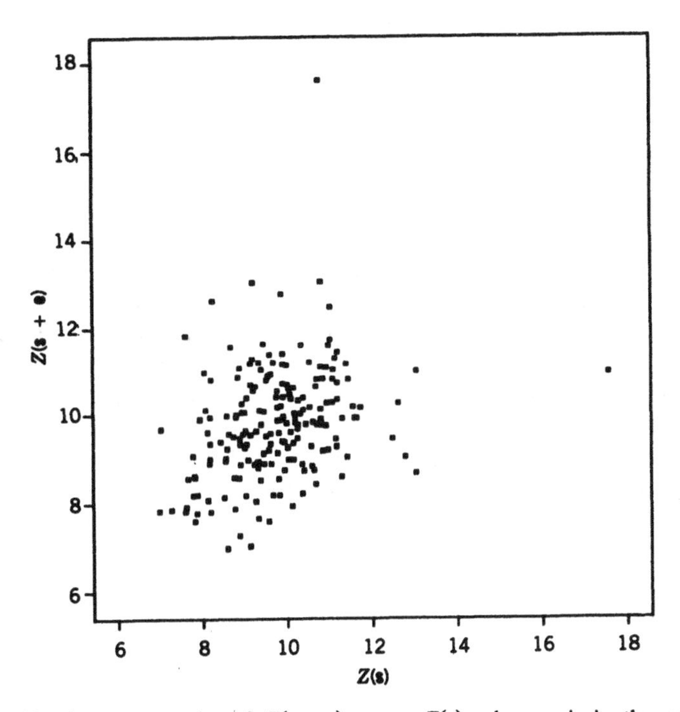

Figure 2.6 Bivariate scatter plot of $Z(\mathbf{s} + \mathbf{e})$ versus $Z(\mathbf{s})$, where $\mathbf{e}$ is in the north–south direction. Units on both axes are % coal ash. [*Source*: Cressie, 1984a]. Reprinted by permission of Kluwer Academic Publishers.

A set of possibly troublesome observations is gradually being accumulated, although an explanation of their "noncompliance" may be global nonstationarity present in the mean of the underlying Gaussian distribution. Whether they do in fact need special attention will ultimately be determined by their influence in variogram estimation (Section 2.4.3) and kriging (Section 3.3).

2.2.1† Intrinsic Stationarity

It is time now to introduce a statistical model and its parameters. Suppose that the grade of an ore body at a point $\mathbf{s}$ (here in $\mathbb{R}^2$) is the realization of a random process $\{Z(\mathbf{s}): \mathbf{s} \in D\}$ and that this is observed at certain points $\{\mathbf{s}_i: i = 1, \ldots, n\}$ (here, an approximately square grid) of the ore body. Then *intrinsic stationarity* is defined through first differences:

$$E(Z(\mathbf{s} + \mathbf{h}) - Z(\mathbf{s})) = 0, \tag{2.2.5}$$

$$\operatorname{var}(Z(\mathbf{s} + \mathbf{h}) - Z(\mathbf{s})) = 2\gamma(\mathbf{h}). \tag{2.2.6}$$

The quantity $2\gamma(\mathbf{h})$ is known as the variogram and is the crucial parameter of geostatistics; for more details, see Matheron (1963b) and Section 2.3. The classical estimator of the variogram proposed by Matheron (1962) is

$$2\hat{\gamma}(\mathbf{h}) \equiv \frac{1}{|N(\mathbf{h})|} \sum_{N(\mathbf{h})} \left(Z(\mathbf{s}_i) - Z(\mathbf{s}_j)\right)^2, \tag{2.2.7}$$

where the sum is over $N(\mathbf{h}) \equiv \{(i, j): \mathbf{s}_i - \mathbf{s}_j = \mathbf{h}\}$ and $|N(\mathbf{h})|$ is the number of distinct elements of $N(\mathbf{h})$. It is unbiased; however, it possesses very poor resistance properties. It is badly affected by atypical observations due to the $(\cdot)^2$ term in the summand of (2.2.7), and its cousin, the variogram cloud (see Figure 2.7), confounds skewness with atypical behavior.

Cressie and Hawkins (1980) present a more robust approach to the estimation of the variogram by transforming the problem to location estimation for an approximately symmetric distribution (Section 2.4.3). Now, for a Gaussian process $Z(\cdot)$, $(Z(\mathbf{s} + \mathbf{h}) - Z(\mathbf{s}))^2$ is distributed as $2\gamma(\mathbf{h}) \cdot \chi_1^2$, where χ_1^2 is a chi-squared random variable on 1 degree of freedom. Thus, $2\gamma(\mathbf{h})$ is the first moment of a *highly skewed* random variable. Using the set of power transformations proposed by Box and Cox (1964), Cressie and Hawkins (1980) found that the fourth-root of χ_1^2 has a skewness of 0.08 and a kurtosis of 2.48 (compared with 0 and 3 for the Gaussian distribution). Estimates of location, such as the mean and the median, can then be applied to the $|N(\mathbf{h})|$ *transformed differences* $\{|Z(\mathbf{s}_i) - Z(\mathbf{s}_j)|^{1/2}: (i, j) \in N(\mathbf{h})\}$. Finally, these estimates are raised to the fourth power, to bring them back to the correct scale, and adjusted for bias. This results in the variogram estimator

$$2\bar{\gamma}(\mathbf{h}) \equiv \left\{\frac{1}{|N(\mathbf{h})|} \sum_{N(\mathbf{h})} |Z(\mathbf{s}_i) - Z(\mathbf{s}_j)|^{1/2}\right\}^4 \Big/ \left(0.457 + \frac{0.494}{|N(\mathbf{h})|}\right). \tag{2.2.8}$$

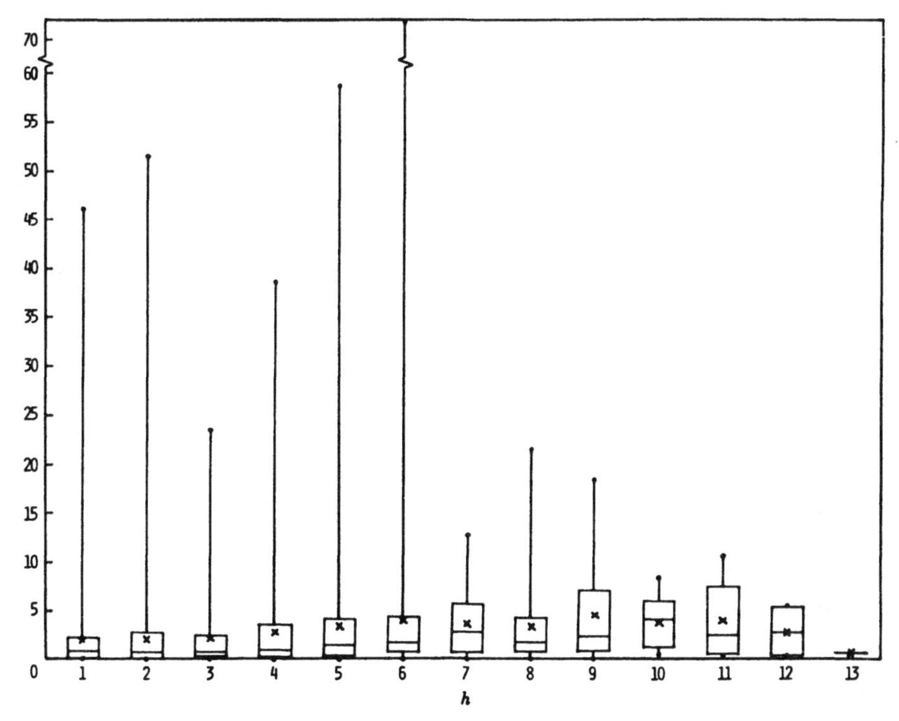

Figure 2.7 Box plots of the variogram cloud in the east–west direction. Units on the vertical axis are (% coal ash)^2. [*Source*: Cressie, 1984a]. Reprinted by permission of Kluwer Academic Publishers.

The properties of this estimator are compared to those of the classical estimator in Hawkins and Cressie (1984); see also Section 2.4.3.

2.2.2† Square-Root-Differences Cloud

Consider the coal-ash data and, for illustrative purposes, suppose that differences are taken in the east–west direction. It has been proposed by Chauvet (1982) that the variogram cloud, cousin to (2.2.7), is a useful diagnostic tool. I propose that a *square-root-differences cloud*, cousin to (2.2.8), is a much more powerful exploratory tool. The variogram cloud in direction $\mathbf{e}$ is simply an *x*-*y* plot of the *y* values $\{(Z(\mathbf{s}_i + h\mathbf{e}) - Z(\mathbf{s}_i))^2: \mathbf{s}_i + h\mathbf{e} = \mathbf{s}_j, (i, j) \in N(h\mathbf{e})\}$ at the *x* value h, as h varies. Figure 2.7 shows box plots (see, e.g., Velleman and Hoaglin, 1981, Chapter 3, for an explanation of the box plot) of the cloud as h varies. Figure 2.8 shows the square-root-differences cloud as an *x*-*y* plot obtained by using $\{|Z(\mathbf{s}_i + h\mathbf{e}) - Z(\mathbf{s}_i)|^{1/2}: \mathbf{s}_i + h\mathbf{e} = \mathbf{s}_j, (i, j) \in N(h\mathbf{e})\}$ as *y* values, at the *x* value h, $h = 1, 2, \ldots, 13$.

Compare these two plots. It is well known what the box plot of a roughly *symmetric* batch of numbers should look like (also the mean should be

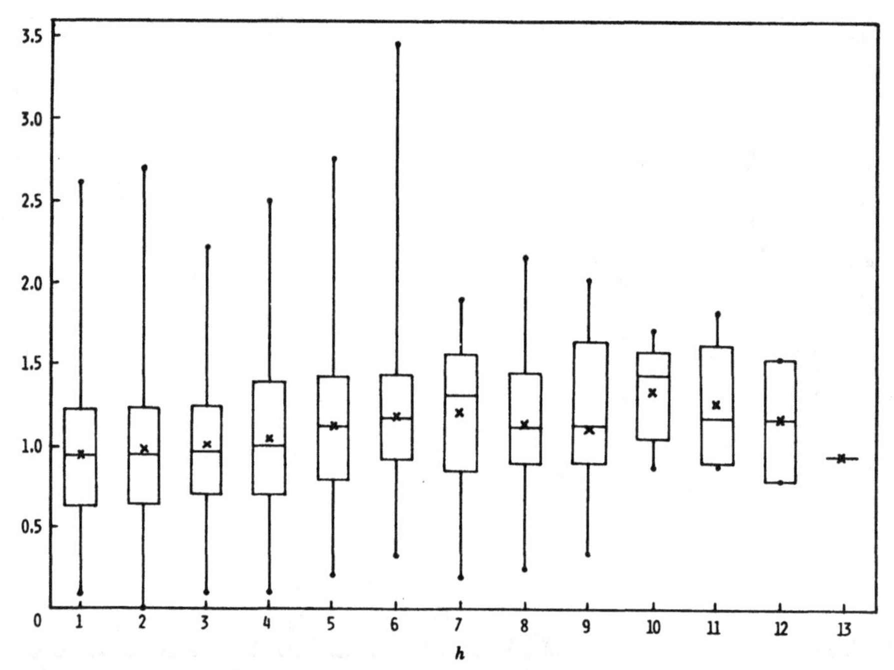

Figure 2.8 Box plots of the square-root-differences cloud in the east–west direction. Units on the vertical axis are (% coal ash)$^{1/2}$. [*Source*: Cressie, 1984a]. Reprinted by permission of Kluwer Academic Publishers.

approximately equal to the median). Therefore, any departures from this are easy to detect by eye. On the contrary, when the batches are inherently skewed, as they are in the variogram cloud, it is difficult to gauge whether a large value is the result of the skewness or of an atypical observation. It is also clear now, when comparing Figures 2.7 and 2.8, how the square-root differences lead to a more resistant variogram estimator; compare the influence of the outlier 17.61 for $h = 6$ in the variogram cloud (Figure 2.7) and in the square-root-differences cloud (Figure 2.8).

2.2.3† The Pocket Plot

The data analytic techniques presented thus far have been successful at detecting either gross trends or isolated outliers. A technique is needed that will identify a localized area as being atypical with respect to a stationary model. As before, this is done by exploiting the spatial nature of the data through row and column coordinates.

A primary aim of the geostatistician is to estimate the spatial relationship (i.e., variogram) between data points. Then, this estimate is used to predict the grade of mining blocks and estimate the predictor's variability. Although (2.2.8) offers a robust approach to variogram estimation, there is still the concern that a nonnegligible fraction of differences $\{Z(\mathbf{s}_i + h\mathbf{e}) - Z(\mathbf{s}_i)\}$ may be inappropriate in the estimation of $2\gamma(h\mathbf{e})$. Locations on the grid that exhibit measurements different from the rest need to be identified. These pockets of nonstationarity, once discovered, may be removed from variogram estimation, but of course must eventually be modeled and incorporated into final resource appraisals. The *pocket plot* (so-called because of its use in detecting pockets of nonstationarity) is a simple idea that I shall illustrate on north–south differences of the coal-ash data. Concentrate on row j of the grid. For any other row, k say, there are a certain number (N_{jk} say) of data differences defined, whose locations are distance $h \equiv |j - k|$ apart in the north–south direction. Let $\bar{Y}_{jk}$ denote the mean of these $|\text{difference}|^{1/2}$, averaged over the N_{jk} terms, and define [see (2.2.8)]

$$\bar{\bar{Y}}_h \equiv \frac{1}{|N(h\mathbf{e})|} \sum_{N(h\mathbf{e})} |Z(\mathbf{s}_i + h\mathbf{e}) - Z(\mathbf{s}_i)|^{1/2},$$

where $\mathbf{e}$ is the unit vector in the north–south direction. Alternatively, $\bar{\bar{Y}}_h$ is a weighted mean of the $\bar{Y}_{jk}$s such that $|j - k| = h$. Then define

$$P_{jk} \equiv \bar{Y}_{jk} - \bar{\bar{Y}}_h; \tag{2.2.9}$$

$\{P_{jk}\colon k = 1,2,\ldots\}$ is just the residual contribution of row j to the (fourth-root-scale) variogram estimator at differing lags. Ideally, these points will be scattered either side of zero, but if there is something unusual about row j, then it will give an unusual contribution at all lags and will typically show a scatter of points above the zero level. Now vary the row j and put the point scatters beside each other; this forms the *pocket plot*, illustrated in Figure 2.9*a*, where the central part of the scatter is replaced by the box of a box plot (e.g., Velleman and Hoaglin, 1981, Chapter 3). The numbers beside the point scatters denote the row numbers k corresponding to the observations $\{P_{jk}\}$.

Clearly rows 2, 6, and 8 are atypical in that the values $\{P_{2,k}\}$, $\{P_{6,k}\}$, and $\{P_{8,k}\}$ are scattered about a level above zero. Notice that the second, sixth, and eighth rows are also identified as extreme in the scatter of points $\{P_{jk}\colon k = 1,2,\ldots\}$, for most j. This serves as verification that indeed rows 2, 6, and 8 are potentially problematic. Trouble with row 6 could be expected because of the two values $z(5,6) = 17.61$ and $z(8,6) = 13.06$ and trouble with row 8 could be expected because of the value $z(6,8) = 13.07$, *but trouble with row 2 is a surprise*. Looking at row 2 one realizes that its observations are consistently lower than anything around (see the row summary in Fig. 2.4), and so there appears to be a pocket of nonstationarity. That we have discovered more interesting features in the data with the pocket plot is a good recommendation for its routine use. But the added bonus is that, at least as far as estimating variograms is concerned, we do not have to worry

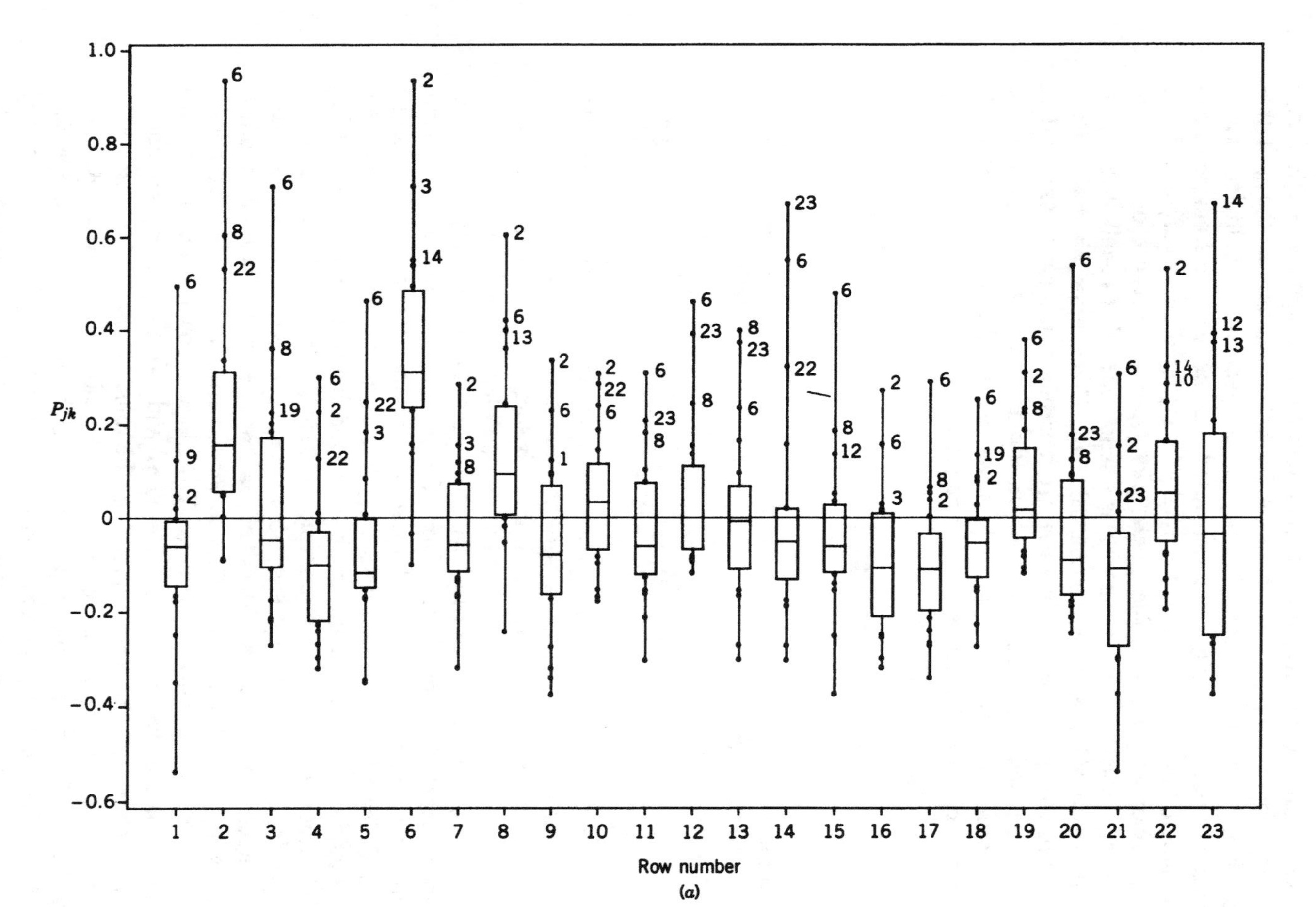

Figure 2.9 (*a*) Pocket plot in the north–south direction. Units on the vertical axis are (% coal ash)$^{1/2}$. [*Source*: Cressie, 1984a]. Reprinted by permission of Kluwer Academic Publishers.

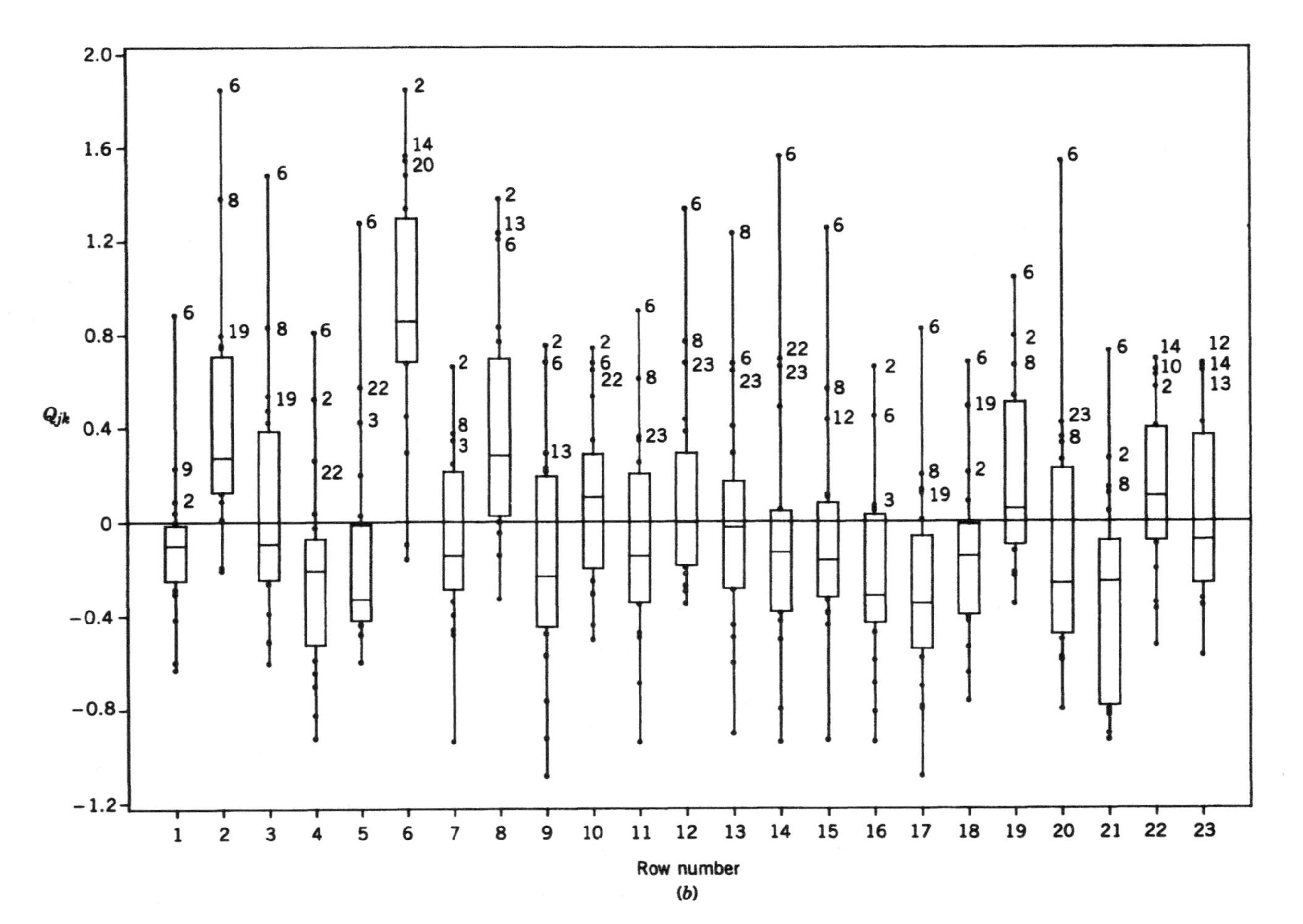

Figure 2.9 (*b*) Variance-standardized pocket plot in the north–south direction.

about other previously identified suspicious values, such as $z(7,2) = 12.65$, $z(3,13) = 12.50$, and $z(5,19) = 12.80$. These data may need further attention when kriging; see Section 3.3.

A further modification of the pocket plot would be to plot *normalized* values of P_{jk}, to ensure an approximately constant variance across k (for fixed j); thus plot instead

$$Q_{jk} \equiv N_{jk}^{1/2}\left\{\left(\bar{Y}_{jk}/\bar{\bar{Y}}_h\right) - 1\right\}.$$

From results in Cressie (1985a), $\mathrm{var}(P_{jk}) \simeq (2\gamma(h))^{1/2}/N_{jk}$, which justifies computing Q_{jk}. This modification will only affect the spread of the points and not the overall picture of being above the zero level, as is illustrated in Figure 2.9*b*.

So far, it has been assumed that data or differences of data come from a process that is stationary, apart from some small atypical pockets. It is unrealistic to expect all data to behave this way; the next subsections will be concerned with trying to find stationary components in spatial data, where the large-scale (mean) effects are nonstationary.

2.2.4† Decomposing the Data into Large- and Small-Scale Variation

The idea of analyzing gridded data as if they were a two-way (or higher-way) table is not new. Miller and Kahn (1962, p. 411) propose a formal two-way analysis of variance and claim to test for nonstationarity by performing the within-rows and within-columns F tests. Unfortunately, the F tests are invalid because the data are in general correlated; however, underlying the two-way analysis of variance is an additive decomposition,

$$\text{data} = \text{all} + \text{row} + \text{column} + \text{residual}, \tag{2.2.10}$$

which is extremely useful. When there are r rows and c columns, and each grid location has an observation Y_{ij}, this decomposition is

$$Y_{ij} = Y_{\cdot\cdot} + (Y_{i\cdot} - Y_{\cdot\cdot}) + (Y_{\cdot j} - Y_{\cdot\cdot}) + (Y_{ij} - Y_{i\cdot} - Y_{\cdot j} + Y_{\cdot\cdot}),$$

where a subscripted dot $(\cdot)$ denotes averaging over that subscript. The formulas are slightly more complicated when there are grid locations with observations missing, but the decomposition (2.2.10) remains simple. In the case of the coal-ash data, $r = 23$ and $c = 16$, although, as is shown in Figure 2.2*a*, there are many empty locations. The resistant way to effect (2.2.10) is not by means, but by medians. This leads to the EDA technique known as *median polishing* (Tukey, 1977; Emerson and Hoaglin, 1983). Put quite simply, the algorithm successively sweeps medians out of rows, then columns, then rows, then columns, and so on, accumulating them in "row," "column,"

																Row
									−0.20	0.00	3.29	0.00	0.51			0.00
						1.35	1.15	−1.32	−0.00	0.00	−1.05	0.64				0.10
				0.10	0.01	0.06	−0.20	0.56	0.00	−1.32	−0.85	−0.00	−1.00			0.12
			0.01	−0.40	0.89	1.74	0.03	0.00	0.11	−0.41	−0.25	−0.00	0.00			−0.71
			−0.00	2.38	−0.74	−1.06	−1.34	−1.22	0.00	−1.97	0.76	0.53	0.23	1.53		0.25
			0.90	−0.10	0.00	−1.80	0.34	0.26	−0.03	0.07	0.25	−1.69	−0.49	0.00		−0.18
		0.27	−0.46	0.00	0.10	−0.00	0.68	−0.64	−0.78	0.58	−0.41	0.56		−0.17	0.00	−0.33
0.53	−0.67	−0.14	−0.26	0.00	0.53	0.42	−0.88	0.16	0.51	−0.43						0.04
−0.54	0.19	1.77	−1.39	−0.15	−1.48	0.15	−0.18	0.72	0.41							−0.14
0.00	0.35	−0.18	−0.75	−0.00	0.25	0.00	−0.36	−0.37	3.03							−0.39
		2.74	−0.59	0.92	−0.13	−0.50	0.09	1.28	1.40	−0.89	−0.09					−0.27
	−0.09	1.78	0.38	2.05	0.65	−0.96	0.00	−1.15	−0.07	−1.24	0.09					−0.76
	0.93	−0.23	−0.73	1.43	−0.20	0.92	0.00	−0.31	1.73	0.00	0.60					−0.39
	0.00	−0.50	−0.70	0.00	0.71	−1.39	0.00	0.83	−1.07	1.92						0.38
		0.59	1.06	−0.59	0.28	−0.58	0.33	−1.03	−0.84	0.00						0.20
	−0.96	−2.00	0.35	−0.09	2.35	0.09	1.72	−0.72	−0.45	1.67						0.20
		−0.20	0.06	0.23	−0.25	−0.64	−1.53	0.63	0.04	0.00						0.56
		−0.39	−0.45	6.66	−0.42		2.62	0.66	0.39	−0.63						0.78
		0.14	1.01	1.03	−1.25	−0.29	−0.04	0.04	−0.19							−0.40
		0.89	0.44	−0.65	0.08	0.08	−0.08	−0.17	−0.99							0.01
			−0.39	−0.19	0.00	2.93	0.43									−0.46
			0.32	0.10	−0.04	−0.38	0.00									−1.52
			0.00	0.16	−1.30											0.10
0.78	0.95	0.21	0.67	0.35	0.70	0.36	−0.16	0.16	−1.03	−0.82	−1.25	−0.91	−0.34	−1.69	−0.42	9.82
							Column									All

Figure 2.10 Results from median polish of coal-ash data. Units are % coal ash. [*Source*: Cressie, 1984a]. Reprinted by permission of Kluwer Academic Publishers.

and "all" registers, and leaves behind the table of residuals. Crucially, at each step of the algorithm, relation (2.2.10) is preserved; see Section 3.5.1 for more details. Figure 2.10 shows the result of median polish on the coal-ash data; ambiguous medians (over an even number of observations) are resolved by averaging the middle two values.

Details on when median polish converges and when its solution minimizes the sum of absolute residuals [i.e., is an L_1 solution to (2.2.10)] can be found in Kemperman (1984) and Section 3.5.1. I am using it here as my *definition* of the large-scale variation; then the median-polish residuals constitute the small-scale variation.

A plot of the row values and the column values yielded a picture almost identical with Figure 2.4 (where the raw medians of rows and columns were considered), showing a definite trend in the east–west direction and very little trend in the north–south direction. But, most importantly, the table of residuals (and their spatial locations) is readily available and can be analyzed in its own right. Furthermore, this table is the result of a *resistant* analysis. Had the data been analyzed by a nonresistant *mean* polish, the analogous figure to Figure 2.10 would have shown the unusualness of the outliers diluted, causing them to appear (confusingly) elsewhere in the table (Emerson and Hoaglin, 1983, pp. 178, 179).

Although the data analyzed here are two dimensional, the median-polish idea carries over to three- and higher-dimensional spatial data on a regular grid. Cook (1985) explores three-way median polish.

One final comment is that this technique for removing trend is very dependent on grid orientation. What is special about the grid directions chosen for exploration drilling? How can one be sure that "all + row + column" captures all the nonstationarity (in the mean)? To check for possible interaction between row and column, one can allow for an extra (interaction) parameter g (Cressie, 1986, p. 629) as follows:

$$\begin{aligned}\text{data} &= \text{all} + \text{row} + \text{column} + g \cdot \{\text{row no.} - \text{ave(row no.)}\}\\ &\quad \{\text{column no.} - \text{ave(column no.)}\} + \text{residual}.\end{aligned}$$

A diagnostic plot of {data − all − row − column} versus {row no. − ave(row no.)}{column no. − ave(column no.)} for the coal-ash data indicated that $g = 0$, so that the purely additive decomposition of (2.2.10) is reasonable here.

It has long been thought that working with residuals introduces bias into variogram estimation. The next section discusses this interesting and complex issue.

2.2.5[†] Analysis of Residuals

I shall now compare estimation techniques to show that the residuals from (2.2.10) can have small bias. When the effective number of observations (Section 1.3) is small, the bias squared is typically much larger than the variance, so that it will dominate the mean-squared error (cf. density estimation).

The comparison of biases will be made for the special case of a process with constant mean and stationary covariance function $C(\mathbf{h}) \equiv \text{cov}(Z(\mathbf{s}+\mathbf{h}), Z(\mathbf{s}))$. Consider its estimator,

$$C^{\#}(\mathbf{h}) \equiv \frac{1}{|N(\mathbf{h})|} \sum_{N(\mathbf{h})} \left(Z(\mathbf{s}_i) - Z^{\#}\right)\left(Z(\mathbf{s}_j) - Z^{\#}\right), \qquad (2.2.11)$$

where $Z^{\#}$ is some estimator of the stationary mean of $Z(\cdot)$; see (2.2.7) for the necessary explanation of notation. More general results for polynomial trend and variogram estimation are presented in Section 3.4.3.

For simplicity, suppose that the n data are equally spaced on a transect. Then, for h fixed (see Section 3.5.4 for the details),

$$\begin{aligned} E(\bar{C}(h)) &= E\left\{\frac{1}{n-h}\sum_{s=1}^{n-h}(Z(s) - \bar{Z})(Z(s+h) - \bar{Z})\right\} \\ &= C(h) - \left\{C(0) + 2\sum_{k=1}^{m} C(k)\right\}/n + O(1/n^2), \qquad (2.2.12) \end{aligned}$$

where m is the order of dependence in $Z(\cdot)$ and $\bar{Z} \equiv (1/n)\sum_{s=1}^{n} Z(s)$. Notice that when the Cs are positive (as they are for the most widely used geostatistical models) the $O(1/n)$ bias is negative; moreover, it does not depend on h.

If a different estimator of the unknown constant mean were used, say $\tilde{Z} \equiv \text{median}\{Z(s): s = 1, \ldots, n\}$ for symmetric Zs, then a different bias may result from using the residuals $\{Z(s) - \tilde{Z}: s = 1, \ldots, n\}$ to estimate $C(h)$. Intuitively, the median-based estimator should have less of a bias problem because there is no longer a linear constraint on the residuals: The mean-based residuals satisfy $\sum_{s=1}^{n}(Z(s) - \bar{Z}) = 0$.

Now the estimator of $C(\cdot)$, based on the residuals $\{Z(s) - \tilde{Z}: s = 1, \ldots, n\}$, is

$$\tilde{C}(h) = \frac{1}{n-h}\sum_{s=1}^{n-h}(Z(s) - \tilde{Z})(Z(s+h) - \tilde{Z}). \qquad (2.2.13)$$

From Section 3.5.4, for $Z(\cdot)$ a Gaussian process,

$$\begin{aligned} E(\tilde{C}(h)) = C(h) - \Bigg\{&\frac{(4-\pi)}{2}C(0) + 4\sum_{k=1}^{m} C(k) \\ &- 2C(0)\sum_{k=1}^{m}\arcsin\left(\frac{C(k)}{C(0)}\right)\Bigg\}\Big/n + O\left(\frac{1}{n^2}\right). \qquad (2.2.14) \end{aligned}$$

Comparing (2.2.14) with (2.2.12) is illuminating. If the Cs are all positive (as they are for the most widely used geostatistical models) and because $x \le \arcsin(x) \le (\pi/2)x$, it follows that (Cressie and Glonek, 1984)

$$|O(1/n) \text{ bias of } (2.2.14)| \le \sum_{k=1}^{m} C(k) + \frac{(4-\pi)}{2} C(0)$$
$$< |O(1/n) \text{ bias of } (2.2.12)|.$$

A more general condition on the Cs is given in Section 3.5.4.

Thus, the estimator $\tilde{C}(\cdot)$, based on resistant estimation of the mean, shows less of a bias problem. Using medians not only ensures resistance to outliers, but also results in superior bias properties.

In the spatial case, analogous mean- and median-based methods of fitting are, respectively, analysis of variance and median polish. Results like (2.2.12) and (2.2.14) are not available, but it is expected that the lessons learned from dealing with transect data will also apply to data on higher-dimensional grids.

2.2.6† Variogram of Residuals from Median Polish

My recommendation now is to proceed with the median-polish residuals (given in Figure 2.10) as if they were a fresh data set; that is, repeat the exploratory data analytic steps outlined in Sections 2.2.1 through 2.2.3 and compute estimates of the variogram. Cressie and Hawkins' (1980) robust estimator of the variogram is designed to deal with atypical points *automatically*, rather than deleting them in some *ad hoc* way. Nevertheless, *both* the classical variogram estimator $2\hat{\gamma}$ [given by (2.2.7)] and the robust estimator $2\bar{\gamma}$ [given by (2.2.8)] were computed on the original data and on the median-polish residuals. The east–west direction was chosen for illustration because of the trend in that direction (see Figure 2.4).

Figure 2.11 shows the east–west robust and classical variogram estimators for the original data. The effect of the trend is plainly obvious, leading to a steadily increasing estimated variogram. However, when the residuals from median polish are used, most of the correlation apparent in Figure 2.11 seems to be due to the trend (see Figure 2.12 and Starks and Fang, 1982a). The robust variogram estimator typically gives lower values than the classical one, which is to be expected because $2\bar{\gamma}$ downweights potentially atypical observations (Cressie and Hawkins, 1980).

The next stage of the analysis, namely model fitting (Section 2.6), can be carried out with confidence after a thorough exploratory look at the data. An effort should also be made to check, in various ways, the fit of a proposed model to the data. Cross-validation is one such check that will be considered

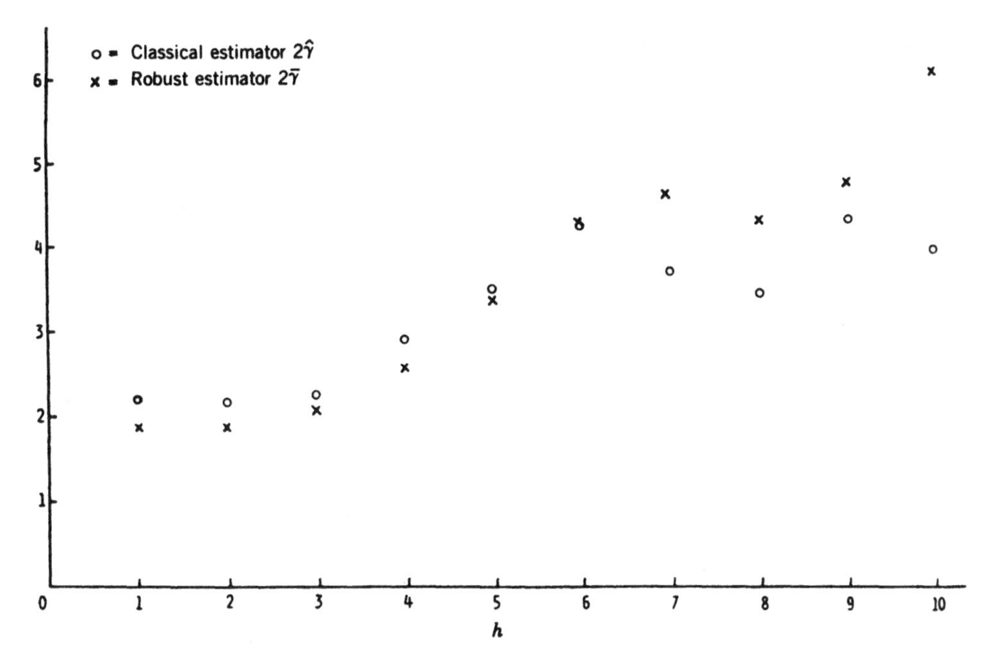

Figure 2.11 Estimated east–west variogram of coal-ash data. Units on the vertical axis are (% coal ash)^2; on the horizontal axis, 1 unit = 2500 ft. [*Source*: Cressie, 1984a]. Reprinted by permission of Kluwer Academic Publishers.

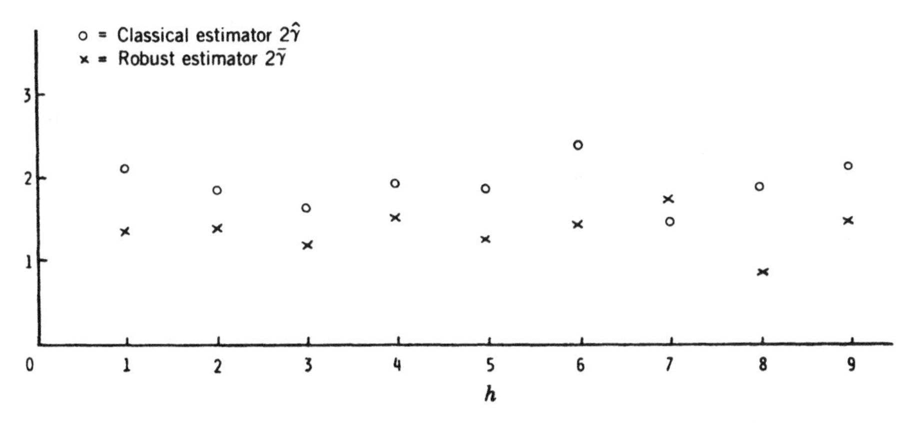

Figure 2.12 Estimated east–west variogram of coal-ash residuals (from median polish). Units are the same as in Figure 2.11. [*Source*: Cressie, 1984a]. Reprinted by permission of Kluwer Academic Publishers.

in Section 2.6.4. Finally, the nonidentifiability of the components of the decomposition

$$\text{data} = \text{large-scale variation} + \text{small-scale variation}$$

leads to many different possible models; Sections 3.2, 3.4, 3.5, and 5.4 explore the various options.

2.3* STATIONARY PROCESSES

The coal-ash data $\{z(x, y): x = 1, \ldots, 16; y = 1, \ldots, 23\}$ of Section 2.2 are a sample from a potentially infinite number of measurements $\{z(\mathbf{s}): \mathbf{s} \in D\}$ that could have been taken throughout the ore body D. Most of the modeling developed in this book is based on the assumption that $\{z(\mathbf{s}): \mathbf{s} \in D\}$ is a *realization* of a *random* process (i.e., random function or random field or stochastic process)

$$\{Z(\mathbf{s}): \mathbf{s} \in D\}, \qquad (2.3.1)$$

where D is a fixed subset of $\mathbb{R}^d$ with positive d-dimensional volume. To emphasize the source of randomness, (2.3.1) is sometimes written as $\{Z(\mathbf{s}; \omega): \mathbf{s} \in D; \omega \in \Omega\}$, where $(\Omega, \mathcal{F}, P)$ is a probability space; the realization $\{z(\mathbf{s}): \mathbf{s} \in D\}$ would correspond to a particular value of ω, say $\omega = \omega_0$. Under the assumption that the *random* process $\{Z(\mathbf{s}): \mathbf{s} \in D\}$ is stationary over D (see subsequent text for different notions of stationarity), preferential spatial clustering of data does *not* make estimators of random-process parameters unrepresentative. The model-based approach that I take throughout this book allows for inhomogeneities in the physical process, most simply through the mean of the random process.

Matheron (1962) calls the quantity $z(\cdot)$ a *regionalized variable* in order to emphasize the continuous spatial nature of the index set D. Indeed, there have been some attempts (Matheron, 1965; Journel, 1985) to build a prediction theory assuming the data $\{z(\mathbf{s}_i): i = 1, \ldots, n\}$ are a probability sampling from a fixed but unknown function $\{z(\mathbf{s}): \mathbf{s} \in D\}$. This idea is not developed any further here, although the interested reader might look at the discussion and a rejoinder following the article by Journel (1985). There is a parallel controversy in classical sampling theory, between model-based and design-based methods of inference; see Hansen et al. (1983) and the discussion following their article.

In problems encountered in Part I of this book, it is not typical to find data locations $\{\mathbf{s}_1, \ldots, \mathbf{s}_n\}$ on a regular grid nor is n often very large. Thus, most modeling of geostatistical data occurs in the spatial domain rather than the frequency domain. Nevertheless, some spectral theory is presented in Section 2.5.

The random process (2.3.1) is usually defined through the *finite-dimensional distributions*

$$F_{\mathbf{s}_1, \ldots, \mathbf{s}_m}(z_1, \ldots, z_m) \equiv P\{Z(\mathbf{s}_1) \le z_1, \ldots, Z(\mathbf{s}_m) \le z_m\}, \qquad m \ge 1,$$

which must satisfy Kolmogorov's conditions of symmetry (i.e., F remains invariant when z_j and $\mathbf{s}_j$ are subjected to the same permutation) and consistency [i.e., $F_{\mathbf{s}_1, \ldots, \mathbf{s}_{k+l}}(z_1, \ldots, z_k, \infty, \ldots, \infty) = F_{\mathbf{s}_1, \ldots, \mathbf{s}_k}(z_1, \ldots, z_k)$]. Furthermore, (2.3.1) is said to be *separable* if there exists a countable set D_0

(e.g., all vectors in $\mathbb{R}^d$ with rational coordinates) such that, for any closed interval B and any open d-dimensional cube $A \subset D$, the two sets $\{\omega: Z(\mathbf{s}; \omega) \in B;\ \mathbf{s} \in A\}$ and $\{\omega: Z(\mathbf{s}; \omega) \in B;\ \mathbf{s} \in A \cap D_0\}$ differ only on a set of probability 0. In this book, Z will usually have continuous sample surfaces and, when possible, a separable version will be chosen to model the data. Adler (1981) is a good source for these theoretical issues.

Suppose $\mu(\mathbf{s}) \equiv E(Z(\mathbf{s}))$ exists for all $\mathbf{s} \in D$; call $\mu(\cdot)$ the *trend* (sometimes called drift, but not in this book). The existence of $\mathrm{var}(Z(\mathbf{s}))$ for all $\mathbf{s} \in D$ allows definition of second-order stationary and intrinsically stationary processes.

Second-Order Stationarity

Statistically speaking, some further assumptions about Z have to be made. Otherwise, the data represent an *incomplete* sampling of a *single* realization, making inference impossible. For example, one might assume that

$$E(Z(\mathbf{s})) = \mu, \quad \text{for all } \mathbf{s} \in D, \tag{2.3.2}$$

or that $F_{\mathbf{s}}(z) \equiv \Pr(Z(\mathbf{s}) \leq z)$ does not depend on $\mathbf{s}$ (an *invariant* distribution for Z is then said to exist). In order to *estimate* optimal linear predictors (Chapter 3), an additional assumption, such as

$$\mathrm{cov}(Z(\mathbf{s}_1), Z(\mathbf{s}_2)) = C(\mathbf{s}_1 - \mathbf{s}_2), \quad \text{for all } \mathbf{s}_1, \mathbf{s}_2 \in D, \tag{2.3.3}$$

is needed, although even this is sometimes not enough (see subsequent text). The function $C(\cdot)$ is called a *covariogram* or a stationary covariance function.

Definition. A random function $Z(\cdot)$ satisfying (2.3.2) and (2.3.3) is defined to be *second-order* (or weak or wide-sense) *stationary*. Furthermore, if $C(\mathbf{s}_1 - \mathbf{s}_2)$ is a function only of $\|\mathbf{s}_1 - \mathbf{s}_2\|$, then $C(\cdot)$ is called *isotropic*. ■

There is a stronger type of stationarity, called strong (or strict) stationarity, that is defined via the conditions $F_{\mathbf{s}_1+\mathbf{h},\ldots,\mathbf{s}_m+\mathbf{h}}(z_1, \ldots, z_m) \equiv F_{\mathbf{s}_1,\ldots,\mathbf{s}_m}(z_1, \ldots, z_m)$, for all $m \geq 1$ and all $\mathbf{h} \in \mathbb{R}^d$. Of course, strong stationarity implies second-order stationarity whenever $F_{\mathbf{s}}(z)$ yields a finite second moment.

Ergodicity

There is a subset of the second-order stationary random functions that possesses a crucial property known as *ergodicity*. Its definition (e.g., Adler, 1981, p. 143, and subsequent text here) does not betray its importance but, simply put, this property allows expectations over the Ω space to be estimated by spatial averages.

To gain some understanding of ergodicity, consider for the moment the simpler setting of data regularly sequenced in time. Let $Z(1), Z(2), Z(3), \ldots$ be an infinite time series of random variables, from which n pieces of data $z(1), \ldots, z(n)$ are observed. In the absence of other information from similarly observed series, a statistician hoping to make inferences about parameters of the time series, from the n data, has to make some assumptions about the series. One assumption that is frequently made, yet not well understood, is that of ergodicity. Blum (1982) is a good source for clarification and explanation of this important concept.

An *ergodic* (from the Greek for wandering) assumption allows for consistent estimation of the joint probability law of various subsets $Z(t_1), \ldots, Z(t_p)$ of the time series. Hence, from observations $Z(n-p+1), \ldots, Z(n)$, prediction of $Z(n+1)$ from *estimated* joint covariation laws can proceed in a straightforward manner. Moreover, mean-squared errors of prediction can be estimated consistently under an ergodic assumption. Thus, without ergodicity, the prediction equations based on an estimated covariogram simply yield a prediction algorithm having no particular *statistical* optimality properties.

In order to talk sensibly about "wandering" time series, it is necessary to introduce a *translation* operator T acting on the time series $Z(\cdot) \equiv \{Z(t): t = 1, 2, \ldots\}$;

$$T(\{Z(t): t = 1, 2, \ldots\}) = \{Y(t): t = 1, 2, \ldots\},$$

where $Y(t) \equiv Z(t+1)$, $t = 1, 2, \ldots$. When the probabilistic properties of $T^k(Z(\cdot))$ are the same for all nonnegative integers k, then $Z(\cdot)$ is said to be (strongly) *stationary*; $T^k(\cdot)$ is interpreted as $T(T(\cdots T(\cdot)\cdots))$.

Contained within the class of stationary time series is the class of ergodic time series, that I shall now define and describe. Let P denote the probability measure for the time series $Z(\cdot)$. Let A be a measurable set (e.g., $A = \{z(\cdot): z(1) \le z_1, \ldots, z(p) \le z_p\}$) and define $T^{-1}(A) = \{b: T(b) \in A\}$. Then $Z(\cdot)$ is stationary if and only if $P(T^{-1}(A)) = P(A)$, for every measurable set A. Finally, a stationary time series is defined to be *ergodic* if $T^{-1}(A) = A$ (i.e., A is *invariant* with respect to T) implies $P(A) = 0$ or $P(A) = 1$.

But what does this mean? Forget for the moment about the invariant sets for which $P(A) = 0$; they appear in the definition for the same reason probabilists make statements only "almost surely." Then, the definition says that almost every realization of the time series $\{Z(t): t = 1, 2, \ldots\}$, when successively translated, completely fills up the space of possible trajectories. Therefore, an ergodic time series is quite irregular because the future holds in reserve any type of behavior allowable under its probability measure P. Now the adjective *ergodic* (wandering) makes sense.

There are a number of consequences of this definition. First, T mixes the space of realizations in the sense that if $P(A) > 0$, then there exists a positive integer k such that $P(T^k(A) \cap B) > 0$; that is, any realization of the time series belonging to A will eventually hit *every* set B of positive measure. In fact, a necessary and sufficient condition for a stationary time series to be ergodic is

$$\lim_{n\to\infty}(1/n)\sum_{j=1}^{n}P(T^j(A)\cap B) = P(A)P(B);$$

hence, for $P(A) > 0$, $P(B) > 0$, it follows that $P(T^k(A) \cap B) > 0$ for *infinitely many* k. Other mixing properties (e.g., Ibragimov and Linnik, 1971) are

weak mixing:

$$\lim_{n\to\infty}(1/n)\sum_{j=1}^{n}|P(T^j(A)\cap B) - P(A)P(B)| = 0;$$

strong mixing: $$\lim_{n\to\infty}P(T^n(A)\cap B) = P(A)P(B).$$

All these mixing conditions are versions of asymptotic independence: The class of strong-mixing time series is contained within the class of weak-mixing time series, which is contained within the class of ergodic time series (which in turn is contained within the class of strongly stationary time series).

Suppose g is a measurable function that is integrable; that is, $\int g\,dP$ exists [e.g., $g = I_A$; then $\int g\,dP = P(A)$].

Ergodic Theorem (Birkhoff, 1931) Let $\{Z(t)\colon t = 1,2,\ldots\}$ be an ergodic time series and suppose that the measurable function g is integrable. Then, for almost every realization $\{z(t)\colon t = 1,2,\ldots\}$,

$$\lim_{n\to\infty}(1/n)\sum_{j=0}^{n-1}g(T^j(\{z(t)\colon t = 1,2,\ldots\})) = \int g\,dP. \qquad \blacksquare$$

Using the Ergodic Theorem

Time-series analysts implicitly make use of this theorem right from the outset of their analysis. Suppose they plot an empirical distribution function (or, equivalently, a histogram) of their data and, based on this picture, they declare, say, a lognormal distribution. Implicitly they are choosing $g(z(\cdot)) = I_{\{z(1)\le z\}}$, the indicator function equal to 1 when $z(1) \le z$ and equal to 0 otherwise. The empirical distribution function formed from data $\{z(t)\colon t = 1,\ldots,n\}$ is

$$F_n(z) = (1/n)\sum_{t=1}^{n}I_{\{z(t)\le z\}},$$

which, according to the Ergodic Theorem, converges to $P(Z(1) \le z)$, the distribution function of any individual member of the stationary ergodic time series. Furthermore, and crucial for covariogram estimation, the choice of both $g_1(z(\cdot)) = (z(h+1) - \mu)(z(1) - \mu)$ and $g_2(z(\cdot)) = z(1)$ in the Ergodic Theorem justifies

$$\hat{C}(h) \equiv \sum_{t=1}^{n-h} (Z(t+h) - \bar{Z})(Z(t) - \bar{Z})/(n-h)$$

as an estimator of the lag-h covariogram $C(h) \equiv \mathrm{cov}(Z(t+h), Z(t))$. The sample mean $\bar{Z}$ is an estimator of the mean μ of the stationary time series.

Choice of $g(z(\cdot)) = (z(h+1) - z(1))^2$ in the Ergodic Theorem justifies

$$2\hat{\gamma}(h) \equiv \sum_{t=1}^{n-h} (Z(t+h) - Z(t))^2/(n-h),$$

as an estimator of the lag-h variogram $2\gamma(h) \equiv \mathrm{var}(Z(t+h) - Z(t))$. The variogram is a less familiar parameter to time-series analysts than the covariogram, but it will be shown in this section to have some attractive properties.

Put succinctly, the Ergodic Theorem says that, for any ergodic time series, time averages over one realization approximate expectations over the Ω space. Notice that ergodicity is an *assumption* of the theorem.

Examples of Ergodic Time Series

The most obvious ergodic time series occurs for $Z(1), Z(2), \ldots,$ independent and identically distributed. But notice that if $W(\cdot)$ is defined as

$$W(t) = \mu t + Z(t),$$

then $W(\cdot)$ is not ergodic (it is not stationary). Stationary time series that are m-dependent [i.e., $Z(j)$ and $Z(j+m+1)$ are independent], weak mixing, or strong mixing are all ergodic.

A stationary *Gaussian* time series is characterized by its mean μ and its covariogram $C(h) = \mathrm{cov}(Z(t+h), Z(t))$. A sufficient condition for the series to be ergodic is $\lim_{h \to \infty} C(h) = 0$ (e.g., Blum, 1982).

An example of a nonergodic but stationary time series (that is not necessarily Gaussian) is $U(t) = A$, $t = 1, \ldots,$ where A is a random variable. Another example is $U(t) = A \cdot \cos(wt + \Phi)$, $t = 1, \ldots,$ where w is a fixed constant, A is a random variable, and Φ is an independent uniform random variable on $(-\pi, \pi]$. A single realization of either of these time series gives no information on the probability law of A, and hence nothing can be gleaned about, say, $P(U(1) \leq z)$, from a time average based on $u(1), \ldots, u(n)$.

Checking for Ergodicity

Sufficient conditions for ergodicity are given in terms of the joint probability laws of the time series. For example, it has been mentioned already that if

$\lim_{h\to\infty} C(h) = 0$, then a stationary Gaussian time series is ergodic. To make use of this result, one might take data $\{z(t): t = 1,\ldots,n\}$ that are thought to come from a Gaussian time series and estimate C with $\hat{C}$. Then, by graphing $\hat{C}(h)$ versus h, a rough "eyeball" check could be made to determine whether $\hat{C}(h)$ approaches zero as h becomes large, and consequently whether the ergodic assumption is reasonable. This graphical check can be comforting, but it is not logically tenable. In order for $\hat{C}(h)$ to be a consistent estimator of $C(h)$, one must first make an assumption of ergodicity in the Ergodic Theorem.

Ergodicity is an assumption made to allow inference to proceed for a series of nonindependent observations. It might only be verifiable in the sense that one fails to reject it. This should not be too worrisome because scientific discovery generally proceeds in this way.

Ergodicity and Second-Order Stationarity

Ergodic time series are strongly stationary. Often only second-order stationarity for $\{Z(t): t = 1,2,\ldots\}$ is assumed; that is, $E(Z(t))$ is constant and $\text{cov}(Z(t+h), Z(t))$ is a function only of h. But, if a second-order stationary time series is also assumed to be ergodic, then it *has* to be strongly stationary. In other words, by working with ergodic processes there is no extra generality in assuming only second-order stationarity (Cressie, 1989b). It seems that statisticians are using only the part of the ergodicity assumption that guarantees the sample mean and covariances converge to their population counterparts. Gardiner (1983, Section 3.7.1) formalizes this into a notion of ergodicity in mean and ergodicity in covariance by specifying that the sample quantities converge in L_2. [The sequence of random variables $\{X_n\}$ is said to converge to X in L_2 if $E(X_n - X)^2 \to 0$, as $n \to \infty$.] He also gives sufficient conditions for convergence that depend on fourth-order moments of the process.

Mixing Ergodicity and Geostatistics

Geostatisticians are concerned with spatial processes, not time series. Generalization of the notion of ergodicity to such processes in $\mathbb{R}^d$ can be found, for example, in Nguyen (1979), Adler (1981), or Rosenblatt (1985). There, the time series definition is modified to processes with continuous spatial index. Data $\{z(1),\ldots,z(n)\}$ are replaced with data $\{z(\mathbf{s}): 0 \le \|\mathbf{s}\| \le H\}$ and H is allowed to tend to infinity. But, more realistically, geostatisticians deal with a *finite sample* $\{z(\mathbf{s}_1),\ldots,z(\mathbf{s}_n)\}$ of a realization $\{z(\mathbf{s}): \mathbf{s} \in D\}$ of a spatial process. No one to my knowledge has generalized ergodicity to this case of a finite sample embedded in a continuous realization. In $\mathbb{R}^1$, Gaposhkin (1988) shows that ergodicity in mean of a continuous process does not imply ergodicity in mean of a consequent discrete process obtained by regular sampling (nor conversely). In such situations, it appears that something more has to be assumed, like strong mixing at a very small scale of spacing between observations or asymptotics that choose more and more sample points in a finite region (called infill asymptotics in Section 5.8).

In the time-series setting it was seen that, although statisticians are assuming ergodicity, they might as well make the weaker ergodic assumption that $\bar{Z}$ and $\hat{C}(h)$ [or $2\hat{\gamma}(h)$] converge in L_2 to μ and $C(h)$ [or $2\gamma(h)$]. The same recommendation is made in the geostatistics setting, with the added complication that the sample $\mathbf{Z}$ consists of a finite number of observations from a continuous process $Z(\cdot)$. It is an open problem to find a convenient condition, say, on the moments of $Z(\cdot)$, that implies this weaker ergodic assumption is satisfied.

Gaussian Processes

When the random processes are Gaussian, second-order stationarity and strong stationarity coincide, because a Gaussian process is characterized by its mean and its covariance function. A sufficient condition for ergodicity is $C(\mathbf{h}) \to 0$, as $\|\mathbf{h}\| \to \infty$ (Adler, 1981, p. 145), and the various forms of mixing are characterized by the rate of this convergence to zero. Gaussian processes are important for two reasons. The first is a pragmatic reason that recognizes that, upon assumption of a Gaussian process, virtually all prediction, estimation, and distribution theory are considerably easier. The second reason comes from asymptotic considerations where the net result of many small-order (possibly non-Gaussian) effects is approximately Gaussian (central limit theorem; see, e.g., Lindgren, 1976, p. 157).

In Chapter 3, it will become apparent that spatial prediction (kriging) can be carried out under slightly weaker assumptions than second-order stationarity. In fact, random processes whose *increments* are second-order stationary (e.g., Yaglom, 1962, pp. 86–93) provide a very useful class for prediction purposes: It is not necessary to estimate the unknown $\mu \equiv E(Z(\mathbf{s}))$, and the class actually contains all second-order stationary random functions. (This approach is generalized to higher-order increments in Section 5.4.)

2.3.1 Variogram

Suppose,

$$\operatorname{var}(Z(\mathbf{s}_1) - Z(\mathbf{s}_2)) = 2\gamma(\mathbf{s}_1 - \mathbf{s}_2), \quad \text{for all } \mathbf{s}_1, \mathbf{s}_2 \in D. \tag{2.3.4}$$

The quantity $2\gamma(\cdot)$, which is a function only of the increment $\mathbf{s}_1 - \mathbf{s}_2$, has been called a *variogram* [and $\gamma(\cdot)$ has been called a *semivariogram*] by Matheron (1962), although earlier appearances of it can be found in the scientific literature. Cressie (1988a) can be consulted for some of the details, but briefly it has been called a *structure function* by Yaglom (1957) in probability and by Gandin (1963) in meteorology, and a *mean-squared difference* by Jowett (1952) in time series. Kolmogorov (1941a) introduced it to study the local structure of turbulence in a fluid and Matern (1960, p. 51) makes incidental note of a Swedish forester, A. Langsaeter, who used this way of expressing variation when dealing with systematic sampling in forest surveys. Nevertheless, it has been Matheron's mining terminology that has persisted.

Throughout this book, the function 2γ (assuming it exists) will be treated as a *parameter* of the random process $Z(\cdot)$. Usually, parameters are real- or vector-valued and are restricted to belong to certain natural parameter spaces (e.g., a variance is always nonnegative, and a probability is always in the interval $[0, 1]$). The variogram, too, is restricted (to be conditionally negative-definite), as is demonstrated in Section 2.5.2.

It will be seen in Section 2.4 that averaged squared differences estimate the parameter $2\gamma(\cdot)$. However, for modeling (see following text) and kriging (Chapter 3), $\gamma(\cdot)$ is all that is needed. Therefore, some authors have called $\gamma(\cdot)$ a variogram. This is a dangerous practice; there is too much to lose from missing 2s.

Nugget Effect

Clearly, $\gamma(-\mathbf{h}) = \gamma(\mathbf{h})$ and $\gamma(\mathbf{0}) = 0$. If $\gamma(\mathbf{h}) \to c_0 > 0$, as $\mathbf{h} \to \mathbf{0}$, then c_0 has been called the *nugget effect* by Matheron (1962). This is because it is believed that microscale variation (small nuggets) is causing a discontinuity at the origin. Mathematically, this *cannot* happen for L_2-continuous processes [i.e., processes $Y(\cdot)$ for which $E(Y(\mathbf{s} + \mathbf{h}) - Y(\mathbf{s}))^2 \to 0$, as $\|\mathbf{h}\| \to 0$]. Hence, if continuity of the phenomenon is expected at the microscale, the only possible reason for $c_0 > 0$ is measurement error; that is, if the measurement were made several times, the results would fluctuate around the true value. Call the measurement-error variance c_{ME}.

How can Matheron's nugget effect be included in the mathematical model? In practice, only data $\{z(\mathbf{s}_i): i = 1, \ldots, n\}$ are available and nothing can be said about the variogram at lag distances smaller than $\min\{\|\mathbf{s}_i - \mathbf{s}_j\|: 1 \le i < j \le n\}$. Therefore, it is not known whether the microscale variation is continuous or not, but Matheron typically makes the assumption that it is *not* (Section 3.2.1). Mathematically speaking, to model the process at very small scales, he adds a white-noise process (i.e., zero mean, constant variance, zero covariance) to a process with continuous sample paths. This is purely an assumption, unverifiable without samples that are very close together. Call the variance of this white-noise process c_{MS}, which represents the nugget effect of the *microscale* process. Thus,

$$c_0 = c_{MS} + c_{ME}. \tag{2.3.5}$$

In practice, there are problems determining c_0 from data whose separations are too large to give accurate microscale information; typically, it is determined by extrapolating variogram estimates from lags closest to zero. Laslett et al. (1987) show, in their studies of soil pH, how crucial it is to model the small-scale and microscale variations correctly. A major problem in prediction that has scarcely been addressed is determining the relative importance of the components in (2.3.5). Most often, kriging equations implicitly assume

$c_{ME} = 0$, without investigating the measuring process through, for example, replicate assays. Failure to recognize the decomposition (2.3.5) is the source of the "kriging is/is not an exact interpolator" controversy; see Cressie (1986) and Section 3.2.1 for further discussion.

Further Properties

The behavior of the variogram near the origin is very informative about the continuity properties of the random process $Z(\cdot)$. Similar results hold for the covariance function (e.g., Loeve, 1963, Section 34.2, discusses the one-dimensional case). The most common types are categorized by Matheron (1971b, p. 58) as follows:

i. $2\gamma(\cdot)$ is continuous at the origin. Then $Z(\cdot)$ is L_2-continuous. [Clearly, $E(Z(\mathbf{s}+\mathbf{h}) - Z(\mathbf{s}))^2 \to 0$ if and only if $2\gamma(\mathbf{h}) \to 0$, as $\|\mathbf{h}\| \to 0$.]

ii. $2\gamma(\mathbf{h})$ does not approach 0 as $\mathbf{h}$ approaches the origin. Then $Z(\cdot)$ is not even L_2-continuous and is highly irregular. This *discontinuity* of γ at the origin is the *nugget effect* discussed previously.

iii. $2\gamma(\cdot)$ is a positive constant (except at the origin where it is zero). Then $Z(\mathbf{s}_1)$ and $Z(\mathbf{s}_2)$ are uncorrelated for any $\mathbf{s}_1 \neq \mathbf{s}_2$, regardless of how close they are; $Z(\cdot)$ is often called *white noise*.

Notice that L_2-continuity of $Z(\cdot)$ does not mean that the realizations are almost surely continuous; for the latter property to hold, stronger conditions than (i) are needed (Kent, 1989).

A random process $Z(\cdot)$ is said to be L_2*-differentiable* at $\mathbf{s}$ if, as $h_j \to 0$, $\{Z(\mathbf{s} + h_j\mathbf{e}_j) - Z(\mathbf{s})\}/h_j$ converges in L_2, $j = 1, \ldots, d$, where $\{\mathbf{e}_j\colon j = 1, \ldots, d\}$ is the natural basis of $\mathbb{R}^d$ (e.g., Adler, 1981, p. 27). In terms of the variogram $2\gamma(\cdot)$ of $Z(\cdot)$, consider the following property:

iv. $\partial^2(2\gamma(\mathbf{h}))/\partial h_1^2, \ldots, \partial^2(2\gamma(\mathbf{h}))/\partial h_d^2$ exist and are finite at $\mathbf{h} = \mathbf{0}$.

Then, as a consequence of iv, $Z(\cdot)$ is L_2-differentiable at all $\mathbf{s} \in \mathbb{R}^d$.

The variogram $2\gamma(\cdot)$ must satisfy a property called *conditional negative-definiteness* (Matheron, 1971b), namely,

$$\sum_{i=1}^{m} \sum_{j=1}^{m} a_i a_j 2\gamma(\mathbf{s}_i - \mathbf{s}_j) \leq 0, \tag{2.3.6}$$

for any finite number of spatial locations $\{\mathbf{s}_i\colon i = 1, \ldots, m\}$ and real numbers $\{a_i\colon i = 1, \ldots, m\}$ satisfying $\sum_{i=1}^{m} a_i = 0$. More details are given in Section 2.5.2.

Definition. Suppose $\{Z(\mathbf{s})\colon \mathbf{s} \in D\}$ satisfies (2.3.2) and (2.3.4) [so (2.3.4) can be rewritten as $E(Z(\mathbf{s}_1) - Z(\mathbf{s}_2))^2 = 2\gamma(\mathbf{s}_1 - \mathbf{s}_2)$]. Then $Z(\cdot)$ is said to be *intrinsically stationary* (or, equivalently, to satisfy the intrinsic hypothesis). Furthermore, if $2\gamma(\mathbf{s}_1 - \mathbf{s}_2) = 2\gamma^o(\|\mathbf{s}_1 - \mathbf{s}_2\|)$, a function only of $\|\mathbf{s}_1 - \mathbf{s}_2\|$, then $2\gamma(\cdot)$ is called *isotropic*. ■

By characterizing first differences of the random process $Z(\cdot)$, one loses the ability to distinguish between the two processes, $Z(\cdot) = \mu + \delta(\cdot)$ and $Z(\cdot) = M + \delta(\cdot)$, where M is *any* random variable. Pathologies of the latter type will be avoided by assuming that the process is at least ergodic in mean. Section 5.4 generalizes intrinsic stationarity to higher-order versions, where the existence of a generalized covariance function is assumed. Again, pathologies like random-coefficient additive polynomials will be avoided.

Some Isotropic Variogram Models

Various parametric variogram models are presented in Journel and Huijbregts (1978, pp. 161–195) and in Section 2.5. Consider the three basic isotropic models given here in terms of the semivariogram: linear, spherical, and exponential.

Linear model (valid in $\mathbb{R}^d$, $d \geq 1$):

$$\gamma(\mathbf{h};\boldsymbol{\theta}) = \begin{cases} 0, & \mathbf{h} = \mathbf{0}, \\ c_0 + b_l\|\mathbf{h}\|, & \mathbf{h} \neq \mathbf{0}, \end{cases} \tag{2.3.7}$$

$\boldsymbol{\theta} = (c_0, b_l)'$, where $c_0 \geq 0$ and $b_l \geq 0$.

Spherical model (valid in $\mathbb{R}^1$, $\mathbb{R}^2$, and $\mathbb{R}^3$):

$$\gamma(\mathbf{h};\boldsymbol{\theta}) = \begin{cases} 0, & \mathbf{h} = \mathbf{0}, \\ c_0 + c_s\{(3/2)(\|\mathbf{h}\|/a_s) - (1/2)(\|\mathbf{h}\|/a_s)^3\}, & 0 < \|\mathbf{h}\| \leq a_s, \\ c_0 + c_s, & \|\mathbf{h}\| \geq a_s, \end{cases} \tag{2.3.8}$$

$\boldsymbol{\theta} = (c_0, c_s, a_s)'$, where $c_0 \geq 0$, $c_s \geq 0$, and $a_s \geq 0$.

Exponential model (valid in $\mathbb{R}^d$, $d \geq 1$):

$$\gamma(\mathbf{h};\boldsymbol{\theta}) = \begin{cases} 0, & \mathbf{h} = \mathbf{0}, \\ c_0 + c_e\{1 - \exp(-\|\mathbf{h}\|/a_e)\}, & \mathbf{h} \neq \mathbf{0}, \end{cases} \tag{2.3.9}$$

$\boldsymbol{\theta} = (c_0, c_e, a_e)'$, where $c_0 \geq 0$, $c_e \geq 0$, and $a_e \geq 0$.

Another semivariogram model is the *rational quadratic model* (valid in $\mathbb{R}^d$, $d \geq 1$):

$$\gamma(\mathbf{h};\boldsymbol{\theta}) = \begin{cases} 0, & \mathbf{h} = \mathbf{0}, \\ c_0 + c_r\|\mathbf{h}\|^2/(1 + \|\mathbf{h}\|^2/a_r), & \mathbf{h} \neq \mathbf{0}, \end{cases} \tag{2.3.10}$$

$\boldsymbol{\theta} = (c_0, c_r, a_r)'$, where $c_0 \geq 0$, $c_r \geq 0$, and $a_r \geq 0$.

A semivariogram model that exhibits negative correlations caused by periodicity of the process is the *wave* (or *hole-effect*) *model* (valid in $\mathbb{R}^1$, $\mathbb{R}^2$, and $\mathbb{R}^3$):

$$\gamma(\mathbf{h};\boldsymbol{\theta}) = \begin{cases} 0, & \mathbf{h} = \mathbf{0}, \\ c_0 + c_w\{1 - a_w \sin(\|\mathbf{h}\|/a_w)/\|\mathbf{h}\|\}, & \mathbf{h} \neq \mathbf{0}, \end{cases} \quad (2.3.11)$$

$\boldsymbol{\theta} = (c_0, c_w, a_w)'$, where $c_0 \geq 0$, $c_w \geq 0$, and $a_w \geq 0$. Further models are given in Section 2.5.

Figure 2.13 shows various variogram shapes for representative values of $\boldsymbol{\theta}$. The linear model is linear and differentiable in its parameters; the spherical, exponential, rational quadratic, and wave models are not linear in their parameters, although they are differentiable.

A further condition that a variogram model *must* satisfy is (Matheron, 1971b)

$$2\gamma(\mathbf{h})/\|\mathbf{h}\|^2 \to 0, \quad \text{as } \|\mathbf{h}\| \to \infty.$$

In fact the *power model*,

$$\gamma(\mathbf{h};\boldsymbol{\theta}) = \begin{cases} 0, & \mathbf{h} = \mathbf{0}, \\ c_0 + b_p\|\mathbf{h}\|^{\lambda}, & \mathbf{h} \neq \mathbf{0}, \end{cases} \quad (2.3.12)$$

$\boldsymbol{\theta} = (c_0, b_p, \lambda)'$, where $c_0 \geq 0$, $b_p \geq 0$, and $0 \leq \lambda < 2$, is a valid semivariogram model in $\mathbb{R}^d$, $d \geq 1$.

An example of a random process in $\mathbb{R}^1$ (the $\mathbb{R}^d$ version is given in Section 5.5) that has variogram $2\gamma(h) = |h|^{\lambda}$, $0 < \lambda < 2$, is fractional Brownian motion with parameter $H = \lambda/2$ (e.g., Mandelbrot and Van Ness, 1968). This is a zero-mean Gaussian process $\{B_H(s): s \geq 0\}$ with $B_H(0) = 0$, stationary increments, and $\text{cov}(B_H(u), B_H(v)) = (1/2)\{|u|^{2H} + |v|^{2H} - |u - v|^{2H}\}$, $0 < H < 1$. Notice that $H = 1/2$ (i.e., $\lambda = 1$) corresponds to the usual Brownian motion (or Wiener process).

Anisotropy

When the process Z is anisotropic [i.e., dependence between $Z(\mathbf{s})$ and $Z(\mathbf{s} + \mathbf{h})$ is a function of both the magnitude *and* the direction of $\mathbf{h}$], the variogram (of the intrinsically stationary process) is no longer purely a function of distance between two spatial locations. Anisotropies are caused by the underlying physical process evolving differentially in space. Most commonly, the process in the vertical direction (where there is a gravity field) will be different from that in the horizontal directions. In geology, if a mineralization in rock occurs in oblong lenses, then the variogram will be

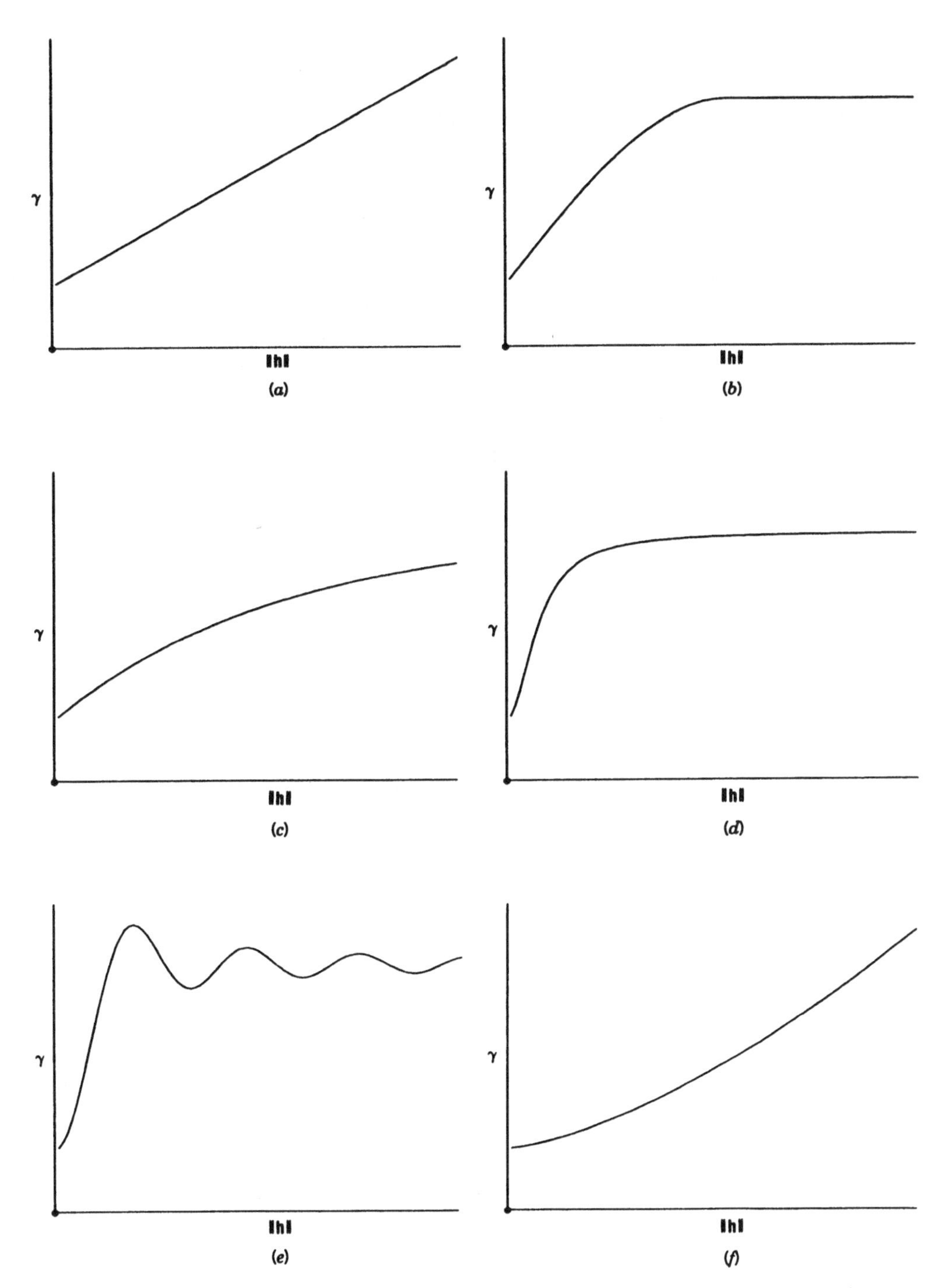

Figure 2.13 (*a*) Linear semivariogram given by (2.3.7). (*b*) Spherical semivariogram given by (2.3.8). (*c*) Exponential semivariogram given by (2.3.9). (*d*) Rational quadratic semivariogram given by (2.3.10). (*e*) Wave semivariogram given by (2.3.11). (*f*) Power semivariogram given by (2.3.12).

different in various horizontal directions as well. Sometimes the anisotropy can be corrected by a linear transformation of the lag vector $\mathbf{h}$. That is, the anisotropic variogram of Z is *geometrically anisotropic*, namely,

$$2\gamma(\mathbf{h}) = 2\gamma^o(\|A\mathbf{h}\|), \qquad \mathbf{h} \in \mathbb{R}^d, \tag{2.3.13}$$

where A is a $d \times d$ matrix and $2\gamma^o$ is a function of a real variable. Here, the Euclidean space is not appropriate for measuring distance between locations, but a linear transformation of it is. One could also imagine ore deposits where lag distances need to be measured on buckled manifolds (e.g., Dagbert et al., 1984).

Dealing with anisotropy becomes very difficult when the overall process decomposes into independent subprocesses,

$$Z(\mathbf{s}) = Z_1(\mathbf{s}) + Z_2(\mathbf{s}) + \cdots + Z_p(\mathbf{s}), \qquad \mathbf{s} \in D,$$

with variogram decomposition

$$2\gamma(\mathbf{h}) = 2\gamma_1(\mathbf{h}) + 2\gamma_2(\mathbf{h}) + \cdots + 2\gamma_p(\mathbf{h}), \qquad \mathbf{h} \in \mathbb{R}^d.$$

Each subprocess may have its own straightforward geometric anisotropy [see (2.3.13)], yet it may be impossible to identify such structure from the observed process $Z(\cdot)$. *A priori* information about the subprocesses is needed to handle this type of anisotropy.

Relative Variogram

Suppose that the region

$$D = \bigcup_{j=1}^{k} D_j \tag{2.3.14}$$

is a disjoint union of subregions $\{D_j: j = 1, \ldots, k\}$ and that within the jth subregion the process Z is intrinsically stationary with mean μ_j and variogram $2\gamma_Z^{(j)}(\cdot)$, $j = 1, \ldots, k$. Clearly, Z is not, in general, intrinsically stationary across the whole region. However, if a very simple relationship of mean to variogram exists, then estimates from all subregions can be combined. Consider

$$Y(\mathbf{s}) \equiv g(Z(\mathbf{s})), \tag{2.3.15}$$

where $g(\cdot)$ is a sufficiently smooth transformation that possesses at least two continuous derivatives. The δ-method (e.g., Kendall and Stuart, 1969, pp. 231, 244) expands $g(Z(\mathbf{s}))$ in a Taylor series about $E(Z(\mathbf{s}))$. Specifically,

$$Y(\mathbf{s}) = g(\mu_j) + g'(\mu_j)(Z(\mathbf{s}) - \mu_j) + g''(\mu_j)(Z(\mathbf{s}) - \mu_j)^2/2! + \cdots, \qquad \mathbf{s} \in D_j,$$

from which,

$$Y(\mathbf{s}+\mathbf{h}) - Y(\mathbf{s}) = g'(\mu_j)(Z(\mathbf{s}+\mathbf{h}) - Z(\mathbf{s})) + \cdots, \qquad \mathbf{s}, \mathbf{s}+\mathbf{h} \in D_j.$$

Hence,

$$2\gamma_Y^{(j)}(\mathbf{h}) \simeq (g'(\mu_j))^2 2\gamma_Z^{(j)}(\mathbf{h}), \qquad j = 1, \ldots, k. \tag{2.3.16}$$

It has been suggested (e.g., Journel and Huijbregts, 1978, p. 187ff) that some forms of nonstationarity can be modeled by a *proportional effect*; that is, $2\gamma_Z^{(j)}(\mathbf{h})/f(\mu_j)$, for some positive function f, is independent of j and hence its estimation can be combined over the various subregions of D. The most common case is $f(\mu_j) = \mu_j^2$. Then the *relative variogram*

$$2\gamma_Z^{(j)}(\mathbf{h})/\mu_j^2 \tag{2.3.17}$$

is independent of j. Equating (2.3.17) with (2.3.16), it is seen that when $g'(\mu_j) = 1/\mu_j$, the process Y is (approximately) intrinsically stationary. That is, when

$$g(x) = \log x, \tag{2.3.18}$$

the process Y defined by (2.3.15) has approximate variogram

$$2\gamma_Y(\mathbf{h}) \simeq \mathrm{var}(Z(\mathbf{s}+\mathbf{h}) - Z(\mathbf{s}))/(E(Z(\mathbf{s})))^2, \qquad \mathbf{h} \in \mathbb{R}^d.$$

Thus, the lack of intrinsic stationarity of the Z process over D can be recovered either by working on the log scale or by estimating the relative variograms (2.3.17).

Similarly, should the *scaled variograms* $2\gamma_Z^{(j)}(\mathbf{h}) \cdot (\mu_j^{\lambda-1})^2$ be independent of j, then the process Y, given by (2.3.15) and $g(x) = x^\lambda$ is approximately intrinsically stationary. Cressie (1985b) analyzes part of a soil–water-infiltration data set using $\lambda = 1/2$ (i.e., the square-root scale); see Section 4.3 for a more complete analysis of these data.

Often the transformation that is chosen so that Z exhibits a proportional effect also transforms the data so that they have an approximately Gaussian (i.e., normal) distribution. This principal of a universal transformation

(Cressie, 1985b) is often seen in data (Box and Cox, 1964). It says that additivity of small effects (Gaussianity), additivity of small effects to large effects (homogeneity of variograms), and additivity of large effects (no interaction) *all* tend to occur on the same scale $g(\cdot)$. The latter will be important in Section 3.5, where an additive decomposition is proposed for the large-scale variation of Z.

Aggregation

Suppose the random process $\{Z(\mathbf{s}): \mathbf{s} \in D\}$ is integrable in quadratic mean. That is, if B is a bounded integrable subset of $\mathbb{R}^d$ such that $|B| \equiv \int_B d\mathbf{s} > 0$, the *aggregated* random variable $Z(B) \equiv \int_B Z(\mathbf{s})\,d\mathbf{s}/\,|B|$ is well defined as a limit in quadratic mean of approximating Riemann sums (e. g., Yaglom, 1962, p. 23).

Suppose further that the random process possesses a variogram $2\gamma(\cdot)$. In Chapter 3, it is seen that the variances of random variables like $(Z(B_1) - Z(B_2))$, $B_1, B_2 \subset D$, are needed for kriging. When $|B_1| = |B_2| = b_0 > 0$, the variances can be interpreted as the variogram of the aggregated process $\{Z(B): |B| = b_0,\ B \subset D\}$. A convenient expression for the variance exists in terms of the (point) variogram $2\gamma(\cdot)$. For general $B_1, B_2 \subset D$,

$$\begin{aligned}\operatorname{var}(Z(B_1) - Z(B_2)) = &-\int_{B_1}\int_{B_1}\gamma(\mathbf{s}-\mathbf{u})\,d\mathbf{s}\,d\mathbf{u}/\,|B_1|^2\\ &-\int_{B_2}\int_{B_2}\gamma(\mathbf{s}-\mathbf{u})\,d\mathbf{s}\,d\mathbf{u}/\,|B_2|^2\\ &+\int_{B_1}\int_{B_2}2\gamma(\mathbf{s}-\mathbf{u})\,d\mathbf{s}\,d\mathbf{u}/(|B_1|\cdot|B_2|).\end{aligned}$$

Thus, an aggregated variogram model is easily calculated from the point variogram model; these change-of-support considerations are discussed further in Section 5.2.

Cross-Variograms

Let $\mathbf{Z}(\mathbf{s}) \equiv (Z_1(\mathbf{s}),\ldots,Z_k(\mathbf{s}))'$, $\mathbf{s} \in D$, be a multivariate spatial process, where each component process is assumed to possess a variogram:

$$2\gamma_{ii}(\mathbf{h}) \equiv \operatorname{var}(Z_i(\mathbf{s}+\mathbf{h}) - Z_i(\mathbf{s})), \qquad \mathbf{h} \in \mathbb{R}^d,\ i = 1,\ldots,k.$$

There are two ways to generalize this notion to account for cross-dependence between $Z_i(\cdot)$ and $Z_j(\cdot)$. The most natural one for multivariate spatial prediction (cokriging) is shown in Section 3.2.3 to be

$$2\gamma_{ij}(\mathbf{h}) \equiv \operatorname{var}(Z_i(\mathbf{s}+\mathbf{h}) - Z_j(\mathbf{s})), \qquad \mathbf{h} \in \mathbb{R}^d. \qquad (2.3.19)$$

The other generalization,

$$2\nu_{ij}(\mathbf{h}) \equiv \text{cov}\{(Z_i(\mathbf{s}+\mathbf{h}) - Z_i(\mathbf{s})),(Z_j(\mathbf{s}+\mathbf{h}) - Z_j(\mathbf{s}))\}, \qquad \mathbf{h} \in \mathbb{R}^d, \tag{2.3.20}$$

can only be used for cokriging under special conditions (Journel and Huijbregts, 1978, p. 326). It has, however, been the generalization traditionally recommended (e.g., Journel and Huijbregts, 1978, p. 324). In Section 3.2.3, the use of $2\nu_{ij}$ in cokriging is contrasted with the use of $2\gamma_{ij}$ and $C_{ij}(\mathbf{h}) \equiv \text{cov}(Z_i(\mathbf{s}+\mathbf{h}), Z_j(\mathbf{s}))$, $\mathbf{h} \in \mathbb{R}^d$.

2.3.2 Covariogram and Correlogram

Call the function $C(\cdot)$, given by (2.3.3) (provided it is well defined), a *covariogram*. (Notice that it has also been called an autocovariance function by time-series analysts.) Likewise, provided $C(\mathbf{0}) > 0$, call

$$\rho(\mathbf{h}) \equiv C(\mathbf{h})/C(\mathbf{0}) \tag{2.3.21}$$

a *correlogram* (also known as an autocorrelation function). It is this quantity that is traditionally used by time-series analysts for diagnosing nonstationarity, for determining the type of stationary dependence, for model fitting, and so forth. It is easy to see that $C(\mathbf{h}) = C(-\mathbf{h})$, $\rho(\mathbf{h}) = \rho(-\mathbf{h})$, and $\rho(\mathbf{0}) = 1$.

Consider the relation

$$\text{var}(Z(\mathbf{s}_1) - Z(\mathbf{s}_2)) = \text{var}(Z(\mathbf{s}_1)) + \text{var}(Z(\mathbf{s}_2)) - 2\,\text{cov}(Z(\mathbf{s}_1), Z(\mathbf{s}_2)); \tag{2.3.22}$$

thus, a model of the second-moment structure of the Z process (i.e., variances and covariances) automatically yields a model of $\text{var}(Z(\mathbf{s}_1) - Z(\mathbf{s}_2))$. Conversely, a model of the variogram and $\text{var}(Z(\mathbf{s}))$ yields a model of $\text{cov}(Z(\mathbf{s}_1), Z(\mathbf{s}_2))$. If $Z(\cdot)$ is second-order stationary, from (2.3.22), $\text{var}(Z(\mathbf{s}_1) - Z(\mathbf{s}_2)) = 2\{C(\mathbf{0}) - C(\mathbf{s}_1 - \mathbf{s}_2)\}$, which implies that $Z(\cdot)$ is intrinsically stationary with

$$2\gamma(\mathbf{h}) = 2(C(\mathbf{0}) - C(\mathbf{h})). \tag{2.3.23}$$

If $C(\mathbf{h}) \to 0$, as $\|\mathbf{h}\| \to \infty$ [e.g., when $Z(\cdot)$ is a stationary ergodic Gaussian process], then $2\gamma(\mathbf{h}) \to 2C(\mathbf{0})$; the quantity $C(\mathbf{0})$ is called the *sill* of the semivariogram. The *partial sill* is defined as $C(\mathbf{0}) - c_0$, where c_0 is the nugget effect defined in Section 2.3.1. The smallest value of $\|\mathbf{r}_0\|$ for which

$2\gamma(\mathbf{r}_0(1+\epsilon)) = 2C(\mathbf{0})$, for any $\epsilon > 0$, is called the *range* of the variogram in the direction $\mathbf{r}_0/\|\mathbf{r}_0\|$.

The class of all second-order stationary processes is strictly contained in the class of all intrinsically stationary processes. An example of a process for which $2\gamma(\cdot)$ is defined but $C(\cdot)$ is not, is d-dimensional isotropic Brownian motion (Levy, 1954, p. 71). If $\{W(\mathbf{s}): \mathbf{s} \in \mathbb{R}^d\}$ is such a process, then $\text{var}(W(\mathbf{s}+\mathbf{h}) - W(\mathbf{s})) = \|\mathbf{h}\|$, $\mathbf{h} \in \mathbb{R}^d$. However, $\text{cov}(W(\mathbf{u}), W(\mathbf{v})) = (1/2)(\|\mathbf{u}\| + \|\mathbf{v}\| - \|\mathbf{u} - \mathbf{v}\|)$, $\mathbf{u}, \mathbf{v} \in \mathbb{R}^d$, which is *not* a function of $\mathbf{u} - \mathbf{v}$.

Of course, $C(\cdot)$ must be *positive-definite*, namely,

$$\sum_{i=1}^{m} \sum_{j=1}^{m} a_i a_j C(\mathbf{s}_i - \mathbf{s}_j) \geq 0,$$

for any finite number of spatial locations $\{\mathbf{s}_i: i = 1, \ldots, m\}$ and real numbers $\{a_i: i = 1, \ldots, m\}$. More details are given in Section 2.5, including the spectral theory of second-order stationary and intrinsically stationary processes.

Recall that the covariogram is called *isotropic* when $C(\cdot)$ can be expressed as

$$C(\mathbf{h}) = C^o(\|\mathbf{h}\|),$$

where $C^o(\cdot)$ is a function of only one real variable. The covariances between aggregated random variables $Z(B_1)$ and $Z(B_2)$ can be expressed in terms of the (point) covariogram (cf. Section 2.3.1)

$$\text{cov}(Z(B_1), Z(B_2)) = \int_{B_1} \int_{B_2} C(\mathbf{s} - \mathbf{u})\, d\mathbf{s}\, d\mathbf{u} / (|B_1| \cdot |B_2|).$$

Cross-covariograms can be defined when $\mathbf{Z}(\cdot)$ is multivariate, in an analogous fashion to cross-variograms; see Section 3.2.3.

Separable covariograms in $\mathbb{R}^d$ are those for which $C(h) = \prod_{i=1}^{d} C_i(h_i)$, $\mathbf{h} = (h_1, \ldots, h_d)' \in \mathbb{R}^d$, where each component in the product is a covariogram in $\mathbb{R}^1$ (Section 2.5.1). For example, in $\mathbb{R}^2$, a separable covariogram is $C(h_1, h_2) \equiv C(\mathbf{0}) \cdot \exp\{-\theta_1|h_1| - \theta_2|h_2|\}$, where $C(\mathbf{0})$, θ_1, and θ_2 are positive.

What is to be gained by postulating a variogram rather than a covariogram? It has already been demonstrated that the class of intrinsically stationary processes is more general. However, it is not *too* general; spatial prediction (i.e., kriging) is easy to carry out on intrinsically stationary processes (Chapter 3). Moreover, the next section demonstrates that the variogram can be estimated more reliably than the covariogram.

2.4 ESTIMATION OF THE VARIOGRAM

The variogram $2\gamma(\cdot)$ is defined in (2.3.4) as

$$2\gamma(\mathbf{s}_1 - \mathbf{s}_2) \equiv \text{var}(Z(\mathbf{s}_1) - Z(\mathbf{s}_2)),$$

although some have mistakenly defined it as $E(Z(\mathbf{s}_1) - Z(\mathbf{s}_2))^2$. When the process $Z(\cdot)$ is intrinsically stationary [i.e., satisfies (2.3.4) *and* the constant mean assumption (2.3.2)], these two definitions coincide. However, should the process $Z(\cdot)$ be represented as

$$Z(\mathbf{s}) = \mu(\mathbf{s}) + \delta(\mathbf{s}), \qquad \mathbf{s} \in D, \tag{2.4.1}$$

where $\delta(\cdot)$ is a zero-mean intrinsically stationary stochastic process with variogram 2γ, and the mean $\mu(\cdot)$ is not constant, then

$$E(Z(\mathbf{s}_1) - Z(\mathbf{s}_2))^2 = 2\gamma(\mathbf{s}_1 - \mathbf{s}_2) + (\mu(\mathbf{s}_1) - \mu(\mathbf{s}_2))^2,$$

which is not in general a function of $\mathbf{s}_1 - \mathbf{s}_2$. Nor will it necessarily satisfy the condition

$$E(Z(\mathbf{s}_1) - Z(\mathbf{s}_2))^2 / \|\mathbf{s}_1 - \mathbf{s}_2\|^2 \to 0, \quad \text{as } \|\mathbf{s}_1 - \mathbf{s}_2\| \to \infty,$$

that all variograms must satisfy (Section 2.5.2). For the rest of this section, I shall assume that data $\{Z(\mathbf{s}_i): i = 1, \ldots, n\}$ can be modeled with an intrinsically stationary process; that is, a process satisfying (2.3.2) and (2.3.4). More general processes are considered in Sections 3.4, 3.5, and 5.4.

Method-of-Moments Estimator

Under the constant-mean assumption (2.3.2), a natural estimator based on the method-of-moments, due to Matheron (1962), is

$$2\hat{\gamma}(\mathbf{h}) \equiv \frac{1}{|N(\mathbf{h})|} \sum_{N(\mathbf{h})} (Z(\mathbf{s}_i) - Z(\mathbf{s}_j))^2, \qquad \mathbf{h} \in \mathbb{R}^d, \tag{2.4.2}$$

where

$$N(\mathbf{h}) \equiv \{(\mathbf{s}_i, \mathbf{s}_j): \mathbf{s}_i - \mathbf{s}_j = \mathbf{h}; i, j = 1, \ldots, n\} \tag{2.4.3}$$

and $|N(\mathbf{h})|$ is the number of distinct pairs in $N(\mathbf{h})$. Notice that $N(-\mathbf{h}) \neq N(\mathbf{h})$, although $2\hat{\gamma}(-\mathbf{h}) = 2\hat{\gamma}(\mathbf{h})$, thus preserving a property of the (theoretical) variogram. Notice also that the mean μ does not have to be estimated. Henceforth, (2.4.2) will be referred to as the *classical* variogram estimator.

When data are irregularly spaced in $\mathbb{R}^d$, the variogram estimator (2.4.2) is usually smoothed by using instead

$$2\gamma^{+}(\mathbf{h}(l)) \equiv \text{ave}\left\{\left(Z(\mathbf{s}_i) - Z(\mathbf{s}_j)\right)^2 : (\mathbf{s}_i, \mathbf{s}_j) \in N(\mathbf{h}); \mathbf{h} \in T(\mathbf{h}(l))\right\};$$

the region $T(\mathbf{h}(l))$ is some specified "tolerance" region in $\mathbb{R}^d$ around $\mathbf{h}(l)$, $l = 1, \ldots, K$, and ave$\{\cdot\}$ denotes a possibly weighted average over the elements in $\{\cdot\}$. (Omre, 1984, gives a method of choosing data-dependent weights.) Tolerance regions should be as small as possible to retain spatial resolution, yet large enough so that the estimator $2\gamma^{+}(\cdot)$ is stable. Journel and Huijbregts (1978, p. 194) recommend that the number of distinct pairs $|\cup\{N(\mathbf{h}): \mathbf{h} \in T(\mathbf{h}(l))\}|$ in $T(\mathbf{h}(l))$ be at least 30; one should choose tolerance regions so that most of them satisfy this condition.

Often the regions $\{T(\mathbf{h}(l)): l = 1, \ldots, K\}$ are chosen to be disjoint; this is directly analogous to histogram smoothing of a univariate batch of data. It is natural then to think about moving-window variogram estimators, analogous to kernel density estimators (e.g., Rosenblatt, 1985, p. 192). Remember though, variograms have to be conditionally negative-definite, a property that the variogram estimate will almost certainly not possess. This is remedied by fitting a valid parametric model to the smoothed estimate; see Section 2.6.

It is often a matter of taste whether the statistical modeler chooses to assume intrinsic stationarity or second-order stationarity. Time-series analysts usually assume the latter and have their own version of a method-of-moments estimator of the covariogram (autocovariance function). For spatial data, this is

$$\hat{C}(\mathbf{h}) \equiv \frac{1}{|N(\mathbf{h})|} \sum_{N(\mathbf{h})} \left(Z(\mathbf{s}_i) - \bar{Z}\right)\left(Z(\mathbf{s}_j) - \bar{Z}\right), \qquad (2.4.4)$$

where

$$\bar{Z} = \sum_{i=1}^{n} Z(\mathbf{s}_i)/n \qquad (2.4.5)$$

is an estimator of the mean μ and $N(\mathbf{h})$ is given by (2.4.3). Notice that $2\hat{\gamma}(\mathbf{h}) \neq 2(\hat{C}(\mathbf{0}) - \hat{C}(\mathbf{h}))$ [i.e., the estimators do not preserve the property (2.3.23)]; however, for $|N(\mathbf{h})|/n$ near 1, the difference between the two will be small.

2.4.1 Comparison of Variogram and Covariogram Estimation

I shall make the case that variogram estimation is to be preferred to covariogram estimation. It has been established in Section 2.3.2 that, should the process be second-order stationary, there is a simple relationship between the covariogram and the variogram. Moreover, the variogram is defined in cases when the covariogram is not; in those cases, $\hat{C}(\cdot)$ is estimating a nonexistent parameter.

For present purposes, suppose the data come equally spaced on $\mathbb{R}^1$ and are represented as $\{Z(s): s = 1, \ldots, n\}$. The comparisons are easier to make in this circumstance, but the conclusions are general.

Bias

The classical variogram estimator (2.4.2) is unbiased for $2\gamma(\cdot)$ when $Z(\cdot)$ is intrinsically stationary. However, when $Z(\cdot)$ is second-order stationary, $\hat{C}$ given by (2.4.4) has $O(1/n)$ bias: $E(\hat{C}(h)) = C(h) + O(1/n)$; see, for example, Fuller (1976, Section 6.2). The bias can contribute substantially to the mean-squared error when n is small.

This bias result can be extended to a polynomial trend model $Z(s) = \sum_{j=0}^{p} \beta_j s^j + \delta(s)$, $s \in \mathbb{R}^1$. Suppose ordinary least squares is used to obtain an estimate of $\boldsymbol{\beta} \equiv (\beta_0, \ldots, \beta_p)'$ [i.e., $\hat{\boldsymbol{\beta}} = (X'X)^{-1}X'\mathbf{Z}$, where X is an $n \times (p+1)$ matrix whose (s, j)th element is s^{j-1}, $s = 1, \ldots, n$, $j = 1, \ldots, p+1$]. Then the residuals

$$\hat{\delta}(s) \equiv Z(s) - (1, s, \ldots, s^p)\hat{\boldsymbol{\beta}}, \qquad s = 1, \ldots, n, \tag{2.4.6}$$

can be used in place of $\{Z(s): s = 1, \ldots, n\}$ in (2.4.2) and (2.4.4) to obtain estimates $2\hat{\gamma}(\cdot)$ and $\hat{C}(\cdot)$ of an assumed variogram and an assumed covariogram, respectively. Cressie and Grondona (1992) show that, for $\delta(\cdot)$ a stationary autoregressive-moving-average process, the bias of $\hat{C}$ is $-(p+1)\{C(0) + 2\sum_{h=1}^{\infty} C(h)\}/n + O((\log n)^{1+\alpha}/n^2)$, $\alpha > 0$. However, the bias of $2\hat{\gamma}$ is $O((\log n)^{1+\alpha}/n^2)$, $\alpha > 0$. That is, although the variogram estimator based on the residuals is a biased estimator of the error variogram, this bias is of smaller order than the corresponding covariogram estimator's bias. Further details are given in Section 3.4.3.

Asymptotic Distribution

The asymptotic joint Gaussianity of the estimators $\{\hat{C}(h)\}$ and of $\{2\hat{\gamma}(h)\}$ has been established under appropriate mixing conditions (i.e., conditions that ensure the dependence in the process dies off sufficiently quickly as lag distance increases); see, for example, Fuller (1976, Section 6.2) and Davis and Borgman (1982). Furthermore, the results of Abril (1987) can be used to show that the joint density of $\{\hat{C}(h)\}$ (and of $\{2\hat{\gamma}(h)\}$) admits an Edgeworth expansion. Indeed, Baczkowski and Mardia's (1987) finite-sample study of the distribution of $2\hat{\gamma}(h)$, with h fixed, shows log Gaussianity to be a more appropriate distributional approximation than Gaussianity. Variances and covariances of $\{2\hat{\gamma}(h): h = 1, \ldots, H\}$, for a fixed H, are given by Cressie (1985a), where they are shown to be $O(1/n)$. These will be used in a generalized-least-squares fit of variogram models to estimated variograms that is described in Section 2.6.2. Variances and covariances of $\{\hat{C}(h): h = 1, \ldots, H\}$, for fixed H, can be found, for example, in Fuller (1976, Section 6.2), and are also $O(1/n)$.

Intuitively, the rate of convergence of these estimators to their limiting distribution will be slower when the correlations between the data are stronger. Sharma (1986) quantifies this intuition for the estimator $\hat{C}(0)$ of $C(0)$.

Trend Contamination

Let $\{S(s): s = 1,2,\ldots\}$ be a zero-mean, unit variance, second-order stationary process in $\mathbb{R}^1$ and define

$$Z(s) \equiv S(s) + \epsilon \cdot (s - (n+1)/2), \qquad s = 1,2,\ldots,n.$$

Then Z is not second-order stationary because it is contaminated with linear trend, ϵ being the degree of contamination (Cox, 1981). The goal is to estimate the variogram $2\gamma_S$ or the covariogram C_S of the random process $S(\cdot)$. Suppose the contamination has not been detected and $\hat{C}$ is computed using the data $\{Z(s): s = 1,\ldots,n\}$. Then

$$\hat{C}(h) = \sum_{s=1}^{n-h} \left(Z(s+h) - \bar{Z}\right)\left(Z(s) - \bar{Z}\right)/(n-h)$$

has leading term $\sum_{s=1}^{n-h} Z(s+h)Z(s)/(n-h)$ (because, as $n \to \infty$, $\bar{Z} = \bar{S}$ converges in probability to 0 at the rate $n^{-1/2}$, provided S is uniformly mixing; Ibragimov and Linnik, 1971, Chapter 18). Now

$$\sum_{s=1}^{n-h} Z(s+h)Z(s)/(n-h) = \left\{\sum_{s=1}^{n-h} \left(S(s+h) - \bar{S}\right)\left(S(s) - \bar{S}\right)/(n-h)\right\} + \epsilon^2 n^2/12 + \text{remainder}.$$

The remainder is a stochastic term that converges in probability to 0 at the rate $n^{-1/2}$. Thus, $\hat{C}(h) \simeq \hat{C}_S(h) + \epsilon^2 n^2/12$, in probability. Clearly, as n becomes large, the second term dominates and information on the underlying S process is lost. For n large, $\hat{\rho}(h) \equiv \hat{C}(h)/\hat{C}(0) \simeq 1$, in probability. Thus, even a small amount of trend contamination can have a disastrous effect on attempts to estimate $C_S(\cdot)$.

However, for the variogram,

$$2\hat{\gamma}(h) = \sum_{s=1}^{n-h} \left(Z(s+h) - Z(s)\right)^2/(n-h)$$
$$= 2\hat{\gamma}_S(h) + \epsilon^2 h^2 + \text{remainder},$$

where the remainder term has similar properties to the one in the previous paragraph. Thus, the effect of a linear-trend contamination term (of magnitude ϵ) is that of a small upward shift (of magnitude $\epsilon^2 h^2$) of the desired variogram estimator $2\hat{\gamma}_S$; that is, $2\hat{\gamma}(h) \simeq 2\hat{\gamma}_S(h) + \epsilon^2 h^2$, in probability.

Some Final Remarks

I believe that a large part of the reason why statisticians prefer the covariogram and ignore the variogram as a way of characterizing dependence is unfamiliarity. The variogram does not require estimation of a constant mean μ, Section 2.3 demonstrates the extra generality that variogram models possess, and Section 2.5 gives a spectral theory very similar to that for the covariogram.

2.4.2 Exact Distribution Theory for the Variogram Estimator

Assume, for the moment, that the data $\{Z(s): s = 1, \ldots, n\}$ are equally spaced on a transect in $\mathbb{R}^1$. The mean-squared successive difference $\Sigma_{s=1}^{n-1}(Z(s+1) - Z(s))^2/(n-1)$ was of particular interest to von Neumann (1941), although not for detecting statistical dependence; see the next paragraph. It is easily recognized as $2\hat{\gamma}(1)$, the lag-1 classical variogram estimator.

Data are Independent

Von Neumann et al. (1941) use $2\hat{\gamma}(1)$ to test whether $\mu(s) = E(Z(s))$ is constant or not. Their distribution theory assumes independence of the data $\{Z(s): s = 1, \ldots, n\}$. Durbin and Watson (1950, 1951) use a variant of $2\hat{\gamma}(1)$ to test for serial correlation, although once again distribution theory is derived under the assumption of independence (the null hypothesis). Vinod (1973) recognized the potential of extending the test for first-order serial correlation based on $2\hat{\gamma}(1)$ to a test for higher-order correlation based on the test statistic $2\hat{\gamma}(h)$, $h \geq 1$. Indeed, Ali (1987) has an accurate small-sample beta approximation for the distribution of $2\hat{\gamma}(h)$ under the hypothesis that the data are i.i.d. Gaussian random variables. He points out that the exact distribution is, in principle, available from an inversion of its characteristic function. A numerical-inversion technique due to Imhof (1961) could then be used to give exact percentage points of $2\hat{\gamma}(h)$ for Gaussian data.

Data are Dependent

In geostatistics, it is rare that the independence assumption is tenable, although it provides a natural null hypothesis for testing purposes. Davis and Borgman (1979) tabulate the sampling distribution of $2\hat{\gamma}(h)$, by Fourier inversion, for equally spaced (in $\mathbb{R}^1$) Gaussian data with true semivariogram,

$$\gamma(h) = \begin{cases} |h|/a, & 0 \leq |h| < a, \\ 1, & |h| \geq a. \end{cases}$$

In principle, the same inversion procedure, applied to multidimensional Fourier transforms, could be used to obtain the exact sampling distribution of $2\hat{\gamma}(\mathbf{h})$, $\mathbf{h} \in \mathbb{R}^d$, for Gaussian data.

Regardless of the dimension of the spatial index, $2\hat{\gamma}(\mathbf{h})$ given by (2.4.2) can be written as a quadratic form in $\mathbf{Z} \equiv (Z(\mathbf{s}_1), \ldots, Z(\mathbf{s}_n))'$:

$$2\hat{\gamma}(\mathbf{h}) = \mathbf{Z}'(Q(\mathbf{h})'Q(\mathbf{h}))\mathbf{Z}/|N(\mathbf{h})| \equiv \mathbf{Z}'A(\mathbf{h})\mathbf{Z}, \qquad (2.4.7)$$

where $Q(\mathbf{h})$ is an $|N(\mathbf{h})| \times n$ matrix with entries that are 1, -1, or 0. Hence, $A(\mathbf{h}) \equiv Q(\mathbf{h})'Q(\mathbf{h})/|N(\mathbf{h})|$ is a symmetric $n \times n$ matrix. Recall that $\mu(\mathbf{s}) = E(Z(\mathbf{s}))$ is assumed to be constant over $\mathbf{s}$, and because $\mathbf{1} \equiv (1, \ldots, 1)'$ is an eigenvector of $A(\mathbf{h})$, without loss of generality put $E(\mathbf{Z}) = \mathbf{0}$.

Define

$$\operatorname{var}(\mathbf{Z}) \equiv \Sigma. \tag{2.4.8}$$

Then, if $\mathbf{Z}$ is Gaussian, it is straightforward to show that

$$2\hat{\gamma}(\mathbf{h}) = \sum_{i=1}^{n} \lambda_i(\mathbf{h}) \chi^2_{1,i}, \tag{2.4.9}$$

a linear combination of independent chi-squared random variables, each on 1 degree of freedom. The coefficients $\{\lambda_i(\mathbf{h}): i = 1, \ldots, n\}$ are eigenvalues of the nonnegative-definite matrix $A(\mathbf{h})\Sigma$. When $\Sigma = \sigma^2 I$ (i.e., data are i.i.d. Gaussian) and the data are equally spaced in $\mathbb{R}^1$, Ali (1987, Appendix) has calculated the eigenvalues of $A(h)\Sigma = A(h) \cdot \sigma^2 = Q(h)'Q(h) \cdot \sigma^2/(n - h)$, where the (i, j)th element of $Q(h)$ is given by

$$q_{ij}(h) = \begin{cases} -1, & j = i, i = 1, \ldots, n - h, \\ 1, & j = i + h, i = 1, \ldots, n - h, \\ 0, & \text{otherwise}. \end{cases} \tag{2.4.10}$$

The first two moments are easily obtainable directly from (2.4.7) or from the representation (2.4.9):

$$\begin{aligned} E(2\hat{\gamma}(\mathbf{h})) &= \operatorname{trace}(A(\mathbf{h})\Sigma), \\ \operatorname{var}(2\hat{\gamma}(\mathbf{h})) &= 2\operatorname{trace}(A(\mathbf{h})\Sigma A(\mathbf{h})\Sigma). \end{aligned} \tag{2.4.11}$$

In order to work with these results, a suitable model for Σ and suitable sample spacings $\{\mathbf{s}_i: i = 1, \ldots, n\}$ have to be chosen. Davis and Borgman (1979) effectively do this; see also Ali (1984).

Of course, when the trend $\mu(\mathbf{s})$ is not constant, $2\hat{\gamma}(\mathbf{h})$ is a poor estimator of the variogram and should not be used until the data are detrended. Section 2.4.1 discusses the bias of the resulting variogram estimator based on the residuals from the estimated trend.

2.4.3 Robust Estimation of the Variogram

The word "robust" in this section heading (and, for that matter, throughout the book) is used to describe inference procedures that are stable when model assumptions depart from those of a central model. (Here, departures will be in the form of a small amount of independent contamination of a Gaussian process.) The word "resistant," on the other hand, refers to a statistic that is arithmetically stable under gross contamination of the data values.

Therefore, there may be many different claims of a procedure's robustness, and sometimes they may conflict. Before claiming robustness, a careful statement needs to be made of the class of models over which the procedure is robust. Moreover, if the procedure is estimation, it should be made clear which theoretical quantity is to be estimated.

Recall from (2.4.2) the classical estimator of the variogram (Matheron, 1962):

$$2\hat{\gamma}(\mathbf{h}) \equiv \frac{1}{|N(\mathbf{h})|} \sum_{N(\mathbf{h})} \left(Z(\mathbf{s}_i) - Z(\mathbf{s}_j)\right)^2, \qquad \mathbf{h} \in \mathbb{R}^d,$$

where $N(\mathbf{h})$ is given by (2.4.3) and $|N(\mathbf{h})|$ is the number of distinct sample pairs lagged by the vector $\mathbf{h}$. In order to use the available knowledge of robust *location* estimation, Cressie and Hawkins (1980) take fourth roots of squared differences, yielding robust (to contamination by outliers; see Hawkins and Cressie, 1984) estimators

$$2\bar{\gamma}(\mathbf{h}) \equiv \left\{ \frac{1}{|N(\mathbf{h})|} \sum_{N(\mathbf{h})} |Z(\mathbf{s}_i) - Z(\mathbf{s}_j)|^{1/2} \right\}^4 \Big/ \left(0.457 + 0.494/|N(\mathbf{h})|\right) \tag{2.4.12}$$

and

$$2\tilde{\gamma}(\mathbf{h}) \equiv \left[\mathrm{med}\left\{|Z(\mathbf{s}_i) - Z(\mathbf{s}_j)|^{1/2} : (\mathbf{s}_i, \mathbf{s}_j) \in N(\mathbf{h})\right\}\right]^4 \Big/ B(\mathbf{h}), \tag{2.4.13}$$

where med$\{\cdot\}$ denotes the median of the sequence $\{\cdot\}$ and $B(\mathbf{h})$ corrects for bias [asymptotically, $B(\mathbf{h}) = 0.457$]. The reasoning behind (2.4.12) is that, for Gaussian data, $(Z(\mathbf{s}_i) - Z(\mathbf{s}_j))^2$ is a chi-squared random variable on 1 degree of freedom. The power transformation that makes this most Gaussian-like is the fourth root (Cressie and Hawkins, 1980), namely, $|Z(\mathbf{s}_i) - Z(\mathbf{s}_j)|^{1/2}$, the square root of the absolute difference. Thus, various location estimators can be applied to $\{|Z(\mathbf{s}_i) - Z(\mathbf{s}_j)|^{1/2} : (\mathbf{s}_i, \mathbf{s}_j) \in N(\mathbf{h})\}$, which, when untransformed and normalized for bias, yield robust variogram estimators.

Using the square root of an estimate of scale to test for homogeneity of variances in a one-way analysis of variance was suggested by Levene (1960). He also suggested the square, the absolute value, and the log transformation, but finally settled on the absolute value. Cressie and Hawkins (1980) show that this transformation, as well as the 2/3 root (Wilson and Hilferty, 1931), does not go far enough. Taking logs goes too far; the square root is just right.

There is another advantage to using $|Z(\mathbf{s}_i) - Z(\mathbf{s}_j)|^{1/2}$ over $(Z(\mathbf{s}_i) - Z(\mathbf{s}_j))^2$: Notice that the summands in each of (2.4.12) and (2.4.2) are not independent, and the more dependent they are, the less efficient is their average in estimating the variogram. Suppose X_1, X_2 are joint Gaussian random variables with zero means, unit variances, and $\mathrm{corr}(X_1, X_2) \equiv \mathrm{cov}(X_1, X_2)/\{\mathrm{var}(X_1)\mathrm{var}(X_2)\}^{1/2} \equiv \rho$. Then it is easily shown that $\mathrm{corr}(X_1^2, X_2^2) = \rho^2$, and Cressie (1985a) gives the formula,

$$\mathrm{corr}\left(|X_1|^{1/2}, |X_2|^{1/2}\right)$$
$$= \Gamma^2(3/4)\left\{\pi^{1/2} - \Gamma^2(3/4)\right\}^{-1}\left\{(1-\rho^2)F(3/4, 3/4; 1/2; \rho^2) - 1\right\},$$

where F is the hypergeometric function (Abramowitz and Stegun, 1965, p. 556).

For example, when $\rho = 0.5$, $\mathrm{corr}(X_1^2, X_2^2) = 0.25$, whereas $\mathrm{corr}(|X_1|^{1/2}, |X_2|^{1/2}) = 0.182$. For ρ small, $\mathrm{corr}(|X_1|^{1/2}, |X_2|^{1/2}) \simeq (5/8)\rho^2 < \rho^2$. In general, the summands $\{|Z(\mathbf{s}_i) - Z(\mathbf{s}_j)|^{1/2}\}$ in (2.4.12) are less correlated than the summands $\{(Z(\mathbf{s}_i) - Z(\mathbf{s}_j))^2\}$ in (2.4.2).

Scale and Quantile Estimators

Other types of estimators, based on robust estimation of scale and on quantiles, have been considered. Armstrong and Delfiner (1980) pursued Cressie and Hawkins' (1980) suggestion that, because the variogram is $\mathrm{var}(Z(\mathbf{s}+\mathbf{h}) - Z(\mathbf{s}))$ under intrinsic stationarity, a "robustification" of (2.4.2) could be achieved by using a robust estimator of scale on the differences $\{Z(\mathbf{s}_i) - Z(\mathbf{s}_j): (i, j) \in N(\mathbf{h})\}$.

Provided stationarity holds, $Z(\mathbf{s}_i) - Z(\mathbf{s}_j)$ is a symmetric random variable with mean 0, a property that ameliorates considerably the scale estimation problem. Armstrong and Delfiner defined "Huberized" variograms (i.e., they used the scale estimator of Huber, 1964, rather than the sample variance) and quantile variograms. The Huberized variograms are lengthy to compute, requiring iteration at every lag. The square of the interquartile range of the differences, however, is a resistant, quick, and easy alternative. Consider then,

$$\left[\mathrm{UQ}\{Z(\mathbf{s}_i) - Z(\mathbf{s}_j)\} - \mathrm{LQ}\{Z(\mathbf{s}_i) - Z(\mathbf{s}_j)\}\right]^2,$$

where UQ (LQ) stands for upper (lower) quartile, the 75th (25th) percentile of the differences $\{Z(\mathbf{s}_i) - Z(\mathbf{s}_j): (i, j) \in N(\mathbf{h})\}$.

Consider also the quantile variogram, based on a sample quantile of $\{(Z(\mathbf{s}_i) - Z(\mathbf{s}_j))^2: (i, j) \in N(\mathbf{h})\}$. The idea is that quantiles are more resistant to outliers than the mean. The most obvious choice is

$$\mathrm{med}\left\{\left(Z(\mathbf{s}_i) - Z(\mathbf{s}_j)\right)^2: (i, j) \in N(\mathbf{h})\right\}.$$

Both of the preceding approaches need some normalization to make them unbiased. However, leaving this aside, it is now shown that both are *equivalent* to $2\tilde{\gamma}(\mathbf{h})$ in (2.4.13), the estimator based on the median of the fourth root of the squared differences. Clearly $\text{med}\{(Z(\mathbf{s}_i) - Z(\mathbf{s}_j))^2\} = [\text{med}\{|Z(\mathbf{s}_i) - Z(\mathbf{s}_j)|^{1/2}\}]^4$, because for $x \geq 0$, $f(x) = x^{1/4}$ is a monotonic function. Also, asymptotically,

$$
\begin{aligned}
\text{UQ}\{Z(\mathbf{s}_i) - Z(\mathbf{s}_j)\} &- \text{LQ}\{Z(\mathbf{s}_i) - Z(\mathbf{s}_j)\} \\
&= 2\,\text{med}\{|Z(\mathbf{s}_i) - Z(\mathbf{s}_j)|\} \\
&= 2\Big[\text{med}\Big\{|Z(\mathbf{s}_i) - Z(\mathbf{s}_j)|^{1/2}\Big\}\Big]^2.
\end{aligned}
$$

Hence, the estimator $2\tilde{\gamma}(\mathbf{h})$, based on fourth roots of squared differences, simultaneously captures the essence of a robust-scale approach and a quantile approach.

Which of $2\bar{\gamma}(\cdot)$ and $2\tilde{\gamma}(\cdot)$ is to be preferred? Cressie and Hawkins (1980) and, in a more general context, Taylor (1987) show through simulation that $2\bar{\gamma}$ is a more efficient estimator than $2\tilde{\gamma}$.

Slope Estimator

Under the assumption of intrinsic stationarity,

$$\text{cov}(Z(\mathbf{s} + \mathbf{h}) - Z(\mathbf{s} + \mathbf{l}), Z(\mathbf{s} + \mathbf{h} + \mathbf{l}) - Z(\mathbf{s})) = 2\gamma(\mathbf{h}) - 2\gamma(\mathbf{l}), \qquad \mathbf{h}, \mathbf{l} \in \mathbb{R}^d.$$

Then, if $Z(\cdot)$ is Gaussian,

$$
\begin{aligned}
E(Z(\mathbf{s} + \mathbf{h} + \mathbf{l}) &- Z(\mathbf{s}) | Z(\mathbf{s} + \mathbf{h}) - Z(\mathbf{s} + \mathbf{l})) \\
&= \{(\gamma(\mathbf{h}) - \gamma(\mathbf{l}))/\gamma(\mathbf{h} - \mathbf{l})\} \cdot (Z(\mathbf{s} + \mathbf{h}) - Z(\mathbf{s} + \mathbf{l})), \qquad (2.4.14)
\end{aligned}
$$

which is a linear regression through the origin. For simplicity, assume that the data are on a transect and are equally spaced, namely $\{Z(s): s = 1, \ldots, n\}$. Put $l = 1$ in (2.4.14) and define

$$C_{h-1} \equiv \{(\gamma(h) - \gamma(1))/\gamma(h - 1)\},$$

the slope of the linear regression of $(Z(s + h + 1) - Z(s))$ on $(Z(s + h) - Z(s + 1))$. Then define the robust estimator

$$2\gamma^{\dagger}(h) \equiv \left(1 + \tilde{C}_{h-1} + \tilde{C}_{h-1}\tilde{C}_{h-2} + \cdots + \tilde{C}_{h-1}\tilde{C}_{h-2} \cdots \tilde{C}_1\right) 2\gamma^{\dagger}(1), \qquad h = 2, 3, \ldots, \qquad (2.4.15)$$

where $2\gamma^{\dagger}(1)$ is any one of the robust (lag-1) variogram estimators referred to previously, and $\tilde{C}_{h-1}$ is a robust slope estimator calculated from the pairs of points $\{(Z(s+h)-Z(s+1), Z(s+h+1)-Z(s)): s=1,\ldots,n-h-1\}$. One such estimator, calculated from pairs of points $\{(X_i, Y_i): i=1,\ldots,m\}$, is the weighted median

$$\text{wtd med}\{Y_i/X_i; |X_i|\} \tag{2.4.16}$$

(Scholz, 1978; Sievers, 1978; Cressie and Keightley, 1981). For the purposes of definition of (2.4.16), assume that the sequence of values $\{Y_i/X_i: i = 1,\ldots,m\}$ is ordered from smallest to largest. Define $w_i^* \equiv |X_i|/\Sigma_{j=1}^m |X_j|$, $i=1,\ldots,m$. Then obtain an interval (or a point)

$$[Y_{l+1}/X_{l+1}, Y_{l+k+1}/X_{l+k+1}], \qquad k \geq 0,$$

such that

$$\sum_{i=1}^{l} w_i^* < \sum_{i=1}^{l+1} w_i^* = \cdots = \sum_{i=1}^{l+k} w_i^* = 1/2 < \sum_{i=1}^{l+k+1} w_i^*;$$

the $k=0$ condition is interpreted as $\Sigma_{i=1}^{l} w_i^* < 1/2 < \Sigma_{i=1}^{l+1} w_i^*$. When $k \geq 1$, the weighted median is defined to be the midpoint of the interval, or when $k=0$ it is simply Y_{l+1}/X_{l+1}.

Suppose then that (2.4.16) defines the robust slope estimator $\tilde{C}_{h-1}$, which from (2.4.15) allows construction of $2\gamma^\dagger(h)$. Although the estimator $2\gamma^\dagger(1)$ is a crucial part of (2.4.15), it is easy to see that its value makes no difference to the kriging weights (and hence the kriging predictor) described in Section 3.2. However, the mean-squared prediction errors do depend (multiplicatively) on the choice of $2\gamma^\dagger(1)$.

A comparison of (2.4.15) along with (2.4.13) is given in Figure 2.14 for the coal-ash data of Section 2.2. The robust slope estimator $2\gamma^\dagger(h)$ was originally suggested by Cressie (1979a), but its statistical properties (bias, variance, etc.) are not well understood. One disadvantageous property of $2\gamma^\dagger$ is the possibility that it could occasionally go negative. The estimators $2\gamma^\dagger(h)$, $h=1,2,\ldots,$ each rely on previous values $2\gamma^\dagger(h-k)$, $k=1,\ldots,h-1$, which give the variogram plot (against lag h) a pleasing smoothness; see Figure 2.14. At the same time, this strong dependence makes distribution theory difficult.

Other Approaches

More traditionally, the covariogram $C(\mathbf{h})$ or correlogram $\rho(\mathbf{h}) = C(\mathbf{h})/C(\mathbf{0})$ might be estimated. However, some caution is needed; $C(\cdot)$ and $\rho(\cdot)$ are well defined when $Z(\cdot)$ is second-order stationary, but they may not exist when $Z(\cdot)$ is intrinsically stationary (Section 2.3).

Robust estimators of covariance could be used to obtain an estimator of $\text{cov}(Z(\mathbf{s}), Z(\mathbf{s}+\mathbf{h}))$. For example, Devlin et al. (1975), Huber (1981, Chapter 8), and Li and Chen (1985) discuss various approaches to robust estimation of

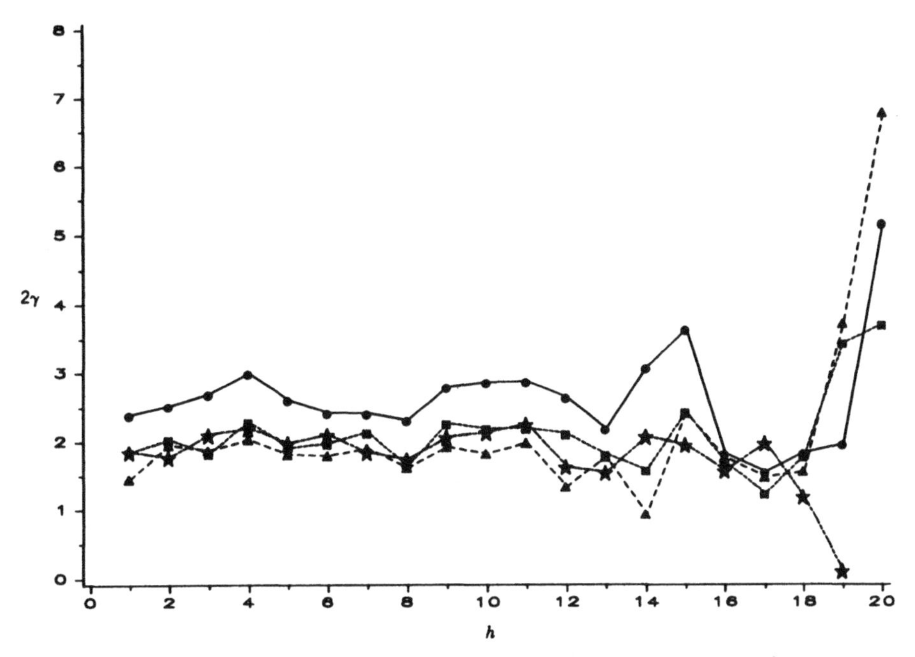

Figure 2.14 Estimated north–south variograms of coal-ash data. Circles denote $2\hat{\gamma}$ the classical estimator (2.4.2); squares denote $2\bar{\gamma}$ the robust estimator (2.4.12); triangles denote $2\tilde{\gamma}$ the robust estimator (2.4.13); stars denote $2\gamma^{\dagger}$ the robust estimator (2.4.15). See Table 2.3 for the plotted values. Units are the same as in Figure 2.11.

$\mathrm{cov}(X, Y)$, the covariance between two random variables X and Y. For that matter, if a robust estimator of scale is available, the obvious relation

$$\mathrm{cov}(X, Y) = \{\mathrm{var}(X + Y) - \mathrm{var}(X - Y)\}/4$$

yields a robust covariance estimator when robust scale estimators are used on the right-hand side. Scale estimators, such as the interquantile range previously suggested, the median of the absolute deviations (MAD), or the bisquare-based estimator, are all possibilities (Lax, 1985).

A Robust Estimator of First-Order Serial Correlation

When the data $\{Z(s): s = 1, \ldots, n\}$ are equally spaced on a transect, a quick way to see whether they are white noise is to compute

$$\frac{\sum_{s=1}^{n-1}(Z(s+1) - Z(s))^2}{\sum_{s=1}^{n}(Z(s) - \bar{Z})^2},$$

which has been presented in Section 2.4.2 as a Durbin–Watson-type statistic.

After noting a small edge effect, it is approximately equal to

$$2\left\{1-\frac{\sum_{s=1}^{n-1}(Z(s+1)-\bar{Z})(Z(s)-\bar{Z})/(n-1)}{\sum_{s=1}^{n}(Z(s)-\bar{Z})^2/n}\right\};$$

the second term inside the braces is an estimator of the (assumed) first-order serial correlation ρ. A way of turning this relationship into a robust estimator for ρ is by working with square-root absolute differences [see (2.4.12)]:

$$\tilde{\rho} \equiv 1-\left(\frac{1}{2}\right)\left\{\frac{\sum_{s=1}^{n-1}|Z(s+1)-Z(s)|^{1/2}/(n-1)}{\sum_{s=1}^{n}|Z(s)-\tilde{Z}|^{1/2}/n}\right\}^4, \tag{2.4.17}$$

where $\tilde{Z} \equiv \text{med}\{Z(1),\ldots,Z(n)\}$, the median of the data.

Statistical Properties of $2\bar{\gamma}$

The estimator $2\bar{\gamma}(\cdot)$, based on the average of the square-root differences, is given by (2.4.12). Under the same mixing conditions referred to in Section 2.4.1, $2\bar{\gamma}(\mathbf{h})$ is asymptotically the fourth power of a Gaussian random variable (Davis and Borgman, 1982; Cressie, 1985a). This is also verified in finite samples by Baczkowski and Mardia (1987, Table II), although there is evidence of a small amount of excess skewness to the right. The quantity $(0.457 + 0.494/|N(\mathbf{h})|)$ in the denominator of (2.4.12) is chosen to achieve approximate unbiasedness for the estimator; that is, $E(2\bar{\gamma}(\mathbf{h})) \simeq 2\gamma(\mathbf{h})$. Cressie and Hawkins (1980) present details of the calculation. The variances and covariances of $\{2\bar{\gamma}(\mathbf{h}(j)): j = 1,\ldots\}$ are given in Cressie (1985a).

The robustness of $2\bar{\gamma}$ compared to $2\hat{\gamma}$ will now be investigated. Robustness is a model-based concept; the model I shall consider here is the same one used by Hawkins and Cressie (1984) and in Section 3.3, namely,

$$Z(\mathbf{s}) = \mu + W(\mathbf{s}) + E(\mathbf{s}),$$

where $\mu \in \mathbb{R}$, $W(\cdot)$ is a zero-mean intrinsically stationary Gaussian process whose variogram is continuous at the origin (i.e., W is continuous in mean square; see Section 2.3.1), and $E(\cdot)$ is a zero-mean white-noise process, independent of $W(\cdot)$, whose departure from Gaussianity is modeled by

$$E(\mathbf{s}) \sim \begin{cases} \text{Gau}(0, c_0), & \text{with probability } 1-\epsilon, \\ \text{Gau}(0, k^2 c_0), & \text{with probability } \epsilon, \end{cases} \tag{2.4.18}$$

where $k^2 \gg 1$. This distribution is a contaminated Gaussian with ϵ measuring the amount of contamination; it quantifies Winsor's principle that "all data are Gaussian in the middle" (Mosteller and Tukey, 1977, p. 12). In a time-series context, Fox (1972) calls it the *additive outliers model*, and Abraham and Box (1979) call it the *aberrant observation model.*

Now

$$2\gamma(\mathbf{h}) = 2\gamma_W(\mathbf{h}) + 2c_0\{1 + \epsilon(k^2 - 1)\}, \qquad (2.4.19)$$

where $2\gamma_W$ is the variogram of the W process. When $\epsilon = 0$,

$$2\gamma(\mathbf{h}) = 2\gamma_W(\mathbf{h}) + 2c_0. \qquad (2.4.20)$$

The goal of a robust estimator of the variogram is to estimate the uncontaminated part, namely, $2\gamma_W(\mathbf{h}) + 2c_0$. That is not to say that unusually high values are not of interest (in mining problems, they are of paramount interest), but rather that if they are unusual in relation to the surrounding samples, they should be set aside for special consideration. They should not be allowed to influence estimation of the underlying spatial dependence.

To illustrate the comparison of $2\bar{\gamma}$ with $2\hat{\gamma}$, the values $\epsilon = 0.05$, $k^2 = 9$, and $c_0 = 1$ were chosen. Let $g = \gamma_W(\mathbf{h})/c_0$; then Table 2.2 shows the respective asymptotic biases caused by the contamination model (2.4.18). Clearly, the bias of $2\bar{\gamma}$ is less than that of $2\hat{\gamma}$, although proportionately less so as g increases; g is a measure of the relative importance of the nugget effect c_0 (Section 2.3.1).

The behavior of the standardized variances in parentheses in Table 2.2 is more complex than that of the biases. It can be seen that

$$\begin{aligned} \mathrm{var}(2\bar{\gamma}) &< \mathrm{var}(2\hat{\gamma}), \quad \text{for } g \text{ small},\\ \mathrm{var}(2\bar{\gamma}) &> \mathrm{var}(2\hat{\gamma}), \quad \text{for } g \text{ large}. \end{aligned}$$

Because g is typically small for small lags, it is clear that $2\bar{\gamma}$ is to be preferred, at least close to the origin. In fact, Table 2.2 guarantees the superiority of $2\bar{\gamma}$ if the nugget effect (c_0) is greater than 50% of any semivariogram value ($\gamma_W(\mathbf{h}) + c_0$). The simulations of McBratney and Webster (1986) show that a blanket recommendation for either $2\bar{\gamma}$ or $2\hat{\gamma}$ is not possible. However, if contamination of Gaussian data occurs, $2\bar{\gamma}$ handles that contamination much better, and so I recommend computation and comparison of *both* estimates.

One further comparison between $2\bar{\gamma}$ [defined by (2.4.12)] and $2\hat{\gamma}$ [defined by (2.4.2)] is made on the coal-ash data analyzed in an exploratory manner in Section 2.2. For completeness, I have also computed the median-based estimates $2\tilde{\gamma}$, given by (2.4.13), with $B(\mathbf{h}) = 0.457 + 0.494/\,|N(\mathbf{h})|$, and the robust-slope estimate $2\gamma^{\dagger}$, given by (2.4.15), with $2\gamma^{\dagger}(1) \equiv 2\bar{\gamma}(1)$. The four estimates were computed in the north-south direction; exploratory spatial data analysis in Section 2.2 indicated the data were trend-free in this direction. The variogram estimates are given in Table 2.3 and plotted in Figure 2.14.

Table 2.2 The Large-Sample Values of $2\bar{\gamma}$ and $2\hat{\gamma}$ under Model (2.4.18) with $\epsilon = 0.05$, $k^2 = 9$, and $c_0 = 1$[a]

	g					
	0	1/4	1/2	1	2	4
$2\bar{\gamma}$	2.42 (19.2)	2.95 (27.8)	3.48 (37.9)	4.52 (62.5)	6.57 (129.2)	10.63 (332.4)
$2\hat{\gamma}$	2.80 (33.9)	3.30 (40.0)	3.80 (47.1)	4.80 (64.3)	6.80 (110.7)	10.80 (251.5)
$2(\gamma_W + c_0)$	2.00	2.50	3.00	4.00	6.00	10.00

Source: Hawkins and Cressie (1984).

[a]Limiting expected values (with n-standardized variances in parentheses) are given.

There is just one large outlier in the coal-ash data, and its effect on the classical estimate is not large. Therefore, the estimated variogram shapes in Figure 2.14 are very similar. As expected, from theoretical considerations (see Table 2.2), the robust variogram estimates are generally smaller than the classical estimates.

Some Final Remarks

Instead of estimating the variogram robustly, one might declare the variogram an unstable parameter and estimate, instead, a more stable parameter

Table 2.3 Variogram Estimates and Number of Pairs Used to Compute Them

		Variogram Estimator			
h	$\lvert N(h)\rvert$	$2\hat{\gamma}$	$2\bar{\gamma}$	$2\tilde{\gamma}$	$2\gamma^{\dagger a}$
1	186	2.40	1.86	1.46	1.86
2	171	2.53	2.04	1.96	1.78 (148)
3	155	2.70	1.84	1.88	2.11 (136)
4	145	3.00	2.30	2.04	2.20 (127)
5	134	2.62	1.92	1.82	1.99 (117)
6	123	2.43	1.99	1.80	2.11 (106)
7	111	2.42	2.14	1.91	1.85 (97)
8	102	2.32	1.65	1.62	1.75 (90)
9	94	2.80	2.27	1.93	2.07 (83)
10	87	2.87	2.20	1.83	2.14 (76)
11	77	2.88	2.20	1.99	2.26 (66)
12	67	2.65	2.12	1.34	1.64 (56)
13	57	2.19	1.82	1.78	1.55 (47)
14	48	3.08	1.59	0.93	2.08 (39)
15	40	3.65	2.44	2.42	1.96 (31)
16	32	1.83	1.70	1.77	1.58 (24)
17	25	1.56	1.24	1.48	1.98 (17)
18	17	1.84	1.79	1.56	1.19 (10)
19	10	1.97	3.45	3.73	0.12 (4)

[a]Number of pairs in $\bar{\bar{C}}_{h-1}$ is given in parentheses.

of the constant-mean Z process, such as

$$E(|Z(\mathbf{s}+\mathbf{h}) - Z(\mathbf{s})|^{w}), \qquad 0 < w \le 2; \tag{2.4.21}$$

the usual variogram is obtained when $w = 2$. For *w-stationary processes* (Miamee and Pourahmadi, 1988), (2.4.21) will depend only on $\mathbf{h}$, but these are not the only processes that have this property. Journel (1988a) obtains kriging predictors based on (2.4.21), analogous to those developed in Chapter 3. However, these predictors lose their attractive interpretation of minimizing an expected convex function of prediction error.

The behavior of robust estimators for dependent data has been investigated by Gastwirth and Rubin (1975). They consider the special case of location estimation in the presence of serial correlation and show that robust estimators obtain higher efficiency relative to the mean for Gaussian processes with *positive* correlation than for independent *Gaussian* observations. For *long-tailed* symmetric distributions, positive correlation acts as an equalizer, in that the mean does not lose as much efficiency as for independence. This helps to justify the emphasis placed on $2\bar{\gamma}$ given by (2.4.12) over $2\tilde{\gamma}$ given by (2.4.13).

2.5* SPECTRAL REPRESENTATIONS

It is well known in the analysis of time series that covariograms fitted to data can only take special forms. In general, these functions $C(\cdot)$ in $\mathbb{R}^d$ must be positive-definite; that is,

$$\sum_{i=1}^{m}\sum_{j=1}^{m} a_i a_j C(\mathbf{s}_i - \mathbf{s}_j) \ge 0, \tag{2.5.1}$$

for any finite number of spatial locations $\{\mathbf{s}_i: i = 1,\ldots,m\}$ and real numbers $\{a_i: i = 1,\ldots,m\}$. It will be shown subsequently that this property characterizes the covariogram of a second-order stationary process.

Section 2.3 shows that second-order stationary models are also intrinsically stationary with variogram

$$2\gamma(\mathbf{h}) = 2(C(\mathbf{0}) - C(\mathbf{h})). \tag{2.5.2}$$

In Section 2.5.1, emphasis will be on constructing valid Cs; (2.5.2) can then be used to convert them to variograms. In Section 2.5.2, the class of valid 2γs will be constructed.

2.5.1* Valid Covariograms

It was stated in Section 2.3 that $C(\mathbf{h}) \equiv \text{cov}(Z(\mathbf{s}+\mathbf{h}), Z(\mathbf{s}))$ of a second-order stationary stochastic process $\{Z(\mathbf{s}): \mathbf{s} \in D\}$ is positive-definite. This is simply expressing the necessary requirement of the model that $\text{var}(\Sigma_{i=1}^m a_i Z(\mathbf{s}_i)) = \Sigma_{i=1}^m \Sigma_{j=1}^m a_i a_j C(\mathbf{s}_i - \mathbf{s}_j)$ be nonnegative.

Is it true that any positive-definite function $C(\cdot)$ corresponds to the covariogram of a second-order stationary (possibly complex-valued) stochastic process? The answer is basically "Yes," and the key result is Bochner's Theorem (see, e.g., Bochner, 1955), which says that a continuous function $C(\cdot)$ is positive-definite if and only if it has a spectral representation

$$C(\mathbf{h}) = \int_{-\infty}^{\infty} \cdots \int_{-\infty}^{\infty} \cos(\boldsymbol{\omega}'\mathbf{h}) G(d\boldsymbol{\omega}), \tag{2.5.3}$$

where G is a positive bounded symmetric measure; $G/C(\mathbf{0})$ is often referred to as the *spectral distribution function*. If $G(d\boldsymbol{\omega})$ can be written as $g(\boldsymbol{\omega})\, d\boldsymbol{\omega}$, where $g \geq 0$ (this happens when $\int_{-\infty}^{\infty} \cdots \int_{-\infty}^{\infty} |C(\mathbf{h})|\, d\mathbf{h} < \infty$), then $g/C(\mathbf{0})$ is called the *spectral density*, where $g(\boldsymbol{\omega}) = g(-\boldsymbol{\omega})$. Now suppose $\{W(\boldsymbol{\omega}): \boldsymbol{\omega} \in \mathbb{R}^d\}$ is a complex-valued d-dimensional zero-mean stochastic process of independent increments, where $E(|W(d\boldsymbol{\omega})|^2) = G(d\boldsymbol{\omega})$ and $\int_{-\infty}^{\infty} \cdots \int_{-\infty}^{\infty} G(d\boldsymbol{\omega}) < \infty$. Then the stochastic process

$$Z(\mathbf{s}) \equiv \int_{-\infty}^{\infty} \cdots \int_{-\infty}^{\infty} e^{i\boldsymbol{\omega}'\mathbf{s}} W(d\boldsymbol{\omega}) \tag{2.5.4}$$

has covariogram $C(\mathbf{h}) = E(Z(\mathbf{s}+\mathbf{h}) \cdot \bar{Z}(\mathbf{s}))$ given by (2.5.3), where here $\bar{z}$ denotes the complex conjugate of z. The integral in (2.5.4) is defined as a limit (in quadratic mean) of finite approximating sums. Yaglom (1957) gives a rigorous treatment of the spectral representations (2.5.3) and (2.5.4) in $\mathbb{R}^d$. In the rest of this section, $C(\cdot)$ is assumed continuous unless otherwise specified.

Construction of Covariogram Models in $\mathbb{R}^d$

First, a word of caution. A valid isotropic covariogram in $\mathbb{R}^{d_1}$ may not be a valid isotropic covariogram in $\mathbb{R}^{d_2}$, where $d_2 > d_1$. For example, the tent covariogram

$$C_T(h) \equiv \begin{cases} \sigma^2(1 - |h|/a), & 0 \leq |h| \leq a, \\ 0, & |h| > a \end{cases}$$

is valid in $\mathbb{R}^1$. But, for $\mathbf{h} \in \mathbb{R}^2$,

$$C(\mathbf{h}) \equiv C_T(\|\mathbf{h}\|)$$

is *not* valid (e.g., Christakos, 1984). A simple counterexample can be obtained from an 8×8 square grid $\{\mathbf{s}_{ij}: i = 1, \ldots, 8,\ j = 1, \ldots, 8\}$ of spacing

$a/2^{1/2}$, with $\{a_{ij}\}$ in (2.5.1) alternately $+1$ and -1. However, it is always true that a valid isotropic covariogram in $\mathbb{R}^{d_2}$ is also valid in $\mathbb{R}^{d_1}$, where $d_2 > d_1$ (Matern, 1960, p. 13). Blanc-Lapierre and Faure (1965) discuss these validity issues in some detail.

If $C_1(\cdot)$ and $C_2(\cdot)$ are two valid covariograms in $\mathbb{R}^d$, then it is obvious from the representation (2.5.3) that $C_1(\cdot) + C_2(\cdot)$ is a valid covariogram in $\mathbb{R}^d$. Likewise, for $b > 0$, $b \cdot C_1(\cdot)$ is a valid covariogram in $\mathbb{R}^d$. Together, these two results imply that the family of valid covariograms is a convex cone.

Separable covariograms in $\mathbb{R}^d$ are those for which $C(\mathbf{h}) = \prod_{i=1}^{d} C_i(h_i)$, $\mathbf{h} = (h_1, \ldots, h_d)' \in \mathbb{R}^d$, where the components in the product are all covariograms in $\mathbb{R}^1$. It is easy to see that if $G_i(\cdot)$ denotes the spectral measure of $C_i(\cdot)$ in the representation (2.5.3), $i = 1, \ldots, d$, then $C(\cdot)$ also has a representation given by (2.5.3), where $G(d\boldsymbol{\omega}) = \prod_{i=1}^{d} G_i(d\omega_i)$. Thus $C(\cdot)$ is a valid covariogram in $\mathbb{R}^d$.

Isotropic covariograms will now be constructed; Section 2.3.1 contains details of how these can in turn be used to develop more general covariograms. Write the covariogram as

$$C(\mathbf{h}) = C^o(\|\mathbf{h}\|), \qquad \mathbf{h} \in \mathbb{R}^d. \tag{2.5.5}$$

In this case, the spectral representation (2.5.3) becomes

$$C^o(h) = \int_0^\infty Y_d(\omega h)\, \Phi(d\omega),$$

where $Y_d(t) \equiv (2/t)^{(d-2)/2}\Gamma(d/2)J_{(d-2)/2}(t)$, $J_\nu(\cdot)$ is the Bessel function of the first kind of order ν, and Φ is a nondecreasing function on $[0, \infty)$ such that $\int_0^\infty \Phi(d\omega) < \infty$; see Yaglom (1957, p. 299).

Let I_d denote the class of all valid isotropic covariograms in $\mathbb{R}^d$ [i.e., C is given by (2.5.5)]. Then $I_1 \supset I_2 \supset I_3 \supset \cdots$. Let $I'_d \subset I_d$ be the set of all isotropic covariograms continuous everywhere except at the origin and let I''_d be the set of all isotropic covariograms continuous everywhere. It is easy to see that any $C \in I'_d$ can be written as $C = a \cdot \delta + b \cdot C_1$, where $a > 0$, $b > 0$, $C_1 \in I''_d$, and $\delta(\mathbf{h}) = 0$ unless $\mathbf{h} = \mathbf{0}$, in which case $\delta(\mathbf{0}) = 1$. Matern (1960) derives important properties of these classes, which in turn enables him to construct valid covariogram models in $\mathbb{R}^d$. Several members of I''_d are:

$$C(\mathbf{h}) = \sigma^2 \exp(-a^2\|\mathbf{h}\|^2), \qquad \mathbf{h} \in \mathbb{R}^d,$$

$$C(\mathbf{h}) = \sigma^2\left\{1 + \left(\|\mathbf{h}\|^2/b^2\right)\right\}^{-\beta}, \qquad \mathbf{h} \in \mathbb{R}^d, \beta > 0,$$

$$C(\mathbf{h}) = \frac{\sigma^2(a^2\|\mathbf{h}\|/2)^\nu 2K_\nu(a^2\|\mathbf{h}\|)}{\Gamma(\nu)}, \qquad \mathbf{h} \in \mathbb{R}^d, \nu > 0,$$

where K_ν is the modified Bessel function of the second kind, $\nu = 1/2$ yields

$$C(\mathbf{h}) = \sigma^2 \exp(-a^2\|\mathbf{h}\|), \qquad \mathbf{h} \in \mathbb{R}^d,$$

and $\nu = 1$ yields

$$C(\mathbf{h}) = \sigma^2 a^2\|\mathbf{h}\| K_1(a^2\|\mathbf{h}\|), \qquad \mathbf{h} \in \mathbb{R}^d.$$

Whittle (1954) proposes that, in $\mathbb{R}^2$, the case $\nu = 1$ is more appropriate than the case $\nu = 1/2$ (although both are valid covariograms).

Sources for these and other valid covariogram models are Whittle (1954, 1962), Heine (1955), Matern (1960), Pearce (1976), Mantoglou and Wilson (1982), Christakos (1984), and Vecchia (1985).

Checking for Validity

How can a proposed covariogram model be checked for validity (i.e., for positive-definiteness)? When $C(\cdot)$ is continuous, it is equivalent that $C(\cdot)/C(\mathbf{0})$ be a characteristic function. In $\mathbb{R}^1$, sufficient conditions for this are due to Polya: $C(h) \geq 0$, $C(h) = C(-h)$, and C is decreasing and convex in $[0, \infty)$ (see Chung, 1968, Section 6.5).

A real function $f(t)$ is said to be completely monotone for $t \geq 0$ if $f(0) = f(0+)$ and $(-1)^n d^n f(t)/dt^n \geq 0$, for $0 < t < \infty$ and $n = 0, 1, \ldots$. Then Schoenberg (1938, p. 821) establishes that $C(\mathbf{h}) \equiv C^o(\|\mathbf{h}\|)$, $\mathbf{h} \in \mathbb{R}^d$, is positive-definite for all $d = 1, 2, \ldots$ if and only if $f(t) \equiv C^o(t^{1/2})$ is completely monotone for $t \geq 0$.

Furthermore, Christakos (1984) shows that the following three conditions are sufficient for positive-definiteness of $C(\mathbf{h}) \equiv C^o(\|\mathbf{h}\|)$, $\mathbf{h} \in \mathbb{R}^d$:

i. At $h = 0$, $dC^o(h)/dh < 0$.

ii. $\lim_{h \to \infty} C^o(h)/h^{(1-d)/2} = 0$.

iii. $d^2C^o(h)/dh^2 \geq 0$, in $\mathbb{R}^1$; $\int_h^\infty u(u^2 - r^2)^{-1/2}(d^3C^o(h)/dh^3)\,dh \geq 0$, in $\mathbb{R}^2$; $d^2C^o(h)/dh^2 - h(d^3C^o(h)/dh^3) \geq 0$, in $\mathbb{R}^3$.

Derivates at $h = 0$ are one-sided, taken from the right.

2.5.2* Valid Variograms

Just as covariograms can only take special forms, variograms $2\gamma(\cdot)$ in $\mathbb{R}^d$ must be *conditionally negative-definite*; that is,

$$\sum_{i=1}^{m} \sum_{j=1}^{m} a_i a_j 2\gamma(\mathbf{s}_i - \mathbf{s}_j) \leq 0, \tag{2.5.6}$$

for any finite number of spatial locations $\{\mathbf{s}_i: i = 1, \ldots, m\}$ and real numbers $\{a_i: i = 1, \ldots, m\}$ satisfying $\sum_{i=1}^m a_i = 0$.

Suppose for the moment that $Z(\cdot)$ is an intrinsically stationary process [i.e., it has constant mean and possesses a variogram $2\gamma(\cdot)$ defined by (2.3.4)] and that $\{\mathbf{s}_i\}$ and $\{a_i\}$ are defined as before. Then

$$\left\{\sum_{i=1}^{m} a_i Z(\mathbf{s}_i)\right\}^2 = -\left(\frac{1}{2}\right)\left\{\sum_{i=1}^{m}\sum_{j=1}^{m} a_i a_j \left(Z(\mathbf{s}_i) - Z(\mathbf{s}_j)\right)^2\right\},$$

because $\sum_{i=1}^{m} a_i = 0$. Upon taking expectations, one obtains

$$\sum_{i=1}^{m}\sum_{j=1}^{m} a_i a_j 2\gamma(\mathbf{s}_i - \mathbf{s}_j) = -2\,\mathrm{var}\left(\sum_{i=1}^{m} a_i Z(\mathbf{s}_i)\right) \leq 0.$$

Is it true that any conditionally negative-definite function $2\gamma(\cdot)$ corresponds to an intrinsically stationary (possibly complex-valued) stochastic process? The answer is basically "Yes"; just as for covariograms the result is established through a spectral representation. Results in Schoenberg (1938, p. 828ff) and Yaglom (1957, p. 289) can be combined to prove the following theorem (see also Johansen, 1966, p. 305).

Theorem If $2\gamma(\cdot)$ is a continuous function on $\mathbb{R}^d$ satisfying $\gamma(\mathbf{0}) = 0$, then the following three properties are equivalent:

i. $2\gamma(\cdot)$ is conditionally negative-definite.
ii. For all $a > 0$, $\exp(-a\gamma(\cdot))$ is positive-definite.
iii. $2\gamma(\cdot)$ is of the form

$$2\gamma(\mathbf{h}) = Q(\mathbf{h}) + \int_{-\infty}^{\infty} \cdots \int_{-\infty}^{\infty} \frac{1 - \cos(\boldsymbol{\omega}'\mathbf{h})}{\|\boldsymbol{\omega}\|^2} G(d\boldsymbol{\omega}), \quad (2.5.7)$$

where $Q(\cdot) \geq 0$ is a quadratic form and $G(\cdot)$ is a positive, symmetric measure continuous at the origin that satisfies $\int_{-\infty}^{\infty} \cdots \int_{-\infty}^{\infty}(1 + \|\boldsymbol{\omega}\|^2)^{-1} G(d\boldsymbol{\omega}) < \infty$. ■

Now define a stochastic process $Z(\cdot)$ by

$$Z(\mathbf{s}) \equiv \int_{-\infty}^{\infty} \cdots \int_{-\infty}^{\infty} \frac{e^{i\boldsymbol{\omega}'\mathbf{s}} - 1}{\|\boldsymbol{\omega}\|} W(d\boldsymbol{\omega}), \quad (2.5.8)$$

where $\{W(\mathbf{s}): \mathbf{s} \in \mathbb{R}^d\}$ is a complex-valued d-dimensional zero-mean stochastic process of independent increments such that $E(|W(d\boldsymbol{\omega})|^2) = G(d\boldsymbol{\omega})/2$. Then

$$Z(\mathbf{s} + \mathbf{h}) - Z(\mathbf{s}) = \int_{-\infty}^{\infty} \cdots \int_{-\infty}^{\infty} e^{i\boldsymbol{\omega}'\mathbf{s}} W_{\mathbf{h}}^*(d\boldsymbol{\omega}),$$

where $W_{\mathbf{h}}^*$ is a process of independent increments such that

$$E\left(|W_{\mathbf{h}}^*(d\boldsymbol{\omega})|^2\right) = G_{\mathbf{h}}^*(d\boldsymbol{\omega})$$

and

$$G_{\mathbf{h}}^*(\boldsymbol{\omega}) = \int_{-\infty}^{\omega_1} \cdots \int_{-\infty}^{\omega_d} \frac{1 - \cos(\boldsymbol{\eta}'\mathbf{h})}{\|\boldsymbol{\eta}\|^2} G(d\boldsymbol{\eta}).$$

Thus, a stochastic process defined by (2.5.8) has variogram $2\gamma(\mathbf{h}) \equiv E(|Z(\mathbf{s} + \mathbf{h}) - Z(\mathbf{s})|^2)$ given by

$$2\gamma(\mathbf{h}) = \int_{-\infty}^{\infty} \cdots \int_{-\infty}^{\infty} \frac{1 - \cos(\boldsymbol{\omega}'\mathbf{h})}{\|\boldsymbol{\omega}\|^2} G(d\boldsymbol{\omega}),$$

which is of the form (2.5.7) with $Q(\mathbf{h}) = 0$.

Notice that, in general, I define the variogram as $\mathrm{var}(Z(\mathbf{s} + \mathbf{h}) - Z(\mathbf{s}))$; unfortunately, it is sometimes defined as $E(Z(\mathbf{s} + \mathbf{h}) - Z(\mathbf{s}))^2$, and in this case $Q(\mathbf{h})$ could be positive. To summarize, any conditionally negative-definite function with $Q(\mathbf{h}) = 0$ in (2.5.7) is a valid variogram of an intrinsically stationary stochastic process.

Construction of Variogram Models in $\mathbb{R}^d$

First, notice that a valid isotropic variogram in $\mathbb{R}^{d_2}$ is also valid in $\mathbb{R}^{d_1}$, where $d_2 > d_1$, but not conversely. Second, if $2\gamma_1(\cdot)$ and $2\gamma_2(\cdot)$ are valid variograms in $\mathbb{R}^d$, then $2\gamma(\cdot) \equiv 2\gamma_1(\cdot) + 2\gamma_2(\cdot)$ is a valid variogram in $\mathbb{R}^d$. Likewise, for $b > 0$, $b\{2\gamma_1(\cdot)\}$ is a valid variogram in $\mathbb{R}^d$. Thus, the family of valid variograms is a convex cone.

I shall now concentrate on constructing isotropic variograms given by

$$2\gamma(\mathbf{h}) = 2\gamma^o(\|\mathbf{h}\|), \qquad \mathbf{h} \in \mathbb{R}^d, \tag{2.5.9}$$

where $2\gamma^o(\cdot)$ is a function in $\mathbb{R}^1$. (Section 2.3.1 shows how more general variograms follow.) In this case, the spectral representation (2.5.7) becomes

$$2\gamma^o(h) = \int_0^{\infty} \frac{(1 - Y_d(\omega h))}{\omega^2} \Phi(d\omega), \tag{2.5.10}$$

where Φ is a nondecreasing function on $(0, \infty)$ such that

$$\int_0^{\infty} (1 + \omega^2)^{-1} \Phi(d\omega) < \infty$$

and recall $Y_d(t) = (2/t)^{(d-2)/2}\Gamma(d/2)J_{(d-2)/2}(t)$; $J_\nu(\cdot)$ is the Bessel function of the first kind of order ν. For more details see Yaglom (1957, p. 310).

The models for covariograms $C(\cdot)$ constructed in Section 2.5.1 also generate models for variograms by using the relation $2\gamma(\mathbf{h}) = 2(C(\mathbf{0}) - C(\mathbf{h}))$.

However, more general models are possible for which $2\gamma(\cdot)$ is unbounded. A study of Yaglom (1957, Section 5) and Christakos (1984) yields the family of valid semivariogram models

$$\gamma(\mathbf{h}) = a^2\|\mathbf{h}\|^\alpha, \qquad \mathbf{h} \in \mathbb{R}^d, 0 \leq \alpha < 2,$$

which is the semivariogram of a fractional isotropic Brownian motion in $\mathbb{R}^d$ (Section 5.5). It is these that Whittle (1962) argues are natural to use in analyzing agricultural field trials (in $\mathbb{R}^2$).

Schoenberg (1938) establishes that

$$\gamma(\mathbf{h}) = a^2\|\mathbf{h}\|^2 \big/ \left(1 + \|\mathbf{h}\|^2/b^2\right), \qquad \mathbf{h} \in \mathbb{R}^d,$$

is a valid semivariogram model, which is called the rational quadratic model in Section 2.3.1. This is derivable from one of the families of covariance models given in Section 2.5.1. Another model derivable from Section 2.5.1 is the so-called Gaussian semivariogram model

$$\gamma(\mathbf{h}) = a^2\left\{1 - \exp\left(-\|\mathbf{h}\|^2/b^2\right)\right\}, \qquad \mathbf{h} \in \mathbb{R}^d.$$

The corresponding stochastic process has very smooth sample paths (derivatives of all orders almost surely exist), which are unlikely to be found in natural phenomena.

Checking for Validity

Christakos (1984) establishes the result that $2\gamma(\cdot)$ is a valid variogram if and only if G in (2.5.7) is continuous at the origin, $G(d\boldsymbol{\omega}) \geq 0$, and $\lim_{\|\mathbf{h}\| \to \infty} \gamma(\mathbf{h})/\|\mathbf{h}\|^2 = 0$ (see also Armstrong and Diamond, 1984, for a similar approach). Thus, a necessary condition for $2\gamma(\cdot)$ to be a variogram is that it must grow more slowly than $\|\mathbf{h}\|^2$.

Another way of checking the validity of $2\gamma(\cdot)$ is to see whether $\exp(-a\gamma(\cdot))$ is positive-definite, for all $a > 0$. The results of the previous section (Section 2.5.1) can be used to do this.

The Spectral Approach

Many of the validity results for variograms are easy to obtain when using the spectral representation (2.5.7); Christakos (1984) is an excellent source for further results. For example, he demonstrates that the (isotropic) spherical model given by (2.3.8) (often used in geostatistical applications) is valid in $\mathbb{R}^3$ (and hence in $\mathbb{R}^2$ and $\mathbb{R}^1$), he finds conditions that ensure variograms such as $2\gamma(\mathbf{h}) = 2\sigma^2\{1 - \exp(-\|\mathbf{h}\|/b^2)\cos(a^2\|\mathbf{h}\|)\}$ are valid (e.g., in $\mathbb{R}^3$ it is necessary that $a^2b^2 < 0.577$), and so on.

There are stochastic processes that are even more general than intrinsically stationary processes. Matheron (1973) has called them intrinsic random functions of order k (IRFks); an intrinsically stationary process is an IRF0. These are considered in more detail in Section 5.4; let it suffice to say here that the variogram can be generalized to a quantity called the generalized covariance of order k. It too has a spectral representation, which can be found in Christakos (1984). Construction of valid generalized covariances then proceeds in the same manner as for valid variograms.

Some Final Remarks

The importance of this section is brought out by scanning the scientific literature and noting that, occasionally, otherwise well reasoned scientific articles have been marred by fitting invalid covariance models. Calculations based on these models run the risk of yielding embarrassing *negative variances*.

A number of researchers have stated that the (square root of the) variogram can be thought of as a distance measure between two locations $\mathbf{s}$, $\mathbf{u}$ in $\mathbb{R}^d$, in place of the Euclidean distance. This intuitive notion can be made rigorous (von Neumann and Schoenberg, 1941), due principally to the conditional negative-definiteness property that characterizes a variogram.

In time-series analysis, debate has also focused on whether the raw covariogram estimates (see Section 2.4) should also be positive-definite. The estimate $\{\Sigma_{i=1}^{n-h}(Z(i) - \bar{Z})(Z(i+h) - \bar{Z})/c(n,h)\colon\ h = 1, 2, \ldots, n-1\}$ *is* positive-definite when $c(n,h) = n$, but it *is not* when $c(n,h) = n - h$. In a parametric setting (e.g., the use of autoregressive-moving-average models), this disagreement is irrelevant; what is important is that the *fitted model* be positive-definite (which, e.g., implies restrictions on the autoregressive-moving-average coefficients). The analogous observation applies to variograms; the next section discusses various ways to fit conditionally negative-definite models to spatial data.

2.6 VARIOGRAM MODEL FITTING

The various variogram estimators $2\hat{\gamma}(\cdot)$, $2\gamma^{+}(\cdot)$, $2\bar{\gamma}(\cdot)$, $2\tilde{\gamma}(\cdot)$, $2\gamma^{\dagger}(\cdot)$, and so on presented in Section 2.4 cannot be used directly for spatial prediction (kriging) described in Chapter 3. They are not necessarily conditionally negative-definite; Section 2.5.2 explains why the absence of this property can result in embarrassing negative mean-squared errors of prediction.

The idea then is to search for a valid variogram that, as a measure of spatial dependence, is closest to the spatial dependence present in the data $\mathbf{Z} = (Z(\mathbf{s}_1), \ldots, Z(\mathbf{s}_n))'$. The space of all valid variograms is a large set over which to search, so usually a parametric family of variograms is chosen. For example, suppose the family of isotropic linear variograms is chosen, which from (2.3.7) is

$$\{2\gamma\colon \gamma(\mathbf{h}) = c_0 + b_l\|\mathbf{h}\|;\ c_0 \geq 0,\ b_l \geq 0\}. \tag{2.6.1}$$

Then an element of (2.6.1) is sought that is closest to the data, in some sense.

2.6.1 Criteria for Fitting a Variogram Model

Suppose that a (parametric) subset of valid variograms can be written as

$$P = \{2\gamma : 2\gamma(\cdot) = 2\gamma(\cdot\,;\boldsymbol{\theta});\ \boldsymbol{\theta} \in \Theta\}; \qquad (2.6.2)$$

here, $2\gamma(\mathbf{h};\boldsymbol{\theta})$ is a valid variogram as described in Section 2.5.2. The subset (2.6.1) is an example of such a parametric family of variograms.

Several goodness-of-fit criteria for finding the best element of P have been proposed. There is one that I prefer because it requires the fewest distributional assumptions about $\mathbf{Z}$: Section 2.6.2 discusses various least-squares criteria based on estimators of the variogram presented in Section 2.4.

Although some use of the Gaussian assumption was made in Section 2.4.3, the variogram estimators presented there rely on it only superficially. Each estimator is trying in its own way to capture the spatial dependence; the Gaussian assumption is used simply to determine a constant that ensures approximate unbiasedness of the variogram estimators. For example, the robust variogram estimator $2\bar{\gamma}$ given by (2.4.12) was constructed by Cressie and Hawkins (1980) assuming that the data were Gaussian or contaminated Gaussian. If $Z(\cdot)$ has an invariant distribution (Section 2.3) other than the Gaussian, $2\bar{\gamma}(\cdot)$ in (2.4.12) can be modified; the only term that changes in (2.4.12) is the bias correction in the denominator, namely, $0.457 + 0.494/|N(\mathbf{h})|$.

I shall now discuss the more parametric methods that have been proposed for fitting a variogram model. In practice, care must be taken to ensure that the parametric assumptions are reasonable.

Maximum Likelihood

Estimation procedures that rely crucially on the Gaussian assumption are maximum likelihood (m.l.) and restricted maximum likelihood (REML) estimation of $\boldsymbol{\theta}$ in (2.6.2). The problem with m.l. estimation is that the estimators of $\boldsymbol{\theta}$ are biased, often prohibitively so in small to moderate samples (Matheron, 1971b; Mardia and Marshall, 1984). The simple case when the data $\mathbf{Z}$ are in fact independent multivariate Gaussian, $\mathrm{Gau}(X\boldsymbol{\beta}, \sigma^2 I)$, yields just one small-scale-variation parameter $\theta = \sigma^2$. The m.l. estimator is $\hat{\sigma}^2 = \sum_{i=1}^{n}(Z(\mathbf{s}_i) - X\hat{\boldsymbol{\beta}})^2/n$, where $\hat{\boldsymbol{\beta}}$ is the ordinary-least-squares estimator of the $q \times 1$ vector $\boldsymbol{\beta}$. It is well known that $\hat{\sigma}^2$ is biased and that $(n/(n-q))\hat{\sigma}^2$ is unbiased; the bias-correction factor $(n/(n-q))$ can be appreciable when q is large relative to n.

More generally, suppose that the data $\mathbf{Z}$ are multivariate Gaussian $\mathrm{Gau}(X\boldsymbol{\beta}, \Sigma(\boldsymbol{\theta}))$, where X is an $n \times q$ matrix of rank $q < n$, and that the $n \times n$ matrix $\Sigma(\boldsymbol{\theta}) = (\mathrm{cov}(Z(\mathbf{s}_i), Z(\mathbf{s}_j)))$ depends on $\boldsymbol{\theta}$ through (2.6.2). Then the negative loglikelihood is

$$L(\boldsymbol{\beta},\boldsymbol{\theta}) = (n/2)\log(2\pi) + (1/2)\log|\Sigma(\boldsymbol{\theta})| + (1/2)(\mathbf{Z} - X\boldsymbol{\beta})'\Sigma(\boldsymbol{\theta})^{-1}(\mathbf{Z} - X\boldsymbol{\beta}), \quad \boldsymbol{\beta}\in\mathbb{R}^q, \boldsymbol{\theta}\in\Theta, \quad (2.6.3)$$

and m.l. estimators $\hat{\boldsymbol{\beta}}$ and $\hat{\boldsymbol{\theta}}$ satisfy

$$L(\hat{\boldsymbol{\beta}},\hat{\boldsymbol{\theta}}) = \inf\{L(\boldsymbol{\beta},\boldsymbol{\theta}): \boldsymbol{\beta} \in \mathbb{R}^q, \boldsymbol{\theta} \in \Theta\}.$$

Computational details of solving for $\hat{\boldsymbol{\beta}}$ and $\hat{\boldsymbol{\theta}}$ are given in Section 7.2.3. Many authors have noted that the m.l. estimator of $\boldsymbol{\theta}$ is seriously biased (see Section 2.6.3). Jackknifing $\hat{\boldsymbol{\theta}}$ is one possible remedy (e.g., Miller, 1974), although this could increase computational time by up to a factor of n. Another option, described in the succeeding text, is to filter the data so that the joint distribution no longer depends on $\boldsymbol{\beta}$.

Restricted Maximum Likelihood

In the case of a one-dimensional process with equally spaced data, Kitanidis and Vomvoris (1983) propose that the likelihood function of the data $\mathbf{W} \equiv (Z(1) - Z(2), Z(2) - Z(3), \ldots, Z(n-1) - Z(n))'$ be maximized. Equivalently, minimize

$$L_W(\boldsymbol{\beta},\boldsymbol{\theta}) = ((n-1)/2)\log(2\pi) + (1/2)\log|A'\Sigma(\boldsymbol{\theta})A| + (1/2)(\mathbf{W} - A'X\boldsymbol{\beta})'(A'\Sigma(\boldsymbol{\theta})A)^{-1}(\mathbf{W} - A'X\boldsymbol{\beta}), \quad (2.6.4)$$

where $A = (a_{ij})$ is an $(n-1) \times n$ matrix whose elements are

$$a_{ij} = \begin{cases} 1, & \text{for } i = j,\ j = 1,\ldots,n-1, \\ -1, & \text{for } i = j+1,\ j = 1,\ldots,n-1, \\ 0, & \text{elsewhere.} \end{cases}$$

Assuming that the process $Z(\cdot)$ has constant mean μ [i.e., $X\boldsymbol{\beta} = (1,\ldots,1)'\mu$], then $A'X\boldsymbol{\beta} = \mathbf{0}$ and hence *L_W does not depend on* $\boldsymbol{\beta}$. The hope is that by sacrificing one observation [$\mathbf{W}$ has $(n-1)$ elements, whereas $\mathbf{Z}$ has n], the estimator of $\boldsymbol{\theta}$ based on L_W will have better bias properties. As a matter of fact, when the elements of $\mathbf{Z}$ are i.i.d. Gau(μ, σ^2), the value of σ^2 that minimizes L_W turns out to be $(n/(n-1))\hat{\sigma}^2$, the bias-corrected m.l. estimator.

Minimizing (2.6.4) to estimate $\boldsymbol{\theta}$ is a special case of a method known as *restricted maximum likelihood* (REML), which was developed by Patterson and Thompson (1971, 1974). REML estimators are obtained by applying maximum likelihood to error contrasts rather than the data themselves. [Rao, 1979, calls this method marginal maximum likelihood (MML) in the context of estimation of variance components. Recently, some authors have also called it residual maximum likelihood, although they have retained the acronym REML.] A linear combination $\mathbf{a}'\mathbf{Z}$ is called an error contrast if $E(\mathbf{a}'\mathbf{Z}) = 0$, for all $\boldsymbol{\beta} \in \mathbb{R}^q$ and $\boldsymbol{\theta} \in \Theta$; thus $\mathbf{a}'\mathbf{Z}$ is an error contrast if and only if $\mathbf{a}'X = \mathbf{0}'$.

Let $\mathbf{W} = A'\mathbf{Z}$ represent a vector of $n - q$ [recall $q = \text{rank}(X)$] linearly independent error contrasts; that is, the $(n - q)$ columns of A are linearly independent and $A'X = 0$. Under the Gaussian assumption (2.6.3), $\mathbf{W} \sim \text{Gau}(\mathbf{0}, A'\Sigma(\boldsymbol{\theta})A)$, which does not depend on $\boldsymbol{\beta}$. Thus, the negative loglikelihood function is

$$L_W(\boldsymbol{\theta}) = ((n - q)/2)\log(2\pi) + (1/2)\log|A'\Sigma(\boldsymbol{\theta})A| + (1/2)\mathbf{W}'(A'\Sigma(\boldsymbol{\theta})A)^{-1}\mathbf{W},$$

which generalizes (2.6.4). If another set of $(n - q)$ linearly independent contrasts were used to define $\mathbf{W}$, the new negative loglikelihood function would differ from $L_W(\boldsymbol{\theta})$ only by an additive constant (Harville, 1974). Indeed, for the A that satisfies $AA' = I - X(X'X)^{-1}X'$ and $A'A = I$,

$$L_W(\boldsymbol{\theta}) = ((n - q)/2)\log(2\pi) - (1/2)\log|X'X| + (1/2)\log|\Sigma(\boldsymbol{\theta})| + (1/2)\log\left|X'\Sigma(\boldsymbol{\theta})^{-1}X\right| + (1/2)\mathbf{Z}'\Pi(\boldsymbol{\theta})\mathbf{Z}, \qquad (2.6.5)$$

where $\Pi(\boldsymbol{\theta}) \equiv \Sigma(\boldsymbol{\theta})^{-1} - \Sigma(\boldsymbol{\theta})^{-1}X(X'\Sigma(\boldsymbol{\theta})^{-1}X)^{-1}X'\Sigma(\boldsymbol{\theta})^{-1}$. A REML estimate of $\boldsymbol{\theta}$ is obtained by minimizing (2.6.5) with respect to $\boldsymbol{\theta}$. The distinction between REML and m.l. estimation becomes important when q is large relative to n; REML does not suffer from the often severe underestimation of $\boldsymbol{\theta}$ that m.l. does. Consequently, kriging variances of universal kriging predictors, obtained by substituting in REML estimators of $\boldsymbol{\theta}$, are often less biased (Section 5.3.3).

The REML method was originally proposed to estimate variance-component parameters: Numerical algorithms (Harville, 1977) and robust adaptations (Fellner, 1986) have been developed in this context, although distribution theory is lacking. (Miller, 1977, has distribution theory for m.l. estimators.) Kitanidis (1983) and Zimmerman (1989a) give computational details for producing an iterative minimization of (2.6.5). Harville (1974) provides a Bayesian justification for REML by assuming a noninformative prior for $\boldsymbol{\beta}$, which is statistically independent of $\boldsymbol{\theta}$, and showing that the marginal posterior density of $\boldsymbol{\theta}$ is proportional to (2.6.5) multiplied by the prior for $\boldsymbol{\theta}$. Thus, when a noninformative prior for $\boldsymbol{\theta}$ is used, REML estimates correspond to marginal maximum *a posteriori* estimates.

MINQ Estimation

Minimum norm quadratic (MINQ) estimation has been developed for the special case where the variance matrix of the data is linear in its parameters:

$$\Sigma(\boldsymbol{\theta}) = \theta_1\Sigma_1 + \cdots + \theta_m\Sigma_m; \qquad (2.6.6)$$

see Rao (1972, 1979). This amounts to searching for an estimator of θ_j

amongst those that can be written as $\hat{\theta}_j = \mathbf{W}'F_j\mathbf{W}$, where $\mathbf{W}$ is the same as that used in (2.6.5). The minimum norm estimator is obtained by minimizing $E(\hat{\theta}_j - \theta_j)^2$, usually subject to unbiasedness or invariance restrictions. Kitanidis (1985) gives formulas for the estimators in a spatial setting, where the data are a sampling from a random process in $\mathbb{R}^d$. Note that the norm he uses is mean-squared error; comparisons with other norms can be found in Kitanidis (1987).

The MINQ approach is particularly appropriate for a variance-components model, but in a spatial setting $\Sigma(\boldsymbol{\theta})$ may be a highly nonlinear function of the small-scale variation parameters $\boldsymbol{\theta}$; for example, let $\mathbf{Z} = (Z(1), \ldots, Z(n))'$ in $\mathbb{R}^1$ covary according to the spherical semivariogram model (2.3.8), where $\boldsymbol{\theta} = (c_0, c_s, a_s)'$. In this case, $\Sigma(\boldsymbol{\theta}) = (c_{ij}(\boldsymbol{\theta}))$, where

$$c_{ij}(\boldsymbol{\theta}) = \begin{cases} c_0 + c_s, & |i-j| = 0, \\ c_s\{1 - (3/2)\,(|i-j|/a_s) \\ \qquad + (1/2)(|i-j|/a_s)^3\}, & 0 < |i-j| < a_s, \\ 0, & |i-j| \geq a_s. \end{cases}$$

In the spatial setting, many authors (Delfiner, 1976; Kitanidis, 1985; Marshall and Mardia, 1985) have hoped that a linear combination of certain (polynomial and spline) generalized covariance functions (Section 5.4) is a sufficiently flexible family to approximate *any* $\Sigma(\boldsymbol{\theta})$. This optimism is unfounded, even in the simplest case of the 0th order generalized covariance function $K(\mathbf{h}) = -\|\mathbf{h}\|$, which cannot approximate, say, the spherical semivariogram given by (2.3.8), for $\|\mathbf{h}\| \geq a_s$.

In order to keep the covariance matrix $\Sigma(\boldsymbol{\theta})$ positive-definite, the parameters $\boldsymbol{\theta}$ often have to satisfy restrictions, some of which are peculiar to fitting linear combinations of generalized covariance functions (see, e.g., Delfiner, 1976, p. 59). More research is needed on how these restrictions can be accounted for efficiently in optimal fitting procedures.

2.6.2 Least Squares

The highly parametric variogram-fitting methods proposed so far ignore the visual appeal of plotting, say, $\{(h, 2\hat{\gamma}(h\mathbf{e})): h = h(1), h(2), \ldots, h(K)\}$, and finding a theoretical variogram curve that is "close" to it. (Regardless of the method of fitting, such a comparison plot of experimental variogram and fitted theoretical variogram values is an invaluable diagnostic tool and highly recommended; see Figure 2.15.) It has been proposed (David, 1977, Sections 6.1 and 6.2; Journel and Huijbregts, 1978, Section III.C.6 and Chapter IV; Clark, 1979, Chapter 2) to measure the closeness by the sum of squares of the differences between a generic variogram estimator $2\gamma^{\#}(h\mathbf{e})$ and a model $2\gamma(h\mathbf{e}; \boldsymbol{\theta})$.

The method of ordinary least squares specifies that $\boldsymbol{\theta}$ is estimated by minimizing

$$\sum_{j=1}^{K} \{2\gamma^{\#}(h(j)\mathbf{e}) - 2\gamma(h(j)\mathbf{e};\boldsymbol{\theta})\}^2, \tag{2.6.7}$$

for some direction $\mathbf{e}$. Multiple directions could also be accounted for in (2.6.7) by adding the appropriate squared differences. Eventually, an *ordinary-least-squares* (o.l.s.) estimator of $\boldsymbol{\theta}$ is obtained. Although (2.6.7) has geometric appeal, it takes no cognizance of the distributional variation and covariation of the generic estimator $2\gamma^{\#}$.

Generalized-Least-Squares Fitting

Suppose that a variogram estimator $2\gamma^{\#}(\cdot)$ is obtained at K lags $\mathbf{h}(1),\ldots,\mathbf{h}(K)$, where K is fixed and the amount of data contributing to the estimator at each lag is large (at least 30 pairs according to Journel and Huijbregts, 1978, p. 194). Let $2\gamma(\mathbf{h};\boldsymbol{\theta})$ be a variogram model whose exact form is known except for the unknown parameters $\boldsymbol{\theta}$ [e.g., the spherical model (2.3.8)].

The method of ordinary least squares is purely a numerical procedure that has an attractive geometric interpretation. To retain the geometry but also to introduce the concept of covariation into the procedure, consider the *generalized-least-squares* (g.l.s.) criterion. Suppose $2\boldsymbol{\gamma}^{\#} \equiv (2\gamma^{\#}(\mathbf{h}(1)),\ldots, 2\gamma^{\#}(\mathbf{h}(K)))'$, a $K \times 1$ vector of random variables, has variance matrix $\mathrm{var}(2\boldsymbol{\gamma}^{\#}) = V$ (which may depend on $\boldsymbol{\theta}$). Then choose the value of $\boldsymbol{\theta}$ that minimizes

$$(2\boldsymbol{\gamma}^{\#} - 2\boldsymbol{\gamma}(\boldsymbol{\theta}))'V^{-1}(2\boldsymbol{\gamma}^{\#} - 2\boldsymbol{\gamma}(\boldsymbol{\theta})), \tag{2.6.8}$$

where $2\gamma(\boldsymbol{\theta}) \equiv (2\gamma(\mathbf{h}(1);\boldsymbol{\theta}),\ldots,2\gamma(\mathbf{h}(K);\boldsymbol{\theta}))'$ is the theoretical model evaluated at the lags $\mathbf{h}(1),\ldots,\mathbf{h}(K)$. Call the estimator $\boldsymbol{\theta}_V^{\#}$.

Along with the ordinary-least-squares estimator $\boldsymbol{\theta}_I^{\#}$ and the generalized-least-squares estimator $\boldsymbol{\theta}_V^{\#}$, there is the *weighted-least-squares* (w.l.s) estimator $\boldsymbol{\theta}_\Delta^{\#}$, where

$$\Delta \equiv \mathrm{diag}\{\mathrm{var}(2\gamma^{\#}(\mathbf{h}(1))),\ldots,\mathrm{var}(2\gamma^{\#}(\mathbf{h}(K)))\}, \tag{2.6.9}$$

a $K \times K$ diagonal matrix with the specified variances along the diagonal.

Generalized least squares uses just the (asymptotic) second-order structure of the variogram estimator and does not make assumptions about the whole distribution of the data. Carroll and Ruppert (1982) show it to possess superior robustness properties over m.l. estimation when the distribution of $\mathbf{Z}$ is misspecified.

Calculation of V in (2.6.8) is not always easy. Cressie (1985a) has the details for $2\hat{\gamma}$ [classical estimator (2.4.2)] and $2\bar{\gamma}$ [robust estimator (2.4.12)], and shows how they simplify for data on a one-dimensional transect. A summary of these results is now presented.

Classical Variogram Estimator $2\hat{\gamma}$

Recall the definition of $2\hat{\gamma}$ given by (2.4.2). Assuming a Gaussian model (this will be relaxed later),

$$\{Z(\mathbf{s}+\mathbf{h}) - Z(\mathbf{s})\}^2 \sim 2\gamma(\mathbf{h}) \cdot \chi_1^2,$$

where χ_1^2 denotes a chi-squared random variable on 1 degree of freedom. Now

$$E\left(\{Z(\mathbf{s}+\mathbf{h}) - Z(\mathbf{s})\}^2\right) = 2\gamma(\mathbf{h}),$$

$$\operatorname{var}\left(\{Z(\mathbf{s}+\mathbf{h}) - Z(\mathbf{s})\}^2\right) = 2(2\gamma(\mathbf{h}))^2,$$

$$\operatorname{corr}\left(\{Z(\mathbf{s}_1+\mathbf{h}_1) - Z(\mathbf{s}_1)\}^2, \{Z(\mathbf{s}_2+\mathbf{h}_2) - Z(\mathbf{s}_2)\}^2\right) \tag{2.6.10}$$

$$= \left[\operatorname{corr}(\{Z(\mathbf{s}_1+\mathbf{h}_1) - Z(\mathbf{s}_1)\}, \{Z(\mathbf{s}_2+\mathbf{h}_2) - Z(\mathbf{s}_2)\})\right]^2$$

$$= \frac{\{\gamma(\mathbf{s}_1 - \mathbf{s}_2 + \mathbf{h}_1) + \gamma(\mathbf{s}_1 - \mathbf{s}_2 - \mathbf{h}_2) - \gamma(\mathbf{s}_1 - \mathbf{s}_2 + \mathbf{h}_1 - \mathbf{h}_2) - \gamma(\mathbf{s}_1 - \mathbf{s}_2)\}^2}{2\gamma(\mathbf{h}_1) \cdot 2\gamma(\mathbf{h}_2)}.$$

From these expressions, $\operatorname{var}(2\hat{\gamma}(\mathbf{h}(j)))$ and $\operatorname{cov}(2\hat{\gamma}(\mathbf{h}(i)), 2\hat{\gamma}(\mathbf{h}(j)))$ can be calculated, which allows V to be written as $V(\boldsymbol{\theta})$ under the assumption that the true variogram expressed in (2.6.10) belongs to $\{2\gamma\colon 2\gamma(\cdot) = 2\gamma(\cdot;\boldsymbol{\theta});\ \boldsymbol{\theta} \in \Theta\}$. From (2.6.8), the generalized-least-squares criterion becomes

$$(2\hat{\boldsymbol{\gamma}} - 2\boldsymbol{\gamma}(\boldsymbol{\theta}))'V(\boldsymbol{\theta})^{-1}(2\hat{\boldsymbol{\gamma}} - 2\boldsymbol{\gamma}(\boldsymbol{\theta})),$$

which can be a complicated function of $\boldsymbol{\theta}$ to minimize. Cressie (1985a) gives asymptotic expressions for $V(\boldsymbol{\theta})$ when the n data are indexed in $\mathbb{R}^1$ and equally spaced. Using a simple model for illustrative purposes, he shows that the off-diagonal terms of $V(\boldsymbol{\theta})$ can be appreciable, but that there is little loss in efficiency in approximating the diagonal terms by

$$\operatorname{var}(2\hat{\gamma}(\mathbf{h}(j))) \simeq 2\{2\gamma(\mathbf{h}(j);\boldsymbol{\theta})\}^2 / |N(\mathbf{h}(j))|. \tag{2.6.11}$$

Thus, w.l.s. is not a good approximation to g.l.s., but minimizing

$$\sum_{j=1}^{K} |N(\mathbf{h}(j))| \left\{\frac{\hat{\gamma}(\mathbf{h}(j))}{\gamma(\mathbf{h}(j);\boldsymbol{\theta})} - 1\right\}^2 \tag{2.6.12}$$

is a good approximation to w.l.s. The fitting criterion (2.6.12) is sensible from the viewpoint that the more pairs of observations $|N(\mathbf{h}(j))|$ there are the more weight the residual at lag $\mathbf{h}(j)$ receives in the overall fit. Also, the smaller the value of the theoretical variogram, the more weight the residual receives at that lag (i.e., lags closest to $\mathbf{h} = \mathbf{0}$ typically get more weight, which is an attractive property because it is important to obtain a good fit of the variogram near the origin; see Stein, 1988).

Criterion (2.6.12) could be seen as a pragmatic compromise between efficiency [generalized least squares, (2.6.8)] and simplicity [ordinary least squares, (2.6.7)] or it could be seen as an initial step in computing the g.l.s. estimator. Call the parameter values obtained from (2.6.12) $\hat{\boldsymbol{\theta}}^{(0)}$. Use these values to calculate $V(\hat{\boldsymbol{\theta}}^{(0)})$, which is used in (2.6.8). Now, minimizing (2.6.8) is relatively straightforward, because the weights do not depend on the parameter being estimated; this produces $\hat{\boldsymbol{\theta}}^{(1)}$. Calculate $V(\hat{\boldsymbol{\theta}}^{(1)})$ and use it in (2.6.8) and so forth until convergence.

Robust Variogram Estimator $2\bar{\gamma}$

Recall the definition of $2\bar{\gamma}$ given by (2.4.12). Consider for the moment the quantity

$$\bar{A}(\mathbf{h}) \equiv \sum_{N(\mathbf{h})} |Z(\mathbf{s}_i) - Z(\mathbf{s}_j)|^{1/2} / |N(\mathbf{h})|.$$

Under Gaussian assumptions (this will be relaxed later), Cressie (1985a) evaluated $\text{var}(\bar{A}(\mathbf{h}))$ and $\text{cov}(\bar{A}(\mathbf{h}_1), \bar{A}(\mathbf{h}_2))$ exactly, and used the δ-method to obtain $\text{cov}(2\bar{\gamma}(\mathbf{h}(i)), 2\bar{\gamma}(\mathbf{h}(j)))$. He was able to show once again that w.l.s. estimation, based on $2\bar{\gamma}$ in (2.6.9), can be approximated (even better than for $2\hat{\gamma}$) by minimizing

$$\sum_{j=1}^{K} |N(\mathbf{h}(j))| \left\{ \frac{\bar{\gamma}(\mathbf{h}(j))}{\gamma(\mathbf{h}(j); \boldsymbol{\theta})} - 1 \right\}^2 \tag{2.6.13}$$

over $\boldsymbol{\theta} \in \Theta$. [As before, (2.6.13) does not approximate well the g.l.s. criterion (2.6.8), although the efficiency loss is less for $2\bar{\gamma}$ than it is for $2\hat{\gamma}$.]

Figure 2.15 shows the fit of a spherical variogram model (2.3.8) to the coal-ash data of Section 2.2, based on the approximate w.l.s. criterion (2.6.13). The north–south direction was chosen and the number of lags was $K = 10$. This respects a recommendation made by Journel and Huijbregts (1978, p. 194) that the fit should be only up to half the maximum possible lag and then only using lags for which $|N(\mathbf{h}(j))| > 30$. Table 2.3 shows the estimates used in the fit. The fitted model is

$$2\gamma(h; \bar{\boldsymbol{\theta}}) = \begin{cases} 0, & h = 0, \\ 2\bar{c}_0 + 2\bar{c}_s\{(3/2)(h/\bar{a}_s) - (1/2)(h/\bar{a}_s)^3\}, & 0 < h < \bar{a}_s, \\ 2\bar{c}_0 + 2\bar{c}_s, & h \geq \bar{a}_s, \end{cases}$$

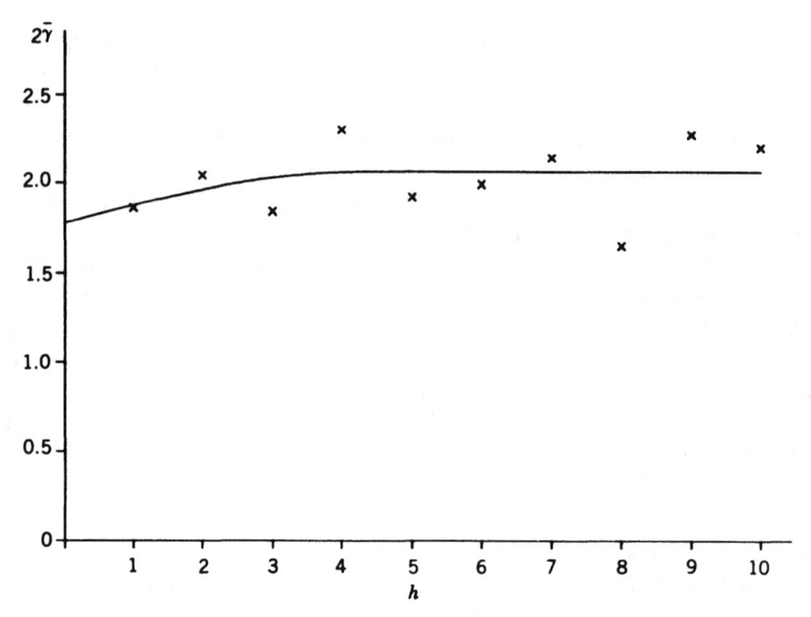

Figure 2.15 Fit of spherical variogram (solid line) to estimated north–south variogram $2\bar{\gamma}$ (crosses) up to lag 10, for the coal-ash data. Units are the same as in Figure 2.11.

where the w.l.s. estimate is

$$\bar{\boldsymbol{\theta}} = (\bar{c}_0, \bar{c}_s, \bar{a}_s)' \equiv (0.89, 0.14, 4.31). \qquad (2.6.14)$$

Non-Gaussian Data

Although the formulas are derived for Gaussian data, they can be easily generalized for transformed Gaussian data. Consider the following approximation (g is a continuous function differentiable in a neighborhood of μ):

$$\begin{aligned}\{g(Z(\mathbf{s}+\mathbf{h})) - g(Z(\mathbf{s}))\}^2 &= \{(\mu + (Z(\mathbf{s}+\mathbf{h}) - \mu)g'(\mu) + \cdots) \\ &\quad -(\mu + (Z(\mathbf{s}) - \mu)g'(\mu) + \cdots)\}^2 \\ &\simeq (g'(\mu))^2\{Z(\mathbf{s}+\mathbf{h}) - Z(\mathbf{s})\}^2;\end{aligned}$$

this expansion is usually called the δ-method (Kendall and Stuart, 1969, pp. 231, 244). Hence, these fitting procedures are also valid for transformed Gaussian data.

Examples and Methods of Weighted-Least-Squares Fitting

Cressie (1985a) fits spherical and linear models to mining data (both original and detrended) by the weighted-least-squares criterion (2.6.13). A combination of differential calculus and gridding was used to find the optimal $\hat{\boldsymbol{\theta}}$. Taylor and Burrough (1986) fit a linear variogram to soils data by the generalized-least-squares criterion (2.6.8) using the iterative approach proposed by Cressie (1985a). They compare their fits to those obtained by weighted and ordinary least squares and find that in most cases w.l.s. is adequate (but o.l.s. is not). McBratney and Webster (1986) and Laslett et al. (1987) fit variograms of various soils data sets by w.l.s.

The w.l.s. method of variogram-model fitting can be implemented through any number of nonlinear estimation algorithms. Finding variogram-model parameters $\boldsymbol{\theta}$ that minimize (2.6.12) is equivalent to finding $\boldsymbol{\theta}$ that minimizes

$$\sum_{j=1}^{K} \frac{|N(\mathbf{h}(j))|}{\{\gamma(\mathbf{h}(j);\boldsymbol{\theta})\}^2}\{\hat{\gamma}(\mathbf{h}(j)) - \gamma(\mathbf{h}(j);\boldsymbol{\theta})\}^2.$$

Viewed as a fitting criterion for weighted nonlinear regression with weights $\{|N(\mathbf{h}(j))|/\{\gamma(\mathbf{h}(j);\boldsymbol{\theta})\}^2: j = 1,\dots,K\}$, the w.l.s. estimates $\hat{\boldsymbol{\theta}}$ can be obtained from an iterative, nonlinear estimation routine, such as the method of steepest descent or the Gauss–Newton algorithm (the NLIN procedure, SAS Institute Inc., 1985, Chapter 25, offers several such routines). See Gotway (1991) for further details. Of course, proper care should be taken to check that the global minimum, and not a local minimum, is attained.

On occasions, when more weight is needed for variogram estimates near the origin, I have carried out a weighted-least-squares fit of a *linear* variogram [see (2.3.7)] to those $\{2\hat{\gamma}(\mathbf{h}(j))\}$ or $\{2\bar{\gamma}(\mathbf{h}(j))\}$ near the origin. This yields estimates of nugget effect and variogram slope at the origin that can either be used as starting values or final values in the full w.l.s. estimation of $\boldsymbol{\theta}$ over Θ.

2.6.3 Properties of Variogram-Parameter Estimators

Distributional properties of $\hat{\boldsymbol{\theta}}$ are not well understood for any of the methods proposed in Sections 2.6.1 and 2.6.2. Asymptotic properties need to be classified according to whether they refer to increasing-domain asymptotics or to infill asymptotics; some discussion of each will follow.

Warnes and Ripley (1987) illustrate potential problems with m.l. estimation based on simulations of some simple spatial processes; however, Mardia and Watkins' (1989) simulations indicate that these problems are caused by likelihoods that are not twice differentiable in $\boldsymbol{\theta}$. In Monte Carlo comparisons of m.l. and REML estimation, Swallow and Monahan (1984), Zimmerman (1986, Section 2.4), Wilson (1988), and McGilchrist (1989) demonstrate that REML estimators can have substantially smaller bias. Tunicliffe-Wilson (1989) concludes that the use of REML in time-series regression generally leads to improvement in inferences over m.l. estimation.

For two intrinsically stationary isotropic Gaussian random processes in $\mathbb{R}^2$ and for various sampling intensities, Zimmerman and Zimmerman (1991) present a Monte Carlo comparison of m.l., REML, MINQ, and w.l.s. estimation. (They included another estimator, due to Delfiner, 1976, but its performance was so poor that it is not considered.) They found that, in terms of bias, mean-squared error, and coverage probability of the classical 95% prediction interval, w.l.s. is sometimes the best procedure and never does badly. Although the comparisons show that likelihood methods can do very well, I doubt their robustness to departures from the Gaussian assumption to which they are so closely tied. [A diagnostic plot of, say, $2\bar{\gamma}(\mathbf{h})$ versus $2\gamma(\mathbf{h}; \hat{\boldsymbol{\theta}})$, where $\hat{\boldsymbol{\theta}}$ is the likelihood-based estimator, would be a way to see whether the estimate is capturing the spatial dependence accurately.] On balance, w.l.s. estimation seems to be preferable because it does well under the Gaussian model, yet it is equally appropriate for non-Gaussian models.

In a spatial setting, there are (at least) two types of asymptotic theories. When n, the number of data in $\mathbf{Z}$, tends to infinity, either $\max\{\|\mathbf{s}_i\|: i = 1, \ldots, n\}$ tends to infinity (increasing-domain asymptotics) or it remains bounded (infill asymptotics). In the context of time-series analysis, most problems require increasing-domain asymptotics; however, which is appropriate in a spatial context depends on the problem being addressed. In this part of the book (Geostatistical Data), infill asymptotics is probably preferable, whereas in Part II (Lattice Data) increasing-domain asymptotics are often more appropriate.

Increasing-Domain Asymptotics

The increasing-domain asymptotic properties of the m.l. estimator of $\boldsymbol{\theta}$ are given in Section 7.3.1. Unfortunately, little is known about REML and MINQ in a spatial setting. Generalized-least-squares or, for that matter, weighted-least-squares estimations yield an asymptotically Gaussian estimator of $\boldsymbol{\theta}$ (Jennrich, 1969; Carroll and Ruppert, 1988, Section 2.6), from which asymptotic efficiency comparisons with m.l. estimation could be made.

The bootstrap technique for estimating standard errors and confidence intervals (Efron, 1979, 1982) has been offered as a panacea in complex estimation problems. But the assumption of independence between the data is crucial to its success. Thus, a straightforward application of the bootstrap to estimate the variability of the variogram parameter estimator $\hat{\boldsymbol{\theta}}$ is not appropriate (Singh, 1981, Remark 2.1). A *spatial bootstrap* would be a very powerful tool, but it is not yet clear how to construct one. One suggestion is to divide the space into regions such that one can assume approximate independence between them. Then the all-important resampling needed to define the bootstrap can be carried out on the regions and not directly on the observations. Hall (1985a) does this but warns that the resampling distorts some of the interactions in the problem. Another possibility is to analyze the problem instead in the spectral domain, where the estimated Fourier coefficients are uncorrelated and, hence, one might hope, approximately independent. For more details on spatial bootstrapping, see Section 7.3.2.

Infill Asymptotics

In mining, soil science, hydrology, and so on, the domain D should often be thought of as bounded; extra units of information come from observations taken between those already observed. I shall borrow a mining term and call this means of obtaining more and more data *infill asymptotics*. Under infill asymptotics, Stein (1987b) considers MINQ estimation of variogram parameters for simple (Gaussian) examples in $\mathbb{R}^1$. Not surprisingly, he finds that the derivative of the variogram at the origin is the only quantity that can be estimated consistently. He goes on to show that the parameters that can be estimated well are the only ones that are important for the kriging predictor (kriging is developed in Chapter 3).

No general results are available for m.l., REML, MINQ, and w.l.s. estimators under infill asymptotics; some ideas may be gained from Zimmerman and Zimmerman's (1991) study, because they increased sample size by infill sampling. For obvious reasons, spatial bootstrapping appears even more difficult under infill asymptotics than under increasing-domain asymptotics. Further discussion of infill asymptotics can be found in Yakowitz and Szidarovsky (1985), Stein (1988, 1989a), and Section 5.8.

The Effect of Variogram Estimation on Inference

The variogram model, fitted by any of the methods described in Sections 2.6.1 and 2.6.2, is useful as a quantification of the spatial dependence in the data $\mathbf{Z}$ or in the process $Z(\cdot)$. More importantly, it is the cornerstone of kriging (Chapter 3). For the calculation of kriging weights and kriging variances, it is often assumed that the fitted $\hat{\boldsymbol{\theta}}$ is the true one. Experience with prediction in random-effects models (Kackar and Harville, 1984; Reinsel, 1984) indicates that estimated kriging variances should be augmented by a term corresponding to the estimation of $\boldsymbol{\theta}$ by $\hat{\boldsymbol{\theta}}$. Zimmerman and Cressie (1991) generalize some of the results from the random-effects model to a spatial setting. More details can be found in Section 5.3.

2.6.4 Cross-Validating the Fitted Variogram

Suppose that a variogram model $2\gamma(\mathbf{h}; \hat{\boldsymbol{\theta}})$, $\mathbf{h} \in \mathbb{R}^d$, has been fitted to data $\{Z(\mathbf{s}_i): i = 1, \ldots, n\}$. A way to diagnose any problems with the fit obtained is to *cross-validate* the variogram model.

Cross-validation has been a popular means of assessing statistical estimation and prediction since the articles by Stone (1974) and Geisser (1975). The basic idea is to delete some of the data and use the remaining data to predict the deleted observations. Then the prediction error can be inferred from the predicted-minus-actual values. Repeating this over many deleted subsets allows an assessment of the variability of prediction error.

In an estimation context, the deletion of observations to improve inference for an estimable parameter was called *jackknifing* by Tukey (1958). It was proposed by Quenouille (1949a) to reduce bias in the original estimator. In jackknifing, the manufacture of pseudo values (e.g., Miller, 1974) makes it different from the cross-validation approach taken here for spatial prediction.

In Chapter 3, the technique of spatial prediction known as *kriging* is presented. For the purposes of this section, it is enough to assume simply that a predictor $\hat{Z}(\mathbf{s}_0)$ [based on the fitted variogram and the data $\{Z(\mathbf{s}_i): i = 1,\ldots,n\}$] of the value of $Z(\cdot)$ at location $\mathbf{s}_0 \in D$ is available, along with a measure of its mean-squared prediction error $\sigma_k^2(\mathbf{s}_0)$. [In general, $\sigma_k^2(\mathbf{s}_0)$ may depend on the fitted variogram, the data, and the spatial locations $\mathbf{s}_0, \mathbf{s}_1, \ldots, \mathbf{s}_n$.]

If the variogram model describes adequately the spatial dependencies implicit in the data set, then the predicted value $\hat{Z}(\mathbf{s}_0)$ should be close to the true value $Z(\mathbf{s}_0)$. Ideally, additional observations on $Z(\cdot)$ could be taken to check this, or initially some of the data might be set aside to validate the spatial predictor. More likely, *all* of the data are used to fit the variogram and build the spatial predictor, and there is *no* possibility of taking more observations. In this case, the cross-validation approach can be used. Let $2\gamma(\mathbf{h}; \hat{\boldsymbol{\theta}})$ be the fitted variogram model (obtained from all the data); now delete a datum $Z(\mathbf{s}_j)$ and predict it with $\hat{Z}_{-j}(\mathbf{s}_j)$ [based on $2\gamma(\mathbf{h}; \hat{\boldsymbol{\theta}})$ and the data $\mathbf{Z}$ *without* $Z(\mathbf{s}_j)$]. Its associated mean-squared prediction error is $\sigma_{-j}^2(\mathbf{s}_j)$, which depends *inter alia* on the fitted variogram model.

The closeness of the predicted values to the true values can be characterized in various ways; for example, consider

$$(1/n)\sum_{j=1}^{n}\left\{\left(Z(\mathbf{s}_j) - \hat{Z}_{-j}(\mathbf{s}_j)\right)\Big/\sigma_{-j}(\mathbf{s}_j)\right\}, \qquad (2.6.15)$$

$$\left[(1/n)\sum_{j=1}^{n}\left\{\left(Z(\mathbf{s}_j) - \hat{Z}_{-j}(\mathbf{s}_j)\right)\Big/\sigma_{-j}(\mathbf{s}_j)\right\}^2\right]^{1/2}, \qquad (2.6.16)$$

$$\text{stem-and-leaf plot of } \left\{\left(Z(\mathbf{s}_j) - \hat{Z}_{-j}(\mathbf{s}_j)\right)\Big/\sigma_{-j}(\mathbf{s}_j) : j = 1,\ldots,n\right\}. \qquad (2.6.17)$$

In all of these summaries, the standardized prediction residuals are used to obtain a diagnostic check of the fit of the variogram model $2\gamma(\cdot; \hat{\boldsymbol{\theta}})$. The mean in (2.6.15) should be approximately *0*, the root-mean-square in (2.6.16) should be approximately *1*, and the histogram in (2.6.17) can be perused for *outliers* whose absolute values are large. [Hawkins and Cressie (1984) use this latter idea to construct a robust kriging predictor; see Section 3.3.]

It is worth noting that the prediction sum of squares (PRESS) statistic for variable subset selection in regression (Allen, 1971; Draper and Smith, 1981, p. 325) is defined in a similar way to (2.6.16), except that no normalization with respect to $\sigma_{-j}(\mathbf{s}_j)$ is used. Following on from a suggestion by Bastin and Gevers (1985), Samper and Neuman (1989) assume that the prediction errors $\{Z(\mathbf{s}_j) - \hat{Z}_{-j}(\mathbf{s}_j; \boldsymbol{\theta}): j = 1,\ldots,n\}$ are Gaussian with negligible correlations; this allows construction of a likelihood (based on the prediction errors), to be

Stem	Leaf
−1	887
−1	333222100
−0	98888887777777666665
−0	44444443333322100000
0	112223333333444444
0	556677788899
1	0000334
1	556677889
2	2
2	
3	1
3	
4	
4	
5	
5	
6	
6	7

Figure 2.16 Stem-and-leaf plot of standardized prediction errors $\{(Z(\mathbf{s}_j) - \hat{Z}_{-j}(\mathbf{s}_j))/\sigma_{-j}(\mathbf{s}_j)\colon j = 1, \ldots, 100\}$, for the coal-ash data. 3|1 denotes 3.1.

maximized with respect to $\boldsymbol{\theta}$. However, even if the correlations between prediction errors are properly recognized in the joint distribution, it is unwise to think of it as a likelihood because unknown parameters are used to define the "data" (namely, the prediction errors). This approach should be treated with caution.

Recall the coal-ash data of Section 2.2; using the weighted-least-squares criterion (2.6.13), a spherical variogram (2.3.8) was fitted with parameters given by (2.6.14). Cressie (1986) establishes a neighborhood of coal-ash data locations from which a deleted value is predicted using (universal) kriging. Ignoring missing values and values near the edges, he is able to delete and predict 100 $Z(\mathbf{s}_j)$ values, all of which have the same neighborhood configuration (see also Section 3.4).

For these 100 data, the fitted spherical variogram given by (2.3.8) and (2.6.14) can be cross-validated; specifically,

$$\text{quantity (2.6.15)} = 0.147,$$
$$\text{quantity (2.6.16)} = 1.167,$$

and (2.6.17) is shown in Figure 2.16. Notice that one value is not very well predicted; this corresponds to the outlier $z(5,6) = 17.61\%$, detected in the exploratory data analysis reported in Section 2.2. When this is deleted from (2.6.17) and the mean and root-mean-square are recomputed, the cross-validation is more satisfactory, namely,

$$\text{quantity (2.6.15)} = 0.081,$$
$$\text{quantity (2.6.16)} = 0.957.$$

I do not advocate cross-validation for confirmatory data analysis (e.g., hypothesis testing, standard error estimation, etc.) at this time, because the correlations between data give rise to complicated distribution theory that needs further study. Consider, for example, the inferences claimed in Carr and Roberts (1989); there, cross-validation-based test statistics are used to define hypothesis tests. Unfortunately, their inferences are invalid. The complicated correlation structure in predictors and prediction residuals also casts doubt on cross-validation measures that try to capture anticipated dependencies; for example, the sample correlation between $\hat{Z}_{-j}(\mathbf{s}_j)$ and $\{Z(\mathbf{s}_j) - \hat{Z}_{-j}(\mathbf{s}_j)\}/\sigma_{-j}(\mathbf{s}_j)$.

After cross-validating the variogram successfully, one can feel confident that prediction based on the fitted variogram is approximately unbiased and that the mean-squared prediction error is about right. Its role is one of model checking, to prevent blunders and to highlight potentially troublesome prediction points. It cannot *prove* that the fitted model is correct, merely that it is not grossly incorrect. Davis (1987) gives more details on most of these issues.

Cross-validation as a tool to *estimate* $\boldsymbol{\theta}$ (Stone, 1974; Geisser, 1975) is computationally prohibitive in most geostatistical problems. The same can be said for its use in selecting the *type* of variogram model [e.g., the linear variogram (2.3.7) versus the spherical variogram (2.3.8)] that will be fitted. Besides, why should this continued sample reuse provide a meaningful way of preferring one $\boldsymbol{\theta}$-value or one class of models over another? The interdependencies introduced into a problem where data are already spatially dependent make confirmatory techniques questionable. It is a difficult problem to adapt the cross-validation, the jackknife, and the bootstrap statistical techniques from independent and identically distributed observations (e.g., Efron, 1982, Chapter 7) to the geostatistical context. For lattice data, there is more hope; see Section 7.3.2.

For time series, model selection and parameter estimation by cross-validation has been called into question by Hjorth (1982), Dawid (1984), and Rissanen (1984, 1987). However, *sequencing* of the observations is very important for these authors' approach of minimizing accumulated prediction errors, and this may be difficult to generalize to the spatial context. Through the notions of stochastic complexity and minimum description length, Rissanen eschews the idea that there is a true model. Instead he bases model choice on parsimony of description of the data given the model, and of the model itself [not unlike the way AIC (Akaike Information Criterion) owing to Akaike, 1973, and BIC (Bayes Information Criterion) owing to Schwarz, 1978, operate in practice]. Theoretical and practical implementation of such ideas to spatial data is an area worthy of considerable further study.

CHAPTER 3

Spatial Prediction and Kriging

The need to obtain accurate predictions from observed data can be found in all scientific disciplines. Those that have embraced statistical notions of random variation are able to do this by exploiting the statistical dependence among the data and the variable(s) to be predicted. However, the statistical approach has not been without its detractors; for example, Philip and Watson (1986) argue that geostatistics is unhelpful for solving problems in mining and geology. Their article and the accompanying discussions (see, in particular, a rebuttal article by Journel, 1986) are worth perusing. I hope this book and, in particular, this chapter, brings more light to a debate that has already seen too much heat. I believe that geostatistics is highly relevant to many problems in the earth (and other) sciences, and I have little to agree with in Philip and Watson's long, emotional, and often incorrect discourse against the subject.

It is helpful to explain first the terms used in the title of this chapter. Let $\{Z(\mathbf{s}): \mathbf{s} \in D \subset \mathbb{R}^d\}$ be a random function (or process), as defined in Section 2.1, from which n data $Z(\mathbf{s}_1), \ldots, Z(\mathbf{s}_n)$ are collected. The data are used to perform inference on the process, here, to predict some known functional $g(\{Z(\mathbf{s}): \mathbf{s} \in D\})$ [or, more simply, $g(Z(\cdot))$] of the random function $Z(\cdot)$. For example, point prediction assumes $g(Z(\cdot)) = Z(\mathbf{s}_0)$, where $\mathbf{s}_0$ is a known spatial location. Throughout this chapter, g is mostly real-valued; Section 3.4.5 (James–Stein prediction) has some discussion of how to deal with vector-valued g.

Sometimes interest is not in $Z(\cdot)$, but in a "noiseless" version of it. Suppose that

$$Z(\mathbf{s}) = S(\mathbf{s}) + \epsilon(\mathbf{s}), \qquad \mathbf{s} \in D,$$

where $\epsilon(\cdot)$ is a white-noise measurement-error process. In this case, one is interested in predicting a known functional $g(S(\cdot))$ of the noiseless random function $S(\cdot)$.

Spatial prediction refers to predicting either $g(Z(\cdot))$ or $g(S(\cdot))$ from data $Z(\mathbf{s}_1), \ldots, Z(\mathbf{s}_n)$ observed at known spatial locations $\mathbf{s}_1, \ldots, \mathbf{s}_n$. Notice that my terminology encompasses the temporal notions of *smoothing* (or interpolation), *filtering*, and *prediction* (e.g., Lewis, 1986, p. 36), which rely on time-ordering for their distinction. If temporal data are available from the past up to and including the present, smoothing refers to prediction of $g(S(\cdot))$ at time points in the past, filtering refers to prediction of $g(S(\cdot))$ at the present time, and prediction refers to prediction of $g(S(\cdot))$ at time points in the future. Rather than stretch the analogies in an unnatural way, I have decided to bring everything together under the general heading of spatial prediction. When clarity demands it, care will be taken to specify exactly what is to be predicted and where. In this book, the word "estimation" will be used exclusively for inference on fixed but unknown parameters; "prediction" is reserved for inference on *random* quantities.

Kriging is a minimum-mean-squared-error method of spatial prediction that (usually) depends on the second-order properties of the process $Z(\cdot)$. Matheron (1963b) named this method of optimal spatial linear prediction after D. G. Krige, a South African mining engineer who, in the 1950s, developed empirical methods for determining true ore-grade distributions from distributions based on sampled ore grades (e.g., Krige, 1951). However, the formulation of optimal spatial linear prediction did not come from Krige's work. (See Matheron, 1971b, pp. 117–119, and Cressie, 1990b, for the extent of the early work of Krige.) The contributions of Wold (1938), Kolmogorov (1941b), and Wiener (1949) all contain optimal linear prediction equations that reflect the notion that observations closer to the prediction point (for them, closer in time) should be given more weight in the predictor.

At the same time as geostatistics was developing in mining engineering under G. Matheron in France, the very same ideas developed in meteorology under L. S. Gandin (Gandin, 1963) in the Soviet Union. The original (and simultaneous) contribution of these authors was to put optimal linear prediction (in terms of variograms) into a spatial setting. Gandin's name for his approach was *objective analysis*, and he used the terminology *optimum interpolation* instead of kriging. Details of the origins of kriging are set out in Cressie (1990b).

Applications of kriging are given in Sections 4.1 (Wolfcamp-aquifer data), 4.5 (wheat-yield data), and 4.6 (acid-deposition data). Throughout this chapter, both the Wolfcamp-aquifer data (Section 4.1) and the coal-ash data (Section 2.2) are used to illustrate the various kriging proposals.

Block Average

A common problem is to predict

$$g(Z(\cdot)) = \int_B Z(\mathbf{s})\, d\mathbf{s}/\,|B|, \qquad B \subset D,$$

the average of the process over a block B whose location and geometry are known and whose d-dimensional volume is $|B|$. To define the integral properly, it is first approximated by Riemann sums; for example, in $\mathbb{R}^2$, let

B be the rectangle $[a_1, a_2] \times [b_1, b_2]$ and consider $\sum_{i=1}^{l} \sum_{j=1}^{m} Z(u_i', v_j')(u_i - u_{i-1})(v_j - v_{j-1})$, where $a_1 = u_0 < u_1 < \cdots < u_l = a_2$, $u_{i-1} \le u_i' \le u_i$, $b_1 = v_0 < v_1 < \cdots < v_m = b_2$, $v_{j-1} \le v_j' \le v_j$. Then, the limit in quadratic mean of the approximating sum is the stochastic integral $\int_B Z(\mathbf{s})\, d\mathbf{s}$ (e.g., Yaglom, 1962, p. 23).

Stochastic Approach

This chapter is devoted to the (spatial) prediction of $Z(\mathbf{s}_0)$ and $\int_B Z(\mathbf{s})\, d\mathbf{s}/|B|$ or noiseless versions of them, namely, $S(\mathbf{s}_0)$ and $\int_B S(\mathbf{s})\, d\mathbf{s}/|B|$. It will be assumed that data, both observed and potential, behave according to a stochastic process indexed in $\mathbb{R}^d$. To understand and quantify the errors made in a spatial prediction, it is necessary to know the source of random variation in the random process $Z(\cdot)$.

When there are infinitely many possible surfaces from which to choose, a standard tactic in science is to deal with these statistically. Then, in the light of observations $\mathbf{Z}$ from $Z(\cdot)$, this "prior" can be updated, yielding a "posterior," which amounts to the distribution of $Z(\cdot)$ conditioned on the data $\mathbf{Z}$. Thus, inferences on the process [e.g., prediction of $Z(\mathbf{s}_0)$] should involve this conditional distribution.

Decision-Theoretic Considerations

To simplify the discussion, assume that $Z(\mathbf{s}_0)$ is to be predicted from data $\mathbf{Z} = (Z(\mathbf{s}_1), \ldots, Z(\mathbf{s}_n))'$; the argument changes little for prediction of a noiseless version or a block average. Using a decision-theoretic formalism (see, e.g., Ferguson, 1967), denote $L(Z(\mathbf{s}_0), p(\mathbf{Z}; \mathbf{s}_0))$ as the *loss* incurred when $Z(\mathbf{s}_0)$ is predicted with $p(\mathbf{Z}; \mathbf{s}_0)$. An optimal predictor p is one that minimizes the so-called *Bayes risk* $E\{L(Z(\mathbf{s}_0), p(\mathbf{Z}; \mathbf{s}_0))\}$, where $E(\cdot)$ denotes expectation with respect to the *joint* distribution of $Z(\mathbf{s}_0)$ and $\mathbf{Z}$.

It is a well known result from Bayesian decision theory that the best predictor minimizes $E\{L(Z(\mathbf{s}_0), p(\mathbf{Z}; \mathbf{s}_0))|\mathbf{Z}\}$, where $E\{\cdot|\mathbf{Z}\}$ denotes (the posterior) expectation with respect to the conditional distribution of $Z(\mathbf{s}_0)|\mathbf{Z}$. Thus, an optimal predictor of unobserved parts of $Z(\cdot)$ will be obtained conditionally on the observed part $\mathbf{Z}$, reinforcing earlier discussion regarding the source of the random variation. A conditional measure of prediction loss is $E\{L(Z(\mathbf{s}_0), p)|\mathbf{Z}\}$, in contrast to the unconditional measure $E\{L(Z(\mathbf{s}_0), p)\}$, which is the Bayes risk. It is mostly the unconditional measure that will be considered throughout this chapter.

Prediction Regions

For a given loss function L and predictor $p(\mathbf{Z}; \mathbf{s}_0)$, prediction regions for $Z(\mathbf{s}_0)$ can be defined. The region

$$\{Z(\mathbf{s}_0) : L(Z(\mathbf{s}_0), p(\mathbf{Z}; \mathbf{s}_0)) < k_\alpha\}$$

is a $100(1 - \alpha)\%$ prediction region if a constant k_α (which may depend on $\mathbf{s}_0$) can be chosen so that $\Pr\{L(Z(\mathbf{s}_0), p(\mathbf{Z}; \mathbf{s}_0)) < k_\alpha\} = 1 - \alpha$, $0 \le \alpha \le 1$.

Loss Functions

In many prediction problems, *squared-error loss*

$$L(Z(\mathbf{s}_0), p(\mathbf{Z}; \mathbf{s}_0)) = (Z(\mathbf{s}_0) - p(\mathbf{Z}; \mathbf{s}_0))^2$$

is used. The resulting optimal predictor, which minimizes $E\{(Z(\mathbf{s}_0) - p(\mathbf{Z}; \mathbf{s}_0))^2 | \mathbf{Z}\}$, is

$$p^0(\mathbf{Z}; \mathbf{s}_0) = E(Z(\mathbf{s}_0) | \mathbf{Z}),$$

the conditional expectation. There are attractive interpretive reasons for choosing squared-error loss, because the Bayes risk is then simply the *mean-squared prediction error*, which depends only on first and second moments. Also, the $100(1 - \alpha)\%$ prediction region is the symmetric interval $(p(\mathbf{Z}; \mathbf{s}_0) - k_\alpha^{1/2}, p(\mathbf{Z}; \mathbf{s}_0) + k_\alpha^{1/2})$, where k_α is defined in the preceding paragraph. Derivations of properties are much easier, too. For example, it is easy to show that the Bayes risk satisfies

$$E(Z(\mathbf{s}_0) - p^0(\mathbf{Z}; \mathbf{s}_0))^2 = \mathrm{var}(Z(\mathbf{s}_0)) - \mathrm{var}(p^0(\mathbf{Z}; \mathbf{s}_0)).$$

Perhaps surprisingly, as the variance of the optimal predictor increases, the mean-squared prediction error decreases.

Other loss functions, such as $|Z(\mathbf{s}_0) - p(\mathbf{Z}; \mathbf{s}_0)|^\nu$, $1 \le \nu < 2$ (e.g., for $\nu = 1$, the optimal predictor is the conditional median), and $I(|Z(\mathbf{s}_0) - p(\mathbf{Z}; \mathbf{s}_0)| > \epsilon)$ (e.g., for ϵ small, the optimal predictor is the conditional mode), may be appropriate at times. In particular, in resource evaluation an underprediction is probably not as serious as an overprediction, indicating that *asymmetric* loss functions should be considered. Zellner (1986), in the context of parameter estimation and prediction, explores the consequences of using the *linex* loss function

$$L(Z(\mathbf{s}_0), p(\mathbf{Z}; \mathbf{s}_0)) = b\big[\exp\{a(Z(\mathbf{s}_0) - p(\mathbf{Z}; \mathbf{s}_0))\} - a(Z(\mathbf{s}_0) - p(\mathbf{Z}; \mathbf{s}_0)) - 1\big], \qquad b > 0,$$

which would satisfy the stated asymmetry for $a < 0$. It is not difficult to show that the optimal predictor is

$$p^0(\mathbf{Z}; \mathbf{s}_0) = (1/a)\log\big[E\{\exp(aZ(\mathbf{s}_0)) | \mathbf{Z}\}\big],$$

which is less than $E(Z(\mathbf{s}_0)|\mathbf{Z})$ for $a < 0$. Journel (1984) also considers asymmetric loss functions that yield conditional quantiles as optimal predictors.

The notion of entropy was discussed in Section 1.1. Consider the distribution of $X \equiv Z(\mathbf{s}_0) - p(\mathbf{Z}; \mathbf{s}_0)$, which is denoted by $f(x; p)$. Then, the minimum error entropy predictor is the $p(\cdot)$ that minimizes $-\int\{\log f(x; p)\}f(x; p)\,dx$. Thomopoulos (1985) gives algorithmic solutions to these problems in the discrete case.

Linear and Nonlinear Prediction

The preceding discussion demonstrates that the conditional distribution of $Z(\mathbf{s}_0)|\mathbf{Z}$ (or some summary of it) is needed. This is calculated from the $(n+1)$-dimensional joint distribution of $(Z(\mathbf{s}_0), \mathbf{Z}')$; however, in practice, estimation of such a distribution from the n data available is not possible, unless some simplifying model assumptions are made.

There is quite a large literature on nonparametric regression, where estimation of $E(Y|\mathbf{X})$ (or some other summary of the distribution of $Y|\mathbf{X}$) is obtained from kernel or nearest-neighbor estimates based on observations $(Y_1, \mathbf{X}_1), \ldots, (Y_n, \mathbf{X}_n)$; see, for example, Stone (1977), Georgiev (1984), and Zhao (1987). Versions of these estimators for (temporal) prediction can be derived, but their statistical properties will not be simple analogues of the regression estimators, nor will the asymptotics necessarily be the same; see Section 5.8.

The simplest assumption to make is that $Z(\cdot)$ is a Gaussian random process, because then the joint distribution of $(Z(\mathbf{s}_0), \mathbf{Z}')$ is Gaussian and $E(Z(\mathbf{s}_0)|\mathbf{Z})$ is *linear* in $\mathbf{Z}$, depending only on $\mu(\mathbf{s}_i) = E(Z(\mathbf{s}_i))$, $i = 0, \ldots, n$, and $C(\mathbf{s}_i, \mathbf{s}_j) = \mathrm{cov}(Z(\mathbf{s}_i), Z(\mathbf{s}_j))$, $0 \le i \le j \le n$. Even so, there are potentially $(n+1) + (1/2)(n+1)(n+2)$ parameters to estimate from just n data. Further simplifying assumptions, such as $E(Z(\mathbf{s})) \equiv \mu$ and $C(\mathbf{s}_i, \mathbf{s}_j) \equiv C^*(\mathbf{s}_i - \mathbf{s}_j)$, where $C^*(\cdot)$ is a positive-definite function on $\mathbb{R}^d$, allow inference to proceed.

Best Linear Predictor

It has been seen that, for squared-error loss, the best predictor is $E(Z(\mathbf{s}_0)|\mathbf{Z})$, which is not always linear in $\mathbf{Z}$. Rather than asking for the best predictor, one could ask instead for the best *linear* predictor; that is, find $l_1, \ldots, l_n, k$ in

$$p(\mathbf{Z}; \mathbf{s}_0) = \sum_{i=1}^{n} l_i Z(\mathbf{s}_i) + k,$$

such that $E(Z(\mathbf{s}_0) - p(\mathbf{Z}; \mathbf{s}_0))^2$ is minimized. Equivalently, minimize (over $l_1, \ldots, l_n, k$)

$$E\left(Z(\mathbf{s}_0) - \sum_{i=1}^{n} l_i Z(\mathbf{s}_i) - k\right)^2 = \mathrm{var}\left(Z(\mathbf{s}_0) - \sum_{i=1}^{n} l_i Z(\mathbf{s}_i)\right) + \left(\mu(\mathbf{s}_0) - \sum_{i=1}^{n} l_i \mu(\mathbf{s}_i) - k\right)^2,$$

where $\mu(\mathbf{s}) = E(Z(\mathbf{s}))$, $\mathbf{s} \in D$. Choice of $k = \mu(\mathbf{s}_0) - \Sigma_{i=1}^{n} l_i \mu(\mathbf{s}_i)$ and

$$\mathbf{l}' \equiv (l_1, \cdots, l_n) = \mathbf{c}'\Sigma^{-1},$$

where $\mathbf{c} \equiv (C(\mathbf{s}_0, \mathbf{s}_1), \ldots, C(\mathbf{s}_0, \mathbf{s}_n))'$ and Σ is an $n \times n$ matrix whose (i, j)th element is $C(\mathbf{s}_i, \mathbf{s}_j)$, yields the optimal linear predictor $p^*(\mathbf{Z}; \mathbf{s}_0)$ [or, more simply, $Z^*(\mathbf{s}_0)$]. That is,

$$p^*(\mathbf{Z}; \mathbf{s}_0) = \mathbf{c}'\Sigma^{-1}(\mathbf{Z} - \boldsymbol{\mu}) + \mu(\mathbf{s}_0),$$

where $\boldsymbol{\mu} \equiv (\mu(\mathbf{s}_1), \ldots, \mu(\mathbf{s}_n))'$. The minimized mean-squared prediction error is

$$\sigma_{sk}^2(\mathbf{s}_0) \equiv C(\mathbf{s}_0, \mathbf{s}_0) - \mathbf{c}'\Sigma^{-1}\mathbf{c}.$$

Matheron (1971b) has called such spatial prediction *simple kriging* because it relies on knowing the mean function $\mu(\cdot)$. Ordinary kriging (Section 3.2) and universal kriging (Section 3.4) yield optimal predictors, linear in the data, but where the mean function is linear in a fixed number of unknown parameters. These parameters are estimated optimally, but the price paid is that mean-squared prediction errors are larger than $\sigma_{sk}^2(\mathbf{s}_0)$. See Section 3.4.5 for more details.

Gaussian Data

If $Z(\cdot)$ is a Gaussian process, then the optimal predictor p^0 and the optimal linear predictor p^* are the same (under squared-error loss). However, it should not be forgotten that optimal linear prediction yields tractable predictors that may perform poorly when $Z(\cdot)$ is far from Gaussian.

The Gaussian assumption has one other feature, namely, conditional homoskedasticity; that is, $\text{var}(Z(\mathbf{s}_0)|\mathbf{Z})$ does *not* depend on $\mathbf{Z}$. Intuitively, the conditional mean-squared prediction error $E\{(Z(\mathbf{s}_0) - p(\mathbf{Z}; \mathbf{s}_0))^2|\mathbf{Z}\}$ is a more appropriate measure of predictor variation than the unconditional mean-squared prediction error $E\{(Z(\mathbf{s}_0) - p(\mathbf{Z}; \mathbf{s}_0))^2\}$, except when the goal is spatial sampling design (see Section 5.6.1). But, for $Z(\cdot)$ a Gaussian process, $p^*(\mathbf{Z}; \mathbf{s}_0) = p^0(\mathbf{Z}; \mathbf{s}_0) \equiv E(Z(\mathbf{s}_0)|\mathbf{Z})$, and the conditional measure is indistinguishable from the unconditional measure. [However, if a prior distribution is put on the covariance parameters of the Gaussian process, $\text{var}(Z(\mathbf{s}_0)|\mathbf{Z})$ is no longer independent of $\mathbf{Z}$; see Section 3.4.4.]

Non-Gaussian Data

Not all data behave as realizations from a Gaussian process. Sometimes it is possible to transform the data to another scale so that they do; see Section 3.2.2. The optimal predictor $E(Z(\mathbf{s}_0)|\mathbf{Z})$ under non-Gaussian model assumptions is typically nonlinear. Kitagawa (1987) proposes a state-space method of (temporal) filtering and prediction that is free from Gaussianity and linearity assumptions, and Matheron (1976a) uses isofactorial models (see also Armstrong and Matheron, 1986a, 1986b) to construct certain optimal nonlinear predictors, an approach he calls disjunctive kriging (Section 5.1).

Consider the conditional mean-squared prediction error

$$E\{(Z(\mathbf{s}_0) - p(\mathbf{Z}; \mathbf{s}_0))^2 | \mathbf{Z}\} = \text{var}(Z(\mathbf{s}_0) | \mathbf{Z}) + \{E(Z(\mathbf{s}_0) | \mathbf{Z}) - p(\mathbf{Z}; \mathbf{s}_0)\}^2.$$

Thus, if the first two moments of the conditional distribution are known or can be estimated, then the conditional mean-squared prediction error can be obtained [as well as the optimal predictor $p^0(\mathbf{Z}; \mathbf{s}_0) = E(Z(\mathbf{s}_0)|\mathbf{Z})$]. For $Z(\cdot)$ a Gaussian process, $p^*(\mathbf{Z}; \mathbf{s}_0) = p^0(\mathbf{Z}; \mathbf{s}_0)$ and $E\{(Z(\mathbf{s}_0) - p^*(\mathbf{Z}; \mathbf{s}_0))^2|\mathbf{Z}\} = \text{var}(Z(\mathbf{s}_0)|\mathbf{Z}) = \sigma_{sk}^2(\mathbf{s}_0)$, which is the (unconditional) mean-squared prediction error. Otherwise,

$$E\{(Z(\mathbf{s}_0) - p^*(\mathbf{Z}; \mathbf{s}_0))^2 | \mathbf{Z}\} = \text{var}(Z(\mathbf{s}_0) | \mathbf{Z}) + \{p^0(\mathbf{Z}; \mathbf{s}_0) - p^*(\mathbf{Z}; \mathbf{s}_0)\}^2.$$

In non-Gaussian situations, where a simple-kriging predictor is used, the preceding expression is the conditional mean-squared prediction error, and its expectation yields $\sigma_{sk}^2(\mathbf{s}_0)$.

Processes that are "almost" Gaussian should give rise to "almost" linear predictors; Le Cam (1985) looks at distributional approximations to Gaussian measures in the independent and the (temporal) dependent cases. O'Brien (1987) proposes a test for nonlinearity of prediction in time series. To quantify the lack of explanation in the ordinary kriging predictor $\boldsymbol{\lambda}'\mathbf{Z}$, given by (3.2.12) in Section 3.2, the parameter

$$R_{ok}^2 \equiv 1 - \{E(Z(\mathbf{s}_0) - \boldsymbol{\lambda}'\mathbf{Z})^2 / E(Z(\mathbf{s}_0) - \bar{Z})^2\}$$

is proposed, where $\bar{Z} \equiv \sum_{i=1}^n Z(\mathbf{s}_i)/n$. Now, $0 \le R_{ok}^2 \le 1$; a value of R_{ok}^2 near 1 indicates that a linear predictor is appropriate.

Estimated Optimal Predictor and Total Prediction Error

The error $Z(\mathbf{s}_0) - E(Z(\mathbf{s}_0)|\mathbf{Z})$ can be referred to as the *probabilistic* prediction error. But, often the conditional expectation has to be estimated. For example, suppose the joint distribution of $Z(\mathbf{s}_0)$ and $\mathbf{Z}$ depends on unknown parameters $\boldsymbol{\eta}$. Write the optimal predictor as $E(Z(\mathbf{s}_0)|\mathbf{Z}; \boldsymbol{\eta})$. If the joint distribution of $\mathbf{Z}$ also depends on $\boldsymbol{\eta}$, then the data can be used to obtain an estimator (e.g., maximum likelihood estimator) $\hat{\boldsymbol{\eta}}$. This yields an (maximum likelihood) estimator $E(Z(\mathbf{s}_0)|\mathbf{Z}; \hat{\boldsymbol{\eta}})$ of the optimal predictor.

Let $\hat{E}(Z(\mathbf{s}_0)|\mathbf{Z})$ denote any estimator of the optimal predictor, $E(Z(\mathbf{s}_0)|\mathbf{Z})$. Then the *total* prediction error

$$\begin{aligned} Z(\mathbf{s}_0) - \hat{E}(Z(\mathbf{s}_0)|\mathbf{Z}) &= \{Z(\mathbf{s}_0) - E(Z(\mathbf{s}_0)|\mathbf{Z})\} \\ &\quad + \{E(Z(\mathbf{s}_0)|\mathbf{Z}) - \hat{E}(Z(\mathbf{s}_0)|\mathbf{Z})\} \end{aligned}$$

can be decomposed into the *probabilistic* prediction error plus the *statistical* prediction error (Bosq, 1983). Typically, the mean-squared total prediction error is *greater* than the mean-squared probabilistic prediction error (see Section 5.3). Strictly speaking, the former should be reported, but often lack of a computable expression leads to a reporting of the latter, and a consequent *under*estimation of the mean-squared prediction error for the predictor $\hat{E}(Z(\mathbf{s}_0)|\mathbf{Z})$. This underestimation is potentially more dangerous than an overestimation.

Chapter Summary

Section 3.1 gives a general discussion of spatial scale. Then Sections 3.2 through 3.5 deal with various forms of spatial prediction, respectively, ordinary kriging, robust (ordinary) kriging, universal kriging, and median-polish kriging. Finally, Section 3.6 shows how geostatistical data can be simulated and gives sources for real data.

3.1 SCALE OF VARIATION

Assumptions about the random process $Z(\cdot)$ will be made that are at times unverifiable. In order to make reasonable assumptions, consideration should be given to the *scale* of fluctuations that the process appears to be exhibiting.

Scale can have two different meanings. One is the observational scale of $Z(\mathbf{s})$; that is, the recording instruments are accurate up to a certain level. The other is the spatial scale of $\mathbf{s}$; that is, the observations are based on a certain aggregation and are taken a certain distance apart. The experiment or observational study is carried out at a particular scale in order to answer questions posed at that or a nearby scale. Thus, there is no absolute standard as to what constitutes a large scale, a small scale, or a microscale; it depends on the goal of the study, the accuracy of the data $\mathbf{Z} = (Z(\mathbf{s}_1), \ldots, Z(\mathbf{s}_n))'$, their level of aggregation, and the spatial locations $\{\mathbf{s}_1, \ldots, \mathbf{s}_n\}$.

Observational and Spatial Scale

The following model is useful. Suppose that the data $(Z(\mathbf{s}_1), \ldots, Z(\mathbf{s}_n))$ represent Z values at points of $D \subset \mathbb{R}^d$, and that they are modeled as a partial realization of the random process

$$\{Z(\mathbf{s}) : \mathbf{s} \in D \subset \mathbb{R}^d\}, \tag{3.1.1}$$

which satisfies the decomposition

$$Z(\mathbf{s}) = \mu(\mathbf{s}) + W(\mathbf{s}) + \eta(\mathbf{s}) + \epsilon(\mathbf{s}), \qquad \mathbf{s} \in D, \tag{3.1.2}$$

where:

$\mu(\cdot) \equiv E(Z(\cdot))$ is the deterministic mean structure that will be called *large-scale variation*.

$W(\cdot)$ is a zero-mean, L_2-continuous [i.e., $E(W(\mathbf{s}+\mathbf{h}) - W(\mathbf{s}))^2 \to 0$ as $\|\mathbf{h}\| \to 0$], intrinsically stationary [i.e., satisfies (2.3.2) and (2.3.4)] process whose variogram range (if it exists) is larger than $\min\{\|\mathbf{s}_i - \mathbf{s}_j\|: 1 \leq i < j \leq n\}$. Call $W(\cdot)$ *smooth small-scale variation*.

$\eta(\cdot)$ is a zero-mean, intrinsically stationary process, independent of W, whose variogram range exists and is smaller than $\min\{\|\mathbf{s}_i - \mathbf{s}_j\|: 1 \leq i < j \leq n\}$. Call $\eta(\cdot)$ *microscale variation*.

$\epsilon(\cdot)$ is a zero-mean white-noise process, independent of W and η. Call $\epsilon(\cdot)$ *measurement error* or noise, and denote $\mathrm{var}(\epsilon(\mathbf{s})) = c_{\mathrm{ME}}$. There are occasions when $\epsilon(\cdot)$ may possess more structure than that of white noise (e.g., Laslett and McBratney, 1990).

Then, in obvious notation,

$$2\gamma_Z(\cdot) = 2\gamma_W(\cdot) + 2\gamma_\eta(\cdot) + 2c_{\mathrm{ME}}.$$

The quantities c_{ME} and $\gamma_Z(\mathbf{h})$, $\|\mathbf{h}\|$ large, are pertinent to the observational scale; the other quantities contain information on the spatial scale.

From the decomposition (3.1.2), write

$$Z(\mathbf{s}) = S(\mathbf{s}) + \epsilon(\mathbf{s}), \qquad \mathbf{s} \in D, \tag{3.1.3}$$

where the "signal" or smooth process $S(\cdot)$ is given by $S(\cdot) \equiv \mu(\cdot) + W(\cdot) + \eta(\cdot)$. The S process is often referred to as the noiseless version of the Z process or, in the engineering literature, as the *state* process. Also, write

$$Z(\mathbf{s}) \equiv \mu(\mathbf{s}) + \delta(\mathbf{s}), \qquad \mathbf{s} \in D, \tag{3.1.4}$$

where the correlated error process $\delta(\cdot)$ is given by $\delta(\cdot) \equiv W(\cdot) + \eta(\cdot) + \epsilon(\cdot)$. When the correlation of $\delta(\mathbf{s})$ with $\delta(\mathbf{s}+\mathbf{h})$ can be written as a function of $\mathbf{h}/a(\mathbf{h})$, where $0 < a(\mathbf{h}) < \infty$, then $a(\mathbf{h})$ is sometimes called the *spatial correlation scale* of the process in direction $\mathbf{h}$.

Nonunique Decomposition

The decomposition (3.1.2) is *not unique* and is largely operational in nature. This means that different scientists might reach different conclusions on the same set of data, depending on how much variation they attribute to the components of (3.1.2). Some idea of the measurement error can be obtained from replicated observations at the same spatial location. More observations at new spatial locations might allow estimation at scales below $\min\{\|\mathbf{s}_i - \mathbf{s}_j\|:$

$1 \le i < j \le n\}$. But, in general, the most important decomposition of $Z(\cdot)$ into $\mu(\cdot)$ and $\delta(\cdot)$, given by (3.1.4), cannot be determined uniquely.

Wold Decomposition

In the absence of extra information or assumptions, and in spite of a theorem called the Wold decomposition theorem (e.g., Koopmans, 1974, p. 225), the dilemma of what to call large-scale variation and what to call small-scale variation also exists in time-series analysis (see, e.g., Nelson and Plosser, 1982; Kunsch, 1986; Bell, 1987). The Wold decomposition theorem says that a second-order stationary (Section 2.3) time series $\{Z(t): t = 0, \pm 1, \pm 2, \ldots\}$ can be expressed uniquely as the sum of two uncorrelated second-order stationary time series $Z_1(\cdot)$ and $Z_2(\cdot)$ such that Z_1 is deterministic and Z_2 is purely nondeterministic.

An analogous theorem for the spatial process (3.1.1) would require generalization of the decomposition from the time dimension to d (≥ 2) dimensions, where notions of past, present, and future are not available. In $\mathbb{Z}^2$, which is the two-dimensional spatial lattice with integer site coordinates, various definitions of "the past" have been made (e.g., half planes, augmented half planes, quarter planes, etc.), leading to different Wold-type decomposition theorems; see Korezlioglu and Loubaton (1986) for details. The generalization from a countable index set to continuous spatial index still remains to be established.

Faced with a finite amount of data, regularly or irregularly located in time or space, these decomposition theorems are of no real help. The large-scale (deterministic) variation usually cannot be extracted unambiguously from the process. In other words, one person's deterministic mean structure may be another person's correlated error structure. When the goal of the analysis is spatial prediction (rather than estimation of mean parameters), this ambiguity does not affect predictions as much as it affects standard errors of prediction (see Section 5.3).

There is a further, potentially serious problem that is difficult to solve in practice. In the absence of information beyond the n data and their spatial locations, the behavior of the microscale component $\eta(\cdot)$ in the decomposition (3.1.2) is, by definition, unobtainable. But knowing this behavior is important (for both predictions and mean squared prediction errors) when a predictor location $\mathbf{s}_0$ is close to an observation location $\mathbf{s}_i$.

Example of Two Decompositions of the Same Data

As an example of the ambiguity in (3.1.4), consider the Wolfcamp-aquifer data presented in Section 4.1. Figure 3.1 shows a plot of variogram estimators in the northeast–southwest and northwest–southeast directions, computed using formula (2.4.2) [actually, using its smoothed version $2\gamma^+$ given just following (2.4.2)]. Notice the monotonic increasing nature of the estimated variograms, to which semivariogram power models of the form $\gamma(h; \boldsymbol{\theta}) = c_0 + b_p|h|^\lambda$ are fitted in Section 4.1.

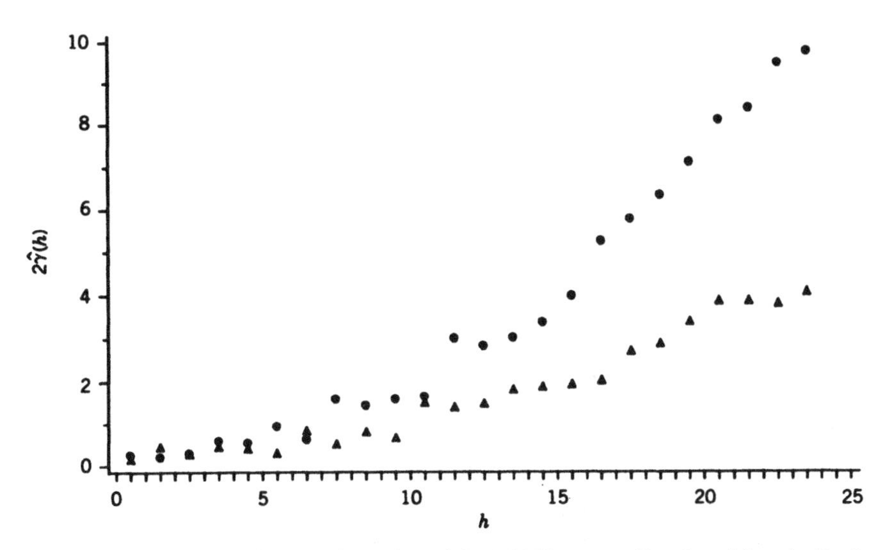

Figure 3.1 Variogram estimators from the original Wolfcamp-aquifer data (given in Section 4.1). Triangles indicate the northwest–southeast direction and circles indicate the northeast–southwest direction. One unit on the horizontal axis is 5 miles and one unit on the vertical axis is 10^5 ft^2. [*Source*: Cressie, 1989c]. Reprinted by permission of the American Statistical Association.

Figure 3.2*a* shows a scatter plot of the data $\{Z(\mathbf{s}_i): i = 1, \ldots, n\}$ and Figure 3.2*b* shows the proposed large-scale variation obtained by median polish (Section 3.5.1). The residuals can then be analyzed as if they were a new spatial data set. Figure 3.3 shows the isotropic variogram estimator based on these residual data.

The first thing to notice when comparing Figures 3.1 and 3.3 is that the observational scale of variation is much larger for the original data. Second, the nature of the spatial dependencies, as pictured by the variogram *shapes*, is very different. For Figure 3.1, the implied model is $Z(\mathbf{s}) = \mu + \delta(\mathbf{s})$, where $\delta(\cdot)$ has zero mean and is intrinsically stationary. For Figure 3.3, the implied model is $Z(\mathbf{s}) = \mu(\mathbf{s}) + \delta(\mathbf{s})$, where $\delta(\cdot)$ has zero mean and is second-order stationary and isotropic. The criterion for choosing one model over another is at present a mixture of scientific context, familiarity, and intuition. Section 2.6.4 contains more on this understudied, but important topic.

Parametric Assumptions on the Mean Function

Of course, the decomposition (3.1.4) can be made unique simply by specifying a smoothing algorithm for $\mu(\cdot)$ based on the data $\{(\mathbf{s}_i, Z(\mathbf{s}_i)): i = 1, \ldots, n\}$. In some circumstances, this may be satisfactory, but it is often the case that as more data are collected the form of $\mu(\cdot)$ changes.

Alternatively, one could parameterize $Z(\cdot)$ or $\mu(\cdot)$. Tamura (1987) surveys various proposals made for time series. In $\mathbb{R}^2$, examples of parameterizations

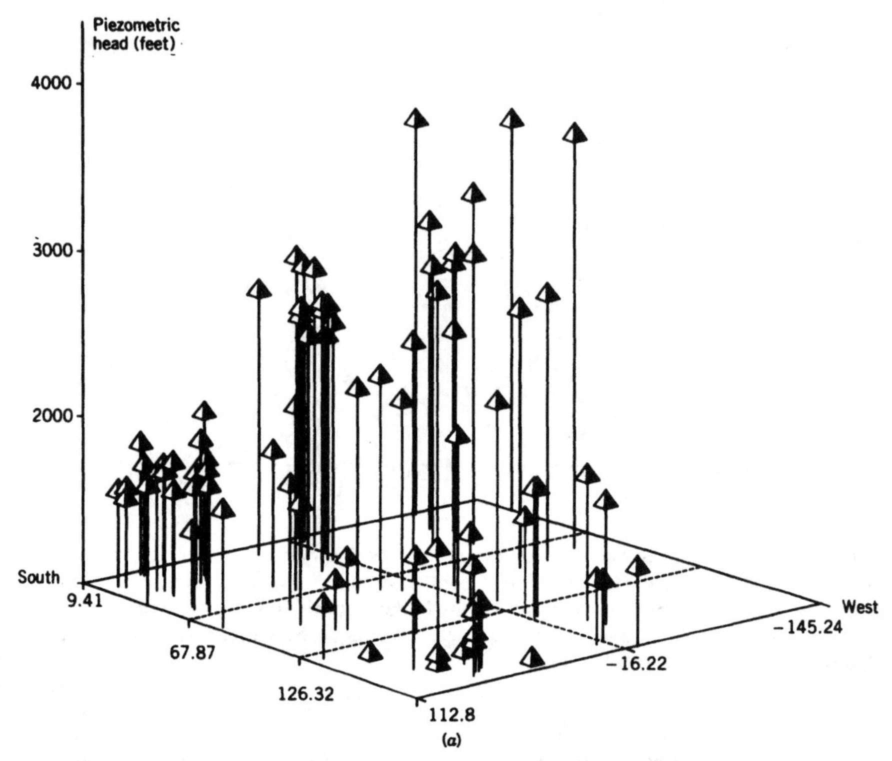

Figure 3.2 (*a*) Scatter plot of the original Wolfcamp-aquifer data (given in Section 4.1). Units on the horizontal axis are displacements in miles and units on the vertical axis are piezometric head in feet above sea level. [*Source*: Cressie, 1989c]. Reprinted by permission of the American Statistical Association.

of $\mu(\cdot)$ are

$$\mu(\mathbf{s}) = a + c(x) + r(y), \qquad \mathbf{s} = (x, y)' \tag{3.1.5}$$

$$\mu(\mathbf{s}) = \beta_0 + \beta_1 x + \beta_2 y + \beta_3 x^2 + \beta_4 xy + \beta_5 y^2, \qquad \mathbf{s} = (x, y)', \tag{3.1.6}$$

$$\mu(\mathbf{s}) = \beta_0 + \beta_1 \cos(\omega_1 x + \omega_2 y), \qquad \mathbf{s} = (x, y)'. \tag{3.1.7}$$

Expression (3.1.5) is the basis of *median-polish kriging* detailed in Section 3.5, expression (3.1.6) is the basis of *universal kriging* detailed in Section 3.4, and expression (3.1.7) is the basis of a two-dimensional *Fourier analysis* where the pair of frequencies (ω_1, ω_2) is considered dominant. Helson and Lowdenslager (1958, 1961) give the theory of spatial Fourier series, and Renshaw and Ford (1983) give examples where a Fourier analysis is useful in obtaining a decomposition of scales of variation.

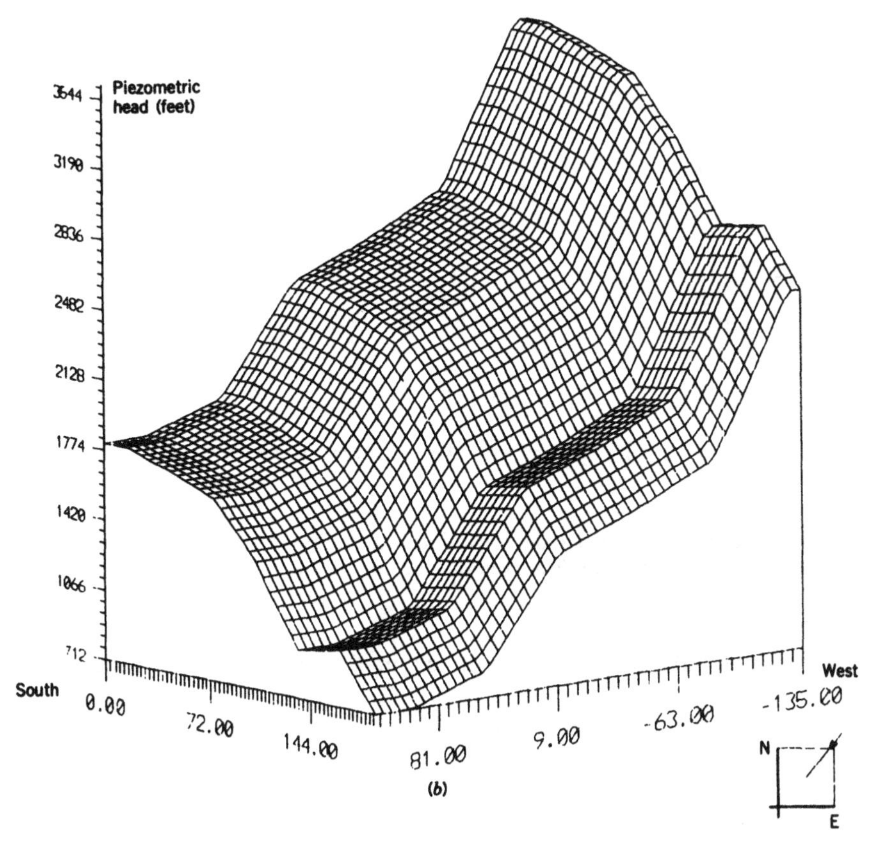

Figure 3.2 (*b*) Estimated mean surface (planar interpolant of median-polish fit) for the Wolfcamp-aquifer data. Units are the same as for (*a*).

Spectral Decomposition

Unless there is a strong reason to believe that the process $Z(\cdot)$ is oscillatory in nature, a Fourier approach may not be sensible. Estimating the spectrum of a two-dimensional process requires a lot of data on a rectangular grid (see, e.g., Ripley, 1981, p. 79), which is an uncommon situation in many geostatistical applications. Shurtz (1985) argues for a fully spectral approach, claiming incorrectly that covariance-based methods of spatial prediction ignore half the information in the data (namely the phase information). It is easy to see from Section 3.2 that, for a minimum-mean-squared-prediction-error criterion, the phase information is irrelevant for linear spatial prediction. However, it is important for conditional simulation (Section 3.6). Sections 6.7.1 and 7.3.2 have more on two-dimensional spectral analysis for lattice data.

Still, the idea of decomposing the data into orthogonal components, the coefficients of which measure the components' importance, is attractive. Large-scale variation could be defined as, say, the first two slowest varying

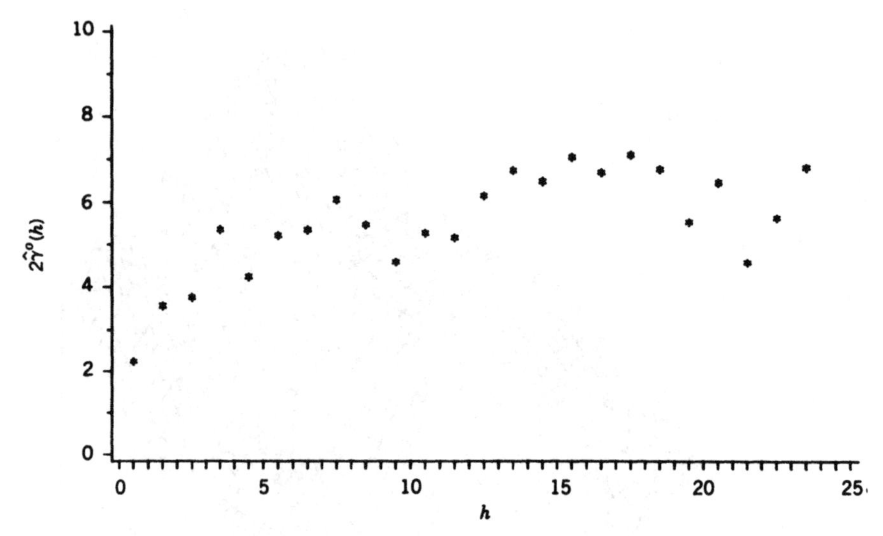

Figure 3.3 Isotropic variogram estimator $2\hat{\gamma}^o$ versus lag, based on median-polish-residual data. One unit on the horizontal axis is 5 miles and one unit on the vertical axis is 10^4 ft^2.

components, and the error would be made up of all the faster varying components. It would not be necessary to adhere to sinusoids, because the analysis could be made using other spectral decompositions. This is a subject of future research.

Scale of Fluctuation

There have been a number of suggestions as to what it is in the process (3.1.1) that embodies spatial scale. Vanmarcke (1983) considers a second-order stationary process $\{Z(s): s \in \mathbb{R}^1\}$ with mean μ, variance σ^2, and autocorrelation function $\rho(\cdot)$. Consider $v(B) \equiv \text{var}(\int_B Z(u)\,du/\,|B|)/\text{var}(Z(s))$, for intervals B such that $|B| \to \infty$ and $|B|v(B) \to \xi$. Then Vanmarcke defines ξ to be the *scale of fluctuation*. In fact, under regularity conditions, $\xi = \int_{-\infty}^{\infty} \rho(h)\,dh$. Also, it is proportional to the zero-frequency ordinate of the spectral density function. Basically, the parameter ξ controls the behavior of $Z(\cdot)$ under long-term averaging: $\text{var}(\int_B Z(u)\,du/\,|B|) \simeq \xi\sigma^2|B|$. Vanmarcke (1983, Chapter 5) is able to obtain asymptotic (as $|B| \to \infty$) results for the mean-squared derivative, the mean threshold crossing rate, and the probability distribution of extreme values of $Z(\cdot)$, in terms of the parameters (μ, σ^2, ξ). Estimation of the parameter ξ is considered by Russo and Jury (1987a, 1987b).

In two dimensions, write $B \equiv B_1 \times B_2$ and define $v(B_1, B_2)$ analogously. Write $v(B_1, B_2) \equiv v(B_2|B_1)v(B_1)$. Then, as $|B_1| \to \infty$ and $|B_2| \to \infty$, suppose $|B_1|\,|B_2|v(B_1, B_2) \to \alpha$, $|B_2|v(B_2|B_1) \to \xi_2(B_1)$, $|B_1|v(B_1) \to \xi_1$, and so forth. The proposed scale of fluctuation is $\kappa_\alpha \equiv \alpha/\xi_1\xi_2$, a dimensionless constant.

This approach of Vanmarcke's concentrates only on the large scale. Black and Freyberg (1987) look at the effect of averaging, for various scales, when the autocorrelation function is $\rho(h) = e^{-h}$.

Analysis of Variation
Batty (1976) has suggested that the various scales exhibited by an observed process $\{z(\mathbf{s}): \mathbf{s} \in D\}$ can be hierarchically decomposed using the notion of entropy of $z(\cdot)$ [actually entropy of the volume proportions p_i of $\mathbf{s} \in D$ that satisfy $z_i < z(\mathbf{s}) \leq z_{i+1}$, where $z_1 < \cdots < z_{m+1}$]. The analysis proceeds much like a nested analysis of variance. Of course, the variogram itself (Section 2.3) is a useful way of looking at the dependence structure of the process at several scales simultaneously (see, e.g., Miesch, 1975; Oliver and Webster, 1986).

Multiple Scales and Fractals
Questions of scale in spatial statistics combine both the issues of level of aggregation of the observations (i.e., the volume of space that an observation represents) and the extent of the observations. As has been demonstrated, these are not unrelated, particularly if the data behave in a self-similar way across a number of scales. Many fractals are self-similar, which is why there has been a great deal of recent interest in their usefulness as models for naturally occurring phenomena. Parameters estimated at one scale are relevant to inference for the process at a smaller, unobserved scale (e.g., microscale); see Sections 5.5 and 9.1.1 for more details. This modeling approach is not confined to fractals. Cushman (1986) gives a stochastic (water) transportation theory that incorporates multiple scales of observations.

Discrete-Index Processes Embedded in a Continuous-Index Process
The development in this part of the book differs from the development in Part II in that, potentially, an observation could be taken at a location arbitrarily close to any other location. That is why spatial generalizations of time-series models (e.g., ARMA models of Box and Jenkins, 1970) are not natural here. However, more useful spatial generalizations are likely to come from statistical methods for (irregularly) sampled continuous-time-indexed stationary processes (e.g., Priestley, 1981; Chan and Tong, 1987).

3.2 ORDINARY KRIGING

The word "kriging" is synonymous with "optimal prediction" (as a noun) or with "optimally predicting" (as the present participle of a verb). In other words, it refers to making inferences on unobserved values of the random process $Z(\cdot)$ given by (3.1.1), or of $S(\cdot)$ given by (3.1.3), from data

$$\mathbf{Z} \equiv (Z(\mathbf{s}_1), \ldots, Z(\mathbf{s}_n))' \tag{3.2.1}$$

observed at known spatial locations

$$\{\mathbf{s}_1, \ldots, \mathbf{s}_n\}. \tag{3.2.2}$$

Denote the generic predictor of $g(Z(\cdot))$ or $g(S(\cdot))$ by

$$p(\mathbf{Z}; g). \tag{3.2.3}$$

Choice of a good predictor will depend on the geometry and location of the region of space where prediction is desired and whether it is the Z process or the S process that is to be predicted. When $g(Z(\cdot)) = Z(B)$ [$\equiv \int_B Z(\mathbf{u})\, d\mathbf{u}/|B|$] or $g(S(\cdot)) = S(B)$, write (3.2.3) as $p(\mathbf{Z}; B)$. In the special case of $B = \{\mathbf{s}_0\}$, write (3.2.3) as $p(\mathbf{Z}; \mathbf{s}_0)$.

Ordinary kriging (Matheron, 1971b; Journel and Huijbregts, 1978, p. 304ff) refers to spatial prediction under the following two assumptions.

Model Assumption. In (3.1.4),

$$Z(\mathbf{s}) = \mu + \delta(\mathbf{s}), \qquad \mathbf{s} \in D, \mu \in \mathbb{R}, \text{ and } \mu \text{ unknown}. \tag{3.2.4}$$

Predictor Assumption.

$$p(\mathbf{Z}; B) = \sum_{i=1}^{n} \lambda_i Z(\mathbf{s}_i), \qquad \sum_{i=1}^{n} \lambda_i = 1. \tag{3.2.5}$$

This latter condition, that the coefficients of the linear predictor sum to 1, guarantees uniform unbiasedness: $E(p(\mathbf{Z}; B)) = \mu = E(Z(B))$, for all $\mu \in \mathbb{R}$. There is a version of kriging called simple kriging, where μ in (3.2.4) is known (Section 3.4.5) and the coefficients are not constrained to sum to 1.

Throughout most of this book, the words "optimal" and "good" will refer to the case of squared-error loss and hence of minimizing mean-squared prediction error, although several other loss functions were suggested in the introduction to this chapter. Hence, if

$$g(Z(\cdot)) = Z(B) \equiv \begin{cases} \int_B Z(\mathbf{u})\, d\mathbf{u}/|B|, & |B| > 0, \\ \text{ave}\{Z(\mathbf{u}): \mathbf{u} \in B\}, & |B| = 0, \end{cases} \tag{3.2.6}$$

then the optimal $p(\cdot; B)$ will minimize the mean-squared prediction error

$$\sigma_e^2 \equiv E(Z(B) - p(\mathbf{Z}; B))^2 \tag{3.2.7}$$

over the class of linear predictors $\sum_{i=1}^{n} \lambda_i Z(\mathbf{s}_i)$ that satisfy $\sum_{i=1}^{n} \lambda_i = 1$.

Optimal Spatial Prediction of the Z Process

For ordinary kriging, the minimization of (3.2.7) is carried out over $(\lambda_1, \ldots, \lambda_n)$, subject to $\sum_{i=1}^{n} \lambda_i = 1$, where the model (3.2.4) is assumed to hold with variogram

$$2\gamma(\mathbf{h}) = \text{var}(Z(\mathbf{s}+\mathbf{h}) - Z(\mathbf{s})), \qquad \mathbf{h} \in \mathbb{R}^d. \tag{3.2.8}$$

Suppose, for the moment, that $B = \{\mathbf{s}_0\}$ in (3.2.6). Then minimize

$$E\left(Z(\mathbf{s}_0) - \sum_{i=1}^{n} \lambda_i Z(\mathbf{s}_i)\right)^2 - 2m\left(\sum_{i=1}^{n} \lambda_i - 1\right) \tag{3.2.9}$$

with respect to $\lambda_1, \ldots, \lambda_n$, and m (the parameter m is a Lagrange multiplier that ensures $\sum_{i=1}^{n}\lambda_i = 1$).

Now the condition $\sum_{i=1}^{n}\lambda_i = 1$ implies that

$$\left(Z(\mathbf{s}_0) - \sum_{i=1}^{n} \lambda_i Z(\mathbf{s}_i)\right)^2 = -\sum_{i=1}^{n}\sum_{j=1}^{n} \lambda_i\lambda_j (Z(\mathbf{s}_i) - Z(\mathbf{s}_j))^2/2$$
$$+ 2\sum_{i=1}^{n} \lambda_i (Z(\mathbf{s}_0) - Z(\mathbf{s}_i))^2/2, \tag{3.2.10}$$

so that under the model (3.2.4), (3.2.9) becomes

$$-\sum_{i=1}^{n}\sum_{j=1}^{n} \lambda_i\lambda_j \gamma(\mathbf{s}_i - \mathbf{s}_j) + 2\sum_{i=1}^{n} \lambda_i \gamma(\mathbf{s}_0 - \mathbf{s}_i) - 2m\left(\sum_{i=1}^{n} \lambda_i - 1\right). \tag{3.2.11}$$

After differentiating (3.2.11) with respect to $\lambda_1, \ldots, \lambda_n$, and m and equating the result to zero, it is seen that the optimal parameters satisfy

$$-\sum_{j=1}^{n} \lambda_j \gamma(\mathbf{s}_i - \mathbf{s}_j) + \gamma(\mathbf{s}_0 - \mathbf{s}_i) - m = 0, \qquad i = 1, \ldots, n,$$

$$\sum_{i=1}^{n} \lambda_i = 1.$$

That is, the optimal $\lambda_1, \ldots, \lambda_n$ can be obtained from

$$\boldsymbol{\lambda}_O = \Gamma_O^{-1} \boldsymbol{\gamma}_O, \tag{3.2.12}$$

where

$$\boldsymbol{\lambda}_O \equiv (\lambda_1, \ldots, \lambda_n, m)', \tag{3.2.13}$$

$$\boldsymbol{\gamma}_O \equiv (\gamma(\mathbf{s}_0 - \mathbf{s}_1), \ldots, \gamma(\mathbf{s}_0 - \mathbf{s}_n), 1)', \tag{3.2.14}$$

$$\Gamma_O \equiv \begin{cases} \gamma(\mathbf{s}_i - \mathbf{s}_j), & i = 1, \ldots, n,\ j = 1, \ldots, n, \\ 1, & i = n+1,\ j = 1, \ldots, n, \\ 0, & i = n+1,\ j = n+1, \end{cases} \tag{3.2.15}$$

and Γ_O is a symmetric $(n + 1) \times (n + 1)$ matrix. It is easy to see that (3.2.12) remains unchanged if $\gamma(\mathbf{h})$ is replaced with $\gamma(\mathbf{h}) + c$, for any $c \in \mathbb{R}$. Such a substitution is sometimes necessary to obtain a numerically stable solution to $\Gamma_O \boldsymbol{\lambda}_O = \boldsymbol{\gamma}_O$.

From (3.2.12), the coefficients $\boldsymbol{\lambda} \equiv (\lambda_1, \ldots, \lambda_n)'$ are given by

$$\boldsymbol{\lambda}' = \left(\boldsymbol{\gamma} + \mathbf{1}\frac{(1 - \mathbf{1}'\Gamma^{-1}\boldsymbol{\gamma})}{\mathbf{1}'\Gamma^{-1}\mathbf{1}}\right)'\Gamma^{-1}$$

and

$$m = -(1 - \mathbf{1}'\Gamma^{-1}\boldsymbol{\gamma})/(\mathbf{1}'\Gamma^{-1}\mathbf{1}),$$

where $\boldsymbol{\gamma} \equiv (\gamma(\mathbf{s}_0 - \mathbf{s}_1), \ldots, \gamma(\mathbf{s}_0 - \mathbf{s}_n))'$ and Γ is the $n \times n$ matrix whose (i, j)th element is $\gamma(\mathbf{s}_i - \mathbf{s}_j)$. Write the optimal (ordinary-kriging) predictor (3.2.5), (3.2.12) as $\hat{p}(\mathbf{Z}; \mathbf{s}_0)$ [or, more simply, as $\hat{Z}(\mathbf{s}_0)$].

The minimized mean-squared prediction error (3.2.7) is sometimes called the *kriging (or prediction) variance*, namely,

$$\sigma_k^2(\mathbf{s}_0) = \boldsymbol{\lambda}_O'\boldsymbol{\gamma}_O = \sum_{i=1}^{n} \lambda_i \gamma(\mathbf{s}_0 - \mathbf{s}_i) + m \tag{3.2.16}$$

$$= \boldsymbol{\gamma}'\Gamma^{-1}\boldsymbol{\gamma} - (\mathbf{1}'\Gamma^{-1}\boldsymbol{\gamma} - 1)^2/(\mathbf{1}'\Gamma^{-1}\mathbf{1}).$$

Also,

$$\sigma_k^2(\mathbf{s}_0) = 2\sum_{i=1}^{n} \lambda_i \gamma(\mathbf{s}_0 - \mathbf{s}_i) - \sum_{i=1}^{n}\sum_{j=1}^{n} \lambda_i \lambda_j \gamma(\mathbf{s}_i - \mathbf{s}_j). \tag{3.2.17}$$

From (3.2.5), (3.2.12), and (3.2.16), *prediction intervals* can be constructed. The interval

$$A \equiv \left(\hat{Z}(\mathbf{s}_0) - 1.96\sigma_k(\mathbf{s}_0), \hat{Z}(\mathbf{s}_0) + 1.96\sigma_k(\mathbf{s}_0)\right)$$

is a nominal 95% prediction interval for $Z(\mathbf{s}_0)$, which, under the assumption that $Z(\cdot)$ is Gaussian, satisfies $\Pr\{Z(\mathbf{s}_0) \in A\} = 95\%$ [the probability calculation is made over the joint distribution of $Z(\mathbf{s}_0), Z(\mathbf{s}_1), \ldots, Z(\mathbf{s}_n)$].

Asymptotic properties of the kriging predictor are considered in Section 5.8. There, infill asymptotics are assumed; that is, n is allowed to tend to infinity by taking more and more observations in a finite domain D.

Finally, it should be observed that stationarity of the variogram is *not* a necessary requirement for kriging; it is assumed for pragmatic reasons, to allow the variogram to be estimated from the data $\mathbf{Z}$ (Section 2.4). Nonstationary versions of the ordinary-kriging equations are given in Section 3.2.4.

Kriging in Terms of the Covariance Function

The squared prediction error in Eq. (3.2.10) can be written more directly as

$$\left(Z(\mathbf{s}_0) - \sum_{i=1}^{n} \lambda_i Z(\mathbf{s}_i)\right)^2 = (Z(\mathbf{s}_0) - \mu)^2 + \sum_{i=1}^{n} \sum_{j=1}^{n} \lambda_i \lambda_j (Z(\mathbf{s}_i) - \mu)(Z(\mathbf{s}_j) - \mu) - 2\sum_{i=1}^{n} \lambda_i (Z(\mathbf{s}_0) - \mu)(Z(\mathbf{s}_i) - \mu),$$

provided $\sum_{i=1}^{n} \lambda_i = 1$. Now suppose the model (3.2.4) holds, with $\delta(\cdot)$ a zero-mean, second-order stationary process (Section 2.3) having covariogram $C(\mathbf{h})$, $\mathbf{h} \in \mathbb{R}^d$. Then (3.2.9) becomes

$$C(\mathbf{0}) + \sum_{i=1}^{n} \sum_{j=1}^{n} \lambda_i \lambda_j C(\mathbf{s}_i - \mathbf{s}_j) - 2\sum_{i=1}^{n} \lambda_i C(\mathbf{s}_0 - \mathbf{s}_i) - 2m\left(\sum_{i=1}^{n} \lambda_i - 1\right).$$

Minimizing this expression with respect to $\lambda_1, \cdots, \lambda_n$, and m yields the ordinary-kriging equations

$$\hat{p}(\mathbf{Z}; \mathbf{s}_0) = \boldsymbol{\lambda}'\mathbf{Z}, \qquad \sigma_k^2(\mathbf{s}_0) = C(\mathbf{0}) - \boldsymbol{\lambda}'\mathbf{c} + m,$$

where

$$\boldsymbol{\lambda}' = \left(\mathbf{c} + \mathbf{1}\frac{(1 - \mathbf{1}'\Sigma^{-1}\mathbf{c})}{\mathbf{1}'\Sigma^{-1}\mathbf{1}}\right)'\Sigma^{-1}, \qquad m = \frac{(1 - \mathbf{1}'\Sigma^{-1}\mathbf{c})}{\mathbf{1}'\Sigma^{-1}\mathbf{1}}.$$

Here, $\mathbf{c} \equiv (C(\mathbf{s}_0 - \mathbf{s}_1), \ldots, C(\mathbf{s}_0 - \mathbf{s}_n))'$ and Σ is an $n \times n$ matrix whose (i, j)th element is $C(\mathbf{s}_i - \mathbf{s}_j)$. These equations require that the process $Z(\cdot)$ be second-order stationary, a stronger assumption than is required for (3.2.12) and (3.2.16). In fact, there is a nonstationary-covariance version of the ordinary-kriging equations, which is given in Section 3.2.4. However, should a stationarity assumption be made in order to estimate $2\gamma(\cdot)$ or $C(\cdot)$, then a larger class of processes can be modeled via $2\gamma(\cdot)$ than via $C(\cdot)$.

The Variogram as a Tool for Forecasting

In time-series analysis, the variogram is a little known tool for forecasting (i.e., kriging), in spite of its extra generality and superior estimation properties (Section 2.4.1). Any expression of the form $\mathrm{var}(\int Z(\mathbf{s})\,F(d\mathbf{s}) - \int Z(\mathbf{s})\,G(d\mathbf{s}))$, where $\int F(d\mathbf{s}) = \int G(d\mathbf{s})$ (= 1, without loss of generality), can *always* be written in terms of the variogram. So far, expressions for $\mathrm{var}(Z(\mathbf{s}_0) - \Sigma_{i=1}^{n}\lambda_i Z(\mathbf{s}_i))$, where $\Sigma_{i=1}^{n}\lambda_i = 1$, have been obtained. In time, where $t_1 < t_2 < \cdots < t_n$ replace the spatial locations and t_{n+1} replaces the prediction location, this quantity is precisely the variance of the forecasting error, based on the linear forecast $\Sigma_{i=1}^{n}\lambda_i Z(t_i)$. More generally,

$$\begin{aligned}\mathrm{var}\left(\int Z(\mathbf{s})\,F(d\mathbf{s}) - \int Z(\mathbf{s})\,G(d\mathbf{s})\right) &= -\int\int \gamma(\mathbf{u}-\mathbf{v})\,F(d\mathbf{u})\,F(d\mathbf{v}) \\ &\quad -\int\int \gamma(\mathbf{u}-\mathbf{v})\,G(d\mathbf{u})\,G(d\mathbf{v}) \\ &\quad +2\int\int \gamma(\mathbf{u}-\mathbf{v})\,F(d\mathbf{u})\,G(d\mathbf{v}).\end{aligned}$$

Thus, one does not need to use the covariogram to obtain expressions for these variances.

Data and Predictor with Different Supports

The ordinary-kriging equations (3.2.12) and (3.2.16) were derived assuming data and predictor have the same (point) support or level of aggregation; the equations are based on the variogram (3.2.8) of the process, also defined on the same (point) support. The physical reality of a point support is an aggregation over a very small volume; to mining engineers, a core sample from a diamond drill bit is a measurement with point support when compared to the thousands of tons of dirt and rock in a mining block. In mining, kriging is often used to predict $Z(B)$, the grade of a block (because that is what is sent to the mill for extraction of minerals), from core-sample assays $\{Z(\mathbf{s}_1), \ldots, Z(\mathbf{s}_n)\}$. However, as the ore body is progressively mined, average grade from previously mined blocks could (and should) be used in the prediction equations. Fortunately, this is easy from knowledge of the variogram (3.2.8).

The general problem just described is to predict $Z(B)$ from data $\{Z(B_1), \ldots, Z(B_n)\}$. The ordinary-kriging equations (3.2.12) and (3.2.16) are easily modified to

$$\boldsymbol{\lambda}_O = \boldsymbol{\Gamma}_O^{-1}\boldsymbol{\gamma}_O(B), \tag{3.2.18}$$

$$\sigma_k^2(B) = \boldsymbol{\lambda}_O'\boldsymbol{\gamma}_O(B) - \gamma(B,B), \tag{3.2.19}$$

where $\boldsymbol{\gamma}_O(B) \equiv (\gamma(B, B_1), \ldots, \gamma(B, B_n), 1)'$, $\gamma(B, B_i) \equiv \int_B\int_{B_i}\gamma(\mathbf{v} - \mathbf{s}_i)\,d\mathbf{s}_i\,d\mathbf{v}/|B_i|\,|B|$, $\boldsymbol{\Gamma}_O$ is the same as in (3.2.12) except that $\gamma(\mathbf{s}_i, \mathbf{s}_j)$ is replaced with $\gamma(B_i, B_j)$, and $\gamma(B, B) \equiv \int_B \int_B \gamma(\mathbf{u} - \mathbf{v})\, d\mathbf{u}\, d\mathbf{v} / |B|^2$. When $B_i = \{\mathbf{s}_i\}$, $\gamma(B, B_i)$ becomes $\gamma(B, \mathbf{s}_i) \equiv \int_B \gamma(\mathbf{v} - \mathbf{s}_i)\, d\mathbf{v} / |B|$, $i = 1, \ldots, n$. In practice, these integrals of the semivariogram can be computed by numerical quadrature.

Wolfcamp-Aquifer Data

Section 4.1 presents piezometric-head data (in feet above sea level) at $n = 85$ distinct locations in the Wolfcamp aquifer of Texas and New Mexico. These data will be used here to illustrate ordinary kriging; a more complete discussion of them can be found in Section 4.1. The estimated variograms are plotted in Figure 3.1, to which an anisotropic power model $\gamma(\mathbf{h}; \boldsymbol{\theta}) = c_0 + \{b_{p,1}^{2/\lambda} r^2 \cos^2((\pi/4) - \phi) + b_{p,2}^{2/\lambda} r^2 \cos^2((\pi/4) + \phi)\}^{\lambda/2}$, $\mathbf{h} = (r \cos \phi, r \sin \phi)'$, $\boldsymbol{\theta} = (c_0, b_{p,1}, b_{p,2}, \lambda)'$, was fitted by weighted least squares in the directions $\pi/4$ and $3\pi/4$ (Section 4.1). The parameter values

$$\hat{\boldsymbol{\theta}} = \left(\hat{c}_0, \hat{b}_{p,1}, \hat{b}_{p,2}, \hat{\lambda}\right)' = (14 \times 10^3\ \mathrm{ft}^2, 38, 15, 1.99)' \qquad (3.2.20)$$

were obtained. These values were used in the ordinary-kriging equations (3.2.12) and (3.2.16) to obtain optimal $\lambda_1, \ldots, \lambda_n$, and m in

$$\hat{p}(\mathbf{Z}; \mathbf{s}_0) \equiv \sum_{i=1}^{n} \lambda_i Z(\mathbf{s}_i), \qquad (3.2.21)$$

$$\sigma_k^2(\mathbf{s}_0) \equiv \sum_{i=1}^{n} \lambda_i \gamma(\mathbf{s}_0 - \mathbf{s}_i) + m. \qquad (3.2.22)$$

Shown in Figure 3.4*a* and *b* are contour maps depicting, respectively, the three-dimensional surfaces

$$\{(\mathbf{s}_0, \hat{p}(\mathbf{Z}; \mathbf{s}_0)) : \mathbf{s}_0 \in D\} \quad \text{and} \quad \{(\mathbf{s}_0, \sigma_k(\mathbf{s}_0)) : \mathbf{s}_0 \in D\}.$$

(In fact, $\mathbf{s}_0$ ranges over a fine grid of points in $D \subset \mathbb{R}^2$. In choosing the grid, the original data locations were avoided.)

Optimal Spatial Sampling Design

Clearly, (3.2.12) and (3.2.16) do *not* depend on the data $\mathbf{Z}$, but they are functions of data location, number of data, and predictor location. (In contrast to popular smoothing algorithms, such as inverse-distance-squared weighting, the ordinary-kriging weight λ_i is *not* solely a function of n and $\|\mathbf{s}_0 - \mathbf{s}_i\|$, $i = 1, \ldots, n$.) For a known variogram, design questions regarding the optimal choice of network size n and locations $\{\mathbf{s}_1, \ldots, \mathbf{s}_n\}$ can be addressed by minimizing *inter alia* functionals of the kriging variances $\{\sigma_k^2(\mathbf{s}_0): \mathbf{s}_0 \in D\}$; see Sections 4.6.2 and 5.6.1 for further details.

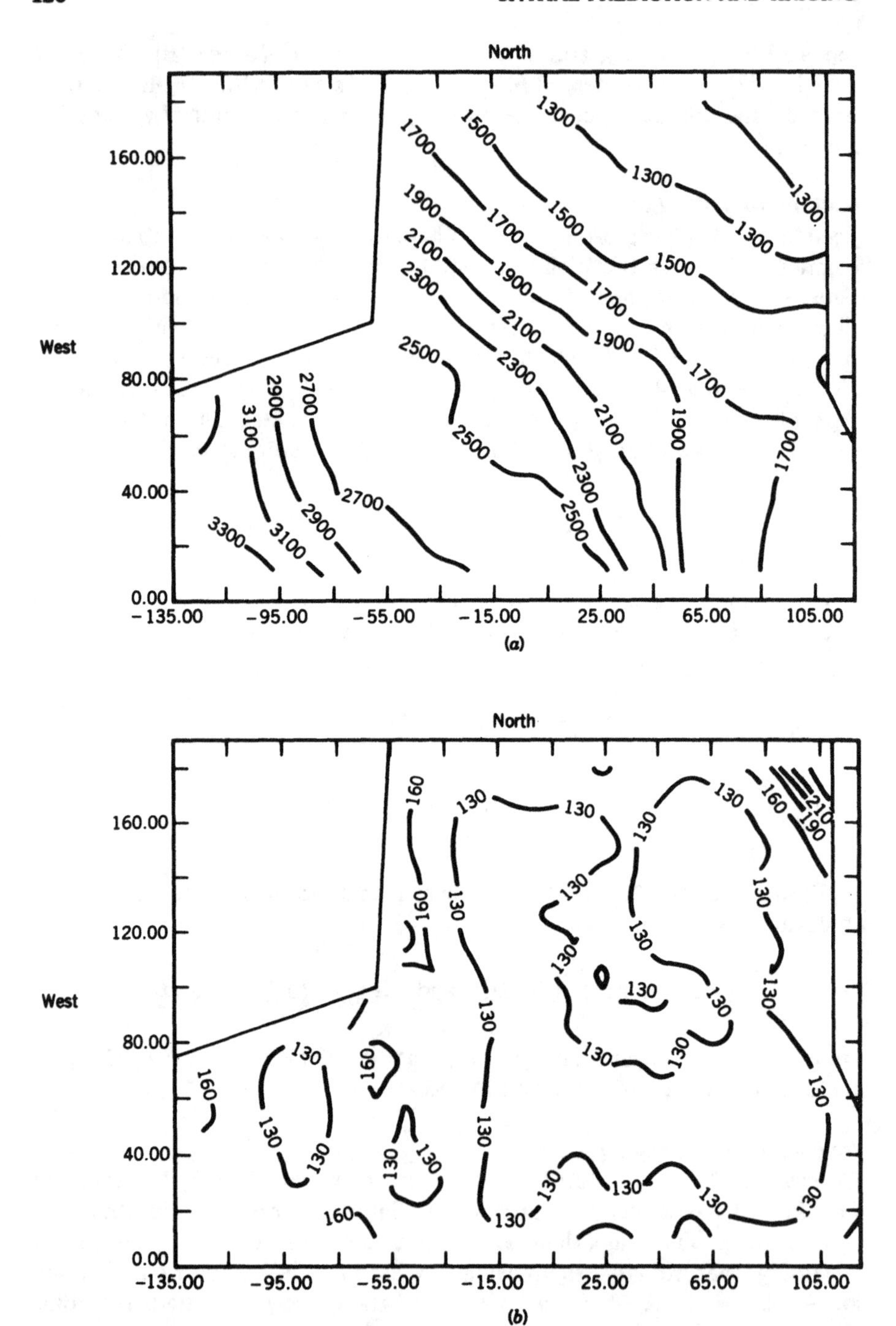

Figure 3.4 (*a*) Predicted surface of piezometric head (using ordinary kriging) for the Wolfcamp-aquifer data. Contour units are in feet above sea level. [*Source*: Cressie, 1989c] (*b*) Kriging (prediction) standard-error surface for the Wolfcamp-aquifer data. Contour units are in feet above sea level. [*Source*: Cressie, 1989c]. Reprinted by permission of the American Statistical Association.

Effect of Variogram Estimation

The kriging equations also depend on the variogram chosen. It is one of the strengths of the geostatistical method that, instead of using a variogram model chosen *a priori* to construct the optimal spatial predictor (3.2.12), there is an initial variography stage consisting of variogram estimation and model fitting. For example, the variogram parameters (3.2.20) were obtained from weighted-least-squares fitting (Section 2.6.2) of variograms to the Wolfcamp-aquifer data.

However, the effect of variogram-parameter estimation on the mean-squared prediction error is not all that well understood. Specifically, $\sigma_k^2(\mathbf{s}_0)$ in (3.2.16) represents a *probabilistic* mean-squared prediction error. Added to this should be a component due to estimation of the variogram. Let $\hat{\hat{p}}(\mathbf{Z}; \mathbf{s}_0)$ denote the predictor obtained from this estimated ordinary kriging; the extra ^ on $\hat{p}(\mathbf{Z}; \mathbf{s}_0)$ is meant to emphasize that $\{\lambda_i: i = 1, \dots, n\}$ in (3.2.21) are obtained from (3.2.12) with an *estimated* variogram. Then one expects

$$\sigma_k^2(\mathbf{s}_0) \le E\left(\hat{\hat{p}}(\mathbf{Z}; \mathbf{s}_0) - Z(\mathbf{s}_0)\right)^2 \tag{3.2.23}$$

and the same inequality to hold with high probability when the left- and right-hand sides are estimated. The (estimated) right-hand side is what should be used to draw maps like Figure 3.4*b*, whereas in practice (an estimate of) $\sigma_k^2(\mathbf{s}_0)$ is typically used. The effect on kriging of not knowing the variogram exactly is discussed further in Section 5.3.

A number of other important remarks on optimal spatial prediction are left to Section 3.2.4. In Section 3.2.1 (following), the effect that various variogram parameters have on ordinary kriging is discussed.

3.2.1 Effect of Variogram Parameters on Kriging

Nugget Effect

Recall from Section 2.3 the definition of the nugget effect as $c_0 = \lim_{\|\mathbf{h}\| \to 0} \gamma(\mathbf{h})$, where $\gamma(\cdot)$ is the semivariogram of the intrinsically stationary process $Z(\cdot)$. The use of this mining terminology conveys the idea that nuggets, or microscale variations, are causing a discontinuity at the origin. Mathematically, this cannot happen for L_2-continuous random processes [i.e., $\mathrm{var}(Y(\mathbf{u}) - Y(\mathbf{v})) \to 0$, as $\mathbf{u} \to \mathbf{v}$]. Hence, if the microscale variation modeled by $\eta(\cdot)$ in (3.1.2) is L_2-continuous, then the only possible reason for c_0 is measurement error. This occurs when a measurement is taken several times (e.g., duplicate assays) and different results are obtained. How can Matheron's nugget effect be interpreted within the decomposition (3.1.2)?

In reality, only data $\{Z(\mathbf{s}_i): i = 1, \dots, n\}$ are available for modeling, and little can be said about the variogram at lag distances smaller than $\min\{\|\mathbf{s}_i - \mathbf{s}_j\|: 1 \le i < j \le n\}$. Unless some $\mathbf{s}_i$s are very close, it is not known whether the microscale variation is continuous or not. Now, assuming that $\lim_{\|\mathbf{h}\| \to 0} \gamma_\eta(\mathbf{h}) = c_{MS}$, that is, the microscale process $\eta(\cdot)$ has nugget effect c_{MS}, we see that

$$c_0 = c_{MS} + c_{ME}, \tag{3.2.24}$$

where c_{ME} is the (measurement-error) variance of the noise process $\epsilon(\cdot)$ in (3.1.2).

Therefore, the so-called nugget effect decomposes into a microscale component plus a measurement-error component. In practice, it is not easy to determine c_0 from data whose spatial separations are large. A value is often obtained by interpolating variogram estimates at lags closest to zero.

An important part of kriging is to determine the relative importance of the components in (3.2.24) (Cressie 1988b). The ordinary-kriging predictor given by Matheron (1962) [see Eqs. (3.2.5) and (3.2.12)] automatically assumes that the value to be predicted is $Z(\mathbf{s}_0)$, which implicitly posits that $\epsilon(\cdot) \equiv 0$ (and $c_{ME} = 0$) in (3.1.2). On the other hand, Yakowitz and Szidarovsky (1985) effectively assume that $\eta(\cdot) \equiv 0$ in (3.1.2) and they develop prediction just for the large-scale and smooth small-scale part of $Z(\mathbf{s}_0)$, namely, $\mu + W(\mathbf{s}_0)$.

More realistically, a *measurement-error-free* (or noiseless) version of $Z(\mathbf{s}_0)$ in (3.1.2), namely $S(\mathbf{s}_0) \equiv \mu + W(\mathbf{s}_0) + \eta(\mathbf{s}_0)$, should be predicted. This is because, when predicting at location $\mathbf{s}_0$, one wants to know what actually exists at that location and not the value distorted by, for example, laboratory measurement error. Then,

$$\hat{S}(\mathbf{s}_0) = \sum_{i=1}^{n} \nu_i Z(\mathbf{s}_i), \tag{3.2.25}$$

where the optimal ν_is [i.e., the ν_is that sum to 1 and minimize the mean-squared-error-prediction error $E(S(\mathbf{s}_0) - \sum_{i=1}^{n} \nu_i Z(\mathbf{s}_i))^2$] satisfy

$$\Gamma_O \boldsymbol{\nu}_O = \boldsymbol{\gamma}_O^*; \tag{3.2.26}$$

here $\boldsymbol{\nu}_O \equiv (\nu_1, \ldots, \nu_n, m)'$, Γ_O is the same as in (3.2.15) (where γ is understood to be γ_Z), and $\boldsymbol{\gamma}_O^* \equiv (\gamma^*(\mathbf{s}_0 - \mathbf{s}_1), \ldots, \gamma^*(\mathbf{s}_0 - \mathbf{s}_n), 1)'$. Notice that the only difference between (3.2.26) and (3.2.12) is the right-hand side, $\boldsymbol{\gamma}_O^*$ instead of $\boldsymbol{\gamma}_O$. When $\mathbf{s}_0 \neq \mathbf{s}_i$, $\gamma^*(\mathbf{s}_0 - \mathbf{s}_i)$ is defined to be $\gamma_Z(\mathbf{s}_0 - \mathbf{s}_i)$, $i = 1, \ldots, n$. However, when $\mathbf{s}_0$ is one of the sample locations, say $\mathbf{s}_1$, $\gamma^*(\mathbf{s}_0 - \mathbf{s}_1)$ is defined to be c_{ME}; that is, the value is not necessarily zero, as it is in (3.2.12). Also, the minimized mean-squared prediction error is given by

$$\tau_k^2(\mathbf{s}_0) = \sum_{i=1}^{n} \nu_i \gamma^*(\mathbf{s}_0 - \mathbf{s}_i) + m - c_{ME}, \tag{3.2.27}$$

which should be compared to $\sigma_k^2(\mathbf{s}_0)$ given by (3.2.16). Analogous expressions to (3.2.26) and (3.2.27), in terms of the covariogram of a second-order stationary process, can be easily derived.

It is important to use (3.2.26) and (3.2.27) when there is measurement error (which is often the case). Most applications of kriging implicitly assume $c_{ME} = 0$ by predicting the Z process rather than the S process.

Comparison of (3.2.12) and (3.2.26) shows, respectively, an exact and a nonexact interpolator. [An exact interpolator, $p(\mathbf{Z}; \mathbf{s}_0)$, is one where $p(\mathbf{Z}; \mathbf{s}_i) = Z(\mathbf{s}_i)$, $i = 1, \ldots, n$.] Thus, kriging can either honor the data, as is sometimes required by mining companies, or smooth the data, as is sensible when it is known (from, say, laboratory experiments) that the data contain measurement error. The larger is c_{ME}, the more one smooths, so that in the extreme case where all the random variation is measurement error [i.e., $Z(\cdot) = \mu + \epsilon(\cdot)$], the ordinary-kriging predictor is

$$\hat{Z}(\mathbf{s}_0) = \begin{cases} \bar{Z}, & \mathbf{s}_0 \notin \{\mathbf{s}_1, \ldots, \mathbf{s}_n\}, \\ Z(\mathbf{s}_j), & \mathbf{s}_0 = \mathbf{s}_j,\ j = 1, \ldots, n, \end{cases} \tag{3.2.28}$$

where $\bar{Z} \equiv \sum_{i=1}^{n} Z(\mathbf{s}_i)/n$. However, from (3.2.26), prediction of its noiseless version at $\mathbf{s}_0$ gives

$$\hat{S}(\mathbf{s}_0) = \sum_{i=1}^{n} \nu_i Z(\mathbf{s}_i) = \bar{Z}, \tag{3.2.29}$$

regardless of whether $\mathbf{s}_0$ is a sample location or not. Equation (3.2.28) shows the extreme lengths the kriging predictor $\hat{Z}(\mathbf{s}_0)$ will go to be an exact interpolator.

In general, *both* $\hat{Z}(\mathbf{s}_0)$ and $\hat{S}(\mathbf{s}_0)$ are discontinuous in $\mathbf{s}_0$, although $\hat{S}$ is less so. When $c_{MS} = 0$, $\hat{S}$ is continuous, but both $c_{MS} = 0$ and $c_{ME} = 0$ are needed for $\hat{Z}$ to be continuous. It is clear from (3.2.12) and (3.2.26) that, when $\mathbf{s}_0 \notin \{\mathbf{s}_1, \ldots, \mathbf{s}_n\}$, $\hat{Z}(\mathbf{s}_0) = \hat{S}(\mathbf{s}_0)$. Therefore, it is tempting to think that, as long as a kriging map is based on prediction locations that do not coincide with the data locations (such as in Section 4.1), the distinction between predicting Z or S is not important. On the contrary, the mean-squared prediction errors $\sigma_k^2(\mathbf{s}_0)$ given by (3.2.16) and $\tau_k^2(\mathbf{s}_0)$ given by (3.2.27) do not coincide unless $c_{ME} = 0$.

This illustration also shows why the literature has been unclear as to whether or not kriging yields exact interpolation. It *is* an exact interpolator if one is willing to assume $c_{ME} = 0$, but in many circumstances this is an unrealistic assumption.

An *incorrect* set of kriging equations is obtained by replacing $\gamma(\mathbf{s}_i - \mathbf{s}_i) = \gamma(\mathbf{0}) = 0$ in (3.2.15) with the nugget effect c_0, $i = 1, \ldots, n$. It is not difficult to show that these would be the equations one would obtain from predicting $U(\mathbf{s}_0)$ from $\{U(\mathbf{s}_i): i = 1, \ldots, n\}$, where $U(\cdot) \equiv Z(\cdot) - \eta(\cdot) - \epsilon(\cdot)$, obtained from (3.1.2). Unless the continuous process $U(\cdot)$ miraculously has

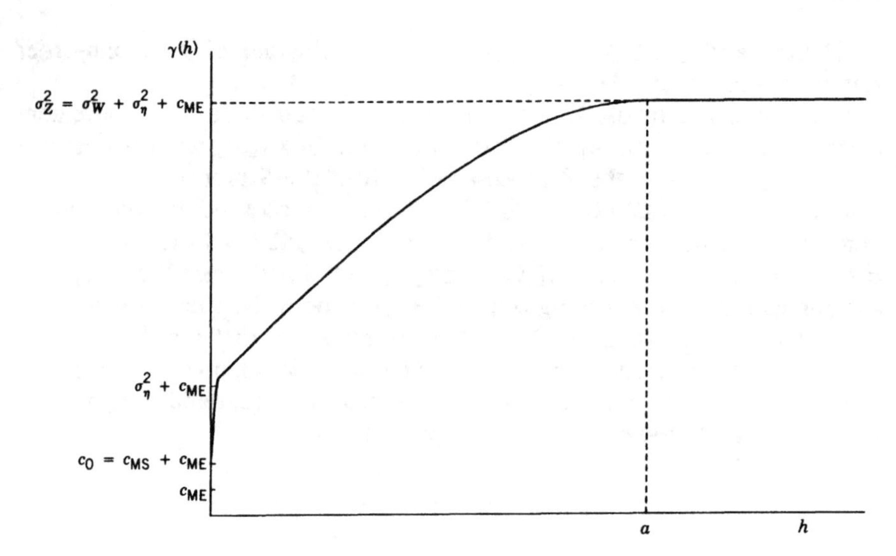

Figure 3.5 A generic semivariogram in $\mathbb{R}^1$, showing sill σ_Z^2, nugget effect c_0, and range a, in terms of the decomposition (3.1.2). [*Source*: Cressie, 1988b]. Reprinted by permission of *Mathematical Geology*.

$U(\mathbf{s}_1), \ldots, U(\mathbf{s}_n)$ available as data, such kriging equations should be religiously avoided.

Sill

Recall from Section 2.3 the definition of the sill as $\sigma_Z^2 \equiv \lim_{\|\mathbf{h}\| \to \infty} \gamma_Z(\mathbf{h})$, provided the limit exists. Assume in the decomposition (3.1.2) that $\mu(\mathbf{s}) \equiv \mu$, $W(\cdot)$ has a sill σ_W^2, $\eta(\cdot)$ has a sill σ_η^2, and $\epsilon(\cdot)$ has variance c_{ME}. Henceforth, refer to σ_W^2, the variance of the smooth small-scale variation, as the *partial sill*. Then $Z(\cdot)$ is intrinsically stationary with sill

$$\sigma_Z^2 = \sigma_W^2 + \sigma_\eta^2 + c_{ME}. \tag{3.2.30}$$

Notice that $\sigma_\eta^2 \geq c_{MS}$ and, hence, $\sigma_Z^2 \geq \sigma_W^2 + c_0$. This inequality may appear strange because the sill is usually thought of as being equal to the partial sill plus the nugget effect; the contradiction is resolved when it is realized that an *estimated* nugget effect, obtained by extrapolating an experimental variogram near the origin, is actually estimating $\sigma_\eta^2 + c_{ME}$ rather than c_0; Figure 3.5 shows the variogram parameters and illustrates the relationships between the sill components in (3.2.30).

Cressie (1988b) shows that, to achieve an unbiased estimator of the total sill, a multiplicative bias adjustment should *not* be applied to the total-sill estimator. Instead, the partial-sill estimator should be adjusted, leaving the nugget-effect estimator alone. This makes sense because, intuitively, such an adjustment is derived from the small- and large-scale components and should not apply to the microscale component.

Range

Recall from Section 2.3.2 the definition of the range in the direction $\mathbf{a}/\|\mathbf{a}\|$. It can be interpreted as the lag beyond which $Z(\mathbf{s})$ and $Z(\mathbf{s}+\mathbf{a})$ [or $Z(\mathbf{s}-\mathbf{a})$] are uncorrelated. It is often thought, mistakenly, that this variogram parameter allows one to determine a kriging neighborhood; for example, one might be tempted to include in the kriging equations (3.2.5), (3.2.12) only those data that are within range of $\mathbf{s}_0$. This rough rule can serve as a guide, but the following considerations show that caution is necessary.

Minimizing the mean-squared prediction error subject to an unbiasedness constraint is equivalent to minimizing (3.2.9) with respect to $\lambda_1, \ldots, \lambda_n$, and m. Under the intrinsic-stationary model (3.2.4), the solution is given by (3.2.12). More generally, by assuming that $C(\mathbf{s}_i, \mathbf{s}_j) \equiv \operatorname{cov}(Z(\mathbf{s}_i), Z(\mathbf{s}_j))$, $0 \le i \le j \le n$, are known, it is shown in Section 3.2.4 that the optimal unbiased predictor can be written as

$$\hat{Z}(\mathbf{s}_0) \equiv \hat{p}(\mathbf{Z}; \mathbf{s}_0) = \boldsymbol{\lambda}'\mathbf{Z}, \tag{3.2.31}$$

where

$$\boldsymbol{\lambda}' = \left(\mathbf{c} + \mathbf{1}\frac{(1 - \mathbf{1}'\Sigma^{-1}\mathbf{c})}{\mathbf{1}'\Sigma^{-1}\mathbf{1}}\right)'\Sigma^{-1}, \tag{3.2.32}$$

Σ is an $n \times n$ matrix whose (i, j)th element is $C(\mathbf{s}_i, \mathbf{s}_j)$, and $\mathbf{c} \equiv (C(\mathbf{s}_0, \mathbf{s}_1), \ldots, C(\mathbf{s}_0, \mathbf{s}_n))'$. [Notice that the expression for $\boldsymbol{\lambda}$ given in Cressie (1988b, eq. 28) is in error, although it is corrected in Cressie (1989d).] Clearly, the behavior of the kriging coefficients $\boldsymbol{\lambda}' = (\lambda_1, \ldots, \lambda_n)$ is strongly influenced by the *inverse* of the covariance matrix Σ.

Write $\Sigma = (c_{ij})$ and $\Sigma^{-1} = (c^{ij})$. Without loss of generality, consider the first two variables $Z(\mathbf{s}_1)$ and $Z(\mathbf{s}_2)$, and partition Σ as

$$\Sigma = \begin{bmatrix} c_{11} & c_{12} & \mathbf{c}'_{13} \\ c_{12} & c_{22} & \mathbf{c}'_{23} \\ \mathbf{c}_{13} & \mathbf{c}_{23} & C_{33} \end{bmatrix}.$$

Then define

$$\begin{bmatrix} e(\mathbf{s}_1) \\ e(\mathbf{s}_2) \end{bmatrix} \equiv \begin{bmatrix} Z(\mathbf{s}_1) - \mu \\ Z(\mathbf{s}_2) - \mu \end{bmatrix} - \begin{bmatrix} \mathbf{c}'_{13} \\ \mathbf{c}'_{23} \end{bmatrix} C_{33}^{-1} \begin{bmatrix} Z(\mathbf{s}_3) - \mu \\ \cdot \\ \cdot \\ \cdot \\ Z(\mathbf{s}_n) - \mu \end{bmatrix},$$

to be the residuals from regression of $Z(\mathbf{s}_1)$ and $Z(\mathbf{s}_2)$ on the other $n - 2$ variables. Then

$$\operatorname{var}((e(\mathbf{s}_1), e(\mathbf{s}_2))') = \begin{bmatrix} c_{11} & c_{12} \\ c_{12} & c_{22} \end{bmatrix} - \begin{bmatrix} \mathbf{c}'_{13} \\ \mathbf{c}'_{23} \end{bmatrix} C_{33}^{-1} [\mathbf{c}_{13} \quad \mathbf{c}_{23}]$$

and it is a straightforward matter to show (e.g., Morrison, 1978, p. 92ff) that

$$\operatorname{cov}(e(\mathbf{s}_1), e(\mathbf{s}_2))/\{\operatorname{var}(e(\mathbf{s}_1))\operatorname{var}(e(\mathbf{s}_2))\}^{1/2} = -c^{12}/\{c^{11}c^{22}\}^{1/2}. \quad (3.2.33)$$

The quantity on the left-hand side of (3.2.33) is often called the *partial correlation* (of the first two variables after fitting the rest). Also, $c^{11} = |C_{22}|/|\Sigma|$, where $C_{22} \equiv \operatorname{var}((Z(\mathbf{s}_2), \ldots, Z(\mathbf{s}_n))')$; a similar expression is available for c^{22}.

When $Z(\cdot)$ is a *Gaussian* random process, the left-hand side of (3.2.33) is actually the *conditional* correlation of $Z(\mathbf{s}_1)$ and $Z(\mathbf{s}_2)$, *given* all other $Z(\mathbf{s}_3), \ldots, Z(\mathbf{s}_n)$, and hence, $c^{12} = 0$ if and only if $Z(\mathbf{s}_1)$ and $Z(\mathbf{s}_2)$ are conditionally independent given $Z(\mathbf{s}_3), \ldots, Z(\mathbf{s}_n)$. Regardless of whether or not the process is Gaussian, it is clear that the (i, j)th entry of Σ^{-1} depends not only of the behavior of the process at $\mathbf{s}_i$ and $\mathbf{s}_j$, but also on its behavior at *every* other location. Thus, it is a mistake to think of the ordinary-kriging weight λ_i as a function only of $\|\mathbf{s}_0 - \mathbf{s}_i\|$, $i = 1, \ldots, n$. For example, suppose $\mathbf{s}_1$ is the only location within range of $\mathbf{s}_0$. Then, from (3.2.32), *no* λ_is are necessarily zero, their being rather complex functions of *all* the covariances $\{c_{ij}: i, j = 1, \ldots, n\}$.

A simple process in $\mathbb{R}^1$, with stationary tent covariogram

$$C(h) = \begin{cases} C(0)(1 - |h|/a), & |h| \le a, \\ 0, & |h| > a, \end{cases}$$

of range a, will serve to illustrate these points. Suppose $s_0 = 0$, $s_1 = -a/2$, and $s_2 = a/2$. Then

$$\Sigma = \begin{bmatrix} 1 & 0 \\ 0 & 1 \end{bmatrix} C(0) \quad \text{and} \quad \mathbf{c} = \begin{bmatrix} 1/2 \\ 1/2 \end{bmatrix} C(0).$$

Hence, $\lambda_1 = \lambda_2 = 1/2$ and $\hat{Z}(s_0) = Z(s_1)/2 + Z(s_2)/2$. Now add a third member to the sample at $s_3 = a/2 + af$, $0 < f \le 1$, and note that when $1/2 < f \le 1$, s_3 is beyond the range of s_0. Then,

$$\Sigma = \begin{bmatrix} 1 & 0 & 0 \\ 0 & 1 & 1-f \\ 0 & 1-f & 1 \end{bmatrix} C(0)$$

and, for $0 < f \le 1/2$,

$$\mathbf{c} = \begin{bmatrix} 1/2 \\ 1/2 \\ 1/2 - f \end{bmatrix} C(0).$$

Hence, from (3.2.32), $\lambda_1 = 2/(4-f)$, $\lambda_2 = (3-f)/(4-f)$, and $\lambda_3 = -1/(4-f)$. Finally,

$$\hat{Z}(s_0) = \{2Z(s_1) + (3-f)Z(s_2) - Z(s_3)\}/(4-f), \qquad 0 < f \le 1/2. \tag{3.2.34}$$

Similarly, for $1/2 < f \le 1$,

$$\hat{Z}(s_0) = \{(5f - 2f^2)Z(s_1) + (2+f)Z(s_2) - 2(1-f)Z(s_3)\}/(8f - 2f^2). \tag{3.2.35}$$

Notice that from (3.2.35), as f ranges over $(1/2, 1]$ ($f = 1$ corresponds to s_3 being the furthest outside the range a), the kriging predictor varies considerably with f. This shows that a simple-minded deletion of values beyond the range (here a), even when those values are "hidden" by closer points (here s_1), is inappropriate. Only for $f = 1$ does (3.2.35) reduce to $\{Z(s_1)/2 + Z(s_2)/2\}$. It is also clear that a restriction on the λ_is to remain positive leads to a nonoptimal predictor; here $\lambda_3 \le 0$, regardless of the value of f (Section 3.2.4).

It is apparent from this example that the influence of a datum is reduced when it is hidden behind another datum. This has sometimes been called a screen effect (see, e.g., Journel and Huijbregts, 1978, p. 346) and makes sense when one realizes that Σ^{-1} is essentially determined by partial correlations; once a closer variable has been "partialled out," the influence of the more distant variable is considerably weakened. An interesting example of this in $\mathbb{R}^1$ is where $C(h) = C(0)\exp(-\lambda|h|)$; notice that the range is infinite. (A Gaussian process with this exponential covariogram is called a Gaussian Markov process.) Suppose the data are at $s_i = i$, $i = 1, \ldots, n$. For $\rho \equiv \exp(-\lambda)$, it can be shown (e.g., Morrison, 1978, p. 339) that

$$\Sigma = \begin{bmatrix} 1 & \rho & \rho^2 & \cdots & \rho^{n-1} \\ \rho & 1 & \rho & \cdots & \rho^{n-2} \\ \rho^2 & \rho & 1 & \cdots & \rho^{n-3} \\ \cdot & & & \cdots & \cdot \\ \cdot & & & & \cdot \\ \cdot & & & & \cdot \\ \rho^{n-1} & \cdot & \cdot & \cdot & 1 \end{bmatrix} \cdot C(0),$$

$$\Sigma^{-1} = \begin{bmatrix} 1 & -\rho & 0 & \cdots & 0 \\ -\rho & 1+\rho^2 & -\rho & \cdots & 0 \\ 0 & -\rho & 1+\rho^2 & \cdots & 0 \\ \cdot & & & & \cdot \\ \cdot & & & & \cdot \\ \cdot & & & 1+\rho^2 & -\rho \\ 0 & \cdot & \cdot & -\rho & 1 \end{bmatrix} \Bigg/ \{C(0)(1-\rho^2)\}.$$

For $s_0 = 0$, (3.2.12) yields ordinary-kriging weights $(\lambda_1, \ldots, \lambda_n) = (\rho + O(1/n), O(1/n), \ldots, O(1/n))$. [The exact formula for the ordinary-kriging predictor is given by (1.3.10).] In other words, the optimal predictor of $Z(s_0)$ is approximately ρ times the (mean-corrected) nearest value $Z(s_1)$, expressing the regression effect caused by the spatial correlation. Similarly, if $s_i < s_0 < s_{i+1}$, then the optimal linear predictor is approximately a weighted combination of the nearest values $Z(s_i)$ and $Z(s_{i+1})$. In general, it is a complicated issue whether a variable receives nontrivial weights in the kriging predictor.

An alternative way of defining a spatial model is to impose a simplifying structure on Σ^{-1} rather than on Σ; that is, assume Σ^{-1} has large blocks of zero elements. Time-series analysts do this when they fit autoregressive processes. Predictors that use coefficients proportional to, for example, inverse distances squared, essentially fall into this category. For Gaussian processes, such models can be seen as special cases of Markov random fields (e.g., Besag, 1974), which are discussed in Chapter 6.

Range and Nugget Effect

Clearly, the range by itself does not provide sufficient guidance to determine the kriging neighborhood. As the nugget effect increases, the screen effect referred to earlier gets weaker. Consider the ordinary kriging predictor (3.2.25) for $f = 1$; it is $\{Z(s_1)/2 + Z(s_2)/2\}$. By adding a nugget effect $c_0 \geq 0$ to the covariance model, the ordinary-kriging predictor becomes $\{(3 + 2c_0)Z(s_1) + (3 + 2c_0)Z(s_2) + 2c_0 Z(s_3)\}/(6 + 6c_0)$. Notice that, as $c_0 \to \infty$, the predictor converges to $\bar{Z}$.

Rivoirard (1987) proposes choosing a neighborhood for which $(1 - \mathbf{1}'\Sigma^{-1}\mathbf{c})/(\mathbf{1}'\Sigma^{-1}\mathbf{1})$ is small. Since m in (3.2.16) [the expression for the kriging variance $\sigma_k^2(\mathbf{s}_0)$] can also be written as $(1 - \mathbf{1}'\Sigma^{-1}\mathbf{c})/(\mathbf{1}'\Sigma^{-1}\mathbf{1})$, it is sensible not only to make it small, but to make it as negative as possible. It is easy to see that a large nugget effect (as a percentage of the sill) makes m large positive; to reduce m, kriging neighborhoods have to be expanded considerably. Conversely, a small nugget effect implies much smaller kriging variances; hence, kriging neighborhoods could be contracted without giving up too much in kriging variance. This compromise between kriging variance and size of kriging neighborhood is worthy of more research; some further discussion can be found in Section 3.4.5.

Conclusion

All three variogram parameters, nugget effect, sill, and range, have an effect on ordinary kriging based on data $Z(\mathbf{s}_1), \ldots, Z(\mathbf{s}_n)$. Under infill asymptotics (see Section 5.8), Stein (1988) shows that, for efficient prediction, it is often the behavior of the variogram near the origin that needs to be captured. These two conclusions are not in conflict, but they do show that care should be taken when judging the relevance of an asymptotic result in a finite-data situation. Because there is no natural scaling for Stein's result, it is hard to judge when the data are sufficiently close to ignore (as his result suggests) all but the nugget effect and the variogram slope at the origin.

3.2.2 Lognormal and Trans-Gaussian Kriging

Lognormal Kriging

A lognormal random process $\{Z(\mathbf{s}): \mathbf{s} \in D\}$ is a positive-valued process such that

$$Y(\mathbf{s}) \equiv \log Z(\mathbf{s}), \qquad \mathbf{s} \in D, \tag{3.2.36}$$

is a Gaussian process [i.e., all finite-dimensional distributions of $Y(\cdot)$ are Gaussian; see Section 2.3]. Sometimes a positive constant is added to $Z(\mathbf{s})$ in (3.2.36) to ensure that $Y(\mathbf{s})$ is well defined. Here, I shall assume $Y(\cdot)$ to be intrinsically stationary. As with ordinary kriging, the goal of lognormal kriging is to predict $Z(B)$ from observations $\mathbf{Z} \equiv (Z(\mathbf{s}_1), \ldots, Z(\mathbf{s}_n))'$, where $Z(B)$ is defined by (3.2.6).

A brief overview of lognormal kriging is presented here, and the case of predicting $Z(\mathbf{s}_0)$ is given prominence. The idea is to transform the problem from the Z scale to the (intrinsically stationary, Gaussian) Y scale. Hence, predict $Y(\mathbf{s}_0)$ with

$$\hat{p}_Y(\mathbf{Z}; \mathbf{s}_0) \equiv \sum_{i=1}^{n} \lambda_i \log Z(\mathbf{s}_i) = \sum_{i=1}^{n} \lambda_i Y(\mathbf{s}_i), \tag{3.2.37}$$

where $\lambda_1, \ldots, \lambda_n$ solve (3.2.12) on the transformed Y scale [i.e., the variogram used in (3.2.12) is $2\gamma_Y(\mathbf{h}) \equiv \mathrm{var}(Y(\mathbf{s} + \mathbf{h}) - Y(\mathbf{s}))$, $\mathbf{h} \in \mathbb{R}^d$]. However, the back-transformed value $\exp(\hat{p}_Y(\mathbf{Z}; \mathbf{s}_0))$ is a *biased* predictor.

If $\mathbf{Y} = (Y_1, Y_2)' \sim \mathrm{Gau}(\boldsymbol{\mu}, \Sigma)$, where $\boldsymbol{\mu} = (\mu_1, \mu_2)'$ and $\Sigma = (\sigma_{ij})$, then the vector $(\exp(Y_1), \exp(Y_2))'$ has mean vector $\boldsymbol{\nu}$ and covariance matrix T, where

$$\boldsymbol{\nu} = (\nu_1, \nu_2)' = \left(\exp\{\mu_1 + \sigma_{11}/2\}, \exp\{\mu_2 + \sigma_{22}/2\}\right)', \tag{3.2.38}$$

$$T = \begin{bmatrix} \nu_1^2(e^{\sigma_{11}} - 1) & \nu_1\nu_2(e^{\sigma_{12}} - 1) \\ \nu_1\nu_2(e^{\sigma_{21}} - 1) & \nu_2^2(e^{\sigma_{22}} - 1) \end{bmatrix} \tag{3.2.39}$$

(e.g., Aitchison and Brown, 1957). Assume that the Gaussian process $Y(\cdot)$ is intrinsically stationary with mean μ_Y and variogram $2\gamma_Y(\cdot)$ and define the variance function $\sigma_Y^2(\mathbf{s}) \equiv \mathrm{var}(Y(\mathbf{s}))$, $\mathbf{s} \in D$. Then, from (3.2.38), an unbiased unbiased predictor is

$$\begin{aligned} \check{p}_Z(\mathbf{Z}; \mathbf{s}_0) &\equiv \exp\{\hat{p}_Y(\mathbf{Z}; \mathbf{s}_0) + \sigma_Y^2(\mathbf{s}_0)/2 - \mathrm{var}(\hat{p}_Y(\mathbf{Z}; \mathbf{s}_0))/2\} \\ &= \exp\{\hat{p}_Y(\mathbf{Z}; \mathbf{s}_0) + \sigma_{Y,k}^2(\mathbf{s}_0)/2 - m_Y\}, \end{aligned} \tag{3.2.40}$$

where $\sigma_{Y,k}^2(\mathbf{s}_0)$ and m_Y are, from (3.2.16), the kriging variance and Lagrange multiplier, respectively, on the Y scale. From (3.2.39), the mean-squared prediction error is

$$E(Z(\mathbf{s}_0) - \check{p}_Z(\mathbf{Z}; \mathbf{s}_0))^2 = \{\exp(2\mu_Y + \sigma_Y^2(\mathbf{s}_0))\} \times \{\exp(\sigma_Y^2(\mathbf{s}_0)) + \exp(\mathrm{var}(\hat{p}_Y(\mathbf{Z}; \mathbf{s}_0))) - 2\exp(\mathrm{cov}(Y(\mathbf{s}_0), \hat{p}_Y(\mathbf{Z}; \mathbf{s}_0)))\}. \quad (3.2.41)$$

The principal feature of (3.2.41) is that, as well as knowledge of the variogram $2\gamma_Y$, the moments μ_Y and $\sigma_Y^2(\cdot)$ are also needed (or have to be estimated). This leads to interesting and more complicated inference problems, which are discussed only briefly at the end of this section.

The situation can be simplified a little by assuming second-order (and hence, strict) stationarity on the Y scale. Then $\sigma_Y^2(\mathbf{s}_0) = C_Y(\mathbf{0})$, which can be estimated from the variogram $2\gamma_Y(\mathbf{h}) = 2(C_Y(\mathbf{0}) - C_Y(\mathbf{h}))$. Furthermore, μ_Y can be estimated by weighted least squares based on the log-transformed data $\mathbf{Y} \equiv (\log Z(\mathbf{s}_1), \ldots, \log Z(\mathbf{s}_n))'$; that is, $\hat{\mu}_Y = \mathbf{1}'\Sigma_Y^{-1}\mathbf{Y}/(\mathbf{1}'\Sigma_Y^{-1}\mathbf{1})$, where $\Sigma_Y \equiv \mathrm{var}(\mathbf{Y})$.

The class of predictors over which the lognormal-kriging predictor (3.2.40) is optimal needs to be mentioned. The predictor $\check{p}_Z(\cdot; \mathbf{s}_0)$ given by (3.2.40) minimizes $E(Y(\mathbf{s}_0) - \log p(\mathbf{Z}; \mathbf{s}_0))^2$, subject to $p(\mathbf{Z}; \mathbf{s}_0) = \exp\{\sum_{i=1}^n \lambda_i \log Z(\mathbf{s}_i) + k\}$, $\sum_{i=1}^n \lambda_i = 1$, and $E(Z(\mathbf{s}_0)) = E(p(\mathbf{Z}; \mathbf{s}_0))$. Notice that (3.2.40) is *not* the best unbiased predictor $E(Z(\mathbf{s}_0)|\mathbf{Z})$ nor is it $E(Z(\mathbf{s}_0)|\mathbf{Z})$ with the unknown μ_Y replaced with its best linear unbiased estimator.

The question of whether μ_Y is assumed known or unknown underlies the difficulty some have with lognormal kriging. Suppose, for the moment, that μ_Y is *known*. Then by transforming to the Y scale, optimally predicting on that scale, and back-transforming *unbiasedly*, the predictor

$$\exp\left\{\mu_Y + \sum_{i=1}^n l_i(Y(\mathbf{s}_i) - \mu_Y) + \sigma_Y^2(\mathbf{s}_0)/2 - \mathbf{c}_Y'\Sigma_Y^{-1}\mathbf{c}_Y/2\right\},$$

where $\mathbf{l}' \equiv (l_1, \cdots, l_n)' = \mathbf{c}_Y'\Sigma_Y^{-1}$, is obtained. It is easy to see that this predictor is precisely $E(Z(\mathbf{s}_0)|\mathbf{Z})$, the optimal predictor that minimizes the mean-squared prediction error. For unknown μ_Y, it is not enough simply to replace μ_Y with $\hat{\mu}_Y = \mathbf{1}'\Sigma_Y^{-1}\mathbf{Y}/(\mathbf{1}'\Sigma_Y^{-1}\mathbf{1})$ in the optimal predictor, because it would then be biased. Upon replacing μ_Y with $\hat{\mu}_Y$ *and* correcting for bias, the proposed lognormal kriging predictor (3.2.40) is obtained.

To predict $Z(B)$, one could use $\check{Z}(B) = \int_B \check{Z}(\mathbf{u})\,d\mathbf{u}/|B|$, where $\check{Z}(\cdot)$ is given by (3.2.40). This predictor handles the change of support from point to block successfully, unlike the proposals of Rendu (1979), Journel (1980), and Dowd (1982), who assume (invalidly) a law of conservation of lognormality from point to block support. The mean-squared prediction error of $\check{Z}(B)$ is obtained by repeated application of the formulas (3.2.38) and (3.2.39). Again, this predictor is not $E(Z(B)|\mathbf{Z})$, but is obtained from optimality properties on the *transformed* Y scale. The approach taken here and elsewhere will be to find a scale where the transformed data are Gaussian (or approximately so), to build optimal predictors on that scale, and finally to back-transform the predictors (and make a bias correction).

Obtaining exact expressions for lognormal kriging (predictors and mean-squared prediction errors) relies on relations (3.2.38) and (3.2.39), which are, respectively, the mean vector and variance matrix of bivariate lognormal variables. Shimizu and Iwase (1987) show that such relations are available for other transformations $\phi(\cdot)$, that satisfy $d^2\phi(y)/dy^2 = a + b\phi(y)$, where a and b are arbitrary real constants. This includes $\phi(y) = y^2$, $\sin^2 y$, $\sinh^2 y$, and e^y; the latter function defines the lognormal distribution. They call any process $Z(\cdot) \equiv \phi(Y(\cdot))$, where $Y(\cdot)$ is a Gaussian process and $d^2\phi(y)/dy^2 = a + b\phi(y)$, a *generalized lognormal process*, and when $Y(\cdot)$ is stationary they obtain exact expressions for $\mu_Z = E(Z(\mathbf{s}))$, $\sigma_Z^2 = \text{var}(Z(\mathbf{s}))$, and $C_Z(\mathbf{h}) = \text{cov}(Z(\mathbf{s}), Z(\mathbf{s} + \mathbf{h}))$. However, for general ϕ, no such exact expressions are available in closed form.

Trans-Gaussian Kriging

Now suppose that the $Z(\cdot)$ process is obtained from

$$Z(\mathbf{s}) = \phi(Y(\mathbf{s})), \qquad \mathbf{s} \in D, \tag{3.2.42}$$

where $Y(\cdot)$ is a Gaussian process that is assumed to be intrinsically stationary and ϕ is a twice-differentiable measurable function. Using the δ-method around $\mu_Y = E(Y(\mathbf{s}))$ to obtain a bias correction, the approximately unbiased predictor recommended is

$$\check{p}_Z(\mathbf{Z}; \mathbf{s}_0) = \phi(\hat{p}_Y(\mathbf{Z}; \mathbf{s}_0)) + \phi''(\hat{\mu}_Y)\{\sigma_{Y,k}^2(\mathbf{s}_0)/2 - m_Y\}, \tag{3.2.43}$$

where the various quantities in (3.2.43) are defined *circa* equation (3.2.40), $\hat{\mu}_Y = \mathbf{1}'\Sigma_Y^{-1}\mathbf{Y}/(\mathbf{1}'\Sigma_Y^{-1}\mathbf{1})$ and $\Sigma_Y = \text{var}(\mathbf{Y})$. The mean-squared prediction error is, approximately,

$$\{\phi'(\mu_Y)\}^2 \sigma_{Y,k}^2(\mathbf{s}_0). \tag{3.2.44}$$

The approximations (3.2.43) and (3.2.44) rely on $\sigma_Y^2(\mathbf{s}_0)$ being small.

Empirical ways of finding a transformation ϕ are discussed by Howarth and Earle (1979). This question is also relevant to nonlinear geostatistics (Section 5.1) and to estimation of an invariant density (Section 5.8).

Verly (1983) uses the same model as (3.2.42) in developing a method to predict block-grade distributions from an assumed invariant distribution (Section 2.3) of $Z(\cdot)$. He calls his method multi-Gaussian kriging.

Notice that for $\sigma_Y^2(\mathbf{s}_0)$ small and $\phi(y) = e^y$, trans-Gaussian-kriging equations (3.2.43) and (3.2.44) are approximately lognormal-kriging equations

(3.2.40) and (3.2.41). Moreover, the same question as to whether μ_Y is assumed known or unknown can be considered for trans-Gaussian kriging. Suppose for the moment that μ_Y is *known*. Then, by transforming to the Y scale, optimally predicting, and back-transforming to achieve approximate unbiasedness, the predictor

$$\phi\left(\mu_Y + \sum_{i=1}^{n} l_i(Y(\mathbf{s}_i) - \mu_Y)\right) + (\phi''(\mu_Y)/2)\left(\sigma_Y^2(\mathbf{s}_0) - \mathbf{c}_Y'\Sigma_Y^{-1}\mathbf{c}_Y\right),$$

where $(l_1, \cdots, l_n) = \mathbf{c}_Y'\Sigma_Y^{-1}$, is obtained. It is not difficult to show that, after using the δ-method around μ_Y, this predictor and the optimal predictor $E(Z(\mathbf{s}_0)|\mathbf{Z})$ are indistinguishable. Again, for unknown μ_Y, replacing it with $\hat{\mu}_Y = \mathbf{1}'\Sigma_Y^{-1}\mathbf{Y}/(\mathbf{1}'\Sigma_Y^{-1}\mathbf{1})$, in the optimal predictor, yields a biased predictor. An (approximately) unbiased predictor is (3.2.43), the proposed trans-Gaussian predictor.

When predicting $Z(B)$, one could use $\check{Z}(B) = \int_B \check{Z}(\mathbf{u})\, d\mathbf{u}/\, |B|$. Alternatively, Stein (1987a) is able to derive an asymptotically optimal predictor for $Z(B)$ as either B shrinks or as the number of data points (in a bounded region) tends to infinity.

I shall close with a suggested way of dealing with large B and finite n. Because $\int_B Z(\mathbf{u})\, d\mathbf{u}/\, |B|$ is an average over a large block, it is already approximately Gaussian distributed even though no transformation has taken place. Then, posit joint Gaussianity for $(Z(B), \phi^{-1}(Z(\mathbf{s}_1)), \ldots, \phi^{-1}(Z(\mathbf{s}_n)))$ and construct a linear predictor based on this assumption, namely, $p(\mathbf{Z}; B) = \sum_{i=1}^{n} \lambda_i \phi^{-1}(Z(\mathbf{s}_i)) + k$, where $\lambda_1, \ldots, \lambda_n, k$ are chosen to minimize $E(Z(B) - p(\mathbf{Z}; B))^2$.

3.2.3 Cokriging

Suppose that the data are $k \times 1$ *vectors* $\mathbf{Z}(\mathbf{s}_1), \ldots, \mathbf{Z}(\mathbf{s}_n)$ and write

$$Z \equiv (\mathbf{Z}(\mathbf{s}_1), \ldots, \mathbf{Z}(\mathbf{s}_n))', \tag{3.2.45}$$

an $n \times k$ matrix with (i, j)th element $Z_j(\mathbf{s}_i)$. It is desired to predict, say, $Z_1(\mathbf{s}_0)$ based not only on $\mathbf{Z}_1 \equiv (Z_1(\mathbf{s}_1), \ldots, Z_1(\mathbf{s}_n))'$, but also based on the covariables $\mathbf{Z}_j \equiv (Z_j(\mathbf{s}_1), \ldots, Z_j(\mathbf{s}_n))'$, $j \neq 1$. More generally, it may be desired to predict $\mathbf{Z}(\mathbf{s}_0) \equiv (Z_1(\mathbf{s}_0), \ldots, Z_k(\mathbf{s}_0))'$, $\mathbf{s}_0 \in D$.

For example, to assess the feasibility of opening a copper mine, mineral exploration is carried out: Samples at known spatial locations $\mathbf{s}_1, \ldots, \mathbf{s}_n$ are taken and assayed. Although the percentage copper is reported, it does not occur in isolation. Percentages of lead and zinc are also typically found, along with other minerals. The data at a single location, reported as fractions of 100%, are compositional in nature (i.e., their sum cannot exceed 100). Assume that this compositional problem has already been resolved (see, e.g.,

Aitchison, 1986), so that the vector-valued stochastic process $\{\mathbf{Z}(\mathbf{s}): \mathbf{s} \in D\}$ is adequately described by its second-order parameters. Other examples can be found in soil science (e.g., Trangmar et al., 1986; Yates and Warrick, 1987), hydrology (e.g., Ahmed and de Marsily, 1987), and geophysics (e.g., Krajewski, 1987).

Let

$$E(\mathbf{Z}(\mathbf{s})) = \boldsymbol{\mu}, \qquad \mathbf{s} \in D, \tag{3.2.46}$$

$$\operatorname{cov}(\mathbf{Z}(\mathbf{s}), \mathbf{Z}(\mathbf{u})) = C(\mathbf{s}, \mathbf{u}), \qquad \mathbf{s}, \mathbf{u} \in D, \tag{3.2.47}$$

where $\boldsymbol{\mu} \equiv (\mu_1, \ldots, \mu_k)'$ and $C(\mathbf{s}, \mathbf{u})$ is a $k \times k$ matrix (not necessarily symmetric). The kriging predictor of $Z_1(\mathbf{s}_0)$ is a linear combination of all the available data values of all the k variables

$$p_1(\mathbf{Z}; \mathbf{s}_0) = \sum_{i=1}^{n} \sum_{j=1}^{k} \lambda_{ji} Z_j(\mathbf{s}_i). \tag{3.2.48}$$

Notice that (3.2.48) assumes that all components of $\mathbf{Z}(\mathbf{s}_i)$ are available at each i. Should this not be the case, an easy modification is possible (e.g., Journel and Huijbregts, 1978, p. 325).

Asking for a predictor that is uniformly unbiased, that is, $E(p_1(\mathbf{Z}; \mathbf{s}_0)) = \mu_1$, for all $\boldsymbol{\mu}$, yields the necessary and sufficient condition

$$\sum_{i=1}^{n} \lambda_{1i} = 1, \qquad \sum_{i=1}^{n} \lambda_{ji} = 0, \quad \text{for } j = 2, \ldots, k. \tag{3.2.49}$$

Therefore, the best linear unbiased predictor is obtained by minimizing

$$E\left(Z_1(\mathbf{s}_0) - \sum_{i=1}^{n} \sum_{j=1}^{k} \lambda_{ji} Z_j(\mathbf{s}_i)\right)^2, \tag{3.2.50}$$

subject to the constraints (3.2.49). In principle, this problem is no more difficult than ordinary kriging, except there are more Lagrange multipliers $m_1, \ldots, m_k$ needed for the extra constraints in (3.2.49).

A covariance-based approach to cokriging [cf. (3.2.31) and (3.2.32)] is straightforward. The cokriging equations are

$$\sum_{i=1}^{n} \sum_{j=1}^{k} \lambda_{ji} C_{jj'}(\mathbf{s}_i, \mathbf{s}_{i'}) - m_{j'} = C_{1j'}(\mathbf{s}_0, \mathbf{s}_{i'}), \qquad i' = 1, \ldots, n,\ j' = 1, \ldots, k,$$

$$\sum_{i=1}^{n} \lambda_{1i} = 1, \qquad \sum_{i=1}^{n} \lambda_{ji} = 0, \quad \text{for } j = 2, \ldots, k. \tag{3.2.51}$$

There are $(n+1)k$ linear equations in $(n+1)k$ unknowns $\{\lambda_{ji}: i = 1,\ldots,n; j = 1,\ldots,k\}$, $m_1,\ldots,m_k$. The minimum mean-squared prediction error, or (co)kriging variance, is

$$\sigma_k^2(\mathbf{s}_0) = C_{11}(\mathbf{s}_0,\mathbf{s}_0) - \sum_{i=1}^{n}\sum_{j=1}^{k}\lambda_{ji}C_{1j}(\mathbf{s}_0,\mathbf{s}_i) + m_1. \qquad (3.2.52)$$

Although the algebra becomes complicated, the principle of exploiting covariation to improve the mean-squared prediction error [going back to Wold (1938), Kolmogorov (1941b), and Wiener (1949)] is the basis of cokriging.

Finally, the covariance-matrix function $\{C(\mathbf{s},\mathbf{u}): \mathbf{s},\mathbf{u} \in \mathbb{R}^d\}$ usually has to be estimated from the data. Assuming stationarity of this covariance-matrix function, that is, $C(\mathbf{s},\mathbf{u}) = C^*(\mathbf{s}-\mathbf{u})$, would allow estimation of $C^*(\cdot)$ from the available data $\mathbf{Z}(\mathbf{s}_1),\ldots,\mathbf{Z}(\mathbf{s}_n)$.

There is no general formulation of cokriging in terms of cross-variograms $2\nu_{jj'}(\cdot)$, where $\nu_{jj'}(\mathbf{h})$ is defined as the (j,j')th element of $G(\mathbf{h}) \equiv (1/2)\mathrm{var}(\mathbf{Z}(\mathbf{s}+\mathbf{h}) - \mathbf{Z}(\mathbf{s}))$. Only under the special condition that the $k \times k$ matrix $C^*(\mathbf{h})$ is symmetric for all $\mathbf{h}$, is one able to formulate cokriging in terms of the $\nu_{jj'}$s (see Journel and Huijbregts, 1978, p. 326).

A more satisfactory definition of the cross-variogram is given by Clark et al. (1989). Modified to account for different means, it is

$$\begin{aligned} 2\gamma_{jj'}(\mathbf{h}) &\equiv \mathrm{var}\big(Z_j(\mathbf{s}+\mathbf{h}) - Z_{j'}(\mathbf{s})\big) \\ &= E\big(Z_j(\mathbf{s}+\mathbf{h}) - Z_{j'}(\mathbf{s})\big)^2 - (\mu_j - \mu_{j'})^2. \end{aligned} \qquad (3.2.53)$$

In general, one is able to formulate cokriging in terms of the $\gamma_{jj'}$s: Simply substitute $-\gamma_{jj'}(\mathbf{s}-\mathbf{u})$ for $C_{jj'}(\mathbf{s},\mathbf{u})$ in cokriging equations (3.2.51). To illustrate why, consider the case of just one extra variable [i.e., $k = 2$ in (3.2.48)]. Then, under the uniform-unbiasedness assumptions $\sum_{i=1}^{n}\lambda_{1i} = 1$, $\sum_{k=1}^{n}\lambda_{2k} = 0$,

$$\begin{aligned} \big(Z_1(\mathbf{s}_0) - p_1(\mathbf{Z};\mathbf{s}_0)\big)^2 = &-\sum_{i=1}^{n}\sum_{j=1}^{n}\lambda_{1i}\lambda_{1j}\big(Z_1(\mathbf{s}_i) - Z_1(\mathbf{s}_j)\big)^2/2 \\ &+ 2\sum_{i=1}^{n}\lambda_{1i}\big(Z_1(\mathbf{s}_0) - Z_1(\mathbf{s}_i)\big)^2/2 \\ &+ 2\sum_{k=1}^{n}\lambda_{2k}\big(Z_1(\mathbf{s}_0) - Z_2(\mathbf{s}_k)\big)^2/2 \\ &- 2\sum_{i=1}^{n}\sum_{k=1}^{n}\lambda_{1i}\lambda_{2k}\big(Z_1(\mathbf{s}_i) - Z_2(\mathbf{s}_k)\big)^2/2 \\ &- \sum_{k=1}^{n}\sum_{l=1}^{n}\lambda_{2k}\lambda_{2l}\big(Z_2(\mathbf{s}_k) - Z_2(\mathbf{s}_l)\big)^2/2. \end{aligned}$$

Take expectations and use (3.2.53) to obtain

$$
\begin{aligned}
(3.2.50) = &-\sum_{i=1}^{n}\sum_{j=1}^{n}\lambda_{1i}\lambda_{1j}\gamma_{11}(\mathbf{s}_i-\mathbf{s}_j)+2\sum_{i=1}^{n}\lambda_{1i}\gamma_{11}(\mathbf{s}_0-\mathbf{s}_i)\\
&+2\sum_{k=1}^{n}\lambda_{2k}\gamma_{12}(\mathbf{s}_0-\mathbf{s}_k)-2\sum_{i=1}^{n}\sum_{k=1}^{n}\lambda_{1i}\lambda_{2k}\gamma_{12}(\mathbf{s}_i-\mathbf{s}_k)\\
&-\sum_{k=1}^{n}\sum_{l=1}^{n}\lambda_{2k}\lambda_{2l}\gamma_{22}(\mathbf{s}_k-\mathbf{s}_l).
\end{aligned}
$$

Now minimize the right-hand side over $\lambda_{11},\ldots,\lambda_{1n},\lambda_{21},\ldots,\lambda_{2n}$, subject to $\Sigma_{i=1}^{n}\lambda_{1i}=1$ and $\Sigma_{k=1}^{n}\lambda_{2k}=0$.

There is a potential problem with using cross-variograms $\{2\gamma_{jj'}(\cdot)\}$ defined by (3.2.53) that can be resolved by rescaling. Notice that in order to perform the subtraction in (3.2.53), a meaningful result is obtained only when $Z_j(\cdot)$ and $Z_{j'}(\cdot)$ are measured in the same units. Therefore, some data preparation is necessary before estimating the cross-variograms [e.g., divide the original jth variable by $\{\hat{\gamma}_{jj}(\mathbf{h}_0)\}^{1/2}$, where $\mathbf{h}_0$ is some fixed nonzero vector in $\mathbb{R}^d$ and $\hat{\gamma}_{jj}(\cdot)$ is a semivariogram estimator calculated from the original observations on the jth variable, $j=1,\ldots,n$].

Although the rescaling is not needed when defining and estimating $\{2\nu_{jj'}(\cdot)\}$, its lack of general applicability in cokriging means that it should be avoided. Moreover, any article that uses cokriging equations with $\{-\nu_{jj'}(\mathbf{s}-\mathbf{u})\}$ replacing $\{C_{jj'}(\mathbf{s},\mathbf{u})\}$ in (3.2.51) should be presumed incorrect, unless the model allows the $k\times k$ matrix $\mathrm{cov}(\mathbf{Z}(\mathbf{s}+\mathbf{h}),\mathbf{Z}(\mathbf{s}))$ to be symmetric and to depend only on $\mathbf{h}$. Wackernagel (1988) assumes simple models for the $\nu_{jj'}$s and shows how they can be fitted. Those of his models for which $C(\mathbf{s},\mathbf{s}+\mathbf{h})=C^*(\mathbf{h})$ are easily shown to result in a symmetric $C^*(\mathbf{h})$, for all $\mathbf{h}\in\mathbb{R}^d$. His other models yield incorrect cokriging equations based on $\{\nu_{jj'}(\cdot)\}$.

Building valid, flexible models for $\{C_{jj'}\}$ or $\{2\gamma_{jj'}\}$ and fitting them to the available data is a problem that requires further research. An important special case is where $Z_j(\mathbf{s}_i)$ is actually $Z(\mathbf{s}_i;\ t_j)$, an observation on the space–time process $Z(\cdot;\ \cdot)$ at location $\mathbf{s}_i$ and time t_j. The goal is prediction of $Z(\mathbf{s}_0;\ t_k)$ [or of $Z(\mathbf{s}_0;\ t_{k+1})$] from data $\{Z(\mathbf{s}_i;\ t_j): i=1,\ldots,n;\ j=1,\ldots,k\}$. In order to do this optimally, equations analogous to (3.2.51) show that knowledge of temporal–temporal, spatial–spatial, and spatial–temporal co-variation is needed. Here, the rescaling on $2\gamma_{jj'}$ is not necessarily needed because all observations arise from one underlying space–time process.

The problem of simultaneously kriging $\mathbf{Z}(\mathbf{s}_0)$ (or, for that matter, some subset of variables at possibly different locations) begs the question of which multivariate criterion will be minimized. Caution is necessary for those who might use

$$\sum_{j=1}^{k} E\left(Z_j(\mathbf{s}_0) - p_j(\mathbf{Z}; \mathbf{s}_0)\right)^2,$$

because $Z_1(\cdot)$, $Z_2(\cdot)$, ... are not necessarily measured in the same units. The following generalized mean-squared prediction error, expressed here for predicting $\mathbf{Z}(\mathbf{s}_0)$, is unitless:

$$E\left[(\mathbf{Z}(\mathbf{s}_0) - \mathbf{p}(\mathbf{Z}; \mathbf{s}_0))' C(\mathbf{s}_0, \mathbf{s}_0)^{-1} (\mathbf{Z}(\mathbf{s}_0) - \mathbf{p}(\mathbf{Z}; \mathbf{s}_0))\right].$$

This criterion is to be preferred to the preceding unweighted sum of mean-squared prediction errors. A matrix criterion is given by Ver Hoef and Cressie (1991).

For a matrix formulation of cokriging, the reader is referred to Myers (1982, 1984) and Ver Hoef and Cressie (1991). Myers does not use the cross-variograms $2\gamma_{jj'}$, although upon replacement of $-\gamma_{jj'}$ for $C_{jj'}$ in those of his equations based on $\{C_{jj'}\}$, an equivalent set of cokriging equations results. The equations he gives in terms of $\{\nu_{jj'}\}$ should be avoided, in general, for reasons given earlier in this discussion.

3.2.4 Some Final Remarks

Kriging Using Nonstationary Covariances and Variograms

Ordinary kriging is usually presented in terms of variograms, namely,

$$\hat{p}(\mathbf{Z}; \mathbf{s}_0) = \boldsymbol{\lambda}'\mathbf{Z}, \qquad \sigma_k^2(\mathbf{s}_0) = \boldsymbol{\lambda}'\boldsymbol{\gamma} + m, \tag{3.2.54}$$

where

$$\boldsymbol{\lambda}' = \left(\boldsymbol{\gamma} + \mathbf{1}\frac{(1 - \mathbf{1}'\Gamma^{-1}\boldsymbol{\gamma})}{\mathbf{1}'\Gamma^{-1}\mathbf{1}}\right)'\Gamma^{-1}, \qquad m = -\frac{(1 - \mathbf{1}'\Gamma^{-1}\boldsymbol{\gamma})}{\mathbf{1}'\Gamma^{-1}\mathbf{1}}, \tag{3.2.55}$$

$\mathbf{1} = (1, \ldots, 1)'$, $\boldsymbol{\gamma} = (\gamma(\mathbf{s}_0 - \mathbf{s}_1), \ldots, \gamma(\mathbf{s}_0 - \mathbf{s}_n))'$, and $\Gamma = (\gamma(\mathbf{s}_i - \mathbf{s}_j))$. However, Eq. (3.2.10) implies that, upon replacing $\gamma(\mathbf{s} - \mathbf{u})$ in (3.2.54) with $\gamma(\mathbf{s}, \mathbf{u}) \equiv (1/2)\mathrm{var}(Z(\mathbf{s}) - Z(\mathbf{u}))$, one obtains the optimal linear unbiased predictor of $Z(\mathbf{s}_0)$ assuming only $\sum_{i=1}^{n} \lambda_i = 1$. That is, it is *not* necessary to assume that $\gamma(\mathbf{s}, \mathbf{u})$ is a function of $\mathbf{s} - \mathbf{u}$. The same remark applies to cokriging, where $\gamma_{jj'}(\mathbf{s}, \mathbf{u}) \equiv (1/2)\mathrm{var}(Z_j(\mathbf{s}) - Z_{j'}(\mathbf{u}))$ can be used in place of $\gamma_{jj'}(\mathbf{s} - \mathbf{u})$.

Equivalently, ordinary kriging in terms of covariograms can be written

$$\hat{p}(\mathbf{Z}; \mathbf{s}_0) = \boldsymbol{\lambda}'\mathbf{Z}, \qquad \sigma_k^2(\mathbf{s}_0) = C(\mathbf{s}_0, \mathbf{s}_0) - \boldsymbol{\lambda}'\mathbf{c} + m, \tag{3.2.56}$$

where

$$\boldsymbol{\lambda}' = \left(\mathbf{c} + \mathbf{1}\frac{(1 - \mathbf{1}'\Sigma^{-1}\mathbf{c})}{\mathbf{1}'\Sigma^{-1}\mathbf{1}}\right)'\Sigma^{-1}, \qquad m = \frac{(1 - \mathbf{1}'\Sigma^{-1}\mathbf{c})}{\mathbf{1}'\Sigma^{-1}\mathbf{1}}, \quad (3.2.57)$$

$\mathbf{c} = (C(\mathbf{s}_0, \mathbf{s}_1), \ldots, C(\mathbf{s}_0, \mathbf{s}_n))'$, and $\Sigma = (C(\mathbf{s}_i, \mathbf{s}_j))$. That is, the kriging equations can be written in terms of covariances that *do not* have to be stationary.

Usually, weak-stationarity or intrinsic-stationarity assumptions are made to allow the variogram or the covariogram to be estimated from data $\mathbf{Z}$ (Section 2.3), but this is not always appropriate if a physical model suggests a particular covariance function $C(\mathbf{s}, \mathbf{u})$ or variogram $\gamma(\mathbf{s}, \mathbf{u})$; see, e.g., Ma et al. (1987) and Arato (1990). Whittle (1954) and Vecchia (1985) show how certain stationary covariance functions result from stochastic differential equations that arise from physical considerations.

Kriging on the Sphere

The calculus of kriging equations (3.2.5) and (3.2.12) relies on a well defined notion of vector addition and subtraction. Young (1987) takes this from the Euclidean setting and proposes analogous definitions on the sphere. A so-called vector variogram is used in the derivation of kriging equations on the sphere.

Kriging with Nonnegative Weights

Because data are never Gaussian and, in particular, are often naturally bounded from above or below (e.g., percentage ore grades are between 0 and 100), practitioners are concerned about prediction techniques that could potentially put the predictor outside the natural boundaries. For example, suppose $Z(\mathbf{s}) \geq 0$, for all $\mathbf{s} \in D$. Then, one way to ensure that the kriging predictor $\hat{Z}(\mathbf{s}_0) = \boldsymbol{\lambda}'\mathbf{Z} \geq 0$ is to specify $\lambda_i \geq 0$, $i = 1, \ldots, n$, as a *further* constraint when minimizing $E(Z(\mathbf{s}_0) - \sum_{i=1}^{n}\lambda_i Z(\mathbf{s}_i))^2$ subject to $\sum_{i=1}^{n}\lambda_i = 1$. Barnes and Johnson (1984), Szidarovsky et al. (1987), and Herzfeld (1989) contain algorithmic details on how to perform this minimization.

Although nonnegative weights imply that $\hat{Z}(\mathbf{s}_0) \geq 0$, it is a constraint that is too heavy-handed unless there are other good reasons to use it; $\boldsymbol{\lambda}'\mathbf{Z}$ can be nonnegative but still have some negative λ_is. The extra constraint leads to a (perhaps unnecessarily large) increase in mean-squared prediction error. In fact, negative kriging weights can be advantageous because they allow $\hat{Z}(\mathbf{s}_0)$ to range outside the limits, $\max\{Z(\mathbf{s}_i): i = 1, \ldots, n\}$ and $\min\{Z(\mathbf{s}_i): i = 1, \ldots, n\}$.

Should it become essential to ensure that a predictor lies between predetermined bounds, a transformation of $Z(\cdot)$ to stretch it to be more Gaussian-like, followed by kriging, followed by back-transforming (Section 3.2.2), may be preferable to kriging with extra constraints.

3.3 ROBUST KRIGING

The optimality of *linear* spatial prediction (or simple kriging) relies on $Z(\cdot)$ being a Gaussian process; recall the discussion at the beginning of this chapter. Even when transformations are used to convert the data to a scale where they are more Gaussian-like, there still might be several observations that are best described as *outliers*. They could be due to, for example, recording errors or instrument failure, or they may be well observed data, reflecting the failure of a Gaussian random process to model the physical process in every possible way. Actual data tend to have density functions whose tails are heavier than the Gaussian density (Huber, 1972; Bartels, 1977), and linear predictors are very sensitive to outlying observations that are a result of heavy tails.

The effect of outliers on inference procedures can be substantial. Deleting outliers when estimating a variogram may be sensible, but when predicting observations, an alternative way of dealing with them is needed. Indeed, in mining, an unusually large assay value may be cause for celebration!

For the purposes of this section, assume that the model (3.2.4) holds; that is, assume that $Z(\cdot)$ is intrinsically stationary (although not necessarily Gaussian). Then, ordinary kriging is *not* resistant to large changes in even a small fraction of the data **Z**. So, for example, an anomalously large gold assay may unduly influence predictions throughout a large portion of the region under consideration. Clearly, that datum still carries information, but caution demands that it not be trusted as much as others, and so should receive less weight. Hawkins and Cressie (1984) propose a way of using neighboring values to determine how much the outlier should be downweighted. This proposal is now summarized; the presentation has a mining flavor to it, but the method is quite general.

Isolated Outliers

Because a Gaussian process is the paradigm for ordinary kriging, it will be assumed that, possibly after data transformation (e.g., gold grades are often transformed to log grades), the data behave as a sampling from an almost Gaussian process. The non-Gaussian behavior is assumed to take the form of isolated outliers that differ markedly from other sampled values nearby. Specifically, in terms of the decomposition (3.1.2), assume that

$$Z(\mathbf{s}) = \mu + W(\mathbf{s}) + \eta(\mathbf{s}) + \epsilon(\mathbf{s}), \qquad \mathbf{s} \in D, \tag{3.3.1}$$

where $W(\cdot)$ is intrinsically stationary and Gaussian. However, $E(\cdot) \equiv \eta(\cdot) + \epsilon(\cdot)$ is not assumed Gaussian. Suppose instead that

$$E(\mathbf{s}) \sim \begin{cases} \mathrm{Gau}(0, c_0), & \text{with probability } 1 - \epsilon, \\ H, & \text{with probability } \epsilon, \end{cases} \tag{3.3.2}$$

where ϵ is thought of as being small (e.g., $\epsilon = 0.05$) and H is some heavy-tailed distribution with mean zero and finite variance. When $\mathbf{s}_1 \neq \mathbf{s}_2$, $E(\mathbf{s}_1)$ and $E(\mathbf{s}_2)$ are assumed independent. A special case of (3.3.2) is given by (2.4.18), which is used to assess robustness of variogram estimators.

It should be noted that the applicability of this model to real data may depend on the scale at which it is applied. Suppose, for example, that a single diamond drill hole intersects a small anomalous area in a deposit; then the model (3.3.2) is applicable. However, if the anomaly is sufficiently large that it covers many mining blocks, then the model's applicability is weakened, though not entirely lost. Because one is more interested in drilling in the richer areas than the poorer areas, at yet a larger scale, the model (3.3.1) should refer to the sample-to-sample variability within a subarea D of above-average grade.

Which Variogram Should be Estimated?

Because the components of (3.3.1) are assumed independent,

$$2\gamma_Z(\mathbf{h}) = 2\gamma_W(\mathbf{h}) + 2\left\{(1-\epsilon)c_0 + \epsilon\int z^2\, H(dz)\right\}, \qquad \mathbf{h} \in \mathbb{R}^d, \quad (3.3.3)$$

where it is assumed that $\int z^2\, H(dz) < \infty$. Further, assuming that any observations from H are mistakes, the process of real interest is

$$Y(\mathbf{s}) \equiv \mu + W(\mathbf{s}) + \tilde{E}(\mathbf{s}), \qquad \mathbf{s} \in D, \quad (3.3.4)$$

where $\tilde{E}(\cdot)$ is the Gaussian white-noise process with marginal distribution $\mathrm{Gau}(0, c_0)$. Thus, the target variogram, to be estimated or used for prediction purposes, is

$$2\gamma_Y(\mathbf{h}) = 2\gamma_W(\mathbf{h}) + 2c_0, \qquad \mathbf{h} \in \mathbb{R}^d.$$

Now specialize (3.3.2) to

$$E(\mathbf{s}) \sim \begin{cases} \mathrm{Gau}(0, c_0), & \text{w.p. } 1-\epsilon, \\ \mathrm{Gau}(0, k^2 c_0), & \text{w.p. } \epsilon, \end{cases} \quad (3.3.5)$$

where k^2 is much larger than 1 (e.g., $k^2 = 9$); compare (2.4.18). As is shown in Tukey (1960) and Kubat (1979), $\epsilon = 0.05$ and $k^2 = 9$ is more than enough contamination to destroy the usual optimality of $\bar{Z} = \sum_{i=1}^{n} Z(\mathbf{s}_i)/n$ in estimating μ in the model $Z(\cdot) = \mu + \epsilon(\cdot)$. Spatial prediction is a slightly different problem, but the literature from robust estimation does offer useful guidelines.

Cressie and Hawkins (1980) and Hawkins and Cressie (1984) show how to estimate $2\gamma_Y(\cdot) = 2\gamma_W(\cdot) + 2c_0$ from data $Z(\mathbf{s}_1), \ldots, Z(\mathbf{s}_n)$ [whose variogram $2\gamma_Z(\cdot) = 2\gamma_Y(\cdot) + \epsilon c_0(k^2 - 1)$ is biased upward from the target variogram $2\gamma_Y(\cdot)$]. Section 2.4.3 covers the main points of this robust estimation of $2\gamma_Y(\cdot)$.

Editing the Data

Turn now to robust kriging and assume the model (3.3.1) and (3.3.5). Hawkins and Cressie (1984, p. 9) give an illustrative example to motivate the desirability of predictors of the form

$$p(\mathbf{Z}; B) = \sum_{i=1}^{n} \lambda_i Z(\mathbf{s}_i) w(Z(\mathbf{s}_i)), \tag{3.3.6}$$

where $0 \le w(\cdot) \le 1$ is a weight function such that $w(Z(\mathbf{s}_i))$ is close to 1 if $Z(\mathbf{s}_i)$ appears to be clean (i.e., appears to come from the uncontaminated distribution) and will decrease to the extent that $Z(\mathbf{s}_i)$ appears to have come from the contaminated distribution.

An equivalent way of looking at (3.3.6) is to write

$$p(\mathbf{Z}; B) = \sum_{i=1}^{n} \lambda_i Z^{(e)}(\mathbf{s}_i), \tag{3.3.7}$$

where $Z^{(e)}(\mathbf{s}_i)$ is an *edited* modification of $Z(\mathbf{s}_i)$ that "behaves" like $Y(\mathbf{s}_i)$. Think of $Z^{(e)}(\mathbf{s}_i)$ as close to $Z(\mathbf{s}_i)$ if $Z(\mathbf{s}_i)$ appears inlying and markedly different if it appears outlying. It is convenient to phrase the robust-kriging proposal in these terms, but the equivalence of (3.3.6) and (3.3.7) should be borne in mind.

The Proposed Method

By analogy with robust smoothing in time-series analysis, the proposal goes as follows:

1. Estimate the variogram using one of the robust estimators of Section 2.4.3. (This represents an attempt to estimate the variogram of the uncontaminated data.) Fit a valid variogram model using, say, the method of Section 2.6.2.
2. Use this variogram to compute the kriging weights for the prediction of each $Z(\mathbf{s}_j)$ from all other data:

$$\hat{Z}_{-j}(\mathbf{s}_j) = \sum_{\substack{i=1 \\ i \ne j}}^{n} \lambda_{ji} Z(\mathbf{s}_i). \tag{3.3.8}$$

Let $\sigma_{-j}^2(\mathbf{s}_j)$ be the associated kriging variance. (Section 3.2 has the details on how to obtain these quantities.)

3. Use the weights in (3.3.8) to obtain a robust prediction of $Z(\mathbf{s}_j)$ from its neighbors:

$$Z^{@}_{-j}(\mathbf{s}_j) = \text{weighted median}\big(\{Z(\mathbf{s}_i): i \neq j\}; \{\lambda_{ji}: i \neq j\}\big), \quad (3.3.9)$$

where the weighted median is defined in Comment 1 in the subsequent text.

4. Edit $Z(\mathbf{s}_j)$ by replacing it with the Winsorized version

$$Z^{(e)}(\mathbf{s}_j) = \begin{cases} Z^{@}_{-j}(\mathbf{s}_j) + c\sigma_{-j}(\mathbf{s}_j), & \text{if } Z(\mathbf{s}_j) - Z^{@}_{-j}(\mathbf{s}_j) > c\sigma_{-j}(\mathbf{s}_j), \\ Z(\mathbf{s}_j), & \text{if } |Z(\mathbf{s}_j) - Z^{@}_{-j}(\mathbf{s}_j)| \leq c\sigma_{-j}(\mathbf{s}_j), \\ Z^{@}_{-j}(\mathbf{s}_j) - c\sigma_{-j}(\mathbf{s}_j), & \text{if } Z(\mathbf{s}_j) - Z^{@}_{-j}(\mathbf{s}_j) < -c\sigma_{-j}(\mathbf{s}_j). \end{cases} \quad (3.3.10)$$

5. Still using the robust variogram, proceed now with predicting $Y(B)$. [Notice that when $|B| > 0$, $Y(B) = Z(B)$.] For any block B, find the weights $\{\lambda_{Bi}: i = 1, \ldots, n\}$ in the ordinary-kriging predictor of $Z(B)$:

$$\hat{p}(\mathbf{Z}; B) = \sum_{i=1}^{n} \lambda_{Bi} Z(\mathbf{s}_i); \quad (3.3.11)$$

see Section 3.2 for details on how to obtain the weights. However, and this is most important, to predict $Y(B)$, do *not* use the original data; instead, use the edited data $\{Z^{(e)}(\mathbf{s}_i): i = 1, \ldots, n\}$:

$$\hat{Y}(B) = \sum_{i=1}^{n} \lambda_{Bi} Z^{(e)}(\mathbf{s}_i). \quad (3.3.12)$$

This will have a kriging variance approximated by the usual formulas (3.2.16) or (3.2.19), but using the robust estimate of the variogram. In view of the symmetric contamination, (3.3.12) will have no bias.

Comments on the Robust Kriging Proposal

1. The weighted median in (3.3.9) is obtained as follows. For the purposes of this definition, assume that the sequence of observations $\{Z(\mathbf{s}_i): i = 1, \ldots, n;\ i \neq j\}$ is ordered from smallest to largest. Define $w^*_{ji} \equiv \lambda_{ji}/\Sigma_{l \neq j}\lambda_{jl}$, if $i \neq j$, and $\equiv 0$, if $i = j$; notice that these weights are not necessarily positive. Then obtain intervals (or points)

$$[Z(\mathbf{s}_{m+1}), Z(\mathbf{s}_{m+k+1})], \qquad k \geq 0, \quad (3.3.13)$$

such that

$$\sum_{i=1}^{m} w_{ji}^* \neq \sum_{i=1}^{m+1} w_{ji}^* = \cdots = \sum_{i=1}^{m+k} w_{ji}^* = 1/2 \neq \sum_{i=1}^{m+k+1} w_{ji}^*.$$

The $k = 0$ condition is interpreted as

$$\sum_{i=1}^{m} w_{ji}^* < 1/2 < \sum_{i=1}^{m+1} w_{ji}^* \quad \text{or} \quad \sum_{i=1}^{m} w_{ji}^* > 1/2 > \sum_{i=1}^{m+1} w_{ji}^*.$$

These intervals and points all qualify to be called a weighted median. To obtain a unique solution, replace all solution intervals (including the degenerate ones) by their midpoints; this leaves as many as l solutions, $\theta_1 < \cdots < \theta_l$, to (3.3.13). The unique solution is declared to be $\theta_{[(l+1)/2]}$, where $[x]$ denotes the integer part of x.

2. Huber (1979) has proposed a robust time-series smoother. Translating it into the language of geostatistics, the procedure is:
   a. Compute a (nonrobust) variogram from $\{Z(\mathbf{s}_i): i = 1, \ldots, n\}$ using the classical estimator (2.4.2).
   b. Use ordinary kriging to predict each $Z(\mathbf{s}_j)$ from the remaining $Z(\mathbf{s}_i)$:

$$\hat{Z}_{-j}(\mathbf{s}_j) = \sum_{\substack{i=1 \\ i \neq j}}^{n} \lambda_{ji} Z(\mathbf{s}_i).$$

   c. Winsorize $Z(\mathbf{s}_j)$. That is, replace it by $\hat{Z}_{-j}(\mathbf{s}_j) \pm c\sigma_{-j}(\mathbf{s}_j)$ if $|Z(\mathbf{s}_j) - \hat{Z}_{-j}(\mathbf{s}_j)| > c\sigma_{-j}(\mathbf{s}_j)$; otherwise, leave it unchanged. Call these edited values $\{Z^{(e)}(\mathbf{s}_j): j = 1, \ldots, n\}$.
   d. Using the edited values, go back to the beginning and repeat the cycle until the edited $Z^{(e)}(\mathbf{s}_j)$ converge.
   e. To predict $Y(B)$, substitute the final edited values in (3.3.12).

   Huber's procedure has the advantage of using a more efficient variogram estimator and a more efficient predictor than (3.3.9), assuming no contamination. However, it has the enormous practical disadvantage of requiring a number of passes through the data to edit them. [Predictor (3.3.12) may be regarded as a single step of the Huber procedure.] Results of Bickel (1975) provide at least a strong hint that, provided the initial estimate is robust, most of the benefit of the editing is obtained in the first one or two cycles through the data, so that the advantage of Huber's procedure may be slight in practice. McLeod et al. (1983) propose a hybrid, of Huber's and

Hawkins and Cressie's methods, for seasonal adjustment of water-quality time series.

3. In the closely related area of time-series smoothing, Masreliez and Martin (1977) have shown that Winsorizing has an optimality property of minimizing the maximum error over all nonparametric neighborhoods of the Gaussian distribution. Thus, although it is known that Winsorizing is not optimal for the contaminated Gaussian model (3.3.5), it may be inferred that Winsorizing has robustness properties because it behaves well regardless of the type of contamination present.
4. In the editing part, namely (3.3.8) through (3.3.10), it is not necessary in practice to include all $\{Z(\mathbf{s}_i): i = 1, \ldots, n\}$ in the kriging calculations; see Section 3.2.1 for discussion of kriging neighborhoods.
5. The constant c controls the amount of editing applied to outlying values. It is under the user's control, but values in the range 1.5 to 2.5 have been found to perform most satisfactorily. A large c implies a small amount of editing or downweighting of data. The value of c might be chosen to match the particular characteristics of the robust variogram estimator used; for example, choose c so that $\text{var}(Z^{(e)}(\mathbf{s}_i))$ is equal to an estimate of $\text{var}(Y(\mathbf{s}_i))$, calculated from, say, a robust estimate of the variogram. Chernick and Murthy's (1983) related approach of using influence functions to replace outliers in the i.i.d. case might be adaptable to this spatial setting.
6. Sometimes one has the sample $\{Z(\mathbf{s}_i): i = 1, \ldots, n\}$ on a grid, but a modest fraction of them are missing. This has unpleasant consequences for block kriging because one should, theoretically, compute fresh kriging weights for each pattern of present/absent sample grades. In the proposed method, the predictor $Z^{@}_{-j}(\mathbf{s}_j)$ given by (3.3.9) could be substituted for such missing grades, so that block kriging could always be carried out using a single standard set of weights. Provided these filled-in values are scattered sparsely, the error of approximation will be small.
7. Like ordinary kriging, the proposed method is an exact interpolator, but now for the *edited* grades $\{Z^{(e)}(\mathbf{s}_i): i = 1, \ldots, n\}$.
8. Under circumstances where there are a large number of irregularly spaced $\{Z(\mathbf{s}_i): i = 1, \ldots, n\}$, the editing of grades may be computationally prohibitive. A less efficient variant of the proposed method is:
   a. Compute the variogram using one of the robust estimators of Section 2.4.3 and fit a valid model.
   b. Use this variogram to compute the kriging weights $\{\lambda_{Bi}: i = 1, \ldots, n\}$ that are used in the predictor

$$\hat{Z}(B) = \sum_{i=1}^{n} \lambda_{Bi} Z(\mathbf{s}_i).$$

c. The predictor of the block grades is,

$$Z^{@}(B) = \text{weighted median}(\{Z(\mathbf{s}_i)\}; \{\lambda_{Bi}\}),$$

which is defined by (3.3.13).

9. Computation of the edited grades (3.3.10) involves use of the cross-validation-like quantities

$$\left\{\left(Z(\mathbf{s}_j) - Z^{@}_{-j}(\mathbf{s}_j)\right)\Big/\sigma_{-j}(\mathbf{s}_j) : j = 1, \ldots, n\right\}.$$

The only difference between these and (2.6.17) is the use of a resistant predictor $Z^{@}_{-j}(\mathbf{s}_j)$, rather than of $\hat{Z}_{-j}(\mathbf{s}_j)$. In effect, robust kriging edits the original data so that it cross-validates successfully; see Section 2.6.4. Mining companies have been doing this with unusually large assay values, on an ad hoc basis, for some time. Overoptimism at the exploration stage can have dire consequences at the production stage, justifying editing of the exploration data $\{Z(\mathbf{s}_1), \ldots, Z(\mathbf{s}_n)\}$.

Comparison with Ordinary Kriging

Robust kriging (3.3.12) differs from ordinary kriging (3.2.18) in the following ways:

- A robust estimator of the variogram is always used in place of the classical (outlier-sensitive) estimator (2.4.2).
- An additional edit phase is introduced before the actual kriging is carried out.
- The edit phase may be used to fill in gaps in a regular grid of sampled grades. In this case, the final kriging may be simplified, as it is no longer necessary to use several different systems of weights depending on which sample grades are missing. [Of course, filling in of missing grades is not novel; Journel and Huijbregts (1978, p. 351) recommend such a preliminary kriging under certain circumstances.]

Implementation of robust kriging is easy, because computational changes can be made by relatively minor additions to, and modifications of, existing geostatistics computer programs and packages.

Further properties of robust kriging can be found in Hawkins and Cressie (1984, pp. 15–17). Research on the biases and efficiencies of the proposed method has yet to be carried out; lack of mathematical tractability will undoubtedly mean that simulation experiments like those of Kunst (1989) (designed for an autoregressive time series) will be necessary.

3.4 UNIVERSAL KRIGING

Recall (from Section 3.2) the constant-mean model given by (3.2.4):

$$Z(\mathbf{s}) = \mu + \delta(\mathbf{s}), \qquad \mathbf{s} \in D, \tag{3.4.1}$$

where $\mu \in \mathbb{R}$ is unknown and $\delta(\cdot)$ is a zero-mean intrinsically stationary random process with variogram $2\gamma(\cdot)$. Prediction of an unknown $Z(B)$ from data $\mathbf{Z}$ is referred to there as *ordinary* kriging. However, it would be folly to fit such a simple model to every data set (see, e.g., the coal-ash data of Section 2.2; Hunt, 1980; Russo and Jury, 1987b).

One possible generalization of (3.4.1) is to assume a more general error process $\delta(\cdot)$; for example, Section 5.4 presents spatial prediction of intrinsic random functions of order k. Instead, in this section it will be assumed that $E(Z(\mathbf{s}))$ [$\equiv \mu(\mathbf{s})$] is no longer constant but is an unknown linear combination of known functions $\{f_0(\mathbf{s}), \ldots, f_p(\mathbf{s})\}$, $\mathbf{s} \in D$. Although each $f_j(\mathbf{s})$ has been written as a function of location $\mathbf{s}$, any one of them could be, say, 1 or a value of an explanatory variable associated with the datum at $\mathbf{s}$, $\mathbf{s} \in D$.

Matheron (1969) set out the details of universal kriging (in French), followed by a shorter version in English (Huijbregts and Matheron, 1971). Others have written variations on the same theme for their own disciplines (e.g., Olea, 1974; Delhomme, 1978).

Model Assumption

Henceforth, in this section, the assumption,

$$Z(\mathbf{s}) = \sum_{j=1}^{p+1} f_{j-1}(\mathbf{s})\beta_{j-1} + \delta(\mathbf{s}), \qquad \mathbf{s} \in D, \tag{3.4.2}$$

is adopted, where $\boldsymbol{\beta} \equiv (\beta_0, \ldots, \beta_p)' \in \mathbb{R}^{p+1}$ is an unknown vector of parameters and $\delta(\cdot)$ is a zero-mean intrinsically stationary random process with variogram $2\gamma(\cdot)$. In obvious notation, data $\mathbf{Z}$ can be written as

$$\mathbf{Z} = X\boldsymbol{\beta} + \boldsymbol{\delta}, \tag{3.4.3}$$

where X is an $n \times (p+1)$ matrix whose (i, j)th element is $f_{j-1}(\mathbf{s}_i)$. Furthermore, from (3.2.6) and (3.4.2),

$$Z(B) = \mathbf{x}'\boldsymbol{\beta} + \delta(B), \tag{3.4.4}$$

where $\mathbf{x} \equiv (f_0(B), \ldots, f_p(B))'$, $f_j(B) \equiv \int_B f_j(\mathbf{u})\, d\mathbf{u}/|B|$, $j = 0, \ldots, p$, and $\delta(B) \equiv \int_B \delta(\mathbf{u})\, d\mathbf{u}/|B|$. When $B = \{\mathbf{s}_0\}$, $\mathbf{x} = (f_0(\mathbf{s}_0), \ldots, f_p(\mathbf{s}_0))'$.

Predictor Assumption

It is desired to predict $Z(B)$ linearly from data $\mathbf{Z}$ using a uniformly unbiased predictor. That is, the predictor is of the form

$$p(\mathbf{Z}; B) = \sum_{i=1}^{n} \lambda_i Z(\mathbf{s}_i), \qquad \text{for } \boldsymbol{\lambda}'X = \mathbf{x}'. \tag{3.4.5}$$

This latter condition on the weights is necessary and sufficient for a uniformly unbiased predictor; that is, $E(p(\mathbf{Z}; B)) = E(\boldsymbol{\lambda}'\mathbf{Z}) = \boldsymbol{\lambda}'X\boldsymbol{\beta}$ is equal to $\mathbf{x}'\boldsymbol{\beta} = E(Z(B))$, for all $\boldsymbol{\beta} \in \mathbb{R}^{p+1}$, if and only if $\boldsymbol{\lambda}'X = \mathbf{x}'$. Notice also that, for $p = 0$ and $f_0(\mathbf{s}) \equiv 1$, one obtains the ordinary-kriging model assumption (3.2.4); in this case, $\boldsymbol{\lambda}'X = \mathbf{x}'$ reduces to $\sum_{i=1}^{n} \lambda_i = 1$.

Optimal Spatial Prediction of the Z Process

For universal kriging, the optimal linear unbiased predictor, which I shall write as $\hat{p}(\mathbf{Z}; B)$ [or, more simply, as $\hat{Z}(B)$], minimizes the mean-squared prediction error

$$\sigma_e^2 = E(Z(B) - p(\mathbf{Z}; B))^2 \tag{3.4.6}$$

over $\lambda_1, \dots, \lambda_n$, subject to $\boldsymbol{\lambda}'X = \mathbf{x}'$. The adjective "universal" was used by Matheron (1969) to refer to the unbiasedness of the predictor when the trend is an unknown linear combination of known functions. Suppose for the moment that $B = \{\mathbf{s}_0\}$. The constrained optimization problem can be written equivalently as the unconstrained minimization of

$$E\left(Z(\mathbf{s}_0) - \sum_{i=1}^{n} \lambda_i Z(\mathbf{s}_i)\right)^2 - 2\sum_{j=1}^{p+1} m_{j-1}\left\{\sum_{i=1}^{n} \lambda_i f_{j-1}(\mathbf{s}_i) - f_{j-1}(\mathbf{s}_0)\right\} \tag{3.4.7}$$

with respect to $\lambda_1, \dots, \lambda_n$, and $m_0, \dots, m_p$ (the latter $p + 1$ coefficients are Lagrange multipliers that ensure $\boldsymbol{\lambda}'X = \mathbf{x}'$). Assume, henceforth, that $f_0(\mathbf{s}) \equiv 1$; this guarantees that $\sum_{i=1}^{n} \lambda_i = 1$ is one of the unbiasedness conditions.

Now, from (3.4.3), (3.4.4), (3.4.5), and using the relation $\sum_{i=1}^{n} \lambda_i = 1$,

$$\begin{aligned}
\left(Z(\mathbf{s}_0) - \sum_{i=1}^{n} \lambda_i Z(\mathbf{s}_i)\right)^2 &= \left(\mathbf{x}'\boldsymbol{\beta} + \delta(\mathbf{s}_0) - \boldsymbol{\lambda}'X\boldsymbol{\beta} - \sum_{i=1}^{n} \lambda_i \delta(\mathbf{s}_i)\right)^2 \\
&= \left(\delta(\mathbf{s}_0) - \sum_{i=1}^{n} \lambda_i \delta(\mathbf{s}_i)\right)^2 \\
&= -\sum_{i=1}^{n}\sum_{j=1}^{n} \lambda_i \lambda_j (\delta(\mathbf{s}_i) - \delta(\mathbf{s}_j))^2/2 \\
&\quad + 2\sum_{i=1}^{n} \lambda_i (\delta(\mathbf{s}_0) - \delta(\mathbf{s}_i))^2/2.
\end{aligned} \tag{3.4.8}$$

Assuming

$$2\gamma(\mathbf{h}) = \text{var}(Z(\mathbf{s}+\mathbf{h}) - Z(\mathbf{s})), \tag{3.4.9}$$

(3.4.7) becomes

$$-\sum_{i=1}^{n}\sum_{j=1}^{n}\lambda_i\lambda_j\gamma(\mathbf{s}_i - \mathbf{s}_j) + 2\sum_{i=1}^{n}\lambda_i\gamma(\mathbf{s}_0 - \mathbf{s}_i)$$
$$-2\sum_{j=1}^{p+1} m_{j-1}\left\{\sum_{i=1}^{n}\lambda_i f_{j-1}(\mathbf{s}_i) - f_{j-1}(\mathbf{s}_0)\right\}. \tag{3.4.10}$$

It is worth emphasizing that if none of the $\{f_{j-1}(\mathbf{s}): j = 1,\ldots,p+1\}$ is identically 1, then minimizing (3.4.6) subject to $\boldsymbol{\lambda}'X = \mathbf{x}'$ is no longer necessarily achieved by minimizing (3.4.10). In that case, the appropriate equations should be written in terms of covariance functions (Section 3.4.5).

Universal-Kriging Equations

Upon differentiating (3.4.10) with respect to $\lambda_1,\ldots,\lambda_n, m_0,\ldots,m_p$, and equating the result to zero, the optimal weights are obtained from

$$\boldsymbol{\lambda}_U = \Gamma_U^{-1}\boldsymbol{\gamma}_U, \tag{3.4.11}$$

where

$$\boldsymbol{\lambda}_U \equiv (\lambda_1,\ldots,\lambda_n, m_0,\ldots,m_p)', \tag{3.4.12}$$

$$\boldsymbol{\gamma}_U \equiv (\gamma(\mathbf{s}_0 - \mathbf{s}_1),\ldots,\gamma(\mathbf{s}_0 - \mathbf{s}_n), 1, f_1(\mathbf{s}_0),\ldots,f_p(\mathbf{s}_0))', \tag{3.4.13}$$

and Γ_U is a *symmetric* $(n+p+1)\times(n+p+1)$ matrix

$$\Gamma_U \equiv \begin{cases} \gamma(\mathbf{s}_i - \mathbf{s}_j), & i = 1,\ldots,n,\ j = 1,\ldots,n, \\ f_{j-1-n}(\mathbf{s}_i), & i = 1,\ldots,n,\ j = n+1,\ldots,n+p+1, \\ 0, & i = n+1,\ldots,n+p+1, \\ & j = n+1,\ldots,n+p+1, \end{cases} \tag{3.4.14}$$

with $f_0(\mathbf{s}) \equiv 1$. Thus, the coefficients $\boldsymbol{\lambda}$ are given by

$$\boldsymbol{\lambda}' = \left\{\boldsymbol{\gamma} + X(X'\Gamma^{-1}X)^{-1}(\mathbf{x} - X'\Gamma^{-1}\boldsymbol{\gamma})\right\}'\Gamma^{-1} \tag{3.4.15}$$

and

$$\mathbf{m}' = -(\mathbf{x} - X'\Gamma^{-1}\boldsymbol{\gamma})'(X'\Gamma^{-1}X)^{-1},$$

where $\boldsymbol{\gamma} \equiv (\gamma(\mathbf{s}_0 - \mathbf{s}_1), \ldots, \gamma(\mathbf{s}_0 - \mathbf{s}_n))'$ and Γ is the $n \times n$ matrix whose (i, j)th element is $\gamma(\mathbf{s}_i - \mathbf{s}_j)$; see, for example, Rao (1973, p. 33, problem 2.7) for the appropriate matrix formula that allows these expressions to be derived from (3.4.11). When predicting a noiseless version of $Z(\cdot)$, generalizations of Eqs. (3.2.25), (3.2.26), and (3.2.27) can be obtained in a straightforward manner.

Now, instead of assuming that $Z(\cdot)$ is intrinsically stationary, strengthen the assumption to second-order stationarity. Then, the covariogram $C(\mathbf{h}) \equiv \text{cov}(Z(\mathbf{s} + \mathbf{h}), Z(\mathbf{s}))$ is well defined and the analogous equation to (3.4.11) is

$$(\lambda_1, \ldots, \lambda_n, -m_0, \ldots, -m_p)' = \Sigma_U^{-1} \mathbf{c}_U,$$

where $\mathbf{c}_U \equiv (C(\mathbf{s}_0 - \mathbf{s}_1), \ldots, C(\mathbf{s}_0 - \mathbf{s}_n), f_0(\mathbf{s}_0), f_1(\mathbf{s}_0), \ldots, f_p(\mathbf{s}_0))'$ and Σ_U is a *symmetric* $(n + p + 1) \times (n + p + 1)$ matrix

$$\Sigma_U \equiv \begin{cases} C(\mathbf{s}_i - \mathbf{s}_j), & i = 1, \ldots, n,\ j = 1, \ldots, n, \\ f_{j-1-n}(\mathbf{s}_i), & i = 1, \ldots, n,\ j = n + 1, \ldots, n + p + 1, \\ 0, & i = n + 1, \ldots, n + p + 1,\ j = n + 1, \ldots, n + p + 1. \end{cases}$$

Notice that it is no longer required that $f_0(\cdot) \equiv 1$.

Hence, the coefficients $\boldsymbol{\lambda}$ are given by

$$\boldsymbol{\lambda}' = \left\{\mathbf{c} + X(X'\Sigma^{-1}X)^{-1}(\mathbf{x} - X'\Sigma^{-1}\mathbf{c})\right\}'\Sigma^{-1},$$

where $\mathbf{c} \equiv (C(\mathbf{s}_0 - \mathbf{s}_1), \ldots, C(\mathbf{s}_0 - \mathbf{s}_n))'$ and Σ is an $n \times n$ matrix whose (i, j)th element is $C(\mathbf{s}_i - \mathbf{s}_j)$. Moreover,

$$\mathbf{m}' = (\mathbf{x} - X'\Sigma^{-1}\mathbf{c})'(X'\Sigma^{-1}X)^{-1}.$$

Kriging Variance and Prediction Intervals

The *kriging (or prediction) variance* (minimum mean-squared prediction error) is

$$\begin{aligned} \sigma_k^2(\mathbf{s}_0) &= \boldsymbol{\lambda}'_U \boldsymbol{\gamma}_U = \sum_{i=1}^{n} \lambda_i \gamma(\mathbf{s}_0 - \mathbf{s}_i) + \sum_{j=1}^{p+1} m_{j-1} f_{j-1}(\mathbf{s}_0) \\ &= \boldsymbol{\gamma}'\Gamma^{-1}\boldsymbol{\gamma} - (\mathbf{x} - X'\Gamma^{-1}\boldsymbol{\gamma})'(X'\Gamma^{-1}X)^{-1}(\mathbf{x} - X'\Gamma^{-1}\boldsymbol{\gamma}). \quad (3.4.16) \end{aligned}$$

Also,

$$\sigma_k^2(\mathbf{s}_0) = 2\sum_{i=1}^{n} \lambda_i \gamma(\mathbf{s}_0 - \mathbf{s}_i) - \sum_{i=1}^{n} \sum_{j=1}^{n} \lambda_i \lambda_j \gamma(\mathbf{s}_i - \mathbf{s}_j). \quad (3.4.17)$$

Assuming the covariogram of $Z(\cdot)$ is well defined, the kriging variance can be written as

$$\sigma_k^2(\mathbf{s}_0) = C(\mathbf{0}) - \sum_{i=1}^{n} \lambda_i C(\mathbf{s}_0 - \mathbf{s}_i) + \sum_{j=1}^{p+1} m_{j-1} f_{j-1}(\mathbf{s}_0)$$

$$= C(\mathbf{0}) - \mathbf{c}'\Sigma^{-1}\mathbf{c} + (\mathbf{x} - X'\Sigma^{-1}\mathbf{c})'(X'\Sigma^{-1}X)^{-1}(\mathbf{x} - X'\Sigma^{-1}\mathbf{c}).$$

Also,

$$\sigma_k^2(\mathbf{s}_0) = C(\mathbf{0}) - 2\sum_{i=1}^{n} \lambda_i C(\mathbf{s}_0 - \mathbf{s}_i) + \sum_{i=1}^{n}\sum_{j=1}^{n} \lambda_i \lambda_j C(\mathbf{s}_i - \mathbf{s}_j).$$

From (3.4.5), (3.4.11), and (3.4.16), *prediction intervals* can be constructed. The interval

$$A \equiv \left(\hat{Z}(\mathbf{s}_0) - 1.96\sigma_k(\mathbf{s}_0), \hat{Z}(\mathbf{s}_0) + 1.96\sigma_k(\mathbf{s}_0)\right)$$

is a nominal 95% prediction interval for $Z(\mathbf{s}_0)$. Under the assumption that $Z(\cdot)$ is Gaussian, $\Pr\{Z(\mathbf{s}_0) \in A\} = 95\%$, where $\Pr\{\cdot\}$ is calculated from the joint distribution of $Z(\mathbf{s}_0), Z(\mathbf{s}_1), \ldots, Z(\mathbf{s}_n)$.

Block Kriging

When $|B| > 0$ in (3.4.6), the universal-kriging equations (3.4.11) and (3.4.16) become

$$\boldsymbol{\lambda}_U = \Gamma_U^{-1}\boldsymbol{\gamma}_U(B), \tag{3.4.18}$$

$$\sigma_k^2(B) = \boldsymbol{\lambda}_U'\boldsymbol{\gamma}_U(B) - \gamma(B, B), \tag{3.4.19}$$

where

$$\boldsymbol{\gamma}_U(B) \equiv \left(\gamma(B, \mathbf{s}_1), \ldots, \gamma(B, \mathbf{s}_n), 1, f_1(B), \ldots, f_p(B)\right)', \tag{3.4.20}$$

$\gamma(B, \mathbf{s}_i) \equiv \int_B \gamma(\mathbf{u} - \mathbf{s}_i)\, d\mathbf{u}/|B|$, $i = 1, \ldots, n$, $f_j(B) \equiv \int_B f_j(\mathbf{u})\, d\mathbf{u}/|B|$, $j = 1, \ldots, p$, and $\gamma(B, B) \equiv \int_B \int_B \gamma(\mathbf{u} - \mathbf{v})\, d\mathbf{u}\, d\mathbf{v}/|B|^2$.

Polynomial Trend

Suppose $\mathbf{s} \in \mathbb{R}^2$. Often $E(Z(\mathbf{s})) = \mu(\mathbf{s})$ is expressed as a linear combination of polynomials in the spatial coordinates $\mathbf{s} = (x, y)'$. A trend surface of degree r is

$$\mu(\mathbf{s}) = \sum\sum_{0 \le k+l \le r} \alpha_{kl} x^k y^l, \qquad \mathbf{s} = (x, y)'. \tag{3.4.21}$$

For example, a *quadratic* trend surface is $\mu(\mathbf{s}) = \alpha_{00} + \alpha_{10}x + \alpha_{01}y + \alpha_{20}x^2 + \alpha_{11}xy + \alpha_{02}y^2$. The polynomial trend surface (3.4.21) has $p+1 = (r+1)(r+2)/2$ terms. It is seen to be a special case of (3.4.2) with

$$\begin{aligned} f_0(\mathbf{s}) &= 1, & \mathbf{s} &= (x, y)' \in D, \\ f_1(\mathbf{s}) &= x, & \mathbf{s} &= (x, y)' \in D, \\ &\vdots \\ f_p(\mathbf{s}) &= y^r, & \mathbf{s} &= (x, y)' \in D, \end{aligned}$$

where $p = \{(r+1)(r+2)/2\} - 1$. For $\mathbf{s} \in \mathbb{R}^d$, $p = \binom{r+d}{d} - 1$.

Estimation of the Mean Parameters

Optimal estimation of the mean parameters $\boldsymbol{\beta}$ in (3.4.2) can be accomplished easily. From (3.4.3), the data $\mathbf{Z}$ are seen to satisfy the general linear model, where $E(\mathbf{Z}) = X\boldsymbol{\beta}$ and $\mathrm{var}(\mathbf{Z}) = \Sigma$. The generalized-least-squares estimator

$$\hat{\boldsymbol{\beta}}_{\mathrm{gls}} = (X'\Sigma^{-1}X)^{-1}X'\Sigma^{-1}\mathbf{Z}$$

has the property that $\mathrm{var}(\hat{\boldsymbol{\beta}}) - \mathrm{var}(\hat{\boldsymbol{\beta}}_{\mathrm{gls}})$ is nonnegative-definite for all linear unbiased estimators $\hat{\boldsymbol{\beta}}$ (e.g., Searle, 1971, Section 3.3). Further details are given in Section 1.3.

To *estimate* the mean parameters optimally, knowledge of $\{\mathrm{cov}(Z(\mathbf{s}_i), Z(\mathbf{s}_j)): 1 \le i \le j \le n\}$ is needed. However, to *predict* an unknown value $Z(\mathbf{s}_0)$ optimally, only knowledge of $\{\mathrm{var}(Z(\mathbf{s}_i) - Z(\mathbf{s}_j)): 0 \le i \le j \le n\}$ is needed (provided X has a column of 1s and there is a 1 in the corresponding entry of $\mathbf{x}$); see (3.4.8). The two notions, of optimal estimation and optimal prediction, are very closely linked (Section 3.4.5).

Some Final Remarks

Universal kriging is an optimal linear spatial prediction procedure that is most simply understood through a linear-model formulation; see (3.4.3) and (3.4.4). Indeed, it was in this guise that Goldberger (1962) gave (kriging) equations for best linear unbiased prediction in the general linear model. His setting was more general than that of spatial prediction; however, he did not consider the implications of the predictand and the predictors arising from a common process $\{Z(\mathbf{s}): \mathbf{s} \in D\}$. Moreover, his prediction equations are given in terms of covariances rather than variograms. Matheron (1963a) and Gandin (1963) used spatial location in their models of large- and small-scale effects; in this setting, stationary variograms are more general and easier to estimate than stationary covariograms (Section 2.3). Nevertheless, the universal-kriging equations can also be written in terms of covariances (Section 3.4.5).

Journel (1977) presents kriging in terms of projections, although the universal-kriging predictor is not a projection with regard to the usual inner product. The universal-kriging predictor $\hat{Z}(\mathbf{s}_0) = \sum_{i=1}^{n} \lambda_i Z(\mathbf{s}_i)$ is *not* orthogonal to (i.e., is correlated with) $Z(\mathbf{s}_0) - \hat{Z}(\mathbf{s}_0)$; see Section 3.4.5 for further details and a sense in which kriging is an orthogonal projection.

Just as ordinary kriging can be generalized to cokriging (Section 3.2.3), it is also possible to derive equations for universal cokriging. The natural generalization of the variogram to a cross-variogram is

$$2\gamma_{jj'}(\mathbf{h}) \equiv \operatorname{var}\left(Z_j(\mathbf{s} + \mathbf{h}) - Z_{j'}(\mathbf{s})\right); \tag{3.4.22}$$

see Section 3.2.3 and Clark et al. (1989).

Finally, it is clear from (3.4.8) that kriging equations (3.4.11) and (3.4.17) remain valid even when the stationarity assumption (3.4.9) does not hold; simply use $\gamma(\mathbf{s}, \mathbf{u}) \equiv (1/2)\operatorname{var}(Z(\mathbf{s}) - Z(\mathbf{u}))$ instead of $\gamma(\mathbf{s} - \mathbf{u})$ in those equations. A similar remark is true for universal cokriging; replace $\gamma_{jj'}(\mathbf{s} - \mathbf{u})$ with $\gamma_{jj'}(\mathbf{s}, \mathbf{u}) \equiv (1/2)\operatorname{var}(Z_j(\mathbf{s}) - Z_{j'}(\mathbf{u}))$. The stationarity assumptions are made in order to motivate various (cross) variogram estimators, but are not needed if $\gamma(\mathbf{s}, \mathbf{u})$ or $\gamma_{jj'}(\mathbf{s}, \mathbf{u})$ are known.

3.4.1 Universal Kriging of Coal-Ash Data

From Sections 2.2, 2.4, and 2.6, the following can be concluded. The coal-ash data (Fig. 2.2*a*) exhibit a strong linear trend in the east–west direction but not in the north–south direction. For the estimated variogram $2\bar{\gamma}$ in the north–south direction (Fig. 2.14), the best fit from the class of spherical variogram models is

$$2\gamma^o(h) = \begin{cases} 0, & h = 0, \\ 1.78 + 0.28\{(3/2)(h/4.31) - (1/2)(h/4.31)^3\}, & \\ & 0 < h \le 4.31, \\ 2.06, & h \ge 4.31, \end{cases} \tag{3.4.23}$$

where "best" is in terms of weighted least squares (see Fig. 2.15). The lag h in (3.4.23) is in units of the square grid spacing (1 unit = 2500 ft).

Because there is no trend in the north–south direction, this variogram model is an accurate representation of the spatial dependence in that direction. There is trend in the east–west direction, but in order to proceed with universal kriging, a two-dimensional variogram model is needed. One way to resolve the problem is to assume isotropy, namely, assume that the one-dimensional variogram (3.4.23) applies in any direction in $\mathbb{R}^2$. Before doing this, care must be taken that the one-dimensional variogram model is conditionally negative-definite as an isotropic model in $\mathbb{R}^2$; in fact, the spherical model is valid in $\mathbb{R}^2$ and $\mathbb{R}^3$ (Section 2.5).

Random-Process Model in $\mathbb{R}^2$

The proposed random-process model for the coal-ash data is

$$Z(\mathbf{s}) = \beta_0 + \beta_1 x + \delta(\mathbf{s}), \qquad \mathbf{s} = (x, y)', \tag{3.4.24}$$

where $\delta(\cdot)$ is a zero-mean random process with stationary, isotropic variogram $2\gamma(\mathbf{h}) = 2\gamma^o(\|\mathbf{h}\|)$, $\mathbf{h} \in \mathbb{R}^2$, and $2\gamma^o$ is given by (3.4.23).

It is fortunate in this case that there is a trend-free direction from which the variogram can be estimated. When there is not, the usual estimators $2\hat{\gamma}$, $2\bar{\gamma}$, and so on can show a confounding of mean effects and the underlying variogram; Section 3.4.3 discusses this more fully.

Choice of Kriging Neighborhood

To predict an unknown value $Z(B)$ involves, in principle, inversion of an $(n+2) \times (n+2)$ matrix; see (3.4.11). Here, $n = 208$, which leads to a computationally prohibitive 210×210 matrix inversion for every prediction value. Many of the 208 points have kriging weights close to zero; data closer to $Z(B)$ will typically have more substantial weights. Because universal-kriging weights necessarily satisfy $\boldsymbol{\lambda}'X = \mathbf{x}'$ [see (3.4.5)], one way to choose a neighborhood of kriging locations is to increase the neighborhood gradually, each time computing the *unconstrained* optimal prediction weights $\mathbf{l}' = \mathbf{c}'\Sigma^{-1}$ [recall $\mathbf{c} = (C(\mathbf{s}_0, \mathbf{s}_1), \ldots, C(\mathbf{s}_0, \mathbf{s}_n))'$, $\Sigma = (C(\mathbf{s}_i, \mathbf{s}_j))$, and $C(\mathbf{s}, \mathbf{u}) \equiv \mathrm{cov}(Z(\mathbf{s}), Z(\mathbf{u}))$]. If it exists, the neighborhood with the fewest points for which $\mathbf{c}'\Sigma^{-1}X \simeq \mathbf{x}'$ is chosen. Section 3.4.5 makes a suggestion that can be implemented more generally, namely, choose a neighborhood that minimizes the value of $\mathbf{m}'\mathbf{x} = (\mathbf{x} - X'\Sigma^{-1}\mathbf{c})'(X'\Sigma^{-1}X)^{-1}\mathbf{x}$.

Figure 3.6 shows various neighborhoods $\{\mathbf{s}_1, \ldots, \mathbf{s}_{n_k}\}$ that were chosen to predict coal-ash value $Z(\mathbf{s}_0)$ located at $\mathbf{s}_0$, where $\mathbf{s}_0$ is the center of the neighborhood. For each neighborhood, $\mathbf{x}'$, $\mathbf{c}'\Sigma^{-1}X$, and $(X'\Sigma^{-1}X)^{-1}$ were computed, where here

$$X = \begin{bmatrix} 1 & x_1 \\ \vdots & \vdots \\ 1 & x_{n_k} \end{bmatrix}, \qquad \mathbf{x} = \begin{bmatrix} 1 \\ x_0 \end{bmatrix}. \tag{3.4.25}$$

For all the neighborhoods under consideration, the symmetry of the kriging locations implies that $\sum_{i=1}^{n_k} l_i x_i = x_0$; hence, the condition $\mathbf{c}'\Sigma^{-1}X \simeq \mathbf{x}'$, reduces to $\mathbf{c}'\Sigma^{-1}\mathbf{1} \equiv \sum_{i=1}^{n_k} l_i \simeq 1$.

However, because the nugget effect is so large $[c_0/(c_0 + c_s) = 0.89/1.03 = 86\%]$, it is likely that *most* of the data locations would be needed to

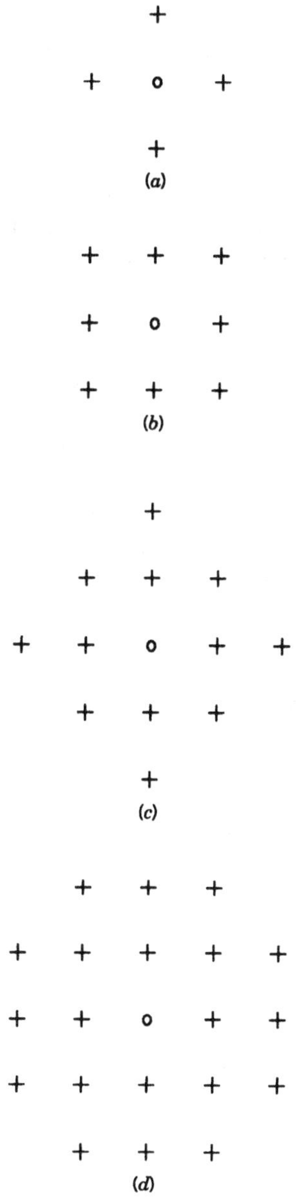

```
        +

    +   o   +

        +
       (a)


    +   +   +

    +   o   +

    +   +   +
       (b)


        +

    +   +   +

+   +   o   +   +

    +   +   +

        +
       (c)


    +   +   +

+   +   +   +   +

+   +   o   +   +

+   +   +   +   +

    +   +   +
       (d)
```

Figure 3.6 Figures (a), (b), (c), and (d) show successively larger neighborhoods about the prediction location (marked as o). Data locations with nonzero kriging weights are marked as +.

ensure $\Sigma l_i = 1$ [100% nugget effect corresponds to uncorrelated data, for which the best predictor is $\bar{Z} = \Sigma_{i=1}^{n} Z(\mathbf{s}_i)/n$], which is computationally undesirable. Here, kriging neighborhood (c) in Figure 3.6 was chosen as a compromise between large unmanageable neighborhoods and small inappropriate ones; for neighborhood (c), $\mathbf{c}'\Sigma^{-1}\mathbf{1} = 0.5424$, with very little increase for neighborhood (d). This compromise choice was corroborated using the values of $\mathbf{m}'\mathbf{x}$: For neighborhood (b), $\mathbf{m}'\mathbf{x} = 0.099$ ($n_k = 8$); for neighborhood (c), $\mathbf{m}'\mathbf{x} = 0.060$ ($n_k = 12$); and for neighborhood (d), $\mathbf{m}'\mathbf{x} = 0.031$ ($n_k = 20$).

Universal Kriging

For cross-validation purposes (Section 2.6.4) and data-editing purposes (Section 3.3), the coal-ash datum at $\mathbf{s}_0$ was *removed* and predicted from its neighbors using neighborhood (c) and universal-kriging equations (3.4.5) and (3.4.11). A prediction location $\mathbf{s}_0$ was used only if all 13 of the locations $\{\mathbf{s}_i: i = 0, \ldots, 12\}$ possessed a datum. This resulted in 100 prediction locations with values $\{Z(\mathbf{s}_{0,i}): i = 1, \ldots, 100\}$ and predictors $\{\hat{Z}_{-i}(\mathbf{s}_{0,i}): i = 1, \ldots, 100\}$; see (3.3.8). The kriging variance was calculated from (3.4.17) to be $1.016(\%)^2$. A plot of $Z(\mathbf{s}_{0,i})$ versus $\hat{Z}_{-i}(\mathbf{s}_{0,i})$ is shown in Figure 3.7a; Figure 3.7b shows a plot of $Z(\mathbf{s}_{0,i}) - \hat{Z}_{-i}(\mathbf{s}_{0,i})$ versus $\hat{Z}_{-i}(\mathbf{s}_{0,i})$. The first plot shows the expected positive dependence and the second shows no obvious pattern, apart from the outlying point caused by the anomalous value $z(5,6) = 17.61\%$. These values were also used for cross-validation purposes in Section 2.6.4.

For universal-kriging equations (3.4.11), the number of nonzero kriging weights in neighborhood (c) is $n_k = 12$ and the number of Lagrange multipliers is $p + 1 = 2$ [corresponding to $\mu(\mathbf{s}) = a_0 + a_1 x$]. Hence, the matrix Γ_U in (3.4.11) is a 14×14 matrix. Notice that the upper-left 12×12 block of Γ_U [consisting of the entries $\gamma(\mathbf{s}_i - \mathbf{s}_j)$] *must* have zeros down its diagonal. Some practitioners make the mistake of putting the nugget effect c_0 along this diagonal, but from the model (3.4.23) and a comment made in Section 3.2.1, this is clearly wrong.

Figure 3.8 shows the kriging weights used to predict $Z(\mathbf{s}_{0,i})$; they are superimposed over their respective neighborhood locations (given by Fig. 3.6c).

Data Editing

Section 3.3 discussed a robust version of ordinary kriging. In fact, the *identical* algorithm can be applied in a universal-kriging context. For example, the comments about predicting missing values with $Z^{@}_{-j}(\mathbf{s}_j)$ given by (3.3.9) and the editing of outlying values using (3.3.10) are equally applicable. Notice from Figure 2.2a that $z(7,1)$ is missing; from (3.3.9), the predicted value is

$$z^{@}(7,1) = 10.27\% \text{ coal ash.}$$

Notice also from Figure 2.2a that $z(5,6) = 17.61\%$ appears to be an outlier; from (3.3.10), the edited value is

$$z^{(e)}(5,6) = 12.84\% \text{ coal ash,}$$

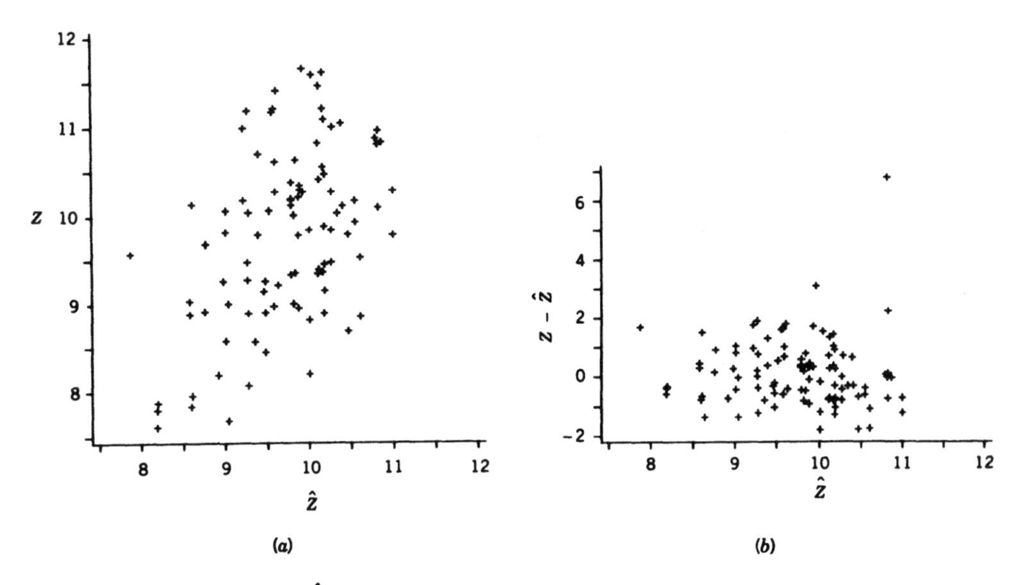

Figure 3.7 (*a*) Plot of $(\hat{Z}_{-i}(\mathbf{s}_{0,i}),\ Z(\mathbf{s}_{0,i}))$. Units on the axes are percent coal ash. [To avoid distortion, the outlying point (10.82, 17.61) is not shown.] (*b*) Plot of $(\hat{Z}_{-i}(\mathbf{s}_{0,i}),\ Z(\mathbf{s}_{0,i}) - \hat{Z}_{-i}(\mathbf{s}_{0,i}))$. Units on the axes are percent coal ash. (All 100 points are shown.)

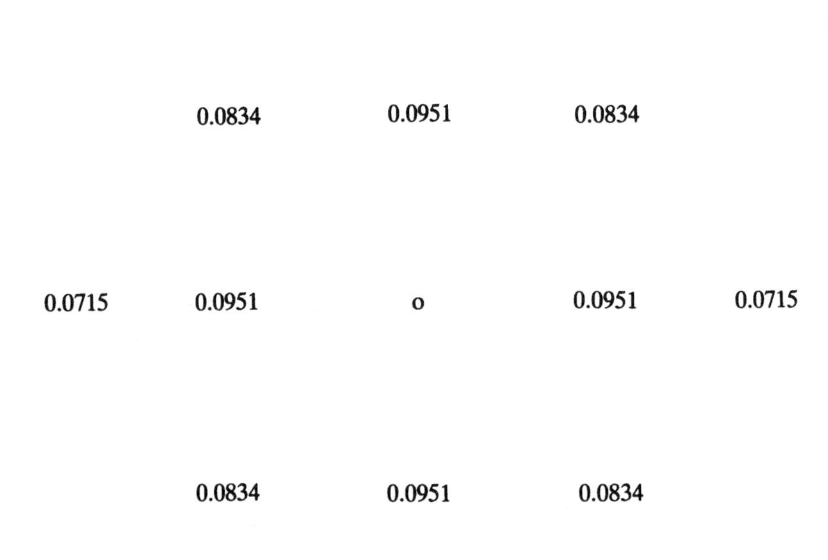

Figure 3.8 Kriging weights for neighborhood (*c*) in Figure 3.6 for the coal-ash data.

where the editing constant $c = 2$ was used. Several of the 100 data (actually 3%) would be similarly edited upon applying (3.3.10).

Predicting Blocks

When the coal is mined, it is mined selectively in large blocks of size $|B|$. Thus, when forecasting mine profitability and production schedules, prediction of $Z(B)$ [$\equiv \int_B Z(\mathbf{u})\, d\mathbf{u}/|B|$] is of more interest than prediction of $Z(\mathbf{s}_0)$, $\mathbf{s}_0 \in D$. A block with a very rich coal seam running through it may be unprofitable to mine if the rest of the block is barren rock. More discussion of this important issue is given in Section 5.2.

To predict $Z(B)$, data both inside and outside the block B should be used. The universal-kriging equations (3.4.18) require computation of $\gamma(B, \mathbf{s})$ and $\gamma(B, B)$, which are normalized integrals of the modeled variogram. In practice, they can be computed by numerical quadrature.

Some Final Remarks

The coal-ash data have been used for illustrative purposes and they will continue to fulfill that purpose in Section 3.5 (median-polish kriging). Although the spatial correlation in this example does not appear to be all that strong, Chapter 4 shows that this can vary a great deal from problem to problem. For a detailed study of the coal-ash data from a mining-engineering point of view, the interested reader is referred to Buxton (1982).

3.4.2 Trend-Surface Prediction

Anyone dealing with time series or spatial data knows that the decomposition (3.1.4),

$$Z(\cdot) = \text{large-scale variation} + \text{smaller-scale variation}, \qquad (3.4.26)$$

cannot be specified uniquely. Recall that throughout this book large-scale variation is synonymous with trend or mean structure, and hence any smaller-scale variation is zero-mean random error. Within disciplines, scientists often have a good idea regarding which part of $Z(\cdot)$ is due to controllable factors and exogenous variables. But even then there is no unanimity, because the decomposition (3.4.26) ultimately depends on personal tastes and preferences.

When the goal of a study is the etiology of a phenomenon, it makes sense to put as much of the variation as can be explained into the mean structure and then to *estimate* the strength of its presence. However, if *prediction* of an observation at a given time or location is the primary goal and the etiology is secondary, then using a trend surface will lead to overfitting and spurious predictions. A parsimoniously parameterized mean structure plus a suitably stationary error structure is probably a preferable model. It is not clear how to quantify "parsimoniously parameterized" and "suitably stationary"; one possibility is a spectral decomposition of the data where the low-order eigenfunctions represent the mean structure. Watson (1972) points out that most geological problems have smaller-scale variation that should be modeled as a stationary, zero-mean random process, with nearby values showing strong positive correlation.

This section gives a method of spatial prediction that uses a rather extreme version of the decomposition (3.4.26). *Trend-surface prediction* relies on an underlying random process

$$Z(\cdot) = \text{large-scale variation} + \text{white noise}. \tag{3.4.27}$$

Thus, all the variation, apart from uncorrelated fluctuations, is absorbed into the mean. In the notation of (3.1.2), the random process is represented as

$$Z(\mathbf{s}) = \mu(\mathbf{s}) + (\eta(\mathbf{s}) + \epsilon(\mathbf{s})), \quad \mathbf{s} \in D, \tag{3.4.28}$$

where the microscale variation $\eta(\cdot)$ is a white-noise process. That is,

$$\begin{aligned} E(Z(\mathbf{s})) &= \mu(\mathbf{s}), \quad \mathbf{s} \in D, \\ \operatorname{cov}(Z(\mathbf{s}), Z(\mathbf{u})) &= \begin{cases} c_0, & \mathbf{s} = \mathbf{u}, \\ 0, & \mathbf{s} \neq \mathbf{u}. \end{cases} \end{aligned} \tag{3.4.29}$$

Trend-Surface Predictor and Mean-Squared Prediction Error

If $\mu(\cdot)$ in (3.4.28) can be written as

$$\mu(\mathbf{s}) = \sum_{j=1}^{p+1} \beta_{j-1} f_{j-1}(\mathbf{s}), \quad \mathbf{s} \in D, \tag{3.4.30}$$

then from (3.1.4) the trend-surface model (3.4.28) is a special case of the universal-kriging model (3.4.2).

Assume the trend-surface model (3.4.28). Then, in the notation of (3.4.2), (3.4.3), and (3.4.4), the best linear unbiased estimator (b.l.u.e.) of $\boldsymbol{\beta}$ is

$$\hat{\boldsymbol{\beta}} = (X'X)^{-1} X'\mathbf{Z}. \tag{3.4.31}$$

Thus, the best linear unbiased predictor (b.l.u.p.) of, say, $Z(\mathbf{s}_0)$ is

$$\hat{Z}(\mathbf{s}_0) = \mathbf{x}'\hat{\boldsymbol{\beta}}, \ \mathbf{s}_0 \notin \{\mathbf{s}_1, \ldots, \mathbf{s}_n\}. \tag{3.4.32}$$

The predictor (3.4.32) has mean-squared prediction error

$$E\big(Z(\mathbf{s}_0) - \hat{Z}(\mathbf{s}_0)\big)^2 = \big(1 + \mathbf{x}'(X'X)^{-1}\mathbf{x}\big)c_0. \tag{3.4.33}$$

Comparison of Kriging and Trend-Surface Prediction

In the 1950s and 1960s, most earth scientists in the United States took a trend-surface approach to their mapping problems, although in the last 15 years the extra advantages in taking a random-field approach (i.e., universal kriging) have been realized. The contributions of Matheron (1967) (written in English) and Watson (1972) were particularly influential.

In one sense, comparison of the two approaches is easy. For the *same* representation of large-scale variation (trend), $\mu(\mathbf{s}) = \sum_{j=1}^{p+1} \beta_{j-1} f_{j-1}(\mathbf{s})$, the trend-surface model (3.4.28) is a special case of the random-field model (3.4.2). Therefore, when the spatial-covariance structure is known, universal kriging generally gives more precise predictions than trend-surface prediction, because universal kriging chooses *optimal* weights to be applied to the data. However, Watson (1971) shows that it is possible for trend-surface prediction to be just as precise as universal kriging even when the error $\delta(\cdot)$ is not pure white noise. Moreover, in practice, there is a price to pay for using universal kriging: One must obtain (efficient) estimators of variogram parameters, whose effects on mean-squared prediction errors should be assessed (Section 5.3).

Gustaffson (1981), Agterberg (1984), and Haining (1987) make comparisons of the two approaches. For example, Agterberg makes an empirical comparison by first sampling from a map and then predicting (known) values using only the sampled data. A 30 × 30 mile area of a contour map of the Arbuckle formation in central Kansas was used; 75 randomly chosen spatial locations yielded $\{\mathbf{s}_1, \ldots, \mathbf{s}_{75}\}$, from which the data $\{Z(\mathbf{s}_1), \ldots, Z(\mathbf{s}_{75})\}$ were obtained. Finally, 50 $\mathbf{s}_0$s were randomly chosen and the (empirical) goodness-of-prediction measure

$$G \equiv \left[1 - \left\{ \sum_{\mathbf{s}_0} \left(Z(\mathbf{s}_0) - \hat{Z}(\mathbf{s}_0)\right)^2 \Big/ \sum_{\mathbf{s}_0} \left(Z(\mathbf{s}_0) - \bar{Z}\right)^2 \right\}\right] 100\% \qquad (3.4.34)$$

was computed for each decomposition (3.4.26), where $\bar{Z} = \sum_{\mathbf{s}_0} Z(\mathbf{s}_0)/50$.

When a quadratic $\mu(\cdot)$ was used for both models, universal kriging ($G = 80.9\%$) dominated trend-surface prediction ($G = 73.8\%$). Now, when a constant mean was used in the random-field approach (i.e., ordinary kriging), the measure G was equal to 76.7%, only slightly better than the quadratic-trend-surface approach ($G = 73.8\%$). Which approach should be preferred? Ordinary kriging has a constant mean and a dependent error structure, whereas the quadratic-trend-surface predictor uses up to five more parameters to describe the mean structure, but needs only one parameter (the variance) for the error structure.

Rather than entering into a fruitless debate over which approach is better, it is more appropriate to address the question through model selection (already discussed briefly in Section 2.6.4), which makes the answer problem-specific. A model with few parameters that predicts well is the ideal.

Methods of cross-validation (Stone, 1974), minimizing an information criterion (Akaike, 1973; Schwarz, 1978), or minimizing description length (Rissanen, 1987) offer ways to discriminate between different models of the same data. Little is known about their properties in the spatial context, a topic of research that needs attention in the future.

3.4.3 Estimating the Variogram for Universal Kriging

Recall, from the universal-kriging equations (3.4.11) through (3.4.20), that the variogram has so far been assumed known. In practice, it must be estimated, which, in the more general situation of nonconstant mean, is not at all straightforward. To use the estimators $2\hat{\gamma}(\cdot)$, $2\bar{\gamma}(\cdot)$, and so on, given in Section 2.4, would be inappropriate because

$$E\big(Z(\mathbf{s}_i) - Z(\mathbf{s}_j)\big)^2 = \operatorname{var}\big(Z(\mathbf{s}_i) - Z(\mathbf{s}_j)\big) + \{\mu(\mathbf{s}_i) - \mu(\mathbf{s}_j)\}^2$$

$$= 2\gamma(\mathbf{s}_i - \mathbf{s}_j) + \left\{\sum_{k=1}^{p+1} \beta_{k-1}\big(f_{k-1}(\mathbf{s}_i) - f_{k-1}(\mathbf{s}_j)\big)\right\}^2.$$

If $\boldsymbol{\beta}$ were known, an estimator of the variogram could be based on $\delta(\cdot) \equiv Z(\cdot) - \sum_{k=1}^{p+1}\beta_{k-1}f_{k-1}(\cdot)$ because

$$E\big(\delta(\mathbf{s}_i) - \delta(\mathbf{s}_j)\big)^2 = 2\gamma(\mathbf{s}_i - \mathbf{s}_j).$$

But $\boldsymbol{\beta}$ is unknown; in order to estimate it efficiently, knowledge of $2\gamma(\cdot)$ [actually the covariance matrix, var($\mathbf{Z}$)] is needed. However, $2\gamma(\cdot)$ is unknown, bringing the discussion right back to where it started. This circularity has led to some dissatisfaction with universal kriging.

A Simple Example

To illustrate the problem, consider the special case where data occur on a transect at locations $\{i: i = 1, \ldots, n\}$ and assume that the true random process is

$$Z(t) = \beta \cdot t + \delta(t), \qquad t \geq 0, \tag{3.4.35}$$

where $\{\delta(t): t \geq 0\}$ is a zero-mean intrinsically stationary random process in $\mathbb{R}^1$, with variogram $2\gamma(\cdot)$.

Notice that $E(Z(t + h) - Z(t))^2 = 2\gamma(h) + \beta^2h^2$, which shows a positive, quadratic bias. A common-sense way to proceed is to estimate the parameter β, compute the residuals, and obtain an estimator of the variogram based on these residuals.

According to (3.4.35), data $\mathbf{Z} = (Z(1), \ldots, Z(n))'$ can be written

$$\mathbf{Z} = \mathbf{p}\beta + \boldsymbol{\delta}, \tag{3.4.36}$$

where $\mathbf{p} \equiv (1, 2, \ldots, n)'$. Strictly speaking, $\text{cov}(\mathbf{Z}) = \Sigma$ is unknown. But, assuming for the moment it is known, it is of interest to examine the bias properties of the residuals based on the most efficient estimator.

The generalized-least-squares (g.l.s.) estimator of β is

$$\hat{\beta}_{\text{gls}} = (\mathbf{p}'\Sigma^{-1}\mathbf{p})^{-1}\mathbf{p}'\Sigma^{-1}\mathbf{Z}. \tag{3.4.37}$$

Its associated residuals are

$$\mathbf{W} \equiv (W(1), \ldots, W(n))' \equiv \mathbf{Z} - \mathbf{p}\hat{\beta}_{\text{gls}} \tag{3.4.38}$$

and $E(\mathbf{W}) = \mathbf{0}$, $\text{var}(\mathbf{W}) = \Sigma - (\mathbf{p}'\Sigma^{-1}\mathbf{p})^{-1}\mathbf{p}\mathbf{p}'$. Thus, $E(W(i) - W(i+h)) = 0$ and

$$E(W(i+h) - W(i))^2 = 2\gamma(h) - h^2\{\mathbf{p}'\Sigma^{-1}\mathbf{p}\}^{-1}. \tag{3.4.39}$$

Notice that the bias is now negative and quadratic in h. Thus, even if Σ were known, residuals based on the most efficient estimator yield a biased estimator of the variogram.

More realistically, Σ is unknown. The ordinary-least-squares (o.l.s.) residuals

$$\mathbf{R} \equiv (R(1), \ldots, R(n))' \equiv \mathbf{Z} - \mathbf{p}\{(\mathbf{p}'\mathbf{p})^{-1}\mathbf{p}'\mathbf{Z}\} \tag{3.4.40}$$

satisfy $E(R(i+h) - R(i)) = 0$, and $E(R(i+h) - R(i))^2$ can be written as $2\gamma(h) + \text{bias}$, where the bias term depends not only on h but also on i (Cressie, 1987). Therefore, every term $(R(i+h) - R(i))^2$ in the estimated variogram $\sum_{i=1}^{n-h}(R(i+h) - R(i))^2/(n-h)$ is estimating a slightly different quantity. Now specialize the model even further by assuming that $\delta(\cdot)$ is a second-order stationary m-dependent process; that is, assume $C(t, u) = C^*(t-u)$ and $C^*(h) = 0$ if $h > m$. Then, provided $i > m$, $i + h < n - m$, and n is large,

$$E(R(i+h) - R(i))^2 \simeq 2\gamma(h) - h^2\left\{C^*(0) + 2\sum_{l=1}^{m} C^*(l)\right\}\Big/(\mathbf{p}'\mathbf{p}). \tag{3.4.41}$$

Thus, if the correlation is positive, the bias is negative and quadratic in h. The following discussion establishes a bias result under more general circumstances.

Variogram Estimation Based on Residuals

A number of authors have proposed solutions to bias problems that occur when residuals are used to estimate the variogram (e.g., Sabourin, 1976; Neuman and Jacobson, 1984). Neuman and Jacobson's approach is to start with an o.l.s. estimator of $\boldsymbol{\beta}$ in (3.4.2), compute a variogram estimator from the residuals, fit a variogram model, obtain a g.l.s. estimator of $\boldsymbol{\beta}$ based on the fitted model, and so forth. This iterative approach makes o.l.s. residuals look more and more like g.l.s. residuals. The simple example above shows that doing this does not solve the bias problem (because even when Σ is known, the bias is not zero), which is at the root of Gambolati and Galeati's (1987) comment on the iterative approach.

Consider the g.l.s. residuals in the general setting of the model (3.4.2). From (3.4.3),

$$\mathbf{Z} = X\boldsymbol{\beta} + \boldsymbol{\delta},$$

the g.l.s. estimator of $\boldsymbol{\beta}$ is

$$\hat{\boldsymbol{\beta}}_{\text{gls}} = \left(X'\Sigma^{-1}X\right)^{-1}X'\Sigma^{-1}\mathbf{Z}, \tag{3.4.42}$$

and the corresponding residuals are

$$\begin{aligned}\mathbf{W} &\equiv \mathbf{Z} - X\hat{\boldsymbol{\beta}}_{\text{gls}} \\ &= \left(\Sigma - X\left(X'\Sigma^{-1}X\right)^{-1}X'\right)\Sigma^{-1}\mathbf{Z}.\end{aligned} \tag{3.4.43}$$

It is not surprising that variogram estimators based on $\mathbf{W}$ are biased, because in addition to the statistical dependence already in the data (expressed through the variance matrix Σ), $\mathbf{W}$ satisfies $(p + 1)$ linear constraints; the projection matrix $(\Sigma - X(X'\Sigma^{-1}X)^{-1}X')\Sigma^{-1}$ in (3.4.43) is of rank $(n - p - 1)$. Intuitively, such *algebraic* constraints lead to residuals $\mathbf{W}$ that exhibit more negative correlations than those of the errors $\boldsymbol{\delta}$. Similar arguments apply to the o.l.s. residuals.

Consequences of the Bias

Any bias is of some concern. If a convenient expression is available for the bias, one could try subtracting, from the original estimator, a consistent estimator of that bias expression. This will improve the order of the bias, perhaps at the expense of the mean-squared error (m.s.e.); recall

$$\text{m.s.e.} = (\text{bias})^2 + \text{variance}.$$

For an estimator based on n observations, typically $(\text{bias})^2 = (O(1/n))^2$ and variance $= O(1/n)$. That is, $(\text{bias})^2$ goes to zero faster than variance, and so one might think that bias can be ignored. However, because n is usually not large in many geostatistical problems, this overreliance on asymptotics can be misleading.

It is generally true that the bias of a residuals-based variogram estimator is small at lags near the origin but more substantial at distant lags. Now, provided a variogram model is fitted by generalized least squares or by weighted least squares (Section 2.6.2), which automatically puts most weight on the estimator at small lags, the effect of the bias should be small. Moreover, because kriging is carried out in local neighborhoods, the fitted variogram is only evaluated at smaller lags, precisely where it has been well fitted.

An estimated kriging variance is more likely to be affected by the bias in the variogram estimator. For simplicity, suppose that the variogram has a sill; that is, suppose $\lim_{\|\mathbf{h}\| \to \infty} 2\gamma(\mathbf{h}) = 2\sigma^2$. A commonly used estimator of σ^2 is $\hat{\sigma}^2 = (\mathbf{W}'\mathbf{W})/(n - p - 1)$. Assuming the covariogram $C^*(\cdot)$ is positive, the bias of $\hat{\sigma}^2$ is typically $O(1/n)$ and *negative* (Section 3.5.4). Because the kriging variance $\sigma_k^2(B)$ is directly proportional to σ^2, a (typically negatively) biased estimate of $\sigma_k^2(B)$ is usually obtained. (A similar result is available when σ^2 is, more generally, a multiplicative semivariogram parameter.)

In conclusion, although the universal-kriging *predictor* may be influenced little by the bias, there is (theoretical and empirical) evidence that the estimated *kriging variance* may be smaller than it should be. Section 5.3 contains a more complete discussion of this issue and gives recommendations regarding when a modified kriging variance estimator should be used.

Bias in Variogram Estimation versus Bias in Covariogram Estimation

Cressie and Grondona (1992) work with the one-dimensional transect model

$$Z(s) = \sum_{j=1}^{p+1} \beta_{j-1} s^{j-1} + \delta(s), \qquad s \geq 0, \tag{3.4.44}$$

where $\delta(\cdot)$ is a second-order stationary process with covariogram $C^*(\cdot)$ that satisfies certain summability conditions given in the following text. Suppose that data $\mathbf{Z} = (Z(1), \ldots, Z(n))'$ are available.

Define the o.l.s. residuals

$$\mathbf{R} \equiv \mathbf{Z} - X(X'X)^{-1}X'\mathbf{Z}, \tag{3.4.45}$$

where X is the $n \times (p + 1)$ matrix whose (i, j)th element is i^{j-1}. Also, define the covariogram estimator

$$\hat{C}(h) \equiv \sum_{i=1}^{n-h} R(i)R(i + h)/(n - h) \tag{3.4.46}$$

and the variogram estimator

$$2\hat{\gamma}(h) = \sum_{i=1}^{n-h} (R(i + h) - R(i))^2/(n - h). \tag{3.4.47}$$

Assume the existence of an infinite sequence $\{q_n\}$ such that, as $n \to \infty$, $q_n \to \infty$, $q_n/n \to 0$, and $\sum_{l=q_n}^{\infty}|C^*(l)| \to 0$. Then Cressie and Grondona

(1992) show that

$$E\big(\hat{C}(h) - C(h)\big) = -(p+1)\left\{C^*(0) + 2\sum_{l=1}^{\infty} C^*(l)\right\}\Big/ n + o(1/n). \tag{3.4.48}$$

That is, $\hat{C}(h)$ has a bias of $O(1/n)$ and each term of the regression contributes the same amount. Under the same assumptions on $C^*(\cdot)$, they then show that

$$E(2\hat{\gamma}(h) - 2\gamma(h)) = o(1/n); \tag{3.4.49}$$

that is, $2\hat{\gamma}(h)$ has a bias of $o(1/n)$.

Notice that the asymptotics used do not pick up a bias effect depending on h, provided h is fixed; compare (3.4.41). Furthermore, when $\delta(\cdot)$ is an autoregressive-moving-average process (e.g., Box and Jenkins, 1970), the $o(1/n)$ terms in (3.4.48) and (3.4.49) become $O((\log n)^{1+\alpha}/n^2)$, $\alpha > 0$ (Cressie and Grondona, 1992).

In the light of these results, is the bias of the estimated variogram important? For n small and h large, it is. Its presence is caused by linear constraints in the residuals $\mathbf{R}$ (or $\mathbf{W}$). One might think then that residuals based on nonlinear estimators of $\boldsymbol{\beta}$ might serve as better proxies for the error vector $\boldsymbol{\delta}$. Section 3.5 uses the residuals from median polish to estimate the spatial dependence.

Some Final Remarks

The substantial bias at large lags of a variogram estimator, based on (ordinary or generalized) least-squares residuals was noted by Matheron (1971b, pp. 152–155). This almost certainly led to a disenchantment with universal kriging and an initiative by Matheron and his group toward kriging with intrinsic random functions (Matheron, 1973; Delfiner, 1976). There the data are differenced, thus filtering out the polynomial trend (rather than estimating it); more details are given in Section 5.4.

It has been demonstrated in the preceding text that trend removal, obtained by subtracting the o.l.s. estimator $X\hat{\boldsymbol{\beta}}_{\text{ols}}$ from the data $\mathbf{Z}$, introduces spurious correlations. For a time series with polynomial trend and autoregressive errors, Kulperger (1987) derives asymptotic results that show the effect of o.l.s. detrending. Tunicliffe-Wilson (1989) proposes restricted (he calls it marginal) maximum likelihood (REML) as a way of differencing out a polynomial regression in a time series and estimating the dependence parameters $\boldsymbol{\theta}$ associated with the error process $\delta(\cdot)$; REML estimation is discussed briefly in Section 2.6.1. Kriging with intrinsic random functions has the same differencing idea as its basis. There has been much concern in the time-series literature with inappropriate fitting of trends to data that have zero mean but where the errors are nonstationary [e.g., the random walk $\delta(t) = \delta(t-1) + \epsilon(t)$, $t = 1, 2, \ldots$]. Articles by Granger and Newbold (1974), Chan et al.

(1977), Nelson and Kang (1981, 1984), and Durlauf and Phillips (1988) show the disastrous consequences of such a misspecification; further comment on this is deferred until Section 5.4.

In conclusion, this section argues that disenchantment with universal kriging has been premature. First, an appropriately weighted fitting of the variogram estimator minimizes the bias effect and second, the kriging predictor typically uses the variogram model at small lags. However, the kriging variance can be susceptible to bias in the estimated variogram; further discussion is given in Section 5.3.

3.4.4 Bayesian Kriging

Recall, from the introductory comments made at the beginning of this chapter, that Bayesian principles can be used to justify modeling an unknown, deterministic process by a random process. The nonstationary-mean model

$$Z(\mathbf{s}) = \mu(\mathbf{s}) + \delta(\mathbf{s}), \qquad \mathbf{s} \in D,$$

where $\delta(\cdot)$ is a zero-mean intrinsically stationary random process, is useful for analyzing physical processes that are spatially heterogeneous.

Often the mean function $\mu(\cdot)$ is not known exactly, leading to the parameterization

$$\mu(\mathbf{s}) = \sum_{j=1}^{p+1} \beta_{j-1} f_{j-1}(\mathbf{s}), \qquad \mathbf{s} \in D, \tag{3.4.50}$$

which is an unknown linear combination of known functions. Thus far, in this section, it has been assumed that $\boldsymbol{\beta} \equiv (\beta_0, \dots, \beta_p)'$ is a vector of fixed but unknown parameters. An alternative approach is once again to express uncertainty (now in the large-scale variation) in a Bayesian way; that is, assume the parameter $\mu(\cdot)$ is a random process independent of $\delta(\cdot)$. Probably the most famous example of this is the Kalman-filter model in time (e.g., Meinhold and Singpurwalla, 1983), where the regression coefficients satisfy a dynamic first-order autoregressive process.

One could assume that $\mu(\cdot)$ is (intrinsically or second-order) stationary or that $\mu(\cdot)$ is a quite general random process with all second-order (prior) parameters known. Omre (1987) takes this latter approach and uses the term Bayesian kriging to describe optimal prediction of $Z(B)$ from data $Z(\mathbf{s}_1), \dots, Z(\mathbf{s}_n)$. The kriging equations are derived from expressions he gives for $E(Z(\mathbf{s}_i))$, $E(Z(B))$, $\mathrm{var}(Z(\mathbf{s}_i) - Z(\mathbf{s}_j))$, and $\mathrm{var}(Z(B) - Z(\mathbf{s}_i))$ in terms of the variogram of $\delta(\cdot)$ and the known mean and variogram functions of $\mu(\cdot)$. Finally, he gives an estimator for the variogram of $\delta(\cdot)$. (Nather, 1985, p. 119, also suggests a hierarchical Bayesian model, but without consideration of the estimation problem.)

The prior model could be specialized by assuming that $\mu(\cdot)$ has the linear form (3.4.50), where now $\boldsymbol{\beta}$ is a *random* vector; see Nather (1985, Chapter 8),

Kitanidis (1986), and Omre and Halvorsen (1989). (The latter show that simple kriging corresponds to a degenerate prior and universal kriging corresponds to a diffuse prior.) In this or the more general setting, one could take an empirical-Bayes approach (e.g., Morris, 1983) and try to *estimate* the parameters of the random process $\mu(\cdot)$ based on the (marginal) distribution of $Z(\mathbf{s}_1), \ldots, Z(\mathbf{s}_n)$. These could then be substituted into the Bayes predictor of $Z(B)$ to yield an *empirical Bayes predictor*.

Initially, one would want to assume that the covariances $\{\text{cov}(\delta(\mathbf{s}), \delta(\mathbf{u})): \mathbf{s}, \mathbf{u} \in D\}$ were known. However, the same (empirical) Bayes approach used for means might also be tried for covariances. In other words, a prior could be put on the space of all positive-definite functions $P_D \equiv \{\{C(\mathbf{s}, \mathbf{u}): \mathbf{s} \in D, \mathbf{u} \in D\}: C \text{ is positive-definite}\}$. (Recall, from Section 2.5, the definition of a positive-definite function.) Alternatively and more simply, a prior could be put on the parameters $\boldsymbol{\theta}$ of a covariance model $C(\mathbf{s}, \mathbf{u}; \boldsymbol{\theta})$. Interestingly, the conditional mean-squared prediction error $\text{var}(Z(\mathbf{s}_0)|\mathbf{Z})$ [of the optimal predictor $E(Z(\mathbf{s}_0)|\mathbf{Z})$] depends on $\mathbf{Z}$, even when the vector $(Z(\mathbf{s}_0), \mathbf{Z}')'$ is, conditional on $\boldsymbol{\theta}$, jointly Gaussian (e.g., Jewell, 1988).

Provided a convenient class of priors can be found, the optimal predictor $E(Z(\mathbf{s}_0)|\mathbf{Z})$ can (in principle) be computed, and would be superior to the universal-kriging predictor. Let $f(\cdot \mid \cdot)$ denote a generic density, whose arguments also represent the random quantities being considered. As well as $Z(\mathbf{s}_0)$ and $\mathbf{Z}$ being random, suppose that the model parameters $\boldsymbol{\theta}$ are random with prior $f(\boldsymbol{\theta})$, $\boldsymbol{\theta} \in \Theta$. Then the *predictive density* is

$$f(Z(\mathbf{s}_0)|\mathbf{Z}) = \int_{\Theta} f(Z(\mathbf{s}_0)|\mathbf{Z}, \boldsymbol{\theta}) f(\boldsymbol{\theta}|\mathbf{Z})\, d\boldsymbol{\theta}.$$

However, if the posterior probabilities $f(\boldsymbol{\theta}|\mathbf{Z})$ and $f(\boldsymbol{\theta}|Z(\mathbf{s}_0), \mathbf{Z})$ are easy to calculate, one might use instead an extension of the formula reported in Besag (1989a):

$$f(Z(\mathbf{s}_0)|\mathbf{Z}) = f(Z(\mathbf{s}_0)|\mathbf{Z}, \boldsymbol{\theta})\{f(\boldsymbol{\theta}|\mathbf{Z})/f(\boldsymbol{\theta}|Z(\mathbf{s}_0), \mathbf{Z})\}.$$

Notice that the first term on the right-hand side is the modeled conditional distribution of $Z(\mathbf{s}_0)$ given $\mathbf{Z}$. In spite of appearances, the right-hand side does *not* depend on $\boldsymbol{\theta}$, but only on prior parameters.

The prior could also be used as a way of quantifying the robustness of various kriging predictors (and their kriging variances) to misspecification of the covariance function. (Different approaches to this quantification are summarized in Section 5.3.) How stable are the kriging weights and kriging variance as $\boldsymbol{\theta}$ ranges over, say, the $100(1 - \alpha)\%$ region of highest prior density?

The discussion above allows $\boldsymbol{\theta}$ to be made up of mean, covariance, or any type of parameters. In general, different priors yield different Bayes (and empirical Bayes) predictors. Bayes predictors based on the Gaussian model are given by Kitanidis (1986).

3.4.5 Kriging Revisited

Kriging, or optimal spatial (linear) prediction, has been presented in terms of changing assumptions about the mean structure $\mu(\cdot) = E(Z(\cdot))$: Simple kriging refers to known mean $\mu(\cdot)$, ordinary kriging to unknown $\mu(\cdot) \equiv \mu$, and universal kriging to unknown $\boldsymbol{\beta}$ in $\mu(\cdot) \equiv \sum_{j=1}^{p+1} \beta_{j-1} f_{j-1}(\cdot)$. (The spatial-dependence structure has been assumed known.)

Assume (for the moment) that the covariance function $C(\cdot, \cdot)$, defined by

$$C(\mathbf{s}, \mathbf{u}) \equiv \operatorname{cov}(Z(\mathbf{s}), Z(\mathbf{u})), \qquad \mathbf{s}, \mathbf{u} \in D, \tag{3.4.51}$$

is known. This implies that the variogram function

$$2\gamma(\mathbf{s}, \mathbf{u}) \equiv \operatorname{var}(Z(\mathbf{s}) - Z(\mathbf{u})), \qquad \mathbf{s}, \mathbf{u} \in D, \tag{3.4.52}$$

is also known because

$$2\gamma(\mathbf{s}, \mathbf{u}) = v(\mathbf{s}) + v(\mathbf{u}) - 2C(\mathbf{s}, \mathbf{u}), \tag{3.4.53}$$

where the variance function $v(\cdot)$ is defined by

$$v(\mathbf{s}) \equiv C(\mathbf{s}, \mathbf{s}), \qquad \mathbf{s} \in D. \tag{3.4.54}$$

Optimal Linear Prediction

Write,

$$Z(\mathbf{s}) = \mu(\mathbf{s}) + \delta(\mathbf{s}), \qquad \mathbf{s} \in D,$$

where $\delta(\cdot)$ is a zero-mean random process with covariance function, variogram function, and variance function given by (3.4.51), (3.4.52), and (3.4.54), respectively. To predict $Z(B)$ from data $\mathbf{Z} = (Z(\mathbf{s}_1), \ldots, Z(\mathbf{s}_n))'$, consider, for the moment, heterogeneously linear predictors (e.g., Toutenburg, 1982, p. 141) of the form

$$\sum_{i=1}^{n} l_i Z(\mathbf{s}_i) + k. \tag{3.4.55}$$

The *best* (heterogeneously) linear predictor minimizes the mean-squared prediction error. Hence, find $l_1, \ldots, l_n$, and k that minimize

$$E\left(Z(B) - \sum_{i=1}^{n} l_i Z(\mathbf{s}_i) - k\right)^2. \tag{3.4.56}$$

It was demonstrated in the introduction to this chapter that the optimal (simple kriging) predictor is

$$Z^*(B) = \sum_{i=1}^{n} l_i Z(\mathbf{s}_i) + \mu(B) - \sum_{i=1}^{n} l_i \mu(\mathbf{s}_i), \tag{3.4.57}$$

where $\mathbf{l}' \equiv (l_1, \cdots, l_n) = \mathbf{c}'\Sigma^{-1}$, $\mathbf{c} \equiv (\int_B C(\mathbf{u}, \mathbf{s}_1)\, d\mathbf{u}/|B|, \ldots, \int_B C(\mathbf{u}, \mathbf{s}_n)\, d\mathbf{u}/|B|)'$, Σ is an $n \times n$ matrix whose (i, j)th element is $C(\mathbf{s}_i, \mathbf{s}_j)$, and $\mu(B) \equiv \int_B \mu(\mathbf{u})\, d\mathbf{u}/|B|$. When $\mu(\cdot)$ is in fact unknown, (3.4.57) is no longer a predictor.

General Linear Model

Henceforth, in this section, parameterize the mean function to be an unknown linear combination of known functions:

$$\mu(\mathbf{s}) = \sum_{j=1}^{p+1} \beta_{j-1} f_{j-1}(\mathbf{s}), \qquad \mathbf{s} \in D. \tag{3.4.58}$$

Then, using the notation of (3.4.3) and (3.4.4), the optimal "predictor" (3.4.57) can be written as

$$Z^*(B) = \mathbf{c}'\Sigma^{-1}\mathbf{Z} + (\mathbf{x} - X'\Sigma^{-1}\mathbf{c})'\boldsymbol{\beta}. \tag{3.4.59}$$

Now, because $\boldsymbol{\beta}$ is unknown, it is natural to estimate it using the data $\mathbf{Z}$, and the natural estimator to use is the generalized-least-squares estimator

$$\hat{\boldsymbol{\beta}}_{\text{gls}} = (X'\Sigma^{-1}X)^{-1} X'\Sigma^{-1}\mathbf{Z}. \tag{3.4.60}$$

It is a simple matter to show that

$$p^{\dagger}(\mathbf{Z}; B) \equiv \mathbf{c}'\Sigma^{-1}\mathbf{Z} + (\mathbf{x} - X'\Sigma^{-1}\mathbf{c})'\hat{\boldsymbol{\beta}}_{\text{gls}} \tag{3.4.61}$$

is unbiased and homogeneously linear (e.g., Toutenburg, 1982, p. 141). Furthermore, it minimizes $E(Z(B) - \sum_{i=1}^{n}\lambda_i Z(\mathbf{s}_i))^2$ over all unbiased homogeneously linear predictors $\sum_{i=1}^{n}\lambda_i Z(\mathbf{s}_i)$ (Goldberger, 1962). Hence, it is identical to the universal kriging predictor $\hat{Z}(B)$.

From (3.4.60) and (3.4.61), the optimal $\boldsymbol{\lambda}$ is

$$\boldsymbol{\lambda}' = \left\{\mathbf{c} + X(X'\Sigma^{-1}X)^{-1}(\mathbf{x} - X'\Sigma^{-1}\mathbf{c})\right\}'\Sigma^{-1}, \tag{3.4.62}$$

which was derived at the beginning of this chapter for the second-order stationary case. Moreover, the minimized value of $E(Z(B) - \sum_{i=1}^{n}\lambda_i Z(\mathbf{s}_i))^2$ is

$$\sigma_k^2(B) = C(B, B) - \boldsymbol{\lambda}'\mathbf{c} + \mathbf{m}'\mathbf{x},$$

where $\mathbf{m}' \equiv (\mathbf{x} - X'\Sigma^{-1}\mathbf{c})'(X'\Sigma^{-1}X)^{-1}$, $C(B, B) \equiv \int_B \int_B C(\mathbf{u}, \mathbf{v})\, d\mathbf{u}\, d\mathbf{v}/\,|B|^2$, and $\mathbf{x} \equiv (f_0(B), \cdots, f_p(B))'$.

Decomposition of Mean-Squared Prediction Error

Because $\hat{\boldsymbol{\beta}}_{gls}$ is unbiased, the unbiasedness of $\hat{Z}(B)$ given by (3.4.61) is automatically guaranteed:

$$E(\hat{Z}(B)) = \mathbf{c}'\Sigma^{-1}X\boldsymbol{\beta} + (\mathbf{x}' - \mathbf{c}'\Sigma^{-1}X)\boldsymbol{\beta} = \mathbf{x}'\boldsymbol{\beta} = E(Z(B)).$$

Thus, by substituting the best linear unbiased estimator of $\boldsymbol{\beta}$ into the simple kriging predictor, one obtains best homogeneous linear unbiased prediction. The following result shows that this is no accident.

Let $\bar{\boldsymbol{\beta}} \equiv A\mathbf{Z}$ denote any linear estimator of $\boldsymbol{\beta}$, and define the predictor

$$\begin{aligned}\bar{Z}(B) &\equiv \mathbf{c}'\Sigma^{-1}\mathbf{Z} + (\mathbf{x} - X'\Sigma^{-1}\mathbf{c})'\bar{\boldsymbol{\beta}} \\ &= Z^*(B) + (\mathbf{x} - X'\Sigma^{-1}\mathbf{c})'(A\mathbf{Z} - \boldsymbol{\beta}).\end{aligned} \tag{3.4.63}$$

Now, $\mathrm{cov}(Z(B) - Z^*(B), A\mathbf{Z}) = \mathbf{0}'$, which implies that

$$\begin{aligned}E(Z(B) - \bar{Z}(B))^2 &= E(Z(B) - Z^*(B))^2 \\ &\quad + (\mathbf{x} - X'\Sigma^{-1}\mathbf{c})'E\{(\bar{\boldsymbol{\beta}} - \boldsymbol{\beta})(\bar{\boldsymbol{\beta}} - \boldsymbol{\beta})'\}(\mathbf{x} - X'\Sigma^{-1}\mathbf{c}).\end{aligned} \tag{3.4.64}$$

The last expression, owing to Harville (1985), shows an attractive decomposition of the mean-squared prediction error. The first term is the simple kriging variance and the second term quantifies the precision of the estimator $\bar{\boldsymbol{\beta}}$. Clearly, among all *unbiased* estimators of $\boldsymbol{\beta}$, $\hat{\boldsymbol{\beta}}_{gls}$ is the most precise. This leaves open the possibility of a biased linear estimator of $\boldsymbol{\beta}$, yielding a biased linear predictor of $Z(B)$ that has smaller mean-squared prediction error than the kriging variance.

Optimal Prediction in Two Stages

The result (3.4.64) has a very useful interpretation. Linear prediction with an unknown mean can be seen as a *two-stage* procedure: First, perform optimal linear prediction with an assumed known mean and, second, estimate the unknown mean. Moreover, linear estimators of $\boldsymbol{\beta}$ that are not unbiased but improve mean-squared error can lead to biased predictors of $Z(B)$ with *smaller* mean-squared prediction error than the universal-kriging predictor (3.4.61); for example, Gotway and Cressie (1992) consider shrinkage prediction and ridge-regression prediction. Because unbiasedness is a global property that has little relevance to the local accuracy of a predictor, there should be little concern that (global) unbiasedness may be lost.

James–Stein Prediction

By giving up both linearity and unbiasedness, improvements in mean-squared prediction error can be obtained. Gotway and Cressie (1992) obtain a

James–Stein predictor that is better than the universal-kriging predictor (3.4.61) for *all* $\boldsymbol{\beta} \in \mathbb{R}^{p+1}$.

Suppose that, instead of predicting a single random variable $Z(B)$ or $Z(\mathbf{s}_0)$, a prediction of the $k \times 1$ random vector $\mathbf{Z}_0$ is needed, where

$$\mathbf{Z}_0 = X_0\boldsymbol{\beta} + \boldsymbol{\delta}_0.$$

To focus the discussion, assume $\mathbf{Z}_0 = (Z(\mathbf{s}_{0,1}), \ldots, Z(\mathbf{s}_{0,k}))'$; hence, X_0 is a $k \times (p+1)$ matrix whose (i, j)th element is $f_{j-1}(\mathbf{s}_{0,i})$ and $\boldsymbol{\delta}_0 \equiv (\delta(\mathbf{s}_{0,1}), \ldots, \delta(\mathbf{s}_{0,k}))'$. The predictor is to be based on the data $\mathbf{Z}$, which from (3.4.3) satisfies the general linear model $\mathbf{Z} = X\boldsymbol{\beta} + \boldsymbol{\delta}$.

Let $\mathbf{p}$ denote a generic predictor of $\mathbf{Z}_0$ and define the loss function as the sum of squared errors

$$L(\mathbf{Z}_0, \mathbf{p}) \equiv \sum_{i=1}^{k} \left(Z(\mathbf{s}_{0,i}) - p_i\right)^2.$$

Then the risk (or sum of mean-squared prediction errors) is

$$r(\mathbf{p}; \boldsymbol{\beta}) = E\big(L(\mathbf{Z}_0, \mathbf{p})\big).$$

Define the *James–Stein predictor* as

$$\mathbf{p}_{\mathrm{js}} \equiv \Sigma_{0Z}\Sigma_{ZZ}^{-1}\mathbf{Z} + \left(X_0 - \Sigma_{0Z}\Sigma_{ZZ}^{-1}X\right)\hat{\boldsymbol{\beta}}_{\mathrm{js}},$$

where

$$\operatorname{var}\left((\mathbf{Z}', \mathbf{Z}_0')'\right) \equiv \begin{bmatrix} \Sigma_{ZZ} & \Sigma_{Z0} \\ \Sigma_{0Z} & \Sigma_{00} \end{bmatrix},$$

$$\hat{\boldsymbol{\beta}}_{\mathrm{js}} \equiv \left[1 - \left\{b\left(\mathbf{Z} - X\hat{\boldsymbol{\beta}}_{\mathrm{gls}}\right)'\Sigma_{ZZ}^{-1}\left(\mathbf{Z} - X\hat{\boldsymbol{\beta}}_{\mathrm{gls}}\right)\right\} \Big/ \left\{\hat{\boldsymbol{\beta}}_{\mathrm{gls}}'X'\Sigma_{ZZ}^{-1}X\hat{\boldsymbol{\beta}}_{\mathrm{gls}}\right\}\right]\hat{\boldsymbol{\beta}}_{\mathrm{gls}},$$

$$\hat{\boldsymbol{\beta}}_{\mathrm{gls}} \equiv \left(X'\Sigma_{ZZ}^{-1}X\right)^{-1}X'\Sigma_{ZZ}^{-1}\mathbf{Z},$$

and b is a positive constant that controls the amount of shrinkage of $\hat{\boldsymbol{\beta}}_{\mathrm{js}}$ away from $\hat{\boldsymbol{\beta}}_{\mathrm{gls}}$ toward $\mathbf{0}$. Assuming data and predictand are jointly Gaussian, Gotway and Cressie (1992) give a range of values of the constant b for which $r(\mathbf{p}_{\mathrm{js}}; \boldsymbol{\beta})$ is *uniformly* (for all $\boldsymbol{\beta} \in \mathbb{R}^{p+1}$) smaller than $r(\mathbf{p}_{\mathrm{uk}}; \boldsymbol{\beta})$, where

$$\mathbf{p}_{\mathrm{uk}} \equiv \Sigma_{0Z}\Sigma_{ZZ}^{-1}\mathbf{Z} + \left(X_0 - \Sigma_{0Z}\Sigma_{ZZ}^{-1}X\right)\hat{\boldsymbol{\beta}}_{\mathrm{gls}},$$

is the universal-kriging predictor. These results can be generalized to the case where data and predictand are jointly elliptically symmetric.

Choosing a Kriging Neighborhood

For every block B to be predicted, $\hat{Z}(B)$ involves inversion of an $n \times n$ matrix. Can a neighborhood of B be chosen so that there is little increase in kriging variance when only data in this neighborhood are used in the kriging predictor? From (3.4.64), it is clear that

$$E\left(Z(B) - \bar{Z}(B)\right)^2 \geq E(Z(B) - Z^*(B))^2, \tag{3.4.65}$$

with equality when $\mathbf{l}'$ $(\equiv \mathbf{c}'\Sigma^{-1})$ in (3.4.57) satisfies

$$\mathbf{l}'X = \mathbf{x}'. \tag{3.4.66}$$

Thus, when a neighborhood around B is expanded until (3.4.66) holds, any predictor of the form of $\bar{Z}(B)$ [not just the universal-kriging predictor $\hat{Z}(B)$] has the same optimal mean-squared prediction error as the simple kriging predictor $Z^*(B)$. Hence, with choice of neighborhood where $\mathbf{c}'\Sigma^{-1}X \simeq \mathbf{x}'$, the universal-kriging predictor has mean-squared prediction error that is as small as that of the simple-kriging predictor.

It is not always possible to choose the neighborhood locations so that (3.4.66) holds. In this case, similar considerations to those given in Section 3.2.1, bring attention to making

$$\mathbf{m}'\mathbf{x} = \sum_{j=1}^{p+1} m_{j-1} f_{j-1}(B) = (\mathbf{x} - X'\Sigma^{-1}\mathbf{c})'\left(X'\Sigma^{-1}X\right)^{-1}\mathbf{x}$$

as small (possibly negative) as possible. Obviously, if there is no spatial dependence (corresponding to 100% nugget effect), $\mathbf{m}'\mathbf{x} = \mathbf{x}'(X'X)^{-1}\mathbf{x} \geq 0$; to reduce it appreciably, kriging neighborhoods have to be expanded considerably.

Another approach to choosing a kriging neighborhood is to view kriging as a two-stage problem. The estimation stage

$$\hat{\boldsymbol{\beta}}_{\text{gls}} = \left(X'\Sigma_n^{-1}X\right)^{-1}X'\Sigma_n^{-1}\mathbf{Z}_n \tag{3.4.67}$$

involves inversion of an $n \times n$ covariance matrix $\Sigma_n \equiv \text{var}(\mathbf{Z}_n)$ [and, far less consequentially, inversion of a $(p+1) \times (p+1)$ matrix] that needs to be carried out just once. Here, $\mathbf{Z}_n \equiv (Z(\mathbf{s}_1), \ldots, Z(\mathbf{s}_n))'$ is the full data vector.

Choice of a neighborhood of B now depends only on the n_k *simple*-kriging weights $\mathbf{l}' = \mathbf{c}'_{n_k}\Sigma_{n_k}^{-1}$, where n_k denotes the number of data locations in the neighborhood, $\mathbf{c}_{n_k}$ is the n_k-dimensional vector $\text{cov}(Z(B), \mathbf{Z}_{n_k})$, Σ_{n_k} is the $n_k \times n_k$ covariance matrix $\text{var}(\mathbf{Z}_{n_k})$, and $\mathbf{Z}_{n_k}$ is the vector of data in the neighborhood. As explained in Section 3.2.1, determination of the smallest neighborhood where all weights in $\mathbf{l}$ are consequential involves consideration of *partial* correlation coefficients of the data. Some models of spatial association work directly with the inverse covariance matrix (e.g., Gaussian Markov random fields, presented in Section 6.6), where the prediction neighborhoods are specified as part of the model.

Therefore, provided $\mathbf{c}'_{n_k}\Sigma_{n_k}^{-1}$ contributes the *only* nonzero elements of the vector $\mathbf{c}'_n\Sigma_n^{-1}$, the optimal kriging predictor is given by

$$\hat{Z}(B) = \mathbf{c}'_{n_k}\Sigma_{n_k}^{-1}\mathbf{Z}_{n_k} + \left(\mathbf{x} - X'_{n_k}\Sigma_{n_k}^{-1}\mathbf{c}_{n_k}\right)'\hat{\boldsymbol{\beta}}_{\text{gls}}, \tag{3.4.68}$$

where $\hat{\boldsymbol{\beta}}_{\text{gls}}$ is given by (3.4.67) and uses the whole data vector $\mathbf{Z}_n$, and X_{n_k} is the $n_k \times (p+1)$ vector such that $E(\mathbf{Z}_{n_k}) = X_{n_k}\boldsymbol{\beta}$. For different B, the neighborhood may change, but in each case one is inverting a matrix of much smaller order than $n \times n$; the $n \times n$ inversion in $\hat{\boldsymbol{\beta}}_{\text{gls}}$ is carried out just once. When all elements of $\mathbf{c}'_n\Sigma_n^{-1}$ are nonzero, small ones could be set to zero or a selection criterion such as AIC or BIC (Schwarz, 1978) might be used to choose an optimal neighborhood.

The final result (3.4.68) shows $\hat{Z}(B)$ to depend on *every* datum in $\mathbf{Z}_n$, but the data in $\mathbf{Z}_{n_k}$ play the most important role in the predictor.

Kriging as an Orthogonal Projection

Return to consideration of the full-data predictor $\hat{Z}(B)$ given by (3.4.61). It has been shown that the simple-kriging predictor $Z^*(B)$ satisfies

$$\operatorname{cov}(Z(B) - Z^*(B), A\mathbf{Z}) = \mathbf{0}', \tag{3.4.69}$$

and hence the simple-kriging prediction error is orthogonal to the subspace spanned by the data $\{Z(\mathbf{s}_1), \ldots, Z(\mathbf{s}_n)\}$. Whereas simple kriging *is* an orthogonal projection, the decomposition (3.4.63) shows that the universal-kriging predictor satisfies

$$\operatorname{cov}\left(Z(B) - \hat{Z}(B), A\mathbf{Z}\right) = (\mathbf{x} - X'\Sigma^{-1}\mathbf{c})'(X'\Sigma^{-1}X)^{-1}X'A', \tag{3.4.70}$$

where A is any $k \times n$ matrix. Choice of $A = \boldsymbol{\lambda}'$ given by (3.4.62) yields $\operatorname{cov}(Z(B) - \hat{Z}(B), \hat{Z}(B))$, which is *not* in general equal to zero.

Thus, unless $\mathbf{c}'\Sigma^{-1}X = \mathbf{x}'$, the universal-kriging prediction error is generally correlated with the universal-kriging predictor. The lack of orthogonality is a reminder once again that even when $Z(\cdot)$ is a Gaussian random process, $\hat{Z}(B) \neq E(Z(B)|\mathbf{Z})$.

However, the prediction error $Z(B) - \hat{Z}(B)$ *is* orthogonal to any $A\mathbf{Z}$, where the row vectors of A are in the orthogonal complement of the column space of X. For example, if $X = \mathbf{1}$, the $n \times 1$ vector of 1s, then $Z(B) - \hat{Z}(B)$ is orthogonal to $Z(\mathbf{s}_0) - \bar{Z}$; see Journel (1977, p. 570).

Homogeneous and Heterogeneous Linear Predictors

Mention was made earlier in this section that the simple-kriging predictor is *heterogeneously* linear, namely, linear in $(\mathbf{Z}', 1)'$, and the ordinary (and universal) kriging predictor is *homogeneously* linear, namely, linear in $\mathbf{Z}$. These notions can be found in Toutenburg (1982, p. 141).

In fact, one could consider various linear predictors, with or without constraints, specifically:

Heterogeneous, Linear. $p_1(\mathbf{Z}; \mathbf{s}_0) = \mathbf{l}'\mathbf{Z} + l_0 1$.

Homogeneous, Linear. $p_2(\mathbf{Z}; \mathbf{s}_0) = \mathbf{l}'\mathbf{Z}$.

Homogeneous, Linear, Weakly Unbiased. $p_3(\mathbf{Z}; \mathbf{s}_0) = \mathbf{l}'\mathbf{Z}$ subject to $\mathbf{l}'X\boldsymbol{\beta} = \mathbf{x}'\boldsymbol{\beta}$.

Homogeneous, Linear, Uniformly Unbiased. $p_4(\mathbf{Z}; \mathbf{s}_0) = \mathbf{l}'\mathbf{Z}$ subject to $\mathbf{l}'X = \mathbf{x}'$.

Then, finding the best predictors within each of these classes yields corresponding mean-squared prediction errors that order smallest (best p_1) to largest (best p_4). In fact, the first mean-squared prediction error corresponds to the *simple*-kriging variance and the fourth mean-squared prediction error corresponds to the *universal*-kriging variance.

Because the best p_1, p_2, and p_3 *all* require knowledge of the mean $\mu(\cdot)$, there seems no reason why one would use p_2 or p_3 (the best p_1 always has the smallest mean-squared prediction error). When the mean is unknown, only p_4 can be used. Perhaps there are situations where partial knowledge of $\mu(\cdot)$ (e.g., in the form of a prior; see Section 3.4.4) or robustness considerations would make the best p_2 or the best p_3 a desirable predictor.

Alternative Formulations of Universal Kriging

For the reader who remains unconvinced that $\boldsymbol{\lambda}$ defined (in terms of covariances) by (3.4.62) does indeed yield the universal-kriging predictor $\boldsymbol{\lambda}'\mathbf{Z}$ given by (3.4.15), consider (3.4.8). Assuming that $\sum_{i=1}^{n}\lambda_i = 1$ is one of the unbiasedness conditions $\boldsymbol{\lambda}'X = \mathbf{x}'$, it can be seen from an equation analogous to (3.4.8) that

$$E\left(Z(B) - \sum_{i=1}^{n}\lambda_i Z(\mathbf{s}_i)\right)^2 = -\sum_{i=1}^{n}\sum_{j=1}^{n}\lambda_i\lambda_j\gamma(\mathbf{s}_i, \mathbf{s}_j) + 2\sum_{i=1}^{n}\lambda_i\gamma(B, \mathbf{s}_i)$$
$$-\int_B\int_B \gamma(\mathbf{u}, \mathbf{v})\, d\mathbf{u}\, d\mathbf{v}/|B|^2, \qquad (3.4.71)$$

where $\gamma(\mathbf{u}, \mathbf{v})$ is given by (3.4.53) and is not necessarily a function of $\mathbf{u} - \mathbf{v}$.

Using the covariance formulation (3.4.62), rather than the variogram formulation (3.4.15), has some advantages when thinking of universal kriging as a two-stage procedure. At the second stage, the generalized-least-squares estimator of $\boldsymbol{\beta}$ *cannot* generally be expressed in terms of Γ alone.

A further formulation is obtained by assuming *only* that $\sum_{i=1}^{n}\lambda_i = 1$. Then it is easy to show that

$$E\left(Z(B) - \sum_{i=1}^{n}\lambda_i Z(\mathbf{s}_i)\right)^2 = -\sum_{i=1}^{n}\sum_{j=1}^{n}\lambda_i\lambda_j E\left((Z(\mathbf{s}_i) - Z(\mathbf{s}_j))^2/2\right) + 2\sum_{i=1}^{n}\lambda_i \int_B E\left((Z(\mathbf{u}) - Z(\mathbf{s}_i))^2/2\right) d\mathbf{u}/|B| - \int_B\int_B E\left((Z(\mathbf{u}) - Z(\mathbf{v}))^2/2\right) d\mathbf{u}\, d\mathbf{v}/|B|^2. \tag{3.4.72}$$

Therefore, optimal homogeneous linear prediction can be expressed in terms of $\{E(Z(\mathbf{u}) - Z(\mathbf{v}))^2 \colon \mathbf{u}, \mathbf{v} \in \mathbb{R}^d\}$ and requires only $\sum_{i=1}^{n}\lambda_i = 1$ rather than $\boldsymbol{\lambda}'X = \mathbf{x}'$. The analogous expression for (3.4.72) in terms of cross-products is

$$E\left(Z(B) - \sum_{i=1}^{n}\lambda_i Z(\mathbf{s}_i)\right)^2 = \sum_{i=1}^{n}\sum_{j=1}^{n}\lambda_i\lambda_j E\left(Z(\mathbf{s}_i)\cdot Z(\mathbf{s}_j)\right) - 2\sum_{i=1}^{n}\lambda_i \int_B E\left(Z(\mathbf{u})\cdot Z(\mathbf{s}_i)\right) d\mathbf{u}/|B| + \int_B\int_B E\left(Z(\mathbf{u})\cdot Z(\mathbf{v})\right) d\mathbf{u}\, d\mathbf{v}/|B|^2. \tag{3.4.73}$$

Dual Formulation of Kriging

As has been shown in the previous paragraphs, there are two equivalent ways to write the universal-kriging predictor of $Z(B)$,

$$\hat{Z}(B) = \boldsymbol{\lambda}'\mathbf{Z};$$

$\boldsymbol{\lambda}$ is given either by (3.4.62) (covariance formulation) or by (3.4.15) (variogram formulation). The equation (3.4.15) for $\boldsymbol{\lambda}$ was in fact obtained by a constrained minimization that yielded the equation (3.4.11):

$$(\lambda_1, \ldots, \lambda_n, m_0, \ldots, m_p)' = \Gamma_U^{-1}\boldsymbol{\gamma}_U,$$

where $\boldsymbol{\gamma}_U$ and Γ_U are given by (3.4.13) and (3.4.14) [notice that $\gamma(\mathbf{s}_i, \mathbf{s}_j)$ can replace $\gamma(\mathbf{s}_i - \mathbf{s}_j)$ in these equations]. The same constrained minimization can be expressed in terms of covariances, yielding

$$(\lambda_1, \ldots, \lambda_n, -m_0, \ldots, -m_p)' = \Sigma_U^{-1}\mathbf{c}_U,$$

where $\mathbf{c}_U$ and Σ_U are the same as (3.4.13) and (3.4.14), respectively, but with $\gamma(\mathbf{s} - \mathbf{u})$ replaced with $C(\mathbf{s}, \mathbf{u})$.

Then,

$$\begin{aligned}\hat{Z}(B) &= \mathbf{Z}'_U \Sigma_U^{-1} \mathbf{c}_U \\ &= \mathbf{v}'_U \mathbf{c}_U \equiv \mathbf{v}'_1 \mathbf{c} + \mathbf{v}'_2 \mathbf{x},\end{aligned} \tag{3.4.74}$$

where $\mathbf{v}_U \equiv \Sigma_U^{-1} \mathbf{Z}_U$ and $\mathbf{Z}_U \equiv (Z(\mathbf{s}_1), \dots, Z(\mathbf{s}_n), 0, \dots, 0)'$, which is an $(n + p + 1) \times 1$ vector. In (3.4.74), write $\mathbf{v}'_U$ as $(\mathbf{v}'_1, \mathbf{v}'_2)$, so that $\mathbf{v}_1$ is $n \times 1$ and $\mathbf{v}_2$ is $(p + 1) \times 1$. Thus, the universal-kriging predictor is given by (3.4.74), where

$$\begin{aligned}\Sigma \mathbf{v}_1 + X \mathbf{v}_2 &= \mathbf{Z}, \\ X' \mathbf{v}_1 \qquad &= \mathbf{0}.\end{aligned} \tag{3.4.75}$$

Equations (3.4.74) and (3.4.75) are known as the *dual-kriging equations*. Similarly, write

$$\hat{Z}(B) = \mathbf{w}'_U \boldsymbol{\gamma}_U \equiv \mathbf{w}'_1 \boldsymbol{\gamma} + \mathbf{w}'_2 \mathbf{x}, \tag{3.4.76}$$

where $\mathbf{w}'_U \equiv (\mathbf{w}'_1, \mathbf{w}'_2)$ solves

$$\begin{aligned}\Gamma \mathbf{w}_1 + X \mathbf{w}_2 &= \mathbf{Z}, \\ X' \mathbf{w}_1 \qquad &= \mathbf{0};\end{aligned} \tag{3.4.77}$$

(3.4.76) and (3.4.77) are also known as the *dual-kriging equations*.

Kriging and Splines

There is a formal connection between these two very important methods of spatial prediction, but there is a large divergence in how they are applied and how their results are interpreted. The comparison between kriging and splines presented here is based on Cressie (1990a).

Dual-kriging equations (3.4.77) are identical in form to smoothing-spline equations presented, for example, in Wahba and Wendelberger (1980) and Section 5.9.2. When the variogram in Γ is replaced with a radial basis function, and the elements of X are polynomials in $\mathbf{s} = (s_1, \dots, s_d)'$, $\mathbf{s} \in \mathbb{R}^d$, the solution of (3.4.77) is a thin-plate spline.

Section 5.4 shows how universal kriging extends to kriging with intrinsic random functions of order k. Instead of using a variogram to evaluate the entries of Γ in (3.4.77), a generalized covariance function of order k can be used without affecting the validity of the dual-kriging equations (Matheron, 1973; Delfiner, 1976). It is precisely here where kriging and splines have their point of contact, because all the commonly used spline basis functions are valid generalized covariances (Micchelli, 1986).

On a theoretical level, the two methods are closely linked through a Bayesian analysis, as was demonstrated by Kimeldorf and Wahba (1970). Other demonstrations of the connection between kriging and splines can be found in Matheron (1981), Salkauskas (1982), and Watson (1984).

For example, consider the two-dimensional thin-plate Laplacian smoothing spline of degree 2:

$$Z^{\&}(\mathbf{s}_0) = \sum_{i=1}^{n} b_i e(\mathbf{s}_0 - \mathbf{s}_i) + a_0 + a_1 x_0 + a_2 y_0, \qquad \mathbf{s}_0 = (x_0, y_0)', \tag{3.4.78}$$

where $e(\mathbf{h}) \equiv \|\mathbf{h}\|^2 \log(\|\mathbf{h}\|^2)/(16\pi)$. In (3.4.78), $\mathbf{a} \equiv (a_0, a_1, a_2)'$ and $\mathbf{b} \equiv (b_1, \ldots, b_n)'$ solve

$$\begin{aligned} (K + n\rho I)\mathbf{b} + X\mathbf{a} &= \mathbf{Z}, \\ X'\mathbf{b} &= \mathbf{0}. \end{aligned} \tag{3.4.79}$$

In (3.4.79), K is the $n \times n$ matrix with (i, j)th entry $e(\mathbf{s}_i - \mathbf{s}_j)$, X is the $n \times 3$ matrix with ith row $(1, x_i, y_i)$, $\mathbf{s}_i = (x_i, y_i)'$, and $0 \le \rho \le \infty$. Comparison of (3.4.79) with (3.4.77) reveals the identical form of kriging and splines mentioned earlier.

From the results of Duchon (1977), the spline $Z^{\&}(\cdot)$ minimizes

$$\sum_{i=1}^{n} \{Z(\mathbf{s}_i) - g(\mathbf{Z}; \mathbf{s}_i)\}^2/n + \rho \int\int \left\{ \left(\frac{\partial^2 g}{\partial x_0^2}\right)^2 + 2\left(\frac{\partial^2 g}{\partial x_0 \, \partial y_0}\right)^2 + \left(\frac{\partial^2 g}{\partial y_0^2}\right)^2 \right\} dx_0 \, dy_0 \tag{3.4.80}$$

with respect to the smoother g. Notice that the preceding criterion is nonstochastic [The general case of a d-dimensional Laplacian smoothing spline of degree m is discussed in Wahba (1990, Section 2.4). In spite of the extra generality, the basic prediction equation (3.4.79) remains unchanged.]

There is just one spline parameter ρ in (3.4.79) that is yet to be specified, and this controls the amount of smoothing of the spline. From a kriging point of view, $n\rho$ is proportional to the nugget effect, which here is made up entirely of measurement error (Section 3.2.1); that is, $c_0 = c_{ME} \propto n\rho$. The goal is to predict the smooth process $S(\cdot)$ from data $\mathbf{Z}$, where $Z(\cdot) = S(\cdot) + \epsilon(\cdot)$ and $\epsilon(\cdot)$ is a white-noise process with $\text{var}(\epsilon(\mathbf{s})) = c_0$.

Notice that in (3.4.78) the covariation of the smooth process $S(\cdot)$ is *determined* by the generalized covariance (Section 5.4)

$$e(\mathbf{h}) = (c_0/n\rho)\|\mathbf{h}\|^2 \log(\|\mathbf{h}\|^2)/(16\pi),$$

which may be in conflict with the actual covariation. Cressie (1987, 1988c) gives a diagnostic procedure for assessing the goodness-of-fit of $e(\cdot)$ (or, for that matter, any other prespecified generalized covariance function) to spatial or temporal data. A strength of the geostatistical approach is that an initial spatial data analysis fits a *model* of the covariation that includes spline basis functions as possible choices. That is, the data suggest the matrix K in (3.4.79), rather than the "spliner" making an *ad hoc* choice.

The spline approach estimates the parameter ρ from cross-validation, which is not in conflict with the geostatistical approach that often uses a type of cross-validation, informally, to check the fit of the model obtained by other means (e.g., Sections 2.6.1 and 2.6.2). Craven and Wahba (1979) developed the generalized cross-validation (GCV) method of estimating ρ, and Wahba (1985) gives results for the consistency of the minimum GCV smoothing spline under mild regularity conditions on the data points and assuming that $Z(\mathbf{s}_1),\ldots,Z(\mathbf{s}_n)$ *are uncorrelated*. Diggle and Hutchinson (1989) show that, in the presence of autocorrelation, the GCV-based smoothing spline is inconsistent. Thus, the spline technology is built for, and is justified by, surfaces that are deterministic or deterministic plus white noise.

There is a fundamental difference between "krigers" and "spliners," which is manifest in how they report their results. Only the spline $\{Z^{\&}(\mathbf{s}_0)$: $\mathbf{s}_0 \in D\}$ is given (through formula or graphic display), without any *local* measure of precision associated with each $Z^{\&}(\mathbf{s}_0)$. In contrast, at the very foundation of kriging is the minimization of a local mean-squared prediction error, whose minimized value (kriging variance) is given by (3.4.16). Thus, accompanying the kriging predictor $\hat{Z}(\mathbf{s}_0)$ is a measure of its uncertainty $\sigma_k^2(\mathbf{s}_0)$, $\mathbf{s}_0 \in D$.

One exception to the essentially deterministic treatment given to splines can be found in Wahba (1983). Adopting the Bayesian approach referred to earlier, she reports posterior prediction intervals that are obtained from a particular prior generated by integrated Brownian motion plus random polynomials. Wahba's prior was chosen because it yielded the required posterior mean $\{Z^{\&}(\mathbf{s}_0)$: $\mathbf{s}_0 \in D\}$, but so do other intrinsic random functions. Unfortunately, these different priors will generally lead to *different* posterior variances, which are the basis of her proposed confidence intervals. Wahba's simulations appear to show accurate intervals because she simulates with additive Gaussian *white noise*.

Finally, the advantage kriging has over splines, in terms of change of support, will be addressed. Equations (3.2.18) and (3.2.19) show how kriging can be adapted easily to handle prediction of $Z(B)$ from data $Z(B_1),\ldots,Z(B_n)$. Then the effect of varying B is easily assessed from the mean-squared prediction error. Wahba (1981) gives splines for predicting $Z(\mathbf{s}_0)$ based on data $Z(B_1),\ldots,Z(B_n)$, but does not give an associated measure of precision nor does she consider the prediction of $Z(B)$ and the effect of varying B.

In conclusion, kriging and splines are formally alike, but practically very different. Both disciplines can benefit from each other's knowledge base. For example, cosplining, formally equivalent to cokriging (Section 3.2.3), could be a way of obtaining a smooth surface based on multivariate spatial data. Kriging can benefit from the stable numerical algorithms that have been developed to solve equations like (3.4.77) (e.g., Wahba, 1990, p. 13). Kriging under inequality constraints (i.e., the predictor belongs to intervals at certain spatial locations) has been implemented from the dual-kriging equations and existing spline results (such as can be found in Laurent, 1972, Chapter 9); see, e.g., Dubrule and Kostov (1986). Nevertheless, between the two methods, my preference is with kriging's obligatory spatial-dependence assessment and its automatic calculation of mean-squared prediction errors (for possibly different supports).

3.5 MEDIAN-POLISH KRIGING

Spatial data can be thought of as a partial sampling of a realization of a random process $\{Z(\mathbf{s}): \mathbf{s} \in D\}$. Recall from (3.1.4) the decomposition

$$Z(\mathbf{s}) = \mu(\mathbf{s}) + \delta(\mathbf{s}), \qquad \mathbf{s} \in D, \tag{3.5.1}$$

where $\mu(\cdot) \equiv E(Z(\cdot))$ is the mean structure and $\delta(\cdot)$ is the error structure. If $\mu(\cdot)$ is known, (3.4.57) gives the optimal linear (or simple kriging) predictor of $Z(\mathbf{s}_0)$ as

$$Z^*(\mathbf{s}_0) = \mu(\mathbf{s}_0) + \sum_{i=1}^{n} l_i(Z(\mathbf{s}_i) - \mu(\mathbf{s}_i)), \tag{3.5.2}$$

where

$$\mathbf{l}' \equiv (l_1,\ldots,l_n) = \mathbf{c}'\Sigma^{-1}, \tag{3.5.3}$$

$$\mathbf{c} \equiv (C(\mathbf{s}_0,\mathbf{s}_1),\ldots,C(\mathbf{s}_0,\mathbf{s}_n))', \tag{3.5.4}$$

$$\Sigma \text{ has } (i,j)\text{th element } C(\mathbf{s}_i,\mathbf{s}_j), \qquad i,j = 1,\ldots,n, \tag{3.5.5}$$

and recall $C(\mathbf{s},\mathbf{u}) \equiv \mathrm{cov}(Z(\mathbf{s}), Z(\mathbf{u}))$.

In reality, $\mu(\cdot)$ is not known. In dimensions higher than one, it is natural to assume $\mu(\cdot)$ decomposes additively into directional components. Here, in

$\mathbb{R}^2$, assume

$$\mu(\mathbf{s}) = a + c(x) + r(y), \qquad \mathbf{s} = (x, y)' \in D. \tag{3.5.6}$$

Furthermore, if $\{\mathbf{s}_i: i = 1, \ldots, n\}$ are actually on a grid $\{(x_l, y_k)': k = 1, \ldots, p;$ $l = 1, \ldots, q\}$, then, in obvious notation, $\mathbf{s}_i = (x_l, y_k)'$ implies that

$$\mu(\mathbf{s}_i) = a + r_k + c_l. \tag{3.5.7}$$

Thus, the row effect r_k can be estimated by exploiting replication in the other dimension; that is, r_k can be estimated from $\{Z(\mathbf{s}_i)$: 2nd coordinate of $\mathbf{s}_i$ is y_k; $i = 1, \ldots, n\}$, where $k = 1, \ldots, p$. Similar considerations allow the column effect c_l to be estimated, where $l = 1, \ldots, q$.

3.5.1 Gridded Data

Gridded spatial data in $\mathbb{R}^2$ can be viewed as a two-way (or higher-way, in $\mathbb{R}^d$) table. Notice that the grid spacings do not have to be equal in either the horizontal direction or the vertical direction. Miller and Kahn (1962, p. 411) propose a formal two-way analysis of variance and claim to test for nonstationarity by performing the within-rows and within-columns F tests. Unfortunately, the tests based on critical values obtained from the usual F tables are incorrect because the data are correlated. Cressie (1984a, 1986) takes a less formal approach, using resistant methods to achieve the decomposition (3.5.1).

Analysis by Means

Assuming precisely one observation at each grid node, ordinary-least-squares (o.l.s) estimators are

$$\begin{aligned} \hat{a} &= \sum_{i=1}^{n} Z(\mathbf{s}_i)/n, \\ \hat{r}_k &= \left\{ \sum_{N(y_k)} Z(\mathbf{s}_i)/q \right\} - \hat{a}, \qquad k = 1, \ldots, p, \\ \hat{c}_l &= \left\{ \sum_{M(x_l)} Z(\mathbf{s}_i)/p \right\} - \hat{a}, \qquad l = 1, \ldots, q, \end{aligned} \tag{3.5.8}$$

where

$$\begin{aligned} N(y_k) &\equiv \{i: \mathbf{s}_i = (\cdot, y_k)'; i = 1, \ldots, n\}, \\ M(x_l) &\equiv \{i: \mathbf{s}_i = (x_l, \cdot)'; i = 1, \ldots, n\}. \end{aligned} \tag{3.5.9}$$

Then, the o.l.s. estimate of $\mu(\mathbf{s}_i)$ is

$$\hat{\mu}(\mathbf{s}_i) = \hat{a} + \hat{r}_k + \hat{c}_l, \qquad \mathbf{s}_i = (x_l, y_k)'. \tag{3.5.10}$$

For $\mathbf{s} = (x, y)'$ in the region bounded by lines joining the four nodes, $(x_l, y_k)'$, $(x_{l+1}, y_k)'$, $(x_l, y_{k+1})'$, and $(x_{l+1}, y_{k+1})'$, where $x_l < x_{l+1}$ and $y_k < y_{k+1}$, define the planar interpolant

$$\begin{aligned}\hat{\mu}(\mathbf{s}) \equiv \hat{a} + \hat{r}_k + &\left(\frac{y - y_k}{y_{k+1} - y_k}\right)(\hat{r}_{k+1} - \hat{r}_k) + \hat{c}_l \\ &+ \left(\frac{x - x_l}{x_{l+1} - x_l}\right)(\hat{c}_{l+1} - \hat{c}_l), \\ &k = 1, \ldots, p-1,\ l = 1, \ldots, q-1.\end{aligned} \tag{3.5.11}$$

The advantages of (3.5.11) as an estimate of $\mu(\cdot)$ are that it is nonparametric (i.e., linear or quadratic trend is not imposed), spatially self-scaling (Section 3.1), and continuous. Its disadvantage is that the residuals $\{Z(\mathbf{s}_i) - \hat{\mu}(\mathbf{s}_i): i = 1, \ldots, n\}$ yield *biased* estimators of the unknown spatial dependence in the error process $\delta(\cdot)$. From the discussion in Section 3.4.3, it is seen that the $p + q - 1$ linear constraints that the residuals must satisfy lead to biased variogram estimators.

The implication of this discussion is that a sensible nonlinear estimator of a, $\{r_k\}$, and $\{c_l\}$ should remove this disadvantage. When the distribution of the error process is symmetric, $E(\text{ave}\{Z(\mathbf{s}_i): i \in A\}) = E(\text{med}\{Z(\mathbf{s}_i): i \in A\})$, where $\text{ave}\{\cdot\}$ denotes the average and $\text{med}\{\cdot\}$ denotes the median over the set of values in $\{\cdot\}$. Further, $\text{med}\{Z(\mathbf{s}_i): i \in A\}$ has attractive outlier-resistance properties (see Section 2.2). So, provided the data are transformed to approximate symmetry, which incidentally makes additivity in (3.5.6) and homoskedasticity more likely (Cressie, 1985b, p. 697), a median analogue to (3.5.8) may yield less-biased residuals (less biased in their estimation of the second-order spatial dependence structure). Section 3.5.4 shows that it does, in $\mathbb{R}^1$. In $\mathbb{R}^2$, *median polish* (Tukey, 1977) allows estimation of the additive effects given by (3.5.7).

The approach taken in this section involves two stages. First, the mean structure is estimated and removed. Then, the spatial-dependence structure is estimated. The procedure could be repeated, where more efficient estimation of the mean structure is implemented the second time around, based on the estimated spatial dependence. Indeed, the two stages could be iterated until there was little change in the results. In problems involving heteroskedastic rather than correlated data, Carroll et al. (1988) demonstrate that only two or three iterations are needed, particularly if a robust estimator of the mean structure is used. Here, the mean structure is estimated robustly by median polish, and only one iteration is used.

Median Polish

The median-polish algorithm that produces the all effect $\tilde{a}$, the row effects $\{\tilde{r}_k: k = 1,\ldots,p\}$, and the column effects $\{\tilde{c}_l: l = 1,\ldots,q\}$ from a $p \times q$ array of numbers $\{Y_{kl}: k = 1,\ldots,p;\ l = 1,\ldots,q\}$, will now be given. In the spatial context, the gridded data $\{Z(\mathbf{s}_i): i = 1,\ldots,n\}$ will play the role of the Ys. At the end, I shall mention how to handle the case where the number of Ys at the (k,l)th node is not always one, but may be zero, one, two, and so on. Emerson and Hoaglin (1983) is a good source for more details on the median-polish algorithm.

Median-Polish Algorithm

For $i = 1,3,5,\ldots,$ define

$$
\begin{aligned}
Y_{kl}^{(i)} &\equiv Y_{kl}^{(i-1)} - \operatorname{med}\{Y_{kl}^{(i-1)}: l = 1,\ldots,q\}, \qquad k = 1,\ldots,p+1,\ l = 1,\ldots,q,\\
Y_{k,q+1}^{(i)} &\equiv Y_{k,q+1}^{(i-1)} + \operatorname{med}\{Y_{kl}^{(i-1)}: l = 1,\ldots,q\}, \qquad k = 1,\ldots,p+1
\end{aligned}
\tag{3.5.12}
$$

and for $i = 2,4,6,\ldots,$ define

$$
\begin{aligned}
Y_{kl}^{(i)} &\equiv Y_{kl}^{(i-1)} - \operatorname{med}\{Y_{kl}^{(i-1)}: k = 1,\ldots,p\}, \qquad k = 1,\ldots,p,\ l = 1,\ldots,q+1,\\
Y_{p+1,l}^{(i)} &\equiv Y_{p+1,l}^{(i-1)} + \operatorname{med}\{Y_{kl}^{(i-1)}: k = 1,\ldots,p\}, \qquad l = 1,\ldots,q+1,
\end{aligned}
\tag{3.5.13}
$$

where $\operatorname{med}\{y_1,\ldots,y_n\}$ is the median of $y_1,\ldots,y_n$ (defined as a number such that half the $y_1,\ldots,y_n$ are to the left of it and half are to the right). To start the algorithm, assume

$$
Y_{kl}^{(0)} = \begin{cases} Y_{kl}, & k = 1,\ldots,p,\ l = 1,\ldots,q,\\ 0, & \text{elsewhere.} \end{cases}
\tag{3.5.14}
$$

In words, start with the $p \cdot q$ data and create $p + q + 1$ extra cells with zero in them. Use (3.5.12) to remove row medians from the data and accumulate the amounts removed in the p extra row cells. Do the same to the columns of the table, removing column medians from not only the data but also the column of accumulated row removals. This last amount removed goes into the extra $(p + 1, q + 1)$th cell. Repeat the process until convergence. Assuming convergence, the estimated effects are

$$
\begin{aligned}
\tilde{a} &\equiv Y_{p+1,q+1}^{(\infty)},\\
\tilde{r}_k &\equiv Y_{k,q+1}^{(\infty)}, \qquad k = 1,\ldots,p,\\
\tilde{c}_l &\equiv Y_{p+1,l}^{(\infty)}, \qquad l = 1,\ldots,q,
\end{aligned}
\tag{3.5.15}
$$

with the property that

$$Y_{kl} = \tilde{a} + \tilde{r}_k + \tilde{c}_l + Y_{kl}^{(\infty)}, \qquad k = 1, \dots, p, l = 1, \dots, q. \quad (3.5.16)$$

Thus, the original $p \times q$ table is replaced with a $p \times q$ table of residuals $\{Y_{kl}^{(\infty)}: k = 1, \dots, p; l = 1, \dots, q\}$, and the $p + q + 1$ extra cells contain the row effects $\{\tilde{r}_k\}$, the column effects $\{\tilde{c}_l\}$, and the all effect $\tilde{a}$.

Should there be an unequal number of observations at each grid node, the only modification needed is notational. The algorithm is unchanged in that it successively removes row medians and column medians from entries in the table. If a whole row (column) of grid nodes has no observations, then that row (column) is ignored. Figure 2.10 shows the results of median polishing the data in Figure 2.2*a*.

The algorithm, as described, starts with row removal, but of course it could have started with column removal. In practice, a stopping criterion needs to be used; for example, when another iteration leaves each entry of the table unchanged within a prespecified tolerance ϵ, median polishing is terminated and the algorithm is said to have converged.

Other Types of Polishing

Tukey originally proposed the median-polish algorithm as a quick, easy, and resistant alternative to a two-way analysis by means, such that the decomposition (3.5.16) is preserved. Notice that, if med$\{\cdot\}$ in (3.5.12) and (3.5.13) is replaced with the arithmetic average ave$\{\cdot\}$, the resulting *mean-polish* algorithm yields o.l.s. estimates (3.5.8). Clearly, any other (weighted) averaging operation, such as the trimmed mean (see, e.g., Hampel et al., 1986, p. 79), will yield its own version of the decomposition (3.5.16).

Properties of Median Polish

Intuitively, the main effects given by (3.5.15) should be like the minimum L_1-norm quantities that minimize

$$\sum_{k=1}^{p} \sum_{l=1}^{q} |Y_{kl} - a - r_k - c_l|. \quad (3.5.17)$$

In fact, it is not hard to see that each iteration of the median polish yields all, row, and column effects that *cannot* increase the L_1 norm (3.5.17) from its value at the previous iteration. However, it is not always true that the median-polish algorithm converges. Moreover, assuming that it does converge, it need not be to an L_1 solution.

Kemperman (1984) considers which pairs of (p, q) will necessarily give convergence to an L_1 solution when the median-polish procedure is applied. Sposito (1987) presents a sufficient condition, based on Kuhn–Tucker optimality conditions, to determine when a median-polish solution is equivalent to the L_1 solution. If, instead of averaging the high median and the low median (in the case where the median of an even number of observations is taken), one decides to take always one or the other, then Fink (1988) shows that the median polish of commensurable data (e.g., data to a fixed number of decimals) always converges. (The high and low median of an even number of observations $y_1 \le y_2 \le \cdots \le y_{2m}$ are defined, respectively, as y_{m+1} and y_m.)

Of course, one could forget about median polish and simply minimize (3.5.17). But the complexities of this minimization and the desire to have an algorithm that any scientist could program, led me to choose median polish as a resistant way of representing the large-scale variation (spatial trend).

Median smoothers have some precedence in the analysis of (one-dimensional) time series and (two-dimensional) image data. They are usually of a moving-window type, taking observations that are temporally or spatially close to allow sufficient "replication." However, such smoothers (see, e.g., Justusson, 1981; Tyan, 1981; Liao et al., 1985) are not based on any special mean structure like (3.5.7).

Median Polish Applied to the Coal-Ash Data

The coal-ash data presented in Section 2.2 occur at locations on an approximately square grid. Figure 2.2*a* shows that not every node of the 23 × 16 grid has a datum. The results of median polish are shown in Figure 2.10. Thus, the data can be expressed as

$$Z(\mathbf{s}_i) = \tilde{a} + \tilde{r}_k + \tilde{c}_l + R(\mathbf{s}_i), \qquad \mathbf{s}_i = (x_l, y_k)', \tag{3.5.18}$$

where $\tilde{\mu}(\mathbf{s}_i) \equiv \tilde{a} + \tilde{r}_k + \tilde{c}_l$ is a flexible estimator of the mean surface that is resistant to outliers (in particular, there is almost no leakage of outliers to other parts of the table). The additivity in $\tilde{\mu}(\mathbf{s}_i)$ allows interpolating planes to be defined uniquely. Use $\tilde{a}$, $\tilde{r}_k$, and $\tilde{c}_l$ in (3.5.11) in place of $\hat{a}$, $\hat{r}_k$, and $\hat{c}_l$ to define the surface

$$\tilde{\mu}(\mathbf{s}) = \tilde{a} + \tilde{r}_k + \left(\frac{y - y_k}{y_{k+1} - y_k}\right)(\tilde{r}_{k+1} - \tilde{r}_k) + \tilde{c}_l + \left(\frac{x - x_l}{x_{l+1} - x_l}\right)(\tilde{c}_{l+1} - \tilde{c}_l), \tag{3.5.19}$$

where $\mathbf{s} = (x, y)'$, $x_l \le x \le x_{l+1}$, and $y_k \le y \le y_{k+1}$, $k = 1, \ldots, 22$, $l = 1, \ldots, 15$. Figure 3.9 shows the surface $\tilde{\mu}(\cdot)$.

Extrapolation Using Median Polish

It is possible also to extrapolate beyond the grid of observations. Suppose $x < x_1$, but $y_1 \le y_k \le y \le y_{k+1} \le y_{23}$. Then, for $\mathbf{s} = (x, y)'$, define

$$\tilde{\mu}(\mathbf{s}) \equiv \tilde{a} + \tilde{r}_k + \left(\frac{y - y_k}{y_{k+1} - y_k}\right)(\tilde{r}_{k+1} - \tilde{r}_k) + \tilde{c}_1 + \left(\frac{x - x_1}{x_2 - x_1}\right)(\tilde{c}_2 - \tilde{c}_1).$$

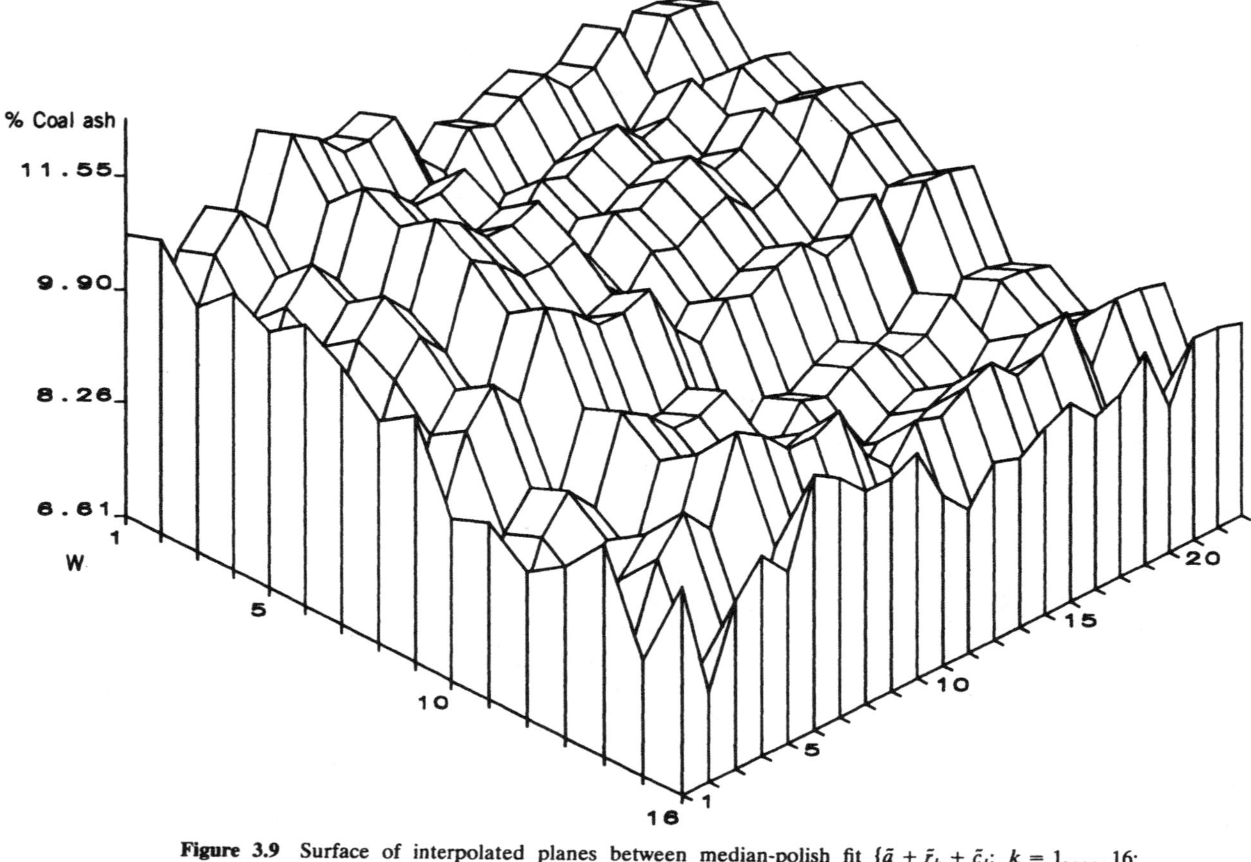

Figure 3.9 Surface of interpolated planes between median-polish fit $\{\tilde{a} + \tilde{r}_k + \tilde{c}_l:\ k = 1, \ldots, 16;\ l = 1, \ldots, 23\}$ for the coal-ash data.

A similar formula is obtained when y is out of range of the y_is or when both x and y are. Therefore, through interpolation and extrapolation, the median-polish surface, $\{(\mathbf{s}, \tilde{\mu}(\mathbf{s})): \mathbf{s} \in \mathbb{R}^2\}$, for all of $\mathbb{R}^2$ can be defined.

Nonadditivity

Median-polish removal of trend depends on the orientation of the grid. Although experimenters often choose their grids in meaningful directions (e.g., the geology dictates the orientation and spacing of the exploration drilling), a check should still be made whether the residuals contain cross-product trend or interaction (in the terminology of contingency tables). The simplest departure from (3.5.6) is through one extra parameter g that accounts for a quadratic term in the fit (Scheffe, 1959, p. 130). This is not the same as Tukey's (1949) 1 degree of freedom for nonadditivity. Define

$$\mu^{\#}(\mathbf{s}_i) \equiv \tilde{a} + \tilde{r}_k + \tilde{c}_l + g \cdot (x_l - \bar{x})(y_k - \bar{y}), \qquad \mathbf{s}_i = (x_l, y_k)', \quad (3.5.20)$$

where $\bar{x} = \text{ave}\{x_1, \dots, x_{16}\}$ and $\bar{y} = \text{ave}\{y_1, \dots, y_{23}\}$. By plotting the median-polish residuals $R(\mathbf{s}_i)$ [defined by (3.5.18)] versus $(x_l - \bar{x})(y_k - \bar{y})$, $i = 1, \dots, n$, a value of g can be diagnosed (Cressie, 1986). For the coal-ash data, the plot supported the additive fit (3.5.18) (i.e., $g = 0$). My experience has been that this is usually the case, provided the data also exhibit homoskedasticity and symmetry about $\mu(\cdot)$. The universal transformation principle (Cressie, 1985b) posits a transformation for which additivity at all scales can be achieved.

If g were found to be nonzero, it could be estimated and $\mu^{\#}(\cdot)$ used in place of $\tilde{\mu}(\cdot)$. Alternatively, on regular rectangular grids, median polish could be carried out first in the row and column directions, followed by a second implementation in the diagonal directions.

Kriging Based on the Residuals

Now think of the median-polish residuals $\{R(\mathbf{s}_i): i = 1, \dots, n\}$ [defined by (3.5.18)] as a new spatial data set, suitably *detrended* to allow *ordinary* kriging (Section 3.2) to be carried out. Time-series analysts take exactly the same approach before they fit stationary models to time series. Furthermore, Cleveland's (1979) robust locally weighted regression (*lowess*) in $\mathbb{R}^1$ is in the same spirit as median polish in $\mathbb{R}^d$, $d \geq 2$.

Cressie's (1986) spatial analysis of $\{R(\mathbf{s}_i): i = 1, \dots, n\}$, for the coal-ash data, yields the estimated variograms shown in Figure 2.12. Upon fitting a spherical model [see (2.3.8)] using weighted least squares, the fit

$$2\gamma^{o}(h; \hat{c}_0, \hat{c}_s, \hat{a}_s) = \begin{cases} 0 & h = 0, \\ 1.50 & h > 0, \end{cases} \quad (3.5.21)$$

is obtained. That is, after the median-polish effects are removed, nothing is left in the small-scale variation but white noise. (This is not a common consequence of median polishing, as is evidenced by the applications in Chapter 4.) More precisely, at the scale of 1 grid spacing $= 2500$ ft, no spatial dependence is detected. The real behavior of the microscale variation [defined in the decomposition (3.1.2)] may be far from white noise, but there is no way to tell without taking more observations closer together.

Based on the fitted variogram (3.5.21) and on the residuals data $\{R(\mathbf{s}_i): i = 1, \ldots, n\}$, the ordinary-kriging equations (3.2.5) and (3.2.12) yield

$$\hat{R}(\mathbf{s}_0) = \sum_{i=1}^{n} \lambda_i R(\mathbf{s}_i), \qquad \mathbf{s}_0 \in \mathbb{R}^2.$$

This kriging based on the residuals is meant to be a proxy for kriging based on the unknown errors $\{\delta(\mathbf{s}_i): i = 1, \ldots, n\}$.

Median-Polish Kriging

From Eq. (3.5.19) and associated extrapolation equations, the median-polish estimate $\tilde{\mu}(\mathbf{s}_0)$ can be defined for *all* $\mathbf{s}_0 \in \mathbb{R}^2$. Furthermore, kriging based on the residuals yields $\hat{R}(\mathbf{s}_0)$ for all $\mathbf{s}_0 \in \mathbb{R}^2$. Finally, then, the median-polish-kriging predictor of $Z(\mathbf{s}_0)$ is defined as

$$\tilde{Z}(\mathbf{s}_0) \equiv \tilde{\mu}(\mathbf{s}_0) + \hat{R}(\mathbf{s}_0), \qquad \mathbf{s}_0 \in \mathbb{R}^2.$$

Notice that $\tilde{Z}(\cdot)$ is an exact interpolator; that is, $\tilde{Z}(\mathbf{s}_i) = Z(\mathbf{s}_i)$, $i = 1, \ldots, n$. To predict $Z(B)$, simply use $\tilde{Z}(B) = \int_B \tilde{Z}(\mathbf{u})\, d\mathbf{u}/|B|$.

Comparing Median-Polish Kriging with Universal Kriging

Repeating the analysis given in Section 3.4.1 [namely, data editing of residuals, using kriging neighborhood (c) of Figure 3.6, and repredicting the 100 residuals for which neighborhood (c) was well defined], 100 values $\{\hat{R}_{-i}(\mathbf{s}_{0,i}): i = 1, \ldots, 100\}$ were computed, with kriging variance $\sigma_m^2 = 0.813(\%)^2$. Define

$$\tilde{Z}_{-i}(\mathbf{s}_{0,i}) \equiv \tilde{\mu}(\mathbf{s}_{0,i}) + \hat{R}_{-i}(\mathbf{s}_{0,i}), \qquad i = 1, \ldots, 100, \qquad (3.5.22)$$

where $\tilde{\mu}$ is given by (3.5.19).

A plot of $Z(\mathbf{s}_{0,i})$ versus $\tilde{Z}_{-i}(\mathbf{s}_{0,i})$ is shown in Figure 3.10; compare this with Figure 3.7a. It appears that the universal kriging of Section 3.4.1 yields predictors almost identical to those of median-polish kriging. This is reinforced by plotting $\hat{Z}_{-i}(\mathbf{s}_{0,i}) - \tilde{Z}_{-i}(\mathbf{s}_{0,i})$ versus $\tilde{Z}_{-i}(\mathbf{s}_{0,i})$; see Figure 3.11.

Median-Polish-Kriging Variance

The kriging variance associated with the median-polish-kriging predictor $\tilde{Z}(\mathbf{s}_0)$ is defined as the ordinary-kriging variance based on the median-polish residuals; call it $\sigma_m^2(\mathbf{s}_0)$. For example, for the coal-ash data, $\sigma_m^2(\mathbf{s}_0)$ is obtained from Eqs. (3.2.16) and (3.2.12), where the fitted residual variogram

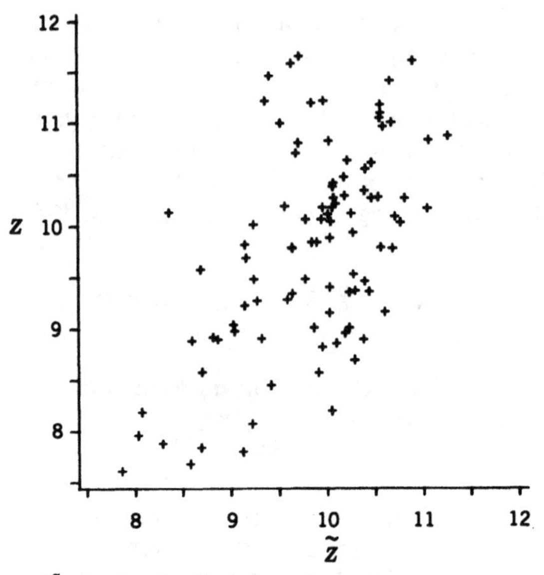

Figure 3.10 Plot of $(\tilde{Z}_{-i}(\mathbf{s}_{0,i}), Z(\mathbf{s}_{0,i}))$. Units on the axes are percent coal ash. [To avoid distortion, the outlying point (10.75, 17.61) is not shown.]

(3.5.21) is used in the equations. The kriging variance associated with $\tilde{Z}_{-i}(\mathbf{s}_{0,i})$ is 0.813, which is close to 1.016, the kriging variance associated with the universal-kriging predictor $\hat{Z}_{-i}(\mathbf{s}_{0,i})$. The difference can be accounted for by the fewer parameters fitted by universal kriging, leaving residuals that are more variable.

Section 3.5.3 justifies using $\sigma_m^2(\mathbf{s}_0)$ as the median-polish kriging variance.

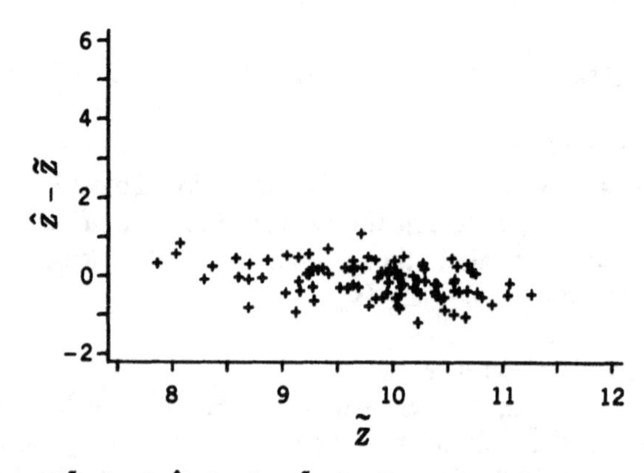

Figure 3.11 Plot of $(\tilde{Z}_{-i}(\mathbf{s}_{0,i}), \hat{Z}_{-i}(\mathbf{s}_{0,i}) - \tilde{Z}_{-i}(\mathbf{s}_{0,i}))$. Units on the axes are percent coal ash. (All 100 points are shown.)

Kriging the Coal-Ash Data: Some Conclusions

For the coal-ash data, it is possible to carry out universal kriging, because there is a trend-free direction that allows estimation of the variogram (although an assumption of isotropy is needed). This section shows that, for the coal-ash data, resistant detrending via median polish gives almost identical predictors and a slightly smaller estimate of mean-squared prediction error (Cressie, 1986, shows the same result for an iron-ore data set).

3.5.2 Nongridded Data

By nongridded data, I mean data whose spatial locations are not designed so that their x or y coordinates are confined to just a few values. Notice that gridded data might have grid lines of nonconstant spacing but are still perfectly amenable to the analyses described in Section 3.5.1.

Low-Resolution Map

Because the median-polish algorithm yields main effects that model large-scale spatial variation, and because these main effects do not necessarily depend on the data's precise spatial locations, a natural way to extend the approach in Section 3.5.1 to nongridded data is to draw a low-resolution map of the spatial locations. The resolution of the spatial coordinates is often chosen in an *ad hoc* way so that each (x_l, y_k) combination has approximately one observation $Z(x_l, y_k)$ at (x_l, y_k). In practice, this is done by overlaying a grid onto the high-resolution map and assigning data locations $\{\mathbf{s}_i: i = 1, \ldots, n\}$ to the nearest nodes of the grid $\{(x_l, y_k)': k = 1, \ldots, p; l = 1, \ldots, q\}$. According to this resolution, $Z(\mathbf{s}_i)$ is written as $Z(x_l, y_k)$, say. Then, median polish is carried out on the data $\{Z(x_l, y_k)\}$.

The resolution of the grid and the grid orientation are choices to be made. Maintaining the same coordinate directions as the original map and obtaining approximately one observation per grid node have worked well on various problems. Sections 4.1, 4.4, and 4.6 all involve nongridded data to which low-resolution mapping has been applied.

Median-Polish Kriging

What should be done with the median-polish residuals? When the data were gridded, the residuals were thought of as a new data set. Ordinary kriging of these residuals yielded predicted values $\{\hat{R}(\mathbf{s}_0): \mathbf{s}_0 \in D\}$ and median-polish kriged values at the original, high resolution:

$$\tilde{Z}(\mathbf{s}_0) \equiv \tilde{\mu}(\mathbf{s}_0) + \hat{R}(\mathbf{s}_0), \qquad \mathbf{s}_0 \in D, \tag{3.5.23}$$

where $\tilde{\mu}(\cdot)$ is given by (3.5.19). Recall that, for the gridded case, $\tilde{Z}(\mathbf{s}_i) = Z(\mathbf{s}_i)$, $i = 1, \ldots, n$, which is the property of exact interpolation. Preservation of this property in the nongridded case is considered important.

The original nongridded data are $\{Z(\mathbf{s}_i): i = 1, \ldots, n\}$. Low-resolution mapping yields an estimate of the large-scale variation $\{\tilde{\mu}(\mathbf{s}): \mathbf{s} \in D\}$ given by (3.5.19). *Define*

$$R(\mathbf{s}_i) \equiv Z(\mathbf{s}_i) - \tilde{\mu}(\mathbf{s}_i), \qquad i = 1, \ldots, n.$$

Now analyze $\{R(\mathbf{s}_i): i = 1, \ldots, n\}$ as a new data set, from which ordinary kriging yields predicted values $\{\hat{R}(\mathbf{s}_0): \mathbf{s}_0 \in D\}$. Finally, use (3.5.23) to define $\{\tilde{Z}(\mathbf{s}_0): \mathbf{s}_0 \in D\}$; notice that the spatial predictor $\tilde{Z}(\cdot)$ is an exact interpolator through the original data. The Wolfcamp-aquifer data analyzed in Section 4.1 gives a good illustration of this method of spatial mapping, adapted for nongridded data.

Augmented Median Polish

J. K. Ord has suggested that one way to decrease the variability in the median-polish main effects is not only to assign $Z(\mathbf{s}_i)$ to the nearest node $(x_l, y_k)'$, but for $\mathbf{s}_i$ in the rectangle $\{(x, y)': x_{l-1} < x \leq x_l, \; y_{k+1} < y \leq y_k\}$, say, assign it to the other three nodes $(x_{l-1}, y_k)'$, $(x_{l-1}, y_{k+1})'$, and $(x_l, y_{k+1})'$, as well. Then there would be about four observations per grid node, to which median polish could be applied. A similar procedure could be carried out on gridded data. Properties of this augmented median polish have not been investigated.

3.5.3 Median Polishing Spatial Data: Inference Results

In Section 2.2, a section on exploratory spatial data analysis, $\tilde{\mu}(\mathbf{s})$ given by (3.5.19) is thought of as *defining* the large-scale variation. Here, think of (3.5.15) as providing *estimators* of the large-scale parameters a, $\{r_k\}$, and $\{c_l\}$ in the decomposition (3.5.7). What then are the distributional properties of these estimators?

Estimation of Main Effects

First, I shall assume that the median-polish estimators (3.5.15) minimize the L_1 norm (3.5.17); Section 3.5.1 has a discussion of when this occurs. Second, write the two-way decomposition (3.5.1) and (3.5.7) as the linear model

$$\mathbf{Z} = X\boldsymbol{\beta} + \boldsymbol{\delta}. \tag{3.5.24}$$

Then, assuming an underlying Gaussian process, the o.l.s. estimator $\hat{\boldsymbol{\beta}} = (X'X)^{-1}X'\mathbf{Z}$ satisfies

$$\sqrt{n}\,(\hat{\boldsymbol{\beta}} - \boldsymbol{\beta}) \,\dot{\sim}\, \mathrm{Gau}\big(\mathbf{0}, n(X'X)^{-1}X'\Sigma X(X'X)^{-1}\big),$$

where $\mathrm{var}(\mathbf{Z}) = \Sigma$. Borrowing from results of Koenker and Bassett (1978) and Babu (1989), it may be possible to obtain an asymptotically Gaussian limit for

$\sqrt{n}(\tilde{\boldsymbol{\beta}} - \boldsymbol{\beta})$, the normalized median-polish estimators. But, for several reasons, the result is not immediate: First, the data are spatially dependent and, second, the number of parameters in the linear model (3.5.7) increases with n.

Spatial Prediction

When spatial prediction is the primary goal, some simplification is possible. From (3.5.23),

$$\left(Z(\mathbf{s}_0) - \tilde{Z}(\mathbf{s}_0)\right) = \left(Z(\mathbf{s}_0) - \tilde{\mu}(\mathbf{s}_0) - \hat{R}(\mathbf{s}_0)\right)$$
$$= \left(R(\mathbf{s}_0) - \hat{R}(\mathbf{s}_0)\right),$$

and hence the mean-squared prediction error of the median-polish-kriging predictor $\tilde{Z}(\mathbf{s}_0)$ is

$$E\left(Z(\mathbf{s}_0) - \tilde{Z}(\mathbf{s}_0)\right)^2 = E\left(R(\mathbf{s}_0) - \hat{R}(\mathbf{s}_0)\right)^2. \tag{3.5.25}$$

Therefore, conditional on $\tilde{\mu}(\cdot)$, the kriging variance is

$$E\left\{\left(R(\mathbf{s}_0) - \hat{R}(\mathbf{s}_0)\right)^2 | \tilde{\mu}(\cdot)\right\} \equiv \sigma_m^2(\mathbf{s}_0),$$

which is an unbiased estimator of $E(Z(\mathbf{s}_0) - \tilde{Z}(\mathbf{s}_0))^2$.

The latter result relies on the belief that the residuals $\{R(\mathbf{s}_i): i = 1, \ldots, n\}$ can be thought of as coming from an intrinsically stationary process. The statistician's skill in stochastic modeling comes in choosing a detrending method for which this belief is tenable. Diagnostic checking of the residuals, as outlined in Section 2.2, and a general desire for the large-scale variation to be as parsimoniously specified as possible plays an important part in the choice.

Median-Polish Kriging in Higher Dimensions

Higher-dimensional median polish (see, e.g., Seheult and Tukey, 1982; Cook, 1985) could be used to solve higher-dimensional spatial prediction problems; the same reasons for which it was used in two dimensions will apply. The additivity assumption needs to be checked, and there is now room for two- and three-way interactions.

Weighted Median Polish

From the model (3.5.24), $\text{var}(\mathbf{Z}) = \Sigma$. It makes sense to transform:

$$\mathbf{W} \equiv \Sigma^{-1/2}\mathbf{Z} = (\Sigma^{-1/2}X)\boldsymbol{\beta} + \boldsymbol{\epsilon}, \tag{3.5.26}$$

where now $\text{var}(\boldsymbol{\epsilon}) = I$. Then, minimizing the L_1 norm

$$\sum_{i=1}^{n} |W_i - (\Sigma^{-1/2}X)_i \boldsymbol{\beta}|$$

would yield *generalized* robust estimators of $\boldsymbol{\beta}$. In practice, this rules out a simple approximating algorithm like median polish, except when Σ is diagonal; then, a *weighted median polish* can be performed by substituting "wt med" for "med" in (3.5.12) and (3.5.13). The weights used are proportional to $(C(\mathbf{s}_i, \mathbf{s}_i))^{-1/2}$; see Sections 6.2.3 and 7.5.1.

3.5.4 Median-Based Covariogram Estimators Are Less Biased

Detrending based on medians rather than means not only avoids leakage of outliers from error terms to unrelated residual terms (as happens when detrending is based on means), but it also avoids linear dependencies in the residuals. Of course, there are still dependencies in the median-based residuals (a number of them are identically zero), but they are of a nonlinear nature. Because the emphasis is on prediction and not on estimation, it is less of a concern that large-scale variation parameters are estimated less efficiently; see (3.5.25).

Dependencies in Residuals

I believe that median-based residuals offer a truer picture of the error correlation structure than do mean-based residuals. Even when $\text{var}(\mathbf{Z}) = \sigma^2 I$ in the model $\mathbf{Z} = X\boldsymbol{\beta} + \boldsymbol{\delta}$, the residuals $\mathbf{Z} - X(X'X)^{-1}X'\mathbf{Z}$ satisfy linear constraints that cause them to exhibit negative dependence. As evidence, I offer a calculation in $\mathbb{R}^1$. The same calculation is not as straightforward in higher dimensions, although I believe the essence of the result still to be true.

Estimating the Covariogram in $\mathbb{R}^1$

It is well known that the usual covariogram estimator of a stationary time series is biased (see, e.g., Fuller, 1976, p. 236). Suppose the second-order stationary time series $\{Z(i): i = 1, \ldots, n\}$ has unknown mean μ and unknown covariogram

$$C(h) = \text{cov}(Z(i), Z(i+h)), \qquad h = 0, 1, 2, \ldots, \tag{3.5.27}$$

independent of i. A frequently used estimator of (3.5.27) is based on its sample version, namely,

$$\bar{C}(h) = \frac{1}{n-h} \sum_{i=1}^{n-h} (Z(i) - \bar{Z})(Z(i+h) - \bar{Z}), \qquad h = 0, 1, 2, \ldots, \tag{3.5.28}$$

where $\bar{Z} = (1/n)\Sigma_{i=1}^{n} Z(i)$. Note that sometimes the divisor in (3.5.28) is n instead of $n - h$.

It is *not* true that $E(\bar{C}(h)) = C(h)$. However, provided the stationary time series has $\Sigma_{i=1}^{n} E\{Z(i+h)Z(i)Z(h+1)Z(1)\}/n$ converging to $C^2(h)$, as $n \to \infty$, it can be shown (Hannan, 1960, p. 32) that, in probability,

$$\bar{C}(h) \to C(h), \quad \text{as } n \to \infty. \tag{3.5.29}$$

Under assumptions made in the succeeding paragraph, this convergence in probability follows easily. The bias of the usual covariogram estimator is $E(\bar{C}(h)) - C(h)$, essentially due to the linear relation $\Sigma_{i=1}^{n}(Z(i) - \bar{Z}) = 0$ among the residuals. The improvement in bias that can be obtained through using the median-based covariogram estimator

$$\tilde{C}(h) \equiv \frac{1}{n-h} \sum_{i=1}^{n-h} (Z(i) - \text{med}(\mathbf{Z}))(Z(i+h) - \text{med}(\mathbf{Z})), \quad h = 0, 1, 2, \ldots, \tag{3.5.30}$$

where med($\mathbf{Z}$) is the median of the observed $Z(1), \ldots, Z(n)$, is investigated in what is to follow. The results are based on those presented in Cressie and Glonek (1984).

The Stationary Model

Assume that $\{Z(s): s = \ldots, -1, 0, 1, \ldots\}$ is a strictly stationary process on the integers in $\mathbb{R}^1$ (Section 2.3) with stationary univariate density $f(\cdot)$ that is symmetric about $\mu = E(Z(s))$, where $f(\mu) \neq 0$ and, for some $\delta > 0$, $f'(z)$ exists and is bounded on the interval $(\mu - \delta, \mu + \delta)$. Let the stationary covariogram $C(h)$ be given by (3.5.27) and assume $C(h) = 0$, for all $h > m$ (i.e., m-dependence).

To obtain a closed-form expression for the bias of $\tilde{C}(\cdot)$, assume further that the data are Gaussian or contaminated Gaussian:

$$Z(s) = S(s) + \epsilon(s), \qquad s = \ldots, -1, 0, 1, \ldots, \tag{3.5.31}$$

where $S(\cdot)$ is a stationary Gaussian process and, independently, $\epsilon(\cdot)$ is a white-noise process (i.i.d. mean zero) with

$$\epsilon(s) \sim \begin{cases} \text{Gau}(0, c_0), & \text{w.p. } 1 - \epsilon, \\ \text{Gau}(0, k^2 c_0), & \text{w.p. } \epsilon, 0 \leq \epsilon \leq 1, k^2 \geq 1. \end{cases} \tag{3.5.32}$$

This is precisely the additive outliers model used in Section 3.3. Putting $\epsilon = 0$ or $k^2 = 1$ in (3.5.32) gives uncontaminated Gaussian white noise. For most of the development that follows, the *uncontaminated* case will be considered.

Bias of the Mean-Based Estimator
Define

$$\bar{B}(h) \equiv \lim_{n\to\infty} \{nE(\bar{C}(h) - C(h))\}, \qquad h = 0, 1, 2, \ldots, \tag{3.5.33}$$

where $\bar{C}(h)$ is given by (3.5.28). Then, for a stationary m-dependent process,

$$\bar{B}(h) = -\left\{C(0) + 2\sum_{j=1}^{m} C(j)\right\}, \tag{3.5.34}$$

which does not depend on h; henceforth, write (3.5.34) as $\bar{B}$.

Bias of the Median-Based Estimator
Define

$$\tilde{B}(h) \equiv \lim_{n\to\infty} \{nE(\tilde{C}(h) - C(h))\}, \qquad h = 0, 1, 2, \ldots, \tag{3.5.35}$$

where $\tilde{C}(h)$ is given by (3.5.30). Its evaluation is trickier and involves writing med(**Z**) in an asymptotically equivalent form:

$$\tilde{Z} \equiv \mu + (1/n)\sum_{i=1}^{n} \operatorname{sgn}(Z(i) - \mu)/2f(\mu), \tag{3.5.36}$$

where $\operatorname{sgn}(z) = 1$ if $z > 0$, $= 0$ if $z = 0$, and $= -1$ if $z < 0$ [i.e., $\operatorname{sgn}(\cdot)$ is the sign function]. Then, results due to Sen (1972) can be invoked to conclude that under a stationary, uniform-mixing process (Ibragimov and Linnik, 1971, p. 312), $\tilde{Z} - \text{med}(\mathbf{Z}) = O(n^{-5/8}\log n)$, with probability 1. Further, $\sup\{\text{var}(n(\tilde{Z} - \text{med}(\mathbf{Z}))^2): n = 1, 2, \ldots\} < \infty$ implies uniform integrability of $\{n(\tilde{Z} - \text{med}(\mathbf{Z}))^2: n = 1, 2, \ldots\}$, which allows calculation of the bias $\tilde{B}(h)$. It is

$$\tilde{B}(h) = -\left\{(2 - (\pi/2))C(0) + 4\sum_{j=1}^{m} C(j) - 2C(0)\sum_{j=1}^{m} \arcsin(C(j)/C(0))\right\}, \tag{3.5.37}$$

under the more specific assumption of a stationary m-dependent *Gaussian* process (Cressie and Glonek, 1984). This also does not depend on h, and will henceforth be written as $\tilde{B}$.

Comparing the Biases

It is easy to show that, for $0 \leq x \leq 1$, $x \leq \arcsin x \leq (\pi/2)x$ and $(\pi/2)(-x) \leq \arcsin(-x) \leq (-x)$. Hence, from (3.5.37),

$$\begin{aligned}(4-\pi)\sum_{j\in P} C(j) + 2\sum_{j\in N} C(j) + C(0)(2-(\pi/2))\\ \leq -\tilde{B} \leq 2\sum_{j\in P} C(j) + (4-\pi)\sum_{j\in N} C(j)\\ + C(0)(2-(\pi/2)),\end{aligned} \qquad (3.5.38)$$

where $N \equiv \{j\colon C(j) < 0;\ j > 0\}$ and $P \equiv \{j\colon C(j) \geq 0;\ j > 0\}$. Thus, it is straightforward from (3.5.34) (the formula for $\bar{B}$) that

$$\begin{aligned}-(\bar{B}-\tilde{B}) &\geq \frac{\pi-2}{2}C(0) + (\pi-2)\sum_{j\in N} C(j),\\ -(\bar{B}+\tilde{B}) &\geq \frac{6-\pi}{2}C(0) + (6-\pi)\sum_{j\in P} C(j) + 4\sum_{j\in N} C(j).\end{aligned} \qquad (3.5.39)$$

Suppose for the moment that $\bar{B} \leq 0$. Then the condition $|\tilde{B}| \leq -\bar{B}$ is equivalent to $\bar{B} - \tilde{B} \leq 0$ and $\bar{B} + \tilde{B} \leq 0$. From (3.5.39), both these conditions are ensured by

$$-\left(\frac{6-\pi}{4}\right)C(0) \leq 2\sum_{j\in N} C(j) \leq 0, \qquad (3.5.40)$$

which incidentally ensures $\bar{B} \leq 0$. Thus, (3.5.40) implies

$$|\tilde{B}| < |\bar{B}|, \qquad h = 0,1,2,\ldots. \qquad (3.5.41)$$

Condition (3.5.40) is not at all strong, because in many applications the negative covariances are inconsequential in comparison to $C(0)$. In particular, when N is empty [true for many variogram models, such as those obtained from (2.3.7) through (2.3.10)], relation (3.5.40) is trivially satisfied. In practice, its validity can be checked by referring to the estimate $\{\tilde{C}(h)\colon h = 0,1,2,\ldots\}$.

The result of the preceding analysis is that, under the very mild condition (3.5.40) that bounds the negative covariances, the median-based estimator of $C(h)$ is less biased than the mean-based estimator, assuming a stationary m-dependent Gaussian process. Cressie and Glonek (1984) show that the *same* result holds true in the m-dependent *contaminated Gaussian* case [i.e., where $S(\cdot)$ in (3.5.31) is a stationary m-dependent Gaussian process independent of the contaminated Gaussian white-noise process (3.5.32)].

Some Final Remarks

In almost all estimation, there is a trade off between bias and variance; choice of interval widths in the classical histogram estimator of a density function is an important example of this trade-off. When sample size is small, the bias term dominates. We have addressed here the problem of covariogram estimation from the point of view of bias and have in mind situations where the sample size is small.

In the purely Gaussian m-dependent case, the condition (3.5.40) can be improved slightly. If the more liberal bound

$$\{(6-\pi)/(\pi-2)\}\bar{B} \leq 2\sum_{j\in N} C(j) \leq 0$$

is satisfied, then (3.5.41) holds.

These bias results can be generalized to stationary Gaussian autoregressive-moving-average (ARMA) processes and, in particular, to Gaussian autoregressive processes. Babu (1989) shows that for such (strong-mixing) processes, $\tilde{Z} - \text{med}(\mathbf{Z}) = O(n^{-5/8}(\log n)^{3/4})$, with probability 1. As an example, consider the AR(1) process

$$Z(i) = \rho Z(i-1) + \epsilon(i), \qquad |\rho| < 1.$$

This is a case where the covariogram, given by

$$C(h) = \rho^h/(1-\rho^2), \qquad h = 0, 1, 2, \ldots,$$

is highly oscillatory for $-1 < \rho < 0$. For $\rho \leq 0$, the condition (3.5.40) becomes $2|\rho|/(1-\rho^2)^2 \leq ((6-\pi)/4)/(1-\rho^2)$; that is, $-0.32 \leq \rho \leq 0$. The more liberal bound referred to in the preceding text translates into $-0.42 \leq \rho \leq 0$. For $\rho > 0$, (3.5.40) is trivially satisfied.

3.6 GEOSTATISTICAL DATA, SIMULATED AND REAL

Data sources are important for the development of statistical methods in a relatively young discipline such as geostatistics. Because much of the early research was related to mining, there was a dearth of data sets that were not proprietary. This is less true in recent times, as other disciplines (with a smaller profit motive) have found geostatistics to be of value.

If real data are not available, one could always simulate; indeed, validating a method on data where the true values of the parameters are known can be an important part of establishing its worth. Most of this section is concerned with methods for simulating spatial processes and their extension to *conditional simulation*. A discussion of real-data sources is left to Section 3.6.3 and Chapter 4.

3.6.1 Simulation of Spatial Processes

Recall that, for geostatistical data, the spatial index of the underlying process $\{Z(\mathbf{s}): \mathbf{s} \in D\}$ varies continuously over some d-dimensional subset D of $\mathbb{R}^d$. In actuality, a simulation of the whole process is usually carried out on a discrete (d-dimensional) grid and then approximated to D. Alternatively, one may only wish to simulate $\{Z(\mathbf{s}_1), \ldots, Z(\mathbf{s}_n)\}$ at n known (possibly irregularly spaced) locations $\{\mathbf{s}_1, \ldots, \mathbf{s}_n\}$, which could be done either directly or by choosing appropriate values from a simulation of the whole process. Furthermore, simulation methods for geostatistical data can often be used for simulating lattice data (Section 7.7).

In one dimension, dependent data are often modeled by autoregressive-moving-average processes of finite order (e.g., Box and Jenkins, 1970), which are easy to simulate. Although they are restrictive in the class of covariograms generated, their principal drawback is that they do not generalize naturally to higher dimensions (e.g., Sharp and Aroian, 1985).

Usually, spatial simulation methods are most appropriate for Gaussian processes, although it will be seen that sometimes small modifications can lead to more general random processes. The most obvious comes from transforming the Gaussian process $\{Z(\mathbf{s}): \mathbf{s} \in D\}$ to, for example, a lognormal process $\{\exp(Z(\mathbf{s})): \mathbf{s} \in D\}$ or a chi-squared process $\{(Z(\mathbf{s}))^2: \mathbf{s} \in D\}$, and so on.

A requirement for simulation of any process is that all parameters of the process must be specified. The source of random variation is usually a pseudo-random-number generator, and the choice of a good one for spatial simulation needs considerable care (e.g., Ripley, 1987a, Chapter 6).

Cholesky Decomposition Method

Suppose it is desired to simulate the process at n prespecified locations $\mathbf{s}_1, \ldots, \mathbf{s}_n$. These might form a dense regular grid over D (in which case n is typically very large) or they may be n (possibly irregular) sampling locations of interest.

Suppose the random process $\{Z(\mathbf{s}): \mathbf{s} \in D\}$ has mean

$$\mu(\mathbf{s}) \equiv E(Z(\mathbf{s})), \qquad \mathbf{s} \in D, \tag{3.6.1}$$

and covariance

$$C(\mathbf{s}, \mathbf{u}) \equiv \operatorname{cov}(Z(\mathbf{s}), Z(\mathbf{u})), \qquad \mathbf{s}, \mathbf{u} \in D. \tag{3.6.2}$$

Let

$$\mathbf{Z} \equiv (Z(\mathbf{s}_1), \ldots, Z(\mathbf{s}_n))' \tag{3.6.3}$$

denote the vector of values to be simulated. Then,

$$E(\mathbf{Z}) = (\mu(\mathbf{s}_1), \ldots, \mu(\mathbf{s}_n))' \equiv \boldsymbol{\mu} \tag{3.6.4}$$

and

$$\mathrm{var}(\mathbf{Z}) = \left(C(\mathbf{s}_i, \mathbf{s}_j)\right) \equiv \Sigma, \tag{3.6.5}$$

where Σ is an $n \times n$ symmetric positive-definite matrix whose (i, j)th element is $C(\mathbf{s}_i, \mathbf{s}_j)$.

The Cholesky decomposition (e.g., Golub and Van Loan, 1983, pp. 86–90) allows Σ to be decomposed as the matrix product

$$\Sigma = LL', \tag{3.6.6}$$

where L is a lower triangular $n \times n$ matrix (i.e., all elements above the diagonal are zero). The Cholesky decomposition is standard output from most computer packages with matrix operations.

Then $\mathbf{Z}$ can be simulated so that it satisfies (3.6.4) and (3.6.5) through the relation

$$\mathbf{Z} = \boldsymbol{\mu} + L\boldsymbol{\epsilon}, \tag{3.6.7}$$

where $\boldsymbol{\epsilon} \equiv (\epsilon(\mathbf{s}_1), \ldots, \epsilon(\mathbf{s}_n))'$ is a vector of uncorrelated random variables, each with zero mean and unit variance. Cressie and Laslett (1986), Quimby (1986), and Davis (1987a) suggested the use of (3.6.7) to simulate geostatistical data. A pseudo-random-number generator produces $\boldsymbol{\epsilon}$ cheaply and quickly, allowing fast (about half the entries of L are zero) simulation of many realizations of $\mathbf{Z}$. The only storage needed is of $\boldsymbol{\mu}$ and L, and the lower triangular nature of L gives fast access to $\Sigma^{-1} = (L^{-1})'L^{-1}$, should it be needed.

Usually the $\epsilon(\mathbf{s}_1), \ldots, \epsilon(\mathbf{s}_n)$ are i.i.d. unit Gaussian random variables, implying that $\mathbf{Z}$ is multivariate Gaussian (e.g., Ripley, 1981, p. 17), although they could be uncorrelated random variables, each with zero mean and unit variance, from any joint distribution. For example, consider the zero-mean, spherically symmetric random vector $\mathbf{Y}$ with characteristic function equal to $\psi(\mathbf{t}'\mathbf{t})$, for some function ψ. Then $\boldsymbol{\epsilon} \equiv \mathbf{Y}/\{-2\psi'(0)\}^{1/2}$ has mean $\mathbf{0}$ and variance matrix I (Muirhead, 1982, p. 47).

When n is larger than say, 1000, and Σ is sparse, various numerical inaccuracies may result from an uncritical use of the Cholesky decomposition method. An approximation that is numerically more stable is the lower-triangular approximation suggested by Quimby (1986, Chapter II). Let R denote the correlation matrix with (i, j)th element $C(\mathbf{s}_i, \mathbf{s}_j)/\{C(\mathbf{s}_i, \mathbf{s}_i)C(\mathbf{s}_j, \mathbf{s}_j)\}^{1/2}$. Write $R = M + I + M'$, where M has zeros *on and above* the diagonal. Then approximate R by $R^* \equiv (M + I)(M + I)'$; the error in the approximation is MM'. Quimby (1986) provides computer code to carry out this and other approximations.

Davis (1987b) gives a computationally efficient way of calculating the square root of Σ based on the spectral decomposition

$$\Sigma = Q\,\mathrm{diag}\{\lambda_1, \ldots, \lambda_n\}Q',$$

where $\lambda_1, \ldots, \lambda_n$ are the eigenvalues of Σ and Q is an orthogonal $n \times n$ matrix whose columns are eigenvectors of Σ. The idea is similar to the Cholesky decomposition method in that if a matrix $\Sigma^{1/2}$ can be found such that $\Sigma = \Sigma^{1/2}(\Sigma^{1/2})'$, then $\mathbf{Z} = \boldsymbol{\mu} + \Sigma^{1/2}\boldsymbol{\epsilon}$ will yield a simulated $\mathbf{Z}$ with the correct mean and variance–covariance structure. Such a matrix is

$$\Sigma^{1/2} = Q\,\mathrm{diag}\{\lambda_1^{1/2}, \ldots, \lambda_n^{1/2}\}Q'.$$

Davis (1987b) then gives a minimax polynomial approximation of degree 8 to $f(x) = x^{1/2}$ on $[0, \lambda_u]$, where $\max\{\lambda_i\} \le \lambda_u$. Hence $\Sigma^{1/2}\boldsymbol{\epsilon}$ can be approximated with a known linear combination of $\boldsymbol{\epsilon}, \Sigma\boldsymbol{\epsilon}, \Sigma^2\boldsymbol{\epsilon}, \ldots, \Sigma^8\boldsymbol{\epsilon}$, that does not actually require a spectral decomposition of Σ (although a bound on its largest eigenvalue is needed). The extension to simulating vector-valued data $\{\mathbf{Z}(\mathbf{s}_1), \ldots, \mathbf{Z}(\mathbf{s}_n)\}$ in this way is given by Myers (1989).

Spectral Methods

Suppose for the moment that $d = 1$ (generalizations to $d > 1$ will be given later) and that the process to be simulated is second-order stationary (Section 2.3.2) with zero mean and covariogram $C(h)$, where $C(0) \equiv \sigma^2 > 0$. A process with nonzero mean function $\mu(\cdot)$ is easily obtained by adding the required mean to a simulated zero-mean process. However, the second-order stationarity requirement makes the spectral method that follows less general than the Cholesky decomposition method.

From Section 2.5, if $\int_{-\infty}^{\infty}|C(h)|\,dh < \infty$, then,

$$C(h) = \int_{-\infty}^{\infty} \cos(\omega \cdot h)g(\omega)\,d\omega,$$

where $g(\cdot) \ge 0$ is called the spectral function. Also, $\int_{-\infty}^{\infty} g(\omega)\,d\omega = C(0) = \sigma^2$, the variance of $Z(s)$, and hence $g(\cdot)/\sigma^2$ is a density function. Shinozuka (1971) and Mejia and Rodriguez-Iturbe (1974) suggest the following formula to simulate the process. Define

$$Z_N(s) \equiv \sigma(2/N)^{1/2} \sum_{i=1}^{N} \cos(\omega_i s + \phi_i), \tag{3.6.8}$$

where $\omega_1, \ldots, \omega_N$ are i.i.d. from a distribution with density $g(\cdot)/\sigma^2$, independent from $\phi_1, \ldots, \phi_N$, i.i.d. uniform random variables on $[-\pi, \pi]$. Now,

$$E(Z_N(s)) = \{(2N)^{1/2}/(2\pi\sigma)\} \int_{-\infty}^{\infty} \int_{-\pi}^{\pi} \cos(\omega s + \phi) g(\omega)\, d\phi\, d\omega = 0,$$

because $\int_{-\pi}^{\pi} \cos\phi\, d\phi = \int_{-\pi}^{\pi} \sin\phi\, d\phi = 0$. Furthermore,

$$\begin{aligned} &E(Z_N(s+h) Z_N(s)) \\ &\quad = \{2\sigma^2/N\} \sum_{k=1}^{N} \sum_{l=1}^{N} E\{\cos(\omega_k s + \phi_k)\cos(\omega_l(s+h) + \phi_l)\} \\ &\quad = 2\sigma^2 E\{\cos(\omega_1 s + \phi_1)\cos(\omega_1(s+h) + \phi_1)\} \\ &\quad = \int_{-\infty}^{\infty} \cos(\omega \cdot h) g(\omega)\, d\omega = C(h). \end{aligned}$$

Regardless of the value of N, (3.6.8) is a random process (not Gaussian, in general) that has mean zero and covariogram $C(\cdot)$. As $N \to \infty$, $\{Z_N(s): s \in [a, b]\}$ converges weakly to $\{Z(s): s \in [a, b]\}$, a Gaussian process (e.g., Billingsley, 1968, p. 41ff), and the process (3.6.8) looks more and more ergodic (see Section 2.3, where ergodicity is discussed).

In higher dimensions, where the zero-mean process has covariogram $C(\mathbf{h})$ $[C(\mathbf{0}) = \sigma^2]$ with spectral function $g(\boldsymbol{\omega})$ satisfying

$$C(\mathbf{h}) = \int_{-\infty}^{\infty} \cdots \int_{-\infty}^{\infty} \cos(\boldsymbol{\omega}'\mathbf{h}) g(\boldsymbol{\omega})\, d\boldsymbol{\omega}, \tag{3.6.9}$$

the process (3.6.8) can be generalized to

$$Z_N(\mathbf{s}) \equiv \sigma(2/N)^{1/2} \sum_{i=1}^{N} \cos(\boldsymbol{\omega}_i'\mathbf{s} + \phi_i), \tag{3.6.10}$$

where $\boldsymbol{\omega}_1, \ldots, \boldsymbol{\omega}_N$ are i.i.d. from a distribution with density $g(\cdot)/\sigma^2$, independent from $\phi_1, \ldots, \phi_N$, i.i.d. uniform random variables on $[-\pi, \pi]$. As $N \to \infty$, (3.6.10) converges to a Gaussian ergodic process. Shinozuka (1971) also shows how a vector-valued process can be similarly simulated.

This method requires knowledge of the spectral density, from which i.i.d. observations are to be drawn. There are no large matrices to invert, and the process generated is not grid dependent and can be integrated (numerically or analytically) over different spatial supports B.

A second spectral method, suggested by Rice (1954, p. 180) and later in a geostatistical context by Borgman et al. (1984), discretizes the frequency domain and uses independent Gaussian random variables and the fast Fourier transform (FFT) to simulate Gaussian random processes on a finite grid. (Dudgeon and Mersereau, 1984, p. 76ff, give a clear explanation of the FFT in $\mathbb{R}^2$.) Shinozuka and Jan (1972) suggest essentially the same method and show that in $\mathbb{R}^1$ it is superior to (3.6.8) with regard to accuracy and cost.

The FFT method also offers a way to simulate non-Gaussian processes, as the following explanation (given in one dimension, for clarity) shows. Suppose $\{W(\omega): -\infty < \omega < \infty\}$ is a zero-mean (possibly complex-valued) random process of independent increments, where $E(|W(d\omega)|^2) = g(\omega)\,d\omega$. Then

$$Z(s) \equiv \int_{-\infty}^{\infty} e^{i\omega s} W(d\omega), \tag{3.6.11}$$

is a zero-mean random process with covariogram $E(Z(s+h)\cdot\bar{Z}(s)) = C(h) = \int_{-\infty}^{\infty}\cos(\omega\cdot h)g(\omega)\,d\omega$, where here $\bar{z}$ denotes the complex conjugate of z (see Section 2.5.1). Discretizing $\mathbb{R}^1$ and the frequency domain $[-\pi,\pi]$ leads to the following approximation of $Z(\cdot)$ on a grid with unit spacing:

$$Z_N(n) = \sum_{m=0}^{N-1} \exp(2\pi i n m/N) J_m, \qquad n = 0,\ldots,N-1, \tag{3.6.12}$$

where $\{J_m: m = 0,\ldots,N-1\}$ are independent, zero-mean, complex-valued random variables for which

$$\mathrm{var}(|J_m|) = (2\pi/N)\sum_{k=-\infty}^{\infty} g((2\pi m/N) + 2k\pi), \qquad m = 0,\ldots,N-1. \tag{3.6.13}$$

Recall that $g(\cdot)$ is the spectral function corresponding to the covariogram $C(\cdot)$ of the process $Z(\cdot)$. Writing $J_m = A_m - iB_m$, it is easily seen that Gaussian $\{A_m\}$ and $\{B_m\}$ yield a Gaussian process $Z_N(\cdot)$. However, a more general result is possible: By writing $J_m = |J_m|e^{i\Phi_m}$, it follows that the simulated values (3.6.12) yield (approximately, for large N) the required second-order structure as long as $\{J_m\}$ are independent, have zero mean, and satisfy (3.6.13). There is no other requirement on the distribution of $|J_m|$ and none at all on the distribution of Φ_m.

To simulate a Gaussian process, generate $|J_m|^2$ to be a scaled χ_2^2 random variable and Φ_m an independent uniform random variable on $[-\pi,\pi]$, independently for each $m = 0,\ldots,N-1$. Elliptically symmetric processes can be simulated from independent, bivariate spherically symmetric random variables (A_m, B_m). In this case, the $\{\Phi_m\}$ are still i.i.d. uniform (Muirhead, 1982, p. 36ff). Clearly, when the phase random variables $\{\Phi_m\}$ are nonuniform, there is information in the simulated data that a purely covariance-based analysis would miss (cf. Shurtz, 1985).

Turning-Bands Method

Of all the methods for Gaussian simulation so far proposed in this section, the spectral methods seem the most appealing. However, Mantoglou and Wilson (1982) demonstrate that the turning-bands method, due to Matheron (1973), converges to a stationary ergodic Gaussian process in $\mathbb{R}^d$, $d > 1$, faster than the FFT method.

The idea behind the turning-bands method is as follows. Suppose $\{Z(\mathbf{s}): \mathbf{s} \in D\}$ is a zero-mean, second-order stationary, isotropic process in $\mathbb{R}^d$, with covariogram $C_d(\mathbf{h}) \equiv C_d^o(\|\mathbf{h}\|)$. When $Z(\cdot)$ is projected onto a one-dimen- sional subspace defined by unit vector $\mathbf{e}$, the projected process $Y_{\mathbf{e}}(\cdot)$ is also second-order stationary with a covariance function $C_1(h)$. The relationship between $C_d^o(\cdot)$ and $C_1(\cdot)$ is given by

$$C_d^o(h) = \frac{2\Gamma(d/2)}{\pi^{1/2}\Gamma((d-1)/2)h} \int_0^h C_1(u)\left(1 - \frac{u^2}{h^2}\right)^{(d-3)/2} du.$$

In particular,

$$C_2^o(h) = (2/\pi)\int_0^h \left\{C_1(u)/(h^2 - u^2)^{1/2}\right\} du. \tag{3.6.14}$$

$$C_3^o(h) = (1/h)\int_0^h C_1(u)\, du. \tag{3.6.15}$$

The latter expression is much easier to solve; that is,

$$C_1(h) = d(hC_3^o(h))/dh. \tag{3.6.16}$$

The turning-bands simulation method reverses the order given in the preceding paragraph. Suppose that, on an arbitrary line of direction $\mathbf{e}$ in three-dimensional space, the process $\{Y_{\mathbf{e}}(t): -\infty < t < \infty\}$ with covariance satisfying (3.6.16) is simulated. This could be carried out by (3.6.7), (3.6.8), (3.6.12), or any other method the yields a zero-mean Gaussian process with covariance $C_1(h)$. for any $\mathbf{s} \in \mathbb{R}^3$, $\mathbf{s}'\mathbf{e}$ is the projection of $\mathbf{s}$ on a line of direction $\mathbf{e}$. Define

$$Z_{\mathbf{e}}(\mathbf{s}) \equiv Y_{\mathbf{e}}(\mathbf{s}'\mathbf{e}), \tag{3.6.17}$$

where $\mathbf{e}$ is a vector on the unit sphere.

Now consider N different unit vectors $\mathbf{e}_1, \ldots, \mathbf{e}_N$, uniformly distributed on the unit sphere in $\mathbb{R}^3$, and define

$$Z_N(\mathbf{s}) \equiv N^{-1/2} \sum_{i=1}^{N} Z_{\mathbf{e}_i}(\mathbf{s}), \qquad \mathbf{s} \in \mathbb{R}^3. \tag{3.6.18}$$

Then, for N large, the simulated process approximates the process $Z(\cdot)$ with covariance function $C_d^o(\cdot)$. Tompson et al. (1989) recommend using $N \geq 100$ to simulate problems characterized by full three-dimensionality.

Often, the 15 axes of a regular icosahedron, rather than uniformly distributed lines, are used in (3.6.18) (Journel and Huijbregts, 1978, p. 503), although this can introduce line-like patterns in the simulated process. The name "turning bands" comes from the bands that a simulation of $Y(\cdot)$ on $\mathbf{e}$ produces through the relation (3.6.17); these bands are then turned through angles to obtain (3.6.18).

Simulation in $\mathbb{R}^2$ looks more problematic because (3.6.14) is not easily inverted; an integral equation is given by Brooker (1985). Mantoglou and Wilson (1982) circumvent this by using a spectral method for simulating $Y(\cdot)$ and show that the spectral function $g_1(\cdot)$ from $C_1(\cdot)$ is one-half the radial spectral function $g_2^o(\cdot)$ from $C_2^o(\cdot)$. They find that the turning-bands method (with the one-dimensional simulations obtained by the FFT method) is as accurate and is much less expensive than multidimensional spectral methods. Mantoglou (1987) shows how it can be implemented for a nonisotropic (but still second-order stationary) covariogram, as does Christakos (1987), who uses a more general approach of Radon projections and space transformations. Also, Mantoglou (1987) gives a multivariate version of the turning-bands method.

Matheron (1973, Section 4) proposes a turning-bands method for simulating the more general intrinsically stationary processes (Sections 2.3 and 5.4) and Cressie (1987) shows how the Cholesky decomposition can be used for simulating such processes. Furthermore, from Section 2.5.2, it should be possible to develop a spectral method for simulating an intrinsic random function.

Conclusion

For nonstationary-covariance problems, where the number of simulated-data locations is fewer than 1000, the Cholesky-decomposition method can be used. For stationary-covariance problems, the turning-bands approach accompanied by the FFT spectral method for the one-dimensional component simulations is recommended.

3.6.2 Conditional Simulation

For the purposes of this section, a notation will be established to distinguish between a physical process $Z(\cdot)$, a (nonconditionally) simulated process $Z_{NS}(\cdot)$, and a conditionally simulated process $Z_{CS}(\cdot)$. Also, recall that $Z^*(\mathbf{s}_0)$, given by (3.4.57), is the simple-kriging predictor of $Z(\mathbf{s}_0)$ based on observations $\mathbf{Z} \equiv (Z(\mathbf{s}_1), \ldots, Z(\mathbf{s}_n))'$.

Should inference on the process proceed conditionally on the data (i.e., in Bayesian terminology, proceed according to the posterior distribution), then conditional properties of $Z(\cdot)$ may be needed. When these cannot be obtained analytically, conditional simulations may be a practical alternative. Even one conditional simulation may be useful, simply to obtain some idea of the amount of variability remaining in the model of the physical process after conditioning with respect to the observations (Journel, 1974).

The building blocks of a conditional simulation are the mean function $\mu(\cdot)$, the covariance function $C(\cdot, \cdot)$, and most importantly the data $\mathbf{Z}$. It is required that the conditionally simulated process $\{Z_{CS}(\mathbf{s}): \mathbf{s} \in D\}$ pass through the data $\mathbf{Z}$ [i.e., $Z_{CS}(\mathbf{s}_i) = Z(\mathbf{s}_i)$, $i = 1, \ldots, n$] and have *unconditional* mean $\mu(\cdot)$ and covariance $C(\cdot, \cdot)$. One might think initially that the kriging predictors $\{\hat{Z}(\mathbf{s}): \mathbf{s} \in D\}$ would satisfy the requirement because kriging does interpolate the data exactly and it is unbiased. However, it does not possess enough variability because it is a smoother as well as an interpolator and it is highly nonstationary.

Consider the decomposition of the process into the simple-kriging predictor and the residual:

$$Z(\mathbf{s}) = Z^*(\mathbf{s}) + (Z(\mathbf{s}) - Z^*(\mathbf{s})), \qquad \mathbf{s} \in D, \qquad (3.6.19)$$

where, by (3.4.69), the two components are orthogonal [i.e., $\text{cov}(Z^*(\mathbf{s}), Z(\mathbf{s}) - Z^*(\mathbf{s})) = 0$]. Now, replace the second component in (3.6.19) with $(Z_{NS}(\mathbf{s}) - Z^*_{NS}(\mathbf{s}))$, which is based on a nonconditional simulation of a process with mean zero and covariance $C(\cdot, \cdot)$. That is, define the *conditional simulation* (Journel, 1974) $Z_{CS}(\cdot)$ by

$$Z_{CS}(\mathbf{s}) \equiv Z^*(\mathbf{s}) + (Z_{NS}(\mathbf{s}) - Z^*_{NS}(\mathbf{s})), \qquad \mathbf{s} \in D, \qquad (3.6.20)$$

where $Z^*_{NS}(\mathbf{s})$ is the simple-kriging predictor whose formula is identical to that of $Z^*(\mathbf{s})$ given by (3.4.57), except that $Z(\mathbf{s}_i)$ is replaced with $Z_{NS}(\mathbf{s}_i)$, $i = 1, \ldots, n$. It is easily verified that the interpolation and mean requirements are satisfied. Furthermore, the orthogonality of the two components of (3.6.20) implies that $Z_{CS}(\cdot)$ has the same (unconditional) covariance as $Z(\cdot)$, namely $C(\cdot, \cdot)$. Notice from (3.6.20) that $E(Z_{CS}(\mathbf{s}) - Z(\mathbf{s}))^2$ is twice the simple-kriging variance at $\mathbf{s}$, reinforcing the idea that prediction and simulation address two different problems.

Journel and Huijbregts (1978, p. 494ff) define conditional simulation in terms of ordinary kriging $\hat{Z}(\mathbf{s}_0)$ (see Section 3.2) rather than simple kriging $Z^*(\mathbf{s}_0)$, in which case they do not have to know the mean (provided it is assumed to be constant). Although the variogram $2\gamma(\cdot, \cdot)$ of the conditional simulation based on $\hat{Z}(\cdot)$ will be the same as the variogram of $Z(\cdot)$, the covariances will not necessarily be the same.

Suppose that the process $\{Z_{NS}(\mathbf{s}): \mathbf{s} \in D\}$ is simulated and define $\mathbf{Z}_{NS} \equiv (Z_{NS}(\mathbf{s}_1), \ldots, Z_{NS}(\mathbf{s}_n))'$. Then, it is a simple matter to establish from (3.6.20) that

$$Z_{CS}(\mathbf{s}) = Z_{NS}(\mathbf{s}) + \mathbf{c}(\mathbf{s})'\Sigma^{-1}(\mathbf{Z} - \mathbf{Z}_{NS}), \qquad \mathbf{s} \in D, \qquad (3.6.21)$$

where $\mathbf{c}(\mathbf{s}) \equiv (C(\mathbf{s}, \mathbf{s}_1), \cdots, C(\mathbf{s}, \mathbf{s}_n))'$ and $\Sigma \equiv \text{var}(\mathbf{Z})$. It is apparent then that when $Z_{NS}(\cdot)$ is Gaussian, so too is $Z_{CS}(\cdot)$. Furthermore, the Cholesky-decomposition method for $Z_{NS}(\cdot)$ is easily generalized to $Z_{CS}(\cdot)$.

Conditional simulation carries with it a danger hinted at before. The conditional surface may be nowhere like reality if there is nontrivial phase information in the data $\mathbf{Z}$ [see Section 3.6.1, circa (3.6.12)]. By simulating a Gaussian process, one makes the implicit assumption that there is no such information. Thus, unless phase information is at hand, conditional-simulation-based solutions to problems like change of support (see Section 5.2) may be far from correct. [Notice, however, that $Z^*(\mathbf{s}_0)$ and $\hat{Z}(\mathbf{s}_0)$ are optimal linear predictors of $Z(\mathbf{s}_0)$, regardless of whether the process is Gaussian or not; that is, the phase information is irrelevant to kriging.]

Conditional simulation is highly dependent on the speed and accuracy of the nonconditional simulation, as can be seen from (3.6.20) or (3.6.21). Nonconditional simulations are discussed extensively in Section 3.6.1.

3.6.3 Geostatistical Data

The analysis of geostatistical data not only requires the observations $\{z(\mathbf{s}_1), \ldots, z(\mathbf{s}_n)\}$, but also knowledge of their spatial locations $\{\mathbf{s}_1, \ldots, \mathbf{s}_n\}$. Very little mining geostatistical data has been published, there being several exceptions. A coal-ash data set is given in Gomez and Hazen (1970), a subset of which is given in Section 2.2. A small copper–zinc data set appears in O'Connor and Leach (1979). Mendelsohn (1980) has collected eight case studies of base mineral deposits and has given the original data. Matheron and Armstrong (1987) have edited a volume of 12 geostatistical case studies, but data are given in only one study (results from 8 out of 37 vertical holes in a uranium deposit are presented). Hohn (1988, Tables 1.1, 2.1, 2.6, 4.4, and 6.1) has presented several geostatistical data sets from petroleum geology.

Laslett et al. (1987) give data on soil surface pH over a regular grid and Zirschky and Harris (1986) give data pertaining to a dioxin-contaminated area surrounding a Missouri state highway. Furthermore, Chapter 4 provides spatial data on a variety of problems; the variables studied include piezometric head, soil–water tension, soil–water infiltration, sudden-infant-death rates, yield of wheat, and acid deposition. Geostatistics has potential applications in a multitude of disciplines.

CHAPTER 4

Applications of Geostatistics

In Chapters 2 and 3, the emphasis has been on applications of geostatistics in the earth sciences; the principal example used to illustrate various spatial statistical methods has been the coal-ash data set (see, in particular, Sections 2.2, 2.6, 3.4, and 3.5). Applications are found in a variety of other areas, for example, with rainfall data (e.g., Ord and Rees, 1979), atmospheric data (e.g., Thiebaux and Pedder, 1987), forestry data (e.g., Samra et al., 1989), soils data (e.g., Burgess and Webster, 1980), regional groundwater geochemical data (e.g., Myers et al., 1982), groundwater contaminant concentrations (e.g., Istok and Cooper, 1988), disease incidences (e.g., Cressie and Read, 1989), and so on.

Chapter Summary

In the sections that follow, I shall present various problems in Statistics for spatial data, whose solutions have required geostatistical tools. Section 4.1 analyzes piezometric head in the Wolfcamp aquifer, in a region originally targeted for nuclear waste disposal. The goal is mapping, to predict the flow of possible radionuclide contamination in the groundwater. Section 4.2 analyzes soil–water tension data, where the purpose of the study is to determine the spatial dependence structure of different tillage treatments. Section 4.3 analyzes water-infiltration properties under different treatments; here the emphasis is on testing for treatment differences in the presence of spatial dependence. Sections 4.4 and 4.5 deal with lattice data (see Chapter 6) but use the tools of geostatistics to characterize the spatial dependence. Section 4.4 considers the incidence of Sudden Infant Deaths in the counties of North Carolina, whereas Section 4.5 presents a geostatistical analysis of the famous Mercer and Hall (1911) wheat-yield data. Finally, Section 4.6 analyzes acid-rain data, from irregularly located monitoring stations in the eastern United States, over different time periods; the goal is to use the (estimated) spatial dependence to carry out optimal network design.

The choice of applications in this chapter is a personal one, because it represents a sample of various collaborative projects in which I have been involved over past years. It is my intention here to demonstrate the flexibility of a geostatistical approach in analyzing spatial data sets.

4.1[†] WOLFCAMP-AQUIFER DATA

This section is based on, but goes beyond, a spatial analysis presented in Cressie (1989c). Here, additionally, a robust/resistant method of spatial-trend removal is compared to ordinary kriging.

Several years ago, there were three potential high-level nuclear waste repository sites being proposed in the United States; the sites were in Nevada, Texas, and Washington state. The chosen repository site will eventually contain more than 68,000 high-level waste canisters placed underground, about 30 feet apart, in holes or trenches surrounded by salt, and will cover an area of about 2 square miles. The U.S. Department of Energy has stipulated that the site chosen must isolate the wastes for 10,000 years. However, leaks could occur, or the radioactive heat could cause the tiny quantities of water in the salt to migrate toward the heat until eventually each canister is surrounded by water (about 6 gallons). The chemical reaction of salt and water would create hydrochloric acid that could slowly corrode the canisters.

Using geostatistics, I shall address the question of where the radionuclide contamination would flow from the site in Deaf Smith County, Texas. Beneath Deaf Smith County is a deep brine aquifer known as the Wolfcamp aquifer, a potential pathway for any radionuclides leaking from the repository. The predicted direction of flow is used to determine locations of downgradient and upgradient wells for a groundwater monitoring system. (As a matter of interest, a decision was made in December 1987 by the U.S. Congress to locate the high-level nuclear waste repository site in Nevada, probably at Yucca Mountain.)

Some Background

Figure 4.1 shows the piezometric-head data (in principle, obtained by drilling a narrow pipe into the aquifer and letting the water find its own level in the pipe) and their locations. The measurements are in units of feet above sea level. The head measurements are from drill stem tests conducted by the Department of Energy. After rigorous screening of unsuitable wells, 85 remained. Figure 4.1 is essentially that found in Harper and Furr (1986), which was my source for the spatial data. For completeness, piezometric heads, along with their spatial coordinates, are repeated here in Table 4.1.

The proposed nuclear waste site was to be in Deaf Smith County, bordering with New Mexico in the Texas panhandle; see Figure 4.1. For country-music listeners who would like to get their bearings, Amarillo, Texas, is also shown on the map.

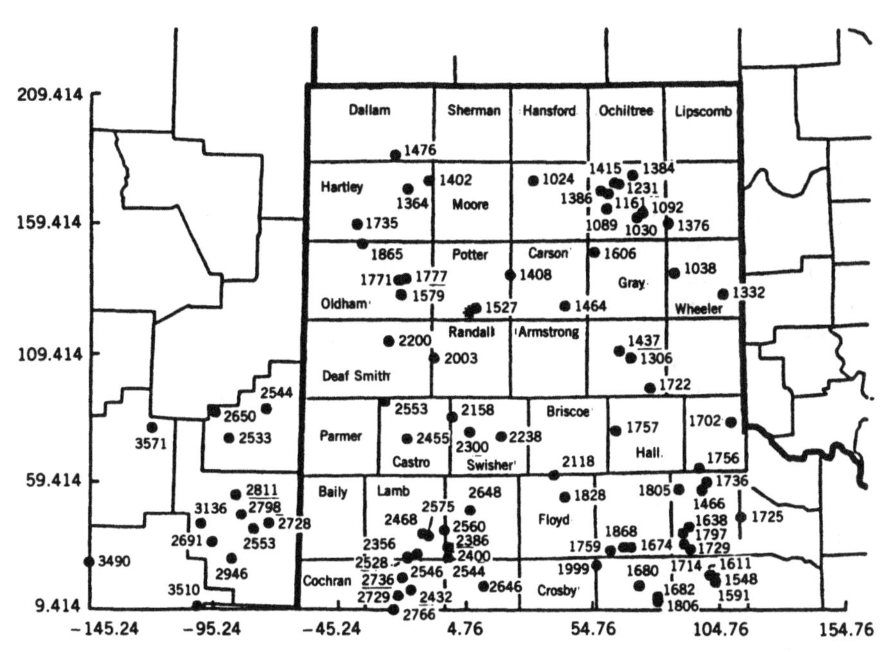

Figure 4.1 Locations and levels (in feet above sea level) of piezometric head for the Wolfcamp aquifer. The asterisk (*) denotes the location of Amarillo, Texas. [*Source:* Harper and Furr, 1986]

Prediction and Prediction-Error Maps

The goal in analyzing this (purely) spatial data set is to draw a map of a predicted surface based on the (irregularly located) 85 data available. An advantage of using a stochastic method of prediction (kriging) is that a map of root-mean-squared prediction error can also be drawn, quantifying the uncertainty in the predicted piezometric-head surface.

In the notation of Chapter 3, it is desired to use a kriging predictor $\hat{Z}(\mathbf{s}_0)$, along with its root-mean-squared prediction error $\sigma_k(\mathbf{s}_0)$, as $\mathbf{s}_0$ varies over the extent of the Wolfcamp aquifer. The kriging equations are straightforward to solve once the variogram is known.

4.1.1† Intrinsic-Stationarity Assumption

Assume that the data $\{Z(\mathbf{s}_1), \ldots, Z(\mathbf{s}_{85})\}$ given in Table 4.1 are observed without measurement error and that they are a sampling from an intrinsically stationary stochastic process. Estimation of the variogram defined by (2.3.4) allows the ordinary-kriging equations (3.2.12) to be solved, from which maps based on (3.2.5) and (3.2.16) can be drawn. (Another analysis of these data is performed in Section 4.1.2, assuming a nonconstant mean.)

Table 4.1 Wolfcamp-Aquifer Data[a]

x	y	$Z(x, y)$	x	y	$Z(x, y)$
42.78275	127.62282	1464	103.26625	20.34239	1591
−27.39691	90.78732	2553	−14.31073	31.26545	2540
−1.16289	84.89600	2158	−18.13447	30.18118	2352
−18.61823	76.45199	2455	−18.12151	29.53241	2528
96.46549	64.58058	1756	−9.88796	38.14483	2575
108.56243	82.92325	1702	−12.16336	39.11081	2468
88.36356	56.45348	1805	11.65754	18.73347	2646
90.04213	39.25820	1797	61.69122	32.49406	1739
93.17269	33.05852	1714	69.57896	33.80841	1674
97.61099	56.27887	1466	66.72205	33.93264	1868
90.62946	35.08169	1729	−36.65446	150.91456	1865
92.55262	41.75238	1638	−19.55102	137.78404	1777
99.48996	59.15785	1736	−21.29791	131.82542	1579
−24.06744	184.76636	1476	−22.36166	137.13680	1771
−26.06285	114.07479	2200	21.14719	139.26199	1408
56.27842	26.84826	1999	7.68461	126.83751	1527
73.03881	18.88140	1680	−8.33227	107.77691	2003
80.26679	12.61593	1806	56.70724	171.26443	1386
80.23009	14.61795	1682	59.00052	164.54863	1089
68.83845	107.77423	1306	68.96893	177.24820	1384
76.39921	95.99380	1722	70.90225	161.38136	1030
64.46148	110.39641	1437	73.00243	162.98960	1092
43.39657	53.61499	1828	59.66237	170.10544	1161
39.07769	61.99805	2118	61.87429	174.30178	1415
112.80450	45.54766	1725	63.70810	173.91453	1231
54.25899	147.81987	1606	5.62706	79.08730	2300
6.13202	48.32772	2648	18.24739	77.39191	2238
−3.80469	40.40450	2560	85.68824	139.81701	1038
−2.23054	29.91113	2544	105.07646	132.03181	1332
−2.36177	33.82002	2386	−101.64278	10.65106	3510
−2.18890	33.68207	2400	−145.23654	28.02333	3490
63.22428	79.49924	1757	−73.99313	87.97270	2594
−10.77860	175.11346	1402	−94.48182	86.62606	2650
−18.98889	171.91694	1364	−88.84983	76.70991	2533
−38.57884	158.52742	1735	−120.25898	80.76485	3571
83.14496	159.11558	1376	−86.02454	54.36334	2811
−21.80248	15.02551	2729	−72.79097	43.09215	2728
−23.56457	9.41441	2766	−100.17372	42.89881	3136
−20.11299	22.09269	2736	−78.83539	40.82141	2553
−16.62654	17.25621	2432	−83.69063	46.50482	2798
29.90748	175.12875	1024	−95.61661	35.82183	2691
100.91568	22.97808	1611	−87.55480	29.39267	2946
101.29544	22.96385	1548			

Source: Harper and Furr (1986).

[a] The First Two Columns are Data Location $\mathbf{s}_i = (x_i, y_i)'$, in Miles from an Arbitrary Origin. The Third Column is Piezometric Head $Z(\mathbf{s}_i)$, in Feet above Sea Level.

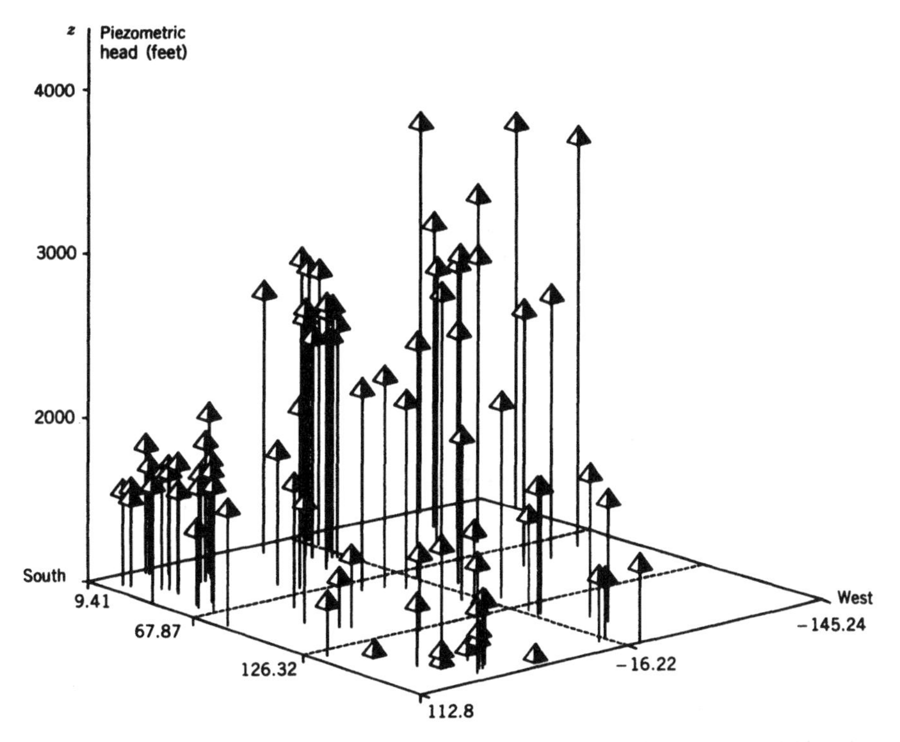

Figure 4.2 Scatter plot of piezometric-head data. Vertical lines are sited at the spatial locations and their heights are equal to the corresponding piezometric heads (in feet above sea level). [*Source:* Cressie, 1989c]. Reprinted by permission of the American Statistical Association.

Anisotropy

It is clear from the data shown in Figure 4.2 that the behavior of the process in the northeast–southwest direction is different from that in the northwest–southeast direction. For this reason, variogram estimators were calculated in these two directions.

Figure 4.3*a* shows $2\hat{\gamma}(\mathbf{h})$ given by (2.4.2) [actually, a smoothed version $2\gamma^+(\mathbf{h})$, as is explained in the following text], for $\mathbf{h} \in \{h(1)\mathbf{e}, \ldots, h(24)\mathbf{e}\}$; $\mathbf{e}$ is a vector in $\mathbb{R}^2$ of length $\|\mathbf{e}\| = 5$ miles and direction $\pi/4$; $h(k) = k - (1/2)$, for $k = 1, \ldots, 24$. Figure 4.3*b* shows a similar plot but where $\mathbf{e}$ has direction $3\pi/4$. Because the data are not on a regular grid, some tolerance regions around $\mathbf{h}$ in (2.4.2) have to be declared. The pair of locations $(\mathbf{s}, \mathbf{u}) \in N(\mathbf{h})$ if $\|\mathbf{s} - \mathbf{u}\| - \|\mathbf{h}\|$ is bounded between ± 2.5 miles and the direction of $(\mathbf{s} - \mathbf{u})$ minus the direction of $\mathbf{h}$ is bounded between $\pm\pi/4$ radians; see Section 2.4 for some guidelines on the choice of tolerance regions.

Superimposed on Figure 4.3*a* and *b* are weighted-least-squares fits (Section 2.6.2) of the power variogram model (Section 2.3.1)

$$2\gamma(h; \boldsymbol{\theta}) = \begin{cases} 0, & h = 0, \\ 2\left\{c_0 + b_p|h|^{\lambda}\right\}, & h \neq 0, \end{cases} \tag{4.1.1}$$

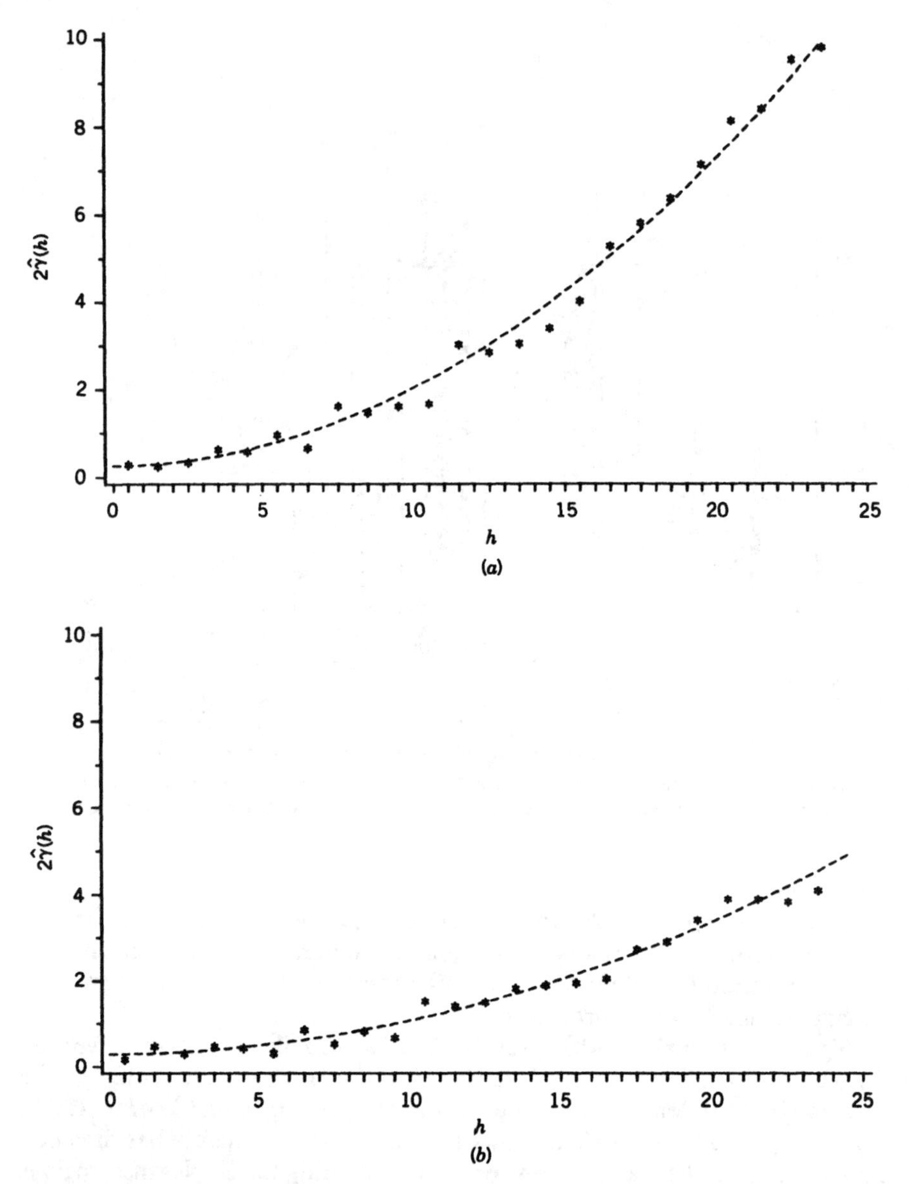

Figure 4.3 (a) Estimated variogram in the northeast–southwest direction. The superimposed dashed line is the weighted-least-squares fit of a power model $2\gamma(h; c_0, b_p, \lambda) = 2\{c_0 + b_p|h|^{\lambda}\}$, $h > 0$. On the horizontal axis, 1 unit = 5 miles; on the vertical axis, 1 unit = 10^5 (ft)^2. [*Source:* Cressie, 1989c] (b) Same as in (a), except that estimated variograms are computed in the northwest–southeast direction. [*Source:* Cressie, 1989c]. Reprinted by permission of the American Statistical Association.

where h is in miles. In Figure 4.3*a* (northeast–southwest direction),

$$\hat{\boldsymbol{\theta}}' = \left(\hat{c}_0, \hat{b}_{p,1}, \hat{\lambda}\right) = (1.4 \times 10^4 \text{ ft}^2, 38, 1.99),$$

whereas in Figure 4.3*b* (northwest–southeast direction),

$$\hat{\boldsymbol{\theta}}' = \left(\hat{c}_0, \hat{b}_{p,2}, \hat{\lambda}\right) = (1.4 \times 10^4 \text{ ft}^2, 15, 1.99).$$

(In fact, the fits in each direction showed slightly different values of $\hat{c}_0$ and $\hat{\lambda}$; the estimates given are the averages of the northeast–southwest and northwest–southeast values.) Thus the process is exhibiting *anisotropy*; the fitted two-dimensional anisotropic semivariogram model is, therefore,

$$\gamma(\mathbf{h}; \hat{\boldsymbol{\theta}}) = \hat{c}_0 + \left\{\hat{b}_{p,1}^{2/\hat{\lambda}} r^2 \cos^2\left(\frac{\pi}{4} - \phi\right) + \hat{b}_{p,2}^{2/\hat{\lambda}} r^2 \cos^2\left(\frac{\pi}{4} + \phi\right)\right\}^{\hat{\lambda}/2}, \tag{4.1.2}$$

where $\mathbf{h}' = (h_1, h_2) \equiv (r \cos \phi, r \sin \phi)$. Geometrical anisotropy [see relation (2.3.13), and Journel and Huijbregts, 1978, p. 179] was assumed in deriving (4.1.2).

Kriging Maps

From the (fitted) variogram (4.1.2), contour maps of the kriging predictor $\hat{Z}(\mathbf{s}_0)$ and the kriging standard error $\sigma_k(\mathbf{s}_0)$ [see Eqs. (3.2.5), (3.2.12), and (3.2.16)] can be drawn. The package Toolkit, by Geostokos Ltd., London was used; the results are displayed in Figure 4.4*a* and *b*. The predictor location $\mathbf{s}_0$ was chosen to range over a fine grid of points, and in choosing the grid the original data locations were avoided.

It can be concluded from these maps that contaminated groundwater from Deaf Smith County, Texas, would flow more or less directly downhill to Amarillo, Texas.

4.1.2† Nonconstant-Mean Assumption

Intrinsic stationarity of the underlying process assumes $E(Z(\mathbf{s})) = \mu$, for all $\mathbf{s} \in D$, which is dubious for the data in Table 4.1. The scatter plot, Figure 4.2, shows a strong trend in the northeast–southwest direction.

Write

$$Z(\mathbf{s}) = \mu(\mathbf{s}) + \delta(\mathbf{s}), \qquad \mathbf{s} \in \mathbb{R}^2, \tag{4.1.3}$$

where $E(Z(\mathbf{s})) = \mu(\mathbf{s})$ and $\delta(\cdot)$ is a zero-mean intrinsically stationary stochastic process with $\text{var}(\delta(\mathbf{s} + \mathbf{h}) - \delta(\mathbf{s})) = \text{var}(Z(\mathbf{s} + \mathbf{h}) - Z(\mathbf{s})) = 2\gamma(\mathbf{h})$. The large-scale variation $\mu(\cdot)$ and the small-scale variation $\delta(\cdot)$ are modeled, respectively, as deterministic and stochastic processes, but there is no way of making the decomposition identifiable.

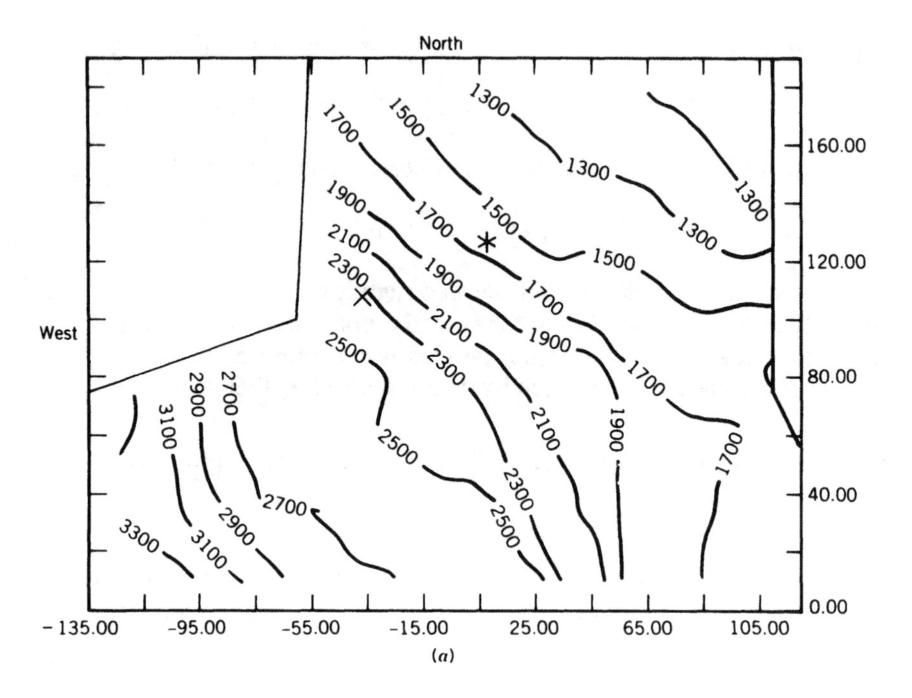

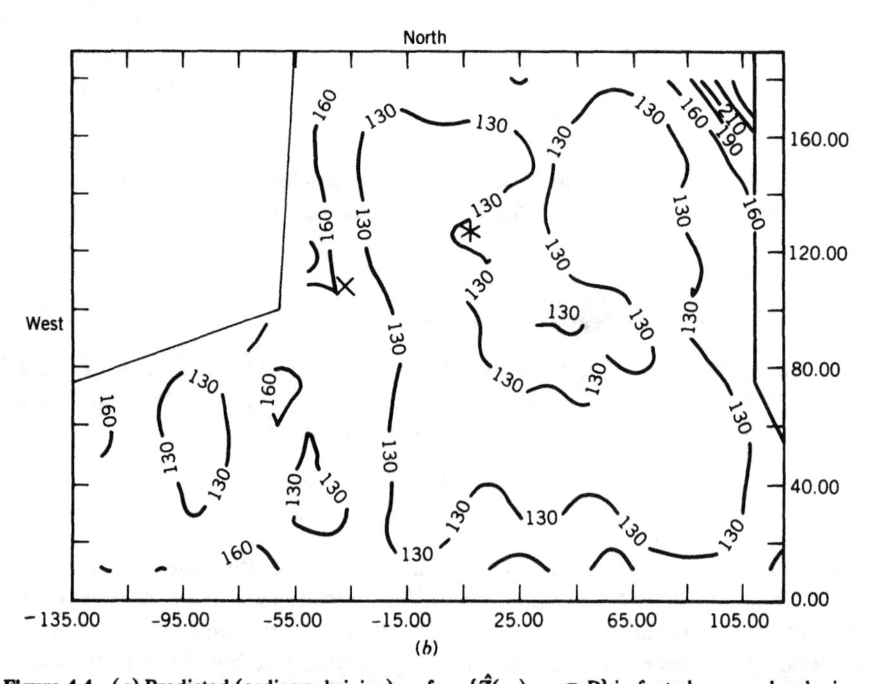

Figure 4.4 (*a*) Predicted (ordinary-kriging) surface $\{\hat{Z}(\mathbf{s}_0): \mathbf{s}_0 \in D\}$ in feet above sea level, given by (3.2.5) and (3.2.12). The cross (×) denotes the center of Deaf Smith County; the asterisk (*) denotes the location of Amarillo, Texas. [*Source:* Cressie, 1989c] (*b*) Kriging standard error surface $\{\sigma_k(\mathbf{s}_0): \mathbf{s}_0 \in D\}$, in feet above sea level, given by (3.2.16). [*Source:* Cressie, 1989c]. Reprinted by permission of the American Statistical Association.

Often this problem is resolved in a substantive application by relying on scientific or habitual reasons for determining the mean structure. Cressie (1986) discusses various ways of kriging in the presence of nonconstant mean. One of these, median-polish kriging, will be used here.

Median-Polish Kriging

The details of this approach, for data with irregular spatial locations, are given in Sections 3.5.2 and 3.5.3. Briefly, a low-resolution map is made of D by superimposing a rectangular grid (possibly of unequal spacing) over Figure 4.1. Each datum is then identified with its nearest grid node and the resulting regular two-dimensional array is median polished. A median-polish surface $\tilde{\mu}(\cdot)$ is then constructed (uniquely) by planar interpolation between the grid

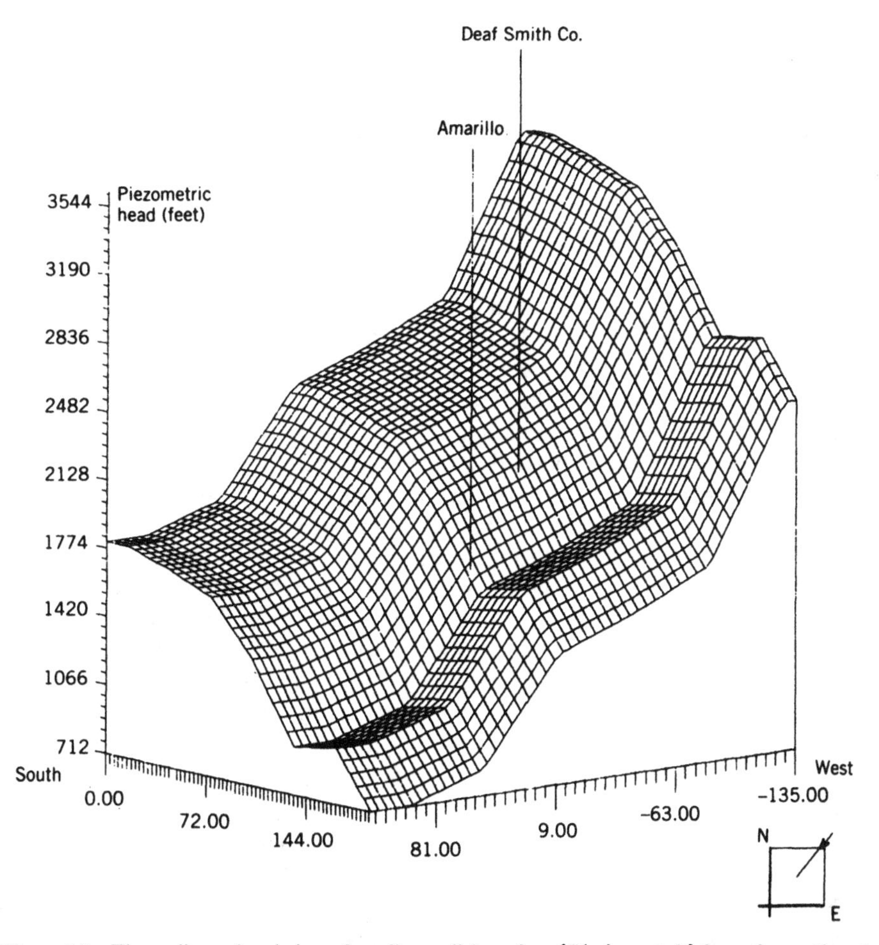

Figure 4.5 Three-dimensional view of median-polish surface $\{\tilde{\mu}(\mathbf{s}_0): \mathbf{s}_0 \in D\}$ from the northeast corner of D; $\tilde{\mu}(\cdot)$ is obtained by planar interpolation and extrapolation. Units on the vertical axis are in feet above sea level.

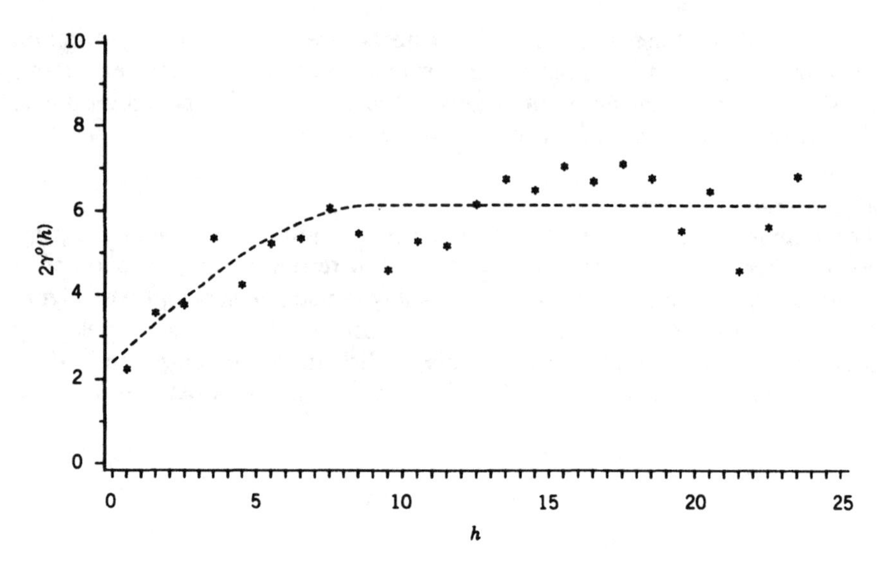

Figure 4.6 Estimated isotropic variogram, of the residuals, based on (2.4.2). The superimposed dashed line is the weighted-least-squares fit of a spherical model $2\gamma^o(h; c_0, c_s, a_s)$ given by (2.3.8). On the horizontal axis, 1 unit = 5 miles; on the vertical axis, 1 unit = 10^4 (ft)^2.

nodes; Figure 4.5 shows the "plates" obtained from the Wolfcamp-aquifer data.

Residuals $\{R(\mathbf{s}_i): i = 1, \ldots, 85\}$ are then obtained by subtracting the surface $\tilde{\mu}(\mathbf{s}_i)$ from $Z(\mathbf{s}_i)$, $i = 1, \ldots, 85$. Treat the residuals as a new data set that is now *intrinsically stationary* and to which a variogram can be fitted. Directional variograms were estimated, but the residuals showed none of the anisotropy encountered in the original data. This is a good illustration of the nonidentifiability of the proposed deterministic and stochastic components of $Z(\cdot)$. Here, anisotropy in one model's error process cannot be distinguished from nonstationarity in another model's mean process.

An isotropic residuals-based variogram estimator was computed, to which was fitted a spherical variogram model $2\gamma^o(h; c_0, c_s, a_s)$, given by (2.3.8). Figure 4.6 shows the variogram estimator $2\hat{\gamma}^o(h)$ given by (2.4.2) (using the same tolerance regions for lag distance described in Section 4.1.1), together with the spherical model fitted by weighted least squares (Section 2.6.2); estimates of $\hat{c}_0 = 1.2 \times 10^4$ (ft)^2, $\hat{c}_s = 1.8714 \times 10^4$ (ft)^2, and $\hat{a}_s = 46$ miles, were obtained. The fitted sill, $\hat{c}_0 + \hat{c}_s$, is slightly smaller than the one fitted by Harper and Furr (1986). This is due to the more flexible large-scale variation posited here (although fitting it by median polish tends to leave more residual variation than if least squares were used).

The ordinary-kriging predictor based on $\{R(\mathbf{s}_i): i = 1, \ldots, n\}$ and isotropic spherical variogram model $2\gamma^o(h; \hat{c}_0, \hat{c}_s, \hat{a}_s)$, yields the prediction surface $\{\hat{R}(\mathbf{s}_0): \mathbf{s}_0 \in D\}$ and the standard error surface $\{\sigma_m(\mathbf{s}_0): \mathbf{s}_0 \in D\}$. These are obtained by applying Eqs. (3.2.5), (3.2.12), and (3.2.16) to the residuals and

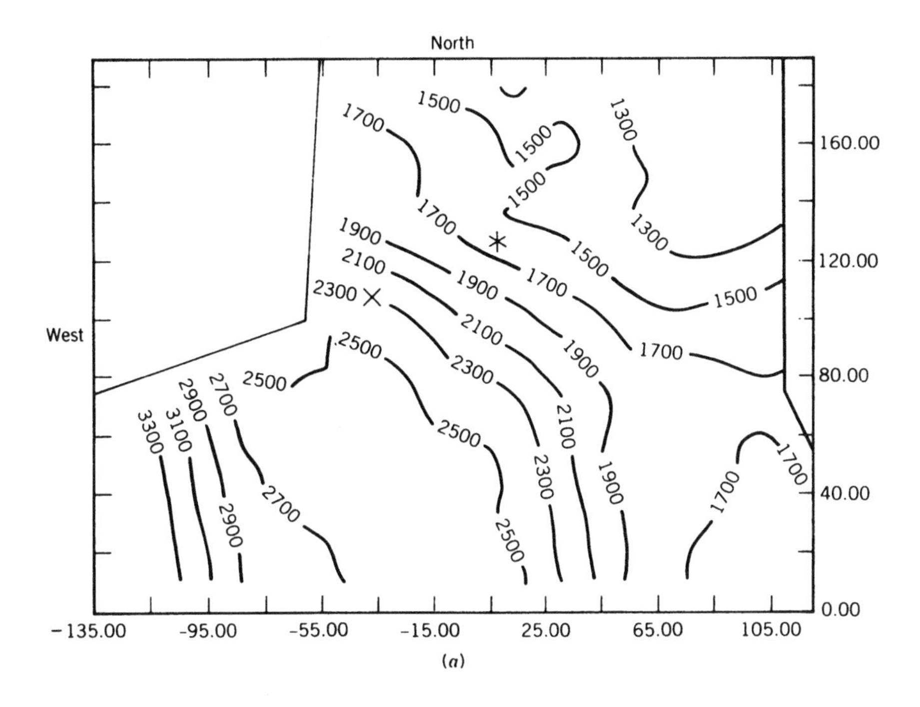

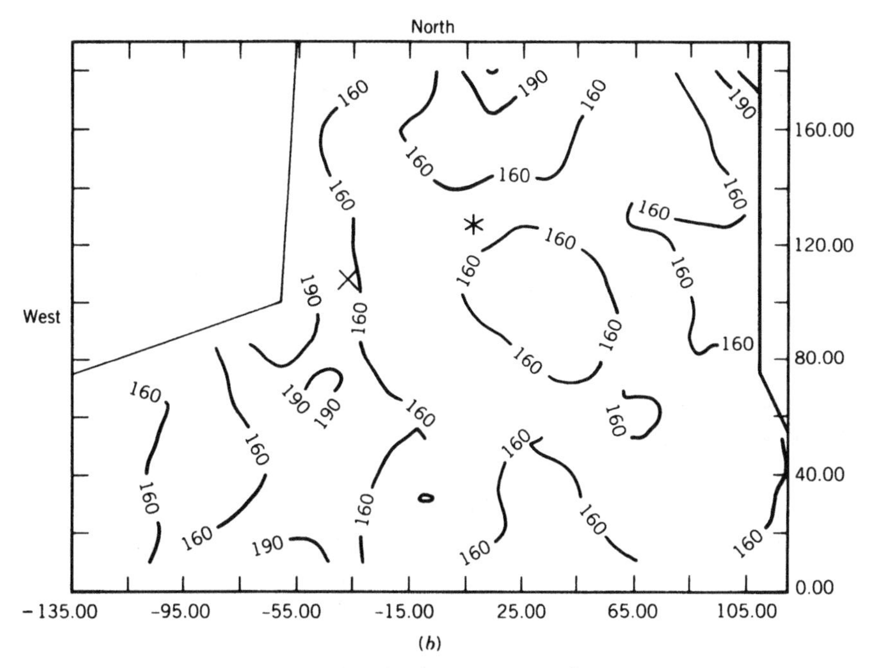

Figure 4.7 (*a*) Predicted (median-polish kriging) surface $\{\tilde{Z}(\mathbf{s}_0): \mathbf{s}_0 \in D\}$, in feet above sea level, given by (4.1.4). (*b*) Median-polish-kriging standard error surface $\{\sigma_m(\mathbf{s}_0): \mathbf{s}_0 \in D\}$, in feet above sea level. The cross ($\times$) denotes the center of Deaf Smith County; the asterisk ($*$) denotes the location of Amarillo, Texas.

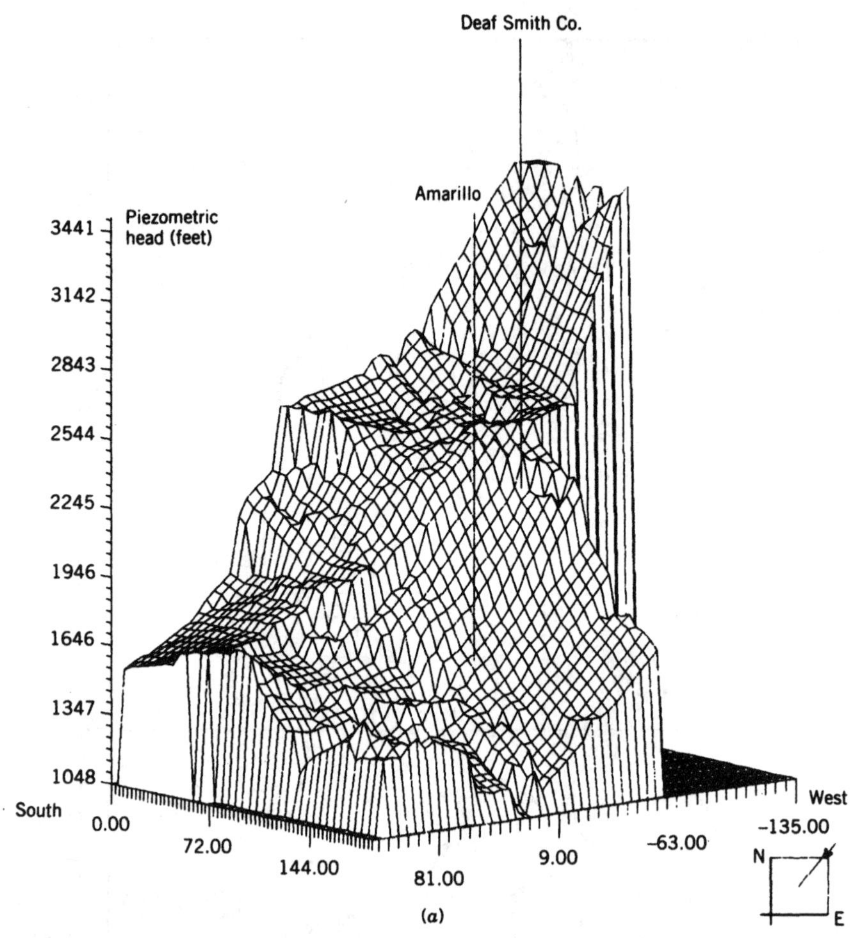

Figure 4.8 (*a*) Three-dimensional view of ordinary-kriging surface $\{\hat{Z}(\mathbf{s}_0): \mathbf{s}_0 \in D\}$, from the northeast corner of D. [*Source:* Cressie, 1989c]. Parts (*a*) and (*b*) reprinted by permission of the American Statistical Association.

the residuals-based fitted variogram. Finally, the median-polish-kriging predictor is defined by

$$\tilde{Z}(\mathbf{s}_0) \equiv \tilde{\mu}(\mathbf{s}_0) + \hat{R}(\mathbf{s}_0), \qquad \mathbf{s}_0 \in D. \tag{4.1.4}$$

Figure 4.7*a* and *b* show the surfaces $\tilde{Z}(\cdot)$ and $\sigma_m(\cdot)$, to be compared with $\hat{Z}(\cdot)$ and $\sigma_k(\cdot)$ presented in Figure 4.4*a* and *b*. The same fine grid of predictor locations was used in each contour plot.

Comparison of Median-Polish Kriging with Ordinary Kriging

The contour plots of $\hat{Z}(\cdot)$ (Fig. 4.4*a*) and $\tilde{Z}(\cdot)$ (Fig. 4.7*a*) look very similar, as expected. However, it is apparent from Figure 4.8*a* and *b* that the fine

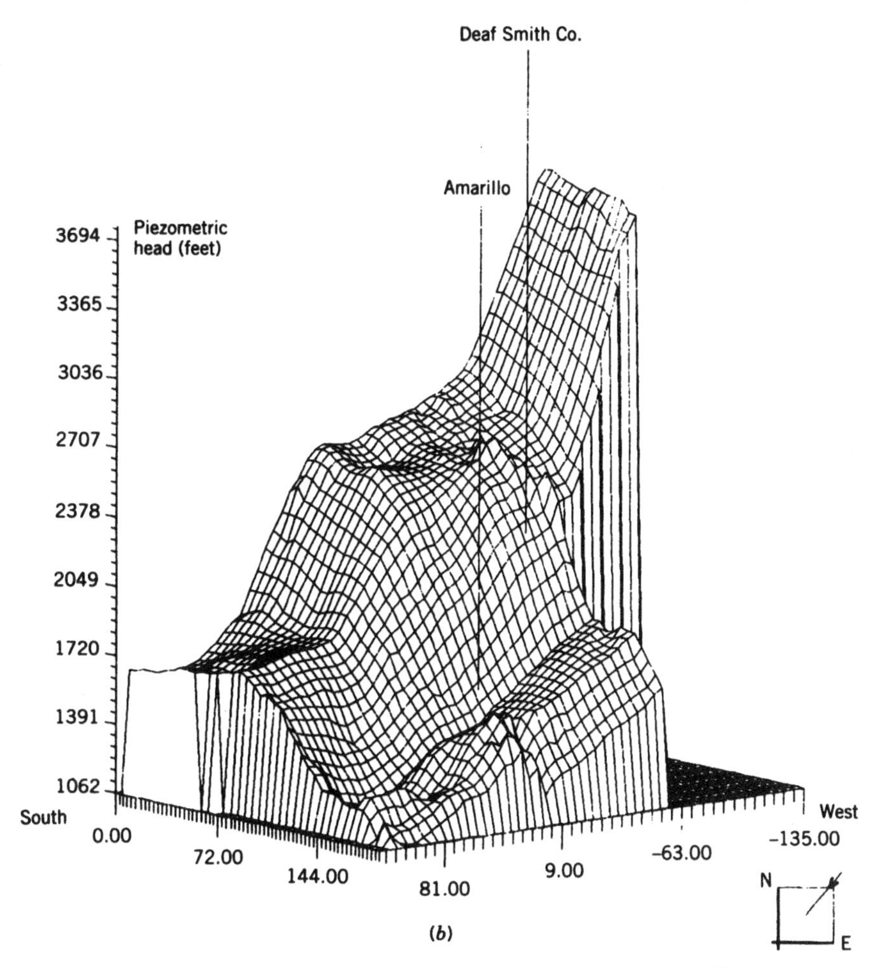

Figure 4.8 (*b*) Same as in (*a*), except the median-polish-kriging surface $\{\tilde{Z}(\mathbf{s}_0): \mathbf{s}_0 \in D\}$ is shown.

features of the two surfaces differ; at a small spatial scale, $\tilde{Z}(\cdot)$ is much smoother.

The contour plots of $\sigma_k(\cdot)$ (Fig. 4.4*b*) and $\sigma_m(\cdot)$ (Fig. 4.7*b*) give qualitatively similar prediction standard errors in the interior of the region of interest D, but differ near the edges. This is also to be expected, because different assumptions about the error process $\delta(\cdot)$ affect how well $Z(\cdot)$ can be predicted. Median-polish kriging assumes that the gross trend is a mean process of "plates", resulting in smaller mean-squared prediction error near the edges. Harper and Furr's (1986) analysis used universal kriging (Section 3.4) to produce maps that were very similar to Figure 4.7*a* and *b*. They used a linear trend after an *ad hoc* analysis of variance indicated that quadratic trend components were not present. However, for reasons given in Section

3.5, I believe that median-polish kriging presents a more flexible and resistant method of spatial prediction than universal kriging.

Although qualitatively similar in the interior of D, the contour maps of $\sigma_k(\cdot)$ and $\sigma_m(\cdot)$ do show quantitative differences that are worthy of comment. Notice that $\sigma_m(\mathbf{s}_0)$ is a little larger in places than $\sigma_k(\mathbf{s}_0)$ (when one might expect the opposite, by comparing variograms in Figs. 4.3 and 4.6). This is due to the relative influence of the nugget effect in each set of kriging equations. Typically, not all n data are used in predicting $Z(\mathbf{s}_0)$; Sections 3.2.1 and 3.4.5 discuss the choice of a neighborhood of $\mathbf{s}_0$ from which to take the n_k $(\leq n)$ data to use in the kriging equations. For the Wolfcamp-aquifer data, all points within a radius of 40 miles determined the n_k data that were used for predicting $Z(\mathbf{s}_0)$. The larger the nugget effect [as a proportion of $\max\{\gamma(\mathbf{s}_i - \mathbf{s}_0): i = 1, \ldots, n_k\}$], the less precise the prediction will be when the neighborhood remains fixed (see Section 3.2.1).

To demonstrate the large-scale similarity between $\hat{Z}(\cdot)$ and $\tilde{Z}(\cdot)$ more graphically, Figure 4.8*a* and *b* shows a three-dimensional view of the contour plots given in Figures 4.4*a* and 4.7*a*, respectively; also compare them to the median-polish surface shown in Figure 4.5. Notice how, at a small-scale, $\tilde{Z}(\cdot)$ has a smoother appearance than $\hat{Z}(\cdot)$.

Regardless of which prediction method is used, a leaking nuclear-waste-repository site in Deaf Smith County would not be good news for regions to its northeast.

4.2† SOIL–WATER TENSION DATA

This section is based on a spatial analysis presented in Hamlett et al. (1986). Resistant and exploratory techniques for a semivariogram analysis are featured.

Researchers, farmers, agribusinesspersons, and others are interested in, and concerned with, crop growth and yield under various tillage systems and configurations. Variation of the physical properties of soil within a given field and under different tillage systems may be sizeable and can affect plant growth and yield. Hence, the quality and form of soil-physics data are important. Classical statistical techniques are based on the assumption that observations are independent in spite of their distribution in space. Past studies (e.g., Campbell, 1978; Burgess and Webster, 1980; Russo and Bresler, 1981; Vieira et al., 1981; Burrough, 1983) imply that an analysis based on *spatial* dependencies should give a more complete understanding of phenomena influencing crop growth and yield.

A useful approach for soil-property investigations is to use statistical concepts to construct a model of the system and then use this model to investigate treatment effects and to predict values at various locations within the system. Such a model will be a function of the distribution of the measured values and the spatial relationships between such values. Geostatistics approaches modeling in two parts. The first part involves estimation of the variogram (see Chapter 2), and the second part involves kriging or spatial prediction of unobserved values from nearby observations (see Chapter 3). The emphasis in this section will be on how to carry out the first part in the presence of outliers and nonstationarity.

Some Background

The soil–water-tension data were collected during the 1983 crop season for the June to October period. Moldboard plow and no-till plot areas were located near Ames, in central Iowa, on a Nicollet clay loam soil (fine–silty, mixed, mesic, Aquic Hapludolls). The plots had 2% land slope and had been in continuous corn growth for the previous eight years. These plots were part of a larger fertility study; because the main interest here is in spatial relationships within a single plot, details of the design are not given. The growing season was somewhat atypical, in that the spring (March through June) season was excessively wet followed by an extremely dry period from the first part of July through mid-September, 1983.

Soil–water tension is the negative of *matric potential*, a commonly measured soil property. Baver et al. (1972, p. 293) define matric potential as "... the amount of work that must be done per unit quantity of pure water in order to transport reversibly and isothermally an infinitesimal quantity of water from a pool containing a solution identical in composition to the soil water at the elevation and the external gas pressure of the point under consideration to the soil water." Because water would flow readily from the reference pool into a dry area with a *release* of energy (in the form of heat), matric potentials are negative (above the water table) or zero (below the water table). It is a soil property associated with the attraction of water molecules to each other (capillary forces) as well as water to soil particles (surface adsorption), and, in general, larger absolute values show the soil is able to retain its moisture better. From its definition it seems likely that soil–water tension will exhibit spatial dependence.

Soil–water-tension data were collected using tensiometers placed at 0.15 m depths within the crop row. Tensiometers were constructed using 100-kPa ceramic cups (2.22 cm diameter by 7.00 cm length) and 2.22-cm-o.d. polyvinyl chloride pipes, similar to those described by Marthaler et al. (1983). A portable pressure transducer, like that described by Marthaler et al. (1983), was used to measure soil–water tensions during the study. A grid network, consisting of three rows by eight columns with spacing of 1.5 m between rows and 3 m between columns, was established on each plot to allow a spatial analysis; see Figure 4.9. Tensiometer readings were generally made once every two or three days.

For illustration and discussion of spatial analysis techniques, the data for the moldboard plot are presented; see Table 4.2. All data are assumed to represent measurements at a point location, a reasonable assumption given the scale of the grid spacings.

↑
N

	Col. 1	Col. 2	Col. 3	Col. 4	Col. 5	Col. 6	Col. 7	Col. 8
Row 3	∘ 17	∘ 18	∘ 19	∘ 20	∘ 21	∘ 22	∘ 23	∘ 24
Row 2	∘ 9	∘ 10	∘ 11	∘ 12	∘ 13	∘ 14	∘ 15	∘ 16
Row 1	∘ 1	∘ 2	∘ 3	∘ 4	∘ 5	∘ 6	∘ 7	∘ 8

Figure 4.9 Spatial locations (row, column, and location number) for the soil–water-tension data presented in Table 4.2. Distance between adjacent columns is 3 m and between adjacent rows is 1.5 m. [*Source:* Hamlett et al., 1986].

Notice that these measurements actually form a space–time data set, although by collapsing them over time, only the spatial component is considered. For further comments on spatiotemporal geostatistical problems, see Section 4.7.

Spatial Dependence Exhibited by Different Tillage Treatments

Because different treatments are being applied, stationarity is not an appropriate assumption for the complete data set. Moreover, spatial trends within a treatment region may be substantial. Exploratory analyses detailed in Hamlett et al. (1986) demonstrate a nonstationarity in the variance, as well. It appears that the variance is proportional to the square of the mean, which is treated by taking *logs* of the original data. Rao et al. (1979) indicates that flow-related soil properties tend to be log Gaussian, whereas capacity-related soil properties tend to be Gaussian.

Write the data of Table 4.2 (for the moldboard-plowed plot) as $\{y(2ja, ia; t): i = 1, 2, 3;\ j = 1, \ldots, 8;\ t = 1, \ldots, 14\}$, where $a = 1.5$ m, the index i refers to data in the ith row, and the index j refers to data in the jth column. Then the analysis will be based on

$$z(2ja, ia; t) \equiv \log y(2ja, ia; t). \tag{4.2.1}$$

Model the Z process by

$$Z(2ja, ia; t) = \mu(2ja, ia; t) + \delta(2ja, ia; t), \tag{4.2.2}$$

where $\delta(\cdot, \cdot\,; t)$ is a zero-mean intrinsically stationary process in $\mathbb{R}^2$ with variogram $2\gamma_t(\mathbf{h})$. Write

$$\gamma_t(\mathbf{h}) = \gamma_{\cdot}(\mathbf{h}) + \big(\gamma_t(\mathbf{h}) - \gamma_{\cdot}(\mathbf{h})\big), \qquad \mathbf{h} \in \mathbb{R}^2, \tag{4.2.3}$$

Table 4.2 Measured Values of Soil–Water Tension at the 0.15 m Depth for the Moldboard-Plowed Plot[a]

Spatial Location			Dates of Measurement (1983)													
Row[b]	Col[b]	Loc[b]	13/6[c]	15/6	16/6	20/6	30/6	06/7	08/7	03/8	25/8	01/9	08/9	07/10	18/10	25/10
1	1	1	170	61	67	61	20	95	186	696	553	235	412	54	54	37
1	2	2	148	59	72	65	30	96	147	642	731	—	—	55	49	42
1	3	3	185	59	71	76	41	95	145	624	671	120	306	60	48	41
1	4	4	193	67	84	65	28	107	176	723	670	110	262	60	52	44
1	5	5	184	49	58	58	29	96	209	376	403	118	402	64	56	42
1	6	6	169	49	55	55	14	70	171	666	714	71	235	—	36	29
1	7	7	169	63	80	73	20	125	264	—	—	102	217	56	49	34
1	8	8	209	62	79	77	37	99	137	421	479	85	—	63	51	—
2	1	9	179	38	50	40	—	79	177	571	421	—	—	51	38	—
2	2	10	270	31	53	37	19	106	247	727	748	171	567	—	47	—
2	3	11	169	44	51	58	18	85	176	599	588	105	350	52	45	32
2	4	12	195	43	75	70	35	107	209	—	300	98	268	63	50	42
2	5	13	161	51	64	57	22	83	143	301	451	133	404	62	45	33
2	6	14	170	38	52	51	11	68	126	514	527	50	275	44	50	35
2	7	15	138	47	54	66	14	88	155	—	561	66	—	—	—	—
2	8	16	145	51	87	77	39	109	167	—	600	114	235	76	68	54
3	1	17	174	74	72	73	26	88	144	363	716	164	360	—	—	—
3	2	18	197	39	51	48	12	89	161	395	501	115	403	—	—	—
3	3	19	252	36	39	22	13	96	227	—	537	61	305	50	35	27
3	4	20	270	33	43	38	18	80	267	303	708	53	361	—	—	—
3	5	21	288	33	57	30	17	89	280	581	750	56	330	47	26	16
3	6	22	210	33	57	42	19	127	371	300	749	46	248	—	—	—
3	7	23	308	48	63	40	39	148	453	333	626	71	310	—	—	—
3	8	24	122	50	85	69	30	99	165	186	630	84	258	73	59	49

Source: Hamlett et al. (1986).

[a] Soil–water tensions (negative potential) are expressed as positive values of 10^{-2} of water. The dashes (—) signify missing values.
[b] See Figure 4.9 for corresponding spatial location.
[c] 13/6, 15/6, etc. represent 13 June, 15 June, etc.

where $\gamma_{.}(\mathbf{h}) \equiv \sum_{t=1}^{14}\gamma_t(\mathbf{h})/14$ is the average semivariogram. Hamlett et al. (1986) demonstrate, graphically, that sample variances are stable over time, indicating that $\gamma_{.}(\mathbf{h})$ is representative of the purely spatial relationships for $t = 1, \ldots, 14$. The goal of this section will be to estimate $2\gamma_{.}(\mathbf{h})$ for the moldboard and no-till treatments.

Trend Removal

Although the ultimate aim of the experiment just described is to compare the soil–water tension of two different tillage treatments, I shall concentrate here on estimating their respective spatial dependencies. This involves removing the mean effect in (4.2.2) and studying the residual variation. The estimated spatial correlations could then be used to obtain efficient estimates of the mean effects, leading to more powerful hypothesis tests than obtained when the spatial correlation is ignored; Section 4.3 has an analysis of this sort. This section is only concerned with estimating the variograms.

Residual data were calculated by subtracting the median for each row, $\tilde{z}(*, ia; t)$, from each (log-transformed) value within the respective row:

$$r_1(2ja, ia; t) \equiv z(2ja, ia; t) - \tilde{z}(*, ia; t). \tag{4.2.4}$$

This was followed with removal of column medians:

$$r_2(2ja, ia; t) \equiv r_1(2ja, ia; t) - \tilde{r}_1(2ja, *; t). \tag{4.2.5}$$

Thus $r_2(\cdot, \cdot\, ; t)$ was the result of one iteration of the median-polish algorithm (see Section 3.5). Stem-and-leaf plots and normal probability plots indicated that the data set $\{r_2(2ja, ia; t)\}$ is from an approximately Gaussian distribution, although independence between the data is not an appropriate assumption.

Because the grid for the moldboard plot was only 3×8, no extensive spatial information can be estimated in the north–south direction (see Figure 4.9); therefore, all variograms were estimated in the east–west direction only. From (2.4.2), define

$$2\hat{\gamma}(ha) = \sum_{t=1}^{14}\sum_{i=1}^{3}\sum_{N_{it}(h)}\left(r_2(2ja, ia; t) - r_2(2la, ia; t)\right)^2 \Bigg/ \sum_{t=1}^{14}\sum_{i=1}^{3}\sum_{N_{it}(h)} 1,$$

$$h = 1, \ldots, 7, \tag{4.2.6}$$

where $N_{it}(h)$, in the innermost summation, is the set of indices $\{(j, l)$: $l - j = h$, and both $r_2(2ja, ia; t)$, $r_2(2la, ia; t)$ are defined$\}$. Figure 4.10a and b presents the estimators $2\hat{\gamma}$ plotted against distance (in meters).

It appears that soil–water tension from the moldboard plot *is* exhibiting spatial dependence, in sharp contrast to the no-till plot. (Can this be tested statistically?) Because, in fact, no-till is a passive treatment where the soil is

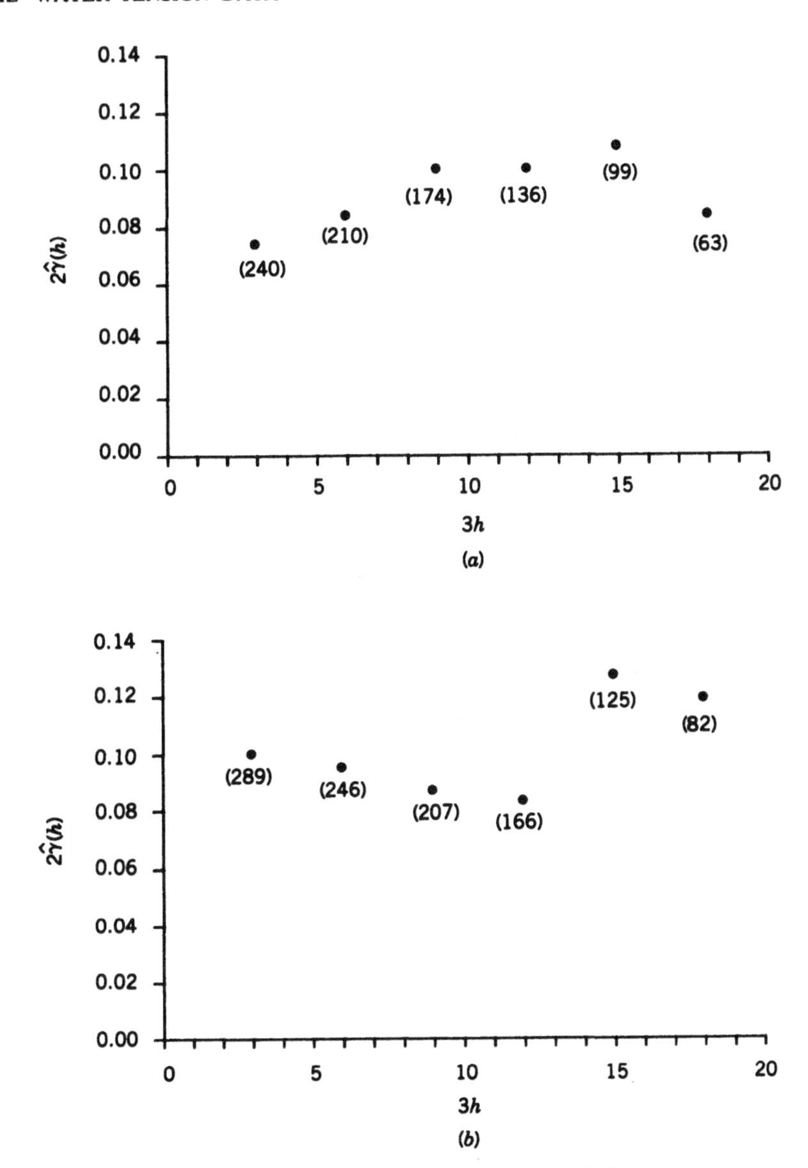

Figure 4.10 (*a*) Plot of east–west variogram estimator $2\hat{\gamma}$ [in $(10^{-2})^2$ of water] versus distance (in meters) for the moldboard plot. The number of pairs of observations, $\Sigma_{l=1}^{14}\Sigma_{i=1}^{3}\Sigma_{N_{il}(h)}1$, is noted in parentheses. [*Source:* Hamlett et al., 1986] (*b*) Same as in (*a*), except for the no-till plot. [*Source:* Hamlett et al., 1986].

not disturbed by plowing, it appears that small-scale soil homogeneity is promoted by tillage. The next section has further evidence of this for a different soil property.

4.3[†] SOIL–WATER-INFILTRATION DATA

This section is based on, but goes beyond, a spatial analysis presented in Cressie and Horton (1987). It is an example of how spatial modeling might be used to analyze a field trial (Section 5.7.2). Here, the effects of different tillage treatments on soil–water infiltration are tested after the spatial dependence has been modeled; more details can be found in Gotway and Cressie (1990).

Soil–water infiltration is one of the most important processes affecting both crop production and the volume, transport route, and water quality of agricultural or vadose zone drainage. Except in regions having excessive precipitation, or where there are storage sites for hazardous materials, or in arid regions having runoff-inducement areas, any practice that increases infiltration and thereby increases water availability to crops is looked upon favorably. Increased utilization of water by crops decreases the total volume of agricultural drainage. Increased infiltration during rainstorms (or snowmelt or irrigation) reduces surface runoff and the potential for water erosion. One hundred years ago, central Iowa's topsoil was 14 to 16 inches deep, but today it is only 5 to 7 inches deep. Unchecked, soil erosion (both water and wind) promises to turn the now-productive farm belt of the United States (from Indiana to Colorado) into a dustbowl. Generally, high intensity rain falling on bare, cultivated soil is the time when the erosion hazard is greatest.

Increased infiltration also reduces the potential for transport of agricultural chemicals with *surface* runoff, not only by decreasing the volume of the carrier, but also by decreasing the concentration of nonadsorbed to moderately adsorbed chemicals in runoff water. However, because increased infiltration usually delays the time during a storm when runoff begins, during this extra time the infiltrating water can carry chemicals downward (see the review by Baker and Laflen, 1983). Thus, the only negative aspect of increased infiltration is that it may increase the potential for transport of agricultural chemicals, deposited wastes, and soluble salts to the groundwater.

At a given location in the field, the ability of water to infiltrate soil depends upon the existing soil–water distribution with depth, the rate of water application to the soil surface, and the soil–pore-structure distribution with depth. As the location varies across the field, that ability will vary spatially so that locations nearby are more alike, with regard to infiltration, than those far apart.

Some Background

Superimposed on the spatial variability of soil–water infiltration were four tillage treatments: moldboard plow (15–20 cm), paraplow (25–30 cm), chisel plow (15–20 cm), and no tillage. These treatments were established in the fall of 1982 at the Agronomy and Agricultural Engineering Research Center near Ames, Iowa. The soil was a Webster silty clay loam (Typic Haplaquoll). All of the tillage plots were chisel plowed in 1981 and left untilled before the fall of 1982 with continuous corn production.

Soil–water infiltration measurements were made at 24 locations (i.e., on a 3×8 grid arrangement) within each of the 4 tillage plots. Two sets of infiltration measurements were obtained, one set in May (where some observations were lost due to compaction of the soil by tractor wheels) and one set in July, 1983. Figure 4.11 gives the data and the spatial locations. Notice that no measurements were taken on the middle of the five plots (where originally another treatment was assigned), because of the necessity to do all measurements on the same day. Conclusions from this study must remain tentative due to the poor design; a randomized block design for the four treatments would have been preferable to considering a single block of the design (as was done here).

Moldboard			Paraplow				Chisel			No-till		
	9.60		**0.64**	**23.87**	**15.24**				**9.32**		**16.28**	**11.43**
31.55	27.90	12.50	7.54	36.64	26.47		10.24	8.93	14.77	4.30	9.75	9.49
	18.84	**4.18**	**2.10**		**23.52**				**1.72**		**8.84**	**14.22**
31.10	35.65	6.84	5.40	38.82	42.02		6.81	8.55	11.84	6.10	13.41	14.84
		4.69	**3.35**	**28.47**	**30.98**				**3.41**		**14.65**	**13.29**
38.05	53.25	13.90	13.43	10.67	20.33		3.99	1.83	7.96	4.48	15.38	10.41
	43.83	**8.02**	**17.40**	**18.44**	**20.79**				**4.14**		**13.46**	**14.48**
17.62	39.04	18.15	26.49	30.28	35.20	→ N	7.10	4.65	5.32	8.67	15.29	12.10
	42.82	**13.29**	**9.71**	**12.54**	**26.44**				**0.50**		**13.98**	**11.20**
8.64	34.14	28.53	39.82	27.52	39.65		2.12	5.29	8.31	3.54	12.56	20.59
	20.64	**8.43**	**3.82**	**28.47**	**24.28**				**6.20**		**10.53**	**18.09**
6.65	23.30	25.97	20.19	25.15	44.42		6.02	3.52	5.84	2.22	15.21	13.12
	22.45	**3.32**	**6.95**	**11.80**	**48.02**				**2.91**		**17.22**	**18.20**
5.78	18.93	38.31	6.48	31.78	60.04		6.33	4.94	8.29	8.58	8.88	18.19
	22.94	**2.99**	**2.79**	**7.65**	**31.10**				**1.33**		**9.84**	**13.00**
22.78	31.29	10.00	16.20	63.32	38.71		8.40	2.53	5.41	10.35	15.32	11.11

Figure 4.11 Thirty-minute cumulative soil–water-infiltration data (in centimeters) and their spatial locations, together with tillage treatments. Distance between readings is 3 m in the east–west direction and 1.5 m in the north–south direction within tillage treatments and 3 m between adjacent treatments. 9 m separates the closest readings associated with paraplow and chisel treatments. Top numbers (in boldface type) are the May data and bottom numbers are the July data. [*Source:* Cressie and Horton, 1987, © The American Geophysical Union.]

Double-ring infiltrometers (Bertrand, 1965) were used to measure ponded infiltration volumes, and water stage recorders were used to record the subsidence of water in the inner ring as a function of time (details can be found in Mukhtar et al., 1985). Similar to Gish and Starr (1983), only the 30-min cumulative infiltration values were used in this analysis (the values are presented in Figure 4.11).

Use of Geostatistics

Several authors (e.g., Sharma et al., 1980; Luxmoore et al., 1981) have used variograms to evaluate the spatial correlation of soil–water infiltration measurements. They estimated variograms using the (method-of-moments) estimator (2.4.2) proposed by Matheron (1962), although Cressie and Hawkins (1980) show that if contaminated Gaussian data are present, a more robust estimator is (2.4.12).

In what follows, a robust/resistant geostatistical analysis of the July soil–water-infiltration data will be carried out. The robust part of the analysis refers to using statistical inference procedures that are little affected by departures from model assumptions, and the resistant part refers to using data analytic techniques that are little affected by outlying or unusual observations.

4.3.1[†] Estimating and Modeling the Spatial Dependence

Although only the July data will be analyzed here, Cressie and Horton (1987) have details of the analysis of the whole data set. A précis of their findings is now presented.

Stationarity is not an appropriate assumption to impose on the (assumed) stochastic mechanism that generated the data. First, different treatments are being applied to different plots resulting in perhaps different means across the plots. Second, spatial variability within a plot may be substantial. Cressie and Horton (1987) find not only nonstationarity in the mean, but in the variance as well. It appears (from resistant plots of interquartile-range-squared versus median, calculated over the columns shown in Fig. 4.11) that the variance is proportional to the mean, which suggests taking *square roots* of the original data. The "universal transformation principle" (Cressie, 1985b) suggests that, on this square-root scale, data can be written as a mean effect (made up of additive components of northing, easting, and treatment effects) *plus* Gaussian error. A thorough spatial analysis looks beyond the mean effects to possible spatial dependence in the error.

Detrending

To estimate the spatial dependence, the mean structure will be (temporarily) removed. This is achieved, resistantly, by subtracting column medians from

Stem	Leaf
−2	
−2	31
−1	9855
−1	443111
−0	9888876665555
−0	4443333322222111110000
0	0000011111111222333334
0	5566666677789
1	00111333
1	567
2	23
2	5

(*a*)

Stem	Leaf
−2	
−2	31
−1	9855
−1	44311
−0	9865
−0	43332110
0	011223334
0	5567
1	111333
1	567
2	23
2	5

(*b*)

Stem	Leaf
−11	1
−10	
−9	
−8	400
−7	
−6	60
−5	742
−4	731
−3	21
−2	422
−1	821
−0	7533
0	33357
1	123348
2	
3	04
4	
5	9
6	0445
7	24
8	0
9	299

(*c*)

Figure 4.12 Stem-and-leaf diagrams of $\{r_{ijk}\}$, the square-root transformed July data after subtraction of column medians: (*a*) all data ($k = 1, 2, 3, 4$), (*b*) moldboard and paraplow ($k = 1, 2$), and (*c*) chisel and no-till ($k = 3, 4$). In (*a*) and (*b*), 0|1 means 0.1 $\text{cm}^{1/2}$; in (*c*), 0|1 means 0.01 $\text{cm}^{1/2}$. [*Source:* Cressie and Horton, 1987, © The American Geophysical Union.]

their respective columns. Write the data of Figure 4.11 as

$$\{y_{ijk}: i = 1, \ldots, 8;\ j = 1, 2, 3;\ k = 1, 2, 3, 4\}, \tag{4.3.1}$$

where the index k refers to the treatments moldboard ($k = 1$), paraplow ($k = 2$), chisel ($k = 3$) and no-till ($k = 4$). Define

$$z_{ijk} \equiv (y_{ijk})^{1/2}, \tag{4.3.2}$$

$$r_{ijk} \equiv z_{ijk} - \tilde{z}_{\ast jk}, \tag{4.3.3}$$

where $\tilde{z}_{\ast jk} \equiv \text{med}\{z_{ijk}: i = 1, \ldots, 8\}$. These residuals data (4.3.3) now appear to have come from a stationary process (plots of row medians versus row numbers showed no trend) that will be used as a surrogate for the unobserved error. Cressie and Glonek (1984) show that median-based removal of trend gives a less-biased estimator of the stationary error covariance than a mean-based removal based on $\bar{z}_{\cdot jk} = \Sigma_{i=1}^{8} z_{ijk}/8$; see also Section 3.5.4. Figure 4.12 shows stem-and-leaf plots of the residuals $\{r_{ijk}\}$. Although the error appears Gaussian, it is clearly not homoskedastic across plots, even after the square-root transformation; treatments $k = 1$ and 2 show much more error variation than treatments $k = 3$ and 4.

Spatial Dependence Exhibited by Different Treatments

To conform with the spatial notation used in this book, write

$$\begin{aligned} z(ja, 2(9-i)a) &\equiv z_{ij1}, \\ z((4+j)a, 2(9-i)a) &\equiv z_{ij2}, \\ z((12+j)a, 2(9-i)a) &\equiv z_{ij3}, \\ z((16+j)a, 2(9-i)a) &\equiv z_{ij4}, \qquad i = 1,\ldots,8,\ j = 1,2,3, \end{aligned} \tag{4.3.4}$$

where $a = 1.5$ m. Define $\{r(x, y)\}$ from $\{r_{ijk}\}$ in an identical manner.

Suppose that data (4.3.4) comes from a process $Z(\cdot)$ modeled by

$$Z(\mathbf{s}) = \mu(\mathbf{s}) + \delta(\mathbf{s}); \qquad \mathbf{s} \in D, \tag{4.3.5}$$

where $\mu(\cdot)$ obtains its structure from the spatial design of the experiment; see Section 4.3.2. In this subsection, it is the structure of the error process $\delta(\cdot)$ that is of interest; that structure will be estimated in the east–west direction from the median-based residuals $\{r_{ijk}\}$. (In fact, a little thought reveals that in this particular case it does not matter whether medians or means are used to define the residuals.)

Variogram Estimation and Fitting

From (2.4.2), define the variogram estimator (for the July data) in the east–west direction:

$$2\hat{\gamma}_k(2ha) \equiv \sum_{j=1}^{3} \sum_{i=1}^{8-h} (r_{i+h,j,k} - r_{i,j,k})^2 \Big/ |N(h)|, \qquad h = 1,\ldots,7, \tag{4.3.6}$$

where $|N(h)| \equiv \Sigma_{j=1}^{3} \Sigma_{i=1}^{8-h} 1 = 3(8-h)$. The robust version due to Cressie and Hawkins (1980) is

$$2\bar{\gamma}_k(2ha) \equiv \frac{\left\{\sum_{j=1}^{3}\sum_{i=1}^{8-h}|r_{i+h,j,k} - r_{i,j,k}|^{1/2}/\,|N(h)|\right\}^4}{0.457 + 0.494/|N(h)|}, \qquad h = 1,\ldots,7, \tag{4.3.7}$$

given by (2.4.12). Outliers in spatial data can be difficult to detect both because of the spatial setting and because it is usually not feasible to check every datum for aberrant behavior. The estimator (4.3.7) automatically downweights contaminated data, whereas in (4.3.6) the squared terms exacerbate them (Cressie, 1985a).

Both variogram estimators were computed for the soil–water infiltration data, but only those for $2\bar{\gamma}_k(\cdot)$ are presented here. It will be seen in the following text that there is indeed a large *spatial* outlier in the data. Figure 4.13 gives variogram plots for each of the treatments up to a lag distance of 15 m. Spatial dependence, as summarized by the estimated variograms, clearly changes with treatment.

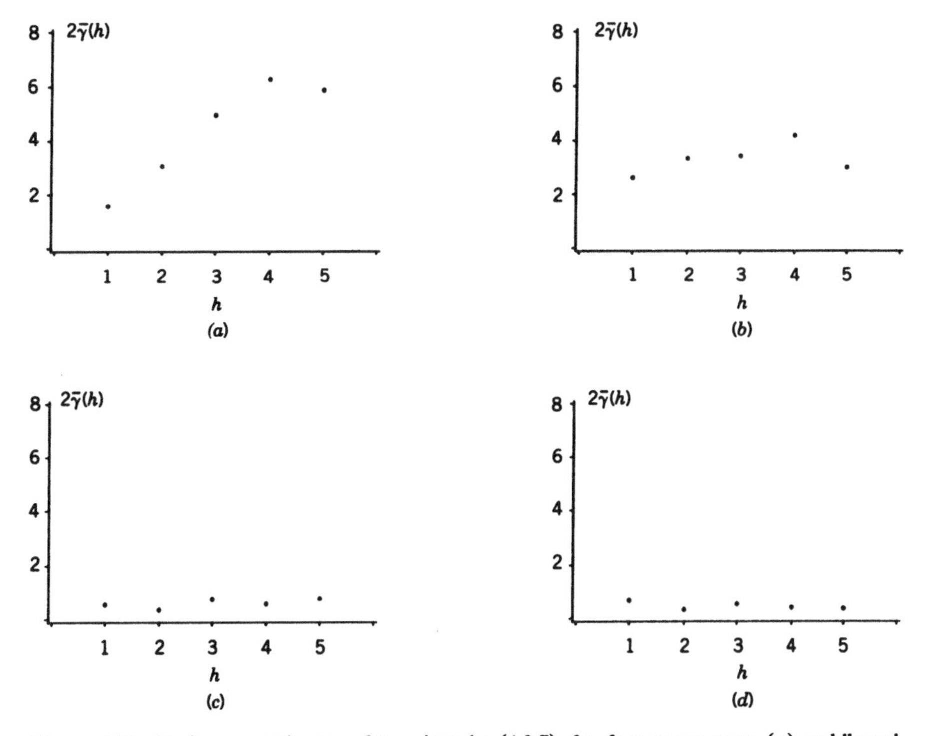

Figure 4.13 Variogram estimators $2\bar{\gamma}_k$, given by (4.3.7), for four treatments: (*a*) moldboard ($k = 1$), (*b*) paraplow ($k = 2$), (*c*) chisel ($k = 3$), and (*d*) no-till ($k = 4$). Units on the vertical axes are in square feet; on the horizontal axis, 1 unit = 3 m. [*Source:* Cressie and Horton, 1987, © The American Geophysical Union.]

The same general behavior reported in Section 4.2 (there, soil–water tension was being measured) occurs again. Specifically, the more the soil is disturbed by the tillage treatment (moldboard plowing causes the most disturbance), the more apparent is the spatial dependence. This corroborating evidence indicates that the noticeable differences in Figure 4.13 are *not* due to plot effects (here, confounded by the design with treatment effects).

Variogram models were fitted to the estimators using the weighted-least-squares method described in Section 2.6.2. Because chisel and no-till variogram estimators were so strikingly similar, they were pooled and the same dependence model was fitted to both treatments. The variogram models

$$2\gamma_1^o(h) = \begin{cases} 0, & h = 0, \\ 6.0616\{(3/2)(h/17.2980) - (1/2)(h/17.2980)^3\}, & 0 < h < 17.29890, \\ 6.0616, & h \geq 17.2980, \end{cases} \quad \text{(moldboard)}, \tag{4.3.8}$$

$$2\gamma_2^o(h) = \begin{cases} 0, & h = 0, \\ 3.3240, & h > 0, \end{cases} \quad \text{(paraplow)}, \tag{4.3.9}$$

$$2\gamma_3^o(h) = 2\gamma_4^o(h) = \begin{cases} 0, & h = 0, \\ 0.5762, & h > 0, \end{cases} \quad \text{(chisel and no-till)}, \tag{4.3.10}$$

were fitted. Because each model has a sill, the spatial dependence could equally be described through a covariogram

$$C_k^o(h) = \gamma_k^0(\infty) - \gamma_k^o(h), \qquad h \geq 0, \tag{4.3.11}$$

where $\gamma_k^o(\infty) = \lim_{h \to \infty} \gamma_k^o(h)$ is the sill of the kth model, $k = 1, 2, 3, 4$.

Building the Complete Covariance Matrix

Very few lags were available to allow estimates of the variograms in the north–south direction. Nevertheless, the estimates indicated that an isotropy assumption for the spatial dependence within each plot would be appropriate. In other words, within the kth plot,

$$2\gamma_k(\mathbf{h}) = 2\gamma_k^o(\|\mathbf{h}\|), \qquad k = 1, 2, 3, 4. \tag{4.3.12}$$

To examine the spatial dependence between neighboring plots, sample correlation coefficients were computed. They indicated a lack of dependence, leading to a block-diagonal covariance model for the data: Let $\mathbf{z}' = (\mathbf{z}_1', \mathbf{z}_2', \mathbf{z}_3', \mathbf{z}_4')$, where $\mathbf{z}_k' = (z_{1,1,k}, z_{2,1,k}, \ldots, z_{8,3,k})$. Then, if $\mathbf{Z}$ is the random vector from which $\mathbf{z}$ is a realization, the preceding discussion has established that

$$\mathrm{var}(\mathbf{Z}) \equiv \sigma^2\Sigma = \sigma^2\begin{bmatrix} \Sigma_1 & 0 & 0 & 0 \\ 0 & \Sigma_2 & 0 & 0 \\ 0 & 0 & \Sigma_3 & 0 \\ 0 & 0 & 0 & \Sigma_4 \end{bmatrix} \tag{4.3.13}$$

is a sensible model. The matrix Σ is a 96×96 block-diagonal matrix where each block Σ_i, $i = 1, 2, 3, 4$, is 24×24. Matrices Σ_2, Σ_3, and Σ_4 are proportional to the identity matrix I: $\Sigma_2 = (1.662)I$, $\Sigma_3 = \Sigma_4 = (0.2881)I$. Only the matrix Σ_1 shows spatial dependence: $\sigma^2\Sigma_1$ is made up of elements $\mathrm{cov}(Z(\mathbf{s}_i), Z(\mathbf{s}_j)) = \sigma^2\{3.0308 - \gamma_1^o(\|\mathbf{s}_i - \mathbf{s}_j\|)\}$, where γ_1^o is given by (4.3.8). If the variogram models (4.3.8), (4.3.9), and (4.3.10) fit the data well, the proportionality constant σ^2 in (4.3.13) should be approximately equal to 1.

Cross-Validation of the Fitted Variogram Models

One way to gauge model fit informally is to check various aspects of the data (including $\sigma^2 = 1$) through cross-validation. For illustrative purposes, attention is directed to the moldboard data.

Section 2.6.4 describes how, after a point $\mathbf{s}_i$ is deleted, $Z(\mathbf{s}_i)$ is predicted from surrounding data yielding predictor $\hat{Z}_{-i}(\mathbf{s}_i)$ and kriging standard error $\sigma_{-i}(\mathbf{s}_i)$. Figure 4.14 shows a normal probability plot of the standardized prediction errors $\{(\hat{Z}_{-i}(\mathbf{s}_i) - Z(\mathbf{s}_i))/\sigma_{-i}(\mathbf{s}_i)\colon i = 1, \ldots, 24\}$, where $\{\mathbf{s}_i\colon i = 1, \ldots, 24\}$ are the locations of the 24 moldboard data; a deleted datum is predicted from the remaining 23.

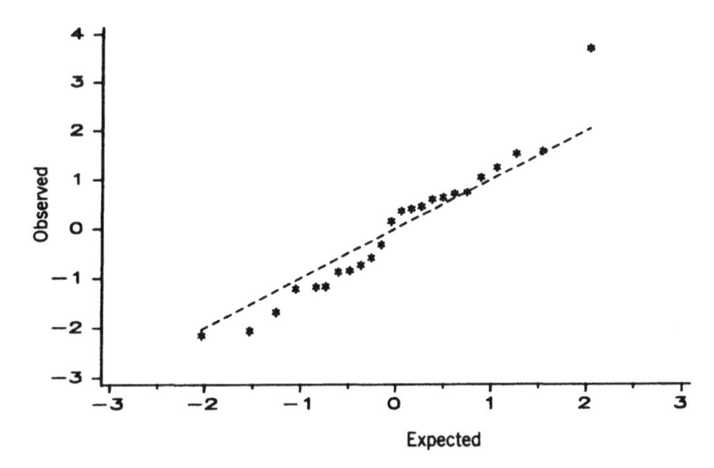

Figure 4.14 Normal probability plot of standardized cross-validation errors for moldboard data. The horizontal axis denotes the expected quantile from a sample of size 24 from a standard Gaussian distribution; the vertical axis denotes the observed quantile. [*Source:* Gotway and Cressie, 1990, © The American Geophysical Union.]

If the data are from a Gaussian model, the points should lie along a 45° line through the origin. Clearly, they do except for the outlying point corresponding to original datum $y_{3,7,1} = 38.31$ cm. Instead of deleting it, Gotway and Cressie (1990) argue for Winsorizing (or editing) it in a manner described in Section 3.3. Using editing constant $c = 2.5$, they obtained an edited value $y_{3,7,1}^{(e)} = 28.22$ cm. From checking the experimental records, no reason for this outlier was apparent (a large subsurface crack might account for the higher than expected infiltration rate).

Spatial outliers such as $y_{3,7,1}$ are large relative to their neighbors. They are hard to detect from a stem-and-leaf plot (or histogram) of the data because such a plot ignores the relative spatial locations of the observations. In the analysis to follow, the edited data will be used, although for notational convenience the superscript (e) will be dropped.

4.3.2† Inference on Mean Effects (Spatial Analysis of Variance)

It has been established that $\delta(\cdot)$ in (4.3.5) is a heteroskedastic dependent error process with covariances described through (4.3.12). The mean function has already been implicitly defined as a function of only the north–south coordinate. Before a formal analysis is carried out (testing whether spatial trend is present, testing whether treatment effects are present, etc.), it is worth emphasizing that the variance matrix (4.3.12) has been *estimated* (from a data set that is not large). This estimate might be seen as the first stage of a procedure that reestimates the parameters of the error process from generalized-least-squares residuals; those residuals are obtained using the initial estimate of the variance matrix. Carroll et al. (1988) give evidence to support a small number of iterations (here, 1) and an initial robust estimator of the large-scale effects (here, the column medians). More research is needed to determine the effects of using estimated variance-matrix parameters as if they were known; Section 5.3 has some discussion of this with regard to kriging.

Linear Model

Model the (square-root) data as a collection of random variables that satisfy

$$Z_{ijk} = \mu + t_k + \beta_{jk} + \delta_{ijk}, \qquad (4.3.14)$$

where t_k is an effect due to treatment k, β_{jk} is an effect associated with the jth column of treatment k, and $\{\delta_{ijk}\}$ is random error with mean $\mathbf{0}$ and variance matrix $\sigma^2\Sigma$ given by (4.3.13). Write (4.3.14) as

$$\mathbf{Z} = X\boldsymbol{\beta} + \boldsymbol{\delta}. \qquad (4.3.15)$$

Hypotheses tested on the transformed data $\{Z_{ijk}\}$ can be formulated and interpreted in terms of the model (4.3.14). However, ultimately one wishes to interpret them in terms of the original scientific problem posed. For example, if the null hypothesis of equal treatment means (of the square-root data) is not rejected, this can be interpreted as inferring no large-scale treatment differences.

Estimation of Mean Effects

The first step toward inference on $X\boldsymbol{\beta}$ is to specify an estimation procedure. If spatial dependence is ignored, or overlooked (as is often the case), the ordinary-least-squares (o.l.s.) estimator of $X\boldsymbol{\beta}$, namely,

$$X\hat{\boldsymbol{\beta}}_{\text{ols}} = X(X'X)^{-}X'\mathbf{Z}, \tag{4.3.16}$$

might be used (e.g., Rao, 1973, Section 4a). For the soil–water infiltration data, because of the heteroskedasticity and spatial dependence, a generalized-least-squares (g.l.s.) estimator of $X\boldsymbol{\beta}$, namely,

$$X\hat{\boldsymbol{\beta}}_{\text{gls}} = X(X'\Sigma^{-1}X)^{-}X'\Sigma^{-1}\mathbf{Z}, \tag{4.3.17}$$

is more appropriate (e.g., Rao, 1973, Section 4a).

Figure 4.15*a* and *b* shows residual plots of estimated mean versus residual for o.l.s. and g.l.s., respectively. Without the weighting, residuals are clearly heteroskedastic, and differences in the estimated treatment means are not as obvious.

A Spatial Analysis of Variance

Let V be an $n \times n$ symmetric positive-definite weight matrix. (In the analysis to follow, V will be either Σ, diag(Σ), or the identity matrix.) A general analysis of variance can be written as follows.

Source of Variation	Degrees of Freedom	Sum of Squares
Model	$KJ-1$	$\hat{\boldsymbol{\beta}}'_V X'V^{-1}\mathbf{Z} - m_V$
Treatments	$K-1$	SS_V(treatments)
Columns within treatments	$K(J-1)$	(obtained by subtraction)
Residual	$n-KJ$	$\mathbf{Z}'V^{-1}\mathbf{Z} - \hat{\boldsymbol{\beta}}'_V X'V^{-1}\mathbf{Z}$
Corrected total	$n-1$	$\mathbf{Z}'V^{-1}\mathbf{Z} - m_V$

The estimator $X\hat{\boldsymbol{\beta}}_V$ is given by (4.3.17) with weight matrix V replacing Σ; $m_V \equiv (\mathbf{1}'V^{-1}\mathbf{1})^{-1}(\mathbf{Z}'V^{-1}\mathbf{1}\mathbf{1}'V^{-1}\mathbf{Z})$. The sum of squares, SS_V(treatments), is obtained from a model-sum-of-squares calculation where the model is $Z_{ijk} = \mu + t_k + \delta_{ijk}$, fitted by generalized least squares using weight matrix V; K (= 4) is the number of treatments, J (= 3) is the number of columns within each treatment, and n (= 96) is the total number of observations. These formulas can be found in Gotway and Cressie (1990) and, in a more general setting, in Wong (1989).

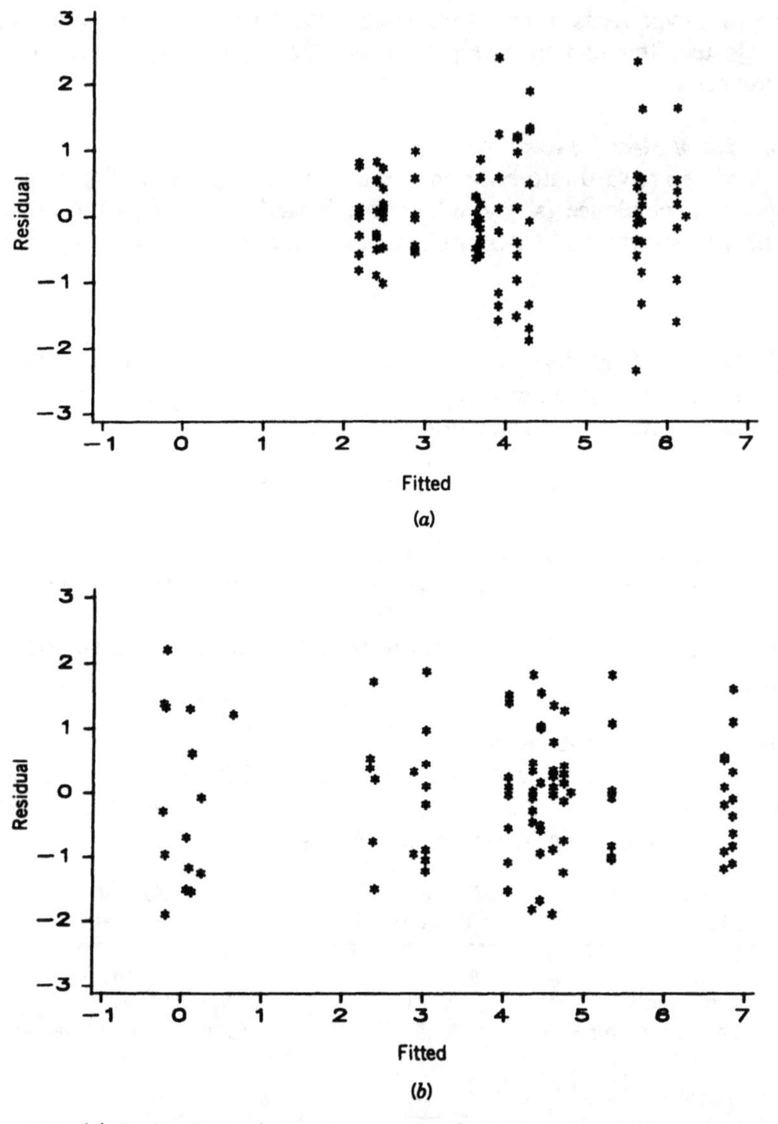

Figure 4.15 (*a*) Residual plot (residual versus fitted) obtained from ordinary-least-squares fitting of model (4.3.14). Units on both axes are in $cm^{1/2}$. (*b*) Same as in (*a*), except the residual plot is obtained from generalized-least-squares fitting. [*Source:* for (*a*) and (*b*); Gotway and Cressie, 1990, © The American Geophysical Union.]

Table 4.3 General Analysis of Variance for Soil–Water-Infiltration Data, using Weight Matrix V

Source of Variation	Degrees of Freedom	Sum of Squares
	$(a)\ V = \Sigma$	
Model	11	183.23
Treatments	3	95.40
Columns in treatments	8	87.83
Residual	84	94.12
Corrected total	95	277.34
	$(b)\ V = \text{diag}(\Sigma)$	
Model	11	168.16
Treatments	3	115.52
Columns in treatments	8	52.64
Residual	84	69.99
Corrected total	95	238.15
	$(c)\ V = I$	
Model	11	157.83
Treatments	3	114.31
Columns in treatments	8	43.52
Residual	84	72.83
Corrected total	95	230.66

Source: Gotway and Cressie (1990) © The American Geophysical Union.

Table 4.3 shows how different the analyses of variance are for $V = \Sigma$ (*full spatial model*), for $V = \text{diag}(\Sigma)$ (*heteroskedastic model*), and for $V = I$ (*classical model*). The correct analysis of variance is given by Table 4.3(a), which is the full spatial model with $V = \Sigma$.

In a pilot study of soil properties, Miller et al. (1988) found evidence of spatial dependence. They then resampled at select sites so that all distances between sites were larger than the variogram ranges; from these new data they correctly performed a classical analysis of variance ($V = I$). The full spatial analysis of variance ($V = \Sigma$) presented in the preceding text, has the advantage that it can be performed in the majority of cases when resampling for independent data cannot be implemented.

Testing Equality of Columns within Treatments

In this spatial context, testing the equality of treatments effects should only be performed after a test for absence of spatial trend (within treatments) is performed. This latter test is equivalent to testing for constant column means (i.e., test H_0: $\beta_{11} = \beta_{21} = \beta_{31}$; $\beta_{12} = \beta_{22} = \beta_{32}$; $\beta_{13} = \beta_{23} = \beta_{33}$;

$\beta_{14} = \beta_{24} = \beta_{34}$) against the model (4.3.14). To do this, compute the ratio

$$F_V \equiv \frac{\{\hat{\boldsymbol{\beta}}'_V X' V^{-1}\mathbf{Z} - m_V - \mathrm{SS}_V(\text{treatments})\}/K(J-1)}{\{\mathbf{Z}'V^{-1}\mathbf{Z} - \hat{\boldsymbol{\beta}}'_V X'V^{-1}\mathbf{Z}\}/(n-KJ)}. \quad (4.3.18)$$

Note that if heteroskedasticity and spatial correlation are present, neither F_I nor $F_{\mathrm{diag}(\Sigma)}$ have an F distribution.

From the soil–water-infiltration data,

$$\begin{aligned} F_\Sigma &= 9.80, \\ F_{\mathrm{diag}(\Sigma)} &= 7.90, \\ F_I &= 6.27. \end{aligned} \quad (4.3.19)$$

Comparing $F_\Sigma = 9.80$ to an F distribution (the only comparison that is asymptotically valid) with 8 and 84 degrees of freedom, it is seen that spatial trend is significant at the 0.1% level. Notice that F_I, the F ratio for the classical model, is smaller than F_Σ.

Testing Equality of Average Treatment Effects

Now that the hypothesis of constant column mean within a treatment has been rejected, the treatment-plot effects can be compared by testing

$$H_0: t_1 + \tfrac{1}{3}\sum_{j=1}^{3}\beta_{j1} = t_2 + \tfrac{1}{3}\sum_{j=1}^{3}\beta_{j2} = t_3 + \tfrac{1}{3}\sum_{j=1}^{3}\beta_{j3} = t_4 + \tfrac{1}{3}\sum_{j=1}^{3}\beta_{j4} \quad (4.3.20)$$

against the general alternative (4.3.14). Because treatment effects are confounded with location, rejection of H_0 may be due to a difference in treatment effects or a difference in spatial locations; this location effect is expressed through average column means added to respective treatment effects in (4.3.20).

To test (4.3.20), compute the ratio

$$G_V \equiv \frac{\mathrm{SS}_V(\text{treatments})/(K-1)}{\{\mathbf{Z}'V^{-1}\mathbf{Z} - \hat{\boldsymbol{\beta}}_V X'V^{-1}\mathbf{Z}\}/(n-KJ)}. \quad (4.3.21)$$

Note that if heteroskedasticity and spatial correlation are present, neither G_I nor $G_{\mathrm{diag}(\Sigma)}$ have an F distribution.

From the soil–water-infiltration data,

$$\begin{aligned} G_\Sigma &= 28.05, \\ G_{\mathrm{diag}(\Sigma)} &= 44.95, \\ G_I &= 43.95. \end{aligned} \quad (4.3.22)$$

Comparing $G_{\Sigma} = 28.05$ to an F distribution (the only comparison that is asymptotically valid) with 3 and 84 degrees of freedom, it is seen that average treatment effects are significantly different at the 0.1% level.

Notice in (4.3.22) that G_I and $G_{\mathrm{diag}(\Sigma)}$ are much larger than G_{Σ}. In general, when data exhibit positive spatial dependence, use of G_I results in more frequent declarations of significant treatment differences than the data warrant.

Pairwise Contrasts

Gotway and Cressie (1990) give the six possible pairwise comparisons between the four treatments and conclude that paraplow ($k = 2$), chisel ($k = 3$), and no-till ($k = 4$) are all different from each other. But, because of the strong positive spatial dependence, moldboard ($k = 1$) cannot be declared significantly different (at the 5% level) from any of the other treatments. Positive spatial dependence reduces the "equivalent number of independent observations" (Section 1.3). Thus, although the estimate of the moldboard effect looks different from other treatment estimates, it has lower precision than the others.

The g.l.s. estimates of the effects given in (4.3.20) are $\{\hat{t}_k + (1/3)\Sigma_{j=1}^{3}\hat{\beta}_{jk}: k = 1,\ldots,4\}$, where the individual treatment and column effects are estimated by (4.3.17). They are, 2.022 $\mathrm{cm}^{1/2}$ for moldboard ($k = 1$), 2.478 $\mathrm{cm}^{1/2}$ for paraplow ($k = 2$), -0.228 $\mathrm{cm}^{1/2}$ for chisel ($k = 3$), and 0.494 $\mathrm{cm}^{1/2}$ for no-till ($k = 4$). Furthermore, $\hat{\mu} = 2.730$ $\mathrm{cm}^{1/2}$, which from (4.3.14) implies that the g.l.s. estimates of the mean soil–water infiltration (on the square-root scale) are

$$\hat{E}(Z_{\cdot\cdot 1}) = 4.752\ \mathrm{cm}^{1/2} \quad (\text{moldboard}),$$

$$\hat{E}(Z_{\cdot\cdot 2}) = 5.208\ \mathrm{cm}^{1/2} \quad (\text{paraplow}),$$

$$\hat{E}(Z_{\cdot\cdot 3}) = 2.502\ \mathrm{cm}^{1/2} \quad (\text{chisel}),$$

$$\hat{E}(Z_{\cdot\cdot 4}) = 3.224\ \mathrm{cm}^{1/2} \quad (\text{no-till}).$$

From the results of Gotway and Cressie (1990), and assuming minimal plot effects, paraplow is declared the superior treatment, followed by no-till and then chisel. Although moldboard looks to be an excellent treatment, there is not enough evidence (due to the presence of spatial correlation) to declare it different from any of the other three treatments.

The multiple comparison of pairwise contrasts needs care when many treatments are involved. Hayter (1984) discusses a heteroskedastic version of the Tukey–Kramer multiple comparison procedure (caused by unequal sample sizes). In situations where correlation is present, Hayter (1989) gives a sufficient condition (whose applicability to spatial correlation models seems unlikely) that guarantees that the usual Tukey–Kramer confidence intervals are conservative.

4.4[†] SUDDEN-INFANT-DEATH-SYNDROME DATA

This section presents the geostatistical part of an exploratory spatial data analysis of sudden-infant-death counts in the 100 counties of North Carolina from 1974–1978; the rest of the analysis is given in Section 6.2. Cressie and Read (1989) and Cressie and Chan (1989) (see also Section 7.6) give a more complete picture of both exploratory and confirmatory approaches to the analysis of these data.

Counts data from spatially contiguous regions offer a challenge to the statistician. Although not apparently of the type of problem considered in Part I, it will be seen how such data can be investigated geostatistically. Typical applications are in epidemiology (e.g., cancer mortality over the counties of the United States) and census surveys (e.g., undercount over census blocks of an urban area).

Time-series analysts have long recognized that data close together in time usually exhibit higher dependence than those far apart. Time-series data analysis relies on methods such as data transformation, detrending, and autocorrelation plotting. This approach is generalized to the spatial setting here and in Section 6.2; the main emphasis in this section is on variogram plotting, to summarize the spatial dependence in the data. This could be followed up with model fitting and kriging for cross-validation or smoothing purposes.

Some Background

Sudden infant death syndrome (SIDS) is currently a leading category of postneonatal death, yet its cause is still a mystery. SIDS is defined as the sudden death of any infant up to 12 months old, that is unexpected by medical history and where the death remains inexplicable after performance of an adequate postmortem examination. It accounts for about 7000 deaths a year in the United States, taking the lives of about two infants per 1000 live births. In contrast to the usual pathologic and physiologic studies of SIDS (see Section 6.2 for details), an epidemiologic approach is taken here and in Sections 6.2 and 7.6.

The 1974–1978 SIDS data can be found in Table 6.1, where the number of SIDS and live births are presented for each of the 100 counties of North Carolina, along with the county seat locations. A map of the counties is given in Figure 6.1.

Section 6.2 details an exploratory spatial data analysis. In the absence of any known cause for SIDS, it is sensible to take this "univariate" look at the data. Much as time-series modeling looks to see how a particular value is influenced by its past values, spatial-process modeling looks to see how a particular value is influenced by its neighboring values. However, the construction of spatial models is more complicated than temporal models, because the space index does not possess the natural ordering of the time index (see Cressie and Chan, 1989, and Section 7.6 for Markov random-field modeling of the SIDS data).

Summary of an Exploratory Spatial Analysis

In order to construct variogram estimators, it is necessary to present, briefly, the conclusions of the exploratory spatial analysis detailed in Section 6.2.

Let $\{S_i: i = 1,\ldots,100\}$, $\{n_i: i = 1,\ldots,100\}$ denote, respectively, the number of SIDS and the number of live births in the 100 counties of North Carolina, 1974–1978. Section 6.2 concludes that the Freeman–Tukey (square-root) transformation

$$Z_i = (1000(S_i)/n_i)^{1/2} + (1000(S_i + 1)/n_i)^{1/2}, \qquad (4.4.1)$$

is a variance-controling transformation; that is,

$$\text{var}(Z_i) \simeq \tau^2/n_i, \qquad (4.4.2)$$

not depending on $E(S_i/n_i)$. However, the variances are still vastly different from county to county because $248 \leq n_i \leq 21{,}588$, $i = 1,\ldots,100$. The transformation (4.4.1) also symmetrizes the data, an effect that is retained even after the spatial trend is removed.

Think of the data as being made up of large-scale variation (spatial trend) plus small-scale variation (spatially dependent error). The principle of transforming to achieve additivity over all scales of variation (Cressie, 1985b) and the two-dimensional (spatial) coordinate representation of the counties' locations, leads naturally to a two-way additive decomposition for the spatial trend of the transformed data. For irregularly located data, such as the counties of North Carolina, the decomposition can be achieved by using a low-resolution map. Cressie and Read (1989) overlay a 9×24 rectangular grid with 20×20 mile grid spacing onto North Carolina; a county is identified with the node of the grid that is closest to its county seat. This particular choice of a regularly spaced rectangular grid gives roughly one county at each grid node; on another occasion (Cressie and Guo, 1987), an irregularly spaced rectangular grid has been used to achieve the same effect.

Moving the datum Z_i to the nearest node of the grid, it is written as $Z(l(i), 10 - k(i))$, where $(k(i), l(i)) \in \{(k,l): k = 1,\ldots,9, \ l = 1,\ldots,24\}$. Decompose

$$Z(l(i), 10 - k(i)) = m + r_{k(i)} + c_{l(i)} + \nu(x_i, y_i), \qquad (4.4.3)$$

where (x_i, y_i) are the Cartesian coordinates (given in Table 6.1) of the ith county seat. Section 6.2 carries out a *weighted median polish* (see, e.g., Section 3.5.1 or Emerson and Hoaglin, 1983, for details on median polish), weighting each Y_i proportional to its $(n_i)^{1/2}$. Then the presence of an interaction term in (4.4.3) was investigated and none was found. The map of the median-polish fitted values, $\{m + \tilde{r}_{k(i)} + \tilde{c}_{l(i)}: i = 1,\ldots,100\}$, is a crude smoothed map that takes into account the unequal variation due to unequal n_is (see Fig. 6.5*a*).

Define the residuals

$$R(x_i, y_i) \equiv Z_i - \tilde{m} - \tilde{r}_{k(i)} - \tilde{c}_{l(i)}, \tag{4.4.4}$$

which, when normalized, yield data

$$\{n_i^{1/2} R(x_i, y_i) : i = 1, \dots, 100\}. \tag{4.4.5}$$

However, the normalized residual of Anson County (county 4) is unacceptably high; it is shown in Section 7.6 that not even a significant covariate can explain it, and its cause remains a mystery. Henceforth, county 4 is deleted from the spatial analysis.

Variogram Estimation

The spatial data analysis just described has a statistical model behind it, which is

$$Z(l(i), 10 - k(i)) = m + r_{k(i)} + c_{l(i)} + n_i^{-1/2}\delta(x_i, y_i), \tag{4.4.6}$$

where $\delta(\cdot)$ is a zero-mean intrinsically stationary random process with variogram

$$2\gamma(\mathbf{h}) \equiv \operatorname{var}(\delta(\mathbf{s} + \mathbf{h}) - \delta(\mathbf{s})). \tag{4.4.7}$$

Further exploratory analysis indicated that $2\gamma(\cdot)$ is isotropic, namely,

$$2\gamma(\mathbf{h}) = 2\gamma^o(\|\mathbf{h}\|), \qquad \mathbf{h} \in \mathbb{R}^2. \tag{4.4.8}$$

The normalized residuals (4.4.5) estimate the errors $\{\delta(x_i, y_i)\}$ whose dependence structure is summarized by the variogram $2\gamma^o(\cdot)$. Thus, it is natural to estimate the spatial dependence of the model (4.4.6) from the normalized residuals $\{n_i^{1/2} R(x_i, y_i)\}$. Combine (4.4.7) and (4.4.8) to obtain $2\gamma^o(d_{ij}) = \operatorname{var}(\delta(x_i, y_i) - \delta(x_j, y_j))$, where $d_{ij} \equiv \{(x_i - x_j)^2 + (y_i - y_j)^2\}^{1/2}$, the Euclidean distance of the ith county seat from the jth county seat. Notice that it is equally possible to model (isotropic) intrinsic stationarity using a distance measure that is not Euclidean, but involves say differences in levels of urbanization, distance by road, and so forth.

In estimating the variogram, the values $\{n_i^{1/2} R(x_i, y_i) : i \neq 4\}$ are associated with the coordinates of the corresponding county seats. However, due to their irregular locations, some lag grouping is necessary in order to obtain multiple contributions to the estimate at each lag.

Define

$$2\hat{\gamma}^o(h) \equiv \sum_{N(h)} \{n_i^{1/2} R(x_i, y_i) - n_j^{1/2} R(x_j, y_j)\}^2 \Big/ |N(h)|, \tag{4.4.9}$$

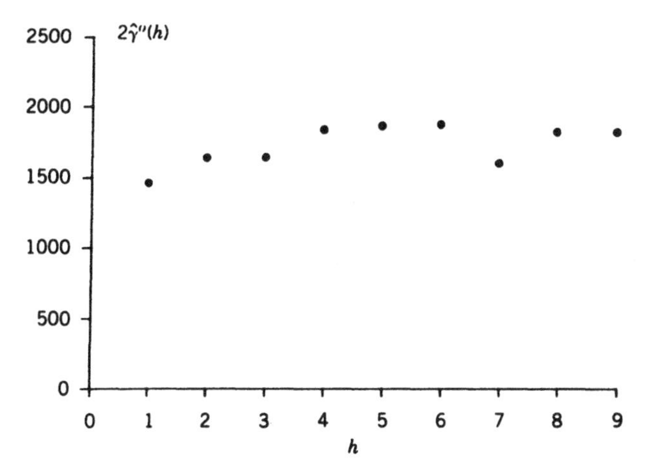

Figure 4.16 Variogram estimator $2\hat{\gamma}^o(\cdot)$ based on standardized residuals (4.4.5), with county 4 deleted. One lag unit on the horizontal axis is 20 miles. [*Source:* Cressie and Read, 1989]

where $N(h) \equiv \{(i,j): d_{ij} \in (20h - 10, 20h + 10]; i \neq 4, j \neq 4\}$, $h = 1, 2, \ldots$. This is the estimator (2.4.2) (actually it is the smoothed version $2\gamma^+$; see Section 2.4) owing to Matheron (1962). There are more robust estimators of $2\gamma(\cdot)$ available [such as (2.4.12)], but with the outlier removed, these should not be necessary. Figure 4.16 shows the plot of $2\hat{\gamma}^o(h)$ versus lag distance h.

Because the variogram appears to asymptote to $2\gamma^o(\infty)$, an isotropic covariogram $C^o(h)$ between data distance h apart can be assumed to exist. Then

$$C^o(h) = \gamma^o(\infty) - \gamma^o(h).$$

Thus, Figure 4.16 can be interpreted as showing positive correlation for lag 1 (10–30 miles), whereas at larger lags independence is indicated. The marked dip at $h = 7$ (i.e., the 130–150 mile range) is interesting; it appears to correspond to the separation of a cluster of high-SID-rate counties in the northeast from a similar high-SID cluster in the south.

Some Final Remarks

The use of median polish to remove spatial trend when the locations are *irregular* shows the versatility of this approach to spatial analysis (viz. resistant detrending followed by variogram estimation). Because the data here are aggregates over counties, precise spatial locations are not really available, making the low-resolution map implied by median polish a sensible goal.

When irregularly located data are not aggregates, such as for the acid-rain data in Section 4.6, it is important to respect the spatial locations more closely. There, interpolating planes are used as estimates of the large-scale variation, from which residuals are computed and used to characterize the spatial dependence.

The next step in a geostatistical analysis of the SIDS data is to fit a variogram model with nugget effect (that accounts for measurement error and location error) to the estimated variogram shown in Figure 4.16. Kriging can then be carried out to cross-validate the model (Section 2.6) and to provide a spatially smooth county map of (Freeman–Tukey-transformed) SID rates. Focus attention for the moment on the ith county. The normalized residuals for the neighboring counties (and for the ith county itself) could be used in the kriging equations (3.2.5) and (3.2.12) to find $\hat{\delta}(x_i, y_i)$, the optimal linear combination of values that predict the stationary error $\delta(x_i, y_i)$. Then the smoothed (predicted) value of Z_i is $\tilde{Z}_i = \tilde{m} + \tilde{r}_{k(i)} + \tilde{c}_{l(i)} + n_i^{-1/2}\hat{\delta}(x_i, y_i)$. Because spatial locations are of low resolution, when the (normalized residual of the) ith county is included in the predictor $\hat{\delta}(x_i, y_i)$, care must be taken in (3.2.12) to offset the prediction location $\mathbf{s}_{0,i}$ slightly from the ith county seat's location (x_i, y_i). This same offsetting technique is employed in Section 4.5 and is necessitated by imprecise location information on the counties.

The spatially smoothed (Freeman–Tukey-transformed) SID rates are $\{\tilde{Z}_i: i = 1, \ldots, 100\}$. From these values, a choropleth map could be drawn that could be used to search for spatial patterns. The choropleth map of unsmoothed (transformed) rates gives a poor picture due to large heteroskedastic variation; see Section 6.2 for further discussion of spatial smoothing and Section 7.6 for a spatial predictor based on a Gaussian Markov-random-field model.

4.5† WHEAT-YIELD DATA

Some of the exploratory spatial data analyses of this section can be found in Cressie (1985c); there the beginnings of a geostatistical analysis of the Mercer and Hall (1911) wheat-yield data were presented. A more complete analysis is given here, together with spatial maps of wheat-yield contours throughout the 1-acre region of interest.

Agriculture and Statistics go back a long way. The origins of Statistical Science can be found in the design and analysis of agricultural experiments that began in earnest around the turn of this century (e.g., Gower, 1988). R. A. Fisher obtained much of his statistical inspiration (and data) from experiments at the Rothamsted Experimental Station in England, as is evidenced in his famous book *Statistical Methods for Research Workers* (Fisher, 1925). In the United States, the Statistical Laboratory's work for the Agricultural Experimental Station at Iowa State College (now Iowa State University) is plainly seen in G. W. Snedecor's book *Statistical Methods* (Snedecor, 1937).

To determine conditions under which a crop will achieve high yields, experiments are usually designed based on three basic concepts: randomization, blocking, and replication. These allow efficient comparison of treatment effects $\boldsymbol{\alpha}$. Specifically, randomization attempts to neutralize the effects of spatial correlation, blocking helps to reduce the residual variation, and replication allows precise estimation of $\boldsymbol{\alpha}$ (see, e.g., Kempthorne, 1952). If $\mathbf{Z}$ is the $n \times 1$ vector of experimental responses, then the model

$$\mathbf{Z} = \mathbf{1}\mu + X_1\boldsymbol{\alpha} + X_2\boldsymbol{\beta} + \boldsymbol{\epsilon}, \tag{4.5.1}$$

is usually fitted, where μ is an overall mean effect, $\boldsymbol{\alpha}$ contains the treatment effects, $\boldsymbol{\beta}$ contains the block effects, and $\boldsymbol{\epsilon}$ is a vector of i.i.d. zero-mean errors.

Classical experimental design of agricultural field trials ignores the spatial position of the treatment in the design; however, recently it has been realized that more efficient treatment-contrast estimators can be obtained by exploiting spatial variation (e.g., Duby, et al., 1977; Wilkinson et al., 1983; Martin, 1986a; Besag and Kempton, 1986; Grondona and Cressie, 1991). Here, $\boldsymbol{\epsilon}$ in the model (4.5.1) is replaced with a spatially dependent error term. Not surprisingly then, expected residual mean squares under classical randomization schemes can be decreased by using systematic designs or restricted randomization schemes. These issues of experimental design in a spatial setting are discussed in some detail in Section 5.6.2.

Some Background

A uniformity trial is an experiment where *all* the plots are treated with the control treatment. The data consist of plot yields for one specific crop, like wheat. By aggregating adjacent plots into new plots and computing mean-squared error as a function of new-plot size, an optimal plot size can be determined for future wheat-yield trials. Such was the goal of Mercer and Hall (1911), who reported the results of mangold and wheat uniformity trials carried out at Rothamsted Experimental Station in 1910. The wheat-yield data are considered here.

The data given by Mercer and Hall (1911) consist of yields on a 20×25 lattice of plots approximately 1 acre in total area. Each of 20 rows runs in the east–west direction and each of 25 columns runs in the north–south direction. In the past, authors have used this data set to fit lattice models, such as those described in Chapter 6. A review of these inference procedures is given in Section 7.1 and a data map is given in Figure 7.1.

Although the data are given on a spatial lattice, it *is* possible to think about wheat yields on potential plots located *between* existing plots. In other words, the spatial index could run continuously over the 1-acre region.

Suppose a stationary variogram or covariogram is needed to characterize the spatial dependence, or suppose a map of wheat yields over the entire area would be useful. Geostatistical methods can provide good answers to these problems.

Spatial Configuration

The dimensions of the plots should determine their spatial locations, but closer inspection of Mercer and Hall's (1911) article reveals some ambiguity as regards the plot size. They state clearly that the plot length (in the east–west direction) is 10.82 ft. The plot width, however, is given as 11 furrows with a guard furrow between each plot; the distance between furrows is not given. A look through the literature shows a great variety of plot sizes assumed:

Fairfield Smith (1938)	10.82 × 8.05 ft
Whittle (1954); Besag (1974)	10.82 × 11 ft
Ripley (1981)	~ 11 ft square
McBratney and Webster (1981)	10.82 × 8.5 ft
Wilkinson et al. (1983)	10.8 × 8.25 ft
Kunsch (1985)	3.3 × 2.5 m

Fairfield Smith's dimensions (actually he worked with an *area* of 87 ft^2) are obtained by dividing 1 acre by 500 plots. It appears that Whittle (and then Besag and Ripley) mistook furrows for feet. The last three articles all use about the same plot dimensions. Because there is no way to tell which is correct, the dimensions of Wilkinson et al. are used; converted to meters, they are 3.30 m (*east–west*) × 2.51 m (*north–south*).

From the brief description of the spatial configuration of plots in Mercer and Hall (1911), it appears that they were not always contiguous nor of exactly the same area. Thus, there are small errors in the spatial locations of the data, which contribute to the nugget effect of the estimated variogram.

4.5.1† Presence of Trend in the Data

From a two-dimensional spectral analysis, McBratney and Webster (1981) found an obvious peak in the east–west spectrum corresponding to a period three plots long. Ripley (1981) found the same peak, guessing that it might be due to variation in soil fertility caused by layers in the outcropping rocks. However, McBratney and Webster found evidence of an earlier ridge and furrow pattern of plowing on the field in question, which seems a more plausible explanation. Earlier, Patankar (1954) had tried to fit a linear east–west trend, but it appears from the plot of column means versus column number, given by Mercer and Hall (1911), that neither a linear trend nor a periodic component is flexible enough to capture the large-scale variation in these data.

No transformation of the wheat yields seems necessary, so all analyses are carried out in the original scale of pounds (of grain). Write the data of Figure 7.1 as $\{z(j(3.30), (21 - i)(2.51)): i = 1, \ldots, 20;\ j = 1, \ldots, 25\}$, where $i = 1$ corresponds to the most northerly row and $j = 1$ corresponds to the most westerly column. Respecting the two-way layout of the plots, I propose the

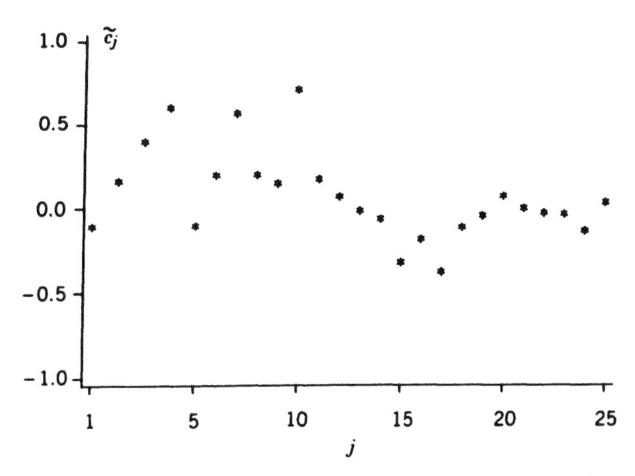

Figure 4.17 Plot of median-polish column effects versus column number. An irregular east–west trend in the wheat-yield data is indicated. Units on the vertical axis are in pounds of wheat.

stochastic model

$$Z(j(3.30),(21-i)(2.51)) = a + r_i + c_j + \delta(j(3.30),(21-i)(2.51)). \tag{4.5.2}$$

The estimated all effect $\tilde{a}$, the estimated row effects $\{\tilde{r}_i: i = 1,\ldots,20\}$, and the estimated column effects $\{\tilde{c}_j: j = 1,\ldots,25\}$ are obtained by median polish (see Section 3.5).

Figure 4.17 shows a plot of $\tilde{c}_j$ versus j, $j = 1,\ldots,25$. Clearly the effects are large, but do not show a regular pattern. Not presented here, the plot of $\tilde{r}_i$ versus i, $i = 1,\ldots,20$, shows very little variation.

Figure 4.18 shows a three-dimensional plot of the surface defined by $\{\tilde{a} + \tilde{r}_i + \tilde{c}_j: i = 1,\ldots,20;\ j = 1,\ldots,25\}$; interpolating planes connect the values at the grid nodes. The variability in the east–west direction is striking.

4.5.2† Intrinsic Stationarity

It is clear that the wheat-yield data should be detrended before variograms can be estimated and modeled. Define

$$R(j(3.30),(21-i)(2.51)) \equiv Z(j(3.30),(21-i)(2.51)) - \tilde{a} - \tilde{r}_i - \tilde{c}_j. \tag{4.5.3}$$

The next step is to analyze the median-polish residuals $\{R(j(3.30),(21-i)(2.51)): i = 1,\ldots,20;\ j = 1,\ldots,25\}$ as if they are a sampling from an intrinsically stationary random process. Estimated variograms based on these

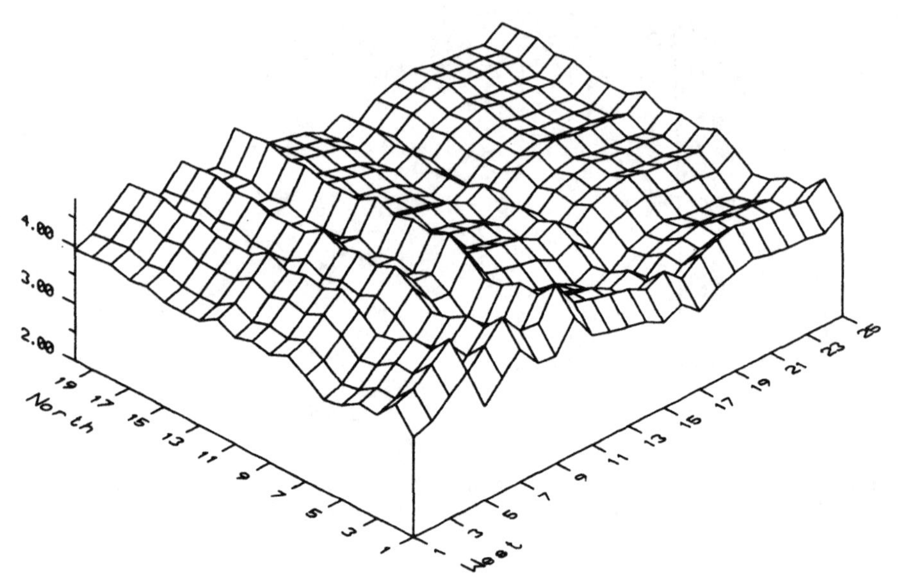

Figure 4.18 Three-dimensional perspective of the median-polish surface given by $\{\tilde{a} + \tilde{r}_i + \tilde{c}_j$: $i = 1, \ldots, 20$; $j = 1, \ldots, 25\}$ and interpolating planes. Units on the vertical axis are in pounds of wheat.

residuals will be used to estimate

$$2\gamma(\mathbf{h}) \equiv \text{var}(\delta(\mathbf{s} + \mathbf{h}) - \delta(\mathbf{s})), \qquad \mathbf{h} \in \mathbb{R}^2. \tag{4.5.4}$$

Section 2.4 offers various alternative estimators of (4.5.4). The two that I routinely compute and compare are

$$2\hat{\gamma}(\mathbf{h}) = \sum_{N(\mathbf{h})} \left(R(\mathbf{s}_i) - R(\mathbf{s}_j)\right)^2 / |N(\mathbf{h})|,$$

$$2\bar{\gamma}(\mathbf{h}) = \left\{ \sum_{N(\mathbf{h})} |R(\mathbf{s}_i) - R(\mathbf{s}_j)|^{1/2} / |N(\mathbf{h})| \right\}^4 \Big/ \{0.457 + 0.494/|N(\mathbf{h})|\};$$

see (2.4.2) and (2.4.12), respectively. The latter estimator automatically downweights outlying data. It will be seen that there *are* a few wheat-yield values that do not fit the Gaussian model very well.

Estimating and Fitting the Variogram

Estimated variograms, based on median-polish residuals, were computed in both directions and indicated that an *isotropic* intrinsic-stationarity assumption would be appropriate for the error process $\delta(\cdot)$ in (4.5.2). Thus,

$$2\gamma(\mathbf{h}) = 2\gamma^o(\|\mathbf{h}\|), \qquad \mathbf{h} \in \mathbb{R}^2, \tag{4.5.5}$$

and $2\gamma^o(\cdot)$ is estimated by

$$2\bar{\gamma}^o(h) \equiv \left\{ \sum_{N(h)} |R(\mathbf{s}_i) - R(\mathbf{s}_j)|^{1/2} / \, |N(h)| \right\}^4 \Big/ \{0.457 + 0.494/\, |N(h)|\}, \tag{4.5.6}$$

where $\mathbf{s}_i, \mathbf{s}_j \in \{(l(3.30), (21 - k)(2.51)) \colon k = 1, \ldots, 20;\ l = 1, \ldots, 25\}$,

$$N(h) \equiv \{(\mathbf{s}_i, \mathbf{s}_j) \colon \|\mathbf{s}_i - \mathbf{s}_j\| \in (h - 4, h]\}, \tag{4.5.7}$$

and $h = 4, 8, 12, \ldots, 52$ m. Furthermore, define

$$d(h) \equiv \sum_{N(h)} \|\mathbf{s}_i - \mathbf{s}_j\| / \, |N(h)|, \qquad h = 4, 8, \ldots, 52. \tag{4.5.8}$$

Figure 4.19 shows a plot of $2\bar{\gamma}^o(h)$ versus $d(h)$. Rather than plot against $d(h)$, one could alternatively use the midpoint $h - 2$ of the interval. It matters little here which is used because tolerance intervals are small and spatial locations are regular. The average distance was used in this section for illustrative purposes. In other examples in this book, I have used mostly the midpoint, in line with various kernel methods of smoothing.

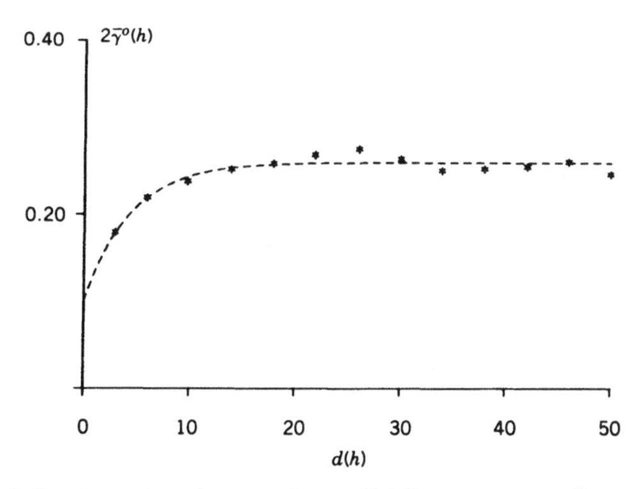

Figure 4.19 Robust isotropic variogram estimator (4.5.6) versus average distance (4.5.8). The superimposed dashed line shows the weighted-least-squares fit of an exponential variogram model $2\gamma(h; c_0, c_e, a_e) = 2\{c_0 + c_e(1 - \exp(-h/a_e))\}$, $h > 0$. Units on the vertical axis are in pounds squared; units on the horizontal axis are in meters.

Superimposed on Figure 4.19 is the weighted-least-squares fit (Section 2.6.2) of the exponential variogram

$$2\gamma^{o}(h;\hat{\boldsymbol{\theta}}) \equiv \begin{cases} 0, & h = 0, \\ 0.10 + 0.16(1 - \exp(-h/4.387)), & h > 0, \end{cases} \quad (4.5.9)$$

where $\hat{\boldsymbol{\theta}} = (0.050, 0.080, 4.387)'$; see (2.3.9). Other models such as the spherical were tried but none fitted as well.

The appearance of the nugget effect $c_0 = 0.050$ in (4.5.9) may look strange at first because the data are aggregations over a 3.30×2.51 m plot. Its presence in the model is largely due to imprecision in the spatial locations of the data; more is made of this in Section 4.5.3.

Cross-Validation of the Data

Section 2.6.4 explains how, by deleting a datum and using the fitted variogram to predict it from surrounding data, a set of standardized prediction residuals

$$\left\{ \frac{Z(\mathbf{s}_j) - \hat{Z}_{-j}(\mathbf{s}_j)}{\sigma_{-j}(\mathbf{s}_j)} : j = 1, \ldots, n \right\} \quad (4.5.10)$$

can be computed. These are highly dependent values; however, marginally each should have mean 0 and variance 1 (and come from a Gaussian distribution). An average and an average sum of squares of (4.5.10) provide a rough check on any bias in fitting the variogram.

Hawkins and Cressie (1984) describe how data can be edited for outliers, essentially by Winsorizing the absolute standardized residuals in (4.5.10) down to 2, 2.5, or 3 (see also Section 3.3). To see whether any data points are unusually large, the values in (4.5.10) are plotted on a normal probability plot; see Figure 4.20.

The three unusual values correspond to locations $\mathbf{s}_i = (3.3, 12.55)'$, $(49.5, 22.59)'$, and $(16.5, 7.53)'$; one is on the edge and two are in the middle of the field. No previous analyses of the wheat-yield data have deemed these values unusual because, when a normal probability plot of the original data (or for that matter the median-polish residuals) is constructed, there are no apparent outliers. This is because spatial information is not used in these latter plots. Nevertheless, the three values seen in Figure 4.20 are unusual *in relation to their surrounding values*.

Rather than edit the data by Winsorizing according to a fixed constant (2, 2.5, or 3 say), I Winsorized each datum such that it fell on the 45° line in Figure 4.20. Thus, I replaced

$$Z(3.3, 12.55) = 5.13 \text{ lb} \quad \text{with} \quad Z^{(e)}(3.3, 12.55) = 4.80 \text{ lb},$$
$$Z(49.5, 22.59) = 4.86 \text{ lb} \quad \text{with} \quad Z^{(e)}(49.5, 22.59) = 4.40 \text{ lb},$$
$$Z(16.5, 7.53) = 4.84 \text{ lb} \quad \text{with} \quad Z^{(e)}(16.5, 7.53) = 4.52 \text{ lb}.$$

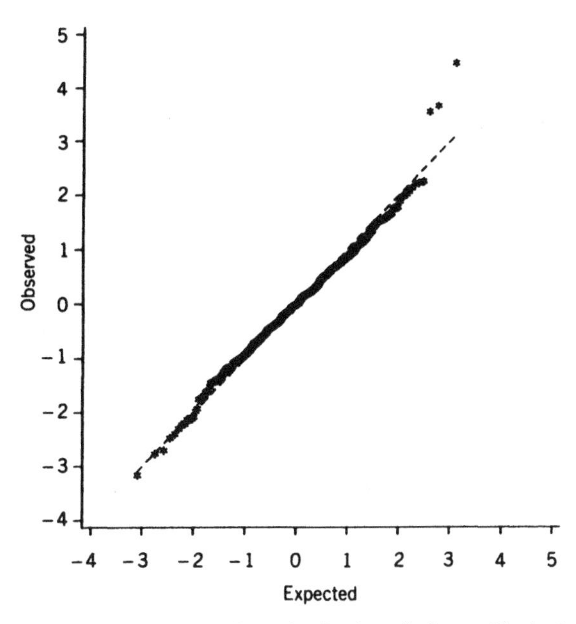

Figure 4.20 Normal probability plot of standardized prediction residuals given by (4.5.10).

Henceforth, all analyses will be carried out on the edited data, which has the effect of downweighting the influence of unusual values. The suggested Winsorization is a compromise between deleting what looks to be an outlier and leaving it in to wreak possible havoc on all the linear predictors. Hawkins and Cressie (1984) propose this as a way of robustifying kriging; see also Section 3.3. Care is needed when making inferences, depending on the question being answered. It should not be forgotten that certain data have been edited, and it may be more appropriate to return to the original values in some circumstances.

4.5.3† Median-Polish (Robust) Kriging

Using the techniques described in Section 3.5 (i.e., krige with the median-polish residuals and then add back the median-polish estimate of trend, pictured in Fig. 4.18), a contour map of predicted wheat yields can be drawn (Fig. 4.21*a*); a three-dimensional view of this surface is given in Figure 4.21*b*. A contour map of its prediction standard errors is given in Figure 4.21*c*.

It is worth emphasizing that $Z(\mathbf{s})$ is itself an aggregation over a plot 3.30×2.51 m, so that the contour map represents predicted wheat yields of identically sized plots located throughout the 1-acre region. As was explained in the introductory paragraphs of this section, the spatial information in Mercer and Hall's experiment is highly ambiguous. A little thought reveals that small errors in the spatial locations lead to a nugget effect in the

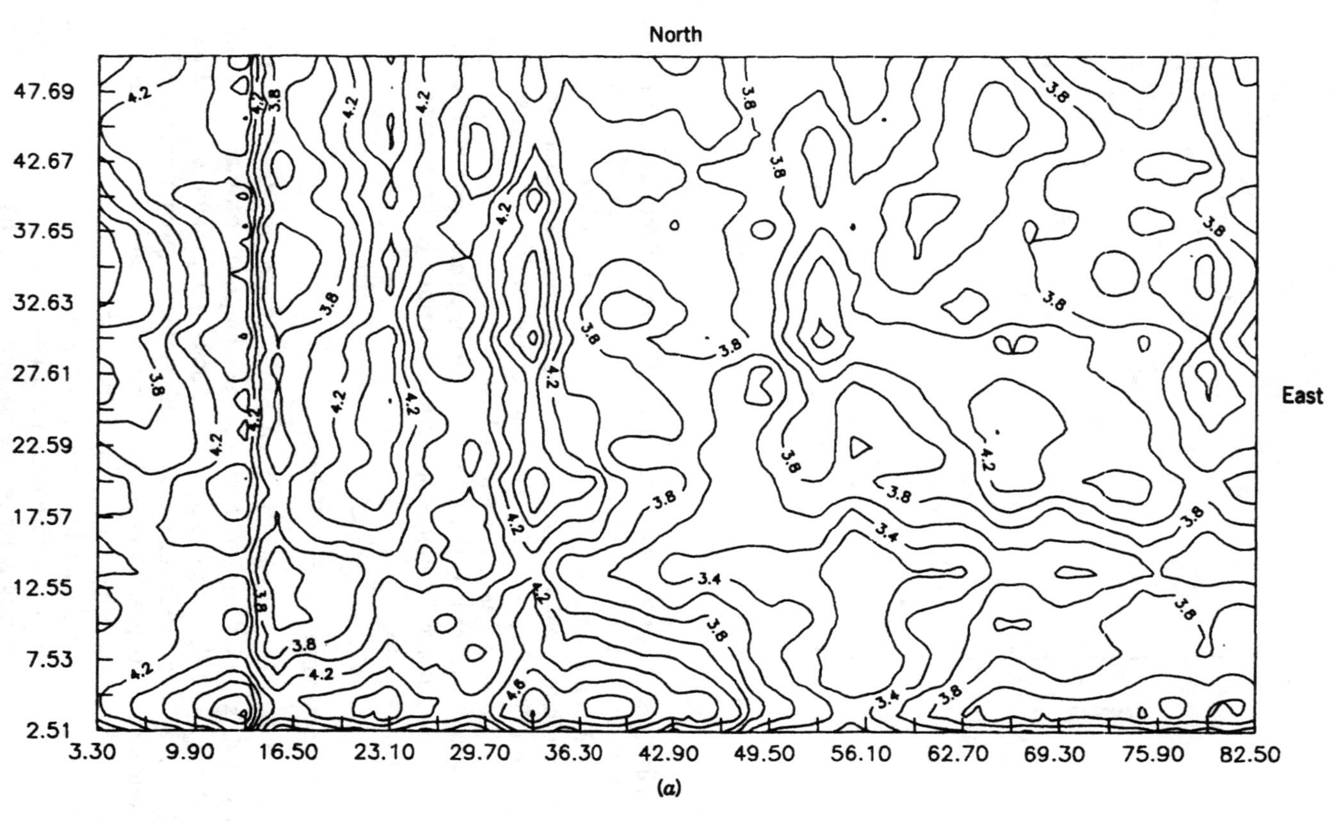

Figure 4.21 (*a*) Contour map obtained by median-polish (robust) kriging. Units of contours are in pounds of wheat.

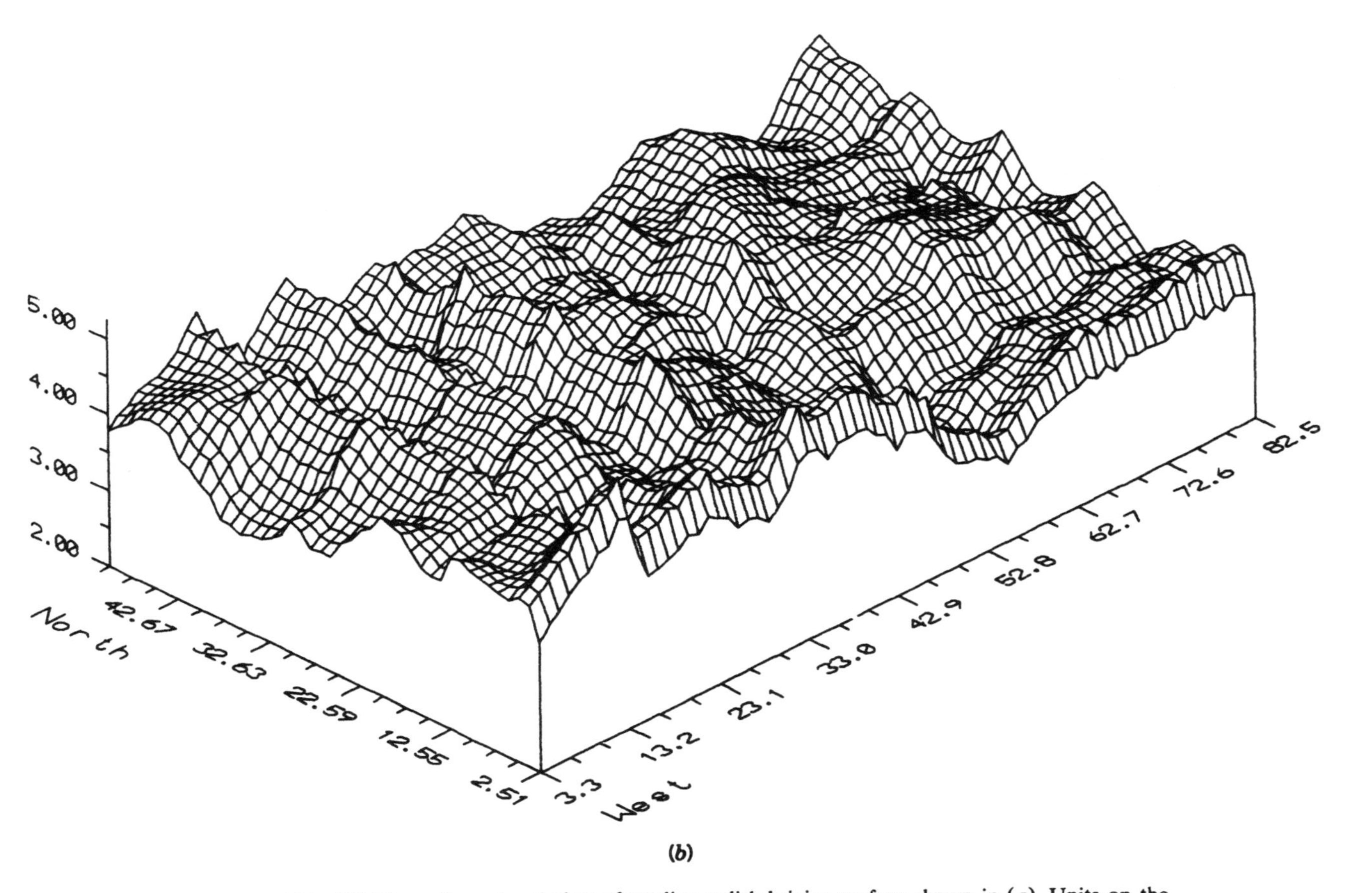

(*b*)

Figure 4.21 (*b*) Three-dimensional view of median-polish-kriging surface shown in (*a*). Units on the horizontal axis are in meters; units on the vertical axis are in pounds of wheat.

Figure 4.21 (*c*) Standard-error map associated with median-polish kriging. Only the southwest corner is shown. Units of contours are in pounds of wheat.

variogram. Equations (3.2.5), (3.2.12), and (3.2.16) are used to draw the maps of Figure 4.21, except that when $\mathbf{s}_0 = \mathbf{s}_i$, a datum location, $\mathbf{s}_0$ is moved infinitesimally away from $\mathbf{s}_i$. Hence, the predictor of Figure 4.21*a* is not (and should not be) an exact interpolator. These equations for the optimal predictor and the kriging standard error are used because the source of the location errors is the imprecision in the original data map.

The inhomogeneity present in this and other uniformity trials shows that spatial considerations are an important part of agricultural field experiments. The traditional approach is to neutralize spatial dependence by randomizing allocation of treatments to plots. Recent research indicates that treatment effects can be estimated more efficiently by modeling the dependence; see Sections 5.6.2 and 5.7.

4.6† ACID-DEPOSITION DATA

This section is based on a spatial analysis presented in Cressie et al. (1990). From an irregular network of monitoring stations in the eastern half of the United States, both median-polish kriging and universal kriging will be used to predict yearly total acid deposition at unobserved locations. Optimal locations of network monitoring sites will also be discussed.

Acid deposition has been (and probably will remain) one of the most controversial environmental issues of recent times. Many of the causes and effects of acid deposition are not well understood and are debated by scientists of all disciplines. It has generally been accepted, though, that an important factor in its increase in recent times is the emission of industrial by-products into the atmosphere; the study by Vong et al. (1988) presents convincing evidence. Government policy regarding the control of such emissions can have a significant impact on the economy of industrial regions, but failure to do anything could lead to international and regional crises. At the root of these economic and political issues is the potentially disastrous consequences of acid deposition on aquatic and terrestrial ecosystems.

It is well known that most fish populations in freshwater lakes are very sensitive to changes in their pH (European Inland Fisheries Advisory Commission, 1969). More fundamentally, such changes could also adversely affect most other aquatic organisms and plants, resulting in a disruption of the food chain. Acid deposition has also been closely connected with forest decline (e.g., Pitelka and Raynal, 1989) in both Europe and North America.

Changes in the delicate balance of the earth's ecosystems could have far-reaching, long-term implications. Thus, an understanding of acid deposition (the causes, the processes, and the effects), will be necessary for the protection, maintenance, and utilization of the environment.

Some Background

In North America, acid deposition results mainly from the atmospheric alteration of sulfur and nitrogen air pollutants produced by industrial processes, combustion, and transportation sources. Total acid deposition includes acid compounds in both wet and dry form. Dry deposition is the removal of gaseous pollutants, aerosols, and large particles from the air by direct contact with the earth (National Acid Precipitation Assessment Program, 1988). Because dry deposition is difficult to monitor, and attempts at any such monitoring are relatively new, this section will focus on wet deposition.

Wet deposition, or acid precipitation as it is commonly called, is defined as the total hydrogen-ion concentration in all forms of water that condense from the atmosphere and fall to the ground (Record et al., 1982). Operationally, acid precipitation has been defined as rain or snow with an annual average pH of less than 5.0 (National Acid Precipitation Assessment Program, 1988). The hydrogen ions present in precipitation are typically due to the strong acids H_2SO_4 and HNO_3.

Depending on a variety of meteorological conditions, pollutants released into the atmosphere may remain there for up to several days and be transported large distances. While they are in the atmosphere, the pollutants are chemically altered, then deposited on the ground via rain, snow, or fog (Mueller et al., 1984). Detailed scientific models and submodels have been established for acid rain (e.g., Dixon, 1989), although their stochastic components are typically rudimentary. The emphasis in this section is on empirical modeling with spatial stochastic processes.

The variable of interest here is the total yearly hydrogen ion (H^+) in wet deposition. (This is nothing more than precipitation-weighted hydrogen-ion concentration cumulated yearly.) The data were collected in 1982 and 1983 from the Utility Acid Precipitation Study Program (UAPSP) network of 19 sites in the eastern and midwestern United States. Stein (1985) constructed a total-yearly-hydrogen-ion variable from Mueller et al. (1984), repeated here in Table 4.4.

For 1982 and 1983, each observation is indexed by $\mathbf{s} =$ (degrees latitude, degrees longitude)$'$. For example, at site 1, Turner Falls, MA, $\mathbf{s}_1 = (42°37', 72°34')'$ and $Z(\mathbf{s}_1; 1982) = 6.13\ \mu\text{mol H}^+/\text{cm}^2$. The network map of all the sites is given in Figure 4.22.

4.6.1† Spatial Modeling and Prediction

An exploratory spatial data analysis was carried out separately for each year; Section 2.2 covers all the techniques used, adapted here for spatial data at irregular locations. This section will present only the conclusions, because details can be found in Cressie et al. (1990).

Random-Field Model

Behind the exploratory spatial data analysis is an assumption that data $\mathbf{Z} = (Z(\mathbf{s}_1), \ldots, Z(\mathbf{s}_n))'$ observed at known spatial locations $\{\mathbf{s}_1, \ldots, \mathbf{s}_n\}$ are a

Table 4.4 Annual Acid Deposition Levels (in μmol H^+/cm^2) for the 19 Sites of the UAPSP Network, for 1982 and 1983

Site	Latitude	Longitude	Annual Acid Deposition (μmol H^+/cm^2) 1982	1983
1. Turner Falls, MA	42°37′	72°34′	6.13	6.21
2. Tunkhannock, PA	41°32′	75°46′	4.80	5.22
3. Zanesville, OH	39°92′	82°02′	6.02	5.92
4. Rockport, IN	37°53′	87°04′	5.84	6.09
5. Fort Wayne, IN	41°05′	85°08′	5.83	4.02
6. Raleigh, NC	35°42′	78°40′	4.69	4.17
7. Gaylord, MI	45°02′	84°41′	2.48	3.39
8. Clearfield, KY	38°15′	83°38′	7.64	5.77
9. Alamo, TN	35°47′	89°09′	2.39	1.94
10. Winterport, ME	44°38′	68°53′	2.93	3.61
11. Uvalda, GA	32°02′	82°30′	2.41	2.83
12. Selma, AL	32°24′	87°01′	4.33	4.22
13. Clinton, MS	32°21′	90°20′	3.81	4.43
14. Marshall, TX	32°33′	94°22′	2.69	2.26
15. Lancaster, KS	39°51′	95°31′	2.25	2.18
16. Brookings, SD	44°19′	96°47′	0.31	0.29
17. Underhill, VT	44°38′	73°13′	5.39	6.70
18. Big Moose, NY	43°49′	74°57′	7.01	6.69
19. McArthur, OH	39°14′	82°29′	7.93	5.38

Source: Cressie et al. (1990).

sampling from a random field $\{Z(\mathbf{s}): \mathbf{s} \in D\}$ that is written

$$Z(\mathbf{s}) = \mu(\mathbf{s}) + \delta(\mathbf{s}), \qquad \mathbf{s} \in D. \tag{4.6.1}$$

Here $\mu(\cdot)$ is the large-scale deterministic mean structure of the process and $\delta(\cdot)$ is the small-scale stochastic error structure that models spatial statistical dependence:

$$\begin{aligned} E(\delta(\mathbf{s})) &= 0, \qquad \mathbf{s} \in D, \\ \operatorname{cov}(\delta(\mathbf{s}), \delta(\mathbf{u})) &\equiv C(\mathbf{s}, \mathbf{u}), \qquad \mathbf{s}, \mathbf{u} \in D. \end{aligned} \tag{4.6.2}$$

The representation (4.6.1) is precisely the sort of approach time-series analysts take; after detrending, they model the small-scale variation with, for example, autoregressive-moving-average processes. Also, as with time-series analysis, the separation of large-scale variation from small-scale variation is somewhat subjective, but exploratory data analysis and model diagnostics can serve as useful tools for determining a plausible decomposition.

Figure 4.22 Monitoring sites of the UAPSP network (1982, 1983). The square denotes an optimally located additional site (Baltimore, MD), discussed in Section 4.6.2. [*Source*: Cressie et al., 1990]

Given the genesis of acid deposition, a constant-mean assumption $\mu(\mathbf{s}) \equiv \mu$ is unlikely to be even approximately true. In (4.6.1), a nonconstant $\mu(\cdot)$ is assumed and $\delta(\cdot)$ is assumed to be *intrinsically stationary*; that is,

$$\operatorname{var}(\delta(\mathbf{s}+\mathbf{h}) - \delta(\mathbf{s})) = 2\gamma(\mathbf{h}), \qquad \mathbf{s}, \mathbf{s}+\mathbf{h} \in D, \qquad (4.6.3)$$

which includes, as a special case, second-order stationarity (see Section 2.3). The goal of this part of the study is to determine the spatial dependence structure through estimation of the variogram $2\gamma(\cdot)$.

Large-Scale Variation

As in the previous sections of this chapter, it is helpful to think of spatial data in two dimensions as a two-way array from which row and column effects (large-scale variation) can be swept. Section 3.5 explains how a low-resolution map is made by superimposing a grid over Figure 4.22. Each datum is then identified with its nearest grid node, and the resulting two-way array is available for row and column effects to be estimated and removed.

A resistant (to outliers) way of removing this large-scale variation is by Tukey's median polish (Tukey, 1977; Emerson and Hoaglin, 1983), which produces an all effect $\tilde{a}$, row effects $\{\tilde{r}_k\}$, and column effects $\{\tilde{c}_l\}$. Suppose that $\mathbf{s} = (x, y)'$ is in the region bounded by lines joining the four grid nodes $(x_l, y_k)'$, $(x_{l+1}, y_k)'$, $(x_l, y_{k+1})'$, and $(x_{l+1}, y_{k+1})'$. Section 3.5.2 adapts Cressie's (1986) median-polish surface for regular grids to the irregular network. Define the median-polish surface by interpolated planes

$$\mu(x, y) \equiv \tilde{a} + \tilde{r}_k + \left(\frac{y - y_k}{y_{k+1} - y_k}\right)(\tilde{r}_{k+1} - \tilde{r}_k) + \tilde{c}_l + \left(\frac{x - x_l}{x_{l+1} - x_l}\right)(\tilde{c}_{l+1} - \tilde{c}_l). \tag{4.6.4}$$

Figure 4.23 is a plot of the surface $\{\tilde{\mu}(\mathbf{s}): \mathbf{s} = (x, y)'; 32° \leq x \leq 44°, 74° \leq y \leq 94°\}$, the (estimated) large-scale variation, for 1982 and 1983. In the east–west direction there appears to be a positive linear trend, reflecting

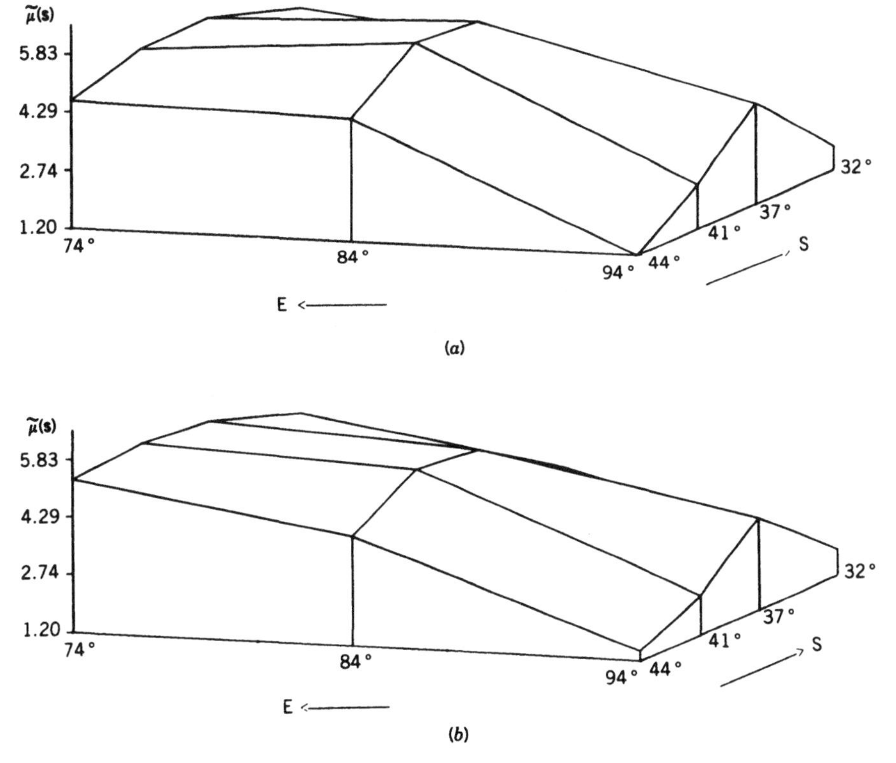

Figure 4.23 Interpolated median-polish surfaces defined by (4.6.4). Units on the vertical axis are in μmol H^+/cm^2. (*a*) 1982; (*b*) 1983. [*Source*: Cressie et al., 1990]

low acid-deposition levels in the west and higher levels in the east. What is more striking, however, is the quadratic trend in the north–south direction, with higher levels in the central region. These general conclusions are also supported by Mueller et al. (1984) and Bilonick (1985); Bilonick analyzes the sulfate-deposition variable, which is highly correlated with acid precipitation. Moreover, there is a consistency of large-scale variation from year to year.

Small-Scale Variation

The variable

$$R(\mathbf{s}) \equiv Z(\mathbf{s}) - \tilde{\mu}(\mathbf{s}), \qquad \mathbf{s} \in D \tag{4.6.5}$$

is an estimate of the error process $\delta(\cdot)$, which has attractive bias properties when compared to least-squares residuals (see Section 3.5.4). The process $R(\cdot)$ will be considered a proxy for $\delta(\cdot)$, and the variogram $2\gamma(\cdot)$, given by (4.6.3), will be estimated from

$$2\hat{\gamma}(\mathbf{h}) = \sum_{N(\mathbf{h})} \left(R(\mathbf{s}_i) - R(\mathbf{s}_j)\right)^2 / \, |N(\mathbf{h})|$$

or from

$$2\bar{\gamma}(\mathbf{h}) = \frac{\left\{\sum_{N(\mathbf{h})} |R(\mathbf{s}_i) - R(\mathbf{s}_j)|^{1/2} / \, |N(\mathbf{h})|\right\}^4}{0.457 + 0.494 / \, |N(\mathbf{h})|};$$

see Section 2.4.3 for a discussion of the merits of these estimators. The latter is robust to isolated outliers, and henceforth in this section $2\bar{\gamma}$ will be the estimator of choice.

The geostatistical method of spatial prediction is very simple. Estimate the variogram with a nonparametric estimator $\{2\bar{\gamma}(\mathbf{h})\colon \mathbf{h} = \mathbf{h}(1), \dots, \mathbf{h}(K)\}$, fit a model $2\gamma(\mathbf{h}; \boldsymbol{\theta})$ whose form is known up to the specification of a few parameters, and finally use the fitted variogram in the best linear unbiased predictor (i.e., kriging predictor) as if it were the true one. This method is now applied to the residuals data

$$\{R(\mathbf{s}_i)\colon i = 1, \dots, n\}, \tag{4.6.6}$$

defined by (4.6.5).

Estimated Isotropic Variograms

The variogram estimator $2\bar{\gamma}$ was computed in various directions to check for anisotropy (wind currents and precipitation patterns may lead to different east–west variograms). It was concluded (albeit from only 19 monitoring stations) that the directional effects were all captured by the large-scale variation $\tilde{\mu}(\cdot)$ and that the small-scale variation $R(\cdot)$ was isotropic. Therefore, an estimator $\{2\bar{\gamma}^o(h): h = 100, 300, \ldots, 1100 \text{ miles}\}$ of the isotropic variogram $2\gamma^o(\cdot)$, where $2\gamma(\mathbf{h}) \equiv 2\gamma^o(\|\mathbf{h}\|)$, was computed for 1982 and 1983. Here the distance $\|\mathbf{s}_1 - \mathbf{s}_2\|$ between $\mathbf{s}_1$ and $\mathbf{s}_2$ is defined to be the length of the great arc (on the earth's surface) joining $\mathbf{s}_1$ and $\mathbf{s}_2$. Write the latitude and longitude of location $\mathbf{s}_1$ as x_1 and y_1, respectively. Similarly, x_2 and y_2 are the latitude and longitude of location $\mathbf{s}_2$. Then the great-arc distance, in miles, between $\mathbf{s}_1$ and $\mathbf{s}_2$ is

$$\begin{aligned}\|\mathbf{s}_1 - \mathbf{s}_2\| = (69.0825)\cos^{-1}\{&\sin(90^\circ - x_1)\sin(90^\circ - x_2)\cos(y_1)\cos(y_2)\\ &+\sin(90^\circ - x_1)\sin(90^\circ - x_2)\sin(y_1)\sin(y_2)\\ &+\cos(90^\circ - x_1)\cos(90^\circ - x_2)\}.\end{aligned}$$

Figure 4.24a and b shows the plot of $2\bar{\gamma}^o(h)$ versus h, for 1982 and 1983, respectively. Superimposed on the estimates is the weighted-least-squares fitted (see Section 2.6.2) spherical variogram model obtained from (2.3.8) with fitted parameters $\bar{\boldsymbol{\theta}} = (\bar{c}_0, \bar{c}_s, \bar{a}_s)'$. These are

$$\begin{aligned}&\text{for 1982: } \bar{c}_0 = 0.608, \quad \bar{c}_s = 2.041, \quad \bar{a}_s = 361.210;\\ &\text{for 1983: } \bar{c}_0 = 0.000, \quad \bar{c}_s = 1.923, \quad \bar{a}_s = 285.260.\end{aligned} \tag{4.6.7}$$

Although the broad structure of spatial dependence evident in $2\bar{\gamma}^o(h)$ is captured, the fitted spherical variograms given by (4.6.7) do not reproduce the experimental variograms all that well. However, I think it would be a mistake to fit more sophisticated variogram models based on only 19 observations.

Duplicate measurements given in Mueller et al. (1984) show that there is almost no measurement error, and hence any nugget effect is due entirely to microscale variation. Consequently, prediction of $Z(\cdot)$, and not a noiseless version of it, is needed (Section 3.2.1 has a full discussion of the role of the nugget effect c_0 in spatial prediction).

A Suggested Model

What can be said about the 1982 process and the 1983 process? First, the median-polish surfaces presented in Figure 4.23 suggest a quadratic mean $\mu(\cdot)$. Second, the variogram for 1982 shows those data to be more variable than for 1983, but the shapes of the variograms appear to be very similar. Consequently, the following simple model for acid precipitation seems reasonable:

$$\begin{aligned}Z(\mathbf{s}; t) &= \mu(\mathbf{s}; t) + \delta(\mathbf{s}; t), \qquad t = 1982, 1983,\\ \delta(\mathbf{s}; 1982) &= \delta'(\mathbf{s}; 1983) + \nu(\mathbf{s}),\end{aligned}$$

where $\delta'(\cdot; 1983)$ is identically distributed to $\delta(\cdot; 1983)$, $\nu(\cdot)$ is an indepen-

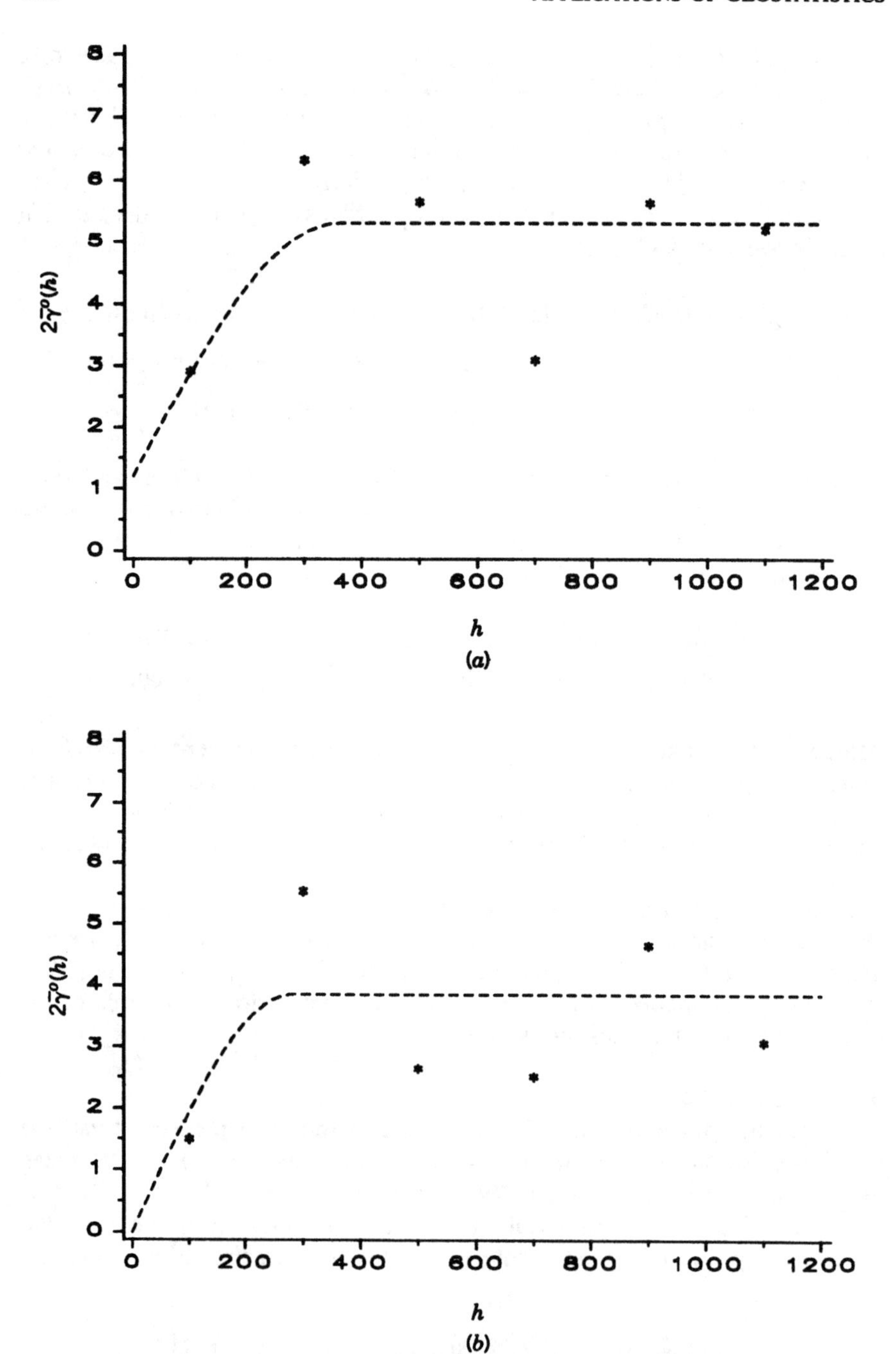

Figure 4.24 Robust empirical variograms for median-polish residuals. The superimposed dashed line shows the weighted-least-squares fit. Units on the horizontal axis are in miles; units on the vertical axis are in $(\mu\text{mol H}^+/\text{cm}^2)^2$. (*a*) 1982; (*b*) 1983. [*Source*: Cressie et al., 1990]

dent white-noise process, and $\mu(\mathbf{s}; 1982)$ and $\mu(\mathbf{s}; 1983)$ are quadratic functions of the coordinates of $\mathbf{s} = (x, y)'$.

For each year then, write

$$Z(\mathbf{s}) = \beta_0 + \beta_1 x + \beta_2 y + \beta_3 x^2 + \beta_4 xy + \beta_5 y^2 + \delta(\mathbf{s}), \quad (4.6.8)$$

where the coordinates of $\mathbf{s} = (x, y)'$ are expressed in radians and $\delta(\cdot)$ is an intrinsically stationary zero-mean process whose variogram is given by (4.6.7). In all that is to follow in this section, 1982 and 1983 will be analyzed separately and the results compared; Section 4.7 discusses briefly how data from both years might be combined to give more precise predictions.

Universal Kriging

The model (4.6.8) is of the form (3.4.2), and hence universal-kriging equations (3.4.11) and (3.4.16) based on fitted variogram (4.6.7) can be used to find the optimal linear predictor $\hat{Z}_U(\mathbf{s}_0) \equiv \sum_{i=1}^{n} \lambda_i Z(\mathbf{s}_i)$ and the associated prediction (kriging) standard error $\sigma_k(\mathbf{s}_0)$.

Alternatively, median-polish kriging (Section 3.5) could be obtained by ordinary kriging of the residuals data (4.6.6), giving rise to an optimal linear predictor $\hat{R}(\mathbf{s}_0) \equiv \sum_{i=1}^{n} \lambda_i R(\mathbf{s}_i)$ and prediction standard error $\sigma_m(\mathbf{s}_0)$ [these are obtained from Eqs. (3.2.12) and (3.2.16)]. Then $Z(\mathbf{s}_0)$ is predicted by

$$\tilde{Z}_{\mathrm{MP}}(\mathbf{s}_0) = \tilde{\mu}(\mathbf{s}_0) + \hat{R}(\mathbf{s}_0), \quad (4.6.9)$$

where $\tilde{\mu}(\cdot)$ is given by (4.6.4) (Cressie, 1986). Table 4.5 shows the results of these predictions for Baltimore, MD, where $\mathbf{s}_0 = (39°17', 76°37')'$.

Apart from Table 4.5, all kriging will be universal kriging based on the model (4.6.8), illustrating the general-linear-model approach of Section 3.4. Here, the role of median polish has been to identify the form of the trend and to provide residuals (4.6.6) from which variograms can be estimated.

Cross-Validation

The proposed model (4.6.8) was checked by cross-validation (see Section 2.6.4). That is, standardized prediction residuals were obtained by deleting an observation and using the remaining 18 observations to predict it (and to

Table 4.5 Universal Kriging and Median-Polish Kriging Predictors and Prediction Standard Errors, for Baltimore, MD[a]

	$\hat{Z}_U(\mathbf{s}_0)$	$\sigma_k(\mathbf{s}_0)$	$\tilde{Z}_{\mathrm{MP}}(\mathbf{s}_0)$	$\sigma_m(\mathbf{s}_0)$
1982	5.1826	1.6777	5.0663	1.5759
1983	5.0309	1.4842	5.4887	1.3723

[a]Units are μmol H^+/cm^2.

provide a prediction standard error). The difference between the actual observation and the predicted observation, divided by the prediction standard error, provides a standardized prediction residual. The ensemble of 19 such residuals should have sample mean and sample standard deviation approximately equal to 0 and 1, respectively, which it did.

4.6.2† Sampling Design

When the data are correlated, the problem of selecting optimal sampling designs to predict points and integrals, to estimate regression parameters, or to detect signals, has received considerable attention, particularly in the time domain (see, e.g., Sacks and Ylvisaker, 1966, 1968, 1970; Ylvisaker, 1975; Schoenfelder and Cambanis, 1982; Cambanis, 1985; Bucklew and Cambanis, 1988). Nather (1985) and Section 5.6.1 have a detailed discussion of these problems in the spatial domain. The statistical problem is part of a much bigger picture that includes such considerations as current pollution patterns, meteorology, cost, accessibility and security of sampling sites, and politics. There is some discussion of these cold, hard realities in Section 5.6.1.

Optimal Deletion of a Site from the Network

For operational, economic, or political reasons suppose that a site is to be *deleted* from the current network of sites $S \equiv \{\mathbf{s}_1, \mathbf{s}_2, \ldots, \mathbf{s}_n\}$. Define $S_{-i} \equiv S - \{\mathbf{s}_i\}$ to be the network without the ith site. Let $\sigma_k^2(\mathbf{s}_i; S_{-i})$ be the kriging variance for predicting the value at the ith site using the network S_{-i}. A sensible statistical criterion for the deletion of a site is to choose the site that achieves $\min\{\sigma_k^2(\mathbf{s}_i; S_{-i}): i = 1, \ldots, n\}$; that is, delete the site that can be predicted best from the remaining $n - 1$ sites. [Other criteria that include the mean level $\mu(\cdot)$, will be discussed.]

Table 4.6 gives the values of $\{\sigma_k^2(\mathbf{s}_i; S_{-i}): i = 1, \ldots, n\}$ based on universal kriging of the acid-deposition data. According to the criterion, McArthur, OH is the city to be deleted in both years. We note that this is also the city suggested by Mueller et al. (1984) in their study of the UAPSP network (their reasons for this recommendation are not clear).

Optimal Addition of a Site to the Network

Now suppose that it is desired to *add* an $(n + 1)$th site. Where is the optimal location $\mathbf{s}_{n+1}$?

"Hot spots" with large $\mu(\cdot)$ need to be featured when the goal is to study the consequences of high acidity levels on the environment. This is different from a longer-term need to study the acid-deposition process from a global point of view, where both low and high levels are of interest.

Define

$$V_n \equiv \int_D \sigma_k^2(\mathbf{s}_0; S) \cdot w(\mathbf{s}_0; \mu(\mathbf{s}_0); \sigma_k^2(\mathbf{s}_0; S))\, d\mathbf{s}_0 \qquad (4.6.10)$$

Table 4.6 Values of $\sigma_k^2(\mathbf{s}_i; S_{-i})$ for 1982 and 1983 Acid-Deposition Data[a]

City	1982	1983
Turner Falls, MA	2.46	1.93
Tunkhannock, PA	2.66	2.00
Zanesville, OH	1.72	0.92
Rockport, IN	2.59	2.08
Fort Wayne, IN	2.73	2.22
Raleigh, NC	4.38	3.19
Gaylord, MI	6.81	5.16
Clearfield, KY	1.99	1.37
Alamo, TN	2.81	2.17
Winterport, ME	6.25	4.79
Uvalda, GA	5.58	4.20
Selma, AL	2.98	2.36
Clinton, MS	2.95	2.40
Marshall, TX	5.43	4.04
Lancaster, KS	4.62	3.36
Brookings, SD	14.01	10.19
Underhill, VT	2.21	1.57
Big Moose, NY	2.01	1.40
McArthur, OH	1.47	0.71

Source: Cressie et al. (1990).
[a]Units are $(\mu\text{mol H}^+/\text{cm}^2)^2$.

and

$$M_n \equiv \sup_{\mathbf{s}_0 \in D} \left\{ \sigma_k^2(\mathbf{s}_0; S) \cdot w\big(\mathbf{s}_0; \mu(\mathbf{s}_0); \sigma_k^2(\mathbf{s}_0; S)\big) \right\} \tag{4.6.11}$$

to be measures of the average and maximum prediction variance, where $\sigma_k^2(\mathbf{s}_0; S)$ denotes the kriging variance at $\mathbf{s}_0$ using the network $S = \{\mathbf{s}_1, \mathbf{s}_2, \ldots, \mathbf{s}_n\}$ and $w(\cdot)$ is a suitably chosen weight function. For example, suppose

$$w\big(\mathbf{s}_0; \mu(\mathbf{s}_0); \sigma_k^2(\mathbf{s}_0; S)\big) = I(\mathbf{s}_0 \in B) \cdot I\big(\mu(\mathbf{s}_0) > K\big) \cdot I\big(\sigma_k^2(\mathbf{s}_0; S) > L^2\big), \tag{4.6.12}$$

where $I(A)$ is an indicator function that is equal to 1 if A is satisfied and is equal to 0 otherwise. The weight function (4.6.12) focuses attention on the prediction variance in a subregion B of D, where the means of the deposition process are larger than K and the prediction variances are beyond a threshold L^2. The subregion B and the constants K and L are exogenous to the sampling design problem.

Adding another site at $\mathbf{s}_{n+1}$ will decrease the average prediction variance, provided new kriging weights are solved for in (3.4.11). To emphasize the addition of this new point, denote the resulting average prediction variance in (4.6.10) as $V_{n+1}(\mathbf{s}_{n+1})$; similarly for $M_{n+1}(\mathbf{s}_{n+1})$ in (4.6.11). Then minimize these quantities over the new location $\mathbf{s}_{n+1} \in D$.

In fact, for reasons such as network coverage, the logistics associated with setting up and then monitoring, the suitability of local conditions, and so on, there probably would be just a small number of potential sites. Therefore, assume that there are $m \geq 2$ potential sites $S_P \equiv \{\mathbf{s}_{n+1}, \mathbf{s}_{n+2}, \ldots, \mathbf{s}_{n+m}\}$, from which the additional one shall be chosen. Define $S_{+j} \equiv S \cup \{\mathbf{s}_j\}$ to be an augmented network, $j = n+1, n+2, \ldots, n+m$. Let $\sigma_k^2(\mathbf{s}_0, S_{+j})$ be the prediction variance for predicting the value at $\mathbf{s}_0$ using the augmented network S_{+j}, and put

$$w\left(\mathbf{s}_0; \mu(\mathbf{s}_0); \sigma_k^2(\mathbf{s}_0; S_{+j})\right) = \frac{I(\mu(\mathbf{s}_0) > K) \cdot I(\mathbf{s}_0 \in S_P - \{\mathbf{s}_j\})}{\sum_{\mathbf{s}_l \in S_P - \{\mathbf{s}_j\}} I(\mu(\mathbf{s}_l) > K)}.$$

Then, according to the criterion (4.6.10), the objective function to be minimized is

$$V_{n+1}(\mathbf{s}_j) = \frac{\sum_{l=n+1, l \neq j}^{n+m} I(\mu(\mathbf{s}_l) > K) \cdot \sigma_k^2(\mathbf{s}_l; S_{+j})}{\sum_{l=n+1, l \neq j}^{n+m} I(\mu(\mathbf{s}_l) > K)},$$

$$j = n+1, \ldots, n+m. \quad (4.6.13)$$

Then $V_{n+1}(\mathbf{s}_j)$ is the average (over the remaining potential sites with large mean) prediction variance using the augmented network S_{+j}. This criterion selects the site in S_P that best predicts the remaining sites in S_P (on the average).

Similarly, according to criterion (4.6.11), an alternative objective function to be minimized is

$$M_{n+1}(\mathbf{s}_j) = \max_{\mathbf{s}_l \in S_P - \{\mathbf{s}_j\}} \left\{ I(\mu(\mathbf{s}_l) > K) \cdot \sigma_k^2(\mathbf{s}_l; S_{+j}) \right\},$$

$$j = n+1, \ldots, n+m. \quad (4.6.14)$$

This criterion selects the site in S_P that minimizes the maximum prediction variance of the remaining sites with large mean.

These ideas are easily generalized to the optimal addition of two (or more) sites from a finite number of potential sites. If the number of potential sites gets *very* large, numerical optimization algorithms of the type discussed in Section 5.6.1 are needed.

Table 4.7 gives the values of the objective functions $V_{20}(\mathbf{s}_j)$ and $M_{20}(\mathbf{s}_j)$ for $m = 11$ potential sites in S_P. The 11 sites were chosen with the idea of

Table 4.7 Potential Cities, Coordinates, and Objective Functions $V_{20}(\mathbf{s}_j)$ and $M_{20}(\mathbf{s}_j)$[a]

			1982		1983	
City	Latitude	Longitude	$V_{20}(\mathbf{s}_j)$	$M_{20}(\mathbf{s}_j)$	$V_{20}(\mathbf{s}_j)$	$M_{20}(\mathbf{s}_j)$
Minneapolis, MN	44°59′	93°16′	2.434	2.806	1.816	2.195
Des Moines, IA	41°35′	93°37′	2.436	2.814	1.819	2.202
Jefferson City, MO	38°34′	92°10′	2.433	2.810	1.815	2.197
Madison, WI	43°04′	89°23′	2.436	2.815	1.819	2.203
Springfield, IL	39°48′	89°38′	2.432	2.809	1.814	2.197
Altoona, PA	40°31′	78°23′	2.287	2.544	1.655	1.871
Charlottesville, VA	38°02′	78°30′	2.522	2.787	1.923	2.187
Charleston, WV	38°20′	81°38′	2.219	2.547	1.565	1.745
Baltimore, MD	39°17′	76°37′	2.103	2.541	1.447	1.614
Trenton, NJ	40°13′	74°46′	2.301	2.547	1.715	2.014
Knoxville, TN	35°58′	83°55′	2.400	2.799	2.254	2.178

Source: Cressie et al. (1990).

[a] Units in the objective-function columns are $(\mu\text{mol H}^+/\text{cm}^2)^2$.

improving geographic coverage of the network. The constant $K = 5$ μmol H^+/cm^2 was chosen to illustrate the optimization method (for a location whose yearly precipitation is 105 cm, this corresponds to a pH of 4.3). For both years, Baltimore, MD is the city that minimizes both objective functions and, according to our statistical criteria, is the city to be added.

Iterated Optimal Design

Notice that

$$\frac{V_n(\mathbf{s}_n)}{\sigma^2} = \frac{V_n(\mathbf{s}_n)}{V_{n-1}(\mathbf{s}_{n-1})}\,\frac{V_{n-1}(\mathbf{s}_{n-1})}{V_{n-2}(\mathbf{s}_{n-2})}\cdots\frac{V_1(\mathbf{s}_1)}{\sigma^2}. \tag{4.6.15}$$

Therefore, by starting on the far right of (4.6.15) and optimizing successively over each factor, a design could be found by adding one point at a time; the $(i+1)$th site is placed in an optimal location having already fixed the i previous sites. Is this iterated design optimal?

The answer to the above question is "No." A very simple counterexample is given by Cressie (1978a) for $Z(\cdot)$ a process of independent increments on the real line. However, if one defines the design measure to be the limiting $(n \to \infty)$ density of design points over the design space, then does the asymptotic *iterated* design measure equal the asymptotic *optimal* design measure? This is an open problem, but the answer is probably "Yes"; see Wu and Wynn (1978).

Although the iterated design may produce design points $S = \{\mathbf{s}_1, \ldots, \mathbf{s}_n\}$ that are quite different from optimal design points when n is finite, it still offers the possibility of providing a suitable yardstick for other, more computer-intensive algorithms, such as those presented in Section 5.6.1.

Equivalent Number of Independent Observations

To see the effect of spatial correlation on the size of the network, the notion of "*equivalent* number of independent observations" can be defined. First take a predictor and calculate its average mean-squared prediction error (4.6.10), under the special case of the model (4.6.8), where data $Z(\mathbf{s}_1), \ldots, Z(\mathbf{s}_{n'})$ are *independent*. Then do the same under the model (4.6.8), where data $Z(\mathbf{s}_1), \ldots, Z(\mathbf{s}_n)$ are *dependent*. Call the former $V_{n', \epsilon}$ and the latter $V_{n, \delta}$. Then the equivalent number of independent observations n' is obtained by solving

$$V_{n', \epsilon} = V_{n, \delta}. \tag{4.6.16}$$

Thus the network designer could think about having n' independent observations and make decisions about precision and cost as if the data were independent. When an n' has been chosen, it can be converted back to the proper "currency" using (4.6.16).

A simple example would be to use the predictor $\bar{Z}$ in (4.6.16). Suppose there is an $\mathbf{s}_0$ of particular interest. Then the preceding considerations lead to

$$n' = n\left\{(1/n) \sum_{i=1}^{n} \sum_{j=1}^{n} \operatorname{corr}(Z(\mathbf{s}_i), Z(\mathbf{s}_j)) - 2 \sum_{i=1}^{n} \operatorname{corr}(Z(\mathbf{s}_i), Z(\mathbf{s}_0))\right\}^{-1},$$

where $\operatorname{corr}(Z(\mathbf{s}), Z(\mathbf{u})) \equiv \operatorname{cov}(Z(\mathbf{s}), Z(\mathbf{u}))/\{\operatorname{var}(Z(\mathbf{s}))\operatorname{var}(Z(\mathbf{u}))\}^{1/2}$. Alternatively, for a second-order stationary process,

$$n' = n\sigma^2\left\{2 \sum_{i=1}^{n} \gamma(\mathbf{s}_i - \mathbf{s}_0) - \frac{1}{n} \sum_{i=1}^{n} \sum_{j=1}^{n} \gamma(\mathbf{s}_i - \mathbf{s}_j) - n\sigma^2\right\}^{-1},$$

where $\sigma^2 \equiv \operatorname{var}(Z(\mathbf{s}))$.

It would be dangerous to demand too much from this rather heuristic approach, because n' very much depends on $\mathbf{s}_0$ and on the criterion chosen in (4.6.16). In other words, an n' calculated according to one criterion, but used for a purpose based on a different criterion, may yield nonsensical results.

Using the acid-deposition data and model (4.6.8), where $n = 19$, the equivalent number of independent observations n' for two representative cities are

Baltimore, MD	28 (1982)	20 (1983)
Rapid City, SD	12 (1982)	13 (1983).

Notice that when $\mathbf{s}_0$ is internal to the network, for example, Baltimore, MD, the equivalent independent random field needs more observations to achieve the same prediction mean-squared error. Instead, when $\mathbf{s}_0$ is external to the network, for example, Rapid City, SD, the equivalent independent random field needs less observations.

Some Final Remarks

Nineteen sites is not a lot from which to build a spatial model. Other acid-precipitation networks, such as the National Coal Association Precipitation Quality Network has about 50 sites. From this network, Bilonick (1988) builds a space–time model based on monthly observations. Nevertheless, from a spatial point of view, the coverage of the eastern United States is still rather sparse.

Although the acid-deposition data came with a space–time index, here they were analyzed separately for each year. Clearly, information gleaned from the 1982 data could improve prediction at unobserved locations in 1983; see Section 3.2.3 on cokriging. The next section discusses briefly how space–time geostatistical data might be analyzed.

4.7 SPACE–TIME GEOSTATISTICAL DATA

Data usually have *both* a spatial and a temporal label associated with them, the recognition of which may be useful in the construction of (deterministic or stochastic) generating mechanisms. For example, the acid-deposition data presented in the previous section were gathered from a network of monitoring sites over the eastern half of the United States during 1982 and 1983. As well as spatial prediction of acid deposition at locations where there is no monitoring site, there is a need to make temporal predictions or forecasts of acid deposition.

The generic problem is to predict $\{Z(\mathbf{s}; t_0): \mathbf{s} \in D(t_0)\}$ from data

$$\left\{\left(Z(\mathbf{s}_{1,i}; t_i), \ldots, Z(\mathbf{s}_{n_i,i}; t_i)\right): i = 1, \ldots, m\right\}, \tag{4.7.1}$$

where $t_1 < t_2 < \cdots < t_m \leq t_0$. The data are assumed to be an incomplete sampling of the stochastic process

$$\{Z(\mathbf{s}; t): \mathbf{s} \in D(t); t \in T\}, \tag{4.7.2}$$

where the domain $D(t) \subset \mathbb{R}^d$ may vary with time; see (1.1.3). Most commonly, $D(t) \equiv D$, and usually $T = \{1, 2, \ldots\}$, which enables (4.7.2) to be viewed as a time series of spatial processes, each process occurring at equally spaced time points. If the temporal correlation is weak, then sampling across time leads to approximate replication of the spatial error process and more precise inferences.

Space–Time Modeling

In order to estimate model parameters with acceptable precision, there has to be repeatability (or more formally, ergodicity, as explained in Section 2.3) across space or time. Also, there is usually an assumption somewhere in the model that the large-scale spatial and temporal components do not interact or that the small-scale components are homogeneous with respect to space or time. Such assumptions allow more precise estimates of model parameters.

For example, Switzer (1989) and Sampson and Guttorp (1992) average over time to obtain estimates of *nonstationary* spatial covariances. These are then used to predict $Z(\mathbf{s}_0; t_i)$ from contemporaneous data $\mathbf{Z}(t_i) \equiv (Z(\mathbf{s}_{1,i}; t_i), \ldots, Z(\mathbf{s}_{n_i,i}; t_i))'$.

Atmospheric Pollution

A problem of immediate national and international concern is the large-scale pollution of our atmosphere with chemical wastes. Not only are ecosystems disrupted, but excess carbon monoxide causes a warming of the planet (greenhouse effect) and chlorofluorocarbons deplete the radiation-protective layer of ozone in the upper atmosphere.

These problems are spatiotemporal in nature. But whereas data are collected in *both* space and time, spatially speaking, they are often rather sparse. There is an ever increasing number of articles being published about space–time models and analysis, and it is clear that many interesting problems remain open. A sampling of such articles in the areas of acid deposition, carbon-monoxide pollution, and ozone depletion, includes Tiao et al. (1975), Bilonick (1983), Eynon and Switzer (1983), Bilonick (1985), Zeger (1985), Egbert and Lettenmaier (1986), Stein (1986), Bilonick (1988), and Le and Petkau (1988).

Other Applications

Haslett and Raftery (1989) assess Ireland's wind resource through a detailed space–time analysis of hourly wind-energy data from 12 synoptic stations during the period 1961–1978. Ranneby (1982) reviews various space–time models as applied to forest inventory. Water resources research into rainfall and runoff is inherently spatiotemporal; the location, time of occurrence, and duration of storms are very important components of (stochastic) models (see, e.g., Gupta and Waymire, 1979; Eagleson, 1984; Bras and Rodriguez-Iturbe, 1985, Chapter 8; Smith and Karr, 1985).

Spatiotemporal Prediction

From data (4.7.1), one wishes to predict $\{Z(\mathbf{s}; t_0): \mathbf{s} \in D\}$, where $t_0 \geq t_m$ and t_m is the last time at which data were taken. When $t_0 > t_m$, which is usually the case, simplifying assumptions (of the type referred to previously) about the appearance of space and time in the models are typically made to allow precise estimation of model parameters. Then optimal (linear) predictors can be developed, much as they were in Chapter 3. If one were not willing to make such simplifying assumptions, the stochastic process $\{Z(\mathbf{s}; t)\}$ could be viewed as having its index vary over a subset of $\mathbb{R}^{d+1}$. Then the methods of Chapters 2 and 3 could be used directly.

When $t_0 = t_m$, the data at each site could be viewed as a multivariate vector. Write the collection of all such vectors as $\{\mathbf{Z}(\mathbf{s}_i): i = 1, \ldots, n\}$. Then, to predict $\{Z(\mathbf{s}; t_m): \mathbf{s} \in D\}$, the methods of cokriging (Section 3.2.3) could be used. For example, the acid-deposition data of Section 4.6 would be amenable to this type of analysis.

CHAPTER 5

Special Topics in Statistics for Spatial Data

Chapter 5 is different from all of the other chapters in this book. Each section covers a topic that bears no particular relationship to topics in the other sections. The choice of topics is very much a personal one, whereas in the rest of the book an attempt has been made to make the coverage complete (albeit colored by my own preferences and prejudices). In some cases, the material in Chapter 5 could be described as peripheral to the main body of knowledge, but it is of interest to me. In other cases, the material is at the leading edge of an area that has generated considerable activity. In all cases, the exposition will avoid the many details that an in-depth discussion would demand.

My goal in writing this chapter is to capture the basic ideas and leave readers with key references, should the material prove interesting to them. Thus, at the very least, a basic review is presented and occasionally more detail is given when I have written something previously on the topic. Each section could itself be a chapter in a book and, in a world of unlimited time, energy, and patience, they would have been.

It is no accident that this chapter appears at the end of Part I; there is often a geostatistical model behind the special topics presented. In order to read on, it will be necessary to have read Chapters 2 and 3.

Chapter Summary

Sections 5.1 and 5.2 consider the closely related topics of, respectively, nonlinear geostatistics and change-of-support. In geostatistics, a number of pragmatic approximations are made when predicting unknown values from observed data, yet the predictions are often very good; Section 5.3 discusses why this is so. However, if the model has highly nonstationary errors, different kriging predictors must be developed. Kriging for intrinsic random functions of order k is presented in Section 5.4. Section 5.5 investigates how results from the theory of random processes can be implemented in Statistics for spatial data.

Spatial sampling designs and spatial experimental designs are considered in Section 5.6, and nearest-neighbor and other methods of analyzing field trials are discussed in Section 5.7. Asymptotics in space can be of the increasing-domain type or of the infill type; Section 5.8 discusses infill asymptotics. Finally, Section 5.9 brings together a large number of stochastic and nonstochastic spatial predictors in a unified notation, and a brief comparison between predictors is made.

5.1* NONLINEAR GEOSTATISTICS

In the introduction to Chapter 3, it is demonstrated that minimizing the mean-squared prediction error $E(Z(\mathbf{s}_0) - p(\mathbf{Z}; \mathbf{s}_0))^2$ yields the optimal predictor

$$p^0(\mathbf{Z}; \mathbf{s}_0) \equiv E(Z(\mathbf{s}_0)|\mathbf{Z}), \tag{5.1.1}$$

where $\mathbf{Z} \equiv (Z(\mathbf{s}_1), \ldots, Z(\mathbf{s}_n))'$ are spatial data observed at spatial locations $\{\mathbf{s}_1, \ldots, \mathbf{s}_n\}$. Aside from a few important exceptions (e.g., the Gaussian model), this conditional expectation is nonlinear in $\mathbf{Z}$. Calculation of (5.1.1) requires knowledge of the $(n+1)$-dimensional distribution of $(Z(\mathbf{s}_0), \mathbf{Z}')$, which is rarely, if ever, available. However, linear approximations may be too smooth and too inaccurate.

Nonlinear geostatistics encompasses techniques that use nonlinear functions of the data to obtain (or approximate) $E(Z(\mathbf{s}_0)|\mathbf{Z})$ and $\Pr(Z(\mathbf{s}_0) \geq z_0|\mathbf{Z})$. Whereas linear geostatistics (Chapter 3) is relatively straightforward to apply, nonlinear techniques often require assumptions for which no methods of verification are currently available and they can yield solutions that are computationally complex.

The quantity $\Pr(Z(\mathbf{s}_0) \geq z_0|\mathbf{Z})$ is particularly important in the monitoring of prespecified standards, where a large value of $Z(\cdot)$ at $\mathbf{s}_0$ indicates that the process there is out of specification or, in an environmental application, is a "hot spot." Under squared–error loss, $\Pr(Z(\mathbf{s}_0) \geq z_0|\mathbf{Z})$ is the optimal predictor of $I(Z(\mathbf{s}_0) \geq z_0)$. Given the data $\mathbf{Z}$, it is *not* appropriate to replace $\Pr(Z(\mathbf{s}_0) \geq z_0|\mathbf{Z})$ with $\Pr(Z(\mathbf{s}_0) \geq z_0)$ (and then to estimate it), because marginally $Z(\mathbf{s}_0)$ has a distribution with more variability than the conditional distribution of $Z(\mathbf{s}_0)|\mathbf{Z}$. Nonlinear approximations to the conditional probability are given in succeeding text.

Disjunctive Kriging

Instead of obtaining a predictor that is linear in the data, disjunctive kriging (Matheron, 1976a) looks for an optimal predictor among the class of all linear combinations of univariate *functions* of the data. More precisely, the mean-squared prediction error is minimized over measurable square-integrable functions $\{f_i: i = 1, \ldots, n\}$ that define the predictor

$$p(\mathbf{Z}; \mathbf{s}_0) \equiv \sum_{i=1}^{n} f_i(Z(\mathbf{s}_i)). \tag{5.1.2}$$

From Hilbert-space theory, the optimal predictor satisfies the orthogonality property

$$E\left[\left\{Z(\mathbf{s}_0) - \sum_{i=1}^{n} f_i(Z(\mathbf{s}_i))\right\} h_j(Z(\mathbf{s}_j))\right] = 0, \qquad j = 1, \ldots, n, \tag{5.1.3}$$

for all measurable functions $\{h_j: j = 1, \ldots, n\}$. Hence, the optimal $\{f_i: i = 1, \ldots, n\}$ satisfy

$$E(Z(\mathbf{s}_0)|Z(\mathbf{s}_j)) = \sum_{i=1}^{n} E(f_i(Z(\mathbf{s}_i))|Z(\mathbf{s}_j)), \qquad j = 1, \ldots, n. \tag{5.1.4}$$

An implicit solution is given by the integral equations (5.1.4), called here disjunctive-kriging equations.

The disjunctive-kriging equations only require knowledge of *bivariate* distributions of $(Z(\mathbf{s}_i), Z(\mathbf{s}_j))$, $0 \leq i < j \leq n$. This is in contrast to the predictor (5.1.1), which requires knowledge of the $(n + 1)$-dimensional distribution of $(Z(\mathbf{s}_0), \mathbf{Z}')$.

Solutions to (5.1.4) can be obtained when the spatial process $Z(\cdot)$ is assumed to follow an isofactorial model (Matheron, 1976a), which is a special case of a bivariate-distribution decomposition introduced by Lancaster (1958). Assuming the existence of random variables Z_1, Z_2 with joint c.d.f. $F_{1,2}(z_1, z_2)$ and marginals $F_1(z_1)$ and $F_2(z_2)$, Lancaster gives a formal expression for $F_{1,2}$ in terms of F_1 and F_2. In the invariant-distribution case, where $F_1 = F_2 = F$, Lancaster's expression coincides with Matheron's (1984) general form of the isofactorial model:

$$F_{1,2}(dz_1 \times dz_2) = \sum_{k=0}^{\infty} \nu_k(1,2)\chi_k(z_1)\chi_k(z_2)F(dz_1)F(dz_2), \tag{5.1.5}$$

where $\{\chi_k: k = 0, 1, \ldots\}$ are complete and orthonormal: $\chi_0(z) \equiv 1$, $\int \chi_k(z)F(dz) = 0$, $\int \chi_k^2(z)F(dz) = 1$, $k = 1, \ldots, \infty$, and $\int \chi_k(z)\chi_l(z)F(dz) = 0$, $k \neq l = 0, \ldots, \infty$.

Hence, $E(\chi_k(Z_1)\chi_l(Z_2)) = 0$, $k \neq l$, $\nu_0(1, 2) = 1$, and $\nu_k(1, 2) = \mathrm{cov}(\chi_k(Z_1), \chi_k(Z_2)) = \mathrm{corr}(\chi_k(Z_1), \chi_k(Z_2))$, $k = 1, 2, \ldots$. Upon extending (5.1.5) to bivariate c.d.f.s of all pairs $(Z(\mathbf{s}_i), Z(\mathbf{s}_j))$, $i \neq j = 0, 1, \ldots,$ the disjunctive-kriging equations (5.1.4) can be written as

$$\sum_{k=0}^{\infty} \nu_k(0, j)\chi_k(Z(\mathbf{s}_j))\int z_0\chi_k(z_0)F(dz_0)$$
$$= \sum_{i=1}^{n}\sum_{k=0}^{\infty} \nu_k(i, j)\chi_k(Z(\mathbf{s}_j))\int f_i(z_i)\chi_k(z_i)F(dz_i), \qquad j = 1,\ldots,n. \tag{5.1.6}$$

Define

$$a_{ik} \equiv \int f_i(z_i)\chi_k(z_i)F(dz_i), \qquad i = 1,\ldots,n,$$
$$b_k \equiv \int z_0\chi_k(z_0)F(dz_0).$$

Completeness and orthonormality of $\{\chi_k\colon k = 0, 1, \ldots\}$ imply

$$f_i(Z(\mathbf{s}_i)) = \sum_{k=0}^{\infty} a_{ik}\chi_k(Z(\mathbf{s}_i)), \qquad i = 1,\ldots,n,$$
$$Z(\mathbf{s}_0) = \sum_{k=0}^{\infty} b_k\chi_k(Z(\mathbf{s}_0)).$$

Thus, the disjunctive-kriging predictor can be written as

$$p_{\mathrm{dk}}(\mathbf{Z}; \mathbf{s}_0) = \sum_{i=1}^{n}\sum_{k=0}^{\infty} a_{ik}\chi_k(Z(\mathbf{s}_i)), \tag{5.1.7}$$

where $\{a_{ik}\}$ solve

$$\nu_k(0, j)b_k = \sum_{i=1}^{n} \nu_k(i, j)a_{ik}, \qquad k = 0, 1, \ldots;\ j = 1, \ldots, n. \tag{5.1.8}$$

Notice that, in (5.1.8), $\{b_k\colon k = 0, 1, \ldots\}$ are properties of the univariate F and $\{\nu_k(i, j)\}$ are properties of the bivariate distributions. At this point, disjunctive kriging should be compared with simple kriging; upon estimation of the $\{b_k\}$ and $\{\nu_k(i, j)\}$ from data $\mathbf{Z}$, it can be compared to ordinary kriging. From (5.1.3), the mean-squared prediction error is

$$E(Z(\mathbf{s}_0) - p_{\mathrm{dk}}(\mathbf{Z}; \mathbf{s}_0))^2 = E(Z(\mathbf{s}_0)^2) - E\left(Z(\mathbf{s}_0)\sum_{i=1}^{n} f_i(Z(\mathbf{s}_i))\right)$$
$$= \sum_{k=0}^{\infty} b_k^2 - \sum_{k=0}^{\infty} b_k \sum_{i=1}^{n} \nu_k(0, i)a_{ik}. \tag{5.1.9}$$

In practice, (5.1.8), the infinite set of n linear equations in n unknowns, is truncated at some large finite integer K [e.g., choose K so that $\text{var}(Z(\mathbf{s}_0)) \simeq \sum_{k=0}^{K} b_k^2$].

Having a well defined isofactorial model relies on the validity of (5.1.5). Matheron (1976a) shows that, for $\{\chi_k: k = 0, 1, \ldots\}$ the set of normalized *Hermite* polynomials (e.g., Beckmann, 1973, Chapter 3), $\nu_k(1,2) = \rho^k$, $-1 < \rho < 1$. Further, for $F(\cdot)$ the c.d.f. of the standard Gaussian distribution, $F_{1,2}$ given by (5.1.5) defines a bivariate Gaussian distribution with standard Gaussian marginals and correlation coefficient ρ. When the invariant c.d.f. F is non-Gaussian, Matheron (1976a) proposes transforming the data first to achieve at least marginal Gaussianity. (The transformation has been given the ferocious name of *anamorphosis*.) Section 5.9.1 contains a brief discussion of disjunctive kriging in this case.

However, it is *not* true in general that (5.1.5) *defines* a valid model, because the function $F_{1,2}$ has to be nonnegative, at least. This important condition is checked by Armstrong and Matheron (1986a) for the bivariate gamma distribution used to define an isofactorial model. [The same bivariate gamma distribution can be found in Mardia (1970, p. 53) and Beckmann (1973, p. 184).] More generally, one has to be very careful to use bivariate distributions that are compatible with an underlying joint distribution.

Although the development here has been for optimal prediction of $Z(\mathbf{s}_0)$, the same expressions hold for $g(Z(\mathbf{s}_0))$, where g is any measurable function of $Z(\mathbf{s}_0)$. The only difference occurs in (5.1.8), where the $\{b_k: k = 0, 1, \ldots\}$ are now the coefficients in the expansion $g(Z(\mathbf{s}_0)) = \sum_{k=0}^{\infty} b_k \chi_k(Z(\mathbf{s}_0))$. For example, $g(Z(\mathbf{s}_0)) = I(Z(\mathbf{s}_0) \geq z_0)$ yields a disjunctive-kriging predictor that approximates the conditional expectation $E(g(Z(\mathbf{s}_0))|\mathbf{Z}) = \Pr(Z(\mathbf{s}_0) \geq z_0|\mathbf{Z})$. Indeed, knowledge of this conditional probability, for all z_0, allows calculation of $E(g(Z(\mathbf{s}_0))|\mathbf{Z})$ for any g. An important example in mining is $g(Z(\mathbf{s}_0)) = Z(\mathbf{s}_0)I(Z(\mathbf{s}_0) \geq z_0)$, which is used in predicting the *quantity of metal* above the cutoff grade z_0 (Section 5.2).

Two other ways of approximating $\Pr(Z(\mathbf{s}_0) \geq z_0|\mathbf{Z})$ will now be presented.

Indicator Kriging

Indicator kriging (Journel, 1983) is the application of kriging (Chapter 3) to indicator functions of the data. There are no assumptions made about the underlying invariant distribution, and the 0–1 indicator transformations of the data make the predictor robust to outliers. However, these advantages are not without a price: Theoretically, indicator kriging gives a worse approximation to the conditional expectation than disjunctive kriging, it requires estimation and modeling of many (indicator) variograms, and the resulting system of kriging equations is very large. Nevertheless, indicator kriging has found applications in estimation of recoverable reserves (Lemmer, 1984), classification schemes for map analysis (Solow, 1986), estimation of spatiotemporal distributions of hydrogen-ion deposition (Bilonick, 1988), and risk assessment in environmental applications (Journel, 1988b).

Define the indicator random variables

$$I(\mathbf{s}, z) \equiv \begin{cases} 1, & \text{if } Z(\mathbf{s}) \le z, \\ 0, & \text{otherwise}, \quad \mathbf{s} \in D,\ z \in \mathbb{R}. \end{cases} \tag{5.1.10}$$

Assume that the process has second-order stationary indicators. That is, independent of $\mathbf{s}$,

$$\begin{aligned} F(z) &\equiv E(I(\mathbf{s}, z)), \qquad z \in \mathbb{R} \\ 2\gamma_z(\mathbf{h}) &\equiv \operatorname{var}(I(\mathbf{s}+\mathbf{h}, z) - I(\mathbf{s}, z)), \qquad \mathbf{h} \in \mathbb{R}^d, \quad z \in \mathbb{R}. \end{aligned} \tag{5.1.11}$$

The *indicator variogram* $2\gamma_z(\cdot)$ is equal to $2F(z) - 2\Pr(Z(\mathbf{s}+\mathbf{h}) \le z,\ Z(\mathbf{s}) \le z)$, so that knowledge of $2\gamma_z(\cdot)$ can be obtained from the bivariate distribution of $Z(\mathbf{s}+\mathbf{h})$ and $Z(\mathbf{s})$. However, the converse is not true because, for $z_1 \ne z_2$, $\Pr(Z(\mathbf{s}+\mathbf{h}) \le z_1,\ Z(\mathbf{s}) \le z_2)$ is not derivable from the indicator variograms. Finally, notice that $E(|Z(\mathbf{s}+\mathbf{h}) - Z(\mathbf{s})|) = \int 2\gamma_z(\mathbf{h})dz$.

Clearly then, a Hilbert-space approximation to $E(g(Z(\mathbf{s}_0))|\mathbf{Z})$ using indicator kriging will be poorer than that for disjunctive kriging, which uses the full bivariate distribution. In practice, a number of approximations are made in order to implement each method, which may make their difference unimportant.

The (ordinary) indicator-kriging predictor of $I(\mathbf{s}_0, z)$, based on data $(I(\mathbf{s}_1, z), \ldots, I(\mathbf{s}_n, z))$, is

$$\hat{I}(\mathbf{s}_0, z) = \sum_{i=1}^{n} \lambda_i(z) I(\mathbf{s}_i, z), \tag{5.1.12}$$

where

$$\sum_{i=1}^{n} \lambda_i(z) = 1 \tag{5.1.13}$$

and $\{\lambda_i(z): i = 1, \ldots, n\}$ satisfy

$$\sum_{i=1}^{n} \lambda_i(z)\gamma_z(\mathbf{s}_i - \mathbf{s}_j) + m(z) = \gamma_z(\mathbf{s}_0 - \mathbf{s}_j), \qquad j = 1, \ldots, n,\ z \in \mathbb{R}. \tag{5.1.14}$$

Equations (5.1.13) and (5.1.14) are $(n+1)$ linear equations in $(n+1)$ unknowns $\{\lambda_1(z), \ldots, \lambda_n(z), m(z)\}$, which should be compared to the ordinary kriging equations (3.2.12). The mean-squared prediction error is, $\sum_{i=1}^{n} \lambda_i(z)\gamma_z(\mathbf{s}_0 - \mathbf{s}_i) + m(z)$.

The expression (5.1.12) that defines $\hat{I}(\mathbf{s}_0, z)$ is actually an estimate of $\Pr(Z(\mathbf{s}_0) \le z | I(\mathbf{s}_1, z), \ldots, I(\mathbf{s}_n, z))$, based only on knowledge of $2\gamma_z(\cdot)$. If, in addition, knowledge of $F(z)$ were assumed, simple kriging (Section 3.4.5) of indicators would yield a better estimator of this conditional probability. It should be noted that $\Pr(Z(\mathbf{s}_0) \le z | I(\mathbf{s}_1, z), \dots, I(\mathbf{s}_n, z))$ is different from the more relevant conditional probability $\Pr(Z(\mathbf{s}_0) \le z | \mathbf{Z})$, so that $\hat{I}(\mathbf{s}_0, z)$ may be a poor estimator of the latter quantity.

Solution of the indicator-kriging equations to obtain (5.1.12) is carried out at K levels, $z_1 < \cdots < z_K$, which requires K variograms, $2\gamma_{z_1}, \dots, 2\gamma_{z_K}$, to be estimated and modeled. Then, $\{\hat{I}(\mathbf{s}_0, z_k): k = 1, \dots, K\}$ is used to estimate $\{\Pr(Z(\mathbf{s}_0) \le z_k | \mathbf{Z}): k = 1, \dots, K\}$, respectively.

Sometimes, the estimates are not monotonic in z_k. Modifications that ensure monotonicity involve using a single variogram $2\gamma_{\tilde{z}}$ in (5.1.14) (where, e.g., $\tilde{z} = \text{med}\{z_1, \dots, z_K\}$) or replacing an offending pair $\hat{I}(\mathbf{s}_0, z_k) > \hat{I}(\mathbf{s}_0, z_{k+1})$, each with the average $(1/2)\{\hat{I}(\mathbf{s}_0, z_k) + \hat{I}(\mathbf{s}_0, z_{k+1})\}$. Also, truncation could be used to deal with predictors that are outside $[0, 1]$. These *ad hoc* modifications to indicator kriging are needed to correct for important consistency conditions, but there is still *no* guarantee that the estimated conditional probabilities are compatible with an underlying joint distribution.

Probability Kriging

In an attempt to incorporate extra information, Sullivan (1984) suggests cokriging (Section 3.2.3) using "data" $\{I(\mathbf{s}_i, z): i = 1, \dots, n\}$ and $\{U(\mathbf{s}_i): i = 1, \dots, n\}$, where $U(\mathbf{s}) \equiv \hat{F}(Z(\mathbf{s}))$ and $\hat{F}$ is an estimate of the marginal c.d.f. F. That is, find an optimal predictor

$$I^{\#}(\mathbf{s}_0, z) \equiv \sum_{i=1}^{n} \lambda_{1i}(z) I(\mathbf{s}_i, z) + \sum_{i=1}^{n} \lambda_{2i}(z) U(\mathbf{s}_i), \qquad (5.1.15)$$

where the coefficients satisfy $\sum_{i=1}^{n} \lambda_{1i}(z) = 1$, $\sum_{i=1}^{n} \lambda_{2i}(z) = 0$, and solve the cokriging equations given by (3.2.51). The mean-squared prediction error is given by (3.2.52). Just as for indicator kriging, the values $\{I^{\#}(\mathbf{s}_0, z_k): k = 1, \dots, K\}$ are used to estimate $\{\Pr(Z(\mathbf{s}_0) \le z_k | \mathbf{Z}): k = 1, \dots, K\}$, respectively.

Some Final Remarks

The preceding nonlinear geostatistical techniques give estimators of the conditional-probability distribution $\Pr(Z(\mathbf{s}_0) \le z | \mathbf{Z})$. [From this, $E(g(Z(\mathbf{s}_0)) | \mathbf{Z})$, where g is any given measurable function, can be obtained.] Disjunctive kriging chooses an approximation family of additive measurable functions that is very rich in comparison to the family of indicator and probability-transform functions of the data. If the indicator functions used for obtaining $\hat{I}(\mathbf{s}_0, z)$ were to include all of $\{I(\mathbf{s}_i, z_k): i = 1, \dots, n;\ k = 1, \dots, K\}$, then indicator (co)kriging would yield a better estimator. Indeed, because the approximating functions in disjunctive kriging are measurable, they in turn can be approximated by linear combinations of indicator functions. Thus, *indicator cokriging* is equivalent to (discretized) disjunctive kriging. To make progress mathematically, disjunctive kriging makes parametric assumptions about bivariate distributions of the random process. To make progress computationally, indicator (and probability) kriging has purged the approximation family of many functions. Although indicator kriging is a nonparametric method, it has been demonstrated that $\hat{I}(s_0, z)$ is more appropriate for $\Pr(Z(\mathbf{s}_0) \leq z | I(\mathbf{s}_1, z), \cdots, I(\mathbf{s}_n, z))$ than for $\Pr(Z(\mathbf{s}_0) \leq z | \mathbf{Z})$. Journel (1983) claims that there is not enough improvement in indicator *co*kriging to make it worthwhile, but a comparison in Lajaunie (1990) suggests otherwise.

The appearance of the disjunctive-kriging predictor (5.1.2) in a generalized-additive form bears a strong resemblance to generalized-additive *regression* models (e.g., Friedman and Stuetzle, 1981; Breiman and Friedman, 1985; Hastie and Tibshirani, 1990). It would be worth investigating whether this regression technology could be adapted to spatial prediction.

Nonlinear prediction is a vast subject; this section has hardly scratched the surface. Several methods from geostatistics have been presented, but the technology could certainly be augmented by borrowing and adapting from other subject areas (see, e.g., Kitagawa, 1987; Priestley, 1988; Christakos, 1989; Gotway and Cressie, 1992).

5.2 CHANGE OF SUPPORT

Consider the random process $\{Z(\mathbf{s}): \mathbf{s} \in D\}$, where $D \subset \mathbb{R}^d$, and assume $\text{var}(Z(\mathbf{s})) < \infty$, for all $\mathbf{s} \in D$. Define the block average

$$Z(B) \equiv \begin{cases} \int_B Z(\mathbf{u})\, d\mathbf{u} / |B|, & |B| > 0, \\ \text{ave}\{Z(\mathbf{u}) : \mathbf{u} \in B\}, & |B| = 0, \end{cases} \tag{5.2.1}$$

where $|B| \equiv \int_B d\mathbf{u}$. Recall from Chapter 3 that the stochastic integral in (5.2.1) is defined as a limit (in mean square) of approximating sums. The *support* of $Z(B)$ is defined to be B, the region over which $Z(\cdot)$ is averaged. The change-of-support problem refers to making inference on block averages whose supports are different from those of the data $\mathbf{Z} \equiv (Z(B_1), \ldots, Z(B_n))'$. Often the data have point support.

It is basic to many statistical methods that averaging reduces variances. For example, if $X_1, \ldots, X_n$ are uncorrelated with $E(X_i) = \mu$ and $\text{var}(X_i) = \sigma^2$, $i = 1, \ldots, n \geq 2$, then $E(\Sigma_{i=1}^n X_i/n) = \mu$ and $\text{var}(\Sigma_{i=1}^n X_i/n) = \sigma^2/n < \sigma^2$. For geostatistical problems, the data are usually correlated; nevertheless, the idea that more averaging produces lower variances is observed empirically (e.g., Mercer and Hall, 1911; Fairfield Smith, 1938) and is verifiable from models (e.g., Whittle, 1962; Modjeska and Rawlings, 1983). (Notice that averaging does not affect the mean μ.)

Modifiable Areal Unit Problem and Ecological Fallacy

Averaging or aggregation of regionalized variables also has an effect on correlations (Yule and Kendall, 1950, pp. 310–313). Typically, the greater the aggregation, the further the sample correlation between the aggregated variables is from zero. This dependence of correlation on support has sometimes been called the modifiable areal unit problem. The ecological fallacy is the presence of a relationship between two variables, at an aggregated level, that is due simply to the aggregation rather than to any real link. Openshaw and Taylor (1979) performed several experiments to show that, by aggregating (in various ways) fixed observations on small areas, a large range of positive and negative correlation coefficients could be produced.

Conditional Expectation and Linear Predictands

In the change-of-support problem, one is concerned with making predictions about $Z(B)$ from data $\mathbf{Z}$, where the supports of predictand and data are not the same. If the proposed predictor is a mean or a conditional mean of a distribution, then the change-of-support problem is relatively easy to solve. The predictor of $Z(B)$, based on data $\mathbf{Z} \equiv (Z(B_1), \dots, Z(B_n))'$, that minimizes mean-squared prediction error is

$$p^0(\mathbf{Z}; B) = E(Z(B)|\mathbf{Z}) = \int_B E(Z(\mathbf{u})|\mathbf{Z})\, d\mathbf{u} / |B|. \qquad (5.2.2)$$

If the integrand of (5.2.2) is approximated by, say, $p_{\mathrm{dk}}(\mathbf{Z}; \mathbf{u})$ given by (5.1.7), then a disjunctive-kriging type predictor of $Z(B)$ is obtained.

Now suppose that $Z(\cdot)$ is a Gaussian process or that a linear predictor is specified. Assume for simplicity that $E(Z(\mathbf{s})) \equiv \mu$. Then the optimal (simple kriging) predictor is

$$p^*(\mathbf{Z}; B) = \sum_{i=1}^{n} l_i(Z(B_i) - \mu) + \mu, \qquad (5.2.3)$$

where $\mathbf{l}' \equiv (l_1, \cdots, l_n) = \mathbf{c}'\Sigma^{-1}$, $\Sigma \equiv \mathrm{var}(\mathbf{Z})$, and $\mathbf{c}' \equiv \mathrm{cov}(Z(B), \mathbf{Z})$; see the introduction to Chapter 3. Importantly, the quantities $l_1, \dots, l_n$, and μ used to compute (5.2.3) can all be obtained from parameters of the point-support process:

$$\mu = E(Z(\mathbf{s})), \qquad \mathbf{s} \in D,$$

and

$$\mathrm{cov}\big(Z(B_i), Z(B_j)\big) = \int_{B_i}\int_{B_j} C(\mathbf{u}, \mathbf{v})\, d\mathbf{u}\, d\mathbf{v} / |B_i||B_j|, \qquad (5.2.4)$$

where

$$C(\mathbf{u}, \mathbf{v}) \equiv \mathrm{cov}(Z(\mathbf{u}), Z(\mathbf{v})), \qquad \mathbf{u}, \mathbf{v} \in D.$$

Should μ or $C(\cdot, \cdot)$ be unknown, they could, in principle, be estimated from point-support data (Chapter 2), which would allow (an estimate of) the predictor (5.2.3) to be computed (Chapter 3).

If point-support data were not available, one might assume a (semi) parametric model for data $Z(B_1), \ldots, Z(B_n)$ and estimate mean and covariance parameters by maximum likelihood, generalized least squares, method of moments, and so forth. In particular, when μ is unknown, the ordinary kriging predictor is

$$\hat{p}(\mathbf{Z}; B) = \sum_{i=1}^{n} \lambda_i Z(B_i), \tag{5.2.5}$$

where $\lambda_1, \ldots, \lambda_n$ satisfy (3.2.18). Thus, change-of-support can be handled fairly easily when the predictands are linear in $\mathbf{Z}$.

General Predictands

For reasons that will become obvious later, consider instead a *direct* disjunctive-kriging approximation to (5.2.2):

$$p(Z; B) = \sum_{i=1}^{n} f_i(Z(\mathbf{s}_i)). \tag{5.2.6}$$

The same approach as in Section 5.1 can be applied, except that now assumptions have to be made about sample–sample and sample–block bivariate distributions. Matheron (1976b) assumes that $Z(\mathbf{s}) = \phi(Y(\mathbf{s}))$, $\mathbf{s} \in D$, where $Y(\cdot)$ is a stationary process whose univariate and bivariate distributions are Gaussian and ϕ is known. (In practice, ϕ is inferred from the data.) Furthermore, he assumes that $Z(B) = \phi_B(X)$, for some ϕ_B, where X is unit Gaussian and $(X, Y(\mathbf{s}))$ are bivariate Gaussian. Now, trivially,

$$\phi_B(X) = Z(B) = \int_B E(\phi(Y(\mathbf{u}))|X)\, d\mathbf{u}/\, |B|. \tag{5.2.7}$$

From the Hermite polynomial expansion of ϕ given by (5.9.38), it is seen from (5.2.7) that ϕ_B can also be expanded in terms of Hermite polynomials. But the coefficients of the expansion depend on the correlation between X and $Y(\mathbf{u})$, $\mathbf{u} \in B$, which typically cannot be inferred for lack of suitable data. Such correlations are also needed in the disjunctive-kriging equations (5.9.38); they yield the optimal disjunctive-kriging predictor

$$p_{\mathrm{dk}}(\mathbf{Z}; B) = \sum_{i=1}^{n} \sum_{k=0}^{\infty} a_{ik} \eta_k(\phi^{-1}(Z(\mathbf{s}_i))), \tag{5.2.8}$$

where $\{\eta_k\colon\ k = 0, 1, \ldots\}$ are normalized Hermite polynomials given in (5.9.38).

The (direct) disjunctive-kriging predictor (5.2.6) was attempted because the orthogonal expansion of (5.2.7) can be generalized to an orthogonal expansion of $g(Z(B))$ for any measurable function g. For example, $g(Z(B)) = I(Z(B) \geq z_0)$ yields a disjunctive-kriging predictor that estimates $\Pr(Z(B) \geq z_0 | \mathbf{Z})$.

In spite of the formula (5.2.8), progress on the change-of-support problem is illusory. Without knowledge of sample–block parameters, (5.2.8) is not a true predictor in the sense that it is not a function only of the data. Neither are there data typically available that allow inference on such parameters, unless a highly specialized model for $Z(\cdot)$ is assumed. Nor is indicator kriging any help, because to construct a predictor like (5.1.12) one needs cross-variograms between indicator functions for block values and indicator functions for point values; again, such parameters typically cannot be inferred from the available point-support data. [Notice that $I(Z(B) \geq z_0) \neq \int_B I(Z(\mathbf{u}) \geq z_0)\, d\mathbf{u}/|B|$; that is, the relation (5.2.2) does not generalize.]

An Application to Resource Appraisal

Change-of-support is an important practical problem. Suppose $Z(\cdot)$ represents a resource (measured in "quantity" per unit volume), but that resource can only be recovered in blocks of volume $|B|$. Processing of the resource is expensive so that only rich blocks should be recovered, that is, blocks B for which $Z(B) \geq z_0$. However, for both financial and physical reasons, only data with point support can be obtained. Before deciding whether to exploit a resource field, the data are used to determine what proportion of blocks of volume $|B|$ are rich enough for recovery.

Consider a block B and its translates $B(\mathbf{s}) \equiv B \oplus \{\mathbf{s}\}$, $\mathbf{s} \in D$. Then the minimum-mean-squared-error predictor of $\sum_{\mathbf{s}} I(Z(B(\mathbf{s})) \geq z_0)/\sum_{\mathbf{s}} 1$, the proportion of rich blocks, is $\sum_{\mathbf{s}} \Pr(Z(B(\mathbf{s})) \geq z_0 | \mathbf{Z})/\sum_{\mathbf{s}} 1$. Without loss of generality, the subsequent discussion will focus on the conditional probability

$$\Pr(Z(B) \geq z_0 | \mathbf{Z}). \tag{5.2.9}$$

It would be wrong to use the histogram of the data $\mathbf{Z} \equiv (Z(\mathbf{s}_1), \ldots, Z(\mathbf{s}_n))'$ to estimate the conditional probability (5.2.9), for a number of reasons. The histogram estimates an assumed invariant distribution, but *not* the conditional distribution. Even if there were no change-of-support (i.e., even if B were $\{\mathbf{s}_0\}$), the density of $Z(\mathbf{s}_0)|\mathbf{Z}$ would be very different from the density of $Z(\mathbf{s}_0)$. In particular, $E(Z(\mathbf{s}_0)|\mathbf{Z}) \neq E(Z(\mathbf{s}_0)) = \mu$ and $E(\mathrm{var}(Z(\mathbf{s}_0)|\mathbf{Z})) \leq \mathrm{var}(Z(\mathbf{s}_0))$; that is, the conditional distribution tends to have less probability in the tails. This effect is exacerbated by making inference on $Z(B)$ (not on $Z(\mathbf{s}_0)$), which is an average of point-support predictands over the block B (see the introductory remarks to this section). In conclusion, it would be to a company's financial ruin to use a *point*-support histogram to determine the proportion of rich *blocks* available. The resulting proportion would typically be far too *optimistic*.

One might hope that knowing (estimating) both the invariant density function *and* the covariance function would resolve the change-of-support problem. Lantuejoul (1988) gives an example of three very different random processes that all have the same density and covariance function. Although very different distributions of $Z(B)$ result, he has not considered the (more relevant) conditional distribution of $Z(B)|\mathbf{Z}$.

From $\Pr(Z(B) \geq z|\mathbf{Z})$, the whole conditional distribution is available and hence $E(g(Z(B))|\mathbf{Z})$ can be calculated for any measurable g. An important example in resource assessment is $g(Z(B)) = Z(B)I(Z(B) \geq z_0)$, which corresponds to predicting $T \equiv \Sigma_{\mathbf{s}} Z(B(\mathbf{s}))I(Z(B(\mathbf{s})) \geq z_0)$, the total quantity of recoverable resource above the (financially determined) threshold z_0. The average value of recoverable resource in a block above the threshold z_0 is, $g(Z(\cdot)) \equiv |B|T/\Sigma_{\mathbf{s}} I(Z(B(\mathbf{s})) \geq z_0)$, which some have rather cavalierly predicted using $E\{Z(B)I(Z(B) \geq z_0)|\mathbf{Z}\}/\Pr(Z(B) \geq z_0|\mathbf{Z})$; the optimal predictor is, instead, $E(g(Z(\cdot))|\mathbf{Z})$. These predictors form the basis of the financial calculations referred to earlier. For example, as world metal prices go up, so the total monetary value of the recoverable quantity of metal goes up, for a fixed z_0. Therefore, in this circumstance, it may be possible to reduce z_0 and still maintain profitability; that is, taking into account the usual operational costs of recovering and processing blocks, smaller quantities of metal in blocks may still be sufficiently valuable to justify their recovery.

These financial considerations do not relegate the problem of prediction of $Z(B)$ itself. An initial analysis might simply involve drawing a map of the predictor $\{E(Z(B(\mathbf{s}_0))|\mathbf{Z}): \mathbf{s}_0 \in D\}$ (or an approximation to it), along with a map of its prediction standard errors.

Affine Correction

The previous paragraphs have discussed the importance (and the difficulty) of the change-of-support problem and some geostatistical approaches were given, although none was all that satisfactory. The simplest approach has not yet been mentioned, namely, *affine correction*. Assume that, conditional on $\mathbf{Z}$, the following equality holds in distribution:

$$\frac{Z(B) - E(Z(B)|\mathbf{Z})}{\{\text{var}(Z(B)|\mathbf{Z})\}^{1/2}} = \frac{Z(\mathbf{s}_0) - E(Z(\mathbf{s}_0)|\mathbf{Z})}{\{\text{var}(Z(\mathbf{s}_0)|\mathbf{Z})\}^{1/2}}$$

This will certainly be true for $Z(\cdot)$ a Gaussian process. Because $E(Z(B)|\mathbf{Z})$ and $\text{var}(Z(B)|\mathbf{Z})$ can be expressed in terms of $E(Z(\mathbf{u})|\mathbf{Z})$ and $\text{cov}(Z(\mathbf{u}), Z(\mathbf{v})|\mathbf{Z})$, quantiles of the conditional distribution of $Z(B)|\mathbf{Z}$ can be expressed as a linear combination of quantiles of $Z(\mathbf{s}_0)|\mathbf{Z}$, whose coefficients can be estimated from data with point support. Matheron (1985) develops diffusion models where exact answers are available, and finds that for such models affine corrections for unconditional distributions do poorly. [Lantuejoul (1988) reaches the same conclusion based on a case study.] However, conditioning on the data will undoubtedly mitigate the circumstances described by Matheron and Lantuejoul.

Another approach, as yet untried in geostatistical applications, is to look for a linear predictor $p^a(\mathbf{Z}; B)$ that minimizes the mean-squared prediction error over all linear predictors $\mathbf{a}'\mathbf{Z}$ such that $E(\mathbf{a}'\mathbf{Z}) = E(Z(B))$ *and* $\text{var}(\mathbf{a}'\mathbf{Z}) = \text{var}(Z(B))$. It is this last constraint that is new and, for $Z(\cdot)$ a Gaussian process, it guarantees that $p^a(\mathbf{Z}; B) = Z(B)$ *in distribution*. In this case, if one wishes to predict $g(Z(B))$, for *any* measurable function $g(\cdot)$, an *unbiased* predictor is $g(p^a(\mathbf{Z}; B))$. Call $p^a(\mathbf{Z}; B)$ the constrained kriging predictor of $Z(B)$.

Assume $E(Z(\mathbf{s})) = \mu(\mathbf{s}) \equiv \mu$ (unknown), and hence $E(Z(B)) = \mu$; generalization to $\mu(\mathbf{s}) = x(\mathbf{s})'\boldsymbol{\beta}$ is straightforward. Further, assume that the covariance function $C(\mathbf{u}, \mathbf{v}) \equiv \text{cov}(Z(\mathbf{u}), Z(\mathbf{v}))$ is known. Then, the appropriate objective function to minimize is

$$E(\mathbf{a}'\mathbf{Z} - Z(B))^2 + 2m_1(\mathbf{a}'\mathbf{1} - 1) + (m_2 - 1)(\mathbf{a}'\Sigma\mathbf{a} - C(B, B)),$$

where $\Sigma \equiv \text{var}(\mathbf{Z})$ and $2m_1, (m_2 - 1)$ are Lagrange multipliers. Straightforward differentiation and simplification yields the optimal predictor $p^a(\mathbf{Z}; B) = \mathbf{a}'\mathbf{Z}$, where

$$\mathbf{a}' = \{(1/m_2)\mathbf{c}'\Sigma^{-1}\} - \{(m_1/m_2)\mathbf{1}'\Sigma^{-1}\},$$

$$m_1 = \{-m_2/(\mathbf{1}'\Sigma^{-1}\mathbf{1})\} + \{(\mathbf{c}'\Sigma^{-1}\mathbf{1})/(\mathbf{1}'\Sigma^{-1}\mathbf{1})\},$$

$$m_2 = \left\{(\mathbf{c}'\Sigma^{-1}\mathbf{c})(\mathbf{1}'\Sigma^{-1}\mathbf{1}) - (\mathbf{c}'\Sigma^{-1}\mathbf{1})^2\right\}^{1/2} / \{C(B, B)(\mathbf{1}'\Sigma^{-1}\mathbf{1}) - 1\}^{1/2},$$

and $\mathbf{c} \equiv (C(B, \mathbf{s}_1), \ldots, C(B, \mathbf{s}_n))'$. Notice that m_2 is well defined provided $C(B, B)(\mathbf{1}'\Sigma^{-1}\mathbf{1}) > 1$ [which is equivalent to $\text{var}(Z(B)) > \text{var}(\hat{\mu}_{\text{gls}})$, where $\hat{\mu}_{\text{gls}}$ is the generalized-least-squares estimator of μ]. Clearly, $g(p^a(\mathbf{Z}; B))$ has computational advantages over $E(g(Z(B))|\mathbf{Z})$, but it has larger (or possibly equal) mean-squared prediction error. For $Z(\cdot)$ a Gaussian process, $g(p^a(\mathbf{Z}; B))$ is an unbiased predictor of $g(Z(B))$ for any choice of g.

From a statistical point of view, current solutions to the change-of-support problem are unsatisfactory; I believe that further progress will have to be model-based. One would like to make as few assumptions as possible and to be able to verify (from data) the assumptions that are made. One also needs to be able to estimate model parameters from available data. Suppose $Z(\cdot)$ is a second-order stationary isotropic random field that is L_2-continuous (Section 2.3.1). Through its spectral representation (Section 2.5.1; Yadrenko, 1983, p. 5), there may be a way to address the change-of-support problem with relatively few assumptions and relatively few parameters to estimate (e.g., Yadrenko, 1983, p. 24ff).

5.3 STABILITY OF THE GEOSTATISTICAL METHOD

The geostatistical method has been set out in considerable detail in Chapters 2 and 3. Briefly, it consists of performing an exploratory spatial data analysis, positing a model of (nonstationary) mean plus (intrinsically stationary) error, nonparametrically estimating the variogram or covariogram of the error, fitting a valid model to the estimate, and kriging (i.e., predicting) unobserved parts of the process from the available data. This last step yields not only a predictor, but a mean-squared prediction error (m.s.p.e.).

Notice that the data are being used twice: first to obtain a model of the spatial dependence and then to predict unobserved parts of the process. A generally held belief is that predictors at points within the network of data locations (interpolation) are more precise than predictors at points outside the network (extrapolation). One reason for this is extrapolation's demonstrably larger m.s.p.e. [provided the decomposition (3.1.2) contains a component of smooth small-scale variation]. A second reason, often forgotten, is the double use of the data referred to previously: With interpolation, the model has been built from surrounding data, but with extrapolation, the fitted model may be inappropriate beyond the network of data locations.

Within the geostatistical method, there are traps for the unwary, which may leave one with the impression that it would be safer to use deterministic methods of smoothing (e.g., Section 5.9.2). I believe this to be a false impression, because deterministic methods are not based on any statistical optimality criterion, and hence their choice of "smoothing constants" is usually an *ad hoc* one. Moreover, there is no way, *a priori*, to quantify whether one smoother is better than another, nor do such methods easily adapt to the change-of-support problem (Section 5.2).

In this section, which is based largely on Cressie and Zimmerman (1992), a mixture of formal and informal arguments exposes the potential pitfalls of the geostatistical method and indicates where it can be expected to be stable or unstable. In general terms, if model misspecification or parameter estimation has little effect on subsequent stages of a statistical method, then the method is said to be stable. Of course, an exploratory spatial data analysis (e.g., Section 2.2) is an important component that, if done properly, will give a solid foundation to the rest of the method.

Using whatever means are available, suppose the following statistical model for $\{Z(\mathbf{s}): \mathbf{s} \in D\}$ is posited:

$$Z(\mathbf{s}) = \mu(\mathbf{s}) + \delta(\mathbf{s}), \qquad \mathbf{s} \in D \subset \mathbb{R}^d. \tag{5.3.1}$$

In (5.3.1), the mean $E(Z(\cdot)) \equiv \mu(\cdot)$ is usually known, except for several large-scale-variation parameters $\boldsymbol{\beta}$ [e.g., $\mu(\mathbf{s}) = \mathbf{x}(\mathbf{s})'\boldsymbol{\beta}$, $\mathbf{s} \in \mathbb{R}^d$, and $\mathbf{x}(\cdot)$ known]. Furthermore, some form of stationarity for the zero-mean error process $\delta(\cdot)$ is usually assumed, to allow estimation of (generalized) covariance or variogram parameters $\boldsymbol{\theta}$.

A number of issues will be addressed in this section. First, there is the estimation of the parameters of the model; second, there is the effect of estimated or misspecified spatial-dependence parameters on kriging predictors; third, there is the same concern for mean-squared prediction errors.

5.3.1 Estimation of Spatial-Dependence Parameters

The estimation of spatial-dependence parameters $\boldsymbol{\theta}$ is usually less stable than the estimation of trend parameters $\boldsymbol{\beta}$. Zimmerman and Harville (1989) prove that the estimated generalized-least-squares (e.g.l.s.) estimator (1.3.26) is unbiased, assuming only that the distribution of the data is symmetric about its mean, and that the estimate of $\boldsymbol{\theta}$ is an even, translation-invariant function of the data.

If there is an overspecification of trend in (5.3.1), estimation of $\boldsymbol{\theta}$ is more stable when based on the variogram than when based on the covariogram. Specifically, suppose that $s \in \mathbb{R}$ and that the specified trend is

$$\mu(s) = \sum_{l=0}^{p+1} \beta_l s^l, \qquad s \in \mathbb{R}. \tag{5.3.2}$$

Suppose further that, unknown to the modeler, $\beta_{p+1} = 0$. Let $\hat{\mu}(s) = \sum_{l=0}^{p+1} \hat{\beta}_l s^l$, where $\hat{\beta}_0, \ldots, \hat{\beta}_{p+1}$ are obtained by an ordinary least-squares (o.l.s.) fit of (5.3.2) to the data $\mathbf{Z} \equiv (Z(1), \ldots, Z(n))'$. Then the results of Cressie and Grondona (1992) show that the variogram estimator (2.4.2) based on the o.l.s. residuals $\{Z(s) - \hat{\mu}(s): s = 1, \ldots, n\}$ has bias of $o(1/n)$, whereas the bias of the analogous covariogram estimator (2.4.4) is of $O(1/n)$ and proportional to $(p + 2)$, the number of terms fitted in (5.3.2). Thus, the greater the overspecification, the greater the bias in the covariogram estimator.

Also, repeating a similar argument to that given in Section 2.4.1, the variogram estimator (2.4.2) is more robust than the covariogram estimator (2.4.4) to small underspecification of trend. That is, suppose $\mu(s) = \sum_{l=0}^{p-1} \beta_l s^l$, $s \in \mathbb{R}$, is posited, but the true mean is $E(Z(s)) = \sum_{l=0}^{p-1} \beta_l s^l + \epsilon s^p$, where ϵ is small. Then the variogram estimator (2.4.2), based on the o.l.s. residuals $\{Z(s) - \hat{\mu}(s): s = 1, \ldots, n\}$, where $\hat{\mu}(s) \equiv \sum_{l=0}^{p-1} \hat{\beta}_l s^l$, is perturbed by a term that increases with ϵ, but the analogous covariogram estimator based on (2.4.4) is perturbed by a term that increases with $n\epsilon$. In conclusion, not only does the variogram exist when the covariogram might not (Section 2.3.1), but, when they both exist, its nonparametric estimator is more stable than that of the covariogram.

If the error process $\delta(\cdot)$ is contaminated by outliers, one should base estimation of the variogram on residuals $\{Z(\mathbf{s}_i) - \tilde{\mu}(\mathbf{s}_i): i = 1, \ldots, n\}$, from a robust estimator $\tilde{\mu}(\cdot)$ of $\mu(\cdot)$, and modify the method-of-moments estimator (2.4.2) to a robust estimator. [The robust estimator (2.4.12) is compared to (2.4.2) in Table 2.2.] Carroll et al. (1988) show that the use of $\tilde{\mu}(\cdot)$ has a bonus effect, at least in the heteroskedastic model for $\delta(\cdot)$. Their problem is one of estimating $\boldsymbol{\beta}$ based on reweighted least squares. When the initial estimator is $\tilde{\mu}(\mathbf{s}) = \mathbf{x}(\mathbf{s})'\tilde{\boldsymbol{\beta}}$, where $\tilde{\boldsymbol{\beta}}$ is a robust estimator of $\boldsymbol{\beta}$, then the weighted-least-squares estimator of $\boldsymbol{\beta}$ after one iteration is usually good enough to obviate further iterations. Discussion of estimated generalized-least-squares (e.g.l.s.) estimation of $\boldsymbol{\beta}$ is given in Section 1.3.

The geostatistical method takes the variogram estimates based on (2.4.2) [or (2.4.12)] and fits a valid variogram model (Section 2.5.2) to them. Section 2.6.2 gives a weighted-least-squares fitting criterion that puts more weight on the variogram estimate at small lags and at lags where larger numbers of data pairs are used. Other, more parametric, methods of estimating $\boldsymbol{\theta}$ are discussed in Section 2.6.1.

For example, consider the method of maximum likelihood (m.l.), where a misspecified family of covariance functions for the Gaussian process $Z(\cdot)$ may have been used in the likelihood. (Indeed, without any initial computation of a nonparametric estimator, a correctly specified likelihood may be more the exception than the rule.) Using increasing-domain asymptotics (Section 7.3.1), Watkins and Al-Boutiahi (1990) show that the maximum (misspecified) likelihood estimator converges to a value that *minimizes* the Kullback–Leibler distance between the posited Gaussian probability density and the true Gaussian probability density.

For geostatisticians, estimation of the parameters $\boldsymbol{\beta}$ and $\boldsymbol{\theta}$ is often a less important problem than obtaining precise spatial predictors and accurate estimators of their m.s.p.e.s. In what follows, it will be seen that, rather than being detrimental, biased estimation of $\boldsymbol{\theta}$ may prove beneficial in solving these geostatistical problems.

5.3.2 Stability of the Kriging Predictor

Suppose that the random process $Z(\cdot)$ satisfies the general linear model

$$Z(\mathbf{s}) = \mathbf{x}(\mathbf{s})'\boldsymbol{\beta} + \delta(\mathbf{s}), \qquad \mathbf{s} \in D \subset \mathbb{R}^d, \tag{5.3.3}$$

where $\delta(\cdot)$ is a zero-mean error process such that $E(\delta(\mathbf{s})^2) < \infty$, for all $\mathbf{s} \in D$. For consistency with Section 3.4, suppose $\mathbf{x}(\mathbf{s}) = (f_0(\mathbf{s}), \ldots, f_p(\mathbf{s}))'$, which is a $(p+1) \times 1$ vector. Define

$$C(\mathbf{s}, \mathbf{u}) \equiv \operatorname{cov}(\delta(\mathbf{s}), \delta(\mathbf{u})), \qquad \mathbf{s}, \mathbf{u} \in D. \tag{5.3.4}$$

Based on data $\mathbf{Z} \equiv (Z(\mathbf{s}_1), \ldots, Z(\mathbf{s}_n))'$, a predictor of $Z(\mathbf{s}_0)$ is sought. Section 3.4.5 shows that, when the vector of large-scale-variation parameters $\boldsymbol{\beta}$ is known, the optimal (heterogeneously) linear predictor (i.e., the linear predictor that minimizes m.s.p.e.) is the simple kriging predictor

$$Z^*(\mathbf{s}_0) = \mathbf{c}'\Sigma^{-1}(\mathbf{Z} - X\boldsymbol{\beta}) + \mathbf{x}'\boldsymbol{\beta}. \tag{5.3.5}$$

The notation in (5.3.5) is explained *circa* (3.4.3) and (3.4.4); recall that $\mathbf{x}(\mathbf{s}_0)$ is abbreviated to $\mathbf{x}$.

In reality, $\boldsymbol{\beta}$ is usually unknown, so an optimal *homogeneously linear unbiased* predictor is sought. Then, from (3.4.61), the optimal predictor is the universal-kriging predictor

$$\hat{Z}(\mathbf{s}_0) = \left\{\mathbf{c} + X(X'\Sigma^{-1}X)^{-1}(\mathbf{x} - X'\Sigma^{-1}\mathbf{c})\right\}'\Sigma^{-1}\mathbf{Z}. \qquad (5.3.6)$$

Clearly, overfitting the trend $\mu(\cdot)$ does not affect the unbiasedness of $\hat{Z}(\mathbf{s}_0)$, but it will make the m.s.p.e. larger than necessary, because the data are being used to estimate parameters that are in fact equal to zero. Underfitting the trend appears to cause more serious problems, because it leads to bias in $\hat{Z}(\mathbf{s}_0)$. However, in practice, underfitting is not as serious as one might first think, because the covariance function is usually unknown and has to be estimated; further discussion follows.

Assume that the trend $\mu(\cdot)$ has been correctly specified and that a model for the spatial dependence is given by

$$\operatorname{cov}(Z(\mathbf{s}), Z(\mathbf{u})) = C(\mathbf{s}, \mathbf{u}; \boldsymbol{\theta}) \quad \text{or} \quad \operatorname{var}(Z(\mathbf{s}) - Z(\mathbf{u})) = 2\gamma(\mathbf{s}, \mathbf{u}; \boldsymbol{\theta}),$$
$$\mathbf{s}, \mathbf{u} \in D, \boldsymbol{\theta} \in \Theta. \qquad (5.3.7)$$

The optimality of $\hat{Z}(\mathbf{s}_0)$ given by (5.3.6) relies not only on the assumption that the trend and the class of covariance or variogram models is correctly specified by (5.3.7), but also that the true value $\boldsymbol{\theta} = \boldsymbol{\theta}_0$ is known and used in (5.3.6).

Studies on the stability of kriging, when the trend has been correctly specified, have usually taken one of two approaches. They have either concentrated on *mathematical stability*, where a known covariance function or variogram is perturbed, or on *statistical stability*, where $\boldsymbol{\theta}$ in (5.3.7) is unknown and has to be estimated. In practice, the latter problem seems to be of more importance than the former, because it is rare that a fully specified covariance function is chosen from among several (or many) without looking at the data. Typically, a parametric family like (5.3.7) is specified and a "best" $\hat{\boldsymbol{\theta}} \in \Theta$ is estimated from the data. Nevertheless, mathematical stability results can indicate the sample size needed to ensure that the estimator $\hat{\boldsymbol{\theta}}$ is "sufficiently close" to the true value $\boldsymbol{\theta}_0$.

Mathematical Stability

Using the isotropic variogram $2\gamma_1^o$, write the ordinary kriging equations (3.2.12) as

$$\Gamma_1 \boldsymbol{\lambda}_1 = \boldsymbol{\gamma}_1. \qquad (5.3.8)$$

Similarly, the ordinary kriging equations based on the isotropic variogram $2\gamma_2^o$ are written as

$$\Gamma_2 \boldsymbol{\lambda}_2 = \boldsymbol{\gamma}_2. \qquad (5.3.9)$$

Define $\Delta\boldsymbol{\lambda} \equiv \boldsymbol{\lambda}_2 - \boldsymbol{\lambda}_1$, $\Delta\Gamma \equiv \Gamma_2 - \Gamma_1$, and $\Delta\boldsymbol{\gamma} \equiv \boldsymbol{\gamma}_2 - \boldsymbol{\gamma}_1$. Diamond and Armstrong (1984) show that if

$$\sup\{|(\gamma_2^o(h)/\gamma_1^o(h)) - 1|: h > 0\} < r, \tag{5.3.10}$$

then $\|\Delta\Gamma\| \le r\|\Gamma_1\|$ and $\|\Delta\boldsymbol{\gamma}\| \le r\|\boldsymbol{\gamma}_1\|$. Hence,

$$\|\Delta\boldsymbol{\lambda}\| \le 2r\|\boldsymbol{\lambda}_1\|C(\Gamma_1)/\{1 - rC(\Gamma_1)\}, \tag{5.3.11}$$

provided $rC(\Gamma_1) < 1$. In (5.3.11), $C(\Gamma_1) \equiv \|\Gamma_1\|\,\|\Gamma_1^{-1}\|$ is the *condition number* of the matrix Γ_1. In the preceding inequalities, the vector norm $\|\mathbf{a}\|$ could be *any* L_p norm, $p \ge 1$. The matrix norm is defined as $\|A\| \equiv \sup\{\|A\mathbf{x}\|: \|\mathbf{x}\| = 1\}$, and so for the L_2 vector norm the condition number $C(A)$ is simply the largest absolute eigenvalue of A divided by the smallest.

From (5.3.11), the relative error satisfies

$$\|\Delta\boldsymbol{\lambda}\|/\|\boldsymbol{\lambda}_1\| \le 2rC(\Gamma_1)/\{1 - rC(\Gamma_1)\}, \tag{5.3.12}$$

provided $rC(\Gamma_1) < 1$. Now suppose $2\gamma_1^o$ is the true variogram, but that any $2\gamma_2^o$ contained in the neighborhood defined by (5.3.10) is used to carry out kriging. Diamond and Armstrong (1984) show that if $rC(\Gamma_1) < 1$ and $r \le \min[1/C(\Gamma_1), \epsilon/\{(2 + \epsilon)C(\Gamma_1)\}]$, then $\|\Delta\boldsymbol{\lambda}\|/\|\boldsymbol{\lambda}_1\| \le \epsilon$. Therefore, the larger the condition number $C(\Gamma_1)$, the less stable is the kriging predictor.

Instead of (5.3.10), Yakowitz and Szidarovsky (1985) choose a neighborhood defined by

$$\sup\{|\gamma_1^o(h) - \gamma_2^o(h)|: h > 0\} < r \tag{5.3.13}$$

and give an analogous inequality to (5.3.11). Warnes (1986) takes an infinitesimal approach by kriging with a perturbed covariogram $C_1^o + \delta C$ and calculating the first- and second-order perturbations in the kriging weights. Not surprisingly, the larger perturbations in $\hat{Z}(\mathbf{s}_0)$ occur where the kriging variance $\sigma_k^2(\mathbf{s}_0)$ is larger. To illustrate his results, Warnes uses two isotropic covariograms (in $\mathbb{R}^2$): $C_1^o(h;\theta) = \sigma^2 \exp\{-(h/\theta)\}$ and $C_2^o(h;\theta) = \sigma^2 \exp\{-(h/\theta)^2\}$, for $h \ge 0$. The latter covariogram is infinitely differentiable at the origin and imposes far more smoothness on $Z(\cdot)$ than would be seen in any physical process. Using the notion of covariogram compatibility (Section 5.8), Stein and Handcock (1989) explain why the extreme sensitivity of $\hat{Z}(\mathbf{s}_0)$ to perturbation of the parameter θ in $C_2^o(h;\theta)$, demonstrated by Warnes (1986), should be discounted: The stability of $\hat{Z}(\mathbf{s}_0)$ to perturbations in the exponential covariogram C_1^o, found by Warnes, is indicative of what can be expected for most other covariogram models (Armstrong and Wackernagel, 1988).

Another possible stability approach is to express uncertainty in model parameters by putting a prior distribution on them. One is still interested in the predictive distribution of $Z(\mathbf{s}_0)$ given $\mathbf{Z}$, but it has to be obtained by first conditioning on $\boldsymbol{\beta}$ and $\boldsymbol{\theta}$, and then integrating out their presence with respect to the prior. More details are given in Section 3.4.4.

An asymptotic stability property of the kriging predictor was established by Yakowitz and Szidarovsky (1985). Let $D \subset \mathbb{R}^d$ be bounded and suppose $\{\mathbf{s}_i: i = 1, 2, \ldots\}$ is a sequence of points in D with limit point $\mathbf{s}_0$. Further, suppose $Z(\cdot)$ is second-order stationary. If a technical condition, on the rate the spectral density $g(\boldsymbol{\omega})$ of $C(\cdot)$ converges to 0 as $|\boldsymbol{\omega}| \to \infty$, holds, then Yakowitz and Szidarovsky show that the ordinary kriging predictor $\hat{Z}(\mathbf{s}_0)$, given by (3.2.12), converges almost surely to $Z(\mathbf{s}_0)$, even should the covariogram $C(\cdot)$ be incorrectly specified.

Statistical Stability

Suppose that $Z(\cdot)$ is a Gaussian process satisfying (5.3.3). Then the predictor with minimum m.s.p.e. is $E(Z(\mathbf{s}_0)|\mathbf{Z}) = \mathbf{c}(\boldsymbol{\theta})'\Sigma(\boldsymbol{\theta})^{-1}(\mathbf{Z} - X\boldsymbol{\beta}) + \mathbf{x}'\boldsymbol{\beta}$, where $\boldsymbol{\theta}$ is a finite vector of small-scale-variation parameters given by the model (5.3.7). If $\boldsymbol{\beta}$ is unknown but $\boldsymbol{\theta}$ is known, then the m.l. estimator of $E(Z(\mathbf{s}_0)|\mathbf{Z})$ is the universal-kriging predictor $\hat{Z}(\mathbf{s}_0)$. Now, if both $\boldsymbol{\beta}$ and $\boldsymbol{\theta}$ are unknown, the m.l. estimator is

$$\hat{E}(Z(\mathbf{s}_0)|\mathbf{Z}) = \mathbf{c}(\hat{\boldsymbol{\theta}})'\Sigma(\hat{\boldsymbol{\theta}})^{-1}(\mathbf{Z} - X\hat{\boldsymbol{\beta}}(\hat{\boldsymbol{\theta}})) + \mathbf{x}'\hat{\boldsymbol{\beta}}(\hat{\boldsymbol{\theta}}), \qquad (5.3.14)$$

where $\hat{\boldsymbol{\beta}}(\hat{\boldsymbol{\theta}}) \equiv (X'\Sigma(\hat{\boldsymbol{\theta}})^{-1}X)^{-1}X'\Sigma(\hat{\boldsymbol{\theta}})^{-1}\mathbf{Z}$ and $\hat{\boldsymbol{\theta}}$ maximizes the profile likelihood

$$(2\pi)^{-n/2}|\Sigma(\boldsymbol{\theta})|^{-1/2}\exp\{-(1/2)(\mathbf{Z} - X\hat{\boldsymbol{\beta}}(\boldsymbol{\theta}))'\Sigma(\boldsymbol{\theta})^{-1}(\mathbf{Z} - X\hat{\boldsymbol{\beta}}(\boldsymbol{\theta}))\},$$

over $\boldsymbol{\theta} \in \Theta$.

It is not necessary that the process be assumed Gaussian for (5.3.14) to be worthy of study; more generally, one takes $\hat{\boldsymbol{\theta}}$ to be any sensible estimator of $\boldsymbol{\theta}$. Indeed, Zimmerman and Cressie (1992) show that (5.3.14) is an unbiased estimator of $E(Z(\mathbf{s}_0)) = \mathbf{x}'\boldsymbol{\beta}$, by assuming simply that the distribution of $(Z(\mathbf{s}_0), \mathbf{Z}')'$ is symmetric about its mean and that $\hat{\boldsymbol{\theta}}$ is an even and translation-invariant function of $\mathbf{Z}$. Thus, $\hat{\boldsymbol{\theta}}$ need not be an unbiased estimator of $\boldsymbol{\theta}$ for (5.3.14) to be an unbiased predictor. In fact, unbiasedness of $\hat{\boldsymbol{\theta}}$ is not always appropriate; it will be seen in Section 5.3.3 that an unbiased estimator of $\Sigma(\boldsymbol{\theta})$ in (5.3.7) can lead to a *negatively* biased estimator of the m.s.p.e. of (5.3.14).

When the data $\mathbf{Z}$ contain outliers, robust methods of estimating $\boldsymbol{\theta}$ should be used. Even so, the kriging predictor will be nonrobust (it is linear in $\mathbf{Z}$); Section 3.3 describes a suggestion for robust kriging based on editing outliers in $\mathbf{Z}$. Both the declaration and editing of outliers use the notion of cross-validation (Section 2.6.4).

Intuitively, decomposing the model into large- and small-scale-variation components, as in (5.3.3), engenders stability in the predictor. Section 3.1 has gone to some lengths to explain the nonuniqueness of this decomposition, but provided $\boldsymbol{\beta}$ and $\boldsymbol{\theta}$ are sensibly estimated this should *not* affect the predictor $\hat{Z}(\mathbf{s}_0)$ very much. For example, if the large-scale variation $E(Z(\cdot))$ is underfitted, then provided an additive decomposition is retained, the small-scale variation is automatically overfitted (and vice versa). Therefore, any misspecifications tend to be compensated for, because the goal is prediction of $Z(\mathbf{s}_0)$. [However, suppose the process is observed with measurement error $c_{ME} > 0$. Then the kriging predictor (3.2.25) of the noiseless value $S(\mathbf{s}_0)$ is *not* stable to the misspecification of c_{ME}.]

5.3.3 Stability of the Kriging Variance

Although the kriging predictor is generally stable, the m.s.p.e. depends heavily on $\text{cov}(Z(\mathbf{s}), Z(\mathbf{u}))$, $\mathbf{s}, \mathbf{u} \in D$, as is illustrated in Section 4.1 [where two very different decompositions (5.3.3), of the same data, are proposed]; compare Figures 4.4*b* and 4.7*b*. Indeed, one might argue that the m.s.p.e. itself is the most important measure of kriging's stability.

The m.s.p.e. (or kriging variance) of (5.3.6) is, from (3.4.64),

$$\sigma_k^2(\mathbf{s}_0) \equiv E\left(Z(\mathbf{s}_0) - \hat{Z}(\mathbf{s}_0)\right)^2$$
$$= C(\mathbf{s}_0, \mathbf{s}_0) - \mathbf{c}'\Sigma^{-1}\mathbf{c} + (\mathbf{x} - X'\Sigma^{-1}\mathbf{c})'\left(X'\Sigma^{-1}X\right)^{-1}(\mathbf{x} - X'\Sigma^{-1}\mathbf{c}). \tag{5.3.15}$$

The statistical stability of (universal) kriging can be quantified by (5.3.15). When the nugget effect of the process (Section 3.2.1) does not dominate the small-scale variation, (5.3.15) is typically small for $\mathbf{s}_0$ surrounded by observations from $\mathbf{Z}$ (interpolation) and large for $\mathbf{s}_0$ away from the data locations (extrapolation).

Mathematical Stability

The kriging predictor minimizes the m.s.p.e. for a given covariance function C or a given variogram 2γ. When $2\gamma_2^o$ is in a neighborhood of $2\gamma_1^o$, defined by (5.3.10), a similar inequality to (5.3.11), for the change in $\sigma_k^2(\mathbf{s}_0)$, is derived by Diamond and Armstrong (1984). Simpler inequalities for the change in $\sigma_k^2(\mathbf{s}_0)$ caused by a change in the nugget effect are derived by Bardossy (1988), and exact results for $\sigma_k^2(\mathbf{s}_0)$, based on a spherical variogram with changing range and changing relative nugget effect, are computed by Brooker (1986). Brooker finds $\sigma_k^2(\mathbf{s}_0)$ to be stable, except if one misspecifies the nugget effect to be close to zero when its true value is much larger; then $\sigma_k^2(\mathbf{s}_0)$ can be grossly understated.

Another approach is given by Stein (1989b), who assumes the decomposition (5.3.3). Suppose that the covariance function in (5.3.7) is

$$C(\mathbf{s},\mathbf{u};\boldsymbol{\theta}) = \theta_1 C_1(\mathbf{s},\mathbf{u}) + \theta_2 C_2(\mathbf{s},\mathbf{u}), \qquad \boldsymbol{\theta} \in (0,\infty)^2, \quad (5.3.16)$$

where C_1 and C_2 are known covariance functions. Let $\boldsymbol{\theta}_0 \equiv (\theta_{01}, \theta_{02})'$ be the true value and $\boldsymbol{\theta}_1 \equiv (\theta_{11}, \theta_{12})'$ be any other value used. Then, for $R \equiv \theta_{01}\theta_{12}/\theta_{11}\theta_{02}$, Stein proves a Kantorovich-type inequality:

$$1 \le \frac{E\left\{\left(Z(\mathbf{s}_0) - \hat{Z}(\mathbf{s}_0;\boldsymbol{\theta})\right)^2\right\}}{E\left\{\left(Z(\mathbf{s}_0) - \hat{Z}(\mathbf{s}_0;\boldsymbol{\theta}_0)\right)^2\right\}} \le 1 + \frac{(R-1)^2}{4R}, \quad (5.3.17)$$

where $\hat{Z}(\mathbf{s}_0;\cdot)$ is the universal-kriging predictor (5.3.6) with covariances defined by (5.3.16), and both expectations in (5.3.17) are taken with $\boldsymbol{\theta} = \boldsymbol{\theta}_0$. Thus, (5.3.17) is more a result about efficiency than stability. However, it does offer guidelines on how far $\boldsymbol{\theta}$ might be allowed to deviate from $\boldsymbol{\theta}_0$ with little attendant increase in true m.s.p.e. [The same ideas in time series can be found in Cleveland (1971).] Under infill asymptotics, Stein (1988) shows that the efficiency ratio in (5.3.17) tends to 1, provided $C(\mathbf{s},\mathbf{u};\boldsymbol{\theta}_1)$ and $C(\mathbf{s},\mathbf{u};\boldsymbol{\theta}_0)$ are *compatible* on a bounded $D \subset \mathbb{R}^d$; see (5.8.6).

To guarantee stability of the kriging variance $\sigma_k^2(\mathbf{s}_0)$ against the choice of C (or 2γ), one might try to obtain a minimax linear predictor. Consider a linear predictor of $Z(\mathbf{s}_0)$, $p(\mathbf{Z};\mathbf{s}_0) = \boldsymbol{\lambda}'\mathbf{Z}$, and its m.s.p.e. $\sigma_e^2 \equiv E(Z(\mathbf{s}_0) - p(\mathbf{Z};\mathbf{s}_0))^2$. Then, choose $\boldsymbol{\lambda}_M$ to satisfy

$$E(Z(\mathbf{s}_0) - \boldsymbol{\lambda}'_M\mathbf{Z})^2 = \inf\{\sup\{\sigma_e^2 : C \in \Delta\} : \boldsymbol{\lambda} \in \mathbb{R}^n\}, \quad (5.3.18)$$

where Δ is a given class of valid covariograms containing at least one member for which σ_e^2 is finite. Call $\boldsymbol{\lambda}'_M\mathbf{Z}$ the *minimax linear predictor* of $Z(\mathbf{s}_0)$. Clearly,

$$\sigma_k^2(\mathbf{s}_0) \le E(Z(\mathbf{s}_0) - \boldsymbol{\lambda}'_M\mathbf{Z})^2,$$

where the left-hand side is calculated for *any* $C \in \Delta$. This approach has received some attention in the time-series literature (e.g., Franke, 1984; Tsaknakis et al., 1986).

Statistical Stability

To emphasize the presence of spatial dependence parameters $\boldsymbol{\theta}$, write (5.3.15) as

$$\begin{aligned}\sigma_k^2(\mathbf{s}_0;\boldsymbol{\theta}) &= C(\mathbf{s}_0,\mathbf{s}_0;\boldsymbol{\theta}) - \mathbf{c}(\boldsymbol{\theta})'\Sigma(\boldsymbol{\theta})^{-1}\mathbf{c}(\boldsymbol{\theta}) \\ &\quad + \left(\mathbf{x} - X'\Sigma(\boldsymbol{\theta})^{-1}\mathbf{c}(\boldsymbol{\theta})\right)'\left(X'\Sigma(\boldsymbol{\theta})^{-1}X\right)^{-1}\left(\mathbf{x} - X'\Sigma(\boldsymbol{\theta})^{-1}\mathbf{c}(\boldsymbol{\theta})\right).\end{aligned} \quad (5.3.19)$$

Notice that (5.3.19) does not depend on the trend parameters $\boldsymbol{\beta}$ nor does it depend on the data. Then the *estimated* m.s.p.e. is

$$\hat{\sigma}_k^2(\mathbf{s}_0) \equiv \sigma_k^2(\mathbf{s}_0; \hat{\boldsymbol{\theta}}), \tag{5.3.20}$$

where $\hat{\boldsymbol{\theta}}$ is an estimator of $\boldsymbol{\theta}$ based on the data $\mathbf{Z}$ [e.g., for $Z(\cdot)$ Gaussian, $\hat{\boldsymbol{\theta}}$ might be the m.l. estimator used in (5.3.14)]. In practice, it is (5.3.20) that is used in the geostatistical method, but its statistical properties can be difficult to obtain.

Now consider the three quantities:

- $m_2(\boldsymbol{\theta}) \equiv E(Z(\mathbf{s}_0) - p_2(\mathbf{Z}; \mathbf{s}_0))^2$, where $p_2(\mathbf{Z}; \mathbf{s}_0)$ is given by (5.3.14). This is the actual m.s.p.e. of the *estimated* kriging predictor.
- $m_1(\boldsymbol{\theta}) \equiv E(Z(\mathbf{s}_0) - \hat{Z}(\mathbf{s}_0))^2$, where $\hat{Z}(\mathbf{s}_0)$ is given by (5.3.6). This is the idealized m.s.p.e. of the kriging predictor with known $\boldsymbol{\theta}$. [The notation $\sigma_k^2(\mathbf{s}_0; \boldsymbol{\theta})$ was used in (5.3.19), but it is not well suited to the comparisons to be presented in the following text.]
- $E(m_1(\hat{\boldsymbol{\theta}}))$, where $m_1(\hat{\boldsymbol{\theta}})$ is given by (5.3.20). This is the expectation of the estimated m.s.p.e. and is also a function of the parameter $\boldsymbol{\theta}$.

In practice, there are almost always spatial-dependence parameters $\boldsymbol{\theta}$ to estimate. Hence, the goal should be to estimate $m_2(\boldsymbol{\theta})$ well; so,

$$E\left(m_1(\hat{\boldsymbol{\theta}})\right) - m_2(\boldsymbol{\theta}) \tag{5.3.21}$$

represents the bias of real interest, and not $E(m_1(\hat{\boldsymbol{\theta}})) - m_1(\boldsymbol{\theta})$, as some have maintained. These issues are discussed in more detail, for the general linear model, by Harville (1988) and Harville and Jeske (1992).

Under the assumption that $Z(\cdot)$ is Gaussian, Zimmerman and Cressie (1992) show that $m_2(\boldsymbol{\theta}) \geq m_1(\boldsymbol{\theta})$, with equality holding if and only if $p_2(\mathbf{Z}; \mathbf{s}_0) = \hat{Z}(\mathbf{s}_0)$, almost surely. They also use a multivariate version of Jensen's inequality to establish that $E(m_1(\hat{\boldsymbol{\theta}})) \leq m_1(\boldsymbol{\theta})$, provided $\Sigma(\boldsymbol{\theta}) - E(\Sigma(\hat{\boldsymbol{\theta}}))$ is nonnegative-definite; the process $Z(\cdot)$ need not be Gaussian. If, in addition, the process is Gaussian, then the two results may be combined to yield:

$$E\left(m_1(\hat{\boldsymbol{\theta}})\right) \leq m_1(\boldsymbol{\theta}) \leq m_2(\boldsymbol{\theta}). \tag{5.3.22}$$

Writing (5.3.21) as $\{E(m_1(\hat{\boldsymbol{\theta}})) - m_1(\boldsymbol{\theta})\} + \{m_1(\boldsymbol{\theta}) - m_2(\boldsymbol{\theta})\}$, it is easily seen that the bias of the estimated m.s.p.e. is *negative* and comes from two sources. (The negative bias should not be surprising in light of the decomposition, total prediction error equals probabilistic prediction error plus statistical prediction error, given in the introductory remarks to Chapter 3.) A further implication is that, to obtain an accurate estimator of $m_2(\boldsymbol{\theta})$, one should look for a $\hat{\boldsymbol{\theta}}$ that yields an *overestimate* of the covariance function C.

Zimmerman and Cressie (1992) actually present these bias results for the more general problem of kriging with intrinsic random functions (Section 5.4).

A number of authors have established that the bias of the estimated m.s.p.e. is asymptotically negligible, for the general linear model (e.g., Toyooka, 1982; Prasad and Rao, 1990) and for time-series models (e.g., Fuller and Hasza, 1981). An open problem remains to establish a similar result in a spatial context (using either infill or increasing-domain asymptotics).

In small samples, the bias (5.3.21) can be appreciable, as is demonstrated in the various examples given by Zimmerman and Cressie (1992); an approximately unbiased estimator of $m_2(\boldsymbol{\theta})$, based on a suggestion by Prasad and Rao (1990), is compared to $m_1(\hat{\boldsymbol{\theta}})$. From their study, Zimmerman and Cressie conclude that Prasad and Rao's estimator is preferable only when $\Sigma(\boldsymbol{\theta}) - E(\Sigma(\hat{\boldsymbol{\theta}}))$ is nonnegative-definite *and* the spatial correlation is weak. Otherwise, the continued use of $m_1(\hat{\boldsymbol{\theta}})$ is recommended. The general conclusion is that moderate-to-strong spatial correlation compensates for the inherent negative bias of $m_1(\hat{\boldsymbol{\theta}})$.

Another potential compensating factor for the negative bias is a tendency to underfit the trend (Starks and Fang, 1982a). Thus, there are at least two self-correcting elements to the estimation of m.s.p.e. that enhance its stability.

In the presence of outliers, the robust-kriging proposal of Hawkins and Cressie (1984) (see Section 3.3) gives the stable predictor (3.3.12). However, its m.s.p.e. is presently unknown. In the meantime, one could use $\sigma_k^2(\mathbf{s}_0; \bar{\boldsymbol{\theta}})$, where $\bar{\boldsymbol{\theta}}$ is based on a *robust* estimator of the variogram. Indeed, Hawkins and Cressie suggest that the editing constant c in (3.3.10) might be chosen so that the mean-squared prediction error of (3.3.12) matches $\sigma_k^2(\mathbf{s}_0; \bar{\boldsymbol{\theta}})$.

5.4 INTRINSIC RANDOM FUNCTIONS OF ORDER k

Recall the model (3.4.2), which expresses a spatial random process as a linear mean function plus an intrinsically stationary error process. Retain the linear model, but suppose instead

$$Z(\mathbf{s}) = \sum_{j=1}^{p+1} f_{j-1}(\mathbf{s})\beta_{j-1} + \delta(\mathbf{s}), \qquad \mathbf{s} \in D, \tag{5.4.1}$$

where $\delta(\cdot)$ is a zero-mean, intrinsic random function of order k (see following text). The presentation of universal kriging in Section 3.4 assumed $\delta(\cdot)$ to be intrinsically stationary with a variogram $2\gamma(\cdot)$; the potential bias in estimating $2\gamma(\cdot)$ from the o.l.s. residuals (Section 3.4.3) led Matheron (1973) to advance an alternative approach. The basic idea is to filter, linearly, the data $\mathbf{Z} \equiv (Z(\mathbf{s}_1), \ldots, Z(\mathbf{s}_n))'$ to remove dependence on the nuisance parameters $\boldsymbol{\beta} \equiv (\beta_0, \ldots, \beta_p)'$.

Consider a linear filter (matrix) Q with the property that $QX = 0$, where X is the $n \times (p+1)$ matrix whose (i, j)th element is $f_{j-1}(\mathbf{s}_i)$. [For example, if $E(Z(\mathbf{s})) \equiv \mu$, then $Q = I - (1/n)\mathbf{1}\mathbf{1}'$ satisfies $Q\mathbf{1} = 0$.] Hence,

$$Q\mathbf{Z} = Q\boldsymbol{\delta}, \tag{5.4.2}$$

where $\boldsymbol{\delta} \equiv (\delta(\mathbf{s}_1), \ldots, \delta(\mathbf{s}_n))'$. Therefore, one option is to apply the various methods of spatial prediction given in Chapter 3 to the transformed data $Q\mathbf{Z}$; from (5.4.2), the second-order properties of $Q\mathbf{Z}$ depend only on small-scale variation parameters. However, the approach taken here is more subtle.

The key to generalizing the model (3.4.2) is to realize that *any* process for which the *filtered* errors are second-order stationary is perfectly suited for optimal spatial prediction. Such a class of processes *includes* all second-order stationary processes and, provided $f_0(\mathbf{s}) \equiv 1$, all intrinsically stationary processes.

Henceforth, in this section, suppose that the explanatory variables $\{f_{j-1}(\mathbf{s}): j = 1, \ldots, p+1\}$ is the set of mixed monomials $x_1^{i_1} \cdots x_d^{i_d}$, where $\mathbf{s} = (x_1, \ldots, x_d)'$ and $i_1, \ldots, i_d$ are nonnegative integers such that $i_1 + \cdots i_d \leq k$; k is a given nonnegative integer. Then the set includes the explanatory variable $f_0(\mathbf{s}) \equiv 1$, and $p + 1 = \binom{k+d}{d}$. Student (1914) noticed that $(k+1)$th order differences of temporal data annihilate the mean effect in (5.4.1). Matheron (1973) went further, by exploiting the theory developed by Yaglom (1955) and Gel'fand and Vilenkin (1964) for higher-order stationary increments and generalized functions. Thus, he defined a rather general class of processes, which he called intrinsic random functions of order k. [In the time domain, they are known as integrated processes; e.g., Box and Jenkins (1970).]

Definition. Suppose k is a nonnegative integer. Then, an *intrinsic random function of order* k (IRFk) is a random process $Z(\cdot)$ for which

$$V(\mathbf{u}) \equiv \sum_{i=1}^{m} \nu_i Z(\mathbf{s}_i + \mathbf{u}), \qquad \mathbf{u} \in D \subset \mathbb{R}^d, \tag{5.4.3}$$

is second-order stationary. (Second-order stationarity is defined in Section 2.3.) In (5.4.3), m is any positive integer, $\{\mathbf{s}_i: i = 1, \ldots, m\}$ are any spatial locations in D, and $\boldsymbol{\nu} \equiv (\nu_1, \ldots, \nu_m)'$ is a *generalized-increment vector* of real numbers for which $(f_j(\mathbf{s}_1), \ldots, f_j(\mathbf{s}_m))\boldsymbol{\nu} = 0$, $j = 1, \ldots, p+1$. ∎

Using results from Gel'fand and Vilenkin (1964), Matheron (1973, Section 2.1) proves that an IRFk possesses a *generalized covariance* $\{K(\mathbf{h}): \mathbf{h} \in \mathbb{R}^d\}$, such that for generalized-increment vectors $\boldsymbol{\nu}$ and $\boldsymbol{\kappa}$ (of order k),

$$\text{cov}(\boldsymbol{\nu}'\mathbf{Z}_m, \boldsymbol{\kappa}'\mathbf{Z}_m) = \sum_{i=1}^{m}\sum_{j=1}^{m} \nu_i \kappa_j K(\mathbf{s}_i - \mathbf{s}_j), \tag{5.4.4}$$

where $\mathbf{Z}_m \equiv (Z(\mathbf{s}_1), \ldots, Z(\mathbf{s}_m))'$. The generalized covariance satisfies $K(-\mathbf{h}) = K(\mathbf{h})$ and is unique up to an even polynomial of degree $2k$.

Now, let $P_1 \equiv X(X'X)^{-1}X'$ be the projection matrix onto the column space of X, $P_2 \equiv I - P_1$ be the projection onto its orthogonal complement, and K be the $n \times n$ matrix whose (i, j)th element is $K(\mathbf{s}_i - \mathbf{s}_j)$. Then $P_2 K P_2$ is positive-semidefinite. To see this, consider the arbitrary $n \times 1$ vector $\mathbf{a}$; then $\mathbf{a}'P_2 K P_2 \mathbf{a} = \boldsymbol{\nu}' K \boldsymbol{\nu} = \mathrm{var}(\boldsymbol{\nu}'\mathbf{Z}) \geq 0$, where $\boldsymbol{\nu} = \mathbf{a}'P_2$ is (by definition of P_2) a generalized-increment vector. A $K(\cdot)$ for which $P_2 K P_2$ is not positive-semidefinite, is not a valid generalized covariance (see following text).

It is worth noticing that an IRF0 is nothing more than an intrinsically stationary process (Section 2.3.1), and the generalized covariance of order 0 can be written $K(\mathbf{h}) = -\gamma(\mathbf{h}) + c$, where $\gamma(\cdot)$ is the semivariogram and c is an arbitrary constant. Because an IRFk is trivially an IRF$(k+1)$, it is clear that there is a nesting of processes: The second-order stationary processes are contained within the intrinsically stationary processes, which are contained within the class of IRF1s, and so forth. From this point of view, it would be logical to refer to a stationary process as an IRF(-1). Indeed, it is my opinion that some confusion would have been avoided had Matheron started numbering the intrinsic random functions with 1 instead of 0. Then, the notation would have been consistent with the order of an integrated process in time series. As it is, a time series such as $(1 - B)^l Z(t) = \epsilon(t)$, $t = 1, 2, \ldots$, where $BZ(t) \equiv Z(t-1)$ and $\epsilon(\cdot)$ is a white-noise process, is an integrated time series of order l, but is (a restriction to the integers of) an IRF$(l - 1)$.

Examples of IRFks

In $\mathbb{R}^1$, let $\{B(s): s \geq 0\}$ be a zero-mean Brownian motion, standardized by specifying $\mathrm{var}(B(1)) = 2$. As was demonstrated in Section 2.3.1, this process does not possess a covariogram, but it has a generalized covariance (of order 0) $K(h) = -|h|$. Matheron (1973, Section 2.3) demonstrates that integrated Brownian motion,

$$B^{(k)}(s) \equiv \int_0^s (s-u)^{k-1} B(u)\, du/(k-1)!, \qquad k = 1, 2, \ldots, \tag{5.4.5}$$

is an IRFk with generalized covariance (of order k) $K(h) = (-1)^{k+1}|h|^{2k+1}/(2k+1)!$. The result is more general: For $\{Y(s): s \geq 0\}$ any second-order stationary process

$$Y^{(k)}(s) \equiv \int_0^s (s-u)^k Y(u)\, du/k! \tag{5.4.6}$$

is an IRFk (Matheron, 1973, p. 453); $k = 0, 1, \ldots$.

In $\mathbb{R}^d$, (Lévy) fractional isotropic Brownian motion $B_H(\cdot)$, with covariance function

$$\text{cov}(B_H(\mathbf{u}), B_H(\mathbf{v})) = \{\|\mathbf{u}\|^{2H} + \|\mathbf{v}\|^{2H} - \|\mathbf{u} - \mathbf{v}\|^{2H}\}, \qquad 0 < H < 1, \tag{5.4.7}$$

is an IRF0 with generalized covariance

$$K(\mathbf{h}) = K^o(\|\mathbf{h}\|) = -\|\mathbf{h}\|^{2H}, \qquad \mathbf{h} \in \mathbb{R}^d, \tag{5.4.8}$$

and semivariogram $\gamma(\mathbf{h}) = \|\mathbf{h}\|^{2H}$ [i.e., the power model given by (2.3.12)]. Notice that $H = 1/2$ corresponds to the usual (Lévy) isotropic Brownian motion in $\mathbb{R}^d$. In $\mathbb{R}^1$, $B_H(\cdot)$, $H > 1/2$ is an example of $1/f$ noise and indicates long-range dependence (e.g., Mandelbrot and Van Ness, 1968; Barton and Poor, 1988). More details are given in Section 5.5.

Let $\delta(\cdot)$ be an IRFk whose mean exists and is everywhere zero. Then

$$Z(\mathbf{s}) \equiv \sum_{j=1}^{p+1} f_{j-1}(\mathbf{s}) B_{j-1} + \delta(\mathbf{s}), \qquad \mathbf{s} \in D \subset \mathbb{R}^d, \tag{5.4.9}$$

where $\mathbf{B} \equiv (B_0, \ldots, B_p)'$ is an arbitrary *random* vector, is also an IRFk. Based on (5.4.9), Cressie (1987) shows how to simulate an IRFk. Depending on the behavior of $\mathbf{B}$, it is possible that $Z(\cdot)$ possesses no moments. Such extreme behavior is usually avoided by assuming that $\mathbf{B} \equiv \boldsymbol{\beta}$, a deterministic vector of (unknown) coefficients. Regardless, when the generalized covariance is known, optimal linear spatial prediction is straightforward for the process $Z(\cdot)$ given by (5.4.9).

IRFk Kriging

It has already been demonstrated in Section 3.2 that optimal linear spatial prediction (kriging), for an intrinsically stationary process, can be obtained via the variogram. An analogous result holds for an IRFk and its generalized covariance $K(\mathbf{h})$, $\mathbf{h} \in \mathbb{R}^d$.

Suppose $\mathbf{Z} \equiv (Z(\mathbf{s}_1), \ldots, Z(\mathbf{s}_n))'$ are observations on the IRFk (5.4.9) at spatial locations $\{\mathbf{s}_1, \ldots, \mathbf{s}_n\}$, where it is assumed that $E(\mathbf{B}) = \boldsymbol{\beta}$. The datum $Z(\mathbf{s}_0)$ at $\mathbf{s}_0$ is to be predicted using an optimal linear predictor

$$\hat{Z}(\mathbf{s}_0) \equiv \boldsymbol{\lambda}'\mathbf{Z}, \tag{5.4.10}$$

such that $E(\hat{Z}(\mathbf{s}_0)) = E(Z(\mathbf{s}_0))$, for all $\boldsymbol{\beta} \in \mathbb{R}^{p+1}$, and $\sigma_e^2 \equiv E(\hat{Z}(\mathbf{s}_0) - Z(\mathbf{s}_0))^2$ is minimized with respect to $\boldsymbol{\lambda}$. The uniform unbiasedness condition is equivalent to $\boldsymbol{\lambda}'X = (f_0(\mathbf{s}_0), \ldots, f_p(\mathbf{s}_0))$, which in turn guarantees that the coefficients of the prediction error, $\lambda_1 Z(\mathbf{s}_1) + \cdots + \lambda_n Z(\mathbf{s}_n) - Z(\mathbf{s}_0)$, constitute a generalized-increment vector. Then,

$$\sigma_e^2 = \mathrm{var}\left(\sum_{i=0}^{n} \nu_i Z(\mathbf{s}_i) \right) = \sum_{i=0}^{n} \sum_{j=0}^{n} \nu_i \nu_j K(\mathbf{s}_i - \mathbf{s}_j),$$

where $\nu_0 = -1$ and $\nu_i = \lambda_i$, $i = 1, \ldots, n$. Because $K(\cdot)$ is assumed known, the optimal linear predictor is obtained from a straightforward minimization of

$$\sigma_e^2 = K(\mathbf{0}) - 2 \sum_{i=1}^{n} \lambda_i K(\mathbf{s}_0 - \mathbf{s}_i) + \sum_{i=1}^{n} \sum_{j=1}^{n} \lambda_i \lambda_j K(\mathbf{s}_i - \mathbf{s}_j), \quad (5.4.11)$$

subject to the constraint $\boldsymbol{\lambda}' X = \mathbf{x}'$, where $\mathbf{x} \equiv (f_0(\mathbf{s}_0), \ldots, f_p(\mathbf{s}_0))'$. This should be recognizable as the universal-kriging objective function (Section 3.4), with generalized covariance $K(\cdot)$ replacing covariogram $C(\cdot)$.

Hence, the *IRFk kriging predictor* is given by (5.4.10), where

$$\boldsymbol{\lambda}' = \left\{ \mathbf{k} + X(X'K^{-1}X)^{-1}(\mathbf{x} - X'K^{-1}\mathbf{k}) \right\}' K^{-1}; \quad (5.4.12)$$

in (5.4.12), $\mathbf{k} \equiv (K(\mathbf{s}_0 - \mathbf{s}_1), \ldots, K(\mathbf{s}_0 - \mathbf{s}_n))'$ and K is the $n \times n$ matrix with (i, j)th element $K(\mathbf{s}_i - \mathbf{s}_j)$. Moreover, the *kriging variance* (minimum mean-squared prediction error) is

$$\sigma_k^2(\mathbf{s}_0) = K(\mathbf{0}) - \mathbf{k}'K^{-1}\mathbf{k} + (\mathbf{x} - X'K^{-1}\mathbf{k})'(X'K^{-1}X)^{-1}(\mathbf{x} - X'K^{-1}\mathbf{k}). \quad (5.4.13)$$

There are several things to notice about (5.4.12) and (5.4.13). First, they are invariant to the choice of generalized covariance $K(\cdot)$ within the equivalence class of generalized covariances plus any even polynomial of degree $2k$. This explains why one can choose $K(\mathbf{0}) = 0$ without loss of generality. Second, if an IRFk $Z(\cdot)$, with generalized covariance $K(\mathbf{h})$, also possesses a covariance function $C(\mathbf{s}, \mathbf{u}) \equiv \mathrm{cov}(Z(\mathbf{s}), Z(\mathbf{u}))$, then the universal-kriging equations based on $C(\cdot, \cdot)$ yield an identical result to the IRFk kriging equations based on $K(\cdot)$.

IRFk Kriging and Splines

An equivalent way to write (5.4.10) is in terms of the dual-kriging equations

$$\hat{Z}(\mathbf{s}_0) = \mathbf{v}_1'\mathbf{k} + \mathbf{v}_2'\mathbf{x}, \quad (5.4.14)$$

where

$$\begin{aligned} K\mathbf{v}_1 + X\mathbf{v}_2 &= \mathbf{Z}, \\ X'\mathbf{v}_1 \qquad &= \mathbf{0}; \end{aligned} \tag{5.4.15}$$

cf. equations (3.4.74) and (3.4.75). But these are precisely the same equations used to obtain Laplacian smoothing splines [cf. equation (3.4.79)]. Indeed, all the commonly used spline basis functions are valid generalized covariances (Micchelli, 1986). The formal equivalence between IRFk kriging and splines is very useful, because each can borrow from each other's knowledge base. Nevertheless, in practice, the two approaches diverge markedly, most notably because the geostatistician attempts to *estimate* the generalized covariance function from the data $\mathbf{Z}$, whereas the "spliner" prespecifies the basis function. A detailed comparison of kriging and splines is given in Section 3.4.5.

Generalized Covariance of Order k

It has already been noted that a generalized covariance $K(\cdot)$, of order k, satisfies $K(\mathbf{h}) = K(-\mathbf{h})$, and is unique up to an even polynomial of degree $2k$. Moreover, a valid generalized covariance must be conditionally positive-definite in the sense that $P_2 K P_2$ is positive-semidefinite for *every* $\{\mathbf{s}_1, \ldots, \mathbf{s}_n\}$ and n, where $P_2 \equiv I - X(X'X)^{-1}X'$. Christakos (1984) and Micchelli (1986) give various results that enable valid generalized covariances to be identified [e.g., if $K(\mathbf{h}) = K^o(\|\mathbf{h}\|)$, K^o is continuous on $[0, \infty)$ and $(-1)^{k+1}\partial^{k+1}K^o(h)/\partial h^{k+1}$ is completely monotone on $(0, \infty)$, then $K(\cdot)$ is a valid generalized covariance function of order k in $\mathbb{R}^d$, for all $d \geq 1$].

An important property to consider is the large-$\|\mathbf{h}\|$ behavior of the generalized covariance $K(\mathbf{h})$ of an IRFk. Matheron (1973, Theorem 2.2) proves that, under mild regularity conditions,

$$K(\mathbf{h})/\|\mathbf{h}\|^{2k+2} \to 0, \quad \text{as } \|\mathbf{h}\| \to \infty. \tag{5.4.16}$$

Furthermore, from his Theorem 2.4, should the IRFk actually be an IRF$(k-1)$, the generalized covariance satisfies an inequality of the form

$$|K(\mathbf{h})| \leq a + b\|\mathbf{h}\|^{2k}, \tag{5.4.17}$$

where a and b are positive constants. Cressie (1987, 1988c) shows how inspection of an estimator of $K(\mathbf{h})/\|\mathbf{h}\|^{2k}$, $k = 0, 1, \ldots,$ for $\|\mathbf{h}\|$ large, can [through use of (5.4.16) and (5.4.17)] indicate the order of the intrinsic random function. Further details are presented in succeeding text.

Parametric models for generalized covariances have been given by Matheron (1973) and Delfiner (1976). They feature the isotropic models

$$K(\mathbf{h}) \equiv K^o(h) = \sum_{j=0}^{k} (-1)^{j+1}\theta_{j+1}h^{2j+1}, \qquad h \equiv \|\mathbf{h}\| \geq 0,$$

where the coefficients $\theta_1, \ldots, \theta_{k+1}$ must satisfy

$$\sum_{j=0}^{k} \frac{\theta_{j+1}\Gamma\{(2j+1+d)/2\}}{\pi^{2j+2+(d/2)}\Gamma\{1+(1/2)(2j+1)\}}\rho^{-d-2j+1} \geq 0,$$

for any $\rho \geq 0$. For example,

$$K^o(h) = \begin{cases} -\theta_1 h, & k=0, \\ -\theta_1 h + \theta_2 h^3, & k=1, \\ -\theta_1 h + \theta_2 h^3 - \theta_3 h^5, & k=2,\ h \geq 0, \end{cases}$$

where $\theta_1 \geq 0$, $\theta_3 \geq 0$, $\theta_2 \geq -10(\theta_1\theta_3/3)^{1/2}$ in $\mathbb{R}^2$, and $\theta_2 \geq -(10\theta_1\theta_3)^{1/2}$ in $\mathbb{R}^3$. Because spline basis functions are also valid, consider the following generalized covariance of order 1, in $\mathbb{R}^2$: $K(\mathbf{h}) = K^o(h) = -\theta_1 h + \theta_2 h^2 \log h + \theta_3 h^3$, $h \equiv \|\mathbf{h}\| \geq 0$, which is valid if $\theta_1 \geq 0$, $\theta_3 \geq 0$, and $\theta_2 \geq -(3/2)(\theta_1\theta_3)^{1/2}$.

Just as for the covariogram, a nugget-effect term, which represents microscale variation and measurement error, can be added to a generalized covariance $K_0(\cdot)$, where $K_0(0) = 0$:

$$K_1(\mathbf{h}) \equiv c_0\delta(\mathbf{h}) + K_0(\mathbf{h}), \qquad c_0 \geq 0,$$

where $\delta(\mathbf{h}) = 0$ unless $\mathbf{h} = \mathbf{0}$, in which case $\delta(\mathbf{0}) = 1$.

Estimation of Generalized Covariance Parameters

One of the strengths of the geostatistical method is the initial fitting of spatial dependence parameters, upon which subsequent spatial predictions are based. Estimation of $\boldsymbol{\theta}$ in the generalized covariance $K(\mathbf{h}; \boldsymbol{\theta})$, from data $\mathbf{Z}$, requires extra assumptions about the process $Z(\cdot)$.

Suppose $Z(\cdot)$ is a Gaussian IRFk. Then a REML estimator of $\boldsymbol{\theta}$ can be obtained by maximizing (2.6.5), where $\Sigma(\boldsymbol{\theta})$ is replaced with $K(\boldsymbol{\theta})$, the $n \times n$ matrix whose (i, j)th element is $K(\mathbf{s}_i - \mathbf{s}_j; \boldsymbol{\theta})$ (Kitanidis, 1983; Zimmerman,

1989a). Maximum likelihood estimation is not possible unless $Z(\cdot)$ possesses a covariance function $C(\mathbf{s}, \mathbf{u}; \boldsymbol{\theta})$ of known form.

The geostatistical package BLUEPACK-3D (e.g., Delfiner et al., 1978) uses a method of estimation advocated by Delfiner (1976), but it can be numerically unstable (Starks and Fang, 1982b) and statistically unstable (Zimmerman and Zimmerman, 1991). In the special case of the time dimension, and for equally spaced observations, estimation of parameters in autoregressive, integrated, moving-average processes have received particular attention in the time-series literature (e.g., Kawashima, 1980; Tsay and Tiao, 1984). Kunsch (1987) extends these time-series models to intrinsic autoregressions on a regular two-dimensional lattice and shows them to be IRF0s. He suggests an estimator of $\boldsymbol{\theta}$ based on the spectral representation of the semivariogram (i.e., of the generalized covariance of order 0) that is analogous to the Whittle–Guyon estimator given by (7.2.41).

Apart from Delfiner's (1976) method (which does not work well), these estimation methods usually require $Z(\cdot)$ to be a Gaussian process. Is there a nonparametric estimator $\hat{K}(\mathbf{h})$, to which a model $K(\mathbf{h}; \boldsymbol{\theta})$, $\mathbf{h} \in \mathbb{R}^d$, can be fitted?

A Nonparametric View of Generalized Covariances

There is a very important component of the geostatistical method that is missing from standard analyses of IRFks, namely *graphing* the spatial dependence. For example, for an intrinsically stationary process (IRF0), a plot of $2\hat{\gamma}(h(l)\mathbf{e})$ versus $h(l)$, $l = 1, \ldots, H$, where $2\hat{\gamma}(h(l)\mathbf{e})$ is a (nonparametric) estimator of the variogram at lag $h(l)$ in direction $\mathbf{e}$, is very useful. By choosing special generalized increments, Cressie and Laslett (1986) and Cressie (1987) show how generalized covariances can be viewed nonparametrically.

In all that follows, a particular direction $\mathbf{e} \in \mathbb{R}^d$ is chosen, and estimates are computed in that direction. By varying $\mathbf{e}$, anisotropies can be investigated. For ease of notation, write $K(h)$ for $K(h\mathbf{e})$, $h \in \mathbb{R}$, and $Z(s)$ for $Z(s\mathbf{e})$, $s \in \mathbb{R}$. Then a *k-primary increment* is defined to be

$$pi_k(s; h) \equiv \begin{cases} Z(s+h) - Z(s), & k = 0, \\ |h|Z(s) - (|h|+1)Z(s+1) + Z(s+|h|+1), & k = 1, \\ -|h|(|h|+1)Z(s) + 2|h|(|h|+2)Z(s+1) & \\ -(|h|+2)(|h|+1)Z(s+2) + 2Z(s+|h|+2), & k = 2. \end{cases} \tag{5.4.18}$$

An expression for general k would have terms in $Z(s), \cdots, Z(s+k)$, and $Z(s+|h|+k)$.

Clearly, $pi_k(s;h)$ is a generalized increment of order k, and hence:

$$2\gamma(h) \equiv \mathrm{var}(pi_0(s;h)) = -2\{K(h) - K(0)\}, \qquad h = 0,1,\ldots, \quad (5.4.19)$$

$$\alpha(h) \equiv \mathrm{var}(pi_1(s;h)) = 2h(h+1)\{\lambda(h+1) - \lambda(h) - K(1)\}, \qquad h = 1,2,\ldots, \quad (5.4.20)$$

$$\beta(h) \equiv \mathrm{var}(pi_2(s;h)) = \left[\frac{\{2\lambda(h+1) - \lambda(h) - \lambda(h+2)\}}{h+1} + \frac{K(2)}{2} - 2K(1)\right], \qquad h = 1,2,\ldots, \quad (5.4.21)$$

where $\lambda(h) \equiv K(h)/h$. Each of (5.4.19) through (5.4.21) is a difference equation that can be solved for $K(h)$. Respectively,

$$-K(h) + K(0) = \gamma(h), \qquad h = 0,1,\ldots, \quad (5.4.22)$$

$$\frac{K(h)}{h^2} - K(1) = \left\{\sum_{j=1}^{h-1} \frac{\alpha(j)}{2j(j+1)}\right\}/h \equiv \gamma_1(h), \qquad h = 2,3,\ldots, \quad (5.4.23)$$

$$-\frac{6K(h)}{h^4} - \left\{2K(1) - \frac{K(2)}{2}\right\} + \left\{\frac{16K(1) - K(2)}{2h^2}\right\} = 6\left\{\sum_{j=1}^{h-2} \frac{(h-1-j)\beta(j)}{4j(j+1)(j+2)}\right\}\bigg/h^3 \equiv \gamma_2(h), \qquad h = 3,4,\ldots, \quad (5.4.24)$$

where $\gamma(\cdot)$, $\gamma_1(\cdot)$, and $\gamma_2(\cdot)$ are called the semivariogram, the *linvariogram*, and the *quadvariogram*, respectively.

The most important thing to notice about (5.4.22) through (5.4.24) is that their right-hand sides can each be estimated *unbiasedly*, based on, respectively, $\hat{\gamma}(h) \equiv (1/2)\mathrm{ave}[\{pi_0(s;h)\}^2]$, $\hat{\alpha}(h) \equiv \mathrm{ave}[\{pi_1(s;h)\}^2]$, and $\hat{\beta}(h) \equiv \mathrm{ave}[\{pi_2(s;h)\}^2]$. (Here, $\mathrm{ave}[\cdot]$ denotes an arithmetic average over all data locations for which a primary increment is defined.) Call the resulting estimators $\hat{\gamma}(h)$, $\hat{\gamma}_1(h)$, and $\hat{\gamma}_2(h)$, which are obtained by substituting $\hat{\gamma}$, $\hat{\alpha}$, and $\hat{\beta}$ for γ, α, and β in (5.4.22), (5.4.23), and (5.4.24), respectively.

The next important thing to notice is that, for IRF0 kriging, $-\gamma(h)$ could be used instead of $K(h)$, for IRF1 kriging, $h^2\gamma_1(h)$ could be used instead of $K(h)$, and for IRF2 kriging, $-h^4\gamma_2(h)/6$ could be used instead of $K(h)$. (Recall that generalized covariances are unique up to even polynomials.) Now, it would be useful to characterize all those $\gamma_1(h)$ for which $K(h) \equiv h^2\gamma_1(h)$ is a valid generalized covariance of order 1; similarly for $\gamma_2(h)$, and so forth.

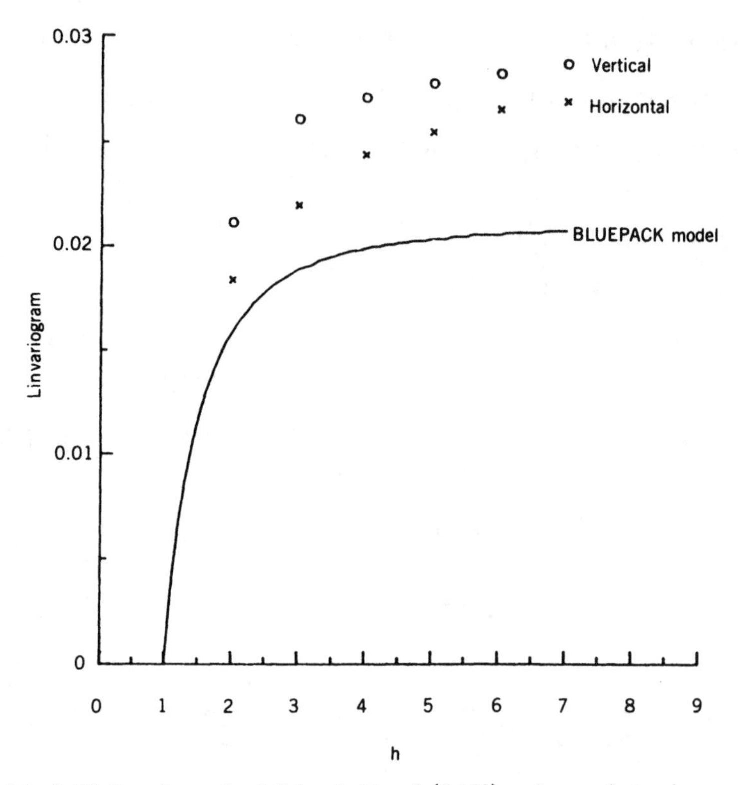

Figure 5.1 Solid line shows the left-hand side of (5.4.23), using an isotropic generalized covariance of order 1, fitted by BLUEPACK-3D to spatial data (soil pH in calcium chloride; see Laslett et al., 1987). Circles (crosses) show $2\hat{\gamma}_1(h)$ versus lag h, in the vertical (horizontal) direction. Units on the vertical axis are $(\text{pH})^2$, and 1 unit on the horizontal axis is 10 m. [*Source:* Cressie, 1987]

Just as a semivariogram model $\gamma(\cdot;\boldsymbol{\theta})$ can be fitted (e.g., by weighted least squares; see Section 2.6.2) to $\hat{\gamma}(\cdot)$, so a generalized covariance of order $k = 1$, $K(\cdot;\boldsymbol{\theta})$, can be fitted by matching the left-hand side of (5.4.23) with $\hat{\gamma}_1(\cdot)$. The same holds true for $k = 2$, where the left-hand side of (5.4.24) can be matched with $\hat{\gamma}_2(\cdot)$.

If, instead, some other technique is used to obtain $\hat{\boldsymbol{\theta}}$ (e.g., REML), the left-hand sides of (5.4.22) through (5.4.24) can be computed using $K(\cdot;\hat{\boldsymbol{\theta}})$ and compared to the nonparametric estimators $\hat{\gamma}(\cdot)$, $\hat{\gamma}_1(\cdot)$, and $\hat{\gamma}_2(\cdot)$, respectively; that is, a *diagnostic* method of assessing goodness-of-fit can be implemented. Cressie (1987) did precisely this on an IRF1 fit, by BLUEPACK-3D (Delfiner et al., 1978), to soil pH data presented in Laslett et al. (1987). In both the vertical direction and the horizontal direction, the estimated model appears to fit poorly; see Figure 5.1.

In practice, the order k of the IRFk has to be determined. Cressie (1988c) shows that this can be achieved by looking for the first graph where one of the functions $\hat{\gamma}(h)$, $\hat{\gamma}_1(h)$, $\hat{\gamma}_2(h)$, and so on "levels off" for large h. For example, if $\hat{\gamma}(\cdot)$ and $\hat{\gamma}_1(\cdot)$ are increasing, but $\hat{\gamma}_2(\cdot)$ levels off, then the properties (5.4.16) and (5.4.17) indicate that the data come from an IRF1 process.

Other methods of determining k, when $Z(\cdot)$ is a time series, are reviewed by Cressie (1988c). Those who analyze economic time series are very concerned about the appropriate degree of differencing needed to achieve a second-order stationary series. The approach given in this section does not distinguish between differencing to annihilate a polynomial trend and differencing to deal with a nonstationary error process, nor, from the point of view of optimal linear prediction, does it matter.

Some Final Remarks

Chauvet (1989) has expressed pessimism that kriging with IRFks can be properly implemented in problems with real data. The results of this section show that his concerns are unfounded.

In practice, there is still an indeterminacy as to what is large-scale variation and what is small-scale variation. Many favor detrending the data and kriging the residuals (e.g., median polish kriging in Section 3.5). But, if the trend is misspecified, considerable asymptotic damage may be done to inference procedures; Durlauf and Phillips (1988) demonstrate this for integrated time series that are spuriously detrended. Nevertheless, it is hard to see how this matters when the number of spatial data are fixed and there is no possibility of obtaining more. (Asymptotics for spatial data are discussed in Sections 5.8 and 7.3.1.)

Taking differences of the data is an elegant way of dealing with the indeterminacy problem because, even if the error process $\delta(\cdot)$ were second-order stationary, the estimated $K(h; \hat{\boldsymbol{\theta}})$ would still yield valid kriging equations. However, the applicability of IRFk kriging is limited, in that polynomial spatial trends are not always appropriate.

5.5* APPLICATIONS OF THE THEORY OF RANDOM PROCESSES

There is a large literature on the *theory* of random processes (or random fields), whose roots can often be found in (spatial) statistical problems. The approach taken in this book has been to stay close to statistical issues; hence, readers interested in more probabilistic questions (and answers) should consult such books as Yaglom (1962), Adler (1981), Vanmarcke (1983), Yadrenko (1983), and Rosenblatt (1985).

To illustrate that there is a healthy symbiosis between Probability and Statistics in problems of a spatial nature, I shall mention briefly some applications of the theory of random processes to Statistics for spatial data. The coverage is by no means complete.

Markov Random Fields

The theory of Markov random fields is presented briefly in Section 6.4. More probabilistically complete presentations can be found in Ruelle (1969), Preston (1976), and Kindermann and Snell (1980). Theoretical results such as the Hammersley–Clifford Theorem (Hammersley and Clifford, 1971; Section 6.4.1) have had an enormous impact on spatial statistical modeling (e.g., Besag, 1974; Geman and Geman, 1984; Cressie and Lele, 1992).

To obtain statistical properties of estimators of model parameters, one needs a law of large numbers and a central limit theorem for spatially dependent data. A number of such results are summarized in Section 7.3.

Optimal spatial prediction of $Z(\mathbf{s}_0)$ from data $\mathbf{Z}$ is discussed in Section 3.1. If mean-squared prediction error is the criterion to be minimized, then the optimal predictor is $E(Z(\mathbf{s}_0)|\mathbf{Z})$. For a Markov random field that is also Gaussian, this predictor is an immediate consequence of the fitted model. [More details are given in Section 5.9.1; in particular, the optimal predictor is given by (5.9.29).] The geostatistical approach to spatial prediction (Chapter 3) models the spatial dependence directly through the covariance function (or the variogram), from which the optimal predictor is derived. Both approaches have their advantages. It would be of great interest to construct a spatial model partly from local specifications of the conditional probabilities (Markov random-field approach) and partly from global specifications of joint (bivariate, trivariate, etc.) probabilities (geostatistical approach). Finding ways to make such specifications, so that a random process satisfying them exists, is an important open problem.

Random Processes Whose Sample Paths are Fractals

Fractals are defined in Section 9.1.1, where they are seen to have an "irregularity index" called the fractal dimension D. The fractal dimension of a fractal surface is strictly greater than its Euclidean dimension (for nonfractal surfaces, the two dimensions are the same).

Many (although not all) fractals possess a self-similarity property (i.e., a small part of the surface, magnified, resembles a larger part of the surface). If the surface of a random process in $\mathbb{R}^d$ is a self-similar fractal, then inference about the process at the spatial scale of the data can yield answers to questions about the process at a smaller spatial scale. Mandelbrot (1982), Pentland (1984), and Goodchild and Mark (1987) have examples where random processes, whose sample paths are fractals, are used to model natural scenes and geographic phenomena. Other applications are to spatial variation of soil properties (e.g., Armstrong, 1986), dispersivity in heterogeneous aquifers (e.g., Wheatcraft and Tyler, 1988), and porosity/permeability variation in potential petroleum reservoirs [the fractal approach is compared to other approaches in Dubrule (1989)].

Consider the surface $\{(t, Z(t)): 0 \leq t \leq T\}$ generated by a fractional Brownian motion $Z(t)$, where $\mathrm{var}(Z(t + \Delta t) - Z(t))$ is proportional to $|\Delta t|^{2H}$, $0 < H < 1$. Mandelbrot and Van Ness (1968) show how such a process can be constructed; the fractal dimension of its surface in $\mathbb{R}^2$ is $D = 2 - H$. Thus, Brownian motion, which corresponds to $H = 1/2$, has fractal dimension $D = 1.5$. For $H > 1/2$, the process exhibits long-range dependence, so that classical tests and confidence intervals based on data sampled from such processes are no longer valid (e.g., Beran, 1989). It is worth remarking that first differences, obtained from equispaced sampling of $\{Z(t)\colon t \geq 0\}$, are equivalent to an integrated Gaussian time series $X(\cdot)$ defined by $(1 - B)^{H-(1/2)}X(j) = \epsilon(j)$, where $\epsilon(\cdot)$ is zero-mean white noise (e.g., Granger and Joyeaux, 1980).

More generally, a fractional Brownian motion in $\mathbb{R}^d$ is a Gaussian process $Z(\cdot)$ characterized by a variogram of the form

$$\mathrm{var}(Z(\mathbf{s} + \mathbf{h}) - Z(\mathbf{s})) \propto \|\mathbf{h}\|^{2H}, \qquad 0 < H < 1. \tag{5.5.1}$$

Its corresponding covariance function $C(\mathbf{s}, \mathbf{u})$ is proportional to $\{\|\mathbf{s}\|^{2H} + \|\mathbf{u}\|^{2H} - \|\mathbf{s} - \mathbf{u}\|^{2H}\}$. The usual isotropic d-dimensional Brownian motion (Lévy, 1954, p. 71) is obtained when $H = 1/2$. Fractional Brownian motions in $\mathbb{R}^2$ are used by Mandelbrot (1982) to simulate topography that has a natural appearance; he uses values of H larger than 1/2, expressing the long-range dependence of phenomena often seen in nature. In general, a fractional-Brownian-motion surface $\{(\mathbf{s}, Z(\mathbf{s}))\colon \mathbf{s} \in \mathbb{R}^d\}$ in $\mathbb{R}^{d+1}$ has fractal dimension, $D = d + 1 - H$, which is greater than its Euclidean dimension d. Taylor (1986) gives the appropriate measure theory for fractals generated by random processes.

In $\mathbb{R}^1$, the class of fractional-Brownian-motion processes has been extended to non-Gaussian (stable) self-similar processes with possibly dependent increments; see Taqqu (1988) and Takashima (1989). In $\mathbb{R}^d$, $d \geq 2$, mathematical physicists (e.g., Major, 1981; Sinai, 1982) have considered construction of self-similar random processes from limits of averages over a regular lattice.

Trying to model a physical process using a random fractal with fractal dimension D has obvious drawbacks. Even after admitting extra mean and variance parameters, it would be overly optimistic to expect that a complex physical process could be summarized by just the fractal dimension D. However, the fractal dimension (or its estimate) may be useful for crude comparisons and categorizations, or for monitoring change of a spatial process over time. For more precise information over different spatial scales, one could use the variogram (a function, which can be thought of as a parameter of the random field)

$$2\gamma(\mathbf{h}) \equiv \mathrm{var}(Z(\mathbf{s} + \mathbf{h}) - Z(\mathbf{s})), \qquad \mathbf{s}, \mathbf{h} \in \mathbb{R}^d, \tag{5.5.2}$$

assuming it exists. From (5.5.1), a fractional Brownian motion (with fractal dimension $D = d + 1 - H$) has variogram that satisfies $\log(2\gamma(\mathbf{h})) = a + (2H)\log(\|\mathbf{h}\|)$, where a is a constant. However, estimated variograms usually demonstrate a much richer structure than a linear one on the log–log scale (e.g., Mark and Aronson, 1984).

Recalling the definition of fractal dimension D given by (9.1.1), it should not be surprising that the value of D corresponds to the behavior of the variogram at the origin (micro spatial scale): If $0 \leq \lambda \leq 2$ is the slope in a graph of $\log(2\gamma(\mathbf{h}))$ versus $\log(\|\mathbf{h}\|)$, for $\|\mathbf{h}\|$ small, then sample paths of the process have fractal dimension $D = \{d + 1 - (\lambda/2)\}$, in $\mathbb{R}^{d+1}$. To look for spatial heterogeneity of D, one could divide the region of interest (large spatial scale) up into smaller, more homogeneous regions (moderate spatial scale) and use the regional data to estimate possibly different fractal dimensions (the estimate is appropriate down to small spatial scales). Typically, the paucity of data will prevent a fine subdivision.

Although it has often been thought sufficient to use D to characterize a spatial process [e.g., sea-floor mapping; see Malinverno (1989)], this micro-spatial-scale parameter may not be relevant to questions concerning larger spatial scales. Under infill asymptotics (Section 5.8), spatial prediction or kriging will depend on the behavior of the variogram at the origin and, by implication, on the fractal dimension D (if it exists). However, spatial data are typically sparse, making the behavior of the process at larger spatial scales (as exhibited by the variogram) important for spatial prediction.

There are practical problems associated with estimating a micro-spatial-scale parameter D from data at only small and moderate spatial scales. First, extrapolation down to the micro scale requires an *assumption* that the process is self-similar. Second, estimation of D by fitting a straight line through a log–log plot (Section 9.1.1) gives notoriously unstable estimates; see Longley and Batty (1989). For further details on the estimation of fractal dimension, see Cutler and Dawson (1990), Ramsay and Yuan (1990), and Theiler (1990).

Level Sets and Suprema

A random process $\{Z(\mathbf{s}): \mathbf{s} \in \mathbb{R}^d\}$ can be studied via the geometry of its level sets:

$$X(\alpha) \equiv \{\mathbf{s} \in \mathbb{R}^d: Z(\mathbf{s}) \geq \alpha\}, \qquad \alpha \in \mathbb{R}.$$

For a nonfractal process with continuous sample paths, $X(\alpha)$ has a smooth boundary with finite length per unit area. The theory for $\mathbb{R}^1$ is presented by Cramér and Leadbetter (1967), and in higher dimensions by Adler (1981).

For a temporal random process, knowledge of the first-passage-time (to level α) distribution, $\alpha \in R$, is equivalent to knowledge of the supremum distribution. Suprema are often useful *test* statistics, which makes their (null) distribution of importance. For example, suppose $\{\mathbf{s}_1, \mathbf{s}_2, \dots\}$ are the random locations of a point process in $A \subset \mathbb{R}^d$, and define

$$Z(\mathbf{s}; \mathbf{h}) \equiv \sum_{i \geq 1} I\{\mathbf{s}_i \in (\mathbf{s}, \mathbf{s} + \mathbf{h}]\}, \tag{5.5.3}$$

where $(\mathbf{s}, \mathbf{s} + \mathbf{h}]$ is the d-dimensional cube $\{\mathbf{u}: s_j < u_j \le s_j + h_j; j = 1, \ldots, d\}$. The random process $Z(\cdot; \mathbf{h})$, in $\mathbb{R}^d$, is called the *scan process*; its properties are best known in $\mathbb{R}^1$ (e.g., Cressie, 1977). To test the null hypothesis of a d-dimensional *homogeneous* Poisson point process of constant intensity (Section 8.4.1), the *scan statistic*

$$N(\mathbf{h}) \equiv \sup\{Z(\mathbf{s}; \mathbf{h}) : \mathbf{s} \in A \cap A_{-\mathbf{h}}\} \tag{5.5.4}$$

was proposed by Naus (1965). In (5.5.4), $A_{-\mathbf{h}} \equiv \{\mathbf{a} - \mathbf{h}: \mathbf{a} \in A\}$.

In $\mathbb{R}^1$ and under the null hypothesis, Cressie (1980b) establishes weak convergence of $Z(\cdot; \mathbf{h})$ to a Gaussian process, as $\lambda \to \infty$. Then the distribution of the scan statistic $N(\mathbf{h})$ converges to the supremum distribution of this Gaussian process, a distribution that is derived by Cressie (1980b). In higher dimensions, some limited approximations to the null distribution of $N(\mathbf{h})$ are available (Naus, 1965; Adler, 1984); exact or asymptotic results, that are sufficiently extensive for hypothesis testing, have proved elusive.

Some Final Remarks

I have not been able to do justice to the wealth of applications of random-process theory to Statistics for spatial data. It is hoped that the three examples given will not only prove to be illustrative of the title of this section, but will also be of interest in their own right. Conversely, I believe that the contents of this book may suggest new problems in the theory of random processes, problems whose solutions will likely advance the research frontiers of Statistics for spatial data.

5.6 SPATIAL DESIGN

In this section, two themes will be developed, namely, *spatial sampling design* and *spatial experimental design*. In the latter, treatments are superimposed onto a random process $\{Z(\mathbf{s}): \mathbf{s} \in D\}$; the goal is to estimate the treatment effects from responses that are spatially correlated. Thus, design here refers to specifying which treatments are applied to which experimental units.

However, in spatial sampling design, potential data would be observations from just the random process $Z(\cdot)$; the goal is to determine sample size n and sample locations $\{\mathbf{s}_1, \ldots, \mathbf{s}_n\}$ from which data $\mathbf{Z} \equiv (Z(\mathbf{s}_1), \ldots, Z(\mathbf{s}_n))'$ can be used to predict $g(Z(\cdot))$. For example, one might be interested in predicting $g(Z(\cdot)) = \int_B Z(\mathbf{u})\, d\mathbf{u} / |B|$, where B is a given block (of volume $|B|$), or in predicting $g(Z(\cdot)) = \sup\{Z(\mathbf{u}): \mathbf{u} \in B\}$.

At first glance, the two themes appear disjoint, apart from having *design* as their motivation. However, Fienberg and Tanur (1987) point out general parallels between sampling and experimental structures that may prove fruitful in a spatial setting.

5.6.1 Spatial Sampling Design

When a resource or a pollutant warrants investigation, it will usually be impossible (financially and operationally) to obtain data at all spatial locations in the domain of interest D. Good spatial sampling design recognizes the spatial inhomogeneities in the process and accounts for them when choosing sample size n and sample locations $\{\mathbf{s}_1, \ldots, \mathbf{s}_n\} \subset D$. This section reviews various approaches to spatial sampling, without becoming problem- or discipline-specific. An important application is to environmental pollution monitoring. [Readers interested in this area might consult Gilbert (1987).] A novel application is to the design of computer experiments (e.g., Sacks et al., 1989), where the deterministic output of computer code (with various inputs) is modeled as a realization of a random process indexed by the input values.

Variogram-model fitting and kriging are very useful for spatial prediction. Recently, they have also been useful in spatial sampling design, in determining the location of the sites of a sampling network for the purpose of point or block prediction.

Throughout this section, the random-field model

$$Z(\mathbf{s}) = \mu(\mathbf{s}) + \delta(\mathbf{s}), \qquad \mathbf{s} \in D, \tag{5.6.1}$$

where $E(Z(\cdot)) = \mu(\cdot)$, is assumed. To simplify matters, it will be assumed further that only one potential observation is available at any spatial location. Thus, the amount of replication at any site will not be considered here as a design parameter.

In *spatial sampling design* the notion of a *best* sampling plan (or network) is central. It is important to specify the optimality criterion (or objective function) to be minimized, which in the statistical context is usually some measure of closeness of estimator (or predictor) to unknown parameter (or datum). Thus, at the very basis of optimal statistical design is first the choice of what is to be estimated or predicted, second the choice of the estimator or predictor, and finally the measure of closeness that is to be minimized. Once the first choice is made, the second is usually one of the standard statistical procedures, such as the best linear unbiased estimator, the best linear unbiased predictor, the generalized-least-squares estimator, and so on. Often, the mean-squared error, the variance, the generalized variance (in a multivariate setting), or the mean-squared prediction error plays a central role in the criterion to be minimized.

Criteria

Assume for the moment that, in the model (5.6.1), $Z(\cdot)$ satisfies the intrinsic-stationarity assumptions (2.3.2) and (2.3.4). Suppose that $Z(\mathbf{s}_0)$ is to be predicted and that the ordinary kriging equations (3.2.12) are used to obtain the best linear unbiased predictor. Then the mean-squared-prediction-error criterion, minimized, is simply the kriging (or prediction) variance given by (3.2.16), and is written here as

$$\sigma_k^2(\mathbf{s}_0) = 2\sum_{i=1}^{n} \lambda_i \gamma(\mathbf{s}_0 - \mathbf{s}_i) - \sum_{i=1}^{n}\sum_{j=1}^{n} \lambda_i \lambda_j \gamma(\mathbf{s}_i - \mathbf{s}_j). \tag{5.6.2}$$

It is clear from (5.6.2) that $\sigma_k^2(\mathbf{s}_0)$ depends on the *number* of points n used to solve the kriging equations, on the *sample locations* of these points, and on the *variogram*. Notice that $\sigma_k^2(\mathbf{s}_0)$ does not depend on the actual values at the observed points; this property makes kriging very useful for designing spatial sampling plans.

More generally, assume that $\mu(\cdot)$ in (5.6.1) is not constant, but is a linear combination of explanatory variables; that is, assume $Z(\cdot)$ satisfies (3.4.2). Let $S \equiv \{\mathbf{s}_1, \ldots, \mathbf{s}_n\}$ denote a sampling plan. Minimize over S, either of

$$V_n(S) \equiv \int_D \sigma_k^2(B(\mathbf{s}_0); S) \cdot w[\mathbf{s}_0; \mu(B(\mathbf{s}_0)); \sigma_k^2(B(\mathbf{s}_0); S)]\, d\mathbf{s}_0 \tag{5.6.3}$$

or

$$M_n(S) = \sup\{\sigma_k^2(B(\mathbf{s}_0); S) \cdot w[\mathbf{s}_0; \mu(B(\mathbf{s}_0)); \sigma_k^2(B(\mathbf{s}_0); S)] : \mathbf{s}_0 \in D\}, \tag{5.6.4}$$

which are measures of the weighted average and weighted maximum prediction variance, respectively. In (5.6.3) and (5.6.4), $\sigma_k^2(B(\mathbf{s}_0); S)$ denotes the kriging variance for the block $B(\mathbf{s}_0)$, located at $\mathbf{s}_0$, using the plan S. Further, $w[\cdot]$ is a suitably chosen nonnegative weight function whose integral (over D) is equal to 1. For example, suppose

$$w[\mathbf{s}_0; \mu(B(\mathbf{s}_0)); \sigma_k^2(B(\mathbf{s}_0); S)] \propto I(\mathbf{s}_0 \in A) \cdot I(\mu(B(\mathbf{s}_0)) > K) \cdot I(\sigma_k^2(B(\mathbf{s}_0); S) > L^2), \tag{5.6.5}$$

where $I(C)$ is an indicator function that is equal to 1 if the event C occurs and is equal to 0 otherwise. The weight function (5.6.5) focuses attention on the prediction variance in a subregion A of D, where the means are larger than K, and the prediction variances are beyond a threshold L^2. The subregion A and the constants K and L are exogenous to the sampling design problem. When $w[\cdot] \propto I(\mathbf{s}_0 \in A)$, (5.6.3) and (5.6.4) are, respectively, the average and maximum kriging variances in A.

The criteria (5.6.3) and (5.6.4) are quite general and, through the presence of $\mu(B(\mathbf{s}_0))$, allow "hot spots" to be featured. Sometimes, such as in the study of pollution levels, the goal is to determine the consequences of high levels of the spatial process $Z(\mathbf{s})$ (e.g., Cressie et al., 1990).

For simplicity, assume henceforth that a point value $Z(\mathbf{s}_0)$ is to be predicted. In practice, criteria (5.6.3) and (5.6.4) are very difficult to evaluate because D is a continuous region. Discrete versions of (5.6.3) and (5.6.4) are given by, respectively,

$$\frac{1}{N}\sum_{j=1}^{N}\sigma_k^2(B(\mathbf{s}_{0j});S)w\left[\mathbf{s}_{0j};\mu(B(\mathbf{s}_{0j}));\sigma_k^2(B(\mathbf{s}_{0j});S)\right] \tag{5.6.6}$$

and

$$\max\left\{\sigma_k^2(B(\mathbf{s}_{0j});S)w\left[\mathbf{s}_{0j};\mu(B(\mathbf{s}_{0j}));\sigma_k^2(B(\mathbf{s}_{0j});S)\right]: j=1,\ldots,N\right\}, \tag{5.6.7}$$

where $\{\mathbf{s}_{01}, \mathbf{s}_{02}, \ldots, \mathbf{s}_{0N}\}$ is a discrete region of interest (usually the nodes of a fine grid on D). The goal is to minimize (5.6.6) or (5.6.7).

Theoretical comparison of and optimal design of sampling plans are usually made assuming that the large-scale variation $\mu(\cdot)$ in (5.6.1) is an unknown linear combination of known variables and that the small-scale variation $\delta(\cdot)$ in (5.6.1) is a zero-mean process with known spatial-dependence structure. In practice, this knowledge is not usually available, in which case a pilot sampling study might be necessary; see Yfantis et al. (1987) for some recommendations on this.

Despite the optimal properties of kriging predictors, some simpler suboptimal predictors (that do not require the inversion of a covariance or variogram matrix) are sometimes used. The most common is the sample mean; see Ripley (1981, p. 19) and Matern (1960, p. 68). Other predictors include the sample median, triangulation, moving average, and so forth; Laslett et al. (1987) and Section 5.9 can be consulted for a comparison of a number of such spatial prediction techniques. Comparisons of different spatial sampling plans should be made conditionally on the predictor employed. Ideally, an optimal plan is designed for an optimal prediction procedure. However, it is also legitimate to look for the optimal plan designed for a nonoptimal procedure that might be used for pragmatic reasons.

Spatial Sampling Plans

There are various classical sampling plans that could be used to select n sites within a region D: Simple random sampling, stratified random sampling, cluster random sampling, regular (or systematic) random sampling, and regular nonrandom sampling are the most commonly used.

In *simple random sampling*, sites are chosen independently, each with a uniform distribution over D.

In *stratified random sampling*, the region D is divided into nonoverlapping strata such that within each stratum a simple random sample is chosen.

Considerations of stratum size and shape and sample size need to be made. Under the assumption of an isotropic stationary process with decreasing correlation function, and using the sample mean $\bar{Z}_n \equiv (1/n)\Sigma_{i=1}^{n} Z(\mathbf{s}_i)$ to predict $Z(D) \equiv \int_D Z(\mathbf{s})\, d\mathbf{s}/|D|$, Matern (1960, p. 72) studies the asymptotic performance of different strata shapes. Applying increasing-domain asymptotics (i.e., as n increases, the region D increases with fixed sampling intensity; see Section 7.3.1), he concludes that for strata shapes with equal area the one with the shortest boundary will yield the smallest "variance per sample point," defined as $\sigma_P^2 \equiv \lim_{n\to\infty} nE(\bar{Z}_n - Z(D))^2$. His conclusion is based on a mixture of formal asymptotics and numerical evaluations for particular covariograms. Among those that yield a nonoverlapping mosaic covering of $\mathbb{R}^2$, the regular hexagon is the stratum with the shortest boundary. (Of course, the hexagon requires more sample points than the square or the triangle; sampling-density factors are given in succeeding text.)

Sometimes, local knowledge of the underlying process could suggest the shape of the strata. For example, consider the problem of estimating the amount of pollutant concentrated around a source (e.g., chemical plant, reactor, etc.). It would be expected that the pollutant concentration decreases with increasing distance from the source. Then concentric circular or concentric rectangular strata of increasing size could be a reasonable choice. In a study where the data are simulated, McArthur (1987) compares several sampling plans and confirms this intuition.

Systematic random sampling consists of choosing an initial site at random and specifying the remaining $(n - 1)$ sites so that all n are located according to some regular pattern. If the initial site is not chosen at random, the resulting plan is called deterministic or regular. This distinction is not important when mean-squared prediction errors are computed using just the superpopulation model (5.6.1), but it is important if expectations are taken with respect to the sampling distribution. The most common regular plans (or grids) are the equilateral triangular grid, the rectangular (and square) grid, and the hexagonal grid; see Figure 5.2. It is easy to see that the regular hexagonal grid can be obtained from the equilateral triangular grid by deleting appropriate points. Conversely, the equilateral triangular grid is obtained from the hexagonal grid simply by adding the center points of the hexagons. Yfantis et al. (1987) discuss the geometrical relationships and properties of these regular patterns. Ignoring edge effects and considering regular grids whose smallest polygons all have sides of the same length, the equilateral triangular grid and the regular hexagonal grid require, respectively, 1.15 and 0.77 times the number of sampling sites required by a square grid to cover the same area $|D|$.

Cluster random sampling consists of the random selection of groups of sites where sites are spatially "close" within groups. This sampling plan has not received much attention in the spatial literature because it is often poorly adapted to this context. Olea (1984) considers cluster sampling in a comparison of several spatial sampling plans.

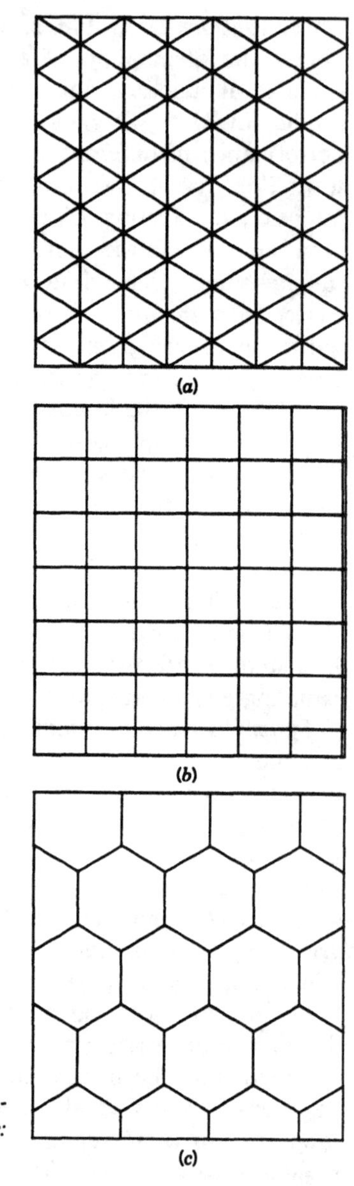

Figure 5.2 Regular sampling grids. (*a*) Equilateral triangular grid. (*b*) Square grid. (*c*) Hexagonal grid. [*Source:* Cressie et al., 1990]

A Comparison of Sampling Plans

Regular random sampling has an advantage over simple and stratified random sampling in that it is usually easier to implement. It provides well-defined directional classes for which reliable variograms can be estimated (Section 2.4). Among the three patterns, the equilateral triangular grid allows investigation of spatial dependence in three basic directions, but with more replication at spatial lags than the hexagonal grid.

Also note that regular sampling plans could yield large computational savings over the randomized plans. First, variogram estimators are evaluated at a small number of lags rather than the many an irregular sampling would engender. In practice, these many are pooled into few, a not altogether satisfactory solution because this gives another source of imprecision in the estimation of the variogram. Second, a "screen effect" could reduce the number of points needed to solve the kriging equations; this effect will be more pronounced in regular patterns of the form discussed here. For a detailed discussion of the screen effect, the reader is referred to Section 3.2.1 or Journel and Huijbregts (1978, p. 346).

Regular (random and nonrandom) sampling plans are usually more efficient than simple and stratified random sampling (Olea, 1984), where efficiency is measured here in terms of the average and maximum kriging variances. Among the regular sampling plans, the equilateral triangular plan is frequently the most efficient (McBratney et al., 1981; Olea, 1984; Yfantis et al., 1987). When the sample mean is used to predict $\int_D Z(\mathbf{u})\, d\mathbf{u}/ |D|$, the regular sampling plans are the most efficient (Matern, 1960, Chapter 5; Ripley, 1981, Section 3.2).

Discrete Design Regions

Suppose now that the continuous design region $D \subset \mathbb{R}^d$ is approximated by a finite set of N sites or that N potential sites at which sampling could take place are selected from D. Denote the finite set of sites by $\{\mathbf{s}_{01}, \ldots, \mathbf{s}_{0N}\}$.

Thus, the construction of an optimal spatial sampling design of size n reduces to finding the "best" n sites from among all $\binom{N}{n}$ possible sampling plans. This is the approach taken in Section 4.6.2, where one additional site ($n = 1$) is chosen from among $N = 11$ potential sites; discrete versions of (5.6.3) and (5.6.4) are given by (4.6.13) and (4.6.14), respectively, and are applied to acid-deposition data. A similar approach is given by Barnes (1989), who points out that when N is large and n is moderate, the naive optimization can be computationally prohibitive. For example, when $N = 100$ and $n = 10$, $\binom{N}{n}$ is more than 17×10^{12}.

Choosing n Sampling Sites from N Possible Sites

Naive optimization involves an exhaustive enumeration of all possible $\binom{N}{n}$ combinations. Instead, computational savings can be obtained by clever search algorithms, such as integer programming branch and bound (e.g., Garside, 1971), discretized partial gradient (e.g., Fedorov, 1972), the annealing algorithm (e.g., Sacks and Schiller, 1988), and the genetic algorithm (e.g., Goldberg, 1989).

The *annealing algorithm* was developed in statistical mechanics and motivated by an analogy to the behavior of physical systems in the presence of a heat bath (Metropolis et al., 1953; Section 7.4.3). Suppose n is fixed and the problem is to find the best sampling plan for predicting $Z(\mathbf{s}_0)$. Recall the definition $V_n(S)$ given by (5.6.3); for simplicity, dependence on n will be dropped and the criterion to be minimized will be denoted $V(S)$.

The annealing algorithm assumes a starting value $S^{(0)}$. At the jth stage, the current value is

$$S^{(j)} \equiv \{\mathbf{s}_1^{(j)}, \ldots, \mathbf{s}_n^{(j)}\}. \tag{5.6.8}$$

The algorithm is based on Markov-chain Monte Carlo simulation (Section 7.7.1) and proceeds as follows:

- Choose $S^{(j+1)}$ in some random or systematic manner (e.g., single-point replacement).
- Replace $S^{(j)}$ by $S^{(j+1)}$ with probability π_j, where

$$\pi_j = \begin{cases} 1, & \text{if } V(S^{(j+1)}) - V(S^{(j)}) \leq 0, \\ \exp\{-(V(S^{(j+1)}) - V(S^{(j)}))/\gamma_j\}, & \text{otherwise}. \end{cases}$$

Notice that the algorithm always demands a change provided there continues to be an improvement in the objective function V. The control parameter γ_j tends to zero as the algorithm proceeds and its influence is analogous to that of temperature in the statistical-mechanics context; for example, $\gamma_j = \gamma/\log j$. For suitable choices of γ_j, the algorithm avoids, at least in theory, entrapment in local minima. This attractive property is obtained by allowing a positive (ever-decreasing) probability of changing the sampling plan, even when no improvement in the objective function is achieved. In practice, it may be impossible to obtain an optimal sampling plan in a reasonable amount of time (Sacks and Schiller, 1988). Geman and Geman (1984) is a good source for the properties of this stochastic-relaxation approach.

Another algorithm with very interesting properties, but as yet untried in the selection of optimal sampling designs, is the *genetic algorithm*, based on the mechanics of natural genetics. It combines a survival-of-the-fittest among string structures with a structured, yet randomized, information exchange. This algorithm has been successfully applied to optimization in engineering problems. For more details, see the book by Goldberg (1989).

Asymptotic Optimal Spatial Sampling

The problem of selecting optimal sampling designs for estimating integrals, estimating regression parameters, or detecting signals has received some attention in the literature, especially in the time domain. This problem, which consists of finding an optimal sampling plan and optimal kriging weights, is very difficult to solve exactly.

Asymptotic approaches are taken to compare different sampling plans, but even here simplifications are needed. For reasons of tractability, various sampling plans are compared via mean-squared prediction errors of predictors with so-called simple weights rather than kriging weights (e.g., the equally weighted sample mean is one such predictor with simple weights).

Assume that $Z(\cdot)$ is an isotropic stationary process with decreasing correlation function; use the sample mean $\bar{Z}_n$ to predict $Z(D) \equiv \int_D Z(\mathbf{u})\,d\mathbf{u}/|D|$. By applying increasing-domain asymptotics, Matern (1960, p. 72) compares the different sampling plans and finds that for grid shapes of equal area the regular hexagonal plan has the smallest value of $\sigma_P^2 = \lim_{n\to\infty} nE\{\bar{Z}_n - Z(D)\}^2$.

Cambanis (1985) considers the rate of convergence of the mean-squared prediction errors to zero, as the number of sites in the sampling plan increases throughout the fixed region of interest $D \subset \mathbb{R}$ (i.e., infill asymptotics). The faster the rate of convergence, the better the sampling plan. Conditions under which these sequences of sampling plans are asymptotically equivalent to the corresponding ones, based on the optimal weights, are given by Schoenfelder and Cambanis (1982) and Cambanis (1985). Further results can be found in Bucklew and Cambanis (1988).

For optimal designs in $\mathbb{R}^d$, $d \geq 2$, few results are available. Ylvisaker (1975) approaches the problem of optimal design for regression estimation through optimal kriging weights and product sampling; that is, sampling each coordinate separately. He finds that asymptotically optimal designs within this class converge with rate $O(n^{-2/d})$.

An important result regarding estimation of multidimensional integrals of a random process was given by Tubilla (1975). Using $\bar{Z}_n$ to estimate the integral of $Z(\mathbf{s})$ over a d-dimensional unit hypercube, Tubilla compared the rate of convergence to zero of the mean-squared prediction errors of simple random sampling, stratified random sampling, and systematic random sampling. The rate of convergence for the last two designs decreased as d increased. For $d \geq 2$, the mean-squared prediction error of stratified random sampling decreased faster than those of the other two sampling plans.

Surprisingly, the rate of convergence for systematic random sampling was slower than or the same as that of simple random sampling. This result appears to contradict Matern's (1960) findings; however, a natural explanation is given by edge effects. By using increasing-domain asymptotics, Matern ignores edge effects. On the other hand, when infill asymptotics are used within the fixed region D, edge effects are present and their influence increases as the dimension d increases. For example, consider 64 sampling points regularly spaced on an interval, square, cubic, etcetera, grid: In $\mathbb{R}^1$, 3.125% of the points are edges; in $\mathbb{R}^2$, 43.75% of the points are edges; in $\mathbb{R}^3$, 87.5% of the points are edges.

Ylvisaker (1987) presents the problem of prediction and design for a class of processes called Gaussian–Markov associated processes (G-MAP), from a nonasymptotic viewpoint. Let $Z(\mathbf{s})$ be a Gaussian process whose index $\mathbf{s}$ ranges over the set D and that has mean zero and covariance function C. Then $Z(\cdot)$ is a G-MAP if the conditional expectation of $Z(\mathbf{s})$ given $Z(\mathbf{s}')$, where $\mathbf{s}' \in D_0 \subset D$, is a linear combination of $\{Z(\mathbf{s}'): \mathbf{s}' \in D_0\}$, such that the coefficients of the linear combination are *nonnegative* with sum less than or equal to 1. It is shown that an important subset of the G-MAP processes can be generated from a continuous Markov process X on D. Conversely, under regularity conditions, any G-MAP process Z on a countable D can be generated from a Markov process X on D. This relationship allows design criteria for G-MAP processes to be expressed in terms of expected functionals of the X process. Thus, optimal designs can be determined empirically by averaging over simulations of the associated Markov process X. Development of algorithms to compare large numbers of competing plans is the next stage of this research.

Finite Population Sampling and Simple Estimators

Assume a finite population of size N and consider the problem of finding an optimal sampling plan of size n to estimate the population mean $\bar{Z}_N$ with the sample mean $\bar{Z}_n$. Furthermore, assume that the values of the population N are a (partial) realization of a stochastic process (i.e., assume a superpopulation model). Cochran (1946) shows that when the population is a realization of a stationary time series with nonnegative and nonincreasing correlogram $\rho(\cdot)$,

$$E_Z\left[\mathrm{var}_p\left(\bar{Z}_{\mathrm{sys},n}\right)\right] \le E_Z\left[\mathrm{var}_p\left(\bar{Z}_{\mathrm{strs},n}\right)\right] \le E_Z\left[\mathrm{var}_p\left(\bar{Z}_{\mathrm{srs},n}\right)\right], \quad (5.6.9)$$

where E_Z denotes the expectation over the stochastic process, var_p denotes the variance over repeated replications of the sampling procedure, and the subindices sys, strs, and srs denote systematic sampling, stratified random sampling, and simple random sampling, respectively. Extensions of the inequality (5.6.9) under less restrictive assumptions for $\rho(\cdot)$ are given by Gautschi (1957) and Iachan (1983).

Consider now a finite population defined by a two-dimensional array of $RC = N$ units arranged in R rows and C columns. Unfortunately, the result (5.6.9) cannot be extended to two-dimensional arrays; Bellhouse (1977) shows the nonexistence of optimal sampling plans for a general class of correlation functions. Nevertheless, in various subclasses of isotropic correlograms, optimal sampling plans do exist, although they are not always equilateral triangular (Dalenius et al., 1960).

Restrict the two-dimensional sampling plans of size $n = rc$ to the following three types (Quenouille, 1949b): plans in which the units are aligned in both row and column directions, plans in which the units are aligned in one direction and unaligned in the other direction, and plans in which the units are unaligned in either direction. That is, in sampling plans where the units are aligned in both directions, the number of sampled elements in any row (length C) of the population is either 0 or c, and the number of sampled elements in any column (length R) of the population is either 0 or r; where the units are aligned in one direction, such as rows, the number of sampled elements in any row of the population is 0 or c and the number of sampled elements in any column is at most r; and where the units are unaligned in both directions, the number of sampled units is at most c in any row and at most r in any column.

Under regularity conditions on the two-dimensional correlation function, Bellhouse (1977) shows that systematic-type designs have smallest average variance (as defined in (5.6.9)) within each of the three types of sampling plans. See also Iachan (1985), where a two-dimensional version of (5.6.9) is shown to hold in the limit. Iachan's asymptotics are similar to Matern's (1960) increasing-domain asymptotics; n and N both increase to infinity at the same rate.

Gy's Theory of Particulate Material Sampling

Thus far, the description given of spatial sampling has been particularly appropriate for the sampling of domains, where any part of the domain is equally accessible. A brief discussion is now given of the problem of sampling particulate solids in one-dimensional "streams" and two- or three-dimensional "piles."

Gy (1982) provides a comprehensive theory based on an analysis of the errors involved in sampling particulate material. Using Gy's approach, it is shown why grab sampling, judgement sampling, and other *ad hoc* sampling methods should be avoided. Gy argues that when the material is not moving in a stream (or cannot be easily split), accessibility to any part of the "pile" is compromised, and this introduces uncontrollable sampling biases.

The variogram is used to characterize the spatial dependence between samples taken at fixed increments, and is used to compute the appropriate measures of sample precision. It is shown that random sampling can be inferior to stratified random sampling and systematic sampling. Systematic sampling provides the smaller sampling error, but biases arise when the sampling period coincides with regular fluctuations in the material. On balance, stratified random sampling is the safest sampling procedure because the variance of its sampling error is not much larger than that of systematic sampling.

Objective Functions with an Economic Component

The sampling optimality criteria presented so far, have all been based on statistical considerations. Frequently, cost or economic considerations provide a very important limitation, which should be introduced into the objective function.

Bras and Rodriguez-Iturbe (1976) propose the objective function

$$\delta(n, S) + \beta\kappa(n, S), \tag{5.6.10}$$

where $\delta(n, S)$ is a measure of statistical accuracy, $\kappa(n, S)$ is the cost of sampling, β is a measure of accuracy obtained from a unit increase in cost, n is the sample size, and $S \equiv \{\mathbf{s}_1, \ldots, \mathbf{s}_n\}$ is the set of sample locations. Bras and Rodriguez-Iturbe suggest how (5.6.10) can be optimized numerically over sampling locations S and sample size n.

Bogardi et al. (1985) consider the problem of optimal spatial sampling design as one of multicriterion decision making. A composite objective function, measuring statistical and economic tradeoffs, is proposed. The optimal *rectangular* network is achieved through compromise programming (Zeleny, 1982).

5.6.2* Spatial Experimental Design

Classical experimental design (e.g., Kempthorne, 1952) is based on the three concepts of randomization, blocking, and replication. Randomization neutralizes the effect of spatial correlation and yields a valid (see following text) analysis of variance (Yates, 1938). Blocking helps to reduce the residual variation and replication allows precise parameter estimation. Given the blocking and the replication, the efficiencies of the treatment-contrast estimators depend solely on the error variation. However, in this classical approach, the spatial positions of the treatments in the design are ignored; in a spatial setting, the efficiency will typically depend not only on the error variation but also on these positions. It is natural then to look for spatial experimental designs and spatial methods of analysis.

Optimal Experimental Design in the Presence of Spatial Dependence

Assuming the covariances for the observations in an experiment are known, the optimal design of the experiment depends on three factors: the estimand [i.e., the parameter(s) to be estimated], the estimator, and the optimality criterion (i.e., a measure of closeness of the estimator to the estimand). In experimental design, one of the principal objectives is the estimation of treatment contrasts $\boldsymbol{\tau}$, and it is this objective that will be considered in the following sections.

Traditionally, two estimators have been discussed in the experimental-design literature when the observations are correlated, namely, the ordinary-least-squares (o.l.s.) estimator and the generalized-least-squares (g.l.s.) estimator. The o.l.s. estimator is usually chosen because of practical considerations; it is easier to compute than the g.l.s. estimator since it does not involve the covariance matrix Σ. (However, because Σ is needed to obtain the variance of the estimators, the merits of using o.l.s., a less-efficient estimator, are not as compelling.)

Consider the following model for responses to an experiment:

$$\mathbf{Z} = X_1\boldsymbol{\alpha} + X_2\boldsymbol{\zeta} + \boldsymbol{\delta}, \tag{5.6.11}$$

where $\boldsymbol{\alpha}$ contains the t treatment effects and $\boldsymbol{\zeta}$ contains the $(p - t)$ local-control parameters (e.g., block effects, row and column effects, etc.), including an overall mean effect. Let $\boldsymbol{\tau} \equiv \Delta\boldsymbol{\alpha}$ be the vector of the t estimable centered treatment contrasts; that is, $\Delta = I - J/t$, where $J = \mathbf{1}\mathbf{1}'$ is a $t \times t$ matrix all of whose elements are 1. Let $\hat{\boldsymbol{\tau}}_{\text{ols}}$ be the o.l.s. estimator of $\boldsymbol{\tau}$, namely,

$$\hat{\boldsymbol{\tau}}_{\text{ols}} = \Delta C(X, I)^{-} X_1'(I - X_2(X_2'X_2)^{-} X_2')\mathbf{Z}, \tag{5.6.12}$$

where

$$C(X, I) = X_1'(I - X_2(X_2'X_2)^{-} X_2')X_1 \tag{5.6.13}$$

and $X \equiv (X_1, X_2)$. Notice that a generalized inverse A^{-}, of A, is used, because the matrices may not be of full rank (e.g., Rao, 1973, p. 24).

The g.l.s. estimator of $\boldsymbol{\tau}$ is given by

$$\hat{\boldsymbol{\tau}}_{\text{gls}} = \Delta C(X, \Sigma)^{-} X_1'\left(\Sigma^{-1} - \Sigma^{-1}X_2(X_2'\Sigma^{-1}X_2)^{-} X_2'\Sigma^{-1}\right)\mathbf{Z}, \tag{5.6.14}$$

where

$$C(X, \Sigma) = X_1'\left(\Sigma^{-1} - \Sigma^{-1}X_2(X_2'\Sigma^{-1}X_2)^{-} X_2'\Sigma^{-1}\right)X_1. \tag{5.6.15}$$

Notice that when $\Sigma = I$, (5.6.14) reduces to (5.6.12).

Let $D(\hat{\boldsymbol{\tau}}, X, \Sigma)$ be the variance of the estimator $\hat{\boldsymbol{\tau}}$ of $\boldsymbol{\tau}$. The matrix $D(\hat{\boldsymbol{\tau}}, X, \Sigma)$ has zero row and column sums because Δ has zero row and column sums. Define $C(\hat{\boldsymbol{\tau}}, X, \Sigma) \equiv D(\hat{\boldsymbol{\tau}}, X, \Sigma)^{+}$ to be the *information matrix* for $\hat{\boldsymbol{\tau}}$, where A^{+} denotes the Moore–Penrose inverse of the real matrix A (e.g., Rao, 1973, p. 26). This too has zero row and column sums because $A^{+} = A^{+}AA^{+} = A'(A^{+})'A^{+} = A^{+}(A^{+})'A'$. Moreover, when $\hat{\boldsymbol{\tau}} = \hat{\boldsymbol{\tau}}_{\text{gls}}$, $C(\hat{\boldsymbol{\tau}}_{\text{gls}}, X, \Sigma) = C(X, \Sigma)$. Thus, in the correlated-errors model, the information matrix $C(X, \Sigma)$ is the natural extension of the C matrix used, for example, by Kiefer (1980), in the uncorrelated-errors model.

Given the estimator, the natural optimality criterion by which a design could be chosen is the "smallest" covariance matrix of the estimator of $\boldsymbol{\tau}$ [i.e., $\text{var}(\hat{\boldsymbol{\tau}})$ is "smaller than" $\text{var}(\tilde{\boldsymbol{\tau}})$ if the matrix $\{\text{var}(\tilde{\boldsymbol{\tau}}) - \text{var}(\hat{\boldsymbol{\tau}})\}$ is nonnull and nonnegative-definite]. However, the minimization of some *function* of the covariance matrix of the estimator is often all that can be achieved.

A large class of optimality criteria can be obtained by minimizing, for a fixed λ, the following functional over a class of designs $\mathscr{X}$ (Kiefer, 1975):

$$\psi_\lambda(C(\hat{\boldsymbol{\tau}}, X, \Sigma)) \equiv \left\{\frac{1}{t-1}\sum_{k=1}^{t-1} \nu_k^{-\lambda}\right\}^{1/\lambda}, \qquad 0 \leq \lambda \leq \infty, \tag{5.6.16}$$

where for $\lambda = 0$ and $\lambda = \infty$ the functional is defined in the limit and $\{\nu_k: k = 1, 2, \ldots, t-1\}$ are the $t-1$ positive eigenvalues of $C(\hat{\tau}, X, \Sigma)$. Thus, ψ_λ-optimality includes the popular A-optimality ($\lambda = 1$), D-optimality ($\lambda = 0$), and E-optimality ($\lambda = \infty$) criteria.

Universal Optimality and Weak Universal Optimality

A stronger optimality criterion, that includes (5.6.16) for all values of λ, is described by Kiefer (1975). Let $\mathscr{B}_{t,0}$ be the class of nonnegative–definite $t \times t$ matrices with zero column sum and zero row sum. Let Ψ be the class of all functions ψ: $\mathscr{B}_{t,0} \to (-\infty, \infty]$, that satisfy the following conditions:

1. ψ is convex; that is, if $B_1, B_2 \in \mathscr{B}_{t,0}$ and $0 \le \omega \le 1$, then $\psi(\omega B_1 + (1-\omega)B_2) \le \omega\psi(B_1) + (1-\omega)\psi(B_2)$.
2. ψ is nonincreasing; that is, if $B_1 - B_2$ is nonnegative-definite, then $\psi(B_1) \le \psi(B_2)$.
3. ψ is invariant under permutation of coordinates; that is, if $B \in \mathscr{B}_{t,0}$ and ΠB is obtained from B by permuting rows and columns according to the permutation matrix Π, then $\psi(\Pi B) = \psi(B)$.

Definition. A design $X^* \in \mathscr{X}$ is said to be *universally optimal* over the class $\mathscr{X}$, for the estimator $\hat{\tau}$, if

$$\psi(C(\hat{\tau}, X^*, \Sigma)) = \min\{\psi(C(\hat{\tau}, X, \Sigma)) : X \in \mathscr{X}\}, \quad \text{for all } \psi \in \Psi. \quad \blacksquare$$

The importance of the conditions 1, 2, and 3 is now clear. The convexity condition 1 ensures that a local minimum is in fact optimal, condition 2 ensures that more informative designs are preferred, and condition 3 simply indicates invariance with respect to treatment ordering. It is shown by Kiefer (1974) that the functional ψ_λ defined in (5.6.16) belongs to the class Ψ for all values of λ; hence, a universally optimal design, within the class of designs $\mathscr{X}$, is ψ_λ-optimal for all values of λ.

A weaker optimality criterion was introduced by Kiefer and Wynn (1981). Although it was originally presented for the o.l.s. estimator of the treatment contrast τ, it can be extended to any estimator $\hat{\tau}$. Let Φ be the class of all functions ϕ: $\mathscr{B}_{t,0} \to (-\infty, \infty]$, that are convex, *nondecreasing*, and invariant under permutation of coordinates.

Definition. A design X^* is said to be *weakly universally optimal* over the class $\mathscr{X}$, for the estimator $\hat{\tau}$, if

$$\phi(D(\hat{\tau}, X^*, \Sigma)) = \min\{\phi(D(\hat{\tau}, X, \Sigma)) : X \in \mathscr{X}\}, \quad \text{for all } \phi \in \Phi,$$

where recall that $D(\cdot)$ is the variance of $\hat{\tau}$. $\blacksquare$

It is pointed out by Kiefer (1975) and Kiefer and Wynn (1981) that the class of functionals Ψ on the set of information matrices $C(\hat{\tau}, X, \Sigma)$ is more general than the class of functionals Φ on the set of variance matrices

$D(\hat{\boldsymbol{\tau}}, X, \Sigma)$. To illustrate the strict inclusion, consider the functional

$$\phi_\lambda(D(\hat{\boldsymbol{\tau}}, X, \Sigma)) \equiv \left\{ \frac{1}{t-1} \sum_{k=1}^{t-1} \xi_k^\lambda \right\}^{1/\lambda}, \quad 0 \le \lambda \le \infty, \quad (5.6.17)$$

where $\{\xi_k : k = 1, \dots, t-1\}$ are the $t-1$ positive eigenvalues of $D(\hat{\boldsymbol{\tau}}, X, \Sigma)$. It is shown by Kiefer (1974) that this functional is convex only for $\lambda \ge 1$.

Kiefer's Theorem (Kiefer, 1975) A design $X^* \in \mathscr{X}$ is universally optimal if it satisfies the following two conditions: (i) $C(\hat{\boldsymbol{\tau}}, X^*, \Sigma)$ is completely symmetric; that is, is of the form $aI + bJ$, where $J = \mathbf{11}'$, and (ii) $\mathrm{tr}(C(\hat{\boldsymbol{\tau}}, X^*, \Sigma))$ is the maximum trace over all $X \in \mathscr{X}$. ∎

Conditions (i) and (ii) are sufficient, but not necessary for universal optimality. It is easy to see that (i) and (ii) can be expressed alternatively in terms of the eigenvalues of the information matrix $C(\hat{\boldsymbol{\tau}}, X, \Sigma)$. That is, a design $X^* \in \mathscr{X}$, whose information matrix $C(\hat{\boldsymbol{\tau}}, X^*, \Sigma)$ has positive eigenvalues $\{\nu_k^* : k = 1, \dots, t-1\}$, is universally optimal if $\nu_1^* = \nu_2^* = \cdots = \nu_{t-1}^*$ and $\sum_{k=1}^{t-1} \nu_k^*$ is maximum over all $X \in \mathscr{X}$. This equivalent representation follows directly from the result that B is a completely symmetric matrix if and only if it has only one positive eigenvalue with multiplicity $t-1$ (J. A. Thompson, 1956). The alternative representation is very helpful in studying the efficiency of designs relative to hypothetical universally optimal designs (see following text).

Similarly, a design $X^* \in \mathscr{X}$ is weakly universally optimal if $D(\hat{\boldsymbol{\tau}}, X^*, \Sigma)$ is completely symmetric and $\mathrm{tr}(D(\hat{\boldsymbol{\tau}}, X^*, \Sigma))$ is the minimum trace over all $X \in \mathscr{X}$. The equivalent representation in terms of the positive eigenvalues $\{\xi_k^* : k = 1, \dots, t-1\}$ of the covariance matrix $D(\hat{\boldsymbol{\tau}}, X^*, \Sigma)$ is $\xi_1^* = \xi_2^* = \cdots = \xi_{t-1}^*$ and $\sum_{k=1}^{t-1} \xi_k^*$ is minimum over all $X \in \mathscr{X}$.

The universal optimality criterion is usually used together with the g.l.s. estimator $\hat{\boldsymbol{\tau}}_{\mathrm{gls}}$ (e.g., Gill and Shukla, 1985a, 1985b; Kunert, 1985, 1987), and the weak universal optimality criterion is usually used together with the o.l.s. estimator $\hat{\boldsymbol{\tau}}_{\mathrm{ols}}$ (e.g., Kiefer and Wynn, 1981; Russell and Eccleston, 1987; Morgan and Chakravarti, 1988). Which criterion is used is usually a matter of mathematical tractability. When the variance matrix Σ has a simple form, as for geostatistical models with small ranges, it is usually easier to find weakly universally optimal designs for the o.l.s. estimator of $\boldsymbol{\tau}$; compare the variance matrix of $\hat{\boldsymbol{\tau}}_{\mathrm{ols}}$ [given by (5.6.12)] to that of $\hat{\boldsymbol{\tau}}_{\mathrm{gls}}$ [given by (5.6.14)]. Conversely, when Σ^{-1} has a simple form (e.g., autoregressive error processes and Gaussian Markov random fields), the information matrix (5.6.15) of $\hat{\boldsymbol{\tau}}_{\mathrm{gls}}$ is usually easier to find; in this case, conditions for universally optimal designs are usually given.

An excellent discussion of spatial experimental designs can be found in Martin (1986a). He gives a thorough treatment of two-dimensional experimental arrays, applying the notions of optimality presented in the preceding text. Similar discussion of one-dimensional experimental arrays can be found in Kunert (1987) and Morgan and Chakravarti (1988). Rather than cover the same ground here, the rest of this section will present selected topics, largely from Grondona and Cressie (1991, 1992), to illustrate the type of results one might expect.

The Covariance Function and the Variogram in Experimental Designs

Clearly, a spatial approach to the analysis and design of experiments depends on how well the spatial covariance structure can be modeled. An assumption often made is the existence of a covariogram, $C(\mathbf{s}_1 - \mathbf{s}_2) \equiv \mathrm{cov}(Z(\mathbf{s}_1), Z(\mathbf{s}_2))$, $\mathbf{s}_1, \mathbf{s}_2 \in D \subset \mathbb{R}^d$ (Section 2.3.2). An alternative way of characterizing the statistical dependence between spatial observations is through the variogram, $2\gamma(\mathbf{s}_1 - \mathbf{s}_2) \equiv \mathrm{var}(Z(\mathbf{s}_1) - Z(\mathbf{s}_2))$, $\mathbf{s}_1, \mathbf{s}_2 \in D \subset \mathbb{R}^d$ (Section 2.3.1). In spatial experimental design, the variogram arises naturally from the expectation of the residual mean square of randomized experimental designs (Grondona and Cressie, 1991) and from expressions for best linear unbiased estimators of treatment contrasts:

Proposition Assume the model (5.6.11), where $X \equiv (X_1, X_2)$ satisfies $X\mathbf{1} \propto \mathbf{1}$. Let $\tau \equiv \boldsymbol{\lambda}'\boldsymbol{\beta}$ be an estimable function of $\boldsymbol{\beta} \equiv (\boldsymbol{\alpha}', \boldsymbol{\zeta}')'$, such that $\boldsymbol{\lambda}'\mathbf{1} = 0$. Then the best linear unbiased estimator of τ is given by

$$\hat{\tau} = \boldsymbol{\lambda}'(X'\Gamma^{-1}X)^{-}X'\Gamma^{-1}\mathbf{Z} \tag{5.6.18}$$

and its variance is

$$\mathrm{var}(\hat{\tau}) = -\boldsymbol{\lambda}'(X'\Gamma^{-1}X)^{-}\boldsymbol{\lambda}, \tag{5.6.19}$$

where Γ is the semivariogram matrix; that is, if the spatial locations of the data are emphasized by writing $\mathbf{Z} = (Z(\mathbf{s}_1), Z(\mathbf{s}_2), \ldots, Z(\mathbf{s}_n))'$, then the (i, j)th element of Γ is $\gamma(\mathbf{s}_i - \mathbf{s}_j)$; and A^- is a generalized inverse of A.

Proof. Let $\mathbf{l}'\mathbf{Z}$ be an unbiased estimator of τ; the unbiasedness condition implies that $\mathbf{l}'X = \boldsymbol{\lambda}'$. Also, $\boldsymbol{\lambda}'\mathbf{1} = 0$ and $X\mathbf{1} \propto \mathbf{1}$ imply $\mathbf{l}'\mathbf{1} = 0$, which in turn implies that

$$(\mathbf{l}'\mathbf{Z} - \mathbf{l}'X\boldsymbol{\beta})^2 = (\mathbf{l}'\boldsymbol{\delta})^2 = \left(\sum_{i=1}^{n} l_i\delta_i\right)^2 = -(1/2)\sum_{i=i}^{n}\sum_{j=1}^{n} l_i l_j(\delta_i - \delta_j)^2, \tag{5.6.20}$$

where $\boldsymbol{\delta} \equiv \mathbf{Z} - X\boldsymbol{\beta}$. Thus,

$$\operatorname{var}(\mathbf{l}'\mathbf{Z}) = E(\mathbf{l}'\mathbf{Z} - \mathbf{l}'X\boldsymbol{\beta})^2 = -(1/2)\sum_{i=1}^{n}\sum_{j=1}^{n} l_i l_j E(\delta_i - \delta_j)^2 = -\mathbf{l}'\Gamma\mathbf{l}. \tag{5.6.21}$$

Therefore, the problem of finding the best linear unbiased estimator can be expressed as one of minimizing $-\mathbf{l}'\Gamma\mathbf{l}$ with respect to $\mathbf{l}$, subject to the restriction $\mathbf{l}'X = \boldsymbol{\lambda}'$. The result (5.6.18) follows immediately from Rao (1973, p. 60) or Section 3.4. Now, $\operatorname{var}(\hat{\tau}) = \operatorname{var}(\mathbf{l}'\mathbf{Z}) = -\mathbf{l}'\Gamma\mathbf{l} = \boldsymbol{\lambda}'(X'\Gamma^{-1}X)^{-}\boldsymbol{\lambda}$, because $\mathbf{l}' = \boldsymbol{\lambda}'(X'\Gamma^{-1}X)^{-}X'\Gamma^{-1}$. ■

Notice that in the proof of the proposition only knowledge of $\operatorname{var}(Z(\mathbf{s}_i) - Z(\mathbf{s}_j))$, $i, j = 1, \ldots, n$, is required. Thus, more generally, Γ could be the $n \times n$ matrix whose (i, j)th element is $\operatorname{var}(Z(\mathbf{s}_i) - Z(\mathbf{s}_j))/2$. Because an experimental design matrix usually satisfies the condition $X\mathbf{1} \propto \mathbf{1}$ and a treatment contrast $\tau \equiv \boldsymbol{\lambda}'\boldsymbol{\beta}$ satisfies the condition $\boldsymbol{\lambda}'\mathbf{1} = 0$, the proposition holds for the g.l.s. estimators of treatment contrasts. Similarly, because $\mathbf{l}'X\mathbf{1} = 0$ for $\mathbf{l}' = \boldsymbol{\lambda}'(X'X)^{-}X'$, the variance of the o.l.s. estimator of τ can also be expressed in terms of the variogram function; that is, $\operatorname{var}(\hat{\tau}_{\text{ols}}) = -\boldsymbol{\lambda}'(X'X)^{-}X'\Gamma X(X'X)^{-}\boldsymbol{\lambda}$.

Randomization for Complete Block Designs

Assume that the model (5.6.11) holds and suppose that the experiment is laid out in $b = r$ blocks each of size t. (For incomplete block designs, the number of treatment replications r is smaller than the number of blocks b.) Write the yield on the ith plot, in the jth block, and receiving the kth treatment as

$$Z_{ij,k} = \mu_{ij} + \alpha_k + W_{ij}, \qquad i = 1, \ldots, t,\ j = 1, \ldots, r,\ k = 1, \ldots, t, \tag{5.6.22}$$

where $\{\alpha_k\colon k = 1, \ldots, t\}$ is an additive treatment effect, $\{\mu_{ij}\}$ constitutes a spatial trend, and $\{W_{ij}\}$ is a collection of spatially correlated zero-mean errors. Equivalently,

$$\begin{aligned} Z_{ij,k} = {} & \{\mu_{..} + W_{..}\} + \{(\mu_{.j} - \mu_{..}) + (W_{.j} - W_{..})\} + \alpha_k \\ & + \{(\mu_{ij} - \mu_{.j}) + (W_{ij} - W_{.j})\}, \end{aligned} \tag{5.6.23}$$

where the dot indicates an average over the omitted index. Then, components μ, $\{\beta_j\}$, $\{\alpha_k\}$, and $\{e_{ij}\}$ are defined from (5.6.23) by identifying terms; that is,

$$Z_{ij,k} = \mu + \beta_j + \alpha_k + e_{ij}. \tag{5.6.24}$$

Notice that the block term β_j is somewhat artificial unless $E(e_{ij}) = 0$; that is, unless $\mu_{ij} = \mu_{\cdot j}$, $i = 1, \ldots, t$, $j = 1, \ldots, r$. An error with zero expectation can be ensured by randomizing within each block, as is now shown.

Because, upon randomization, each plot can receive only one treatment, the yield of treatment k from block j is

$$
\begin{aligned}
Z_{jk} &\equiv \sum_{i=1}^{t} \delta_{ij,k} Z_{ij,k} = \mu + \beta_j + \alpha_k + \sum_{i=1}^{t} \delta_{ij,k} e_{ij} \\
&\equiv \mu + \beta_j + \alpha_k + \omega_{jk},
\end{aligned}
\tag{5.6.25}
$$

where

$$
\delta_{ij,k} \equiv \begin{cases} 1, & \text{if plot } i \text{ in block } j \text{ receives treatment } k, \\ 0, & \text{otherwise.} \end{cases}
$$

Relation (5.6.25) describes the response to treatment k in block j, and $\{\delta_{ij,k}\}$ are random variables whose joint probability distribution is induced by the randomization (Kempthorne, 1952, Chapter 8). It can be easily shown that:

1. $E(\omega_{jk}) = E_W E_R(\omega_{jk} | \{\delta_{ij}: i = 1, \ldots, t\}) = 0,$

2. $\operatorname{cov}(\omega_{jk}, \omega_{j'k'}) = \begin{cases} \sigma_{\omega_j}^2, & j = j', k = k', \\ -\sigma_{\omega_j}^2/(t-1), & j = j', k \neq k', \\ 0, & \text{otherwise}, \end{cases}$

3. $\operatorname{cov}(\beta_j, \omega_{jk}) = 0,$

where $E_W(\cdot)$ denotes expectation with respect to the probability distribution of the random error process $\{W_{ij}\}$; $E_R(\cdot)$ denotes expectation with respect to the randomization distribution; and

$$
\sigma_{\omega_j}^2 \equiv (1/t) \sum_{i=1}^{t} E_W\left(e_{ij}^2\right),
\tag{5.6.26}
$$

where

$$
E_W\left(e_{ij}^2\right) = (\mu_{ij} - \mu_{\cdot j})^2 + E_W(\delta_{ij} - \delta_{\cdot j})^2.
\tag{5.6.27}
$$

Clearly, from item 2, randomization does *not* eliminate spatial correlation; rather, it neutralizes it by changing the problem to one where there is small negative intrablock correlation. Let $\mathbf{Z}$ denote the responses in (5.6.25), and define $\Sigma \equiv \operatorname{var}(\mathbf{Z})$. Although Σ is *not* proportional to the identity matrix, it is

Table 5.1 Analysis of Variance for the Randomized Complete Block Design[a]

Source	Sum of Squares	Expected Mean Square
Block	$t\sum_{j=1}^{r}(Z_{.j}-Z_{..})^2$	$\frac{t}{r-1}\sum_{j=1}^{r}E_W(\beta_j^2)+\sigma_\epsilon^2$
Treatment	$r\sum_{k=1}^{t}(Z_{k.}-Z_{..})^2$	$\frac{r}{t-1}\sum_{k=1}^{t}\alpha_k^2+\frac{t}{r(t-1)}\sum_{j=1}^{r}\sigma_{\omega_j}^2+\sigma_\epsilon^2$
Residual	$\sum_{k=1}^{t}\sum_{j=1}^{r}(Z_{kj}-Z_{k.}-Z_{.j}+Z_{..})^2$	$\frac{t}{r(t-1)}\sum_{j=1}^{r}\sigma_{\omega_j}^2+\sigma_\epsilon^2$
Total	$\sum_{k=1}^{t}\sum_{j=1}^{r}(Z_{kj}-Z_{..})^2$	

Source: Grondona and Cressie (1991).
[a]Based on the randomized spatial experiment whose responses are given by (5.6.28).

not difficult to see that Zyskind's (1967) conditions are satisfied, and hence the o.l.s. estimator of any treatment contrast τ is also the best linear unbiased estimator. As was pointed out in Section 1.3, $\text{var}(\hat{\tau}_{\text{ols}})$ will still depend on Σ. However, in this instance, it can be shown that $\widehat{\text{var}}(\hat{\tau}_{\text{ols}})$ is identical to the analogous expression obtained when $\Sigma = I\sigma^2$. One cannot count on this fortunate circumstance in general (Grondona, 1989, p. 8).

If measurement errors are present, then the responses $\{Z_{jk}\}$ satisfy

$$Z_{jk} = \mu + \beta_j + \alpha_k + \omega_{jk} + \epsilon_{jk}, \tag{5.6.28}$$

where $\{\epsilon_{jk}\}$ is a collection of i.i.d. random variables with variance σ_ϵ^2, independent of the $\{\omega_{jk}\}$. This linear model for randomized complete block designs has a very convenient form for the analysis of variance, as is demonstrated in Table 5.1. Expectations of the mean squares are computed over the probability distribution induced by the randomization, the distribution of the small-scale component $\{W_{ij}\}$, and the distribution of the measurement error $\{\epsilon_{jk}\}$. (The usual randomization approach considers the data as fixed and computes expectations under the induced randomization distribution; see Kempthorne, 1952, Chapter 8. In either case, under the null hypothesis of equal treatment effects, the expectation of the treatment mean square is equal to the expectation of the residual mean square; that is, randomization provides a *valid* test for the hypothesis of equal treatment effect.)

Because the expected residual mean square (r.m.s.) in Table 5.1 is an average over all possible permutations generated by the randomization procedure, it should be possible to find systematic designs or restricted sets of randomized designs with either smaller expected r.m.s. than under the unrestricted procedure or with the same expected r.m.s. but with less variability. That is, in the presence of spatial correlation, randomized complete block designs may be inferior to other designs; an illustrative example is now given.

Efficiency of Block Designs under Second-Order Autoregressive Errors

The results presented in this section are taken from Grondona and Cressie (1992): Under a stationary second-order autoregressive error process, sufficient conditions are presented for the *universal optimality* of block designs for estimation of treatment contrasts via g.l.s. Efficiency and robustness implications of the results are investigated.

Let $\mathscr{X}_{t,b,k}$ denote the class of *incomplete block designs* with t equireplicated treatments and b blocks of size k such that each treatment appears once in each of r blocks; that is, $r = bk/t$. Consider the model

$$\mathbf{Z} = \mu \mathbf{1}_n + X_a \boldsymbol{\alpha} + X_b \boldsymbol{\beta} + \boldsymbol{\delta}, \qquad (5.6.29)$$

where,

1. $\mathbf{Z} \equiv (Z_{11}, Z_{12}, \ldots, Z_{1k}, Z_{21}, \ldots, Z_{bk})'$ is the data vector of dimension $n = b \cdot k$ and Z_{ij} is the response of the ith unit in the jth block.
2. μ represents the overall mean.
3. $\boldsymbol{\alpha}$ represents the t unknown fixed treatment effects.
4. $\boldsymbol{\beta}$ represents the b unknown fixed block effects.
5. $\boldsymbol{\delta}$ is a random vector with zero mean and dependence structure described in (5.6.30).
6. $\mathbf{1}_n$ is a vector of dimension n with all elements equal to 1, X_a is the treatment incidence matrix of dimension $n \times t$, and $X_b \equiv I_b \otimes \mathbf{1}_k$ is the block incidence matrix of dimension $n \times b$ (I_b is the identity matrix of order b and $\otimes$ denotes the Kronecker product); $X \equiv (\mathbf{1}, X_a, X_b)$ is defined to be the design matrix, where for ease of notation the subscript on $\mathbf{1}_n$ is dropped.

We assume that the observations from different blocks are uncorrelated and that within each block j the error process $\{\delta_{ij}: i = 1, \ldots, k\}$ is a partial realization from a second-order autoregression [AR(2)] defined as

$$\delta_{ij} - \rho_1 \delta_{i-1,j} - \rho_2 \delta_{i-2,j} = \epsilon_{ij}, \qquad i = 0, \pm 1, \pm 2, \ldots, \pm \infty, \quad (5.6.30)$$

where the ϵ_{ij}s are independently and identically distributed zero-mean random variables with constant variance σ^2. Thus, (5.6.30) specifies $\operatorname{var}(\mathbf{Z}) \equiv \Sigma$.

The reduced generalized-least-squares normal equations for the vector of treatments effects $\boldsymbol{\alpha}$ are

$$X_a' W X_a \boldsymbol{\alpha} = X_a' W \mathbf{Z},$$

where

$$W = I_b \otimes \left(\Sigma^{-1} - (\mathbf{1}'\Sigma^{-1}\mathbf{1})^{-1}\Sigma^{-1}\mathbf{1}\mathbf{1}'\Sigma^{-1}\right). \tag{5.6.31}$$

Consider the estimable treatment contrasts, $\boldsymbol{\tau} = (I - \mathbf{1}\mathbf{1}'/t)\boldsymbol{\alpha}$; that is, $\tau_i = \alpha_i - (1/t)(\alpha_1 + \alpha_2 + \cdots + \alpha_t)$. Let $(X_a'WX_a)^+$ be the Moore–Penrose inverse of $(X_a'WX_a)$ (e.g., Rao, 1973, p. 26). Then

$$\hat{\boldsymbol{\tau}}_{\text{gls}} = (I - \mathbf{1}\mathbf{1}'/t)(X_a'WX_a)^+ X_a'W\mathbf{Z} = (X_a'WX_a)^+ X_a'W\mathbf{Z}. \tag{5.6.32}$$

Furthermore,

$$\text{var}(\hat{\boldsymbol{\tau}}_{\text{gls}}) = (X_a'WX_a)^+ \tag{5.6.33}$$

and hence the quantity

$$C(\hat{\boldsymbol{\tau}}_{\text{gls}}, X, \Sigma) \equiv X_a'WX_a \tag{5.6.34}$$

is the information matrix for $\hat{\boldsymbol{\tau}}_{\text{gls}}$.

Because $C(\hat{\boldsymbol{\tau}}_{\text{gls}}, X, \Sigma)$ has zero row and column sums and $\text{tr}(C(\hat{\boldsymbol{\tau}}_{\text{gls}}, X, \Sigma))$ is constant for every design in $\mathscr{X}_{t,b,k}$, a sufficient condition for a design X to be universally optimal over $\mathscr{X}_{t,b,k}$ is for the corresponding information matrix to be completely symmetric (Kiefer's Theorem). Grondona and Cressie (1992) derive sufficient conditions that guarantee a completely symmetric information matrix under the model (5.6.29) and (5.6.30). For simplicity, consider $k > 4$ and let $l \neq m = 1, 2, \ldots, t$. Then the conditions are:

1. $$\lambda_{l,m} = \lambda \equiv \frac{r(k-1)}{t-1},$$

2. $$N_{l,m}^{(1)} = \frac{2\lambda}{k},$$

3. $$N_{l,m}^{(2)} = \frac{2b(k-2)}{t(t-1)},$$

4. $$e_{l,m}^* = f_{l,m}^* = \frac{4(k-1)b}{t(t-1)},$$

5. $$\sum_{j=1}^{b} f_{j,l}f_{j,m} = \sum_{j=1}^{b} e_{j,l}e_{j,m} = \frac{2b}{t(t-1)},$$

6. $$\sum_{j=1}^{b} N_{j,l,m}^{(1)}(e_{j,l}f_{j,m} + e_{j,m}f_{j,l})$$
$$= 1/2\sum_{j=1}^{b}(e_{j,l}f_{j,m} + e_{j,m}f_{j,l}) = \frac{4b}{t(t-1)},$$

Table 5.2 A Universally Optimal Complete Block Design for Five Treatments

Block	1	2	3	4	5	6	7	8	9	10
	5	2	3	1	4	5	3	4	2	1
	4	5	2	3	1	1	5	3	4	2
	1	4	5	2	3	2	1	5	3	4
	3	1	4	5	2	4	2	1	5	3
	2	3	1	4	5	3	4	2	1	5

Source: Grondona and Cressie (1992).

where

$$N_{j,l,m}^{(1)} = \begin{cases} 1, & \text{if treatments } l \text{ and } m \text{ are adjacent in block } j, \\ 0, & \text{otherwise,} \end{cases}$$

$$e_{j,l} = \begin{cases} 1, & \text{if treatment } l \text{ is applied to an edge plot in block } j, \\ 0, & \text{otherwise,} \end{cases}$$

$$f_{j,l} = \begin{cases} 1, & \text{if treatment } l \text{ is applied to a next-to-edge plot in block } j, \\ 0, & \text{otherwise,} \end{cases}$$

$N_{l,m}^{(h)}$ is the number of blocks in which treatments l and m occur as hth nearest neighbors,

$\lambda_{l,m}$ is the number of blocks in which treatments l and m occur together,

$e_{l,m}^*$ is the number of blocks in which treatments l and m occur together *and* l or m are applied to an edge plot, and

$f_{l,m}^*$ is the number of blocks in which treatments l and m occur together *and* l or m are applied to a next-to-edge plot.

An example of a universally optimal complete block design for five treatments (i.e., $t = k = 5$) is presented in Table 5.2. When Type I or Type II orthogonal arrays of strength 2 exist (e.g., Rao, 1961; Morgan and Chakravarti, 1988), incomplete block designs satisfying conditions 1 through 6 can be easily constructed: The columns of the array are used as blocks.

First-Order NN Balanced Block Designs

Universally optimal block designs for an AR(2) error process may not exist for particular choices of t, b, and k. In the sufficient condition (5), $2b/t(t-1)$ has to be an integer. Thus, the number of blocks has to be a multiple of $t(t-1)/2$, which can be large. However, the first-order NN balanced block designs typically require a smaller number of blocks. [A block design is called first-order NN balanced if the number of times that two treatments appear adjacent in a block is constant for every pair of treatments, that is, condition (2) of the preceding sufficient conditions.] Construction of these block designs is presented in Kiefer and Wynn (1981) and Cheng (1983).

Here, for an AR(2) error process, we compare the efficiency of first-order NN balanced complete block designs relative to "hypothetical universally optimal" designs and relative to randomized complete block designs; Gill and Shukla (1985a) give a similar comparison for a first-order autoregressive [AR(1)] error process. Suppose that for any block design $X \in \mathscr{X}_{t,b,k}$, with information matrix $C(\hat{\tau}, X, \Sigma)$, the trace of $C(\hat{\tau}, X, \Sigma)$ does not depend on X. Then the design $X^* \in \mathscr{X}_{t,b,k}$, with all positive eigenvalues equal to

$$\nu \equiv \text{tr}(C(\hat{\tau}, X, \Sigma))/(t-1), \tag{5.6.35}$$

will be universally optimal. Because such a design X^* may not exist, we refer to it as the *hypothetical* universally optimal design. For the purposes of comparison, the first-order NN balanced complete block design presented in Table 1.1 will be used.

Recall the definitions of the A-optimality criterion, D-optimality criterion, and E-optimality criterion, given as special cases of (5.6.16). Then, write the A-efficiency, D-efficiency, and E-efficiency of the first-order NN balanced complete block design as, respectively,

$$A_O/A_N \equiv \nu^{-1} \Big/ \left\{ \sum_{l=1}^{t-1} \nu_l^{-1}/(t-1) \right\},$$

$$D_O/D_N \equiv \nu^{-1} \Big/ \left\{ \prod_{l=1}^{t-1} \nu_l^{-1} \right\}^{1/(t-1)},$$

$$E_O/E_N \equiv \nu^{-1} / \{\max(\nu_1^{-1}, \ldots, \nu_{t-1}^{-1})\},$$

where $\{\nu_1, \ldots, \nu_{t-1}\}$ are the $(t-1)$ positive eigenvalues of $C(\hat{\tau}_{\text{gls}}, X, \Sigma)$ given by (5.6.34), and X is the incidence matrix of 0s and 1s defined by Table 1.1. Table 5.3 shows the three efficiencies for those combinations of (ρ_1, ρ_2) that yield a stationary AR(2) error process. Notice that, when $\rho_2 = 0$, an AR(1) error process is obtained; these results agree with Gill and Shukla (1985a) and are used to form Table 1.2. Clearly, the first-order NN balanced complete block design considered is very A- and D-efficient for an AR(2) error model that exhibits small to moderate departures from the AR(1) model. E-efficiency is the most sensitive to small departures; that is, variances of individual contrasts can be very different.

Table 5.3 ***A*-Efficiency (A_O/A_N), *D*-Efficiency (D_O/D_N), and *E*-Efficiency (E_O/E_N) of a First-Order NN Balanced Complete Block Design**

		ρ_2						
ρ_1		−0.75	−0.50	−0.25	0.00	0.25	0.50	0.75
	A	0.96						
−1.50	*D*	0.95						
	E	0.77						
	A	0.93	0.96					
−1.25	*D*	0.92	0.96					
	E	0.67	0.78					
	A	0.87	0.94	0.98				
−1.00	*D*	0.86	0.93	0.98				
	E	0.55	0.68	0.85				
	A	0.78	0.89	0.97	1.00			
−0.75	*D*	0.76	0.88	0.97	1.00			
	E	0.40	0.57	0.78	0.99			
	A	0.65	0.82	0.95	1.00	0.96		
−0.50	*D*	0.64	0.81	0.95	1.00	0.96		
	E	0.26	0.45	0.72	0.99	0.73		
	A	0.51	0.76	0.94	1.00	0.95	0.82	
−0.25	*D*	0.52	0.74	0.93	1.00	0.94	0.81	
	E	0.15	0.36	0.66	1.00	0.68	0.45	
	A	0.45	0.73	0.93	1.00	0.94	0.78	0.61
0.00	*D*	0.47	0.72	0.92	1.00	0.93	0.78	0.62
	E	0.12	0.33	0.64	1.00	0.66	0.41	0.25
	A	0.51	0.76	0.94	1.00	0.94	0.79	
0.25	*D*	0.52	0.74	0.93	1.00	0.93	0.78	
	E	0.15	0.36	0.66	1.00	0.67	0.41	
	A	0.63	0.81	0.95	1.00	0.95		
0.50	*D*	0.62	0.80	0.94	1.00	0.94		
	E	0.24	0.43	0.70	1.00	0.70		
	A	0.73	0.86	0.96	1.00			
0.75	*D*	0.71	0.85	0.96	1.00			
	E	0.33	0.51	0.74	1.00			
	A	0.80	0.90	0.97				
1.00	*D*	0.79	0.89	0.97				
	E	0.43	0.59	0.70				
	A	0.86	0.93					
1.25	*D*	0.85	0.92					
	E	0.51	0.65					
	A	0.90						
1.50	*D*	0.89						
	E	0.58						

Source: Grondona and Cressie (1992).

The efficiencies of the randomized complete block design (relative to the hypothetical optimal design) can be very poor; see Table 1.2 for selected results of how low they can go. More complete results are presented in Grondona and Cressie (1992). Generally, any small departure from the uncorrelated-errors model causes the randomized complete block design to be highly inefficient.

So far, the efficiencies of first-order NN balanced complete block designs, assuming a known AR(1) or AR(2) error process, have been considered. AR(1) processes have been used as a component of error models in the analysis of field trials with a block design structure (e.g., Patterson and Hunter, 1983; Gleeson and Cullis, 1987; Grondona and Cressie, 1991). However, one might wonder whether the efficient first-order NN balanced block designs are robust to misspecification of the AR(1) error structure. This question can be answered by assuming the correct model to be an AR(2) process with parameters ρ_1 and ρ_2, but, when estimating the treatment contrasts, an AR(1) model with parameter ρ_1 is used (incorrectly). For the first-order NN complete block design used earlier (Table 1.1), Grondona and Cressie (1992) computed the relative A-, D-, and E-efficiencies of the misspecified g.l.s. estimator, relative to the correctly specified g.l.s. estimator. Most were above 90%, indicating that first-order NN complete block designs are robust to a misspecified AR(1) process that is in fact a nearby AR(2) process.

It is sensible to select a restricted design (e.g., first-order NN balanced) at random from the class of optimal (or near optimal) restricted designs: Randomization within this restricted subset will likely yield either a smaller expected residual mean square or the same expected residual mean square but with less variability. Moreover, if the true covariance model is different from the assumed model, then randomization will give some robustness to the analysis.

Valid randomization schemes are needed, for which the expected treatment mean square is, under no treatment effects, equal to the expected residual mean square; here, expectation refers to the (restricted) randomization distribution. Bailey (1985) gives valid randomization schemes for first-order NN balanced complete block designs for the cases $b = t/2$ and $b = t$. (She also gives schemes for first-order partial NN balanced designs when $b = t - 1$ and $b = t + 1$.)

Spatial Analysis of a Nonspatial Design

Randomized complete block designs have an enormous practical advantage over other designs: They are very, very easy to construct. Can the inefficiency of using this simple design be mitigated by using a spatial *analysis* on the responses? The answer is "Yes," provided the *location* of the treatments within blocks are available for use in the analysis. Ways to deal with the analysis of experiments, in the presence of spatial dependence, are discussed in Section 5.7.

5.7 FIELD TRIALS

Although originally employed in an agricultural context, the general purpose of field trials is to compare the effects of a number of treatments applied to a collection of experimental units. (The union of proximate experimental units can be called a field.) For example, in the manufacture of integrated-circuit chips from silicon wafers, treatments might be different types of silicon production, wafer fabrication, or chip fabrication, in various combinations. In crop production, treatments might be fertilizers on a particular crop variety or different varieties of the same crop.

Obtaining precise estimators of treatment effects is of primary concern, so that it becomes important to reduce or control the residual variation not due to treatment effects. The latter can be handled in at least two ways. The first and most common approach is to use designs of increasing complexity, e.g., complete block designs, incomplete block designs, lattice designs, and so forth (Section 5.6.2). The second approach, which is not incompatible with the first, is to use alternative models and analyses that account for spatial heterogeneity. This is the topic that will be addressed in this section. Although analyses based on spatial considerations (Papadakis, 1937; Bartlett, 1938) date from around the same time as the incomplete block designs and lattice designs introduced by Yates (1936), it has only been relatively recently that the spatial analyses have been well understood.

The goal is efficient estimation of treatment effects. This section reviews methods that use the spatial location of treatments to achieve that goal, even when a nonoptimal design is used. Ideally, an optimal spatial design is used in the first place (Section 5.6.2), but in practice this has rarely been the case.

5.7.1 Nearest-Neighbor Analyses

Papadakis (1937) was the first to introduce a nearest-neighbor (NN) method for the analysis of field trials, although an investigation of its theoretical properties was not initiated until over 30 years later (Atkinson, 1969). Further understanding of and contributions to the method can be found in Bartlett (1978), Wilkinson et al. (1983), Green et al. (1985), Besag and Kempton (1986), Gleeson and Cullis (1987), and Zimmerman and Harville (1989). Indeed, Papadakis himself continued to use and modify his method over the decades since his original 1937 proposal (Papadakis, 1984).

Estimation of Treatment Effects

Let $\mathbf{Z}$ denote the vector of responses to a field trial and suppose that the experimental design employed allows one to write

$$\mathbf{Z} = X_1\boldsymbol{\alpha} + \mathbf{W} + \boldsymbol{\epsilon}, \tag{5.7.1}$$

where $\boldsymbol{\alpha}$ is a vector of treatment contrasts and $(\mathbf{W} + \boldsymbol{\epsilon})$ is the underlying spatial variation; $\mathbf{W}$ is a zero-mean (spatially dependent) vector representing the smooth small-scale variation (e.g., fertility variation) and is independent of $\boldsymbol{\epsilon}$, which is a zero-mean vector of identically distributed uncorrelated errors. [Green (1985) considers the model, $\mathbf{Z} = X_1\boldsymbol{\alpha} + X_2\boldsymbol{\zeta} + \mathbf{W} + \boldsymbol{\epsilon}$, where $X_2\boldsymbol{\zeta}$ represents large-scale spatial inhomogeneities; e.g., the elements of $\boldsymbol{\zeta}$ are block effects. Although the rest of this section deals with the simpler model (5.7.1), analogous results are true for the more general block designs that are orthogonal.]

Now suppose $\mathbf{W}^{(l)}$ is a current indication of the value of $\mathbf{W}$. Then, an ordinary least-squares (o.l.s.) estimator of $\boldsymbol{\alpha}$ is

$$\boldsymbol{\alpha}^{(l+1)} = (X_1'X_1)^{-1}X_1'\{\mathbf{Z} - \mathbf{W}^{(l)}\}. \tag{5.7.2}$$

Thus, from $\mathbf{Z} - X_1\boldsymbol{\alpha}^{(l+1)}$, a subsequent indication of $\mathbf{W}$ can be found: Papadakis suggested solving

$$\mathbf{W}^{(l+1)} = C(\boldsymbol{\gamma})\left(\mathbf{Z} - X_1\boldsymbol{\alpha}^{(l+1)}\right) \tag{5.7.3}$$

and using as starting value, say, $\mathbf{W}^{(0)} = (0, \ldots, 0)'$. In (5.7.3), the $n \times n$ matrix $C(\boldsymbol{\gamma})$ reflects similarity in spatially close experimental units; for example, in a one-dimensional layout, $C(\gamma)$ might average the residuals in nearest-neighboring units (except at the edges, where just a single nearest neighbor is used) and multiply the result by $\gamma \in \mathbb{R}$. Iterating (5.7.2) and (5.7.3) yields Papadakis' *iterated* NN method (Papadakis, 1970; Bartlett, 1978). (Papadakis' 1937 proposal stopped at $\boldsymbol{\alpha}^{(2)}$.) Consequently, the final estimate, $\hat{\boldsymbol{\alpha}} = \boldsymbol{\alpha}^{(\infty)}$, satisfies

$$X_1'(I - C(\boldsymbol{\gamma}))X_1\hat{\boldsymbol{\alpha}} = X_1'(I - C(\boldsymbol{\gamma}))\mathbf{Z}. \tag{5.7.4}$$

Assuming the $k \times 1$ vector of parameters $\boldsymbol{\gamma}$ in $C(\boldsymbol{\gamma})$ is known, (5.7.4) is clearly the generalized-least-squares (g.l.s.) estimator of $\boldsymbol{\alpha}$ in the linear model

$$E(\mathbf{Z}) = X_1\boldsymbol{\alpha}, \qquad \operatorname{var}(\mathbf{Z}) = (I - C(\boldsymbol{\gamma}))^{-1}\tau^2. \tag{5.7.5}$$

Thus, if the error component in (5.7.1),

$$\boldsymbol{\delta} \equiv \mathbf{W} + \boldsymbol{\epsilon}, \tag{5.7.6}$$

can be identified with a zero-mean conditionally specified spatial Gaussian model (CG; Section 6.6) with spatial dependence matrix $C(\boldsymbol{\gamma})$, the iterated Papadakis NN estimator is just the maximum likelihood (m.l.) estimator of $\boldsymbol{\alpha}$ (Ripley, 1981, p. 97; Martin, 1982; Draper and Faraggi, 1985).

The vector $\boldsymbol{\gamma}$ can now be interpreted as a set of spatial, small-scale-variation parameters; for the purpose of derivation of (5.7.4), $\boldsymbol{\gamma}$ was assumed known. Discussion of the estimation of $\boldsymbol{\gamma}$ will be given later.

One-Dimensional Case

Because the field of units is often a long narrow strip of contiguous plots or is made up of adjoining blocks whose plots down the blocks have more horizontal length than vertical height, the case of *one-dimensional* spatial dependence has received much attention. Indeed, most of the authors mentioned previously have concentrated on this case. The spatial modeling approach in Section 5.7.2 has a broader perspective that allows spatial analyses to be carried out in $\mathbb{R}^d$, $d \geq 1$.

Gleeson and Cullis (1987) have brought all the one-dimensional approaches together by assuming that the dependence in $\mathbf{W}$ follows an autoregressive-integrated-moving-average time series, abbreviated as ARIMA (p, m, q). Suppose $\{W(s): s \in \mathbb{Z}\}$ is a zero-mean random process on the integers, where the index s denotes the spatial location of an experimental unit in a one-dimensional strip. Define the backshift operator B by $B\delta(s) \equiv \delta(s-1)$, and hence $B^2\delta(s) = \delta(s-2)$ and so on. Let $\{a(s): s \in \mathbb{Z}\}$ denote a sequence of independent and identically distributed (i.i.d.) random variables with zero mean and variance σ_a^2. Then $W(\cdot)$ is an ARIMA (p, m, q) if it can be represented as

$$\{1 - \phi_1 B - \cdots - \phi_p B^p\}(1 - B)^m W(s) = (1 - \theta_1 B - \cdots - \theta_q B^q)a(s), \tag{5.7.7}$$

where the roots of the equations $1 - \phi_1 y - \cdots - \phi_p y^p = 0$ and $1 - \theta_1 y - \cdots - \theta_q y^q = 0$ are outside the unit circle and the two equations have no roots in common. Because the elements of $\boldsymbol{\epsilon}$ in (5.7.6) are assumed i.i.d., then $(1 - B)^m(W(s) + \epsilon(s))$ is an ARIMA $(p, 0, Q)$, where $Q \equiv \max(p + q, m)$ (Box and Jenkins, 1970, Appendix 4.4).

Empirical evidence from agricultural field trials indicates that differenced data usually permit the use of a more parsimonious model than undifferenced data. So, although Papadakis' original proposal effectively assumed $m = 0$ (see following text), many of the more recent NN analyses have included some order of differencing. Green et al. (1985) assume that $W(\cdot)$ is an ARIMA $(0, 2, 0)$, so that $\{(1 - B)^2(W(s) + \epsilon(s)): s \in \mathbb{Z}\}$ is an ARIMA $(0, 0, 2)$ with moving average coefficients restricted to satisfy $\theta_2 = -\theta_1(1 - \theta_1)/4$. Wilkinson et al. (1983), Besag and Kempton (1986), and Williams (1986) assume that $W(\cdot)$ is an ARIMA $(0, 1, 0)$, so that $\{(1 - B)(W(s) + \epsilon(s)): s \in \mathbb{Z}\}$ is an ARIMA $(0, 0, 1)$. Differencing the data has an added bonus: For $m = 1$, Besag and Kempton show that $\hat{\boldsymbol{\alpha}}$ depends on the second-differenced data (apart from edge effects) and hence the estimator is invariant to a linear trend $\{\beta_0 + \beta_1 s: s \in \mathbb{Z}\}$ that can be added to $W(\cdot)$ without loss of generality. Gleeson and Cullis (1987) assume that $W(\cdot)$ is an ARIMA $(1, m, 0)$, $m = 1$ or 2, although they point out that more general ARIMA processes could be fitted.

Papadakis (1937), Bartlett (1938, 1978), and Atkinson (1969) all effectively assume that $W(\cdot)$ is an ARIMA $(1,0,0)$ and that $\boldsymbol{\epsilon} = \mathbf{0}$; that is, no differencing is carried out. The ARIMA $(1,0,0)$ satisfies

$$E(\delta(s)|\{\delta(u): u \neq s\}) = \gamma\{\delta(s-1) + \delta(s+1)\}, \qquad 2|\gamma| < 1, \quad (5.7.8)$$

which, together with a Gaussian assumption on the innovation process $a(\cdot)$ in (5.7.7), implies that (Section 6.6)

$$\boldsymbol{\delta} \sim \text{Gau}\left(\mathbf{0}, (I - \gamma H)^{-1}\tau^2\right). \qquad (5.7.9)$$

In (5.7.9), $\boldsymbol{\delta} \equiv (\delta(1), \dots, \delta(n))'$ and H is a symmetric $n \times n$ matrix with (i, j)th element h_{ij}, where $h_{i,i-1} = 1$ $(i = 2, \dots, n)$, $h_{i,i+1} = 1$ $(i = 1, \dots, n-1)$, and $h_{ij} = 0$ elsewhere. [Notice that to handle edge effects, $\delta(0)$ and $\delta(n+1)$ in (5.7.8) have been set equal to zero, their expected values.] From (5.7.1) and (5.7.9),

$$\mathbf{Z} \sim \text{Gau}\left(X_1\boldsymbol{\alpha}, (I - \gamma H)^{-1}\tau^2\right), \qquad (5.7.10)$$

and hence from (5.7.4) the iterated Papadakis NN method yields

$$\hat{\boldsymbol{\alpha}} = \left(X_1'(I - \gamma H)X_1\right)^{-1}X_1'(I - \gamma H)\mathbf{Z}. \qquad (5.7.11)$$

In terms of the iterations (5.7.2) and (5.7.3), $\boldsymbol{\alpha}^{(1)} = (X_1'X_1)^{-1}X_1'\mathbf{Z}$, which is the o.l.s. estimate of $\boldsymbol{\alpha}$. The o.l.s. residuals are $\mathbf{Z} - X_1\boldsymbol{\alpha}^{(1)}$, and hence $\mathbf{W}^{(1)} = \gamma H(\mathbf{Z} - \mathbf{X}_1\boldsymbol{\alpha}^{(1)})$. In words, the ith element of $\mathbf{W}^{(1)}$ is obtained by adding the neighboring $(i-1)$th and $(i+1)$th o.l.s. residuals and multiplying by the regression coefficient γ. One then uses as "data," $\mathbf{Z} - \mathbf{W}^{(1)}$, from which an o.l.s. estimate $\boldsymbol{\alpha}^{(2)} = (X_1'X_1)^{-1}X_1'\{\mathbf{Z} - \mathbf{W}^{(1)}\}$ is obtained. In practice, γ could be estimated from an analysis of covariance on the data $\mathbf{Z}$, where the covariate is the sum of o.l.s. residuals on adjacent experimental units. This was precisely the approach taken by Papadakis (1937), but the preceding discussion shows that the estimator of $\boldsymbol{\alpha}$ is an inefficient approximation to the m.l. estimator. Also, an m.l. estimator of γ would be more efficient.

Finally, notice that, as $\gamma \to (1/2)$ in (5.7.9), $\delta(\cdot)$ becomes an ARIMA $(0,1,0)$. In this case, the data $\{Z(i)\}$ are differenced once and $\hat{\boldsymbol{\alpha}}$ is a function of the adjusted, second-differenced data $\{Z(i) - (Z(i-1) + Z(i+1))/2\}$ (Besag and Kempton, 1986).

Two-Dimensional Case

In two dimensions, one could assume a spatial model analogous to (5.7.8):

$$\begin{aligned} E(\delta(u,v)|\{\delta(k,l)&: (k,l) \neq (u,v)\}) \\ &= \gamma_1\{\delta(u-1,v) + \delta(u+1,v)\} + \gamma_2\{\delta(u,v-1) \\ &\qquad + \delta(u,v+1)\}, \quad 2|\gamma_1| + 2|\gamma_2| < 1. \quad (5.7.12) \end{aligned}$$

Then a homoskedastic Gaussian assumption like (6.6.3), on the conditional distribution of $\{\delta(u, v)\}$, implies that $\delta(\cdot)$ is a Gaussian Markov random field and that

$$\boldsymbol{\delta} \sim \text{Gau}\big(\mathbf{0}, (I - C(\boldsymbol{\gamma}))^{-1}\tau^2\big), \tag{5.7.13}$$

where $C(\boldsymbol{\gamma})$ is defined by comparing (6.6.2) with (5.7.12) and (6.6.4) with (5.7.13). From the iterations (5.7.2) and (5.7.3), an analysis of covariance is suggested, where there are now *two* covariates, namely, the sum of o.l.s. residuals on horizontally adjacent experimental units and the sum of o.l.s. residuals on vertically adjacent experimental units. Edge effects have been ignored in this discussion, but they do require careful consideration; see, for example, Martin (1982), Wilkinson et al. (1983), and Besag and Kempton (1986).

Should $\gamma_1 = \gamma_2 = \gamma$ be assumed, then a single covariate, the sum of all adjacent o.l.s. residuals, is suggested. As $\gamma \to (1/4)$, $\delta(\cdot)$ becomes an intrinsic autoregression (Kunsch, 1987), which is an IRF0 on a square lattice in $\mathbb{R}^2$ (Section 5.4).

Martin (1990) and Cullis and Gleeson (1991) suggest using separable processes (Section 2.5.1) to model the spatial variation in **W**. They fit ARIMA models in each direction to two-way data from field trials.

Estimation of Small-Scale Variation Parameters $\boldsymbol{\gamma}$

In the one-dimensional case, Papadakis (1937) proposed a formal analysis of covariance on data **Z** with covariate $H(\mathbf{Z} - X_1(X_1'X_1)^{-1}X_1'\mathbf{Z})$, where H is previously defined. The small-scale variation parameter γ represents the regression coefficient of data on the covariate, and hence

$$\hat{\gamma} = \{\mathbf{Z}'H(I - P_1)\mathbf{Z}\}/\{\mathbf{Z}'(I - P_1)H'H(I - P_1)\mathbf{Z}\}, \tag{5.7.14}$$

where $P_1 \equiv X_1(X_1'X_1)^{-1}X_1'$. This estimator is inefficient when compared to the m.l. estimator resulting from the model

$$\mathbf{Z} \sim \text{Gau}\big(X_1\boldsymbol{\alpha}, (I - \gamma H)^{-1}\tau^2\big). \tag{5.7.15}$$

Computational details for obtaining m.l. estimates of $\boldsymbol{\alpha}$, τ^2, and γ can be found in Section 7.2.3.

In the one-dimensional case, Wilkinson et al. (1983) and Besag and Kempton (1986) assume away the need for estimating γ by setting it equal to (1/2). This imposes a spatial model on the data that may not be realistic. Formal inference to justify their claim that most data from field trials should only be differenced, has not been forthcoming. Gleeson and Cullis (1987) also impose some degree of differencing, but allow autoregressive dependence in **Y**, the differenced data. Under a Gaussian assumption, they obtain,

$$\mathbf{Y} \sim \text{Gau}\left(A_1\boldsymbol{\alpha}, V(\boldsymbol{\gamma})\tau^2\right), \tag{5.7.16}$$

where A_1 and $V(\boldsymbol{\gamma})$ are easily calculated. Instead of estimating $\boldsymbol{\gamma}$ and τ^2 by m.l., they propose the use of restricted maximum likelihood (REML); for further discussion of REML, see Section 2.6.1.

In the two-dimensional case, Cullis and Gleeson (1991) use separable ARIMA processes to model var(**Z**). Again, the variance–covariance parameters are fitted using REML.

Which analysis one uses on field-trials data should be determined by the underlying spatial model; in general,

$$\mathbf{Z} \sim \text{Gau}\left(X_1\boldsymbol{\alpha}, \Sigma(\boldsymbol{\gamma})\tau^2\right). \tag{5.7.17}$$

The estimated g.l.s. estimator of $\boldsymbol{\alpha}$ is

$$\hat{\boldsymbol{\alpha}}_{\text{egls}} = \left(X_1'\Sigma(\hat{\boldsymbol{\gamma}})^{-1}X_1\right)^{-1}X_1'\Sigma(\hat{\boldsymbol{\gamma}})^{-1}\mathbf{Z}, \tag{5.7.18}$$

where $\hat{\boldsymbol{\gamma}}$ is an estimator of $\boldsymbol{\gamma}$ (m.l. is more efficient than REML, but the m.l. estimator has large bias when the number of data is small in relation to the number of treatments). Notice that, when interest is centered on treatment contrasts, it is not necessary to model the covariance of the data, as in (5.7.17), but only the variogram; see (5.6.18). When working with the variogram, the form of $\Sigma(\boldsymbol{\gamma})$ may not be known, and hence the likelihood may be unavailable. However, $\boldsymbol{\gamma}$ can still be estimated by REML (Kitanidis, 1983).

Section 5.7.2 gives an alternative way of estimating $\boldsymbol{\gamma}$ and τ^2. There, a spatial variogram model is fitted to a nonparametric variogram estimator based on nonlinear residuals $\mathbf{Z} - X_1\tilde{\boldsymbol{\alpha}}$, where $\tilde{\boldsymbol{\alpha}}$ is a robust/resistant estimator of $\boldsymbol{\alpha}$ in $E(\mathbf{Z}) = X_1\boldsymbol{\alpha}$.

Some Final Remarks

This section, on NN analyses of field trials, has concentrated on presenting those spatial models from which (approximate) maximum likelihood estimation amounts to a NN analysis. However, some of the recent statistical literature on field trials has been skeptical of the value of fitting models; it is often claimed that standard errors of estimators are overreliant on model assumptions. It is my belief that, provided the modeling is done well (with diagnostics playing an important role), there is much to gain from spatial modeling. Problems of edge effects, parameter estimation, and adjustment of degrees of freedom now take on their proper perspective: A discussion of how to deal with edge effects is given in Section 7.3. For spatially correlated data, one cannot expect an easy modification of the usual analysis of variance or covariance; this is discussed further in Section 4.3.2, where a spatial analysis of variance is presented. Accounting for the effect (on standard errors) of estimation of small-scale-variation parameters $\boldsymbol{\gamma}$, by a simple adjustment of the degrees of freedom, is unlikely to be satisfactory. Section 5.3.2 shows that, for spatial prediction, the problem is complex and requires careful analysis.

The nonuniqueness of the decomposition (3.1.2), into large- and smaller-scale variation, affects how the spatial process $\delta(\cdot)$ is modeled. By removing row or column effects and doing a conventional (nonspatial) analysis, one may on occasions be able to reduce the treatment-estimate standard errors as much as if one were to carry out a (one-dimensional) NN analysis (Kempton and Howes, 1981). However, the two-dimensional spatial analyses seem to have a clear advantage over a conventional row-column analysis (Cullis and Gleeson, 1991).

There have been a large number of NN analyses proposed. One could legitimately ask which is best. Besag and Kempton (1986, Appendix 2) present evidence that their method of first differences does well on various *agricultural* field trials, but there is no guarantee that the same recommendation holds for "industrial" field trials, for example, a field trial to determine which of several procedures would improve integrated-circuit manufacturing. Spatial modeling allows the responses (data) from the field trial to have an influence on which model will be fitted, and hence which NN analysis will be used to estimate the treatment effects. That is not to say that previous experience with similar types of field trials should be ignored; such experience could be used to define a class of potential spatial models or to put a prior distribution on a specified class of spatial models.

5.7.2 Analyses Based on Spatial Modeling

All the NN analyses presented in Section 5.7.1 can be based on a spatial linear model. Consider a generalization of the model (5.7.1), which allows large-scale spatial inhomogeneities. Suppose

$$\mathbf{Z} = X\boldsymbol{\beta} + \boldsymbol{\delta}, \tag{5.7.19}$$

where a subset of the elements of the fixed effects $\boldsymbol{\beta}$ are treatment effects $\boldsymbol{\alpha}$ and $\boldsymbol{\delta}$ is a zero-mean error vector that is spatially correlated. Assume

$$\text{var}(\boldsymbol{\delta}) = \Sigma(\boldsymbol{\gamma}). \tag{5.7.20}$$

In (5.7.20), $\Sigma(\boldsymbol{\gamma})$ is a variance matrix of regional lattice data. Brewer and Mead (1986) discuss the types of regional models that result from integrating point-support models with continuous spatial index.

By further assuming $\mathbf{Z} \sim \text{Gau}(X\boldsymbol{\beta}, \Sigma(\boldsymbol{\gamma}))$, m.l. estimates of treatment effects can be obtained from $\hat{\boldsymbol{\beta}}_{\text{egls}}$, which is defined analogously to (5.7.18) with $\hat{\boldsymbol{\gamma}}$ the m.l. estimate.

A commonly used estimate of the variance of the estimated effects is given by

$$\widehat{\text{var}}\left(\hat{\boldsymbol{\beta}}_{\text{egls}}\right) = \left(X'\Sigma(\hat{\boldsymbol{\gamma}})^{-1}X\right)^{-1}, \tag{5.7.21}$$

but this does not take into account the estimation of $\boldsymbol{\gamma}$ in $\hat{\boldsymbol{\beta}}_{\text{egls}}$. A careful consideration of this extra source of variability should allay concerns expressed in the literature regarding the inadequacy of model-based standard errors. (Similar concerns, for mean-squared prediction errors of spatial predictors, are discussed in Section 5.3.3.)

REML estimators of small-scale-variation parameters $\boldsymbol{\gamma}$ have superior small-sample biases and seem to yield a valid F test for treatments (Lill et al., 1988). (REML is defined in Section 2.6.1.) Gleeson and Cullis (1987), Cullis and Gleeson (1991), and Zimmerman and Harville (1991) posit various parametric families of spatial models for $\boldsymbol{\delta}$, whose parameters $\boldsymbol{\gamma}$ they recommend be estimated by REML.

An analysis of field-trials data based on spatial modeling adapts to the quantity and quality of the spatial dependence demonstrated by the data, and exploits that dependence to obtain more precise estimators of large-scale-variation parameters, such as treatment effects (Besag and Kempton, 1986, Appendix 2; Cullis and Gleeson, 1991; Grondona and Cressie, 1991; Zimmerman and Harville, 1991). This is another example of why Statistics for spatial data deserves a place in the statistician's repertoire; further examples are given in Section 1.3.

Spatial Modeling After Robust / Resistant Detrending

Residuals $\mathbf{R}$, from a robust/resistant fit to the large-scale variation, estimate well the (spatially dependent) error vector $\boldsymbol{\delta}$ (see Section 3.5). Based on $\mathbf{R}$, one could then fit an intrinsically stationary model (Sections 2.3, 2.4, and 2.6) of the spatial dependence and use this model to obtain more precise (estimated) g.l.s. estimates of treatment contrasts. Does the approach work? Grondona and Cressie (1991) show that gains in efficiency of over 30% are possible; a summary of their results is given in the following text.

An artificial experiment was conducted on the wheat uniformity trials of Mercer and Hall (1911); the data are discussed in Section 7.1 and presented in Figure 7.1. After computing a nonparametric variogram estimator from resistantly detrended residuals $\mathbf{R}$, the following isotropic covariogram model was fitted:

$$\text{cov}(Z(\mathbf{s}), Z(\mathbf{s}+\mathbf{h})) = C^{o}(\|\mathbf{h}\|) = 0.1457\exp\{-\|\mathbf{h}\|/3.1227\}, \qquad \mathbf{s}, \mathbf{h} \in \mathbb{R}^2, \tag{5.7.22}$$

where $\|\mathbf{h}\|$ is in units of meters. For the purposes of this demonstration, it is assumed that the nugget effect is zero, although there is reason to believe (Section 4.5) that spatial location error may contribute to a nonzero jump in the covariogram at the origin. The expression (5.7.22) will henceforth be referred to as the true model.

Six randomized complete block designs, each with 20 treatments and 4 blocks, were applied to the data in Figure 7.1. The blocks ran from north to south and each contained 20 experimental units. A guard column was left between blocks. Treatments 1 to 20 were assigned at random within each block and treatment effects were (arbitrarily) set proportional to the treatment number. The observation in unit (i, j) that received treatment k was artificially generated using the expression

$$y_{ij,k} = z_{ij} + (k - 10.5)(0.14)^{1/2}/5, \tag{5.7.23}$$

where z_{ij} is the original response (wheat yield in pounds) and $0.14 \cong \sigma^2 = 0.1457$, the fitted error variance. Formula (5.7.23) ensures that treatment effects lie between $\pm 2\sigma$.

Because a guard column was left between blocks, the underlying error covariance (5.7.22) implies that observations from different blocks are essentially uncorrelated. After removing block effects and treatment effects from the six artificial experiments by median polish (Section 3.5.1), robust variogram estimators for the individual experiments were computed in the north–south direction (i.e., down the blocks) using the expression (2.4.12).

Let $2\bar{\gamma}_j(h)$ denote the robust variogram estimator, at lag h, for the jth experiment. Then, the combined variogram estimator from J experiments (here $J = 6$) is

$$2\bar{\gamma}(h) \equiv \left[\frac{1}{|N(h)|} \sum_{j \in J(h)} |N_j(h)| \left\{2\bar{\gamma}_j(h)\left(0.457 + \frac{0.494}{|N_j(h)|}\right)\right\}^{1/4}\right]^4 \Bigg/ \left[0.457 + \frac{0.494}{|N(h)|}\right], \tag{5.7.24}$$

where $J(h)$ is the set of experiments with a variogram estimator at lag h, $|N_j(h)|$ denotes the number of data pairs, at lag h, for the jth experiment, and $|N(h)| \equiv \Sigma_{j \in J(h)} |N_j(h)|$ is the total number of data pairs at lag h.

A parametric model was fitted to (5.7.24) using weighted least squares (Section 2.6.2), and the individual experiments were analyzed using this fitted model as if it were the true model. Some justification for this is given by Martin (1986a): A well estimated covariance matrix $\hat{\Sigma}$ is closer to the true covariance matrix Σ than to the identity matrix. Therefore, inferences based on the estimated covariance matrix are expected to be more accurate than those based on the identity matrix; the analysis that follows reinforces this viewpoint.

A weighted-least-squares fit to (5.7.24) yielded the variogram model (in the north–south direction)

$$2\gamma^{\#}(h) = 0.2312\{1 - \exp(-h/2.3893)\}, \qquad h \in \mathbb{R}. \qquad (5.7.25)$$

Equivalently, the spatial dependence can be quantified in terms of the covariogram

$$C^{\#}(h) = 0.1156 \exp(-h/2.3893), \qquad h \in \mathbb{R}. \qquad (5.7.26)$$

Notice that both parameters in (5.7.26) are underestimated when compared to the true model (5.7.22). Other classes of models were tried, but none achieved a better fit (according to the weighted-least-squares criterion) than the exponential model. Also, because the covariogram model is exponential, residuals within any block can be interpreted as a realization of a first-order autoregressive process.

Each of the six experiments was analyzed initially using the usual mean structure for a complete block design; that is,

$$E(\mathbf{Y}) = \mathbf{1}\mu + X_a\boldsymbol{\alpha} + X_b\boldsymbol{\beta}, \qquad (5.7.27)$$

where $\mathbf{Y}$ is the 80×1 data vector, μ is the overall mean, $\boldsymbol{\beta}_{4\times 1}$ contains the unknown block effects, and $\boldsymbol{\alpha}_{20\times 1}$ contains the unknown treatment effects. Write $X \equiv (\mathbf{1}, X_a, X_b)$.

Three different error structures $\Sigma_m = I_4 \otimes U_m$, $m = 1, 2, 3$, are used. The symbol $\otimes$ denotes the Kronecker matrix product and U_1, U_2, and U_3 are 20×20 matrices defined as follows.

Model 1. U_1 is the covariance matrix whose elements are obtained from expression (5.7.22), which is assumed to be the true covariogram.

Model 2. $U_2 = I\sigma^2$; that is, the errors are uncorrelated with equal variances.

Model 3. U_3 is the covariance matrix whose elements are obtained from expression (5.7.26), which is the covariogram fitted after detrending.

Under model 2, o.l.s. estimation is used, and under models 1 and 3, g.l.s. estimation is used. Further, because $C^{\#}(\cdot)$ is estimated from the combined residuals, its variation is assumed to be negligible.

Table 5.4 Analysis of the Six Artificial Experiments, for g.l.s. ($m = 1$), o.l.s. ($m = 2$), and Estimated g.l.s. ($m = 3$)

Estimation		Experiment 1	2	3	4	5	6
		Model 1					
g.l.s.	RMS_1	1.3996	1.0040	1.1217	0.9725	1.4652	0.9411
($m = 1$)	A_1	0.0504	0.0510	0.0510	0.0506	0.0506	0.0512
	$\hat{A}_1$	0.0705	0.0512	0.0572	0.0492	0.0741	0.0482
		Model 2					
o.l.s.	RMS_2	0.1746	0.1216	0.1424	0.1234	0.2184	0.1193
($m = 2$)	A_2	0.0672	0.0672	0.0672	0.0672	0.0672	0.0672
	$\hat{A}_2$	0.0873	0.0608	0.0712	0.0617	0.1092	0.0597
	A_2/A_1	1.3333	1.3176	1.3176	1.3281	1.3281	1.3125
		Model 3					
e.g.l.s.	RMS_3	1.6052	1.1398	1.2935	1.1151	1.7586	1.0805
($m = 3$)	A_3	0.0511	0.0515	0.0515	0.0511	0.0512	0.0517
	$\hat{A}_3$	0.0742	0.0529	0.0600	0.0515	0.0812	0.0501
	A_3/A_1	1.0139	1.0098	1.0098	1.0099	1.0119	1.0098

Source: Grondona and Cressie (1991).

A solution to the reduced normal equations for treatment effects $\boldsymbol{\alpha}$, under the mth model, is given by

$$\hat{\boldsymbol{\alpha}}_m = (X_a'W_mX_a)^{-}X_a'W_m\mathbf{y}, \tag{5.7.28}$$

where $W_m = \Sigma_m^{-1} - \Sigma_m^{-1}X_b(X_b'\Sigma_m^{-1}X_b)^{-}X_b'\Sigma_m^{-1}$ and G^{-} denotes a generalized inverse of G (e.g., Rao, 1973, p. 24). The true average variance of pairwise treatment differences is obtained through the expression

$$\begin{aligned} A_m &\equiv \frac{1}{t(t-1)}\sum_{i=1}^{t}\sum_{k=1}^{t}\operatorname{var}(\hat{\alpha}_{i,m} - \hat{\alpha}_{k,m}) \\ &= \frac{2}{t-1}\operatorname{tr}\{\Delta(X_a'W_mX_a)^{-}X_a'W_m\Sigma_1W_mX_a(X_a'W_mX_a)^{-}\Delta\}, \end{aligned} \tag{5.7.29}$$

where $\hat{\alpha}_{i,m}$ denotes the ith element of the vector $\hat{\boldsymbol{\alpha}}_m$ and $\Delta \equiv I - (1/t)\mathbf{1}\mathbf{1}'$. The expression (5.7.29) provides a measure of efficiency for evaluation and comparison of the different approaches.

Table 5.4 presents a summary of the analysis under the o.l.s. and g.l.s. approaches implied by models 1, 2, and 3. The estimated average variance $\hat{A}_m$ is given by

$$\hat{A}_m \equiv \frac{1}{t(t-1)} \sum_{i=1}^{t} \sum_{k=1}^{t} \widehat{\text{var}}(\hat{\alpha}_{i,m} - \hat{\alpha}_{k,m})$$
$$= \frac{2}{t-1} \text{tr}\{\Delta(X_a' W_m X_a)^- \Delta\} \text{RMS}_m, \quad (5.7.30)$$

where

$$\text{RMS}_m \equiv \frac{\mathbf{Y}'\left(\Sigma_m^{-1} - \Sigma_m^{-1} X (X' \Sigma_m^{-1} X)^- X' \Sigma_m^{-1}\right)\mathbf{Y}}{n - \text{rank}(X)} \quad (5.7.31)$$

is the residual mean square after fitting (5.7.27) using the mth model ($m = 1, 2, 3$) and $n = 80$. That is, the measure $\hat{A}_m$ is the average variance estimated under the assumption that the mth model yields the true covariance matrix up to an unknown multiplicative constant σ^2, and σ^2 is estimated by the residual mean square. Under o.l.s. estimation ($m = 2$), (5.7.30) reduces to $\hat{A}_2 = \widehat{\text{var}}(\hat{\alpha}_{i,2} - \hat{\alpha}_{k,2}) = (2/r)\text{RMS}_2$, where $r = 4$ is the number of blocks.

The true average variance of pairwise treatment contrasts is A_m, $m = 1, 2, 3$. Notice from Table 5.4 that the true average variance A_2 of the o.l.s. estimator is always larger, about 30% larger, than the true average variance A_3 of the (estimated) g.l.s. estimator. The true average variance A_3 is very close (within 1%) to the true average variance A_1 of the g.l.s. estimator obtained from model 1. Using a randomization criterion, Zimmerman and Harville (1991) also show that considerable efficiency gains are possible by using (REML) estimated g.l.s. estimators of treatment effects (for various agricultural field trials).

The results in Table 5.4 indicate that estimated g.l.s. estimators based on model 3 are as efficient as those based on the true model 1 and 30% more efficient than the o.l.s. estimators based on model 2. That is, an analysis based on an approximate covariance structure is more efficient than one based on independence.

Grondona and Cressie (1991) also give various ways to check the fit of the spatial models. They find that the constant-mean assumption within blocks does not appear to hold for experiments 1 and 5, and suggest that postblocking could be used to remedy this. That is, the use of spatial correlation is not always an alternative to the use of greater local controls, but can be a complement.

5.8 INFILL ASYMPTOTICS

Exact inference results based on a sample $\mathbf{Z} \equiv (Z(\mathbf{s}_1), \ldots, Z(\mathbf{s}_n))'$, of spatial data located at $\{\mathbf{s}_1, \ldots, \mathbf{s}_n\}$, are often not available. Simulation remains a possibility, but many times the statistics in question are not pivotal nor easily made pivotal, so the extent of any simulated results may be too narrow. A common approach taken by statisticians is to derive *asymptotic* inference results by letting sample size n tend to infinity.

When the model has a component of replication in it, the asymptotic calculations involve allowing that component to replicate an infinite number of times. When the data are a time series, an infinite number of observations are obtained by observing the process indefinitely. However, when the data are spatial, there are at least two ways n could tend to infinity.

The first is to allow more and more observations to be taken by increasing the domain of observation $D \subset \mathbb{R}^d$, so that $|D| \to \infty$. For example, if $\inf\{\|\mathbf{s}_i - \mathbf{s}_j\|: 1 \leq i < j \leq n\} > \Delta > 0$, then $n \to \infty$ implies $|D| \to \infty$. Or, one could assume that the number of sample locations per unit area tends to a finite value, as $n \to \infty$. In this book, such asymptotics are called *increasing-domain asymptotics*, and they are the spatial analogue of the usual asymptotics seen in time-series analysis. It is often (although not always) the case that asymptotic inference results based on lattice data (Part II) employ increasing-domain asymptotics; see Section 7.3.1 for further details.

On the other hand, for geostatistical data, the spatial index $\mathbf{s}$ ranges continuously over $D \subset \mathbb{R}^d$, where D is the extent of the phenomenon under study. If one views D as a bounded domain, an obvious way to increase n is to take observations at locations between the existing ones. In mining, this practice is called infill sampling; here, any asymptotics where $n \to \infty$, but $0 < |D| < \infty$, will be called *infill asymptotics*.

In practice, the n data in $\mathbf{Z}$ always come from a finite domain D; how one decides to carry out the asymptotic calculations depends very much on the spatial scale (Section 3.1) of the phenomenon under study. Thus, one could imagine not being concerned about the finite extent of the phenomenon (e.g., D an oil field) when the subject of study is at the microscale (e.g., fluid flow through fractured rock). Nevertheless, infill asymptotics is usually of most relevance to inference for models of geostatistical data.

Estimation of an Invariant Density

Recall from Section 2.3 that if

$$\Pr(Z(\mathbf{s}) \leq z) \equiv F(z), \qquad z \in \mathbb{R}, \tag{5.8.1}$$

is independent of $\mathbf{s}$, then F is called the invariant cumulative distribution function. For example, if the process $Z(\cdot)$ is strongly stationary (Section 2.3), then (5.8.1) holds. Assume henceforth that its density $f(z) = dF(z)/dz$ exists for all $z \in \mathbb{R}$; then $f(\cdot)$ is called the *invariant density*.

It is worth stating explicitly that the goal here is to estimate $F(\cdot)$ or $f(\cdot)$, a model-based quantity. Then, a transformation of the data based on this estimator may follow, such as in trans-Gaussian kriging (Section 3.2.2) or disjunctive kriging (Section 5.1). If one views $z(\cdot)$, the process *observed*, as deterministic, then the goal would be to estimate

$$\bar{F}(z) \equiv \int_D I(z(\mathbf{u}) \leq z)\, d\mathbf{u}/|D|$$

from incomplete knowledge of $z(\cdot)$, namely, $z(\mathbf{s}_1), \ldots, z(\mathbf{s}_n)$. This is a laudable goal that can sometimes be accomplished using a design-based sampling approach. (Section 2.3 briefly compares it to the model-based approach used throughout this book.) However, the design-based approach is not well adapted to Statistics for spatial data, where more complex inferences, concerning direction and strength of spatial dependence, are needed.

Unfortunately, most of the literature on density estimation assumes that the data are independent and identically distributed (i.i.d.) or they are Markov in time or they satisfy weak-dependence (e.g., mixing) conditions. Using increasing-domain asymptotics in the latter two cases, one obtains asymptotic results that are indistinguishable from the independence case (e.g., Roussas, 1969; Bosq, 1973; Ahmad, 1979; Masry, 1983; Rosenblatt, 1985, Chapter VII; Castellana and Leadbetter, 1986). Only Masry (1983) makes a brief comment about how his results are no longer valid if $\inf\{|t_{i+1} - t_i|: i = 1, \ldots, n-1\}$ tends to zero.

To my knowledge, the literature on asymptotics for invariant density estimation has yet to consider, specifically, *efficient* estimation. Clearly, the equal weighting of data (that is natural in the i.i.d. case) is no longer necessarily appropriate for the spatial-dependence case: Imagine a positively spatially dependent process $Z(\cdot)$ on $[0,1] \times [0,1] \subset \mathbb{R}^2$; $(n-1)$ of the observations are located in a tight cluster near the origin, but the remaining observation is near $(1,1)$ and statistically independent of the other $(n-1)$ observations. Intuitively, the positive spatial dependence and tight clustering makes the $(n-1)$ data near $\mathbf{0}$ "worth" about as much as the isolated datum.

One may wish to attach a weight w_i to $Z(\mathbf{s}_i)$, where $0 \leq w_i \leq 1$ and $\sum_{i=1}^n w_i = 1$; the weights are to be used in calculations like estimation of F or f. For example,

$$\hat{F}_{\mathbf{w}}(z) \equiv \sum_{i=1}^{n} w_i I(Z(\mathbf{s}_i) \leq z), \qquad z \in \mathbb{R}, \tag{5.8.2}$$

is an estimator of F. The weights $\mathbf{w} \equiv (w_1, \ldots, w_n)'$ will at least depend on $\{\mathbf{s}_i: i = 1, \ldots, n\}$ and may also depend on $\mathbf{Z}$ and z.

A number of suggestions for choice of weights have been made. Switzer (1977) proposes to minimize $E(\hat{F}_{\mathbf{w}}(\text{med}(Z)) - F(\text{med}(Z)))^2$, where $\text{med}(Z)$ denotes the median of the invariant F. When $Z(\cdot)$ is a Gaussian process, this yields

$$\mathbf{w} \equiv A^{-1}\mathbf{1}/(\mathbf{1}'A^{-1}\mathbf{1}), \tag{5.8.3}$$

where $\mathbf{1} \equiv (1,\cdots,1)'$, A is an $n \times n$ matrix with (i,j)th element equal to $\arcsin(\rho(\mathbf{s}_i, \mathbf{s}_j))$, and $\rho(\mathbf{s}_i, \mathbf{s}_j) \equiv \mathrm{cov}(Z(\mathbf{s}_i), Z(\mathbf{s}_j))/\{\mathrm{var}(Z(\mathbf{s}_i))\mathrm{var}(Z(\mathbf{s}_j))\}^{1/2}$. Another possibility is

$$\mathbf{w} \equiv \Sigma^{-1}\mathbf{1}/(\mathbf{1}'\Sigma^{-1}\mathbf{1}), \tag{5.8.4}$$

which are the weights used in best linear unbiased estimation of the mean $\mu \equiv \int_{-\infty}^{\infty} z\, dF(z)$; Σ has (i,j)th element equal to $\mathrm{cov}(Z(\mathbf{s}_i), Z(\mathbf{s}_j))$.

An *ad hoc*, but sensible proposal, called the cell-declusterizing technique, has been made by Journel (1983); see also Isaaks and Srivastava (1989, pp. 421–428). Divide D into c mutually exclusive, d-dimensional rectangles $R_1, \ldots, R_c$, each of equal volume. Define

$$w_i \propto \left\{ \sum_{k=1}^{c} \sum_{l=1}^{n} I(\mathbf{s}_i \in R_k) I(\mathbf{s}_l \in R_k) \right\}^{-1}, \qquad i = 1, \ldots, n. \tag{5.8.5}$$

Thus, as more observation locations are shared by $\mathbf{s}_i$ in R_k, so the weight decreases; w_i is inversely proportional to the number of such locations in R_k. Deutsch (1989) provides a computer program for finding the weights.

In a small, finite-sample investigation, Switzer's weights (5.8.3) were found to perform better than the others, even when the underlying process was not Gaussian. A study of *asymptotically optimal* weighting would likely resolve the issue of which weighting scheme is to be preferred. However, the asymptotics should probably be more delicate than infill asymptotics: Suppose $n \to \infty$ in such a way that

$$(1/n) \sum_{i=1}^{n} I(\mathbf{s}_i \in A) \to \int_A w(\mathbf{u})\, d\mathbf{u}, \quad \text{for all } A \subset D,$$

where $w(\cdot)$ is a sampling intensity (i.e., density) function. Now, in (5.8.5), suppose that the number of rectangles $c = O(n)$, as $n \to \infty$. Therefore, for n large, (5.8.5) becomes

$$w_i \simeq w(\mathbf{s}_i)^{-1}/c, \qquad i = 1, \ldots, n,$$

and from (5.8.2),

$$\begin{aligned} \hat{F}_{\mathbf{w}}(z) &\simeq \sum_{i=1}^{n} \{w(\mathbf{s}_i)^{-1}/c\} I(Z(\mathbf{s}_i) \le z) \\ &\simeq \int_D I(Z(\mathbf{u}) \le z)\, d\mathbf{u}/|D| = \bar{F}(z). \end{aligned}$$

Indeed, as $n \to \infty$, $\hat{F}_{\mathbf{w}}(z)$ converges in quadratic mean to the *random variable* $\bar{F}(z)$, whose expectation is $E(\bar{F}(z)) = F(z)$. Thus, Journel's weights (5.8.5) yield an approximation to $\bar{F}(z)$, but not necessarily to $F(z)$. When $Z(\cdot)$ is ergodic and D is large, $\bar{F}(z) \simeq F(z)$; the weaker the spatial dependence, the better the approximation.

If more delicate asymptotics are found so that $\hat{F}_{\mathbf{w}}(z) \to F(z)$, then weights $\mathbf{w}$ in (5.8.2) could be sought to optimize the rate of convergence. [The asymptotics in Hardle and Tuan (1986), for the time domain, suggest that a compromise between infill asymptotics and increasing-domain asymptotics may be appropriate.] Finite-sample performance would still have to be investigated.

Estimation of a Conditional Expectation

There is a large literature in nonparametric regression that involves estimation of $E(Y|\mathbf{X} = \mathbf{x})$ based on independent data $\{(\mathbf{X}_i, Y_i): i = 1, \ldots, n\}$ (e.g., Stone, 1977; Georgiev, 1984; Silverman, 1984; Zhao, 1987). It is relatively easy then to adapt these procedures to obtain estimators of the conditional density of $Z(\mathbf{s}_0)$ given $\mathbf{Z}$ and of $E(Z(\mathbf{s}_0)|\mathbf{Z})$, for cases when the process $Z(\cdot)$ is a weakly dependent stationary time series (e.g., Collomb, 1983, 1985; Robinson, 1983; Yakowitz, 1985; Bosq, 1989). The literature on estimation and prediction for dependent data is almost exclusively for time series, and hence little can be found on differential weighting caused by irregular data locations or on infill asymptotics.

Kriging and Infill Asymptotics

Yakowitz and Szidarovsky (1985) consider a random process $Z(\cdot)$ in $\mathbb{R}^d$ that is intrinsically stationary (Section 2.3.1), and hence the variogram $2\gamma(\cdot)$ is well defined. The problem they consider is prediction of $Z(\mathbf{s}_0)$, from data $\mathbf{Z}_n \equiv (Z(\mathbf{s}_1), \ldots, Z(\mathbf{s}_n))'$, using the ordinary kriging predictor $\hat{p}(\mathbf{Z}_n; \mathbf{s}_0)$ given by (3.2.5) and (3.2.12). They show that if $2\gamma(\cdot)$ is known and continuous in a neighborhood of $\mathbf{0}$ and if a limit point of $\{\mathbf{s}_1, \mathbf{s}_2, \ldots\}$ is $\mathbf{s}_0$, then $\hat{p}(\mathbf{Z}_n; \mathbf{s}_0)$ converges in quadratic mean (i.e., in L_2) to $Z(\mathbf{s}_0)$, as $n \to \infty$. Under suitable conditions on the functions $\{f_{j-1}: j = 1, \ldots, p + 1\}$ in (3.4.2), they generalize this result to universal kriging.

The effect of misspecification of the variogram on kriging is considered by Yakowitz and Szidarovsky (1985) and Stein (1988). Stein assumes that $\mathbf{s}_0$ is a limit point of $\{\mathbf{s}_1, \mathbf{s}_2, \ldots\}$, that the random process $Z(\cdot)$ has a trend term given by (3.4.2), and the existence of a (generally nonstationary) covariance function $C(\mathbf{s}, \mathbf{u}) \equiv \mathrm{cov}(Z(\mathbf{s}), Z(\mathbf{u}))$. He then considers two compatible covariance functions C_0 and C_1 [i.e., their respective Gaussian measures are mutually absolutely continuous; see, e.g., Ibragimov and Rozanov (1978), Chapter III].

Let $\hat{p}_0(\mathbf{Z}_n; \mathbf{s}_0)$ denote the universal-kriging predictor $\boldsymbol{\lambda}'\mathbf{Z}_n$, given by (3.4.62), using covariance function C_0; define $\hat{p}_1(\mathbf{Z}_n; \mathbf{s}_0)$ similarly. Assuming C_j to be the true covariance function, the mean-squared prediction errors are

$E_j(\hat{p}_0(\mathbf{Z}_n; \mathbf{s}_0) - Z(\mathbf{s}_0))^2$ and $E_j(\hat{p}_1(\mathbf{Z}_n; \mathbf{s}_0) - Z(\mathbf{s}_0))^2$, respectively, $j = 0, 1$. Stein (1988) proves that if $E_0(\hat{p}_0(\mathbf{Z}_n; \mathbf{s}_0) - Z(\mathbf{s}_0))^2 \to 0$, then

$$\frac{E_0(\hat{p}_0(\mathbf{Z}_n; \mathbf{s}_0) - Z(\mathbf{s}_0))^2}{E_0(\hat{p}_1(\mathbf{Z}_n; \mathbf{s}_0) - Z(\mathbf{s}_0))^2} \to 1, \qquad \frac{E_0(\hat{p}_0(\mathbf{Z}_n; \mathbf{s}_0) - Z(\mathbf{s}_0))^2}{E_1(\hat{p}_0(\mathbf{Z}_n; \mathbf{s}_0) - Z(\mathbf{s}_0))^2} \to 1, \tag{5.8.6}$$

as $n \to \infty$. That is, within equivalence classes of compatible covariances, the universal-kriging predictor is stable with regard to the mean-squared prediction error, provided $E_0(\hat{p}_0(\mathbf{Z}_n; \mathbf{s}_0) - Z(\mathbf{s}_0))^2 \to 0$. This latter condition is equivalent to $\hat{p}_0(\mathbf{Z}_n; \mathbf{s}_0) \to Z(\mathbf{s}_0)$ in L_2, as $n \to \infty$; conditions for this to hold were discussed previously.

It is an open problem to identify all the equivalence classes of compatible covariances. For stationary covariances $C^*(\mathbf{h})$, $\mathbf{h} \in \mathbb{R}^d$, Stein (1988) gives a *necessary* condition in terms of the derivatives of C^* at the origin $\mathbf{0}$. For stationary isotropic covariances $C^o(\|\mathbf{h}\|)$, $\mathbf{h} \in \mathbb{R}^d$, Stein and Handcock (1989) give a sufficient condition in terms of the spectral density. Suppose that covariance functions C_0 and C_1 are obtained from Gaussian random fields $Z_0(\cdot)$ and $Z_1(\cdot)$ in $\mathbb{R}^d$, defined by

$$P_j Z_j(\cdot) = \epsilon_j(\cdot), \qquad j = 0, 1,$$

where $\epsilon_0(\cdot)$ and $\epsilon_1(\cdot)$ are each Gaussian white-noise processes in $\mathbb{R}^d$, and P_0 and P_1 are each elliptic differential operators of order $2p$. Then Inoue (1976) and Arato (1989) give sufficient and necessary conditions, on the order of $P_0 - P_1$, for compatibility of C_0 and C_1.

It should be noted that, although Stein's stability result is comforting, its practical import is limited for two reasons: The result gives no rate-of-convergence result to gauge when n is "large enough." The second (not unrelated) reason is that infill sampling is often not concentrated around just one or even several prediction locations.

Estimation of Covariance Parameters and Infill Asymptotics

In general, the covariance function cannot be determined completely from knowledge of sample functions in a finite region (Doob, 1953, p. 531). Thus, for infill asymptotics, where more and more locations are sampled in a finite region, one might ask which parameters of the covariance function can be estimated consistently and which cannot? Stein (1987b, 1989a) considers a Gaussian process $Z(\cdot)$ with covariance function of the form

$$\mathrm{cov}(Z(\mathbf{s}), Z(\mathbf{u})) \equiv \sum_{j=1}^{k} \theta_j C_j(\mathbf{s}, \mathbf{u}), \tag{5.8.7}$$

where $C_1, \ldots, C_k$ are known covariance functions and $\boldsymbol{\theta} \equiv (\theta_1, \ldots, \theta_k)'$ are parameters to be estimated. He considers minimum norm quadratic (MINQ) estimators of $\boldsymbol{\theta}$ (Section 2.6.1) based on spatial data $Z(\mathbf{s}_1), \ldots, Z(\mathbf{s}_n)$ observed at locations $\{\mathbf{s}_1, \ldots, \mathbf{s}_n\}$. He then proves that, provided $\{C_j: j = 1, \ldots, k\}$ are compatible and $\{\mathbf{s}_1, \mathbf{s}_2, \ldots\}$ is dense in D, the MINQ estimator of $\Sigma_{i=1}^k \theta_i$ is consistent and asymptotically normal with asymptotic variance of order n^{-1}. However, the MINQ estimator of any other linear combination of $\boldsymbol{\theta}$ is not consistent. Interestingly, Stein (1989a) shows that the only linear combination of elements of $\boldsymbol{\theta}$ that can have a nonnegligible (infill) asymptotic impact on kriging is $\Sigma_{i=1}^k \theta_i$. That is, using infill asymptotics, the parameter that can be estimated well is the parameter that it is important to know for spatial prediction.

Under quite general conditions on D, which allow D to be *inter alia* a bounded subset of $\mathbb{R}^d$, Beder (1988) uses compatibility of covariance functions to find conditions under which parameters $\boldsymbol{\theta}$ in

$$C(\mathbf{s}, \mathbf{u}; \boldsymbol{\theta}) \equiv \operatorname{cov}(Z(\mathbf{s}), Z(\mathbf{u})) \tag{5.8.8}$$

can be estimated from data $\{Z(\mathbf{s}): \mathbf{s} \in D\}$. Thus, $\boldsymbol{\theta}$ appears in (5.8.8) in quite a general way, in contrast to (5.8.7). His results should then also hold using infill asymptotics, where $\{\mathbf{s}_1, \mathbf{s}_2, \ldots\}$ is a sequence of sampling locations dense in D.

There are no infill asymptotic results for the important problem where an estimator $\hat{\boldsymbol{\theta}}$ is substituted into the kriging predictor, yielding what might be called an estimated kriging predictor. Some finite-sample results and simulation studies are given by Zimmerman and Cressie (1992); see also Section 5.3.3.

Some Final Remarks

In $\mathbb{R}^1$, *regular* infill sampling in an interval can be achieved by dividing the interval into $(n - 1)$ equal-length subintervals. In $\mathbb{R}^2$, regular infill sampling in a square can be obtained by using quadtrees (e.g., Samet et al., 1984); this generalizes to octrees in $\mathbb{R}^3$ (e.g., Dunstan and Mill, 1989) and so forth.

Although the model (3.4.2) decomposes the data into trend plus error, all of the infill-asymptotic considerations have concentrated on prediction of $Z(\mathbf{s}_0)$ and estimation of covariance parameters $\boldsymbol{\theta}$. Estimation of trend parameters $\boldsymbol{\beta}$ will be inherently problematic [just as the estimation of the invariant cumulative distribution function $F(\cdot)$ is problematic]. This is demonstrated by Morris and Ebey (1984) in the simple case of constant trend μ and infill sampling on the interval $D = [0, 1]$, at locations $t_i = (i - 1)/(n - 1)$, $i = 1, \ldots, n$. When $\operatorname{cov}(Z(t), Z(t + h)) = \theta_1 e^{-\theta_2 |h|}$, $\theta_1 > 0$, $\theta_2 > 0$ (i.e., an exponential covariogram), they show that the estimator $\hat{\mu}_n \equiv \Sigma_{i=1}^n Z(t_i)/n$ deteriorates as $n \to \infty$ [$\operatorname{var}(\hat{\mu}_n) = O(1)$ and is minimized for some finite value n']. This type of result will be unavoidable unless a more delicate type of asymptotics is chosen, such as that used by Hardle and Tuan (1986). (Recall from Section 1.3 that, *for increasing-domain asymptotics*, $\hat{\mu}_n$ is an asymptotically efficient estimator of μ.)

Under infill asymptotics, edge effects are far more worrisome than under increasing-domain asymptotics. Tubilla (1975) gave systematic and stratified spatial sampling results using infill asymptotics, which are different from Matern's (1960) results that were obtained using increasing-domain asymptotics.

For infill asymptotics, the combination of large edge effects and strong spatial dependence between nearby data makes the uncritical use of resampling methods, such as the bootstrap (Section 7.3.2), highly suspect. In classical statistical problems, resampling has proved to be extremely powerful for finite-sample inference. Considerably more research is needed before it can be adapted to the present context.

5.9 THE MANY FACES OF SPATIAL PREDICTION

Basic to geostatistics is prediction of an unknown value (or average of values) of a random function from (possibly noisy) observations on the same random function at known locations. The geostatistical method models the covariation of the random function and then uses it to predict, either linearly or nonlinearly, the unknown value. This (kriging) predictor has an advantage over deterministic predictors in that it adapts to the quantity and quality of spatial dependence demonstrated by the data. Moreover, the mean-squared prediction error (i.e., the kriging variance) quantifies the faith that can be put in the predictor.

Kriging presupposes that the data can be modeled as a random function; however, such an assumption is not universally accepted. Here, I present both stochastic and nonstochastic methods of spatial prediction, and various properties of the methods are discussed briefly. Some overlap with other parts of the book can be expected, but I wanted to bring together the many faces in one place and in a unified notation. A shorter version of this section can be found in Cressie (1989a). Davis (1986, Chapter 5) also reviews various methods of spatial prediction (which he calls mapping).

For the purposes of this section, *spatial prediction* is defined as the prediction of unobserved values from observed data for which the only exogenous variables are their spatial locations. Specifically, suppose that measurements, both actual and potential, are denoted

$$\{z(\mathbf{s}): \mathbf{s} \in D\}, \tag{5.9.1}$$

where $\mathbf{s}$ is a spatial location vector in $D \subset \mathbb{R}^2$ (for the purposes of this section). The index set D gives the extent of the region of interest and $\mathbf{s}$

varies continuously over it. Suppose that data

$$\mathbf{z} \equiv (z(\mathbf{s}_1), \ldots, z(\mathbf{s}_n))' \tag{5.9.2}$$

are observed at known sites

$$\{\mathbf{s}_1, \ldots, \mathbf{s}_n\}. \tag{5.9.3}$$

This section is concerned with the prediction of $z(\mathbf{s}_0)$, an unknown value at a known location $\mathbf{s}_0$. I shall present various suggestions for the predictor

$$p(\mathbf{z}; \mathbf{s}_0), \tag{5.9.4}$$

and briefly critique them. Section 5.9.1 presents methods that are derived from stochastic considerations, whereas Section 5.9.2 presents deterministic-based methods. Some comparisons and final remarks are given in Section 5.9.3.

5.9.1 Stochastic Methods of Spatial Prediction

When studying one particular oilfield or one particular aquifer, it has been argued (Matheron, 1965; Journel, 1985) that there is no random variation in $\{z(\mathbf{s}): \mathbf{s} \in D\}$ at all (apart from measurement error). Then, prediction could be carried out assuming the data $\mathbf{z}$ are a probability sampling from a fixed but unknown $z(\cdot)$ defined by (5.9.1), eschewing a stochastic model for the process. A similar choice between model-based and design-based methods of inference occurs in classical sampling theory [see, e.g., Hansen et al. (1983) and discussion subsequent to it].

The presentation here will be completely model-based, where (5.9.1) will be thought of as a realization from a random process

$$\{Z(\mathbf{s}): \mathbf{s} \in D\}. \tag{5.9.5}$$

The source of the randomness in (5.9.5) is discussed in the introduction to Chapter 3. Inferences should involve the conditional probability measure generated by

$$\{Z(\mathbf{s}): \mathbf{s} \in D\} \quad \text{given} \quad (Z(\mathbf{s}_1), \ldots, Z(\mathbf{s}_n)) = \mathbf{z}'. \tag{5.9.6}$$

The error of the predictor $p(\mathbf{z}; \mathbf{s}_0)$ is

$$z(\mathbf{s}_0) - p(\mathbf{z}; \mathbf{s}_0) \tag{5.9.7}$$

and the mean-squared prediction error is

$$\text{mspe}(p; \mathbf{s}_0) \equiv E(Z(\mathbf{s}_0) - p(\mathbf{Z}; \mathbf{s}_0))^2, \tag{5.9.8}$$

where the expectation is taken over both the random variable $Z(\mathbf{s}_0)$ and the random vector of data

$$\mathbf{Z} \equiv (Z(\mathbf{s}_1), \ldots, Z(\mathbf{s}_n))'. \tag{5.9.9}$$

Suppose that *minimizing* (5.9.8) with respect to predictor p defines the *best* predictor.

It is seen in Chapter 3 that

$$\text{mspe}(p; \mathbf{s}_0) \geq \text{mspe}(p^0; \mathbf{s}_0), \tag{5.9.10}$$

for all measurable p, where

$$p^0(\mathbf{z}; \mathbf{s}_0) \equiv E(Z(\mathbf{s}_0)|\mathbf{z}). \tag{5.9.11}$$

In other words, the optimal predictor is the conditional expectation, which is calculated from the $(n + 1)$-dimensional joint distribution of $Z(\mathbf{s}_0)$, $Z(\mathbf{s}_1), \ldots,$ and $Z(\mathbf{s}_n)$. Unfortunately, it is not possible to estimate this from just n pieces of data, unless some simplifying model assumptions are made. Specifically, if the stochastic process $Z(\cdot)$ is written as

$$Z(\mathbf{s}) = \mu(\mathbf{s}) + \delta(\mathbf{s}), \qquad \mathbf{s} \in D, \tag{5.9.12}$$

where $\delta(\cdot)$ is a zero-mean random process, then these assumptions might refer to the form of $\mu(\cdot)$, the finite-dimensional distributions of $\delta(\cdot)$, or a stationarity condition on $\delta(\cdot)$. Also, simplifying assumptions are often made about the *form* of $p(\mathbf{z}; \mathbf{s}_0)$; for example, unbiased, linear, and so on. Optimal (i.e., minimum mean-squared prediction error) predictors are then sought from the restricted class satisfying these predictor assumptions.

In what is to follow, various suggestions for the predictor p will be made and critiqued. Coupled with the statistical criterion of making (5.9.8) small, p will be assessed according to various pragmatic criteria such as simplicity, ease of computation, resistance, and so forth. Notice that exact interpolation is a desirable property of the spatial predictor when the data exhibit no measurement error (Cressie, 1988b). Throughout this section, the goal will be to predict $Z(\mathbf{s}_0)$, although if measurement error is present one should predict the noiseless version of $Z(\mathbf{s}_0)$.

For each predictor, a statement will be made regarding what parameters have to be known. In practice, they will be estimated, usually from the original data $\mathbf{z}$. Only scant details on their estimation will be given here; further information is available from the sources referenced.

Simple Kriging (Wold, 1938; Kolmogorov, 1941b; Wiener, 1949; Matheron, 1962; Section 3.4.5)

- *Model Assumptions.* In (5.9.12), $\mu(\mathbf{s})$ is known, as is

$$C(\mathbf{s},\mathbf{u}) \equiv \text{cov}(Z(\mathbf{s}), Z(\mathbf{u})),\ \mathbf{s},\mathbf{u} \in D.$$

- *Predictor Assumptions.* p given by (5.9.4) is heterogeneously linear in $\mathbf{z}$; that is,

$$p(\mathbf{z};\mathbf{s}_0) = \sum_{i=1}^{n} l_i z(\mathbf{s}_i) + k.$$

- *Optimal Predictor.*

$$p(\mathbf{z};\mathbf{s}_0) = \mu(\mathbf{s}_0) + \mathbf{c}'\Sigma^{-1}(\mathbf{z} - \boldsymbol{\mu}), \qquad (5.9.13)$$

where $\mathbf{c} \equiv (C(\mathbf{s}_0,\mathbf{s}_1),\ldots,C(\mathbf{s}_0,\mathbf{s}_n))'$, Σ is a symmetric $n \times n$ matrix whose (i,j)th element is $C(\mathbf{s}_i,\mathbf{s}_j)$, and $\boldsymbol{\mu} \equiv (\mu(\mathbf{s}_1),\ldots,\mu(\mathbf{s}_n))'$. Notice that (5.9.13) is unbiased, although that was not a predictor assumption.

- *Properties.*

1. $$\text{mspe}(p;\mathbf{s}_0) = C(\mathbf{s}_0,\mathbf{s}_0) - \sum_{i=1}^{n} \lambda_i C(\mathbf{s}_0,\mathbf{s}_i), \qquad (5.9.14)$$

   where $(\lambda_1,\ldots,\lambda_n) = \mathbf{c}'\Sigma^{-1}$.
2. Knowledge of the mean function $\mu(\cdot)$ and the covariance function $C(\cdot,\cdot)$ is needed.
3. Prediction at $\mathbf{s}_0$ requires inversion of an $n \times n$ matrix.
4. The predictor p is optimal among all heterogeneously linear unbiased predictors. When $\delta(\cdot)$ is assumed to be a Gaussian process, p coincides with $E(Z(\mathbf{s}_0)|\mathbf{z})$, so in this case p is best among all predictors.
5. Because the data appear linearly in (5.9.13), the predictor p is not resistant to outliers (e.g., caused by independent contamination of assumed Gaussian errors).
6. $p(\mathbf{z};\mathbf{s}_i) = z(\mathbf{s}_i)$; that is, p is an exact interpolator.

In general, the mean and covariance structure are not known, in which case the simple kriging predictor is not practically useful. Some simplifying assumptions can be made that bring it into the realm of applicability.

Ordinary Kriging (Matheron, 1962; Gandin, 1963; Section 3.2)

- *Model Assumptions.* In (5.9.12), $\mu(\mathbf{s}) \equiv \mu$, where μ is unknown and $\delta(\cdot)$ is intrinsically stationary; that is,

$$\text{var}(\delta(\mathbf{s}+\mathbf{h}) - \delta(\mathbf{s})) = 2\gamma(\mathbf{h}), \quad \text{for all } \mathbf{s}, \mathbf{s}+\mathbf{h} \in \mathbf{D}. \quad (5.9.15)$$

  The quantity $2\gamma(\cdot)$ is known as the variogram (Section 2.3.1).
- *Predictor Assumptions.* p given by (5.9.4) is homogeneously linear in $\mathbf{z}$ and is uniformly unbiased; that is,

$$E(p(\mathbf{Z};\mathbf{s}_0)) = \mu, \quad \text{for all } \mu \in \mathbb{R}.$$

  This latter assumption can be replaced by uniform equivariance under location and scale change; that is, $p(a\mathbf{1} + b\mathbf{z};\mathbf{s}_0) = a + bp(\mathbf{z};\mathbf{s}_0)$, for all $a \in \mathbb{R}$ and all $b > 0$.
- *Optimal Predictor.*

$$p(\mathbf{z};\mathbf{s}_0) = \sum_{i=1}^{n} \lambda_i z(\mathbf{s}_i), \quad (5.9.16)$$

  where coefficients $\boldsymbol{\lambda} \equiv (\lambda_1, \ldots, \lambda_n)'$ are given by

$$\boldsymbol{\lambda}' = \left(\boldsymbol{\gamma} + \mathbf{1}\frac{(1 - \mathbf{1}'\Gamma^{-1}\boldsymbol{\gamma})}{\mathbf{1}'\Gamma^{-1}\mathbf{1}}\right)'\Gamma^{-1}; \quad (5.9.17)$$

  Γ is an $n \times n$ matrix whose (i,j)th element is $\gamma(\mathbf{s}_i - \mathbf{s}_j)$, $\boldsymbol{\gamma} \equiv (\gamma(\mathbf{s}_0 - \mathbf{s}_1), \ldots, \gamma(\mathbf{s}_0 - \mathbf{s}_n))'$, and $\mathbf{1} \equiv (1, \cdots, 1)'$.
- *Properties.*

1. $$\text{mspe}(p;\mathbf{s}_0) = \sum_{i=1}^{n} \lambda_i \gamma(\mathbf{s}_0 - \mathbf{s}_i) + m; \quad (5.9.18)$$

   m is a Lagrange multiplier that ensures uniform unbiasedness through $\sum_{i=1}^{n}\lambda_i = 1$. Specifically, $m = -(1 - \mathbf{1}'\Gamma^{-1}\boldsymbol{\gamma})/(\mathbf{1}'\Gamma^{-1}\mathbf{1})$.
2. Knowledge of the variogram $2\gamma(\cdot)$ is needed; under an ergodic assumption (Section 2.3), this can be estimated from one realization of $Z(\cdot)$.
3. Prediction at $\mathbf{s}_0$ requires inversion of an $n \times n$ matrix.
4. The predictor p is optimal among all homogeneously linear unbiased predictors.

5–6. Properties are the same as for simple kriging. With regard to Property 5, Hawkins and Cressie (1984) have proposed a way of robustifying ordinary kriging (Section 3.3).

Notice that the constant mean μ need not be estimated. Sometimes, however, it is more realistic to assume that the mean is an unknown linear combination of known functions.

Universal Kriging (Goldberger, 1962; Matheron, 1969; Section 3.4)

- *Model Assumptions.* In (5.9.12),

$$\mu(\mathbf{s}) = \sum_{j=1}^{p+1} \beta_{j-1} f_{j-1}(\mathbf{s}), \tag{5.9.19}$$

$f_0(\cdot) \equiv 1$, and $\delta(\cdot)$ is intrinsically stationary, as defined by (5.9.15).
- *Predictor Assumptions.* p given by (5.9.4) is linear in $\mathbf{z}$ and is uniformly unbiased for $\boldsymbol{\beta} \equiv (\beta_0, \dots, \beta_p)'$ in a nontrivial region of $\mathbb{R}^{p+1}$; that is,

$$E(p(\mathbf{Z}; \mathbf{s}_0)) \equiv \sum_{j=1}^{p+1} \beta_{j-1} f_{j-1}(\mathbf{s}_0). \tag{5.9.20}$$

- *Optimal Predictor.*

$$p(\mathbf{z}; \mathbf{s}_0) = \sum_{i=1}^{n} \lambda_i z(\mathbf{s}_i), \tag{5.9.21}$$

where coefficients $\boldsymbol{\lambda} \equiv (\lambda_1, \dots, \lambda_n)'$ are given by

$$\boldsymbol{\lambda}' = \left\{\boldsymbol{\gamma} + X(X'\Gamma^{-1}X)^{-1}(\mathbf{x} - X'\Gamma^{-1}\boldsymbol{\gamma})\right\}'\Gamma^{-1}. \tag{5.9.22}$$

In (5.9.22), $2\gamma(\cdot)$ is the variogram of $\delta(\cdot)$ [and hence of $Z(\cdot)$], X is an $n \times p$ matrix whose (i, j)th element is $f_{j-1}(\mathbf{s}_i)$, and $\mathbf{x} \equiv (f_0(\mathbf{s}_0), \dots, f_p(\mathbf{s}_0))'$.
- *Properties.*

1. $$\text{mspe}(p; \mathbf{s}_0) = \sum_{i=1}^{n} \lambda_i \gamma(\mathbf{s}_0 - \mathbf{s}_i) + \sum_{j=1}^{p+1} m_{j-1} f_{j-1}(\mathbf{s}_0); \tag{5.9.23}$$

   $\mathbf{m} \equiv (m_0, \dots, m_p)'$ are Lagrange multipliers that ensure uniform unbiasedness through

$$\sum_{i=1}^{n} \lambda_i f_{j-1}(\mathbf{s}_i) = f_{j-1}(\mathbf{s}_0), \qquad j = 1, \dots, p+1. \tag{5.9.24}$$

Specifically,

$$\mathbf{m}' = -(\mathbf{x} - X'\Gamma^{-1}\boldsymbol{\gamma})'(X'\Gamma^{-1}X)^{-1}.$$

2. Knowledge of the variogram of $\delta(\cdot)$ is needed.

3–6. Properties are the same as for ordinary kriging.

If $\delta(\cdot)$ is in fact a white-noise process, then $2\gamma(\mathbf{h})$ is constant for $\mathbf{h} \neq \mathbf{0}$, and the best linear unbiased predictor (5.9.21) reduces to

$$p(\mathbf{z}; \mathbf{s}_0) = \mathbf{x}'(X'X)^{-1}X'\mathbf{z}.$$

By introducing a nonstationary mean, a larger class of models is obtained at the price of not being able to estimate $2\gamma(\cdot)$ directly. Because $\beta_0, \dots, \beta_p$ are unknown parameters, their estimates yield residuals $\hat{\delta}(\cdot)$ with different correlation structure than the errors $\delta(\cdot)$; Cressie (1986) has a discussion of this problem of biased estimation of the variogram and suggests preliminary estimation of $\mu(\cdot)$ using a median-based technique.

Median-Polish Kriging (Cressie, 1986; Section 3.5)

- *Model Assumptions.* In (5.9.12), $\delta(\cdot)$ is intrinsically stationary, as defined by (5.9.15).
- *Predictor Assumptions.* p given by (5.9.4) is nonlinear in $\mathbf{z}$. Specifically,

$$p(\mathbf{z}; \mathbf{s}_0) = \tilde{\mu}(\mathbf{s}_0) + \sum_{i=1}^{n} \lambda_i R(\mathbf{s}_i),$$

where $\tilde{\mu}(\cdot)$ is the median-polish surface given by the right-hand side of (5.9.50) and $R(\cdot) \equiv Z(\cdot) - \tilde{\mu}(\cdot)$.

- *"Optimal" Predictor.*

$$p(\mathbf{z}; \mathbf{s}_0) = \tilde{\mu}(\mathbf{s}_0) + \hat{R}(\mathbf{s}_0), \qquad (5.9.25)$$

where $\hat{R}(\mathbf{s}_0)$ is the ordinary kriging predictor (5.9.16) based on the residuals $R(\mathbf{s}_1), \dots, R(\mathbf{s}_n)$. [Use the fitted variogram, obtained from $\{R(\mathbf{s}_i)\}$, to compute the "optimal" $\lambda_1, \dots, \lambda_n$.]

- *Properties.*

1.

$$\text{mspe}(p; \mathbf{s}_0) \simeq \sum_{i=1}^{n} \lambda_i \gamma(\mathbf{s}_0 - \mathbf{s}_i) + m;$$

see (5.9.18) for the definition of $\lambda_1, \dots, \lambda_n$, and m.

2. Knowledge of the grid nodes in $\tilde{\mu}(\cdot)$ is needed.

3. Prediction at $\mathbf{s}_0$ requires inversion of an $n \times n$ matrix.
4. When $\tilde{\mu}(\cdot) \simeq \mu(\cdot)$, the predictor p is approximately optimal.
5. Although $\tilde{\mu}(\cdot)$ is resistant to outliers, the predictor p is not, unless ordinary kriging of $R(\cdot)$ is robustified (Section 3.3).
6. p is an exact interpolator.

The median-polish surface $\tilde{\mu}(\cdot)$ is a very flexible continuous surface consisting of intersecting planes (see, e.g., Figure 4.5). It is based upon assigning nearby data to nodes of a (possibly irregularly spaced) rectangular grid. Thus, the data themselves do *not* have to be located on the grid. The choice of grid will have some effect on the predictor; as always, there is a trade-off between bias and variance, because the closer $\tilde{\mu}(\cdot)$ is to $\mu(\cdot)$, the more the median-polish-kriging predictor looks like the simple kriging predictor.

Rather than estimating $\mu(\cdot)$ and spatially predicting the residuals $R(\cdot)$, one could model the error $\delta(\cdot)$ as an intrinsic random function.

Kriging with Intrinsic Random Functions (Matheron, 1973; Delfiner, 1976; Cressie, 1987; Section 5.4)

- *Model Assumptions.* In (5.9.12), $\delta(\cdot)$ is a zero-mean intrinsic random function of order k (IRFk) with generalized covariance $K(\mathbf{h})$, $\mathbf{h} \in \mathbb{R}^2$, and $\mu(\cdot)$ is given by (5.9.19), where $f_0, \ldots, f_p$ are mixed monomials of degree less than or equal to k.
- *Predictor Assumptions.* p given by (5.9.4) is linear in $\mathbf{z}$ and its coefficients satisfy (5.9.24), so guaranteeing unbiasedness.
- *Optimal Predictor.*

$$p(\mathbf{z}; \mathbf{s}_0) = \sum_{i=1}^{n} \lambda_i z(\mathbf{s}_i), \tag{5.9.26}$$

where coefficients $\boldsymbol{\lambda} \equiv (\lambda_1, \ldots, \lambda_n)'$ solve (5.9.22), but with $-\gamma(\cdot)$ in (5.9.22) replaced with $K(\cdot)$, the generalized covariance.

- *Properties.*

1. $$\operatorname{mspe}(p; \mathbf{s}_0) = K(\mathbf{0}) - \sum_{i=1}^{n} \lambda_i K(\mathbf{s}_0 - \mathbf{s}_i) + \sum_{j=1}^{p+1} m_{j-1} f_{j-1}(\mathbf{s}_0), \tag{5.9.27}$$

where $\mathbf{m} \equiv (m_0, \ldots, m_p)'$ are Lagrange multipliers that ensure uniform unbiasedness. Specifically,

$$\mathbf{m}' = (\mathbf{x} - X'K^{-1}\mathbf{k})'(X'K^{-1}X)^{-1},$$

where K and $\mathbf{k}$ are analogous to Γ and $\boldsymbol{\gamma}$, that is, with entries $K(\mathbf{h})$ in place of $-\gamma(\mathbf{h})$.

2. Knowledge of the generalized covariance is needed. Usually, a parametric form is assumed (Delfiner, 1976), although nonparametric estimation is possible (Cressie, 1987).

3–6. Properties are the same as for universal kriging.

The generalized covariance is more general than the variogram and allows spatial prediction to be carried out for processes that do not possess a variogram. One property that illustrates this is

$$K(\mathbf{h}) = o(\|\mathbf{h}\|^{2k+2}), \qquad \|\mathbf{h}\| \text{ large.} \tag{5.9.28}$$

Notice that $\gamma(\mathbf{h}) = o(\|\mathbf{h}\|^2)$. In fact an IRF0 is precisely an intrinsically stationary process with $K(\mathbf{h}) = -\gamma(\mathbf{h})$.

Markov-Random-Field Prediction (Besag, 1974; Cressie and Chan, 1989; Haining et al., 1989)

- *Model Assumptions.* In (5.9.12), $\delta(\cdot)$ is assumed to be a Gaussian process such that $(Z(\mathbf{s}_0), Z(\mathbf{s}_1), \ldots, Z(\mathbf{s}_n))$ constitutes a Markov random field with distance-based neighborhood structure (see following text), known mean function $\mu(\cdot)$, and homogeneous conditional variance σ^2.
- *Predictor Assumptions.* From the Gaussian model assumption, p given by (5.9.4) is linear in $\mathbf{z}$. Assume that the coefficient of $Z(\mathbf{s}_i)$ is proportional to $h(d_{0,i})$, a known decreasing function of distance between $\mathbf{s}_0$ and $\mathbf{s}_i$.
- *Optimal Predictor.*

$$p(\mathbf{z}; \mathbf{s}_0) = \mu(\mathbf{s}_0) + \theta \sum_{i=1}^{n} h(d_{0,i})\{z(\mathbf{s}_i) - \mu(\mathbf{s}_i)\}, \tag{5.9.29}$$

where θ is a spatial dependence parameter, $\mu(\cdot)$ is given in (5.9.12), and $d_{0,i} \equiv \|\mathbf{s}_0 - \mathbf{s}_i\|$ is the distance between locations $\mathbf{s}_0$ and $\mathbf{s}_i$. Some common examples of $h(\cdot)$ are

$$h(d_{0,i}) = d_{0,i}^{-k} I(d_{0,i} \le r), \qquad k = 0, 1, 2, \tag{5.9.30}$$

where the radius r defines the neighborhood structure for the Gaussian Markov random field.

- *Properties.*

1. $$\text{mspe}(p; \mathbf{s}_0) = \sigma^2 \mathbf{l}'(I - \theta H)^{-1}\mathbf{l}, \tag{5.9.31}$$

   where $\mathbf{l} \equiv (-1, \theta h(d_{0,1}), \ldots, \theta h(d_{0,n}))'$ and the (i, j)th element of H is $h(d_{i-1,j-1})$, $i, j = 1, \ldots, n+1$.

2. Knowledge of $\mu(\cdot)$ is needed. Also, knowledge of small-scale-variation parameters θ and σ^2 is needed. Under the assumption that predictand and data $(Z(\mathbf{s}_0), \ldots, Z(\mathbf{s}_n))$ together constitute a Gaussian Markov random field with $\text{var}(Z(\mathbf{s}_i)|\{Z(\mathbf{s}_j)\colon j \neq i\}) = \sigma^2$, for $i = 0, \ldots, n$, the joint distribution of $\mathbf{Z} = (Z(\mathbf{s}_1), \ldots, Z(\mathbf{s}_n))'$ is Gaussian with mean vector $\boldsymbol{\mu} \equiv (\mu(\mathbf{s}_1), \ldots, \mu(\mathbf{s}_n))'$ and variance matrix $\Sigma(\theta, \sigma^2)$, whose (i, j)th element is

   $$\text{cov}\big(Z(\mathbf{s}_i), Z(\mathbf{s}_j)\big) = \sigma^2 (I - \theta H)^{i+1, j+1}, \quad i, j = 1, \ldots, n, \tag{5.9.32}$$

   and A^{kl} denotes the (k, l)th element of A^{-1}. Thus,

   $$\mathbf{Z} \sim \text{Gau}\big(\boldsymbol{\mu}, \Sigma(\theta, \sigma^2)\big), \tag{5.9.33}$$

   which means that θ and σ^2 can be estimated by maximizing the likelihood obtained from (5.9.33). Another, simpler way of estimating θ and σ^2 is by conditional least squares (e.g., Klimko and Nelson, 1978).

3. Assuming θ is known, prediction at $\mathbf{s}_0$ requires no inversion of large-order matrices; coefficients are proportional to a known function of distance. When θ is unknown, its maximum likelihood estimation requires one $n \times n$ matrix inversion (e.g., Rao, 1973, p. 33) for each evaluation of $\Sigma(\theta, \sigma^2)^{-1}$ or $\Sigma(\theta, \sigma^2)$.

4. Because $\delta(\cdot)$ is assumed to be a Gaussian process, the predictor p is best among all predictors.

5–6. Properties are the same as for simple kriging.

This method, where conditional distributions are modeled, contrasts with the previous methods where models are built for joint distributions and the parameters of the conditional distributions are derived by inversion. However, (5.9.29) is not well adapted to geostatistical data, where there are potentially many $Z(\mathbf{s}_{0,1}), Z(\mathbf{s}_{0,2}), \ldots$ to predict. Then, a Gaussian Markov random-field assumption needs to be made about the joint distribution of $\mathbf{Z}_0 \equiv (Z(\mathbf{s}_{0,1}), Z(\mathbf{s}_{0,2}), \ldots)'$ and $Z(\mathbf{s}_1), \ldots, Z(\mathbf{s}_n)$, from which $E(\mathbf{Z}_0|\{Z(\mathbf{s}_i)\colon i = 1, \ldots, n\})$ and the conditional variance matrix can be computed. Haining et al. (1989) apply the prediction method to find missing data on a lattice.

Trans-Gaussian Kriging (Section 3.2.2)

- *Model Assumptions.* $Z(\mathbf{s}) = \phi(G(\mathbf{s}))$, $\mathbf{s} \in D$, where $G(\cdot)$ is a Gaussian process satisfying (5.9.12). To keep the development simple, assume that $G(\cdot) = \mu + \delta(\cdot)$, an intrinsically stationary random function with $\text{var}(G(\mathbf{s}) - G(\mathbf{u})) = 2\gamma(\mathbf{s} - \mathbf{u})$, $\mathbf{s}, \mathbf{u} \in D$, known.
- *Predictor Assumptions.* p given by (5.9.4) is an approximately unbiased, back-transformed predictor that is linear in the transformed process $G(\cdot)$.
- *"Optimal" Predictor.*

$$p(\mathbf{z}; \mathbf{s}_0) = \phi\left(\sum_{i=1}^{n} \lambda_i \phi^{-1}(z(\mathbf{s}_i))\right) + \phi''(\hat{\mu})\left\{(\sigma_k^2(\mathbf{s}_0)/2) - m\right\}, \quad (5.9.34)$$

where $\lambda_1, \ldots, \lambda_n$, and m are given in (5.9.17) and (5.9.18), $\sigma_k^2(\mathbf{s}_0)$ is the kriging variance given by (5.9.18), $\hat{\mu} = \mathbf{1}'\Sigma^{-1}\mathbf{z}/(\mathbf{1}'\Sigma^{-1}\mathbf{1})$, and $\Sigma = \text{var}(\mathbf{Z})$.

- *Properties.*

1. $$\text{mspe}(p; \mathbf{s}_0) \simeq \left\{\sum_{i=1}^{n} \lambda_i \gamma(\mathbf{s}_0 - \mathbf{s}_i) + m\right\}\{\phi'(\mu)\}^2. \quad (5.9.35)$$
2. Knowledge of the transformation ϕ is needed, as well as the variogram $2\gamma(\cdot)$ of the *transformed* process $\{\phi^{-1}(Z(\mathbf{s})): \mathbf{s} \in D\}$.
3. Prediction at $\mathbf{s}_0$ requires inversion of an $n \times n$ matrix.
4. Because $E(Z(\mathbf{s}_0)|Z(\mathbf{s}_1), \ldots, Z(\mathbf{s}_n))$ depends on the unknown μ, the predictor p is only an approximation to the optimal predictor.
5. The predictor p is not, in general, resistant to outliers, unless the transformation ϕ^{-1} has a dampening effect on unusually large or unusually small observations.
6. p is an exact interpolator, because the bias correction in (5.9.34) is zero when $\mathbf{s}_0 = \mathbf{s}_i$.

The special case $\phi(x) = \exp(x)$, $-\infty < x < \infty$, yields the familiar lognormal kriging for which the expressions (5.9.34) and (5.9.35) have exact versions (Section 3.2.2).

Disjunctive Kriging (Matheron, 1976a; Section 5.1)

- *Model Assumptions.* $Z(\mathbf{s}) = \phi(Y(\mathbf{s}))$, $\mathbf{s} \in D$, where $Y(\cdot)$ is a stationary process whose univariate distribution is unit Gaussian and whose bivariate distributions are Gaussian.

- *Predictor Assumptions.* p given by (5.9.4) is a sum of measurable univariate functions; that is,

$$p(\mathbf{z};\mathbf{s}_0) = \sum_{i=1}^{n} f_i(z(\mathbf{s}_i)),$$

where $E(f_i(Z(\mathbf{s}_i))^2) < \infty$, $i = 1,\ldots,n$.

- *Optimal Predictor.*

$$p(\mathbf{z};\mathbf{s}_0) = \sum_{i=1}^{n} f_i(z(\mathbf{s}_i)), \tag{5.9.36}$$

where $\{f_i\colon i = 1,\ldots,n\}$ satisfy the system of equations

$$\sum_{i=1}^{n} E\big(f_i(Z(\mathbf{s}_i))|Z(\mathbf{s}_j)\big) = E\big(Z(\mathbf{s}_0)|Z(\mathbf{s}_j)\big), \qquad j = 1,\ldots,n. \tag{5.9.37}$$

Equations (5.9.37) cannot be solved in general. Instead use Hermite polynomial expansions (see, e.g., Beckmann, 1973, Chapter 3) for $h_i(\cdot) \equiv f_i(\phi(\cdot))$ and for $\phi(\cdot)$, and truncate at the Kth term:

$$\begin{aligned} h_i(y) &= \sum_{k=0}^{K} a_{ik}\eta_k(y), \qquad i = 1,\ldots,n, \\ \phi(y) &= \sum_{k=0}^{K} b_k\eta_k(y). \end{aligned} \tag{5.9.38}$$

In (5.9.38), $\{\eta_k(\cdot)\colon k = 0,1,2,\ldots\}$ are the Hermite polynomials, having the orthonormality property that

$$\int_{-\infty}^{\infty} \eta_k(y)\eta_j(y)(2\pi)^{-1/2}e^{-y^2/2}\,dy = 0, \qquad j \neq k = 0,1,2,\ldots,$$

and

$$\int_{-\infty}^{\infty} \eta_k^2(y)(2\pi)^{-1/2}e^{-y^2/2}\,dy = 1, \qquad k = 0,1,2,\ldots.$$

That is,

$$\eta_k(y) = (k!)^{-1/2}\exp(y^2/2)d^k\{\exp(-y^2/2)\}/dy^k.$$

Equations (5.9.37) then reduce to solving K sets of n equations in n unknowns: For $k = 1,\ldots,K$, solve for $\{a_{ik}\colon i = 1,\ldots,n\}$ in

$$\sum_{i=1}^{n} a_{ik}\rho_{i,j}^k = b_k\rho_{0,j}^k, \qquad j = 1,\ldots,n, \tag{5.9.39}$$

where $\rho_{i,j} \equiv \operatorname{cov}(Y(\mathbf{s}_i), Y(\mathbf{s}_j))/\{\operatorname{var}(Y(\mathbf{s}_i))\operatorname{var}(Y(\mathbf{s}_j))\}^{1/2}$, which here is simply $E(Y(\mathbf{s}_i)Y(\mathbf{s}_j))$. The approximate optimal predictor is

$$p(\mathbf{z};\mathbf{s}_0) = \sum_{i=1}^{n}\sum_{k=0}^{K} a_{ik}\eta_k(\phi^{-1}(z(\mathbf{s}_i))). \tag{5.9.40}$$

- *Properties.*
  1. For p given by (5.9.40),

$$\text{mspe}(p;\mathbf{s}_0) \simeq \sum_{k=0}^{K} b_k^2 - \sum_{k=0}^{K} b_k \sum_{i=1}^{n} a_{ik}\rho_{0,i}^k. \tag{5.9.41}$$

  2. Knowledge of the transformation ϕ is needed [recall $Y(\mathbf{s})$ is a *unit* Gaussian random variable]. One also needs to know the pairwise correlation coefficients $\{\rho_{i,j}: 0 \le i < j \le n\}$ of the $Y(\cdot)$ process, and the truncation value K has to be specified [e.g., choose a K so that $\sum_{k=0}^{K} b_k^2 \simeq \text{var}(Z(\mathbf{s}))$].
  3. Prediction at $\mathbf{s}_0$ requires inversion of $K\ n \times n$ matrices.
  4. When $Z(\cdot)$ is Gaussian, so that $\phi(y) = \mu + \sigma y$, the predictor p coincides with $E(Z(\mathbf{s}_0)|\mathbf{Z})$, and so it is the optimal predictor.
  5. The predictor p is not, in general, resistant to outliers, unless the transformation ϕ^{-1} has a dampening effect on unusually large or unusually small observations.
  6. p is an exact interpolator.

Approximations to the equations (5.9.37) are not limited to Hermitian expansions; provided appropriate univariate and bivariate assumptions are made about $Y(\cdot)$, the larger class of *isofactorial models* can be used [see, e.g., Armstrong and Matheron (1986a, 1986b) and Section 5.1]. Regardless, practical difficulties with prediction of $Z(\mathbf{s}_0)$ by disjunctive kriging still remain, in particular with the estimation of ϕ. Not only is an assumption of an invariant distribution required [i.e., $\Pr(Z(\mathbf{s}) \le z)$ does not depend on $\mathbf{s}$], but the distribution has then to be estimated from *dependent* observations $\mathbf{z}$. Some discussion of this difficult problem is given in Section 5.8. [Notice that the mean-squared prediction error in (5.9.41) does not take the estimation of ϕ into account.]

Bayesian Nonparametric Smoothing (Weerahandi and Zidek, 1988; Cressie, 1989a)

- *Model Assumptions.* For every $\mathbf{s}$, the second-order stationary process $Z(\mathbf{s})$ is locally expandable, about predictor location $\mathbf{s}_0$, in a Taylor series up to order k. For the purpose of presentation, assume $k = 2$. Then,

$$\mathbf{Z} = X\boldsymbol{\beta} + \boldsymbol{\nu}, \tag{5.9.42}$$

where $\boldsymbol{\beta}$ is a 6×1 matrix with $\beta_0 = Z(\mathbf{s}_0)$, $\beta_1 = \partial Z(\mathbf{s}_0)/\partial x_0$, $\beta_2 = \partial Z(\mathbf{s}_0)/\partial y_0$, $\beta_3 = \partial^2 Z(\mathbf{s}_0)/\partial x_0^2$, $\beta_4 = \partial^2 Z(\mathbf{s}_0)/\partial x_0\, \partial y_0$, and $\beta_5 = \partial^2 Z(\mathbf{s}_0)/\partial y_0^2$; X is an $n \times 6$ matrix whose ith row is $(1, x_i - x_0, y_i - y_0, (x_i - x_0)^2/2, (x_i - x_0)(y_i - y_0), (y_i - y_0)^2/2)$; and $\boldsymbol{\nu} \equiv (\nu(\mathbf{s}_1), \ldots, \nu(\mathbf{s}_n))'$ is an $n \times 1$ vector whose entries are the remainder terms in the Taylor series expansions of $Z(\mathbf{s}_i)$ about $\mathbf{s}_0$. Now, put a prior distribution on $\boldsymbol{\beta}$ and $\boldsymbol{\nu}$:

$$\boldsymbol{\beta} \sim \text{Gau}(\boldsymbol{\beta}_0, U), \quad \text{independent from } \boldsymbol{\nu} \sim \text{Gau}\big(\mathbf{0}, \sigma^2 H(\mathbf{s}_0)\big), \qquad (5.9.43)$$

where the elements of U come from the relation,

$$\text{cov}\left(\frac{\partial^{i+j} Z(\mathbf{s}_0)}{\partial x_0^i\, \partial y_0^j}, \frac{\partial^{k+l} Z(\mathbf{s}_0)}{\partial x_0^k\, \partial y_0^l} \right) = (-1)^{k+l} \left. \frac{\partial^{i+j+k+l} C(h_1, h_2)}{\partial h_1^{i+k}\, \partial h_2^{j+l}} \right|_{h_1=0,\, h_2=0},$$

$C(\mathbf{h})$ is the covariogram of $Z(\cdot)$, and $H(\mathbf{s}_0) \equiv \text{diag}(1 + \tau\|\mathbf{s}_0 - \mathbf{s}_1\|^6, \ldots, 1 + \tau\|\mathbf{s}_0 - \mathbf{s}_n\|^6)$, for some $\tau \in [0, \infty]$.

- *Predictor Assumptions.* Under the model (5.9.42) and (5.9.43), the posterior distribution of $\boldsymbol{\beta}$ given $\mathbf{Z}$ is Gaussian. Denote the posterior expectation $E(\boldsymbol{\beta}|\mathbf{Z})$ as the predictor, $\mathbf{p}(\mathbf{z}; \mathbf{s}_0)$. Assume $U^{-1} \to 0$, so that the prior for $\boldsymbol{\beta}$ is diffuse.
- *Predictor.*

$$p(\mathbf{z}; \mathbf{s}_0) = \text{first element of } \mathbf{p}(\mathbf{z}; \mathbf{s}_0), \qquad (5.9.44)$$

where $\mathbf{p}(\mathbf{z}; \mathbf{s}_0) = (X'H(\mathbf{s}_0)^{-1}X)^{-1}X'H(\mathbf{s}_0)^{-1}\mathbf{z}$; the second, third, and so forth, elements of $\mathbf{p}(\mathbf{z}; \mathbf{s}_0)$ yield predictors of the derivatives of $Z(\cdot)$ at $\mathbf{s}_0$.

- *Properties.*
  1. The mean-squared prediction error is the $(1,1)$th element of $\sigma^2(X'H(\mathbf{s}_0)^{-1}X)^{-1}$, and the posterior variance $\text{var}(\boldsymbol{\beta}|\mathbf{Z}) = \sigma^2(X'H(\mathbf{s}_0)^{-1}X)^{-1}$ allows construction of credibility regions for $\boldsymbol{\beta}$.
  2. Knowledge of k, σ^2, and τ is needed. Weerahandi and Zidek (1988) give a number of suggestions for estimation of σ^2 and τ.
  3. Prediction at $\mathbf{s}_0$ requires inversion of an $n \times n$ matrix.
  4. Although the predictor p is optimal under the assumptions, the existence of at least $2k$ derivatives of $C(\mathbf{h})$ with respect to h_1 and h_2, at the origin, the assumption that $U^{-1} \to 0$, and the diagonal-matrix assumption for $H(\mathbf{s}_0)$, are difficult to verify.
  5. Because the data appear linearly, the predictor p is not resistant to outliers.
  6. The parameter τ represents a trade-off between smoothness ($\tau = 0$) and goodness-of-fit to the data ($\tau = \infty$). When $\tau = \infty$, p is an exact

interpolator, because the first element of $\mathbf{p}(\mathbf{z}; \mathbf{s}_i)$ is $z(\mathbf{s}_i)$; but, when $0 \leq \tau < \infty$, p is not an exact interpolator. When $\tau = 0$, $\mathbf{p}(\mathbf{z}; \mathbf{s}_0) = (X'X)^{-1}X'\mathbf{z}$, for all $\mathbf{s}_0 \in D$.

5.9.2 Nonstochastic Methods of Spatial Prediction

The nonstochastic methods available in the literature are mostly sensible, *ad hoc* approaches to spatial interpolation or smoothing (in contrast to prediction). A stochastic model is not assumed, save occasionally for additive measurement error. Assuming that measurement error is zero, predictors in this section do not possess a mean-squared prediction error. [Of course, the various predictors that follow *do* have a mean-squared prediction error with respect to the model (5.9.12), but it will not be calculated.]

Global Measure of Central Tendency

- *Predictor Assumptions.* It is thought that the surface $Z(\cdot)$ is fluctuating "equally" either side of a central tendency; the predictor is chosen to reflect this.
- *Predictor.*

$$p(\mathbf{z}; \mathbf{s}_0) = \sum_{i=1}^{n} z(\mathbf{s}_i)/n \tag{5.9.45}$$

or

$$p(\mathbf{z}; \mathbf{s}_0) = \text{med}\{z(\mathbf{s}_1), \ldots, z(\mathbf{s}_n)\}. \tag{5.9.46}$$

- *Properties.*
  1. These predictors are simple, requiring no knowledge of spatial model parameters, and hence no preliminary estimation is needed.
  2. They are computationally fast, although computing the median in (5.9.46) takes more time than computing the mean in (5.9.45).
  3. Predictor (5.9.45) is not resistant to outliers, whereas (5.9.46) is.
  4. In general, $p(\mathbf{z}; \mathbf{s}_i) \neq z(\mathbf{s}_i)$; that is, p is not an exact interpolator.

This predictor is often too smooth, because it does not allow for local fluctuations of $Z(\cdot)$. In the parlance of Section 5.9.1, it could also be viewed as an estimator of a constant unknown mean μ in (5.9.12).

Simple Moving Average

- *Predictor Assumptions.* It is thought that the surface $Z(\cdot)$ is "smooth," and the predictor is chosen to reflect this.

- *Predictor.*

$$p(\mathbf{z}; \mathbf{s}_0) = \sum_{i=1}^{n} z(\mathbf{s}_i) I(d_{0,i} \le r) \Big/ \sum_{i=1}^{n} I(d_{0,i} \le r), \qquad (5.9.47)$$

where $I(A)$ denotes the indicator function of the event A, $d_{0,i} \equiv \|\mathbf{s}_0 - \mathbf{s}_i\|$, and r defines a fixed-neighborhood radius. Alternatively, let $d_{[0,1]} \le d_{[0,2]} \le \cdots \le d_{[0,n]}$ denote the ordered Euclidean distances. Then the kth nearest-neighbor predictor is

$$p(\mathbf{z}; \mathbf{s}_0) = \sum_{i=1}^{n} z(\mathbf{s}_i) I(d_{0,i} \le d_{[0,k]})/k. \qquad (5.9.48)$$

- *Properties.*
  1. The predictors do not depend formally on model parameters, although choice of r in (5.9.47) or of k in (5.9.48) has to be made.
  2. Sorting and testing distances $\{d_{0,1}, \ldots, d_{0,n}\}$ can slow computations.
  3. The moving means are not resistant to outliers; however, moving-median analogues of (5.9.47) and (5.9.48) would be resistant.
  4. In general, $p(\mathbf{z}; \mathbf{s}_i) \ne z(\mathbf{s}_i)$; that is, p is not an exact interpolator. Here, p is acting as a smoother.

The property of exact interpolation is often important to retain. For some purposes, a prediction surface that does not reproduce the original data is thought to be too smooth.

Inverse-Distance-Squared Weighted Average

- *Predictor Assumptions.* Instead of allowing the data values to contribute equally to the average, they are weighted according to how far they are from $\mathbf{s}_0$.
- *Predictor.*

$$p(\mathbf{z}; \mathbf{s}_0) = \sum_{i=1}^{n} d_{0,i}^{-2} z(\mathbf{s}_i) \Big/ \sum_{i=1}^{n} d_{0,i}^{-2}. \qquad (5.9.49)$$

[It is also possible to define a weighted-median version of this predictor:

$$p(\mathbf{z}; \mathbf{s}_0) = \text{wt med}\{\mathbf{z}; d_{0,1}^{-2}, \ldots, d_{0,n}^{-2}\},$$

where the right-hand side is defined in the same way as (2.4.16).]

- *Properties.*
  1–3. Properties are the same as for the global measure of central tendency.
  4. p is an exact interpolator.

Moving-average versions of (5.9.49) could also be defined by modifying (5.9.47) and (5.9.48) to include weights $d_{0,i}^{-2}$ in their summands. Notice also that any positive decreasing function of $d_{0,i}$ could replace $d_{0,i}^{-2}$ in (5.9.49) and achieve the same effect of giving less weight to observations further away from the predictor location. For example, weights $\{d_{0,i}^{-\alpha}: i = 1, \ldots, n\}$, $\alpha > 0$, yield predictors whose properties 1, 2, 3, and 4 are the same as the preceding for $\alpha = 2$. However, weights $\{\exp(-\beta d_{0,i}): i = 1, \ldots, n\}$, $\beta > 0$, do not yield exact interpolators. How can α or β be chosen "optimally"? Other *ad hoc* weights that express the "neighborliness" of surrounding data have been proposed (e.g., Cliff et al., 1975, Section 10.2; Tobler and Kennedy, 1985).

Moving-median versions of the weighted median, or indeed those based on other resistant summaries, have been considered in one dimension by Cleveland (1979). He has developed a locally weighted regression (*lowess*) smoother that is a resistant weighted moving average based on the kth nearest neighbor form, (5.9.48) [rather than the fixed neighborhood form, (5.9.47)]. Cleveland and Devlin (1988) adapt lowess to two or more dimensions.

Median-Polish Plating (Section 3.5)

- *Predictor Assumptions.* Low-resolution mapping creates replication in two (orthogonal) directions. Each observation is assigned to a row and a column, from which row and column effects can be estimated and plates defined.
- *Predictor.* Suppose that a (possibly irregularly spaced) rectangular grid $\{(x_l, y_k)': k = 1, \ldots, p; l = 1, \ldots, q\}$ is overlaid onto the domain D, and data at locations $\{\mathbf{s}_i: i = 1, \ldots, n\}$ are assigned to the nearest nodes of the grid. If $\mathbf{s}_i$ is closest to the node $(x_l, y_k)'$, rewrite $Z(\mathbf{s}_i)$ as Y_{kl}. (Notice that there may be no observations at a node, or more than one observation at a node; in the latter case, the observations are written as $Y_{kl1}, Y_{kl2}, \ldots$.) Then, the data are median polished (Tukey, 1977; Section 3.5.1) to obtain the all effect $\tilde{a}$, the row effects $\{\tilde{r}_k\}$, and the column effects $\{\tilde{c}_l\}$ in the decomposition (3.5.16). The predicted value at $\mathbf{s}_0 = (x_l, y_k)'$ is

$$p(\mathbf{z}; (x_l, y_k)') = \tilde{a} + \tilde{r}_k + \tilde{c}_l, \qquad k = 1, \ldots, p; l = 1, \ldots, q.$$

For $\mathbf{s}_0 = (x_0, y_0)'$, where $x_l \le x_0 \le x_{l+1}$ and $y_k \le y_0 \le y_{k+1}$, the *me-*

dian-polish plates are

$$p(\mathbf{z};\mathbf{s}_0) = \tilde{a} + \tilde{r}_k + \left(\frac{y_0 - y_k}{y_{k+1} - y_k}\right)(\tilde{r}_{k+1} - \tilde{r}_k)$$
$$+ \tilde{c}_l + \left(\frac{x_0 - x_l}{x_{l+1} - x_l}\right)(\tilde{c}_{l+1} - \tilde{c}_l),$$
$$k = 1,\ldots,p-1;\ l = 1,\ldots,q-1. \quad (5.9.50)$$

Thus, the predictor consists of piecewise interpolating planes, or plates, that interpolate between the values $\{\tilde{a} + \tilde{r}_k + \tilde{c}_l\colon k = 1,\ldots,p;\ l = 1,\ldots,q\}$ located at the grid nodes. (Figure 4.5 shows median-polish plating for the Wolfcamp-aquifer data.) Extrapolation is discussed in Section 3.5.1.

- *Properties.*
  1. A choice of grid nodes $\{(x_k, y_l)'\colon k = 1,\ldots,p;\ l = 1,\ldots,q\}$ must be made. A recommendation is given in Section 3.5.2 that the grid nodes be chosen so that each receives approximately one observation from the low-resolution mapping.
  2. The predictor p is computationally fast, provided the median-polish algorithm converges; see Section 3.5.1 for a summary of numerical convergence results.
  3. The predictor p is resistant to outliers.
  4. In general, p is not an exact interpolator.

A number of suggestions for embellishing the basic median-polish plating, defined by (5.9.50), are given in Section 3.5. A priori, it is not clear that a plane will interpolate between the *four* values, $\tilde{a} + \tilde{r}_k + \tilde{c}_l$, $\tilde{a} + \tilde{r}_{k+1} + \tilde{c}_l$, $\tilde{a} + \tilde{r}_k + \tilde{c}_{l+1}$, and $\tilde{a} + \tilde{r}_{k+1} + \tilde{c}_{l+1}$, located at nodes $(x_k, y_l)'$, $(x_{k+1}, y_l)'$, $(x_k, y_{l+1})'$, and $(x_{k+1}, y_{l+1})'$, respectively. However, the additive structure of the main effects guarantees it. Thus, the prediction surface is continuous, and it could be thought of as a resistant version of a linear smoothing spline.

When the data locations are used to define a triangulation of the plane, it is straightforward to use planar interpolation between the *data* to define a predictor.

Delauney Triangulation

- *Predictor Assumptions.* The domain D can be triangulated according to the locations of the data $\{\mathbf{s}_1,\ldots,\mathbf{s}_n\}$ and, on any triangle, the predictor is a planar interpolant of the surrounding three data values.
- *Predictor.* The Delauney triangulation will be used (see following text), because the greatest distances over which interpolations must be carried out are smaller than for any other triangulation. Consider the locus of

points V_i that are closer to $\mathbf{s}_i$ than any other data location. The union of all such loci partition D into Voronoi polygons, the ith polygon referring to the data location $\mathbf{s}_i$. If the jth polygon shares a common boundary with the ith polygon, join $\mathbf{s}_i$ and $\mathbf{s}_j$ with a straight line. The set of all such joins defines the Delauney triangulation. Then consider,

$$p(\mathbf{z}; \mathbf{s}_0) = T\big(z(\mathbf{s}_i), z(\mathbf{s}_j), z(\mathbf{s}_k); \mathbf{s}_0, \mathbf{s}_i, \mathbf{s}_j, \mathbf{s}_k\big), \qquad (5.9.51)$$

where $\mathbf{s}_0$ is contained in the Delauney triangle defined by vertices $\mathbf{s}_i$, $\mathbf{s}_j$, and $\mathbf{s}_k$, and the right-hand side of (5.9.51) is the planar interpolant through the coordinates $(\mathbf{s}_i, z(\mathbf{s}_i))$, $(\mathbf{s}_j, z(\mathbf{s}_j))$, and $(\mathbf{s}_k, z(\mathbf{s}_k))$, at $\mathbf{s}_0$. By joining $\mathbf{s}_0$ to $\mathbf{s}_i, \mathbf{s}_j$, and $\mathbf{s}_k$, three subtriangles are formed. Let A_i denote the area of the subtriangle opposite vertex i, and so on. Then $\mathbf{s}_0 = (x_0, y_0)'$ has the property that $\mathbf{s}_0 = (A_i\mathbf{s}_i + A_j\mathbf{s}_j + A_k\mathbf{s}_k)/(A_i + A_j + A_k)$, and the predictor in (5.9.51) is

$$p(\mathbf{z}; \mathbf{s}_0) = \big(A_i z(\mathbf{s}_i) + A_j z(\mathbf{s}_j) + A_k z(\mathbf{s}_k)\big)/\big(A_i + A_j + A_k\big). \qquad (5.9.52)$$

- *Properties.*
  1. The predictor is unique and requires no *ad hoc* choices, with two exceptions. There are ambiguities at the edges of the spatial domain $D \subset \mathbb{R}^2$. Kenkel et al. (1989) propose to delete all affected polygons. Moreover, in the special case of a regular rectangular grid, there are two choices as to how each rectangle is triangulated.
  2. It is computationally much slower than the nonstochastic predictors mentioned previously, although there are algorithms available that produce almost the optimal (Delauney) triangulation on the first pass (e.g., Gold et al., 1977; McCullagh and Ross, 1980).
  3. The predictor p is resistant in the sense that any one outlier affects it only locally.
  4. p is an exact interpolator.

The surface produced is continuous but not differentiable, due to abrupt changes of slope at the edges of the triangulation.

Natural Neighbor Interpolation (Sibson, 1981)

- *Predictor Assumptions.* The domain D is tesselated according to the Voronoi polygons based on the predictor locations and data locations $\{\mathbf{s}_0, \mathbf{s}_1, \ldots, \mathbf{s}_n\}$; see immediately preceding text. The predictor based on these Voronoi polygons will interpolate the data and be continuously differentiable.
- *Predictor.* Let V_i denote the Voronoi polygon around $\mathbf{s}_i$, when only $\{\mathbf{s}_1, \ldots, \mathbf{s}_n\}$ are used to tesselate D; $V_{i,j}$ is that part of V_i that is closest

to $\mathbf{s}_j$ in the event that $\mathbf{s}_i$ is removed. Define $V_0(\mathbf{s}_0)$ and $V_{0,j}(\mathbf{s}_0)$ similarly, when $\{\mathbf{s}_0, \mathbf{s}_1, \ldots, \mathbf{s}_n\}$ is used to tesselate D. A continuous, but not everywhere differentiable, predictor is

$$p^{(0)}(\mathbf{z}; \mathbf{s}_0) = \sum_{j=1}^{n} \lambda_{0,j}(\mathbf{s}_0) z(\mathbf{s}_j), \tag{5.9.53}$$

where

$$\lambda_{0,j}(\mathbf{s}_0) \equiv |V_{0,j}(\mathbf{s}_0)| / |V_0(\mathbf{s}_0)|, \quad j = 1, \ldots, n, \tag{5.9.54}$$

a ratio of areas. Notice that all but neighboring locations $\mathbf{s}_j$ of $\mathbf{s}_0$ yield zero coefficients. Now define

$$\mathbf{b}_i \equiv H_i^{-1} \mathbf{g}_i, \quad i = 1, \ldots, n, \tag{5.9.55}$$

where

$$H_i \equiv \sum_{j=1}^{n} \lambda_{i,j} (\mathbf{s}_i - \mathbf{s}_j)(\mathbf{s}_i - \mathbf{s}_j)' / \|\mathbf{s}_i - \mathbf{s}_j\|^2,$$

$$\mathbf{g}_i \equiv \sum_{j=1}^{n} \lambda_{i,j} (\mathbf{s}_i - \mathbf{s}_j)\big(z(\mathbf{s}_i) - z(\mathbf{s}_j)\big) / \|\mathbf{s}_i - \mathbf{s}_j\|^2,$$

and $\lambda_{i,j} \equiv |V_{i,j}| / |V_i|$. Also define

$$\zeta(\mathbf{s}_0) \equiv \left\{ \sum_{j=1}^{n} \lambda_{0,j}(\mathbf{s}_0) \zeta_j(\mathbf{s}_0) / d_{0,j}(\mathbf{s}_0) \right\} \Bigg/ \left\{ \sum_{j=1}^{n} \lambda_{0,j}(\mathbf{s}_0) / d_{0,j}(\mathbf{s}_0) \right\}, \tag{5.9.56}$$

where

$$\zeta_j(\mathbf{s}_0) \equiv z(\mathbf{s}_j) + \mathbf{b}_j'(\mathbf{s}_0 - \mathbf{s}_j) \tag{5.9.57}$$

and

$$d_{0,j}(\mathbf{s}_0) \equiv \|\mathbf{s}_0 - \mathbf{s}_j\|, \quad j = 1, \ldots, n. \tag{5.9.58}$$

Finally, the natural neighbor interpolant is

$$p(\mathbf{z}; \mathbf{s}_0) = \{a_1(\mathbf{s}_0) p^{(0)}(\mathbf{s}_0) + a_2(\mathbf{s}_0) \zeta(\mathbf{s}_0)\} / \{a_1(\mathbf{s}_0) + a_2(\mathbf{s}_0)\}, \tag{5.9.59}$$

where

$$a_1(\mathbf{s}_0) \equiv \left\{ \sum_{j=1}^{n} \lambda_{0,j}(\mathbf{s}_0) d_{0,j}(\mathbf{s}_0) \right\} \bigg/ \left\{ \sum_{j=1}^{n} \lambda_{0,j}(\mathbf{s}_0) / d_{0,j}(\mathbf{s}_0) \right\}$$

and

$$a_2(\mathbf{s}_0) \equiv \sum_{j=1}^{n} \lambda_{0,j}(\mathbf{s}_0) d_{0,j}^2(\mathbf{s}_0).$$

- *Properties.*
  1. The predictor (5.9.59) is unique and requires no *ad hoc* choices.
  2–4. Properties are the same as for the Delauney triangulation. Add to this the property of continuous differentiability and the ability to reproduce spherical quadratics exactly, except near the edges of D.

This nearest-neighbor method is in the spirit of (5.9.48) rather than of (5.9.47) (a fixed neighborhood method), so that in the case of irregularly spaced data, points very far from $\mathbf{s}_0$ can receive considerable weight in (5.9.59).

Splines (Whittaker, 1923; Henderson, 1924; Duchon, 1977; Wahba, 1978; Section 3.4.5)

- *Predictor Assumptions.* Predictors $p(\cdot;\mathbf{s}_0)$ will be sufficiently smooth so that $\partial^2 p/\partial x_0^2$, $\partial^2 p/\partial y_0^2$, and $\partial^2 p/\partial x_0\,\partial y_0$ exist [where $\mathbf{s}_0 = (x_0, y_0)'$].
- *Predictor.* The two-dimensional Laplacian smoothing spline of degree 2 is

$$p(\mathbf{z};\mathbf{s}_0) = a_0 + a_1 x_0 + a_2 y_0 + \sum_{i=1}^{n} b_i e(\mathbf{s}_0 - \mathbf{s}_i), \qquad (5.9.60)$$

where $e(\mathbf{s}) \equiv \|\mathbf{s}\|^2 \log(\|\mathbf{s}\|^2)/(16\pi)$. In (5.9.60), $\mathbf{a} \equiv (a_0, a_1, a_2)'$ and $\mathbf{b} \equiv (b_1, \ldots, b_n)'$ solve

$$\begin{aligned} (K + n\rho I)\mathbf{b} + X\mathbf{a} &= \mathbf{z}, \\ X'\mathbf{b} &= \mathbf{0}, \end{aligned} \qquad (5.9.61)$$

where K is the $n \times n$ matrix with (i, j)th entry $e(\mathbf{s}_i - \mathbf{s}_j)$, X is the $n \times 3$ matrix with ith row $(1, x_i, y_i)$, and $0 \le \rho \le \infty$.

- *Properties.*
  1. Knowledge of the parameter ρ is needed; Wahba and Wendelberger (1980) recommend its estimation by cross-validation.

2. Prediction at $\mathbf{s}_0$ requires inversion of an $n \times n$ matrix.
3. The predictor p is not resistant to outliers, in that $\mathbf{z}$ appears linearly in the defining equations (5.9.61).
4. When $\rho = 0$, $p(\mathbf{z}; \mathbf{s}_i) = z(\mathbf{s}_i)$, but for $0 < \rho \leq \infty$, p is not an exact interpolator. When $\rho = \infty$, $p(\mathbf{z}; \mathbf{s}_0) = a_0 + a_1 x_0 + a_2 y_0$, for all $\mathbf{s}_0 \in D$.
5. Duchon (1977) demonstrates that $p(\mathbf{z}; \mathbf{s}_0)$ minimizes

$$\sum_{i=1}^{n} \{z(\mathbf{s}_i) - p(\mathbf{z}; \mathbf{s}_i)\}^2 / n + \rho \iint \left\{ \left(\partial^2 p / \partial x_0^2\right)^2 + 2\left(\partial^2 p / \partial x_0 \, \partial y_0\right)^2 + \left(\partial^2 p / \partial y_0^2\right)^2 \right\} dx_0 \, dy_0,$$

which is a trade-off between goodness-of-fit to the data (measured by the first term) and surface roughness (measured by the second term); the parameter ρ measures the trade-off.

Should more smoothness be required, higher-degree Laplacian smoothing splines are available. Then, (5.9.60) becomes the sum of a $(k-1)$th order polynomial trend surface plus the linear combinations of terms in e, where $e(\mathbf{s}) = \|\mathbf{s}\|^{2(k-1)} \log(\|\mathbf{s}\|^2)/(16\pi)$, $k \in \{3, 4, \ldots\}$ (e.g., Wahba, 1990, p. 31).

Multiquadric-Biharmonic Interpolation (Hardy, 1971; Hardy and Nelson, 1986)

- *Predictor Assumptions.* The predictor is a smooth finite-sum approximation to a biharmonic representation of disturbing potential.
- *Predictor.*

$$p(\mathbf{z}; \mathbf{s}_0) = a + \sum_{i=1}^{n} b_i l(\mathbf{s}_0 - \mathbf{s}_i), \tag{5.9.62}$$

where $l(\mathbf{s}) \equiv \{\delta + \|\mathbf{s}\|^2\}^{1/2}$, for δ a positive constant. Now, a and $\mathbf{b} \equiv (b_1, \ldots, b_n)'$ solve

$$\begin{aligned} L\mathbf{b} + \mathbf{1}a &= \mathbf{z}, \\ \mathbf{1}'\mathbf{b} &= 0, \end{aligned} \tag{5.9.63}$$

where L is the $n \times n$ matrix with (i, j)th entry $l(\mathbf{s}_i - \mathbf{s}_j)$ and $\mathbf{1} \equiv (1, \cdots, 1)'$.
- *Properties.*
  1. Knowledge of δ in $l(\mathbf{s})$ is needed.
  2. Prediction at each point $\mathbf{s}_0$ requires inversion of an $n \times n$ matrix.
  3. As for splines, the predictor p is not resistant to outliers.

4. Solution of (5.9.63) with $l(\mathbf{s}) = \{\delta + \|\mathbf{s}\|^2\}^{1/2} - \delta^{1/2}$ yields the same predictor p and hence (5.9.61) is an exact interpolator. It is also continuous and differentiable.

5.9.3 Comparisons and Some Final Remarks

Franke (1982) compares several of the nonstochastic methods presented in Section 5.9.2. Notice that the nonstochastic method of inverse-distance-squared weighting, used extensively by practitioners, is seen to be very much like Markov-random-field prediction in Section 5.9.1. Splines and multiquadric biharmonic interpolation (Section 5.9.2) are simply intrinsic random-function-kriging predictors (Section 5.9.1) using prespecified generalized covariance functions (e.g., Micchelli, 1986). A detailed comparison of kriging and splines can be found in Section 3.4.5 and Cressie (1990a).

Among the stochastic methods of Section 5.9.1, all of the kriging methods (except disjunctive kriging) bear a strong resemblance to each other. There is some similarity between disjunctive kriging and generalized-additive models, which is discussed briefly in Section 5.1.

Confronted with a set of data, which is the best method to apply? Depending upon which model is true, the answer will change markedly. For this reason, several authors have conducted investigations where the values to be predicted were actually known (but were not used in the prediction). They then compared the performance of various methods by determining the closeness of predicted to actual. The interested reader can consult the articles of Creutin and Obled (1982), Agterberg (1984), Bardossy and Bardossy (1984), Puente and Bras (1986), Bardossy et al. (1987), and Laslett et al. (1987). In all of these comparisons, on real and simulated data, universal kriging generally did as well or better than the other methods. Laslett et al. (1987) found interpolating methods in general to be poor, and Laplacian smoothing splines to be as good as kriging (for predicting soil pH).

These studies reinforce the intuition that the prediction method has to be flexible, according to the underlying spatial variation in the data. Unlike splines, kriging has this flexibility because the spatial dependence structure is first gauged from an initial data analysis before the kriging equations are solved.

A number of the methods proposed are incomplete in the sense that unknown model parameters or unknown constants have to be chosen before the algorithm for $p(\mathbf{z}; \mathbf{s}_0)$ can be computed. Often they are chosen (estimated) adaptively, according to the data at hand, and this leads to another source of error that has not been considered in this section. In calculating the mean-squared prediction errors in Section 5.9.1, it was assumed that such parameters were fixed and known. Thus, the calculations actually give the means of the square of the *probabilistic* prediction error. By adding to this probabilistic error the statistical error due to parameter estimation, the total prediction error is obtained. It is the mean of this squared total prediction error that, strictly speaking, should be reported (Section 5.3.3).

Throughout this section, it was assumed that there was no measurement error in the process $\{Z(\mathbf{s}): \mathbf{s} \in D\}$. This is not always realistic and, under the circumstance where there is measurement error, exact interpolation of the spatial predictor is *not* a desirable property. For example, the ordinary kriging equations have to be modified to predict the noiseless (or "smooth") process $S(\mathbf{s}_0)$; the optimal predictor is given by (3.2.25) and (3.2.26). Some of the methods presented, such as splines, already have a trade-off constant that can be changed according to the amount of measurement error in the data $\mathbf{Z}$.

Suppose that somehow a predictor $\{p(\mathbf{z}; \mathbf{s}_0): \mathbf{s}_0 \in D\}$ is chosen, based (perhaps) on a combination of context and prejudice. One is then faced with presenting the surface, defined over $D \subset \mathbb{R}^2$, for easy interpretation. Three-dimensional views (e.g., Figure 4.8) from different angles, or contour maps (e.g., Figure 4.7) can be computer drawn. A number of such methods are reviewed in Jones et al. (1986).

PART II

Lattice Data

CHAPTER 6

Spatial Models on Lattices

In Chapter 1, a categorization was made of Statistics for spatial data into three broad types. Assuming data can be thought of as a (partial) realization of a random phenomenon (i.e., a stochastic process) $\{Z(\mathbf{s}): \mathbf{s} \in D\}$, this second part of the book is concerned with the situation where the index set D is a countable collection of spatial sites at which data are observed. The collection D of such sites is called a lattice, which is then supplemented with neighborhood information. Mathematically speaking, the sites become vertices, which are connected by edges (this is the neighborhood structure). The graph-theoretic formalism is scarcely needed here, though some discussion of it is given in Section 6.1.

Chapter Summary

The development of spatial *models* is initiated in Section 6.2 through an *exploratory* spatial data analysis of Sudden Infant Death Syndrome (SIDS) in North Carolina, 1974–1978. (These preliminary exploratory passes through the data are essential for effective model building.) Section 6.3 discusses the two modeling approaches usually used, namely, the conditional approach and the simultaneous approach. I concentrate in subsequent sections on the conditional approach, giving consistency conditions for the conditional probabilities in Section 6.4. Models for discrete and continuous data on regular and irregular lattices are given in Sections 6.5 and 6.6. Section 6.7 discusses various other models for lattice data, including regression models with spatial autoregressive-moving-average errors. Finally, Section 6.8 contains a brief presentation of space–time models for lattice data.

6.1 LATTICES

The purpose of this section is to show how the lattice of sites that indexes $Z(\cdot)$ is supplemented (either implicitly or explicitly) with neighborhood information about the sites. Here "lattice" refers to a countable collection of

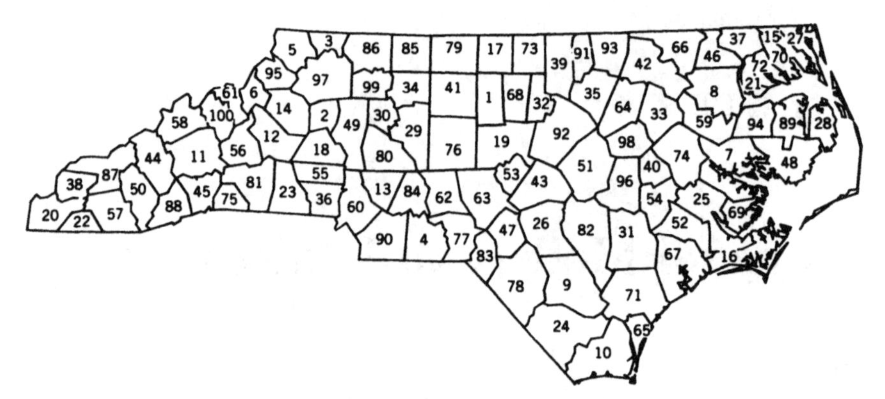

Figure 6.1 Map showing the 100 counties of North Carolina, numbered in alphabetical order. County names are given in Table 6.1.

(spatial) sites, to be distinguished in subsequent text as being either spatially regular or irregular; that is, the word is being used here in neither of its mathematical senses (e.g., Birkhoff, 1967; Hammersley and Mazzarino, 1983). Notice that this usage is at slight variance with Besag (1974, 1975), who reserves the word lattice for spatially *regular* sites.

Figure 6.1 shows a map of the 100 counties of North Carolina. The numbering of the counties is in alphabetical order, county 1 being Alamance and county 100 being Yancey. However, *without* this map the index set $\{1, 2, \ldots, 100\}$ shows no spatial information and so does not qualify to be a spatial lattice.

A direct way of forming a lattice is to specify an identifying feature common to each county, for example, the county seat, and to record its "longitude" (x) and "latitude" (y). Let

$$D \equiv \{(i; x_i, y_i): i = 1, \ldots, 100\},$$

which is usually abbreviated to

$$D \equiv \{(x_i, y_i): i = 1, \ldots, 100\}. \tag{6.1.1}$$

Therefore, Alamance has site (1; 278, 151), where the 278, 151 indicates that Alamance's county seat, Graham, is 278 miles east and 151 miles north of an arbitrarily chosen origin (southwest of North Carolina).

Neighborhoods

From the spatial lattice D, neighborhood information (based on, e.g., Euclidean distance) can be specified. For example, call any county (excluding the ith) whose county seat is within 30 miles of the ith county seat, a neighbor of the ith county. Let

$$N_i \equiv \{k: k \text{ is a neighbor of } i\}, \qquad i = 1,\ldots,100 \tag{6.1.2}$$

$$D_N \equiv \{(i; N_i): i = 1,\ldots,100\}. \tag{6.1.3}$$

Therefore, in the preceding example, the first site, Alamance, has neighbors $\{17, 19, 32, 41, 68\}$; see Table 6.1.

The set D_N will also be called a spatial lattice; although it does not contain exact site-location information, there is typically enough to build a model of spatial dependence (albeit at low spatial resolution) between counties. In mathematical parlance, D_N is a *graph*.

Another way of defining N_i is to take those counties sharing a common boundary with the ith county, $i = 1,\ldots,n$. This could be extended to include all second-generation counties sharing a boundary with the original first-generation neighbors.

Table 6.1 sets out a neighborhood system for the counties of North Carolina, where counties whose seats are closer than 30 miles are deemed neighbors. A neighborhood system for North Carolina's counties based on common county boundaries is given in Cressie and Read (1985).

Distance Measures

At times a site may be close "as the crow flies" to another site, but a spatial analysis based on this distance may be strictly for our feathered friends. For example, a spatial study over a highly urbanized area might use the city-block distance $|x_i - x_j| + |y_i - y_j|$ between two site locations (x_i, y_i) and (x_j, y_j), rather than the Euclidean distance $\{(x_i - x_j)^2 + (y_i - y_j)^2\}^{1/2}$. Urban geographical studies on commuting habits should probably record distance between work and home in units of travel time or a combination of time and cost. The closeness of two census enumeration districts might be a function of both distance and, say, percent urbanization.

6.2† SPATIAL DATA ANALYSIS OF SUDDEN INFANT DEATHS IN NORTH CAROLINA

Before giving a more formal presentation of models for lattice data, namely, parametric forms, parameter estimation, precision of parameter estimators, goodness-of-fit, and so on, an example is introduced and its data analyzed in an exploratory manner. Much of what is presented in this section comes from Cressie and Read (1985, 1989). Section 7.6, Cressie and Chan (1989), and Cressie (1992) continue the analysis into the parametric and confirmatory stages of the problem.

In any statistical analysis of data on a spatial lattice, it should be determined whether the locations (1) are regular or irregular, (2) represent points

Table 6.1 For each county entries show: name, number, county seat coordinates, total number of live births and SIDs, number of nonwhite live births, county numbers for counties with seats within 30 miles of the county seat

Name	i	Coordinates		1974–1978 Births	1974–1978 SIDs	1974–1978 Nonwhite Births	1979–1984 Births	1979–1984 SIDs	1979–1984 Nonwhite Births	County and Its Neighbors[a]
Alamance	1	278	151	4672	13	1243	5767	11	1397	1 17 19 32 41 68
Alexander	2	179	142	1333	0	128	1683	2	150	2 14 18 49 97
Alleghany	3	183	182	487	0	10	542	3	12	3 5 86 97
Anson	4	240	75	1570	15	952	1875	4	1161	4 62 77 84 90
Ashe	5	164	176	1091	1	10	1364	0	19	3 5 95 97
Avery	6	138	154	781	0	4	977	0	5	6 12 14 56 61 95 100
Beaufort	7	406	118	2692	7	1131	2909	4	1163	7 59 74 94
Bertie	8	411	148	1324	6	921	1616	5	1161	8 21 46 59 94
Bladen	9	321	53	1782	8	818	2052	5	1023	9 24 78
Brunswick	10	353	6	2181	5	659	2655	6	841	10 65
Buncombe	11	104	121	7515	9	930	9956	18	1206	11 44 45 56 58 88 100
Burke	12	150	130	3573	5	326	4314	15	407	6 12 14 18 56 61 81
Cabarrus	13	211	105	4099	3	856	5669	20	1203	13 60 80 84 90
Caldwell	14	158	142	3609	6	309	4249	9	360	2 6 12 14 18 56 95 97
Camden	15	453	173	286	0	115	350	2	139	15 27 70 72 89
Carteret	16	429	62	2414	5	341	3339	4	487	16 69
Caswell	17	281	175	1035	2	550	1253	2	597	1 17 68 73 79
Catawba	18	177	125	5754	5	790	6883	21	914	2 12 14 18 36 49 55
Chatham	19	291	127	1646	2	591	2398	3	687	1 19 32 43 53 63 68 92
Cherokee	20	19	92	1027	2	32	1173	1	42	20 22 38
Chowan	21	428	154	751	1	386	899	1	491	8 21 37 46 59 70 72 89 94
Clay	22	31	88	284	0	1	419	0	5	20 22 38 57

Cleveland	23	158	99	4866	10	1491	5526	21	1729	23 36 55 81
Columbus	24	316	33	3350	15	1431	4144	17	1832	9 24 78
Craven	25	407	88	5868	13	1744	7595	18	2342	25 52 69
Cumberland	26	308	83	20366	38	7043	26370	57	10614	26 43 47 82
Currituck	27	461	182	508	1	123	830	2	145	15 27 70
Dare	28	482	145	521	0	43	1059	1	73	28
Davidson	29	231	133	5509	8	736	7143	8	941	29 30 34 76 80
Davie	30	213	139	1207	1	148	1438	3	177	29 30 34 49 80 99
Duplin	31	355	76	2483	4	1061	2777	7	1227	31 71 82
Durham	32	306	146	7970	16	3732	10432	22	4948	1 19 32 39 68 73 92
Edgecome	33	378	142	3657	10	2186	4359	9	2696	33 42 59 64 74 98
Forsyth	34	233	153	11858	10	3919	15704	18	5031	29 30 34 41 85 99
Franklin	35	337	155	1399	2	736	1863	0	950	35 39 64 91 92 93
Gaston	36	178	96	9014	11	1523	11455	26	2194	18 23 36 55 60
Gates	37	421	177	420	0	254	594	2	371	21 37 46 72
Graham	38	32	107	415	0	40	488	1	45	20 22 38 57 87
Granville	39	322	171	1671	4	930	2074	4	1058	32 35 39 73 91 93
Greene	40	371	111	870	4	534	1178	4	664	40 54 74 96 98
Guilford	41	258	150	16184	23	5483	20543	38	7089	1 34 41 76 79
Halifax	42	376	171	3608	18	2365	4463	17	2980	33 42 66
Harnett	43	309	105	3776	6	1051	4789	10	1453	19 26 43 51 53 92
Haywood	44	80	116	2110	2	57	2463	8	62	11 44 50 58 87 88
Henderson	45	109	101	2574	5	158	3679	8	264	11 45 75 81 88
Hertford	46	411	176	1452	7	954	1838	5	1237	8 21 37 46 66 72
Hoke	47	289	75	1494	7	987	1706	6	1172	26 47 63 78 83
Hyde	48	446	110	338	0	134	427	0	169	48
Iredell	49	195	132	4139	4	1144	5400	5	1305	2 18 30 49 55 80 97 99
Jackson	50	67	108	1143	2	215	1504	5	307	44 50 57 87 88

Table 6.1 *(Continued)*

Name	i	Coordinates		1974–1978 Births	1974–1978 SIDs	1974–1978 Nonwhite Births	1979–1984 Births	1979–1984 SIDs	1979–1984 Nonwhite Births	County and Its Neighbors[a]
Johnston	51	335	114	3999	6	1165	4780	13	1349	43 51 92 96 98
Jones	52	389	84	578	1	297	650	2	305	25 52 54 67
Lee	53	292	109	2252	5	736	2949	6	905	19 43 53 63
Lenoir	54	377	98	3589	10	1826	4225	14	2047	40 52 54 74 96
Lincoln	55	174	111	2216	8	302	2817	7	350	18 23 36 49 55 60
McDowell	56	133	127	1946	5	134	2215	5	128	6 11 12 14 56 61 81 100
Macon	57	55	96	797	0	9	1157	3	22	22 38 50 57 87
Madison	58	98	135	765	2	5	926	2	3	11 44 58 100
Martin	59	405	139	1549	2	883	1849	1	1033	7 8 21 33 59 74 94
Mecklenberg	60	198	95	21588	44	8027	30757	35	11631	13 36 55 60 90
Mitchell	61	128	148	671	0	1	919	2	4	6 12 56 61 95 100
Montgomery	62	251	102	1258	3	472	1598	8	588	4 62 63 76 84
Moore	63	278	101	2648	5	844	3534	5	1151	19 47 53 62 63
Nash	64	354	146	4021	8	1851	5189	7	2274	33 35 64 98
New Hanover	65	357	26	5526	12	1633	6917	9	2100	10 65 71
Northampton	66	385	176	1421	9	1066	1606	3	1197	42 46 66
Onslow	67	384	66	11158	29	2217	14655	23	3568	52 67
Orange	68	294	151	3164	4	776	4478	6	1086	1 17 19 32 68 73
Pamlico	69	423	90	542	1	222	631	1	277	16 25 69
Pasquotank	70	451	172	1638	3	622	2275	4	933	15 21 27 70 72 89
Pender	71	357	48	1228	4	580	1602	3	763	31 65 71
Perquimans	72	437	163	484	1	230	676	0	310	15 21 37 46 70 72 89 94
Person	73	302	173	1556	4	613	1790	4	650	17 32 39 68 73
Pitt	74	387	121	5094	14	2620	6635	11	3059	7 33 40 54 59 74

Polk	75	123	97	533	1	95	673	0	79	45 75 81
Randolph	76	256	126	4456	7	384	5711	12	483	29 41 62 76
Richmond	77	259	72	2756	4	1043	3108	7	1218	4 77 83
Robeson	78	302	51	7889	31	5904	9087	26	6899	9 24 47 78 83
Rockingham	79	257	173	4449	16	1243	5386	5	1369	17 41 79 85
Rowan	80	218	124	4606	3	1057	6427	8	1504	13 29 30 49 80 84
Rutherford	81	136	104	2992	12	495	3543	8	576	12 23 45 56 75 81
Sampson	82	336	79	3025	4	1396	3447	4	1524	26 31 82
Scotland	83	276	61	2255	8	1206	2617	16	1436	47 77 78 83
Stanly	84	234	103	2356	5	370	3039	7	528	4 13 62 80 84
Stokes	85	233	175	1612	1	160	2038	5	176	34 79 85 86
Surry	86	204	174	3188	5	208	3616	6	260	3 85 86 97 99
Swain	87	53	113	675	3	281	883	2	406	38 44 50 57 87
Transylvania	88	93	97	1173	3	92	1401	4	104	11 44 45 50 88
Tyrrell	89	450	144	248	0	116	319	0	141	15 21 70 72 89 94
Union	90	214	76	3915	4	1034	5273	9	1348	4 13 60 90
Vance	91	332	172	2180	4	1179	2753	6	1492	35 39 91 93
Wake	92	319	133	14484	16	4397	20857	31	6221	19 32 35 43 51 92
Warren	93	344	177	968	4	748	1190	2	844	35 39 91 93
Washington	94	422	139	990	5	521	1141	0	651	7 8 21 59 72 89 94
Watauga	95	151	162	1323	1	17	1775	1	33	5 6 14 61 95
Wayne	96	353	106	6638	18	2593	8227	23	3073	40 51 54 96 98
Wilkes	97	181	157	3146	4	200	3725	7	222	2 3 5 14 49 86 97 99
Wilson	98	358	128	3702	11	1827	4706	13	2330	33 40 51 64 96 98
Yadkin	99	208	156	1269	1	65	1568	1	76	30 34 49 86 97 99
Yancey	100	120	142	770	0	12	869	1	10	6 11 56 58 61 100

Source: Cressie and Chan (1989). Reprinted by permission of the American Statistical Association.

[a]Counties with seats within 30 miles.

or regions, (3) are indices for continuous or discrete random variables. By nature or by design, many spatial problems are "regular," often "points," and less often "continuous." Counts data from neighboring geopolitical regions offer a particular challenge to the statistician because the spatial problem is "irregular," "regions," and "discrete." Important applications include epidemiological studies (e.g., cancer mortality over the counties of the United States) and Census Bureau surveys (e.g., undercount over the census blocks of an urban area).

It has long been recognized by time-series analysts that data close together in time usually exhibit higher dependence than those far apart. Time-series data analysis relies on methods of data transformation, detrending, and autocorrelation plotting. This section shows how a similar approach can be taken in a spatial setting.

Some Background

Here, sudden-infant-death-syndrome (SIDS) data for the 100 counties of North Carolina are analyzed. SIDS is currently a leading category of post-neonatal death, yet its cause is still a mystery. It accounts for about 7000 deaths per year in the United States, taking the lives of about two infants per 1000 live births.

SIDS is defined as the sudden death of any infant up to 12 months old, that is unexpected by history and where the death remains inexplicable after performance of an adequate postmortem examination. Atkinson (1978) provides an early review of the SIDS literature. Much attention has been directed toward the relationship of apnea (cessation of respiratory air flow) and SIDS (see, e.g., Steinschneider, 1972); Goldberg and Stein (1978) notice excess mortality in the winter months. Fogerty et al. (1984) present evidence that poor nutrition, indicated by high liver fatty acids, has an effect. Giulian et al. (1987) conclude that a high level of fetal hemoglobin is a useful postmortem marker for a SIDS infant, which may have value as a prospective marker for some infants at risk for SIDS. Much of the current research into SIDS has been physiologic, specifically, in the areas of developmental neurophysiology; autonomic disturbances and sleep state; respiratory, laryngeal, and cardiac functions; and immunology and infection.

From an epidemiologic point of view, certain risk factors for SIDS (summarized and compared by Peters and Golding, 1986) have been identified. Symons et al. (1983) find geographical clustering in a North Carolina (1974–1978) data set, which could be due to some surrogate variable such as race; indeed, Cressie and Chan (1989) find that the baby's race is an important explanatory variable for the SIDS data presented in this section.

In the 100 counties of North Carolina, the number of sudden infant deaths (SIDs) and live births from 1974–1978 were presented by Atkinson (1978) and augmented by Symons et al. (1983), Cressie and Read (1985, 1989), and Cressie and Chan (1989). The 1974–1978 data are analyzed in this section, whereas Section 7.6 gives a more formal analysis of both the 1974–1978 data and data from a later period (1979–1984); Table 6.1 contains the full data set.

A primary component of any spatial analysis is a *map*. A complete spatial data set includes not only "how much" or "how many," but also "where." Figure 6.1 shows a map of the counties numbered alphabetically, as given in Table 6.1.

6.2.1[†] Nonspatial Data Analysis

For the period 1974–1978, define

$$\begin{aligned} S_i &\equiv \text{number of SIDS in county } i, \\ n_i &\equiv \text{number of live births in county } i, \\ W_i &\equiv 1000(S_i + 1)/n_i, \qquad i = 1, \cdots, 100. \end{aligned} \tag{6.2.1}$$

The definition of W_i facilitates discrimination between the 13 counties with zero SIDs but different numbers of live births. A choropleth map (i.e., a map with different shadings, according to the magnitude of the variable being mapped) of the *raw* death rates $\{W_i: i = 1, \ldots, 100\}$ can be misleading, because counties with small n_is typically are far more variable than those with large n_is; see Section 6.2.2.

Figure 6.2 shows a stem-and-leaf plot of the rates $\{W_i\}$ defined in the preceding text. It is tempting to think about the stem-and-leaf plot as estimating a true density, but this would not be correct because the observed

```
 0 | 7899
 1 | 0122222233444
 1 | 55556666778888889999
 2 | 0001122223333444
 2 | 56666788999999
 3 | 000122444
 3 | 5678999
 4 | 000013
 4 | 7
 5 | 01223
 5 | 579
 6 | 0
 6 |
 7 | 0
 7 |
 8 |
 8 |
 9 |
 9 |
10 | 1
```

Figure 6.2 Stem-and-leaf plot of $\{W_i: i = 1, \ldots, 100\}$, the 1974–1978 North Carolina SID rates. 1|2 represents 1.2. [*Source:* Cressie and Read, 1989]

rates cannot be thought of as a batch of identically distributed random variables. Figure 6.2 must be interpreted simply as a concise summary of the 100 SID rates.

Probability Map

Cressie and Read (1985) draw a *probability map* (Choynowski, 1959) based on the data (6.2.1). In particular, define

$$E_i \equiv n_i \hat{p}, \tag{6.2.2}$$

where $\hat{p} = \Sigma_{i=1}^{100} S_i / \Sigma_{i=1}^{100} n_i$. Under the assumption that $\{S_i: i = 1, \ldots, 100\}$ are independent Poisson random variables with means $\{\lambda_i: i = 1, \ldots, 100\}$, respectively, and $\lambda_1/n_1 = \cdots = \lambda_{100}/n_{100} = p$, an index of deviation from equal $\{\lambda_i/n_i\}$ can be defined as

$$\rho_i \equiv \begin{cases} \sum_{x \geq S_i} \exp(-E_i)(E_i)^x/x!, & S_i \geq E_i, \\ \sum_{x \leq S_i} \exp(-E_i)(E_i)^x/x!, & S_i < E_i. \end{cases} \tag{6.2.3}$$

A small value of ρ_i indicates that county i's SID rate is unusually high or low.

A choropleth map based on $\{\rho_i: i = 1, \ldots, 100\}$ is called a *probability map*. One such map is presented in Cressie and Read (1985), where counties with small (< 0.05) values of ρ_i are highlighted; it indicates definite clustering of counties with unusually high SIDs in the northeast and south of the state. This clustering is in close agreement with that observed by Symons et al. (1983) through their more complicated risk classification model. However, the independent-Poisson model assumed in preceding text and by Symons et al., is not really appropriate in this situation due to the small-area spatial correlations that will be illustrated in Section 6.2.4.

Although the idea of the probability map is to standardize rates onto a probability scale for proper comparison, it does not achieve this for widely varying n_is. It is well known that goodness-of-fit tests always reject the null hypothesis when sample size is large enough. Thus, an extreme value of ρ_i defined by (6.2.3) may be more due to its lack of fit to the Poisson model than to its deviation from the constant rate assumption in (6.2.2).

Generalized Linear Models

Suppose the SIDS data are assumed to be independently Poisson distributed and are fitted to a generalized linear model (e.g., McCullagh and Nelder, 1983) with link function $g(\cdot)$. The special case of $g(x) = x^{1/2}$ is considered by Cochran (1940), who derives the likelihood equations for estimating a no-interaction decomposition for the square roots of the Poisson means. Other examples of link functions are $g(x) = x$ and $g(x) = \log(x)$.

I wish to emphasize here that without exploratory data analyses, such as those described in the subsequent paragraphs, there is little guidance in the choice of joint distribution and link function. Section 6.2.2 indicates a preference for the square-root link function. However, the Poisson assumption probably should be replaced by a model with extra-Poisson variation due to microscale spatial correlations; and, at a larger scale, Section 6.2.4 indicates that the independence assumption does not hold between neighboring counties. The analysis of deviance (McCullagh and Nelder, 1983) can be carried out by comparing likelihoods of the various models being fitted. For example, the homogeneity model, which asserts that the Poisson mean of S_i is $\lambda_i = n_i p$, can be tested against the fully saturated model $\lambda_i = n_i p_i$. This leads to a deviance of 203.3, which is to be compared to a chi-squared random variable on 99 degrees of freedom. The homogeneity model is hence rejected (p value < 0.0001); however, due to doubts about the model assumptions, the quoted p value may not be all that accurate.

A standard generalized linear analysis could proceed by writing

$$g(\lambda_i) = \sum_{j=1}^{q} x_{ij}\beta_j,$$

where the xs are explanatory variables. The significance of the βs could be gauged by a further analysis of deviance. I shall not pursue this confirmatory, hypothesis-testing approach here because the necessary model assumptions may well be violated; in particular, the data may bc spatially dependent.

6.2.2† Spatial Data Analysis

Attempts to recognize the spatial component in *confirmatory* data analyses have appeared in recent articles. Carter and Rolph (1974) obtain James–Stein estimators of the fire-alarm probabilities in the Bronx using geographical groupings into neighborhoods, but their assumption of independent observations within neighborhoods is probably not appropriate. Cook and Pocock (1983) fit a parametric (linear) model to the trend, followed by an investigation of residual spatial dependence. However, their fitting techniques lead to residuals that give biased estimates of spatial correlation. A more sophisticated (but more difficult to implement) set of spatial models can be constructed using the Markov random-field approach (Besag, 1974), as outlined in the following sections of this chapter. Hill et al. (1984) and Clayton and Kaldor (1987) assume conditional independence of the data given the parameters, and then the *parameters* are assumed to follow a Markov random field. They use empirical-Bayes shrinkage to predict each of the parameters.

Exploratory Spatial Data Analysis

Section 6.2.1 described a (nonspatial) parametric approach to fitting and smoothing counts data. However, when a significant deviance is obtained, it is unclear whether the parametric assumption itself is causing the lack of fit. Exploratory data analysis (Tukey, 1977) aims at making the initial stage of looking at the data resistant to the presence of unusual observations, often by graphical methods. In what follows, I describe an exploratory analysis of the SIDS data that recognizes its obvious spatial component.

Implications of Unequal Variances on Mapping

A simple way to illustrate the relative magnitudes of the number of SIDs associated with each county is to draw a choropleth map on which different shading patterns are used to compare the levels of the SID *rates*. From (6.2.1), the SID rate per thousand for county i is $1000(S_i)/n_i$. We use the (transformed) rates $W_i = 1000(S_i + 1)/n_i$ to facilitate discrimination between the 13 counties with zero SIDs, but different numbers of live births. Why is a map of the $\{W_i: i = 1, \ldots, 100\}$ potentially misleading?

The main reason is that the counties have substantially *different* numbers of live births. In general, counties with smaller n_is will have larger variances for their estimated SID rates. Consequently, they are more likely to exhibit SID rates that fluctuate greatly from the true (unknown) rate. The Poisson-based probability map of Section 6.2.1 transforms the SID rates into probabilities of deviation from homogeneity. However, as was explained there, this map still retains dependence on the n_is.

Spatial Smoothing

When the data are on a regular grid and come from an underlying continuous population, Cressie (1984a) has proposed a resistant spatial data analysis (i.e., an analysis resistant to outlying data values). Consider a two-dimensional $r \times s$ grid with data $\{Z_{kl}: k = 1, \ldots, r; l = 1, \ldots, s\}$. A first decomposition might be

$$Z_{kl} = \mu + \alpha_k + \beta_l + \delta_{kl}, \tag{6.2.4}$$

where $\{\alpha_k\}$ and $\{\beta_l\}$ are fixed row and column effects, respectively, μ is a fixed overall effect, and $\{\delta_{kl}\}$ represent random errors. Cressie (1986) shows how the assumption in (6.2.4) of no interaction can be checked, using a diagnostic plot to detect cross-product drift terms, and analyzes two geological data sets using (6.2.4) with stationary errors. The resistant method proposed to obtain the decomposition (6.2.4) is *median polish*, which will be described in subsequent text.

For the SIDS data, this approach needs some modification to take into account the irregular spacing of the counties. First, the data represent regions. Therefore, for the purposes of a spatial analysis, I shall realize the value W_i at the location of the county seat given by the Cartesian coordinates (x_i, y_i) from some arbitrary (fixed) origin; see Table 6.1. From here, a decomposition such as $W_i = \mu_i + \delta_i$ can be attempted, where μ_i is a function of the location (x_i, y_i).

However, due to the nature of the data (they are counts), one should consider using a transformation to remove possible dependence of the variance on the mean. That is, consider the more general decomposition

$$g(W_i) = \mu_i + \nu_i. \tag{6.2.5}$$

Transformations

The data $\{S_i: i = 1, \ldots, 100\}$ are counts. It follows from a binomial model that

$$E(S_i/n_i) = p_i \quad \text{and} \quad \text{var}(S_i/n_i) = p_i(1 - p_i)/n_i. \tag{6.2.6}$$

For small p_i, the variance is proportional to the mean, suggesting a square-root transformation to remove this mean-variance dependence. However, it is more likely that the $\{S_i\}$ are sums of spatially *dependent* Bernoulli random variables, in which case a more severe transformation may be necessary. With this in mind, the following transformations were considered:

$$\begin{aligned} W_i &\equiv 1000(S_i + 1)/n_i, \\ Z_i &\equiv (1000(S_i)/n_i)^{1/2} + (1000(S_i + 1)/n_i)^{1/2}, \\ V_i &\equiv \log W_i. \end{aligned} \tag{6.2.7}$$

I used the Freeman–Tukey square-root transformation Z_i, because it shows more stability than the usual square-root transformation $(W_i)^{1/2}$ (Freeman and Tukey, 1950).

In order to compare the stability of the variance associated with each transform, it is necessary to eliminate the effect of the n_is in the variance formula (6.2.6). This was accomplished by partitioning the set of 100 counties into 6 similarly sized subsets with roughly equal $(1/n_i)$ values. For each group, medians and interquartile ranges (IQs) were calculated for the W_is, Z_is, and V_is, together with the group averages of the $(1/n_i)$s. Plots of $(\text{IQ})^2/\{\text{ave}(1/n_i)\}$ versus median, for the different transformations, indicated that by using the Freeman–Tukey transformation Z_i the variance was no longer a function of the mean (Cressie and Read, 1989). The unequal numbers of live births $\{n_i\}$ make it impossible to assume homoskedasticity; however, it appeared that $\{n_i^{1/2}Z_i\}$ did have roughly equal variances. A further piece of evidence that Z_i is the "natural" transformation to use when analyzing these SIDs data is found in Figure 6.3. Here, the stem-and-leaf plot of the Z_is indicates that the Freeman–Tukey transformed SID rates are more symmetrically distributed (albeit with unequal variances) than the untransformed W_is plotted in Figure 6.2. (Now that the data have been symmetrized, there is a hint that county 4, corresponding to the outlying value $Z_4 = 6.2$ in Figure 6.3, may cause future trouble in the analysis.)

Stem	Leaf
0	8
1	111234
1	5577888899
2	000111112233344444
2	55556666677888899999
3	0000011112223333333333444
3	5588899
4	033344
4	55557
5	1
5	
6	2

Figure 6.3 Stem-and-leaf plot of $\{Z_i: i = 1, \ldots, 100\}$, the Freeman–Tukey transformed SID rates. 1|3 represents 1.3. [*Source:* Cressie and Read, 1989]

The idea that there is often a natural reexpression for data is discussed by Cressie (1985b). There, a universal-transformation principle is proposed that interprets symmetry, homoskedasticity, and lack of interaction in decompositions like (6.2.4), as the property of additivity at all scales. Data often, although not always, exhibit this multiscale additivity after a suitable transformation.

6.2.3† Trend Removal

Several different approaches (Symons et al., 1983; Cressie and Read, 1985) have shown clustering among the North Carolina-county SID rates. However, once this obvious trend in the rates (or rather the transformed rates Z_i) has been accounted for, are there residual, smaller-scale variations due to (stationary) spatial dependence? (Symons et al. have essentially no spatial component in their analysis.) The variety of problems analyzed in Chapter 4 offers evidence that this is a sensible question, often with a positive answer.

Weighted Median Polish

This approach to trend removal involves overlaying a rectangular grid onto North Carolina and placing nodes at 20 mile intervals, with 8 north–south intervals and 23 east–west intervals. Counties are then identified with the node closest to their county seat, resulting in a 9×24 matrix of nodes: some have no counties, most have one county, and some have more than one county. An *ad hoc* guideline is to choose a grid spacing that has one county at each node; Cressie and Guo (1987) chose their rectangular grid with unequal spacing to achieve this more closely. Figure 6.4 shows the Freeman–Tukey transformed rates $\{Z_i\}$ associated with each of the nodes. One can think of this figure as a map of the county seats of North Carolina using a low resolution of 20 miles (cf. the Coordinates columns in Table 6.1 for a higher resolution of 1 mile).

A median polish (Tukey, 1977, Chapter 11, or Section 3.5.1) can now be performed on this matrix of nodes in order to remove the row and column

	1	2	3	4	5	6	7	8	9	10	11	12	13	14	15	16	17	18	19	20	21	22	23	24
1	·	·	·	·	·	·	·	2.31	1.43	2.62	·	1.90	3.85	3.09	3.40	3.28	2.87 4.31	·	4.53 5.17	·	1.54 4.54	·	1.87 3.39 2.92	·
2	·	·	·	·	·	·	1.13 1.22	2.10	2.39	·	2.14	1.88	2.41	3.40	2.38	·	2.66	·	·	·	·	2.79 3.47	·	·
3	·	·	·	·	3.60	1.14	3.36	2.48 2.68	0.87	2.08	2.20	2.48	·	·	2.88	2.13	·	2.91	3.39 3.52	2.53	4.43	·	2.01	1.39 4.71
4	·	1.55	2.94 4.54	2.17	2.25	·	·	·	1.95 3.92	·	1.74	·	2.59	·	2.45 3.12	·	2.55	·	4.54 3.37	3.34	·	1.72	·	·
5	3.10	1.88	1.12	·	3.45	2.92 3.31	4.09	2.94	2.26	2.87	1.84	3.05	3.33	2.88	·	2.62	·	3.34	3.42	·	3.28	·	·	·
6	·	·	·	·	·	·	·	·	·	·	2.14	6.28	2.55	4.48	2.75	·	2.44	2.69	3.18	3.03	·	·	·	·
7	·	·	·	·	·	·	·	·	·	·	·	·	·	3.88	4.00	4.37	·	3.82	3.25	·	3.02	·	·	·
8	·	·	·	·	·	·	·	·	·	·	·	·	·	·	·	4.30	·	·	·	·	·	·	·	·
9	·	·	·	·	·	·	·	·	·	·	·	·	·	·	·	·	·	3.17 3.01	·	·	·	·	·	·

Figure 6.4 Low-resolution map of the Freeman–Tukey transformed SID rates associated with each of the 9 × 24 nodes of a grid overlaid on the county map of North Carolina.

effects, hopefully leaving a set of stationary residuals; see (6.2.4). However, due to the dependence of var(Z_i) on n_i, Cressie and Read (1989) polish using *weighted* medians, where each county rate is weighted by $(n_i)^{1/2}$; see Cressie (1980a) for justification of this weighting.

A brief explanation of *median polish* is as follows (Section 3.5.1 can be consulted for more details). The idea is to obtain an additive decomposition like (6.2.4), of a two-way table into an "all" effect plus a row effect plus a column effect, that is resistant to outliers. Each polish of the table involves the subtraction of each row (column) median from every member of that row (column), the median being added into an extra cell created for the purpose of accumulating the row (column) effect. Polishing continues from rows to columns and back again until there is no change in the table. For the data in Figure 6.4, the weighted-median polish started with columns and each polish involved the subtraction of *weighted* medians (with weights equal to the square roots of the number of live births); see (3.3.13) for the exact definition of the weighted median.

To obtain residuals of roughly equal variances, the median-polish residuals R_i can then be multiplied by $(n_i)^{1/2}$ (see Section 6.2.2).

Smoothed Map of SIDS Data

Move the datum Z_i to the nearest node of the grid and write it as $Z(l(i), 10 - k(i))$, where $(k(i), l(i)) \in \{(k, l)\colon\ k = 1, \ldots, 9,\ l = 1, \ldots, 24\}$. Decompose

$$Z(l(i), 10 - k(i)) = \tilde{a} + \tilde{r}_{k(i)} + \tilde{c}_{l(i)} + R(x_i, y_i), \qquad (6.2.8)$$

where $\tilde{a}$, $\tilde{r}_{k(i)}$, and $\tilde{c}_{l(i)}$ are the all, the row, and the column effects of weighted-median polish, and (x_i, y_i) is the spatial location of the ith county seat (see Table 6.1). Then, a choropleth map of $\{\tilde{a} + \tilde{r}_{k(i)} + \tilde{c}_{l(i)}\colon\ i = 1, \ldots, 100\}$ is a smoothed map of the contiguous regions and accounts for unequal variation. This is shown in Figure 6.5*a* for the 1974–1978 SIDS data. For comparison, the analogous map for the 1979–1984 SIDS data (which are also given in Table 6.1) is also shown (Fig. 6.5*b*). The difference, which shows up very clearly at this exploratory stage, is investigated further in Section 7.6.

The clustering observable in Figure 6.5*a* is similar to that seen previously in a probability map (Cressie and Read, 1985), namely, a clustering of high SID rates in the northeast and south and a clustering of low SID rates in the near west. However, Figure 6.5 is the preferred map because it does not rely on parametric model assumptions nor are the unequal n_is confusing the overall picture.

Another method of producing a smoothed map, by weighted moving averages of neighboring counties, is discussed by Cressie and Read (1989). Although it is an attractive way of estimating trend, its residuals do not contain accurate information on spatial dependence.

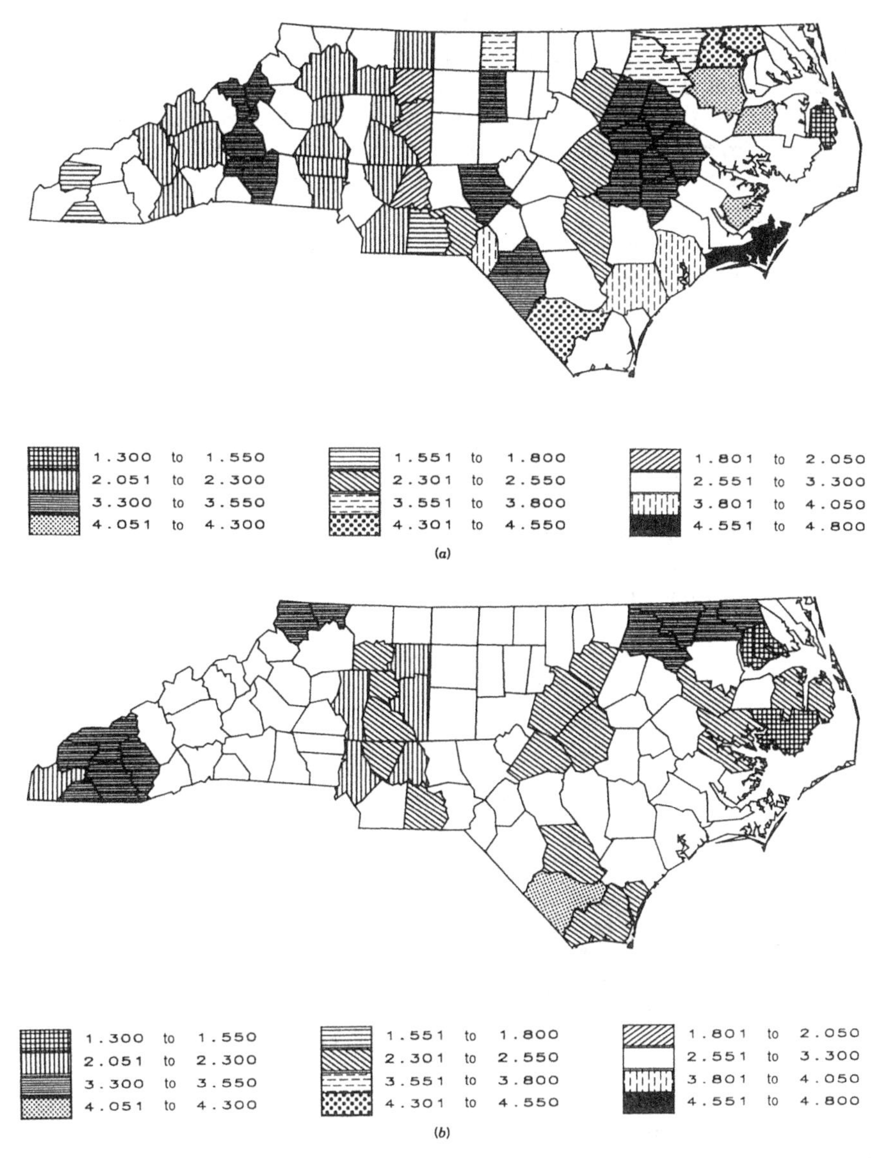

Figure 6.5 (*a*) Choropleth map of weighted-median-polish fit $\{\tilde{a} + \tilde{r}_{k(i)} + \tilde{c}_{l(i)}: i = 1, \ldots, 100\}$ for the 1974–1978 North Carolina (Freeman–Tukey transformed) SIDS data. (*b*) Same as in (*a*) except for 1979–1984.

Median-Polish Residuals

Attention is now turned to characterizing the small-scale variation by performing a spatial analysis on the weighted-median-polish residuals. The stem-and-leaf plot of $\{n_i^{1/2}R(x_i, y_i)\}$ presented in Figure 6.6, which for the first time can be thought of as indicating the stationary density (previously there was confounding with trend), now shows county 4 as a definite outlier. A more formal test in Section 7.6 confirms this. Therefore, county 4 has been set aside from further consideration in this section; the rest of the exploratory spatial data analysis is aimed at characterizing the small-scale variation of the other 99 counties.

A weighted-median polish was repeated on the 99 counties, and in the reanalysis no unusual counties appeared. Similar figures gave the same message with regard to distribution shape and dependence, but without the distraction of the outlier. This is to be expected from a resistant analysis.

Stationarity of the weighted residuals $\{n_i^{1/2}R(x_i, y_i)\}$ was checked by calculating the median and squared interquartile range across each row and plotting median versus row number and squared interquartile range versus row number. Similar calculations and plots were made for columns. The plots should be roughly level, which they were found to be (apart from columns or rows containing only a few entries). Moreover, a diagnostic plot for nonadditivity (Cressie, 1986; compare McNeil and Tukey, 1975) was made and it showed no obvious pattern. Thus, following (6.2.5), the model

$$Z_i = \mu_i + n_i^{-1/2}\delta_i, \qquad i \neq 4,$$

is suggested, where the $\{\delta_i\}$ are stationary errors whose correlation structure can be estimated from the weighted residuals.

```
−0 | 8
−0 | 66
−0 | 55555544
−0 | 33322222222222
−0 | 11111000000000000000000
+0 | 0000000000000000000000000001111111111
+0 | 22333
+0 | 44444555
+0 | 67
+0 |
 1 |
 1 |
 1 | 5
```

Figure 6.6 Stem-and-leaf plot of weighted-median-polish residuals $\{n_i^{1/2}R(x_i, y_i): i = 1, \ldots, 100\}$. 1|5 represents 150. [*Source:* Cressie and Read, 1989]

The Variogram

Although usually defined for spatial processes whose index varies continuously over $\mathbb{R}^d$, this measure of spatial dependence can also be useful when working with lattice data. Consider a random process $\{\delta(\mathbf{s}): \mathbf{s} \in D\}$ with constant mean μ and

$$\mathrm{var}(\delta(\mathbf{s}_1) - \delta(\mathbf{s}_2)) = 2\gamma(\mathbf{s}_1 - \mathbf{s}_2), \qquad \mathbf{s}_1, \mathbf{s}_2 \in D,$$

a function only of the vector difference $\mathbf{s}_1 - \mathbf{s}_2$. Then $2\gamma(\cdot)$ is called the variogram of the random process δ (see Section 2.3 for more details). Because $E(\delta(\mathbf{s})) = \mu$, $2\gamma(\cdot)$ can be estimated by

$$2\hat{\gamma}(\mathbf{h}) = \sum_{N(\mathbf{h})} \left(\delta(\mathbf{s}_i) - \delta(\mathbf{s}_j)\right)^2 / |N(\mathbf{h})|,$$

where $N(\mathbf{h}) = \{(\mathbf{s}_i, \mathbf{s}_j): \mathbf{s}_i - \mathbf{s}_j = \mathbf{h}\}$. This and other estimators are discussed at some length in Section 2.4. If the process δ is isotropic, then $2\gamma(\mathbf{h}) \equiv 2\gamma^o(\|\mathbf{h}\|)$.

Based on data $\{n_i^{1/2}R(x_i, y_i): i = 1, \ldots, 100, i \neq 4\}$, an isotropic variogram estimator $2\hat{\gamma}^o(h)$ can be calculated and plotted as a function of distance $h \equiv \|\mathbf{h}\|$. Details of this part of the exploratory spatial analysis are given in Section 4.4, where Figure 4.16 can be interpreted as showing positive spatial dependence at distances up to 30 miles, whereas at greater distances zero spatial correlation is indicated.

6.2.4† Some Final Remarks

Based on a univariate exploratory data analysis, there appears to be an important spatial component to SIDS. Symons et al. (1983) notice it on a large scale (i.e., excesses seem to occur in the east, deficits in the west), and now it appears to be present on a small scale as well. This may be saying nothing more than neighboring counties are alike, but, assuming spatial continuity of county profiles throughout North Carolina, that in itself indicates certain *types* of counties (and by implication certain types of families) have higher SIDS risks associated with them than others. Another possible cause for county differences may be reporting bias.

Determining what characterizes high-risk counties is the next stage of the analysis. For the epidemiologist seeking a clue to the mystery of SIDS, it is the dependence of the SID rate on race, health care, and socioeconomic variables that is most interesting. Indeed, these "explanatory" variables may themselves have a strong spatial behavior that could explain completely the spatial behavior of the SID rate. Nevertheless, it would be wrong to fit a regression and assume that the errors from the model are statistically independent. Econometricians are aware of this when fitting their models to temporal data; epidemiologists should have the same concerns about their spatial data.

This section has developed exploratory spatial data analysis techniques that give the statistical modeler ideas about the (spatial) behavior of the data. Data transformation, trend removal, estimation of correlation, and so forth, are natural concepts for a time-series analyst to use, and should be equally so for a spatial data analyst. From these explorations, spatial models can be built, such as those presented in Sections 7.5 and 7.6. Section 7.6 gives a confirmatory analysis of the SIDS data and addresses such topics as efficient parameter estimation, model selection, prediction, and interpretation of the results back on the untransformed scale.

6.3 CONDITIONALLY AND SIMULTANEOUSLY SPECIFIED SPATIAL GAUSSIAN MODELS

A spatial process $\{\mathbf{Z}(\mathbf{s}): \mathbf{s} \in D\}$ on a lattice D is thought about differently from the continuous-spatial-index processes in Part I. When building models for data on a specific lattice, no possibility is given to a realization occurring between locations. Alamance county (county 1) borders Guilford county (county 41) in North Carolina, and no county is between them. Equally, there is no possibility of a measurement (of yield, say) in between adjacent trees in an orchard. These spatial models on lattices are analogues of time-series models (where, e.g., for monthly series, July 1989 immediately precedes August 1989; at the monthly level an intermediate measurement does not exist). This observation essentially resolves the question of how to develop an asymptotic theory. Here, it is more appropriate to think of the (finite) data set of $\mathbf{Z}$s as part of an infinite lattice whose sites tend out to infinity in at least one direction of Euclidean space. However, in Part I, where the spatial index $\mathbf{s}$ varies continuously over a domain D, it is often more reasonable to think of the locations of observations to be part of a finer and finer sampling within a fixed and bounded D (called infill asymptotics in Section 5.8); spatial analogues to time-series analysis may not be appropriate there.

The main objective in this section is to provide a couple of simple methods that use the spatial locations of sites to model the probability distribution of $\{\mathbf{Z}(\mathbf{s}): \mathbf{s} \in D\}$. In Chapter 7, I treat in some detail estimation of model parameters and the properties of those estimators.

Unless stated otherwise, suppose henceforth that the process $\{\mathbf{Z}(\mathbf{s}): \mathbf{s} \in D\}$ is a univariate process that is written as $\{Z(\mathbf{s}): \mathbf{s} \in D\}$. Because time-series techniques and models are familiar to many, the link between space and time will be emphasized as often as possible.

Markov chain

Suppose that $\{Z(t): t = 0, 1, \ldots\}$ is a random process in time and that the data are observed at the times $\{0, 1, \ldots, n\}$. For illustrative purposes, let the Zs be discrete-valued variables.

Notation. Write $\Pr(Z(t_1) = z(t_1), \ldots, Z(t_I) = z(t_I) | Z(t_{I+1}) = z(t_{I+1}), \ldots, Z(t_{I+J}) = z(t_{I+J}))$, a conditional probability of the process, as $\Pr(z(t_1), \ldots, z(t_I) | z(t_{I+1}), \ldots, z(t_{I+J}))$. ■

The most obvious departure from the independence model is to assume that $\{Z(t): t = 0, 1, \ldots\}$ is a Markov chain. One definition of this is through joint probabilities:

$$\Pr(z(1), \ldots, z(i) | z(0)) = \prod_{t=1}^{i} Q_t(z(t); z(t-1)), \quad \text{for all } i \geq 1, \quad (6.3.1)$$

where Q_t is some function of $z(t)$ and $z(t-1)$, generic realizations of $Z(t)$ and $Z(t-1)$. Notice that the Zs are independent if and only if $Q_t(z(t); z(t-1)) = Q_t^*(z(t))$, a function only of $z(t)$.

There is an equivalent, and probably more familiar, definition of a Markov chain based on conditional probabilities:

$$\Pr(z(i) | z(0), \ldots, z(i-1)) = \Pr(z(i) | z(i-1)), \quad \text{for all } i \geq 1. \quad (6.3.2)$$

This expresses the "lack of memory" property that says only the most recent past determines conditional probabilities about the present given the past.

That (6.3.2) implies (6.3.1) is easy to demonstrate:

$$\begin{aligned} \Pr(z(1), \ldots, z(i) | z(0)) &= \prod_{t=1}^{i} \Pr(z(t) | z(0), \ldots, z(t-1)) \\ &= \prod_{t=1}^{i} \Pr(z(t) | z(t-1)). \end{aligned}$$

The converse is obtained by writing

$$\begin{aligned} \Pr(z(i) | z(0), \ldots, z(i-1)) &= \frac{\Pr(z(1), \ldots, z(i) | z(0))}{\Pr(z(1), \ldots, z(i-1) | z(0))} \\ &= \frac{\prod_{t=1}^{i} Q_t(z(t); z(t-1))}{\sum_{z(i)} \prod_{t=1}^{i} Q_t(z(t); z(t-1))} \\ &= \frac{Q_i(z(i); z(i-1))}{\sum_{z(i)} Q_i(z(i); z(i-1))}, \end{aligned}$$

which does not depend on $z(0), \ldots, z(i-2)$.

Spatial Dependence

The equivalence of joint-probability and conditional-probability definitions in the time domain relies upon the unidirectional flow of time. The two

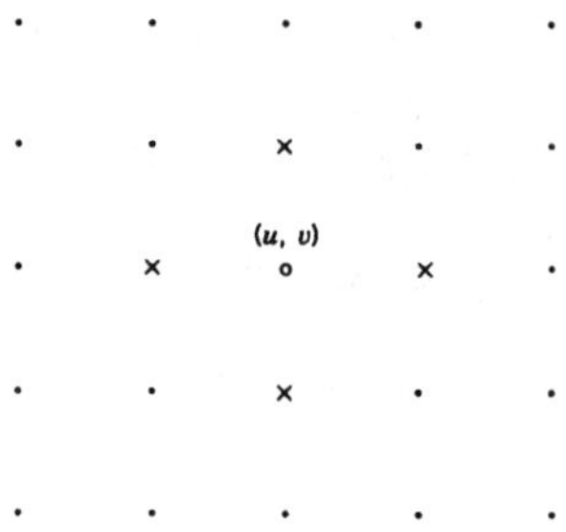

Figure 6.7 Square lattice showing central site (marked as ∘) and its four nearest neighbors (marked as ×).

approaches, however, give different models in the spatial domain. For simplicity, consider the square lattice in the plane $\mathbb{R}^2$:

$$D = \{\mathbf{s} = (u, v)' : u = \ldots, -2, -1, 0, 1, 2, \ldots; v = \ldots, -2, -1, 0, 1, 2, \ldots\}.$$

An obvious analogue to the joint approach of (6.3.1) is to assume

$$\Pr(\mathbf{z}) = \prod_{(u,v)\in D} Q_{uv}(z(u,v); z(u-1,v), z(u+1,v), z(u,v-1), z(u,v+1)), \quad (6.3.3)$$

which expresses the probability of $\{Z(u,v): (u,v)' \in D\}$ as a product of functions each depending on a site $(u,v)'$ and its four nearest-neighbor sites; Figure 6.7 illustrates this spatial structure. Models of the type (6.3.3) have received little attention, except in the article by Whittle (1963).

Conditional Approach

The spatial conditional approach assumes

$$\begin{aligned} &\Pr(z(u,v) \mid \{z(k,l): (k,l) \neq (u,v)\}) \\ &\quad = \Pr(z(u,v) \mid z(u-1,v), z(u+1,v), z(u,v-1), z(u,v+1)), \\ &\qquad \text{for all } (u,v)' \in D, \quad (6.3.4) \end{aligned}$$

using a similar notational convention as for the temporal case. This expresses dependence of $Z(u,v)$ on the four nearest-neighbor sites shown in Figure 6.7, in a different way to (6.3.3). Early references to assumption (6.3.4) include Bartlett (1955, Section 2.2; 1967; 1968), Dobrushin (1968), and Besag (1974).

The joint approach given by (6.3.3) is not, in general, equivalent to the conditional approach given by (6.3.4). In terms of building spatial models for Markov-type departures from independence, it is the conditional approach that turns out to be more natural. In particular, for Gaussian data, the conditionally specified model has a particularly simple joint distribution; see Section 6.3.2. The purpose of this section is to contrast this spatial model with another Gaussian model that is based on a *simultaneous* specification of the spatial dependencies (Whittle, 1954). In general, the models are different, and there has been some debate as to which approach is preferable (Besag, 1974; Cliff and Ord, 1975; Haining, 1979).

6.3.1 Simultaneously Specified Spatial Gaussian Models

Whittle (1954) introduced the following prescription for a class of stationary processes in the plane. Suppose $\{\epsilon(u,v)\colon u = \ldots,-1,0,1,\ldots;\ v = \ldots,-1,0,1,\ldots\}$ is a process of independent-and-identically distributed random variables. Then, define the process $\{Z(u,v)\}$ by

$$\phi(T_1,T_2)Z(u,v) = \epsilon(u,v), \tag{6.3.5}$$

where T_1 and T_2 are translation operators defined by

$$T_1Z(u,v) = Z(u+1,v), \qquad T_1^{-1}Z(u,v) = Z(u-1,v),$$
$$T_2Z(u,v) = Z(u,v+1), \qquad T_2^{-1}Z(u,v) = Z(u,v-1)$$

and ϕ is given by

$$\phi(T_1,T_2) = \sum_i \sum_j a_{ij}T_1^iT_2^j, \tag{6.3.6}$$

where the summations are potentially over all integers. This simultaneous specification of variables $\{Z(u,v)\}$ in (6.3.5) is analogous to the way autoregressive models are specified in time series.

The range of the summation in (6.3.6) determines the degree of departure from independence. An example is the nearest-neighbor dependence described by Figure 6.7:

$$\phi(T_1,T_2) = 1 - \xi_1(T_1 + T_1^{-1}) - \xi_2(T_2 + T_2^{-1}).$$

In order for a stationary spatial autoregressive process of the form (6.3.5) to exist [with a finite range of summation in (6.3.6)], it is necessary and sufficient that $\phi(c_1,c_2)$, as a function of the two complex variables c_1, c_2, not be zero for any c_1, c_2 that simultaneously satisfy $|c_1| = 1$, $|c_2| = 1$ (Whittle, 1954; Rosanov, 1967).

Gaussian (Normal) Model

For the moment assume that $\{Z(\mathbf{s})\colon \mathbf{s} \in D\} = \{Z(\mathbf{s}_i)\colon i = 1,\ldots,n\}$ is defined on a finite subset of the integer lattice in the plane. Let $\boldsymbol{\epsilon} \sim \text{Gau}(\mathbf{0},\Lambda)$, be a joint ($n$-dimensional) Gaussian distribution with mean $\mathbf{0}$ and diagonal variance matrix Λ (e.g., $\Lambda = \sigma^2I$). The elements of $\boldsymbol{\epsilon}$ are also indexed by their locations $\{\mathbf{s}_i\colon i = 1,\ldots,n\}$ on the lattice.

Let $B = (b_{ij})$ denote a matrix to be interpreted as the spatial-dependence matrix for this simultaneous modeling approach. Intuitively, if $Z(\mathbf{s}_1)$ is thought to depend positively on, say, $Z(\mathbf{s}_2)$, then specify $b_{12} > 0$; and if $Z(\mathbf{s}_1)$ is thought not to depend on, say, $Z(\mathbf{s}_6)$, then specify $b_{16} = 0$. (It will be seen below that this intuition is oversimplistic.) It is assumed that $b_{ii} = 0$, for $i = 1, \ldots, n$, and that $(I - B)^{-1}$ exists (Ripley, 1981, p. 89); it is *not* a requirement of the model that $b_{ij} = b_{ji}$. Then, *define* $\mathbf{Z} = (Z(\mathbf{s}_1), \ldots, Z(\mathbf{s}_n))'$ by

$$(I - B)(\mathbf{Z} - \boldsymbol{\mu}) = \boldsymbol{\epsilon}. \tag{6.3.7}$$

Clearly, $E(\mathbf{Z}) = \boldsymbol{\mu}$ and $\mathrm{var}(\mathbf{Z}) \equiv E\{(\mathbf{Z} - \boldsymbol{\mu})(\mathbf{Z} - \boldsymbol{\mu})'\} = (I - B)^{-1}\Lambda(I - B')^{-1}$. Also, $(\mathbf{Z} - \boldsymbol{\mu})$ is a linear combination of $\boldsymbol{\epsilon}$, which is jointly Gaussian, so that

$$\mathbf{Z} \sim \mathrm{Gau}\left(\boldsymbol{\mu}, (I - B)^{-1}\Lambda(I - B')^{-1}\right). \tag{6.3.8}$$

Spatial Autoregression

Equation (6.3.7) can be written equivalently as

$$Z(\mathbf{s}_i) = \mu_i + \sum_{j=1}^{n} b_{ij}\left(Z(\mathbf{s}_j) - \mu_j\right) + \epsilon_i, \qquad i = 1, \ldots, n, \tag{6.3.9}$$

which shows more transparently how (6.3.7) is the spatial analogue of the autoregressive time-series model (recall $b_{ii} = 0$, for all i). For example, data $(Z(\mathbf{s}_1), \ldots, Z(\mathbf{s}_n))$ from the nearest-neighbor model given by (6.3.5) with $\phi(T_1, T_2) = 1 - \xi_1(T_1 + T_1^{-1}) - \xi_2(T_2 + T_2^{-1})$, can be generated via (6.3.7) as follows: Suppose $\mathbf{s}_1 = (u, v)'$, $\mathbf{s}_2 = (u - 1, v)'$, $\mathbf{s}_3 = (u + 1, v)'$, $\mathbf{s}_4 = (u, v - 1)'$, and $\mathbf{s}_5 = (u, v + 1)'$. Then $b_{12} = b_{13} = \xi_1$, $b_{14} = b_{15} = \xi_2$, and all other $b_{1j} = 0$. Similarly, the site at $\mathbf{s}_2$ is dependent on its four nearest neighbors, and so forth. [Edge effects are a problem that I discuss in Section 6.7. Clearly, any site in the data set can be made to depend in various ways on any other site. The problem is to specify the $\{b_{ij}: j = 1, \ldots, n, j \neq i\}$ when $\mathbf{s}_i$ is located on the edge of D, so that $\mathbf{Z} = (Z(\mathbf{s}_1), \ldots, Z(\mathbf{s}_n))'$ is a part of the planar process defined by (6.3.5).]

In order to estimate the parameters $\boldsymbol{\mu}$, B, and Λ of (6.3.7), the likelihood,

$$(2\pi)^{-n/2}|\Lambda|^{-1/2}|I - B|\exp\{-(1/2)(\mathbf{z} - \boldsymbol{\mu})'(I - B') \times \Lambda^{-1}(I - B)(\mathbf{z} - \boldsymbol{\mu})\}, \tag{6.3.10}$$

is maximized. Computational details are given in Section 7.2.

Notice that $\mathrm{cov}(\boldsymbol{\epsilon}, \mathbf{Z}) = E(\boldsymbol{\epsilon}\mathbf{Z}') = \Lambda(I - B')^{-1}$, which is *not* diagonal; that is, the error is *not* independent of the autoregressive variables, unlike in time-series autoregression. Consequently, least-squares estimators of spatial-dependence parameters are not guaranteed to be consistent (Whittle, 1954).

In Section 6.7, the simultaneously specified Gaussian model (6.3.7) will be extended to spatial-moving-average models and to regression models with spatially correlated errors. The intervening sections of this chapter concentrate on another type of model, obtained through specification of conditional-probability distributions. Section 6.3.2 introduces a conditionally specified Gaussian model that is compared to (6.3.7) in Section 6.3.3.

6.3.2 Conditionally Specified Spatial Gaussian Models

Earlier in this section, I indicated that spatial Markov-type departures from independence were more satisfactorily modeled via a conditional approach. It will be shown in Section 6.4 that, for Gaussian data, (6.3.4) can be written as

$$f\left(z(\mathbf{s}_i)|\{z(\mathbf{s}_j): j \neq i\}\right)$$
$$= \left(2\pi\tau_i^2\right)^{-1/2} \exp\left[-\left\{z(\mathbf{s}_i) - \theta_i\left(\{z(\mathbf{s}_j): j \neq i\}\right)\right\}^2 \Big/ 2\tau_i^2\right],$$
$$i = 1,\ldots,n, \quad (6.3.11)$$

where f denotes the conditional density of $Z(\mathbf{s}_i)$ given $\{Z(\mathbf{s}_j) = z(\mathbf{s}_j): j = 1,\ldots,n;\ j \neq i\}$ and $\theta_i(\cdot)$ and τ_i^2 are its conditional mean and variance, respectively. Under a regularity condition of "pairwise-only dependence" between sites

$$\theta_i\left(\{z(\mathbf{s}_j): j \neq i\}\right) = \mu_i + \sum_{j=1}^{n} c_{ij}\left(z(\mathbf{s}_j) - \mu_j\right), \qquad i = 1,\ldots,n, \quad (6.3.12)$$

where $c_{ij}\tau_j^2 = c_{ji}\tau_i^2$, $c_{ii} = 0$, and $c_{ik} = 0$ unless there is pairwise dependence between site i and site k. This result is established in Section 6.4, along with the theorem

$$\mathbf{Z} \sim \text{Gau}\left(\boldsymbol{\mu}, (I - C)^{-1}M\right), \qquad (6.3.13)$$

provided $(I - C)$ is invertible and $(I - C)^{-1}M$ is symmetric and positive-definite; here, $\mathbf{Z} \equiv (Z(\mathbf{s}_1),\ldots,Z(\mathbf{s}_n))'$, $\boldsymbol{\mu} \equiv (\mu_1,\ldots,\mu_n)'$, $C \equiv (c_{ij})$ is an $n \times n$ matrix whose (i, j)th element is c_{ij}, and $M \equiv \text{diag}(\tau_1^2,\ldots,\tau_n^2)$ is an $n \times n$ diagonal matrix.

Specification of the spatial dependencies $\{c_{ij}\}$ near the edge of the finite lattice D can be problematic; discussion is left to Section 6.6. Estimation of parameters $\boldsymbol{\mu}$, C, and M of (6.3.13) is usually achieved through maximizing the likelihood

$$(2\pi)^{-n/2}|M|^{-1/2}|I - C|^{1/2} \exp\{-(1/2)(\mathbf{z} - \boldsymbol{\mu})' M^{-1}(I - C)(\mathbf{z} - \boldsymbol{\mu})\}. \tag{6.3.14}$$

Computational details are given in Section 7.2.

The variance matrix in (6.3.13) is in direct contrast with the variance matrix in (6.3.8). Of course, when $(I - C)^{-1}M = (I - B)^{-1}\Lambda(I - B')^{-1}$ they yield the same model, but clearly $\{b_{ij}\}$ and $\{c_{ij}\}$ *cannot* have the same interpretation. (Although, they are both measures of spatial dependence.) Because the form of the spatial dependence is usually chosen by the modeler, the conclusions will differ markedly according to whether the modeler thinks conditionally or thinks simultaneously.

Spatial Autoregression

That (6.3.12) qualifies more than (6.3.9) to be called the spatial analogue of an autoregressive time series, can be seen if pseudoerrors $\boldsymbol{\nu}$ are *defined* via $\boldsymbol{\nu} \equiv (I - C)(\mathbf{Z} - \boldsymbol{\mu})$. That is,

$$Z(\mathbf{s}_i) - \mu_i \equiv \sum_{j=1}^{n} c_{ij}\big(Z(\mathbf{s}_j) - \mu_j\big) + \nu_i, \qquad i = 1, \cdots, n. \tag{6.3.15}$$

Although $\text{var}(\boldsymbol{\nu}) = M(I - C')$, which is not diagonal, it is easy to see that $E(\boldsymbol{\nu}\mathbf{Z}') = M$. That is, ν_i is independent of $\{Z(\mathbf{s}_j): j \neq i\}$, in contrast with the errors of the simultaneously specified Gaussian model.

6.3.3 Comparison

Brook (1964) was the first to make the distinction between the simultaneous specification and the conditional specification of spatial models. Here, for the Gaussian case, the merits of each will be discussed.

Henceforth, I shall use SG to denote the *simultaneously specified Gaussian* model (6.3.7) or (6.3.9) and CG to denote the *conditionally specified Gaussian* model (6.3.11) or (6.3.15). It has been noted that for SG, ϵ_i is correlated with $\{Z(\mathbf{s}_j): j \neq i\}$, but for CG, ν_i is uncorrelated with $\{Z(\mathbf{s}_j): j \neq i\}$. Thus, in terms of estimation and interpretation of model parameters, CG is to be preferred. For time-series autoregressions, parameters are estimated consistently and efficiently by least squares; however, Whittle (1954) shows that least-squares estimators of SG parameters are inconsistent. Finally, CG defined through (6.3.11) and (6.3.12) immediately gives the best (minimum mean squared prediction error) interpolation of $Z(\mathbf{s}_i)$ based on $\{z(\mathbf{s}_j): j \neq i\}$, namely, the right-hand side of (6.3.12).

Equivalence of SG and CG

Assuming that the mean level μ has been successfully modeled (by no means a trivial task, but one that has no effect on choice between SG and CG), the two models are equivalent if and only if their variance matrices are equal:

$$(I - C)^{-1}M = (I - B)^{-1}\Lambda(I - B')^{-1}. \tag{6.3.16}$$

Because M is diagonal, it is clear that *any* SG can be represented as a CG, but not necessarily *vice versa*. A simple example is provided by starting with a first-order SG on a $U \times V$ rectangular lattice, with $\Lambda = \sigma^2 I$. For the purposes of this example, the regular lattice D is such that $4 < U, V < \infty$ and is wrapped on a torus (the two-dimensional equivalent of a circle). This can be formalized by imagining extra sites $\{(1, V + 1), \dots, (U, V + 1), (U + 1, 1), \dots, (U + 1, V)\}$, but then identifying them with the left-hand and bottom sites of D, so that $(u, V + l) \equiv (u, l)$ and $(U + k, v) \equiv (k, v)$. The first-order SG, which involves first-nearest neighbors, is

$$\begin{aligned} Z(u,v) = {} & \xi_1(Z(u-1,v) + Z(u+1,v)) \\ & + \xi_2(Z(u,v-1) + Z(u,v+1)) + \epsilon(u,v). \end{aligned} \tag{6.3.17}$$

Order the $n = UV$ sites as $(1,1), (1,2), \dots, (1,V), (2,1), \dots, (U,V)$ and write (6.3.17) as $(I - B)\mathbf{Z} = \boldsymbol{\epsilon}$, where entries of B are mostly zero except for two ξ_1s and two ξ_2s in each row and $\boldsymbol{\epsilon} \sim \text{Gau}(\mathbf{0}, \sigma^2 I)$. Now $(I - B)(I - B')/\sigma^2 = (I - 2B + B^2)/\sigma^2$, because B is symmetric in this example. Hence from (6.3.16) an equivalent CG model is found by solving $M^{-1}(I - C) = (I - 2B + B^2)/\sigma^2$. Matching matrix entries yields $M = \{\sigma^2/(1 - 2\xi_1^2 - 2\xi_2^2)\}I$ [i.e., $\tau^2 = \sigma^2/(1 - 2\xi_1^2 - 2\xi_2^2)$] and a CG model that involves third-nearest neighbors:

$$\begin{aligned} & E(Z(u,v) \mid \{z(k,l) : (k,l) \neq (u,v)\}) \\ & = \{1 - 2\xi_1^2 - 2\xi_2^2\}^{-1}\{2\xi_1(z(u-1,v) + z(u+1,v)) \\ & \qquad + 2\xi_2(z(u,v-1) + z(u,v+1)) \\ & \qquad + 2\xi_1\xi_2(z(u-1,v-1) + z(u+1,v-1) \\ & \qquad\qquad + z(u-1,v+1) + z(u+1,v+1)) \\ & \qquad + \xi_1^2(z(u-2,v) + z(u+2,v)) \\ & \qquad + \xi_2^2(z(u,v-2) + z(u,v+2))\}. \end{aligned}$$

Consequently, a CG model of *third order* results from a symmetric SG of first order. Thus, a first- or second-order CG model generally has no equivalent SG model, aside from special cases.

Further Comparisons

Notice that CG must have a *symmetric* spatial dependence matrix C, whereas the spatial dependence matrix B in SG need not be symmetric. At first sight this may appear to be an advantage for SG, but inspection of (6.3.10) reveals that the spatial dependence parameters $\{b_{ij}\}$ are estimated through $(I - B')\Lambda^{-1}(I - B)$, leading to possible nonidentifiability of these parameters. For those modelers who do not want site i to depend on site j in a symmetric way, a *space–time* model (see Section 6.8) is probably more appropriate; cause–effect components of this model naturally give rise to asymmetries. However, a purely spatial model should be used for data that have achieved temporal stationarity; then, symmetric dependencies are usually a natural consequence.

Consider the set of all stationary models with a given finite set of covariances (including the variances). Then the CG model achieves maximum entropy among all such models (Kunsch, 1981), a property not satisfied by the SG models.

Finding the maximum likelihood estimates of CG parameters involves the same numerical difficulties as for SG parameters. An iterative solution is slowed considerably by the need to evaluate the determinant $|I - C|$ in (6.3.14) at each stage of the iteration. Inspection of (6.3.10) shows that the same problem occurs with $|I - B|$. More details are given in Section 7.2.

Econometricians usually build their models according to the SG approach. In the model $(I - B)(\mathbf{Z} - \boldsymbol{\mu}) = \boldsymbol{\epsilon}$, the vector $\boldsymbol{\mu}$ could be called a vector of exogenous variables and $\mathbf{Z}$ the endogenous variables. (Of course, econometricians use systems that also develop over time, but that does not affect the general issues here.) The simultaneous econometric method can be viewed as a perfectly valid way of making deterministic relationships stochastic. But it seems to me that the parameters estimated from such equations are often interpreted incorrectly, as if they come from a conditional-expectation structure. If a conditional-expectation interpretation is more natural, then the CG approach should be adopted at the outset.

6.4* MARKOV RANDOM FIELDS

This section is largely review, giving the results needed for construction of the likelihood from a conditional specification. The fundamental contribution of Besag (1974) should be noted, from which much of the material presented here was drawn.

For Gaussian data, Section 6.3.2 states that, under certain conditions, the conditional (auto-Gaussian) specification (6.3.11) implies the joint Gaussian distribution (6.3.13). Thus, from (6.3.14), the negative loglikelihood is given by

$$L(\boldsymbol{\mu}, M, C) = (n/2)\log(2\pi) + (1/2)\log|M| - (1/2)\log|I - C| + (1/2)(\mathbf{z} - \boldsymbol{\mu})' M^{-1}(I - C)(\mathbf{z} - \boldsymbol{\mu}), \qquad (6.4.1)$$

which is to be *minimized* with respect to mean parameters $\boldsymbol{\mu}$, variance parameters $M = \text{diag}(\tau_1^2, \ldots, \tau_n^2)$, and spatial dependence parameters C.

This section develops, in the general case, the links between the conditional specification and the likelihood for a finite collection of sites. Care must be taken when specifying a model at the local level (via conditioning) to ensure that the pieces fit together in a noncontradictory, consistent way. Under conditions in the following text, the construction is straightforward.

Not so straightforward is the development of the local Markov models on a *countable* collection of sites, such as the integer coordinates in $\mathbb{R}^d$. Albeverio et al. (1981) give results for the case when these models also possess a global Markov property and so have a transition matrix for the states of the lattice. Questions of uniqueness and identifiability are deferred until Section 6.4.3.

Positivity Condition

Let $\{\mathbf{s}_i: i = 1, \ldots, n\}$ be the locations on a lattice at which a discrete variable Z is observed. Define $\zeta \equiv \{\mathbf{z}: \Pr(\mathbf{z}) > 0\}$ and $\zeta_i \equiv \{z(\mathbf{s}_i): \Pr(z(\mathbf{s}_i)) > 0\}$, $i = 1, \ldots, n$. Then the *positivity condition* is said to be satisfied if $\zeta = \zeta_1 \times \cdots \times \zeta_n$. For a continuous variable Z, the same definition applies except $\Pr(\cdot)$ is replaced with $f(\cdot)$, representing the joint and marginal densities.

The example that follows, where the positivity condition does *not* hold, is due to J. E. Besag. Consider a *contact* model of infection, and associate with each site a variable $z(\cdot)$ that is 1 if the site is infected and 0 if it is healthy. Consider the evolution of a site from healthy to infected to be a one-way process and suppose the sites are on the square lattice in $\mathbb{R}^2$. Let $\Pr(z(u, v)|\{z(k, l): (k, l) \neq (u, v)\})$, the conditional probability of observing $z(u, v)$ at (u, v), depend only on $z(u - 1, v)$, $z(u + 1, v)$, $z(u, v - 1)$, and $z(u, v + 1)$, the four nearest neighbors. Because of the one-way contact nature of infection, an event like

$$\begin{matrix} & 0 & \\ 0 & 1 & 0 \\ & 0 & \end{matrix} \qquad (6.4.2)$$

must have probability 0.

Suppose that the infection is started with two consecutive sites infected, at $(0, 0)$ and $(1, 0)$. After evolving for a fixed time, the pattern of 0s and 1s is taken as the data and analyzed via a spatial model. It is now shown that the positivity condition (and hence the Markov random-field approach to be given) is inappropriate for these data.

Consider the random variable $Z(u, v)$ with neighboring values and extra values, appearing as

$$\begin{matrix} z(u-2, v+1) & 0 & z(u, v+1) & z(u+1, v+1) \\ 0 & 1 & Z(u, v) & z(u+1, v) \\ z(u-2, v-1) & 0 & z(u, v-1) & z(u+1, v-1) \end{matrix} .$$

$\Pr\{Z(u, v) = 0|$"neighboring values" and "extra values"$\}$ should be 0, because otherwise it would result in the excluded event (6.4.2). But, given only the neighboring values of $Z(u, v)$,

$$\begin{matrix} & z(u, v+1) & \\ 1 & Z(u, v) & z(u+1, v), \\ & z(u, v-1) & \end{matrix}$$

$\Pr\{Z(u, v) = 0|$"neighboring values"$\}$ will not be zero. Thus, conditional probability models for data of this kind cannot be of the simple nearest-neighbor variety. The positivity condition fails here because each of the individual site realizations has positive probability, even though the realization (6.4.2) has zero probability.

Notation. If the data are discrete, then $\Pr(\cdot)$ will be used to denote (conditional) probabilities of events; for continuous data, $f(\cdot)$ will be used to denote (conditional) densities. For example, $\Pr(\{z(\mathbf{s}_i): i \in E_1\}|\{z(\mathbf{s}_j): j \in E_2\})$ denotes the conditional probability that $\{Z(\mathbf{s}_i) = z(\mathbf{s}_i): i \in E_1\}$, for given $\{Z(\mathbf{s}_j) = z(\mathbf{s}_j): j \in E_2\}$, for $E_1, E_2 \subset D$. ■

Write the data as $\mathbf{Z} \equiv (Z(\mathbf{s}_1), \dots, Z(\mathbf{s}_n))'$, where $\{\mathbf{s}_1, \dots, \mathbf{s}_n\} \subset D$ denotes the spatial locations of the n sites at which observations are available.

Consistency of Conditional Probabilities

From the joint probabilities $\{\Pr(\mathbf{z}): \mathbf{z} \in \zeta\}$, the conditional probabilities at each site $\{\Pr(z(\mathbf{s}_i)|\{z(\mathbf{s}_j): j \neq i\}): i = 1, \dots, n\}$ can be calculated. Conversely, any conditional specification must satisfy the following theorem.

Factorization Theorem (Besag, 1974) Suppose the variables $\{Z(\mathbf{s}_i): i = 1, \dots, n\}$ have joint probability mass function $\Pr(\cdot)$, whose support ζ satisfies the positivity condition. Then,

$$\frac{\Pr(\mathbf{z})}{\Pr(\mathbf{y})} = \prod_{i=1}^{n} \frac{\Pr(z(\mathbf{s}_i)|z(\mathbf{s}_1), \dots, z(\mathbf{s}_{i-1}), y(\mathbf{s}_{i+1}), \dots, y(\mathbf{s}_n))}{\Pr(y(\mathbf{s}_i)|z(\mathbf{s}_1), \dots, z(\mathbf{s}_{i-1}), y(\mathbf{s}_{i+1}), \dots, y(\mathbf{s}_n))}, \quad \mathbf{z}, \mathbf{y} \in \zeta, \tag{6.4.3}$$

where $\mathbf{y} \equiv (y(\mathbf{s}_1), \ldots, y(\mathbf{s}_n))'$, $\mathbf{z} \equiv (z(\mathbf{s}_1), \ldots, z(\mathbf{s}_n))'$ are possible realizations of $\mathbf{Z}$.

Proof

$$\Pr(\mathbf{z}) = \Pr\left(z(\mathbf{s}_n) | \{z(\mathbf{s}_j) : j \neq n\}\right) \cdot \Pr\left(\{z(\mathbf{s}_j) : j \neq n\}\right)$$

$$= \frac{\Pr\left(z(\mathbf{s}_n) | \{z(\mathbf{s}_j) : j \neq n\}\right) \cdot \Pr\left(\{z(\mathbf{s}_j) : j \neq n\}, y(\mathbf{s}_n)\right)}{\Pr\left(y(\mathbf{s}_n) | \{z(\mathbf{s}_j) : j \neq n\}\right)},$$

for some $y(\mathbf{s}_n) \in \zeta_n$. Under the positivity condition, the denominator of the last expression is positive.

Now,

$$\Pr(\{z(\mathbf{s}_j) : j \neq n\}, y(\mathbf{s}_n))$$

$$= \Pr(z(\mathbf{s}_{n-1}) | \{z(\mathbf{s}_i) : i \neq n-1, n\}, y(\mathbf{s}_n)) \cdot \Pr(\{z(\mathbf{s}_i) : i \neq n-1, n\}, y(\mathbf{s}_n))$$

$$= \frac{\Pr(z(\mathbf{s}_{n-1}) | z(\mathbf{s}_1), \ldots, z(\mathbf{s}_{n-2}), y(\mathbf{s}_n)) \cdot \Pr(z(\mathbf{s}_1), \ldots, z(\mathbf{s}_{n-2}), y(\mathbf{s}_{n-1}), y(\mathbf{s}_n))}{\Pr(y(\mathbf{s}_{n-1}) | z(\mathbf{s}_1), \ldots, z(\mathbf{s}_{n-2}), y(\mathbf{s}_n))},$$

for some $y(\mathbf{s}_{n-1}) \in \zeta_{n-1}$. Again the positivity condition is used to guarantee that the last expression is well defined. By continuing in the same way, the theorem is proved. ■

From (6.4.3), the joint probability is in principle available from the conditional probabilities by recalling that $\Sigma_{\mathbf{y} \in \zeta} \Pr(\mathbf{y}) = 1$. But the normalizing constant, which depends on the model parameters to be estimated, is not usually available in closed form. An exception is the CG model specified by (6.3.11).

The factorization theorem has a version for continuous data, where $\Pr(\cdot)$ is replaced with $f(\cdot)$; then the normalizing condition is $\int_\zeta f(\mathbf{y})\, d\mathbf{y} = 1$. Through this, a result claimed in Section 6.3.2 will now be proved.

Proposition The conditional (auto-Gaussian) specification (6.3.11) and (6.3.12) implies that

$$\mathbf{Z} \sim \text{Gau}\left(\boldsymbol{\mu}, (I - C)^{-1} M\right),$$

provided $(I - C)$ is invertible and $(I - C)^{-1}M$ is symmetric and positive-definite.

Proof. Use the factorization theorem for densities and take $\mathbf{y} = \boldsymbol{\mu}$ in (6.4.3). Therefore,

$$\begin{aligned}
&\log(f(\mathbf{z})/f(\boldsymbol{\mu}))\\
&\quad = -(1/2)\sum_{i=1}^{n}\left\{z(\mathbf{s}_i) - \mu(\mathbf{s}_i) - \sum_{j=1}^{i-1} c_{ij}(z(\mathbf{s}_j) - \mu(\mathbf{s}_j))\right\}^2 \Big/ \tau_i^2\\
&\qquad + (1/2)\sum_{i=1}^{n}\left\{\sum_{j=1}^{i-1} c_{ij}(z(\mathbf{s}_j) - \mu(\mathbf{s}_j))\right\}^2 \Big/ \tau_i^2\\
&\quad = -(1/2)\sum_{i=1}^{n}(z(\mathbf{s}_i) - \mu(\mathbf{s}_i))^2/\tau_i^2\\
&\qquad + \sum_{i=1}^{n}\sum_{j=1}^{i-1} c_{ij}(z(\mathbf{s}_i) - \mu(\mathbf{s}_i))(z(\mathbf{s}_j) - \mu(\mathbf{s}_j))/\tau_i^2\\
&\quad = -(1/2)(\mathbf{z} - \boldsymbol{\mu})'M^{-1}(I - C)(\mathbf{z} - \boldsymbol{\mu}).
\end{aligned}$$

The right-hand side is the exponent of an n-dimensional Gaussian distribution with mean $\boldsymbol{\mu}$ and variance matrix $(I - C)^{-1}M$. ■

The Factorization Theorem shows how severe the conditional-probability consistency conditions can be. Because there are $n!$ ways of ordering the sites, there are $n!$ factorizations of $\Pr(\mathbf{z})/\Pr(\mathbf{y})$, which must all be equal. More discussion of such conditions is delayed until some important definitions and results are established.

6.4.1* Neighbors, Cliques, and the Negpotential Function Q

Definition. A site k is defined to be a *neighbor* of site i if the conditional distribution of $Z(\mathbf{s}_i)$, given all other site values, depends functionally on $z(\mathbf{s}_k)$, for $k \neq i$. Also define

$$N_i \equiv \{k : k \text{ is a neighbor of } i\} \tag{6.4.4}$$

to be the *neighborhood set* of site i. ■

For example, the neighborhood structure of Table 6.1 for the counties of North Carolina is defined by distance between county seats. Thus, $N_1 = \{17, 19, 32, 41, 68\}$, $N_2 = \{14, 18, 49, 97\}$, and so forth. It is not difficult to show that $k \in N_i$ if and only if $i \in N_k$.

Definition. A *clique* is defined to be a set of sites that consists either of a single site or of sites that are all neighbors of each other. ■

For example, a quick check of Table 6.1 shows that $\{1, 17, 68\}$ is a clique.

Definition. Any probability measure whose conditional distributions define a neighborhood structure $\{N_i: i = 1, \ldots, n\}$ through (6.4.4) is defined to be a *Markov random field*. ■

Without loss of generality, assume that zero can occur at each site; that is, $\mathbf{0} \in \zeta$. *Define*

$$Q(\mathbf{z}) \equiv \log\{\Pr(\mathbf{z})/\Pr(\mathbf{0})\}, \qquad \mathbf{z} \in \zeta. \tag{6.4.5}$$

Then knowledge of $Q(\cdot)$ is equivalent to knowledge of $\Pr(\cdot)$, because

$$\Pr(\mathbf{z}) = \exp(Q(\mathbf{z})) \Big/ \sum_{\mathbf{y} \in \zeta} \exp(Q(\mathbf{y}))$$

in the discrete case. An analogous Q function in the continuous case can be defined by replacing $\Pr(\cdot)$ with $f(\cdot)$ and Σ with $\int$. In the statistical mechanics literature (e.g., Ruelle, 1969), $-Q$ plays the role of a potential energy function, and the normalizing term $\Sigma_{\mathbf{y} \in \zeta} \exp(Q(\mathbf{y}))$ is called the partition function. Hereafter, Q will be referred to as the *negpotential function*.

Proposition The negpotential function Q satisfies the following two properties:

i.
$$\frac{\Pr\left(z(\mathbf{s}_i)|\{z(\mathbf{s}_j): j \neq i\}\right)}{\Pr\left(0(\mathbf{s}_i)|\{z(\mathbf{s}_j): j \neq i\}\right)} = \frac{\Pr(\mathbf{z})}{\Pr(\mathbf{z}_i)} = \exp(Q(\mathbf{z}) - Q(\mathbf{z}_i)), \tag{6.4.6}$$

where $0(\mathbf{s}_i)$ denotes the event $Z(\mathbf{s}_i) = 0$ and $\mathbf{z}_i \equiv (z(\mathbf{s}_1), \ldots, z(\mathbf{s}_{i-1}), 0, z(\mathbf{s}_{i+1}), \ldots, z(\mathbf{s}_n))'$.

ii. Q can be expanded uniquely on ζ as

$$\begin{aligned} Q(\mathbf{z}) = &\sum_{1 \le i \le n} z(\mathbf{s}_i) G_i(z(\mathbf{s}_i)) + \sum\sum_{1 \le i < j \le n} z(\mathbf{s}_i) z(\mathbf{s}_j) G_{ij}(z(\mathbf{s}_i), z(\mathbf{s}_j)) \\ &+ \sum\sum\sum_{1 \le i < j < k \le n} z(\mathbf{s}_i) z(\mathbf{s}_j) z(\mathbf{s}_k) G_{ijk}(z(\mathbf{s}_i), z(\mathbf{s}_j), z(\mathbf{s}_k)) + \cdots \\ &+ z(\mathbf{s}_1) \cdots z(\mathbf{s}_n) G_{1 \cdots n}(z(\mathbf{s}_1), \ldots, z(\mathbf{s}_n)), \qquad \mathbf{z} \in \zeta. \end{aligned} \tag{6.4.7}$$

Proof. Proof of (6.4.6) is straightforward and (6.4.7) follows by defining

$$z(\mathbf{s}_i)G_i(z(\mathbf{s}_i)) \equiv Q(0,\ldots,0,z(\mathbf{s}_i),0,\ldots,0),$$
$$\begin{aligned} z(\mathbf{s}_i)z(\mathbf{s}_j)G_{ij}(z(\mathbf{s}_i),z(\mathbf{s}_j)) \equiv{} & Q(0,\ldots,0,z(\mathbf{s}_i),0,\ldots,0,z(\mathbf{s}_j),0,\ldots,0) \\ & - Q(0,\ldots,0,z(\mathbf{s}_i),0,\ldots,0,0,0,\ldots,0) \\ & - Q(0,\ldots,0,0,0,\ldots,0,z(\mathbf{s}_j),0,\ldots,0) \end{aligned}$$

and similar higher-order difference formulas for the rest of the Gs. ■

Although the expansion (6.4.7) is unique, the functions $\{G_{ij\ldots}\}$ are not uniquely specified. By defining $G_{ij\ldots}(z(\mathbf{s}_i), z(\mathbf{s}_j),\ldots) \equiv 0$, whenever $z(\mathbf{s}_i) = 0$, or $z(\mathbf{s}_j) = 0$, or $\ldots$, uniqueness is obtained.

The implication of properties (i) and (ii) together is that the expansion (6.4.7) for $Q(\mathbf{z})$ is actually made up of *conditional* probabilities. For example,

$$z(\mathbf{s}_i)G_i(z(\mathbf{s}_i)) = \log\left[\Pr\left(z(\mathbf{s}_i)|\{0(\mathbf{s}_j): j \neq i\}\right)/\Pr\left(0(\mathbf{s}_i)|\{0(\mathbf{s}_j): j \neq i\}\right)\right]$$

is obtained from the expression for $z(\mathbf{s}_i)G(z(\mathbf{s}_i))$ in terms of differences of appropriate Qs; see (6.4.6). The pairwise interaction term is likewise

$$\begin{aligned} & z(\mathbf{s}_i)z(\mathbf{s}_j)G_{ij}(z(\mathbf{s}_i),z(\mathbf{s}_j)) \\ & \quad = \log\left[\frac{\Pr(z(\mathbf{s}_i)|z(\mathbf{s}_j),\{0(\mathbf{s}_k): k \neq i,j\})}{\Pr(0(\mathbf{s}_i)|z(\mathbf{s}_j),\{0(\mathbf{s}_k): k \neq i,j\})} \cdot \frac{\Pr(0(\mathbf{s}_i)|\{0(\mathbf{s}_k): k \neq i\})}{\Pr(z(\mathbf{s}_i)|\{0(\mathbf{s}_k): k \neq i\})}\right]. \end{aligned}$$

An example of a three-way interaction term is

$$\begin{aligned} & z(\mathbf{s}_1)z(\mathbf{s}_2)z(\mathbf{s}_3)G_{123}(z(\mathbf{s}_1),z(\mathbf{s}_2),z(\mathbf{s}_3)) \\ & \quad = \log\left[\frac{\Pr(z(\mathbf{s}_1)|z(\mathbf{s}_2),z(\mathbf{s}_3),0(\mathbf{s}_4),\ldots,0(\mathbf{s}_n))}{\Pr(0(\mathbf{s}_1)|z(\mathbf{s}_2),z(\mathbf{s}_3),0(\mathbf{s}_4),\ldots,0(\mathbf{s}_n))}\right. \\ & \qquad \cdot \frac{\Pr(0(\mathbf{s}_1)|z(\mathbf{s}_2),0(\mathbf{s}_3),\ldots,0(\mathbf{s}_n))}{\Pr(z(\mathbf{s}_1)|z(\mathbf{s}_2),0(\mathbf{s}_3),\ldots,0(\mathbf{s}_n))} \\ & \qquad \cdot \frac{\Pr(0(\mathbf{s}_1)|0(\mathbf{s}_2),z(\mathbf{s}_3),0(\mathbf{s}_4),\ldots,0(\mathbf{s}_n))}{\Pr(z(\mathbf{s}_1)|0(\mathbf{s}_2),z(\mathbf{s}_3),0(\mathbf{s}_4),\ldots,0(\mathbf{s}_n))} \\ & \qquad \left. \cdot \frac{\Pr(z(\mathbf{s}_1)|0(\mathbf{s}_2),\ldots,0(\mathbf{s}_n))}{\Pr(0(\mathbf{s}_1)|0(\mathbf{s}_2),\ldots,0(\mathbf{s}_n))}\right]. \end{aligned}$$

Higher-way interaction terms follow similarly.

Well Defined G Functions

The consistency conditions on the conditional probabilities (needed for reconstruction of a joint probability) can then be expressed as those conditions needed to yield well defined G functions. That is, in the preceding expression for the pairwise interaction term (for example), the right-hand side must be invariant to whether the ith site's or the jth site's conditional probabilities are featured. Similar invariances must also hold for the right-hand sides of the expressions for three- and higher-way interaction terms.

From Conditional Specifications to the Likelihood

Recall that $\Pr(\mathbf{z})$ [or $f(\mathbf{z})$] is proportional to $\exp(Q(\mathbf{z}))$. Because the likelihood is $\Pr(\mathbf{z})$, expressed as a function of the random-process parameters, it is clear that finding the proportionality constant as a (closed-form) function of those parameters is important. This is not always possible. However, there is a powerful result available regarding the form that the Q function must take:

Hammersley–Clifford Theorem (Hammersley and Clifford, 1971) Suppose that $\mathbf{Z}$ is distributed according to a Markov random field on ζ that satisfies the positivity condition. Then, the negpotential function $Q(\cdot)$ given by (6.4.7) must satisfy the property

$$\text{if sites } i, j, \ldots, s \text{ do not form a clique, then } G_{ij \cdots s}(\cdot) \equiv 0, \quad (6.4.8)$$

where the cliques are defined by the neighborhood structure $\{N_i : i = 1, \ldots, n\}$.

Proof (Besag, 1974). Without loss of generality, consider site 1 in detail: The expansion (6.4.7) implies that

$$\begin{aligned} Q(\mathbf{z}) - Q(\mathbf{z}_1) &= z(\mathbf{s}_1)\Big\{G_1(z(\mathbf{s}_1)) + \sum_{2 \le j \le n} z(\mathbf{s}_j) G_{1j}(z(\mathbf{s}_1), z(\mathbf{s}_j)) \\ &\quad + \sum\sum_{2 \le j < k \le n} z(\mathbf{s}_j) z(\mathbf{s}_k) G_{1jk}(z(\mathbf{s}_1), z(\mathbf{s}_j), z(\mathbf{s}_k)) + \cdots \\ &\quad + z(\mathbf{s}_2) \cdots z(\mathbf{s}_n) G_{12 \cdots n}(z(\mathbf{s}_1), z(\mathbf{s}_2), \ldots, z(\mathbf{s}_n))\Big\}. \end{aligned}$$

Now suppose that site l is not a neighbor of site 1. Then, from (6.4.6), $Q(\mathbf{z}) - Q(\mathbf{z}_1)$ does *not* depend on $z(\mathbf{s}_l)$. Putting $z(\mathbf{s}_2) = \cdots = z(\mathbf{s}_{l-1}) = z(\mathbf{s}_{l+1}) = \cdots = z(\mathbf{s}_n) = 0$ in the preceding expression yields the right-hand side equal to $z(\mathbf{s}_1)G_1(z(\mathbf{s}_1)) + z(\mathbf{s}_1)z(\mathbf{s}_l)G_{1l}(z(\mathbf{s}_1), z(\mathbf{s}_l))$, which must not depend on $z(\mathbf{s}_l)$. Hence $G_{1l}(\cdot) \equiv 0$. [The other possibility, namely,

$z(\mathbf{s}_l)G_{1l}(z(\mathbf{s}_1), z(\mathbf{s}_l)) = c \cdot g(z(\mathbf{s}_1))$, cannot occur because it would have already been absorbed into $G_1(z(\mathbf{s}_1))$.]

Now put $z(\mathbf{s}_2) = \cdots = z(\mathbf{s}_{j-1}) = z(\mathbf{s}_{j+1}) = \cdots = z(\mathbf{s}_{l-1}) = z(\mathbf{s}_{l+1}) = \cdots = z(\mathbf{s}_n) = 0$ in the preceding expression; similar reasoning yields $G_{1jl}(\cdot) \equiv 0$. By continuing in the same manner, it can be seen that all functions G in (6.4.7) that involve both $z(\mathbf{s}_1)$ and $z(\mathbf{s}_l)$ *must* be identically zero. Further, the same result must hold for any pair of sites that are not neighbors; that is, if $i, j, \ldots, s$ is not a clique, then $G_{ij \cdots s}(\cdot) \equiv 0$. ■

From the symmetric appearance of pairs of sites in Q, it can be seen that $k \in N_i$ if and only if $i \in N_k$ (which can also be proved directly from the conditional probabilities). For further discussion of the theorem, see Spitzer (1971) and Kindermann and Snell (1980).

Let κ be a clique and define $\mathbf{z}_\kappa \equiv (z(\mathbf{s}_i): i \in \kappa)'$. Also define $V_\kappa(\mathbf{z}_\kappa) \equiv \{\prod_{i \in \kappa} z(\mathbf{s}_i)\} G_\kappa(\mathbf{z}_\kappa)$. Then,

$$Q(\mathbf{z}) = \sum_{\kappa \in \mathscr{C}} V_\kappa(\mathbf{z}_\kappa),$$

where $\mathscr{C}$ is the set of all cliques. Thus, from (6.4.6), one obtains

$$\Pr\left(z(\mathbf{s}_i) | \{z(\mathbf{s}_j): j \neq i\}\right) \propto \exp\left(\sum_{\kappa: i \in \kappa} V_\kappa(\mathbf{z}_\kappa)\right), \qquad i = 1, \ldots, n.$$

The importance of the Hammersley–Clifford theorem to the spatial modeler is that a conditional specification typically involves just a few nonzero functions $\{G_{ij \cdots s}(\cdot)\}$. The converse is also important.

Corollary Suppose ζ is countable in $\mathbb{R}^n$ (or ζ is a Lebesgue measurable subset of $\mathbb{R}^n$) and that a well defined set of G functions can be obtained from specified conditional probabilities and neighborhoods $\{N_i: i = 1, \ldots, n\}$. Then, the resulting negpotential function Q defined by (6.4.7) yields a unique well defined joint probability function proportional to $\exp(Q(\cdot))$, provided the summability condition

$$\sum_{\mathbf{z} \in \zeta} \exp(Q(\mathbf{z})) < \infty \quad \left[\text{or} \int_\zeta \exp(Q(\mathbf{z}))\, d\mathbf{z} < \infty\right]$$

holds.

Proof. The proof is by construction. Define the function $p(\cdot)$ on ζ as

$$p(\mathbf{z}) \equiv \exp(Q(\mathbf{z})) \Big/ \sum_{\mathbf{y} \in \zeta} \exp(Q(\mathbf{y})), \qquad \mathbf{z} \in \zeta.$$

Then $p(\cdot)$ is nonnegative and sums to 1. Furthermore, because the G functions were constructed from the conditional probabilities, the uniqueness of the expansion (6.4.7) guarantees that $p(\cdot)$ is a joint probability with conditional probabilities $\Pr(z(\mathbf{s}_i)|\{z(\mathbf{s}_j): j \neq i\})$, $i = 1, \ldots, n$. ■

The theorem has been proved independently in the statistical-mechanics literature; see, for example, Preston (1974). The corollary shows that, apart from a normalizing constant, the joint probability can be written explicitly in terms of the nonnull G functions:

$$\Pr(\mathbf{z}) = \exp(Q(\mathbf{z})) \Big/ \sum_{\mathbf{y} \in \zeta} \exp(Q(\mathbf{y})), \qquad \mathbf{z} \in \zeta,$$

where Q is expressed in terms of the G functions through (6.4.7). However, it is precisely this normalizing constant that causes difficulty when interpreting $\Pr(\mathbf{z})$ as a likelihood. Unless $\Sigma_{\mathbf{y} \in \zeta} \exp(Q(\mathbf{y}))$ can be written as a function of the parameters in closed form (which is more the exception than the rule), maximization of the exact likelihood can be computationally prohibitive. See Section 7.2.1 for alternative inference procedures.

The corollary is precisely the result needed by modelers. It gives them a way of seeing the consequences (on the joint probability) of various assumptions about the neighborhood structure $\{N_i: i = 1, \ldots, n\}$. Furthermore, it says that provided the specified conditional probabilities yield well defined G functions (of which, only those defined on the cliques are nonzero) and a summability condition is satisfied, then there is a unique Markov random field over $\{Z(\mathbf{s}_i): i = 1, \ldots, n\}$.

6.4.2* Pairwise-Only Dependence and Conditional Exponential Distributions

Having specified the neighbors of each site, there still remains the choice of a specific form for the conditional distribution at each site. Suppose the exponential family of distributions (a flexible family that includes the binomial, Poisson, Gaussian, gamma, etc., distributions) is used to model the conditional distributions. In the discrete case,

$$\Pr\left(z(\mathbf{s}_i)|\{z(\mathbf{s}_j): j \neq i\}\right) = \exp\left[A_i(\{z(\mathbf{s}_j): j \neq i\})B_i(z(\mathbf{s}_i)) + C_i(z(\mathbf{s}_i)) + D_i(\{z(\mathbf{s}_j): j \neq i\})\right], \qquad i = 1, \ldots, n, \quad (6.4.9)$$

where $\{B_i(\cdot)\}$ and $\{C_i(\cdot)\}$ have specified forms and $\{A_i(\cdot)\}$ and $\{D_i(\cdot)\}$ are functions of the values observed at neighboring sites of i. In the continuous case, $\Pr(\cdot)$ is replaced by $f(\cdot)$, and (6.4.9) defines conditional densities.

Theorem (Besag, 1974) Assume (6.4.9) and pairwise-only dependence between sites [i.e., all $G_A(\cdot) \equiv 0$ for any A whose number of distinct elements is 3 or more]. Then,

$$A_i(\{z(\mathbf{s}_j): j \neq i\}) = \alpha_i + \sum_{j=1}^{n} \theta_{ij} B_j(z(\mathbf{s}_j)), \qquad i = 1, \ldots, n, \quad (6.4.10)$$

where $\theta_{ji} = \theta_{ij}$, $\theta_{ii} = 0$, and $\theta_{ik} = 0$ for $k \notin N_i$.

Proof. The proof assumes that if $z(\mathbf{s}_i) = 0$, then (6.4.9) is strictly positive. Besag (1974) shows that the case where (6.4.9) is zero can be handled by transforming $z(\mathbf{s}_i)$. For convenience, write A_i and D_i as functions of $\mathbf{z}_i = (z(\mathbf{s}_1), \ldots, z(\mathbf{s}_{i-1}), 0, z(\mathbf{s}_{i+1}), \ldots, z(\mathbf{s}_n))'$, although, in fact, they depend only on $\{z(\mathbf{s}_k): k \in N_i\}$. By assumption,

$$Q(\mathbf{z}) = \sum_{i=1}^{n} z(\mathbf{s}_i) G_i(z(\mathbf{s}_i)) + \sum\sum_{1 \le i < j \le n} z(\mathbf{s}_i) z(\mathbf{s}_j) G_{ij}(z(\mathbf{s}_i), z(\mathbf{s}_j)) \quad (6.4.11)$$

and

$$\log[\Pr(z(\mathbf{s}_i) | \{z(\mathbf{s}_j): j \neq i\})] = A_i(\{z(\mathbf{s}_j): j \neq i\}) B_i(z(\mathbf{s}_i)) + C_i(z(\mathbf{s}_i)) + D_i(\{z(\mathbf{s}_j): j \neq i\}).$$

Putting these into, respectively, the left- and right-hand sides of

$$\exp(Q(\mathbf{z}) - Q(\mathbf{z}_i)) = \Pr(z(\mathbf{s}_i) | \{z(\mathbf{s}_j): j \neq i\}) / \Pr(0(\mathbf{s}_i) | \{z(\mathbf{s}_j): j \neq i\})$$

and putting $z(\mathbf{s}_1) = \cdots = z(\mathbf{s}_{i-1}) = z(\mathbf{s}_{i+1}) = \cdots = z(\mathbf{s}_n) = 0$, the relation

$$z(\mathbf{s}_i) G_i(z(\mathbf{s}_i)) = A_i(\mathbf{0}) \{B_i(z(\mathbf{s}_i)) - B_i(0)\} + C_i(z(\mathbf{s}_i)) - C_i(0)$$

is obtained.

Moreover, supposing sites 1 and 2 are neighbors and putting $z(\mathbf{s}_3) = z(\mathbf{s}_4) = \cdots = z(\mathbf{s}_n) = 0$, the relations

$$z(\mathbf{s}_1) G_1(z(\mathbf{s}_1)) + z(\mathbf{s}_1) z(\mathbf{s}_2) G_{1,2}(z(\mathbf{s}_1), z(\mathbf{s}_2)) = A_1(z(\mathbf{s}_2), 0, \ldots, 0) \{B_1(z(\mathbf{s}_1)) - B_1(0)\} + C_1(z(\mathbf{s}_1)) - C_1(0),$$

$$z(\mathbf{s}_2) G_2(z(\mathbf{s}_2)) + z(\mathbf{s}_1) z(\mathbf{s}_2) G_{1,2}(z(\mathbf{s}_1), z(\mathbf{s}_2)) = A_2(z(\mathbf{s}_1), 0, \ldots, 0) \{B_2(z(\mathbf{s}_2)) - B_2(0)\} + C_2(z(\mathbf{s}_2)) - C_2(0)$$

are obtained. Thus,

$$\begin{aligned} z(\mathbf{s}_1)z(\mathbf{s}_2)G_{1,2}(z(\mathbf{s}_1), z(\mathbf{s}_2)) \\ &= \{B_1(z(\mathbf{s}_1)) - B_1(0)\}\{A_1(z(\mathbf{s}_2), 0, \dots, 0) - A_1(\mathbf{0})\} \\ &= \{B_2(z(\mathbf{s}_2)) - B_2(0)\}\{A_2(z(\mathbf{s}_1), 0, \dots, 0) - A_2(\mathbf{0})\}, \end{aligned}$$

which implies

$$\begin{aligned} \frac{A_2(z(\mathbf{s}_1), 0, \dots, 0) - A_2(\mathbf{0})}{B_1(z(\mathbf{s}_1)) - B_1(0)} &= \frac{A_1(z(\mathbf{s}_2), 0, \dots, 0) - A_1(\mathbf{0})}{B_2(z(\mathbf{s}_2)) - B_2(0)} \\ &\equiv \theta_{12}, \end{aligned}$$

a constant independent of both $z(\mathbf{s}_1)$ and $z(\mathbf{s}_2)$. A similar argument for any neighboring pair i and j yields

$$z(\mathbf{s}_i)z(\mathbf{s}_j)G_{ij}(z(\mathbf{s}_i), z(\mathbf{s}_j)) = \theta_{ij}\{B_i(z(\mathbf{s}_i)) - B_i(0)\}\{B_j(z(\mathbf{s}_j)) - B_j(0)\},$$

where $\theta_{ij} = \theta_{ji}$, $i \neq j$. Therefore,

$$\begin{aligned} Q(\mathbf{z}) - Q(\mathbf{z}_1) = \alpha_1 B_1(z(\mathbf{s}_1)) + \left\{ \sum_{2 \le j \le n} \theta_{1j} B_j(z(\mathbf{s}_j)) \right\} B_1(z(\mathbf{s}_1)) \\ + \text{terms not depending on } B_1(z(\mathbf{s}_1)), \end{aligned}$$

for some constant α_1. Because $A_1(\{z(\mathbf{s}_j): j \neq 1\})$ is the coefficient of $B_1(z(\mathbf{s}_1))$, the result follows. ■

An easy consequence is that, under the same conditions as those stated in the theorem, $Q(\mathbf{z})$ defined by (6.4.5) is (up to an additive constant)

$$\sum_{i=1}^{n} \{\alpha_i B_i(z(\mathbf{s}_i)) + C_i(z(\mathbf{s}_i))\} + \sum\sum_{1 \le i < j \le n} \theta_{ij} B_i(z(\mathbf{s}_i)) B_j(z(\mathbf{s}_j)),$$

where $\theta_{ji} = \theta_{ij}$, $\theta_{ii} = 0$, and $\theta_{ik} = 0$ for $k \notin N_i$. Furthermore, from the proof of the theorem, it is easily seen that $\theta_{ij} = \theta_{ji}$ guarantees that the pairwise G functions are well defined.

The pairwise-only-dependence assumption does not say that all cliques must have two or fewer sites; it merely says that some of the potentially nonzero G functions in (6.4.7) turn out to be zero. The exponential family (6.4.9) is univariate in the natural parameter $A_i(\{z(\mathbf{s}_j): j \neq i\})$. A useful extension of the theorem would be to prove a similar result in a multivariate-parameter situation. Any extension would have to ensure well defined G functions; see Cressie and Lele (1992).

This theorem narrows down the search for sensible and interpretable models: Equation (6.4.10) shows that spatial dependence is *necessarily* embodied in the parameters $\{\theta_{ij}: i = 1, \ldots, n;\ j = 1, \ldots, n\}$, because $\theta_{ik} = 0$ if and only if $k \notin N_i$. Their (efficient) estimation is discussed in Sections 7.2 and 7.3.

6.4.3* Some Final Remarks

Suppose that the lattice D is countably infinite, but that each of the neighborhood sets N_i, $i = 1, 2, \ldots$, is finite. It is not obvious that the conditional probabilities

$$\Pr\left(z(\mathbf{s}_i)|\{z(\mathbf{s}_j): j \in N_i\}; \boldsymbol{\theta}\right), \qquad i = 1, 2, \ldots,$$

give rise to a unique and identifiable probability measure $P_{\boldsymbol{\theta}}$; $\boldsymbol{\theta} \in \Theta$, for the random process $\{Z(\mathbf{s}): \mathbf{s} \in D\}$.

Uniqueness is characterized by the property, if $\boldsymbol{\theta}_1 = \boldsymbol{\theta}_2$, then $P_{\boldsymbol{\theta}_1} = P_{\boldsymbol{\theta}_2}$. For example, for binary lattices where $Z(\mathbf{s}_i) = 0$ or 1, physicists have called a lack of uniqueness *phase transition*. For $D = \mathbb{Z}^d$, the lattice of d-dimensional integer coordinates, conditions under which uniqueness is guaranteed are given by Dobrushin (1968) and Follmer (1982). Identifiability is characterized by the property, if $\boldsymbol{\theta}_1 \neq \boldsymbol{\theta}_2$, then $P_{\boldsymbol{\theta}_1} \neq P_{\boldsymbol{\theta}_2}$. For translation-invariant processes on $D = \mathbb{Z}^d$, Gidas (1988) gives a necessary and sufficient condition for identifiability.

Edge Sites

What to do with edge sites is a thorny problem that does not go away in dimensions greater than 1; in fact, it gets worse because proportionately more sites are on the edge.

When D is a countable lattice from which a sampling of sites $\{\mathbf{s}_1, \ldots, \mathbf{s}_n\}$ is taken, the conditional specification (and consequently the likelihood) should respect the fact that the finite data $\mathbf{Z}$ are embedded in a process $\{Z(\mathbf{s}): \mathbf{s} \in D\}$ over a potentially infinite set of sites. Nearest-neighbor assumptions leave all but the edge sites of $\{\mathbf{s}_1, \ldots, \mathbf{s}_n\}$ well specified in terms of conditional probabilities.

Conceptually, conditional probabilities at edge sites that are conditional on unobserved data can be handled by integrating out those data from the conditioning event. However, this can result in a complicated likelihood. Another possibility is to perform all inferences conditional on the edge sites, which is equivalent to forming a "guard region" inside the perimeter of sites at which observations are taken. Observations at sites in this guard region contribute to the likelihood only through their neighborhood relations with internal sites not in the guard region. These and other possibilities are discussed for conditionally specified Gaussian models in Section 6.6 and for simultaneously specified Gaussian models in Section 6.7.1. The effect of edges on asymptotic distribution theory for parameter estimators is considered in Section 7.3.

In summary, this section has presented a number of theoretical results that allow the construction of (conditionally specified) Markov random fields in a rather general setting. Strauss (1986) shows that such an approach can be taken in a variety of other situations. Perhaps the best known is modeling of point processes (Section 8.5.5). The next two sections are devoted to applying the results of this section to various models for discrete lattice data (Section 6.5) and continuous lattice data (Section 6.6).

6.5 CONDITIONALLY SPECIFIED SPATIAL MODELS FOR DISCRETE DATA

This section makes use of the Hammersley–Clifford theorem and its consequences (presented in Section 6.4) to build classes of models for discrete data. When $Z(\mathbf{s}_i)$ is binary, a class known as the autologistic models is defined in a natural way. Auto-Poisson and other models for *counts* data are then defined; an important application is to the modeling of regional incidences of a particular disease.

6.5.1 Binary Data

Assume that data are either 0 (white) or 1 (black), which often derives from the absence or presence of a characteristic. For example, Besag (1974) analyzed presence/absence of *Plantago lanceolata* plants in a 10 × 940 grid over an apparently homogeneous area of lead–zinc tailings in defunct mine workings at Treloggan, Flintshire (U.K.).

Under the assumption of pairwise-only dependence, the conditional distribution of binary variable $Z(\mathbf{s}_i)$ is now shown to be necessarily of the logistic form.

Autologistic Model

Because of the binary nature of the data, the only important values of the G functions in (6.4.11) are, assuming pairwise-only dependence between sites, $G_i(1) \equiv \alpha_i$ and $G_{ij}(1,1) \equiv \theta_{ij}$. Thus,

$$Q(\mathbf{z}) = \sum_{i=1}^{n} \alpha_i z(\mathbf{s}_i) + \sum\sum_{1 \le i < j \le n} \theta_{ij} z(\mathbf{s}_i) z(\mathbf{s}_j), \qquad (6.5.1)$$

where $\theta_{ij} = 0$ unless sites i and j are mutual neighbors. Then, $Q(\mathbf{z}) - Q(\mathbf{z}_i)$

$= \alpha_i z(\mathbf{s}_i) + \Sigma_{j=1}^n \theta_{ij} z(\mathbf{s}_i) z(\mathbf{s}_j)$, where $\theta_{ji} = \theta_{ij}$. To maintain identifiability of the parameters, define $\theta_{ii} \equiv 0$. Therefore, from (6.4.6),

$$\frac{\Pr\left(z(\mathbf{s}_i)|\{z(\mathbf{s}_j): j \neq i\}\right)}{\Pr\left(0(\mathbf{s}_i)|\{z(\mathbf{s}_j): j \neq i\}\right)} = \exp\left\{\alpha_i z(\mathbf{s}_i) + \sum_{j=1}^{n} \theta_{ij} z(\mathbf{s}_i) z(\mathbf{s}_j)\right\}. \quad (6.5.2)$$

Because $z(\mathbf{s}_i) = 0$ or 1,

$$\frac{1 - \Pr\left(0(\mathbf{s}_i)|\{z(\mathbf{s}_j): j \neq i\}\right)}{\Pr\left(0(\mathbf{s}_i)|\{z(\mathbf{s}_j): j \neq i\}\right)} = \exp\left\{\alpha_i + \sum_{j=1}^{n} \theta_{ij} z(\mathbf{s}_j)\right\}$$

and, hence,

$$\Pr\left(z(\mathbf{s}_i)|\{z(\mathbf{s}_j): j \neq i\}\right) = \frac{\exp\left\{\alpha_i z(\mathbf{s}_i) + \Sigma_{j=1}^n \theta_{ij} z(\mathbf{s}_i) z(\mathbf{s}_j)\right\}}{1 + \exp\left\{\alpha_i + \Sigma_{j=1}^n \theta_{ij} z(\mathbf{s}_j)\right\}},$$

$$z(\mathbf{s}_i) = 0, 1. \quad (6.5.3)$$

Here, the exponential form in (6.5.3) is a *consequence* of the algebra and is not an assumption, in contrast to the theorem in Section 6.4.2 that derives (6.4.10). The model specified by (6.5.3) is called an *autologistic model.*

When $\theta_{ij} = 0$, for all $i, j = 1, \ldots, n$, the variables $\{Z(\mathbf{s}_i): i = 1, \ldots, n\}$ are independent. Moreover, $\theta_{ik} = 0$ if and only if $k \notin N_i$, the neighborhood set of site i. Thus the $\{\theta_{ij}\}$ can be thought of as spatial-dependence parameters.

How are the parameters in (6.5.3) to be estimated? Section 7.2 can be consulted for a detailed discussion of parameter estimation. Suppose maximum likelihood estimates are required. The likelihood is simply

$$\Pr(\mathbf{z}) = \exp(Q(\mathbf{z})) \Big/ \sum_{\mathbf{y} \in \zeta} \exp(Q(\mathbf{y})), \quad (6.5.4)$$

considered as a function of the parameters. Here $Q(\mathbf{z})$ is given by (6.5.1), showing explicitly how the spatial-trend parameters $\{\alpha_i\}$ and the spatial-dependence parameters $\{\theta_{ij}\}$ enter. When the lattice is infinite, the summation in (6.5.4) is infinite, but no closed-form expression is available. The intractability of this normalizing constant makes *exact* maximum likelihood estimation cumbersome.

Modeling the Large-Scale Variation

Because there are n sites and n parameters $\{\alpha_i\}$ (not counting the, at most, $n(n-1)/2$ parameters $\{\theta_{ij}\}$), there are at least as many parameters as data. This is a common situation in statistical modeling and it is usually resolved by regressing (or some other form of smoothing) a dependent variable on q independent variables, *where* $q < n$. Thus there are q regression parameters to be estimated, allowing $n - q$ degrees of freedom with which to estimate the small-scale-variation parameters.

Although the usual linear model is not to be recommended here, the same ideas can be invoked by modeling the large-scale-variation parameters as $\boldsymbol{\alpha} = X\boldsymbol{\beta}$, where $\boldsymbol{\beta}$ is a $q \times 1$ vector of regression coefficients. Often, the most important explanations of the data are to be found in $\boldsymbol{\beta}$. However, it may not be appropriate to model spatial data as independent observations. The small-scale variations then are left to be explained by spatial dependence parameters $\{\theta_{ij}\}$, which in turn allow $\boldsymbol{\beta}$ to be estimated efficiently.

Ising Model in Statistical Mechanics

Markov random fields of binary data have been of interest to physicists for some time. The classical *Ising model* (e.g., Ruelle, 1969) is simply a homogeneous first-order autologistic on a countable regular lattice $D = \{(u, v): u = \ldots, -1, 0, 1, \ldots; v = \ldots, -1, 0, 1, \ldots\}$, which reduces to

$$\Pr(z(u,v)|\{z(k,l):(k,l) \neq (u,v)\}) = \exp(z(u,v)g)/\{1 + \exp(g)\},$$
$$u, v = \ldots, -1, 0, 1, \ldots, \quad (6.5.5)$$

where

$$g \equiv \{\alpha + \gamma_1(z(u-1,v) + z(u+1,v)) + \gamma_2(z(u,v-1) + z(u,v+1))\}.$$

The model was originally presented by Ising (1925) under the added assumption that $\gamma_1 = \gamma_2$ (sometimes mistakenly called isotropy). It was used as a model for ferromagnetism; each site is thought of as possessing a spin, which may be either an up spin or a down spin. See Grimmett (1987) for a review of the properties of the model and Kunsch (1984) for a discussion of how an interactive particle system, allowed to evolve over a long period of time, can generate the Ising model.

There are still delicate questions to resolve, namely, whether (6.5.5) is a sensible definition (because the lattice is infinite). Pickard (1987) shows that, provided the boundary of an increasing sequence (increasing to D) of finite lattices always has dimension 1, then (6.5.5) can be interpreted as a limit of conditional probabilities over the finite lattices. Moreover, a limiting Markov random field exists; however, there are regions of the parameter space where two different probability measures give rise to the same conditional probabilities (6.5.5) (physicists call this phenomenon phase transition). Further, for $\gamma_1 = \gamma_2 = \gamma$, critical values of γ exist, below which the limiting nearest-neighbor correlation is zero, but above which the process exhibits long-range dependence (Pickard, 1987).

Instead of increasing-domain asymptotics, one might think about a bounded region A within which a grid is made finer and finer. At the nodes of the grid, a binary Markov random field (whose neighborhoods may go beyond the nearest neighbors) is realized. Assuming pairwise-only dependence between nodes, Besag et al. (1982) show that this type of infill asymptotics yields a Markov point process (Section 8.5.5) in the limit.

Hajek and Berger (1987) show that a binary Markov random field can be represented as the nodewise modulo-2 sum of two independent binary random fields, one of which is "white binary noise" with success probability θ. Under modest assumptions, they prove that the largest such θ is strictly positive.

Random Media

Sites of a (possibly irregular) lattice are supposed connected or not, from which $\binom{n}{2}$ connection values of 1 (connected) or 0 (not connected) can be defined. Suppose the medium is connected according to a given set of connection values. Then, the percolation model is obtained by converting independently each connection value of 1 to a 0 with probability $1 - p$. Combinatorial theorists have studied the process asymptotically by assuming that all connection values are initially 1, allowing p to depend on n, and letting $n \to \infty$.

In the case of a regular lattice $\mathbb{Z}^d$ $(d \geq 2)$, where connections are formed only with nearest neighbors and conversions occur independently with fixed probability p, there is a critical probability $\pi(d)$ such that: If $p < \pi(d)$, only finite clusters of connected sites occur, and if $p > \pi(d)$, an infinite cluster occurs. Spatial dependence can be introduced through a binary Markov random field on $\mathbb{Z}^d$. For example, if nearest neighbors are both black, then they are deemed connected; otherwise, any pair of sites is deemed not connected (e.g., Gandolfi, 1989). Grimmett (1987, 1989) gives an overview of percolation and related processes.

Multicolored Data

Binary data can be viewed as a black-and-white picture, where $z(\mathbf{s}_i) = 1$ corresponds to the ith site being colored black and $z(\mathbf{s}_i) = 0$ corresponds to the ith site being colored white. When more than two colors (or categories) are allowed, the Markov random field (6.5.1) can be generalized.

Let $z(\mathbf{s}_i) \in \{0, 1, \ldots, c\}$, the set of $c + 1$ colors available. Assuming homogeneity and pairwise-only dependence between sites, Strauss (1977) shows that

$$\begin{aligned} Q(\mathbf{z}) &= \sum_{i=1}^{n} z(\mathbf{s}_i) G_1(z(\mathbf{s}_i)) + \sum\sum_{1 \leq i < j \leq n} z(\mathbf{s}_i) z(\mathbf{s}_j) G_{ij}\left(z(\mathbf{s}_i), z(\mathbf{s}_j)\right) \\ &= \sum_{r=1}^{c} m_r u_r + \sum_{r=1}^{c} \sum_{s=1}^{c} n_{rs} v_{rs}, \end{aligned}$$

where $u_r \equiv r G_1(r)$, $G_{ij} \equiv G_2$ if i and j are neighbors and $\equiv 0$ otherwise, and $v_{rs} \equiv$

$rsG_2(r, s)$, $r, s = 1, \ldots, c$. The data appear through $m_r \equiv$ number of sites having color r and $n_{rs} \equiv$ number of pairs of neighboring sites where one has color r and the other has color s. Under two further assumptions (color indifference and equal strength of attraction for all colors), Strauss is able to find approximate maximum likelihood estimates; the approximation is necessary due to the unwieldly normalizing constant $\Sigma_{\mathbf{y} \in \zeta} \exp(Q(\mathbf{y}))$.

Testing for Spatial Dependence

From data $\mathbf{Z} = (Z(\mathbf{s}_1), \ldots, Z(\mathbf{s}_n))'$, whose components take the values 0 or 1, consider the following statistic,

$$I \equiv (\mathbf{Z}'W\mathbf{Z})/(\mathbf{Z}'\mathbf{Z}),$$

where W is an $n \times n$ matrix whose (i, j)th element is

$$w_{ij} = \begin{cases} 1 \Big/ \sum_{k \in N_i} 1, & j \in N_i, \\ 0, & j \notin N_i. \end{cases}$$

The statistic I is known as Moran's contiguity ratio, due to earlier related work by Moran (1948, 1950); as a test statistic, it tests for random spatial scatter of, say, black sites in black/white images by counting the number of black–black and black–white edges (or joins). The image is deemed to exhibit spatial dependence if I is large. Cliff and Ord (1973) can be consulted for the details, including the generalization to multicolored images. Howe and Webb (1983) discuss computational aspects and apply the test to pollen data.

Image Analysis

The conditionally specified spatial models presented in the preceding section have proved particularly useful in the analysis of images, pattern recognition, and classification from satellite data. Geman and Geman's (1984) maximum a posteriori (MAP) estimate of an image, given the degraded observations, represents a major advance in the subject. Section 7.4 contains a brief review of the field of statistical image analysis and remote sensing.

6.5.2 Counts Data

When spatial data arise as counts, the natural model that comes to mind is one based on the Poisson distribution. The *auto-Poisson* conditional specification (assuming pairwise-only dependence between sites) is

$$\Pr\left(z(\mathbf{s}_i) | \{z(\mathbf{s}_j): j \neq i\}\right)$$
$$= \exp\left(-\lambda_i(\{z(\mathbf{s}_j): j \neq i\})\right)\left(\lambda_i(\{z(\mathbf{s}_j): j \neq i\})\right)^{z(\mathbf{s}_i)} \Big/ z(\mathbf{s}_i)!, \quad (6.5.6)$$

where from (6.4.10),

$$\lambda_i(\{z(\mathbf{s}_j): j \neq i\}) = \exp\left\{\alpha_i + \sum_{j=1}^{n} \theta_{ij} z(\mathbf{s}_j)\right\}, \tag{6.5.7}$$

$\theta_{ij} = \theta_{ji}$, $\theta_{ii} = 0$, and $\theta_{ik} = 0$ for $k \notin N_i$.

Now, from (6.4.6),

$$Q(\mathbf{z}) - Q(\mathbf{z}_i) = \log\left[\Pr(z(\mathbf{s}_i) | \{z(\mathbf{s}_j): j \neq i\}) / \Pr(0(\mathbf{s}_i) | \{z(\mathbf{s}_j): j \neq i\})\right];$$

that is, from (6.4.7) and (6.5.6),

$$\begin{aligned} z(\mathbf{s}_i) G_i(z(\mathbf{s}_i)) &+ \sum_{j \neq i} z(\mathbf{s}_i) z(\mathbf{s}_j) G_{ij}(z(\mathbf{s}_i), z(\mathbf{s}_j)) \\ &= \alpha_i z(\mathbf{s}_i) - \log(z(\mathbf{s}_i)!) + \sum_{j \neq i} \theta_{ij} z(\mathbf{s}_i) z(\mathbf{s}_j). \end{aligned}$$

Hence, $G_{ij}(z(\mathbf{s}_i), z(\mathbf{s}_j)) = \theta_{ij}$, $i = 1, \ldots, n$, $j = 1, \ldots, n$, and

$$Q(\mathbf{z}) = \sum_{i=1}^{n} \alpha_i z(\mathbf{s}_i) + \sum_{1 \leq i < j \leq n}\sum \theta_{ij} z(\mathbf{s}_i) z(\mathbf{s}_j) - \sum_{i=1}^{n} \log(z(\mathbf{s}_i)!). \tag{6.5.8}$$

Finally, from (6.4.5), the joint distribution is

$$\Pr(\mathbf{z}) = \exp(Q(\mathbf{z})) \Big/ \sum_{\mathbf{y} \in \zeta} \exp(Q(\mathbf{y})),$$

where $Q(\mathbf{z})$ is given by (6.5.8). The normalizing constant $\Sigma_{\mathbf{y} \in \zeta} \exp(Q(\mathbf{y}))$ is a sequence of n infinite sums, because the support of each Z variable is $\{0, 1, 2, \ldots\}$. Although the constant does not depend upon the data $\mathbf{z}$, it is a function of the parameters and so is crucial in determining the exact likelihood. The parameter space is $\{\boldsymbol{\alpha}, (\theta_{ij}): \Sigma_{\mathbf{z} \in \zeta} \exp(Q(\mathbf{z})) < \infty\}$. Care must be taken to ensure that parameter estimates are indeed in the parameter space. For the auto-Poisson model (6.5.6), the parameter space is $\{\boldsymbol{\alpha}, (\theta_{ij}): \boldsymbol{\alpha} \in \mathbb{R}^n;\ \theta_{ij} \leq 0$ for all $i, j = 1, \ldots, n\}$ (Besag, 1974).

SIDS Data

The sudden infant death syndrome (SIDS) data of Section 6.2 can be modeled according to (6.5.6). Assuming that the number of live births $\{n_i: i = 1, \ldots, 100\}$ in each country is large and that a sudden infant death is a rare occurrence, a Poisson approximation to the counts $\{S_i: i = 1, \ldots, 100\}$ is appropriate (a more exact analysis might involve an autobinomial model; see succeeding text). Moreover, if the actual location of the SIDs were thought to follow a Markov point process (Section 8.5.5), then the results of Besag et al. (1982) indicate that an auto-Poisson model for the county numbers of SIDs is a reasonable one. It is apparent, from Figure 6.5, that there are spatial clusters of high SID rate counties. In other words, there is a spatial trend (large-scale variation) across counties that has to be accounted for; and then there is spatial dependence (small-scale variation) between neighboring countries to be considered.

Large-Scale Variation

In Section 7.6, three levels of large-scale variation are fitted. They are (1) $\exp(\alpha_i) = n_i\rho$, (2) $\exp(\alpha_i) = n_i\rho_{R(i)}$, and (3) $\exp(\alpha_i) = n_i\rho_{M(i)}$. The first level uses only one parameter and corresponds to a homogeneous SID rate (certainly unrealistic for these data). The second level uses 12 parameters, each corresponding to one of 12 artificially created *regions* of North Carolina, which partition the state into contiguous parcels of counties. Thus $R: \{1,2,\ldots,100\} \to \{1,2,\ldots,12\}$.

The third level uses 32 parameters, obtained by overlaying the 9×24 grid, referred to in Section 6.2, onto the country map of North Carolina. For $i = 1,\ldots,100$, county i is assigned to the nearest node $(k(i), l(i))$ of the grid and its rate is modeled by

$$\exp(\alpha_i) = n_i \cdot \rho_{M(i)} = n_i \cdot \exp(\mu + \omega_{k(i)} + \nu_{l(i)}),$$

a *multiplicative* two-way fit to the SID rates. With restrictions on the parameters to make the model identifiable, there are $9 + 24 - 1 = 32$ parameters to be estimated.

Notice that a purely spatial trend has been proposed; however, the use of county-level explanatory variables (such as nonwhite live birth rate, percent urban, etc.) could also be used. In general, the large-scale variation might be modeled by

$$\exp(\alpha_i) = n_i \cdot \exp(\mathbf{x}_i'\boldsymbol{\beta}).$$

Small-Scale Variation

Now there remains the statistical modeling of the small-scale variation. Spatial-dependence parameters $\{\theta_{ij}\}$ in (6.5.7) make the model (6.5.6) more general than the classical model, where $\Pr(z(\mathbf{s}_i)) = \Pr(z(\mathbf{s}_i)|\{z(\mathbf{s}_j): j \neq i\})$. There, the likelihood is simply the product of these probabilities.

There are 100 observations (or 99 if, as is recommended in Section 6.2, county 4 is omitted) and as many as 32 parameters already used for modeling spatial trend. Potentially, the number of spatial-dependence parameters $\{\theta_{ij}\}$ is large; it is the task of the modeler to reduce this without losing interpretability of the model. If the sites were regular, a homogeneous first- or second-order nearest-neighbor model could be tried. With irregularly located sites, a very simple model to start with is

$$\theta_{ij} \propto \begin{cases} \gamma d_{ij}^{-k}, & j \in N_i, \\ 0, & j \notin N_i. \end{cases}$$

A more complicated model might introduce two parameters, γ_1 and γ_2, for modeling spatial dependence with nearest and second-nearest neighbors, respectively.

Here, the single parameter γ captures the spatial dependence, and the distance d_{ij} between neighboring counties i and j is exogenous. The idea is that counties further apart should have weaker dependence; the exponent k controls the rate of decline of this dependence, and it may be prespecified (e.g., $k = 0$, $k = 1$, and $k = 2$ are common choices) or left as a parameter to be estimated. In order to be able to interpret γ, it needs to be dimensionless. This can be achieved by normalizing d_{ij}^{-k} with a quantity of the same dimension. A reasonable choice is

$$\theta_{ij} = \begin{cases} \gamma d_{ij}^{-k} \big/ \max\{d_{ij}^{-k} : j \in N_i, i = 1, \ldots, n\}, & j \in N_i, \\ 0, & j \notin N_i, \end{cases} \tag{6.5.9}$$

ensuring comparability of models across different values of k. If it is desired that the spatial-dependence parameter lies between -1 and $+1$ (like Pearson's product-moment correlation coefficient), the following model could be fitted:

$$\theta_{ij} = \begin{cases} \log\left(\dfrac{1+\gamma}{1-\gamma}\right) d_{ij}^{-k} \Big/ \max\{d_{ij}^{-k} : j \in N_i, i = 1, \ldots, n\}, & j \in N_i, \\ 0, & j \notin N_i, \end{cases} \tag{6.5.10}$$

where the single spatial-dependence parameter $\gamma \in (-1, 1)$.

Due to the unwieldly normalizing constant, maximum likelihood estimation of the 32 + 1 parameters [assuming a two-way multiplicative trend and spatial dependence given by (6.5.9)] is not possible, and hence an approximation based on what Besag (1975) calls the *pseudolikelihood* (Section 7.2) is computed for the SIDS data; details can be found in Section 7.6.

Spatial Model for Latent Poisson Parameters

Another way to model spatial counts data is to assume that the *intensities* of the Poisson variables $\mathbf{Z} \equiv (Z(\mathbf{s}_1), \ldots, Z(\mathbf{s}_n))'$ are random and behave according to a Markov random field, but that conditional on those intensities $Z(\mathbf{s}_1), \ldots, Z(\mathbf{s}_n)$ are independent. (This is the type of approach often taken in remote sensing, where it is assumed that the pixel data are observed with measurement error, but that the noiseless mean process varies according to a spatial process; see Section 7.4.) For the case where the Poisson intensities are assumed multivariate log Gaussian (not necessarily according to a spatial model), Aitchison and Ho (1989) give the properties of the resulting discrete multivariate distribution for **Z**.

Clayton and Kaldor (1987) analyze lip-cancer incidence in the local districts of Scotland this way: Conditional on the intensities, the counts are assumed Poisson, but the logarithm of the intensity is modeled as an auto-Gaussian process. It is not clear that *Poisson* measurement error is an appropriate model for disease incidence; however, it does allow positive and negative dependence between sites (recall that for an auto-Poisson model the dependence is restricted to be negative; i.e., $\theta_{ij} \le 0$, for all $i, j = 1, \ldots, n$). Section 7.5.2 contains more details on this type of spatial model for counts data.

Autobinomial Model

Often the Poisson distribution is an approximation to the binomial distribution, a finite sum of independent 0–1 outcomes. Therefore, disease-incidence data might be modeled more accurately using an *autobinomial model*.

In the case of SIDS in North Carolina, condition on the number of live births in county i, assume independence and homogeneity *within* any county, and write

$$\Pr\left(z(\mathbf{s}_i) \mid \{z(\mathbf{s}_j): j \neq i\}\right)$$
$$\equiv \binom{n_i}{z(\mathbf{s}_i)} p_i\left(\{z(\mathbf{s}_j): j \neq i\}\right)^{z(\mathbf{s}_i)} \left(1 - p_i\left(\{z(\mathbf{s}_j): j \neq i\}\right)\right)^{n_i - z(\mathbf{s}_i)},$$
$$z(\mathbf{s}_i) = 0, 1, \ldots, n_i, \quad (6.5.11)$$

where $z(\mathbf{s}_i)$ here represents the number of sudden infant deaths in county i (elsewhere denoted as S_i). Then, because (6.5.11) is of exponential family form, and assuming pairwise-only dependence between sites,

$$p_i\left(\{z(\mathbf{s}_j): j \neq i\}\right) = \frac{\exp\left\{\alpha_i + \sum_{j=1}^{n} \theta_{ij} z(\mathbf{s}_j)\right\}}{1 + \exp\left\{\alpha_i + \sum_{j=1}^{n} \theta_{ij} z(\mathbf{s}_j)\right\}}.$$

Hence, up to an additive constant, $Q(\mathbf{z})$ is

$$\sum_{i=1}^{n} \alpha_i z(\mathbf{s}_i) + \sum\sum_{1 \le i < j \le n} \theta_{ij} z(\mathbf{s}_i) z(\mathbf{s}_j) + \sum_{i=1}^{n} \log\left(\binom{n_i}{z(\mathbf{s}_i)}\right), \quad (6.5.12)$$

where $\theta_{ij} = \theta_{ji}$, $\theta_{ii} = 0$, and $\theta_{ik} = 0$ for $k \notin N_i$. When $n_1 = n_2 = \cdots = 1$ in (6.5.11) and (6.5.12), the relevant expressions are obtained for the autologistic model for binary data (Section 6.5.1).

Modeling of the large-scale variation can be handled by $\boldsymbol{\alpha} = X\boldsymbol{\beta}$; for the small-scale variation one might try something like (6.5.9) or (6.5.10). Once again, the intractability of the normalizing constant $\Sigma_{\mathbf{y}\in\zeta}\exp(Q(\mathbf{y}))$ in the likelihood makes exact maximum likelihood estimation cumbersome; see Section 7.2 for other approaches.

Auto Negative Binomial Model

A distribution that has more variation than the Poisson (the variance of a Poisson variable is equal to its mean) is the negative binomial distribution. Consider a Poisson distribution with mean λ, which is itself a gamma random variable with parameters $\beta > 0$, $\mu > 0$. Then the marginal distribution, after mixing over λ, is

$$\Pr(z) = \int_0^\infty \frac{e^{-\lambda}\lambda^z}{z!}\frac{\beta^\mu \lambda^{\mu-1}e^{-\beta\lambda}}{\Gamma(\mu)}\,d\lambda$$

$$= \frac{\Gamma(z+\mu)}{z!\Gamma(\mu)}\left(\frac{\beta}{\beta+1}\right)^\mu\left(1-\frac{\beta}{\beta+1}\right)^z, \qquad z = 0,1,\ldots.$$

When $\mu = r$, an integer,

$$\Pr(z) = \binom{z+r-1}{r-1}\left(\frac{\beta}{\beta+1}\right)^r\left(\frac{1}{\beta+1}\right)^z, \qquad z = 0,1,\ldots,$$

which is the *negative binomial* probability mass function. A negative binomial random variable is usually defined as the number of successes needed before the rth failure occurs in a (potentially infinite) sequence of independent Bernoulli trials, where Pr(success) $= 1/(\beta+1) = 1 -$ Pr(failure). Also, for such a distribution, the mean is r/β and the variance is $(r/\beta)(1+\beta^{-1})$, which is greater than r/β.

Therefore an *auto negative binomial* spatial model for counts data can be defined as

$$\Pr\left(z(\mathbf{s}_i)\middle|\{z(\mathbf{s}_j): j \neq i\}\right)$$

$$= \frac{\Gamma(z(\mathbf{s}_i)+\mu_i)}{z(\mathbf{s}_i)!\Gamma(\mu_i)}\left(\frac{\beta_i(\{z(\mathbf{s}_j): j\neq i\})}{\beta_i(\{z(\mathbf{s}_j): j\neq i\})+1}\right)^{\mu_i}$$

$$\times\left(\frac{1}{\beta_i(\{z(\mathbf{s}_j): j\neq i\})+1}\right)^{z(\mathbf{s}_i)}, \qquad z(\mathbf{s}_i) = 0,1,2,\ldots. \qquad (6.5.13)$$

If pairwise-only dependence between sites is assumed, then

$$\beta_i\big(\{z(\mathbf{s}_j): j \neq i\}\big) = \exp\left\{-\alpha_i - \sum_{j=1}^{n} \theta_{ij} z(\mathbf{s}_j)\right\} - 1, \quad \text{where } \theta_{ij} = \theta_{ji}, \theta_{ii} = 0.$$

Some Final Remarks

The range of models that can be specified at the local level, via conditioning, and that have easily interpretable trend and dependence parameters, is really only limited by the extent of the (univariate) exponential family of models. However, it is almost never possible to construct a closed-form likelihood at the global level. Parameter estimation for these spatial models is discussed in Section 7.2.

6.6 CONDITIONALLY SPECIFIED SPATIAL MODELS FOR CONTINUOUS DATA

Auto-Gaussian (or CG) models

By far the most useful class of Markov random-field models for continuous data is the class of auto-Gaussian (or CG) models. They have already been discussed briefly in Section 6.3, but are presented here from the advanced standpoint of the theoretical development in Section 6.4.

Let the data $\mathbf{Z} = (Z(\mathbf{s}_1), \ldots, Z(\mathbf{s}_n))'$ be observed at site locations $\{\mathbf{s}_1, \ldots, \mathbf{s}_n\}$. By substituting f for Pr in (6.4.9) and assuming that the conditional density has the Gaussian form (6.3.11), it is easy to show that

$$\begin{aligned} &f\big(z(\mathbf{s}_i) | \{z(\mathbf{s}_j): j \neq i\}\big) \\ &\quad = \exp\Big[\theta_i\big(\{z(\mathbf{s}_j): j \neq i\}\big)\tau_i^{-2} z(\mathbf{s}_i) + C\big(z(\mathbf{s}_i)\big) + D\big(\{z(\mathbf{s}_j): j \neq i\}\big)\Big], \\ &\qquad\qquad i = 1, \ldots, n. \quad (6.6.1) \end{aligned}$$

Assuming pairwise-only dependence between sites, the conditional expectation $E(Z(\mathbf{s}_i)|\{z(\mathbf{s}_j): j \neq i\}) \equiv \theta_i(\{z(\mathbf{s}_j): j \neq i\})$ can be written as

$$\theta_i\big(\{z(\mathbf{s}_j): j \neq i\}\big) = \mu_i + \sum_{j=1}^{n} c_{ij}\big(z(\mathbf{s}_j) - \mu_j\big), \qquad (6.6.2)$$

where $c_{ij}\tau_j^2 = c_{ji}\tau_i^2$, $c_{ii} = 0$, and $c_{ik} = 0$ unless i and k are neighbors. Therefore, from (6.3.11) and (6.6.2), the conditional distribution is given by

$$Z(\mathbf{s}_i) | \{z(\mathbf{s}_j): j \neq i\} \sim \text{Gau}\left(\mu_i + \sum_{j=1}^{n} c_{ij}\big(z(\mathbf{s}_j) - \mu_j\big), \tau_i^2\right). \quad (6.6.3)$$

As a direct consequence of the Factorization Theorem, it is shown in Section 6.4 that

$$\mathbf{Z} \sim \text{Gau}\big(\boldsymbol{\mu}, (I - C)^{-1} M\big), \tag{6.6.4}$$

where C is an $n \times n$ matrix whose (i, j)th element is c_{ij}, $c_{ij}\tau_j^2 = c_{ji}\tau_i^2$, and $c_{ii} = 0$; $M = \text{diag}(\tau_1^2, \ldots, \tau_n^2)$, and $\mu_i = E(Z(\mathbf{s}_i))$. Thus,

$$\begin{aligned} Q(\mathbf{z}) = \log(f(\mathbf{z})/f(\mathbf{0})) &= -(1/2)(\mathbf{z} - \boldsymbol{\mu})' M^{-1}(I - C)(\mathbf{z} - \boldsymbol{\mu}) \\ &\quad + (1/2)\boldsymbol{\mu}' M^{-1}(I - C)\boldsymbol{\mu}. \end{aligned}$$

It is now apparent why the normalizing constant $\int_{\mathbf{R}^n} \exp\{-(1/2)(\mathbf{z} - \boldsymbol{\mu})'\mathrm{M}^{-1}(\mathrm{I} - \mathrm{C})(\mathbf{z} - \boldsymbol{\mu})\}\, d\mathbf{z}$ can be evaluated in the auto-Gaussian case. Its integrand is recognizable as (constant) $\times$ (Gaussian density). By definition, the integral of a density is equal to 1; thus, the constant equals $(2\pi)^{n/2}|M|^{1/2}/\,|I - C|^{1/2}$.

Notice that *any* Gaussian distribution on a finite set of sites $\mathbf{Z} \sim \text{Gau}(\boldsymbol{\mu}, \Sigma)$ can be expressed as a CG process (6.6.4). This is easily established by writing $M = \{\text{diag}(\sigma^{(11)}, \cdots, \sigma^{(nn)})\}^{-1}$, a diagonal matrix defined by the inverse of the diagonal terms of Σ^{-1}, and $C = I - M\Sigma^{-1}$.

Mardia (1988) develops multivariate versions of (6.6.3) for a vector process $\{\mathbf{Z}(\mathbf{s}): \mathbf{s} \in D\}$. In a similar manner, the likelihood can be constructed and maximum likelihood estimates can be obtained.

SIDS Data

The fortuitous recognition of a known density in $\exp(Q(\mathbf{z}))$ is very much the exception when building models from conditional specifications. For this reason, as well as for reasons of statistical tradition and available software, data that are not Gaussian (e.g., continuous but skewed, or discrete) might be transformed to approximate Gaussianity and analyzed spatially in the transformed scale. The sudden infant death rates in the counties of North Carolina were subjected to an exploratory analysis in Section 6.2, and a Freeman–Tukey transformation given by (6.2.7) was indicated. Notice that the (transformed) variable $Z(\mathbf{s}_i)$ need not have homogeneous conditional variances. For the SIDS data it seems sensible to model τ_i^2 proportional to $1/n_i$ (n_i is the number of live births in county i). Section 7.6 contains the auto-Gaussian-based analysis of these Freeman–Tukey transformed data, including details on choice of mean structure, conditional variances, and spatial dependencies.

Wheat-Yield Data

The Mercer and Hall (1911) wheat-yield data consist of the weight (in pounds) of the wheat grain harvested from plots (approximately) 3.30×2.51 m, layed out on a regular grid. There were 500 plots with 20 rows, each row running east to west, and 25 columns, each column running north to south (see Section 4.5 for more details on the spatial configuration of the plots). Whittle (1954) analyzes these data using a simultaneously specified Gaussian (SG) model, and Besag (1974) fits a CG model. The analyses are given in Section 7.1. The conclusion to be reached from them is that neither model fits well (as judged by comparing theoretical spatial covariances to sample spatial covariances). I want to say briefly why.

The data $\{Z(u,v): u = 1,\ldots,25,\ v = 1,\ldots,20\}$ are on a regular lattice. A histogram of the 500 Zs presented by Mercer and Hall (1911), shows a remarkably close agreement with a Gaussian distribution. Thus the following model, with first-order dependence structure, does not seem unreasonable (Besag, 1974):

$$\begin{aligned} E(Z(u,v)|\{z(k,l)\colon (k,l) \neq (u,v)\}) &= \mu + \gamma_1(z(u-1,v) + z(u+1,v)) \\ &\quad + \gamma_2(z(u,v-1) + z(u,v+1)). \end{aligned} \tag{6.6.5}$$

Unfortunately, (6.6.5) does not fit, nor does the model with second-order dependence structure, nor do similar models with measurement error in the Zs (Besag, 1977a), nor do the analogous SG models. The reason is simple. The process is not homogeneous; that is, μ in (6.6.5) really should be $\mu(u,v)$, which in turn might be expressed as $\boldsymbol{\mu} = X\boldsymbol{\beta}$ (X is a matrix whose columns are explanatory variables). Often, some of the explanatory variables are spatial variables; indeed, Section 4.5 shows that the decomposition of $\mu(u,v)$ into a function of u plus a function of v fits well. It is clear that there is large-scale variation (spatial trend) in the wheat-yield data that must be taken into account before the parameters of the small-scale variation (spatial dependence) can be interpreted meaningfully.

CG Regression Models

Statisticians prepared to use spatial models need to keep the role of the models in perspective. When scientific interest centers on the large-scale effects, the idea is to use a few extra small-scale (spatial dependence) parameters so that the large-scale parameters are estimated more efficiently. Ideally, the estimated spatial dependence parameters also have a scientific interpretation; certainly, when studying a homogeneous medium, they can be interpreted as measuring local interaction between sites. When prediction is the primary goal, spatial dependence parameters are important because the dependence can be exploited to improve predictions (Chapter 3).

Suppose that the data $\mathbf{Z}$ are modeled according to a conditionally specified Gaussian (CG) model and that the large-scale variation is modeled as

$$\boldsymbol{\mu} = X\boldsymbol{\beta}, \tag{6.6.6}$$

where X is an $n \times q$ matrix whose columns are q explanatory variables. In other words, the response $Z(\mathbf{s}_i)$ at the ith site is accompanied by q explanatory variables $\{x_j(\mathbf{s}_i): j = 1, \ldots, q\}$, such that $E(Z(\mathbf{s}_i)) = \sum_{j=1}^{q} x_j(\mathbf{s}_i)\beta_j$. So far, there is nothing spatial about this model, except that some of the explanatory variables may depend on the site locations $\{\mathbf{s}_1, \ldots, \mathbf{s}_n\}$ [e.g., choosing $x_1(\mathbf{s}_i) = x_i$, $x_2(\mathbf{s}_i) = y_i$, where $\mathbf{s}_i = (x_i, y_i)'$, yields a linear trend in either direction].

The usual Gaussian regression model is $\mathbf{Z} \sim \text{Gau}(X\boldsymbol{\beta}, \tau^2 I)$. Maximum likelihood estimation of parameters $\boldsymbol{\beta}$ and τ^2 yields $\hat{\boldsymbol{\beta}} = (X'X)^{-1}X'\mathbf{Z}$ and $\hat{\tau}^2 = (\mathbf{Z} - X\hat{\boldsymbol{\beta}})'(\mathbf{Z} - X\hat{\boldsymbol{\beta}})/n$. Inference about the β_is can proceed using standard F tests and the variance matrix estimate $\widehat{\text{var}}(\hat{\boldsymbol{\beta}}) = (X'X)^{-1}\hat{\tau}^2$. However, when (6.6.4) holds, these (ordinary-least-squares) estimators are no longer appropriate.

Together, (6.6.4) and (6.6.6) allow one to write

$$\mathbf{Z} = X\boldsymbol{\beta} + \boldsymbol{\delta}, \qquad (6.6.7)$$

where the error $\boldsymbol{\delta}$ satisfies

$$\boldsymbol{\delta} \sim \text{Gau}\left(\mathbf{0}, (I - C)^{-1}M\right).$$

When $C = 0$ and $M = \tau^2 I$, the CG regression model (6.6.7) reduces to the usual Gaussian regression model. Assuming C and M are known, the generalized-least-squares estimator is

$$\boldsymbol{\beta}^* = \left(X'M^{-1}(I - C)X\right)^{-1} X'M^{-1}(I - C)\mathbf{Z}, \qquad (6.6.8)$$

which is also the maximum likelihood estimator. Section 1.3 compares these two estimators $\hat{\boldsymbol{\beta}}$ and $\boldsymbol{\beta}^*$; under the CG regression model (6.6.7), $\boldsymbol{\beta}^*$ is uniformly superior in that any linear combination of $\hat{\boldsymbol{\beta}}$ always has variance at least as large as the same linear combination of $\boldsymbol{\beta}^*$.

The poor performance of various classical statistical inference procedures when $C \neq 0$ (e.g., Cliff and Ord, 1981, Chapter 7; Anselin and Griffith, 1988) should convince scientists analyzing spatial data that their classical regression models need to be generalized to include spatial dependence terms. If $C = 0$ is inferred from the data, the classical regression model is fitted. If it appears that $C \neq 0$, there may be a missing x variable (varying spatially) that is causing it. By (unknowingly) modeling its presence through spatial dependence parameters, the spatial model is more resistent to misspecification errors. Another possible cause of a nonzero C is a spatially dependent error due to the cumulative effect of a lot of small-scale, spatially varying components.

When M and C are unknown, as is usually the case, they are often expressed in terms of a few (small-scale) parameters, at least one of which captures the spatial dependence and ensures that, in general, $C \neq 0$. Inference on $\boldsymbol{\beta}$ (and these parameters) usually proceeds through maximizing the likelihood:

$$(2\pi)^{-n/2}|M|^{-1/2}|I - C|^{1/2} \exp\{-(1/2)(\mathbf{z} - X\boldsymbol{\beta})' M^{-1}(I - C)(\mathbf{z} - X\boldsymbol{\beta})\}; \tag{6.6.9}$$

computational details are given in Section 7.2. To illustrate these inferential procedures, a confirmatory analysis of the (suitably transformed) SIDS data is given in Section 7.6.

The CG regression model (6.6.7) can be compared with the SG regression model (and more general variants) presented in detail in Section 6.7. There is no difference in computational complexity, but there is a difference in interpretation of the spatial-dependence parameters. Section 6.3.3 contains a comparison of the two approaches.

Small-Scale Variation (Spatial Dependence)

For the CG model, spatial dependence is characterized through the conditional expectation. On a regular lattice, spatial dependence in a homogeneous process is expressed as

$$E(Z(u,v)|\{z(k,l): (k,l) \neq (u,v)\}) = \mu + \sum_i \sum_j \gamma(i,j) z(u-i, v-j),$$

where $\gamma(i,j) = \gamma(-i,-j)$ and $\gamma(0,0) = 0$; Section 7.2.2 discusses estimation of the spatial-dependence parameters $\{\gamma(i,j)\}$. The first-order dependence model (6.6.5) is the special case: $\gamma(1,0) = \gamma(-1,0) = \gamma_1$, $\gamma(0,1) = \gamma(0,-1) = \gamma_2$, and all other $\gamma(i,j)$s are zero. Besag (1981) investigates the covariance properties of this simple CG model; he shows *inter alia* that the correlation between two observations decays rather slowly with increasing distance between their lattice sites.

For the irregular lattice, one could model the c_{ij} in (6.6.2) as some function of distance d_{ij} between site i and site j. For example, c_{ij} might be given by (6.5.9); see Section 7.6, where a transformed SID rate is modeled according to a heteroskedastic CG model. Notice that, although $C = (c_{ij})$ depends only on $\{d_{ij}: 1 \le i < j \le n\}$, the covariance $\text{cov}(Z(\mathbf{s}_i), Z(\mathbf{s}_j))$ does *not* in general depend on $\mathbf{s}_i - \mathbf{s}_j$, and hence the finite-lattice process is generally not stationary.

Care must be taken with parameterizing the spatial-dependence matrix C. It is not difficult to show that $\text{corr}^2(Z(\mathbf{s}_i), Z(\mathbf{s}_j)|\{z(\mathbf{s}_k): k \neq i, j\}) = c_{ij}c_{ji}$. Thus, $0 \le c_{ij}c_{ji} \le 1$. Indeed, Speed and Kiiveri (1986) show how zeros in the off-diagonal elements in C correspond to conditional independence statements; they go on to characterize all such statements that arise from a given pattern of zeros.

Clearly, for auto-Gaussian (or CG) processes, modeling is in terms of the *inverse* covariance matrix, which is in contrast to Part I (Geostatistical Data) where the covariance matrix is modeled directly (often through the variogram).

An Initial Check for Spatial Dependence
Write $C = \gamma H$, H a known symmetric matrix, and $M = \tau^2 I$, and assume the CG regression model (6.6.7). Then, a statistic for testing

$$H_0: \gamma = \gamma_0 \quad \text{versus} \quad H_A: \gamma \neq \gamma_0$$

has been proposed by Singh and Shukla (1983). Reject H_0 for large values of the test statistic

$$O \equiv \{\mathbf{e}'(I - \gamma_0 H) H \mathbf{e}\} / \{\mathbf{e}'(I - \gamma_0 H)\mathbf{e}\}, \qquad (6.6.10)$$

where $\mathbf{e} \equiv \mathbf{Z} - X\hat{\boldsymbol{\beta}}$ and $\hat{\boldsymbol{\beta}}$ is the usual ordinary-least-squares estimator $(X'X)^{-1}X'\mathbf{Z}$. Under suitable regularity conditions, (6.6.10) is asymptotically Gaussian, and hence confidence intervals for γ can be constructed. When the confidence interval contains zero, the data support the classical model with no spatial dependence.

Infinite Lattices and Edge Sites
When data in $\mathbb{R}^2$ occur at a potentially infinite number of sites, construction of a Gaussian Markov random field proceeds in the same manner as Moran's (1973a) construction. There, an initial process is constructed on a torus (a donut-shaped surface that is the two-dimensional analogue of the circle) lattice, which is then allowed to tend to infinity. Moran (1973a) gives such a construction on a square lattice for the model (6.6.5), a stationary Gaussian process with first-order neighborhood structure. Moran (1973b) extends the approach to certain non-Gaussian processes, and Martin (1986b) considers it for d-dimensional lattices.

Suppose the data are $\{Z(u, v): u = 1, \ldots, U;\ v = 1, \ldots, V\}$, a sampling of an infinite process on $\mathbb{Z}^2$. Then, what should be done with the conditional specification (6.6.5) at the edge sites where $u = 1, U$ or $v = 1, V$? One possibility is to condition on the edge sites when constructing the likelihood (although they do appear as neighbors of internal sites); see, for example, Gleeson and McGilchrist (1980). For a $U \times V$ lattice, this amounts to setting aside $\{2(U + V - 2)/UV\}100\%$ of the sites, which can be a substantial amount (e.g., $U = V = 10$ yields a loss of 36% of the sites). Another possibility is to wrap the $U \times V$ lattice on a torus; however, the donut-shaped space may be distasteful to those who object to having, for example, $(1, v)$ and (U, v) as nearest neighbors.

A third possibility is to recognize that there are unobserved sites bordering the edge sites and work with them in some way. One way would be to assume that they take whatever values are needed for (6.6.5) not to depend on the offending edge site. For example, if $(u, v) = (u, V)$, where $1 < u < U$, then for the relation

$$E(Z(u,V)|\{z(k,l):(k,l) \neq (u,V)\})$$
$$= \mu + \gamma_1(z(u-1,V) + z(u+1,V)) + \gamma_2 z(u,V-1)$$

to hold, one must assume $z(u, V+1) = 0$. This approach was taken by Haining (1978a) for simulating CGs on small lattices, and seems as artificial as wrapping the lattice on a torus.

A more rigorous approach would be to integrate out the presence of the variable $z(u, V+1)$ in the conditional distribution. Thus,

$$E(Z(u,V))|\{z(k,l):(k,l) \neq (u,V),(u,V+1)\}$$
$$= \mu + \gamma_1(z(u-1,V) + z(u+1,V)) + \gamma_2 z(u,V-1),$$

as desired. However, if conditional homoskedastic variances were assumed; that is, $\mathrm{var}(Z(u,v)|\{z(k,l)\colon (k,l) \neq (u,v)\}) = \tau^2$, then $\mathrm{var}(Z(u,V)|\{z(k,l)\colon (k,l) \neq (u,V),(u,V+1)\})$ would be heteroskedastic through marginalization with respect to the unobserved border values. For the autoregressive-moving-average process in time series, Ljung and Box (1979) give the full unconditional likelihood. Haining (1978b) claims to give the likelihood for a spatial moving average, but fails to handle the edge sites properly.

There is room for further research on the problem of how to deal with edge sites. Of the options presented, the assumption that edge sites are zero (or, more generally, equal to their respective means) is usually favored because the likelihood does not change from the finite-site case defined by (6.6.3).

Some Final Remarks

When the model is no longer CG, the normalizing constant $\int_{\zeta} \exp(Q(\mathbf{y}))\,d\mathbf{y}$ is typically intractable. Dobrushin (1988) considers Markov random fields where $Q(\cdot)$ is perturbed slightly from its Gaussian form by additive (non-Gaussian) interactions. He discusses approximation of the normalizing constant under this perturbation model.

A large class of non-Gaussian models is generated by members of the exponential family whose distributions are absolutely continuous. For example, assuming pairwise-only dependence between sites, an *autogamma* model is defined by

$$f\big(z(\mathbf{s}_i)|\{z(\mathbf{s}_j): j \neq i\}\big) = \big(\Gamma(\lambda_i)\big)^{-1}\beta_i^{\lambda_i} z(\mathbf{s}_i)^{\lambda_i - 1} e^{-\beta_i z(\mathbf{s}_i)} I\big(z(\mathbf{s}_i) > 0\big), \tag{6.6.11}$$

where dependence of λ_i on the rest of the observations $\{z(\mathbf{s}_j): j \neq i\}$ is momentarily suppressed. Indeed,

$$\lambda_i(\{z(\mathbf{s}_j): j \neq i\}) = \alpha_i + \sum_{j=1}^{n} \theta_{ij} \log(z(\mathbf{s}_j)),$$

where $\theta_{ij} = \theta_{ji}$, $\theta_{ii} = 0$, and $\theta_{ik} = 0$ for $k \notin N_i$. From (6.6.11) and the Corollary in Section 6.4.2, $Q(\mathbf{z})$ is (up to an additive constant)

$$\sum_{i=1}^{n} \alpha_i \log(z(\mathbf{s}_i)) + \sum\sum_{1 \leq i < j \leq n} \theta_{ij} \log(z(\mathbf{s}_i))\log(z(\mathbf{s}_j))$$
$$- \sum_{i=1}^{n} (\beta_i z(\mathbf{s}_i) + \log z(\mathbf{s}_i)).$$

The real hurdle to obtaining the likelihood (from which inference on the parameters can proceed) remains the normalizing constant, $\int_{\zeta} \exp(Q(\mathbf{y}))\, d\mathbf{y}$.

The autogamma model expresses neighborhood dependence through the shape parameter λ_i. Is it possible to express such dependence also through the scale parameter β_i? The question becomes particularly relevant should one wish to construct an "autobeta" process; there, both parameters are shape parameters. Cressie and Lele (1992) provide some answers.

6.7 SIMULTANEOUSLY SPECIFIED AND OTHER SPATIAL MODELS

The emphasis in this chapter has been on conditionally specified models, in particular using them to define a (parametric) Markov random field. Section 6.3 introduced another approach, based on the specification of how data at the various sites of the process interact simultaneously. This section develops these models further, allowing for regression (large-scale) effects, along with spatial autoregressive-moving-average (small-scale) effects.

6.7.1 Simultaneously Specified Spatial Models

Whittle's (1954) prescription for simultaneously specified stationary processes in the plane is given by (6.3.5). For a (finite) data set $\mathbf{Z} \equiv (Z(\mathbf{s}_1), \ldots, Z(\mathbf{s}_n))'$ at locations $\mathbf{s}_1, \ldots, \mathbf{s}_n$, the analogous specification is

$$(I - B)\mathbf{Z} = \boldsymbol{\epsilon}, \tag{6.7.1}$$

where $\boldsymbol{\epsilon} \equiv (\epsilon(\mathbf{s}_1), \ldots, \epsilon(\mathbf{s}_n))'$ is a vector of i.i.d. zero-mean errors and B is a matrix whose diagonal elements $\{b_{ii}\}$ are zero. Another way to write (6.7.1) is

$$Z(\mathbf{s}_i) = \sum_{j=1}^{n} b_{ij} Z(\mathbf{s}_j) + \epsilon(\mathbf{s}_i), \qquad (6.7.2)$$

which is why it is sometimes referred to as a *spatial autoregressive* (SAR) process. Notice that (6.7.2) does *not* define a stationary process in general, even when the $\{b_{ij}\}$ are purely functions of distance between their respective sites.

When $\boldsymbol{\epsilon}$ is Gaussian,

$$\mathbf{Z} \sim \mathrm{Gau}\big(\mathbf{0}, (I - B)^{-1}(I - B')^{-1}\sigma^2\big), \qquad (6.7.3)$$

a special case of (6.3.8). From (6.7.3), the (small-scale) parameters σ^2 and B can be estimated by maximum likelihood estimation; see Section 7.2.

To include large-scale regression parameters in the model, there are two possibilities, although they have sometimes been confused in the literature. I now give the model I prefer, and make brief mention of the other.

Spatial Autoregressive Regression Model

If it is desired to interpret large-scale effects $\boldsymbol{\beta}$ through $E(\mathbf{Z}) = X\boldsymbol{\beta}$ (where the columns of X might be treatments, spatial trends, factors, etc.), then from (6.3.7), (6.7.1) should be modified to

$$(I - B)(\mathbf{Z} - X\boldsymbol{\beta}) = \boldsymbol{\epsilon}. \qquad (6.7.4)$$

When $\boldsymbol{\epsilon}$ is Gaussian,

$$\mathbf{Z} \sim \mathrm{Gau}\big(X\boldsymbol{\beta}, (I - B)^{-1}(I - B')^{-1}\sigma^2\big), \qquad (6.7.5)$$

a special case of (6.3.8) with $\boldsymbol{\mu} = X\boldsymbol{\beta}$. From (6.7.5), the large-scale parameters $\boldsymbol{\beta}$ and the small-scale parameters σ^2 and B can be estimated by maximum likelihood; see Section 7.2.

Define $\boldsymbol{\delta} \equiv \mathbf{Z} - X\boldsymbol{\beta}$, which is the error process. Then,

$$\mathbf{Z} = X\boldsymbol{\beta} + \boldsymbol{\delta}, \qquad (6.7.6)$$

where $\boldsymbol{\delta}$ satisfies

$$\boldsymbol{\delta} = B\boldsymbol{\delta} + \boldsymbol{\epsilon}. \qquad (6.7.7)$$

Time series typically exhibit *serial* correlation, which can be modeled via the error process $\boldsymbol{\delta}$.

An Initial Check for Spatial Dependence

Durbin and Watson (1950) present a test for first-order serial correlation in a time series, and by analogy Cliff and Ord (1973) develop a likelihood-ratio test for

$$H_0: \xi = 0 \quad \text{versus} \quad H_A: \xi \neq 0,$$

where $B = \xi W$ in (6.7.4) and $W = (w_{ij})$ is a known $n \times n$ matrix. Write $\mathbf{e} = \mathbf{Z} - X\hat{\boldsymbol{\beta}}$, where $\hat{\boldsymbol{\beta}}$ is the usual ordinary-least-squares estimate $(X'X)^{-1}X'\mathbf{Z}$. Then the test statistic is (cf. Moran's contiguity ratio I in Section 6.5.1)

$$J \equiv \mathbf{e}'W\mathbf{e}/\mathbf{e}'\mathbf{e}, \tag{6.7.8}$$

where H_0 is rejected for large values of J. Burridge (1980) shows this to be the same as the Lagrange multiplier test (Silvey, 1959) of the two hypotheses and demonstrates that under H_0,

$$nJ\{\text{tr}(W^2 + W'W)\}^{-1/2}$$

is distributed asymptotically according to a standard Gaussian distribution. Furthermore, King (1981) shows these tests to be locally best invariant not only for the autoregressive model $(I - \xi W)\boldsymbol{\delta} = \boldsymbol{\epsilon}$, but also for the moving average model $\boldsymbol{\delta} = (I - \xi W)\boldsymbol{\epsilon}$. If H_0 is not rejected, then the data support the classical model with no spatial dependence. By replacing W in (6.7.8) with $\text{diag}\{\Sigma_{j=1}^{n}(w_{ij} + w_{ji})/2\} - W$, a regression analogue to Geary's contiguity ratio is obtained (Geary, 1954).

Spatial Autoregressive-Moving-Average Regression Model

First, the spatial-moving-average regression model is defined by (6.7.6), where

$$\boldsymbol{\delta} = (I - E)\boldsymbol{\epsilon}. \tag{6.7.9}$$

Haining (1978b) and Moore (1988) give detailed discussions of this model for $X\boldsymbol{\beta} = \mathbf{0}$ and $X\boldsymbol{\beta} = \mathbf{1}\mu$, respectively.

The spatial autoregressive-moving-average regression (SARMAX) model is defined by (6.7.6), where

$$(I - B)\boldsymbol{\delta} = (I - E)\boldsymbol{\epsilon}; \tag{6.7.10}$$

Huang (1984) has suggested parametrizations of B and E of the form

$$B = \sum_{j=1}^{p} \xi_j W^j, \qquad E = \sum_{j=1}^{q} \phi_j W^j, \tag{6.7.11}$$

where W is a prespecified spatial dependence matrix. When $\boldsymbol{\epsilon}$ is Gaussian,

$$\mathbf{Z} \sim \operatorname{Gau}\left(X\boldsymbol{\beta}, (I-B)^{-1}(I-E)(I-E')(I-B')^{-1}\sigma^2\right), \quad (6.7.12)$$

from which the likelihood for $\boldsymbol{\beta}$, B, and E is,

$$(2\pi)^{-n/2}\left\{\frac{\det(I-E)}{\det(I-B)}\right\}\exp\left\{-\left(\frac{1}{2}\right)(\mathbf{z}-X\boldsymbol{\beta})'(I-B')\right.$$
$$\left.\times(I-E')^{-1}(I-E)^{-1}(I-B)(\mathbf{z}-X\boldsymbol{\beta})\right\}.$$

An Alternative SARMAX Model

The model

$$(I-B)\mathbf{Z} = X\boldsymbol{\beta} + (I-E)\boldsymbol{\epsilon} \quad (6.7.13)$$

has the same variance matrix as the model given by (6.7.6) and (6.7.10), but

$$E(\mathbf{Z}) = (I-B)^{-1}X\boldsymbol{\beta};$$

thus, the regression is with respect to explanatory variables $(I - B)^{-1}X$, so confounding large- and small-scale effects. Authors who have taken this approach include Ord (1975), Doreian (1980), Gleeson and McGilchrist (1980), Huang (1984), and Anselin (1988). Anselin has used the term "spatial econometrics" to describe his work, because he is adapting econometric methods to address spatial and spatiotemporal problems deriving from human phenomena. From a statistician's perspective, the distinction is not particularly helpful.

Spectral Analysis of Lattice Data

Although this book is written very much from a space-domain point of view, the reader should also be aware of the spectral-domain approach [more details can be found in Helson and Lowdenslager (1958, 1961) and Bartlett (1975)].

Suppose $\{Z(\mathbf{s}): \mathbf{s} \in D \subset \mathbb{R}^2\}$ is a realization of a two-dimensional second-order stationary process in a domain D. Assume that the process has zero mean; if not, then substitute the mean-corrected process $\{Z(\mathbf{s}) - \mu(\mathbf{s}): \mathbf{s} \in D\}$ for $\{Z(\mathbf{s}): \mathbf{s} \in D\}$ in the formulas that follow. Define the covariogram and correlogram, $C(\cdot)$ and $\rho(\cdot)$, respectively, as

$$\begin{aligned} C(\mathbf{h}) &\equiv \sigma^2\rho(\mathbf{h}) \equiv E\{Z(\mathbf{s}+\mathbf{h})Z(\mathbf{s})\}, \qquad \mathbf{h} \in \mathbb{R}^2, \\ \sigma^2 &\equiv C(\mathbf{0}), \end{aligned} \quad (6.7.14)$$

which are functions only of the relative spatial separation $\mathbf{h}$. The spectral distribution function $G_0(\boldsymbol{\omega})$, satisfying $\int_{\mathbb{R}^2} G_0(d\boldsymbol{\omega}) = 1$, is related to $\rho(\mathbf{h})$ by

$$\rho(\mathbf{h}) = \int_{\mathbb{R}^2} \exp(ih_1\omega_1 + ih_2\omega_2) G_0(d\boldsymbol{\omega}), \qquad (6.7.15)$$

where $\mathbf{h} = (h_1, h_2)'$ and $\boldsymbol{\omega} = (\omega_1, \omega_2)'$. When a spectral density $g_0(\boldsymbol{\omega})$ exists,

$$\rho(\mathbf{h}) = \int_{\mathbb{R}^2} \exp(ih_1\omega_1 + ih_2\omega_2) g_0(\boldsymbol{\omega})\, d\boldsymbol{\omega}. \qquad (6.7.16)$$

If in fact $Z(\mathbf{s})$ is not observed continuously in space, but over a square lattice of unit spacing,

$$\rho(\mathbf{h}) = \int_{-\pi}^{\pi}\int_{-\pi}^{\pi} \exp(ih_1\omega_1 + ih_2\omega_2) G(d\boldsymbol{\omega}), \qquad h_1, h_2 = \ldots, -1, 0, 1, \ldots, \qquad (6.7.17)$$

where, due to aliasing, $G(\boldsymbol{\omega}) = \sum^{\infty}\sum^{\infty}_{k,l=-\infty} G_0(\omega_1 + 2k\pi, \omega_2 + 2l\pi)$. Then write its spectral density as $g(\boldsymbol{\omega})$ or as $g_Z(\boldsymbol{\omega})$.

Define the linear transformation

$$Y(u,v) \equiv \sum_{k,l=-\infty}^{\infty}\sum^{\infty} a_{u-k,v-l} Z(k,l), \qquad (6.7.18)$$

where the coefficients are assumed to be absolutely summable. Then, by defining

$$a(\boldsymbol{\omega}) \equiv \sum_{k,l=-\infty}^{\infty}\sum^{\infty} \exp(-ik\omega_1 - il\omega_2) a_{k,l}, \qquad (6.7.19)$$

it is easy to see that,

$$\sigma_Y^2 g_Y(\boldsymbol{\omega}) = |a(\boldsymbol{\omega})|^2 \sigma_Z^2 g_Z(\boldsymbol{\omega}), \qquad \boldsymbol{\omega} \in (-\pi, \pi]^2, \qquad (6.7.20)$$

where the Y and Z subscripts on g and σ^2 refer, respectively, to the variance and spectral density of those processes.

Spatial Autoregressive Model: Spectral Analysis

The first-order dependent, zero-mean, simultaneously specified (or spatial autoregressive) model on a square lattice is

$$Z(u,v) = \xi_1(Z(u-1,v) + Z(u+1,v)) + \xi_2(Z(u,v-1) + Z(u,v+1)) + \epsilon(u,v), \qquad (6.7.21)$$

where $|\xi_1| + |\xi_2| < 1/2$ and the $\epsilon(u, v)$s are zero-mean independent-and-identically distributed random variables. But this is a linear transformation of the form of (6.7.18), where

$$a(\boldsymbol{\omega}) = 1 - 2(\xi_1 \cos \omega_1 + \xi_2 \cos \omega_2). \tag{6.7.22}$$

Because the $\epsilon(\cdot)$ process is white noise, $g_\epsilon(\boldsymbol{\omega}) = (1/2\pi)^2$, $\boldsymbol{\omega} \in (-\pi, \pi]^2$, and hence it follows from (6.7.20) that

$$g_Z(\boldsymbol{\omega}) \propto \{1 - 2(\xi_1 \cos \omega_1 + \xi_2 \cos \omega_2)\}^{-2}, \qquad \boldsymbol{\omega} \in (-\pi, \pi]^2. \tag{6.7.23}$$

The expression (6.7.23) does *not* involve a Gaussian assumption, although Gaussianity is usually assumed in order to find efficient estimators of ξ_1 and ξ_2 (e.g., by maximum likelihood). A maximum entropy method of spectral estimation based on Gaussianity is another possible approach (Newman, 1977; Lim and Malik, 1981; McClellan, 1982; Dudgeon and Mersereau, 1984).

It has already been mentioned in Section 6.3 that, for the Gaussian model, transforming from the (independent) ϵs to the observed Zs involves a complicated Jacobian (i.e., normalizing constant) that prevents consistent estimation of ξ_1 and ξ_2 by ordinary least squares. Whittle (1954) derives an asymptotic expression for this Jacobian in terms of the spectral density $g_Z(\boldsymbol{\omega})$. Suppose the data are observed on a $U \times V$ regular lattice. For U and V large, the approximate negative loglikelihood (up to an additive constant), after minimization with respect to σ_ϵ^2, reduces to

$$\log\left[\sum_u \sum_v \{\phi(T_1, T_2) z(u, v)\}^2 / UV\right]$$
$$+ \int_{-\pi}^{\pi} \int_{-\pi}^{\pi} \log\{\phi(e^{i\omega_1}, e^{i\omega_2}) \phi(e^{-i\omega_1}, e^{-i\omega_2})\}\, d\boldsymbol{\omega}/(2\pi)^2, \tag{6.7.24}$$

where $\phi(T_1, T_2)$ is the operator in (6.3.5). Whittle (1954) gives an approximation to this general expression based on (biased) empirical autocovariances, which is shown by Guyon (1982) to lead to inconsistent estimators; see Sections 7.2.3 and 7.3 for further details.

For the first-order process being considered here, the second term of (6.7.24) is

$$(2\pi)^{-2} \int_{-\pi}^{\pi} \int_{-\pi}^{\pi} (-2)\log(1 - 2\xi_1 \cos \omega_1 - 2\xi_2 \cos \omega_2)\, d\boldsymbol{\omega}$$
$$= \sum_{j=1}^{\infty} \sum_{k=0}^{j} \frac{(2j)!}{j\{k!(j-k)!\}^2} (\xi_1)^{2k} (\xi_2)^{2j-2k},$$

and, of course, the first term of (6.7.24) involves ξ_1, ξ_2 through $\phi(T_1, T_2) = 1 - \xi_1(T_1 + T_1^{-1}) - \xi_2(T_2 + T_2^{-1})$.

A Comparison of the CG Model to the SG Model: Spectral Analysis

Spectral analysis for the CG model is presented here to provide a comparison with the formulas obtained previously for the SG model. Consider the conditionally specified (zero-mean) Gaussian process

$$E(Z(u,v)|\{z(k,l):(k,l) \neq (u,v)\})$$
$$= \gamma_1\{z(u-1,v) + z(u+1,v)\} + \gamma_2\{z(u,v-1) + z(u,v+1)\},$$
$$\text{var}(Z(u,v)|\{z(k,l):(k,l) \neq (u,v)\}) = \tau^2.$$

Bartlett (1975, p. 23) has shown that

$$g_Z(\boldsymbol{\omega}) \propto \{1 - 2\gamma_1 \cos\omega_1 - 2\gamma_2 \cos\omega_2\}^{-1}, \quad \boldsymbol{\omega} \in (-\pi, \pi]^2, \tag{6.7.25}$$

and hence the maximum likelihood estimates of γ_1, γ_2 are obtained by minimizing

$$\log\left[\sum_u \sum_v \{z(u,v) - 2\gamma_1(z(u-1,v) + z(u+1,v)) - 2\gamma_2(z(u,v-1) + z(u,v+1))\}^2/UV\right]$$
$$+ (1/2)\sum_{j=1}^{\infty}\sum_{k=0}^{j} \frac{(2j)!}{j[k!(j-k)!]^2}(\gamma_1)^{2k}(\gamma_2)^{2j-2k}. \tag{6.7.26}$$

The only difference between (6.7.24) and (6.7.26) is the factor (1/2) in the second term.

From (6.7.17) and (6.7.25), the correlogram for $\gamma_1 = \gamma_2 = \gamma$ is given by

$$\rho(\mathbf{h}) = \int_{-\pi}^{\pi}\int_{-\pi}^{\pi} \cos(h_1\omega_1 + h_2\omega_2)\{1 - 2\gamma(\cos\omega_1 + \cos\omega_2)\}^{-1}\,d\omega_1\,d\omega_2,$$
$$|\gamma| \leq 1/4.$$

Consider, for example, the spatial correlation $\rho(1,0)$ $(= \rho(0,1))$ as a function of γ. Bartlett (1975) and Besag (1981) show that this function approaches 1 *very* slowly, as γ approaches 0.25; for example, when $\gamma = 0.24$, $\rho(1,0) = 0.4341$.

McBratney and Webster (1981) perform a spectral analysis on the Mercer and Hall (1911) wheat-yield data, referred to in Section 6.6. They are able to detect a periodicity that they attribute to an earlier ridge-and-furrow pattern; Sections 4.5 and 7.1 have more details on these data.

Infinite Lattices and Edge Sites

When data in $\mathbb{R}^2$ occur at a potentially infinite number of sites, a simultaneously specified autoregressive-moving-average model is defined under conditions given by Whittle (1954), Rosanov (1967), Soltani (1984), and Moore (1988). Rosenblatt (1985, Chapter VIII) specifically considers non-Gaussian linear processes.

From a finite data set, the objective is to make inference on the (infinite) model parameters, usually via the likelihood. For example, suppose $\mathbf{Z}$ is modeled as being generated by a (first-order, say) simultaneously specified Gaussian process (6.3.6), observed on a $U \times V$ square lattice. How should a datum $Z(u, v)$ at $u = 1, U$, or $v = 1, V$, appear in the likelihood? According to the model, it depends on neighbors $Z(u + 1, v)$, $Z(u - 1, v)$, $Z(u, v - 1)$, and $Z(u, v + 1)$, some of which are not observed. In Section 6.6, four possible solutions were suggested to solve this problem for the conditionally specified model. One of those, marginalization with respect to the unobserved neighbors, is very difficult here due to the simultaneous specification of the model. One could, however, condition on the unobserved values in (6.7.4) being equal to their respective means, causing those values to drop from sight in the likelihood. [Haining, (1978b) does this for the first-order spatial-moving-average model.]

The other possibilities, of wrapping the lattice onto a torus or conditioning on edge sites, have the same advantages and disadvantages as were discussed in Section 6.6. Griffith (1983) and Griffith and Amrhein (1983) compare different methods for handling edge sites; the general message is that no method is all that satisfactory.

Even if the full likelihood were available, maximum likelihood estimators of spatial-dependence parameters may be badly biased; restricted maximum likelihood (REML) appears to have better bias properties (Section 2.6.1).

6.7.2 Other Spatial Models

This section describes a few spatial models that I have come across, but which are not covered in previous sections of this chapter. Johnson and Kotz (1972) can be consulted for details on general multivariate models, most of which would have a spatial version by allowing dependence parameters to be functions of data locations.

Covariogram Modeling

Covariogram models from Part I of this book can be used to represent the spatial dependence between data on a lattice. Because the models in Part I are constructed to be valid throughout $\mathbb{R}^d$, they will automatically be valid for data located on some countable subset $D \equiv \{\mathbf{s}_i: i = 1, 2, \dots\}$. For example, let $\{Z(\mathbf{s}): \mathbf{s} \in \mathbb{R}^d\}$ be a second-order stationary process on $\mathbb{R}^d$ with covariogram $C(\cdot)$ and suppose $Z(\cdot)$ is observed on the lattice D. Then $\text{cov}(Z(\mathbf{s}_i), Z(\mathbf{s}_j)) = C(\mathbf{s}_i - \mathbf{s}_j)$, $\mathbf{s}_i, \mathbf{s}_j \in D$. Valid covariogram models are given in Section 2.5.1. An important example is the separable class of models (Section 2.5.1) applied to regular lattice data in $\mathbb{R}^2$; Martin (1979, 1990) has successfully used time-series models for the component covariograms.

Cepstrum Modeling

Solo (1986) has proposed a model based on the spectral density $g(\boldsymbol{\omega})$, defined in Section 6.7.1. The *cepstrum* is the array $\{\Psi_{kl}: k, l = \dots, -1, 0, 1, \dots\}$ in

$$\log g(\boldsymbol{\omega}) \equiv \sum_{-\infty}^{\infty} \sum_{-\infty}^{\infty} \Psi_{kl} \exp(-i\omega_1 k - i\omega_2 l).$$

Modeling the $\{\Psi_{kl}\}$ is often easier because they take simple forms for commonly used random fields.

Unilateral Models

Rather than look for dependence on sites in all directions, Tjostheim (1978, 1983) defines processes for which $Z(u, v)$ depends on Zs to the southwest (quarter-plane dependence) or to the west (half-plane dependence). Any stationary process that has a continuous and positive spectral density can be approximated arbitrarily closely, in quadratic mean, by a quarter-plane unilateral process (Tjostheim, 1978); see Korezlioglu and Loubaton (1986) for further results in the half-plane case.

These models have an evolutionary flavor to them, although applications appear to be scarce [Bronars and Jansen (1987) fit such a model to unemployment rates in the United States]. This is because, even if an accurate representation of the data can be found, it lacks interpretability in a purely spatial problem. However, the likelihoods of unilateral models are much easier to work with than those of Section 6.7.1.

Models for Spatial Flows

Geographers and urban planners have developed models for flows (of people, for example) between a set of origins and a set of destinations (e.g., Bennett and Haining, 1985; Baxter, 1986; Brouwer et al., 1988; Upton and Fingleton, 1989, Chapter 8). The spatial interaction (SPIN) model of flows, the pure-competition model (essentially a Markov random field), the supply–demand model (a space–time model), and the hierarchial spatial model (based on central-place theory), are all discussed by Bennett and Haining (1985).

6.8 SPACE–TIME MODELS

In this chapter I have concentrated on the conditionally specified spatial models. In the Gaussian case, Section 6.3 contrasts them with simultaneously specified models. One of the necessities of the conditional approach is that the dependence between data at any two sites is *symmetric*, which is a reasonable property if the spatial process is in equilibrium. However, if for good physical reasons $Z(\mathbf{s}_i)$ is thought to depend on $Z(\mathbf{s}_j)$ differently from the dependence of $Z(\mathbf{s}_j)$ on $Z(\mathbf{s}_i)$, probably a space–time process that models the asymmetry with *causative* parameters should be fitted. For example, a model of the spatial diffusion of the Human Immunodeficiency Virus (HIV) epidemic should also be temporal. Gardner et al. (1989) present a geographic analysis of the HIV epidemic in the United States and provide compelling evidence for the existence of sub–state-level epidemics growing spatially and temporally into areas some distance from the original HIV epicenters.

Space–Time Autoregressive-Moving-Average

Suppose $Z(\mathbf{s}_i; t)$ is an observation taken at spatial location $\mathbf{s}_i$ and time t. Then the space–time autoregressive-moving-average (STARMA) models (Cliff et al., 1975) offer a way of generalizing both the autoregressive-moving-average (ARMA) time-series models and the simultaneously specified spatial models. They are characterized by linear dependence lagged in both space and time. Let $\{Z(\mathbf{s}_i; t): i = 1, \ldots, n;\ t = 1, \ldots, T_0\}$ denote the collection of data and $\mathbf{Z}(t) \equiv (Z(\mathbf{s}_1; t), \ldots, Z(s_n; t))'$. Then the STARMA model in its most general form is defined by

$$\mathbf{Z}(t) = \sum_{k=0}^{p} \sum_{j=1}^{\lambda_k} \xi_{kj} W_{kj} \mathbf{Z}(t-k) - \sum_{l=0}^{q} \sum_{j=1}^{\mu_l} \phi_{lj} V_{lj} \boldsymbol{\epsilon}(t-l) + \boldsymbol{\epsilon}(t), \quad (6.8.1)$$

where W_{kj} and V_{lj} are given weight matrices, λ_k is the extent of the spatial lagging on the autoregressive component, μ_l is the extent of the spatial lagging on the moving-average component, for each t, $\epsilon(t) \equiv (\epsilon(\mathbf{s}_1; t), \ldots, \epsilon(\mathbf{s}_n; t))'$ is a vector of independent-and-identically-distributed zero-mean error terms at spatial locations $\{\mathbf{s}_i: i = 1, \ldots, n\}$, and $\{\xi_{kj}\}, \{\phi_{lj}\}$ are the STARMA parameters to be estimated (restrictions are needed on the Ws and Vs to ensure these parameters are identifiable). In order to see more simply what is happening, if the matrix $\sum_{j=1}^{\lambda_k} \xi_{kj} W_{kj}$ is written as B_k and the matrix $\sum_{j=1}^{\mu_l} \phi_{lj} V_{lj}$ is written as E_l, (6.8.1) is written as

$$\mathbf{Z}(t) - \sum_{k=0}^{p} B_k \mathbf{Z}(t-k) = \boldsymbol{\epsilon}(t) - \sum_{l=0}^{q} E_l \boldsymbol{\epsilon}(t-l), \quad (6.8.2)$$

where B_0, E_0 necessarily have zeros down their diagonals.

Special Cases

Ali (1979) considers the case $q = 0$ and, for certain simple $\{W_{kj}\}$, gives computational formulas to obtain m.l. estimators (assuming Gaussianity) for the parameters $\{\xi_{kj}\}$. When $p = q = 0$, (6.8.2) becomes

$$(I - B_0)\mathbf{Z}(t) = (I - E_0)\boldsymbol{\epsilon}(t), \qquad t = 1,\ldots,T_0, \tag{6.8.3}$$

which at any time point t is simply the spatial ARMA process defined in Section 6.7. When $B_0 = E_0 = 0$, the zero matrix, $B_k = \text{diag}(b_{k1},\ldots,b_{kn})$, $k = 1,\ldots,p$, and $E_l = \text{diag}(e_{l1},\ldots,e_{ln})$, $l = 1,\ldots,q$, (6.8.2) becomes

$$Z(\mathbf{s}_i;t) - \sum_{k=1}^{p} b_{ki}Z(\mathbf{s}_i;t-k) = \epsilon(\mathbf{s}_i;t) - \sum_{l=1}^{q} e_{li}\epsilon(\mathbf{s}_i;t-l),$$

$$i = 1,\ldots,n,$$

which models the $n \cdot T_0$ data as n independent purely temporal ARMA processes, each process being associated with a site.

When $\lambda_k = 1$, the spatial dependence matrix $B_k = \xi_{k1}W_{k1}$, which depends on only one parameter. When $\lambda_k = n$ and for a particular choice of $\{W_{kj}: j = 1,\ldots,n\}$, B_k can be written as $\text{diag}(\xi_{k1},\ldots,\xi_{kn})H_k$, where H_k is a known matrix.

Instantaneous Spatial Dependence

The STARMA models *for the case* $B_0 = E_0 = 0$ have been considered by Martin and Oeppen (1975), Aroian (1980), Pfeifer and Deutsch (1980a), and Abraham (1983). These authors discuss questions of stationarity and invertibility, and Abraham gives the exact likelihood (assuming Gaussianity) for the parameters of (6.8.1) (with $\xi_{0j} = 0$, $j \geq 0$, and $\phi_{0j} = 0$, $j \geq 0$). I think it is a weakness of these models that there is no instantaneous spatial component; effectively, they assume

$$\mathbf{Z}(t) - \sum_{k=1}^{p} B_k\mathbf{Z}(t-k) = \boldsymbol{\epsilon}(t) - \sum_{l=1}^{q} E_l\boldsymbol{\epsilon}(t-l);$$

that is, summations have a lower limit of 1 rather than 0.

Although it is difficult to imagine how spatial dependence can arise from any other source than integration of causations over time, it is the *time scale* that is crucial in determining whether any space–time model should have a purely spatial component. If the generating mechanism of the data occurs over a scale of minutes, but data are *observed* over a scale of days, then it is not wise to assume that $B_0 = E_0 = 0$ in (6.8.2). It has already been mentioned in Section 6.3.1 that parameter estimation in the (spatial) simultaneously specified Gaussian model needs considerable care. Clearly, there remains important work to do in estimating, consistently and efficiently, the parameters of the *full* STARMA model (6.8.1).

STARIMA and STARMAX Models

The STARMA model (6.8.1) can be generalized further to include temporal differences (STARIMA models; see Pfeifer and Deutsch, 1980b) and a nonstationary mean function of independent X variables (STARMAX models; see, e.g., Stoffer, 1985, 1986). Again these authors arbitrarily assume $B_0 = E_0 = 0$.

Large-Scale Variation

In general, the mean function $\boldsymbol{\mu}(t) \equiv E(\mathbf{Z}(t))$ will include temporal trend, spatial trend, and explanatory variables. Concentrate for the moment on site i. Suppose

$$\mu(\mathbf{s}_i; t) = \alpha(t) + \nu(\mathbf{s}_i) + \mathbf{X}_i'\boldsymbol{\beta}, \tag{6.8.4}$$

where the first term is the temporal trend, appearing additively with the spatial and explanatory components. Differencing over time yields

$$E(\mathbf{Z}(t) - \mathbf{Z}(t-1)) = \{\alpha(t) - \alpha(t-1)\}(1,\ldots,1)',$$

independent of everything but t. If $\alpha(t)$ is linear in t, then $\mathbf{Y}(t) \equiv \mathbf{Z}(t) - \mathbf{Z}(t-1)$ is mean stationary and can be analyzed using (6.8.1) (after a mean correction). Higher-order differencing can be used when $\alpha(t)$ is a higher-order polynomial in t.

The assumption of additivity in (6.8.4) is crucial to this approach. Interaction of t and $\mathbf{s}$ in $\mu(\mathbf{s}; t)$ requires a more careful analysis.

Conditionally Specified Space–Time Models

The interpretability of the parameters of the conditionally specified spatial models, as well as its flexibility with discrete data, skewed continuous data, and so forth (see Sections 6.5 and 6.6), makes it an attractive class of models to study in space *and* time. Consider

$$dZ(\mathbf{s}_i; t) = -\lambda\phi(T_1, T_2)Z(\mathbf{s}_i; t)\,dt + dY(\mathbf{s}_i; t), \qquad i = 1,\ldots,n,\ t \geq 0,$$

where $\phi(T_1, T_2)$ is any linear spatial displacement operator acting on $Z(\mathbf{s}_i; t)$ and $\{dY(\mathbf{s}_i; t)\}$ are zero-mean independent-and-identically-distributed error terms. For data on a regular grid and Gaussian Zs, the equilibrium model is the auto-Gaussian (Bartlett, 1975, p. 23). For binary data, Bartlett (1971) shows that only for the one-dimensional lattice is there an equilibrium model that is linear; nonlinear equilibrium models on two- and three-dimensional lattices are presented. Further space–time Markov models , namely, binomial and Poisson, are given by Bartlett (1975, pp. 43–44).

Durrett (1981) gives an excellent survey of binary interacting particle systems on $\mathbb{R}^d$, which evolve as follows: If at time t the configuration of the system is $\mathbf{Z}$, an infinite collection of 0s and 1s, then during the time interval

$(t, t + dt]$ the particle at the ith site $\mathbf{s}_i$ changes its state with probability $p_i(\mathbf{Z})dt$. Typically $p_i(\mathbf{Z})$ depends only on $Z(\mathbf{s}_i)$ and the $Z(\mathbf{s}_j)$s at neighboring sites $\{\mathbf{s}_j\}$. The equilibria and time evolution of these systems is of interest, because they offer crude models of phase transitions in magnets and superconductors (e.g., Holley, 1971), the spread of infectious diseases (e.g. Bailey, 1980), and tumor growth (Section 9.7.1).

Tumor growth is sometimes modeled very simply through a particular interacting-particle system called the *contact process* (Harris, 1974), which is discussed more fully in Section 9.7.1.

Competition Models

Growth over time of entities occupying the same domain, but with fixed location (that forces them to share resources), leads naturally to space–time *competition models*. Here the usual premise of spatial dependence—that neighboring observations tend to be alike—gives way to one of negative association. For example, in a forest a big tree is likely to be surrounded by unusually small trees because of the competition for light and soil nutrients. Gates and Westcott (1981) build a space–time model for the development of plant and forest communities based on the notion of a "zone of influence," and prove that, during the early stages of competition, the sizes of neighboring entities are negatively correlated and the size distribution is negatively skewed (something that is observed in data).

CHAPTER 7

Inference for Lattice Models

Chapter 6 concentrated on lattice models based on Markov random fields and gave the appropriate conditions for such models to be well defined and identifiable. The next step of a statistical analysis is to estimate parameters of the model from data and, hopefully, to obtain expressions for estimation variances that can themselves be estimated. This last part, which allows statistical inferences to be made, is crucial to interpretation of the data through the model's parameters.

For nonstandard problems (i.e., non-i.i.d. problems), such as are encountered in Statistics for spatial data, making the leap from indication to estimation (see Section 1.1) is often difficult. This chapter reviews various ways to perform inference for lattice models and includes discussion of computational aspects.

Chapter Summary

Section 7.1 presents the Mercer and Hall (1911) wheat-yield data and discusses various attempts to fit lattice models to them. Sections 7.2 and 7.3 consider parameter estimation and associated statistical properties, giving the basic tools of inference for lattice models. An important application of Markov-random–field models is to image analysis and remote sensing; a review is presented in Section 7.4. Counts from a known base for spatially irregular lattices are commonly encountered in, for example, epidemiology. Sections 7.5 and 7.6 give ways of handling such data by, respectively, building a spatial mixture model or by transforming rates to be modeled as continuous variables. The former analyzes lip-cancer incidences in the districts of Scotland and offers comparisons of different methods of regional mapping. The latter analyzes sudden-infant-death syndrome in the counties of North Carolina. Section 7.7 discusses simulated and real spatial lattice data.

7.1[†] INFERENCE FOR THE MERCER AND HALL WHEAT-YIELD DATA

This section uses spatial autoregressive models to determine the degree of spatial dependence in wheat yields over a 20×25 configuration of plots. It will be seen that Mercer and Hall (1911) were rather vague in their description of the spatial aspect of the experiment, requiring some guesswork as to the exact lattice dimensions and configurations.

7.1.1[†] Data Description

The data given by Mercer and Hall (1911) consist of a 20×25 lattice of plots with 20 rows of plots running east to west and 25 columns of plots running north to south. Presented in Figure 7.1 is the yield of wheat grain (in pounds) for the 500 plots, as well as the lattice configuration.

The object of their study was to determine the plot size that would "reduce the inevitable error within working limits." As a matter of interest, they concluded that 5 plots, or 1/40 acre, should give adequate precision.

A uniform area of 1 acre was harvested in separate plots, each 1/500 acre in area. The wheat sheaves were then threshed out by hand and weighed.

However, closer investigation of their experiment reveals that the field was not all that homogeneous, but was "laid up in lands 15–17 yards broad." There were three or four rows of plots on each land and a strip of varying breadth containing a furrow was left uncut between each row of plots.

The dimension of each plot was given as 10.82 ft by 11 furrows, with a "guard furrow" between each plot. Assuming each plot occupied 1/500 acre, the other dimension can be easily calculated as 8.05 ft (a little less than 3 yd). However, it seems clear that with, say, four plots on a "land," there would be only 11 or 12 yd of breadth accounted for. This would give rise to precisely the divider strips of varying breadth referred to earlier.

Mercer and Hall's choice of 10.82 ft for plot length looks mysterious. I believe it was a rough calculation obtained by dividing 1 acre by 500 and then dividing the result by the average width of a plot (a figure that is not given, so my conjecture cannot be verified). Moreover, because of the presence of the lands, 1 acre is probably an approximation to the area studied, anyway.

Clearly, the spatial information is not very precise, but for *lattice* models it does not always have to be; see Section 6.2, where the location of a county was arbitrarily specified to be at its county seat. However, it is seen in Section 4.5, where a *geostatistical* analysis of these data is carried out, that the lack of precise spatial locations is a hindrance to spatial mapping.

Exploratory Data Analysis

Mercer and Hall carried out some preliminary analyses of the data; further exploratory spatial analyses are presented in Section 4.5. In summary, a stem-and-leaf plot of the 500 data indicate that the data are Gaussian, although plots of column median versus column number and row median versus row number show a disquieting spatial periodicity. Further discussion of this is given in McBratney and Webster (1981) and in Section 4.5.

North

West

3.63	4.15	4.06	5.13	3.04	4.48	4.75	4.04	4.14	4.00	4.37	4.02	4.58	3.92	3.64	3.66	3.57	3.51	4.27	3.72	3.36	3.17	2.97	4.23	4.53
4.07	4.21	4.15	4.64	4.03	3.74	4.56	4.27	4.03	4.50	3.97	4.19	4.05	3.97	3.61	3.82	3.44	3.92	4.26	4.36	3.69	3.53	3.14	4.09	3.94
4.51	4.29	4.40	4.69	3.77	4.46	4.76	3.76	3.30	3.67	3.94	4.07	3.73	4.58	3.64	4.07	3.44	3.53	4.20	4.31	4.33	3.66	3.59	3.97	4.38
3.90	4.64	4.05	4.04	3.49	3.91	4.52	4.52	3.05	4.59	4.01	3.34	4.06	3.19	3.75	4.54	3.97	3.77	4.30	4.10	3.81	3.89	3.32	3.46	3.64
3.63	4.27	4.92	4.64	3.76	4.10	4.40	4.17	3.67	5.07	3.83	3.63	3.74	4.14	3.70	3.92	3.79	4.29	4.22	3.74	3.55	3.67	3.57	3.96	4.31
3.16	3.55	4.08	4.73	3.61	3.66	4.39	3.84	4.26	4.36	3.79	4.09	3.72	3.76	3.37	4.01	3.87	4.35	4.24	3.58	4.20	3.94	4.24	3.75	4.29
3.18	3.50	4.23	4.39	3.28	3.56	4.94	4.06	4.32	4.86	3.96	3.74	4.33	3.77	3.71	4.59	3.97	4.38	3.81	4.06	3.42	3.05	3.44	2.78	3.44
3.42	3.35	4.07	4.66	3.72	3.84	4.44	3.40	4.07	4.93	3.93	3.04	3.72	3.93	3.71	4.76	3.83	3.71	3.54	3.66	3.95	3.84	3.76	3.47	4.24
3.97	3.61	4.67	4.49	3.75	4.11	4.64	2.99	4.37	5.02	3.56	3.59	4.05	3.96	3.75	4.73	4.24	4.21	3.85	4.41	4.21	3.63	4.17	3.44	4.55
3.40	3.71	4.27	4.42	4.13	4.20	4.66	3.61	3.99	4.44	3.86	3.99	3.37	3.47	3.09	4.20	4.09	4.07	4.09	3.95	4.08	4.03	3.97	2.84	3.91
3.39	3.64	3.84	4.51	4.01	4.21	4.77	3.95	4.17	4.39	4.17	4.17	4.09	3.29	3.37	3.74	3.41	3.86	4.36	4.54	4.24	4.08	3.89	3.47	3.29
4.43	3.70	3.82	4.45	3.59	4.37	4.45	4.08	3.72	4.56	4.10	3.07	3.99	3.14	4.86	4.36	3.51	3.47	3.94	4.47	4.11	3.97	4.07	3.56	3.83
4.52	3.79	4.41	4.57	3.94	4.47	4.42	3.92	3.86	4.77	4.99	3.91	4.09	3.05	3.39	3.60	4.13	3.89	3.67	4.54	4.11	4.58	4.02	3.93	4.33
4.46	4.09	4.39	4.31	4.29	4.47	4.37	3.44	3.82	4.63	4.36	3.79	3.56	3.29	3.64	3.60	3.19	3.80	3.72	3.91	3.35	4.11	4.39	3.47	3.93
3.46	4.42	4.29	4.08	3.96	3.96	3.89	4.11	3.73	4.03	4.09	3.82	3.57	3.43	3.73	3.39	3.08	3.48	3.05	3.65	3.71	3.25	3.69	3.43	3.38
5.13	3.89	4.26	4.32	3.78	3.54	4.27	4.12	4.13	4.47	3.41	3.55	3.16	3.47	3.30	3.39	2.92	3.23	3.25	3.86	3.22	3.69	3.80	3.79	3.63
4.23	3.87	4.23	4.58	3.19	3.49	3.91	4.41	4.21	4.61	4.27	4.06	3.75	3.91	3.51	3.45	3.05	3.68	3.52	3.91	3.87	3.87	4.21	3.68	4.06
4.38	4.12	4.39	3.92	4.84	3.94	4.38	4.24	3.96	4.29	4.52	4.19	4.49	3.82	3.60	3.14	2.73	3.09	3.66	3.77	3.48	3.76	3.69	3.84	3.67
3.85	4.28	4.69	5.16	4.46	4.41	4.68	4.37	4.15	4.91	4.68	5.13	4.19	4.41	3.54	3.01	2.85	3.36	3.85	4.15	3.93	3.91	4.33	4.21	4.19
3.61	4.22	4.42	5.09	3.66	4.22	4.06	3.97	3.89	4.46	4.44	4.52	3.70	4.28	3.24	3.29	3.48	3.49	3.68	3.36	3.71	3.54	3.59	3.76	3.36

Figure 7.1 Mercer and Hall (1911) wheat-yield data. Shown are yields of grain (in pounds) for the 20×25 plots and the two principal directions of the lattice.

7.1.2† Spatial Lattice Models

In spite of the heterogeneities in the field (e.g., presence of "lands," thistles in 187 out of 500 plots) and periodicities in the trend, most lattice models fitted to the data in Figure 7.1 assume constant large-scale variation, hoping that (homogeneous) spatial dependence parameters can explain all of the spatial variation. In fact, commenting on their frequency curve of all 500 data, Mercer and Hall (1911) say that "since the [frequency] curve fits the [Gaussian] one as well as may be expected ... we may conclude that the material is fairly homogeneous." From the analysis given in Section 4.5, it is clear that their conclusion is not a proper one. Nevertheless, I shall use these data, and the Gaussian models fitted to them, to *illustrate* maximum likelihood estimation and likelihood ratio testing.

Simultaneous Gaussian Model

Recall from Section 6.3.1 the simultaneously specified Gaussian model (SG). Here, assume

$$Z(u,v) = \xi_1(Z(u+1,v) + Z(u-1,v)) + \xi_2(Z(u,v-1) + Z(u,v+1)) + \epsilon(u,v), \quad (7.1.1)$$

where $\{\epsilon(u,v)\}$ are i.i.d. zero-mean Gaussian random variables. This is a special case of the model (6.3.9) whose likelihood is given by (6.3.10). Whittle (1954) approximates the determinant $|I - B|$ of the spatial dependence matrix in (6.3.10) using a spectral representation. This results in (approximate) maximum likelihood estimates

$$\hat{\xi}_1 = 0.213, \qquad \hat{\xi}_2 = 0.102; \quad (7.1.2)$$

no estimated variances are given by Whittle.

Other models fitted were

$$Z(u,v) = \nu_1 Z(u+1,v) + \nu_2 Z(u,v+1) + \epsilon(u,v) \quad (7.1.3)$$

and

$$Z(u,v) = \xi_{11}Z(u-1,v) + \xi_{12}Z(u+1,v) + \xi_{21}Z(u,v-1) + \xi_{22}Z(u,v+1) + \epsilon(u,v). \quad (7.1.4)$$

Whittle (1954) also gives details of a likelihood-ratio test and, using the same likelihood approximation referred to in the preceding text, concludes that the unilateral model (7.1.3) fits the data better than the spatial models (7.1.1) and (7.1.4).

However, Guyon (1982) shows that Whittle's approximation leads to an estimator whose asymptotic distribution is noncentral Gaussian and to likelihood-ratio statistics whose asymptotic distributions are noncentral chi-squareds. Guyon modifies Whittle's approximation to obtain central asymptotic distributions; in Section 7.2, compare (7.2.42) with (7.2.29).

In my opinion, Whittle's conclusion about the poor fit of spatial models to the Mercer and Hall data is less due to the inconsistent likelihood approximation chosen and more due to his not accounting for the large-scale variation (trend) in the data. In fact, (7.1.1), (7.1.3), and (7.1.4) assume not just that the mean is constant, but that *it is zero*. This is clearly inappropriate.

Conditional Gaussian Model

Recall from Section 6.3.2 the conditionally specified Gaussian model (CG). Here, assume the first-order model

$$\begin{aligned} E(Z(u,v)|\{Z(k,l):(k,l)\neq(u,v)\}) &\\ = \alpha + \gamma_1(Z(u-1,v)+Z(u+1,v)) &\\ + \gamma_2(Z(u,v-1)+Z(u,v+1)), & \qquad (7.1.5)\end{aligned}$$

conditional variances are constant, and the conditional distribution is Gaussian. Using the same type of determinant approximation as Whittle (1954), Besag (1974) obtains the (approximate) maximum likelihood estimates

$$\hat{\gamma}_1 = 0.368, \qquad \hat{\gamma}_2 = 0.107; \qquad (7.1.6)$$

no estimated variances are given. Although he does not give a value for $\hat{\alpha}$, he implies that he has fitted a model with nonzero (but constant) mean.

Other models fitted were the second-order model

$$\begin{aligned} E(Z(u,v)|\{Z(k,l):(k,l)\neq(u,v)\}) &\\ = \alpha + \gamma_1(Z(u-1,v)+Z(u+1,v)) &\\ + \gamma_2(Z(u,v-1)+Z(u,v+1)) &\\ + \lambda_1(Z(u-1,v-1)+Z(u+1,v+1)) &\\ + \lambda_2(Z(u-1,v+1)+Z(u+1,v-1)) & \qquad (7.1.7)\end{aligned}$$

and (7.1.7) where α is replaced with $\tau \cdot v$, a linear trend term. Whittle's approximate likelihood-ratio test between (7.1.5) and (7.1.7) led Besag (1974) to accept the simpler first-order model (7.1.5).

As noted previously, Guyon (1982) gives an improved approximation to the likelihood, yielding maximum likelihood estimators that have a *central* asymptotic Gaussian distribution and likelihood-ratio statistics distributed according to *central* chi-squared random variables. This is achieved very simply by replacing the non–edge-corrected estimator of the autocovariance function by its edge-corrected version; see (7.2.42). It is easily seen that such edge corrections do not matter in $\mathbb{R}^1$, but that they are crucial in $\mathbb{R}^d$, $d \geq 2$. This is discussed further in Sections 7.2.3 and 7.3.

Although the spatial models are different, both Whittle (1954) and Besag (1974) express dissatisfaction with the fit they achieve, and Besag's (1977a) suggestion to add a measurement-error term does not help very much. Cressie (1985c) and Kunsch (1985) observe large-scale variation (trend) in the east–west direction that is not as simple as a linear trend; see Figure 4.17. After detrending the data by removing column means, Kunsch found that a first-order CG fit well. A geostatistical analysis of the Mercer and Hall wheat-yield data, including model fitting and spatial mapping (kriging), is given in Section 4.5.

7.2 PARAMETER ESTIMATION FOR LATTICE MODELS

When data are modeled as a random sample from a distribution of known form but with unknown parameters, then under mild regularity conditions maximum likelihood (m.l.) estimation of the parameters yields consistent, asymptotically Gaussian, and asymptotically efficient estimators (e.g., Cox and Hinkley, 1974, Chapter 9). In nonstandard situations, such as those presented by the spatial models of Chapter 6, estimation of model parameters is not always so straightforward. Of course, finite-sample properties, such as sufficiency, completeness, ancillarity, unbiasedness, minimum mean-squared error (e.g., Cox and Hinkley, 1974, Bickel and Doksum, 1977), are still desirable; however, they are even more elusive than for the i.i.d. paradigm. Thus, methods of estimation are usually assessed via their asymptotic properties.

7.2.1 Estimation Criteria

For both theoretical and computational reasons, the method of maximum likelihood (m.l.) for estimation of lattice-model parameters is no longer automatically the method of choice. When statistical independence of observations is not guaranteed, the negative loglikelihood can no longer be written as a sum of *independent* random variables. Thus, a more general form of the central limit theorem, assuming dependence between variables, is needed to obtain asymptotic properties of the m.l. estimator. Moreover, it is not obvious in these more general situations how to obtain a global maximum on the likelihood surface.

When the data are indexed by time, there are a number of results available that establish consistency, asymptotic Gaussianity, and efficiency of the m.l. estimator. Roussas (1972), Bhat (1974), Basawa and Prakasa Rao (1980, Sections 2.2, 2.3, 2.4), Hall and Heyde (1980), Heijmans and Magnus (1986a), Moore (1987), and the references therein, provide a sampling of results available for m.l. estimation of parameters of temporal random processes.

In what is to follow, various criteria for estimation of lattice-model parameters are presented. Maximum likelihood remains the most popular, although it is often difficult to prove asymptotic Gaussianity and asymptotic efficiency in the spatial context.

Likelihood-Based Estimation

To achieve unity and to avoid arbitrariness, Cox (1985) calls for Statistics to concentrate on likelihood-based procedures. Of course, when the exact likelihood is unavailable (as it is for many spatial models), modified methods are used, allowing some of that arbitrariness (in the choice of modification) back in. It is important, therefore, to look at the losses of efficiency caused by the various modifications and to use this together with computational considerations to choose an estimation procedure.

Recall that the data $\{Z(\mathbf{s}_i): i = 1, \ldots, n\}$ are observed on a lattice D (with a given neighborhood structure). Suppose a model of one of the types proposed in Sections 6.5, 6.6, or 6.7 is to be fitted. Call the collection of large-scale variation (trend) parameters together with small-scale variation (spatial dependence) parameters, $\boldsymbol{\eta}$. The likelihood $l(\boldsymbol{\eta})$ is defined by (using obvious notation)

$$l(\boldsymbol{\eta}) \equiv \Pr(Z(\mathbf{s}_1) = z(\mathbf{s}_1), \ldots, Z(\mathbf{s}_n) = z(\mathbf{s}_n); \boldsymbol{\eta}), \tag{7.2.1}$$

for discrete data, and for continuous data it is defined to be the joint *density* of $Z(\mathbf{s}_1), \ldots, Z(\mathbf{s}_n)$. The maximum likelihood estimator is obtained by maximizing $l(\boldsymbol{\eta})$ with respect to $\boldsymbol{\eta}$, fixing the observations $\mathbf{z} \equiv (z(\mathbf{s}_1), \ldots, z(\mathbf{s}_n))'$. Of course, the maximized value $\hat{\boldsymbol{\eta}}$ is a function of the data and hence is a random variable whose sampling distribution is needed, for example, for variance and confidence-interval calculations.

Estimation Based on the Exact Likelihood

It is clear from Sections 6.5 and 6.6 that many of the conditionally specified models have intractable likelihoods: $Q(\mathbf{z}) \equiv \log(\Pr(\mathbf{z})/\Pr(\mathbf{0}))$ has a simple form, but the normalizing constant, $\Sigma_{\mathbf{z} \in \zeta} \exp(Q(\mathbf{z}))$, which is a function of unknown parameters, usually has no closed-form expression.

Consider the class of (spatial) models defined by $Q(\mathbf{z}) = \boldsymbol{\eta}'\mathbf{w}(\mathbf{z})$, or equivalently (without loss of generality assume the data are continuous),

$$f(\mathbf{z}; \boldsymbol{\eta}) \equiv l(\boldsymbol{\eta}) = \exp(\boldsymbol{\eta}'\mathbf{w}(\mathbf{z}))/K(\boldsymbol{\eta}), \tag{7.2.2}$$

where $\mathbf{W} \equiv \mathbf{w}(\mathbf{Z})$ is a minimally sufficient and complete statistic for the parametric class and, of course, the evasive normalizing constant is $K(\boldsymbol{\eta}) = \int_\zeta \exp(\boldsymbol{\eta}'\mathbf{w}(\mathbf{z}))\, d\mathbf{z}$. Now the *moment generating function* of a random vector $\mathbf{W}$

is given as

$$M_{\mathbf{W}}(\mathbf{t}) \equiv E(\exp(\mathbf{t}'\mathbf{W})) \tag{7.2.3}$$

and the *cumulant generating function* is $\log M_{\mathbf{W}}(\mathbf{t})$. From (7.2.2) and (7.2.3), using obvious notation,

$$M_{\mathbf{W}}(\mathbf{t}; \boldsymbol{\eta}) = K(\boldsymbol{\eta} + \mathbf{t})/K(\boldsymbol{\eta}), \tag{7.2.4}$$

provided $\boldsymbol{\eta} + \mathbf{t}$ is in the parameter space [which is $\{\boldsymbol{\eta}: K(\boldsymbol{\eta}) < \infty\}$].

Conversely, for any $\boldsymbol{\eta}_0$ and $\boldsymbol{\eta}_1$ in the parameter space,

$$K(\boldsymbol{\eta}_1) = K(\boldsymbol{\eta}_0) M_{\mathbf{W}}(\boldsymbol{\eta}_1 - \boldsymbol{\eta}_0; \boldsymbol{\eta}_0). \tag{7.2.5}$$

Thus, for the class of models given by (7.2.2), if the (joint) distributional properties of $\mathbf{Z}$ are known at just one $\boldsymbol{\eta}_0$, the normalizing constant $K(\boldsymbol{\eta})$ is known for all $\boldsymbol{\eta}$. In particular, if $\boldsymbol{\eta}_0 = \mathbf{0}$ is in the parameter space and represents complete randomness, then (7.2.5) shows that $K(\boldsymbol{\eta}_1)$ is determined by the properties of a completely random system. Therefore, calculations (or simulations) to determine the normalizing constant under $\boldsymbol{\eta}_1$ can be carried out assuming a much simpler system.

Maximizing the likelihood (7.2.2) is achieved by differentiating the loglikelihood with respect to $\boldsymbol{\eta}$ and equating to $\mathbf{0}$. That is, solve for $\boldsymbol{\eta}$ in

$$\mathbf{w}(\mathbf{z}) = \frac{\partial \log K(\boldsymbol{\eta})}{\partial \boldsymbol{\eta}}. \tag{7.2.6}$$

Now, $(\partial/\partial\boldsymbol{\eta})\int\{\exp(\boldsymbol{\eta}'\mathbf{w})/K(\boldsymbol{\eta})\}\, d\mathbf{w} = \mathbf{0}$, so, assuming the interchange of differentiation and integration is valid,

$$\int\left[\left\{\frac{\mathbf{w}\cdot\exp(\boldsymbol{\eta}'\mathbf{w})}{K(\boldsymbol{\eta})}\right\} - \left\{\frac{\exp(\boldsymbol{\eta}'\mathbf{w})}{K(\boldsymbol{\eta})}\right\}\left\{\frac{\partial K(\boldsymbol{\eta})}{\partial\boldsymbol{\eta}}\Big/K(\boldsymbol{\eta})\right\}\right] d\mathbf{w} = \mathbf{0}.$$

That is, in obvious notation,

$$E(\mathbf{W}; \boldsymbol{\eta}) = \frac{\partial \log K(\boldsymbol{\eta})}{\partial \boldsymbol{\eta}}. \tag{7.2.7}$$

Combining (7.2.6) and (7.2.7), it is seen that the m.l. estimator of $\boldsymbol{\eta}$ in (7.2.2) is obtained by solving

$$\mathbf{w}(\mathbf{z}) = E(\mathbf{W}; \boldsymbol{\eta}), \tag{7.2.8}$$

for $\boldsymbol{\eta}$. Thus the problem now reduces to determining the dependence of just the first moment $E(\mathbf{W}; \boldsymbol{\eta})$, on $\boldsymbol{\eta}$.

Strauss (1977) considers a finite, multicolored random variable whose values $\mathbf{Z}$ on a regular two-dimensional lattice (under assumptions of homogeneity, pairwise-only dependence between sites, color indifference, and the same strength of attraction for all colors) follow the distribution

$$\Pr(\mathbf{z}; \boldsymbol{\eta}) \propto \exp(\eta \cdot w),$$

where w is the number of adjacencies of like color and η is a real parameter that measures the spatial dependence between adjacent sites. (The binary case was considered earlier by Strauss, 1975b.) The likelihood equation (7.2.8) can be solved approximately by replacing the random variable W, under $\eta = 0$, with a linearly transformed chi-squared random variable (the approximation is achieved by matching their first three moments). This random variable gives computable expressions for the right-hand side of (7.2.5) with $\eta_0 = 0$ and $\eta_1 = \eta$, which allows evaluation of (7.2.7) and hence solution of (7.2.8); see Strauss (1977). For data where $Z(\mathbf{s})$ can only be one of a finite number of possibilities, Younes (1988) shows how a Gibbs sampler can be used to obtain m.l. estimates of $\boldsymbol{\eta}$ in (7.2.2). He gives a stochastic gradient algorithm that solves (7.2.8) by simulating the Markov random field without having to know $K(\boldsymbol{\eta})$; see Section 7.4.3.

In the Gaussian case, the normalizing constant is known and the likelihood can be written out exactly. Details of the maximization of this likelihood are given in Sections 7.2.2 and 7.2.3; it should be noted that, even when the normalizing constant is known, approximations to the likelihood are often used for computational convenience.

Estimation Based on the Pseudolikelihood

In the general case (for conditionally specified models), a number of modified likelihood-based estimation procedures have been proposed. They trade away efficiency in exchange for closed-form expressions that avoid working with the exact likelihood's unwieldy normalizing constant. The efficiency loss can occur because their maximization does not always yield functions of a minimal sufficient statistic, unlike for the maximum likelihood estimator.

Besag (1975) coined the term *pseudolikelihood* for the function

$$p(\boldsymbol{\eta}) \equiv \prod_{i=1}^{n} \Pr\left(z(\mathbf{s}_i) | \{z(\mathbf{s}_j) : j \neq i\}; \boldsymbol{\eta}\right). \qquad (7.2.9)$$

Also denote $P(\boldsymbol{\eta}) \equiv -\log p(\boldsymbol{\eta})$; that is,

$$P(\boldsymbol{\eta}) = -\sum_{i=1}^{n} \log \Pr\left(z(\mathbf{s}_i) | \{z(\mathbf{s}_j) : j \neq i\}; \boldsymbol{\eta}\right). \qquad (7.2.10)$$

The pseudolikelihood (7.2.9) is to be maximized [or equivalently (7.2.10) is to be minimized] with respect to $\boldsymbol{\eta}$, to yield the maximum pseudolikelihood estimator $\hat{\boldsymbol{\eta}}_P$. For the models of Sections 6.5 and 6.6, $\boldsymbol{\eta}$ consists of the large-scale variation parameters $\{\alpha_i: i = 1, \ldots, n\}$ and the small-scale (spatial dependence) parameters $\{\theta_{ij}: \theta_{ii} = 0, \theta_{ij} = \theta_{ji}; i, j = 1, \ldots, n\}$. It is easy to see that, for the conditionally specified Gaussian model (discussed in Section 6.6), maximum pseudolikelihood estimators of these parameters are simply the ordinary–least-squares estimators. (Note that Gong and Samaniego, 1981, define an approximate likelihood for estimation in the presence of nuisance parameters, which they also call a pseudolikelihood. Theirs is formed by substitution of an estimated value of the nuisance parameter into the likelihood, and should *not* be confused with (7.2.9).)

For the autologistic model defined in Section 6.5.1, the normalizing constant in (6.5.4) cannot be written in closed form. However, the conditional distribution (6.5.3) can be expressed as

$$\operatorname{logit}\left[\Pr\left(Z(\mathbf{s}_i) = 1 \middle| \{z(\mathbf{s}_j): j \neq i\}\right)\right] = \alpha_i + \sum_{j=1}^{n} \theta_{ij} z(\mathbf{s}_j), \qquad i = 1, \ldots, n, \tag{7.2.11}$$

where $\operatorname{logit}(p) \equiv \log(p/(1-p))$. Assuming a so-called isotropic Ising model on $\mathbb{Z}^2$ [i.e., assume (6.5.5) with $\gamma_1 = \gamma_2 = \gamma$], the right-hand side of (7.2.11) can be written as

$$\alpha + \gamma \cdot n_i, \tag{7.2.12}$$

where n_i is the number of the four nearest-neighbor sites of site i that have a value of 1. Thus, maximum pseudolikelihood estimation of α and γ is equivalent to formal maximum likelihood estimation of α and γ from the logistic-regression model (7.2.11) and (7.2.12), based on dependent and independent variables $\{(z(\mathbf{s}_i), n_i): i = 1, \ldots, n\}$; see, for example, Cox (1970, Section 6.4). This observation, which can be found in Possolo (1986) and Strauss and Ikeda (1990), allows one to use logistic-regression options in standard computer packages such as BMDP (BMDP Statistical Software Inc., 1988, Vol. 2, pp. 941–969) to fit Ising models by maximum pseudolikelihood (although for this autologistic application the package's standard errors should be ignored).

For the auto-Poisson model defined in Section 6.5.2, the normalizing constant is intractable. However, the pseudolikelihood given by (7.2.10) has a very simple form; from (6.5.6),

$$P(\{\alpha_i\}, \{\theta_{ij}\}) = \sum_{i=1}^{n} \Big\{\lambda_i\big(\{z(\mathbf{s}_j): j \neq i\}\big) - z(\mathbf{s}_i) \cdot \log \lambda_i\big(\{z(\mathbf{s}_j): j \neq i\}\big) + \log z(\mathbf{s}_i)!\Big\}, \tag{7.2.13}$$

where $\lambda_i(\{z(\mathbf{s}_j): j \neq i\}) = \exp\{\alpha_i + \Sigma_{j=1}^n \theta_{ij} z(\mathbf{s}_j)\}$. Apart from symmetry and zero diagonals for the $n \times n$ matrix (θ_{ij}), the only other restriction on the parameter space is that $\theta_{ij} \leq 0$, for all $i, j = 1, \ldots, n$ (see Section 6.5.2). Of course, there are far too many parameters to fit, for the data available, and some initial statistical modeling has to be done to reduce this number to a fraction of the number of data. The example in Section 6.2 on SIDS counts in the counties of North Carolina is analyzed in Section 7.6 using (among others) a pseudolikelihood approach.

For temporal models that satisfy the Markov property (6.3.2), $\Pr(z(t)|\{z(i): i \leq t-1\}) = \Pr(z(t)|z(t-1))$. Thus, even in this simple case, the pseudolikelihood differs from the exact likelihood. For spatial auto-Gaussian (or CG) models with equal conditional variances, maximum pseudolikelihood is clearly equivalent to ordinary least squares.

Just as for exact-likelihood optimization, numerical stability of pseudolikelihood optimization is a separate issue to be addressed. Although the pseudolikelihood usually has a closed-form expression, its maximization needs careful consideration (Section 7.2.3).

Inference procedures require (asymptotic) distribution theory for the maximum pseudolikelihood estimator. Section 7.3.1 gives a number of asymptotic results.

Estimation Based on Coding

For conditionally specified models with near-neighbor dependence, Besag (1974) gives a method of estimation he calls *coding*; again, efficiency is traded for tractability. The idea is to divide up the lattice D into two disjoint sublattices D_0 and D_1: $D = D_0 \cup D_1$, where the neighborhood structure of D_0 is the trivial one of no two sites being neighbors of each other. Now the elements of $\{Z(\mathbf{s}_i): \mathbf{s}_i \in D_0\}$, given $\{Z(\mathbf{s}_i): \mathbf{s}_i \in D_1\}$, are mutually independent. Coding estimators are obtained by minimizing

$$C(\boldsymbol{\eta}) \equiv - \sum_{\mathbf{s}_i \in D_0} \log \Pr\left(z(\mathbf{s}_i) \middle| \{z(\mathbf{s}_j): j \neq i\}; \boldsymbol{\eta}\right), \qquad (7.2.14)$$

which is the *conditional* likelihood of $\{Z(\mathbf{s}_i): \mathbf{s}_i \in D_0\}$ *given* $\{Z(\mathbf{s}_i): \mathbf{s}_i \in D_1\}$. For example, for the first-order dependence model on $\mathbb{Z}^2$, the neighborhoods are given by

$$N_{(u,v)} = \{(u-1, v), (u+1, v), (u, v-1), (u, v+1)\}; \quad (7.2.15)$$

then D_0 can be constructed by deleting every other point from D. This is clearly an inefficient way to use the data (half are being deleted), but note that in this example the other half could be used in the same way. Thus, there are two possible coding estimators, $\hat{\boldsymbol{\eta}}_C^{(0)}$ and $\hat{\boldsymbol{\eta}}_C^{(1)}$, which could then be (say) averaged to provide a better estimator than each of the individual ones; more efficient ways of combining coding estimators are needed.

Figure 7.2 Crosses show the sublattice D_0, which together with (7.2.14) defines a coding estimator for the spatial-dependence parameters implied by the neighborhoods (7.2.16).

For a model with the more extended neighborhood structure given by

$$N_{(u,v)} = \{(u-1,v),(u+1,v),(u-1,v-1),(u,v-1),$$
$$(u+1,v-1),(u-1,v+1),(u,v+1),(u+1,v+1)\}, \quad (7.2.16)$$

Figure 7.2 shows the sublattice D_0. There are approximately $n/4$ data points in D_0. Three other coding estimators are possible in this example by performing shifts of D_0, and once again the four estimators could be combined.

Estimation Based on Other Types of Likelihoods

There have been other attempts in the literature to deal with complicated problems by means of modified likelihoods. Maximum quasilikelihood (Wedderburn, 1972; McCullagh, 1983) is a special case of generalized least squares where the variance matrix is modeled as a function of the mean; spatial models are not usually of this type. The partial likelihood (Cox, 1975) (a generalization of marginal and conditional likelihoods) is designed for problems where the observations have a natural sequencing; again spatial models are not usually of this type.

Restricted maximum likelihood (REML) first filters the data by taking a sufficiently high order of differencing to remove the large-scale variation parameters from the model (of the differenced data). Then, maximum likelihood is applied to the filtered data to estimate the small-scale-variation parameters in isolation. Section 2.6.1 has a discussion of this approach for purely spatial Gaussian data, although the general principle is applicable to nonspatial models provided the appropriate filtering operation can be found.

Alternative Estimation Criteria

Estimating equations offer a general approach to the estimation of parameters from data and include the previous likelihood-based approaches as special cases. Any real function g of the data $\mathbf{Z} = (Z(\mathbf{s}_1), \ldots, Z(\mathbf{s}_n))'$ and the parameter η (here real-valued) is called a regular unbiased estimating function if $E_F(g(\mathbf{Z}; \eta)) = 0$, for all F in a family of distributions that define the stochastic model for $\mathbf{Z}$. Then solve for η in the estimating equation

$$g(\mathbf{Z}; \eta) = 0. \tag{7.2.17}$$

Godambe (1985) shows how to find the optimal g when the index set D is finite, although this involves perhaps an unnatural sequencing of the spatial indices $\mathbf{s}_1, \ldots, \mathbf{s}_n$. He shows that maximum likelihood estimation is not optimal when the model is misspecified. For various examples, he also shows ordinary least squares, the method of moments, and conditional least squares (Klimko and Nelson, 1978) to be special cases of estimating equations and generally inferior to the optimal estimating equation. In practice, these less-than-optimal methods may be used for reasons of simplicity and computing time; for example, in image analysis, n may be 512×512, making straightforward Gaussian maximum likelihood estimation prohibitive, but something like the method-of-moments approach of Derin and Elliott (1987) can be implemented.

Consider a general linear model where the expected value of the data is linear in large-scale-variation parameters and the covariance matrix Σ depends on small-scale-variation parameters. Green (1985) proposes a generalized-least-squares estimator for the regression parameters and cross-validation to choose the spatial dependence parameters; however, his goal was efficient estimation of only the large-scale variation in the presence of small-scale nuisance parameters. A number of the estimation methods suggested in Section 2.6 for geostatistical data could be applied equally well to lattice data, assuming the (possibly transformed) data follow a general linear model. For example, MINQ estimation, generalized-least-squares estimation of variogram parameters, and so on, could be implemented most easily in cases where the lattice model is expressed in terms of Σ rather than Σ^{-1}.

7.2.2 Gaussian Maximum Likelihood Estimation

Under the assumption

$$\mathbf{Z} \sim \text{Gau}(\boldsymbol{\mu}, \Sigma),$$

the negative loglikelihood $L(\boldsymbol{\eta})$ is given by

$$L(\boldsymbol{\eta}) = (n/2)\log(2\pi) + (1/2)\log(|\Sigma|) + (1/2)(\mathbf{z} - \boldsymbol{\mu})'\Sigma^{-1}(\mathbf{z} - \boldsymbol{\mu}), \tag{7.2.18}$$

where the parameter $\boldsymbol{\eta}$ is made up of functions of the mean vector $\boldsymbol{\mu}$ and the variance matrix Σ. The m.l. estimator of $\boldsymbol{\eta}$ minimizes (7.2.18). Typically, $\boldsymbol{\mu} = X\boldsymbol{\beta}$, which models the large-scale variation of the data with a small number of parameters that appear as linear combinations of independent variables X: These variables may be functions of spatial location (trend-surface analysis) or concomitant data associated with each site (purported causal variables) or a mixture of both. Usually $X\boldsymbol{\beta}$ is of most interest in the model, but in order to estimate the coefficients $\boldsymbol{\beta}$ efficiently, the small-scale variation in Σ has to be recognized (Section 1.3). If $\boldsymbol{\mu}$ and Σ are generated by recursive relations in the time dimension, Kalman filtering can be an effective way of both maximizing the likelihood and optimally predicting the next observation (e.g., Anderson and Moore, 1979). However, for two-way *spatial* dependencies in Σ, this filtering approach is not natural.

Recall from Section 6.3 that the simultaneously specified (SG) and conditionally specified (CG) Gaussian models have variance matrices given by

$$\text{SG:} \qquad \Sigma = (I - B)^{-1}\Lambda(I - B')^{-1}, \tag{7.2.19}$$

$$\text{CG:} \qquad \Sigma = (I - C)^{-1}M, \tag{7.2.20}$$

where $B = (b_{ij})$ and $C = (c_{ij})$ are spatial-dependence parameters and $\Lambda = \text{diag}\{\sigma_1^2, \ldots, \sigma_n^2\}$ and $M = \text{diag}\{\tau_1^2, \ldots, \tau_n^2\}$ are (simultaneous and conditional, respectively) variances at the n sites of the lattice. Substitution of these Σs into (7.2.18) shows that the spatial-dependence parameters appear in *both* the log determinant and the generalized sum of squares. In time-series analysis, $\log(|\Sigma|)$ is often ignored, yielding the so-called conditional estimates; Box and Jenkins (1976, Section 7.1) argue that for large n its influence on the likelihood is small. This is not true in the spatial context, where Whittle (1954) shows that ignoring it can result in inconsistent estimators.

Spatial-dependence parameters of geostatistical models (Part I) appear directly in Σ rather than in Σ^{-1}. Under certain regular data configurations, Zimmerman (1989b) shows that Σ possesses patterned structure that can greatly reduce the computational burden of maximizing the likelihood. In what is to follow in Section 7.2.2, spatial dependence will be modeled through Σ^{-1} and homoskedasticity will be assumed; that is, $\Lambda = \sigma^2 I$ and $M = \tau^2 I$. Appropriate modifications for heteroskedasticity can be made to the m.l. estimating equations; see Section 7.6.3, where it is necessary to do so for the auto-Gaussian analysis of the transformed SIDS data. The development that follows is taken principally from Besag (1974).

From (7.2.18) and (7.2.20), the CG model yields

$$L(\boldsymbol{\beta}, \tau^2, C) = (n/2)\log(2\pi\tau^2) - (1/2)\log(|I - C|) + (1/2)(\mathbf{z} - X\boldsymbol{\beta})'(I - C)(\mathbf{z} - X\boldsymbol{\beta})/\tau^2, \quad (7.2.21)$$

where $I - C$ is a symmetric, positive–definite matrix. This can be minimized in stages: For $I - C$ fixed,

$$\hat{\boldsymbol{\beta}} = (X'(I - C)X)^{-1}X'(I - C)\mathbf{z},$$
$$\hat{\tau}^2 = (\mathbf{z} - X\hat{\boldsymbol{\beta}})'(I - C)(\mathbf{z} - X\hat{\boldsymbol{\beta}})/n \quad (7.2.22)$$

are the m.l. estimators of $\boldsymbol{\beta}$ and τ^2. Substituting back into (7.2.21), the m.l. estimator of C can be obtained by minimizing the negative log *profile likelihood*,

$$L^*(C) \equiv (n/2)(\log(2\pi) + 1) - (1/2)\log(|I - C|) + (n/2)\log\left[\mathbf{z}'(I - C)\{I - X(X'(I - C)X)^{-1}X'(I - C)\}\mathbf{z}/n\right], \quad (7.2.23)$$

with respect to the unknown c_{ij}s. Moreover, interval estimates for these spatial-dependence parameters can be obtained from (7.2.23) (Critchley et al., 1988), as is illustrated in Section 7.6. An analogous expression to (7.2.23) can be easily derived for the SG model.

In time series, n is often large and $|\Sigma|$ close to 1, so that the log-determinant term of (7.2.23) has often been ignored. However, work by McLeod (1977), Dent and Min (1978), and Ansley (1979) demonstrate the importance of the $\log(|\Sigma|)$ term. When the spectral decomposition $\Sigma = P\Delta P'$ can be written so that the small-scale variation parameters only appear in Δ, the diagonal matrix of eigenvalues, minimization of (7.2.23) is often relatively straightforward, as is illustrated by the examples that follow and those in Kiiveri and Campbell (1989).

Determinant of (I − C): Regular Lattice and Stationary Model

Construction of spatial models for which the eigenvalues $\{\lambda_i: i = 1, \ldots, n\}$ of Σ are available as closed-form expressions of the spatial-dependence parameters allows easy evaluation of (7.2.18) through $\log(|\Sigma|) = \sum_{i=1}^n \log \lambda_i$. Suppose that the data $\mathbf{Z}$ are observed on a regular two-dimensional lattice $D = \{(u, v): u = 1, \ldots, U;\ v = 1, \ldots, V\}$ wrapped onto a torus. Consider the zero-mean CG model where

$$E(Z(u,v)|\{z(k,l): (k,l) \neq (u,v)\}) = \sum_i \sum_j \gamma(i,j) z(u - i, v - j),$$

$$\operatorname{var}(Z(u,v)|\{z(k,l): (k,l) \neq (u,v)\}) = \tau^2, \quad (7.2.24)$$

where $\gamma(0,0) = 0$, $\gamma(i,j) = \gamma(-i,-j)$, and $\gamma(i,j) = 0$ for all $|i| + |j| > K$. For example, the nearest-neighborhood structure given by (7.2.15) and shown in Figure 6.7 yields just two spatial-dependence parameters, $\gamma(1,0)$ and $\gamma(0,1)$. If the $n = UV$ sites are ordered $(1,1), (1,2), \ldots, (1,V)$, $(2,1), \ldots, (2,V), \ldots, (U,V)$ and the matrix C formed according to the neighborhood structure defined by (7.2.24) [i.e., each row of C has mostly zero entries except for entries $\gamma(i,j)$, $|i| + |j| \le K$], then it can be shown (Moran, 1973a; Besag and Moran, 1975) that C has eigenvalues

$$\left\{\sum_k \sum_l \gamma(k,l)\cos(\omega_p k + \eta_q l): p = 1,\ldots,U; q = 1,\ldots,V\right\}, \tag{7.2.25}$$

where $\omega_p \equiv 2\pi p/U$, $p = 1,\ldots,U$, and $\eta_q \equiv 2\pi q/V$, $q = 1,\ldots,V$. Then,

$$|I - C| = \prod_{p=1}^{U} \prod_{q=1}^{V} \left\{1 - \sum_k \sum_l \gamma(k,l)\cos(\omega_p k + \eta_q l)\right\}, \tag{7.2.26}$$

where now it is clear that the choice of $\{\gamma(k,l)\}$ in (7.2.24) must also ensure the positivity of each term in braces in (7.2.26) (i.e., $I - C$ must be positive-definite).

Gleeson and McGilchrist (1980) assume that $\gamma(k,l)$ can be written as a product of a parameter depending only on k and a parameter depending only on l. This results in Toeplitz matrices in the likelihood function, whose determinants and inverses are easily obtained using an algorithm of Trench (1964). They then use the Nelder–Mead simplex minimization method to obtain m.l. estimates. (They are actually dealing with SG, but the principle is the same for CG.)

Stationary Model

Suppose that the process on the torus is stationary with unknown mean μ; that is,

$$\begin{aligned} E(Z(u,v)|\{z(k,l):(k,l) \ne (u,v)\}) \\ = \mu + \sum_i \sum_j \gamma(i,j)(z(u-i,v-j) - \mu), \\ \mathrm{var}(Z(u,v)|\{z(k,l):(k,l) \ne (u,v)\}) = \tau^2. \end{aligned} \tag{7.2.27}$$

Then, from (7.2.22),

$$\begin{aligned} \hat{\mu} &= \sum_{u=1}^{U} \sum_{v=1}^{V} z(u,v)/UV \equiv \bar{z}, \\ \hat{\tau}^2 &= (\mathbf{z} - \bar{z}\mathbf{1})'(I - C)(\mathbf{z} - \bar{z}\mathbf{1})/UV, \end{aligned} \tag{7.2.28}$$

where $\mathbf{1} \equiv (1,\ldots,1)'$. Define

$$\bar{C}(h_1,h_2) \equiv \sum_{u=1}^{U-h_1}\sum_{v=1}^{V-h_2}(z(u,v)-\bar{z})(z(u+h_1,v+h_2)-\bar{z})/UV,$$
$$h_1 = 0,1,\ldots,U-1,\ h_2 = 0,1,\ldots,V-1, \quad (7.2.29)$$

the empirical covariogram. Then, from (7.2.28),

$$\hat{\tau}^2 = \bar{C}(0,0) - \sum_k\sum_l \gamma(k,l)\bar{C}(k,l),$$

because $\bar{C}(-h_1,-h_2) = \bar{C}(h_1,h_2)$. Substituting this into (7.2.21) yields the negative log profile likelihood

$$(UV/2)(\log(2\pi)+1) + (UV/2)\log\left\{\bar{C}(0,0) - \sum_k\sum_l\gamma(k,l)\bar{C}(k,l)\right\}$$
$$-(1/2)\sum_{p=1}^{U}\sum_{q=1}^{V}\log\left\{1-\sum_k\sum_l\gamma(k,l)\cos(\omega_p k+\eta_q l)\right\}, \quad (7.2.30)$$

which is to be minimized with respect to the unknown $\gamma(k,l)$s. Differentiating (7.2.30) and putting the result equal to zero yields

$$\frac{\bar{C}(k,l)}{\bar{C}(0,0)-\Sigma_i\Sigma_j\hat{\gamma}(i,j)\bar{C}(i,j)} = \frac{1}{UV}\sum_{p=1}^{U}\sum_{q=1}^{V}\frac{\cos(\omega_p k+\eta_q l)}{1-\Sigma_i\Sigma_j\hat{\gamma}(i,j)\cos(\omega_p i+\eta_q j)};$$

that is,

$$\frac{\bar{C}(k,l)}{\hat{\tau}^2} = \frac{\hat{C}(k,l)}{\hat{\tau}^2}, \qquad k,l = \ldots,-1,0,1,\ldots, \quad (7.2.31)$$

where $\{\hat{C}(k,l)\}$ denotes the m.l. estimate of the covariogram $C(h_1,h_2) \equiv \text{cov}(Z(u,v),Z(u+h_1,v+h_2))$ under the stationary model (7.2.27). The justification for (7.2.31) comes from realizing that

$$\left[1-\sum_k\sum_l\gamma(k,l)\cos(\omega_p k+\eta_q l)\right]^{-1}, \qquad p=1,\ldots,U,\ q=1,\ldots,V$$

are the eigenvalues of $(I-C)^{-1} = (\text{cov}(Z(\mathbf{s}_i),Z(\mathbf{s}_j))/\tau^2)$ for the CG model (7.2.27) on the toroidal $U \times V$ regular lattice. Diagonalization of this variance matrix in an analogous way to the time-series case (see, e.g., Fuller,

1976, p. 135) yields

$$\frac{1}{UV}\sum_{p=1}^{U}\sum_{q=1}^{V}\cos(\omega_p k+\eta_q l)\left\{1-\sum_i\sum_j\gamma(i,j)\cos(\omega_p i+\eta_q j)\right\}^{-1}$$
$$=C(k,l)/\tau^2.$$

Thus, for data on a torus, the m.l. estimates of $\{\gamma(k,l)\colon\ k,l=\ldots,-1,0,1,\ldots\}$ are given by solving (7.2.31), so that the empirical covariograms and the model-based [see (7.2.27)] covariograms match. In time-series analysis, the Yule–Walker equations are the direct analogy of the estimating equations (7.2.31).

Stationary Model: First-Order Dependence

This is a special case of the preceding model where $\gamma(1,0)$ and $\gamma(0,1)$ are typically nonzero and all other $\gamma(k,l)$s are zero. The restriction $|\gamma(1,0)|+|\gamma(0,1)|<1/2$ is imposed to ensure that $(I-C)$ is positive-definite. The particular model considered here is the so-called isotropic one, where $\gamma(1,0)=\gamma(0,1)=\gamma$.

Now, $\hat{\mu}=\bar{z}$ and from (7.2.30), $\hat{\gamma}$ is obtained by minimizing

$$\log\{\overline{C}(0,0)-2\gamma(\overline{C}(1,0)+\overline{C}(0,1))\}$$
$$-(1/UV)\sum_{p=1}^{U}\sum_{q=1}^{V}\log\{1-2\gamma(\cos\omega_p+\cos\eta_q)\}$$

with respect to γ. It is shown in Besag and Moran (1975) that $\hat{\gamma}$ is, approximately (replace $\Sigma\Sigma$ by $\int\int$), the solution of

$$\frac{\overline{C}(0,0)}{\overline{C}(0,0)-2\gamma(\overline{C}(1,0)+\overline{C}(0,1))}=\frac{1}{2\pi}K_1(16\gamma^2), \qquad (7.2.32)$$

where $K_1(\cdot)$ is the complete elliptic integral of the first kind and is widely tabulated (see, e.g., Abramowitz and Stegun, 1965, Table 17.1). The approximation is shown to be excellent when $|4\gamma|<0.9$.

Consider for the moment the distributional properties of the m.l. estimators for this special model (more details are given in Section 7.3.1). Besag and Moran (1975) show that $(\hat{\mu},\hat{\tau}^2,\hat{\gamma})$ is asymptotically Gaussian,

$$\text{var}(\hat{\mu})=\text{var}(\overline{Z})=\tau^2/\{UV(1-4\gamma)\}, \qquad (7.2.33)$$

and $\overline{Z}$ is asymptotically independent of $(\hat{\tau}^2,\hat{\gamma})$. They give $\text{var}\{(\hat{\tau}^2,\hat{\gamma})'\}$ in terms of various special functions; when $\gamma = 0$,

$$\text{var}(\hat{\gamma}) = 1/(2UV), \tag{7.2.34}$$

which enables construction of a test of

> H_0: The data come from a random sample from a Gaussian distribution

versus

> H_A: The data come from a stationary, first-order, isotropic CG.

The test agrees with that of Moran (1950) and its more general version (6.6.10), due to Singh and Shukla (1983), and is analogous to the Durbin–Watson (Durbin and Watson, 1950) test in time-series analysis.

When the $U \times V$ data are viewed as imbedded in a sequence of problems that allows $U \to \infty$ and $V \to \infty$, then modifications to the likelihood equations (7.2.31) have to be made to guarantee estimators with asymptotic central Gaussian distributions (Guyon, 1982). Section 7.3 has a discussion of these corrections for edge effects.

Determinant of $(I - C)$: Possibly Irregular Lattice

Suppose it is possible to express $C = \gamma H$, H a known symmetric matrix with zero diagonal elements, so that there is only one real-valued spatial-dependence parameter to be estimated. Let $h_1 \le h_2 \le \cdots \le h_n$ be the ordered eigenvalues of H. Because trace$(H) = 0$, $h_1 < 0 < h_n$. Then,

$$|I - C| = \prod_{i=1}^{n} (1 - \gamma h_i), \tag{7.2.35}$$

where now it is clear that the choice of γ must ensure the positivity of each term in the product. Because $h_1 < 0 < h_n$, choose $h_1^{-1} < \gamma\ h_n^{-1}$. Thus (7.2.23) can be minimized relatively easily with respect to γ by using a grid search or a Newton–Raphson algorithm. Eigenvalues are evaluated *once* at the beginning of the optimization and not at every iteration. To begin the iterative procedure (described in Section 7.2.3), the starting value $\gamma = (\mathbf{z} - X\hat{\boldsymbol{\beta}})'H(\mathbf{z} - X\hat{\boldsymbol{\beta}})/(\mathbf{z} - X\hat{\boldsymbol{\beta}})'(\mathbf{z} - X\hat{\boldsymbol{\beta}})$ might be tried [it solves $\hat{\tau}^2 = 0$ in (7.2.22)].

Cliff and Ord (1981, p. 160) give details for ordinary-least-squares (o.l.s.) estimation of $\gamma_1, \ldots, \gamma_k$ in the slightly more general situation where $I - C = I - \sum_{j=1}^{k} \gamma_j H_j$. The o.l.s. estimators are obtained by writing $\mathbf{Y}_j \equiv H_j \mathbf{Z}$ and solving

$$(\mathbf{Z} - \gamma_1 \mathbf{Y}_1 - \cdots - \gamma_k \mathbf{Y}_k)' \mathbf{Y}_j = 0, \qquad j = 1, \ldots, k. \qquad (7.2.36)$$

However, the solution is not necessarily constrained to the parameter space (i.e., the estimate of $I - C$ may not be positive-definite), and the parameter estimates are not fully efficient (in fact they are maximum pseudolikelihood estimators).

For the simultaneously specified Gaussian (SG) processes and $I - B \equiv I - \Sigma_{j=1}^{k} \phi_j W_j$, ordinary-least-squares estimators are not even consistent. Maximum likelihood estimation for the case $k = 2$ has been studied by Hepple (1976) and Brandsma and Ketellapper (1979), and for the particular form $W_j = W^j$, $j = 1, \ldots, k$, by Huang (1984).

7.2.3 Some Computational Details

Various computational aspects of parameter estimation have already received some attention in Sections 7.2.1 and 7.2.2. Most of the development in this section concerns computational algorithms for Gaussian m.l. estimation. Non-Gaussian m.l. estimation usually involves a normalizing constant that cannot be written in closed form. For a Markov random field whose realizations at a site take only a finite number of colors, Younes (1988) introduces a stochastic gradient algorithm based on the Gibbs sampler that avoids the normalizing constant when computing m.l. estimates for the spatial-dependence parameters; see Section 7.4.3.

More generally, maximum pseudolikelihood (m.p.l.) estimation yields a less unwieldy objective function (7.2.9). Nevertheless, there still may be trouble with writing or finding an appropriate optimization routine. Many of the computer packages have well tested algorithms where the objective function is a generalized sum of squares. This is suitable for the Gaussian case where maximum pseudolikelihood is equivalent to least squares; however, for the non-Gaussian case, (7.2.10) does not necessarily have this structure. The IMSL subroutine UMINF (IMSL Inc., 1987, Chapter 8, pp. 802–806) optimizes general functions using a quasi-Newton method; see Section 7.6.2 where it is used to fit an auto-Poisson model to the SIDS data of Section 6.2. When an m.p.l. estimator is easily computable, it could then be used as a starting value in an iteration that yields a more efficient estimator (such as the m.l. estimator).

Algorithms for m.l. Estimation of Gaussian Parameters

Consider the Gaussian model, in particular the CG model with negative log profile likelihood given by (7.2.23). Computational goals for minimizing (7.2.23) are to construct algorithms that converge quickly and to the m.l. estimator. Suppose C, the matrix of spatial dependencies, is a function of k parameters $\boldsymbol{\gamma} \equiv (\gamma_1, \ldots, \gamma_k)'$. Write $C = C(\boldsymbol{\gamma})$. Then $L^*(C(\boldsymbol{\gamma}))$, given by

(7.2.23), must be minimized with respect to $\boldsymbol{\gamma}$; the Newton–Raphson algorithm yields

$$\boldsymbol{\gamma}^{(l+1)} = \boldsymbol{\gamma}^{(l)} - (B^{(l)})^{-1}\partial L^*(C(\boldsymbol{\gamma}))/\partial\boldsymbol{\gamma}\big|_{\boldsymbol{\gamma}=\boldsymbol{\gamma}^{(l)}}, \qquad l = 0, 1, \cdots, \quad (7.2.37)$$

where $B^{(l)}$ is the matrix whose (i, j)th element is $(\partial^2 L^*/\partial\gamma_i\,\partial\gamma_j)|_{\boldsymbol{\gamma}=\boldsymbol{\gamma}^{(l)}}$. Sometimes the second term in (7.2.37) is multiplied by a scalar step size $\alpha^{(l)}$ to improve the speed of convergence (e.g., Kitanidis and Lane, 1985).

The spatial-dependence parameter space should be restricted to ensure positive-definiteness of $I - C(\boldsymbol{\gamma})$. Unconstrained minimization of (7.2.23) may yield a maximum likelihood estimator $\hat{\boldsymbol{\gamma}}$ outside the parameter space. Most commonly, these constraints are positivity restrictions, which can be handled by reparameterization. If γ_i must be positive, one could put $\gamma_i = \nu_i^2$ and minimize (7.2.23) over ν_i, unconstrained (the trivial constraint $\nu_i \neq 0$ is ordinarily satisfied in practice). Other possible approaches include partial stepping strategies (e.g., Jennrich and Sampson, 1976), interior penalty techniques such as that of Carroll (1961), and the gradient projection method (e.g., Harville, 1977).

An alternative to minimizing (7.2.23) is to minimize the negative loglikelihood (7.2.18) simultaneously over all unknown parameters. Kitanidis and Vomvoris (1983), Mardia and Marshall (1984), and Kitanidis and Lane (1985) show how the scoring (Gauss–Newton) algorithm proceeds.

Kitanidis and coauthors have a more general perspective than Mardia and Marshall. That is, suppose some parameters are common to both the mean and the variance, and write $\mu(\boldsymbol{\eta})$ and $\Sigma(\boldsymbol{\eta})$ for μ and Σ in (7.2.18). Then the scoring (Gauss–Newton) algorithm for minimizing the negative loglikelihood $L(\boldsymbol{\eta})$, given by (7.2.18), is

$$\boldsymbol{\eta}^{(l+1)} = \boldsymbol{\eta}^{(l)} - (A^{(l)})^{-1}\partial L(\boldsymbol{\eta})/\partial\boldsymbol{\eta}\big|_{\boldsymbol{\eta}=\boldsymbol{\eta}^{(l)}}, \qquad l = 0, 1, \cdots, \quad (7.2.38)$$

where $A^{(l)}$ is the matrix whose (i, j)th element is $E(\partial^2 L(\boldsymbol{\eta})/\partial\eta_i\,\partial\eta_j)|_{\boldsymbol{\eta}=\boldsymbol{\eta}^{(l)}}$. Sometimes the second term in (7.2.38) is multiplied by a scalar step size $\alpha^{(l)}$ to improve the speed of convergence (e.g., Kitanidis and Lane, 1985).

The vector $\partial L(\boldsymbol{\eta})/\partial\boldsymbol{\eta}$ in (7.2.38) has ith term

$$\begin{aligned}\partial L(\boldsymbol{\eta})/\partial\eta_i = {} & (1/2)\mathrm{tr}\{\Sigma(\boldsymbol{\eta})^{-1}(\partial\Sigma(\boldsymbol{\eta})/\partial\eta_i)\} \\ & -(\mathbf{z} - \boldsymbol{\mu}(\boldsymbol{\eta}))'\Sigma(\boldsymbol{\eta})^{-1}\partial\boldsymbol{\mu}(\boldsymbol{\eta})/\partial\eta_i \\ & -(1/2)(\mathbf{z} - \boldsymbol{\mu}(\boldsymbol{\eta}))'\Sigma(\boldsymbol{\eta})^{-1} \\ & \times(\partial\Sigma(\boldsymbol{\eta})/\partial\eta_i)\Sigma(\boldsymbol{\eta})^{-1}(\mathbf{z} - \boldsymbol{\mu}(\boldsymbol{\eta})),\end{aligned}$$

where the matrix operator $\mathrm{tr}(G)$, called the trace, sums the diagonal elements of the square matrix G. Furthermore, the matrix $A \equiv E(\partial^2 L(\boldsymbol{\eta})/\partial\boldsymbol{\eta}^2)$, from which $A^{(l)} \equiv A|_{\boldsymbol{\eta}=\boldsymbol{\eta}^{(l)}}$ is obtained, has (i, j)th term

$$E\left(\partial^2 L(\boldsymbol{\eta})/\partial\eta_i\,\partial\eta_j\right) = (1/2)\mathrm{tr}\left\{\Sigma(\boldsymbol{\eta})^{-1}\left(\partial\Sigma(\boldsymbol{\eta})/\partial\eta_i\right)\Sigma(\boldsymbol{\eta})^{-1}\left(\partial\Sigma(\boldsymbol{\eta})/\partial\eta_j\right)\right\}$$
$$+\left(\partial\boldsymbol{\mu}(\boldsymbol{\eta})/\partial\eta_i\right)'\Sigma(\boldsymbol{\eta})^{-1}\left(\partial\boldsymbol{\mu}(\boldsymbol{\eta})/\partial\eta_j\right).$$

These expressions are obtained by noting that

$$\partial \log(|\Sigma(\boldsymbol{\eta})|)/\partial\eta_i = \mathrm{tr}\left(\Sigma(\boldsymbol{\eta})^{-1}\,\partial\Sigma(\boldsymbol{\eta})/\partial\eta_i\right),$$
$$\partial\Sigma(\boldsymbol{\eta})^{-1}/\partial\eta_i = -\Sigma(\boldsymbol{\eta})^{-1}\left(\partial\Sigma(\boldsymbol{\eta})/\partial\eta_i\right)\Sigma(\boldsymbol{\eta})^{-1}$$

(Magnus and Neudecker, 1988, Chapter 8). The scoring algorithm is generally preferable to the Newton–Raphson algorithm (where the expectation in A is not taken).

For spatial regression, $\boldsymbol{\eta} = (\boldsymbol{\beta}', \boldsymbol{\gamma}')'$, $\boldsymbol{\mu}(\boldsymbol{\eta}) = X\boldsymbol{\beta}$, and $\Sigma(\boldsymbol{\eta}) = \Sigma(\boldsymbol{\gamma})$. Then it is easy to see that the (information) matrix A is block diagonal, because $E(\partial^2 L(\boldsymbol{\beta}, \boldsymbol{\gamma})/\partial\boldsymbol{\beta}\,\partial\boldsymbol{\gamma}) = 0$. Hence, (7.2.38) becomes

$$\boldsymbol{\beta}^{(l)} = \left(X'\Sigma(\boldsymbol{\gamma}^{(l)})^{-1}X\right)^{-1}X'\Sigma(\boldsymbol{\gamma}^{(l)})^{-1}\mathbf{Z},$$
$$\boldsymbol{\gamma}^{(l+1)} = \boldsymbol{\gamma}^{(l)} - \left(J_{\gamma}^{(l)}\right)^{-1}\mathbf{L}_{\gamma}^{(l)}, \qquad l = 0, 1, \cdots,$$

where $\mathbf{L}_{\gamma}^{(l)}$ and $J_{\gamma}^{(l)}$ are given in Section 7.3.1 by (7.3.9) and (7.3.15), respectively, evaluated at $\boldsymbol{\beta} = \boldsymbol{\beta}^{(l)}$ and $\boldsymbol{\gamma} = \boldsymbol{\gamma}^{(l)}$.

Further details on implementation of these algorithms can be found in Mardia and Marshall (1984) and Kitanidis and Lane (1985). Kiiveri and Campbell (1989) show that for a large class of models, including those given as examples in Section 7.2.2, the scoring algorithm (7.2.38) can be written as an iteratively reweighted least-squares algorithm.

When the likelihood is multimodal, these algorithms may converge to a local minimum or maximum. Mardia and Watkins (1989) argue, through simulations, that multimodality seems to be caused by likelihoods that are not twice differentiable in their spatial-dependence parameters. Therefore, for such nonsmooth spatial covariances, attempting to minimize the negative loglikelihood by scoring may not succeed.

Vecchia's Approximation

For stationary Gaussian models in $\mathbb{R}^2$, whose covariogram has a rational spectral density function, Vecchia (1988) has given an approximation to the likelihood. Write (7.2.1) as $\prod_{i=1}^{n} \Pr(z(\mathbf{s}_i)|\{z(\mathbf{s}_j): 1 \le j \le i-1\}; \boldsymbol{\eta})$; approximate it by

$$l^{(m)}(\boldsymbol{\eta}) \equiv \prod_{i=1}^{n} \Pr\left(z(\mathbf{s}_i)\middle|\left\{z(\mathbf{s}_k): k \in T_i^{(m)}\right\}; \boldsymbol{\eta}\right),$$

where $T_i^{(m)}$ is a subset of $\{1, 2, \ldots, i-1\}$ containing indices k such that $\|\mathbf{s}_k - \mathbf{s}_i\|$ is one of the m smallest distances out of the $i-1$ possible distances $\{\|\mathbf{s}_j - \mathbf{s}_i\|: j = 1, \ldots, i-1\}$. Maximizing $l^{(m)}(\boldsymbol{\eta})$ is computationally much easier than maximizing $l(\boldsymbol{\eta})$; it is also seen to compromise between maximizing the pseudolikelihood $p(\boldsymbol{\eta})$ and the exact likelihood $l(\boldsymbol{\eta})$.

As m increases, the approximation improves; Vecchia recommends increasing m iteratively up to $m = 10$. An unattractive feature of the method is that it requires an arbitrary ordering of the spatial locations $\{\mathbf{s}_i: i = 1, \ldots, n\}$ to define $l^{(m)}(\boldsymbol{\eta})$.

Whittle's Approximation

When the eigenvalues of $(I - C)$ are not known in closed form, computation of its m.l. estimate [obtained by minimizing (7.2.23)] can be prohibitive. Whittle (1954) considers a stationary Gaussian *planar* process and approximates the likelihood estimating equations as follows. Because he is working with a stationary SG (rather than a CG), his equations involve variance σ^2 (rather than τ^2) and spatial-dependence parameters $\xi(i, j)$ [rather than $\gamma(i, j)$]. Then an integral approximation to the likelihood estimating equations [cf. Eqs. (7.2.31)] yields

$$\begin{aligned}\overline{C}(k, l) = \frac{\hat{\sigma}^2}{(2\pi)^2} \int_{-\pi}^{\pi} \int_{-\pi}^{\pi} &\cos(\omega k + \eta l)\\ &\times \left\{1 - \sum_i \sum_j \xi(i, j) \cos(\omega i + \eta j)\right\}^{-2} d\omega\, d\eta,\\ &k, l = \ldots, -1, 0, 1, \ldots, \qquad (7.2.39)\end{aligned}$$

and hence the (approximate) maximum likelihood estimator of $\{\xi(i, j)\}$ solves (7.2.39). The analogous estimating equations for a stationary planar CG have τ^2 instead of σ^2, $\gamma(i, j)$ instead of $\xi(i, j)$, and, most importantly, a power of -1 instead of -2 in the right-hand side of (7.2.39). That is, solve for $\{\gamma(i, j)\}$ in

$$\begin{aligned}\overline{C}(k, l) = \frac{\hat{\tau}^2}{(2\pi)^2} \int_{-\pi}^{\pi} \int_{-\pi}^{\pi} &\cos(\omega k + \eta l)\\ &\times \left\{1 - \sum_i \sum_j \gamma(i, j) \cos(\omega i + \eta j)\right\}^{-1} d\omega\, d\eta,\\ &k, l = \ldots, -1, 0, 1, \ldots, \qquad (7.2.40)\end{aligned}$$

where $\hat{\tau}^2$ and $\overline{C}$ are given by (7.2.28) and (7.2.29), respectively.

However, Guyon (1982) shows that using (7.2.40) leads to estimators with asymptotic noncentral Gaussian distributions. Is it the likelihood approach itself or the approximation that is causing the trouble? Guyon (1982) modifies the approximate maximum likelihood equations (7.2.40) to

$$C^+(k,l) = \frac{\hat{\tau}^2}{(2\pi)^2}\int_{-\pi}^{\pi}\int_{-\pi}^{\pi}\cos(\omega k + \eta l) \times\left\{1 - \sum_i\sum_j \gamma(i,j)\cos(\omega i + \eta j)\right\}^{-1} d\omega\, d\eta, \quad (7.2.41)$$

where

$$C^+(h_1,h_2) \equiv \sum_u\sum_v (z(u,v) - \bar{z})(z(u+h_1, v+h_2) - \bar{z}) \Big/ \sum_u\sum_v 1, \quad h_1, h_2 = 0,1,\ldots, \quad (7.2.42)$$

is less biased than $\bar{C}$. [Actually, Guyon assumes $Z(\cdot)$ has zero mean; the mean correction $\bar{z}$ in $\bar{C}$ and C^+ introduces a negligible effect on his asymptotic results.] Now, (approximate) m.l. estimators of $\{\gamma(i,j)\}$ obtained by solving (7.2.41) are consistent and have a central asymptotic Gaussian distribution. Section 7.3 presents further discussion of and improvements to Whittle's approximation.

Consider an example from Section 7.2.2, but in the plane rather than on the torus. Suppose the data $\{Z(u,v)\}$ are modeled as coming from a CG on a regular $U \times V$ lattice with no periodic boundary conditions. Suppose further that the only nonzero spatial-dependence parameters are $\gamma(1,0) = \gamma(0,1) = \gamma$. From (7.2.28) and (7.2.41), the (approximate) maximum likelihood estimators of parameters μ, τ^2, and γ are obtained from

$$\hat{\mu} = \sum_{u=1}^{U}\sum_{v=1}^{V}\frac{z(u,v)}{UV} \equiv \bar{z},$$

$$\hat{\tau}^2 = C^+(0,0) - 2\hat{\gamma}(C^+(1,0) + C^+(0,1)), \quad (7.2.43)$$

$$\frac{C^+(1,0) + C^+(0,1)}{2} = \frac{\hat{\tau}^2}{2(2\pi)^2}\int_{-\pi}^{\pi}\int_{-\pi}^{\pi}\frac{\cos\omega + \cos\eta}{\{1 - 2\gamma(\cos\omega + \cos\eta)\}}\,d\omega\, d\eta.$$

Parenthetically, it should be noted that Whittle (1954) is able to write an exact m.l. estimation equation for this first-order process: A solution for γ can be obtained by minimizing

$$(1/2)\sum_{j=1}^{\infty}\frac{1}{j}\binom{2j}{j}^2\gamma^{2j} + \log\hat{\tau}^2; \quad (7.2.44)$$

however, Whittle points out that this approach is useful only when γ is small. (In practice, a finite truncation of the infinite series will be made.) Therefore (7.2.43) or, indeed, iterative solutions of exact maximum likelihood estimators are preferred. Asymptotic distribution theory and finite-sample implications (e.g., edge effects) are considered in the next section.

7.3 PROPERTIES OF ESTIMATORS

Having decided on an estimation criterion and having successfully implemented an algorithm to compute a parameter estimate from the data, the inference problem is not solved until some quantification of the estimator's random variation is obtained (and itself estimated). This necessarily requires an assumption that the spatial process in question possesses ergodic-like properties to guarantee that moments (of measurable functionals of the process) can be approximated by appropriate spatial averages; see Nguyen (1979) and Section 2.3. The scientist's nightmare is that nothing is repeatable or predictable; small doses of ergodicity help alleviate this fear.

In practice, inference always relates to a finite amount of data. Thus, exact finite-sample distribution theory for estimators, for their variance estimators, for test statistics, and so on, is the ultimate goal. Such results are available, but only in special circumstances (e.g., Jensen, 1988). Simulation offers a realistic alternative to intractable distribution theory.

Simulation Results

Several small simulation studies have been carried out to examine the finite-sample properties of different types of estimators of $\boldsymbol{\eta} = (\boldsymbol{\beta}', \boldsymbol{\gamma}')'$, including m.l. estimation; see, for example, Haining (1978c), Mardia and Marshall (1984), Swallow and Monahan (1984), Haining et al. (1989), and Zimmerman and Zimmerman (1991). These have all been for Gaussian models (7.2.18), and have mostly been concerned with modeling the covariance matrix Σ rather than $\Sigma^{-1} = \tau^{-2}(I - C)$. However, a few general conclusions emerge from these simulations. When the spatial dependence is positive, the bias of its m.l. estimator tends to be negative, especially so for small sample sizes. Section 2.6 suggests REML or jackknifing the m.l. estimator as ways of reducing this bias. Furthermore, asymptotic variances of small-scale variation parameters are good approximations to exact variances only when the spatial dependence is weak: The large-scale variation parameters $\boldsymbol{\beta}$ are estimated via m.l. with little bias, and asymptotic variances for the m.l. estimators correspond closely to the exact variances.

Edge Effects and Asymptotic Inference

Asymptotic inferences for parameters of random processes indexed in time rely heavily on martingale convergence theorems (e.g., Hall and Heyde, 1980). For processes indexed in space, the usual models do not generate martingales, because increasing sequences of σ algebras, representing accumulated knowledge about the process, cannot be defined naturally.

Asymptotic results will be given in Section 7.3.1. In this part of the book (Part II), the lattice is assumed fixed and more observations can only be obtained by increasing the spatial domain. Sometimes this assumption is realistic, sometimes not. For example, Haining (1978c) studies the spatial pattern of corn and wheat yields in 45 contiguous counties in southwestern Nebraska and northwestern Kansas. The counties that border the study region make up an augmented region that can be augmented again by adding a further layer of bordering counties, and so on. Thus, in this case, the data are obtained by placing an artificial mask over the counties of the midwest.

However, a study of the yearly agricultural production for the island of Puerto Rico, divided into its 72 agricultural administrative regions (Griffith, 1979), cannot be embedded naturally into a sequence of problems of increasing spatial domain. Geographical barriers such as oceans and mountains make asymptotic inference for fixed-lattice problems simply a convenient approximation to finite-sample inference. The SIDS study for the 100 countries of North Carolina, introduced in Section 6.2, presents a mix of the two situations just presented (contrast Robeson County, a county bordering with South Carolina, and the ocean county of Dare).

At the very least, asymptotic inferences applied to the Nebraska–Kansas data, the Puerto Rico data, and the North Carolina data, are meant to approximate finite-sample inferences. Until Guyon's (1982) article, it was assumed that concerns about edge sites could be ignored, in the limit, allowing the same asymptotic distribution theory to be applied to either of these three types of problems. More recently, it has been realized that edge effects sometimes cannot be ignored, even asymptotically, although scant attention has been given to differences caused by *type* of edge site. Asymptotic theory should allow the domain to increase, but still respect the geometric configuration of lattice sites and the types of edge sites present in the problem under consideration.

As the spatial dimension d increases, edge effects clearly become more important. Consider 100 consecutive observations sited at the integers of $\mathbb{R}^1$; only 2% are on the edge ($d = 1$). But of 100 observations sited on a 10×10 regular lattice in $\mathbb{R}^2$, 36% are on the edge ($d = 2$). For 10,000 observations there are 0.02% edges for $d = 1$ and 3.96% edges for $d = 2$. For $d \geq 1$, this percentage goes to zero like $n^{-1/d}$ as sample size n tends to infinity. Hence, for $d \geq 2$, it does not go to zero fast enough to annihilate $n^{-1/2}$ terms that arise from central-limit theory.

Modifications for Edge Effects

Section 7.2.3 discusses modifications for edge effects from the point of view of approximating the likelihood. Here, the inferential aspects will be emphasized. Guyon (1982) considers a d-dimensional $U \times V \times \cdots \times W$ regular lattice and shows that the d-dimensional equivalent of the approximate likelihood equations (7.2.40) for the CG model leads to a bias in parameter estimators of order $(U \cdot V \cdot \cdots \cdot W)^{-1/d} = n^{-1/d}$. Because an asymptotic central Gaussianity result relies on $n^{1/2}(\hat{\boldsymbol{\eta}} - \boldsymbol{\eta}) \to \text{Gau}(0, \text{var})$ in distribution as $n \to \infty$, it is necessary that $n^{1/2}(\text{bias}) \to 0$, as $n \to \infty$. However, $n^{1/2} \cdot n^{-1/d}$ does *not* satisfy this for dimensions $d \geq 2$. The trouble disappears simply by using the approximate likelihood equations (7.2.41) with the edge-corrected empirical covariogram C^+ given by (7.2.42).

Dahlhaus and Kunsch (1987) observe that the two-way array $\{C^+(h_1, h_2): h_1, h_2 = \ldots, -1, 0, 1, \ldots\}$ is not necessarily positive-definite and propose instead to work with a tapered version of the empirical covariogram. Not only is their modification positive-definite, but it leads to an m.l. estimator that is asymptotically efficient for $d = 1$, 2, and 3. The positive-definiteness guarantees the existence and uniqueness of a solution to the approximate likelihood estimating equations (Kunsch, 1981), and it appears that the tapered estimator possesses superior finite-sample properties, at least for $d = 1$ (Dahlhaus, 1988).

For the autologistic model of binary data on a regular lattice (Section 6.5.1), Pickard (1982) has an asymptotic expression for the m.l. estimating equations, but its solution produces a bias of the same order as the random fluctuations. Kunsch (1983) modifies this expression to produce consistent estimators, in much the same way that Guyon (1982) modifies Whittle's (1954) expression for the Gaussian case.

Edge corrections are not a panacea for model misspecification. In particular, specification of the large-scale variation is of primary importance. To illustrate the edge-corrected likelihood procedure, Guyon (1982) analyzes the Mercer and Hall (1911) wheat-yield data that was previously analyzed *inter alia* by Whittle (1954) and Besag (1974); Section 7.1 presents a summary of their conclusions. Unfortunately, Guyon fails to recognize that a stationary model for these data is not appropriate due to the presence of large-scale (column) variation (see Section 4.5). When there is a nonstationary linear mean effect, $\mu(u, v) = \mathbf{x}(u, v)'\boldsymbol{\beta}$, the estimated covariances given by (7.4.42) are modified to

$$C^+(h_1, h_2) = \sum_u \sum_v \left(z(u, v) - \mathbf{x}(u, v)'\hat{\boldsymbol{\beta}}\right)\left(z(u + h_1, v + h_2) - \mathbf{x}(u, v)'\hat{\boldsymbol{\beta}}\right) \Big/ \sum_u \sum_v 1.$$

Strictly speaking, $\hat{\boldsymbol{\beta}}$ should be the m.l. estimator of $\boldsymbol{\beta}$ given by (7.2.22) [depending itself on $\{\gamma(i, j)\}$]; in practice, one might use the ordinary-least-squares estimator of $\boldsymbol{\beta}$, at least as a starting value.

The effect of edges is felt at the modeling stage and at the (asymptotic) inference stage. Discussion of how to model edge sites is given in Section 6.6 (conditionally specified models) and in Section 6.7.1 (simultaneously specified models). The preceding inference results reinforce the empirical observation: As d increases, infinity "moves further away." This is caused by an increase in the number of edge sites (due to an increase in the number of nearest neighbors of a site) as d increases.

7.3.1* Increasing-Domain Asymptotics

When making inference on a random field $\{Z(\mathbf{s}): \mathbf{s} \in D\}$ from data $\mathbf{Z} = (Z(\mathbf{s}_1), \ldots, Z(\mathbf{s}_n))'$, exact distribution theory of estimators, predictors, test statistics, and so on, is rarely available. Asymptotic distribution theory is the natural place to turn for approximations, although, spatially, there are a number of ways n might tend to infinity. This book addresses two such possibilities.

More and more observations might be sampled in the same finite domain D. I call this *infill asymptotics*, after the mining term "infill drilling," where extra core samples are drilled between existing ones. Geostatistical data, considered in Part I, are most likely to belong to a domain of finite extent (e.g., a basin of oil-bearing rock), so that infill asymptotics are usually most appropriate there. Infill asymptotics are discussed briefly in Section 2.6.3 and as a future direction for research in Section 5.8.

In contrast, more and more observations may be taken by increasing the domain of observation to an infinite subset of $\mathbb{R}^d$. Specifically, suppose $D_n \equiv \{\mathbf{s}_1, \ldots, \mathbf{s}_n\} \subset \mathbb{R}^d$, $n = 1, 2, \ldots$, where $D_n \uparrow D$, such that the volume of the convex hull of D (i.e., the smallest convex set containing D) is infinite. I call this *increasing-domain asymptotics*. When lattice data have a spacing between neighbors that is fixed (e.g., locations of fruit trees in an orchard), increasing-domain asymptotics are more appropriate. Some circumstances may warrant other types of asymptotics for lattice data; however, they will not be considered here. Temporal lattice data (i.e., time series) are most naturally analyzed using increasing-domain asymptotics because the opportunity to "turn back the clock" and take an intermediate sample comes only to the heros of science-fiction writers.

A review of (increasing-domain) asymptotic inference results for lattice models follows. More basic questions of central limit theory for random fields are addressed by, for example, Bolthausen (1982), Takahata (1983), Guyon and Richardson (1984), and Goldie and Morrow (1986). Establishment of weak–convergence results for spatial empirical processes would be of considerable interest.

Hypothesis Testing for Spatial Dependence in a Linear Regression Model

Suppose the data follow a linear model $\mathbf{Z} = X\boldsymbol{\beta} + \boldsymbol{\delta}$, where $\boldsymbol{\delta}$ is a zero-mean conditionally specified Gaussian (CG) process such that $\boldsymbol{\delta} \sim \text{Gau}(\mathbf{0}, (I - \gamma H)^{-1}\tau^2)$, where H is a known symmetric $n \times n$ matrix and γ is the spatial-dependence parameter. A statistic for testing H_0: $\gamma = \gamma_0$, versus H_A: $\gamma \neq \gamma_0$ has been proposed by Singh and Shukla (1983): Define the test statistic $O \equiv \{\mathbf{e}'(I - \gamma_0 H)H\mathbf{e}\}/\{\mathbf{e}'(I - \gamma_0 H)\mathbf{e}\}$, where $\mathbf{e} \equiv \mathbf{Z} - X\hat{\boldsymbol{\beta}}$ and $\hat{\boldsymbol{\beta}}$ is the usual ordinary-least-squares estimator $(X'X)^{-1}X'\mathbf{Z}$. Then, under suitable regularity conditions, O is asymptotically Gaussian, and hence confidence intervals for γ can be constructed.

Suppose instead that $\boldsymbol{\delta}$ is a zero-mean simultaneously specified Gaussian (SG) process such that $\boldsymbol{\delta} \sim \text{Gau}(\mathbf{0}, (I - \xi W)^{-1}(I - \xi W')^{-1}\sigma^2)$, where W is a known $n \times n$ matrix. Recall that $\mathbf{e} = \mathbf{Z} - X\hat{\boldsymbol{\beta}}$. Then a test statistic for testing H_0: $\xi = 0$ versus H_A: $\xi \neq 0$ is $J \equiv \mathbf{e}'W\mathbf{e}/\mathbf{e}'\mathbf{e}$, where H_0 is rejected for large values of J. Burridge (1980) shows this to be the same as the Lagrange multiplier test (Silvey, 1959) of the two hypotheses and demonstrates that $nJ\{\text{tr}(W^2 + W'W)\}^{-1/2}$ is distributed asymptotically as a standard Gaussian under H_0. Furthermore, King (1981) shows the test to be locally best invariant not only for the autoregressive model $(I - \xi W)\boldsymbol{\delta} = \boldsymbol{\epsilon}$, but also for the moving-average model $\boldsymbol{\delta} = (I - \xi W)\boldsymbol{\varepsilon}$. See Section 6.7 for an explanation of how J relates to Moran's and Geary's contiguity ratios.

Asymptotic Results for m.l. Estimation

Sections 6.5, 6.6, and 6.7 discuss various classes of spatial-lattice models for $\mathbf{Z}$, typically on a finite lattice. This usually means that the model is nonstationary. However, for the purposes of asymptotic inference, consider the lattice data as part of an infinite-lattice model with stationary small-scale variation. (Section 2.3 contains the necessary formalism for stationary models.) Increasing domain asymptotic results for m.l. estimation are presented in this section. Under general circumstances of nonidentically distributed, dependent observations, asymptotic likelihood inference results are reviewed by Fahrmeir (1987) and, in the case of Gaussian observations, by Heijmans and Magnus (1986b); mostly the data are assumed sequenced, an unnatural supposition for spatial data.

Now consider the Markov random field with negpotential function

$$Q(\mathbf{z}_n; \boldsymbol{\eta}, D_n) = \sum_{1 \le i \le n} z(\mathbf{s}_i)G_i(z(\mathbf{s}_i)) + \sum\sum_{1 \le i < j \le n} z(\mathbf{s}_i)z(\mathbf{s}_j)G_{ij}(z(\mathbf{s}_i), z(\mathbf{s}_j))$$
$$+ \cdots + z(\mathbf{s}_1) \cdots z(\mathbf{s}_n)G_{1 \cdots n}(z(\mathbf{s}_1), \ldots, z(\mathbf{s}_n)), \quad (7.3.1)$$

where the function $G_{ij \cdots s}$ is the zero function for $\{i, j, \ldots, s\}$ not a clique (see the Hammersley–Clifford Theorem in Section 6.4.1). The finite lattice $D_n \equiv \{\mathbf{s}_i\colon\ i = 1, \ldots, n\}$ and the data $\mathbf{z}_n \equiv (z(\mathbf{s}_1), \ldots, z(\mathbf{s}_n))'$ are, for the moment, subscripted with n. Recall

$$f(\mathbf{z}_n; \boldsymbol{\eta}, D_n) = \frac{\exp(Q(\mathbf{z}_n; \boldsymbol{\eta}, D_n))}{\int_\zeta \exp(Q(\mathbf{y}_n; \boldsymbol{\eta}, D_n))\, d\mathbf{y}_n} \quad (7.3.2)$$

and the maximum likelihood estimator $\hat{\boldsymbol{\eta}}_n$ is obtained by minimizing

$$L_n(\boldsymbol{\eta}) \equiv -\log f(\mathbf{z}_n; \boldsymbol{\eta}, D_n), \quad (7.3.3)$$

with respect to $\boldsymbol{\eta}$.

Gidas (1988) proves the consistency of the m.l. estimator $\hat{\boldsymbol{\eta}}_n$ of a stationary Markov random field on $\mathbb{Z}^d$, whose range of values is either a finite set or a compact metric space. Pickard (1987) establishes consistency and asymptotic normality of a judiciously chosen function of $\boldsymbol{\eta}$ for the classical Ising model of binary lattice data on $\mathbb{Z}^d$ (defined in Section 6.5.1).

Mase (1984) shows that if (1) the sequence of lattices $\{D_n\}$ tends to $\mathbb{Z}^2 = \{(u, v)': u = \ldots, -1, 0, 1, \ldots; v = \ldots, -1, 0, 1, \ldots\}$ monotonically and regularly, (2) the measure defined by (7.3.2) converges to a stationary measure on $\mathbb{Z}^2$ as $\{D_n\}$ tends to $\mathbb{Z}^2$, for each $\boldsymbol{\eta}$, (3) the negpotential function given by (7.3.1) satisfies an asymptotic Gaussian result as $\{D_n\}$ tends to $\mathbb{Z}^2$, for each $\boldsymbol{\eta}$, and (4) a uniform clustering condition on cross moments of $\mathbf{Z}_n$ holds as $\{D_n\}$ tends to $\mathbb{Z}^2$, for each $\boldsymbol{\eta}$, *then* the family of measures defined by (7.3.2) and indexed by $\boldsymbol{\eta}$ is locally asymptotically normal (Le Cam, 1960). Employing known results about these families, Mase (1984) deduces that m.l. estimators are also efficient. However, he notes their computational intractability and goes on to construct (inefficient) method-of-moments estimators.

For zero-mean Gaussian processes on $\mathbb{Z}^2$, Whittle (1954) approximates the m.l. estimating equations by (7.2.40). Guyon (1982) shows that a better (from an asymptotic point of view) approximation is given by the edge-corrected version (7.2.41); results are presented in $\mathbb{Z}^d$ ($d \geq 1$), the lattice whose sites are d tuples of integer coordinates. (His propositions 3 and 4 prove $\sqrt{n}$ consistency, asymptotic central Gaussianity, and asymptotic efficiency of edge-corrected approximate m.l. estimators.) Indeed, Guyon shows that if the Gaussian-based equations (7.2.41) are used, but the process is not Gaussian, then consistency and asymptotic central Gaussianity is preserved provided the process is linear, has fourth moments, and its spectral density satisfies mild regularity conditions.

Consider now the edge-corrected approximate likelihood ratio for testing two embedded hypotheses, where the null hypothesis' parameter space is of dimension p and the alternative hypothesis' parameter space is of dimension $p + r$, where $r > 0$. For a Gaussian process, Guyon (1982) shows that minus twice the log of this ratio converges to a *central* χ_r^2 distribution [and that Whittle's (1954) non–edge-corrected version converges to a noncentral chi-squared distribution].

There is no distinction here between CG and SG for these issues of $\sqrt{n}$ consistency, because both cases satisfy Guyon's (1982, p. 97) condition that the spectral density be rational without zeros or poles. Dahlhaus and Kunsch (1987) obtain similar asymptotic results for a modification to (7.2.40) that involves replacing $\bar{C}$ [given by (7.2.29)] with a tapered version.

It is worth reiterating that all of these results are for spatial-dependence parameters; the mean of the process has been assumed known and hence, without loss of generality, assumed to be zero. Often, most of the interest centers on the unknown mean parameters; the spatial-dependence parameters are important, but only in terms of efficient estimation of the mean parameters (Section 1.3). Asymptotic results for the linear model with spatially dependent errors that extend Guyon's can be found in Chalmond (1986).

Rather than modeling spatial dependence from Σ^{-1}, the partial covariances, Mardia and Marshall (1984) assume that the process $\{Z(\mathbf{s}): \mathbf{s} \in S\}$ on index set S (which is not necessarily a lattice) satisfies

$$E(Z(\mathbf{s})) = \mathbf{x}(\mathbf{s})'\boldsymbol{\beta}, \tag{7.3.4}$$

where $\mathbf{x}(\mathbf{s}) \equiv (x_1(\mathbf{s}), \dots, x_q(\mathbf{s}))'$ is a $q \times 1$ vector of nonrandom regressors, $\boldsymbol{\beta}$ is a vector of mean parameters in some open subset of $\mathbb{R}^q$, and

$$\operatorname{cov}(Z(\mathbf{s}), Z(\mathbf{s}')) = C(\mathbf{s}, \mathbf{s}'; \boldsymbol{\gamma}), \tag{7.3.5}$$

where $\boldsymbol{\gamma}$ is a $k \times 1$ vector of spatial-dependence parameters in some open subset of $\mathbb{R}^k$. Assume that C is twice differentiable with respect to $\boldsymbol{\gamma}$ and is positive-definite in the sense that for every finite subset $D_n = \{\mathbf{s}_1, \dots, \mathbf{s}_n\}$ of S, the matrix $\Sigma_n \equiv (C(\mathbf{s}_i, \mathbf{s}_j; \boldsymbol{\gamma}))$ is positive-definite. Let $\mathbf{Z}_n \equiv (Z(\mathbf{s}_1), \dots, Z(\mathbf{s}_n))'$ denote the data, $X_n \equiv (\mathbf{x}(\mathbf{s}_1), \dots, \mathbf{x}(\mathbf{s}_n))'$ denote the matrix of regressors (assumed to be of rank q), and $\boldsymbol{\eta} \equiv (\boldsymbol{\beta}', \boldsymbol{\gamma}')'$ denote the $(q + k) \times 1$ parameter vector. Then the negative loglikelihood is

$$\begin{aligned} L_n(\boldsymbol{\eta}) = {} & (n/2)\log(2\pi) + (1/2)\log(|\Sigma_n|) \\ & + (1/2)(\mathbf{Z}_n - X_n\boldsymbol{\beta})'\Sigma_n^{-1}(\mathbf{Z}_n - X_n\boldsymbol{\beta}). \end{aligned} \tag{7.3.6}$$

By minimizing (7.3.6), the maximum likelihood estimators $\hat{\boldsymbol{\beta}}_n$ and $\hat{\boldsymbol{\gamma}}_n$ are obtained. For notational convenience, the subscript n on $\mathbf{Z}_n$, X_n, Σ_n, and so forth, is now dropped, although it is retained on $\hat{\boldsymbol{\eta}}_n$.

The derivative vector of L is $(\partial L(\boldsymbol{\eta})/\partial \eta_1, \dots, \partial L(\boldsymbol{\eta})/\partial \eta_{q+k})'$, which can be written as

$$\mathbf{L}^{(1)} \equiv (\mathbf{L}'_\beta, \mathbf{L}'_\gamma)'. \tag{7.3.7}$$

In (7.3.7),

$$\mathbf{L}_\beta = X'\Sigma^{-1}X\boldsymbol{\beta} - X'\Sigma^{-1}\mathbf{Z} \tag{7.3.8}$$

and the ith element of $\mathbf{L}_\gamma$ is

$$(\mathbf{L}_\gamma)_i = (1/2)\operatorname{tr}(\Sigma^{-1}\Sigma_i) + (1/2)\boldsymbol{\delta}'\Sigma^i\boldsymbol{\delta}, \tag{7.3.9}$$

where $\boldsymbol{\delta} \equiv \mathbf{Z} - X\boldsymbol{\beta}$, $\Sigma_i \equiv \partial\Sigma/\partial\gamma_i$, $\partial \log(|\Sigma|)/\partial\gamma_i = \operatorname{tr}(\Sigma^{-1}\Sigma_i)$, and $\Sigma^i \equiv \partial\Sigma^{-1}/\partial\gamma_i = -\Sigma^{-1}\Sigma_i\Sigma^{-1}$, $i = 1, \dots, k$. [The derivative of a matrix is the matrix of elementwise derivatives, and the matrix operator $\operatorname{tr}(G)$, called the trace, sums the diagonal elements of the square matrix G.]

The (i, j)th element of the second-derivative matrix of L is $\partial^2 L(\boldsymbol{\eta})/\partial\eta_i\,\partial\eta_j$; this matrix can be written as

$$L^{(2)} \equiv \begin{bmatrix} L_{\beta\beta} & L_{\beta\gamma} \\ L'_{\beta\gamma} & L_{\gamma\gamma} \end{bmatrix}. \tag{7.3.10}$$

In (7.3.10),

$$L_{\beta\beta} = X'\Sigma^{-1}X, \tag{7.3.11}$$

$L_{\beta\gamma}$ has ith column

$$X'\Sigma^{i}X\boldsymbol{\beta} - X'\Sigma^{i}\mathbf{Z}, \qquad i = 1, \ldots, k, \tag{7.3.12}$$

and $L_{\gamma\gamma}$ has (i, j)th term

$$(1/2)\left\{\mathrm{tr}\left(\Sigma^{-1}\Sigma_{ij} + \Sigma^{i}\Sigma_{j}\right) + \boldsymbol{\delta}'\Sigma^{ij}\boldsymbol{\delta}\right\}, \tag{7.3.13}$$

where $\Sigma_{ij} \equiv \partial^2\Sigma/\partial\gamma_i\,\partial\gamma_j$ and $\Sigma^{ij} \equiv \partial^2\Sigma^{-1}/\partial\gamma_i\,\partial\gamma_j$, $i, j = 1, \ldots, k$. Using $\Sigma^{ij} = \Sigma^{-1}(\Sigma_i\Sigma^{-1}\Sigma_j + \Sigma_j\Sigma^{-1}\Sigma_i - \Sigma_{ij})\Sigma^{-1}$, the (expected) information matrix $E(L^{(2)})$ is given by

$$J \equiv \begin{bmatrix} J_\beta & 0 \\ 0 & J_\gamma \end{bmatrix}, \tag{7.3.14}$$

where $J_\beta = X'\Sigma^{-1}X$, the (i, j)th element of J_γ is $(1/2)t_{ij}$, and

$$(1/2)t_{ij} \equiv (1/2)\mathrm{tr}\left(\Sigma^{-1}\Sigma_i\Sigma^{-1}\Sigma_j\right) = (1/2)\mathrm{tr}(\Sigma\Sigma^{i}\Sigma\Sigma^{j}),$$
$$i, j = 1, \ldots, k. \tag{7.3.15}$$

More general expressions, of which (7.3.7) through (7.3.15) are special cases, can be found in Magnus and Neudecker (1988, Chapter 15).

Section 7.2.3 contains a discussion of the computational aspects involved in minimizing $L(\boldsymbol{\eta})$ with respect to $\boldsymbol{\eta}$. What statistical properties does the resulting maximum likelihood estimator $\hat{\boldsymbol{\eta}}_n \equiv (\hat{\boldsymbol{\beta}}'_n, \hat{\boldsymbol{\gamma}}'_n)'$ possess?

Because $\mathbf{Z}$ is a single observation from an n-dimensional random vector with distribution $\mathrm{Gau}(X\boldsymbol{\beta}, \Sigma)$, it is not obvious that $\hat{\boldsymbol{\beta}}_n$ and $\hat{\boldsymbol{\gamma}}_n$ are consistent and asymptotically Gaussian. Magnus (1978) and Sweeting (1980) have each given a general result for m.l. estimators $\hat{\boldsymbol{\beta}}_n$ and $\hat{\boldsymbol{\gamma}}_n$. Based on Sweeting's result, Mardia and Marshall (1984) prove the following theorem.

Theorem (Mardia and Marshall, 1984) Suppose $\mathbf{Z} \sim \mathrm{Gau}(X\boldsymbol{\beta}, \Sigma)$, where $\boldsymbol{\beta}$ is a $q \times 1$ vector of unknown mean parameters and Σ is a function of $\boldsymbol{\gamma}$, a $k \times 1$ vector of unknown spatial-dependence parameters. Let $\lambda_1 \leq \cdots \leq \lambda_n$

be the eigenvalues of Σ and let those of Σ_i and Σ_{ij} be $\{\lambda_l^i: l = 1, \ldots, n\}$ and $\{\lambda_l^{ij}: l = 1, \ldots, n\}$, with $|\lambda_1^i| \leq \cdots \leq |\lambda_n^i|$ and $|\lambda_1^{ij}| \leq \cdots \leq |\lambda_n^{ij}|$, for $i, j = 1, \ldots, k$. Suppose that, as $n \to \infty$,

i. $\lambda_n \to e < \infty$, $|\lambda_n^i| \to e_i < \infty$, $|\lambda_n^{ij}| \to e_{ij} < \infty$, $i, j = 1, \ldots, k$.
ii. $\|\Sigma_i\|^{-2} = O(n^{-1/2-\delta})$, for some $\delta > 0$, $i = 1, \ldots, k$ ($\|G\|$ denotes the Euclidean matrix norm, $(\Sigma_i \Sigma_j g_{ij}^2)^{1/2} = \{\text{tr}(G'G)\}^{1/2}$).
iii. $t_{ij}/(t_{ii}t_{jj})^{1/2} \to a_{ij}$, for all $i, j = 1, \ldots, k$, where t_{ij} is given by (7.3.15) and $A \equiv (a_{ij})$ is a nonsingular matrix.
iv. $(X'X)^{-1} \to 0$, a $q \times q$ matrix all of whose elements are zero.

Then the m.l. estimator $\hat{\boldsymbol{\eta}}_n$ of $\boldsymbol{\eta} \equiv (\boldsymbol{\beta}', \boldsymbol{\gamma}')'$ satisfies

$$\hat{\boldsymbol{\eta}}_n \to \boldsymbol{\eta}, \quad \text{in probability,}$$
$$J^{1/2}(\hat{\boldsymbol{\eta}}_n - \boldsymbol{\eta}) \to \text{Gau}(0, I), \quad \text{in distribution,} \tag{7.3.16}$$

where J is the information matrix given by (7.3.14). ■

To use this theorem in the spatial context, where $(\beta_1, \ldots, \beta_q)$ are large-scale variation parameters and $(\gamma_1, \ldots, \gamma_k)$ are small-scale variation parameters [e.g., $\boldsymbol{\gamma}$ consists of $\tau_1^2, \ldots, \tau_n^2$ and (c_{ij}) for the CG model], restrict the lattice D_n to satisfy $\|\mathbf{s} - \mathbf{s}'\| \geq a > 0$, for all pairs $\mathbf{s}, \mathbf{s}'$ in D_n. This ensures that the spatial domain increases without bound, as $n \to \infty$. Assume further that $Z(\cdot)$ is covariance stationary in $\mathbb{R}^d$; that is, $C(\mathbf{s}, \mathbf{s} + \mathbf{h}; \boldsymbol{\gamma}) = \sigma^2\rho(\mathbf{h}; \boldsymbol{\gamma})$ with $\rho(\mathbf{0}; \boldsymbol{\gamma}) = 1$. Define $\rho_i \equiv \partial\rho/\partial\gamma_i$, and $\rho_{ij} \equiv \partial^2\rho/\partial\gamma_i\,\partial\gamma_j$.

Corollary (Mardia and Marshall, 1984) Consider a covariance stationary Gaussian process sampled on a $U \times V \times \cdots \times W$ regular lattice D_n. If suppositions (iii) and (iv) of the Theorem hold and ρ, ρ_i, and ρ_{ij} are absolutely summable over the set $\{\mathbf{h}: \mathbf{h} \in \mathbb{Z}^d\}$, for all $i, j = 1, \ldots, k$, then $\hat{\boldsymbol{\eta}}_n$ is consistent and the asymptotic Gaussian result (7.3.16) holds. ■

Notice that $\hat{\boldsymbol{\beta}}_n$ and $\hat{\boldsymbol{\gamma}}_n$ are asymptotically independent and their asymptotic variance matrix is simply the inverse of the usual (expected) Fisher information matrix. This allows construction of confidence regions on the parameter $\boldsymbol{\eta}$ and, using the results of Mase (1984), these regions should be the "tightest" possible. Mardia and Marshall (1984) go on to show that similar asymptotic results are true for covariance stationary Gaussian models with additive measurement error.

These results are general enough to have as a special case the stationary auto-Gaussian model (7.2.27) defined on a regular lattice of the torus. The local periodic conditions of the model imply that a stationary correlogram $\rho(\cdot)$ exists, which satisfies the conditions of the preceding corollary. For example, the stationary, (so-called) isotropic, first-order dependent auto-

Gaussian model (Section 7.2.2) has $\boldsymbol{\beta} = \mu$ and $\boldsymbol{\gamma} = (\tau^2, \gamma)'$. Thus, asymptotic Gaussianity and asymptotic independence of $\hat{\mu}_n$ and $(\hat{\tau}_n^2, \hat{\gamma}_n)'$ follow, as was derived by Besag and Moran (1975) in this particular case.

Inference in Practice

To use Mardia and Marshall's theorem, an expression for the (expected) information matrix given by (7.3.14) is needed. Because this matrix is itself a function of $\boldsymbol{\eta}$, in practice this means evaluating it at $\boldsymbol{\eta} = \hat{\boldsymbol{\eta}}_n$; call the result $\hat{J}$, the estimated information matrix. [Efron and Hinkley (1978) suggest using instead the *observed* Fisher information, the matrix of second-order partial derivatives evaluated at $\hat{\boldsymbol{\eta}}_n$; further justification is presented by Skovgaard (1985).] Therefore, from (7.3.16), an approximate $100(1-\alpha)\%$ confidence ellipsoid for $\boldsymbol{\eta}$ is

$$\left\{\boldsymbol{\eta}: (\hat{\boldsymbol{\eta}}_n - \boldsymbol{\eta})'\hat{J}(\hat{\boldsymbol{\eta}}_n - \boldsymbol{\eta}) \le \chi^2_{q+k}(\alpha)\right\}, \tag{7.3.17}$$

where $\chi_l^2(\alpha)$ denotes the $100(1-\alpha)\%$ quantile of the chi-squared distribution on l degrees of freedom. Similar expressions for confidence ellipsoids for $\boldsymbol{\beta}$ (or $\boldsymbol{\gamma}$) can be obtained by replacing $\hat{\boldsymbol{\eta}}_n$ with $\hat{\boldsymbol{\beta}}_n$ (or $\hat{\boldsymbol{\gamma}}_n$), $\hat{J}$ with $\hat{J}_\beta$ (or $\hat{J}_\gamma$) from (7.3.14), and $\chi^2_{q+k}(\alpha)$ with $\chi^2_q(\alpha)$ [or $\chi^2_k(\alpha)$].

In practice, conditions of theorems may not be easy to check or may not be appropriate to the problem under study. Nevertheless, some measure of variability of the m.l. estimator is needed. When possible, the information matrix should be computed and its inverse used as such a measure. Consider the spatial regression model

$$\mathbf{Z} = X\boldsymbol{\beta} + \boldsymbol{\delta}, \qquad \boldsymbol{\delta} = (I - \phi W)^{-1}\boldsymbol{\epsilon}, \tag{7.3.18}$$

where $\boldsymbol{\epsilon}$ is an $n \times 1$ vector of i.i.d. $\mathrm{Gau}(0, \sigma^2)$ random variables and ϕW is a matrix of spatial autoregressive coefficients (ϕ is an unknown real-valued parameter). In other words, $\boldsymbol{\delta}$ is an SG process (Section 6.3.1) with parameters $\boldsymbol{\gamma} = (\sigma^2, \phi)'$. Then Ord (1975) gives the information matrix

$$J = \sigma^{-4}\begin{bmatrix} \sigma^2 X'F'FX & \mathbf{0} & \mathbf{0} \\ \mathbf{0}' & n/2 & \sigma^2\,\mathrm{tr}(G) \\ \mathbf{0}' & \sigma^2\,\mathrm{tr}(G) & \sigma^4\{\mathrm{tr}(G'G) - \nu\} \end{bmatrix}, \tag{7.3.19}$$

where $F \equiv I - \phi W$, $G \equiv WF^{-1}$, $\nu \equiv -\sum_{i=1}^n w_i^2/(1-\phi w_i)^2$, and $\{w_i: i = 1, \ldots, n\}$ are the eigenvalues of W; see Doreian (1980) for a detailed derivation. The inverse of (7.3.19) is a measure of the variation of the m.l. estimator of $\boldsymbol{\eta} = (\boldsymbol{\beta}', \sigma^2, \phi)'$.

Asymptotic Results for Maximum Pseudolikelihood Estimation

Pseudolikelihoods are most naturally defined for conditionally specified models; see (7.2.9). Hence, the discussion that follows only applies to lattice models where the conditional probabilities (or densities) are modeled.

Section 7.2.1 contains details of the method of maximum pseudolikelihood (m.p.l.) estimation. In short, it looks for an $\boldsymbol{\eta}$ that minimizes

$$P(\boldsymbol{\eta}) \equiv -\sum_{i=1}^{n} \log f\big(z(\mathbf{s}_i)|\{z(\mathbf{s}_j): j \neq i\}; \boldsymbol{\eta}\big). \qquad (7.3.20)$$

Because the conditional densities are specified and not derived, optimization of the objective function $P(\boldsymbol{\eta})$ can be much easier than m.l. estimation, where the objective function often has an unwieldy normalizing constant. But because $P(\boldsymbol{\eta})$ is not necessarily a function of the minimally sufficient statistic, m.p.l. estimation will not be as efficient, in general, as m.l. estimation.

Little is known about (asymptotic) distribution theory for m.p.l. estimators: At the very least, the underlying model should be sufficiently ergodic to allow for consistent estimation of $\boldsymbol{\eta}$ from a single set of observations $\mathbf{Z}$; see succeeding text. [In the less interesting case where the data consist of N independent realizations of spatial data $\mathbf{Z}$, Grenander (1989) proves that the m.p.l. estimator is consistent and asymptotically Gaussian, as $N \to \infty$.] In many applications, the number of components n of $\mathbf{Z}$ is large (e.g., in image analysis, n might be 512×512), in which case asymptotic approximations may be appropriate.

Geman and Graffigne (1987) prove the consistency of the m.p.l. estimator of the parameters of a Markov random field on $\mathbb{Z}^d$, where the range of values of $Z(\mathbf{s})$ is a finite set, the conditional probabilities $\Pr(z(\mathbf{s}_i)|\{z(\mathbf{s}_j): j \neq i\}; \boldsymbol{\eta})$ are homogeneous over (interior) sites, and the neighborhoods are finite. Gidas' (1988) m.p.l consistency result extends this to the case where the range of values of $Z(\mathbf{s})$ is a compact metric space. Guyon (1987) considers a generally irregular lattice of sites in $\mathbb{R}^d$ and a generally nonstationary lattice model. He shows the m.p.l. estimator to be consistent and asymptotically Gaussian under, *inter alia*, a spatial mixing (i.e., weak dependence) assumption that assumes an exponentially decreasing mixing coefficient. The CG model and the Markov random field, whose range of values is a finite set, are considered as examples. Unfortunately, the asymptotic variance of the m.p.l. estimator is not generally available in closed form, although it could be approximated using the Gibbs sampler (Section 7.7.1).

Guyon (1987) has similar results for the estimator based on coding. Recall from Section 7.2.1, the coding estimator is obtained by minimizing (7.2.14). Intuitively, the coding estimator should be least efficient, followed by the m.p.l. estimator, and finally the m.l. estimator should be fully efficient (Mase, 1984).

Table 7.1 Efficiencies of m.p.l. and Coding-Based Estimators of γ, Relative to the m.l. Estimator

True Value 4γ	Pseudolikelihood Efficiency	Coding Efficiency
0.1	0.9975	0.991
0.2	0.987	0.965
0.3	0.971	0.921
0.4	0.946	0.859
0.5	0.912	0.779
0.6	0.864	0.681
0.7	0.807	0.564
0.8	0.709	0.419
0.9	0.553	0.248

Source: Kashyap and Chellappa (1983) © 1983 IEEE.

For the CG model (7.2.27), Kashyap and Chellappa (1983) have established consistency of the m.l., m.p.l., and coding-based estimators and have obtained expressions for their asymptotic variances. In this case, the question of loss of efficiency can be addressed directly. For example, consider the zero-mean, so-called isotropic CG model on a regular toroidal $U \times U$ lattice with first-order neighborhood structure [given by (7.2.15)]. That is, consider the CG model

$$\begin{aligned} E(Z(u,v)|\{z(k,l)\colon (k,l) \neq (u,v)\}) \\ = \gamma\{z(u-1,v) + z(u+1,v) + z(u,v-1) + z(u,v+1)\}, \\ \operatorname{var}(Z(u,v)|\{z(k,l)\colon (k,l) \neq (u,v)\}) = \tau^2, \end{aligned} \tag{7.3.21}$$

where $4|\gamma| < 1$ and $\tau^2 > 0$. Apparently unaware that Besag (1977b) had already tabulated relative efficiencies of the m.p.l. and coding-based estimators of γ, Kashyap and Chellappa (1983) derived a very similar table, part of which is presented in Table 7.1; Besag's results differ occasionally in the third decimal place. Ratios of variances under the different estimation techniques, which serve as measures of their relative efficiencies, are shown.

To read this table, I use as *my* yardstick the "36% rule." How much efficiency loss are *you* willing to tolerate in exchange for other estimation considerations, such as computational ease and robustness? I have resolved this question by tolerating the loss of asymptotic efficiency caused by estimating the population mean of a Gaussian random sample by a sample median instead of the fully efficient sample mean, namely, a loss of $(1 - (2/\pi))100\% \simeq 36\%$. Thus, using my yardstick, m.p.l. estimation is acceptable for 4γ up to and including 0.8 (and is fully efficient for $\gamma = 0$), whereas coding-based estimation is acceptable for 4γ up to and including

0.6. One can conclude in this case, and hopefully others, that m.p.l. estimation is acceptable for the major part of the parameter space.

These inefficient estimators are worth considering because of their computational tractability. But they may also be used as an initial estimator in a scoring algorithm [see (7.2.38)] that, under certain conditions (see Dzhaparidze, 1974), yields an asymptotically fully efficient estimator. This idea warrants further attention in the spatial context.

7.3.2 The Jackknife and Bootstrap for Spatial Lattice Data

When the data are independent, the jackknife (Quenouille, 1949a, 1956; Tukey, 1958) and bootstrap (Efron, 1979, 1982) techniques for estimating a statistic/estimator's distributional properties have proved their worth in situations where very little can be assumed about the underlying distribution of the data. Suppose $Z_1, \ldots, Z_n$ are independent and identically distributed (i.i.d.) random variables with a cumulative distribution function F and let $R(\mathbf{Z}; F)$ denote a random variable (test statistic, estimator, pivotal quantity, etc.) of interest.

The Classical Jackknife

The jackknife deals with the special case

$$R(\mathbf{Z}; F) = \hat{\theta}(\mathbf{Z}) - \theta, \tag{7.3.22}$$

where θ is some functional of F [e.g., $\theta = \int zF(dz)$, the mean of F] and $\hat{\theta}(\mathbf{Z})$ is an estimator of θ. The basic (delete-1) jackknife proceeds as follows:

1. Write $\hat{\theta}_{-i} \equiv \hat{\theta}(\{Z_j: j \neq i\})$, which is the estimator of θ based on the sample [of size $(n-1)$] with the ith observation deleted.
2. Define the pseudovalues

$$\tilde{\theta}_i \equiv n\hat{\theta} - (n-1)\hat{\theta}_{-i}, \qquad i = 1, \ldots, n. \tag{7.3.23}$$

3. Define the jackknife mean and variance

$$\tilde{\theta} \equiv \sum_{i=1}^{n} \tilde{\theta}_i / n, \qquad S_J^2 \equiv \sum_{i=1}^{n} \left(\tilde{\theta}_i - \tilde{\theta}\right)^2 \Big/ (n-1) \tag{7.3.24}$$

and the studentized jackknife statistic

$$T \equiv n^{1/2}(\tilde{\theta} - \theta)/S_J. \tag{7.3.25}$$

Under appropriate conditions on $\hat{\theta}$ and F (see, e.g., Miller, 1974), the bias of $\tilde{\theta}$ is $O(1/n^2)$, in contrast to the $O(1/n)$ bias of $\hat{\theta}$. Furthermore, T given by

(7.3.25) is distributed approximately as a t distribution on $(n - 1)$ degrees of freedom, allowing confidence intervals for θ to be constructed. It was for this reason that Tukey called the method the jackknife. Think of it as a trusty, multipurpose Boy Scout tool that, in the absence of other evidence about the underlying distribution, provides approximate confidence intervals for θ.

Jackknifing in more general, unbalanced situations has been considered by Hinkley (1977) and Wu (1986). Jackknifing is not always appropriate. It does not yield asymptotically valid confidence intervals when $\hat{\theta}$ is the median, for instance, which is a result of the estimator not being locally linear in the data.

The Classical Bootstrap

Jackknifing involves a systematic deletion of each datum (or more generally, subsets of the data). On the other hand, the bootstrap is based on resampling the data. Generally speaking, the applicability of the bootstrap is more comprehensive than that of the jackknife. Its goal is to obtain properties of the sampling distribution of a general random variable $R(\mathbf{Z}; F)$ without having to assume a distributional form for F.

The basic method proceeds as follows:

1. Construct the empirical distribution function

$$\hat{F}(z) \equiv \sum_{i=1}^{n} I(Z_i \leq z)/n. \tag{7.3.26}$$

2. Draw a random sample of size n from $\hat{F}$; suppose $Z_1^*, \ldots, Z_n^*$ are i.i.d. as $\hat{F}$. Thus, $Z_1^*, \ldots, Z_n^*$ are randomly sampled *with replacement* from $\{Z_1, \ldots, Z_n\}$.
3. Compute

$$R^* \equiv R(\mathbf{Z}^*; \hat{F}). \tag{7.3.27}$$

4. Repeat steps 2 and 3 B times (e.g., $B = 500$) and use the empirical distribution of $R_1^*, \ldots, R_B^*$ to approximate the sampling distribution of $R(\mathbf{Z}; F)$.

Notice that if the true distribution function F were to replace $\hat{F}$ in (7.3.27), then $R^* = R(\mathbf{Z}; F)$, in distribution. The attractive feature of the bootstrap is its sample reuse for estimating F, which obviates having to assume a parametric model. It was for this reason that Efron (1979) chose the name "bootstrap," after the expression, "He pulled himself up by his bootstraps."

The bootstrap is a computer-intensive method only in that the bootstrap distribution, namely, the distribution of R^*, usually has to be approximated by Monte Carlo methods. It has also spawned much theoretical research in obtaining conditions on F and R for which the distributions (and moments) of R^* and R are close (and if so, how close), as $n \to \infty$.

Efron and Tibshirani (1986) can be consulted for a general review of the bootstrap, applied in not only the one-sample problem considered here, but also in regression models, proportional hazards models, autoregressive time-series models, and so forth. What is crucial in all these variants of the basic bootstrap method is that somehow the stochastic model can be written in terms of i.i.d. components $\{\epsilon_1, \ldots, \epsilon_n\}$, which can be approximated and then resampled. For spatial processes, the same reasoning can be attempted. It should be noted that the bootstrapped spatial maps presented in Diaconis and Efron (1983) were not obtained in this way, and their validity is questionable. Blind application of the classical jackknife and bootstrap techniques to dependent data can give incorrect answers (e.g., Singh, 1981, Remark 2.1). Moreover, asymptotic validity of these techniques would seem less likely under infill asymptotics than under increasing-domain asymptotics.

Jackknifing Dependent Data

Consider the special case of a regularly spaced stationary time series $\{Z(t): t = 1, 2, \ldots\}$, of which there are n observations $\mathbf{Z} \equiv (Z(1), \ldots, Z(n))'$. Define $\mathbf{Z}_i^m \equiv (Z(i+1), \ldots, Z(i+m))'$ to be a subseries of values of length m. Then $\{\mathbf{Z}_{j \cdot m}^m: j = 0, \ldots, [n/m] - 1\}$ denotes a sequence of subseries, where $[x]$ denotes the integer part of x. If $\hat{\theta}(\mathbf{Z})$ is a statistic that can be written as a function of $\{\mathbf{Z}_k^m: 0 \le k \le n - m\}$, then a jackknife can be defined by deleting each of the subseries

$$\hat{\theta}_{-i} \equiv \hat{\theta}\left(\left\{\mathbf{Z}_{j \cdot m}^m: j \ne i\right\}\right), \qquad i = 0, \ldots, [n/m] - 1,$$

$$\tilde{\theta}_i \equiv [n/m]\hat{\theta} - ([n/m] - 1)\hat{\theta}_{-i},$$

$$\tilde{\theta} \equiv \sum_{i=0}^{[n/m]-1} \tilde{\theta}_i \Big/ [n/m], \qquad S_J^2 \equiv \sum_{i=0}^{[n/m]-1} \left(\tilde{\theta}_i - \tilde{\theta}\right)^2 \Big/ ([n/m] - 1),$$

$$T \equiv [n/m]^{1/2}(\tilde{\theta} - \theta)/S_J.$$

Inference on θ is achieved through T being approximately (as $n \to \infty$, $m \to \infty$, $[n/m] \to \infty$) distributed as a standard Gaussian random variable. Kunsch (1989) gives conditions under which this recipe, modified to leave out (weighted) subseries of overlapping blocks, gives valid results.

Carlstein's (1986, 1988) approach is not in the spirit of the jackknife in the sense that he *selects* (nonoverlapping) blocks rather than deleting them. He computes an empirical variance $\sum_{j=0}^{[n/m]-1}\{\hat{\theta}(\mathbf{Z}_{j \cdot m}^m) - \bar{\theta}(\mathbf{Z})\}^2/[n/m]$, where $\bar{\theta}(\mathbf{Z}) \equiv \sum_{j=0}^{[n/m]-1}\hat{\theta}(\mathbf{Z}_{j \cdot m}^m)/[n/m]$, and gives conditions under which it converges to $\lim_{n \to \infty} \operatorname{var}(n^{1/2}\hat{\theta}(\mathbf{Z}))$. Furthermore, Carlstein's approach can be generalized to the spatial setting. For data $\mathbf{Z} \equiv (Z(\mathbf{s}_1), \ldots, Z(\mathbf{s}_n))'$ from a stationary lattice process defined on $D \subset \mathbb{R}^d$, divide D into mutually exclusive and exhaustive congruent subregions $D_1, \ldots, D_K$ (or *tiles*, according to Hall, 1985a). Let the corresponding sublattice data be $\mathbf{Z}_{D_1}, \ldots, \mathbf{Z}_{D_K}$ from which statistics $\hat{\theta}(\mathbf{Z}_{D_k})$, $k = 1, \ldots, K$, can be computed. One might conjecture that $\sum_{k=1}^{K}(\hat{\theta}(\mathbf{Z}_{D_k}) - \bar{\theta}(\mathbf{Z}))^2/K$, where $\bar{\theta}(\mathbf{Z}) \equiv \sum_{k=1}^{K}\hat{\theta}(\mathbf{Z}_{D_k})/K$, is an appropriate estimator of $\text{var}(n^{1/2}\hat{\theta}(\mathbf{Z}))$, but it remains to be proved. One would use increasing-domain asymptotics and assume that both the subregions and the domain D are increasing without bound, such that the number of observations in each subregion goes to infinity, but at a rate slower than the sample size n.

The analogous spatial jackknife would involve deleting a subregion and closing up the remaining subregions when computing $\hat{\theta}_{-i}$, although it is far from obvious how to do this. In the circumstances where $\hat{\theta}(\mathbf{Z})$ is defined by estimating equations of the form

$$\sum_{i=1}^{n} g_i(\mathbf{Z}; \theta) = 0, \tag{7.3.28}$$

Lele (1991a) proposes deleting the ith summand in (7.3.28). An estimator $\hat{\theta}_{-i}$ is then obtained, from which pseudovalues (7.3.23) can be defined; that is, jackknife the estimating equations. For example, maximum pseudolikelihood estimation of parameters of a Markov random field can be written as (7.3.28) with $g_i(\mathbf{Z}; \theta) \equiv \partial \log f(Z(\mathbf{s}_i)|\{Z(\mathbf{s}_j): j \in N_i\}; \theta)/\partial\theta$, where N_i is the set of neighbors of the ith site. Lele (1991a) gives conditions under which $n^{1/2}(\tilde{\theta} - \theta)/\{\sum_{i=1}^{n}\sum_{j \in N_i}(\tilde{\theta}_i - \tilde{\theta})(\tilde{\theta}_j - \tilde{\theta})/(n-1)\}^{1/2}$, converges to a standard Gaussian distribution, as $n \to \infty$.

Bootstrapping Dependent Data

The bootstrap paradigm involves resampling (groups of) observations rather than deleting them. When D is divided into congruent subregions $D_1, \ldots, D_K$, Hall (1985a) has suggested two types of resampling. One is to assign the data $\mathbf{Z}_k^*$ to the region D_k, $k = 1, \ldots, K$, by random sampling with replacement from the original sequence of data $\{\mathbf{Z}_1, \ldots, \mathbf{Z}_K\}$, where $\mathbf{Z}_k$ is a vector whose components are $\{Z(\mathbf{s}_i): i \in D_k\}$, $k = 1, \ldots, K$.

The other resampling scheme is to choose a region D_k^* uniformly from all possible regions within D that are congruent to D_1 (and hence to $D_2, \ldots, D_K$). Then the data in D_k^* are assigned to the region D_k and denoted Z_k^*, $k = 1, \ldots, K$. Kunsch (1989) has proposed an analogous resampling scheme for time-series data.

Let $\hat{\theta}(\mathbf{Z}) = \hat{\theta}(\mathbf{Z}_1, \ldots, \mathbf{Z}_K)$ denote a statistic of spatial data from a stationary process and suppose $\{\mathbf{Z}_1^{*(j)}, \ldots, Z_K^{*(j)}\}$ is the result of the jth bootstrap, $j = 1, \ldots, B$. Then an estimate of $\text{var}(n^{1/2}\hat{\theta}(\mathbf{Z}))$ is obtained from $\sum_{j=1}^{B}(\hat{\theta}(\mathbf{Z}_1^{*(j)}, \ldots, \mathbf{Z}_K^{*(j)}) - \bar{\theta})^2/B$, where $\bar{\theta} \equiv \sum_{j=1}^{B}\hat{\theta}(\mathbf{Z}_1^{*(j)}, \ldots, \mathbf{Z}_K^{*(j)})/B$.

Hall (1988b) takes a different tack, to avoid the obvious, unnatural pasting together of data whose spatial dependencies across subregion boundaries do not reflect those in the original spatial data set. He bootstraps within groups of (approximately) independent data and uses a Bonferroni inequality to construct conservative confidence intervals.

Semiparametric Bootstrap

If one is willing to make parametric or semiparametric assumptions about the large- and small-scale variation in the data, it is sometimes possible to write the stochastic model in terms of i.i.d. components that can be estimated and then resampled (Freedman and Peters, 1984; Solow, 1985).

Suppose spatial data $\mathbf{Z} \equiv (Z(\mathbf{s}_1), \ldots, Z(\mathbf{s}_n))'$ follow the linear model

$$\mathbf{Z} = X\boldsymbol{\beta} + \boldsymbol{\delta}, \tag{7.3.29}$$

where X is a matrix of explanatory variables that may include trend-surface variables and $\boldsymbol{\delta} \equiv (\delta(\mathbf{s}_1), \ldots, \delta(\mathbf{s}_n))'$ is a vector of spatially dependent error terms with zero mean and positive-definite covariance matrix Σ. Furthermore, suppose that $\boldsymbol{\epsilon} \equiv \Sigma^{-1/2}\boldsymbol{\delta}$ is a vector of i.i.d. random variables distributed according to a cumulative distribution function F.

Sections 3.4.3, 3.5, and 5.4 give different approaches to estimation of the covariance function $C(\cdot, \cdot)$ from the data $\mathbf{Z}$. Henceforth, assume that a positive-definite estimator $\hat{\Sigma}$ of Σ has been obtained from the data. The Cholesky decomposition (e.g., Golub and Van Loan, 1983, pp. 86–90) allows $\hat{\Sigma}$ to be decomposed as the matrix product

$$\hat{\Sigma} = \hat{L}\hat{L}', \tag{7.3.30}$$

where $\hat{L}$ is a lower triangular $n \times n$ matrix (i.e., all elements above the diagonal are zero). Now, from (7.3.29), define

$$(\hat{\epsilon}_1, \ldots, \hat{\epsilon}_n)' \equiv \hat{\boldsymbol{\epsilon}} \equiv \hat{L}^{-1}(\mathbf{Z} - X\hat{\boldsymbol{\beta}}), \tag{7.3.31}$$

$\hat{\boldsymbol{\beta}} \equiv (X'\hat{\Sigma}^{-1}X)^{-1}X'\hat{\Sigma}^{-1}\mathbf{Z}$, and $\tilde{\epsilon}_i \equiv \hat{\epsilon}_i - (\sum_{j=1}^{n}\hat{\epsilon}_j/n)$, $i = 1, \ldots, n$. The spatial bootstrap is obtained by bootstrapping the $\tilde{\epsilon}_1, \ldots, \tilde{\epsilon}_n$ [i.e., sampling according to the empirical distribution function $\hat{F}(\epsilon) \equiv \sum_{i=1}^{n} I(\tilde{\epsilon}_i \leq \epsilon)/n$], to obtain $\boldsymbol{\epsilon}^* \equiv (\epsilon_1^*, \ldots, \epsilon_n^*)'$, and then untransforming according to

$$\mathbf{Z}^* \equiv X\hat{\boldsymbol{\beta}} + \hat{L}\boldsymbol{\epsilon}^*. \tag{7.3.32}$$

In the i.i.d. case, where $\Sigma = \sigma^2 I$, Wu (1986) suggests a small bias-correcting modification to (7.3.32), whose analogue could be sought for the more general case considered here.

Finally, the distribution of any random quantity $R(\mathbf{Z}; F, \boldsymbol{\beta}, \Sigma)$ is approximated with the distribution of $R^* \equiv R(\mathbf{Z}^*; \hat{F}, \hat{\boldsymbol{\beta}}, \hat{\Sigma})$. This is usually estimated empirically from $R_1^*, \ldots, R_B^*$, obtained by repeated samplings with replacement from $\tilde{\epsilon}_1, \ldots, \tilde{\epsilon}_n$.

This bootstrap proposal is very much in the spirit of the classical bootstrap, but its validity is unproved (increasing-domain asymptotics are needed). Moreover, the assumption that the random quantity R depends only on F, $\boldsymbol{\beta}$, and Σ is crucial; spatial dependence in higher-order moments is ignored by this approach. For example, does the bootstrap break down when the independence assumption on $\epsilon_1, \ldots, \epsilon_n$ is replaced with an assumption of zero correlation? It probably does not, as long as R depends only on F and first- and second-order moments (e.g., R is a mean-squared error of prediction). There are many interesting theoretical and methodological problems that remain to be solved.

Although the preceding discussion treats the general linear model, a more general bootstrapping approach emerges. Suppose

$$\mathbf{Z} = \mathbf{h}(\boldsymbol{\epsilon}), \tag{7.3.33}$$

where $\mathbf{h}(\cdot)$ is a one-to-one vector function of vector argument and $\boldsymbol{\epsilon}$ is a vector of i.i.d. random variables distributed according to a cumulative distribution function F [e.g., $Z(\mathbf{s})$ is obtained from a spatial moving average]. Let $\hat{\mathbf{h}}$ denote an estimator of $\mathbf{h}$ that is also one-to-one and define $\hat{\boldsymbol{\epsilon}} \equiv \hat{\mathbf{h}}^{-1}(\mathbf{Z})$. Bootstrapping $\hat{\boldsymbol{\epsilon}}$ yields $\boldsymbol{\epsilon}^*$, which in turn yields

$$\mathbf{Z}^* \equiv \hat{\mathbf{h}}(\boldsymbol{\epsilon}^*). \tag{7.3.34}$$

Finally, the distribution of the random quantity $R(\mathbf{Z}; F, \mathbf{h})$ is approximated by that of $R(\mathbf{Z}^*; \hat{F}, \hat{\mathbf{h}})$. The same comments, about proving the validity of the proposal and about the effect of replacing independent ϵ_is with uncorrelated ϵ_is, are pertinent here.

An example of (7.3.33) is obtained from the spectral representation of $\mathbf{Z}$ and is adapted from a time-series bootstrap suggested by Hurvich and Zeger (1987). Their basic idea was to bootstrap in the frequency domain because discrete Fourier transforms at different frequencies are asymptotically independent (even assuming strong dependence of the stationary time series; see Yajima, 1989). Assume $\mathbf{Z}$ is obtained by sampling a second-order stationary process $\{Z(\mathbf{s})\colon \mathbf{s} \in D\}$, where D is a d-dimensional integer lattice wrapped onto a $U \times V \times \cdots \times W$ torus. Let the process have mean μ, covariogram $C(\cdot)$, and spectral function $g(\cdot)$. For the purposes of presentation, suppose that $d = 2$ and data locations $\{\mathbf{s}_1, \ldots, \mathbf{s}_n\}$ are given by the integer coordinates $\{(u, v)'\colon\ u = 0, 1, \ldots, U - 1;\ v = 0, 1, \ldots, V - 1\}$. Therefore, $C(h_1, h_2) = \int_{-\pi}^{\pi}\int_{-\pi}^{\pi} \exp(ih_1\omega_1 + ih_2\omega_2) g(\boldsymbol{\omega})\, d\boldsymbol{\omega}$, and $g(\boldsymbol{\omega}) = (2\pi)^{-2}\sum_{h_1=-\infty}^{\infty}\sum_{h_2=-\infty}^{\infty} C(h_1, h_2)\exp(-ih_1\omega_1 - ih_2\omega_2)$. Moreover, there is a one-to-one relation between the data $\{Z(u, v)\colon u = 0, 1, \ldots, U - 1;\ v = 0, 1, \ldots, V - 1\}$ and the discrete Fourier transform $\{J(p, q)\colon$

$p = 0, 1, \ldots, U-1;\ q = 0, 1, \ldots, V-1\}$:

$$J(p,q) \equiv (UV)^{-1} \sum_{u=0}^{U-1} \sum_{v=0}^{V-1} Z(u,v) \exp\{-i\omega_p u - i\eta_q v\},$$
$$p = 0, \ldots, U-1,\ q = 0, \ldots, V-1, \quad (7.3.35)$$

$$Z(u,v) = \sum_{p=0}^{U-1} \sum_{q=0}^{V-1} J(p,q) \exp\{i\omega_p u + i\eta_q v\},$$
$$u = 0, \ldots, U-1,\ v = 0, \ldots, V-1, \quad (7.3.36)$$

where $\omega_p \equiv 2\pi p/U$, $p = 0, \ldots, U-1$, and $\eta_q \equiv 2\pi q/V$, $q = 0, \ldots, V-1$. The values (7.3.35) can be computed very quickly using a two-dimensional version of the fast Fourier transform (FFT) (e.g., Dudgeon and Mersereau, 1984, p. 76ff).

Write $J(p,q) = CC(p,q) + SS(p,q) - i\{CS(p,q) + SC(p,q)\}$, where

$$CC(p,q) = (UV)^{-1} \sum_{u=0}^{U-1} \sum_{v=0}^{V-1} Z(u,v) \cos(\omega_p u) \cos(\eta_q v),$$
$$SS(p,q) = -(UV)^{-1} \sum_{u=0}^{U-1} \sum_{v=0}^{V-1} Z(u,v) \sin(\omega_p u) \sin(\eta_q v),$$
$$CS(p,q) = (UV)^{-1} \sum_{u=0}^{U-1} \sum_{v=0}^{V-1} Z(u,v) \cos(\omega_p u) \sin(\eta_q v),$$
$$SC(p,q) = (UV)^{-1} \sum_{u=0}^{U-1} \sum_{v=0}^{V-1} Z(u,v) \sin(\omega_p u) \cos(\eta_q v). \quad (7.3.37)$$

Note that, because of aliasing, only frequencies satisfying $0 \le \omega_p \le \pi$ and $0 \le \eta_q \le \pi$ need to be considered. It can be demonstrated that the bivariate random vectors $\{(CC(p,q), SS(p,q)), (CS(p,q), SC(p,q)) \colon 0 \le \omega_p \le \pi,\ 0 \le \eta_q \le \pi\}$ are all mutually uncorrelated. Furthermore, apart from the special cases $\omega_p = 0, \pi$ or $\eta_q = 0, \pi$, the bivariate random vectors $(CC(p,q), SS(p,q))$ and $(CS(p,q), SC(p,q))$ are asymptotically ($U \to \infty$, $V \to \infty$) independent and identically distributed Gaussian with mean $(0,0)$, asymptotic variances $(4UV)^{-1}\sum_{h_1=-\infty}^{\infty}\sum_{h_2=-\infty}^{\infty} C(h_1, h_2)\cos(\omega_p h_1)\cos(\eta_q h_2)$, and asymptotic covariance $-(4UV)^{-1}\sum_{h_1=-\infty}^{\infty}\sum_{h_2=-\infty}^{\infty} C(h_1, h_2) \times \sin(\omega_p h_1)\sin(\eta_q h_2)$.

Thus, from (7.3.36), one can write $\mathbf{Z} = \mathbf{h}(\boldsymbol{\epsilon})$, where $\boldsymbol{\epsilon}$ consists of bivariate components of the discrete Fourier transform, which, when suitably normalized, are approximately i.i.d. The vector function $\mathbf{h}$ depends on μ and trigonometric transforms of the covariogram $C(\cdot)$; the latter can be computed very quickly using the FFT. [In practice, a tapered estimate of $C(\cdot)$ is used.]

The relationship (7.3.33) is not a necessary requirement for spatial data to be bootstrapped. One simply needs a stochastic model whose generating mechanism can be estimated nonparametrically. Suppose that $f(\cdot)$ is the unknown density of data $\mathbf{Z}$ and that $\hat{f}(\cdot)$ is a nonparametric estimator of f. Then, to approximate the probability $\Pr(\mathbf{Z} \in A) = \int_A f(\mathbf{z})\,d\mathbf{z}$, use the bootstrap distribution

$$P^*(A) \equiv \int_A \hat{f}(\mathbf{z})\,d\mathbf{z}, \qquad (7.3.38)$$

which in turn can be approximated by $\sum_{j=1}^{B} I(\mathbf{Z}_j^* \in A)/B$, where $\mathbf{Z}_1^*, \ldots, \mathbf{Z}_B^*$ are the result of i.i.d. sampling from a distribution with density $\hat{f}(\cdot)$. (The asymptotic validity of the approximation $\Pr(\mathbf{Z} \in A) \simeq P^*(A)$ needs to be established; indeed, much of the recent theoretical research on the classical bootstrap has been devoted to finding conditions under which $\Pr(\mathbf{Z} \in A) - P^*(A)$ decreases rapidly to zero as n increases.) Lele (1989) uses this idea to bootstrap from Markov random fields on regular lattices. He obtains nonparametric estimators of the conditional densities $\{f(z(\mathbf{s}_i)|\{z(\mathbf{s}_j)\colon j \in N_i\})\colon i = 1, \ldots, n\}$ from data $\mathbf{Z}$ and then uses a Gibbs sampler (see, e.g., Section 7.7.1) to generate $\mathbf{Z}^*$ from the joint distribution of the Markov random field. The same approach could be used in (7.3.29), where the ordinary-least-squares residuals $\hat{\boldsymbol{\delta}} \equiv \mathbf{Z} - X\hat{\boldsymbol{\beta}}_{\text{ols}}$, $\hat{\boldsymbol{\beta}}_{\text{ols}} \equiv (X'X)^{-1}X'\mathbf{Z}$, replace $\mathbf{Z}$ in obtaining (7.3.38). From nonparametrically bootstrapped errors $\boldsymbol{\delta}^*$, define $\mathbf{Z}^* \equiv X\hat{\boldsymbol{\beta}}_{\text{ols}} + \boldsymbol{\delta}^*$.

Parametric Bootstrap

The spatial bootstrap proposals given so far all have a nonparametric component to them, for example, the unknown cdf F of the i.i.d. $\epsilon_1, \ldots, \epsilon_n$. This maintains a certain amount of generality and is true to the original reasons why the name "bootstrap" was chosen. Nevertheless, it is possible to define a *parametric bootstrap*, whereby the data $\mathbf{Z}$ are assumed distributed according to a joint density (or probability mass function) known up to the parameter $\boldsymbol{\eta}$; call it $f(\mathbf{z}; \boldsymbol{\eta})$, where $\mathbf{z} \in \zeta$. In fact, the use of simulation, to provide inference in parametric models where analytic results are unobtainable, predates Efron's (1979) bootstrap; for example, the Monte Carlo tests of Hope (1968) and Besag and Diggle (1977) had their origins in a suggestion from Barnard (1963).

Let $\hat{\boldsymbol{\eta}}$ be a consistent estimator of $\boldsymbol{\eta}$ based on the data $\mathbf{Z}$. Then the parametric bootstrap distribution is obtained from the probability measure

$$P^*(A) \equiv \int_A f(\mathbf{z}; \hat{\boldsymbol{\eta}})\,d\mathbf{z}, \qquad (7.3.39)$$

where A is a set in the σ algebra associated with the sample space ζ. When $f(\cdot\,; \boldsymbol{\eta})$ becomes the density of n i.i.d. observations whose cdf F is unknown, one could think of an infinite-dimensional $\boldsymbol{\eta} = F$ and (7.3.39) reduces to the classical bootstrap with $\hat{\boldsymbol{\eta}}$ the empirical distribution function.

Notice that (7.3.39) does not depend on any unknown parameters and can, in principle, be calculated for any measurable set A. The bootstrap approach is to harness fast computers to sample $\mathbf{Z}_1^*, \ldots, \mathbf{Z}_B^*$ repeatedly from $f(\cdot; \hat{\boldsymbol{\eta}})$ and use the approximation

$$P^*(A) \simeq \sum_{j=1}^{B} I(\mathbf{Z}_j^* \in A)/B. \tag{7.3.40}$$

There are ways to achieve more accurate approximations than the straightforward simulations suggested here. The interested reader might consult, for example, Bratley et al. (1987, Chapter 2); in particular, Johns (1988) has shown how importance sampling can increase simulation-size effectiveness.

Now suppose the distribution (or moments) of a complicated statistic $T(\mathbf{Z})$ is needed. The idea behind the parametric bootstrap is to replace $\Pr(T(\mathbf{Z}) \leq t; \boldsymbol{\eta}) = \int_{A_t} f(\mathbf{z}; \boldsymbol{\eta})\, d\mathbf{z}$ with $P^*(A_t)$ given by (7.3.39), and then to approximate it with (7.3.40). The "law of the unconscious statistician" does not require A_t to be calculated; (7.3.40) can be replaced with

$$P^*(A_t) \simeq \sum_{j=1}^{B} I\big(T(\mathbf{Z}_j^*) \leq t\big)/B.$$

Obviously, the parametric bootstrap can be used in spatial problems provided the stochastic model is known up to a set of parameters $\boldsymbol{\eta}$, and bootstrap observations can be generated from that model when $\boldsymbol{\eta} = \hat{\boldsymbol{\eta}}$. What properties must $\hat{\boldsymbol{\eta}}$ satisfy in order for the proposal to be asymptotically valid? [Bickel and Freedman (1981) give answers to this question in the i.i.d. case.] What is the order of the approximation and can it be improved? Moreover, what happens to the parametric bootstrap when the model is misspecified?

7.3.3 Cross-Validation and Model Selection

The conditionally specified models are in a very convenient form for cross-validation. Suppose the observation $Z(\mathbf{s}_i)$ is deleted from the data set and predicted using the other observations $\{Z(\mathbf{s}_j): j \neq i\}$. Depending on the loss function specified for prediction, an optimal predictor $p_i^0(\{Z(\mathbf{s}_j): j \neq i\})$ will be obtained (see the introductory remarks in Chapter 3). For example, for squared-error loss, the best predictor (that is, the predictor that minimizes mean-squared prediction error) is $E(Z(\mathbf{s}_i) | \{Z(\mathbf{s}_j): j \neq i\})$.

Let $L(Z(\mathbf{s}_i), p_i(\{Z(\mathbf{s}_j): j \neq i\}))$ denote the loss incurred when $Z(\mathbf{s}_i)$ is predicted with p_i. Hence, the optimal predictor $p_i^0(\{Z(\mathbf{s}_j): j \neq i\})$ minimizes

$$V_i(p_i|\{Z(\mathbf{s}_j): j \neq i\}) \equiv E\big(L\big(Z(\mathbf{s}_i), p_i(\{Z(\mathbf{s}_j): j \neq i\})\big)\big|\{Z(\mathbf{s}_j): j \neq i\}\big), \tag{7.3.41}$$

the posterior expected loss. Usually, p_i^0 can be calculated easily from the conditionally specified models, as can V_i. Then, if the *fitted* model is correct, each of

$$u_i^2 \equiv L(Z(\mathbf{s}_i), p_i^0)/V_i(p_i^0|\{Z(\mathbf{s}_j): j \neq i\}), \qquad i = 1, \ldots, n, \tag{7.3.42}$$

is estimating the quantity 1. Thus, the summary $(\Sigma_{i=1}^n u_i^2/n) - 1$ or the histogram of $\{u_1^2, \ldots, u_n^2\}$ can indicate whether any data do not fit the model well. For example, squared-error loss given by $L(Z(\mathbf{s}_i), p_i) \equiv (Z(\mathbf{s}_i) - p_i)^2$, yields

$$u_i^2 = \big[Z(\mathbf{s}_i) - E\big(Z(\mathbf{s}_i)|\{Z(\mathbf{s}_j): j \neq i\}\big)\big]^2\big/\mathrm{var}\big(Z(\mathbf{s}_i)|\{Z(\mathbf{s}_j): j \neq i\}\big), \tag{7.3.43}$$

a quantity that was already considered in Section 2.6.4 for cross-validating the variogram model. In the analysis of SID rates for the counties of North Carolina presented in Section 7.6, Anson county's rate in 1974–1978 is diagnosed as unusual using (the square root of) (7.3.43).

The role of cross-validation here is to prevent blunders and to highlight potentially troublesome data. It cannot prove that the fitted model is correct, merely that it is not grossly incorrect.

Use of cross-validation to *estimate* $\boldsymbol{\eta}$ or to *select* a model (Stone, 1974; Geisser, 1975) needs close scrutiny. Why should continued sample reuse provide a meaningful way of preferring one $\boldsymbol{\eta}$ value or one class of models over another? The interdependencies introduced into a problem where data are already spatially dependent make confirmatory techniques questionable. Hjorth (1982), Dawid (1984), and Rissanen (1984, 1987) give alternative approaches for time-series data. However, *sequencing* of the observations is very important for these authors' approach of minimizing accumulated prediction errors, and this may be difficult to generalize to the spatial context.

Rissanen bases model choice on parsimony of description of the data given the model and of the model itself. This is not unlike the way Akaike's information criterion (AIC) owing to Akaike (1973) and the Bayes information criterion (BIC) owing to Schwarz (1978) choose a model (from a set of typically nested models). Kashyap and Chellappa (1983) use BIC to choose the neighborhood of the CG model (7.2.24). A fertile area for future research is in developing sensible model-selection criteria when the data are spatial.

7.4 STATISTICAL IMAGE ANALYSIS AND REMOTE SENSING

Image analysis is concerned with the restoration and interpretation of images that have been contaminated by noise and possibly some (nonlinear) transformation. Applications include remote sensing by satellites, optical astronomy, radar, electron microscope imaging, photography, computer vision and graphics, ultrasound, magnetic resonance imaging, and photon emission tomography. The data are a degraded image obtained via some photochemical or photoelectric sensor and are often multivariate, consisting of the intensity of radiation in different bands (e.g., seven for the orbiting satellite *LANDSAT 4*) of the electromagnetic spectrum. Digitization transforms the data further to gray levels or a finite set of colors on a rectangular lattice of pixels (picture elements). Going from the continuous world to digitized two-dimensional arrays, and back again, leads to difficult mathematical and statistical problems.

The purpose of this section is to review various statistical methods that have been proposed for the analysis of images. In the following, let $\{\theta(\mathbf{s}): \mathbf{s} \in D\}$ denote the true image on a two-dimensional regular lattice D. Values of $\theta(\mathbf{s}_0)$ at intermediate values $\mathbf{s}_0 \notin D$ are usually assumed to be undefined. Images are characterized by the range of possible values for the image function $\theta(\mathbf{s})$ evaluated at $\mathbf{s}$. Types of images include (e.g., Switzer, 1986):

Dichotomous images. $\theta(\mathbf{s}) \in \{0, 1\}$, for all $\mathbf{s} \in D$.
Polychotomous images. $\theta(\mathbf{s}) \in \{1, \ldots, K\}$, for all $\mathbf{s} \in D$ and some integer $K \geq 2$.
Nonnegative images. $\theta(\mathbf{s}) \in [0, \infty)$, for all $\mathbf{s} \in D$.
Multivariate combinations of the previous types.

In the following, let the true image on a finite lattice D (with n sites) be denoted $\boldsymbol{\theta} \equiv (\theta(\mathbf{s}_1), \ldots, \theta(\mathbf{s}_n))'$.

Restoration is concerned with the estimation of the image $\boldsymbol{\theta}$ from the contaminated-image data $\mathbf{Z} \equiv (Z(\mathbf{s}_1), \ldots, Z(\mathbf{s}_n))'$. They are often assumed to be related by

$$\mathbf{Z} = \phi\{H(\boldsymbol{\theta})\} \odot N, \tag{7.4.1}$$

where H is a blurring matrix corresponding to a translation-invariant point-spread function, ϕ is any (possibly nonlinear) function, and N is random noise. The operator $\odot$ is any suitable invertible operation, such as addition or multiplication (Geman and Geman, 1984). For example,

$$Z(u,v) = \phi\left\{ \sum_{(i,j)} H(i,j)\theta(u-i, v-j) \right\} + \epsilon(u,v),$$

where $\{\epsilon(u,v)\}$ is Gaussian white noise. For photoelectric sensors, ϕ is a linear function, but for photochemical sensors, ϕ is often a known nonlinear function.

Often the lattice D can, in principle, be partitioned into regions corresponding to different objects or surface types in the scene. *Segmentation* is concerned with the partitioning of the domain into homogeneous (possibly unlabeled) regions; boundaries separating regions may be modeled through an edge process. *Classification* refers to assigning labels to pixels or regions. Let $\theta(\mathbf{s})$ denote the true (unknown) class to which pixel $\mathbf{s}$ belongs, and assume that the intensity of emitted or reflected radiation $Z(\mathbf{s})$ from a pixel $\mathbf{s}$ is a (usually Poisson or Gaussian) random variable whose parameters are a function of $\theta(\mathbf{s})$. Emitted or reflected radiation may be scattered (by the atmosphere) or distorted by the sensors themselves. As for image restoration, the data often take the form

$$\mathbf{Z} = \phi\{H(\boldsymbol{\theta})\} \odot N.$$

Thus, although the goals of restoration and segmentation are different, these image analysis problems share several common statistical techniques.

In photon emission tomography (Section 7.4.6) and magnetic resonance imaging, the data $\{Z(\mathbf{u}): \mathbf{u} \in D^{(p)}\}$ are line integrals of a corrupted version of the unknown image $\{\theta(\mathbf{s}): \mathbf{s} \in D\}$. Here, there is no one-to-one correspondence between the sites $\mathbf{s} \in D$ and the sites $\mathbf{u} \in D^{(p)}$. Instead, the data are collected by an array of sensors positioned around the region of interest. A given sensor $\mathbf{u} \in D^{(p)}$ detects the intensity of emissions along a line passing through D. *Reconstruction* is concerned with the estimation of $\boldsymbol{\theta}$ from $\mathbf{Z}$.

In what follows, various Bayesian methods of image restoration, segmentation, and reconstruction are considered. Implementation may be either in a joint manner over the whole region D, or pixel by pixel. Let $\pi(\cdot)$ denote the prior distribution of $\boldsymbol{\theta} \in \Theta$ and let $f(\cdot|\boldsymbol{\theta})$ denote the conditional distribution of $\mathbf{Z}$ given $\boldsymbol{\theta}$. Then, the posterior distribution of $\boldsymbol{\theta}$ given $\mathbf{Z}$ is

$$p(\boldsymbol{\theta}|\mathbf{Z}) = \frac{f(\mathbf{Z}|\boldsymbol{\theta})\pi(\boldsymbol{\theta})}{\int_{\Theta} f(\mathbf{Z}|\boldsymbol{\tau})\pi(\boldsymbol{\tau})\, d\boldsymbol{\tau}}. \tag{7.4.2}$$

Assume a 0-1 loss function that assigns zero loss if and only if a correct joint classification is made. Then the joint restoration of $\boldsymbol{\theta}$, given by the $\hat{\boldsymbol{\theta}} \in \Theta$ that maximizes (7.4.2), is the Bayes rule, and is sometimes called the maximum *a posteriori* (MAP) estimator. For pixel-by-pixel restoration, the Bayes rule finds the $\hat{\theta}(\mathbf{s})$ that maximizes the marginal posterior distribution $p(\theta(\mathbf{s})|\mathbf{Z})$. The methods for image restoration that follow differ in their choices of priors $\pi(\cdot)$, in their choices of conditional distributions $f(\cdot|\boldsymbol{\theta})$, and according to whether joint restoration or pixel-by-pixel restoration is implemented.

Section 7.4.1 gives a description of how remote-sensing data are collected and processed. Sections 7.4.2 and 7.4.3 present various Bayesian methods of image restoration and segmentation. Models for edge processes are considered in Section 7.4.4 and methods for segmentation of textured images are presented in Section 7.4.5. Section 7.4.6 discusses the reconstruction of tomographic images. Least squares and image regularization are considered in Section 7.4.7. The method of sieves is presented in Section 7.4.8. Finally, Section 7.4.9 discusses briefly the application of mathematical morphology to image analysis.

The emphasis in this section is on low-level image analysis. The modeling, recognition, and analysis of whole objects or parts of objects from images is not nearly as well developed, although low-level methods are often applied to such problems (with mixed success). Chapter 9 addresses higher levels of image analysis and modeling.

7.4.1 Remote Sensing

Remote sensing by satellites or high-flying aircraft has become an important tool for *inter alia* the inventory of natural resources, providing an attractive alternative to relatively inefficient ground surveys. Remote-sensing data may be used, for example, to estimate crop yields or to monitor the effects of desertification, forest clearing, erosion, and so forth. Statistical techniques are required for the processing and classification of such data.

When surfaces are irradiated, they emit electromagnetic energy at a rate related to their temperature. Let ϵ be the emissivity (a constant depending on the type of object) and let T be the thermodynamic temperature of an object. Then the rate of emission (in emitted photons per unit time) is given by $M = c\epsilon T^4$, where c is a constant (e.g., Lamp, 1984). Solar radiation is absorbed by carbon dioxide, dust, ozone, water, and other constituents of the atmosphere, so passive methods of remote sensing can be used only for certain "windows" in the electromagnetic spectrum; *LANDSAT 1, 2,* and *3* satellites can detect four bands, namely, green, red, and two infrared; *LANDSAT 4* can detect seven bands. The sensors measure the intensity of reflected radiation within each band from each area or pixel. The two scanning sensors in operation are the Multi-Spectral Scanner, with pixel size 80×80m, and the Thematic Mapper, with pixel size 30×30m. For each band and each pixel, intensity of radiation is integrated over recent dates, the area of the pixel, and the width of the band.

Before remote-sensing data can be used, data must be preprocessed. Distortions may result from the sensors, the angle of the satellite platform relative to the surface, and the geometry of the surface (e.g., presence of mountains). If data are collected from more than one pass of the satellite, the scale and projection must be standardized. Noise due to detection, recording, or transmission errors should be removed. Different angles of the sun relative to the sensor and the ground may also cause distortions.

The data are denoted by vectors

$$\mathbf{Z}(\mathbf{s}) \equiv (Z_1(\mathbf{s}), \dots, Z_p(\mathbf{s}))', \qquad \mathbf{s} \in D,$$

where $Z_i(\mathbf{s})$ is the emitted or reflected energy in band i, integrated over a pixel centered at $\mathbf{s} = (u, v)'$. Different surfaces reflect different spectra of electromagnetic radiation, so the data vector might be used to classify surface types. From such a classification, a map of the terrain may be constructed and the total area belonging to each class may be calculated. The classification of pixels into surface types usually requires some calibration from training pixels, where the surface type is known from ground-truth surveys.

Because remote-sensing data sets are frequently very large (a single scene may contain 30 Mbytes of data), computationally efficient statistical techniques are required for pixel classification. One approach has been to extend usual methods of multivariate analysis to the remote-sensing problem; these have included multiple linear discriminant analysis (e.g., Hjort and Mohn, 1984), principal components analysis (e.g., Esbensen and Geladi, 1989), and cluster analysis (e.g., Dubes and Jain, 1976). However, these techniques do not really exploit the spatial component of the data, and so may result in a high error rate. In the last 10 years, a variety of spatial techniques have been developed.

In the exposition to follow, assume that the data are located on a square lattice D with sites $\mathbf{s} \in D$. Let $\mathbf{Z}(\mathbf{s})$ denote the data vector for the pixel located at $\mathbf{s}$ and let $\theta(\mathbf{s})$ denote the true (unknown) ground class to which the pixel $\mathbf{s}$ belongs. Suppose that the surface can be classified into K classes and assume that each pixel belongs to just one of these classes. Although these techniques are illustrated in a remote-sensing context, they can be applied to any segmentation/classification problem.

7.4.2 Ordinary Discriminant Analysis

Suppose that, given $\theta(\mathbf{s})$, the remote-sensing or imagery data $\mathbf{Z}(\mathbf{s})$ are distributed according to a conditional density $f(\cdot\,|\theta(\mathbf{s}))$ that does not depend on $\mathbf{s}$. Let $\{\pi(k)\colon k = 1, \dots, K\}$ be the prior probability distribution of $\theta(\mathbf{s})$. Based only on the realization $\mathbf{z}(\mathbf{s})$ of $\mathbf{Z}(\mathbf{s})$, a Bayes classification rule declares $\hat{\theta}(\mathbf{s}) = k^*$, where k^* maximizes the posterior distribution

$$p(k|\mathbf{z}(\mathbf{s})) = \frac{f(\mathbf{z}(\mathbf{s})|k)\pi(k)}{\sum_{l=1}^{K} f(\mathbf{z}(\mathbf{s})|l)\pi(l)}; \tag{7.4.3}$$

this is a Bayes rule for the loss function,

$$l(\theta(\mathbf{s}), \hat{\theta}(\mathbf{s})) = \begin{cases} 0, & \hat{\theta}(\mathbf{s}) = \theta(\mathbf{s}), \\ 1, & \text{otherwise.} \end{cases}$$

Because the denominator of (7.4.3) is constant in k, it is equivalent to declare $\hat{\theta}(\mathbf{s}) = k^*$, where k^* maximizes $f(\mathbf{z}(\mathbf{s})|k)\pi(k)$. If $f(\cdot|k)$ are normal densities with means $\boldsymbol{\mu}_k$ and identical variance matrices Σ, then this rule is ordinary linear discriminant analysis; that is, declare $\hat{\theta}(\mathbf{s}) = k^*$, where k^* maximizes the score

$$\boldsymbol{\mu}_k'\Sigma^{-1}\mathbf{Z}(\mathbf{s}) - (1/2)\boldsymbol{\mu}_k'\Sigma^{-1}\boldsymbol{\mu}_k + \log \pi(k). \tag{7.4.4}$$

If the variance–covariance matrices are not equal, then quadratic discriminant analysis is obtained; that is, declare $\hat{\theta}(\mathbf{s}) = k^*$, where k^* maximizes the score

$$-(1/2)(\mathbf{Z}(\mathbf{s}) - \boldsymbol{\mu}_k)'\Sigma_k^{-1}(\mathbf{Z}(\mathbf{s}) - \boldsymbol{\mu}_k) - (1/2)\log|\Sigma_k| + \log \pi(k). \tag{7.4.5}$$

In practice, $\boldsymbol{\mu}_k$ and Σ_k are estimated by sample-mean vectors and sample-variance matrices calculated from training data, for which the true classes are known. For example, suppose training data $\{Z(\mathbf{s})\colon \mathbf{s} \in D_0\}$ are available, where $D_0 = \bigcup\{D_k\colon k = 1,\ldots,K\}$, a disjoint union of training pixels whose values are known to belong to $1,\ldots,K$, respectively. It is assumed that each set of pixels in the union is known *a priori*, and hence $\boldsymbol{\mu}_k$ and Σ_k can be estimated by

$$\hat{\boldsymbol{\mu}}_k = \sum_{\mathbf{s}\in D_k} \mathbf{Z}(\mathbf{s}) \Big/ \sum_{\mathbf{s}\in D_k} 1$$

and

$$\hat{\Sigma}_k = \sum_{\mathbf{s}\in D_k} (\mathbf{Z}(\mathbf{s}) - \hat{\boldsymbol{\mu}}_k)(\mathbf{Z}(\mathbf{s}) - \hat{\boldsymbol{\mu}}_k)' \Big/ \left\{\left(\sum_{\mathbf{s}\in D_k} 1\right) - 1\right\}.$$

This ordinary-discriminant-analysis approach to pixel classification effectively assumes that data vectors in neighboring pixels are independent, but clearly this is not so. Spatial dependencies between pixels may be caused by scattering of reflected electromagnetic radiation from the surface of the earth, contamination resulting from oversampling and resampling, or spatial continuity of the ground classes. Moreover, data from neighboring pixels will not be (marginally) independent even if they are independent conditional on the ground classes. Unless training pixels are selected to be sufficiently far apart, these spatial dependencies can result in biased estimates of the covariance matrix Σ (Lawoko and McLachlan, 1983), and hence in an increased classification error rate. Moreover, discriminant analysis acts on pixels independently and so may result in classification maps that are patchier than the true scene.

Several *ad hoc* approaches have been taken for the incorporation of spatial information in the classification of pixels. These include simple augmentation of the data vector by the data vectors of surrounding pixels, smoothing algorithms, and various contextual classification methods.

Simple Augmentation

Switzer (1980) proposes to augment the $p \times 1$ data vector $\mathbf{Z}(\mathbf{s})$ with the average of the data vectors from the four neighboring pixels with common edges. He considers the $2p \times 1$ vector $\mathbf{Z}^*(\mathbf{s}_0) \equiv (\mathbf{Z}(\mathbf{s}_0)', \mathbf{Z}^{(a)}(\mathbf{s}_0)')'$, where $\mathbf{s}_0 = (u, v)'$ and

$$\mathbf{Z}^{(a)}(\mathbf{s}_0) = (1/4)\{\mathbf{Z}(u-1, v) + \mathbf{Z}(u+1, v) + \mathbf{Z}(u, v-1) + \mathbf{Z}(u, v+1)\}. \quad (7.4.6)$$

Then, ordinary linear discriminant analysis is performed using the augmented data vector $\mathbf{Z}^*(\mathbf{s}_0)$. This approach has the advantage of easy implementation through standard discriminant-analysis software.

Mardia (1984) modifies Switzer's method by augmenting $\mathbf{Z}(\mathbf{s})$ with each of the data vectors for neighboring pixels. Let $\mathbf{Z}(\mathbf{s}_0)$ denote the $p \times 1$ data vector for the central pixel to be classified, let $\mathbf{Z}(\mathbf{s}_1), \ldots, \mathbf{Z}(\mathbf{s}_r)$ denote data vectors of its neighbors, and define $\mathbf{Z}^*(\mathbf{s}_0) \equiv (\mathbf{Z}(\mathbf{s}_0)', \mathbf{Z}(\mathbf{s}_1)', \ldots, \mathbf{Z}(\mathbf{s}_r)')'$. Assume that the distribution of $\mathbf{Z}(\mathbf{s})$, conditional on $\theta(\mathbf{s}) = k$, is Gaussian with mean $\boldsymbol{\mu}_k$ and variance Σ, and assume that the $p \times p$ covariance matrix between $\mathbf{Z}(\mathbf{s})$ and $\mathbf{Z}(\mathbf{u})$ can be factored as

$$\operatorname{cov}(\mathbf{Z}(\mathbf{s}), \mathbf{Z}(\mathbf{u})) = \rho^o(\|\mathbf{s} - \mathbf{u}\|)\Sigma, \quad (7.4.7)$$

where ρ^o is an isotropic correlogram and $\rho^o(0) = 1$. Then the variance matrix of $\mathbf{Z}^*(\mathbf{s}_0)$ is given by the Kronecker product

$$\Sigma^* = P \otimes \Sigma \quad (7.4.8)$$

[recall $A_{m \times n} \otimes B_{p \times q} = ((a_{ij}B))_{mp \times nq}$], where P is the $(r+1) \times (r+1)$ spatial correlation matrix whose (i, j)th element is $\rho^o(\|\mathbf{s}_{i-1} - \mathbf{s}_{j-1}\|)$. Assume local spatial homogeneity; that is, if $\theta(\mathbf{s}_0) = k$, then $\theta(\mathbf{s}_i) = k$ with probability close to 1, for $i = 1, 2, \ldots, r$. Then the mean vector for $\mathbf{Z}^*(\mathbf{s}_0)$ is

$$E(\mathbf{Z}^*(\mathbf{s}_0) | \theta(\mathbf{s}_0) = k) \simeq \mathbf{1} \otimes \boldsymbol{\mu}_k, \quad (7.4.9)$$

where $\mathbf{1}$ is an $(r+1) \times 1$ vector of 1s. Replacing $\boldsymbol{\mu}_k$ with $\mathbf{1} \otimes \boldsymbol{\mu}_k$, Σ with $\Sigma^* = P \otimes \Sigma$, and $\mathbf{Z}(\mathbf{s}_0)$ with $\mathbf{Z}^*(\mathbf{s}_0)$ in Eq. (7.4.4), the score to be maximized is

$$\begin{aligned} S_k &= (\mathbf{1}' \otimes \boldsymbol{\mu}'_k)(P^{-1} \otimes \Sigma^{-1})\mathbf{Z}^*(\mathbf{s}_0) \\ &\quad -(1/2)(\mathbf{1}' \otimes \boldsymbol{\mu}'_k)(P^{-1} \otimes \Sigma^{-1})(\mathbf{1} \otimes \boldsymbol{\mu}_k) + \log \pi(k) \\ &= \boldsymbol{\mu}_k\Sigma^{-1} \sum_{i=0}^{r} \eta_i \mathbf{Z}(\mathbf{s}_i) - (1/2)\left(\sum_{i=0}^{r} \eta_i\right)\boldsymbol{\mu}'_k\Sigma^{-1}\boldsymbol{\mu}_k + \log \pi(k), \end{aligned} \quad (7.4.10)$$

where $\boldsymbol{\eta} \equiv (\eta_0, \eta_1, \ldots, \eta_r)' \equiv P^{-1}\mathbf{1}$.

Haslett and Horgan (1987) relax the assumption that $\theta(\mathbf{s}_i) = k$ with probability close to 1, but assume that $\rho^o(h) = 0$, for $h > 0$.

Smoothing

Another approach to incorporating spatial information into discriminant analysis is either to presmooth the input electromagnetic-intensity data or to postsmooth the output of standard classification algorithms (Switzer, 1983). Such smoothers are expected to improve classification in the interior of homogeneous regions, but may also result in blurred edges between regions.

Presmoothing involves filtering the data prior to classification. A filtered version of the data is obtained from the moving average

$$\mathbf{Z}^*(\mathbf{s}_0) \equiv \frac{\sum_{i=1}^{n} w(\mathbf{s}_0 - \mathbf{s}_i)\mathbf{Z}(\mathbf{s}_i)}{\sum_{i=1}^{n} w(\mathbf{s}_0 - \mathbf{s}_i)}, \qquad \mathbf{s}_0 \in D, \quad (7.4.11)$$

where $w(\cdot)$ is a nonnegative weight function. If $w(\mathbf{s}_0 - \mathbf{s}_i)$ is equal to 1 when $\mathbf{s}_i$ is a neighbor of $\mathbf{s}_0$ and is equal to 0 otherwise, then (7.4.11) is a moving-mean filter. Standard pointwise classification procedures can then be applied to the filtered data $\{\mathbf{Z}^*(\mathbf{s}): \mathbf{s} \in D\}$.

Mean filters blur edges between regions and are sensitive to outliers. Median filtering (e.g., Tukey, 1977, pp. 210–211) preserves edges and is not sensitive to outliers. An example of a moving-median filter is

$$Z_i^*(\mathbf{s}_0) = \text{median}\{Z_i(\mathbf{s}): \mathbf{s} \in N_{\mathbf{s}_0} \cup \{\mathbf{s}_0\}\}, \qquad i = 1, 2, \ldots, p, \quad (7.4.12)$$

where $N_{\mathbf{s}_0}$ denotes the neighbors of $\mathbf{s}_0$. Properties of median filters are considered by Justusson (1981) and Tyan (1981).

Median filters round the corners of objects. Narendra (1981) considers a *separable* median filter that does not round corners of objects, provided the corner is oriented in the same direction as the rows and columns of the pixels. For an $M \times M$ neighborhood centered on $\mathbf{s}_0$, a separable median filter is obtained from successive applications of a one-dimensional median filter of size M, first applied down the columns of an image, and then applied along the rows (or vice versa). The properties of separable median filters are considered by Liao et al. (1985).

Postsmoothing operates on the results of pixel-by-pixel classification. Let $\hat{\theta}(\mathbf{s})$ denote the assigned class of the pixel at $\mathbf{s}$. One simple postsmoothing routine would be to replace $\hat{\theta}(\mathbf{s})$ by the class previously assigned to the plurality of pixels in the neighborhood of $\mathbf{s}$.

Switzer et al. (1982) suggest the following empirical Bayes smoother. Assume that $\theta(\mathbf{s})$ is stochastic, taking the value k with probability $\pi(k;\mathbf{s})$, $k = 1,\ldots,K$. If the vector of prior probabilities $\boldsymbol{\pi}(\mathbf{s}) \equiv (\pi(1;\mathbf{s}),\ldots,\pi(K;\mathbf{s}))'$ is approximately the same for all pixels in a fixed neighborhood of $\mathbf{s}$, then it is possible to estimate $\boldsymbol{\pi}(\mathbf{s})$ from the data as follows. The $K \times K$ confusion matrix $F \equiv (f_{kl})$ is calculated from the training pixels, where f_{kl} is the proportion of class k pixels that have been assigned to class l by the ordinary discriminant analysis classifier. For each pixel $\mathbf{s}$, calculate the proportion $p(l;\mathbf{s})$ of pixels in the fixed neighborhood of $\mathbf{s}$ that have been assigned to class l by the ordinary discriminant analysis classifier and define $\mathbf{p}(\mathbf{s}) \equiv (p(1;\mathbf{s}),\ldots,p(K;\mathbf{s}))'$. By the law of total probability, $E(p(l;\mathbf{s})) = E(\sum_{k=1}^{K} f_{kl}\pi(k;\mathbf{s}))$. Then, assuming F is nonsingular, the prior probability vector is estimated by

$$\hat{\boldsymbol{\pi}}(\mathbf{s}) = (F')^{-1}\mathbf{p}(\mathbf{s}). \tag{7.4.13}$$

The smoothed classification of pixel $\mathbf{s}$ is obtained by substituting $\hat{\pi}(k;\mathbf{s})$ for $\pi(k)$ in Eq. (7.4.4) and declaring $\hat{\theta}(\mathbf{s}) = k^*$, where k^* maximizes the resultant score.

Contextual Classification

Contextual classification exploits the tendency for certain ground cover classes to occur more frequently in some contexts than in others. The classification of a given pixel is influenced by the probable classifications of surrounding pixels.

Suppose that the conditional density

$$f(\mathbf{z}(u,v),\mathbf{z}(u\pm 1,v),\mathbf{z}(u,v\pm 1)\,|\,\theta(u,v),\theta(u\pm 1,v),\theta(u,v\pm 1)) \tag{7.4.14}$$

and the joint prior probability distribution $\pi(\theta(u,v),\theta(u\pm 1,v),\theta(u,v\pm 1))$ are known. Then the posterior distribution of the five levels can be obtained from Bayes' Theorem:

$$p(\theta(u,v),\theta(u\pm 1,v),\theta(u,v\pm 1)\,|\,\mathbf{Z}(u,v),\mathbf{Z}(u\pm 1,v),\mathbf{Z}(u,v\pm 1))$$
$$= \frac{f(\mathbf{Z}(u,v),\mathbf{Z}(u\pm 1,v),\mathbf{Z}(u,v\pm 1)\,|\,\theta(u,v),\theta(u\pm 1,v),\theta(u,v\pm 1)) \times\pi(\theta(u,v),\theta(u\pm 1,v),\theta(u,v\pm 1))}{f(\mathbf{Z}(u,v),\mathbf{Z}(u\pm 1,v),\mathbf{Z}(u,v\pm 1))}. \tag{7.4.15}$$

Hence the posterior distribution of $\theta(u, v)$ given the data $\{\mathbf{Z}(u, v), \mathbf{Z}(u \pm 1, v), \mathbf{Z}(u, v \pm 1)\}$ can be obtained by summation:

$$\begin{aligned} p(\theta(u,v) &= k|\mathbf{Z}(u,v), \mathbf{Z}(u \pm 1,v), \mathbf{Z}(u,v \pm 1)) \\ &= \sum_{i=1}^{K} \sum_{j=1}^{K} \sum_{l=1}^{K} \sum_{m=1}^{K} p(\theta(u,v) = k, \theta(u+1,v) = i, \theta(u-1,v) = j, \\ &\qquad \theta(u,v+1) = l, \theta(u,v-1) = m \\ &\qquad |\mathbf{Z}(u,v), \mathbf{Z}(u \pm 1,v), \mathbf{Z}(u,v \pm 1)), \end{aligned} \tag{7.4.16}$$

where it is assumed there are a finite number K of colors. Owen (1984), Hjort and Mohn (1984), and Haslett (1985) declare $\hat{\theta}(u, v) = k^*$, where k^* maximizes (7.4.16). All assume conditional independence of the data vectors in (7.4.14), but they differ in their specifications of the prior.

Haslett (1985) assumes that the θs of nearest-neighbor pixels are conditionally independent given $\theta(u, v)$. Owen (1984) and Hjort and Mohn (1984) assume that pixels are small relative to the grain of the pattern so that the site $(u, v)'$ and its four nearest neighbors are one of three possible patterns:

X pattern	T pattern	L pattern
i	j	j
$i \quad i \quad i$	$i \quad i \quad i$	$j \quad i \quad i$
i	i	i

All rotations are assumed to be equally likely. In practice, it is difficult to check these assumptions, and in many circumstances they would not be appropriate.

Hjort and Mohn (1984) let the distribution of the three pattern types be arbitrary. Owen (1984) assumes further that the region is partitioned into convex regions by lines realized from a Poisson line process. Let β be the probability that a given neighborhood intersects a boundary line (i.e., more than one pixel type is present in the neighborhood) and let α be the conditional probability of an L pattern given that a boundary line intersects the neighborhood. For a Poisson line process, Owen calculates $\alpha \simeq 0.41$, and $1 - \beta$ is estimated by the proportion of training pixels that are of the same type as their neighbors. Let $\pi(i)$ denote the marginal prior probability that $\theta(u, v) = i$. Then, under the assumption that $\boldsymbol{\theta}$ is a stationary process, associated with the three patterns are the probabilities

$$\begin{aligned} &\Pr\{\theta(u+1,v) = i, \theta(u-1,v) = i, \theta(u,v+1) = i, \\ &\qquad \theta(u,v-1) = i|\theta(u,v) = i\} = 1 - \beta + \pi(i)\beta, \\ &\Pr\{\theta(u+1,v) = i, \theta(u-1,v) = i, \theta(u,v+1) = j, \\ &\qquad \theta(u,v-1) = i|\theta(u,v) = i\} = \pi(j)(1-\alpha)\beta/4, \\ &\Pr\{\theta(u+1,v) = i, \theta(u-1,v) = j, \theta(u,v+1) = j, \\ &\qquad \theta(u,v-1) = i|\theta(u,v) = i\} = \pi(j)\alpha\beta/4, \end{aligned} \tag{7.4.17}$$

which can then be substituted into (7.4.16) and maximized, yielding a declaration for $\theta(u, v)$.

Compound-Decision-Theory Approach

Swain et al. (1981) and Tilton et al. (1982) consider an approach to contextual classification based on compound decision theory. Let the classification of pixel $\mathbf{s}$ be denoted by $a(\mathbf{s}; \{\mathbf{Z}(\mathbf{u}): \mathbf{u} \in D\})$, a function of pixel location and data, taking values in $\{1, 2, \ldots, K\}$. Let $l(\theta(\mathbf{s}), a(\mathbf{s}; \{\mathbf{Z}(\mathbf{u}): \mathbf{u} \in D\}))$ be the loss suffered by taking action $a(\mathbf{s}; \{\mathbf{Z}(\mathbf{u}): \mathbf{u} \in D\})$ when the true class is $\theta(\mathbf{s})$. Then, the average loss over the n pixel locations ($n = |D|$, the number of elements in D) is given by

$$\bar{l}(\boldsymbol{\theta}) = \frac{1}{n} \sum_{\mathbf{s} \in D} l(\theta(\mathbf{s}), a(\mathbf{s}; \{\mathbf{Z}(\mathbf{u}): \mathbf{u} \in D\})). \qquad (7.4.18)$$

Compare this loss to the one that leads to the MAP estimator. There, the loss is 0 if *all* the pixels are correctly classified, and 1 otherwise. The expected average loss or risk is

$$R(\boldsymbol{\theta}) = E(\bar{l}(\boldsymbol{\theta})). \qquad (7.4.19)$$

The goal is to find a decision rule $a(\mathbf{s}; \{\mathbf{Z}(\mathbf{u}): \mathbf{u} \in D\})$ such that the risk $R(\boldsymbol{\theta})$ is small for any given classification vector $\boldsymbol{\theta} \in \boldsymbol{\Theta}$.

Swain et al. (1981) take the approach of controlling the risk through $a(\cdot\,;\cdot)$. They restrict the classification function $a(\mathbf{s}; \{\mathbf{Z}(\mathbf{u}): \mathbf{u} \in D\})$ so that it depends only on the data for $\mathbf{s}$ and its neighbors. Thus, $a(\mathbf{s}; \{\mathbf{Z}(\mathbf{u}): \mathbf{u} \in D\}) = d(\mathbf{s}; \mathbf{Z}(\mathbf{s}), \{\mathbf{Z}(\mathbf{u}): \mathbf{u} \in N_{\mathbf{s}}\})$, where $N_{\mathbf{s}}$ denotes the neighbors of $\mathbf{s}$. Let $\boldsymbol{\theta}^* \equiv (\theta(\mathbf{s}_0), \theta(\mathbf{s}_1), \ldots, \theta(\mathbf{s}_r)) \in \boldsymbol{\Theta}^*$ denote an $(r+1) \times 1$ vector containing the true classes of a pixel $\mathbf{s}_0 \in D$ and its neighbors $\mathbf{s}_i \in N_{\mathbf{s}_0}$, $i = 1, \ldots, r$. Assume that $\mathbf{Z}(\mathbf{s}_0), \mathbf{Z}(\mathbf{s}_1), \ldots, \mathbf{Z}(\mathbf{s}_r)$ are conditionally independent given $\boldsymbol{\theta}^*$ and that this distribution does not depend on $\mathbf{s}_0$. Therefore,

$$f(\mathbf{Z}(\mathbf{s}_0), \mathbf{Z}(\mathbf{s}_1), \ldots, \mathbf{Z}(\mathbf{s}_r) | \boldsymbol{\theta}^*) = \prod_{i=0}^{r} g(\mathbf{Z}(\mathbf{s}_i) | \theta(\mathbf{s}_i)). \qquad (7.4.20)$$

Assume that $\boldsymbol{\theta}$ is realized from a stationary process and let $\pi(\boldsymbol{\theta}^*)$ be the context (i.e., the prior) distribution; that is, $\pi(\boldsymbol{\theta}^*)$ is the probability of a given configuration $\boldsymbol{\theta}^*$. Consider the 0-1 loss function

$$l(\theta, a) = \begin{cases} 0, & \text{if } \theta = a, \\ 1, & \text{if } \theta \neq a, \end{cases} \qquad (7.4.21)$$

in (7.4.18). Then the decision rule of Swain et al. (1981) declares $d(\mathbf{s}; \mathbf{Z}(\mathbf{s}),$

$\{\mathbf{Z}(\mathbf{u})\colon \mathbf{u} \in N_s\}) = k^*$, where k^* maximizes

$$\left\{\sum{}^{(k)}\pi(\boldsymbol{\theta}^*)\cdot\prod_{\mathbf{u}\in N_s} g(\mathbf{Z}(\mathbf{u})|\theta(\mathbf{u}))\right\}\cdot g(\mathbf{Z}(\mathbf{s})|k); \tag{7.4.22}$$

the summation $\Sigma^{(k)}$ in (7.4.22) is over all $\boldsymbol{\theta}^* \in \Theta^*$ such that the first element of $\boldsymbol{\theta}^*$ is $\theta(\mathbf{s}) = k$. The implementation of this decision rule requires estimates of $g(\cdot|k)$, for $k = 1,\ldots,K$, and a specification or estimation of $\pi(\boldsymbol{\theta}^*)$ for each $\boldsymbol{\theta}^* \in \Theta^*$. Assuming Gaussianity, the parameters of $g(\cdot|k)$ can be estimated from the sample-mean vector and the sample-variance matrix calculated from the training data. Swain et al. (1981) propose to estimate the context distribution $\pi(\cdot)$ from a preliminary classification using ordinary discriminant analysis; that is, $\hat{\pi}(\boldsymbol{\theta}^*)$ is the relative frequency of occurrence of $\boldsymbol{\theta}^*$ in the array $\hat{\boldsymbol{\theta}}$. Although this approach to estimating $\pi(\boldsymbol{\theta}^*)$ gave good results for simulated data sets, Swain et al. (1981) found that it was not adequate for real data. This poor performance may be due to the obvious bias in $\hat{\pi}$, because it has smaller variance than the true π. If a representative sample of ground truth were available (for this or a similar problem), then $\pi(\cdot)$ could be estimated by relative frequencies. Alternatively, Tilton et al. (1982) give an unbiased estimator for $\pi(\boldsymbol{\theta}^*)$ that is derived from a standard empirical-Bayes technique based on the method of moments.

7.4.3* Markov-Random-Field Models

Suppose that the prior distribution of $\{\theta(\mathbf{s})\colon \mathbf{s} \in D\}$ is a Markov random field whose probability distribution π can be expressed as

$$\pi(\boldsymbol{\theta}) = c^{-1}\exp\{Q(\boldsymbol{\theta})\} = c^{-1}\exp\left\{\sum_{\kappa\in\mathscr{C}} V_\kappa(\boldsymbol{\theta}_\kappa)\right\}, \tag{7.4.23}$$

where $Q(\boldsymbol{\theta})$ is the negative total potential energy (negpotential function), $\boldsymbol{\theta}_\kappa \equiv (\theta(\mathbf{s})\colon \mathbf{s} \in \kappa)'$, κ is a clique, and the summation is over the set of all cliques $\mathscr{C}$ (e.g., Wong, 1968). More details on Markov random fields can be found in Section 6.4. In particular, the functions $V_\kappa(\boldsymbol{\theta}_\kappa)$ correspond to the functions $z(\mathbf{s}_{i_1})\cdots z(\mathbf{s}_{i_s})G_{i_1\cdots i_s}(z(\mathbf{s}_{i_1}),\ldots,z(\mathbf{s}_{i_s}))$ in Eqs. (6.4.7) and (6.4.8), and the normalizing constant is given by

$$c = \sum_{\boldsymbol{\theta}\in\Theta}\exp\left\{\sum_{\kappa\in\mathscr{C}} V_\kappa(\boldsymbol{\theta}_\kappa)\right\}.$$

Several examples of (7.4.23) have been applied to image restoration. For binary images, where $\Theta = \{0,1\}^n$, Greig et al. (1986, 1989) consider the model

$$\pi_{\boldsymbol{\beta}}(\boldsymbol{\theta}) \propto \exp\left[(1/2) \sum_{i=1}^{n} \sum_{j=1}^{n} \beta_{ij}\{\theta(\mathbf{s}_i)\theta(\mathbf{s}_j) + (1-\theta(\mathbf{s}_i))(1-\theta(\mathbf{s}_j))\}\right],$$

where $\beta_{ii} = 0$, $\beta_{ij} = \beta_{ji} \geq 0$, and, if $\beta_{ij} > 0$, then $\mathbf{s}_i$ and $\mathbf{s}_j$ are neighbors. For polychotomous images, where $\Theta = \{1, 2, \ldots, K\}^n$, Besag (1986) considers

$$\pi_{\boldsymbol{\alpha},\boldsymbol{\beta}}(\boldsymbol{\theta}) \propto \exp\left\{ \sum_{k=1}^{K} \alpha_k n_k - \sum\sum_{1 \leq k < l \leq K} \beta_{kl} n_{kl} \right\},$$

where n_k is the number of pixels colored k and n_{kl} is the number of neighboring pairs colored (k, l) (Strauss, 1977). For nonnegative images, where $\Theta = [0, \infty)^n$, the auto-Poisson or auto-Gaussian models (Sections 6.5.2 and 6.6) may be used depending on whether Θ is discrete or continuous.

Consider also the *Markov mesh* random fields (e.g., Abend et al., 1965; Devijver, 1988), which are defined via the conditional probabilities

$$\pi(\theta(u,v) | \{\theta(u',v') : 1 \leq u' \leq u, 1 \leq v' \leq v; (u',v') \neq (u,v)\}).$$

Markov mesh random fields are special cases of Markov random fields (Abend et al., 1965). For example, consider

$$\begin{aligned} &\pi(\theta(u,v) | \{\theta(u',v') : 1 \leq u' \leq u, 1 \leq v' \leq v; (u',v') \neq (u,v)\}) \\ &\quad = \pi(\theta(u,v) | \theta(u-1,v), \theta(u,v-1), \theta(u-1,v-1)). \end{aligned}$$

Abend et al. show that this is also a second-order Markov random field with

$$\begin{aligned} &\pi(\theta(u,v) | \{\theta(u',v') : (u',v') \neq (u,v)\}) \\ &\quad = \pi(\theta(u,v) | \{\theta(u+i, v+j) : i,j = -1,0,1; (i,j) \neq (0,0)\}). \end{aligned}$$

For a Markov mesh random field, the joint distribution of $\boldsymbol{\theta}$ is given by

$$\begin{aligned} \pi(\boldsymbol{\theta}) = \prod_{u=1}^{U} \prod_{v=1}^{V} \pi(\theta(u,v) | \{\theta(u',v') : 1 \leq u' \leq u, 1 \leq v' \leq v; \\ (u',v') \neq (u,v)\}) \end{aligned}$$

(Abend et al., 1965). The special form of this joint probability distribution makes a Markov mesh random field easy to simulate using the conditional probabilities and a raster scan beginning at $(u, v) = (1, 1)$.

Posterior Distribution

Suppose that the image is polychotomous and that the conditional density of $\mathbf{Z}$ given $\boldsymbol{\theta}$ is $f(\mathbf{z}|\boldsymbol{\theta})$. Here, the parameter space is contained in a space of dimension $n = |D|$ (e.g., $n = 512^2$). Then the posterior distribution of $\boldsymbol{\theta}$ given $\mathbf{Z}$ is

$$p(\boldsymbol{\theta}|\mathbf{Z}) \propto f(\mathbf{Z}|\boldsymbol{\theta}) \cdot \pi(\boldsymbol{\theta}). \tag{7.4.24}$$

Geman and Geman (1984) note that for suitable f and π, the posterior distribution (7.4.24) is also a Markov random field. For example, assume pairwise-only dependence between sites for the prior distribution of $\boldsymbol{\theta}$; that is,

$$\pi(\boldsymbol{\theta}) = c^{-1} \exp\left\{ \sum_{i=1}^{n} \theta(\mathbf{s}_i) G_i(\boldsymbol{\theta}(\mathbf{s}_i)) + \sum\sum_{1 \le i < j \le n} \theta(\mathbf{s}_i)\theta(\mathbf{s}_j) G_{ij}\big(\theta(\mathbf{s}_i), \theta(\mathbf{s}_j)\big) \right\}, \tag{7.4.25}$$

where $c > 0$ is a normalizing constant. Suppose further that, given $\boldsymbol{\theta}$, $\mathbf{Z}(\mathbf{s}_1), \mathbf{Z}(\mathbf{s}_2), \ldots, \mathbf{Z}(\mathbf{s}_n)$ are conditionally independent and that each $\mathbf{Z}(\mathbf{s}_i)$ has conditional density function $g_i(\mathbf{Z}(\mathbf{s}_i)|\theta(\mathbf{s}_i))$ that depends on i and $\theta(\mathbf{s}_i)$. Then, the conditional density of $\mathbf{Z}$ given $\boldsymbol{\theta}$ is

$$f(\mathbf{Z}|\boldsymbol{\theta}) = \prod_{i=1}^{n} g_i(\mathbf{Z}(\mathbf{s}_i)|\theta_i(\mathbf{s}_i)). \tag{7.4.26}$$

From (7.4.25) and (7.4.26), the posterior distribution of $\boldsymbol{\theta}$ given $\mathbf{Z}$ is

$$p(\boldsymbol{\theta}|\mathbf{Z}) \propto \exp\left[\sum_{i=1}^{n} \{\theta_i(\mathbf{s}_i) G_i(\boldsymbol{\theta}(\mathbf{s}_i)) + \log g_i(\mathbf{Z}(\mathbf{s}_i)|\theta(\mathbf{s}_i))\} + \sum\sum_{1 \le i < j \le n} \theta(\mathbf{s}_i)\theta(\mathbf{s}_j) G_{ij}\big(\theta(\mathbf{s}_i), \theta(\mathbf{s}_j)\big) \right], \tag{7.4.27}$$

which describes the distribution of a Markov random field with pairwise-only dependence between sites. Because of the local nature of the dependence, images simulated from the posterior distribution do not give realistic restorations near the edges of imaged objects. Nor, for that matter, does the prior distribution (7.4.25) express the presence of edges very well. Remedies are suggested in Section 7.4.4.

Suppose the loss function is

$$l(\boldsymbol{\theta}, \hat{\boldsymbol{\theta}}) = \begin{cases} 0, & \text{if } \hat{\boldsymbol{\theta}} = \boldsymbol{\theta}, \\ 1, & \text{otherwise}. \end{cases} \tag{7.4.28}$$

Then, it is easily seen that the Bayes decision rule is to take $\hat{\boldsymbol{\theta}}$ to be a value of $\boldsymbol{\theta}$ that maximizes the posterior probability $p(\boldsymbol{\theta}|\mathbf{Z})$ given by Eq. (7.4.24); that is, $\hat{\boldsymbol{\theta}}$ is the MAP (maximum *a posteriori*) estimator.

A naive approach to finding $\hat{\boldsymbol{\theta}}$ would involve searching through the K^n possible realizations of $\boldsymbol{\theta}$ to obtain the one with highest posterior probability. However, even for $K = 2$ and n moderate, the number of possible realizations can be astronomical. Efficient optimization algorithms are needed to find the MAP estimator.

Simulated Annealing

Geman and Geman (1984) use simulated annealing to find a $\boldsymbol{\theta} \in \{1, \ldots, K\}^n$ that maximizes the posterior distribution $p(\boldsymbol{\theta}|\mathbf{Z})$. Consider the probability distribution

$$p_T(\boldsymbol{\theta}|\mathbf{Z}) \equiv (c(T))^{-1}\{f(\mathbf{Z}|\boldsymbol{\theta})\pi(\boldsymbol{\theta})\}^{1/T}, \qquad (7.4.29)$$

where $c(T)$ is the normalizing constant. The parameter $T > 0$ corresponds to the absolute temperature of the system. When $T = 1$, (7.4.29) gives (7.4.24), and if (7.4.24) defines a Markov random field, so too does (7.4.29) for all $T > 0$. As $T \to \infty$, $p_T(\cdot|\mathbf{Z})$ approaches a uniform distribution on Θ. Conversely, as $T \to 0$, $p_T(\cdot|\mathbf{Z})$ becomes concentrated on the MAP estimate $\hat{\boldsymbol{\theta}}$, and if multiple global maxima exist, $p_T(\cdot|\mathbf{Z})$ approaches a uniform distribution over these multiple $\hat{\boldsymbol{\theta}}$s.

Simulated annealing is a sequential procedure made up of a Gibbs sampler and an annealing schedule. The *Gibbs sampler* (Section 7.7.1) simulates a realization of the Markov random field with joint probability distribution $p_T(\cdot|\mathbf{Z})$ given by (7.4.29). Let $\hat{\boldsymbol{\theta}}_t \in \Theta$ denote the estimated state at time t; the initial state $\hat{\boldsymbol{\theta}}_0$ might be obtained from, say, ordinary discriminant analysis. All n sites are visited repeatedly in a fixed order $\mathbf{s}_{(1)}, \mathbf{s}_{(2)}, \ldots$ (e.g., a raster scan of sites). Let $\mathbf{s}_{(t)}$ denote the site visited at time t. Then choose $\hat{\theta}_t(\mathbf{s}_{(t)}) = k$ with probability

$$\begin{aligned} p_T\left(\hat{\theta}_t(\mathbf{s}_{(t)}) = k \middle| \left\{\hat{\theta}_{t-1}(\mathbf{u}) : \mathbf{u} \neq \mathbf{s}_{(t)}\right\}; \mathbf{Z}\right) \\ = c^{-1}\Big[g\left(\mathbf{Z}(\mathbf{s}_{(t)}) \middle| \hat{\theta}_t(\mathbf{s}_{(t)}) = k\right) \\ \cdot \pi\left(\hat{\theta}_t(\mathbf{s}_{(t)}) = k \middle| \left\{\hat{\theta}_{t-1}(\mathbf{u}) : \mathbf{u} \in N_{\mathbf{s}_{(t)}}\right\}\right)\Big]^{1/T(t)}, \end{aligned} \qquad (7.4.30)$$

where here g does not depend on the site; the normalizing constant is

$$c = \sum_{l=1}^{K}\left[g\left(\mathbf{Z}(\mathbf{s}_{(t)}) | \hat{\theta}_t(\mathbf{s}_{(t)}) = l\right) \cdot \pi\left(\hat{\theta}_t(\mathbf{s}_{(t)}) = l \middle| \left\{\hat{\theta}_{t-1}(\mathbf{u}) : \mathbf{u} \in N_{\mathbf{s}_{(t)}}\right\}\right)\right]^{1/T(t)};$$

and $N_{\mathbf{s}_{(t)}}$ and $\mathbf{Z}(\mathbf{s}_{(t)})$ are the neighbors and data, respectively, for pixel $\mathbf{s}_{(t)}$. Finally, take $\hat{\theta}_t(\mathbf{s}) = \hat{\theta}_{t-1}(\mathbf{s})$, for $\mathbf{s} \neq \mathbf{s}_{(t)}$.

The sequence $\hat{\boldsymbol{\theta}}_0, \hat{\boldsymbol{\theta}}_1, \hat{\boldsymbol{\theta}}_2, \ldots$ is an inhomogeneous Markov chain with state space $\boldsymbol{\Theta}$, whose transition probabilities are defined by (7.4.30) (Geman and Geman, 1984). Note that the posterior probability of state $\hat{\boldsymbol{\theta}}_t$ is not necessarily larger than the posterior probability of state $\hat{\boldsymbol{\theta}}_{t-1}$. Deterministic algorithms that only allow transitions to states that improve the criterion frequently converge to local maxima.

The *annealing schedule* is simply the rate at which $T(t)$ approaches zero. As $T(t)$ decreases, $\hat{\boldsymbol{\theta}}_t$ is forced to approach the MAP estimator $\hat{\boldsymbol{\theta}}$. If $T(t)$ converges to zero too rapidly, then the chain may get stuck in a local maximum. Conversely, if $T(t)$ converges to zero too slowly, then the convergence of $\hat{\boldsymbol{\theta}}_t$ to $\hat{\boldsymbol{\theta}}$ will be too slow. For an integer j such that $n \cdot (j-1) < t \leq n \cdot j$, Geman and Geman (1984) take

$$T(t) = \frac{\alpha}{\log(1+j)},$$

where α is a tuning constant (chosen to be 3.0 or 4.0). Geman and Geman show that, for this annealing schedule, $\hat{\boldsymbol{\theta}}_t \to \hat{\boldsymbol{\theta}}$ as $t \to \infty$, and for their examples they find good restoration of images upward of $300 \cdot n$ iterations.

Maximum Posterior Marginal Probability Estimator

Instead of maximizing the probability of correct classification of the entire scene, a more relevant requirement may be to maximize the expected proportion of correctly classified pixels; that is, use the loss function

$$L(\boldsymbol{\theta}, \hat{\boldsymbol{\theta}}) \equiv \sum_{\mathbf{s} \in D} l(\theta(\mathbf{s}), \hat{\theta}(\mathbf{s})),$$

where

$$l(j,k) = \begin{cases} 0, & \text{if } j = k, \\ 1, & \text{otherwise;} \end{cases}$$

this is a special case of (7.4.18). Then the Bayes decision rule is obtained, pixel by pixel, by finding $\hat{\theta}(\mathbf{s})$ that maximizes the posterior distribution of $\theta(\mathbf{s})$ given all the data $\mathbf{Z}$:

$$p(\theta(\mathbf{s}) | \mathbf{Z}) = \sum_{\boldsymbol{\theta}_{-\mathbf{s}} \in \boldsymbol{\Theta}_{-\mathbf{s}}} f(\mathbf{Z} | \boldsymbol{\theta}_{-\mathbf{s}}, \theta(\mathbf{s})) \cdot \pi(\boldsymbol{\theta}) \qquad (7.4.31)$$

(Abend, 1968), where $\boldsymbol{\theta}_{-\mathbf{s}} \equiv (\theta(\mathbf{u}): \mathbf{u} \neq \mathbf{s})'$ and $\boldsymbol{\Theta}_{-\mathbf{s}}$ is the space of all possible realizations of $\boldsymbol{\theta}_{-\mathbf{s}}$. A *maximum posterior marginal probability estimator* of $\boldsymbol{\theta}$ is obtained by finding $\hat{\theta}(\mathbf{s})$ that maximizes (7.4.31) for each $\mathbf{s} \in D$. Notice that, in general, this does *not* coincide with the sth element of the MAP estimator.

For most Markov random fields, (7.4.31) is not available in closed form, but Marroquin et al. (1987) show how an approximation to the maximum posterior marginal probability estimator can be obtained using the Gibbs sampler. Suppose a Gibbs sampler is used to simulate realizations of the posterior distribution (7.4.29) with $T(t) \equiv 1$. Then, as the number of iterations increases, the fraction of time that $\theta_t(\mathbf{s})$ spends in a state k will converge to (7.4.31), evaluated at $\theta(\mathbf{s}) = k$. Hence, the conditional probability (7.4.31) evaluated at $\theta(\mathbf{s}) = k$ can be approximated by

$$\sum_{t=t_0}^{\nu} I(\theta_t(\mathbf{s}) = k)/(\nu - t_0 + 1),$$

where t_0 is the time required for the system to reach equilibrium. When this latter expression is maximized over $k = 1, \ldots, K$, an approximate maximum posterior marginal probability estimator is obtained. [In principle, the same type of approximation can be used for any other property of the posterior distribution $\{p(\boldsymbol{\theta}|\mathbf{Z}) \colon \boldsymbol{\theta} \in \boldsymbol{\Theta}\}$.]

Iterated Conditional Modes

In an attempt to approximate the MAP estimator, Besag (1986) introduced the iterated conditional modes algorithm for estimating $\boldsymbol{\theta}$. Instead of a random choice of pixel value in the simulated-annealing equation (7.4.30), the mode is automatically selected. Such a choice guarantees that the posterior distribution never decreases; however, convergence may be to a *local* maximum. The following algorithm departs slightly from the simulated-annealing paradigm by using a synchronous approximation; advantages seem to depend on computing environment (Besag, 1986).

Assume that, given $\boldsymbol{\theta}$, $\mathbf{Z}(\mathbf{s}_1), \ldots, \mathbf{Z}(\mathbf{s}_n)$ are conditionally independent and that each $\mathbf{Z}(\mathbf{s}_i)$ has the same conditional density function $g(\mathbf{Z}(\mathbf{s}_i)|\theta(\mathbf{s}_i))$ that depends only on $\theta(\mathbf{s}_i)$. Let $\hat{\theta}_t(\mathbf{s})$ denote the classification at pixel location $\mathbf{s}$ for the tth iterate. As usual, the initial classification $\hat{\theta}_0(\mathbf{s})$ might be obtained from ordinary discriminant analysis. Then, for each pixel $\mathbf{s}$ at time t, choose a state $\hat{\theta}_t(\mathbf{s}) = k^*$, where k^* maximizes the probability

$$\begin{aligned} &\Pr\left(\hat{\theta}_t(\mathbf{s}) = k \middle| \{\hat{\theta}_{t-1}(\mathbf{u}) \colon \mathbf{u} \neq \mathbf{s}\}; \mathbf{Z}\right) \\ &\qquad = c^{-1} g\left(\mathbf{Z}(\mathbf{s}) | \hat{\theta}_t(\mathbf{s}) = k\right) \cdot \pi\left(\hat{\theta}_t(\mathbf{s}) = k \middle| \{\hat{\theta}_{t-1}(\mathbf{u}) \colon \mathbf{u} \in N_{\mathbf{s}}\}\right), \quad (7.4.32) \end{aligned}$$

and c is the appropriate normalizing constant. Notice that this corresponds to an annealing schedule where $T(t) \equiv 0$. The procedure is continued for a fixed number of cycles or until convergence, to produce an estimate of $\boldsymbol{\theta}$.

Exact Posterior Estimation of Binary Images

Greig et al. (1986, 1989) extend the Ford–Fulkerson algorithm (Ford and Fulkerson, 1962) to obtain exact MAP estimates for binary images $\theta(\mathbf{s}) \in \{0, 1\}$, $\mathbf{s} \in D$. They go on to make comparisons with approximate MAP estimates obtained by simulated annealing.

Assume that the data $\{\mathbf{Z}(\mathbf{s}): \mathbf{s} \in D\}$ (written in vector form as $\mathbf{Z}$) are conditionally independent given $\{\theta(\mathbf{s}): \mathbf{s} \in D\}$ (written in vector form as $\boldsymbol{\theta}$) with the same conditional density g, so that

$$
\begin{aligned}
f(\mathbf{Z}|\boldsymbol{\theta}) &= \prod_{i=1}^{n} g(\mathbf{Z}(\mathbf{s}_i)|\theta(\mathbf{s}_i)) \\
&= \prod_{i=1}^{n} g(\mathbf{Z}(\mathbf{s}_i)|1)^{\theta(\mathbf{s}_i)} g(\mathbf{Z}(\mathbf{s}_i)|0)^{1-\theta(\mathbf{s}_i)}. \qquad (7.4.33)
\end{aligned}
$$

The prior distribution π is modeled as a binary Markov random field with pairwise-only interactions between sites and takes the form

$$
\pi(\boldsymbol{\theta}) = c^{-1} \exp\left[(1/2) \sum_{i=1}^{n} \sum_{j=1}^{n} \beta_{ij} \{\theta(\mathbf{s}_i)\theta(\mathbf{s}_j) + (1 - \theta(\mathbf{s}_i))(1 - \theta(\mathbf{s}_j))\} \right], \qquad (7.4.34)
$$

where $\beta_{ii} = 0$, $\beta_{ij} = \beta_{ji} \geq 0$, and, if $\beta_{ij} > 0$, then $\mathbf{s}_i$ and $\mathbf{s}_j$ are neighbors. Hence, the negative log posterior distribution is

$$
\begin{aligned}
L(\boldsymbol{\theta}|\mathbf{Z}) = a - \sum_{i=1}^{n} \lambda_i \theta(\mathbf{s}_i) - (1/2) \sum_{i=1}^{n} \sum_{j=1}^{n} \beta_{ij} \{\theta(\mathbf{s}_i)\theta(\mathbf{s}_j) \\
+ (1 - \theta(\mathbf{s}_i))(1 - \theta(\mathbf{s}_j))\}, \qquad (7.4.35)
\end{aligned}
$$

where a is an additive constant that does not depend on $\boldsymbol{\theta}$ (although it may depend on $\mathbf{Z}$) and $\lambda_i \equiv \log\{g(\mathbf{Z}(\mathbf{s}_i)|1)/g(\mathbf{Z}(\mathbf{s}_i)|0)\}$. The MAP estimate is the $\hat{\boldsymbol{\theta}}$ that minimizes (7.4.35).

To see the link with the Ford–Fulkerson algorithm, consider a capacitated network consisting of the n pixels $\{\mathbf{s}_i: i = 1, \ldots, n\}$, a source $\mathbf{u}_0$, and a sink $\mathbf{u}_{n+1}$. Define a directed edge from $\mathbf{u}_0$ to pixel $\mathbf{s}_i$ with capacity $c_{0i} = \max\{0, \lambda_i\}$, a directed edge from pixel $\mathbf{s}_i$ to sink $\mathbf{u}_{n+1}$ with capacity $c_{i,n+1} = \max\{0, -\lambda_i\}$, and an undirected edge between pixels $\mathbf{s}_i$ and $\mathbf{s}_j$ with capacity $c_{ij} = \beta_{ij}$. Let $B \equiv \{\mathbf{u}_0\} \cup \{\mathbf{s}_i: \theta(\mathbf{s}_i) = 1; i = 1, \ldots, n\}$ and $W \equiv \{\mathbf{u}_{n+1}\} \cup \{\mathbf{s}_i: \theta(\mathbf{s}_i) = 0; i = 1, \ldots, n\}$ define a partition of the network vertices. The set of edges with one vertex in B and the other vertex in W is called a *cut* and is a function of $\boldsymbol{\theta}$.

The capacity of the cut is

$$\sum_{k \in B} \sum_{l \in W} c_{kl} = \sum_{i=1}^{n} \theta(\mathbf{s}_i)\max\{0, -\lambda_i\} + \sum_{i=1}^{n} (1 - \theta(\mathbf{s}_i))\max\{0, \lambda_i\}$$
$$+ (1/2) \sum_{i=1}^{n} \sum_{j=1}^{n} \beta_{ij}(\theta(\mathbf{s}_i) - \theta(\mathbf{s}_j))^2, \qquad (7.4.36)$$

which differs from $L(\boldsymbol{\theta}|\mathbf{Z})$ [Eq. (7.4.35)] by a constant that does not depend on $\boldsymbol{\theta}$. So, finding a $\hat{\boldsymbol{\theta}}$ that maximizes the posterior distribution is equivalent to finding a $\hat{\boldsymbol{\theta}}$ that minimizes the capacity (7.4.36). Ford and Fulkerson (1962) give an efficient algorithm for solving this problem.

Restoration of a Sequence of Images

Green and Titterington (1987) consider the restoration of a sequence of images $\{\boldsymbol{\theta}_t\colon t = 0, 1, \ldots, T\}$ observed at fixed points of time. Let $\mathbf{Z}_t \equiv (Z_t(\mathbf{s}_1), \ldots, Z_t(\mathbf{s}_n))'$ denote the data recorded at time t. Assume that $\{\boldsymbol{\theta}_t\colon t = 0, 1, \ldots, T\}$ is a Markov chain and that the distribution of $\mathbf{Z}_t$ depends only on $\boldsymbol{\theta}_t$. Then

$$f(\mathbf{Z}_t|\boldsymbol{\theta}_0, \boldsymbol{\theta}_1, \ldots, \boldsymbol{\theta}_t; \mathbf{Z}_0, \mathbf{Z}_1, \ldots, \mathbf{Z}_{t-1}) = f(\mathbf{Z}_t|\boldsymbol{\theta}_t) \qquad (7.4.37)$$

and

$$\Pr(\boldsymbol{\theta}_t = \boldsymbol{\theta}|\boldsymbol{\theta}_0, \boldsymbol{\theta}_1, \ldots, \boldsymbol{\theta}_{t-1}) = \Pr(\boldsymbol{\theta}_t = \boldsymbol{\theta}|\boldsymbol{\theta}_{t-1}), \qquad \boldsymbol{\theta} \in \{1, \ldots, K\}^n. \qquad (7.4.38)$$

Assume that this distribution of $\boldsymbol{\theta}_t$ conditional on $\boldsymbol{\theta}_{t-1}$ takes the form

$$\Pr(\boldsymbol{\theta}_t|\boldsymbol{\theta}_{t-1}) \equiv \pi(\boldsymbol{\theta}_t|\boldsymbol{\theta}_{t-1}) = c(\boldsymbol{\theta}_{t-1})^{-1} \cdot \exp\left\{\sum_{\kappa \in \mathscr{C}} V_\kappa(\boldsymbol{\theta}_{\kappa,t}; \boldsymbol{\theta}_{t-1})\right\}, \qquad (7.4.39)$$

where the summation is over the set of all cliques $\mathscr{C}$. From (7.4.37) and (7.4.38), the conditional distribution of $\boldsymbol{\theta}_t$ given $\{\boldsymbol{\theta}_i\colon i < t\}$ and $\{\mathbf{Z}_i\colon i \le t\}$ is

$$p(\boldsymbol{\theta}_t|\boldsymbol{\theta}_0, \ldots, \boldsymbol{\theta}_{t-1}; \mathbf{Z}_0, \ldots, \mathbf{Z}_t) \propto f(\mathbf{Z}_t|\boldsymbol{\theta}_t)\pi(\boldsymbol{\theta}_t|\boldsymbol{\theta}_{t-1}). \qquad (7.4.40)$$

Green and Titterington (1987) suggest a recursive algorithm for restoring the sequence of images $\{\boldsymbol{\theta}_t\colon t = 0, 1, \ldots, T\}$. Let $\hat{\boldsymbol{\theta}}_t$ denote the restoration of the tth image. Then the $(t + 1)$th image is restored by finding $\hat{\boldsymbol{\theta}}_{t+1}$ that maximizes $p(\boldsymbol{\theta}_{t+1}|\hat{\boldsymbol{\theta}}_0, \ldots, \hat{\boldsymbol{\theta}}_t; \mathbf{Z}_0, \ldots, \mathbf{Z}_{t+1})$, which is simplified using (7.4.40). For each t, this maximization problem can be solved using simulated annealing, or any of the other algorithms previously described. Because the restoration $\hat{\boldsymbol{\theta}}_t$ is "smoother" than the true image $\boldsymbol{\theta}_t$, Green and Titterington's restoration of $\boldsymbol{\theta}_{t+1}$ might be improved by using an estimator of the tth image whose ensemble properties match those of $\boldsymbol{\theta}_t$.

Parameter Estimation

The implementation of Markov random-field methods to image restoration requires estimation of the parameters of the prior probability distribution of $\boldsymbol{\theta}$,

$$\pi_{\boldsymbol{\eta}}(\boldsymbol{\theta}) = c(\boldsymbol{\eta})^{-1} \exp\left\{ \sum_{\kappa \in \mathscr{C}} V_\kappa(\boldsymbol{\theta}_\kappa; \boldsymbol{\eta}) \right\} \tag{7.4.41}$$

[see Eq. (7.4.23)], and of the conditional density $f_{\boldsymbol{\xi}}(\cdot \mid \boldsymbol{\theta})$ of $\mathbf{Z}$ given $\boldsymbol{\theta}$. Suppose there are training pixels D_0 for which data $\{Z(\mathbf{s})\colon \mathbf{s} \in D_0\}$ are available and for which the respective true values $\{\theta(\mathbf{s})\colon \mathbf{s} \in D_0\}$ are known. Assuming conditional independence of the data vector $\mathbf{Z}_0$ (whose elements are $\{\mathbf{Z}(\mathbf{s})\colon \mathbf{s} \in D_0\}$) given $\boldsymbol{\theta}_0$ (whose elements are $\{\theta(\mathbf{s})\colon \mathbf{s} \in D_0\}$), the likelihood of $\boldsymbol{\xi}$ is given by

$$l(\boldsymbol{\xi}) \equiv f_{\boldsymbol{\xi}}(\mathbf{Z}_0 \mid \boldsymbol{\theta}_0) = \prod_{\mathbf{s} \in D_0} g_{\boldsymbol{\xi}}(Z(\mathbf{s}) \mid \theta(\mathbf{s}))$$

[see Eq. (7.4.26)]. This factorization ensures a straightforward calculation of maximum likelihood estimates of the parameter $\boldsymbol{\xi}$ from the training data.

Maximum likelihood estimation of $\boldsymbol{\eta}$ from observed θs on training pixels is usually computationally intensive because the normalizing constant in (7.4.41) depends on $\boldsymbol{\eta}$ and cannot be easily evaluated. Methods of estimating $\boldsymbol{\eta}$ for Markov random fields on lattices are considered in Section 7.2. A number of approaches to parameter estimation have been considered in image analysis. Hassner and Sklansky (1980) and Cross and Jain (1983) use the coding method of Besag (1974), where $\hat{\boldsymbol{\eta}}$ is chosen to maximize the conditional likelihood

$$\prod_{i \in M} \pi\left(\theta(\mathbf{s}_i) \mid \{\theta(\mathbf{u})\colon \mathbf{u} \in N_{\mathbf{s}_i}\}; \boldsymbol{\eta}\right); \tag{7.4.42}$$

the product is over a maximal set M of pixels of D_0 such that no two pixels are neighbors. Depending on the choice of M, different estimates are obtained and there is no known optimal method for combining these different estimates into a single estimate for $\boldsymbol{\eta}$. Further discussion is given in Sections 7.2.1 and 7.3.1.

Besag (1986) uses a maximum pseudolikelihood estimator, where $\hat{\boldsymbol{\eta}}$ is chosen to maximize

$$\prod_{i \in D_0} \pi\left(\theta(\mathbf{s}_i) \mid \{\theta(\mathbf{u})\colon \mathbf{u} \in N_{\mathbf{s}_i}\}; \boldsymbol{\eta}\right) \tag{7.4.43}$$

(cf. Besag, 1975). Discussion of computational aspects and asymptotic properties of maximum pseudolikelihood estimation can be found in Sections 7.2.1 and 7.3.1.

Derin and Elliott (1987) propose a parameter-estimation scheme consisting of histograms and least-squares estimation of the parameters of a Markov random field. Suppose $\boldsymbol{\theta}$ is a Markov random field with finite range space $\{1, \ldots, K\}$ and for which a site $\mathbf{s} = (u, v)'$ has neighbors, $\mathbf{u}_1 = (u+1, v)'$, $\mathbf{u}_2 = (u, v+1)'$, $\mathbf{u}_3 = (u-1, v)'$, and $\mathbf{u}_4 = (u, v-1)'$. Assume

$$\pi(\boldsymbol{\theta}) \propto \exp\left\{ \sum_{i=1}^{n} \theta(\mathbf{s}_i) G_i(\theta(\mathbf{s}_i)) + \sum_{1 \le i < j \le n} \theta(\mathbf{s}_i)\theta(\mathbf{s}_j) G_{ij}\big(\theta(\mathbf{s}_i), \theta(s_j)\big) \right\},$$

where $\theta(\mathbf{s}_i)G_i(\theta(\mathbf{s}_i)) = \alpha_k$ if $\theta(\mathbf{s}_i) = k$,

$$\theta(\mathbf{s}_i)\theta(\mathbf{s}_j) G_{ij}\big(\theta(\mathbf{s}_i), \theta(\mathbf{s}_j)\big)$$
$$= (-1)^{I(\theta(\mathbf{s}_i) = \theta(\mathbf{s}_j))} \begin{cases} \beta_1, & \text{if } \mathbf{s}_j \in \{\mathbf{s}_i + (1,0)', \mathbf{s}_i + (-1,0)'\}, \\ \beta_2, & \text{if } \mathbf{s}_j \in \{\mathbf{s}_i + (0,1)', \mathbf{s}_i + (0,-1)'\}, \\ 0, & \text{otherwise}, \end{cases}$$

and $I(A)$ is the indicator function of the event A. Let $\boldsymbol{\eta} \equiv (\alpha_1, \alpha_2, \ldots, \alpha_K, \beta_1, \beta_2)'$ denote the vector of parameters and define

$$V\big(\theta(\mathbf{s}_0), \{\theta(\mathbf{u}) : \mathbf{u} \in N_{\mathbf{s}_0}\}; \boldsymbol{\eta}\big)$$
$$\equiv \theta(\mathbf{s}_0) G_0(\theta(\mathbf{s}_0)) + \sum_{\mathbf{s}_i \in N_{\mathbf{s}_0}} \theta(\mathbf{s}_0)\theta(\mathbf{s}_i) G_{0i}(\theta(\mathbf{s}_0), \theta(\mathbf{s}_i)). \quad (7.4.44)$$

For neighboring pixels $\mathbf{s}$ and $\mathbf{u}$, let

$$I(\mathbf{s}, \mathbf{u}) = \begin{cases} -1, & \text{if } \theta(\mathbf{s}) = \theta(\mathbf{u}), \\ 1, & \text{otherwise} \end{cases}$$

and

$$J_k(\mathbf{s}) = \begin{cases} 1, & \text{if } \theta(\mathbf{s}) = k, \\ 0, & \text{otherwise}. \end{cases}$$

Then (7.4.44) can be rewritten as

$$V\big(\theta(\mathbf{s}), \{\theta(\mathbf{u}) : \mathbf{u} \in N_{\mathbf{s}}\}; \boldsymbol{\eta}\big) = \boldsymbol{\eta}'\boldsymbol{\phi}\big(\theta(\mathbf{s}), \{\theta(\mathbf{u}) : \mathbf{u} \in N_{\mathbf{s}}\}\big),$$

where

$$\boldsymbol{\phi}(\theta(\mathbf{s}),\{\theta(\mathbf{u}):\mathbf{u}\in N_{\mathbf{s}}\})$$
$$=\left[J_1(\mathbf{s}),\ldots,J_K(\mathbf{s}),(I(\mathbf{s},\mathbf{u}_1)+I(\mathbf{s},\mathbf{u}_3)),(I(\mathbf{s},\mathbf{u}_2)+I(\mathbf{s},\mathbf{u}_4))\right]'.$$

Derin and Elliott (1987) show that

$$\boldsymbol{\eta}'\left[\boldsymbol{\phi}(k,\{\theta(\mathbf{u}):\mathbf{u}\in N_{\mathbf{s}}\})-\boldsymbol{\phi}(j,\{\theta(\mathbf{u}):\mathbf{u}\in N_{\mathbf{s}}\})\right]$$
$$=\log\left[\pi(\theta(\mathbf{s})=j,\{\theta(\mathbf{u}):\mathbf{u}\in N_{\mathbf{s}}\})/\pi(\theta(\mathbf{s})=k,\{\theta(\mathbf{u}):\mathbf{u}\in N_{\mathbf{s}}\})\right]; \tag{7.4.45}$$

compare Eq. (7.2.11). Assuming the right-hand side of (7.4.45) is determined, or estimated, (7.4.45) reduces to a linear system of equations for each pair of $j,k\in\{1,2,\ldots,K\}$. Using training data, Derin and Elliott estimate $\pi(\theta(\mathbf{s}),\{\theta(\mathbf{u}):\ \mathbf{u}\in N_{\mathbf{s}}\})$ from histograms. When these estimates are substituted into (7.4.45), a least-squares estimate $\hat{\boldsymbol{\eta}}$ can be obtained.

Younes (1988) considers a stochastic gradient algorithm for estimating $\boldsymbol{\eta}$. Suppose the prior probability distribution of $\boldsymbol{\theta}$ can be written in the form

$$\pi_{\boldsymbol{\eta}}(\boldsymbol{\theta})=c(\boldsymbol{\eta})^{-1}\exp\left\{-\sum_{i=1}^{p}\eta_iT_i(\boldsymbol{\theta})\right\}, \tag{7.4.46}$$

where $c(\boldsymbol{\eta})$ is a normalizing constant, $\boldsymbol{\eta}\equiv(\eta_1,\ldots,\eta_p)'$ is the vector of prior parameters, and $\mathbf{T}(\boldsymbol{\theta})\equiv(T_1(\boldsymbol{\theta}),\ldots,T_p(\boldsymbol{\theta}))'$ is a known sufficient statistic. Consider, for example, the model (Strauss, 1977)

$$\pi_{\boldsymbol{\eta}}(\boldsymbol{\theta})=c(\boldsymbol{\eta})^{-1}\exp\left\{\sum_{k=1}^{K}\alpha_kn_k-\sum\sum_{1\le k<l\le K}\beta_{kl}n_{kl}\right\},$$

where n_k is the number of pixels colored k and n_{kl} is the number of neighboring pairs of sites colored (k,l). Here, $\boldsymbol{\eta}$ consists of the parameters $\{\alpha_k\}$ and $\{\beta_{kl}\}$, and the sufficient statistics are the corresponding n_ks and n_{kl}s. Let $\{\theta(\mathbf{s}):\ \mathbf{s}\in D_0\}$ be a realization of (7.4.46) on the training pixels D_0. The maximum likelihood estimator of $\boldsymbol{\eta}$ is obtained by finding $\hat{\boldsymbol{\eta}}$ that maximizes (7.4.46). The negative loglikelihood $L(\boldsymbol{\eta})$ $[=-\log\pi_{\boldsymbol{\eta}}(\boldsymbol{\theta})]$ is convex in $\boldsymbol{\eta}$ and its differential is

$$\partial L(\boldsymbol{\eta})/\partial\boldsymbol{\eta}\equiv\mathbf{h}(\boldsymbol{\eta})=\mathbf{T}(\boldsymbol{\theta})-E_{\boldsymbol{\eta}}(\mathbf{T}(\boldsymbol{\theta})).$$

Let $\hat{\boldsymbol{\theta}}_t$ be an estimated state at time t and let $\hat{\boldsymbol{\eta}}_t$ be an estimate of $\boldsymbol{\eta}$ at time t; for the initial state, $\hat{\boldsymbol{\theta}}_0\equiv\boldsymbol{\theta}$ and $\hat{\boldsymbol{\eta}}_0$ is an initial guess for $\boldsymbol{\eta}$. The stochastic gradient algorithm is defined by

$$\hat{\boldsymbol{\eta}}_{t+1} = \hat{\boldsymbol{\eta}}_t + \frac{\mathbf{T}(\hat{\boldsymbol{\theta}}_{t+1}) - \mathbf{T}(\hat{\boldsymbol{\theta}}_0)}{a \cdot (t+1)},$$

where $a > 0$ is a constant and $\hat{\boldsymbol{\theta}}_{t+1}$ is obtained from $\hat{\boldsymbol{\theta}}_t$ using the Gibbs sampler (described earlier in this section and in Section 7.7.1). In the Gibbs sampler, use $\boldsymbol{\eta} = \hat{\boldsymbol{\eta}}_t$ in the conditional probability that chooses the tth site's value randomly from $\{1, \ldots, K\}$. Younes (1988) shows that, as $t \to \infty$, $\hat{\boldsymbol{\eta}}_t$ converges almost surely to the solution of $\mathbf{h}(\boldsymbol{\eta}) = \mathbf{0}$.

Training Pixels Not Available

When $\boldsymbol{\theta}$ is completely unknown, Besag (1986) suggests an iterative procedure for simultaneous parameter estimation and image restoration, based on his iterated conditional modes algorithm for image restoration, and the maximum (pseudo) likelihood method (Section 7.2.1) for parameter estimation. Following stage t of the iterated conditional modes algorithm, maximum likelihood and maximum pseudolikelihood is used to obtain $\boldsymbol{\xi}_t$ and $\boldsymbol{\eta}_t$, estimates of the parameters $\boldsymbol{\xi}$ and $\boldsymbol{\eta}$ that appear in $f_{\boldsymbol{\xi}}(\mathbf{Z}|\hat{\boldsymbol{\theta}}_t)$ and $\pi_{\boldsymbol{\eta}}(\hat{\boldsymbol{\theta}}_t)$, respectively. Then $\hat{\boldsymbol{\theta}}_t$ is updated using the iterated conditional modes algorithm, using $\boldsymbol{\xi}_t$ and $\boldsymbol{\eta}_t$ as estimates of $\boldsymbol{\xi}$ and $\boldsymbol{\eta}$.

Qian and Titterington (1989) suggest using the EM algorithm (Dempster et al., 1977; Section 7.5.2) for finding m.l. estimates of $\boldsymbol{\xi}$ and $\boldsymbol{\eta}$ that appear in $f_{\boldsymbol{\xi}}(\mathbf{Z}|\boldsymbol{\theta})$ and $\pi_{\boldsymbol{\eta}}(\boldsymbol{\theta})$. Suppose that

$$\pi_{\boldsymbol{\eta}}(\boldsymbol{\theta}) = B_1(\boldsymbol{\theta})\exp\{\boldsymbol{\mu}(\boldsymbol{\eta})'\mathbf{T}_1(\boldsymbol{\theta})\}/A_1(\boldsymbol{\eta})$$

and

$$f_{\boldsymbol{\xi}}(\mathbf{Z}|\boldsymbol{\theta}) = B_2(\boldsymbol{\theta}, \mathbf{Z})\exp\{\boldsymbol{\nu}(\boldsymbol{\xi})'\mathbf{T}_2(\boldsymbol{\theta}, \mathbf{Z})\}/A_2(\boldsymbol{\xi}),$$

where $\boldsymbol{\mu}(\boldsymbol{\eta})$, $\mathbf{T}_1(\boldsymbol{\theta})$ and $\boldsymbol{\nu}(\boldsymbol{\xi})$, $\mathbf{T}_2(\boldsymbol{\theta}, \mathbf{Z})$ are vectors with compatible dimensions. (The assumed exponential families make the whole procedure computationally feasible.) Let $\boldsymbol{\xi}_t$ and $\boldsymbol{\eta}_t$ be current estimates of $\boldsymbol{\xi}$ and $\boldsymbol{\eta}$, respectively. Then the E step of the EM algorithm consists of computing

$$E\{\mathbf{T}_1(\boldsymbol{\theta})|\mathbf{Z}, \boldsymbol{\xi}_t, \boldsymbol{\eta}_t\} \equiv \mathbf{T}_1^{(t)}$$

and

$$E\{\mathbf{T}_2(\boldsymbol{\theta}, \mathbf{Z})|\mathbf{Z}, \boldsymbol{\xi}_t, \boldsymbol{\eta}_t\} \equiv \mathbf{T}_2^{(t)},$$

where the expectations are taken with respect to the conditional distribution of $\boldsymbol{\theta}$ given $\mathbf{Z}$, at parameter values $\boldsymbol{\xi} = \boldsymbol{\xi}_t$ and $\boldsymbol{\eta} = \boldsymbol{\eta}_t$. The M step is to find

$\boldsymbol{\eta}_{t+1}$ that maximizes

$$\boldsymbol{\mu}(\boldsymbol{\eta})'\mathbf{T}_1^{(t)} - \log A_1(\boldsymbol{\eta})$$

and to find $\boldsymbol{\xi}_{t+1}$ that maximizes

$$\boldsymbol{\nu}(\boldsymbol{\xi})'\mathbf{T}_2^{(t)} - \log A_2(\boldsymbol{\xi}).$$

Qian and Titterington (1989) show how this algorithm can be implemented when $\pi_{\boldsymbol{\eta}}(\boldsymbol{\theta})$ is a CG process (Section 6.3) and $f_{\boldsymbol{\xi}}(\mathbf{Z}|\boldsymbol{\theta})$ is Gaussian.

Finally, a fully Bayesian approach could be taken by specifying priors and hyperpriors for all unknown parameters. Inference is then based on the joint posterior distribution of $\boldsymbol{\theta}$, $\boldsymbol{\eta}$, and $\boldsymbol{\xi}$, conditional on $\mathbf{Z}$ (e.g., Besag, 1989b).

7.4.4* Edge Processes

In realizations of a Markov random field for the gray level of an image, adjacent pixels tend to have similar gray levels; abrupt changes in level rarely occur. By augmenting such random processes with an edge process, an image can be realized in which adjacent pixels can have sharply different gray levels. This section introduces a Bayesian approach to modeling edge processes. No discussion is given here of edge *detection*, in which the aim is to draw contours (edges) between regions of a realized image with very different gray levels (e.g., Rosenfeld and Kak, 1976; Marr and Hildreth, 1980; Huertas and Medioni, 1986).

Consider the boundaries (or edges) separating the different segments of the image. Establish new sites located midway between each vertical or horizontal pair of pixels as possible locations of edge elements. Thus, the original lattice D of pixel locations is augmented with a lattice $D^{(e)}$ consisting of sites for potential edge elements, as shown in Figure 7.3. Let $L(\mathbf{s}) = 1$ if there is an edge element at $\mathbf{s} \in D^{(e)}$, and $L(\mathbf{s}) = 0$ otherwise. An edge process is a stochastic model for L that locates edges on the lattice $D^{(e)}$.

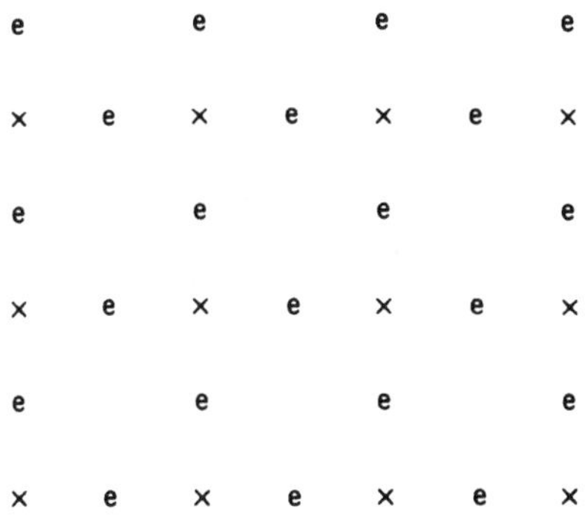

```
e       e       e       e

×   e   ×   e   ×   e   ×

e       e       e       e

×   e   ×   e   ×   e   ×

e       e       e       e

×   e   ×   e   ×   e   ×
```

Figure 7.3 Square lattice showing locations of pixels (×) and locations of edge sites (e).

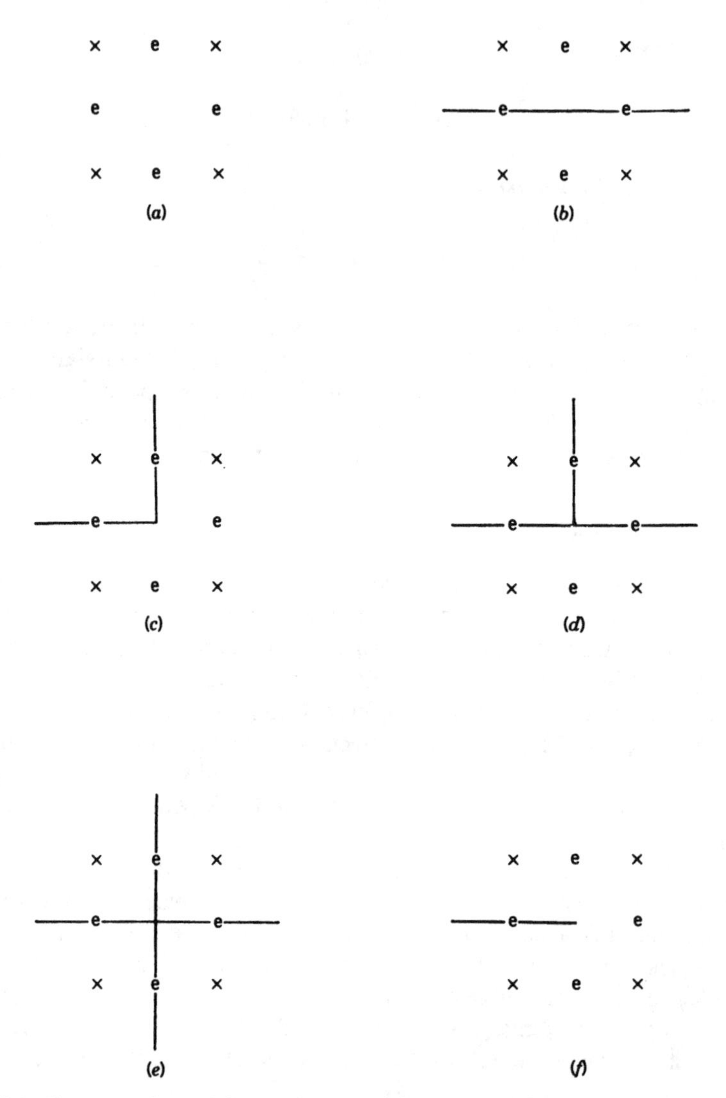

Figure 7.4 Shown are six possible edge patterns. (*a*) No edges. (*b*) Straight line. (*c*) Corner. (*d*) T pattern. (*e*) X pattern. (*f*) Ending.

Edge elements are likely to be located adjacent to one another and adjacent edge elements are more likely to form a straight line than a corner, a T pattern, an X pattern, or an ending; these five possible edge configurations are illustrated in Figure 7.4. Adjacent pixels separated by edge elements are likely to have quite different colors or gray levels. Such spatial dependencies among adjacent pixels and edge sites make the Markov random field a natural model for the joint distribution of edge elements and pixel colors.

(Notice that, in this section, an edge site refers to a site of the lattice $D^{(e)}$ and *not* to a pixel at the edge of the lattice D.)

Suppose that the prior distribution of $\{\theta(\mathbf{s}): \mathbf{s} \in D\}$, $\{L(\mathbf{u}): \mathbf{u} \in D^{(e)}\}$ (written in vector form as $\boldsymbol{\theta}, \mathbf{L}$) is a Markov random field with probability distribution π and that the conditional density of $\mathbf{Z}$ given $\boldsymbol{\theta}$ and $\mathbf{L}$ is $f(\mathbf{Z}|\boldsymbol{\theta})$. Then, a MAP estimator for $(\boldsymbol{\theta}, \mathbf{L})$ is obtained by finding $(\hat{\boldsymbol{\theta}}, \hat{\mathbf{L}})$ that maximizes the posterior distribution

$$p(\boldsymbol{\theta}, \mathbf{L}|\mathbf{Z}) \propto f(\mathbf{Z}|\boldsymbol{\theta}) \cdot \pi(\boldsymbol{\theta}, \mathbf{L}). \qquad (7.4.47)$$

Geman and Geman (1984) first construct a prior distribution for the edge process $\mathbf{L}$ and then model the prior for the pixels conditional on the edge process; that is, the joint prior distribution of $\boldsymbol{\theta}$ and $\mathbf{L}$ takes the form

$$\pi(\boldsymbol{\theta}, \mathbf{L}) = c^{-1} \exp\{Q_{\circ}(\boldsymbol{\theta}; \mathbf{L}) + Q_{\cdot}(\mathbf{L})\}, \qquad (7.4.48)$$

where examples of $Q_{\circ}$ and $Q_{\cdot}$ are given in the following text. Two pixels separated by an edge site $\mathbf{s}^{(e)}$, for which $L(\mathbf{s}^{(e)}) = 1$ (i.e., an edge element is present at the site), are not considered to be neighbors, so they can have quite different colors or gray levels with high probability, according to the prior distribution (7.4.48) (Geman and Geman, 1984). This does not happen for prior models having no edge component.

Geman and Geman (1986) consider the following neighborhood system for a pixel and for an edge site. Each pixel has eight pixel neighbors and four edge neighbors (see Fig. 7.5*a*), and each edge site has six edge neighbors and two pixel neighbors (see Fig. 7.5*b*).

Under the neighborhood structure given by Figure 7.5, Geman and Geman (1986) specify the prior distribution (7.4.48) as follows. Assume

$$Q_{\cdot}(\mathbf{L}) = -\beta_1 \sum_{\kappa} \psi_{\kappa}(\mathbf{L}_{\kappa}), \qquad \beta_1 > 0, \qquad (7.4.49)$$

where $\mathbf{L}_{\kappa} \equiv (L(\mathbf{s}^{(e)}): \mathbf{s}^{(e)} \in \kappa)'$ and the summation is over all cliques κ containing four neighboring edge sites (the largest cliques formed by the neighborhoods in Figure 7.5*b*). The six possible clique states are given in Figures 7.4*a* through 7.4*f*; call them type 1 through type 6, respectively. If clique κ is of type i, let $\psi_{\kappa}(\mathbf{L}_{\kappa}) = \xi_i$, $i = 1, 2, \ldots, 6$. Assuming that most pixels are not next to edges, that edges tend to be straight and continuous, and that edges are not congested, choose $\xi_1 \leq \xi_2 \leq \cdots \leq \xi_6$. Silverman et al. (1990) note that boundaries between regions of an image need not all run vertically or horizontally. They consider a scheme for assigning values to $\{\xi_i\}$, based on desired cost per unit length of boundaries, cost per region of the pattern (a region is a contiguous area of homogeneous color or gray level), and cost per ending.

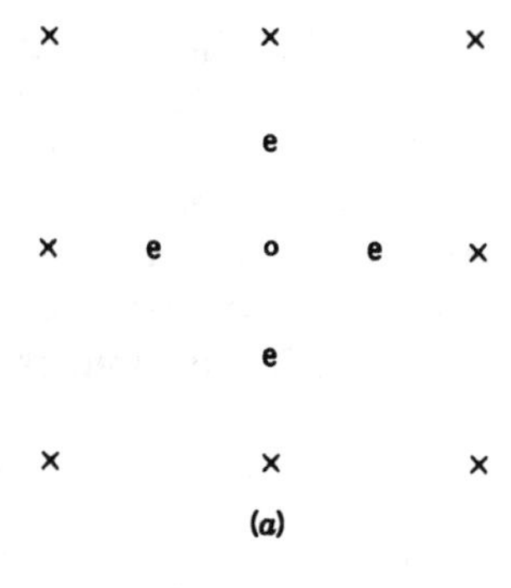

(b)

Figure 7.5 (a) Central pixel ($\circ$) and its neighbors, made up of the eight nearest pixels ($\times$) and the four nearest edges (e). (b) Central edge site ($\cdot$) and its neighbors, made up of the two nearest pixels ($\times$) and the six nearest edges (e).

Let L_{ij} denote the edge value (0 or 1) at the edge site between adjacent pixels $\mathbf{s}_i$ and $\mathbf{s}_j$. Geman and Geman (1986) give an example of $Q_\circ$ in (7.4.48), namely,

$$Q_\circ(\boldsymbol{\theta};\mathbf{L}) = \sum\{\beta_2\phi(\theta(\mathbf{s}_i) - \theta(\mathbf{s}_j)) - \beta_3\}\{1 - L_{ij}\}, \quad (7.4.50)$$

where the summation is over adjacent pixels, $\beta_2 > \beta_3$ are positive parameters, and

$$\phi(u) = 1/\{1 + |u/\alpha_1|^{\alpha_2}\};$$

usually $\alpha_2 = 1.5$ or 2.0 and α_1 depends on the range of the image $\sup_{\mathbf{s}\in D}\theta(\mathbf{s}) - \inf_{\mathbf{s}\in D}\theta(\mathbf{s})$.

Another example of $Q_\circ$ (Geman et al., 1990) is

$$Q_\circ(\boldsymbol{\theta};\mathbf{L}) = -\beta_2\sum\phi(\theta(\mathbf{s}_i) - \theta(\mathbf{s}_j))(1 - L_{ij}),$$

where $\beta_2 > 0$, $\phi(0) = -1$, and $\phi(\cdot)$ is even and nondecreasing on $[0,\infty)$; for example, $\phi(u) = 1 - 2\{1 + (u/\alpha_1)^2\}^{-1}$. Once the posterior distribution (7.4.47) has been specified, the MAP estimate of $(\boldsymbol{\theta},\mathbf{L})$ can be found by simulated annealing as described in Section 7.4.3.

The elegance of these models is very appealing, but of some concern to me is the way nonstationarities in the image are imposed through the edge process. Equations (7.4.49) and (7.4.50) assume a homogeneous process, when prior knowledge of certain images should suggest strong inhomogeneities (e.g., remote sensing of a large city). Clifford and Middleton (1989) use a polygonal Markov random field (suggested by the work of Arak and Surgailis, 1989) as a prior for a process that is expected to vary smoothly over most of D but may exhibit jump discontinuities over line segments.

7.4.5* Textured Images

The intensity of reflected radiation from a homogeneous region may exhibit spatial correlation. This might result from either the texture of the underlying region or from spatially correlated noise. The goal here is to model the texture of a homogeneous region and to segment the domain D into regions having the same texture types. The segmentation/classification of textured images has been considered by Deguchi (1986), Cohen and Cooper (1987), Derin and Elliott (1987), Geman and Graffigne (1987), and Klein and Press (1989).

The conditional distribution of the data $\mathbf{Z}$, given the true image $\boldsymbol{\theta}$, requires a more complicated model for textured image analysis. A brief introduction to a Markov random field model for texture, originally proposed by Hassner and Sklansky (1980), is now given.

Assume that the texture types $\{\theta(\mathbf{s}): \mathbf{s} \in D\}$ and the intensity of reflected radiation $\{Z(\mathbf{s}): \mathbf{s} \in D\}$ given the texture type are each realized from a Markov random field. Suppose that the prior is

$$\pi(\boldsymbol{\theta}) \propto \exp\{Q_1(\boldsymbol{\theta})\}$$

and that the conditional density of $\mathbf{Z}$ given $\boldsymbol{\theta}$ is

$$f(\mathbf{Z}|\boldsymbol{\theta}) \propto \exp\{Q_2(\mathbf{Z};\boldsymbol{\theta})\}.$$

Then the conditional distribution of $\boldsymbol{\theta}$ given $\mathbf{Z}$ is

$$p(\boldsymbol{\theta}|\mathbf{Z}) \propto \exp\{Q_1(\boldsymbol{\theta}) + Q_2(\mathbf{Z};\boldsymbol{\theta})\}. \tag{7.4.51}$$

Thus, the segmented image can be estimated by finding the $\hat{\boldsymbol{\theta}}$ that maximizes the right-hand side of (7.4.51). This can be obtained, for example, by simulated annealing (Section 7.4.3).

7.4.6* Single Photon Emission Tomography

Single photon emission computed tomography (SPECT) is a procedure for determining the distribution of a pharmaceutical in an organ of the body.

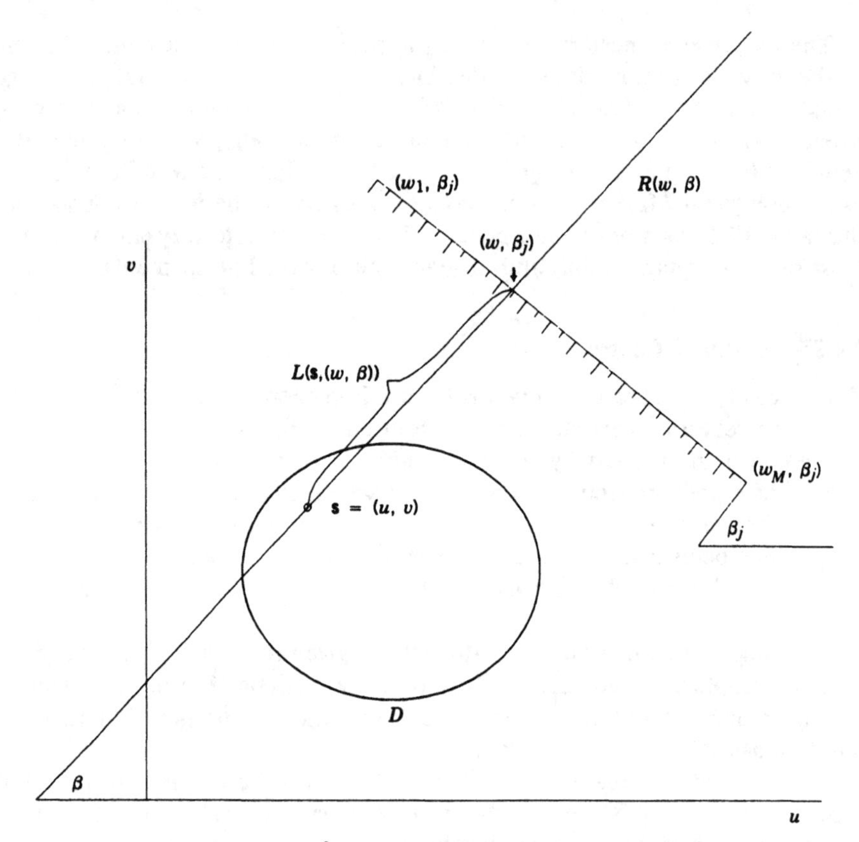

Figure 7.6 A stylized drawing in $\mathbb{R}^2$ of the region under investigation D and the row of sensors $w_1, \dots, w_M$ at an angle β_j to the horizontal axis. A photon is assumed emitted at an angle $\beta \in [\beta_j - \Delta\beta/2,\ \beta_j + \Delta\beta/2]$ from a location $\mathbf{s} = (u, v)'$ in D, and travels along the line segment $L(\mathbf{s}, (w, \beta))$ to the sensor (w, β); $R(w, \beta)$ is the continuation of the line segment.

Following injection or inhalation, a radio-labeled pharmaceutical will tend to become concentrated in regions of high metabolic activity or high blood flow. The radioactive tag emits photons and the pharmaceutical concentration is reconstructed or estimated from counts of the number of photons detected by sensors positioned near the region of interest.

Figure 7.6 illustrates the geometry of photon emission tomography. Let $\theta(\mathbf{s})$ denote the concentration of the pharmaceutical at location $\mathbf{s} \in D$, where D is the region of interest. For the purposes of presentation, assume D is a subset of $\mathbb{R}^2$. Photon emissions are detected by a row of equally spaced sensors $w_1, w_2, \dots, w_M$, which are positioned near the region of interest D and oriented at various angles to the horizontal axis. The sensors are rotated around the region of interest at equally spaced angles β_j, $j = 1, 2, \dots, n$, positioned at each angle for a time interval of duration T. Let the row of sensors, oriented at angle β_j, be denoted as $S(\beta_j) \equiv \{(w_i, \beta_j): i = 1, 2, \dots, M\}$

and let $Z(w_i, \beta_j)$ denote the number of photons detected by sensor w_i at angle β_j. The goal of single photon emission tomography is to reconstruct the concentration of the pharmaceutical $\{\theta(\mathbf{s}): \mathbf{s} \in D\}$ from the data

$$\{Z(w_i, \beta_j): i = 1, 2, \ldots, M;\ j = 1, 2, \ldots, n\}.$$

Photons are emitted from a site $\mathbf{s}$ in all directions, and a sensor only detects those photons arriving within a small angular interval defined by the direction of the vector perpendicular to the sensor $\pm \Delta\beta/2$. Moreover, the ith sensor only detects photons in the interval $[w_i - \Delta w/2, w_i + \Delta w/2]$. However, not all emitted photons arrive at the sensors; some are absorbed by matter encountered along their trajectories. Such attenuation can be described by an attenuation function $\mu(\mathbf{s})$ on D, which is assumed to be known. (In practice, it is measured for different types of matter using transmission tomography.) Consider a photon, traveling in a trajectory with angle β, emitted from a site $\mathbf{s}$. The probability that it survives and strikes the sensor array at the position $(w, \beta) \in S(\beta)$ is given by

$$\exp\left\{ -\int_{L(\mathbf{s},(w,\beta))} \mu(\mathbf{v}) \nu(d\mathbf{v}) \right\},$$

where $L(\mathbf{s}, (w, \beta))$ is the line segment from $\mathbf{s}$ to (w, β) (see Fig. 7.6) and $\nu(\cdot)$ is Lebesgue measure along that segment.

Assume that photons are generated by an inhomogeneous space–time Poisson point process with intensity $\theta(\mathbf{s})$, $\mathbf{s} \in D$, and that the orientations of photon trajectories are uniformly distributed on $[0, 2\pi)$. Let $R(w, \beta)$ be a line with orientation β, passing through a point $(w, \beta) \in S(\beta)$. Then $\{Z(w, \beta): (w, \beta) \in S(\beta)\}$ are the counts from an inhomogeneous Poisson process; the mean of $Z(w, \beta)$ over the interval of duration T is

$$\lambda(w, \beta) = T \int_{R(w,\beta)} \theta(\mathbf{s}) \exp\left\{ -\int_{L(\mathbf{s},(w,\beta))} \mu(\mathbf{v}) \nu(d\mathbf{v}) \right\} \nu(d\mathbf{s}). \quad (7.4.52)$$

Then $Z(w_i, \beta_j)$ is a Poisson random variable with mean

$$\lambda_{ij} \equiv E\big(Z(w_i, \beta_j)\big) = \int_{w_i - \Delta w/2}^{w_i + \Delta w/2} \int_{\beta_j - \Delta\beta/2}^{\beta_j + \Delta\beta/2} \lambda(w, \beta)\, d\beta\, dw. \quad (7.4.53)$$

Geman and McClure (1985, 1987) consider a Markov-random-field approach to the reconstruction of tomographic images. (A nonparametric approach is taken by Jones and Silverman, 1989.) Suppose $\{\theta(\mathbf{s}): \mathbf{s} \in D\}$ is a Markov random field with prior distribution

$$\pi(\boldsymbol{\theta}) \propto \exp\{Q(\boldsymbol{\theta})\}. \tag{7.4.54}$$

Assume that the (Poisson) components of **Z** are conditionally independent. Then, the conditional distribution of **Z** given $\boldsymbol{\theta}$ is

$$\begin{aligned} f(\mathbf{Z}|\boldsymbol{\theta}) &= \prod_{i=1}^{M} \prod_{j=1}^{n} \frac{e^{-\lambda_{ij}} \lambda_{ij}^{Z(w_i, \beta_j)}}{Z(w_i, \beta_j)!} \\ &= \exp\left[\sum_{i=1}^{M} \sum_{j=1}^{n} \{ -\log Z(w_i, \beta_j)! + Z(w_i, \beta_j) \log \lambda_{ij} - \lambda_{ij} \} \right]. \end{aligned} \tag{7.4.55}$$

Hence, from (7.4.54) and (7.4.55), the posterior distribution of $\boldsymbol{\theta}$ given **Z** is

$$p(\boldsymbol{\theta}|\mathbf{Z}) = c(\mathbf{Z})^{-1} \exp\left\{ Q(\boldsymbol{\theta}) + \sum_{i=1}^{M} \sum_{j=1}^{n} [Z(w_i, \beta_j) \log \lambda_{ij} - \lambda_{ij}] \right\}, \tag{7.4.56}$$

where, of course, $\{\lambda_{ij}\}$ depends on $\theta(\cdot)$ through (7.4.52) and (7.4.53).

Recall that maximum *a posteriori* (MAP) restoration finds a $\hat{\boldsymbol{\theta}}$ that maximizes (7.4.56), which in practice can be found, for example, through simulated annealing (Section 7.4.3). Geman and McClure (1985) also consider a maximum likelihood estimator for $\boldsymbol{\theta}$; that is, they maximize (7.4.55) with respect to $\boldsymbol{\theta}$. To solve this optimization problem, they use the EM algorithm (brief details of this algorithm are given in Section 7.5.2).

7.4.7* Least Squares and Image Regularization

Consider degraded images of the form

$$Z(\mathbf{s}) = \sum_{\mathbf{u}} H(\mathbf{u}) \theta(\mathbf{s} - \mathbf{u}) + \sigma \epsilon(\mathbf{s}), \qquad \mathbf{s} \in D, \tag{7.4.57}$$

where $\epsilon(\cdot)$ is a white-noise process with unit variance, σ is a parameter for varying the signal-to-noise ratio, and D is a square lattice. The point-spread function $H(\cdot)$ is assumed to be known and describes the blurring of the image by the recording device. The first term on the right-hand side of (7.4.57) is called the *convolution* of $\theta(\cdot)$ with $H(\cdot)$. In matrix notation, (7.4.57) can be written

$$\mathbf{Z} = H\boldsymbol{\theta} + \sigma\boldsymbol{\epsilon}. \tag{7.4.58}$$

The goal of image restoration is to estimate the image $\boldsymbol{\theta}$ from the data $\mathbf{Z}$. A least-squares or deconvolved estimator of $\boldsymbol{\theta}$ can be obtained by finding $\hat{\boldsymbol{\theta}}_0$ that minimizes

$$(\mathbf{Z} - H\boldsymbol{\theta})'(\mathbf{Z} - H\boldsymbol{\theta}).$$

Standard linear algebra shows that $\hat{\boldsymbol{\theta}}_0$ satisfies

$$H'H\hat{\boldsymbol{\theta}}_0 = H'\mathbf{Z}. \tag{7.4.59}$$

If H is nonsingular, then

$$\hat{\boldsymbol{\theta}}_0 \equiv H^{-1}\mathbf{Z} = \boldsymbol{\theta} + \sigma H^{-1}\boldsymbol{\epsilon}. \tag{7.4.60}$$

Although the least-squares estimator $\hat{\boldsymbol{\theta}}_0$ is unbiased for $\boldsymbol{\theta}$, it does not give a satisfactory reconstruction of the image. First, because n is usually very large, computation of H^{-1} may be very expensive. Second, H is often ill-conditioned, so small changes in $\mathbf{Z}$ can result in large changes in $\hat{\boldsymbol{\theta}}_0$. Thus, in the presence of random noise, least-squares solutions can be quite unstable (Titterington, 1985a).

Efficient algorithms for computing H^{-1}, or an approximation thereof, can be found when H has certain specific structure. For example, Hall and Titterington (1986) find an approximate solution to this matrix inversion problem assuming that H is Toeplitz. Alternatively, iterative solutions to the minimization problem can sometimes be found; see the following text.

Regularization

Instability due to ill-conditioned matrices H may be reduced by some form of regularization, such as smoothing or shrinking the least-squares estimator $\hat{\boldsymbol{\theta}}_0$. Let $\hat{\boldsymbol{\theta}}_\infty$ be an ultrasmooth estimator for $\boldsymbol{\theta}$; $\hat{\boldsymbol{\theta}}_\infty$ may be of uniform intensity [i.e., $\hat{\theta}_\infty(\mathbf{s}) = k$, for all $\mathbf{s} \in D$] or may be a reference map based on prior knowledge or a preliminary analysis of the data. A regularized estimator $\hat{\boldsymbol{\theta}}_\lambda$ can be obtained by solving

$$\inf_{\boldsymbol{\theta} \in \boldsymbol{\Theta}} \Delta_1(\boldsymbol{\theta}, \hat{\boldsymbol{\theta}}_0) + \lambda \Delta_2(\boldsymbol{\theta}, \hat{\boldsymbol{\theta}}_\infty), \tag{7.4.61}$$

where Δ_1 and Δ_2 are distance measures (Titterington, 1985a, 1985b). A Bayesian justification for (7.4.61) is given in Titterington (1985a). Fidelity to the data is obtained by taking $\lambda = 0$ and minimizing $\Delta_1(\boldsymbol{\theta}, \hat{\boldsymbol{\theta}}_0)$. As λ increases, $\hat{\boldsymbol{\theta}}_\lambda$ becomes smoother and smoother and shrinks toward the ultrasmooth estimator $\hat{\boldsymbol{\theta}}_\infty$.

Titterington (1985a) lists some of the available candidates for Δ. These include:

1. *Quadratic Distance.*

$$\Delta_I(\boldsymbol{\theta}_1, \boldsymbol{\theta}_2) \equiv (\boldsymbol{\theta}_1 - \boldsymbol{\theta}_2)'(\boldsymbol{\theta}_1 - \boldsymbol{\theta}_2).$$

2. *Weighted Quadratic Distance.*

$$\Delta_W(\boldsymbol{\theta}_1, \boldsymbol{\theta}_2) \equiv (\boldsymbol{\theta}_1 - \boldsymbol{\theta}_2)'W(\boldsymbol{\theta}_1 - \boldsymbol{\theta}_2),$$

where W is a prespecified symmetric, positive-definite matrix.

3. *Kullback–Leibler Directed Divergence.* If all components of any $\boldsymbol{\theta}$ in Θ are nonnegative, the image can be normalized by taking

$$p(\mathbf{s}) = \frac{\theta(\mathbf{s})}{\Sigma_{\mathbf{u} \in D}\theta(\mathbf{u})}, \qquad \mathbf{s} \in D.$$

Then the Kullback–Leibler directed divergence is

$$\Delta_{\mathrm{KL}}(\boldsymbol{\theta}_1, \boldsymbol{\theta}_2) \equiv \sum_{\mathbf{s} \in D} p_1(\mathbf{s})\log(p_1(\mathbf{s})/p_2(\mathbf{s})),$$

where $p_1(\cdot)$ and $p_2(\cdot)$ are obtained from $\theta_1(\cdot)$ and $\theta_2(\cdot)$, as before, by normalization. [Notice that Δ_{KL} is not symmetric in its two arguments; in general, $\Delta_{\mathrm{KL}}(\boldsymbol{\theta}_1, \boldsymbol{\theta}_2) \neq \Delta_{\mathrm{KL}}(\boldsymbol{\theta}_2, \boldsymbol{\theta}_1)$.]

There is no particular need for Δ_1 and Δ_2 to be the same. Typically, Δ_1 is taken to be either quadratic distance or weighted quadratic distance. If $\hat{\boldsymbol{\theta}}_\infty$ represents a prior image of specific interest, then Δ_2 is often taken to be the Kullback–Leibler directed divergence (Gull and Skilling, 1984).

In many instances, $\Delta_2(\boldsymbol{\theta}, \hat{\boldsymbol{\theta}}_\infty)$ appears in a slightly different form: $\Delta_2(\boldsymbol{\theta}, \hat{\boldsymbol{\theta}}_\infty) = \Phi(\boldsymbol{\theta})$, where Φ is a measure of roughness of $\boldsymbol{\theta}$. Examples of Φ are given by Titterington (1985a):

1. $\Phi_1(\boldsymbol{\theta}) \equiv \Sigma_{\mathbf{s}, \mathbf{u} \in D}(\theta(\mathbf{s}) - \theta(\mathbf{u}))^2$.
2. $\Phi_2(\boldsymbol{\theta}) \equiv \boldsymbol{\theta}'K\boldsymbol{\theta}$, where K is a prespecified, positive-definite matrix, chosen to reflect the local smoothness of the image.
3. $\Phi_3(\boldsymbol{\theta}) \equiv \Sigma_{\mathbf{s} \in D}p(\mathbf{s})\log p(\mathbf{s})$, where $p(\cdot)$ is the normalized version of $\theta(\cdot)$; that is, $p(\cdot) = \theta(\cdot)/\Sigma_{\mathbf{s} \in D}\theta(\mathbf{s})$.
4. $\Phi_4(\boldsymbol{\theta}) \equiv -\Sigma_{\mathbf{s} \in D} \log p(\mathbf{s})$, where $p(\cdot)$ is the normalized version of $\theta(\cdot)$.

Explicit solutions to the image restoration problem are available if Eq. (7.4.61) is a quadratic form. For example, if

$$\begin{aligned}\Delta_1(\boldsymbol{\theta}, \hat{\boldsymbol{\theta}}_0) + \lambda\Delta_2(\boldsymbol{\theta}, \hat{\boldsymbol{\theta}}_\infty) &= \Delta_l(\boldsymbol{\theta}, \hat{\boldsymbol{\theta}}_0) + \lambda\Phi_2(\boldsymbol{\theta}) \\ &= (\mathbf{Z} - H\boldsymbol{\theta})'(\mathbf{Z} - H\boldsymbol{\theta}) + \lambda\boldsymbol{\theta}'K\boldsymbol{\theta}, \quad (7.4.62)\end{aligned}$$

then the minimum penalized distance estimator $\hat{\boldsymbol{\theta}}_\lambda$ satisfies

$$(H'H + \lambda K)\hat{\boldsymbol{\theta}}_\lambda = H'\mathbf{Z}. \quad (7.4.63)$$

Provided $(H'H + \lambda K)$ is nonsingular, then

$$\hat{\boldsymbol{\theta}}_\lambda = (H'H + \lambda K)^{-1}H'\mathbf{Z}, \quad (7.4.64)$$

which is the generalized ridge regression estimator for $\boldsymbol{\theta}$ (Titterington, 1985b). An iterative procedure for solving (7.4.62) is considered by Ripley (1988, p. 84). Equation (7.4.63) can be rewritten as

$$\hat{\boldsymbol{\theta}}_\lambda = \alpha \cdot H'\mathbf{Z} + U\hat{\boldsymbol{\theta}}_\lambda,$$

where $\alpha = (\lambda + 1)^{-1}$ and $U = (1 - \alpha)(I - K) + \alpha(I - H'H)$. Then, the iterative scheme

$$\hat{\boldsymbol{\theta}}_\lambda^{(i+1)} = \alpha \cdot H'\mathbf{Z} + U\hat{\boldsymbol{\theta}}_\lambda^{(i)}$$

converges provided all the eigenvalues of U have modulus strictly less than 1. Solutions to (7.4.63) can have negative entries, an undesirable property because $\boldsymbol{\theta}$ is usually assumed to be nonnegative. Molina and Ripley (1989) consider constrained estimation of $\boldsymbol{\theta}$ under the nonnegativity restriction.

The functions Φ_3 and Φ_4 are negatives of entropy measures (see, e.g., Gull and Daniell, 1978; Burch et al., 1983; Gull and Skilling, 1984). Thus, minimization of functions of the form

$$\Delta_1(\boldsymbol{\theta}, \hat{\boldsymbol{\theta}}_0) + \lambda\Phi_{\mathrm{NE}}(\boldsymbol{\theta}), \quad (7.4.65)$$

where $-\Phi_{\mathrm{NE}}$ is an entropy measure, is called maximum-entropy restoration. Equation (7.4.65) is not a quadratic function of $\boldsymbol{\theta}$ and, in general, there is no explicit solution to this minimization problem. Iterative algorithms for finding the maximum-entropy estimator are given by Burch et al. (1983) and Mohammad-Djafari and Demoment (1988). Because of the appearance of the logarithm in Φ_{NE} (see Φ_3, Φ_4), $\boldsymbol{\theta}$ and its maximum-entropy estimator must be nonnegative.

Consider the relationship between the MAP estimators of Section 7.4.3 and the regularized least-squares estimators of the present section. Suppose that the prior distribution π of $\boldsymbol{\theta}$ follows a CG model Gau$(\mathbf{0}, \tau^2(I - C)^{-1})$ and that $\mathbf{Z}$ given $\boldsymbol{\theta}$ is Gaussian and satisfies Eq. (7.4.58). Then, the posterior distribution of $\boldsymbol{\theta}$ given $\mathbf{Z}$ is

$$p(\boldsymbol{\theta}|\mathbf{Z}) = c(\mathbf{Z})^{-1}\exp\left\{-\frac{1}{2\sigma^2}(\mathbf{Z} - H\boldsymbol{\theta})'(\mathbf{Z} - H\boldsymbol{\theta}) - \frac{1}{2\tau^2}\boldsymbol{\theta}'(I - C)\boldsymbol{\theta}\right\}. \tag{7.4.66}$$

Thus, maximizing (7.4.66) with respect to $\boldsymbol{\theta}$ is the same as minimizing (7.4.62) with respect to $\boldsymbol{\theta}$, where $K = (I - C)$ and $\lambda = \sigma^2/\tau^2$.

Choice of Smoothing Parameter

Consider now choice of the smoothing parameter λ; at least two approaches are available (Titterington, 1985a). The first approach is to base the choice on the fit of $\hat{\boldsymbol{\theta}}_\lambda$ to the data; that is, treat λ as a Lagrange multiplier and minimize $\Delta_2(\boldsymbol{\theta}, \hat{\boldsymbol{\theta}}_\infty)$ subject to the constraint that $\Delta_1(\boldsymbol{\theta}, \hat{\boldsymbol{\theta}}_0) = a$, where a is some appropriately chosen constant. For example, if $\Delta_1(\boldsymbol{\theta}, \hat{\boldsymbol{\theta}}_0) = \Delta_W(\boldsymbol{\theta}, \hat{\boldsymbol{\theta}}_0)$ with $W = \sigma^{-2}H'H$, and $\boldsymbol{\theta}$ is the true image, then $\Delta_1(\boldsymbol{\theta}, \hat{\boldsymbol{\theta}}_0)$ has a chi-squared distribution on n degrees of freedom. Hence, one might take $a = n$ (the expected value) or $a = n + 2\sqrt{2n}$ (the approximate 95th percentile). Titterington (1985a) provides a Bayesian justification for this example; if the prior distribution of $\boldsymbol{\theta}$ is uniform, then the posterior distribution of $\Delta_W(\boldsymbol{\theta}, \hat{\boldsymbol{\theta}}_0)$ is chi-squared on n degrees of freedom.

Titterington's second approach is to choose the λ that minimizes risk. Letting $\boldsymbol{\theta}$ denote the true image and viewing $\Delta(\hat{\boldsymbol{\theta}}(\lambda, \mathbf{Z}), \boldsymbol{\theta})$ as a loss function, the expectation $E_{\boldsymbol{\theta}}\{\Delta(\hat{\boldsymbol{\theta}}(\lambda, \mathbf{Z}), \boldsymbol{\theta})\}$ is a measure of risk for a given $\boldsymbol{\theta}$. If $\Delta = \Delta_I$ is quadratic distance, then $E_{\boldsymbol{\theta}}\{\Delta_I(\hat{\boldsymbol{\theta}}(\lambda, Z), \boldsymbol{\theta})\}$ is the mean-squared error, which can be broken down into the variance plus the bias squared. By minimizing mean-squared error, one obtains a $\hat{\lambda}$ that depends on $\boldsymbol{\theta}$. A value of λ, based only on the data, could be chosen by cross-validation. Let $\hat{\theta}_{\lambda, -i}(\mathbf{s}_j)$ be the estimated value of $\theta(\mathbf{s}_j)$ obtained when pixel $\mathbf{s}_i$ is removed from the data. Then choose λ so that $Z(\mathbf{s}_i)$ is close to $\Sigma_{\mathbf{u}} H(\mathbf{u})\hat{\theta}_{\lambda, -i}(\mathbf{s}_i - \mathbf{u})$ for all i. Alternatively, the risk can be averaged over a prior distribution of $\boldsymbol{\theta}$ to obtain the Bayes risk, a function of λ alone. An optimal value of λ can then be found by minimizing the Bayes risk.

7.4.8* Method of Sieves

Many of the methods of image analysis considered earlier are Bayesian. That is, starting with $\pi(\boldsymbol{\theta})$, a prior distribution on $\boldsymbol{\theta}$, and $f(\mathbf{Z}|\boldsymbol{\theta})$, the conditional distribution of the data $\mathbf{Z}$ given $\boldsymbol{\theta}$, the MAP estimator (or some other estimator) is calculated from the posterior distribution $p(\boldsymbol{\theta}|\mathbf{Z})$. In contrast, a frequentist analysis seeks a maximum likelihood estimator; that is, a $\hat{\boldsymbol{\theta}}$ that maximizes the likelihood $l_n(\boldsymbol{\theta}) \equiv f(\mathbf{Z}|\boldsymbol{\theta})$. With infinite resolution, the image $\theta(\cdot)$ is a function over a continuous domain, say A. However, the data are collected on a spatial lattice $D \subset A$. Because the dimension of the parameter $\{\theta(\mathbf{s}): \mathbf{s} \in A\}$ is greater than the sample size $n = |D|$, a maximum likelihood estimator based on $f(\mathbf{Z}|\{\theta(\mathbf{s}): \mathbf{s} \in A\})$ is not consistent, in general, as $n \to \infty$. To remedy this problem, Grenander (1981) developed the method of sieves. Notice that the asymptotics used here are infill asymptotics; see Section 7.3.1, where they are contrasted with increasing-domain asymptotics.

The method of sieves restricts the index space A to a lattice $A(\alpha_n) \subset A$, where α_n depends on n and represents the distance between adjacent vertices of the lattice $A(\alpha_n)$. Then, by definition, the maximum likelihood estimator $\hat{\boldsymbol{\theta}}_{n,\alpha}$ satisfies,

$$l_n\left(\hat{\boldsymbol{\theta}}_{n,\alpha}\right) \geq l_n(\boldsymbol{\theta}), \quad \text{for all } \boldsymbol{\theta} \in A(\alpha_n),$$

where $l_n(\cdot)$ is the likelihood function. As $n \to \infty$, suppose the lattice spacing $\alpha_n \to 0$, and hence $m(n) \equiv |A(\alpha_n)| \to \infty$. But, to guarantee consistency of the estimator $\hat{\boldsymbol{\theta}}_{n,\alpha_n}$, the rate of decrease of lattice spacing must not be too fast. The sequence $\{A(\alpha_n): n = 1, 2, \ldots\}$ is called a *sieve*.

To illustrate the method of sieves, consider the triangular sieve, where $A(\alpha_n)$ is a triangular lattice for each α_n. At each vertex $\mathbf{s}$ of $A(\alpha_n)$, six equilateral triangles meet, forming a hexagon $G(\mathbf{s})$. Define $\phi_{\mathbf{s}}(\mathbf{u})$ to be the continuous piecewise linear function, vanishing outside $G(\mathbf{s})$, equal to 1 at $\mathbf{s}$, and equal to 0 along the boundary of $G(\mathbf{s})$. Define the set of functions $\theta_{\alpha_n}(\cdot)$ such that

$$\theta_{\alpha_n}(\mathbf{u}) \equiv \sum_{\mathbf{s} \in A(\alpha_n)} b_{\mathbf{s}} \phi_{\mathbf{s}}(\mathbf{u}), \quad \mathbf{u} \in \mathbb{R}^2, \tag{7.4.67}$$

where $\{b_{\mathbf{s}}: \mathbf{s} \in A(\alpha_n)\}$ are parameters to be estimated. Suppose

$$Z(\mathbf{s}) = \theta(\mathbf{s}) + \epsilon(\mathbf{s}), \quad \mathbf{s} \in A, \tag{7.4.68}$$

where $\epsilon(\cdot)$ is a white-noise process. Then the method-of-sieves estimator is obtained by finding $\hat{b}_{\mathbf{s}_1}, \hat{b}_{\mathbf{s}_2}, \ldots, \hat{b}_{\mathbf{s}_{m(n)}}$ that minimize the quadratic form

$$\sum_{\mathbf{u} \in D} \{Z(\mathbf{u}) - \theta_{\alpha_n}(\mathbf{u})\}^2 = \sum_{\mathbf{u} \in D} \left\{Z(\mathbf{u}) - \sum_{\mathbf{s} \in A(\alpha_n)} b_{\mathbf{s}} \phi_{\mathbf{s}}(\mathbf{u})\right\}^2, \tag{7.4.69}$$

where $m(n) < n = |D|$.

7.4.9 Mathematical Morphology

Based on a theory that involves topological spaces whose elements are sets in $\mathbb{R}^d$, mathematical morphology approaches the analysis of images through the forms of objects. Consider a binary image consisting of black $[\theta(\mathbf{s}) = 1]$ and white $[\theta(\mathbf{s}) = 0]$ pixels. The image can be completely described by the set of black pixels. Mathematical morphological operations can be used to simplify image data by smoothing rough edges or eliminating peninsulas and islands. Thus, the general shape of the object is preserved whereas irrelevancies are eliminated. A very brief presentation of various operations is given in the following text; Section 9.3 also discusses mathematical morphology from the point of view of random-set models. Serra's (1982) book is a rich and detailed resource for both the theory and applications of the approach.

Only binary images will be considered here and, for greater generality, a continuous spatial index can be assumed. Let $P \subset A \subset \mathbb{R}^d$ denote the image to be processed; that is, $P = \{\mathbf{s} \in A: \theta(\mathbf{s}) = 1\}$. The two basic mathematical morphological operations are dilation and erosion. Dilation combines two sets using vector addition of set elements. The dilation of an image P by a set B is denoted by $P \oplus \check{B}$ (where $\check{B} \equiv \{-\mathbf{b} : \mathbf{b} \in B\}$) and is defined by

$$P \oplus \check{B} \equiv \{\mathbf{s} + (-\mathbf{b}) : \mathbf{s} \in P \text{ and } \mathbf{b} \in B\}; \qquad (7.4.70)$$

the set B is called a structuring element. Dilation is commutative and associative. Erosion combines two sets using vector subtraction of set elements. The erosion of an image P by a structuring element B is denoted by $P \ominus \check{B}$ and is defined by

$$P \ominus \check{B} \equiv \{\mathbf{s} \in A: \mathbf{s} + \mathbf{b} \in P \text{ for all } \mathbf{b} \in B\}. \qquad (7.4.71)$$

Dilation and erosion can be applied together; that is, either image dilation followed by erosion of the dilated image or image erosion followed by dilation of the eroded image. These two operations are referred to, respectively, as the closing and opening of an image (Matheron, 1975, p. 18). The closing of an image P by a structuring element B is denoted by $P \bullet B$ and is defined by $P \bullet B \equiv (P \oplus \check{B}) \ominus B$. Closing an image with a disk B smooths contours, fuses narrow breaks and long thin gulfs, eliminates small holes, and fills in gaps. The opening of an image P by a structuring element B is denoted by $P \circ B$ and is defined by $P \circ B \equiv (P \ominus \check{B}) \oplus B$. Opening an image with a disk B smooths contours, breaks narrow isthmuses, and eliminates small islands and sharp spikes or capes. Openings and closings are idempotent operations; that is, $(P \circ B) \circ B = P \circ B$ and $(P \bullet B) \bullet B = P \bullet B$. So, repeated applications of either operation cause no further transformation of the image.

Openings and closings are highly nonlinear, deterministically based smoothing operators. For a stochastic approach to mathematical morphology and its subsequent use in higher-level image analysis, see Sections 9.3 through 9.7.

7.5 REGIONAL MAPPING: SCOTLAND LIP-CANCER DATA

The first part of this section is based on a resistant method of regional mapping presented in Cressie and Guo (1987), and the second part on a parametric (Bayesian) spatial analysis presented in Clayton and Kaldor (1987); both are applied to lip-cancer incidence data in the districts of Scotland. The general problem being addressed here is how to deal with a regional variable that is a proportion with respect to a base that varies from region to region.

A map showing (usually political) boundaries of regions is an essential document for maintaining the definition of those regions. A very accurate definition is also available from sequences of latitude–longitude pairs that define lines on the earth's surface, but, from a visual point of view, this definition is sterile. Economists, demographers, epidemiologists, ecologists, geographers, sociologists, and so on, have need of regional information to pinpoint where, and then to explain why, certain phenomena (e.g., cancer incidence, unemployment, crime, etc.) are exhibiting spatial dependence. Political boundaries are not always ideal for, say, an epidemiologic study, but for bureaucratic reasons the data are often collected that way.

Having established a regional data set, it is tempting to present the data on the regional map by color or gray-tone shading or with overlaid symbols proportional to the size of the variable (e.g., Tufte, 1983). Of course, if the variable is, say, the number of unemployed people in the 1980 civilian labor force by states of the United States, then unless it is *standardized* by the states' total civilian labor force, such a regional map does not mean very much. In other words, it is the *percent* unemployed that is the variable of interest, or in the epidemiological context, it is the rate of disease per population-years-at-risk that is of interest.

This standardization is essential for comparison of measures from region to region. A large amount of regional data are counts from a base that is itself variable, so that although the standardization in some sense yields comparability of means, the unequal base from region to region results in unequal variances. Of course, without a statistical model, there is no notion of mean or variance, but I claim that the very act of shading the standardized variable on a regional map and then looking for geographical clusters is an attempt to assimilate a statistical model. Unusually high or low regional values are featured by such a shading. Explanations for them are then sought that can, and usually do, include something about their spatial locations relative to other regions. Nevertheless, this practice of mapping variables can be *misleading*.

"Unusually high" and "unusually low" are terms that, in statistical parlance, mean "in the tails of the distribution." But, if the precision of each datum is very different from one to the other, then the data should not be thought of as coming from a single distribution. For example, a Sudden Infant Death (SID) rate of 0.1% per number of live births is far more reliable from a region whose recorded number of live births is 20,000 than from a region whose recorded number of live births is 2000. To put it another way, an unusually high or low SID rate for a region may be due to very few live births from year to year, so that chance fluctuations may have its SID rate high one year and low the next. Strictly speaking then, any region is probably only comparable to *itself* across time and then only for a base that is roughly constant across time within that region. In order to compare regions spatially, allowance must be made for this nonconstant spatial variability.

These issues are introduced by Cressie and Read (1989), who analyze SID rates for the counties of North Carolina. Wallin (1984) discusses *inter alia* the problem of mapping ratios and identifies a need for an analysis of their uncertainty and variability. However, he does not pursue this, instead suggesting that mapping of ratios should be *avoided*. In Section 6.2 and here, the very opposite approach is taken. I believe that the mapping of standardized (or ratio) variables is so basic to most spatial studies that the problem must be dealt with head on.

Some Background

For the purpose of definition, cancer of the lip is cancer of the vermilion border of the lip that arises in a relatively small anatomical region between the mouth and the hair-bearing skin of the face (and so excludes cancer of the skin of the lip).

Descriptive epidemiologic studies of lip cancer have revealed that its incidence is about 10 times higher in males than in females, that it is far more common in rural areas than in urban areas, and that the cancer almost always arises on the lower lip. Increased exposure to sunlight has been implicated in the excess occurrence of lip cancers among rural populations, producing high rates in people who work outdoors. There is also a smaller, independent contribution to the risk of lip cancer associated with smoking (International Agency for Research on Cancer, 1985, pp. 58, 59).

Table 7.2 gives $\{L_i: i = 1, \ldots, 56\}$, the number of male lip-cancer cases in the districts of Scotland for the six years from 1975–1980 (International Agency for Research on Cancer, 1985). Also given are $\{n_i: i = 1, \ldots, 56\}$, the number of male population-years-at-risk (1975–1980), kindly supplied by J. Kaldor.

Some explanation of the political geography is needed. The Local Government (Scotland) Act, passed in 1973 by the British Parliament, introduced major changes in the system of local government in Scotland. Nine regional authorities were created and divided into a total of 53 districts. Added to this were three single-tier Island Authorities (Western, Orkney, and Shetland Isles). For the purposes of this study, I shall call the Island Authorities, districts, resulting in a regional map of Scotland that consists of *56 districts* (Fig. 7.7).

Table 7.2 Scotland Lip-Cancer Data

District No. and Name	L_i^a	n_i^b	District No. and name	L_i^a	n_i^b
1. Skye-Lochalsh	9	28324	29. Perth-Kinross	16	346041
2. Banff-Buchan	39	231337	30. West Lothian	11	382702
3. Caithness	11	83190	31. Cumnock-Doon	5	139148
4. Berwickshire	9	51710	32. Stewartry	3	65448
5. Ross-Cromarty	15	129271	33. Midlothian	7	249667
6. Orkney	8	53199	34. Stirling	8	233125
7. Moray	26	245513	35. Kyle-Carrick	11	319316
8. Shetland	7	62603	36. Inverclyde	9	296238
9. Lochaber	6	59183	37. Cunninghame	11	391513
10. Gordon	20	165554	38. Monklands	8	319072
11. Western Isles	13	87815	39. Dumbarton	6	231227
12. Sutherland	5	37521	40. Clydebank	4	156924
13. Nairn	3	29374	41. Renfrew	10	617413
14. Wigtown	8	86444	42. Falkirk	8	426519
15. NE Fife	17	185472	43. Clackmannan	2	141294
16. Kincardine	9	111665	44. Motherwell	6	449231
17. Badenoch	2	27075	45. Edinburgh	19	1287561
18. Ettrick	7	94145	46. Kilmarnock	3	238170
19. Inverness	9	162867	47. East Kilbride	2	246849
20. Roxburgh	7	102697	48. Hamilton	3	312103
21. Angus	16	263205	49. Glasgow	28	2316353
22. Aberdeen	31	583327	50. Dundee	6	547016
23. Argyll-Bute	11	190816	51. Cumbernauld	1	179194
24. Clydesdale	7	163818	52. Bearsden	1	110707
25. Kirkcaldy	19	432132	53. Eastwood	1	146112
26. Dunfermline	15	378946	54. Strathkelvin	1	246744
27. Nithsdale	7	163703	55. Tweeddale	0	38704
28. East Lothian	10	231185	56. Annandale	0	103412

Source: Cressie and Guo (1987).
[a]The number of male lip-cancer cases (1975–1980).
[b]The number of male population-years-at-risk (1975–1980).

7.5.1[†] Exploratory Regional Mapping

This section uses resistant, spatial-data–analysis techniques described in Section 6.2 (where exploratory analyses of the North Carolina SIDS data were presented) to map the Scottish lip-cancer rates. A feature of the Scottish data is the highly irregular geographical locations and shapes of districts. In Figure 7.7, compare district 40 (Clydebank) with district 8 (Shetland) with district 29 (Perth-Kinross). The resistant method of disease mapping presented here should be of particular interest to medical geographers.

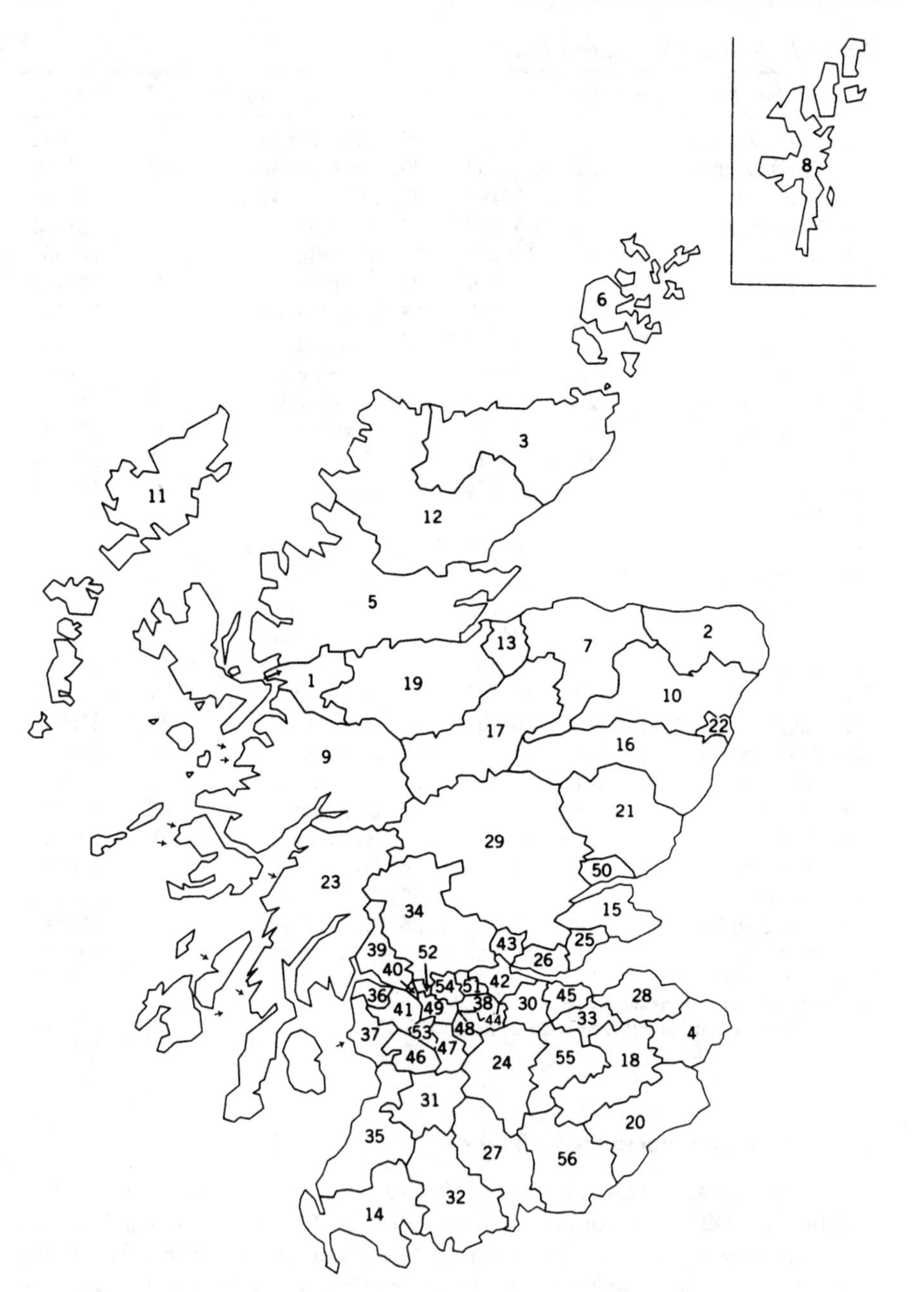

Figure 7.7 Map shows the 56 districts of Scotland, numbered according to Table 7.2. The small arrows indicate to which district the islands belong.

The first and most obvious regional map of Scotland to make is one I cast doubts upon earlier, namely, the raw incidence rates (number of male lip-cancer patients per population-years-at-risk). Cressie and Guo (1987) show a choropleth map of the districts (i.e., a shading of the districts) using the variable

$$A_i \equiv 1000(L_i)/n_i, \qquad i = 1,\ldots,56. \tag{7.5.1}$$

They compare this to another choropleth map using the variable

$$W_i \equiv 1000(L_i + 1)/n_i, \qquad i = 1,\ldots,56, \tag{7.5.2}$$

and note that there is a considerable difference. This is symptomatic of a problem encountered with districts whose n_is are small, namely, high variability. Either map is misleading in that variables of widely different variability are being compared on the same map. For a similar reason, the probability map (Choynowski, 1959) is also avoided; see Section 6.2.1.

My emphasis in this section is to make as few parametric assumptions as possible. Nevertheless, I do have a vague model in mind, which is

$$\text{data} = \text{spatial trend} + \text{error},$$

where the error might hopefully be second-order stationary (i.e., error has zero mean, constant variance, and covariance depending only on the distance between two districts).

Locate in some unique way a center of the district and identify its datum with that center. It could be done by visually assessing a central point (as was done here for Scotland) or by using a regional capital (as was done for the counties of North Carolina in Sections 6.2 and 7.6). These districts and district centers are irregularly located throughout Scotland; inspection of Figure 7.7 shows the compact nature of the districts in the Strathclyde area (Clydebank, Bearsden, Glasgow, Strathkelvin, Cumbernauld, Monklands, etc.) and the expansive nature of the districts in the Highlands area (Caithness, Sutherland, Ross-Cromarty, etc.).

Following Cressie and Read (1989), overlay a grid of rows and columns and identify a district to a particular node of the grid if its center is closest to that node. This low-resolution mapping is an excellent way of looking for spatial trend. There is some subjectivity in the choice of the grid, and what is different in the case of Scotland is that the grid spacing is variable within both its rows and its columns. I chose it this way in order to have roughly one district at each grid node. Set out in Table 7.3 are the locations of the districts within the grid (the variable grid spacing is deemphasized).

Table 7.3 Locations of Districts of Scotland, Discretized onto a 10×7 Grid[a]

	1	2	3	4	5	6	7
1	11,5	12		3	6		8
2	1	19		13	7	10	2
3	9			17		16	22
4				29		21,50	
5	23	39	34	43	25,26	15	
6		36,40	51,52,54	42	45	28	
7		41	49	30,38,44	33		
8		37,46	47,48,53				4
9		35	31	24	55	18,20	
10		14	32	27	56		

Source: Cressie and Guo (1987).
[a]Districts in row 1 are the most northerly districts and districts in column 1 are the most westerly districts.

Transforming the Data
Now *try* to model the data by fitting additive row and column effects:

$$A_i = a + r_{k(i)} + c_{l(i)} + \delta_i, \qquad k(i) \in \{1,\ldots,10\}, l(i) \in \{1,\ldots,7\}, \tag{7.5.3}$$

where the ith district is located at grid node $(k(i), l(i))$, a is the overall mean, r_k is the kth row effect ($k = 1,\ldots,10$), and c_l is the lth column effect ($l = 1,\ldots,7$). Note that if there are more than two districts at the same node (k, l), *each* is assumed to have the same additive mean, $a + r_k + c_l$. One would like to assume that $\{\delta_i\}$ are (second-order stationary) random errors, with $\text{var}(\delta_i) =$ constant and $\text{cov}(\delta_i, \delta_j)$ a function of the Euclidean distance between the ith and jth counties. Unfortunately, $\text{var}(A_i)$ is not constant because it varies both with $E(L_i/n_i)$ and with n_i: The incidence of (albeit locally spatially correlated) rare events might be approximated by a sum of a large number of Bernoulli random variables with very small success probability, conditions that imply $\text{var}(L_i) \propto E(L_i)$; see Section 6.2.2. This variance-mean relation is removed by the Freeman–Tukey transformation (Freeman and Tukey, 1950):

$$Z_i \equiv (1000(L_i)/n_i)^{1/2} + (1000(L_i + 1)/n_i)^{1/2}, \qquad i = 1,\ldots,56, \tag{7.5.4}$$

which can be thought of as a finely tuned *square-root* transformation of the data.

Just as for the SIDS data (see Section 6.2),

$$\text{var}(Z_i) \simeq \sigma^2/n_i. \tag{7.5.5}$$

Therefore, write the Freeman–Tukey transformed rate as

$$Z_i = a + r_{k(i)} + c_{l(i)} + n_i^{-1/2}\delta_i, \qquad i = 1,\ldots,56, \tag{7.5.6}$$

where a, $r_{k(i)}$, $c_{l(i)}$, and δ_i have the same definition as for (7.5.3), and n_i is the number of population-years-at-risk associated with Z_i. Although model (7.5.3) is clearly *not* realistic, model (7.5.6) *is* because the data have been appropriately transformed and the error has been modeled as $\{n_i^{-1/2}\delta_i: i = 1,\ldots,56\}$.

Fitting the Spatial Model

Given in the following text is a (resistant) way of mapping the districts of Scotland, to look for unusually large and unusually small lip-cancer incidence rates, that takes into account unequal variances. From the model (7.5.6), a, $\{r_k\}$, and $\{c_l\}$ can be fitted to the data and substituted into $\{a + r_k + c_l: k = 1,\ldots,10;\ l = 1,\ldots,7\}$; this (fitted) large-scale variation is then used to indicate districts with unusually large or small values (unusual on the Freeman–Tukey-transformed scale defines unusual on the original scale). Fitting should take into account the unequal error variances using weights that depend on the $\{n_i\}$.

Median polish (Tukey, 1977) is the initial method used to fit the spatial model. There are two reasons: (1) Exploratory methods need to be resistant to outlying values and (2) residuals from median polish are a less biased estimate of the error structure (Cressie, 1986). In fact, *weighted* median polish is used with weights equal to $(n_i)^{1/2}$; see Section 6.2.3 for more details. The (weighted) median-polish algorithm gives the decomposition

$$Z_i = \tilde{a} + \tilde{r}_{k(i)} + \tilde{c}_{l(i)} + R_i, \tag{7.5.7}$$

where $\tilde{a}$, $\tilde{r}_{k(i)}$, and $\tilde{c}_{l(i)}$ are the all-, the row-, and the column-effect estimators of a, $r_{k(i)}$, and $c_{l(i)}$, respectively. Table 7.4 gives the estimates derived from weighted (and unweighted) median polish.

Table 7.4 Estimates of All Effect ($\tilde{a}$), Row Effects ($\tilde{r}_1,\ldots,\tilde{r}_{10}$), and Column Effects ($\tilde{c}_1,\ldots,\tilde{c}_7$)[a]

$\tilde{a} = 0.39849\ (0.43732)$	$\tilde{r}_9 = 0.01534\ (-0.00144)$
$\tilde{r}_1 = 0.31531\ (0.30174)$	$\tilde{r}_{10} = 0.01549\ (0.03968)$
$\tilde{r}_2 = 0.27475\ (0.25410)$	$\tilde{c}_1 = 0.07024\ (0.05217)$
$\tilde{r}_3 = 0.08245\ (0.10853)$	$\tilde{c}_2 = 0.00000\ (0.00000)$
$\tilde{r}_4 = 0.00000\ (-0.03561)$	$\tilde{c}_3 = -0.03890\ (-0.06312)$
$\tilde{r}_5 = 0.02214\ (0.00144)$	$\tilde{c}_4 = 0.01387\ (-0.00282)$
$\tilde{r}_6 = -0.13014\ (-0.14907)$	$\tilde{c}_5 = -0.01619\ (-0.03434)$
$\tilde{r}_7 = -0.13775\ (-0.13084)$	$\tilde{c}_6 = 0.10221\ (0.10165)$
$\tilde{r}_8 = -0.14834\ (-0.16961)$	$\tilde{c}_7 = -0.01619\ (0.04390)$

[a]Obtained by weighted median polish of the Scottish (Freeman–Tukey transformed) lip-cancer incidence rates. Estimates obtained by unweighted median polish are given in parentheses.

Define the fitted values

$$\tilde{\mu}_i \equiv \tilde{a} + \tilde{r}_{k(i)} + \tilde{c}_{l(i)}, \qquad i = 1,\ldots,56. \tag{7.5.8}$$

A district map of Scotland with these fitted values used to shade the districts can be found in Cressie and Guo (1987). It is a smoothed map that is intended to capture *spatial trends* (or large-scale spatial variation) in the presence of heteroskedastic small-scale variation.

To untransform the fitted values (7.5.8) back to the original scale, assign the ith district the smoothed value

$$\left\{\tilde{a} + \tilde{r}_{k(i)} + \tilde{c}_{l(i)}\right\}^2 + \left\{\hat{\sigma}_\delta^2/n_i\right\}, \qquad i = 1,\ldots,56, \tag{7.5.9}$$

where $\hat{\sigma}_\delta^2$ is the estimated variance of δ_i defined in (7.5.6); for example, $\hat{\sigma}_\delta^2$ is (a robust version of) the sample variation of the weighted residuals (7.5.10). A choropleth map of these untransformed values yields the final smooth regional map.

These maps reinforce an earlier conjecture that the incidence rates are highest in areas where substantial proportions of the population are employed in outdoor activities such as farming, forestry, and fishing. Lip-cancer incidence disaggregated on age was not available, but if it were, one might discover that the differential lip-cancer rate can be explained by a differential age profile. Because lip cancer is generally not a disease of the young, the flight of youth to the cities in search of employment and social opportunities would leave behind an aging rural population, more susceptible to lip cancer (or most other ailments, for that matter). Without *age-specific* incidence rates, it is difficult to conclude that excessive exposure to the sun leads to a high incidence of lip cancer. Section 7.5.2 looks at this further, and defines standardized ratios of observed numbers to expected numbers, based on national age-specific incidence rates.

Weighted Residuals

There may be information on spatial variability left in the weighted residuals:

$$n_i^{1/2} R_i = n_i^{1/2}\left(Z_i - \tilde{a} - \tilde{r}_{k(i)} - \tilde{c}_{l(i)}\right), \qquad i = 1,\ldots,56. \tag{7.5.10}$$

Figure 7.8 gives a stem-and-leaf plot of the weighted residuals. Notice that there are two outliers from the spatial model (7.5.6), namely, district 4 (Berwickshire), corresponding to the weighted residual 142, and district 50 (Dundee), corresponding to the weighted residual -209. The weighted-median-polish method of spatial smoothing is resistant to these apparent unusual values.

The spatial model (7.5.6) is an *additive* model, which is not always appropriate. However, it is often the case that a transformation to obtain

−20	9
−18	
−16	
−14	
−12	
−10	
−8	6
−6	74
−4	71
−2	83931
−0	97544100000000
0	5880345800000000
2	2377899
4	5918
6	30
8	1
10	
12	
14	2

Figure 7.8 Stem-and-leaf plot of weighted-median-polish residuals $\{n_i^{1/2} R_i: i = 1, \ldots, 56\}$ defined by (7.5.10). 8|1 represents 81. [*Source*: Cressie and Guo, 1987]

variances independent of the mean *also* yields additivity and, incidentally, symmetry (Box and Cox, 1964; Cressie, 1985b). Nevertheless, it is worthwhile to make a diagnostic plot of the type suggested by Cressie (1986) that looks for the presence of *cross-product* trend. For the data set considered here, no such trend was obvious from the diagnostic plot.

Explanatory Variables

The model (7.5.6) is *purely* spatial in that its explanatory variables have entries of 0 and 1, describing the spatial locations of the counties relative to each other. To investigate reasons for the spatial clustering and the incidence rates of districts 4 and 50, the epidemiologist might fit another model:

$$\mathbf{Z} = X\boldsymbol{\beta} + \operatorname{diag}\{n_1^{-1/2}, \ldots, n_{56}^{-1/2}\} \cdot \boldsymbol{\delta}, \tag{7.5.11}$$

where the columns of X represent say urbanicity, health care, and percent males over 50 years of age. Regardless of which model is being fitted, it is advised to include in the analysis resistant procedures and diagnostic checking of the type described in the this section and in Section 6.2.

Spatial Dependence

To obtain more efficiency in the fit of $X\boldsymbol{\beta}$ [see Eq. (7.5.11)] to the data, the spatial dependence in $\boldsymbol{\delta}$ should be used. For the case of the purely spatial model (7.5.6), the weighted-median-polish residuals (7.5.10) can proxy for the (assumed) stationary errors $\boldsymbol{\delta}$. In a similar manner to the geostatistical analysis of the SIDS data presented in Section 4.4, the weighted residuals

(7.5.10) can also be differenced, squared, averaged, and plotted against respective distances to yield an estimated variogram (see also Cook and Pocock, 1983). This plot indicates oscillatory autocorrelation, with maximum negative dependence between 45 and 65 miles (due in part to districts 4 and 50 being considerably different from their neighboring districts). Chapter 4 can be consulted for various examples of how a variogram or a correlogram can be modeled, fitted, and used to obtain efficient estimates of large-scale-variation parameters.

Some Final Remarks

A map of regional variables is usually produced by coloring directly onto the outline of the regions according to a graded color code (or gray scale). For disease mapping, this has the disadvantage that large areas of low population density dominate the map, whereas often most of the interest is in cities, small areas of high population density. One possible alternative is the demographic-base map (e.g., Forster, 1972; Kidron and Segal, 1984), where the area of each region is made proportional to the population-years-at-risk, while contiguity of the geographical boundaries and the relative geographical positions are maintained as far as possible. Although offering interesting possibilities, constructing the required map is difficult and often arbitrary.

Cleveland and McGill (1984) and Cleveland (1985, p. 208) suggest that the traditional choropleth map be replaced with a regional map that has framed rectangles of *equal* size superimposed on, or attached in some way to, the corresponding regions. The framed rectangle is like a chemist's measuring beaker with differing amounts of black liquid in it; its degree of fullness is proportional to the observed (transformed) incidence rate. Dunn (1988) describes an experiment in graphical perception that demonstrates the superiority of the framed-rectangle method of mapping over choropleth mapping. One possible enhancement is to make the width of the ith framed rectangle proportional to $n_i^{1/2}$ (i.e., wider framed rectangles correspond to rates with smaller standard deviations).

Neither method of mapping addresses the question of how to smooth the data or of how to account for the unequal variation inherent in incidence data from unequal bases. A weighted resistant smoother has been presented in this section that responds to both these questions and has weighted residuals that accurately reflect any spatial correlation in the errors. When mapping the smoothed values (7.5.9), either the traditional choropleth map or a map with framed rectangles could be used.

7.5.2 Parametric Empirical Bayes Mapping

Clayton and Kaldor (1987) propose *inter alia* the following parametric spatial model for the lip-cancer data of Table 7.2. They assume that each district has a latent Poisson parameter and that collectively they behave according to a Markov random field whose neighborhoods $\{N_i: i = 1, \ldots, 56\}$ are defined by the sets of district adjacencies. One could call this a spatial Poisson-log-Gaussian model (cf. Aitchison and Ho, 1989). A different type of parametric spatial model is presented in Section 7.6; there the incidence data are modeled directly, without assuming random parameters.

A summary of Clayton and Kaldor's (1987) spatial-latent-parameter approach will be presented in this section. They begin by age-standardizing their data. (The exploratory spatial data analysis in Section 7.5.1 could equally have been carried out on the age-standardized data. Let the population of males be divided into p age groups and suppose that age-specific incidence rates $\{\xi_m: m = 1, \ldots, p\}$ are available (e.g., from national incidence rates in the p age groups). Suppose also that for the ith district, the numbers of male-population-years-at-risk in each age group $\{n_{im}: m = 1, \ldots, p\}$ are available, $i = 1, \ldots, 56$. Then assume that each district has a relative risk $\{\theta_i: i = 1, \ldots, 56\}$ associated with it, such that

$$L_i \sim \text{Po}(\theta_i E_i), \qquad i = 1, \ldots, 56, \tag{7.5.12}$$

where Po(λ) is shorthand for a Poisson distribution with mean λ and

$$E_i \equiv \sum_{m=1}^{p} \xi_m n_{im}, \qquad i = 1, \ldots, 56, \tag{7.5.13}$$

are the expected lip-cancer incidences in the 56 districts. This *age standardization* is based upon an implicit assumption of proportional mortality, which Keiding et al. (1990) show may not always be appropriate.

The vector $\boldsymbol{\theta} \equiv (\theta_1, \ldots, \theta_{56})'$ is a vector of unknown parameters to be estimated. Assuming that all distributions in (7.5.12) are independent, the maximum likelihood (m.l.) estimator is $(L_1/E_1, \ldots, L_{56}/E_{56})'$; the quantity (L_i/E_i) is known as a standardized (mortality) ratio for the ith district. These estimates are not smooth, just as the incidence rates $\{L_i/n_i: i = 1, \ldots, 56\}$ of Section 7.5.1 were not smooth. In fact, the resistant method of regional mapping proposed in Section 7.5.1, there applied to the data $\{(L_i, n_i): i = 1, \ldots, 56\}$, could equally be applied to the age-standardized data $\{(L_i, E_i): i = 1, \ldots, 56\}$. Here, rather, the parameters $\{\theta_i: i = 1, \ldots, 56\}$ are assumed to be a realization from a stochastic process, in particular, a Markov random field.

Assume that

$$\nu_i \equiv \log \theta_i, \qquad i = 1, \ldots, 56, \tag{7.5.14}$$

is a conditionally specified Gaussian (CG) model; that is, the (prior) distribution of $\boldsymbol{\nu} \equiv (\nu_1, \ldots, \nu_{56})'$ is given by:

$$\boldsymbol{\nu} \sim \text{Gau}\big(\boldsymbol{\mu}, \tau^2(I - \gamma H)^{-1}\big), \tag{7.5.15}$$

where H is a 56×56 matrix whose (i, j)th element is

$$h_{ij} = \begin{cases} 1, & \text{if the } i\text{th and } j\text{th districts are adjacent,} \\ 0, & \text{otherwise;} \end{cases}$$

the adjacencies are given in Table 1 of Clayton and Kaldor (1987). Recall from Section 6.4 that the spatial dependence parameter γ ranges over all values for which $(I - \gamma H)$ is positive-definite. The mean parameter $\boldsymbol{\mu}$ in (7.5.15) is typically expressed as

$$\boldsymbol{\mu} = X\boldsymbol{\beta}, \qquad (7.5.16)$$

where the columns of X are explanatory variables that may include functions of site location. Tsutakawa et al. (1985) and Manton et al. (1989) take a similar approach, but their prior distributions do not exhibit any spatial dependence such as in (7.5.15). The prior parameters β, γ, and τ^2 could then be assumed distributed according to some hyperprior, resulting in an hierarchical Bayes approach. Instead, it will be assumed in this section that the prior parameters are fixed but unknown, to be estimated from the data, which is an *empirical Bayes* approach.

Although the spatial model (7.5.12) and (7.5.15) is highly parametric, it accounts for the unequal variation and spatial dependence in a clever way. Through the latent θ variable, the model becomes a continuous one (for which numerical optimization methods are available; see Section 7.2), but with a Poisson-based noise component. A similar model is given by Zeger (1988) for analyzing time series of counts.

The next task is to determine the posterior density of $\boldsymbol{\nu}$ given the data $\mathbf{L} \equiv (L_1, \ldots, L_{56})'$; for the purpose of this presentation, $E_1, \ldots, E_{56}$ are assumed known. Clayton and Kaldor (1987) show that the posterior distribution is approximately multivariate Gaussian with mean and variance:

$$E(\boldsymbol{\nu}|\mathbf{L}) \simeq \hat{\boldsymbol{\nu}} \equiv \left[\{(I - \gamma H)/\tau^2\} + \Delta\right]^{-1}$$
$$\times\left[\{(I - \gamma H)/\tau^2\}X\boldsymbol{\beta} + \Delta\boldsymbol{\nu}^0 - (1/2)\mathbf{1}\right], \qquad (7.5.17)$$

$$\text{var}(\boldsymbol{\nu}|\mathbf{L}) \simeq V \equiv \left[\{(I - \gamma H)/\tau^2\} + \Delta\right]^{-1}, \qquad (7.5.18)$$

where $\boldsymbol{\nu}^0$ is a 56×1 vector whose ith element is $\log\{(L_i + (1/2))/E_i\}$ and Δ is a 56×56 diagonal matrix whose (i, i)th element is $(L_i + (1/2))$. The right-hand side of (7.5.17) shows the familiar compromise one sees in Bayes estimators between the raw estimate $\boldsymbol{\nu}^0$ and the prior mean $X\boldsymbol{\beta}$. Under squared-error loss, the best predictor of $\boldsymbol{\theta}$ is $\hat{\boldsymbol{\theta}} \equiv E(\boldsymbol{\theta}|\mathbf{L}) = (E(e^{\nu_1}|\mathbf{L}), \ldots, E(e^{\nu_{56}}|\mathbf{L}))'$. From (7.5.17), (7.5.18), and standard results for the log Gaussian distribution,

$$\hat{\theta}_i = \exp(\hat{\nu}_i + (1/2)v_{ii}), \qquad i = 1, \ldots, 56, \tag{7.5.19}$$

where v_{ii} is the (i, i)th element of the matrix V defined by (7.5.18).

If $\boldsymbol{\beta}$, τ^2, and γ were known, $\hat{\boldsymbol{\theta}} = (\hat{\theta}_1, \ldots, \hat{\theta}_{56})'$ could be used in a (choropleth or framed-rectangle) map of relative risk, instead of the noisy, heteroskedastic raw incidence rates $(L_1/E_1, \ldots, L_{56}/E_{56})'$. Estimation of these parameters can be achieved by maximum likelihood, using the EM algorithm (Dempster et al., 1977); see following text.

Smoothed regional maps are a very effective way of discovering geographical patterns in highly variable regional data. However, the importance of obtaining and presenting measures of precision for stochastic smoothers has not generally been realized. For example, the smoother $\hat{\boldsymbol{\theta}}$ given by (7.5.19) has an accompanying mean quadratic prediction-error matrix $E\{(\boldsymbol{\theta} - \hat{\boldsymbol{\theta}})(\boldsymbol{\theta} - \hat{\boldsymbol{\theta}})'\}$, where $E\{\cdot\}$ is a joint expectation over $\boldsymbol{\theta}$ and $\mathbf{L}$. It is important to obtain the whole matrix and not just diagonal elements, because regions may be aggregated later. Section 7.6 gives an approximate expression for the mean quadratic prediction-error matrix of a Markov random-field-based smoother of sudden infant death rates in the counties of North Carolina.

For some purposes, the empirical Bayes predictors $\hat{\theta}_1, \ldots, \hat{\theta}_{56}$ are too smooth. For example, to predict the (weighted) proportion of districts whose lip-cancer rate is greater than, say 4×10^{-5}, $\hat{\boldsymbol{\theta}} = E(\boldsymbol{\theta}|\mathbf{L})$ is not appropriate. Because $\text{var}(\boldsymbol{\theta}) - \text{var}(\hat{\boldsymbol{\theta}})$ is always nonnegative-definite, a rougher estimator of $\boldsymbol{\theta}$ is needed; Cressie (1990c) suggests a constrained empirical Bayes predictor.

EM Algorithm

The EM algorithm (Dempster et al., 1977) is designed to compute maximum likelihood estimates of parameters when some of the data are missing. It proceeds in an iterative manner, from the E step to the M step and back to the E step until convergence is attained. The E step (E stands for expectation) evaluates the conditional expectation of the loglikelihood based on both the observed and the missing data, conditional on the data observed; the conditional expectation is calculated using the prameter values from the previous M step. The subsequent M step (M stands for maximization) maximizes this expected loglikelihood over the unknown parameter values, providing an estimate for the next E step. The algorithm converges to a local maximum of the likelihood, albeit rather slowly.

Assuming the model given by (7.5.12), (7.5.15), and (7.5.16), m.l. estimates of $\boldsymbol{\beta}$, τ^2, and γ are sought based on the data $\mathbf{L}$. The EM algorithm can be used once one realizes that $\boldsymbol{\nu}$ can be considered as missing data. Thus, the

E step reduces to evaluating

$$Q(\boldsymbol{\beta}, \tau^2, \gamma) = E\{-(n/2)\log(2\pi) + (1/2)\log(|I - \gamma H|) - (1/2)\log \tau^2 \\ -(1/2)(\boldsymbol{\nu} - X\boldsymbol{\beta})'(I - \gamma H)(\boldsymbol{\nu} - X\boldsymbol{\beta})/\tau^2 | \mathbf{L}\},$$

where the conditional expectation over $\boldsymbol{\nu}$ given $\mathbf{L}$ is calculated using parameter values equal to $\boldsymbol{\beta}^0$, $(\tau^0)^2$, and γ^0 given by the previous M step. The subsequent M step maximizes Q with respect to $\boldsymbol{\beta}$, τ^2, and γ, and the resulting estimates are used in the subsequent E step. This sequence of EM steps is repeated until the estimates converge. (Further details are given in Appendix 1 of Clayton and Kaldor, 1987.)

The EM algorithm yields m.l. estimators $\hat{\boldsymbol{\beta}}$, $\hat{\tau}^2$, and $\hat{\gamma}$, which, upon substitution into (7.5.17) through (7.5.19), yields empirical Bayes predictors of $\theta_1, \ldots, \theta_{56}$. It is these predictors upon which Clayton and Kaldor (1987) propose to base their regional map of relative risks. This parametric approach is in the same spirit as the spatial exploratory map based on the smoothed values (7.5.9), although the latter should be seen as a preliminary, nonparametric analysis.

7.6 SUDDEN-INFANT-DEATH-SYNDROME DATA

This section completes the study of data on Sudden Infant Death Syndrome (SIDS) in North Carolina, begun in Section 6.2 (see also Section 4.4). There, an exploratory spatial data analysis was carried out to gain some understanding of the spatial relationships present in the data. With the modeling, estimation, and inference results of Sections 6.4, 7.2, and 7.3 now in place, a parametric model of the SIDS data can be built and subjected to a confirmatory analysis. Most of the material presented in this section can be found in Cressie and Chan (1989). Further analysis is given by Cressie (1992).

The data analyzed here are the number of sudden infant deaths $\{S_i: i = 1, \ldots, 100\}$ and the number of live births $\{n_i: i = 1, \ldots, 100\}$ in the 100 counties of North Carolina from July 1, 1974 to June 30, 1978. An identical analysis is given of SIDS data from July 1, 1979 to June 30, 1984; data from the intervening one year period was not available to me. Table 6.1 has the original counts and the necessary spatial information.

The very use of the word "sudden" to describe this category of postneonatal death implies that very little is known about it. Some background is given in Section 6.2; the results of this section will show that SIDS appears to have changed drastically from the earlier time period to the later time period. This may be due to reporting bias or may be an indication of the effect of public education programs in infant health.

Given that the race of the baby is thought to be an important factor, a more extensive analysis would obtain SIDS data broken down by race. Then one could either carry out a multivariate spatial analysis on the resulting data or race-standardize the SID rates in each county [see (7.5.12) and (7.5.13)] and analyze the standardized data.

7.6.1† Exploratory Spatial Data Analysis

A summary of the findings of Section 6.2, for the 1974–1978 SIDS data, follows.

1. It is concluded that the Freeman–Tukey (square-root) transformation

$$Z_i \equiv (1000(S_i)/n_i)^{1/2} + (1000(S_i + 1)/n_i)^{1/2} \qquad (7.6.1)$$

   is a variance-controlling transformation; that is,

$$\text{var}(Z_i) \simeq \sigma^2/n_i.$$

   This transformation controls the variance in the sense that it does not depend on $E(S_i/n_i)$. However, the variances are still vastly different from county to county, because $248 \leq n_i \leq 21{,}588$, $i = 1, \ldots, 100$.

   A comparison of stem-and-leaf plots of $\{1000(S_i + 1)/n_i\colon i = 1, \ldots, 100\}$ and $\{Z_i\colon i = 1, \ldots, 100\}$ shows that transformation (7.6.1) also symmetrizes the data, an effect that is retained even after spatial trend is removed.
2. By thinking of the data as being equal to large-scale variation (spatial trend) plus small-scale variation (spatially dependent error), an attempt was made in Section 6.2 to isolate each component. The principle of transforming to achieve additivity over all scales of variation (Cressie, 1985b) and the two-dimensional (spatial) coordinate representation of the counties' locations, leads naturally to a two-way additive decomposition for the spatial trend of the transformed data. A rectangular grid with 20 × 20 mile grid spacing is overlaid onto North Carolina, yielding 9 east–west transects, 24 north–south transects, and 9 × 24 nodes of the grid. The ith county is identified with the node $(k(i), l(i))$ that is closest to its county seat. This particular choice of a regularly spaced rectangular grid gives roughly one county at each grid node; on other occasions (see Section 7.5.1), irregularly spaced rectangular grids have been used to achieve the same effect.
3. From a weighted median polish, the decomposition

$$Z_i = \tilde{m} + \tilde{r}_{k(i)} + \tilde{c}_{l(i)} + R(x_i, y_i), \qquad (7.6.2)$$

   is obtained, where $\tilde{m}$ is an estimated all effect, $\{\tilde{r}_k\colon k = 1, \ldots, 9\}$ are estimated row (or north–south) effects, $\{\tilde{c}_l\colon l = 1, \ldots, 24\}$ are estimated column (or east–west) effects, and $\{R(x_i, y_i): i = 1, \ldots, 100\}$ are the residuals, placed at their respective county-seat locations.

4. From $\{n_i^{1/2} R(x_i, y_i): i = 1, \ldots, 100\}$, it is demonstrated that Anson county (county 4) has an unexplainably high value. Because we do not want county 4 to influence the model fit to the other 99 counties, it is removed from subsequent analysis. However, the fitted spatial model will allow a predicted value to be obtained.
5. Based on data $\{n_i^{1/2} R(x_i, y_i): i = 1, \ldots, 100, i \neq 4\}$, an isotropic variogram estimator $2\hat{\gamma}^o$ is calculated in Section 4.4 and plotted in Figure 4.16. The figure indicates positive spatial dependence at distances up to 30 miles and zero correlations at greater distances.

This univariate spatial data analysis should be accompanied with various scatter plots of Z on potential explanatory variables X. Any relationship observed is usually modeled as large-scale variation, which might also help to explain some of the spatial dependence in the extant small-scale variation. Parameters from both scales of variation can then be estimated, for example, by maximum likelihood.

Figure 7.9 shows plots of the points $\{(X_{5,i}, Z_i): i = 1, \ldots, 100\}$, where $X_{5,i}$ is the (Freeman–Tukey transformed) *nonwhite live birth rate* in county i. That is,

$$X_{5,i} \equiv (1000(\overline{w}_i)/n_i)^{1/2} + (1000(\overline{w}_i + 1)/n_i)^{1/2}, \qquad (7.6.3)$$

where $\overline{w}_i$ is the number of nonwhite live births in the ith county of North Carolina; see Table 6.1. For 1974–1978, Figure 7.9*a* shows a positive relationship; this is to be expected in light of some of the introductory remarks in Section 6.2. Surprisingly, for 1979–1984, Figure 7.9*b* shows that this relationship is not nearly as strong.

Explanatory variables were sought by linearly regressing (Freeman–Tukey transformed) SID rate on (functions of) X_1 = population density, X_2 = percent urban, X_3 = number of hospital beds per 100,000 population, X_4 = median family income, and X_5 = Freeman–Tukey transformed nonwhite live birth rate. Logistic regression based on the five explanatory variables $X_1, \ldots, X_5$ was also carried out on the raw SID rates. In both cases, a diagonal error variance matrix was assumed on this first pass. Regardless of which regression was used, X_5 was the only significant variable for 1974–1978, whereas for 1979–1984 none was significant.

For the earlier period, the fitted value for county 4 was 3.42, obtained from regression of Z on X_5 (fitted by weighted least squares using weights proportional to numbers of live births, and data from all counties except county 4). When compared with the observed (Freeman–Tukey transformed) value of 6.28, a *standardized* difference of 17.29 is obtained. Thus, not even the explanatory variable can account for county 4's outlying value.

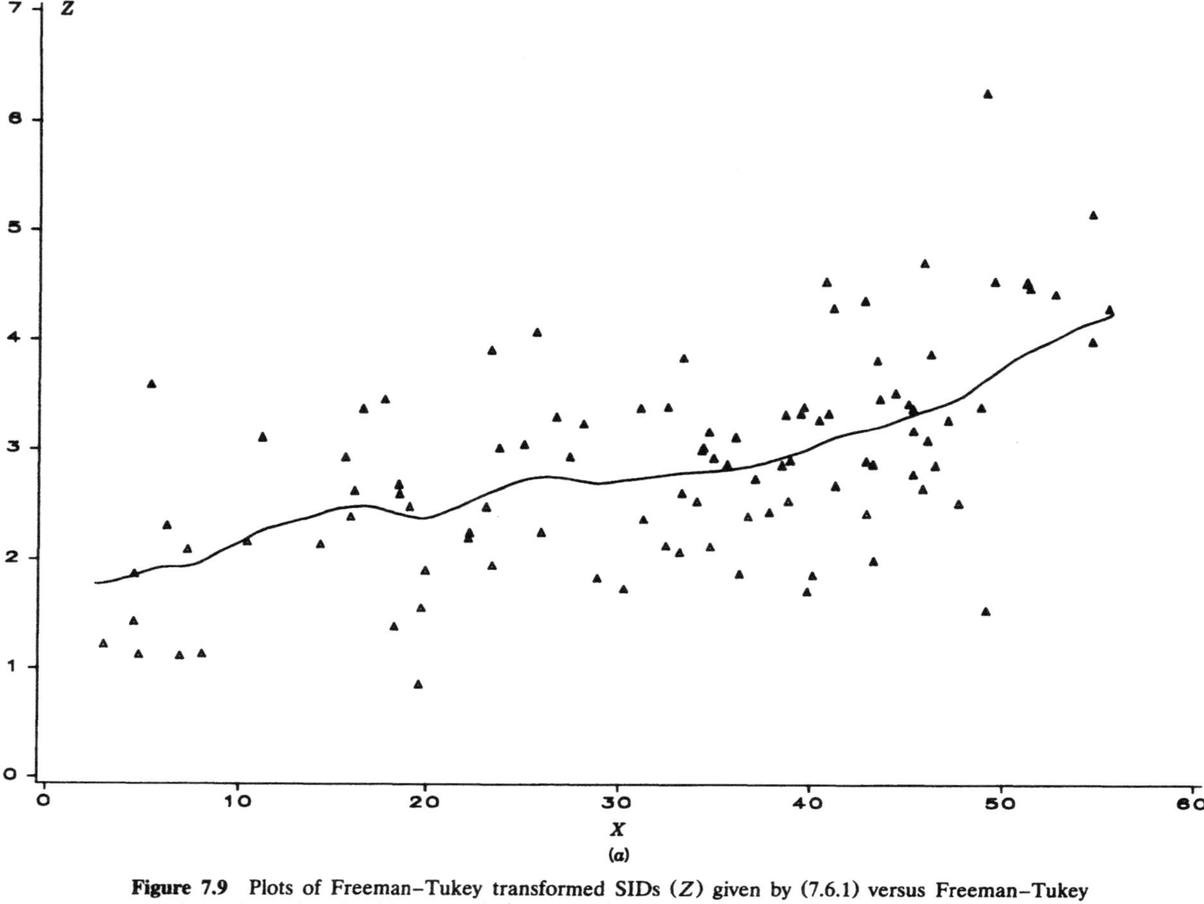

Figure 7.9 Plots of Freeman–Tukey transformed SIDs (Z) given by (7.6.1) versus Freeman–Tukey transformed nonwhite live birth rates (X) given by (7.6.3), for the counties of North Carolina. Triangles denote the pairs $\{(X_i, Z_i): i = 1, \ldots, 100\}$ and the solid line is the smooth, $\hat{Z}(x) = \{\sum_{i=1}^{100} \exp(-|X_i - x|/4)I(|X_i - x| \le 10)Z_i\}/\{\sum_{i=1}^{100} \exp(-|X_i - x|/4)I(|X_i - x| \le 10)\}$. (*a*) 1974–1978; (*b*) 1979–1984.

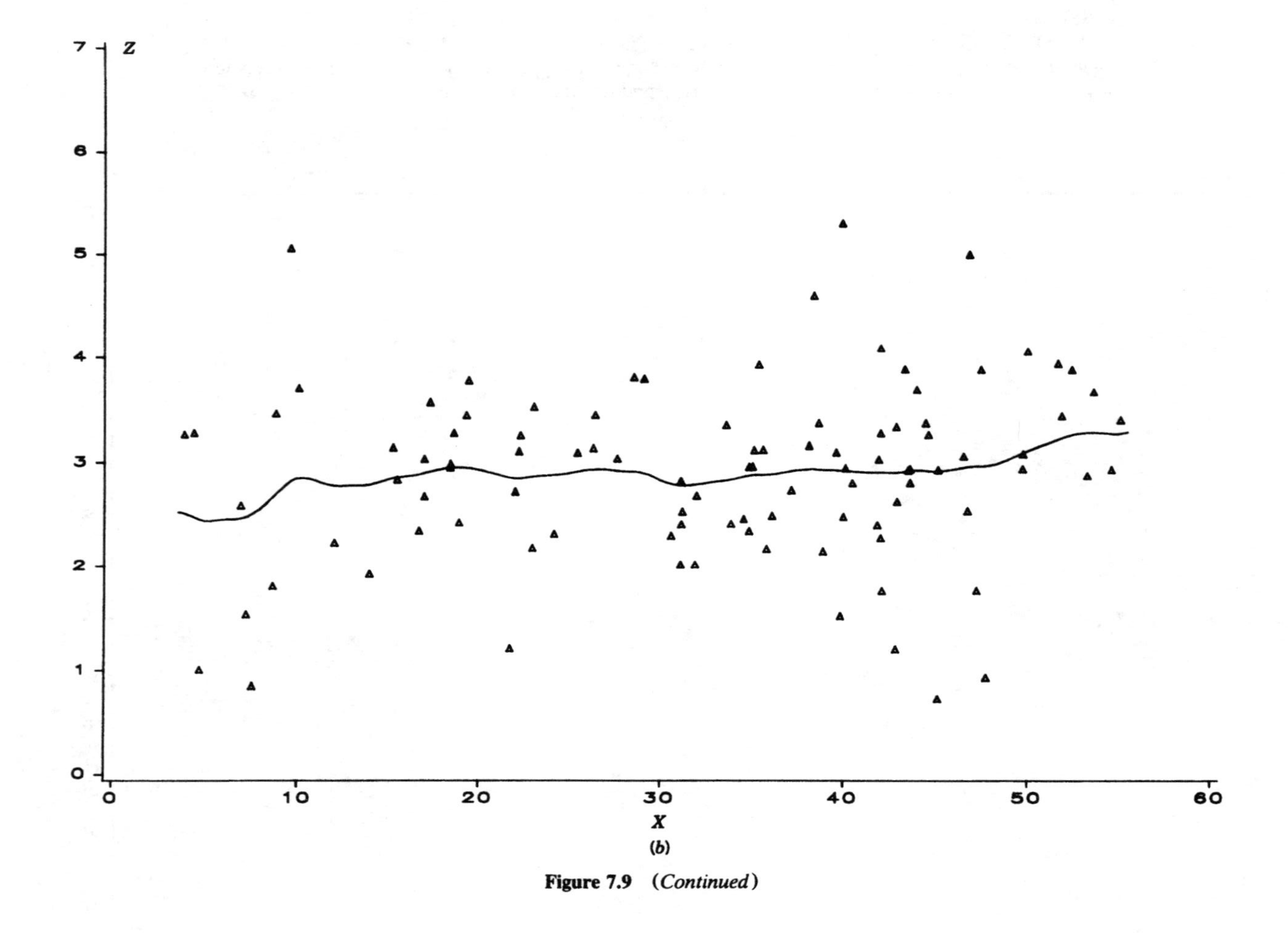

(b)

Figure 7.9 (*Continued*)

7.6.2 Auto-Poisson Model

Recall (from Section 6.5.2) that the auto-Poisson's conditional specification (assuming pairwise-only dependence between sites) is

$$\Pr\left(z(\mathbf{s}_i) \middle| \{z(\mathbf{s}_j): j \in N_i\}\right) = \exp(-\lambda_i)\lambda_i^{z(\mathbf{s}_i)}/z(\mathbf{s}_i)!, \qquad i = 1,\ldots,n, \tag{7.6.4}$$

where $\lambda_i = \lambda_i(\{z(\mathbf{s}_j): j \in N_i\})$ is a function of data observed for the regions N_i that neighbor the region i $(i = 1,\ldots,n)$. Then, from (6.4.10),

$$\lambda_i\left(\{z(\mathbf{s}_j): j \in N_i\}\right) = \exp\left\{\alpha_i + \sum_{j=1}^{n} \theta_{ij} z(\mathbf{s}_j)\right\}, \tag{7.6.5}$$

where $\theta_{ij} = \theta_{ji}$, $\theta_{ii} = 0$, and $\theta_{ik} = 0$ for $k \notin N_i$.

To estimate the parameters $\{\alpha_i\}$ and $\{\theta_{ij}\}$ by maximum likelihood, one needs the *joint* probability mass function implied by (7.6.4) and (7.6.5). It is

$$\Pr(\mathbf{z}) = \exp(Q(\mathbf{z})) \Big/ \sum_{\mathbf{y} \in \zeta} \exp(Q(\mathbf{y})), \tag{7.6.6}$$

where $Q(\mathbf{z})$ is given by (6.5.8). The normalizing constant $\Sigma_{\mathbf{y} \in \zeta}\exp(Q(\mathbf{y}))$ is a sequence of n infinite sums, because the support of each Z variable is $\{0, 1, 2, \ldots\}$. Although the constant does not depend on the data, it is a function of the parameters and so is crucial in determining the likelihood. The parameter space is $\{\boldsymbol{\alpha}, (\theta_{ij}): \Sigma_{\mathbf{y} \in \zeta} \exp(Q(\mathbf{y})) < \infty\}$, which for the auto-Poisson process is $\{\boldsymbol{\alpha}, (\theta_{ij}): \boldsymbol{\alpha} \in \mathbb{R}^n;\ \theta_{ij} \le 0 \text{ for all } i, j = 1,\ldots,n\}$ (Besag, 1974). Thus, spatial dependence parameters must be *negative* for the auto-Poisson process, so that any estimates must also be negative.

Section 6.5.2 shows how the SIDS data can be modeled according to an auto-Poisson process. Here, I shall present the analysis only for the 1974–1978 data (with county 4 removed). Three levels of large-scale variation are fitted. The simplest one is

$$\exp(\alpha_i) = n_i \rho, \qquad i = 1,\ldots,100, \tag{7.6.7}$$

which assumes a homogeneous SID rate throughout the counties of North Carolina.

The second level of large-scale variation is

$$\exp(\alpha_i) = n_i \rho_{R(i)}, \qquad i = 1,\ldots,100, \tag{7.6.8}$$

which assumes that the rates are homogeneous within each of 12 regions of North Carolina, but possibly different across regions. The regions consist of

12 parcels of contiguous counties and are defined by the mapping R:

$$\begin{aligned} R&: \{3,5,6,34,85,86,95,97,99\} \to \{1\},\\ R&: \{1,17,32,35,39,41,64,68,73,79,91,93\} \to \{2\},\\ R&: \{8,15,21,27,37,42,46,66,70,72\} \to \{3\},\\ R&: \{11,44,56,58,61,87,100\} \to \{4\},\\ R&: \{2,12,14,18,29,30,49,80\} \to \{5\},\\ R&: \{19,43,51,53,76,92,96,98\} \to \{6\},\\ R&: \{7,28,33,40,48,59,74,89,94\} \to \{7\},\\ R&: \{20,22,38,45,50,57,75,81,88\} \to \{8\},\\ R&: \{4,13,23,36,55,60,84,90\} \to \{9\},\\ R&: \{26,31,47,62,63,77,82,83\} \to \{10\},\\ R&: \{16,25,52,54,67,69\} \to \{11\},\\ R&: \{9,10,24,65,71,78\} \to \{12\}. \end{aligned} \tag{7.6.9}$$

The third level of large-scale variation is

$$\exp(\alpha_i) = n_i \rho_{M(i)} \equiv n_i \exp(\mu + \omega_{k(i)} + \nu_{l(i)}), \tag{7.6.10}$$

where $(k(i), l(i))$ is the node of the 9×24 grid, referred to in Section 7.6.1, nearest to county $i (i = 1, \dots, 100)$.

The small-scale variation is modeled by

$$\theta_{ij} = \begin{cases} \gamma d_{ij}^{-k} / \max\{d_{il}^{-k} : l \in N_i;\, i = 1, \dots, n\}, & j \in N_i, \\ 0, & j \notin N_i, \end{cases} \tag{7.6.11}$$

where $k = 0$, 1, or 2 and N_i is the set of neighbors of county i, here defined to be the set of counties whose county seat is within 30 miles of the ith county seat, $i = 1, \dots, 100$. See Table 6.1 for the complete set of county neighbors.

Section 6.5.2 has a more detailed discussion that motivates (7.6.11). Briefly, the dimensionless parameter γ captures the spatial dependence, and the normalization by $\max\{\cdot\}$ in (7.6.11) ensures comparability of models across different values of k. Recall that a value of $\gamma > 0$ is not in the parameter space of the auto-Poisson model.

After omitting county 4 (both as a lattice site and as a neighbor of other lattice sites), the parameters were estimated by maximizing the *pseudolikelihood* (see Section 7.2.1), rather than the exact likelihood (7.6.6). Although they are more efficient, m.l. estimators are difficult to find because of a computationally intractable normalizing constant. The negative log pseudolikelihood,

$$P(\boldsymbol{\alpha}, \gamma) \equiv \sum_{i=1}^{n} \{\lambda_i - z(\mathbf{s}_i)\log \lambda_i + \log(z(\mathbf{s}_i)!)\},$$

where λ_i is given by (7.6.5) and (7.6.11), can be maximized using the IMSL subroutine UMINF (IMSL Inc., 1987, Chapter 8, pp. 802–806).

Starting with the very simple two-parameter auto-Poisson model, (7.6.7), (7.6.11), and $k = 1$, the maximum pseudolikelihood (m.p.l.) estimates

$$\hat{\rho} = 1.785 \times 10^{-3}, \qquad \hat{\gamma} = 8.216 \times 10^{-3}$$

were obtained. The positive value of $\hat{\gamma}$ is unfortunate and is also found for the more complicated models (7.6.8), (7.6.11) and (7.6.10), (7.6.11). Even if $\hat{\gamma}$ were negative, there is no way to assess whether this indicates true spatial dependence or whether it is simply a result of sampling from the (spatially independent) case where $\gamma = 0$. Guyon (1987) shows that, under a spatial exponential mixing assumption, the m.p.l. estimator is asymptotically Gaussian, but its asymptotic variance is not available in closed form. Thus, inference on γ based on the m.p.l. estimator is not immediately possible.

The auto-Poisson model is clumsy to work with because of the restricted parameter space, difficulty in computing m.l. estimates, and because few inference results are available. Moreover, the auto-Poisson model (7.6.4) does *not* yield a Poisson marginal distribution for each $Z(\mathbf{s}_i)$, $i = 1, \ldots, 100$. A way around some of these problems is to assume a spatial Gaussian latent process $\{\nu(\mathbf{s})\colon \mathbf{s} \in D\}$. Given $\nu(\cdot)$, assume

$$S_i \sim \text{Po}(n_i \cdot \exp\{\nu(\mathbf{s}_i)\}), \qquad i = 1, \ldots, n \tag{7.6.12}$$

(Clayton and Kaldor, 1987; Section 7.5.2), where Po(λ) is shorthand for a Poisson distribution with mean λ. Another approach, taken in the next section, is to transform the counts data so that they are approximately continuous.

7.6.3 Auto-Gaussian Model

For reasons given in the previous section, the confirmatory analysis of the SIDS data will be based on the auto-Gaussian (or CG) model (Section 6.6) of Freeman–Tukey transformed counts (7.6.1). The logistic transformation was not used because it has small-scale parameters that depend functionally on large-scale parameters.

There are other approaches that could have been taken. Clayton and Kaldor (1987) propose a spatial mixture model where the data $\{S_i\colon i = 1, \ldots, n\}$ are, conditional on their means, independently Poisson distributed. Then the logarithms of the means are assumed to follow an auto-Gaussian process. This was referred to as a Poisson-log-Gaussian model in Section 7.5.2. Alternatively, a *quasilikelihood* (e.g., McCullagh, 1983) could be constructed by modeling the variance matrix of the counts $\{S_i: i = 1, \ldots, n\}$ as a function of its mean vector $\boldsymbol{\mu}$. Spatial covariances between pairs of counts S_i, S_j could be expressed as a (positive-definite) function of the distance d_{ij} between county i and county j. Although spatial dependence is allowed in this approach, it must be in the form of a precisely known function of $\boldsymbol{\mu}$, apart from a possibly unknown multiplicative constant. That is, the correlation between the S_is cannot have any unknown parameters. Not only is this unrealistic, but it is also undesirable in spatial statistical modeling because one would like the data to determine the strength of spatial dependence.

Various auto-Gaussian models suggested by the data analysis described in Section 7.6.1 will be presented. The conclusions of Section 6.2 suggest that the large-scale variation can be successfully captured with a two-way spatial trend. Equally, the significance of the explanatory variable "non-white live birth rate" for the 1974–1978 data (see Section 7.6.1) suggests a regression relationship. We shall try both of these in the presence of small-scale variation by fitting different parameterizations of the auto-Gaussian model. Identical spatial analyses are performed on the data sets for the earlier and later time periods, although the data analysis in Section 7.6.1 indicates that "nonwhite live birth rate" is not an important explanatory variable in the later period.

Modeling Large-Scale Variation

Consider the auto-Gaussian model (6.3.11), where the Freeman–Tukey transformed SID rate Z_i [given by (7.6.1)] plays the role of the $Z(\mathbf{s}_i)$ and $\mathbf{s}_i = (x_i, y_i)'$ are the coordinates of the ith county seat. Then, from (6.6.4),

$$\mathbf{Z} \sim \text{Gau}\big(\boldsymbol{\mu}, (I - C)^{-1}M\big). \tag{7.6.13}$$

The following models for $\boldsymbol{\mu} \equiv (\mu_1, \ldots, \mu_n)'$ were fitted.

Model I. $\quad \mu_i = m, \qquad i = 1, \ldots, n;$

$$\text{that is,} \quad \boldsymbol{\beta} = m. \tag{7.6.14}$$

Model II. $\quad \mu_i = m_{R(i)}, \qquad i = 1, \ldots, n;$

$$\text{that is,} \quad \boldsymbol{\beta} = (m_1, \ldots, m_{12})'. \tag{7.6.15}$$

Model III. $\quad \mu_i = m_{M(i)}, \qquad i = 1, \ldots, n;$

$$\text{that is,} \quad \boldsymbol{\beta} = (m, r_1, \ldots, r_8, c_1, \ldots, c_{23})'. \tag{7.6.16}$$

Model IV. $\quad \mu_i = m + bX_{5,i}, \qquad i = 1, \ldots, n;$

$$\text{that is,} \quad \boldsymbol{\beta} = (m, b)'. \tag{7.6.17}$$

Model II uses 12 parameters, each corresponding to one of 12 parcels of contiguous counties [see (7.6.9)]. Model III uses 32 parameters, which are obtained by overlaying the same 9×24 grid, referred to in Section 7.6.1, onto the county map of North Carolina. County i is assigned to the nearest node $(k(i), l(i))$ of the grid and its mean is modeled as

$$m_{M(i)} = m + r_{k(i)} + c_{l(i)}. \tag{7.6.18}$$

With restrictions on the parameters to make them identifiable, there are $9 + 24 - 1 = 32$ large-scale variation parameters to be estimated.

Model IV uses just two parameters. Assume that $\{X_{5,i}: i = 1, \dots, n\}$, the (Freeman–Tukey transformed) nonwhite live birth rates given by (7.6.3), appear linearly in the mean of the (Freeman–Tukey transformed) SID rates.

Notice that Model I is nested within Model II, Model III and Model IV, and Model II is essentially nested within Model III. In what is to follow, the fits of these various models will be compared.

Modeling Small-Scale Variation

The neighborhood sets (6.4.4) of the Markov random field were defined as

$$j \in N_i \quad \text{if} \quad d_{ij} \equiv \left\{(x_i - x_j)^2 + (y_i - y_j)^2\right\}^{1/2} \le 30 \text{ miles}, \tag{7.6.19}$$

where recall that $(x_i, y_i)'$ is the location of the ith county seat. This 30-mile criterion was conservatively assessed by looking at the estimated variogram in Figure 4.16 and seeing at what distance spatial correlation became approximately zero.

Now, from (6.3.12), any $j \notin N_i$ has $c_{ij} = 0$. For those remaining nonzero c_{ij}s, the spatial dependence is modeled with just one parameter ϕ. Different dependencies between county i and its neighbors are assumed to be functions of distance (Cliff and Ord, 1981, p. 144):

$$c_{ij} = \begin{cases} \phi\left(d_{ij}^{-k}/C(k)\right)\left(n_j/n_i\right)^{1/2}, & j \in N_i, \\ 0, & j \notin N_i, \end{cases} \tag{7.6.20}$$

where $C(k) = \max\{d_{ij}^{-k}: j \in N_i; i = 1, \dots, n\}$ and $k = 0$, 1, or 2. Choice of this $C(k)$ allows comparability of interpretation of ϕ across different values of k. The parameterization (7.6.20) is compatible with

$$\operatorname{var}\left(Z(\mathbf{s}_i) \mid \{Z(\mathbf{s}_j): j \in N_i\}\right) = \tau^2/n_i, \tag{7.6.21}$$

because $c_{ij}\tau_j^2 = c_{ji}\tau_i^2$. Note that the square of the partial correlation of $Z(\mathbf{s}_i)$ and $Z(\mathbf{s}_j)$ is $c_{ij}c_{ji} = \phi^2\{C(k)d_{ij}^{-k}\}^2$, independent of n_i and n_j.

Combining the small-scale variation parameters ϕ and τ^2 with the large-scale ones, it can be seen that the auto-Gaussian Models I, II, III, and IV have 3, 14, 34, and 4 parameters, respectively, to be estimated.

In the analysis to follow, Anson County (county 4) is set aside. Its 1974–1978 SID rate of 9.6 per 1000 is very unlike the other 99 counties. This unusually high rate is not explainable from an unusual value of the covariate, nonwhite live birth rate, in that county. By setting aside county 4, my intention is to avoid one unusual value affecting the fit to the other 99 counties. However, county 4 should not be forgotten; rather it requires separate consideration.

Maximum Likelihood Estimation of Auto-Gaussian Parameters

Now proceed with fitting auto-Gaussian Models I, II, III, and IV to the Freeman–Tukey transformed SID rates for 99 of the 100 counties of North Carolina, 1974–1978, and do the same for the 1979–1984 data. The negative loglikelihood is given by $L(\boldsymbol{\mu}, M, C)$ in (6.4.1). After invoking the parameterizations: $\boldsymbol{\mu} = X\boldsymbol{\beta}$ in (7.6.14), (7.6.15), (7.6.16), and (7.6.17); $M = \tau^2 \cdot \mathrm{diag}\{n_i^{-1}: i = 1, \ldots, 100; i \neq 4\}$ in (7.6.21); and c_{ij} in (7.6.20), L can be written as

$$L(\boldsymbol{\beta}, \tau^2, \phi) = (n/2)\log(2\pi) + (n/2)\log \tau^2 - (1/2)\log|D^{-1}(I - \phi H)| + (1/2)(\mathbf{z} - X\boldsymbol{\beta})'D^{-1}(I - \phi H)(\mathbf{z} - X\boldsymbol{\beta})/\tau^2. \quad (7.6.22)$$

In (7.6.22), $D = \mathrm{diag}\{n_i^{-1}: i \neq 4\}$, $H = (h_{ij})$, with $h_{ij} = (d_{ij}^{-k}/C(k))(n_j/n_i)^{1/2}$ if $j \in N_i$, $h_{ij} = 0$ otherwise, and $n = 99$. Henceforth, if an index i runs from 1 to 99, it is implicit that this refers to all of the counties except county 4.

To minimize (7.6.22), assume for the moment that ϕ is fixed. The maximum likelihood (m.l.) estimators of $\boldsymbol{\beta}$ and τ^2 are

$$\hat{\boldsymbol{\beta}}(\phi) = (X'D^{-1}(I - \phi H)X)^{-1}X'D^{-1}(I - \phi H)\mathbf{z}, \quad (7.6.23)$$

$$\hat{\tau}^2(\phi) = (\mathbf{z} - X\hat{\boldsymbol{\beta}})'D^{-1}(I - \phi H)(\mathbf{z} - X\hat{\boldsymbol{\beta}})/n = \mathbf{z}'D^{-1}(I - \phi H)\{I - X(X'D^{-1}(I - \phi H)X)^{-1} \cdot X'D^{-1}(I - \phi H)\}\mathbf{z}/n. \quad (7.6.24)$$

Substituting (7.6.23) and (7.6.24) back into (7.6.22), the m.l. estimate of ϕ can be obtained by minimizing the negative log profile likelihood $L^*(\phi) \equiv L(\hat{\boldsymbol{\beta}}(\phi), \hat{\tau}^2(\phi), \phi)$, given by

$$L^*(\phi) = (99/2)\log(2\pi) + (99/2) - (1/2)\sum_{i=1}^{99}\log(n_i) + (99/2)\log \hat{\tau}^2(\phi) - (1/2)\sum_{i=1}^{99}\log(1 - \phi\eta_i), \quad (7.6.25)$$

Table 7.5 Maximum Likelihood Estimates of Small-Scale-Variation Parameters ($\hat{\tau}^2, \hat{\phi}$) and 95% Confidence Intervals (*ci*) for Spatial Dependence Parameter ϕ, for the 1974–1978 Data[a]

	k		
	0	1	2
		Model I	
$\hat{\tau}^2$	1230.68	1443.17	1651.48
$\hat{\phi}$	0.173	0.833	0.596
ci	[0.13, 0.18[b]]	[0.53, 0.90[b]]	[−0.73, 0.99[b]]
		Model II	
$\hat{\tau}^2$	884.45	864.61	876.68
$\hat{\phi}$	0.0792	0.710	0.810
ci	[−0.06, 0.17]	[−0.10, 0.89]	[−0.22, 0.99[b]]
		Model III	
$\hat{\tau}^2$	681.86	681.91	682.13
$\hat{\phi}$	0.117	0.081	0.0336
ci	[−0.20, 0.17]	[−0.98, 0.89]	[−0.99[b], 0.99[b]]
		Model IV	
$\hat{\tau}^2$	1138.7	1169.2	1214.6
$\hat{\phi}$	0.113	0.640	0.336
ci	[−0.01, 0.18[b]]	[−0.24, 0.90[b]]	[−0.98, 0.99[b]]

Source: Cressie and Chan (1989). Reprinted by permission of the American Statistical Association.

[a]Upper (ϕ_u) and lower (ϕ_l) permissible values of ϕ are determined so that $(I - \phi D^{-1/2} H D^{1/2})$ is positive-definite. For $k = 0$, $(\phi_l, \phi_u) = (-0.328, 0.190)$; for $k = 1$, $(\phi_l, \phi_u) = (-0.997, 0.902)$; for $k = 2$, $(\phi_l, \phi_u) = (-0.999, 0.998)$.

[b]These limits of the confidence interval are rounded inward to avoid being outside the permissible range (ϕ_l, ϕ_u).

where $\{\eta_i: i = 1, \ldots, 99\}$ are the ordered eigenvalues of the symmetric matrix $D^{-1/2}HD^{1/2}$. Because $\eta_1 < 0 < \eta_n$, then $\phi \in (\eta_1^{-1}, \eta_n^{-1}) \equiv (\phi_l, \phi_u)$; see (7.2.35). The estimates for the 1974–1978 data are summarized in Tables 7.5 and 7.6.

Figure 7.10 shows a plot of $L^*(\phi)$ against ϕ for auto-Gaussian Model I given by (7.6.14), (7.6.20), and (7.6.21), for $k = 1$ (i.e., $c_{ij} \propto d_{ij}^{-1}$). This enables a $100(1 - \alpha)\%$ confidence interval for ϕ to be determined. From Whittle (1954, p. 441),

$$\Pr\left\{L^*(\phi) \le L^*(\hat{\phi}) + (n/(n - q - 2))\chi_1^2(\alpha)/2\right\} \simeq 1 - \alpha, \quad (7.6.26)$$

where $L^*(\phi)$ is given by (7.6.25), $\hat{\phi}$ is the m.l. estimator obtained by

Table 7.6 Estimates $\hat{\boldsymbol{\beta}}$ of $\boldsymbol{\beta}$ in $E(\mathbf{Z}) = X\boldsymbol{\beta}$, for $k = 1$ in (7.6.21), for the 1974–1978 Data

	Model I	
$\hat{m} = 2.8378$		
	Model II	
$\hat{m}_1 = 2.0559$	$\hat{m}_2 = 2.8704$	$\hat{m}_3 = 4.2564$
$\hat{m}_4 = 2.4778$	$\hat{m}_5 = 2.1505$	$\hat{m}_6 = 2.6388$
$\hat{m}_7 = 3.2788$	$\hat{m}_8 = 3.1126$	$\hat{m}_9 = 2.6793$
$\hat{m}_{10} = 2.8369$	$\hat{m}_{11} = 3.1786$	$\hat{m}_{12} = 3.6866$
	Model III	
$\hat{m} = 2.8256$		
$\hat{r}_1 = 0.5332$	$\hat{r}_2 = -0.4364$	$\hat{r}_3 = -0.1554$
$\hat{r}_4 = -0.0551$	$\hat{r}_5 = 0.1798$	$\hat{r}_6 = -0.2261$
$\hat{r}_7 = 0.3464$	$\hat{r}_8 = 1.7900$	$\hat{r}_9 = -0.1751$[a]
$\hat{c}_1 = 0.0994$	$\hat{c}_2 = -1.1723$	$\hat{c}_3 = -0.0449$
$\hat{c}_4 = -0.6183$	$\hat{c}_5 = -0.2885$	$\hat{c}_6 = -0.3209$
$\hat{c}_7 = 0.4522$	$\hat{c}_8 = -0.1693$	$\hat{c}_9 = -0.5585$
$\hat{c}_{10} = -0.2663$	$\hat{c}_{11} = -0.8086$	$\hat{c}_{12} = -0.4334$
$\hat{c}_{13} = 0.0699$	$\hat{c}_{14} = 0.7044$	$\hat{c}_{15} = 0.2574$
$\hat{c}_{16} = -0.3301$	$\hat{c}_{17} = -0.1000$	$\hat{c}_{18} = 0.4031$
$\hat{c}_{19} = 0.5548$	$\hat{c}_{20} = 0.3764$	$\hat{c}_{21} = 0.7087$
$\hat{c}_{22} = 0.2892$	$\hat{c}_{23} = -0.4697$	$\hat{c}_{24} = -1.2848$[a]
	Model IV	
$\hat{m} = 1.644$	$\hat{b} = 0.03457$	

[a]There are 32 independent parameters in Model III. $\hat{r}_9$ and $\hat{c}_{24}$ are obtained from the linear constraints $\Sigma_{i=1}^{9} q_i \hat{r}_i = 0$ and $\Sigma_{j=1}^{24} w_j \hat{c}_j = 0$, where $q_i \equiv \text{sum}\{n_h\colon k(h) = i\}$ and $w_j \equiv \text{sum}\{n_h\colon l(h) = j\}$. $(k(h), l(h))$ is the grid node nearest county h.

minimizing $L^*(\cdot)$, q is the number of *large-scale* parameters fitted, and $\chi_1^2(\alpha)$ is the upper $100(1-\alpha)\%$ point of the chi-squared distribution on 1 degree of freedom. For $n = 99$, $q = 1$, $k = 1$, and $\alpha = 0.05$, the 95% confidence interval becomes $\{\phi\colon L^*(\phi) \le L^*(\hat{\phi}) + 1.98\}$. Now, for auto-Gaussian Model I, $\hat{\phi} = 0.833$, $L^*(\hat{\phi}) = 124.87$, and hence $\{\phi\colon L^*(\phi) \le 126.85\}$ is a 95% confidence interval (shown in Table 7.5 to be [0.53, 0.90]). Because $L^*(0) = 130.26$, the null hypothesis H_0 is rejected when testing

$$H_0\colon \phi = 0 \quad \text{versus} \quad H_1\colon \phi \neq 0. \tag{7.6.27}$$

Thus, the spatial interaction *is* significant.

To see the effect on the models for varying values of k, Figure 7.11 shows a plot of $L^*(\phi)$ against ϕ for auto-Gaussian Model II given by (7.6.15),

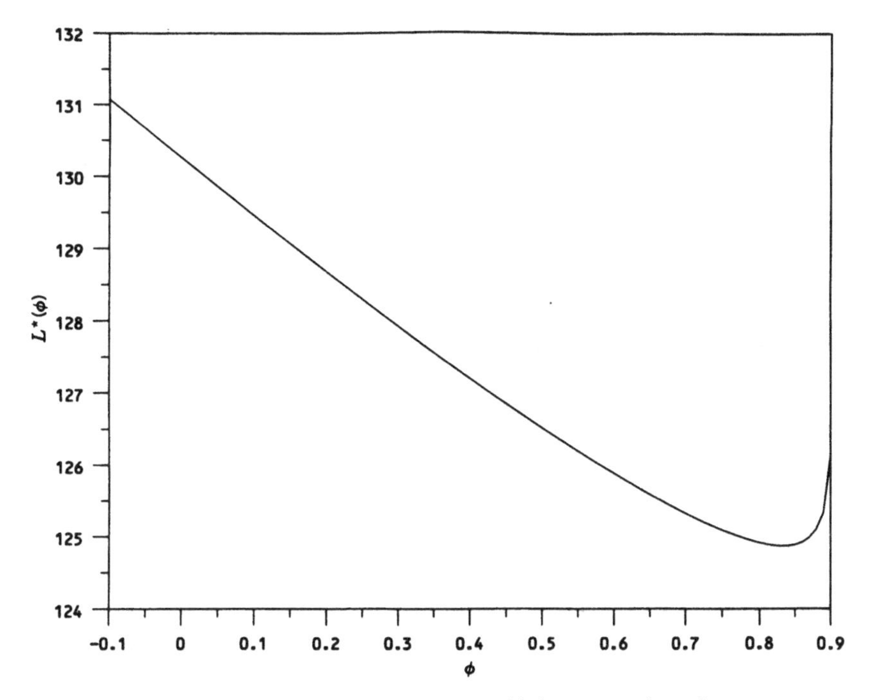

Figure 7.10 Plot of negative log profile likelihood $L^*(\phi)$ given by (7.6.25), as a function of spatial-dependence parameter ϕ; auto-Gaussian Model I with $k = 1$. [*Source*: Cressie and Chan, 1989]. Reprinted by permission of the American Statistical Association.

(7.6.20), and (7.6.21), for $k = 0$ (i.e., $c_{ij} \propto 1$), $k = 1$ (i.e., $c_{ij} \propto d_{ij}^{-1}$), and $k = 2$ (i.e., $c_{ij} \propto d_{ij}^{-2}$) together.

The likelihoods for Model II are considerably flatter than those for Model I. It can be seen from (7.6.26) and Table 7.5 that, for Model II ($q = 12$), $\phi = 0$ is contained in the 95% confidence interval for ϕ (for $k = 0$, 1, and 2). For Model III, Table 7.5 shows that the $q = 32$ large-scale parameters in $\boldsymbol{\beta}$ have removed all spatial dependence (as measured by $\hat{\phi}$).

Model IV has just $q = 2$ large-scale parameters, measuring the linear dependence on nonwhite live births. From Table 7.5, the estimated spatial dependence parameter $\hat{\phi}$ is not significantly different from zero, although there is still some spatial dependence present; see the $k = 0$ case in particular. In conclusion, the explanatory variable "nonwhite live birth rate" has captured much (but not all) of the strong spatial dependence present in Model I.

Model Selection

Models II through IV, given by (7.6.15) through (7.6.17), can each be compared with Model I [given by (7.6.14)], the simplest model. Whittle (1954, p. 441) gives the appropriate statistic to compute. If L_p is the negative

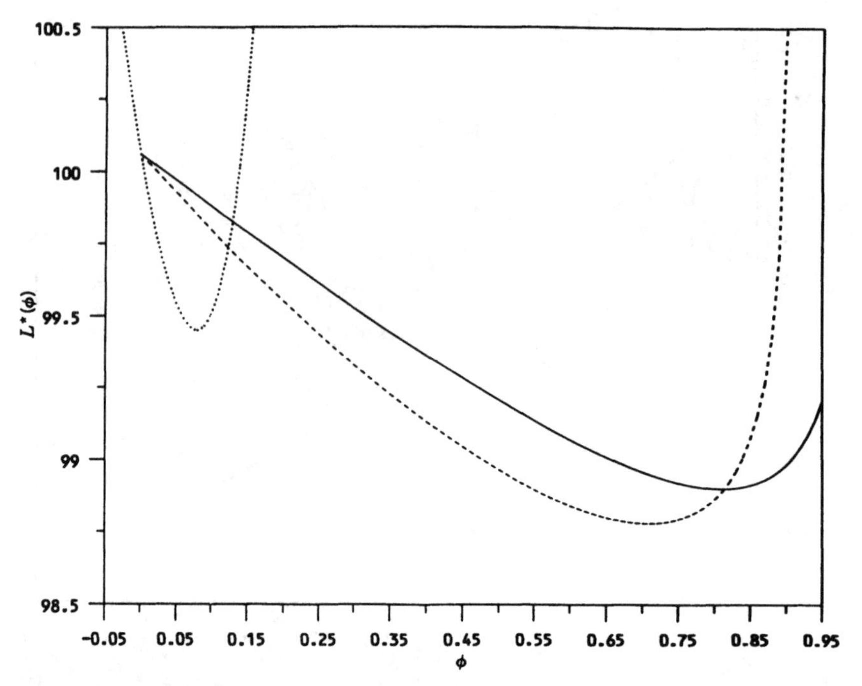

Figure 7.11 Same as in Figure 7.10, except auto-Gaussian Model II is shown for $k = 0$ (dotted line), $k = 1$ (dashed line), and $k = 2$ (solid line).

loglikelihood of a model where p parameters have been fitted by maximum likelihood and L_{p+r} is the same but with r *extra* parameters fitted, then the significance of those extra parameters can be assessed by comparing

$$U^2 \equiv 2\{(n-p-r)/n\}\{L_p - L_{p+r}\} \tag{7.6.28}$$

with a χ_r^2 random variable on r degrees of freedom. Table 7.7 shows the values of U^2 for both periods.

For 1974–1978, the best fit is Model II or Model IV. Within Model II the spatial-dependence parameter ϕ is not significantly different from zero [p value = 0.14, obtained from (7.6.26)]; similarly for Model IV [p value = 0.13, obtained from (7.6.26)]. For reasons of interpretability and parsimony, I prefer Model IV, which says that SID rates depend significantly on the race of the babies born; the estimated linear dependence is given by $\hat{b}$ in Table 7.6. Notice that there is still some evidence of spatial dependence in the model errors, but that it is not significant.

The 1979–1984 data present a different story; see Table 7.7. Model I, which assumes a homogeneous SID rate throughout North Carolina, is the best choice and, within Model I, the spatial-dependence parameter is not significantly different from zero (p value = 0.15). It is intriguing that the

Table 7.7 Goodness-of-Fit Statistic U^2 Defined by (7.6.28), $k = 1$

	U^2	Upper χ^2 Percentiles
	Model I versus Model II	
1974–1978	44.80***	$\chi^2_{11}(0.001) = 31.3$
1979–1984	13.36	$\chi^2_{11}(0.1) = 17.3$
	Model II versus Model III	
1974–1978	16.89	$\chi^2_{20}(0.5) = 19.3$
1979–1984	24.25	$\chi^2_{20}(0.1) = 28.4$
	Model I versus Model IV	
1974–1978	21.92***	$\chi^2_{1}(0.001) = 11.0$
1979–1984	1.83	$\chi^2_{1}(0.1) = 2.71$

Source: Cressie and Chan (1989). Reprinted by permission of the American Statistical Association.

***Significance at the 0.1% level.

nature of SIDS seems to have changed from the earlier period to the later period; this is discussed further in the following text.

Some Consequences of the Fitted Model

The intercounty correlation can be estimated from

$$\widehat{\text{corr}}(Z(\mathbf{s}_i), Z(\mathbf{s}_j)) = a_{ij}/(a_{ii}a_{jj})^{1/2}, \tag{7.6.29}$$

where $(a_{ij}) \equiv (I - \hat{\phi}D^{-1/2}HD^{1/2})^{-1}$. The values from (7.6.29) were used to check the fit of Model IV by comparing them (use Model IV's m.l. estimate $\hat{\phi} = 0.640$) to an empirical correlogram calculated from $D^{-1/2}(\mathbf{Y} - X\hat{\boldsymbol{\beta}})$, the standardized residuals using the m.l. estimate $\hat{\boldsymbol{\beta}}$ for Model IV. On the whole, the two sets of correlations showed good agreement, although there was an indication of a more complex spatial dependence in the residuals than that given by (7.6.29).

The variance matrix of the m.l. estimator $\hat{\boldsymbol{\beta}}$ is estimated by

$$\widehat{\text{var}}(\hat{\boldsymbol{\beta}}) = \left(X'D^{-1}(I - \hat{\phi}H)X\right)^{-1}\hat{\tau}^2. \tag{7.6.30}$$

For example, consider the 1974–1978 analysis. For Model I, $\hat{m} = 2.838$ with $\widehat{\text{var}}(\hat{m}) = 0.006218$, and for Model IV, $\hat{m} = 1.644$ with $\widehat{\text{var}}(\hat{m}) = 0.0557$, $\hat{b} = 0.03457$ with $\widehat{\text{var}}(\hat{b}) = 0.4307 \times 10^{-4}$, and $\widehat{\text{cov}}(\hat{m}, \hat{b}) = -0.001483$.

By conditioning on a county's neighboring values, the Markov random-field model can be used to predict that county's value. When applied to all counties, the result is a spatial smoother. Under the auto-Gaussian model, the minimum mean-squared error predictor is

$$Z_i^P \equiv E\big(Z(\mathbf{s}_i)|\{Z(\mathbf{s}_j): j \in N_i\}\big) = \mu_i + \sum_{j \in N_i} c_{ij}\big(Z(\mathbf{s}_j) - \mu_j\big),$$

$$i = 1, \ldots, n.$$

Therefore, $\mathbf{Z}^P = \boldsymbol{\mu} + C(\mathbf{Z} - \boldsymbol{\mu})$, which is estimated by

$$\hat{\mathbf{Z}}^P = X\hat{\boldsymbol{\beta}} + (\hat{\phi}H)(\mathbf{Z} - X\hat{\boldsymbol{\beta}}). \tag{7.6.31}$$

Ignoring the variation in $\hat{\phi}$, the matrix of mean-quadratic prediction errors is $\Sigma^P \equiv E\{(\hat{\mathbf{Z}}^P - \mathbf{Z})(\hat{\mathbf{Z}}^P - \mathbf{Z})'\} = F\hat{\Sigma}F'$, where $F \equiv (I - \hat{\phi}H) \cdot \{I - X(X'\hat{\Sigma}^{-1}X)^{-1}X'\hat{\Sigma}^{-1}\}$ and $\hat{\Sigma} \equiv (I - \hat{\phi}H)^{-1}D\hat{\tau}^2$. Then Σ_{ii}^P, the (i, i)th element of Σ^P, is the mean-squared prediction error (or kriging variance) of $\hat{Z}_i^P$. Discussion of how the variation in $\hat{\phi}$ should be accounted for can be found in Cressie (1990c) and Prasad and Rao (1990).

Notice that $Z(\mathbf{s}_i)$ does not appear in the predictor Z_i^P. Consider the errors-in-variables model (Besag, 1977a), $Z(\cdot) = S(\cdot) + \epsilon(\cdot)$, where $\epsilon(\cdot)$ is a (heteroskedastic) white-noise process independent of the CG process $S(\cdot)$. Here, $E(S(\mathbf{s}_i)|\mathbf{Z})$ is the predictor of interest and it is a function of all elements of $\mathbf{Z} \equiv (Z(\mathbf{s}_1), \ldots, Z(\mathbf{s}_n))'$. Because of the imprecise spatial locations of the counties and possible measurement error, Cressie (1992) developed an errors-in-variables model. See also Cressie (1990c), who constructs such a spatial model for census undercount and derives empirical Bayes and constrained empirical Bayes predictors of undercount for the states of the United States. (The constrained predictors are used to predict ensemble quantities, such as the weighted proportion of states whose undercount is greater than, say, 3%.)

Figure 7.12 shows choropleth maps of Model IV smoothed values $\hat{\mathbf{Z}}^P$, determined from linear combinations of neighboring values (alternative methods of mapping are discussed in Section 7.5.1). Notice that county 4 has been included; its value was predicted from its neighboring data, $\{Z(\mathbf{s}_j): j \in N_4\}$, using (7.6.31). The statistics $\{(\Sigma_{ii}^P)^{-1/2}(\hat{Z}_i^P - Z(\mathbf{s}_i)): i = 1, \ldots, 100\}$ could also be used to cross-validate the model, similar to the suggestion in Section 2.6.4.

The (confirmatory) maps of Figure 7.12 are directly comparable to the (exploratory) maps of Figure 6.5, obtained from a weighted median polish fit. Qualitatively, they both convey the same message, that the variability has decreased considerably in the later time period. To see this more graphically, as well as the effect of smoothing, Figure 7.13 shows plots of $\{(\hat{Z}_i^P, Z(\mathbf{s}_i)): i = 1, \ldots, 100\}$ for both time periods.

For SIDS, as well as for most other variables, North Carolina is not a closed system. Some North Carolina counties have neighboring counties in Virginia, Tennessee, Georgia, and South Carolina, although the neighbors do not appear in the Markov-random-field analysis because no SIDS data are at

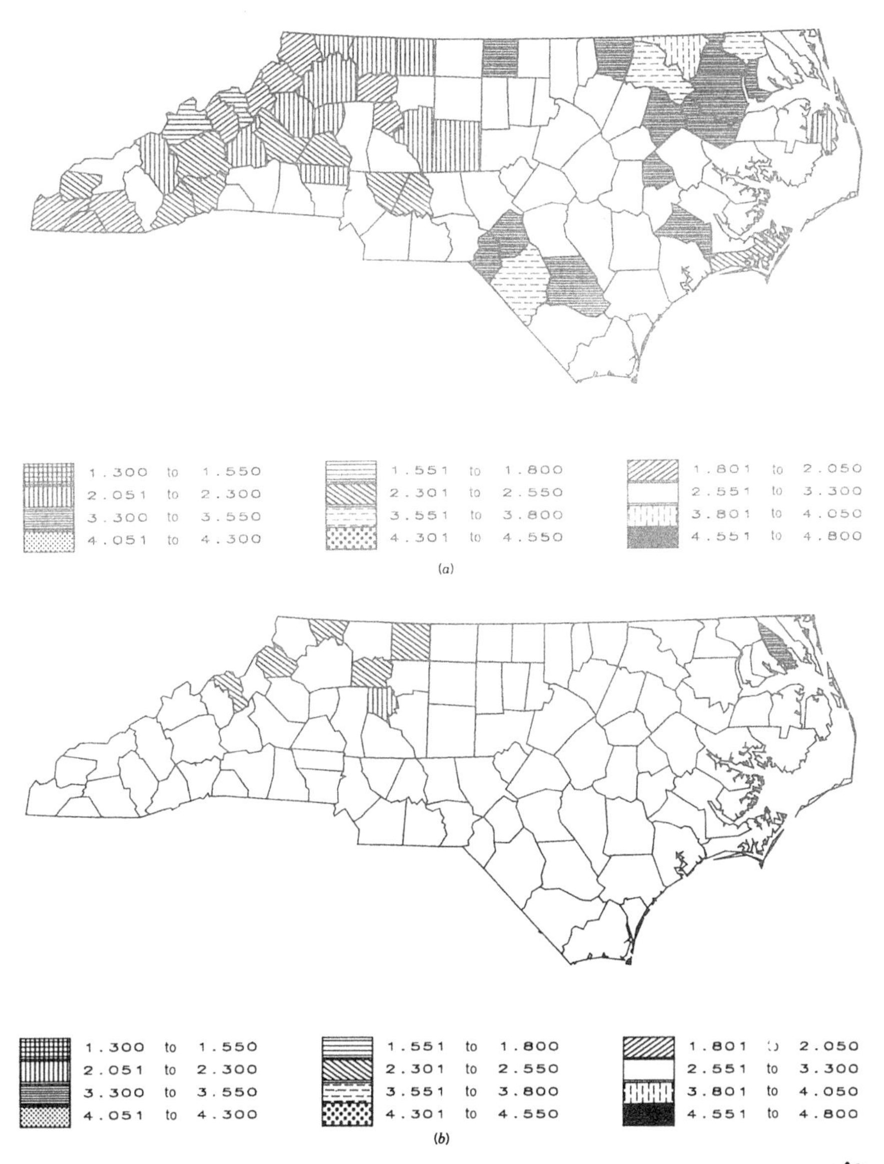

Figure 7.12 Choropleth map of counties of North Carolina showing the predicted values $\{\hat{Z}_i^P: i = 1, \ldots, 100\}$ given by (7.6.31); auto-Gaussian Model IV with $k = 1$. (*a*) 1974–1978; (*b*) 1979–1984. [*Source*: Cressie and Chan, 1989]. Reprinted by permission of the American Statistical Association.

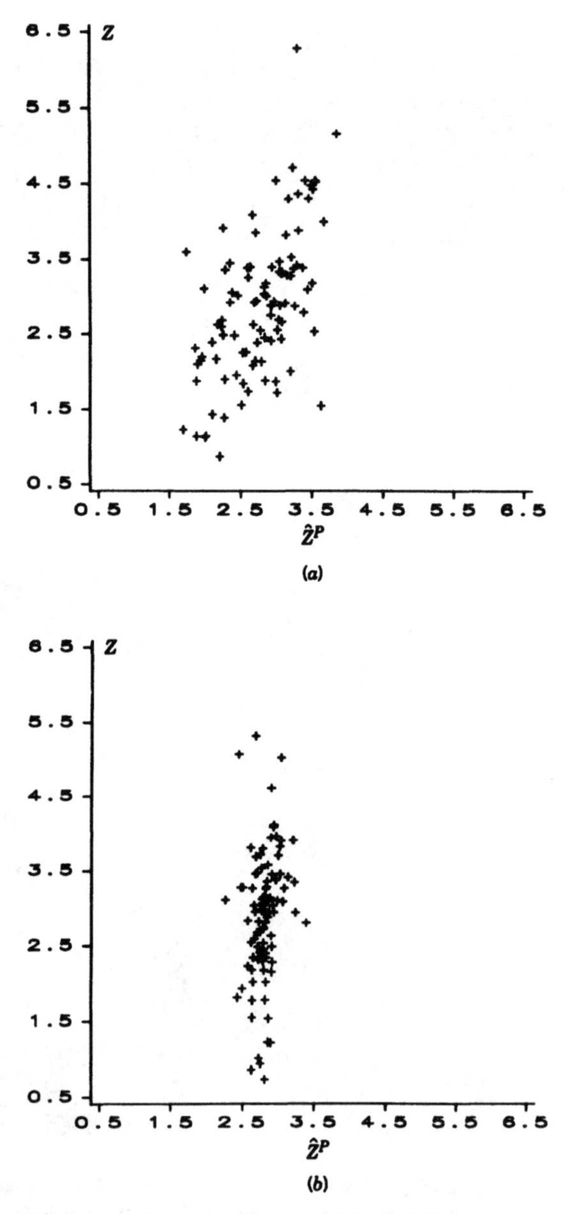

Figure 7.13 Plot of (Freeman–Tukey transformed) SID rate (Z), given by (7.6.1), versus Model IV ($k = 1$) predicted value ($\hat{Z}^P$), given by (7.6.31). (*a*) 1974–1978; (*b*) 1979–1984.

hand. What is the effect of ignoring them in forming the neighborhoods $\{N_i\}$? If the spatial dependence is truly as stated in (7.6.19), namely, a 30-mile distance criterion, then ignoring neighbors in bordering states is equivalent to observing a $Z(\mathbf{s}_j)$ equal to its mean μ_j in the ignored county. This was the approach taken here for the analysis of the SIDS data. Alternative suggestions are given in Section 6.6.

Transforming the data is usually for the statistician's convenience. It is not difficult to transform the results back to the original scale. Recall from Section 6.2 that

$$1000p_i \equiv E(1000(S_i/n_i)), \qquad i = 1,\ldots,100, \qquad (7.6.32)$$

are the *parameters* of interest. In the context of the preceding auto-Gaussian models, it is unwise and statistically inefficient to *estimate* p_i by (S_i/n_i). Besides, what confidence can one put in (i.e., what is the standard error of) such an estimator? In terms of the transformed variable,

$$1000p_i \simeq E\left(Z(\mathbf{s}_i)^2/4\right) = (1/4)\left[\operatorname{var}(Z(\mathbf{s}_i)) + \{E(Z(\mathbf{s}_i))\}^2\right], \qquad i = 1,\ldots,100. \qquad (7.6.33)$$

The quantities in (7.6.33) are each available from the auto-Gaussian model fit, so that the maximum likelihood estimator of p_i is obtained by substituting $\hat{\boldsymbol{\beta}}$, $\hat{\tau}^2$, and $\hat{\phi}$ into the right-hand side of (7.6.33) to yield a maximum likelihood estimator $1000\hat{p}_i$ of the mean SID rate per thousand in county i. This estimator has approximate estimated variance $(1/4)\widehat{\operatorname{var}}(\hat{\mu}_i)(\hat{\mu}_i)^2$. For example, under Model IV in 1974–1978, $1000\hat{p}_{80} = 1.88$ with estimated standard error $= 0.10$. Compare this with the estimator (S_i/n_i) for p_i, whose estimated variance might be computed inappropriately as $(S_i/n_i)(1 - (S_i/n_i))/n_i$; for 1974–1978, $1000(S_{80}/n_{80}) = 0.65$, with (inappropriately) estimated standard error 0.38.

Some Final Remarks

After analyzing SIDS data at the county level in North Carolina over two disjoint time periods, several conclusions may be drawn. The earlier time period (1974–1978) shows a significant dependence of (Freeman–Tukey transformed) SID rate on the (Freeman–Tukey transformed) nonwhite live birth rate. This confirms earlier opinion that the baby's race is an important SIDS risk factor. The errors in this linear regression still exhibit some spatial structure, although it is not significant. A similar fit is obtained using a purely spatial model that requires 12 regression parameters. For reasons of both parsimony and interpretability, the model chosen for the 1974–1978 data is

$$Z(\mathbf{s}_i) = m + bX_{5,i} + \nu(\mathbf{s}_i), \qquad i \neq 4, i = 1,\ldots,100,$$

where $\boldsymbol{\nu} \equiv (\nu(\mathbf{s}_1), \ldots, \nu(\mathbf{s}_{99}))'$ is Gaussian with zero mean and variance matrix $(I - \phi H)^{-1}\tau^2 D$; see (7.6.22). Spatial dependence in the errors is represented by the parameter ϕ; when $\phi = 0$, they are independent.

However, in the subsequent period, 1979–1984, the effect due to race of the baby is no longer significant. This comparison of SIDS over time, using a spatial model, leads to a surprising but statistically verifiable conclusion. The nature of SIDS (or of SIDS reporting) has changed considerably over the 10-year period, 1974–1984.

Raw SID rates by race, available for the state of North Carolina, are (in deaths per 1000 live births),

$$
\begin{aligned}
&1974\text{–}1978: \quad 1.201\ (\text{white});\ 3.806\ (\text{nonwhite})\\
&1979\text{–}1984: \quad 1.512\ (\text{white});\ 2.970\ (\text{nonwhite}).
\end{aligned}
$$

Thus, the white SID rate has increased considerably and the nonwhite SID rate has decreased even more considerably. Whether this is due to reporting bias or the effect of public education programs in infant health is difficult to say. In the later period there is a cluster of counties in the northeast (counties 35, 48, 59, 72, 89, and 94) whose nonwhite birth rate is high but whose SID rate is low. However, their total number of live births is not substantial, so that they alone cannot account for this change. The considerable increase in the white SID rate is of concern and warrants further attention.

The overall SID rate in North Carolina decreased slightly (but significantly) from 2.031 per thousand live births in the earlier period to 1.979 in the later period.

7.7 LATTICE DATA, SIMULATED AND REAL

Real data are important for the development of statistical methods, and, ideally, their analysis also stimulates research in statistical theory.

Simulated data have a different role. They may be used to validate or establish properties of a statistical method under an assumed model, which includes checking the validity of asymptotic properties in finite samples. This role is particularly valuable when several competing methods are available, but there is little or no theory to indicate which is superior. Simulation can also be useful in checking the fit of a model to real data by simulating the model using the estimated parameters. This is done, say, 99 times, and a diagnostic statistic is computed each time. The diagnostic statistic based on the real data is then compared to the 99 statistics based on the simulated data. A further use of simulation is in stochastic optimization techniques such as the annealing algorithm presented in Section 7.4.3.

7.7.1 Simulation of Lattice Processes

Consider the problem of simulating a random field $\{Z(\mathbf{s})\colon \mathbf{s} \in D\}$ with spatial locations on a d-dimensional finite lattice D. If $\{Z(\mathbf{s})\colon \mathbf{s} \in D\}$ is Gaussian, then the problem of its simulation is no different, in principle, from the methods for simulating geostatistical data described in Section 3.6.1. For example, the Cholesky decomposition method can be extended to simulate an auto-Gaussian (or CG) process by instead decomposing the inverse of the variance matrix. Other approaches may be required to simulate non-Gaussian processes.

Rejection Sampling

A simulation method that does not require the process to be Gaussian is rejection sampling (e.g., Ripley, 1987a, pp. 60–63). Suppose $\{Z(\mathbf{s})\colon \mathbf{s} \in D\}$ is a random field with joint probability density $f(\cdot)$. Suppose further that we can write $f(\mathbf{z}) = g(\mathbf{z}) \cdot h(\mathbf{z})$, where $g(\mathbf{z})$ is a probability density that is readily simulated; for example, g is the joint probability density of independent random variables. Let $h_{\max} = \sup_{\mathbf{z} \in \zeta} h(\mathbf{z})$, where ζ is the sample space. Then generate a realization $\mathbf{X}$ from the probability density function g and accept $\mathbf{Z} = \mathbf{X}$ with probability $h(\mathbf{X})/h_{\max}$. If $\mathbf{X}$ is rejected, resample from g until an acceptable $\mathbf{X}$ is obtained.

To see that the preceding prescription yields a random vector with density f, notice that

$$\begin{aligned} \Pr\{\mathbf{X} \in A \text{ and } \mathbf{X} \text{ is accepted}\} &= (h_{\max})^{-1} \int_A h(\mathbf{x}) \cdot g(\mathbf{x})\, d\mathbf{x} \\ &= (h_{\max})^{-1} \int_A f(\mathbf{x})\, d\mathbf{x}, \qquad A \subset \zeta. \quad (7.7.1) \end{aligned}$$

Thus,

$$\Pr\{\mathbf{X} \text{ is accepted}\} = (h_{\max})^{-1} \int_\zeta f(\mathbf{x})\, d\mathbf{x} = (h_{\max})^{-1} \qquad (7.7.2)$$

and

$$\Pr\{\mathbf{X} \in A | \mathbf{X} \text{ is accepted}\} = \int_A f(\mathbf{x})\, d\mathbf{x}. \qquad (7.7.3)$$

For example, consider the auto-Gaussian model given by (6.6.4); that is, $\mathbf{Z} \sim \text{Gau}(\boldsymbol{\mu}, (I - C)^{-1} M)$ and

$$\begin{aligned} f(\mathbf{z}) = (2\pi)^{-n/2} &\left|(I - C)^{-1} M\right|^{-1/2} \\ &\times \exp\{-(1/2)(\mathbf{z} - \boldsymbol{\mu})' M^{-1} (I - C)(\mathbf{z} - \boldsymbol{\mu})\}. \quad (7.7.4) \end{aligned}$$

Let G be any diagonal matrix such that $M^{-1}(I - C) - G^{-1}$ is positive-definite. Then the rejection sampling algorithm can be implemented by taking g to be a Gaussian density with mean $\boldsymbol{\mu}$ and variance G:

$$g(\mathbf{x}) = (2\pi)^{-n/2}|G|^{-1/2}\exp\{-(1/2)(\mathbf{x} - \boldsymbol{\mu})'G^{-1}(\mathbf{x} - \boldsymbol{\mu})\}. \quad (7.7.5)$$

Upon dividing (7.7.4) by (7.7.5),

$$h(\mathbf{x}) = \left|(I - C)^{-1}M\right|^{-1/2}|G|^{1/2} \times\exp\{-(1/2)(\mathbf{x} - \boldsymbol{\mu})'[M^{-1}(I - C) - G^{-1}](\mathbf{x} - \boldsymbol{\mu})\}, \quad (7.7.6)$$

whose maximum is

$$h_{\max} = \left|(I - C)^{-1}M\right|^{-1/2}|G|^{1/2}, \quad (7.7.7)$$

because $M^{-1}(I - C) - G^{-1}$ is positive-definite. Thus, if $\mathbf{X}$ is realized from a Gaussian distribution with mean $\boldsymbol{\mu}$ and diagonal variance matrix G (i.e., the elements of $\mathbf{X}$ are independent), and $\mathbf{Z} = \mathbf{X}$ is accepted with probability

$$\exp\{-(1/2)(\mathbf{X} - \boldsymbol{\mu})'[M^{-1}(I - C) - G^{-1}](\mathbf{X} - \boldsymbol{\mu})\}, \quad (7.7.8)$$

then $\mathbf{Z}$ will be a Gaussian random vector with mean $\boldsymbol{\mu}$ and variance $(I - C)^{-1}M$.

In theory, rejection sampling can be implemented for any lattice process, but in practice it is not always feasible to do so. Unfortunately, h is often too variable, so that the probability of accepting a given realization is very small.

Markov Chain Approach

Markov chains have been used frequently in statistical physics to simulate random fields (e.g., Metropolis et al., 1953; Barker, 1965; Flinn, 1974). Let ζ denote the sample (or state) space for the random field $(Z(\mathbf{s}_1), \ldots, Z(\mathbf{s}_n))$ defined on a finite set of sites $\{\mathbf{s}_1, \ldots, \mathbf{s}_n\}$. For simplicity, assume $Z(\mathbf{s}_i) \in \{1, 2, \ldots, K\}$ so that there are K^n distinct elements in ζ. The simulation algorithms presented in the following text are for large finite state spaces; theory for general state spaces may be found in the stochastic-processes literature (e.g., Doob, 1953, Section 5.5; Orey, 1971).

The idea is to construct a Markov chain with state space ζ and limiting probability mass function f, where f defines the probability distribution from which the random field is to be sampled. Let $\mathbf{Z}_t \equiv (Z_t(\mathbf{s}_1), \ldots, Z_t(\mathbf{s}_n))' \in \zeta$ denote the state of the Markov chain at time t. Then, for properly specified transition probabilities between any two states $\mathbf{z}$ and $\mathbf{x}$ in ζ, the Markov chain $\mathbf{Z}_0, \mathbf{Z}_1, \ldots,$ will converge in distribution to a random field with probability mass function $f(\cdot)$. Thus, by running the chain sufficiently long, one can obtain an observation $\mathbf{Z}_\nu$ that can be considered a sample from $f(\cdot)$ on ζ. The methods described in the subsequent text give various simulation algorithms that have properly specified transition probabilities.

Metropolis Algorithm

The Metropolis algorithm (Metropolis et al., 1953; Geman and Geman, 1984; Gidas, 1985; Ripley, 1987a, pp. 113–114) was developed to compute ensemble averages, $E(\phi(\mathbf{Z})) \equiv \Sigma_{\mathbf{z} \in \zeta} \phi(\mathbf{z}) f(\mathbf{z})$, where ϕ is some variable of interest. A straightforward Monte Carlo approach would sample $\mathbf{z}$ uniformly from ζ and weight by $f(\mathbf{z})$ but, from a simulation point of view, one obtains samples of very low probability just as often as those of high probability. Alternatively, the idea behind the Metropolis algorithm is to run a Markov chain that, in the long run, samples from $f(\cdot)$, and then weight the samples evenly.

A simple description is as follows. As the chain is run, all n sites must be visited infinitely often; in practice, the sites might be visited in a fixed order (e.g., a raster scan of sites) or in a random order that chooses $\mathbf{s}_{(t)}$, the site visited at time t, from say a uniform distribution on $\{1, 2, \dots, n\}$. Recall that $\mathbf{Z}_{t-1}$ denotes the state at time $(t-1)$. At time t, draw Y uniformly from $\{1, 2, \dots, K\}$. (The uniform drawing can be generalized; see, e.g., Ripley, 1987a, p. 113.) Define $\mathbf{y} \equiv (Z_{t-1}(\mathbf{s}_1), \dots, Y, \dots, Z_{t-1}(\mathbf{s}_n))'$, where Y replaces $Z_{t-1}(\mathbf{s}_{(t)})$. Then let $\mathbf{Z}_t = \mathbf{y}$ with probability

$$\alpha(\mathbf{Z}_{t-1}, \mathbf{y}) \equiv \min\{1, f(\mathbf{y})/f(\mathbf{Z}_{t-1})\}, \qquad (7.7.9)$$

and let $\mathbf{Z}_t = \mathbf{Z}_{t-1}$ with probability $1 - \alpha(\mathbf{Z}_{t-1}, \mathbf{y})$. Thus, from time $t-1$ to time t, the state vector is changed in at most one entry, corresponding to the site $\mathbf{s}_{(t)}$. This rejection-sampling approach determines transition probabilities for a Markov chain whose limiting probability mass function is given by $f(\cdot)$ (e.g., Ripley, 1987a, pp. 113–114).

Variations on the Metropolis algorithm have been proposed by Barker (1965) and Hastings (1970). Barker replaces (7.7.9) with the smaller quantity, $\alpha(\mathbf{Z}_{t-1}, \mathbf{y}) = f(\mathbf{y})/\{f(\mathbf{Z}_{t-1}) + f(\mathbf{y})\}$, and Hastings' general algorithm includes those of Metropolis and Barker as special cases. It is intuitively clear that, from the point of view of encouraging a better sampling of states, the Metropolis algorithm is preferable because it makes more transitions than Barker's algorithm. Peskun (1973) formalizes this intuition by demonstrating the Metropolis algorithm's superior efficiency for estimation of $E(\phi(\mathbf{Z}))$.

Gibbs Sampler

The Gibbs sampler is described by Geman and Geman (1984) and is appropriate for simulating *Markov* random fields. As before, all n sites are visited repeatedly, in some order $\mathbf{s}_{(1)}, \mathbf{s}_{(2)}, \dots$. At time t, choose $Z_t(\mathbf{s}_{(t)}) = k$ with probability

$$\Pr\Big(Z(\mathbf{s}_{(t)}) = k \Big| \{Z(\mathbf{u}) = Z_{t-1}(\mathbf{u}) : \mathbf{u} \in N_{\mathbf{s}_{(t)}}\}\Big); \; k = 1, \dots, K, \quad (7.7.10)$$

where $N_{\mathbf{s}_{(t)}}$ is the neighborhood of $\mathbf{s}_{(t)}$ and where the conditional probabilities in (7.7.10) are given by the Markov-random-field model. In addition, for $\mathbf{s} \neq \mathbf{s}_{(t)}$, take $Z_t(\mathbf{s}) = Z_{t-1}(\mathbf{s})$. In the context of the simulated annealing algorithm given in Section 7.4.3, the Gibbs sampler corresponds to an annealing schedule of $T(t) \equiv 1$.

It is not difficult to show that the Gibbs sampler corresponds to a Barker-type variation of the Markov-chain Monte-Carlo algorithm that uses $\alpha(\mathbf{Z}_{t-1}, \mathbf{y}) = f(\mathbf{y})/\Sigma_{k=1}^{K} f((Z_{t-1}(\mathbf{s}_1), \ldots, k, \ldots, Z_{t-1}(\mathbf{s}_n)))$. Thus, a modification of the Gibbs sampler, to a Metropolis-type variation, would likely yield a more efficient algorithm

Notice the power of the Gibbs sampler; a rather complex multivariate probability distribution of a Markov random field can be obtained by successive simulations from univariate (conditional) distributions of the type (7.7.10). When these simulations are performed in parallel within coding sublattices (Section 7.2.1), considerable gains in computing time are possible. Swendsen and Wang (1987) give a simulation algorithm that allows large clusters to be changed in a single move and is potentially more efficient than the site-by-site simulation used in the Gibbs sampler.

Spin Exchange

Introduced by Flinn (1974), the spin-exchange algorithm simulates a Markov random field $Z(\cdot)$ that will have a fixed frequency of discrete gray levels. At time 0, the given gray levels are assigned to sites completely randomly. At time t, a pair of sites $\mathbf{s}_{(t)}$ and $\mathbf{u}_{(t)}$ are selected at random. Suppose $Z_{t-1}(\mathbf{s}_{(t)}) = k$ and $Z_{t-1}(\mathbf{u}_{(t)}) = l$. From the Markov-random-field model, calculate

$$\alpha_t \equiv \frac{\Pr\left(Z(\mathbf{s}_{(t)}) = l,\, Z(\mathbf{u}_{(t)}) = k \,\middle|\, \{Z(\mathbf{v}) = \mathbf{Z}_{t-1}(\mathbf{v})\colon \mathbf{v} \neq \mathbf{s}_{(t)}, \mathbf{u}_{(t)}\}\right)}{\Pr\left(Z(\mathbf{s}_{(t)}) = k,\, Z(\mathbf{u}_{(t)}) = l \,\middle|\, \{Z(\mathbf{v}) = \mathbf{Z}_{t-1}(\mathbf{v})\colon \mathbf{v} \neq \mathbf{s}_{(t)}, \mathbf{u}_{(t)}\}\right)}.$$

Flinn's approach is to exchange the values at the two sites [i.e., take $Z_t(\mathbf{s}_{(t)}) = l$ and $Z_t(\mathbf{u}_{(t)}) = k$] with probability $\alpha_t/(1 + \alpha_t)$; otherwise, the values are left the way they are. As usual, for $\mathbf{s} \neq \mathbf{s}_{(t)}, \mathbf{u}_{(t)}$, take $Z_t(\mathbf{s}) = Z_{t-1}(\mathbf{s})$. Then the Markov chain $\mathbf{Z}_0, \mathbf{Z}_1, \ldots$ converges in distribution to the desired Markov random field $Z(\cdot)$. Cross and Jain (1983) use a Metropolis-type variation that exchanges the values at the two sites with probability $\min\{1, \alpha_t\}$; otherwise, the values are left the way they are.

7.7.2 Lattice Data

Published data sets are not always complete in that spatial information is not always given. Part of the data should be a map (or equivalent) showing site locations, from which neighborhood information can be read. I shall only reference data sets here for which the necessary spatial information is given.

Mercer and Hall (1911) give a 20 × 25 array of wheat yields (see Sections 4.5 and 7.1), although they are not concerned with spatial dependence. Federer and Schlottfeldt (1954) consider the effect of cathode rays on the growth of tobacco plants; an 8 × 7 array of plant-height data (actually plant height totaled over the 20 plants in the plot) is given. They are the result of a randomized block design; nevertheless, by modeling the spatial dependence, one can achieve considerably higher efficiency than the classical analysis that assumes independent data (Section 5.7.2).

Haining (1978a) studied the spatial pattern of corn and wheat yields in 45 contiguous counties in southwestern Nebraska and northwestern Kansas. Data for 1964 and 1969 and a map of the counties are given. Gleeson and McGilchrist (1980) analyze plant weights of individual plants sown on a 7 × 7 square grid; they have three replicates of the experiment that they assume to be independent. Cliff and Ord (1981, pp. 206ff) analyze the economic effects of road accessibility in the 26 counties of Eire (Ireland) and Upton and Fingleton (1985, pp. 267ff) are concerned with people whose blood group is A in those same counties. The latter authors (1985, pp. 203ff) also present data on the flora of the Galapagos Islands. Anselin (1988, Section 12.2) relates crime to measures of income and housing value for 49 contiguous planning neighborhoods in Columbus, Ohio. He also analyzes (Anselin, 1988, Section 12.3) space–time lattice data on wage rates, unemployment, and net migration for 25 southwestern Ohio counties.

Two further sets of lattice data are available in this book. The Sudden-Infant-Death-Syndrome data is presented in Table 6.1 and various analyses are given in Sections 4.4, 6.2, and 7.6. The Scotland lip-cancer data is presented in Table 7.2 and used to illustrate exploratory regional mapping methods in Section 7.5.1.

PART III

Spatial Patterns

CHAPTER 8

Spatial Point Patterns

This is the first of two chapters in Part III of the book, the part devoted to (marked) spatial point patterns. Recall from Chapter 1 that a spatial process is written most generally as

$$\{\mathbf{Z}(\mathbf{s}) : \mathbf{s} \in D\},$$

where both $\mathbf{Z}(\cdot)$ and D are random. In both Chapters 8 and 9, D will be a random set—specifically, a *spatial point process*. Formal definitions are given in Section 8.3; for the moment it suffices to think of the index set D as a collection of random events whose realization is called a *spatial point pattern*. In these two chapters, my objective is to give methods for inferring parameters of the point process (the model) from the observed point pattern (the data). Although, in general, D is contained in d-dimensional Euclidean space $\mathbb{R}^d$, applications are usually for $d = 2$ or $d = 3$.

There is a large literature on processes $\{\mathbf{Z}(t): t \in T\}$, where T is a point process in *time*. Although there is considerable overlap of methods for point processes occurring in space and for those occurring in time, it would be wrong to say that the temporal case can be handled by putting $d = 1$. The unidirectional flow of time forces one to distinguish between the temporal and the one-dimensional spatial cases. The reader interested in point patterns in time can consult Cox and Lewis (1966), Lewis (1972), Snyder (1975), Cox and Isham (1980), and Daley and Vere-Jones (1988).

The hybrid space–time point process

$$\{\mathbf{Z}(\mathbf{s}, t) : \mathbf{s} \in D(t), t \in T\}$$

models data occurring at both random locations and random times. For example, in studying and predicting earthquakes, data on their magnitudes, locations, and times of occurrence would be useful. Space–time processes are considered in Sections 8.8 and 9.7.

A final remark on nomenclature. To avoid confusion with "points of the process" and "points of $\mathbb{R}^d$," I shall refer throughout to *events* of the process (D) or of the pattern (realization of D) and to *points* of $\mathbb{R}^d$. However, I shall resist the temptation to refer to a spatial event process or a spatial event pattern. In their most general form, point processes can have multiple events at the same point. Only *simple* spatial point processes in $\mathbb{R}^d$ (i.e., almost surely at most single events occur) are considered in this book; a comparison with orderly point processes is given by Daley and Vere-Jones (1988, p. 44). Actually, multiple events can be modeled through positive-integer–valued marks of a marked spatial point process (Section 8.7).

Chapter Summary

Section 8.1 introduces the notion of a random spatial index, an example of which is given in Section 8.2. There, a variety of *ad hoc* analysis and inference techniques are implemented on the locations of longleaf pines in a study region of the Wade Tract in Thomas County, Georgia. Section 8.3 gives the theory underlying spatial point processes, which then allows distance functions and second moment measures to be discussed in Section 8.4 with more rigor than in Section 8.2.6. Section 8.5 presents a large number of point-process models and then discusses how their parameters might be fitted from an observed point pattern. Multivariate point processes and marked point processes are presented in Sections 8.6 and 8.7, respectively. Section 8.8 discusses (briefly) space–time point patterns. Finally, algorithms for simulating any of the point processes of Section 8.5 and sources for actual data are given in Section 8.9.

8.1 RANDOM SPATIAL INDEX

There are many examples in science where it is sensible to model the locations of events as random. Indeed, the study of spatial point patterns has a long history in ecology and forestry (Goodall, 1952, 1970; Pielou, 1977; Ripley, 1987b); ecological examples go back to Gleason (1920) and Svedberg (1922). Early studies of spatial point patterns were primarily concerned with comparing area (or quadrat) counts to a Poisson distribution (e.g., Student, 1907); departures indicate that the pattern is not completely spatially random (e.g., Neyman, 1939). The degree of departure was usually measured by an index based on the quadrat counts (e.g., Fisher et al., 1922; David and Moore, 1954; Morisita, 1959; Lloyd, 1967) or based on distance measures between events, points and events, and so on (e.g., Skellam, 1952; Clark and Evans, 1954; Hopkins, 1954). These indices do not show characteristics of the pattern at multiple scales very well. By employing a grid or transect of contiguous quadrats, Greig-Smith (1952) and subsequent authors (e.g., Hill, 1973; Mead, 1974; Goodall, 1974) extended the usual quadrat-counts analysis by examining several scales of pattern simultaneously.

Spatial point patterns have also found application in fields as diverse as archeology (e.g., Hodder and Orton, 1976), cosmology (e.g., Neyman and Scott, 1958), geography (e.g., Glass and Tobler, 1971; Cliff and Ord, 1981,

Chapter 4), and seismology (e.g., Ogata, 1989). Although many of the early techniques associated with analyzing point patterns have been developed outside Statistics, a turning point for the modern development of the field was the article by Ripley (1977). There, the K function, a powerful descriptive and modeling tool, originally suggested by Bartlett (1964), was featured; more details followed in Ripley (1981).

Although quadrat and distance methods have their place in distinguishing complete spatial randomness from spatially regular or clustered patterns in the field, K-function plots give a picture of such behavior, for mapped data, at a multitude of scales. Since the development of the K function, statisticians have used it to fit a variety of mathematically well defined models to data (e.g., Diggle, 1978, 1979; Diggle and Milne, 1983; Harkness and Isham, 1983; Ogata and Tanemura, 1986).

This chapter is meant as a review of an important area of Statistics for spatial data and is written from a data analysis and modeling point of view. For completeness, a summary of the underlying *theory* of spatial point processes can be found in Section 8.3. Because this theory can lead to formidable model definitions, I have included with most models a description of how they can be simulated.

8.2 SPATIAL DATA ANALYSIS OF LONGLEAF PINES (*Pinus palustris*)

Before I give a more formal presentation of point-process theory, models, and model fitting, an example is introduced and analyzed in various (mostly *ad hoc*) ways. Some of the material presented in this section can be found in Rathbun and Cressie (1990b).

8.2.1† Data Description

The longleaf-pine data are from the Wade Tract, an old-growth forest in Thomas County, Georgia, and were collected by W. J. Platt and S. L. Rathbun with the support of Tall Timbers Research Station, Tallahassee, Florida. These data consist of the coordinates and diameters (at breast height) of all longleaf pine trees at least 2 cm in diameter at breast height (dbh) in 4 ha of forest in 1979. A map showing the locations and relative diameters is given in Figure 8.1. On occasion, information is required on the locations and sizes of trees outside the 4-ha region. When appropriate I shall refer to the central study region (4 ha) and the extended study region (4 ha *plus* surrounding 12 ha).

Prior to European settlement, forests dominated by longleaf pine (*Pinus palustris*) occurred throughout most of the southern Atlantic and Gulf coastal plains of the United States. The Wade Tract is one of the largest remaining stands and is relatively free of disturbances by humans. There is evidence that this forest was present prior to European settlement. Longleaf pine is a

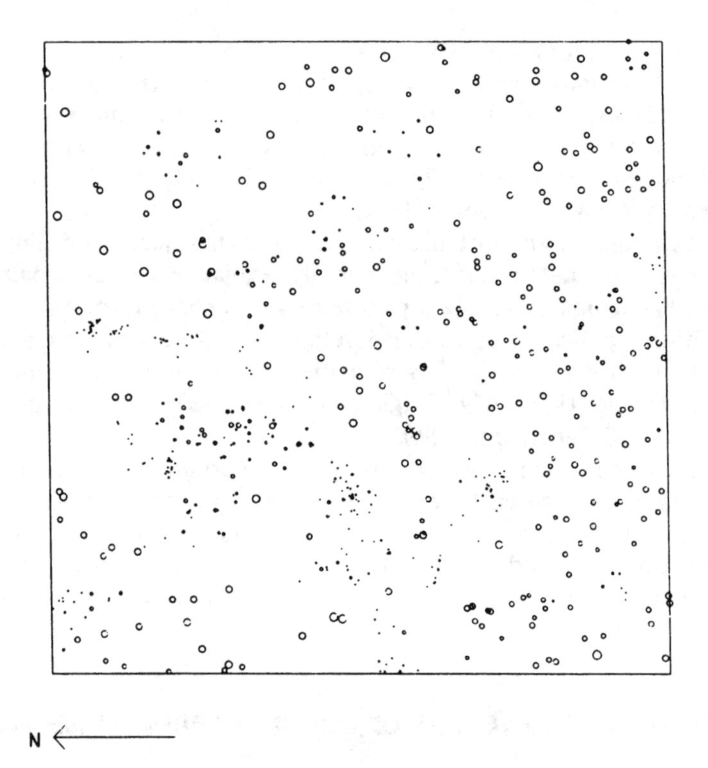

Figure 8.1 Map of locations and relative diameters (at breast height) of all longleaf pines in the 200 × 200-m (4-ha) study region in 1979.

fire-adapted species; ground fires occur frequently in these forests, removing most competing hardwoods. In addition to longleaf pine, there are a number of other tree species, predominantly the scrub oak species *Quercus incana*, *Quercus laevis*, *Quercus marilandica*, and *Quercus margaretta*. These species are few in number and small in stature. In the present study, only the data for longleaf pine are considered. A random sample of 400 tree cores revealed that the Wade Tract contains all ages of trees up to 250 years. For more details, see Platt et al. (1988).

The 4-ha study region illustrated in Figure 8.1 was chosen for its relatively gentle topography, for its absence of notable recent disturbances, and because all sizes of trees are represented. The locations of the observed pattern $D = \{\mathbf{s}_1, \ldots, \mathbf{s}_{584}\}$ and the associated dbhs $\{z(\mathbf{s}_1), \ldots, z(\mathbf{s}_{584})\}$, are given here in Table 8.1.

8.2.2 Complete Spatial Randomness, Regularity, and Clustering

Do the spatial locations of Figure 8.1 appear completely spatially random? Or, are they clustered? Such intuitive notions need quantification because

Table 8.1 Locations and Diameters at Breast Height (dbh, in centimeters) of all 584 Longleaf Pine Trees in the 4-ha Study Region[a]

x	y	dbh	x	y	dbh	x	y	dbh	x	y	dbh
200.0	8.8	32.9	89.7	4.9	72.0	10.4	61.2	19.2	92.8	61.5	11.4
199.3	10.0	53.5	10.8	0.0	31.4	30.9	52.2	43.5	91.3	69.5	33.4
193.6	22.4	68.0	26.4	5.4	55.1	48.9	67.8	33.7	95.9	59.7	35.8
167.7	35.6	17.7	11.0	5.5	36.0	49.5	73.8	43.3	93.4	71.5	54.4
183.9	45.4	36.9	5.1	3.9	28.4	46.3	80.9	36.6	89.6	86.3	33.6
182.5	47.2	51.6	10.1	8.5	24.8	44.1	78.0	46.3	99.5	78.9	35.5
166.1	48.8	66.4	18.9	11.3	44.1	48.5	94.8	48.3	100.6	53.1	7.4
160.7	42.4	17.7	28.4	11.0	50.9	45.9	90.4	20.4	103.5	72.1	36.6
162.9	29.0	21.9	41.1	9.2	47.5	44.2	84.0	40.5	104.7	74.0	19.1
166.4	33.6	25.7	41.2	12.6	58.0	37.0	64.3	44.0	104.0	67.1	34.9
163.0	35.8	25.5	33.9	21.4	36.9	36.3	67.7	40.9	104.2	64.7	37.3
156.1	38.7	28.3	40.8	39.8	65.6	36.7	71.5	51.0	105.0	59.8	16.3
157.6	42.8	11.2	49.7	18.2	52.9	35.3	78.3	36.5	111.8	73.2	39.1
154.4	36.2	33.8	6.7	46.9	39.5	33.5	81.6	42.1	112.4	69.8	36.5
150.8	45.8	2.5	11.6	46.9	42.7	29.3	83.8	15.6	110.0	65.9	25.0
144.6	25.4	4.2	17.2	47.9	44.4	22.4	84.1	18.5	120.4	79.2	46.8
142.7	25.4	2.5	19.4	50.0	40.3	17.1	84.7	43.0	109.4	62.5	18.7
144.0	28.3	31.2	26.9	47.2	53.5	27.3	89.4	28.9	109.7	62.9	23.2
143.5	36.9	16.4	39.6	47.9	44.2	27.9	90.6	21.3	113.3	60.4	20.4
123.1	14.3	53.2	38.0	50.7	53.8	48.4	99.5	30.9	118.0	69.3	42.3
113.9	13.1	67.3	19.1	45.2	38.0	43.6	98.4	42.7	126.5	69.2	38.1
114.9	8.1	37.8	32.1	35.0	48.3	39.0	97.3	37.6	125.1	68.2	17.9
101.4	9.3	49.9	28.4	35.5	42.9	14.9	91.2	47.1	114.2	54.6	39.7
105.7	9.1	46.3	3.8	44.8	40.6	6.1	96.2	44.6	110.6	51.5	14.5
106.9	14.7	40.5	8.5	43.4	34.5	10.7	98.6	44.3	147.3	73.8	33.5
127.0	29.7	57.7	11.2	40.2	45.7	22.2	100.0	26.1	146.7	73.0	56.0
129.8	45.8	58.0	22.4	34.3	51.8	32.7	99.1	25.9	148.1	86.2	66.1
136.3	44.2	54.9	23.8	33.3	52.0	0.9	100.0	41.4	138.2	73.4	26.3
106.7	49.4	25.3	24.9	29.8	44.5	93.5	96.2	59.5	135.7	70.7	44.8
103.4	49.6	18.4	9.0	38.9	35.6	85.1	90.6	26.1	134.9	72.7	24.2

Table 8.1 Locations and Diameters at Breast Height (dbh, in centimeters) of all 584 Longleaf Pine Trees in the 4-ha Study Region[a] *(Continued)*

x	y	dbh	x	y	dbh	x	y	dbh	x	y	dbh
98.0	27.7	39.0	11.0	34.4	44.1	80.4	90.7	21.7	133.4	77.1	35.7
93.5	28.7	15.1	17.5	21.9	51.5	71.0	88.8	42.4	129.9	76.1	12.1
82.3	16.8	35.6	4.3	31.3	51.6	73.0	85.6	40.2	126.5	77.3	35.4
79.2	25.3	21.6	5.9	8.1	33.3	56.7	95.3	37.4	129.1	83.1	32.7
84.2	29.0	17.2	1.9	68.5	13.3	66.5	86.2	40.1	134.4	87.0	30.1
88.8	35.1	22.3	1.8	71.0	5.7	67.0	84.7	39.5	130.7	90.1	28.4
82.5	36.3	18.2	1.1	82.5	3.3	62.9	87.9	32.5	130.9	90.7	16.5
75.6	28.1	55.6	2.4	95.3	45.9	61.8	89.0	39.5	132.0	94.5	12.7
72.9	36.2	23.2	4.6	94.0	32.6	51.9	94.5	35.6	136.8	96.7	5.5
79.1	43.6	27.0	3.1	79.5	11.4	60.9	71.6	44.1	137.7	98.0	2.5
50.0	48.8	50.1	3.9	72.1	9.1	61.0	69.8	42.2	157.8	99.9	3.0
59.9	34.4	45.5	4.1	70.9	5.2	61.7	66.2	39.4	187.1	98.1	3.2
60.5	13.0	47.2	7.9	68.7	4.9	57.3	68.4	35.5	190.6	92.1	3.2
60.2	11.4	37.8	14.8	81.8	42.0	54.2	76.4	39.1	185.4	93.1	4.0
66.5	15.9	31.9	9.4	67.7	32.0	76.1	52.9	9.5	186.6	92.2	3.6
70.4	6.6	38.5	15.9	78.7	32.8	67.2	57.6	48.4	185.9	91.7	3.8
70.7	2.2	23.8	16.6	78.8	22.0	81.9	58.5	31.9	184.3	92.1	4.3
71.7	1.9	46.3	18.2	80.3	20.8	90.1	59.6	30.7	188.2	91.2	3.3
179.5	92.6	2.8	174.1	135.6	7.3	135.3	126.6	15.0	104.4	145.1	6.3
186.1	91.0	3.2	173.0	127.4	3.0	135.0	124.0	24.5	104.9	145.0	18.4
178.3	92.4	5.8	174.0	125.7	2.2	136.2	122.1	15.0	101.5	148.4	5.4
178.6	91.8	3.5	177.3	121.0	2.2	129.7	127.0	22.2	102.4	148.7	5.4
186.2	90.3	2.3	177.6	120.3	2.2	134.8	120.2	27.5	123.4	128.9	26.0
185.2	89.9	3.8	195.7	144.1	59.4	136.9	116.8	10.8	123.8	135.1	22.3
185.5	89.8	3.2	197.0	142.5	48.1	137.0	116.0	26.2	127.0	133.8	35.2
185.8	89.1	4.4	178.2	112.6	51.5	128.9	124.2	10.2	109.6	145.9	24.1
186.5	88.8	3.9	173.8	112.7	50.3	127.5	125.0	18.9	112.4	145.0	6.9
176.7	92.3	7.8	172.8	124.4	2.9	127.6	121.7	44.2	133.1	144.8	61.0
177.7	91.5	4.7	162.7	114.6	19.1	129.7	119.0	13.8	139.4	143.1	20.6
184.0	89.0	4.8	164.6	120.9	15.1	126.6	121.1	16.7	140.4	143.6	6.5

184.1	88.2	2.8	162.9	119.9	29.9	127.1	119.9	14.5	139.7	145.8	20.0
183.5	88.5	4.8	158.4	113.4	14.9	120.7	115.6	12.0	145.5	148.4	8.9
183.0	88.0	5.4	153.9	108.3	38.7	115.3	112.6	2.2	146.4	148.4	27.6
176.1	91.0	4.3	156.1	116.0	31.5	134.1	105.2	2.3	105.8	149.8	4.5
175.6	90.2	4.0	156.5	118.9	27.8	134.6	104.1	3.2	96.7	149.1	9.2
173.8	89.9	3.2	156.8	122.3	28.5	135.6	103.3	3.0	66.5	150.0	2.3
164.9	93.7	2.8	159.0	126.1	21.6	128.9	102.6	50.6	55.7	148.5	5.0
163.0	95.3	4.9	161.0	131.9	2.0	116.3	106.5	2.6	54.7	146.8	4.0
163.2	94.1	3.5	161.3	132.8	2.6	104.3	104.0	50.0	57.1	144.0	21.8
162.4	94.5	2.9	160.6	132.6	2.3	111.5	100.0	52.2	61.7	145.3	10.9
161.5	94.9	2.4	161.3	134.9	3.5	100.5	149.7	5.2	60.1	143.7	14.9
162.2	94.3	3.3	159.7	129.8	3.6	100.0	145.5	5.2	77.7	144.8	45.0
161.0	94.7	2.1	161.7	136.1	2.6	100.8	145.0	6.7	67.2	139.3	16.4
157.7	95.7	2.0	161.1	136.4	2.0	100.9	143.5	14.0	80.7	133.2	43.3
154.9	96.2	3.9	160.1	133.0	2.0	100.3	140.8	12.7	85.1	133.5	55.6
154.6	92.7	5.0	159.0	133.6	2.7	101.5	120.8	59.5	94.7	143.7	10.6
152.9	93.7	2.3	160.0	134.8	2.6	99.3	110.6	52.0	81.2	125.0	45.9
153.2	93.2	2.2	160.2	135.5	2.2	99.2	106.0	45.9	81.9	123.2	45.2
168.2	73.0	67.7	159.1	136.5	2.7	102.0	137.1	18.0	83.8	123.1	35.5
151.6	93.0	2.9	154.7	126.8	30.1	105.4	115.7	43.5	84.8	121.4	43.6
151.4	93.4	2.4	151.9	127.5	16.6	103.6	134.2	3.3	82.9	119.2	44.6
157.6	67.2	56.3	151.3	124.7	10.4	103.9	139.4	4.3	82.1	116.4	38.8
149.4	63.0	39.4	151.0	127.3	11.8	102.6	141.6	7.4	84.3	114.8	34.9
149.4	64.3	59.5	150.4	123.0	32.3	102.0	143.3	10.1	96.7	142.6	17.0
167.3	54.6	42.4	149.6	124.6	33.5	102.1	144.4	23.1	92.0	109.0	50.4
157.4	51.5	63.7	146.2	127.1	30.5	103.5	141.3	8.1	96.1	146.6	2.0
181.5	66.1	66.6	146.1	127.4	10.5	102.9	143.8	5.7	78.5	102.5	33.8
196.5	55.2	69.3	144.4	131.8	13.8	105.7	138.2	13.3	78.7	103.0	51.1
189.9	85.2	56.9	143.3	131.5	22.8	106.6	135.1	12.8	59.5	107.4	21.8
155.1	149.2	23.5	140.6	137.7	31.7	108.5	133.2	11.6	56.5	105.5	46.5
154.5	148.4	9.1	143.2	125.4	10.1	105.2	142.3	6.3	64.3	132.1	5.6

Table 8.1 Locations and Diameters at Breast Height (dbh, in centimeters) of all 584 Longleaf Pine Trees in the 4-ha Study Region[a] *(Continued)*

x	y	dbh	x	y	dbh	x	y	dbh	x	y	dbh
152.7	146.7	19.6	141.0	127.8	7.8	107.5	138.5	17.8	56.8	116.0	46.0
155.8	145.4	32.3	140.1	127.3	17.0	107.9	139.5	3.7	62.2	137.7	7.8
161.2	138.1	3.7	140.9	121.4	36.4	116.5	122.6	19.0	58.2	125.1	54.9
161.0	138.1	2.7	135.0	132.3	19.6	114.5	127.7	11.2	54.1	115.5	45.5
162.1	136.9	2.5	139.3	122.9	15.0	115.3	127.4	27.6	59.5	138.1	9.2
166.2	132.0	2.5	142.0	117.2	28.8	115.3	128.1	14.5	58.6	140.3	13.2
168.7	133.4	2.4	140.4	117.2	20.1	119.0	127.4	34.4	58.8	141.5	15.3
169.3	133.7	7.2	138.5	121.5	39.3	119.4	127.7	20.0	57.9	137.3	8.5
57.9	140.7	7.0	28.7	158.8	37.9	94.7	179.8	2.9	153.5	159.9	2.2
57.5	142.3	11.8	33.7	162.3	40.6	89.3	185.0	7.3	155.9	183.7	58.8
57.3	141.7	8.5	23.1	160.8	33.0	90.8	174.0	52.7	160.4	176.6	47.5
56.0	137.7	9.5	11.3	158.9	35.7	95.3	158.4	8.7	171.3	185.1	52.2
53.4	139.3	7.0	18.2	168.2	20.6	90.9	162.1	3.6	182.8	187.4	56.3
53.1	136.0	10.5	21.5	172.3	22.0	90.2	162.1	4.6	182.5	196.0	39.8
54.0	137.7	6.6	15.9	168.3	16.3	90.2	161.7	11.4	176.3	197.7	38.1
54.5	136.7	6.6	15.4	172.8	5.6	90.6	160.8	11.0	161.9	199.4	38.9
53.3	137.8	8.8	14.0	174.2	7.4	93.0	158.0	18.7	199.5	179.4	9.7
52.1	139.3	11.6	6.8	179.6	42.3	78.4	172.4	5.6	197.6	176.9	7.4
48.0	114.4	48.2	6.0	184.1	43.8	76.2	171.4	2.1	196.3	192.4	22.1
44.2	129.6	36.2	1.6	194.9	53.0	75.8	171.0	3.3	195.7	180.5	16.9
39.4	136.8	44.9	43.6	197.3	48.1	75.7	169.7	11.5	196.2	177.1	5.9
42.7	124.0	43.0	39.4	195.5	41.9	82.7	163.5	2.6	196.3	176.0	10.5
38.1	134.4	37.5	37.1	196.1	48.0	76.7	166.3	4.4	193.7	185.8	9.5
37.1	131.9	31.5	23.7	193.9	75.9	74.7	167.1	18.3	191.7	189.2	45.9
37.6	125.4	39.9	21.5	187.9	40.4	119.4	170.8	7.5	194.5	173.8	11.4
31.2	127.9	35.5	27.7	188.7	40.9	74.2	164.3	17.2	192.7	177.3	7.8
40.1	112.2	51.7	32.3	178.9	39.4	73.9	162.7	4.6	188.9	182.1	14.4
29.3	118.6	36.5	32.6	168.6	40.9	81.7	156.7	32.0	190.1	174.4	8.3
23.8	114.5	40.2	37.7	176.9	17.6	79.5	156.3	56.7	186.9	179.4	30.6

26.9	111.3	44.4	40.4	179.3	37.7	79.2	155.6	27.0	186.9	174.7	23.9
17.9	111.0	38.7	41.0	176.6	36.8	61.6	158.2	2.6	181.2	176.9	5.2
34.4	104.2	41.5	43.9	182.2	33.6	70.3	153.1	4.9	181.1	176.1	7.6
31.9	103.2	34.5	44.7	184.6	47.9	79.8	151.8	35.0	177.2	174.5	27.8
20.6	101.5	31.8	45.6	175.2	32.0	110.1	150.4	23.7	182.8	162.9	49.6
14.1	103.1	39.7	47.5	175.9	40.3	116.1	156.8	42.9	180.0	160.2	51.0
2.9	122.8	23.3	51.2	177.9	42.5	114.0	165.1	14.2	189.1	156.3	50.7
6.4	125.9	37.7	55.0	159.3	59.7	103.2	154.4	3.3	196.9	151.4	43.4
2.2	142.2	43.0	58.0	180.3	44.2	112.3	167.0	28.4	171.4	161.6	55.6
11.7	116.2	39.2	54.6	188.7	30.9	110.4	167.3	10.0	169.1	160.0	4.3
14.2	116.5	40.4	58.9	180.0	39.5	110.6	166.4	6.4	162.5	157.3	2.5
15.6	118.1	36.7	63.9	178.6	48.7	107.0	165.0	22.0	156.7	155.3	23.5
13.6	127.4	48.4	64.3	178.9	32.8	105.6	160.6	4.3	154.1	150.8	8.0
11.1	134.8	27.9	65.6	179.3	47.2	104.0	162.4	10.0	87.7	200.0	11.7
7.2	141.7	46.4	61.0	184.9	42.1	104.0	166.1	9.2			
12.2	140.1	38.5	63.1	183.3	43.8	103.7	167.2	3.7			
23.0	132.7	39.4	86.1	186.9	30.5	108.6	182.1	66.7			
30.2	133.9	50.0	65.8	194.9	28.3	105.7	182.6	68.0			
27.7	136.5	51.6	90.0	195.1	10.4	102.8	169.7	23.1			
3.4	148.8	38.7	94.3	196.1	15.0	101.5	171.8	5.7			
15.4	145.6	39.6	91.9	197.1	7.4	100.4	170.5	11.7			
16.7	146.4	29.1	86.5	197.4	15.3	144.1	199.0	40.4			
24.3	145.7	44.0	87.5	199.3	17.5	138.3	197.9	43.3			
0.4	175.2	50.9	93.9	199.2	5.0	142.7	197.2	60.2			
0.0	177.5	50.8	92.4	199.3	12.2	118.8	188.0	55.5			
7.9	151.0	43.0	81.8	198.9	9.0	142.3	173.3	54.1			
33.2	151.2	44.5	99.0	158.1	2.4	143.8	156.0	22.3			
36.6	150.6	29.8	94.1	187.2	13.7	145.3	155.6	21.4			
42.2	153.7	44.3	95.4	182.9	13.1	151.2	192.2	55.7			
24.5	153.4	51.2	97.1	168.4	12.8	153.7	176.5	51.4			

[a]The x coordinates are distances (in meters) from the tree to the southern boundary. The y coordinates are distances (in meters) from the tree to the eastern boundary. [Source: Dr. W. J. Platt and S. L. Rathbun, Tall timbers Research Station.]

different observers may disagree as to the amount of clustering or randomness in a spatial point pattern. Besides, it will be shown that patterns from completely spatially random processes can appear clustered to the untrained eye.

Complete Spatial Randomness

The standard against which spatial point patterns are often compared is (a realization from) a completely spatially random point process. In this book, *complete spatial randomness* (csr) is synonymous with a *homogeneous* Poisson process in $\mathbb{R}^d$ (although occasionally others have not specified the homogeneity requirement). This process has the property that, conditional on $N(A)$, the number of events in a bounded region $A \subset \mathbb{R}^d$, the events of the process are independently and uniformly distributed over A. That is, given $N(A) = n$, the ordered n tuple of events $(\mathbf{s}_1, \ldots, \mathbf{s}_n)$ in A^n satisfies

$$\Pr(\mathbf{s}_1 \in B_1, \ldots, \mathbf{s}_n \in B_n) = \prod_{i=1}^{n} (|B_i|/|A|), \qquad B_1, \ldots, B_n \subset A, \quad (8.2.1)$$

where $|B| \equiv \int_B d\mathbf{s}$. Intuitively, this says that events are equally likely to occur anywhere within A and that events do not interact with each other, either repulsively (regularity of events) or attractively (clustering of events).

To illustrate these three different types of patterns, Figure 8.2*a*, *b*, and *c* shows realizations from a completely spatially random (csr) process, a regular process, and a cluster process, respectively (each conditioned to have 100 points), in the unit square.

Notice from Figure 8.2*a* that the csr realization seems to exhibit some clustering. This is not an unrepresentative realization, but illustrates a well known property of homogeneous Poisson processes: event-to-nearest-event distances are proportional to χ_2^2 random variables, whose densities put a substantial amount of probability near zero. True clustering is shown in Figure 8.2*c*, which should be compared with Figure 8.2*a*.

Several different approaches will be taken to quantify types of spatial point pattern, and the longleaf-pine data will be used to illustrate them. The general goal in the following subsections is to reduce the spatial data to informative descriptive statistics that can help elucidate models that might be fitted to the complex point pattern.

One type of descriptive statistic is based on quadrats (i.e., well defined areas, often rectangular in shape, in a region of interest A). Usually, quadrats of random location and orientation are sampled, the number of events in the quadrats are counted (here the events are trees), and statistics derived from the counts are computed. As well as a count of trees, the percent of area covered or the biomass in the quadrats might also be recorded.

Another type of statistic is based on distances between events, or between randomly sampled points (of A, not of the point pattern) and events. One such is the K function, discussed more fully in Sections 8.2.6 and 8.4.

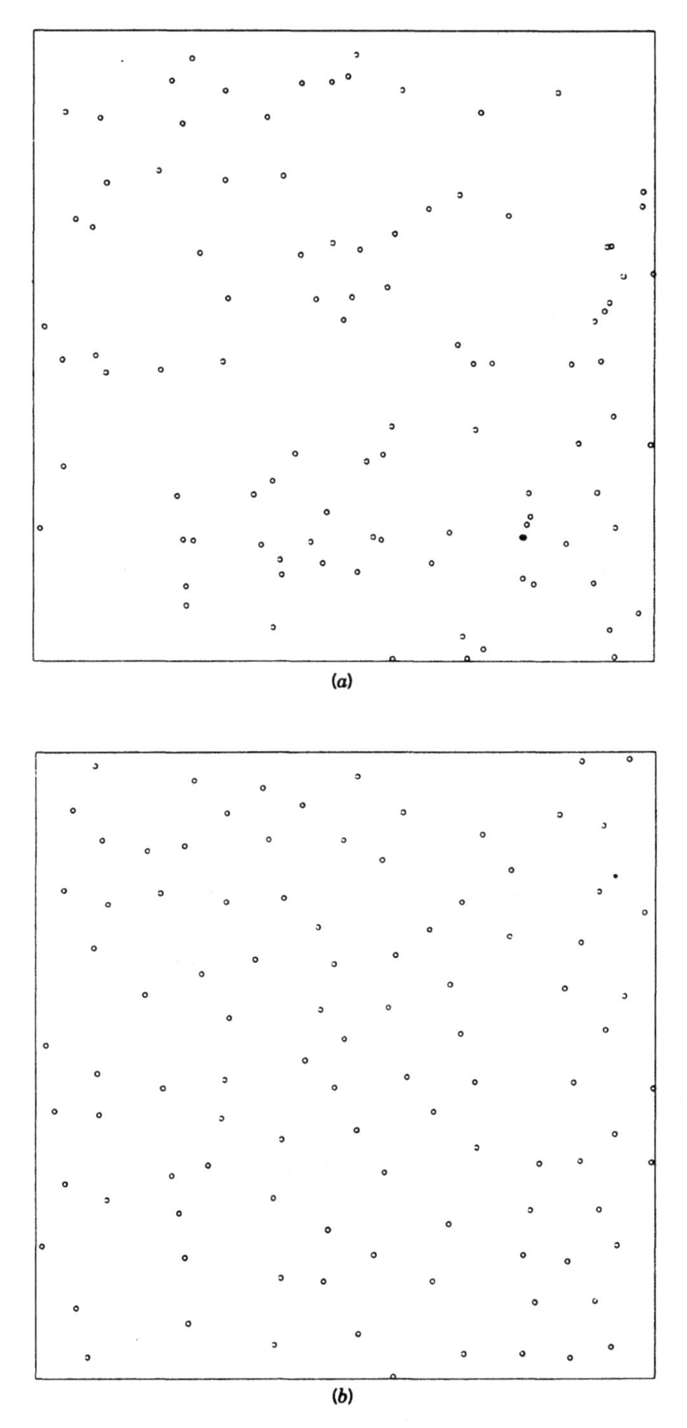

Figure 8.2 (*a*) A realization of a two-dimensional csr point process, conditioned on 100 events in the unit square. (*b*) Same as (*a*) except the point process exhibits regularity.

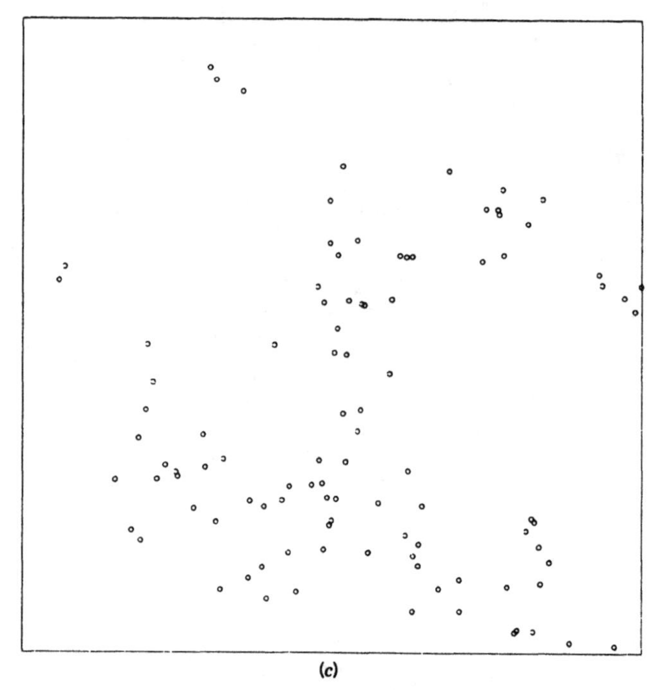

(c)

Figure 8.2 (c) Same as (a) except the point process exhibits clustering.

8.2.3[†] Quadrat Methods

Quadrat sampling involves collecting counts of the number of events in subsets of the study region A. Traditionally, these subsets are rectangular (hence the name quadrats), although any shape is possible. Quadrats are typically placed either randomly or layed out contiguously in A.

Random Quadrats

Although random quadrats are primarily intended for sampling events in the field, I shall illustrate their use on the mapped longleaf-pine data set. Figure 8.3 depicts the locations of 100 circular quadrats, 6 m in radius, in the extended study region (16 ha) of the Wade Tract. Quadrat locations were constrained so that no two quadrats overlap.

The trees in each quadrat were enumerated. Table 8.2 gives the frequency distribution of the number of trees per quadrat. Under complete spatial randomness (csr), the number of trees in a quadrat A_1, of area $|A_1|$, has a Poisson distribution with mean $\lambda|A_1|$, where λ is the intensity of the Poisson process. So, one test for csr is Pearson's X^2 goodness-of-fit test. Table 8.2 also gives the expected frequency distribution of number of trees per quadrat under a Poisson distribution with (estimated) mean 1.43. Here, quadrats with five or more trees were pooled to ensure that the expected number of trees in

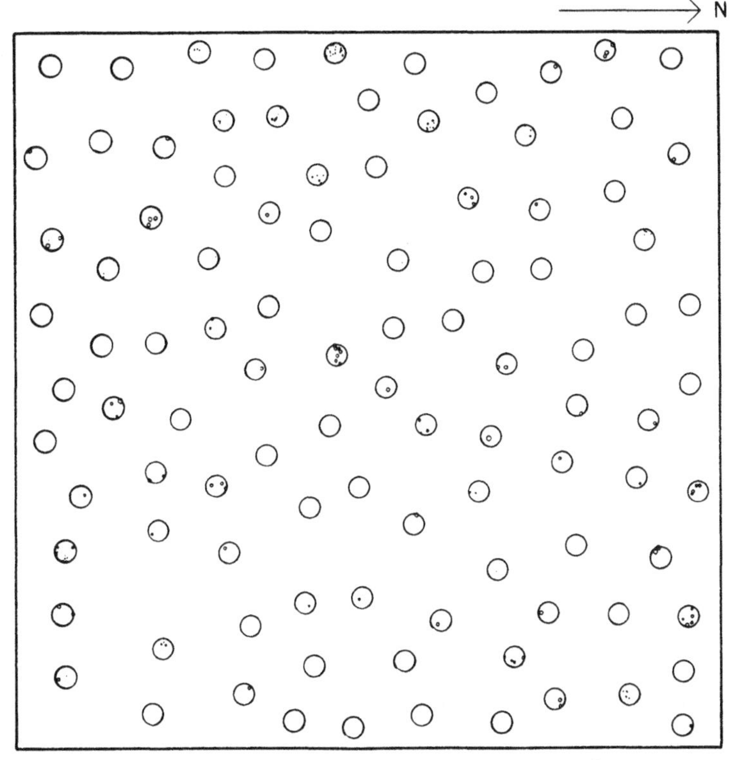

Figure 8.3 Map of the Wade Tract (extended study region of 16 ha) showing 100 randomly placed circular quadrats, each of radius 6 m.

Table 8.2 Frequency Distribution of Number of Trees per Quadrat in a Sample of 100 quadrats Each of Radius 6 m[a]

Trees per Quadrat	Observed Frequency		Expected Frequency	
0	34		23.93	
1	33		34.22	
2	17		24.47	
3	7		11.66	
4	3		4.17	
5	1	6	1.19	1.54
6	1		0.28	
7	2		0.06	
8	1		0.01	
9	0		0.00	
10	1		0.00	

[a]Observed frequencies are compared to expected Poisson (mean = 1.43) frequencies.

Table 8.3 Indices for Quadrat Count Data[a]

Index	Estimator	Realization	Reference
I	$\frac{S^2}{\overline{X}}$	2.335	Fisher et al. (1922)
ICS	$\frac{S^2}{\overline{X}} - 1$	1.335	David and Moore (1954)
ICF	$\frac{\overline{X}^2}{S^2 - \overline{X}}$	1.072	Douglas (1975)
$\overset{*}{X}$	$\overline{X} + \frac{S^2}{\overline{X}} - 1$	2.765	Lloyd (1967)
IP	$\frac{\overset{*}{X}}{\overline{X}}$	1.933	Lloyd (1967)
I_δ	$\frac{n\Sigma_{i=1}^n X_i(X_i - 1)}{n\overline{X}(n\overline{X} - 1)}$	1.930	Morisita (1959)

[a] X_i is the number of trees in the ith quadrat, $\overline{X}$ is the sample mean of the quadrat counts, and S^2 is the sample variance.

each frequency class is sufficiently large for a chi-squared approximation to hold (Cochran, 1954). Pearson's X^2 test indicates significant departure from a Poisson distribution. There are more empty quadrats and quadrats with five or more trees than would be expected under a Poisson distribution. Thus, trees appear to be clustered.

Regularity and Clustering

Once complete spatial randomness (csr) is rejected, the next step in a spatial analysis may be to measure the departure from csr. Various indices that have appeared in the literature were calculated from the sample of 100 quadrats and are presented in Table 8.3. The relative variance (I) and the David–Moore index (ICS) were obviously motivated by the equality of mean and variance of Poisson quadrat counts. Bartko et al. (1968) have shown that the expected value of ICS is 0 for Poisson quadrat counts. Values of I greater than 1 and ICS greater than 0 in Table 8.3 indicate that the longleaf pines are clustered. If ICS were less than 0, then this would indicate a tendency for regular spacings.

Douglas (1975) considers ICS to be an index of cluster size. He shows that if quadrats are large relative to cluster area and if the number of events per cluster has a Poisson distribution with mean μ, then ICS is approximately equal to μ for large sample sizes. Assuming ICS measures cluster size, then the index of cluster frequency ICF $\equiv \overline{X}/\text{ICS}$ should measure the average number of clusters per quadrat. In the present example, quadrats are small, only 6 m in radius, so Douglas' interpretation of ICS and ICF is not valid here.

Lloyd's (1967) "mean crowding" $\overset{*}{X}$ indicates that the average number of events sharing a quadrat with an arbitrary event is approximately 2.765. Lloyd also defines IP as an index of patchiness.

Morisita (1959) assumes that the point process consists of a mosaic of patches of differing intensities of plants with random spacing within patches. If quadrat size is small relative to patch size, so that quadrats tend to lie within only one patch each, then Morisita's index I_{δ} should measure variability between patches.

The reduction of complex point patterns to counts of the number of events in random quadrats and to one-dimensional indices results in a considerable loss of information. Only a single scale of pattern can be measured by a sample of random quadrats, and there is no consideration of quadrat locations or of the relative positions of events within quadrats, so most of the spatial information is lost.

Although I have illustrated the use of random quadrat samples on mapped data, they are primarily intended for use in the field. Limited time and resources require quick and efficient field-sampling techniques. Therefore, only one sample of quadrats of a given size can usually be taken. Moreover, the choice of quadrat size can substantially affect the results of the analysis. Interpretation of ICS and ICF as indices of cluster size requires that quadrats be large relative to the average areal coverage of clusters (which may be impractical in the field). If mean crowding is of interest, the natural choice of quadrat size is the average ambit or zone of influence of an individual (Lloyd, 1967), for example, territory size for motile organisms. Morisita's index I_{δ} requires that quadrats be small relative to the scale of pattern.

Grids of Contiguous Quadrats

Figure 8.4 shows the number of trees in a 32×32 grid of 6.25×6.25-m quadrats located in the 4-ha central study region. The analysis of grids of contiguous quadrats takes advantage of information on quadrat locations. A grid of contiguous quadrats is a spatial lattice, so the data could be analyzed using the methods described in Part II. Techniques used by ecologists to analyze such grids have been inspired by Watt (1947), who suggested that plant communities are comprised of a mosaic pattern of patches consisting of vegetation in differing stages of development. Therefore, it is fundamental to the understanding of plant communities to be able to recognize these patches and then to determine the average patch size.

In ecology, the analysis of contiguous quadrats has been approached in two different ways. In the agglomerative approach, pairs of adjacent quadrats are combined into blocks of size two, which are in turn successively combined, until a hierarchy of block sizes ranging from $2, 4, 8, 16, \ldots, K$ quadrats is obtained; see Figure 8.5. The between-blocks sum of squares from the block counts in pairs of blocks of size r, that comprise a block of size $2r$, is then considered (Greig-Smith, 1952). The second approach is to pair quadrats separated by given distances (Goodall, 1974; Ludwig and Goodall, 1978); variation between quadrat pairs is then considered.

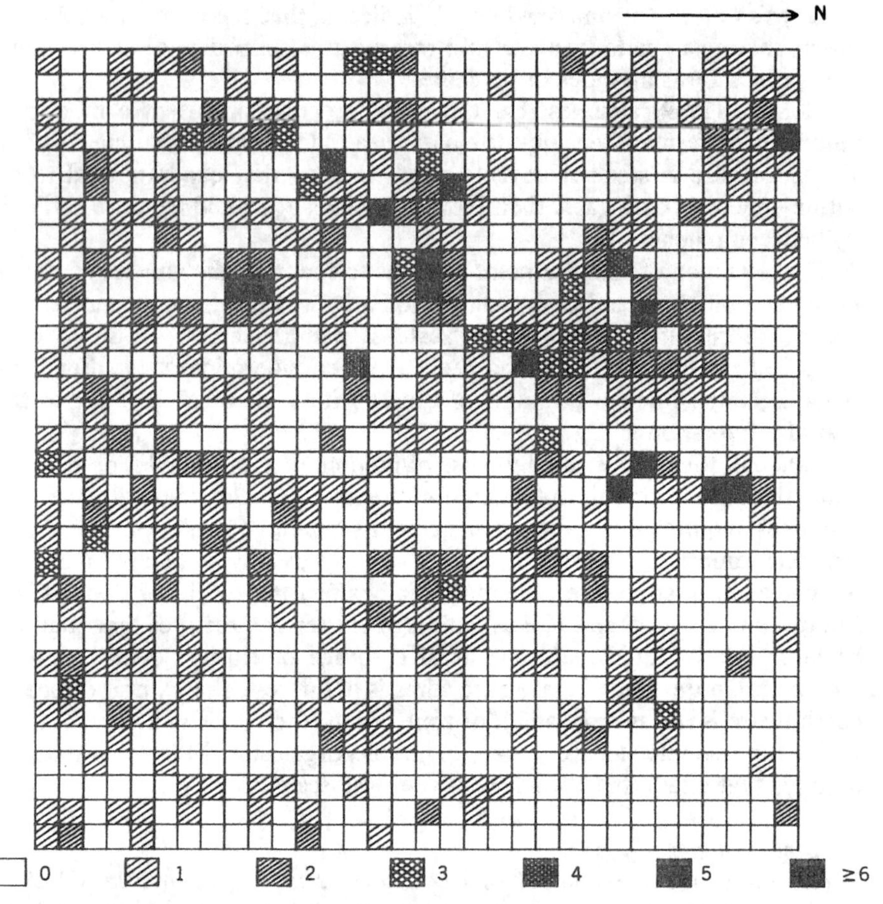

Figure 8.4 Number of trees in a 32 × 32 grid of quadrats (a quadrat is 6.25 × 6.25 m), which divide up the 4-ha central study region.

Agglomerative Approach

The agglomerative approach was introduced by Greig-Smith (1952), who analyzed contiguous-quadrat data using a nested analysis of variance. Let A_i and B_i denote the number of events in the ith pair of blocks, each containing r quadrats. The between-blocks sum of squares for blocks made up of r quadrats is, after suitable normalization,

$$\begin{aligned} SS_r &\equiv \frac{1}{r}\sum_{i=1}^{m}\left(A_i^2 + B_i^2\right) - \frac{1}{2r}\sum_{i=1}^{m}\left(A_i + B_i\right)^2 \\ &= \frac{1}{2r}\sum_{i=1}^{m}\left(A_i - B_i\right)^2, \end{aligned} \tag{8.2.2}$$

where $m = K/2r$. The mean square is $MS_r \equiv SS_r/m$. Traditionally, mean

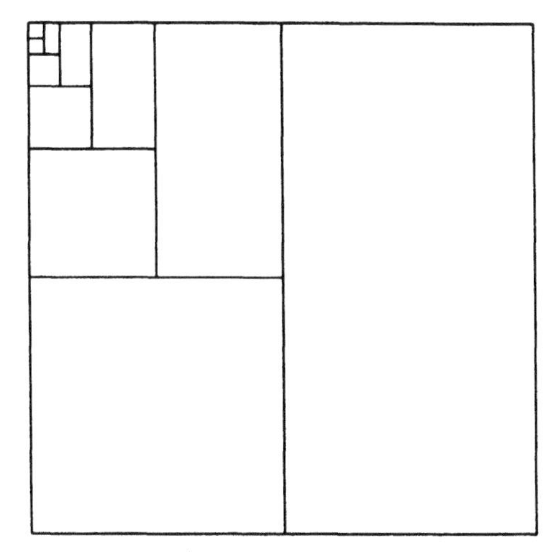

Figure 8.5 Agglomeration method for combining quadrats into blocks.

squares are plotted against block size. Peaks in mean-square plots are considered to be associated with patch size. If one or more peaks are detected, biological interpretations are often attached to each peak. However, such patterns of peaks may not be statistically significant: As a result of small degrees of freedom, variances of mean squares increase with increasing block size, and observed peaks may simply be due to random fluctuation. For Poisson quadrat counts, the mean and variance of MS_r are

$$E(MS_r) = \lambda|Q|, \tag{8.2.3}$$

$$\text{var}(MS_r) = [\lambda|Q|\{1 + 4r\lambda|Q|\}]/K, \tag{8.2.4}$$

where $|Q|$ denotes the quadrat area (Thompson, 1955). In general, the exact (joint) distribution of mean squares is unknown.

An alternative way to obtain acceptance envelopes (based on the null hypothesis of complete spatial randomness) for the MS_r versus r plot is to use randomization of counts among quadrat locations. Figure 8.6 shows a plot of mean square against block size and a 95% simulation envelope obtained from 200 random rearrangements of the original quadrat counts. Mean squares fall below the lower acceptance envelope for the two smallest block sizes, indicating that adjacent quadrat counts are more homogeneous than would be expected from randomly arranged quadrat counts. For most of the larger block sizes, mean squares fall above the upper acceptance envelope. Peak mean squares occur in block sizes of 64 and 256 quadrats (50 × 50-m and 100 × 100-m blocks, respectively) and are above the upper acceptance envelope, indicating patches of size 50 m square and 100 m square.

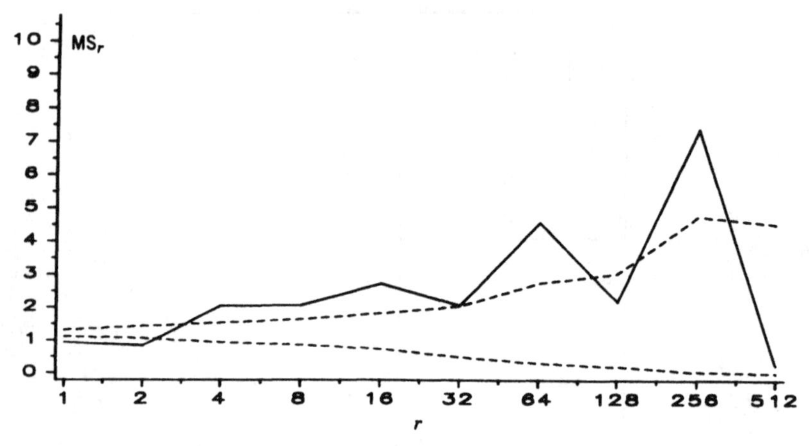

Figure 8.6 Plot of mean square MS$_r$ (solid line) versus block size r, together with 95% simulation envelopes (dashed lines).

Nested Analysis of Variance

Comparisions of successive mean squares are sometimes used to test for clustering at given block sizes. In Table 8.4, mean squares for blocks of r quadrats are compared to 95% acceptance regions obtained assuming normality (the limits of the regions are estimated by $m^{-1}\text{MS}_{r/2} \cdot \chi^2_{m,0.025}$ and $m^{-1}\text{MS}_{r/2} \cdot \chi^2_{m,0.975}$, where $m = K/2r$ and $K = (32)^2$) and from randomization of blocks of $r/2$ quadrats. There is little agreement between acceptance regions obtained assuming normality and those obtained from randomization. This implies that the normal approximation for the acceptance regions is poor for this example.

Table 8.4 Tests for Clustering at Each Block Size Using Greig-Smith's Hierarchical Analysis of Variance[a]

			95% Acceptance Regions	
r	m	MS$_r$	$m^{-1}\text{MS}_{r/2} \cdot \chi^2_m$	Randomization
1	512	0.926		(1.127, 1.322)
2	256	0.834	(0.773, 1.093)	(1.354, 1.701)
4	128	2.039	(0.642, 1.050)	(1.773, 2.527)
8	64	2.049	(1.395, 2.804)	(1.807, 2.894)
16	32	2.713	(1.171, 3.168)	(1.826, 3.822)
32	16	2.043	(1.171, 4.891)	(1.730, 4.253)
64	8	4.549	(0.557, 4.478)	(1.838, 6.109)
128	4	2.182	(0.551, 12.673)	(0.525, 5.494)
256	2	7.361	(0.055, 8.049)	(0.455, 7.361)
512	1	0.250	(0.007, 36.981)	(0.250, 0.250)

[a]Mean squares are compared to 95% acceptance regions obtained from (1) a normal approximation, under the hypothesis $E(\text{MS}_r) = E(\text{MS}_{r/2})$, and (2) randomization of blocks of size r.

Assuming normality, significant clustering is indicated for blocks of 4 and 64 quadrats. However, under the randomization distribution, significant clustering is not indicated at any block size.

Distribution-Free Randomization Test

Mead (1974) suggests a distribution-free randomization test for clustering at each block size. Randomizations are carried out within blocks of four units (a unit is a quadrat or a block of quadrats). Consider a block of four units with counts 4, 7, 3, and 6, paired as

$$\overline{4\quad 7}\qquad \overline{3\quad 6}$$

Consider the set of possible differences between pairs of units within blocks:

$$\{|(4+7)-(3+6)|, |(4+3)-(7+6)|, |(4+6)-(7+3)|\}$$
$$= \{2,6,0\}.$$

The observed pair difference is $|(4+7)-(3+6)| = 2$. The sum of the observed pair differences over all such blocks of four units can then be compared to the randomization distribution of the sum. But, to avoid domination by blocks containing wide ranges of counts, differences are *ranked* to form a distribution-free test. By starting the ranking at 0 and allowing for ties, ranked differences take one of three forms: $\{0,0,2\}$, $\{0,1,2\}$, or $\{0,2,2\}$. Block configurations for which all three pair differences are equal are ignored. For the preceding example, we have $\{0,1,2\}$ with a ranked observed difference of 1. The sum of the ranked observed differences T is compared to the randomization distribution. A normal approximation is convenient for small block sizes (and so a large number of blocks). Let m_1, m_2, and m_3, be the number of blocks coded as $\{0,0,2\}$, $\{0,1,2\}$, and $\{0,2,2\}$, respectively. Then, under the null hypothesis of complete spatial randomness,

$$Z \equiv \frac{3T - (2m_1 + 3m_2 + 4m_3)}{\sqrt{8m_1 + 6m_2 + 8m_3}} \tag{8.2.5}$$

is approximately normally distributed with mean 0 and variance 1 (Upton and Fingleton, 1985, p. 47). Results of this "2 within 4" randomization test, given in Table 8.5, suggest that there is significant clustering when the block size is four quadrats.

Disadvantages

For agglomerative methods, the effect of block size is confounded with average spacing between block centers. Center-to-center distances are doubled with each second doubling of block size. Thus, it is not possible to judge the relative importance of spacing and patch size from mean-square plots

Table 8.5 Results of Mead's "2 within 4" Randomization Test

Block Size[a]	T	$E(T)$	Z	p Value
2	76	81.7	−0.50	0.6915
4	95	63.3	3.15	0.0008
8	42	45.67	−0.75	0.7722
16	39	24.7	1.40	0.0805
32	10	13.3	−0.82	0.7922
64	10	7.3	0.91	0.1832
128	4	3.7	0.16	0.5309
256	3	2.0	0.71	0.3333
512	0	1.0	−1.00	1.0000

[a]For block sizes of less than 128 quadrats, the normal approximation was used to calculate the p-value column. For block sizes of 128 or more quadrats, randomization was used.

(Goodall, 1963). Also, much of the peaked structure of mean-square plots may result from alternation of square and rectangular blocks (Pielou, 1977) and the increase in variance of mean squares as block size increases (Thompson, 1955). In addition, because block size is doubled at each step, only a limited number of block sizes can be considered (Goodall, 1974). Finally, agglomerative methods are sensitive to the starting position of the quadrats (Usher, 1969, 1975; Errington, 1973; Hill, 1973; Galiano, 1983; Upton and Fingleton, 1985, p. 48).

Paired-Quadrat Approach

The paired-quadrat-variance method considers only the effect of spacing between quadrat centers. In his original analysis, Goodall (1974) randomly pairs quadrats separated by a variety of fixed distances, but in Ludwig and Goodall (1978) the random pairing is dropped. Variance within a pair of quadrats is computed and averaged over all quadrat pairs separated by a given distance, say h. Let A_i and B_i be the number of events in the ith pair of (randomly selected) quadrats h grid units apart. The mean square for quadrats separated by h grid units is

$$2V_h \equiv \frac{1}{n_h} \sum_{i=1}^{n_h} (A_i - B_i)^2, \tag{8.2.6}$$

where the sum is over all n_h pairs of (selected) quadrats separated by h units, $h = 1, 2, \cdots$. Note that this is precisely a variogram estimator; see Section 2.4. Under a random arrangement of quadrat counts, the variance estimates $\{V_h\}$ are expected to be roughly constant. Goodall (1974) compares V_h/V_1 to an F distribution on $n_h - 1$ and $n_1 - 1$ degrees of freedom. He also suggests using Bartlett's test for equality of the entire set of variances; but, if quadrats are exhaustively sampled, variance estimators are not independent, so overall significance levels are unknown (Zahl, 1977).

Figure 8.7 shows a plot of paired-quadrat variance (estimated from all possible pairs of quadrats) against distance between quadrat centers. Separate variances were calculated for quadrat pairs oriented from north to south and pairs oriented from east to west. Figure 8.7 also shows 95% simulation envelopes for 200 randomizations of quadrat counts. These plots suggest that variation between adjacent quadrats is smaller than would be expected for randomly arranged quadrat counts. This result is consistent with the earlier result that blocks of two and four quadrats have smaller mean squares than expected from randomized quadrat counts. Thus, there appears to be some small-scale spatial homogeneity in quadrat counts.

Summary

Analysis of plant communities based on grids of contiguous quadrats have found wide favor among ecologists. Significant clustering at a given block size is considered to be evidence that a population consists of a mosaic of patches of differing densities. Observed peaks in mean-square plots are often declared to be real, and then subjected to rather close biological interpretation. Alternative statistical models of spatial pattern are typically not considered.

Although analyses of grids of contiguous quadrats take advantage of relative locations of quadrats, precise spatial information on the locations of events with quadrats is lost. Additionally, although Goodall's paired-quadrat-variance method yields results that depend only on distances between quadrat centers (and are not confounded with block size), information at multiple levels of aggregation is lost because quadrats are not combined into blocks. For contiguous quadrats on a line transect, Ver Hoef et al. (1991) show how both the agglomerative approach and the paired-quadrat approach are based on classical variogram estimators (Section 2.4), that are not only functions of lag h but also of level of aggregation. Ripley (1978) suggests an alternative, spectral approach (to the analysis of quadrat counts on a line transect) that appears to be easier to interpret. However, there are other ways to analyze mapped spatial point patterns that are spatially more informative than quadrat methods. The next sections will present such methods.

8.2.4[†] Kernel Estimators of the Intensity Function

Suppose that the region of interest A is divided into regularly spaced square quadrats of size $a \times a$. It has been argued in the previous subsection that, for large a, the resulting quadrat counts destroy spatial information. However, they do give a global idea of subregions with high or low numbers of events per unit area. For small a, more spatial information is retained, but at the expense of a "busier" picture of numbers of events per unit area; as $a \to 0$, the picture degenerates into a mosaic of 0s and $(1/a^2)$s.

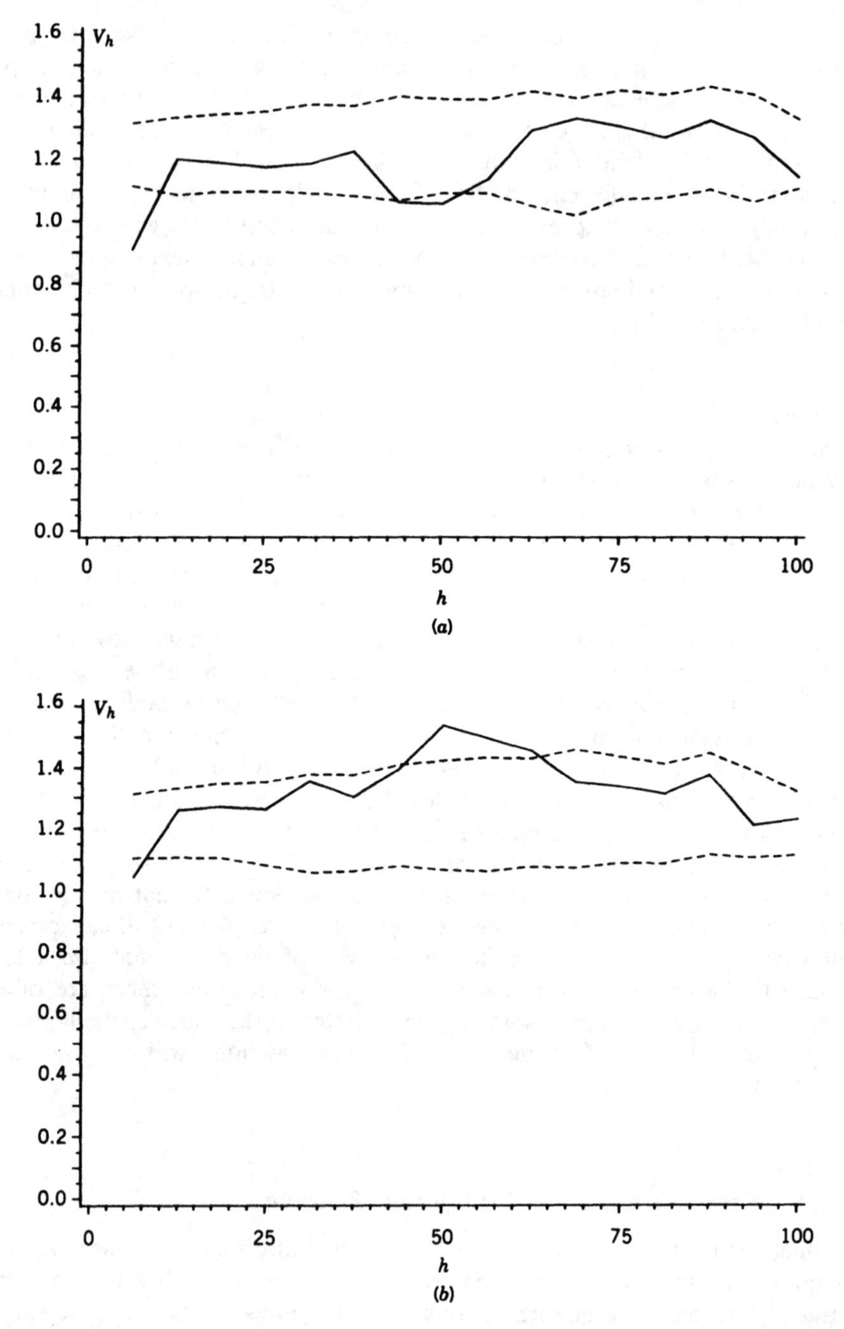

Figure 8.7 Plots of V_h given by (8.2.6) versus h, for the paired-quadrat-variance method. Units on the horizontal axis are in meters. (*a*) North–south; (*b*) east–west; (*c*) combined.

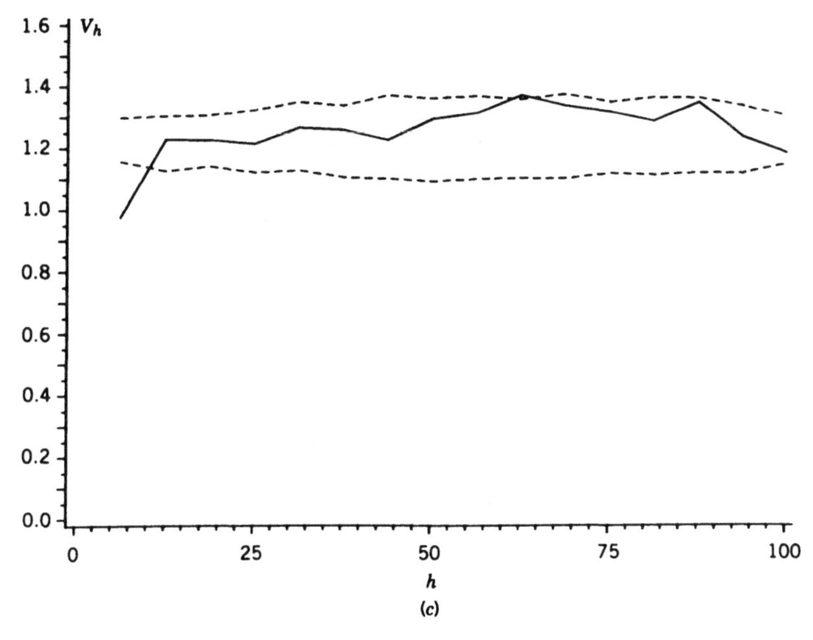

Figure 8.7 (*Continued*)

Intensity Function

In terms of the underlying point process, are the quadrat counts per unit area estimating something? Let $N(\mathbf{s}, a)$ denote the number of events in an $a \times a$ square at location $\mathbf{s}$. Consider

$$\lambda_a(\mathbf{s}) \equiv \Pr(N(\mathbf{s}, a) > 0)/a^2.$$

If $\lambda_a(\mathbf{s}) \to \lambda(\mathbf{s})$ as $a \to 0$, for all $\mathbf{s} \in A$, then call $\{\lambda(\mathbf{s}): \mathbf{s} \in A\}$ the *intensity function* (a more formal definition is given in Section 8.3.1).

For $\mathbf{s}$ a node of the square grid of dimension $a \times a$, the quadrat count per unit area, $N(\mathbf{s}, a)/a^2$, is an unbiased estimator of $\int_0^a \int_0^a \lambda(\mathbf{u} + \mathbf{s})\, d\mathbf{u}/a^2$. When $\lambda(\cdot)$ does not vary much over the $a \times a$ square, this integral is approximately equal to $\lambda(\mathbf{s})$. Thus, the set of quadrat counts per unit area, $\{N(\mathbf{s}, a)/a^2$: $\mathbf{s}$ is a node of the square $a \times a$ grid in $A\}$, estimates $\{\lambda(\mathbf{s})$: $\mathbf{s} \in A\}$.

Kernel Estimators

The reader may notice that this discussion resembles another problem in Statistics, namely the estimation of a (multivariate) density function. In fact, *kernel estimators* of density functions can be extended to obtain nonparametric estimators for $\lambda(\cdot)$ (e.g., Diggle, 1985; Section 8.5.1).

Let $(\mathbf{s}_1, \mathbf{s}_2, \ldots, \mathbf{s}_n)$ be the spatial locations of $n = N(A)$ events in a bounded study region $A \subset \mathbb{R}^d$. Consider estimators of the form

$$\hat{\lambda}_h(\mathbf{s}) \equiv \frac{1}{p_h(\mathbf{s})}\left\{\sum_{i=1}^{n} \kappa_h(\mathbf{s} - \mathbf{s}_i)\right\}, \qquad \mathbf{s} \in A, \tag{8.2.7}$$

where $\kappa_h(\cdot)$ is a probability density (kernel) function symmetric about the origin, $h > 0$ determines the amount of smoothing, and $p_h(\mathbf{s}) \equiv \int_A \kappa_h(\mathbf{s} - \mathbf{u})\, d\mathbf{u}$ is an edge correction (Diggle, 1985). For example, $\kappa_h(\cdot) = h^{-d}\kappa_1(\cdot/h)$; see (8.5.10). The choice of an appropriate smoothing constant (or bandwidth) h is of primary concern when estimating $\lambda(\cdot)$ (Silverman, 1978a). Choice of the kernel function is of secondary importance; any reasonable kernel gives close to optimal results (Epanechnikov, 1969). Consider the simple product kernel $\kappa_h(\mathbf{u}) = \rho_h(u_1) \cdot \rho_h(u_2)$ in two dimensions, where

$$\rho_h(u) = \begin{cases} 0.75h^{-1}\left[1 - (u/h)^2\right], & -h \le u \le h, \\ 0, & \text{otherwise}. \end{cases} \tag{8.2.8}$$

Epanechnikov (1969) has shown that this kernel has certain optimality properties. Diggle (1981a) recommends $h = 0.68n^{-0.2}$ for estimating the intensity on a unit square. Here, $h = 0.715n^{-0.2}$ is chosen to obtain a bandwidth of 40 m (a round number) on a 200 × 200-m square.

Longleaf-Pine Data

For the longleaf-pine data, Figure 8.8 shows an intensity estimator based on (8.2.8), using a 40-m kernel. This figure suggests that the point pattern is nonstationary; there is a clear trend of increasing intensity from the eastern to the western half of the study region. Peak intensities of more than 0.025 m^{-2} are found in a band extending from the west-central to the north-central part of the study region and in the northwestern corner. Intensities in the eastern portion are generally less than 0.015 m^{-2}. Although the kernel estimate of the intensity function shows the overall trend in intensity of the longleaf pine forest, a 40-m kernel is too large to show the fine structure of the spatial pattern evident in Figure 8.1. Several small clusters of densely packed small trees are present, a pattern that is not evident in Figure 8.8. A smaller kernel size could be used, but this would result in smaller bias only at the expense of a large increase in variance.

Some Final Remarks

Looking at the estimated intensity function is just one way of summarizing the spatial point pattern. It is analogous to estimating the mean function $\mu(\cdot)$ for geostatistical or lattice data (see Parts I and II of this book), and in the same way does not address the spatial dependence (or small-scale variations) in the data.

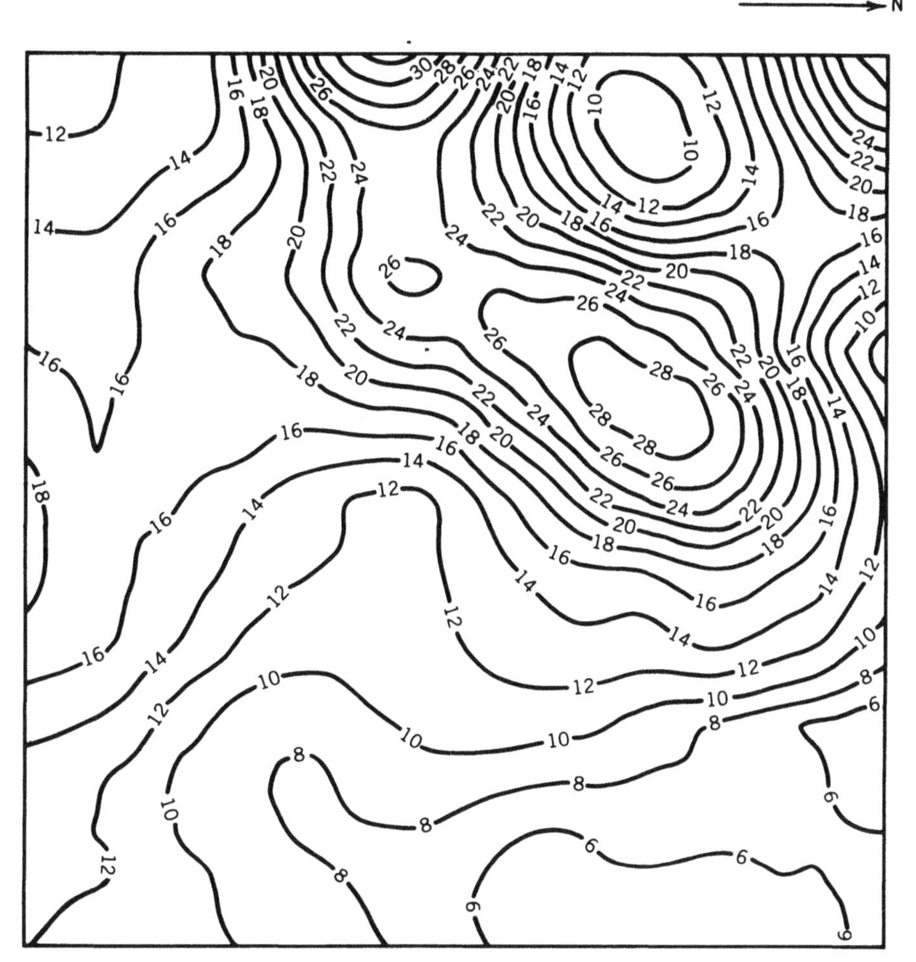

Figure 8.8 Nonparametric kernel estimate of $\{\lambda(\mathbf{s}): \mathbf{s} \in \text{4-ha study region}\}$, using a quadratic kernel of extent 40×40 m. Contour values shown are in units of 10^{-3} m^{-2}.

If the point process were an inhomogeneous Poisson process with intensity function $\lambda(\cdot)$ (see Section 8.3), then there is no spatial dependence component. For example, If $N(\mathbf{s}_1, a_1)$ and $N(\mathbf{s}_2, a_2)$ refer to the random number of events in *disjoint* regions, then $N(\mathbf{s}_1, a_1)$ and $N(\mathbf{s}_2, a_2)$ are *independent* random variables. However, other point processes do not share this property, although they may have the same intensity function. In an analogous role to the variogram and covariogram, second-order intensity functions and K functions help characterize the spatial dependence; see Sections 8.2.6 and 8.4.

8.2.5† Distance Methods

Whereas quadrat methods lend themselves to field sampling, some of the more powerful distance methods rely on having a good map of all the events. Distance methods make use of precise information on the locations of events and have the advantage of not depending on arbitrary choices of quadrat size or shape.

Nearest-Neighbor Methods
Here, event-to-event or point-to-event distances are computed and summarized. Figure 8.9 illustrates various possibilities. Distances may be measured between events and nearest-neighboring events (W) or between sample points and nearest events (X). Sample points usually are located randomly in the study area, but may be placed systematically (Byth and Ripley, 1980).

Distances may be measured also between an event nearest a sample point and the event's nearest neighbor (Y). Distance Y is not independent of distance X because no events can occur in the circle of radius X around the sample point. For that reason, Besag and Gleaves (1973) suggest the T-square sampling method. The T-square method involves dividing the region into two half-planes as illustrated in Figure 8.9. Distance Z is measured between the

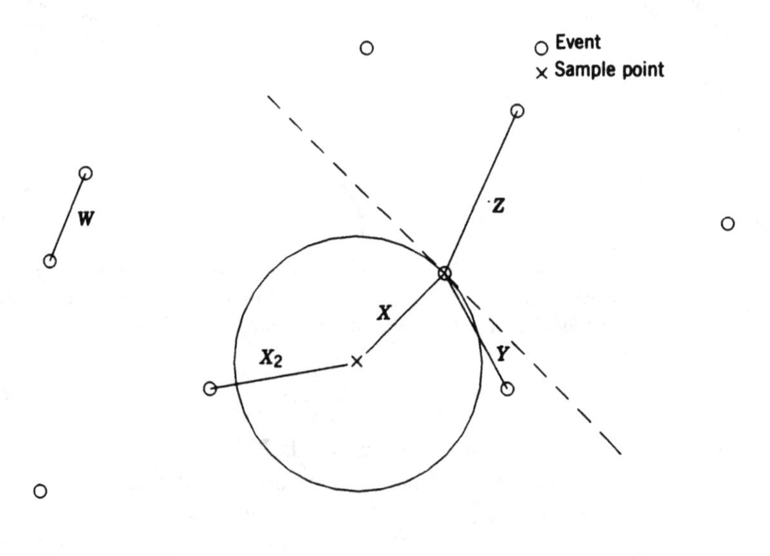

Figure 8.9 Types of nearest-neighbor distances: event to nearest event (W); sample point to nearest event (X); sample point to second-nearest event (X_2); event-nearest-sample point to nearest event (Y); event-nearest-sample point to nearest-event-in-half-plane-not-containing-sample point (Z).

event nearest the sample point and the nearest event in the half-plane not containing the sample point. For a homogeneous Poisson (i.e., completely spatially random) process, distance Z is independent of distance X. Holgate (1965a) considers distances between a sample point and the second nearest event (X_2).

The distribution theory for nearest-neighbor distances W and X under complete spatial randomness (csr) is well known. In $\mathbb{R}^2$, the density of the positive random variable W is

$$g(w) = 2\pi\lambda w \exp(-\pi\lambda w^2), \qquad w > 0; \tag{8.2.9}$$

see Section 8.4.2. The distance from a randomly placed sample point to the nearest event, X, has the same distribution as W. The probability that there are no events within distance z from an arbitrary event with an exclusion angle of 180° is $\exp(-\pi\lambda z^2/2)$. So, the density of Z under csr is

$$h(z) = \pi\lambda z \exp(-\pi\lambda z^2/2), \qquad z > 0. \tag{8.2.10}$$

Hines and Hines (1989) propose a variant of the T-square sampling method based on the wandering-quarter method of Catana (1963). Instead of sampling events in the half-plane, as shown in Figure 8.9, events are sampled in the quarter-plane formed by the subset of $\mathbb{R}^2$ lying within $\pm 45°$ of the direction from sample point to nearest event. The origin of the first quarter-plane is positioned at this nearest event. The nearest event in the quarter-plane is obtained and the interevent distance is measured. The second quarter-plane (of identical orientation) is then positioned at this new event and the procedure is repeated. The resulting interevent distances are computed. Under csr, the interevent distances are i.i.d. with density $h(z) = (\pi\lambda z/2)\exp(-\pi\lambda z^2/4)$, $z > 0$.

Test Statistics

Many statistics have been proposed for testing complete spatial randomness (csr), usually based on a random sample of n points or a random sample of n events. The latter cannot be performed in the field and require mapped data of the type presented in Table 8.1. A summary of test statistics and their asymptotic distributions under csr is presented in Table 8.6, which is a modification of Upton and Fingleton's (1985) Table 1.10.

Distribution theory for these tests is based on independence of n nearest-neighbor measurements randomly sampled from a region A, an assumption that is unlikely to hold when A is intensively sampled. Byth and Ripley (1980) recommend the bound $n \leq N(A)/10$; otherwise, Monte Carlo procedures should be used (see following text and Section 8.4.1).

Distribution Theory

For a sparsely sampled homogeneous Poisson process, the distributions of B, G, I, N, and O are known. Suppose a random sample of n events or n points

Table 8.6 Nearest-Neighbor Statistics and Their Asymptotic Distributions under csr

Basic Measurement		Test Statistic	Asymptotic or Exact Distribution	Reference
W	A	$2(\lambda)^{1/2}\Sigma W_i/n$	$N(1,(4-\pi)/n\pi)$	Clark and Evans (1954)
	B	$2\pi\lambda\Sigma W_i^2$	χ^2_{2n}	Skellam (1952)
X	C	$\pi\lambda\Sigma X_i^2/n$	$N(1,1/n)$	Pielou (1959)
	D	$n(\Sigma X_i^2)/(\Sigma X_i)^2$	By simulation	Eberhardt (1967)
	E	$12n\{n\log(\Sigma X_i^2/n) - \Sigma\log X_i^2\}/(7n+1)$	χ^2_{n-1}	Pollard (1971)
X, X_2	F	$\{\Sigma(X_i^2/X_{2,i}^2)\}/n$	$N(1/2,1/12n)$	Holgate (1965a)
	G	$(\Sigma X_i^2)/(\Sigma X_{2,i}^2)$	beta(n,n)	Holgate (1965a)
X, W	H	$[\Sigma\{X_i^2/(X_i^2+W_i^2)\}]/n$	$N(1/2,1/12n)$	Byth and Ripley (1980)
	I	$(\Sigma X_i^2)/(\Sigma W_i^2)$	$F_{2n,2n}$	Hopkins (1954)
X, Z	J	$2n\Sigma(2X_i^2+Z_i^2)/\{\Sigma(\sqrt{2}X_i+Z_i)\}^2$	By simulation	Hines and Hines (1979)
	K	$48n[n\log\{\Sigma(2X_i^2+Z_i^2)/n\} - \Sigma\log(2X_i^2+Z_i^2)]/(13n+1)$	χ^2_{n-1}	Diggle (1977)
	L	$\Sigma\{2X_i^2/(2X_i^2+Z_i^2)\}/n$	$N(1/2,1/12n)$	Besag and Gleaves (1973)
	M	$2\Sigma\{\min(2X_i^2,Z_i^2)/(2X_i^2+Z_i^2)\}/n$	$N(1/2,1/12n)$	Diggle et al. (1976)
	N	$2(\Sigma X_i^2)/(\Sigma Z_i^2)$	$F_{2n,2n}$	Besag and Gleaves (1973)
	O	$-2\Sigma[\log\{2X_i^2/(2X_i^2+Z_i^2)\}]$	χ^2_{2n}	Cormack (1979)
X, Y	P	$\Sigma R_i/n'$	$N(1/2,1/12n')$	Cox and Lewis (1976)
	Q	$\bar{A}_X - \bar{A}_Y$	$N(0,(\bar{A}_X^2+\bar{A}_Y^2)/n')$	Satyamurthi (1979)

is obtained. The random variables $2\pi\lambda W^2$, $2\pi\lambda X^2$, and $\pi\lambda Z^2$ have chi-squared distributions on 2 degrees of freedom (Skellam, 1952). So Skellam's statistic (B) is chi-squared distributed on $2n$ degrees of freedom, and Hopkins' (I) and Besag and Gleaves' (N) statistics are F distributed on $2n$ and $2n$ degrees of freedom. The random variables $\pi\lambda X^2$ and $\pi\lambda(X_2^2 - X^2)$ have independent and identical exponential distributions with mean 1. So $s = \pi\lambda\Sigma X_i^2$ and $t = \pi\lambda\Sigma(X_{2,i}^2 - X_i^2)$ have independent and identical gamma distributions with parameters 1 and n. Taking $u = s/(s+t)$ and $v = t$, and integrating out v, the distribution of Holgate's (G) statistic is beta(n, n). Taking $p = 2X^2/(2X^2 + Z^2)$ and $q = 2\pi\lambda X^2$ and integrating out q, the distribution of p is uniform on $[0, 1]$. If $r = -2\log(p)$, then r has a chi-squared distribution on 2 degrees of freedom; so Cormack's (O) statistic is chi-squared distributed on $2n$ degrees of freedom (for sparse sampling).

For large samples, A, C, F, H, L, and M are approximately normally distributed. Normal approximations should be good for $n \geq 10$ (Diggle et al., 1976). Neither Clark and Evans (1954) nor Pielou (1959) intended that their statistics (A and C, respectively) be used to *test* for csr using a normal approximation. They were intended as indices for measuring departure from csr. Clark and Evans suggest a test based on Skellam's (B) statistic, and Pielou suggests a test based on $2\pi\lambda\Sigma X_i^2$, both using a χ_{2n}^2 distribution under the null hypothesis of csr.

The null distributions of Eberhardt's (D) and Hines and Hines' (J) statistics are unknown, so inference must be based on simulation. Both statistics are identically distributed under csr (Hines and Hines, 1979); a table of critical values compiled by simulation is presented in Hines and Hines (1979) and reproduced in Upton and Fingleton (1985, Appendix D).

Pollard's (E) and Diggle's (K) statistics are likelihood ratio tests for homogeneity of a Poisson process. Suppose the intensity at the ith sample point is λ_i. Let $u_i = 2\pi X_i^2$. For a homogeneous Poisson process, the maximum likelihood estimate of λ is $\hat{\lambda} = n/\Sigma_{i=1}^n u_i$. For an inhomogeneous Poisson process, the maximum likelihood estimate of λ_i is u_i^{-1}. So, minus twice the log-likelihood ratio is

$$-2\left[n \log\left(\sum_{i=1}^{n} X_i^2/n\right) - \sum_{i=1}^{n} \log(X_i^2)\right].$$

Pollard's statistic (E) is obtained by multiplying by Bartlett's (1937) correction factor. Diggle's (K) statistic is obtained in the same way, taking $u_i = \pi(2X_i^2 + Z_i^2)$.

Cox and Lewis (1976) and Satyamurthi (1979) suggest test statistics based on distances X and Y. The geometric configuration of a point P, event Q nearest P, and event R nearest Q is illustrated in Figure 8.10. Let $B \equiv$

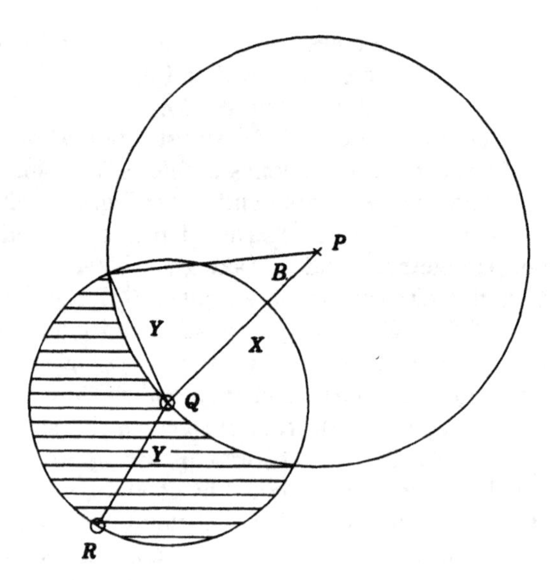

Figure 8.10 Point P, nearest event Q, and event R nearest Q. Distances X and Y and angle B are marked. The area of the circle centered at P is A_X. The shaded area is A_Y.

$\arcsin(Y/2X)$, $Y < 2X$. Cox and Lewis found it easier to consider the joint distribution of X and B rather than X and Y. From this they found that, under csr,

$$R \equiv (4/3)(1 - \pi T), \tag{8.2.11}$$

where $T \equiv (2\pi + \sin B - (\pi + B)\cos B)$, is uniformly distributed on $[0, 1]$. Then, under csr, the mean of R over the n' pairs (X_i, Y_i) that satisfy $Y_i < 2X_i$ is approximately normally distributed with mean $1/2$ and variance $1/(12n')$.

Satyamurthi (1979) considers the null hypothesis that the intensity λ_P of events at a random point P is equal to the intensity λ_Q at event Q nearest P. Define areas A_X and A_Y as in Figure 8.10. Under csr, and provided $Y < 2X$, the joint density of A_X and A_Y is

$$g(A_X, A_Y) = \lambda_P \exp(-\lambda_P A_X) \cdot \lambda_Q \exp(-\lambda_Q A_Y).$$

Let $\theta_P \equiv 1/\lambda_P$ and $\theta_Q \equiv 1/\lambda_Q$. Suppose a sample of n points is taken and, as before, let n' be the number of pairs (X_i, Y_i) that satisfy $Y_i < 2X_i$. Let $\bar{A}_X$ and $\bar{A}_Y$ each denote the mean of n' values obtained from the respective areas. Then, under the null hypothesis that $\theta_P = \theta_Q$, $\bar{A}_X - \bar{A}_Y$ is approximately normally distributed with mean 0 and variance $(\bar{A}_X^2 + \bar{A}_Y^2)/n'$.

Edge Effects

For the study region A, some events or sample points may be closer to the border of A than to their nearest neighbors within A. Because the nearest event may be located outside A, distance to the nearest event is unknown. If the nearest neighbor is taken to be the closest event within the study region, expected nearest-neighbor distances will be greater for events (points) located near the boundary of the study region than for events (points) located near its center. Thus, estimates based on nearest-neighbor statistics are biased unless some method is used to correct for edge effects (e.g., Ripley, 1982). There are at least three general approaches to correcting for edge effects.

The first approach consists of constructing a guard area inside the perimeter of A. Distances are measured only from events (points) located some minimum distance away from the edge of A. Distances are not measured from events (points) within the guard area, but events in the guard area are allowed as nearest neighbors of selected events (points).

The second approach is to regard a rectangular study region as a torus, so that events near opposite edges are considered to be close. Thus, the bottom edge of the plot is joined to the top edge and the left edge is joined to the right edge. An equivalent interpretation is to regard the study region as the center plot of a 3×3 grid of plots each identical to the study region.

The third approach is to obtain finite-sample corrections to the distribution theory for specific test statistics. For example, Donnelly (1978) and Doguwa and Upton (1988) give corrections to the first two moments of, respectively, Clark and Evans' (1954) event-to-event statistic (statistic A in Table 8.6) and its point-to-event analogue. However, their results are empirical and do not generalize to arbitrarily shaped sampling regions nor to nonregular point sampling.

Longleaf-Pine Data

Nearest-neighbor methods are illustrated using the 584 trees in the 4-ha central study region A. Distances were measured from 50 randomly chosen events (trees, chosen without replacement) and 50 randomly located points. Less than 1 in 10 trees were sampled, so the assumed null distributions should be approximately correct (Diggle et al., 1976). To determine how sensitive nearest-neighbor statistics are to edges, distances were calculated with and without using the surrounding 12 ha of the extended study region as a guard area. Results of these analyses are given in Table 8.7. Clark and Evans' (A), Skellam's (B), and Pielou's (C) statistics require an estimate of intensity. Here, I used $N/|A| = 584/(4 \times 10^4) = 0.0146 \text{ m}^{-2}$.

When the 12-ha guard area within the extended study region is used, nearest-neighbor statistics generally show significant (p value < 0.05) departures from complete spatial randomness, indicating that trees are clustered (Table 8.7). Statistics C, I, and K had the smallest p values, suggesting that, for this realization, these statistics are the most sensitive.

Table 8.7 Results of Nearest-Neighbor Analyses With and Without Edge Correction

Test Statistic	Author	With Guard Area		Without Edge Correction	
		Value	p Value	Value	p Value
A	Clark and Evans	0.847	0.01949	0.867	0.03595
B	Skellam	87.073	0.20083	94.817	0.37234
C	Pielou	1.785	< 0.00001	2.474	< 0.00001
D	Eberhardt	1.424	< 0.01	1.684	< 0.005
E	Pollard	75.212	0.00942	98.486	0.00004
F	Holgate	0.557	0.08207	0.542	0.15294
G	Holgate	0.630	0.00413	0.639	0.00229
H	Byth and Ripley	0.614	0.00256	0.619	0.00179
I	Hopkins	2.049	0.00020	2.609	< 0.00001
J	Hines and Hines	1.449	< 0.005	1.636	< 0.005
K	Diggle	91.903	0.00020	126.506	< 0.00001
L	Besag and Gleaves	0.592	0.01211	0.594	0.01093
M	Diggle et al.	0.445	0.08974	0.442	0.07734
N	Besag and Gleaves	1.646	0.00671	2.116	0.00011
O	Cormack	85.612	0.15316	85.304	0.14751
P	Cox and Lewis	0.413	0.02895	0.400	0.01342
Q	Satyamurthi	52.497	0.00417	99.685	0.00413

The sensitivity of Hopkins' (I) test is consistent with findings of Holgate (1965b) and Diggle et al. (1976), which suggest that against clustered alternatives Hopkins' test (I) is more powerful than either of Holgate's tests (F and G) or Besag and Gleaves' T-square tests (L and N). The power of Hopkins' test is a consequence of the clustering simultaneously reducing event-to-event distances while increasing point-to-event distances (Holgate, 1965b). Diggle's statistic (K) has been shown to be powerful against clustering alternatives by Hines and Hines (1979), who also show that Hines and Hines' (J), Cox and Lewis' (P), and Besag and Gleaves' (L) statistics are powerful against a variety of alternative point processes.

Skellam's (B), Holgate's (G), Diggle et al.'s (M), and Cormack's (O) statistics were not significant for the longleaf-pine data (Table 8.7). Diggle et al. (1976) found that both of Holgate's statistics (F, G) were not as powerful as Hopkins' (I) nor as powerful as both of Besag and Gleaves' (L, N) statistics. Likewise, Hines and Hines (1979) found Diggle et al.'s (M) statistic to be weak against a variety of clustered and regular point patterns.

A comparison of values of statistics calculated with and without edge correction suggests that some nearest-neighbor statistics are more sensitive to edge effects than others. Specifically, Table 8.7 suggests that C, D, E, I, J, K, and N are sensitive to edge effects, whereas statistics A, F, G, H, L, M, O, and P are not. Most of the latter (F, H, L, M, and O) are the sum of some function of W_i, X_i, $X_{2,i}$, Y_i, and Z_i. In general, it appears that functions of

Table 8.8 Comparison of Byth and Ripley's (1980) and Hopkins' (1954) Statistics under Random Sampling and Semisystematic Sampling

Test Statistic	Author	Random Sampling Value	Random Sampling p Value	Semisystematic Sampling Value	Semisystematic Sampling p Value
H	Byth and Ripley	0.614	0.00256	0.684	< 0.00001
I	Hopkins	2.049	0.00020	2.797	< 0.00001

sums of distances are more sensitive to edge effects than sums of functions of distances. Satyamurthi's (Q) statistic increased dramatically when edge effects were not corrected, but because there was a corresponding increase in its variance, there was very little change in its p value.

Semisystematic Sampling

Although Hopkins' test is sensitive to departures from complete spatial randomness, the requirement that a random sample of events (trees) be obtained is difficult to implement in the field. Byth and Ripley (1980) suggest a semisystematic sampling scheme that may increase the practicality of Hopkins' (I) test as well as their own (H). Their scheme was applied to the longleaf-pine data to compare results obtained for random samples of trees and points with those for semisystematic samples of trees and points. A grid of 100 points was layed out over the 4-ha central study region. Every other point was used as a sample point for measuring point-to-event distances (X). A 20×20-m quadrat was centered on each of the remaining 50 points. Trees in these quadrats were enumerated and 50 were chosen at random for measuring event-to-event distances. Hopkins' (I) and Byth and Ripley's (H) statistics were then calculated; see Table 8.8. Both statistics showed substantially higher values (and hence smaller p values) under semisystematic sampling than under random sampling. This resulted from both an increase in average point-to-event distances and a decrease in average event-to-event distances.

Intensive Sampling

The distribution theory for nearest-neighbor statistics (see Table 8.6) assumes that nearest-neighbor measurements are independent. If a mapped forest is sampled intensively, nearest-neighbor measurements are *not* independent. To determine how well distribution theory holds for an intensive sampling scheme, each of the nearest-neighbor statistics was calculated from all 584 trees in the central study region A and from 584 randomly located points in A. A 95% acceptance region was calculated for each statistic assuming the distribution theory (given by Table 8.6) holds for independent nearest-neighbor distances. In addition, 200 realizations of a homogeneous Poisson process were obtained. For each realization, events consist of 841 pairs of X, Y coordinates selected from a uniform distribution on a 240×240-m plot.

(The density of 841 events in a 240 × 240-m plot is the same as the density of 584 events in a 200 × 200-m plot. The slightly larger plot allows the introduction of a 20-m-wide guard area.) Distances were measured from 584 randomly sampled points located at least 20 m from the plot edge. Likewise, distances were measured only from events located at least 20 m from the plot edge (and hence not exactly 584 events were chosen for each realization). Thus, a 20-m-wide guard area was provided. Each of the nearest-neighbor statistics was calculated for each realization. The number of realizations falling outside the nominal 95% acceptance region for each statistic was then counted. If the distribution theory for a given statistic holds well, then 10 of the 200 realizations should fall outside the 95% region.

Results in Table 8.9 show that, for most of the nearest-neighbor statistics, more realizations of the completely spatially random point process fell outside the 95% acceptance region than expected. Thus, if distribution theory that assumes independent observations is used, one would tend to reject the null hypothesis of complete spatial randomness too often. This result suggests that acceptance regions obtained from the preceding distribution theory are too narrow, and, unless estimates fall well outside these intervals, Monte Carlo tests should be used for inference (see Section 8.4.1; Besag and Diggle, 1977). Finally, each nearest-neighbor statistic in Table 8.9,

Table 8.9 Results of Nearest-Neighbor Analyses using All 584 Pines and 584 Random Points[a]

Test Statistic	Author	Value	95% Acceptance Region	−	+	T
A	Clark and Evans	0.825	0.985, 1.042	5	10	15
B	Skellam	982.5	1074.7, 1264.1	6	6	12
C	Pielou	1.561	0.919, 1.081	8	7	15
D	Eberhardt	1.366	1.242, 1.306	13	7	20
E	Pollard	726.1	517.5, 651.3	19	5	24
F	Holgate	0.533	0.477, 0.523	6	4	10
G	Holgate	0.534	0.471, 0.528	8	6	14
H	Byth and Ripley	0.638	0.477, 0.523	6	11	17
I	Hopkins	1.855	0.891, 1.122	9	10	19
J	Hines and Hines	1.409	1.242, 1.306	2	6	8
K	Diggle	896.0	517.5, 651.3	15	16	31
L	Besag and Gleaves	0.559	0.477, 0.523	6	9	15
M	Diggle et al.	0.446	0.477, 0.523	11	9	20
N	Besag and Gleaves	1.316	0.891, 1.122	14	17	31
O	Cormack	1021.8	1074.8, 1264.1	8	15	23
P	Cox and Lewis	0.421	0.474, 0.526	13	10	23
Q	Satyamurthi	27.39	−12.23, 12.23	6	2	8

[a]The number of realizations (out of 200) of completely spatially random point processes having values of nearest-neighbor statistics below the theoretical lower 95% acceptance limit (−), above the theoretical upper 95% acceptance limit (+), and their totals (T) are given. Expected total number is 10 if the distribution theory given by Table 8.6 holds.

calculated for the longleaf-pine data, falls well outside its simulation envelope, in a direction indicating that trees are clustered.

Summary of Nearest-Neighbor Methods

The reduction of complex point patterns to a one-dimensional nearest-neighbor summary statistic results in a considerable loss of information. Information on individual nearest-neighbor distances is lost. Because distances are measured only to the closest events, only the smallest scales of pattern are considered, and information on larger scales of pattern is unavailable. Nearest-neighbor statistics indicate only the direction of departure from complete spatial randomness (csr). Little is known about the behavior of these statistics when csr does not hold.

Unlike quadrat methods, these statistics do not depend on some arbitrary choice of quadrat size. However, choosing to measure distances to nearest neighbors as opposed to second, third, seventh, or fiftieth nearest neighbors, for example, is also arbitrary. In conclusion, because much of the spatial information is lost, and because for non-csr models it is debatable what these statistics are measuring, nearest-neighbor statistics for *mapped* data cannot be generally recommended.

In all fairness, nearest-neighbor statistics were primarily intended for use in the *field*. Hopkins' (I) test appears to be the most powerful against a variety of alternatives (Holgate, 1965b; Diggle et al., 1976), but it is difficult to implement in the field because a random sample of events is required. Byth and Ripley's (H) semisystematic sampling technique reduces the effort required to obtain a random sample of events, but their approach still requires considerable effort and may remain impractical. The results of this section and those of Payendeh (1970) suggest that Pielou's (C) test is also powerful against a variety of alternatives, but this test requires an estimate of intensity. Besag and Gleaves' (1973) sampling methods can be readily applied in the field; T-square methods have been shown to be powerful against a variety of alternatives (Diggle et al., 1976; Hines and Hines, 1979). If clustering is suspected, informative results can be obtained using Diggle's (K) and Hines and Hines' (J) statistics (Hines and Hines, 1979).

Extensions of Nearest-Neighbor Methods

There are a number of ways in which distance methods can be extended to use more spatial information. Average distances to the first, second, third,... nearest neighbors may be calculated and compared to expected distances under complete spatial randomness (csr), so combining information on many spatial scales.

Consider the distance W_k from an arbitrary event to the kth nearest event. H. R. Thompson (1956) shows that, under csr,

$$E(W_k) = k(2k)!/\left\{(2^k k!)^2 \lambda^{1/2}\right\}, \qquad k = 1, 2, \cdots. \tag{8.2.12}$$

From a random sample of n events, calculate the kth nearest-neighbor distances $\{W_{ki}: i = 1, \ldots, n\}$, $k = 1, \ldots, K$. Then form the ratio of the sample mean to the expected kth nearest neighbor:

$$R_k \equiv \left\{ \sum_{i=1}^{n} W_{ki}/n \right\} \left\{ (2^k k!)^2 \lambda^{1/2} \right\} / k(2k)!, \qquad k = 1, \ldots, K. \quad (8.2.13)$$

Notice that R_1 is identical to Clark and Evans' (1954) statistic given in Table 8.6. Finally, plot the ratios R_k versus k and look for departures from 1: $R_k > 1$ indicates regularity and $R_k < 1$ indicates clustering, for small k. (If events are randomly spaced within clusters, R_k should be less than 1 and approximately constant for small k and should increase as k approaches and exceeds cluster size.) Distributions are derived in Section 8.4.2.

Distances to the first 100 nearest neighbors were calculated from each of the 584 trees in the 4-ha study region (trees in the surrounding 12 ha were allowed as neighbors, but were not used as originating events for nearest-neighbor measurements). Figure 8.11 shows the plot of R_k versus k, up to $K = 100$, which is clearly more informative than the Clark–Evans statistic (R_1) alone. The shape of the plot suggests that clusters of longleaf pines are small, but that there are large numbers of these small-scale clusters.

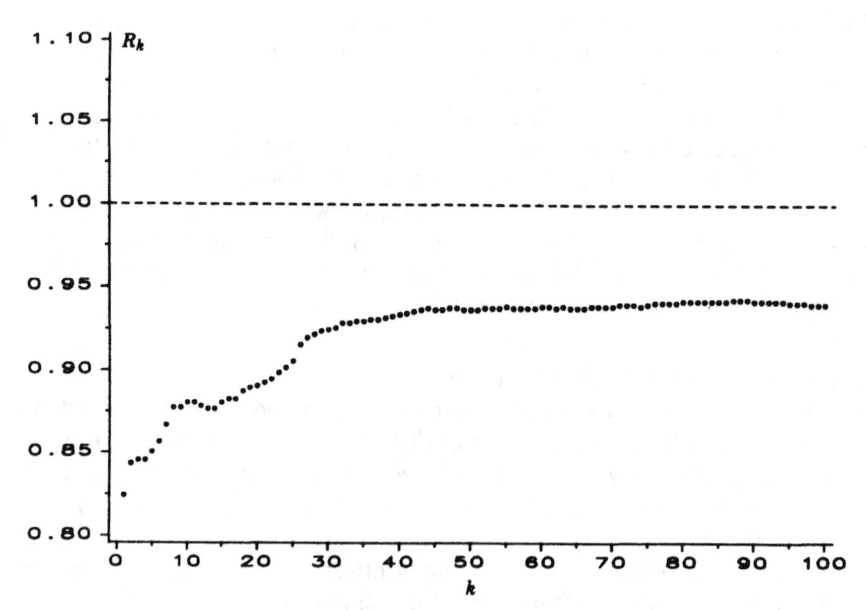

Figure 8.11 Plot of the kth nearest-neighbor ratio R_k versus k, $k = 1, \ldots, 100$, based on distances from all 584 trees in the 4-ha central study region. A 12-ha guard area was used to control for edge effects.

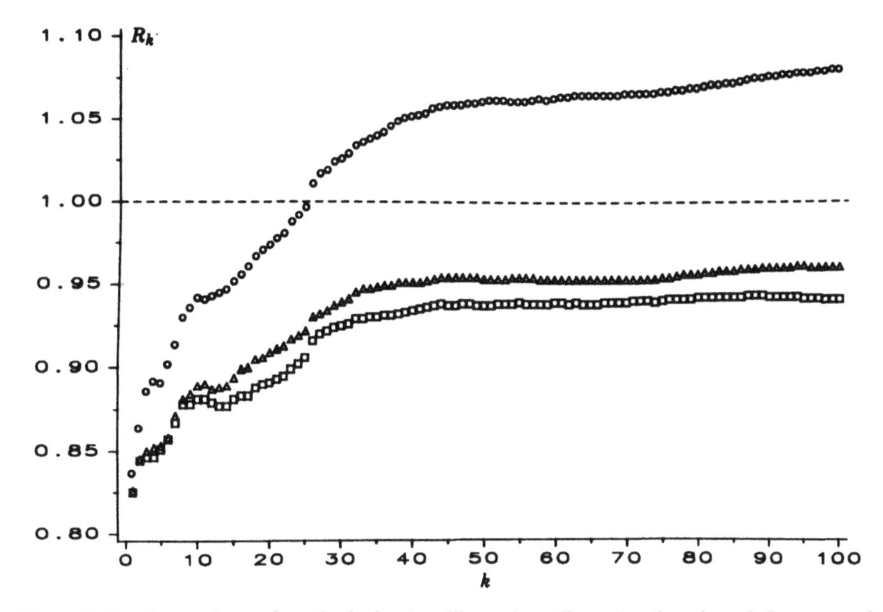

Figure 8.12 Comparison of methods for handling edge effects for the plot of R_k versus k, $k = 1, \ldots, 100$. The methods are: using no edge correction (circles), using a guard area (squares), and putting the 4-ha region on a torus (triangles).

The importance of an edge correction is demonstrated in Figure 8.12, where the plot from Figure 8.11 is compared with a toroidal edge correction, and with no edge correction at all.

8.2.6[†] Nearest-Neighbor Distribution Functions and the K Function

The distance methods presented in the preceding subsection are designed to indicate departures from complete spatial randomness (csr), but do not suggest a way of fitting alternative models. Two methods that do are now presented briefly. Their advantages in data summary are illustrated on the longleaf-pine data. Theoretical details, as well as a discussion of their role in model fitting, are given in Sections 8.3.5 and 8.4.

Nearest-Neighbor Distribution Functions

Define $G(r)$ to be the probability that the distance from a randomly chosen event to its nearest event is less than or equal to r. Likewise, define $F(r)$ to be the probability that the distance from a randomly chosen point to its nearest event is less than or equal to r.

Each of these can be estimated from the observed pattern by their respective empirical distribution functions. For example, from the total number $N(A)$ of events in A, suppose $i = 1, \ldots, n$ were chosen at random (without replacement). Let $r_{i,A}$, $r_{i,E}$ and d_i denote the distance from the ith event to, respectively, the nearest event in A, the nearest event (using a guard area or toroidal edge correction), and the nearest boundary of A. This leads to estimators

$$\hat{G}_1(r) \equiv \sum_{i=1}^{n} I(r_{i,A} \leq r)/n, \qquad r > 0, \tag{8.2.14}$$

$$\hat{G}_2(r) \equiv \sum_{i=1}^{n} I(r_{i,E} \leq r)/n, \qquad r > 0, \tag{8.2.15}$$

$$\hat{G}_3(r) \equiv \sum_{i=1}^{n} I(r_{i,A} \leq r, d_i > r) \Big/ \sum_{i=1}^{n} I(d_i > r), \qquad r > 0, \tag{8.2.16}$$

where $I(A_1)$ is the indicator function of the event A_1: $I(A_1) = 1$ if A_1 is true, $= 0$ otherwise; and $0/0 \equiv 0$. The estimator $\hat{G}_3(\cdot)$ is due to Ripley (1976a).

Theoretical calculations (e.g., Matern, 1971) show that for an inhomogeneous Poisson process in $\mathbb{R}^d$ with intensity function $\lambda(\cdot)$, the distribution function of the distance from an event at $\mathbf{s}_0$ to the nearest event is $1 - \exp\{-\int_{b(\mathbf{s}_0, r)} \lambda(\mathbf{u})\, d\mathbf{u}\}$, where $b(\mathbf{s}_0, r)$ is the closed ball of radius r centered at $\mathbf{s}_0$. Therefore, under csr in $\mathbb{R}^2$,

$$G(r) = 1 - \exp(-\lambda \pi r^2), \qquad r \geq 0, \tag{8.2.17}$$

and the same expression is obtained for $F(r)$ under the same conditions.

For the longleaf-pine data, an exhaustive sample of all 584 pines in the 4-ha study region A was taken. Estimators $\hat{G}_1$, $\hat{G}_2$, and $\hat{G}_3$ were computed but only $\hat{G}_3$ will be presented. Simulation envelopes were also computed for $\hat{G}_3$ based on 100 realizations of $\mathbf{s}_1, \ldots, \mathbf{s}_{584}$ from a uniform distribution on A. The estimator $\hat{G}_3$ was calculated from each realization and for each r the largest and smallest values defined the simulation envelope. The results are given in Figure 8.13, which is a Q-Q (or quantile-quantile) plot of $\hat{G}_3$ versus G, given by (8.2.17), based on the data and on the simulations ($\hat{\lambda} = 0.0146$ m^{-2} was used). Clearly, the data are exhibiting clustering; that is, trees tend to be closer to their neighbors than would be expected under csr. This was confirmed with analogous plots based on $F(\cdot)$, and is formally tested (by a Monte Carlo test) in Section 8.4.

Interestingly, edge-effect corrections were not really needed here (i.e., $\hat{G}_1$ was not very different from $\hat{G}_2$ and $\hat{G}_3$). This is almost certainly a result of there being a large number of trees in A and of using *nearest* neighbors. This is a different manifestation of the same potentially serious drawback of neighbor methods, namely, they do not use information in the pattern over a wide range of scales. The K function seems to address these concerns.

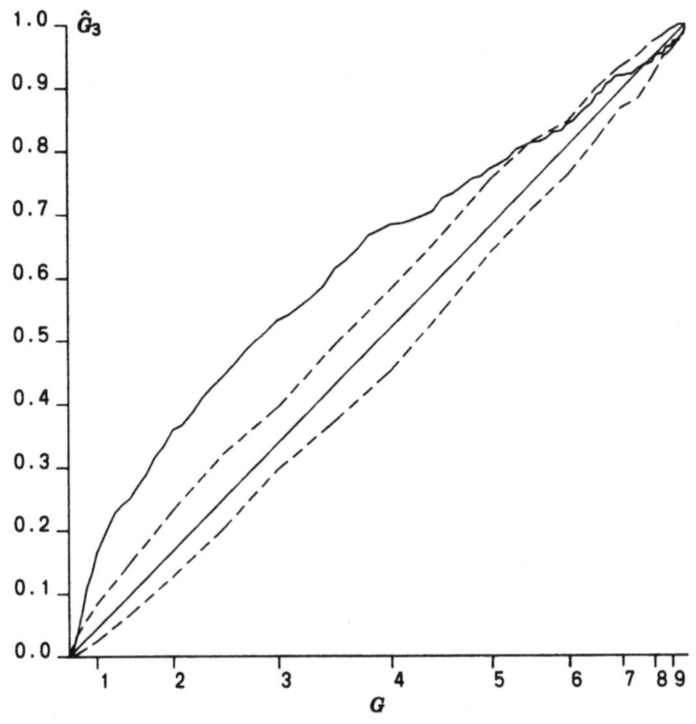

Figure 8.13 A quantile–quantile (*Q-Q*) plot of empirical distribution function $\hat{G}_3(r)$ versus $G(r)$ (solid line), together with upper and lower envelopes from 100 simulations of csr (dashed lines either side of the 45° csr line). The estimate $\hat{G}_3$ is given by (8.2.16). The value $\lambda = 0.0146$ m^{-2} was used to obtain the figure. Units shown on the horizontal axis are units of r (in meters).

The K Function

In Parts I and II of this book, much of the inference was concerned with careful modeling of the mean function $\mu(\cdot)$ and the variogram $2\gamma(\cdot)$ [or covariogram $C(\cdot)$]. Here, in Part III, the intensity $\lambda(\cdot)$ plays the role of $\mu(\cdot)$. What plays the role of $C(\cdot)$? Section 8.4 argues that the K function (Bartlett, 1964; Ripley, 1976a, 1977), for mapped data, captures the spatial dependence between different regions of the point process. Its definition is,

$$K(h) \equiv \lambda^{-1}E(\text{number of extra events within distance } h \text{ of an arbitrary event}), \qquad h \geq 0.$$

It is sometimes called the reduced second moment measure, because it is closely related to the second-order intensity of a stationary isotropic point process (see Sections 8.3.5 and 8.4.3, and Ripley, 1976a).

Its estimation is based on an empirical average replacing the expectation operator. Additionally, because the estimator counts numbers of events within a range of distances, it is not very efficient to work with sampled events. From the complete map of events, let $(\mathbf{s}_1, \ldots, \mathbf{s}_N)$ denote the $N \equiv N(A)$ locations of all events in a bounded study region A and suppose $(d_1, \ldots, d_N)$ are the associated distances from events to the nearest boundary of A. Define

$$\hat{K}_1(h) \equiv \hat{\lambda}^{-1} \sum_{i=1}^{N} \sum_{\substack{j=1 \\ i \neq j}}^{N} I(\|\mathbf{s}_i - \mathbf{s}_j\| \leq h)/N, \qquad h > 0. \tag{8.2.18}$$

There is a version of this that corrects for edge effects by computing the distances $\|\mathbf{s}_i - \mathbf{s}_j\|$ toroidally. Further, define

$$\hat{K}_2(h) \equiv \hat{\lambda}^{-1} \sum_{i=1}^{N^+} \sum_{\substack{j=1 \\ i \neq j}}^{N} I(\|\mathbf{s}_i - \mathbf{s}_j\| \leq h)/N, \qquad h > 0, \tag{8.2.19}$$

where N^+ is the number of events in A *and* a surrounding guard area. This method of edge correction is only valid for h smaller than the width of the guard area. A variable-width edge correction yields the estimator

$$\hat{K}_3(h) \equiv \hat{\lambda}^{-1} \sum_{i=1}^{N} \sum_{\substack{j=1 \\ i \neq j}}^{N} I(\|\mathbf{s}_i - \mathbf{s}_j\| \leq h, d_j > h) \Big/ \sum_{j=1}^{N} I(d_j > h), \qquad h > 0, \tag{8.2.20}$$

where $0/0 \equiv 0$.

Ripley (1976a) has proposed a closely related edge-corrected estimator in $\mathbb{R}^2$ that even uses information on events for which $d_j \leq h$:

$$\hat{K}_4(h) \equiv \hat{\lambda}^{-1} \sum_{i=1}^{N} \sum_{\substack{j=1 \\ i \neq j}}^{N} w(\mathbf{s}_i, \mathbf{s}_j)^{-1} I(\|\mathbf{s}_i - \mathbf{s}_j\| \leq h)/N, \qquad h > 0, \tag{8.2.21}$$

where the weight $w(\mathbf{s}_i, \mathbf{s}_j)$ is the proportion of the circumference of a circle centered at $\mathbf{s}_i$, passing through $\mathbf{s}_j$, and that is inside the study region A.

For all of the estimators of $K(\cdot)$, an estimate of the intensity λ is needed; use $\hat{\lambda} = N/|A|$. Estimators $\hat{K}_2$ and $\hat{K}_3$ (assuming stationarity) and $\hat{K}_4$ (assuming stationarity and isotropy) are unbaised when λ is known. It seems that use of $\hat{\lambda}$ does not upset this unbiasedness too much (Ripley, 1981, p. 159).

Under the assumption of csr in $\mathbb{R}^2$, $K(h) = \pi h^2$, $h \geq 0$. Under *regularity*, $K(h)$ tends to be *less* than πh^2, whereas under *clustering* $K(h)$ tends to be *greater* than πh^2; more details are given in Section 8.4.3.

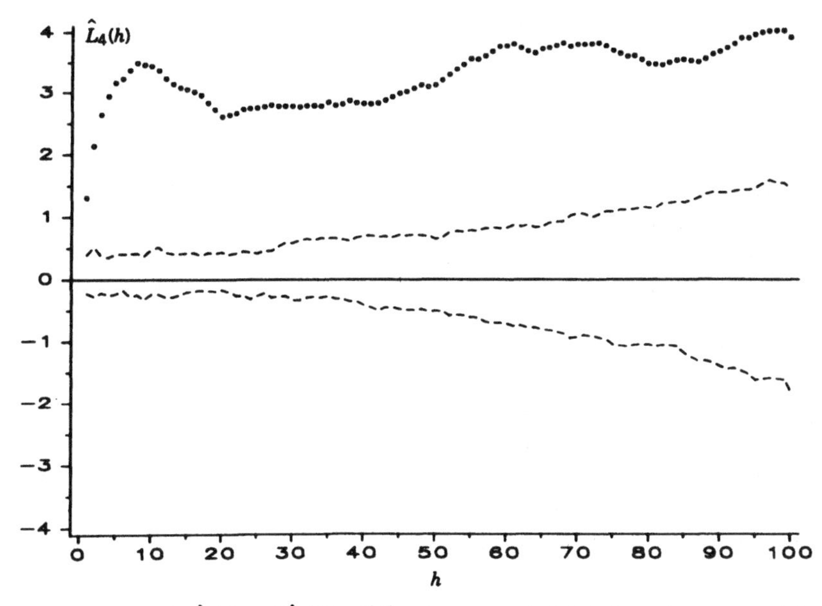

Figure 8.14 Plot of $\hat{L}_4(h) \equiv \{\hat{K}_4(h)/\pi\}^{1/2} - h$, versus h (circles), and upper and lower envelopes from 100 simulations of csr (dashed lines). The estimate $\hat{K}_4$ is given by (8.2.21). Units on the horizontal axis and the vertical axis are in meters.

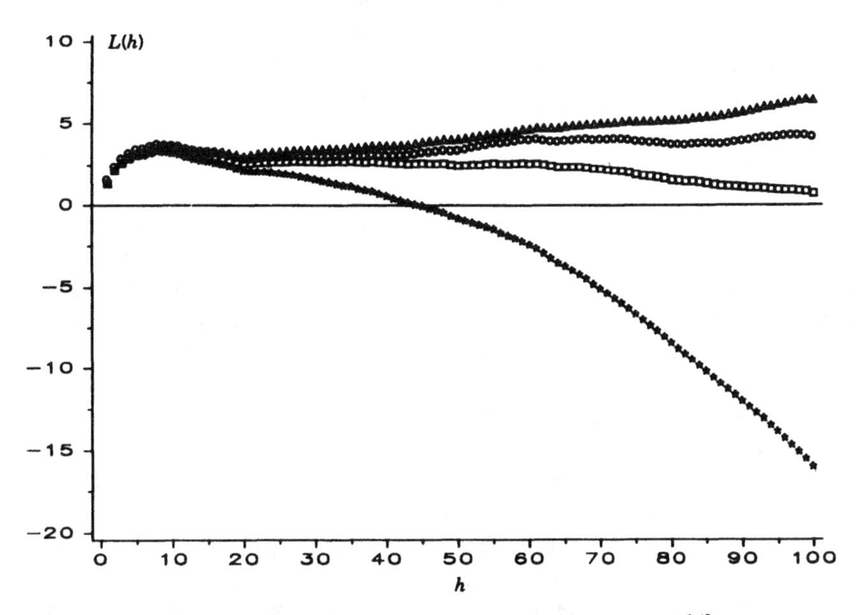

Figure 8.15 Comparison of edge-corrected estimates of $L(h) \equiv \{K(h)/\pi\}^{1/2} - h$. Shown are estimates derived from $\hat{K}_1$ without edge correction (stars), $\hat{K}_1$ based on toroidal distances (squares), $\hat{K}_2$ based on a guard area (triangles), and the same $\hat{K}_4$ given in Figure 8.14 (circles). Units on the horizontal axis and the vertical axis are in meters.

The estimated K function $\hat{K}_4$ given by (8.2.21) was calculated from the 584 locations of pine trees in the 4-ha central study region. Simulation envelopes were also constructed from the same 100 realizations (of csr) referred to earlier in this subsection. Figure 8.14 shows a plot of $\hat{L}_4(h) \equiv \{\hat{K}_4(h)/\pi\}^{1/2} - h$ versus h (following a suggestion by Besag, 1977c), together with the simulation envelopes. The data are clearly demonstrating clustering.

The importance of edge-effect corrections for estimating the K function is shown in Figure 8.15. Because the plots show characteristics of the spatial pattern at a multitude of scales, it is not surprising that, at large distances h, the estimate without edge correction gives a very misleading result. It indicates regularity at large scales, when in reality clustering is maintained.

As a method of data summary, the K function has obvious advantages. Ripley's (1976a, 1977) adaption of Bartlett's (1964) original suggestion ensures that it does not depend on the shape of the study region A. Moreover, it presents spatial information at all scales of pattern, and precise spatial locations of events are used in its estimation. Section 8.4.3 will discuss the K function's role as a modeling tool. It is defined here only for stationary point processes; versions for more general processes are clearly needed.

8.2.7† Some Final Remarks

The longleaf-pine data have now been analyzed in many different ways, and most show a statistically significant departure from complete spatial randomness toward clustering. It will be seen in Section 8.5.3 that a Poisson cluster process (of offspring from parents) also does not provide a very good fit to the data. I suggest that these models fail to fit the data because they ignore the dynamics of the biological processes that generated the spatial point pattern.

If the longleaf-pine population is driven by a cluster process, then such dynamics indicate that, for low mortality, the parent population (offspring of the grandparents) should also be a cluster process. But this contradicts the homogeneous-parent-process assumption of the Poisson cluster process.

In fact, the longleaf-pine data from the Wade Tract are more extensive than Table 8.1 (which gives the locations and dbhs of 584 trees censused in 1979). Annual mortality censuses were also taken in 1983 and 1987, and recruits (trees not in the 1979 census but bigger than 2 cm dbh in 1983 or 1987) were added to the mapped data set. Thus, locations and diameters of all mapped trees are available for 1979, 1983, and 1987.

A proper analysis of these data would involve dynamic modeling of the birth process, the death process, and the growth process, with a view toward integrating these components back into a global model (e.g., Rathbun and Cressie, 1990b). It is more than will be presented in this book on Statistics for spatial data, but some discussion of space–time modeling of spatial point patterns is given in Section 8.8.

8.3* POINT PROCESS THEORY

The development of point-process theory given in this section owes much to the work of Matthes et al. (1978), Kallenberg (1986), Karr (1986), Stoyan et al. (1987), and Daley and Vere-Jones (1988). My intention is to provide a brief overview of the subject, emphasizing those aspects necessary for an understanding of the models developed in Sections 8.5 through 8.8. A more detailed description of the theory of point processes can be found in the texts cited above.

Informally speaking, a point process is a stochastic model governing the location of events $\{\mathbf{s}_i\}$ in some set X. Because my interest is in *spatial* point processes, I shall take X to be a subset of $\mathbb{R}^d$, but more generally X could be any locally compact Hausdorff space whose topology has a countable base (Karr, 1986, p. 4; Kallenberg, 1986, p. 168). If an event located at $\mathbf{s} \in \mathbb{R}^d$ is marked by a quantity $z \in \mathscr{F}$, then $(\mathbf{s}, z)$ is a point in $\mathbb{R}^d \times \mathscr{F}$. Therefore, a *marked* point process could be represented mathematically as a point process on the product space $X = \mathbb{R}^d \times \mathscr{F}$. Similarly, if marked point processes are observed at time points $\{t_1, t_2, \dots\} \equiv T$, then they could be thought of as a point process on the product space $X = \mathbb{R}^d \times \mathscr{F} \times T$. In the following, let $\mathscr{X}$ be the Borel σ algebra of $X \subset \mathbb{R}^d$ and let ν be Lebesgue measure on X [recall $\nu(B) = |B|$, the volume of B].

The most natural way of defining a spatial point pattern (a realization of a spatial point process) is through the spatial locations of events $\mathbf{s}_1, \mathbf{s}_2, \dots$ in X. However, it is often mathematically more convenient to define a point pattern through a counting measure ϕ on X. For each Borel set B, $\phi(B)$ is the number of events in B; so $\phi(B) \in \{0, 1, 2, \dots\}$, for all $B \in \mathscr{X}$. Assume the counting measure ϕ is *locally finite*; that is, $\phi(B) < \infty$, for all bounded sets $B \in \mathscr{X}$ (Kallenberg, 1986, p. 12). Knowing $\phi(B)$ for all $B \in \mathscr{X}$ is equivalent to knowing the spatial locations of all events $\mathbf{s}_1, \mathbf{s}_2, \dots$ in X: Clearly, $\phi(B) = \sum_{i=1}^{\infty} I(\mathbf{s}_i \in B)$, for all $B \in \mathscr{X}$. Conversely, $\mathbf{s}$ is an (possibly multiple) event of the point pattern if $\phi(\{\mathbf{s}\}) > 0$. These equivalent definitions of a spatial point pattern lead to two equivalent characterizations for a spatial point process.

Characterization Based on Random Measures

Consider a characterization of a spatial point process based on random measures. Let $(\Omega, \mathscr{A}, P)$ be a probability space and let Φ be a collection of locally finite counting measures on $X \subset \mathbb{R}^d$. On Φ define $\mathscr{N}$, the smallest σ algebra generated by sets of the form $\{\phi \in \Phi\colon \phi(B) = n\}$, for all $B \in \mathscr{X}$ and all $n \in \{0, 1, 2, \dots\}$. Then, a *spatial point process* N on X is a measurable mapping of $(\Omega, \mathscr{A})$ into $(\Phi, \mathscr{N})$. This random counting measure N on X is analogous to a random variable Z, say, on $\mathbb{R}$, and probabilities are computed from $P(N \in Y) \equiv P(\omega\colon N(\omega) \in Y)$, $Y \in \mathscr{N}$.

A spatial point process defined over $(\Omega, \mathscr{A}, P)$ induces a probability measure $\Pi_N(Y) \equiv P(N \in Y)$ on $(\Phi, \mathscr{N})$, for all $Y \in \mathscr{N}$.

Definition. Two point processes N_1 and N_2 are said to be identically distributed if $\Pi_{N_1}(Y) = \Pi_{N_2}(Y)$, for all $Y \in \mathcal{N}$. ■

Definition. A spatial point process N is said to be *simple* if $\phi(\{s\}) \in \{0, 1\}$, for all $\mathbf{s} \in X$ and almost all $\phi \in \Phi$. ■

In all that is to follow in Part III, only simple spatial point processes in $\mathbb{R}^d$ will be considered. However, patterns with multiple events can still be modeled, through the marked point processes presented in Section 8.7.

Poisson Process

Let μ be any Radon measure on X; that is, $\mu(K) < \infty$ for any compact set $K \in \mathscr{X}$. Then N is a (inhomogeneous) Poisson process with mean measure μ if:

1. For any $B \in \mathscr{X}$, $P(N(B) \in \{0, 1, \dots\}) = 1$, and for any collection of disjoint sets $B_1, B_2, \dots, B_k \in \mathscr{X}$ the random variables $N(B_1)$, $N(B_2), \dots, N(B_k)$ are independent.
2. For all $\mathbf{s} \in X$,

$$\begin{aligned} P\{N(d\mathbf{s}) = 0\} &= 1 - \mu(d\mathbf{s}) + o(\mu(d\mathbf{s})), \\ P\{N(d\mathbf{s}) = 1\} &= \mu(d\mathbf{s}) + o(\mu(d\mathbf{s})), \\ P\{N(d\mathbf{s}) > 1\} &= o(\mu(d\mathbf{s})), \end{aligned}$$

   where $d\mathbf{s}$ is an infinitesimal region located at $\mathbf{s}$.

From these postulates, it can be shown that $N(B)$ has a Poisson distribution with mean $\mu(B)$, for all $B \in \mathscr{X}$ (e.g., Rogers, 1974, pp. 13–14):

$$P\{N(B) = n\} = \frac{(\mu(B))^n e^{-\mu(B)}}{n!}, \qquad n = 0, 1, \cdots. \tag{8.3.1}$$

The *homogeneous* Poisson process is a special case of a (inhomogeneous) Poisson process where $\mu(B) = \lambda \nu(B) = \lambda |B|$, for some $\lambda > 0$ and all $B \in \mathscr{X}$.

In one dimension, alternative characterizations of the Poisson process are given by Renyi (1967) and Huang and Puri (1987). Renyi shows that a Poisson process is characterized by the simple condition

$$P\{N(B) = n\} = \frac{(\mu(B))^n e^{-\mu(B)}}{n!}, \qquad n = 0, 1, 2, \dots,$$

where μ is any Radon measure on $\mathbb{R}$ and $B \subset \mathbb{R}$ is any set consisting of the union of a finite number of half-open intervals. However, it does not suffice for B to be any interval of the form $(a, b]$; for a counterexample, generalized to $\mathbb{R}^2$, see Moran (1976).

If the mean measure $\mu(\cdot)$ has no atoms, then the inhomogeneous Poisson process is simple (Daley and Vere-Jones, 1988, pp. 32, 33, 210–212). Henceforth, only inhomogeneous Poisson processes with nonatomic mean measures will be considered. Although this excludes the compound Poisson process, it is considered in Section 8.7.3 as a marked point process.

Characterization Based on Locations of Events

An alternative characterization for a point process is through the locations of the events in X. Usually X is assumed to be a bounded subset of $\mathbb{R}^d$. Let $X^n \equiv X \times X \times \cdots \times X$ (n times) be the set of collections of n (not necessarily distinct) elements of X; take $X^0 \equiv \emptyset$. If $n \neq k$, then X^n and X^k are disjoint. Define the *exponential* of X by

$$X_e \equiv \bigcup_{n=0}^{\infty} X^n.$$

Every set $B \in X_e$ can be expressed uniquely as the union of disjoint sets; that is, $B = \bigcup_{n=0}^{\infty} B^{(n)}$, where $B^{(n)} \equiv B \cap X^n$. Let $\mathscr{X}^{(n)}$ be the smallest σ algebra of sets in X^n containing all product sets $B_1 \times B_2 \times \cdots \times B_n$, such that $B_i \in \mathscr{X}$, $i = 1, 2, \ldots, n$. Let $\mathscr{X}_e$ be the class of all sets $\bigcup_{n=0}^{\infty} B^{(n)}$ in X_e, such that $B^{(n)} \in \mathscr{X}^{(n)}$. Then $\mathscr{X}_e$ is the smallest σ algebra of sets in X_e generated by sets $B^{(n)} \in \mathscr{X}^{(n)}$, $n = 0, 1, \ldots$ (Moyal, 1962). The pair $(X_e, \mathscr{X}_e)$ is called an *exponential space* (Carter and Prenter, 1972). The probability measure on $(X_e, \mathscr{X}_e)$ can now be defined. Let Π_n be a measure on $(X^n, \mathscr{X}^{(n)})$, $n = 0, 1, \ldots,$ such that $\sum_{n=0}^{\infty} \Pi_n(X^n) = 1$. Then the function Π on $\mathscr{X}_e$ defined by $\Pi(B) \equiv \sum_{n=0}^{\infty} \Pi_n(B^{(n)})$ is the unique probability distribution on $\mathscr{X}_e$ whose restriction to $\mathscr{X}^{(n)}$ agrees with Π_n for all n (Moyal, 1962).

Homogeneous Poisson Process

Through this alternative characterization of a spatial point process by the locations of events, an equivalent definition of the homogeneous Poisson process with intensity λ is

1. The number of events in any bounded region A has a Poisson distribution with mean $\lambda\nu(A)$, where recall that $\nu(\cdot)$ is Lebesgue measure.
2. Given there are n events in A, those events are independent and form a random sample from a uniform distribution on A.

By postulate 2, the conditional density of the ordered n tuple $(\mathbf{s}_1, \mathbf{s}_2, \ldots, \mathbf{s}_n)$ given $N(A) = n$ is

$$f(\mathbf{s}_1, \mathbf{s}_2, \ldots, \mathbf{s}_n) = 1/(\nu(A))^n. \tag{8.3.2}$$

Thus, from postulates 1 and 2, the joint "density" of n and $(\mathbf{s}_1, \mathbf{s}_2, \ldots, \mathbf{s}_n)$ is given by

$$f((\mathbf{s}_1, \ldots, \mathbf{s}_n), n) = \frac{\lambda^n e^{-\lambda\nu(A)}}{n!}. \tag{8.3.3}$$

Notice,

$$\sum_{n=0}^{\infty} \frac{\lambda^n e^{-\lambda\nu(A)}}{n!} \int_{A^n} d\mathbf{s}_1\, d\mathbf{s}_2 \cdots d\mathbf{s}_n = \sum_{n=0}^{\infty} \frac{(\lambda\nu(A))^n e^{-\lambda\nu(A)}}{n!} = 1,$$

as it should be.

Exponential spaces need not be formulated in terms of collections X^n of n points from X (Carter and Prenter, 1972). Instead, the members of X_e may be regarded as collections of finite counting measures on X. That is, let $X^n = \{\phi\colon \phi(X) = n\}$. Theorem 3.1 of Moyal (1962) shows that there is a one-to-one correspondence between these two characterizations of a point process, allowing one to choose between the two based on computational convenience.

8.3.1* Moment Measures

The moment measures of a spatial point process are analogous to the moments of a random variable or random vector. The increasing complexity from random variables through random vectors to point processes is paralleled in the increasing complexity of the definition of their moments. Moments of random variables are scalar quantities, moments of random vectors are vectors, matrices, or higher-order arrays, and moments of point processes are measures defined on the space $(X, \mathscr{X})$ and on cross-product spaces.

For a point process N and Borel set B, the number of points $N(B)$ in B is a random variable with first moment

$$\mu_N(B) \equiv E(N(B)) = \int_{\Phi} \phi(B) \Pi_N(d\phi), \tag{8.3.4}$$

a measure on $(X, \mathscr{X})$. Recall that N induces a probability measure Π_N on $(\Phi, \mathscr{N})$. The measure μ_N is called the *mean measure* or *first moment measure* of N. Higher-order moments can also be defined. The *kth moment measure* of N is given by

$$\begin{aligned} \mu_N^{(k)}(B_1 \times \cdots \times B_k) &\equiv E(N(B_1)N(B_2) \cdots N(B_k)) \\ &= \int_{\Phi} \phi(B_1)\phi(B_2) \cdots \phi(B_k)\Pi_N(d\phi), \end{aligned} \tag{8.3.5}$$

where $B_1, \ldots, B_k \in \mathscr{X}$. Note that $\mu_N^{(k)}$ is a measure on $(X^k, \mathscr{X}^{(k)})$ and recall that $\mathscr{X}^{(k)}$ is the smallest σ algebra formed by the product sets $B_1 \times \cdots \times B_k$, where $B_i \in \mathscr{X}$, $i = 1, 2, \ldots, k$. Analogous to the covariance of random variables is the *covariance measure* of a point process:

$$C_N(B_1 \times B_2) \equiv \mu_N^{(2)}(B_1 \times B_2) - \mu_N(B_1)\mu_N(B_2), \qquad (8.3.6)$$

where $B_1, B_2 \in \mathscr{X}$. Notice that C_N is a signed measure on $(X^2, \mathscr{X}^{(2)})$.

Just as factorial moments can be defined for random variables, factorial moment measures can be defined analogously for point processes. Ordinary moment measures of point processes are often unsatisfactory because the B_is in $B_1 \times B_2 \times \cdots \times B_k$ are not necessarily disjoint. This means that events of the process can occur in two or more of the B_is, causing a possible redundancy of information and singularity of the moment measures (Karr, 1986, pp. 10–11). Factorial moment measures of simple point processes do not suffer from these problems. Let $\mathbf{s}_1, \mathbf{s}_2, \ldots \in X$ be the events corresponding to the realization ϕ of a point process N. Then the *kth factorial moment measure* of N is defined by

$$\alpha_N^{(k)}(B_1 \times B_2 \times \cdots \times B_k) \equiv \int_\Phi \sum_{\substack{\text{distinct} \\ \mathbf{s}_1, \mathbf{s}_2, \ldots, \mathbf{s}_k \in \phi}} I(\mathbf{s}_1 \in B_1) I(\mathbf{s}_2 \in B_2) \cdots I(\mathbf{s}_k \in B_k) \Pi_N(d\phi). \qquad (8.3.7)$$

Note, $\alpha_N^{(k)}(B^k) = E\{N(B)(N(B) - 1) \cdots (N(B) - k + 1)\}$ is the kth factorial moment of the random variable $N(B)$, so the term "factorial moment measure" is appropriate. For $k = 2$,

$$\alpha_N^{(2)}(B_1 \times B_2) = \mu_N^{(2)}(B_1 \times B_2) - \mu_N(B_1 \cap B_2). \qquad (8.3.8)$$

Consider the behavior of the moment measures as $B \downarrow \{\mathbf{s}\}$, where $\mathbf{s} \in X$. Let $d\mathbf{s}$ and $d\mathbf{u}$ be small regions located at $\mathbf{s}$ and $\mathbf{u}$, respectively. Then the *first-order intensity* is defined by

$$\lambda(\mathbf{s}) \equiv \lim_{\nu(d\mathbf{s}) \to 0} \mu_N(d\mathbf{s})/\nu(d\mathbf{s}), \qquad (8.3.9)$$

provided the limit exists. Similarly, the *second-order intensity* is defined by

$$\lambda_2(\mathbf{s}, \mathbf{u}) \equiv \lim_{\substack{\nu(d\mathbf{s}) \to 0 \\ \nu(d\mathbf{u}) \to 0}} \frac{\mu_N^{(2)}(d\mathbf{s} \times d\mathbf{u})}{\nu(d\mathbf{s})\nu(d\mathbf{u})}, \qquad (8.3.10)$$

provided the limit exists. The *pair-correlation function* is $g(\mathbf{s}, \mathbf{u}) \equiv \lambda_2(\mathbf{s}, \mathbf{u})/(\lambda(\mathbf{s})\lambda(\mathbf{u}))$ and the *covariance density* is $\beta(\mathbf{s}, \mathbf{u}) \equiv \lambda_2(\mathbf{s}, \mathbf{u}) - \lambda(\mathbf{s})\lambda(\mathbf{u})$.

Homogeneous Poisson Process

For a homogeneous Poisson process with intensity λ, we have

$$\mu_N(B) = \lambda\nu(B)$$

and

$$C_N(B_1 \times B_2) = \lambda\nu(B_1 \cap B_2).$$

Because the numbers of events in disjoint regions are independent, $\lambda_2(\mathbf{s}, \mathbf{u}) = \lambda^2$, $g(\mathbf{s}, \mathbf{u}) = 1$, and $\beta(\mathbf{s}, \mathbf{u}) = 0$, for all $\mathbf{s}, \mathbf{u} \in X$.

8.3.2* Generating Functionals

Moment measures provide an incomplete description of point processes. Two point processes can have identical moment measures, even when they are not identically distributed. Zessin (1983) gives conditions under which the moment measures uniquely determine a point process and a counterexample is given by Ruelle (1969, p. 106). This section is concerned with specifications that completely determine the probability measure of a simple point process. In the following, the zero-probability functional and generating functionals will be defined, and each will be shown to determine uniquely the distribution of a simple point process. Finally, it will be shown how the moment measures can be obtained from the generating functionals.

The *zero-probability functional* of a point process N is the mapping z_N: $\mathscr{X} \to [0, 1]$, defined by

$$z_N(B) \equiv P(N(B) = 0), \qquad B \in \mathscr{X}. \tag{8.3.11}$$

The role of Laplace functionals and probability generating functionals for point processes is analogous to the role of moment generating functions and probability generating functions, respectively, for random variables. The moment generating function of a random variable Z is $E(e^{bZ})$, a function of the scalar b. For a random vector $\mathbf{Z}$, bZ is replaced with the vector product $\mathbf{b}'\mathbf{Z}$. For a point process, the role of $\mathbf{b}$ is taken by the set of nonnegative $\mathscr{X}$-measurable functions ζ on X. The Laplace functional of a point process N is defined by

$$L_N(\zeta) \equiv E\left[\exp\left\{-\int_X \zeta(\mathbf{s}) N(d\mathbf{s})\right\}\right] = E\left[\exp\left\{-\sum_{\mathbf{s}_i \in N} \zeta(\mathbf{s}_i)\right\}\right], \tag{8.3.12}$$

where ζ is any nonnegative $\mathscr{X}$-measurable function for which the right-hand side of (8.3.12) exists. Similarly, the probability generating functional of N is defined by

$$G_N(\xi) \equiv E\left[\exp\left\{\int_X \log \xi(\mathbf{s}) N(d\mathbf{s})\right\}\right] = E\left[\prod_{\mathbf{s}_i \in N} \xi(\mathbf{s}_i)\right], \quad (8.3.13)$$

where ξ is $\mathscr{X}$ measurable and $0 \le \xi(\mathbf{s}) \le 1$ for all $\mathbf{s} \in X$. Notice that $G_N(\xi)$ is well defined if $1 - \xi$ vanishes outside some bounded set $B \subset X$. The relationship between the Laplace functional and the probability generating functional is given by $L_N(\zeta) = G_N(e^{-\zeta})$.

The following equivalence theorem shows that the zero-probability functional and the two generating functionals uniquely determine the distribution of a simple point process. A sketch of the proof is provided here. Further details can be found in Grandell (1977), Kallenberg (1986, p. 27), and Karr (1986, pp. 9–10).

Equivalence Theorem For simple point processes (defined in the introduction to this section) N_1 and N_2, the following are equivalent:

i. N_1 and N_2 are identically distributed.
ii. $(N_1(B_1), N_1(B_2), \ldots, N_1(B_k))$ is identically distributed to $(N_2(B_1), N_2(B_2), \ldots, N_2(B_k))$, for all positive integers k and all $B_1, B_2, \ldots, B_k \in \mathscr{B}$, the ring of all bounded sets in $\mathscr{X}$ (see, e.g., Karr, 1986, p. 4).
iii. $z_{N_1}(B) = z_{N_2}(B)$, for all $B \in \mathscr{X}$.
iv. $L_{N_1}(\zeta) = L_{N_2}(\zeta)$, for all nonnegative $\mathscr{X}$-measurable functions ζ.
v. $G_{N_1}(\xi) = G_{N_2}(\xi)$, for all $\mathscr{X}$-measurable functions ξ such that $0 \le \xi(\mathbf{s}) \le 1$.

Proof (Sketch). It is obvious that i implies ii through v, and that iv and v are equivalent from the definition of $L_N(\zeta)$ and $G_N(\xi)$.

Suppose, conversely, that ii is true. Define

$$\mathscr{D} \equiv \{Y \in \mathscr{N}: P(N_1 \in Y) = P(N_2 \in Y)\}.$$

From ii, $\mathscr{D}$ contains the class $\mathscr{C}$ of all sets of the form

$$\{\phi \in \Phi: \phi(B_1) \le n_1, \ldots, \phi(B_k) \le n_k\},$$

where $k, n_1, n_2, \ldots, n_k \in \{0, 1, 2, \ldots\}$, and $B_1, B_2, \ldots, B_k \in \mathscr{B}$. Because $\mathscr{C}$ is closed under finite intersections and because $\mathscr{D}$ is closed under proper set differences and monotone limits and contains Φ, it follows from a monotone-class theorem (Kallenberg, 1986, p. 163) that $\mathscr{D} \supset \sigma(\mathscr{C})$, the smallest σ algebra generated by $\mathscr{C}$. But $\sigma(\mathscr{C}) = \mathscr{N}$ (see Kallenberg, 1986, Lemma 1.4), so $P(N_1 \in Y) = P(N_2 \in Y)$, for all $Y \in \mathscr{N}$, and hence i is true.

Suppose that iii is true. Let $\mathscr{D}$ be defined as before. From iii, $\mathscr{D}$ contains the class of sets $\mathscr{E} = \{\phi \in \Phi: \phi(B) = 0\}$, for all $B \in \mathscr{X}$. Because $\mathscr{E}$ is closed under finite intersections, it follows from a monotone-class theorem (Kallenberg, 1986, p. 163) that $\mathscr{D} \supset \sigma(\mathscr{E})$. But, by Lemma 7 of Grandell (1977), $\sigma(\mathscr{E}) = \mathscr{N}$, so $P(N_1 \in Y) = P(N_2 \in Y)$, for all $Y \in \mathscr{N}$, and hence i is true.

Suppose again that ii is true. Let $\zeta = \sum_{i=1}^{k} b_i I_{B_i}$ be any simple nonnegative function, where $I_B(\mathbf{s})$ is an indicator function taking the value 1 when $\mathbf{s} \in B$, and 0 otherwise. Then

$$\begin{aligned} L_{N_1}(\zeta) &= E\left[\exp\left\{-\sum_{i=1}^{k} b_i N_1(B_i)\right\}\right] \\ &= E\left[\exp\left\{-\sum_{i=1}^{k} b_i N_2(B_i)\right\}\right] = L_{N_2}(\zeta). \end{aligned} \tag{8.3.14}$$

Because the result is true for simple functions, standard measure-theoretic arguments imply it is also true for general nonnegative functions ζ. So ii implies iv. Conversely, suppose iv is true. Then in particular (8.3.14) is true, and ii follows by the uniqueness theorem for Laplace transforms; so, iv implies ii. ■

Note that if the assumption of simple point processes in the Equivalence Theorem is dropped, then it is still true that i, ii, iv, and v are equivalent.

Moment Measures from Generating Functionals

Just as moment generating functions can be used to obtain the moments of a random variable, the Laplace functional can be used to obtain the moment measures of a point process. Provided it exists, the kth moment measure of a point process N can be obtained from the Laplace functional by differentiation:

$$\begin{aligned} &\mu_N^{(k)}(B_1 \times \cdots \times B_k) \\ &\quad = (-1)^k \lim_{\alpha_1, \ldots, \alpha_k \downarrow 0} \frac{\partial^k}{\partial \alpha_1 \cdots \partial \alpha_k} L_N(\alpha_1 I_{B_1} + \cdots + \alpha_k I_{B_k}). \end{aligned} \tag{8.3.15}$$

Moreover, if all moment measures of N are locally finite and the right-hand side of the following equation (8.3.16) converges, then

$$L_N(\zeta) = 1 + \sum_{k=1}^{\infty} \frac{(-1)^k}{k!} \int_{X^k} \zeta(\mathbf{s}_1) \cdots \zeta(\mathbf{s}_k) \mu_N^{(k)}(d\mathbf{s}_1 \times \cdots \times d\mathbf{s}_k), \tag{8.3.16}$$

where ζ is a nonnegative $\mathscr{X}$-measurable function (Stoyan et al., 1987, p. 197).

Likewise, the kth factorial moment of a point process N can be obtained from the probability generating functional by differentiation:

$$\alpha_N^{(k)}(B_1 \times \cdots \times B_k)$$
$$= (-1)^k \lim_{\alpha_1, \ldots, \alpha_k \downarrow 0} \frac{\partial^k}{\partial \alpha_1 \cdots \partial \alpha_k} G_N(1 - \alpha_1 I_{B_1} - \cdots - \alpha_k I_{B_k}). \quad (8.3.17)$$

Moreover,

$$G_N(1 - \xi) = 1 + \sum_{k=1}^{\infty} \frac{(-1)^k}{k!} \int_{X^k} \xi(\mathbf{s}_1) \cdots \xi(\mathbf{s}_k) \alpha_N^{(k)}(d\mathbf{s}_1 \times \cdots \times d\mathbf{s}_k), \quad (8.3.18)$$

provided $\alpha_N^{(k)}$ are Radon measures for all k and provided the right-hand side of (8.3.18) converges (Westcott, 1972; Stoyan et al., 1987, p. 109).

Homogeneous Poisson Process

Let N be a homogeneous Poisson process with intensity λ. From Eq. (8.3.1),

$$z_N(B) = e^{-\lambda \nu(B)}, \qquad B \in \mathscr{X}.$$

The probability generating functional of N is

$$G_N(\xi) = \exp\left\{-\lambda \int_X (1 - \xi(\mathbf{s}))\nu(d\mathbf{s})\right\}. \quad (8.3.19)$$

To prove (8.3.19), consider the function $\xi_1 \equiv 1 - (1 - b)I_{B_1}$, where $b \in [0, 1]$, I_{B_1} is an indicator function, and $B_1 \in \mathscr{X}$ is bounded. Then

$$G_N(\xi_1) = E(b^{N(B_1)}),$$

the probability generating function of a Poisson random variable with mean $\lambda \nu(B_1)$. So

$$G_N(\xi_1) = \exp\{-\lambda \nu(B_1)(1 - b)\}. \quad (8.3.20)$$

Now consider functions $\xi_k \equiv 1 - \Sigma_{i=1}^k (1 - b_i) I_{B_i}$, where $b_i \in [0, 1]$ and $B_1, B_2, \ldots, B_k \in \mathscr{X}$ are disjoint bounded sets. Then, by independence of the

$N(B_i)$ and by Eq. (8.3.20),

$$G_N(\xi_k) = E\left(\prod_{i=1}^{k} b_i^{N(B_i)}\right)$$
$$= \prod_{i=1}^{k} \exp\{-\lambda\nu(B_i)(1-b_i)\}$$
$$= \exp\left\{-\lambda\int_X (1-\xi_k(\mathbf{s}))\nu(d\mathbf{s})\right\}.$$

The result (8.3.19) then follows by approximation of ξ by simple functions using standard measure-theoretic arguments.

From the Equivalence Theorem and Eq. (8.3.19), it is easy to show that the superposition of two independent Poisson processes is a Poisson process. [The superposition $N_1 + N_2$ of two point processes N_1 and N_2 is defined through $(N_1 + N_2)(B) \equiv N_1(B) + N_2(B)$, for all $B \in \mathscr{X}$.] Specifically, let N_1 and N_2 be independent homogeneous Poisson processes with intensities λ_1 and λ_2, respectively. Then, by independence of N_1 and N_2 and from (8.3.19),

$$G_{N_1+N_2}(\xi) = G_{N_1}(\xi)G_{N_2}(\xi) = \exp\left\{-(\lambda_1+\lambda_2)\int_X (1-\xi(\mathbf{s}))\nu(d\mathbf{s})\right\},$$

and hence by the Equivalence Theorem, $N_1 + N_2$ is a homogeneous Poisson process with intensity $(\lambda_1 + \lambda_2)$.

8.3.3* Stationary and Isotropic Point Processes

Because spatial point pattern data are usually a single realization of a spatial point process, the additional assumptions of stationarity, and sometimes isotropy, are often made to reduce the parameter space and to allow parameter estimation. For each $\mathbf{s} \in \mathbb{R}^d$, let $\tau_{\mathbf{s}}: \mathbb{R}^d \to \mathbb{R}^d$ be the translation operator $\tau_{\mathbf{s}}(\mathbf{u}) \equiv \mathbf{u} - \mathbf{s}$. For each $\phi \in \Phi$, let $\phi_{\mathbf{s}}(A) \equiv \phi\tau_{\mathbf{s}}^{-1}(A) = \phi(\{\mathbf{u}: \mathbf{u} + \mathbf{s} \in A\})$ and let $N_{\mathbf{s}} \equiv N\tau_{\mathbf{s}}^{-1}$. A point process N is *stationary* if N and $N_{\mathbf{s}}$ are identically distributed for all $\mathbf{s} \in \mathbb{R}^d$; that is, if $\Pi_N(B) = \Pi_{N_{\mathbf{s}}}(B)$, for all $B \in \mathscr{X}$ and all $\mathbf{s} \in \mathbb{R}^d$. Thus, stationary point processes are invariant under translations $\tau_{\mathbf{s}}$. Similarly, a point process N is *isotropic* if it is invariant under rotations ρ_θ about the origin, that is, if N and $N\rho_\theta^{-1}$ are identically distributed for all $\theta \in [0, 2\pi)$. Stationary and isotropic point processes are said to be *motion invariant* (e.g., Stoyan et al., 1987, p. 98). More generally, stationarity can be defined under a group of $\mathscr{X}$-measurable transformations $\xi_{\mathbf{s}}$, $\mathbf{s} \in X$ (Karr, 1986, pp. 38–39).

If a point process N is stationary, then the mean measure μ_N is translation invariant. So, for all $B \in \mathscr{X}$, $\mu_N(B)$ is proportional to $\nu(B)$, the

Lebesgue measure of B. That is, there exists a constant $\lambda > 0$ such that $\mu_N(B) = \lambda\nu(B)$, for all $B \in \mathscr{X}$. This constant λ is called the *intensity* of N. Therefore, for a stationary process, the intensity λ is just the first-order intensity $\lambda(\cdot)$ defined by (8.3.9). Moreover, the second-order intensity defined by (8.3.10) is $\lambda_2(\mathbf{s}, \mathbf{u}) = \lambda_2^*(\mathbf{s} - \mathbf{u})$, a function of the difference between $\mathbf{s}$ and $\mathbf{u}$ only. If, in addition, the point process N is isotropic, then $\lambda_2(\mathbf{s}, \mathbf{u}) = \lambda_2^o(\|\mathbf{s} - \mathbf{u}\|)$, a function of the distance separating $\mathbf{s}$ and $\mathbf{u}$. Sometimes a weaker form of stationarity is defined based on λ_2: A point process N is said to be second-order stationary if $\lambda(\mathbf{s}) = \lambda$ and $\lambda_2(\mathbf{s}, \mathbf{u}) = \lambda_2^*(\mathbf{s} - \mathbf{u})$, for all $\mathbf{s}, \mathbf{u} \in X$. Obviously stationarity implies second-order stationarity, but not *vice versa*.

Local and Global Intensity

Two additional notions of intensity can be defined (Karr, 1986, p. 40):
(1) the local limit

$$\lambda_l \equiv \lim_{B_n \downarrow \{\mathbf{0}\}} \frac{P\{N(B_n) = 1\}}{\nu(B_n)}; \tag{8.3.21}$$

(2) the global limit

$$\lambda_g \equiv \lim_{B_n \uparrow X} \frac{N(B_n)}{\nu(B_n)}. \tag{8.3.22}$$

For sufficiently well behaved stationary point processes, $\lambda = \lambda_l = \lambda_g$. For example, when the point process N is simple as well as stationary, equivalence of λ and λ_l follows from the Korolyuk–Khinchin Theorem. [See Khinchin (1960), pp. 41–42, where this theorem is proved for $X = \mathbb{R}$. Extensions to $\mathbb{R}^d$ and more general X can be found in Leadbetter (1972), Daley and Vere-Jones (1972), Daley (1974), and Karr (1986), p. 42.]

Ergodicity of Point Processes

Equivalence of λ and λ_g requires an assumption of ergodicity.

Definition. A measurable set $Y \in \mathscr{N}$ is *invariant* under translations $\tau_{\mathbf{s}}$ if $\phi\tau_{\mathbf{s}}^{-1} \in Y$, for all $\mathbf{s} \in X$ and all $\phi \in Y$. ■

Definition. A point process N is said to be *ergodic* if $\Pi_N(Y)$ is 0 or 1 for every invariant measurable set Y. ■

Then, it can be shown that if N is a stationary ergodic point process,

$$\lambda_g \equiv \lim_{B_n \uparrow X} \frac{N(B_n)}{\nu(B_n)} = \lambda \tag{8.3.23}$$

(Karr, 1986, pp. 43–44). The number of events in a bounded region $A \subset X$ is a realization of the random variable $N(A)$. The result (8.3.23) shows that $\hat{\lambda} = N(A)/\nu(A)$ is a consistent (as $A \uparrow X$) estimator of λ provided N is stationary and ergodic. Note that homogeneous Poisson processes are both stationary and ergodic, so that $\hat{\lambda}$ is a consistent estimator of the intensity in this case.

8.3.4* Palm Distributions

The concept of the Palm distribution is one of the most important to the theory of spatial point processes. Originally applied to an investigation of long-distance telephone queues, Palm (1943) defined the function $\psi(t; t_0)$ to be the conditional probability that there are no calls (events) in the interval $(t_0, t_0 + t)$ given that there is a call at time t_0. The first rigorous definition and systematic investigation of Palm distributions on $\mathbb{R}$ is found in Khinchin (1960, pp. 37–40). Jagers (1973) and Papangelou (1974) extended Palm distributions to more general spaces X.

The modern definition of Palm distributions is based on Campbell measures (Kallenberg, 1986, pp. 83–120; Karr, 1986, pp. 33–38; Stoyan et al., 1987, pp. 110–115). Assume that the mean measure μ_N is σ finite and let $\mathscr{X} \times \mathscr{N}$ denote the smallest σ algebra generated by sets of the form $B \times Y (B \in \mathscr{X}, Y \in \mathscr{N})$. The *Campbell measure* of a point process N is a measure on $\mathscr{X} \times \mathscr{N}$ defined by

$$\mathscr{C}_N(B \times Y) \equiv \int_Y \phi(B)\Pi_N(d\phi), \qquad B \in \mathscr{X}, Y \in \mathscr{N}. \quad (8.3.24)$$

Note that $\mathscr{C}_N(B \times \Phi) = \mu_N(B)$, for all $B \in \mathscr{X}$, so $\mathscr{C}_N$ is absolutely continuous with respect to μ_N. By the Radon–Nikodym Theorem, there exist almost everywhere (a.e.) uniquely determined measures $P_{N,s}$ on $(\Phi, \mathscr{N})$ such that

$$\mathscr{C}_N(B \times Y) = \int_B P_{N,s}(Y)\mu_N(ds). \quad (8.3.25)$$

The measure $P_{N,s}$ is called the *Palm distribution* of N with respect to s.

Reduced Palm Distributions

Defined through the reduced Campbell measure, the reduced Palm distribution is of greater importance to the theory of point processes than the (unreduced) Palm distribution. The *reduced Campbell measure* of a point process N is a measure on $\mathscr{X} \times \mathscr{N}$ defined by

$$\mathscr{C}_N^!(B \times Y) \equiv \int_\Phi \int_B I[(\phi - \delta_s) \in Y]\phi(ds)\Pi_N(d\phi), \qquad B \in \mathscr{X}, Y \in \mathscr{N}. \quad (8.3.26)$$

For an event of ϕ at s, $\phi - \delta_s$ represents the point pattern ϕ minus the event at s. The indicator function $I[(\phi - \delta_s) \in Y]$ is equal to 1 if $\phi - \delta_s \in Y$, and is equal to 0 otherwise. For each realization $\phi \in \Phi$, the integral $\int_B I[(\phi - \delta_s) \in Y]\phi(ds)$ counts the number of events s of ϕ, contained in B, and such that $\phi - \delta_s \in Y$. Clearly, for $Y \in \mathcal{N}$, $0 \leq \mathscr{C}_N^!(B \times Y) \leq \mathscr{C}_N(B \times Y) \leq \mu_N(B)$, for all $B \in \mathscr{X}$, so $\mathscr{C}_N^!$ is absolutely continuous with respect to μ_N. By the Radon–Nikodym theorem, there exist (a.e.) uniquely determined measures $P_{N,s}^!$ on $(\Phi, \mathcal{N})$ such that

$$\mathscr{C}_N^!(B \times Y) = \int_B P_{N,s}^!(Y)\mu_N(ds). \tag{8.3.27}$$

The measure $P_{N,s}^!$ is called the *reduced Palm distribution* of N with respect to s. Heuristically, $P_{N,s}^!$ can be interpreted as the conditional distribution of N on the *reduced* set $X - \{s\}$, given that there is an event at $s \in X$. For example, if $Y = \{\phi \in \Phi: \phi(b(s, r)) = 0\}$, then $P_{N,s}^!$ is the conditional probability, given there is an event at s, that the closed ball of radius r centered at s contains no additional events. Notice that, for each $s \in X$, the reduced Palm distribution is a probability measure on $X - \{s\}$. The point process on $X - \{s\}$ with probability measure $P_{N,s}^!$ is called the *reduced Palm process* of N.

For a stationary point process N, the Palm distributions $P_{N,s}$ and $P_{N,s}^!$ do not depend on s; hence write $P_{N,s} \equiv P_{N,0}$ and $P_{N,s}^! \equiv P_{N,0}^!$. Then, by Eqs. (8.3.25) and (8.3.27),

$$\mathscr{C}_N(B \times Y) = \lambda\nu(B)P_{N,0}(Y) \tag{8.3.28}$$

and

$$\mathscr{C}_N^!(B \times Y) = \lambda\nu(B)P_{N,0}^!(Y). \tag{8.3.29}$$

Homogeneous Poisson Process

Let N be a homogeneous Poisson process with intensity λ. In this special case (Stoyan et al., 1987, p. 114), the Palm distribution $P_{N,s} = \Pi_N * \delta_{\delta_s}$, for all $s \in X$, which is a convolution of measures corresponding to the superposition of N and the single event s. (Here, δ_s represents a point pattern consisting of just one event at s, and δ_{δ_s} is the induced measure of a point process that puts all its mass on $\delta_s \in \Phi$.) The reduced Palm distribution is simply $P_{N,s}^! = \Pi_N$, for all $s \in X$ (Stoyan et al., 1987, p. 115). For example, let N be a Poisson process on $\mathbb{R}^2$ with intensity λ and let $Y = \{\phi \in \Phi: \phi(b(s, r)) = 0\}$. Then, $P_{N,s}^!(Y) = \Pi_N(Y) = \exp(-\lambda\pi r^2)$.

8.3.5* Reduced Second Moment Measure

Reduced second moment measures illustrate the usefulness of reduced Palm distributions and are important to the modeling of stationary point processes (see, e.g., Ripley, 1981; Diggle, 1983). In the following, assume that the point process N is stationary with intensity λ. Let $B_{-\mathbf{h}} = B \oplus \{-\mathbf{h}\} \equiv \{\mathbf{s} - \mathbf{h}: \mathbf{s} \in B\}$. From Ripley (1976a) and Stoyan (1983), there exists a measure $\mathscr{K}_N$ on $(X, \mathscr{X})$ such that

$$\alpha_N^{(2)}(A \times B) = \lambda^2 \int_X \nu(A \cap B_{-\mathbf{h}}) \mathscr{K}_N(d\mathbf{h}), \qquad A, B \in \mathscr{X}. \quad (8.3.30)$$

The measure $\mathscr{K}_N$ is called the *reduced second moment measure* of N. An equivalent definition of the reduced second moment measure for a stationary point process is given by

$$\lambda \mathscr{K}_N(B) \equiv \int_\Phi \phi(B) P_{N,\mathbf{0}}^!(d\phi) = \int_\Phi \phi(B - \{\mathbf{0}\}) P_{N,\mathbf{0}}(d\phi), \qquad B \in \mathscr{X}, \quad (8.3.31)$$

where $P_{N,\mathbf{0}}^!$ is the reduced Palm distribution of N (Stoyan et al., 1987, p. 117). If $B = b(\mathbf{0}, h)$, the closed ball of radius h centered at $\mathbf{0}$, define

$$K(h) \equiv \mathscr{K}_N(b(\mathbf{0}, h)), \qquad h \geq 0.$$

The function $K(\cdot)$ is often referred to as the K function (see Sections 8.2.6 and 8.4). Notice that $\phi(b(\mathbf{0}, h) - \{\mathbf{0}\})$ is the number of events of the realization ϕ in $b(\mathbf{0}, h) - \{\mathbf{0}\}$ and $P_{N,\mathbf{0}}^!$ is the conditional probability associated with N minus the event at $\mathbf{0}$, given that there is an event at $\mathbf{0}$. So, $\lambda K(h)$ is the expected number of extra events of N within distance h of an event of N. Reduced second moment measures can also be extended to marked point processes. Let N be a marked stationary point process of events located in $\mathbb{R}^d$ and marks in $\mathscr{F}$. If $B = b(\mathbf{0}, h) \times \mathscr{V}$, where $\mathscr{V}$ is a measurable set of $\mathscr{F}$, then $\lambda K_{\mathscr{V}}(h)$ is defined to be the expected number of extra events $\mathbf{s}$ for which $Z(\mathbf{s}) \in \mathscr{V}$ and $\mathbf{s}$ is within distance h of an arbitrary event of N (Section 8.7).

K Function and Second-Order Intensity

Consider the relationship between the K function and the second-order intensity λ_2^o when N is a simple, stationary, isotropic point process on $\mathbb{R}^d$. Let $a(\mathbf{0}, dh)$ be the annulus of width dh of a sphere of radius h centered at $\mathbf{0}$. Let $d\mathbf{0}$ be a small region located at $\mathbf{0}$. Note that the surface area of a d-dimensional sphere of radius h is $d\pi^{d/2}h^{d-1}/\Gamma(1 + \frac{d}{2})$. Then

$$\begin{aligned}
&\lim_{\substack{dh \to 0 \\ \nu(d\mathbf{0}) \to 0}} \frac{P\{N(a(\mathbf{0}, dh)) > 0 | N(d\mathbf{0}) > 0\}}{dh} \\
&\quad = \lim_{\substack{dh \to 0 \\ \nu(d\mathbf{0}) \to 0}} \frac{P\{N(a(\mathbf{0}, dh)) > 0, N(d\mathbf{0}) > 0\}}{P\{N(d\mathbf{0}) > 0\} dh} \\
&\quad = \frac{d\pi^{d/2}h^{d-1}\lambda_2^o(h)}{\lambda\Gamma(1 + \frac{d}{2})}.
\end{aligned}$$

Integrating with respect to h, the expected number of events within distance h of an arbitrary event is

$$\lambda K(h) = \frac{d\pi^{d/2}}{\lambda \Gamma(1 + \frac{d}{2})} \int_0^h u^{d-1} \lambda_2^o(u)\, du, \qquad h \geq 0. \tag{8.3.32}$$

Then

$$\lambda_2^o(h) = \frac{\lambda^2 \Gamma(1 + \frac{d}{2})}{d\pi^{d/2} h^{d-1}} K'(h), \qquad h > 0. \tag{8.3.33}$$

Homogeneous Poisson Process

Let N be a homogeneous Poisson process on $\mathbb{R}^d$ with intensity λ. Then, by the independence property of a Poisson process, $\lambda_2^o(h) = \lambda^2$. So, from Eq. (8.3.32), the K function of N is

$$K(h) = \frac{\pi^{d/2} h^d}{\Gamma(1 + \frac{d}{2})}, \qquad h \geq 0. \tag{8.3.34}$$

Thus, $L(h) \equiv \{K(h)\Gamma(1 + \frac{d}{2})/\pi^{d/2}\}^{1/d} - h$, $h \geq 0$, can be used to monitor departures from csr.

8.4 COMPLETE SPATIAL RANDOMNESS, DISTANCE FUNCTIONS, AND SECOND MOMENT MEASURES

Each of the topics featured in this section has already been discussed briefly in Section 8.2, where they were used in the analysis of the longleaf-pine data. A more complete discussion will be given here, interspersed with practical recommendations.

8.4.1 Complete Spatial Randomness

Complete spatial randomness (csr) is the "white noise" of spatial point processes. It characterizes the *absence* of structure (or signal) in the data. As such, it is often the null hypothesis in a statistical test to determine whether there is spatial structure in a given point pattern. Because all stochastic processes have components of randomness in them, the term *complete* spatial randomness is used and is synonymous with a *homogeneous* Poisson process. (It should be noted that some authors equate csr with a generally inhomogeneous Poisson process.)

Homogeneous Poisson Process

In Section 8.3, several equivalent definitions of csr were given, via counting measures, the zero-probability functional, the Laplace generating functional, and the probability generating functional. An illuminating way of seeing the lack of spatial structure in csr is a definition using infinitesimal spatial regions.

Let $d\mathbf{s}$ denote an infinitesimal region in $\mathbb{R}^d$, located at $\mathbf{s}$. Suppose $N(B)$ is the number of events of a point process in $B \subset \mathbb{R}^d$ and denote the d-dimensional volume of B as $|B|$ or $\nu(B)$. Because Section 8.3 may prove difficult to the less-theoretically inclined, the (homogeneous) Poisson process is defined once again, although in a way that allows heuristic derivation of some of its properties. The point process N is a homogeneous Poisson process if

i.
$$\lim_{|d\mathbf{s}| \to 0} \frac{1 - \Pr(N(d\mathbf{s}) = 0)}{|d\mathbf{s}|} = \lambda. \tag{8.4.1}$$

ii.
$$\lim_{|d\mathbf{s}| \to 0} \frac{\Pr(N(d\mathbf{s}) = 1)}{|d\mathbf{s}|} = \lambda \tag{8.4.2}$$

[which in turn implies $\Pr(N(d\mathbf{s}) > 1)/\,|d\mathbf{s}| \to 0$].

iii. $N(d\mathbf{s}_1), N(d\mathbf{s}_2), \ldots,$ are statistically independent for any disjoint sequence of regions $d\mathbf{s}_1, d\mathbf{s}_2, \ldots$.

The constant λ in (8.4.1) and (8.4.2) is known as the intensity of the homogeneous Poisson process. [The homogeneous Poisson process becomes inhomogeneous when λ is replaced by $\lambda(\mathbf{s})$, a function of the location $\mathbf{s}$ of the infinitesimal region.]

Consider any (Lebesgue measurable) B and divide B into $|B|/|d\mathbf{s}|$ equal infinitesimal areas each of volume $|d\mathbf{s}|$. By the axioms i, ii, and iii, $N(B)$ is approximately a *binomial* distribution with parameters $n = |B|/\,|d\mathbf{s}|$ and $p = \lambda|d\mathbf{s}|$. Then, a well known result in distribution theory (see, e.g., Hogg and Craig, 1978, p. 189) says that, as $|d\mathbf{s}| \to 0$, $N(B)$ converges in distribution to a Poisson distribution with mean $\lim_{|d\mathbf{s}| \to 0} np = \lambda|B|$. That is, the distribution of $N(B)$ does not depend on the location or shape of B, but just on its d-dimensional volume. Moreover, conditional on $N(B) = n$, events in B occur at i.i.d. uniform (over B) locations and, when B_1 and B_2 are disjoint, $N(B_1)$ and $N(B_2)$ are independently Poisson distributed.

Other properties of the homogeneous Poisson process include the expressions:

$$z_N(B) = \exp(-\lambda|B|) \quad \text{(zero probability functional)}, \tag{8.4.3}$$

$$L_N(\zeta) = \exp\left\{-\lambda \int (1 - \exp(-\zeta(\mathbf{s})))\, d\mathbf{s}\right\} \quad \text{(Laplace functional)}, \tag{8.4.4}$$

$$G_N(\xi) = \exp\left\{-\lambda \int (1 - \xi(\mathbf{s}))\, d\mathbf{s}\right\} \quad \text{(probability generating functional)}, \tag{8.4.5}$$

$$\mu_N(B) = \lambda|B| \quad \text{(mean measure)}, \tag{8.4.6}$$

$$C_N(B_1 \times B_2) = \lambda|B_1 \cap B_2| \quad \text{(covariance measure)}, \tag{8.4.7}$$

$$\lambda_2(\mathbf{s}, \mathbf{u}) = \lambda^2 \quad \text{(second-order intensity)}. \tag{8.4.8}$$

Also, recall that the superposition of two independent Poisson processes with parameters λ_1 and λ_2 is a Poisson process with parameter $\lambda_1 + \lambda_2$.

Simulating a Homogeneous Poisson Process

One approach to simulating a homogeneous Poisson process (intensity λ) in a study region A follows directly from the observation that $N(A)$ has a Poisson distribution with mean $\lambda\nu(A)$, and, given $N(A) = n$, the n events in A form an independent random sample from the uniform distribution on A [see (8.3.2)]. Care must be taken, however, when choosing a pseudo-random-number generator. Although most random-number generators will yield a uniform distribution on the interval $[0, 1)$, many will yield quite regular patterns on the unit square or unit cube (Ripley, 1983, 1987a, p. 23).

Lewis and Shedler (1979) give an alternative method for simulating a homogeneous Poisson process, intensity λ, in a rectangle $A = (0, a_1) \times (0, a_2)$: Consider the x coordinates of events, located in $(0, \infty) \times (0, a_2)$, from a Poisson process with intensity $\lambda \cdot a_2$. Now, the differences between successive x coordinates are i.i.d. exponential random variables with parameter $\lambda \cdot a_2$. Suppose that x coordinates are generated in this way; then the corresponding y coordinates can be generated from a uniform distribution on $(0, a_2)$. This algorithm generates the events in order of increasing x coordinates and is terminated when the latest x coordinate exceeds a_1.

Monte Carlo Tests of csr

Various statistics for testing csr from data were presented in Section 8.2. The distribution theory for complicated functions of the data can be intractible, even under the null hypothesis of csr. The *Monte Carlo* test (Barnard, 1963; Hope, 1968; Birnbaum, 1975) is a way around this problem. In the present context, the test is implemented by simulating values of the test statistic under csr and comparing them to the corresponding statistic calculated from the observed pattern (Besag and Diggle, 1977). It should be noted that the same idea can be applied to any testing problem, spatial or not.

Let U denote a test statistic and let $\{U_i: i = 2, \ldots, k\}$ denote $(k - 1)$ values of the statistic generated by independently simulating data (of the same size) $(k - 1)$ times under the null hypothesis. Call U_1 the observed value of U and ask whether U_1 is different from, or typical of, the other U_is. This is ascertained by ordering $U_{(1)} \leq U_{(2)} \leq \cdots \leq U_{(k)}$ and rejecting the null hypothesis if U_1 is one of the smaller or one of the larger order statistics (depending on the direction of the alternative hypothesis with regard to U).

For example, in a bounded region $A \subset \mathbb{R}^2$, suppose Pielou's (1959) statistic (see Table 8.6) is used for testing csr versus a clustering alternative:

$$U = \pi\hat{\lambda} \sum_{i=1}^{n} X_i^2/n,$$

where $\hat{\lambda} = N(A)/\,|A|$ is an estimator of the global intensity and $X_1, \ldots, X_n$

is a random sample of $n \leq N(A)$ point-to-nearest-event distances. Now the null hypothesis of csr is not simple, because it involves the intensity parameter λ. By conditioning on the sufficient statistic $N(A)$, the number of events in A, the Monte Carlo test can be carried out.

Because large values of U indicate a clustering-type departure from csr, the size-α $(0 < \alpha < 1)$ Monte Carlo test is performed as follows. Suppose $U_1 = U_{(j)}$ for some $j \in \{1, \dots, k\}$. Then reject H_0 if $(k + 1 - j)/k \leq \alpha$. For example, based on 99 simulations (i.e., $k = 100$), rejection at the 5% level occurs if U_1 is $U_{(96)}, \dots, U_{(100)}$. From Table 8.7, the edge-corrected value of $U_1 = 1.785$ was compared to U_is calculated from 99 simulations of 584 longleaf pine locations in the 4-ha central study region. It was found that $U_1 = U_{(100)}$, leading to rejection of csr at the 5% level of significance. (Approximate distribution theory for Pielou's statistic is also available; see Section 8.2.5.)

8.4.2 Distance Functions

Recall from Figure 8.9 the two basic distances W (event-to-event) and X (sample point-to-event). Their distribution theory under csr is well known (e.g., Upton and Fingleton, 1985, p. 56). For a homogeneous Poisson process in $\mathbb{R}^2$, the probability that there are no events within distance x of an arbitrary point is $\exp(-\lambda\pi x^2)$; see (8.4.3). Therefore, the distribution function of the point-to-nearest-event distance is $1 - \exp(-\lambda\pi x^2)$, which has density

$$f(x) = 2\pi\lambda x \exp(-\lambda\pi x^2), \qquad x > 0. \tag{8.4.9}$$

The results of Section 8.3.4 for the homogeneous Poisson process show that the same arguments apply for event-to-event distance W. Its density in $\mathbb{R}^2$ is

$$g(w) = 2\pi\lambda w \exp(-\lambda\pi w^2), \qquad w > 0. \tag{8.4.10}$$

Generalizations of (8.4.9) and (8.4.10) to $\mathbb{R}^d$ are straightforward using (8.4.3).

Extension to kth Nearest Neighbors

H. R. Thompson (1956) considered kth nearest-neighbor distances W_k from an arbitrary event. For a homogeneous Poisson process, the density of W_k given $W_1 = w_1, \dots, W_{k-1} = w_{k-1}$ is

$$f(w_k | w_1, \dots, w_{k-1}) = 2\pi\lambda w_k \exp\{-\pi\lambda(w_k^2 - w_{k-1}^2)\},$$
$$0 \leq w_1 \leq \cdots \leq w_k.$$

Then, the joint density of $W_1, W_2, \ldots, W_k$ is

$$f(w_1, w_2, \ldots, w_k) = f(w_1) f(w_2|w_1) \cdots f(w_k|w_1, w_2, \ldots, w_{k-1})$$
$$= (2\pi\lambda)^k w_1 w_2 \cdots w_k \exp(-\pi\lambda w_k^2),$$
$$0 \le w_1 \le w_2 \le \cdots \le w_k. \quad (8.4.11)$$

Integrating out $w_1, w_2, \ldots, w_{k-1}$, the density of W_k is obtained:

$$g_k(w_k) = 2(\pi\lambda)^k w_k^{2k-1} \exp(-\pi\lambda w_k^2) / (k-1)!, \qquad w_k > 0. \quad (8.4.12)$$

It follows that

$$E(W_k) = k(2k)! / \{(2^k k!)^2 \lambda^{1/2}\}, \quad (8.4.13)$$

a formula already given in (8.2.12). This result was used in Section 8.2.5 to compute observed to expected ratios $\{R_k: k = 1, \ldots, K\}$ defined by (8.2.13). A plot of these ratios against k for the longleaf-pine data is given in Figure 8.11.

Nearest-Neighbor Distribution Functions

The probability measure Π_N, induced on a space Φ of locally finite counting measures (Section 8.3), in turn induces probability measures on event-to-event and point-to-event distances. Let $G(r)[F(r)]$ be the probability that the distance from a *randomly* chosen event [point] to its nearest event is less than or equal to r. The random choice of event or point is an extra source of randomness beyond that obtained from Π_N, ensuring that G or F are functions only of r.

Estimating G

Estimating the cdf G, or its density, from an observed pattern is complicated by edge effects. The emphasis here will be on estimating G; Fiksel (1988a) and Doguwa (1989a) can be consulted for edge-corrected density estimation. Various edge-corrected estimators $\hat{G}_1$, $\hat{G}_2$, and $\hat{G}_3$ are presented in Section 8.2.6 [see Eqs. (8.2.14), (8.2.15), and (8.2.16)]. In this subsection, consider

$$\hat{G}_3(r) = \sum_{i=1}^{n} I(r_i \le r, d_i > r) \Big/ \sum_{i=1}^{n} I(d_i > r), \qquad r > 0, \quad (8.4.14)$$

where $0/0 \equiv 0$ and n events are randomly chosen so that from the ith event the nearest-event distance (r_i) and the nearest-boundary distance (d_i) are measured.

The estimator (8.4.14) can also be computed when there is exhaustive sampling of the region A. Figure 8.13 is based on such sampling for the longleaf-pine data. It shows a quantile-quantile plot of $\hat{G}_3(r)$ versus $G(r) = 1 - \exp(-\lambda\pi r^2)$, $r > 0$, the cumulative distribution function under csr. As well as a visual assessment, it is possible to use a Monte Carlo testing procedure to determine if the point pattern came from a csr process. For given λ, define the Cramér–von Mises-type statistic

$$g \equiv \int_0^\infty \left\{\hat{G}_3(r) - (1 - \exp(-\lambda\pi r^2))\right\}^2 dr. \qquad (8.4.15)$$

For the observed point pattern, use $\lambda = 584/40{,}000 = 0.0146 \text{ m}^{-2}$, which yields $g = 0.1285$. It is well above the largest value 0.0107 for 100 realizations of csr (with $\lambda = 0.0146 \text{ m}^{-2}$). This is consistent with the results in Section 8.2 that suggest the longleaf-pine locations are clustered. In fact, all nearest neighbors were within 16 m and 50% were within 3 m of the 584 trees (under csr, the median nearest-neighbor distance is almost 4 m).

An Alternative Estimator of G

The estimator $\hat{G}_3$ given by (8.4.14), due to Ripley (1976a), effectively uses a guard area of variable width r. The guard area contains events that can be nearest neighbors to an initiating event, but cannot be initiating events themselves. It is easy to see that $\hat{G}_3$ might ignore valuable nearest-neighbor information: Suppose an event is in the guard area of width r, but its nearest-neighbor distance is smaller than r. Another unattractive feature of $\hat{G}_3$ is that it is not necessarily nondecreasing in r; see Ripley (1981, Figure 8.3).

Hanisch (1984) proposes an alternative estimator of G that makes more efficient use of the available nearest-neighbor information and is nondecreasing. It is

$$\hat{G}_4(r) \equiv \sum_{i=1}^{n} I(r_i \le r, d_i > r_i) \Big/ \sum_{i=1}^{n} I(d_i > r_i), \qquad r > 0, \quad (8.4.16)$$

where $0/0 \equiv 0$. Like (8.4.14), the estimator (8.4.16) can also be computed from an exhaustive sampling of the region A.

Estimating F

A similar development to that given for G can also be given for the random-point-to-nearest-event distribution function F. For the longleaf-pine data, 584 points in the 4-ha central study region A were chosen uniformly, from which nearest-event distances r_i^* and nearest-boundary distances d_i^* were found, $i = 1, \ldots, 584$. Then, the estimator

$$\hat{F}(r) \equiv \sum_{i=1}^{584} I(r_i^* \le r, d_i^* > r) \Big/ \sum_{i=1}^{584} I(d_i^* > r), \qquad r > 0, \quad (8.4.17)$$

was computed and compared to the theoretical value under csr, using a Cramér–von Mises-type statistic:

$$f \equiv \int_0^{\infty} \{\hat{F}(r) - (1 - \exp(-\lambda\pi r^2))\}^2 \, dr \tag{8.4.18}$$

(where $\lambda = 584/40{,}000 = 0.0146 \text{ m}^{-2}$ was used). Again, the test statistic $f = 0.1131$ was well above the largest value 0.0117 for 100 realizations of csr.

Under either exhaustive or intensive sampling, the usual distribution theory for a Cramér–von Mises statistic is *not* appropriate for (8.4.15) or (8.4.18) due to spatial dependence in the data. The Monte Carlo test provides a statistically valid alternative.

8.4.3 *K* Function

The origins of the K function can be found in Bartlett (1964). However, its importance as an effective summary of spatial dependence over a wide range of scales was first realized and developed by Ripley (1976a, 1977). The (theoretical) K function of a stationary spatial point process is defined as:

$$K(h) \equiv \lambda^{-1} E(\text{number of extra events within distance } h \text{ of a randomly chosen event}), \quad h \geq 0. \tag{8.4.19}$$

Theoretical considerations given in Section 8.3.5 show that, for a stationary point process N, (8.4.19) is a special case of (8.3.31): Write (8.3.31) as

$$\mathscr{K}_N(B) = \lambda^{-1} \int_{\Phi} \phi(B - \{\mathbf{0}\}) P_{N,\mathbf{0}}(d\phi), \tag{8.4.20}$$

where B is a Borel set in $\mathbb{R}^d$. Then put $B = b(\mathbf{0}, h) \equiv \{\mathbf{s}: \|\mathbf{s}\| \leq h\}$ in (8.4.20) to obtain (8.4.19), where $K(h) \equiv \mathscr{K}_N(b(\mathbf{0}, h))$, $h \geq 0$.

For a nonstationary point process, the expectation on the right-hand side of (8.4.19) still does not depend on the location of the event, but only on h, because the event is *randomly* chosen. This extra source of randomness introduced into the problem allows K functions to be considered for nonstationary processes. The following development, however, assumes stationarity.

Second-Order Properties

The relationship between the second-order intensity $\lambda_2^o(\cdot)$ of a stationary isotropic process in $\mathbb{R}^d$ and the K function is given by (8.3.32). In $\mathbb{R}^2$ it is

$$\lambda K(h) = (2\pi/\lambda) \int_0^h u \lambda_2^o(u) \, du, \qquad h \geq 0. \tag{8.4.21}$$

Then (8.3.33) becomes $\lambda_2^o(h) = \lambda^2 K'(h)/(2\pi h)$, and a nonparametric estimator of the derivative of K yields a nonparametric estimator of λ_2^o (Diggle et al., 1987).

Thus the K function is only a second-order property of a point process. Therefore, two processes having the same intensity function and K function may behave very differently (e.g., Baddeley and Silverman, 1984), just as the mean and covariance structures do not characterize the random processes described in Parts I and II of this book. Nevertheless, there is considerable merit in estimating K functions and fitting models to them (Section 8.5).

Estimating K

Estimating $K(\cdot)$ from an observed pattern in a bounded $A \subset \mathbb{R}^d$ is complicated by edge effects. As d increases, the edge effects become more important. For the purposes of this subsection, assume $d = 2$. Various edge-corrected estimators $\hat{K}_1$, $\hat{K}_2$, $\hat{K}_3$, and $\hat{K}_4$ are presented in Section 8.2.6 [see Eqs. (8.2.18) through (8.2.21)], based on $N(A) = N$ events in A.

In this subsection, consider Ripley's edge-corrected estimator given by (8.2.21), for which the additional assumption of isotropy is made:

$$\hat{K}_4(h) = \hat{\lambda}^{-1} \sum_{i=1}^{N} \sum_{\substack{j=1 \\ i \neq j}}^{N} w(\mathbf{s}_i, \mathbf{s}_j)^{-1} I(\|\mathbf{s}_i - \mathbf{s}_j\| \le h)/N, \qquad h > 0, \quad (8.4.22)$$

where the weight $w(\mathbf{s}_i, \mathbf{s}_j)$ is the proportion of the circumference of a circle centered at $\mathbf{s}_i$, passing through $\mathbf{s}_j$, and that is inside the study region A. Finally, $\hat{\lambda} = N/|A|$, the total number of events in the study region A divided by the volume of A. For a rectangular region A, an explicit formula for $w(\mathbf{s}_i, \mathbf{s}_j)$ is given by Diggle (1983, p. 72): Let $u = \|\mathbf{s}_i - \mathbf{s}_j\|$ and let d_1 and d_2 be, respectively, the distances of $\mathbf{s}_i$ from the nearest vertical and horizontal edges of A. If $u^2 \le d_1^2 + d_2^2$, then

$$w(\mathbf{s}_i, \mathbf{s}_j) = 1 - \pi^{-1}\left[\cos^{-1}\{\min(d_1, u)/u\} + \cos^{-1}\{\min(d_2, u)/u\}\right],$$

or if $u^2 > d_1^2 + d_2^2$,

$$w(\mathbf{s}_i, \mathbf{s}_j) = 3/4 - (2\pi)^{-1}\left\{\cos^{-1}(d_1/u) + \cos^{-1}(d_2/u)\right\}.$$

The estimator $\hat{K}_4$ given by (8.4.22) is approximately unbiased provided $\hat{K}_4(h)$ and N are approximately independent; for small h, this is true (Ripley, 1981, p. 159). Further distribution theory is given subsequently.

Exhaustive Sampling of Events

It is worth noting that in estimating the K function an exhaustive sampling of the bounded region A is usually the easiest way to proceed. If random initiating events were chosen, a complete map of events around each initiating event would be needed in order to compute any of the estimators $\hat{K}_1$ through $\hat{K}_4$. This means that most of the region A would need to be mapped, leaving little reason to take a random subset of events in A.

Empirical K Function for Longleaf-Pine Data

Using the 584 pines in the 4-ha central study region, the empirical K function given by (8.4.22) was calculated. In Figure 8.16, $\{\hat{K}_4(h)/\pi\}^{1/2}$ is plotted against radius h; a square-root transformation stabilizes the variance under csr (Besag, 1977c; Ripley, 1979a). Moreover, under csr $K(h) = \pi h^2$, which implies that $\{\hat{K}_4(h)/\pi\}^{1/2}$ is estimating h under csr. By definition, $\lambda K(h)$ can be interpreted as a measure of mean crowding, in the sense of Lloyd (1967); from the right vertical axis of Figure 8.16, one can read off the estimated expected number of extra events $\hat{\lambda}\hat{K}_4(h)$.

The figure shows the estimated K function (dotted line) to be above and roughly parallel to the 45° solid line (expected under csr). It also shows that, on average, a tree has more than eight neighboring trees within 10 m, and up to 20 m (the extent of the roots of a large tree) the expected number of trees is about 23.

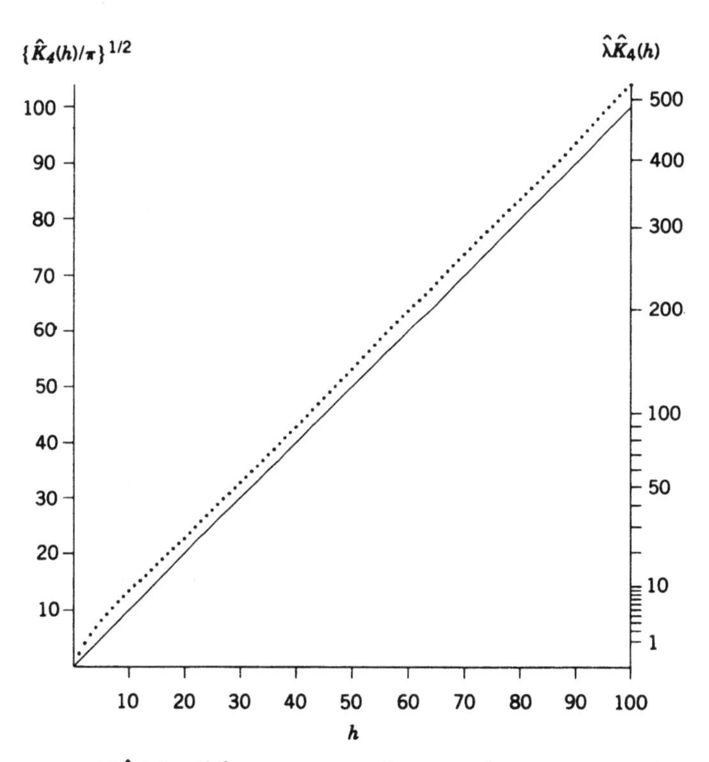

Figure 8.16 Plot of $\{\hat{K}_4(h)/\pi\}^{1/2}$ versus radius h (dotted line), and the expected line under csr (45° solid line). Units on the horizontal axis and the left-hand vertical axis are in meters. The right-hand vertical axis gives the estimated expected number of extra events within distance h of an arbitrary event.

Monte Carlo Test

To test whether the data come from a csr process, a Monte Carlo test based on the Cramér–von Mises-type statistic

$$k \equiv \int_0^\infty \left\{ \left(\hat{K}_4(h) \right)^{1/2} - \pi^{1/2} h \right\}^2 dh \tag{8.4.23}$$

was carried out. Specifically, 100 simulations of a csr process (using $\lambda = 584/40{,}000 = 0.0146 \text{ m}^{-2}$) gave 100 ks, the largest of which was 211.3. This is well below the value of $k = 3504.2$, calculated from the observed pattern, leading to rejection of csr. A further illustration of how the data differ from a csr process is given by Figure 8.14. There, the 100 simulations were used to define an envelope of $\hat{L}_4$s under csr, where $\hat{L}_4(h) \equiv \{\hat{K}_4(h)/\pi\}^{1/2} - h$. It is patently clear from Figure 8.14 how different the observed $\hat{L}_4$ is from these envelopes.

Distribution Theory for $\hat{K}_4$

The Monte Carlo procedure provides a method of inference in a setting where data are dependent and distribution theory is difficult. Nevertheless, a few results in $\mathbb{R}^2$ are available to help with inference.

Under csr, Stoyan et al. (1987, p. 58) give the formula

$$\text{var}\left(\hat{K}_4(h)\right) \simeq 2\lambda^{-2}(h/a)^2\{1 + 0.61(h/a) + 0.083\lambda h^3/a\},$$

where the study region A is a disk of radius a. From this, an approximate acceptance region for csr is easy to construct *for fixed h*. Ripley (1979a) gives simultaneous regions obtained by simulation. For example, the acceptance region of a 5% test for csr of N events in a square $A = [0, a] \times [0, a]$, based on $\hat{L}_4$, is given as

$$\left\{\left(h - (1.42)|A|^{1/2}/N,\ h + (1.42)|A|^{1/2}/N\right): 0 < h \leq a/4\right\}.$$

Silverman (1978b) proves $\hat{K}_4(h)$ to be asymptotically normal under csr as $N \to \infty$; see Ripley (1981, p. 163) for its potential use in testing for csr.

Weak convergence of $\{(\lambda^3|A|)^{1/2}(\hat{K}_1(h) - \pi h^2): 0 < h \leq h_0\}$, as $A \uparrow \mathbb{R}^2$, is proved under csr (and other regularity conditions) by Heinrich (1988). The limiting process $Y(\cdot)$ is Gaussian with zero mean and covariance:

$$\text{cov}(Y(h_1), Y(h_2)) = 2\lambda\pi h_1^2(1 + 2\lambda\pi h_2^2), \qquad h_1 \leq h_2.$$

(Heinrich's, 1988, result is actually given in $\mathbb{R}^d$.) A similar result should be true for $\hat{K}_4$.

An Alternative Estimator of K

Ohser and Stoyan (1981) suggest an (approximately) unbiased estimator of an anisotropic version of the K function; see also Hanisch and Stoyan (1984).

For the (isotropic) K function, defined by (8.4.19), this estimator becomes

$$\hat{K}_5(h) = (\hat{\lambda})^{-2} \sum_{i=1}^{N} \sum_{\substack{j=1 \\ i \neq j}}^{N} \left\{ I(\|\mathbf{s}_i - \mathbf{s}_j\| \le h) \Big/ |(A \oplus \mathbf{s}_i) \cap (A \oplus \mathbf{s}_j)| \right\}, \tag{8.4.24}$$

where $A \oplus \mathbf{s} \equiv \{\mathbf{a} + \mathbf{s}: \mathbf{a} \in A\}$ and $\hat{\lambda} = N/|A|$; see also Ohser (1983), who shows that $\hat{K}_4(h)$, but not $\hat{K}_5(h)$, is a biased estimator of $K(h)$ for large h.

Under csr, for $A = [0, a_1] \times [0, a_2]$ and $\max(a_1, a_2) \to \infty$, Heinrich (1984) shows that

$$\sup\left\{ |\hat{K}_5(h) - \pi h^2| : 0 < h \le h_0 < \infty \right\} \to 0$$

and in distribution

$$(a_1 a_2)^{1/2}\left(\hat{K}_5(h) - \pi h^2 \right) \to \text{Gau}(0, v);$$

he also shows how to estimate v. A multivariate asymptotic Gaussian result for $(\hat{K}_5(h_1), \ldots, \hat{K}_5(h_k))$ should follow from a weak-convergence result similar to the one proved by Heinrich (1988) for $\{(\lambda^3|A|)^{1/2}(\hat{K}_1(h) - \pi h^2): 0 < h \le h_0\}$, as $A \uparrow \mathbb{R}^2$ (see preceding text).

Although these results have been proved under csr, there are undoubtedly more general stationary point processes (e.g., whose spatial dependence between counts in disjoint sets decreases as their distance apart increases) for which similar results hold. For example, under appropriate assumptions of stationarity and ergodicity, the various estimators of $K(\cdot)$ are consistent as $A \uparrow \mathbb{R}^d$ (Daley and Vere-Jones, 1988, pp. 360–363).

Second-Order Neighborhood Analysis

Getis and Franklin's (1987) second-order neighborhood analysis is similar to the second-order analyses involving the K function, except that consideration is given only to pairs of events that include a given event $\mathbf{s}_i$ as one of the pairs. Conditional on an event at $\mathbf{s}_i$, they define

$$\lambda K^{(i)}(h) \equiv E\{\text{number of extra events within distance } h \text{ of event } \mathbf{s}_i\}, \qquad h \ge 0,$$

and estimate it using a variation of Ripley's edge-corrected estimator of $K(h)$:

$$\hat{K}^{(i)}(h) \equiv \frac{|A|}{N(N-1)} \sum_{\substack{j=1 \\ j \neq i}}^{N} w(\mathbf{s}_i, \mathbf{s}_j)^{-1} I(\|\mathbf{s}_i - \mathbf{s}_j\| \le h), \qquad h > 0,$$

where $w(\cdot,\cdot)$ and $I(\cdot)$ are as defined in Eq. (8.4.22). Getis and Franklin (1987) use $\hat{K}^{(i)}(h)$ to measure the crowding of events around the event $\mathbf{s}_i$.

Doguwa (1989b) suggests that $\hat{K}^{(i)}(h)$ can be used to distinguish between two possibly different point processes having identical K functions. He defines the following functions of $\hat{K}^{(i)}(h)$:

$$R(h) \equiv \max_{1 \leq i \leq N} \left\{\hat{K}^{(i)}(h)/\pi\right\}^{1/2},$$

$$G(h) \equiv \frac{1}{N} \sum_{i=1}^{N} \left\{\hat{K}^{(i)}(h)/\pi\right\}^{1/2},$$

and

$$D(h) \equiv \sum_{i=1}^{N} \left(\hat{K}^{(i)}(h)\right)^2 \Big/ N\pi^2.$$

Although the cellular process of Baddeley and Silverman (1984) has the same K function as the homogeneous Poisson process, Doguwa shows that these functions can be used to distinguish between realizations of the two point processes.

8.4.4[†] Animal-Behavior Data

This book is about Statistics for spatial data, but the methods described can often be applied to temporal data. It is important to reemphasize that the time dimension is different from $\mathbb{R}^1$ due to the unidirectional flow of time. The data presented in this subsection are temporal, meaning that some care is needed in adapting the K function and its estimators.

Some Background

Observational techniques identify behavior as discrete acts, but may quantify the acts in various ways (Norton, 1968; Hutt and Hutt, 1970). Usually the acts are simply counted; that is, each initiation of an act is recorded as one occurrence and then occurrences are totaled. The total time spent performing an act, its average duration, or even the time between initiations are quantities of interest in psychiatry (Wolff, 1968; Pohl, 1976; Baumeister, 1978). However, these alone are not enough to describe abnormal behavior. Hyperactive behavior in children is believed to be qualitatively different from normal behavior, so that simple summaries of act initiations may not be sensitive enough (Wender, 1971). Hyperactivity in children is thought to consist of randomly organized behavior that gives only a false impression of increased motor activity (Pontius, 1973). These perceptions of hyperactivity emphasize the need to understand the structure of behavior and to expand quantification beyond measures of the frequency and duration of behavioral acts.

Workers on oil rigs and at oil refineries are exposed to chemicals that may affect them adversely, causing erratic and sometimes violent behavior. To test the effects of certain chemicals, Dr. W. Kernan and Dr. D. Hopper of the Veterinary Diagnostic Laboratory at Iowa State University conducted experiments on adult *Macaca fascicularis* (monkeys) exposed to three levels of amphetamine as well as a placebo exposure.

Six adult male monkeys and six adult female monkeys were each exposed to two replications of placebo, 0.11 mg/kg, 0.33 mg/kg, and 1.0 mg/kg of amphetamine. Thus, the experiment consisted of monkeys (12 levels made up of sex × number) × treatment (4 levels), replicated twice. Within a replication, the order of treatment exposure was balanced across monkeys, and sufficient time was allowed for the drug to pass before the next exposure was started. Beginning 15 min after exposure to the drug, the animal was observed for 13 min at half second (half-s) intervals; in each half-s interval it was classified as performing one out of a set of acts (e.g., sit, quadrupedal stand) described by Kernan et al. (1980). Therefore, the data are categorical, coming as a time series with 1560 observations, each spaced 1 time unit (half-s) apart.

The classification of behavior by a human observer is an arduous, subjective task, although permanent records provided by film and videotape make it less error prone. Kernan et al. (1980, 1981) describe their method of classification using computer pattern recognition to relieve the tedium and to remove observer bias from data collection.

There is obviously a lot of data; in all, there are 96 time series, each of length 1560, and each datum is categorical. How can it be analyzed? Various summaries, based on numbers of acts, initiations of acts, times spent in acts, and so forth, have been tried (see, e.g., Norton, 1973, 1977; Kernan et al., 1981; Mullenix, 1981; Mullenix et al., 1986). But, as was argued in the beginning paragraph, these summaries are not powerful enough to discriminate between complex behavior patterns. It will be shown in succeeding text how to convert the data into temporal point patterns, whose $\hat{K}$ functions give summaries of an aspect missed by previous analyses.

The Data

For ease of presentation, only one act will be analyzed. The act "sit" was chosen because the monkeys were often sitting, drugged or not, and because it illustrates well the K-function approach. The *pattern* of sequences of "sit" and "not sit" will be used to indicate a difference between the monkey's behavior when drugged and when not drugged. Purely to illustrate the potential power of the K function to detect differences, the data from just one monkey will be analyzed. These are presented in Table 8.10. I am grateful to Dr. Kernan and Dr. Hopper of the Veterinary Diagnostic Laboratory for making these data available.

Table 8.10 Entries Either Side of the Word "Sit" Show the First and Last Half-s Time Periods that Define the Intervals During which the Monkey was Sitting[a].

Placebo			0.33 mg/kg
1	624	993	2
Sit	Sit	Sit	Sit
30	626	1012	28
34	629	1014	83
Sit	Sit	Sit	Sit
36	683	1027	134
48	685	1037	137
Sit	Sit	Sit	Sit
48	702	1037	137
50	708	1043	145
Sit	Sit	Sit	Sit
52	709	1124	202
55	740	1131	210
Sit	Sit	Sit	Sit
56	797	1136	328
58	799	1138	439
Sit	Sit	Sit	Sit
62	817	1138	602
66	819	1142	607
Sit	Sit	Sit	Sit
102	821	1147	607
117	826	1150	842
Sit	Sit	Sit	Sit
217	828	1155	868
223	830	1223	1248
Sit	Sit	Sit	Sit
225	833	1254	1260
229	835	1256	1263
Sit	Sit	Sit	Sit
229	836	1318	1265
243	838	1323	1267
Sit	Sit	Sit	Sit
346	840	1323	1274
353	843	1325	1276
Sit	Sit	Sit	Sit
353	844	1325	1384
355	846	1328	1387
Sit	Sit	Sit	Sit
355	846	1343	1417
476	852	1348	
Sit	Sit	Sit	
544	860	1350	
551	883	1352	
Sit	Sit	Sit	
602	883	1499	
614	885	1506	
Sit	Sit	Sit	
616	974	1560	

[a]The two drug levels are: placebo and 0.33 mg/kg amphetamine.

Analysis by K Functions

It is clear from Table 8.10 that under the drug there are fewer "sit" initiations and that the total sitting time is less. Some of the more standard ways of analyzing behavioral data involve comparing such quantities, comparing length-of-sit histograms, as well as summaries based on several acts. None of these methods takes into account the *pattern* of sit initiations (random, clustered, or regular), upon which I shall now concentrate. I would like to emphasize that this analysis is not at all exhaustive and is merely an illustration of how the K function can be used to discriminate between patterns.

First, the data are transformed to remove the intervals of sitting. Because the intention is to analyze the *initiations* of "sit," retaining those intervals might bias even completely random initiations toward a more regular-looking pattern. The transformation is carried out by removing all but the first half second of each "sit" interval and closing up the gap caused by the removal. Figure 8.17 shows how the first 250 half-s time periods for the placebo data are transformed. Theoretical justification for removing and closing up is given in the theorem on marker points in Section 9.5.

The transformed data are now analyzed as a temporal point pattern, where the events are initiations of the act "sit." Distances are defined only from earlier points to later points, preserving the unidirectionality of the time

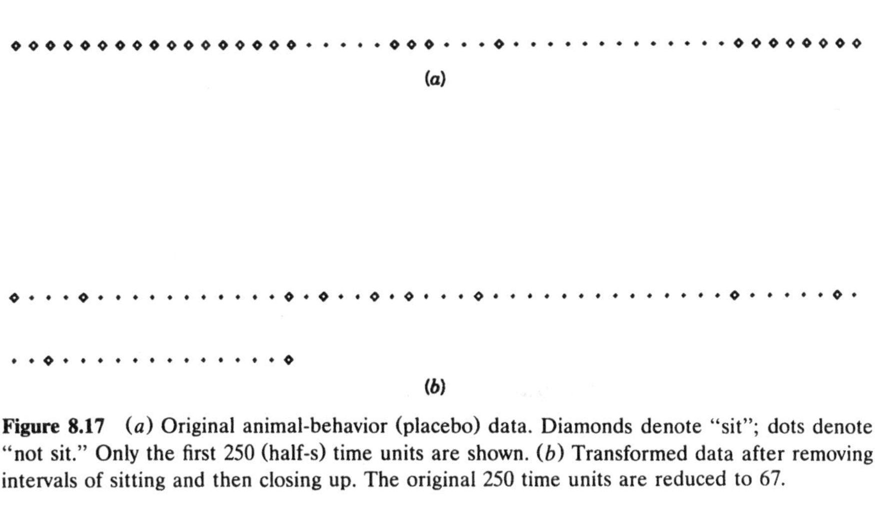

(a)

(b)

Figure 8.17 (a) Original animal-behavior (placebo) data. Diamonds denote "sit"; dots denote "not sit." Only the first 250 (half-s) time units are shown. (b) Transformed data after removing intervals of sitting and then closing up. The original 250 time units are reduced to 67.

dimension. Moreover, because time has been discretized, the distance between a point at t_1 and a point at t_2 ($t_1 < t_2$) is defined in this subsection to be

$$|t_2 - t_1| \equiv t_2 - t_1 - 1, \qquad t_2 > t_1,$$

where t_1, t_2 are positive integers (units of half-s).

The temporal analogue of $\hat{K}_3$ given by (8.2.20) is computed; that is, a variable-width edge correction is used:

$$\hat{K}_3(t) \equiv \frac{T_s}{N_s} \sum_{i=1}^{N_s - 1} \sum_{j=i+1}^{N_s} I(0 < |t_j - t_i| \leq t, d_i > t) \Big/ \sum_{i=1}^{N_s - 1} I(d_i > t), \tag{8.4.25}$$

where T_s is the total *transformed* time (i.e., 1560 minus the total "sit" time), N_s is the number of sit initiations, t_i is the transformed time of the ith initiation, and d_i is the transformed time between t_i and $T_s + 1$. If the events are a realization of a (temporally) stationary point process with intensity λ_s, then $\hat{K}_3(t)$ is estimating

$$K(t) = \lambda_s^{-1} E(\text{no. extra events occurring at times} \leq t \text{ from an arbitrarily chosen event}), \qquad t \geq 0.$$

For a completely temporally random process (i.e., homogeneous Poisson process in time), it is easily demonstrated that $K(t) = t$.

Regardless of whether $\hat{K}_3$ is estimating anything, it still has a role as a summary measure of temporal dependence in the point pattern. Figure 8.18 shows $\hat{K}_3$ for placebo and drug, and the reference line $\hat{K}(t) = t$. It appears as if the drugged monkey's sit initiations are clustered, causing its $\hat{K}_3$ to appear well above the reference line. Under the placebo, the sit initiations appear completely random.

Inference Using K Functions

The obvious question to ask is whether the two K functions in Figure 8.18 are significantly different (assuming, say, a null hypothesis that both are completely temporally random). The answer needs further theoretical research; some adaptation of the spatial asymptotics presented in Section 8.4.3 may be possible.

With more replication of units (monkeys) and assuming small intermonkey variability, the sample variance of response summaries may be adequate for inference, or a bootstrap procedure (see, e.g., Efron, 1982) might be developed.

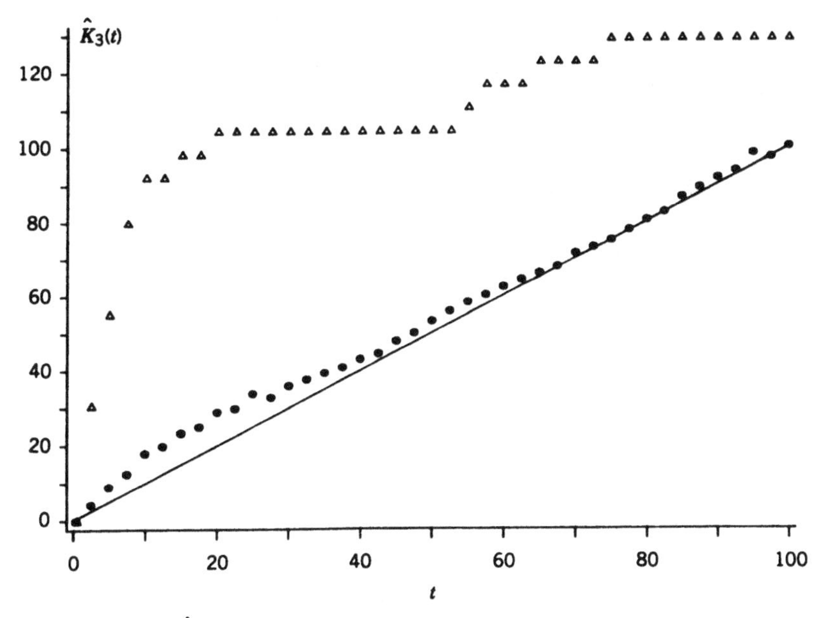

Figure 8.18 Plot of $\hat{K}_3(t)$ versus t. Circles denote placebo and triangles denote drug treatment. Units on the horizontal axis and the vertical axis are in half-s.

Monkeys are expensive to care for but have the advantage of being closer to humans in behavior than other animals. Because any experiment has cost constraints, more units can be obtained by using a less costly animal, like the rat (in a cage). A K-function analysis of experiments of this sort is described in Kernan et al. (1988), and approximate inference is carried out using the bootstrap. Olshen et al. (1989) show how the bootstrap can be used for simultaneous prediction of angular-rotation curves in gait analysis; analogous simultaneous inference may be possible for the K function.

8.4.5 Some Final Remarks

As summaries of spatial point pattern, the nearest-neighbor empirical distribution functions and the empirical K function offer a vast improvement over the indices reviewed in Sections 8.2.2 and 8.2.3. Monte Carlo testing allows these functions to be effective discriminators between csr and various alternative processes.

Only smaller scales of pattern can be modeled using the nearest-neighbor functions; however, the K function can be used to model a wide range of scales of pattern. Its superior performance as a descriptive statistic makes it desirable as a test statistic for formal inference in, say, an analysis of variance. This inference requires finite-region distribution theory that is not yet available. The discussion of the animal-behavior data given in Section

8.4.4 illustrates the power of the K function to discriminate between treatments where the responses are point patterns, and underlines the importance of having such distribution theory.

8.5 MODELS AND MODEL FITTING

Testing for complete spatial randomness (csr) is only the first step in the analysis of a spatial point pattern. If the null hypothesis of csr is rejected, the next obvious step is to fit some alternative (parametric) model to the data. After a model has been fitted, diagnostic tests should be performed to assess its goodness-of-fit. Finally, inference for the estimated parameters is often needed in response to a specific scientific question. The necessary distribution theory for the estimators can be difficult to obtain, in which case approximations may be necessary; at the very least, a bias and variance should be obtained.

In the following, I shall consider a variety of models for spatial point processes. Departure from csr is usually toward either clustering or regularity of events. Clustering can be modeled through an inhomogeneous Poisson process, a Cox process, or a Poisson cluster process. Simple inhibition processes can be used to model regular point patterns. Markov point processes can incorporate both elements through small-scale regularity and large-scale clustering. Thinning and related processes describe the result of randomly removing events from a point process.

For pragmatic reasons, methods used to fit models to data may change according to the class of models under consideration. For some models, it is seen that maximum likelihood estimators are available, but for others the likelihood function is intractable. For stationary point processes, the empirical K function can often be exploited for model fitting. These and other *ad hoc* procedures, devised for specific models, will be discussed in this section.

Methods of inference for various spatial–point-process models are considered in the following text. When asymptotics are needed, I shall consider statistical properties of parameter estimators as the sample window A approaches $\mathbb{R}^d$. Because independent-and-identically-distributed realizations of point processes are not commonly encountered in scientific applications, I shall only consider the case where the data are a single realization of a point process. Karr (1986) gives methods of fitting models to independent-and-identically-distributed realizations.

8.5.1* Inhomogeneous Poisson Process

The inhomogeneous Poisson process is perhaps the simplest alternative to complete spatial randomness (i.e., homogeneous Poisson process). In contrast to the homogeneous Poisson process, the mean measure of an inhomogeneous Poisson process is not proportional to Lebesgue measure, so the first-order intensity is a deterministic function of spatial location.

Let μ be a Radon measure on X; that is, $\mu(B)$ is finite for any compact set $B \in \mathscr{X}$, the Borel σ algebra of $X \subset \mathbb{R}^d$. Recall from Section 8.3 the definition of an *inhomogeneous Poisson process* with mean measure μ. As a consequence, such a point process N satisfies

$$P\{N(B) = n\} = \frac{e^{-\mu(B)}(\mu(B))^n}{n!}, \qquad n = 0, 1, 2, \ldots . \tag{8.5.1}$$

Let $\lambda(\mathbf{s}) \equiv \lim_{\nu(d\mathbf{s}) \to 0} \mu(d\mathbf{s})/\nu(d\mathbf{s})$ be the first-order intensity of N (Section 8.3.1). Here $\nu(B)$ ($= |B|$) is Lebesgue measure (or the d-dimensional volume) of $B \subset \mathbb{R}^d$. Given $N(A) = n$, for a bounded set $A \in \mathscr{X}$, the ordered n tuple of events $(\mathbf{s}_1, \mathbf{s}_2, \ldots, \mathbf{s}_n)$ in A^n is distributed as an independent random sample from the distribution on A with probability density proportional to $\lambda(\mathbf{s})$. That is,

$$f_A(\mathbf{s}) = \frac{\lambda(\mathbf{s})}{\int_A \lambda(\mathbf{u})\nu(d\mathbf{u})}, \qquad \mathbf{s} \in A. \tag{8.5.2}$$

[Note that $\mu(A) = \int_A \lambda(\mathbf{u})\nu(d\mathbf{u})$, the denominator of (8.5.2).] Then, the conditional density of the ordered n tuple $(\mathbf{s}_1, \mathbf{s}_2, \ldots, \mathbf{s}_n) \in A^n$, given $N(A) = n$, is

$$f(\mathbf{s}_1, \mathbf{s}_2, \ldots, \mathbf{s}_n) = \frac{\prod_{i=1}^n \lambda(\mathbf{s}_i)}{(\mu(A))^n}. \tag{8.5.3}$$

From (8.5.1) the joint "density" of $(\mathbf{s}_1, \mathbf{s}_2, \ldots, \mathbf{s}_n)$ and n is given by

$$f((\mathbf{s}_1, \mathbf{s}_2, \ldots, \mathbf{s}_n), n) = \begin{cases} e^{-\mu(A)}, & n = 0, \\ e^{-\mu(A)} \prod_{i=1}^n \lambda(\mathbf{s}_i)/n!, & n \geq 1. \end{cases} \tag{8.5.4}$$

Upon summing and integrating,

$$\begin{aligned} e^{-\mu(A)} &+ \sum_{n=1}^{\infty} \frac{e^{-\mu(A)}}{n!} \int_{A^n} \prod_{i=1}^n \lambda(\mathbf{s}_i)\nu(d\mathbf{s}_1) \cdots \nu(d\mathbf{s}_n) \\ &= \sum_{n=0}^{\infty} \frac{e^{-\mu(A)}}{n!} \left(\int_A \lambda(\mathbf{u})\, d\mathbf{u} \right)^n \\ &= \sum_{n=0}^{\infty} \frac{e^{-\mu(A)}(\mu(A))^n}{n!} = 1, \end{aligned}$$

as it should be.

Generating Functionals

The probability generating functional of an inhomogeneous Poisson process N on X with mean measure μ is given by

$$G_N(\xi) = \exp\left\{-\int_X (1 - \xi(\mathbf{s}))\mu(ds)\right\}. \tag{8.5.5}$$

Proof of (8.5.5) parallels that of (8.3.19) be replacing $\lambda\nu$ with μ in each step of the proof. Because $L_N(\zeta) = G_N(e^{-\zeta})$, (8.5.5) implies that the Laplace functional of N is

$$L_N(\zeta) = \exp\left\{-\int_X (1 - e^{-\zeta(\mathbf{s})})\mu(ds)\right\}. \tag{8.5.6}$$

It is easy to show that the superposition of two independent inhomogeneous Poisson processes is an inhomogeneous Poisson process: Let N_1 and N_2 be independent inhomogeneous Poisson processes with mean measures μ_1 and μ_2, respectively. Then, by the independence of N_1 and N_2 and by (8.5.5),

$$G_{N_1+N_2}(\xi) = G_{N_1}(\xi)G_{N_2}(\xi) = \exp\left\{-\int_X (1 - \xi(\mathbf{s}))(\mu_1 + \mu_2)(ds)\right\},$$

which is the probability generating functional of an inhomogeneous Poisson process with mean measure $\mu_1 + \mu_2$. The result then follows by the Equivalence Theorem given in Section 8.3.2.

Moment Measures

From (8.5.6) and (8.3.17), the second factorial moment of an inhomogeneous Poisson process N is

$$\alpha_N^{(2)}(B_1 \times B_2) = \mu(B_1) \cdot \mu(B_2). \tag{8.5.7}$$

So, from (8.3.8), the second moment measure of N is

$$\mu_N^{(2)}(B_1 \times B_2) = \mu(B_1) \cdot \mu(B_2) + \mu(B_1 \cap B_2) \tag{8.5.8}$$

and the covariance measure is

$$C_N(B_1 \times B_2) = \mu(B_1 \cap B_2). \tag{8.5.9}$$

From equations (8.5.8) and (8.3.10), the second-order intensity of N is $\lambda_2(\mathbf{s}, \mathbf{u}) = \lambda(\mathbf{s}) \cdot \lambda(\mathbf{u})$.

Simulating an Inhomogeneous Poisson Process

An inhomogeneous Poisson process in a sampling window A with mean measure μ may be simulated through Lewis and Shedler's (1979) rejection-sampling algorithm: Let $\lambda_{max} \equiv \sup_{\mathbf{s} \in A} \lambda(\mathbf{s})$, where $\lambda(\cdot)$ is the first-order intensity of the process. First, simulate a homogeneous Poisson process on A with intensity λ_{max}. Then, independently retain each event $\mathbf{s}$ of the homogeneous process with probability $\lambda(\mathbf{s})/\lambda_{max}$. The realization of the inhomogeneous Poisson process consists of the retained events.

Model Fitting

Two approaches can be taken to fitting an inhomogeneous Poisson process to spatial data. Taking a nonparametric approach, methods for multivariate density estimation can be extended to the problem of estimating $\lambda(\cdot)$ (e.g., Diggle, 1985); that is, kernel methods or nearest-neighbor methods may be used to estimate $\lambda(\cdot)$. Alternatively, consider a family of parameterized intensity functions $\{\lambda_{\boldsymbol{\theta}}(\mathbf{s}): \boldsymbol{\theta} \in \Theta\}$. The intensity function $\lambda_{\boldsymbol{\theta}}(\mathbf{s})$ may be a function of some concomitant spatial variable, of spatial location $\mathbf{s}$ alone, or of both (Cox, 1955).

Nonparametric Estimation of Intensity

Equation (8.5.2) shows that there is a close relationship between the first-order intensity of events at a spatial location $\mathbf{s}$ and the corresponding density of events in a sample region A. Therefore, it is reasonable to extend methods for multivariate density estimation to the problem of obtaining nonparametric estimators of the intensity function $\lambda(\cdot)$ on $\mathbb{R}^d$. Two types of estimators, kernel estimators and nearest-neighbor estimators, will be considered briefly; for more details see Silverman (1986).

An edge-corrected kernel estimator for $\lambda(\mathbf{s})$ is given by

$$\hat{\lambda}_h(\mathbf{s}) = \frac{1}{p_h(\mathbf{s})} \sum_{i=1}^{n} h^{-d} \kappa\left\{\frac{\mathbf{s} - \mathbf{s}_i}{h}\right\}, \qquad (8.5.10)$$

where the sum is over all n events in the sample window $A \subset \mathbb{R}^d$. The kernel function $\kappa(\cdot)$ is any probability density function symmetric about the origin. The function $p_h(\mathbf{s}) \equiv \int_A h^{-d} \kappa\{(\mathbf{s} - \mathbf{u})/h\} \nu(d\mathbf{u})$ is an edge correction (Diggle, 1985). The choice of an appropriate smoothing constant (or bandwidth) $h > 0$ is of primary concern when estimating $\lambda(\cdot)$ (Silverman, 1978a). Choice of the kernel function $\kappa(\cdot)$ is of secondary importance; any reasonable kernel gives close to optimal results (Epanechnikov, 1969). Examples of kernels are the multivariate normal density function

$$\kappa(\mathbf{u}) = (2\pi)^{-d/2} \exp(-\mathbf{u}'\mathbf{u}/2), \qquad \mathbf{u} \in \mathbb{R}^d, \qquad (8.5.11)$$

or the Epanechnikov kernel

$$\kappa(\mathbf{u}) = \begin{cases} \dfrac{\Gamma(1 + \frac{d}{2})}{2\pi^{d/2}}(d + 2)(1 - \mathbf{u}'\mathbf{u}), & \text{if } \mathbf{u}'\mathbf{u} < 1, \\ 0, & \text{if } \mathbf{u}'\mathbf{u} \geq 1. \end{cases} \quad (8.5.12)$$

The kernel $\kappa(\cdot)$ and smoothing constant h could be chosen to minimize the mean integrated squared error

$$MISE(\hat{\lambda}) = \mathrm{E}\left\{\int_A \left(\hat{\lambda}(\mathbf{s}) - \lambda(\mathbf{s})\right)^2 \nu(d\mathbf{s})\right\}. \quad (8.5.13)$$

Epanechnikov (1969) gives an equation for h_{opt}, the value of h that minimizes MISE($\hat{\lambda}$), although h_{opt} is a function of the kernel $\kappa(\cdot)$ and the intensity $\lambda(\cdot)$. But $\lambda(\cdot)$ is unknown, so it is not clear what value of h should be chosen; Diggle (1981a) recommends experimentation with a range of values.

Let $r_k(\mathbf{s})$ be the Euclidean distance from $\mathbf{s}$ to the kth nearest event and let $V_k(\mathbf{s}) \equiv (\pi^{d/2}(r_k(\mathbf{s}))^d)/(\Gamma(1 + \frac{d}{2}))$ be the volume of a sphere of radius $r_k(\mathbf{s})$. Then, a nearest-neighbor intensity estimator for $\lambda(\mathbf{s})$ is given by

$$\hat{\lambda}(\mathbf{s}) \equiv \frac{k}{V_k(\mathbf{s})} = \frac{k\Gamma(1 + \frac{d}{2})}{\pi^{d/2}(r_k(\mathbf{s}))^d} \quad (8.5.14)$$

(Silverman, 1986, p. 96). This estimator comes from expecting $\lambda(\mathbf{s})V_k(\mathbf{s})$ events in the ball of radius $r_k(\mathbf{s})$ centered at $\mathbf{s}$. Setting this number equal to k and solving for $\lambda(\mathbf{s})$, we obtain the estimator (8.5.14). An edge-corrected version of (8.5.14) can be obtained by replacing $V_k(\mathbf{s})$ with the volume of $b(\mathbf{s}, r_k(\mathbf{s})) \cap A$, where $b(\mathbf{s}, r)$ is the closed ball of radius r centered at $\mathbf{s}$. The smoothing constant k should be chosen to minimize MISE($\hat{\lambda}$). Again, the optimal value of k depends on $\lambda(\cdot)$, so that experimentation with a range of values of k can be tried.

A combined kernel and nearest-neighbor intensity estimator can also be given:

$$\hat{\lambda}(\mathbf{s}) = \frac{1}{q_k(\mathbf{s})} \sum_{i=1}^{n} r_k(\mathbf{s})^{-d} \kappa\left\{\frac{\mathbf{s} - \mathbf{s}_i}{r_k(\mathbf{s})}\right\} \quad (8.5.15)$$

(Silverman, 1986, p. 21), where $q_k(\mathbf{s}) \equiv \int_A r_k(\mathbf{s})^{-d} \kappa\{(\mathbf{s} - \mathbf{u})/(r_k(\mathbf{s}))\} \nu(d\mathbf{u})$ is an edge correction.

Parametric Estimation of Intensity

Suppose the data consist of the locations $\mathbf{s}_1, \ldots, \mathbf{s}_n$ of n events in a bounded region $A \subset X$. Then the likelihood function is given by the Janossy density (e.g., Daley and Vere-Jones, 1988, p. 122).

For the inhomogeneous Poisson process, whose intensity function belongs to the parametric family $\{\lambda_{\boldsymbol{\theta}}(\mathbf{s}): \boldsymbol{\theta} \in \Theta\}$, the likelihood function is

$$l(\boldsymbol{\theta}; A) \equiv \left\{\prod_{i=1}^{n} \lambda_{\boldsymbol{\theta}}(\mathbf{s}_i)\right\} \exp\left\{-\int_A \lambda_{\boldsymbol{\theta}}(\mathbf{u}) \nu(d\mathbf{u})\right\}, \qquad (8.5.16)$$

where $n = N(A)$ is the random number of events in A. Maximum likelihood estimators of $\boldsymbol{\theta}$ are obtained by finding $\hat{\boldsymbol{\theta}}$ that maximizes Eq. (8.5.16). Usually, this maximization problem will not admit a closed-form solution, so numerical techniques must be applied (e.g., Fletcher and Powell, 1963).

For inhomogeneous Poisson processes on $[0, T]$, Kutoyants (1984, pp. 138–141) considers the asymptotic behavior of the maximum likelihood estimator for $\boldsymbol{\theta}$ under such regularity conditions as the intensity functions $\lambda_{\boldsymbol{\theta}}(t)$ must be strictly positive for all $\boldsymbol{\theta} \in \Theta$, have sufficiently good separability for adjacent values of $\boldsymbol{\theta}$, and be continuously differentiable in $\boldsymbol{\theta}$ for all $t > 0$ (almost everywhere ν). Furthermore, if $J_T(\boldsymbol{\theta})$ is the matrix whose (i, j)th element is

$$\int_0^T \left\{\frac{\partial \lambda_{\boldsymbol{\theta}}(t)}{\partial \theta_i}\right\} \left\{\frac{\partial \lambda_{\boldsymbol{\theta}}(t)}{\partial \theta_j}\right\} \{\lambda_{\boldsymbol{\theta}}(t)\}^{-1} \, dt,$$

then $|J_T(\boldsymbol{\theta})|^{-1}$ is assumed to approach zero at a uniform rate as $T \to \infty$. Also, conditions on the smoothness and rate of increase of $(\partial \lambda_{\boldsymbol{\theta}}(\cdot)/\partial \boldsymbol{\theta})$ are required. Under these regularity conditions, the maximum likelihood estimator $\hat{\boldsymbol{\theta}}$ is consistent, asymptotically Gaussian, and asymptotically efficient as $T \to \infty$. In distribution,

$$(J_T(\boldsymbol{\theta}))^{1/2}(\hat{\boldsymbol{\theta}} - \boldsymbol{\theta}) \to \mathrm{Gau}(\mathbf{0}, I),$$

as $T \to \infty$, where I is the identity matrix. Lin'kov (1985) shows that a test based on the loglikelihood ratio,

$$\log \Lambda(\boldsymbol{\theta}_0, \boldsymbol{\theta}_1) \equiv \log l(\boldsymbol{\theta}_0; [0, T]) - \log l(\boldsymbol{\theta}_1; [0, T]),$$

is asymptotically ($T \to \infty$) most powerful for testing H_0: $\boldsymbol{\theta} = \boldsymbol{\theta}_0$ versus H_1: $\boldsymbol{\theta} = \boldsymbol{\theta}_1$. Kutoyants' results have been extended by Krickeberg (1982) and Rathbun and Cressie (1990a) to inhomogeneous Poisson processes on $\mathbb{R}^d$.

In order to ensure that the intensity function $\lambda_{\boldsymbol{\theta}}$ is nonnegative, restrictions must often be imposed on the parameter space Θ. Ogata and Katsura (1986) use a smooth convex penalty function $R(\boldsymbol{\theta})$ that takes the value zero for all $\boldsymbol{\theta} \in \Theta_+$, the region for which $\lambda_{\boldsymbol{\theta}}$ is nonnegative, but takes large values for $\boldsymbol{\theta} \notin \Theta_+$. They then minimize the function

$$G(\boldsymbol{\theta}; A) \equiv -\log l(\boldsymbol{\theta}; A) + R(\boldsymbol{\theta}). \qquad (8.5.17)$$

The nonnegativity condition on $\lambda_{\boldsymbol{\theta}}$ could also be ensured by taking $\lambda_{\boldsymbol{\theta}}(\mathbf{s}) \equiv \exp\{-\psi_{\boldsymbol{\theta}}(\mathbf{s})\}$ (e.g., Ogata and Tanemura, 1986). Then the negative loglikelihood is

$$-\log l(\boldsymbol{\theta}; A) = \sum_{i=1}^{n} \psi_{\boldsymbol{\theta}}(\mathbf{s}_i) + \int_A \exp\{-\psi_{\boldsymbol{\theta}}(\mathbf{u})\}\nu(d\mathbf{u}). \qquad (8.5.18)$$

Further restrictions on $\boldsymbol{\theta}$ may be required to ensure that $-\log l(\boldsymbol{\theta}; A)$ has at most one minimum and that the Hessian matrix with (i, j)th element $-\partial^2 \log l(\boldsymbol{\theta}; A)/\partial\theta_i\,\partial\theta_j$ is everywhere positive-definite.

Diagnostics

Having fitted a statistical model to data, diagnostic tests are needed to assess the fit of the model. Berman (1983) proposes a method for analyzing the residuals of point patterns on $\mathbb{R}_+$, the nonnegative real numbers. Suppose that the data $\{t_i\}$ are generated by the intensity function $\lambda(t)$. Consider $\Lambda(t) \equiv \int_0^t \lambda(u)\,du$, a monotonically increasing function of t. If $\tau_i \equiv \Lambda(t_i)$, then $\{t_i\}$ is transformed one-to-one into $\{\tau_i\}$, and if $\lambda(\cdot)$ is the true intensity function, the transformed process $\{\tau_i\}$ has the distribution of a homogeneous Poisson process with intensity 1 (Papangelou, 1972). Berman (1983) plots the cumulative number of points $\{\tau_i\}$ as a function of transformed time $\tau \equiv \Lambda(t)$ and applies a Kolmogorov–Smirnov test to assess the goodness-of-fit of the intensity function $\lambda(t)$. Further, let $Y_i \equiv \tau_i - \tau_{i-1} = \Lambda(t_i) - \Lambda(t_{i-1})$. Then, if $\lambda(\cdot)$ is the true intensity function, $U_i \equiv 1 - \exp(-Y_i)$, $i = 1, 2, \cdots$, are independent and identically distributed according to a uniform distribution on $[0, 1]$. A plot of U_i against U_{i-1} could be used to look for serial correlation (Berman, 1983).

This residual analysis can be extended readily to point patterns in $\mathbb{R}^d$. In $\mathbb{R}^2$, for example, let $\{(x_i, y_i)\}$ be events from an inhomogeneous Poisson process with intensity $\lambda(x, y)$, realized on the sample window $A = [0, a] \times [0, b]$. Then, for each event $(x_i, y_i) \in A$, define

$$u_i \equiv \int_0^{x_i}\int_0^b \lambda(x, y)\,dy\,dx$$

and

$$v_i \equiv \int_0^a\int_0^{y_i} \lambda(x, y)\,dy\,dx.$$

Finally, the events $\{u_i\}$ and $\{v_i\}$ are realizations of homogeneous Poisson processes in $\mathbb{R}^1$ with unit intensity, and tests to check for this can be constructed.

Further generalization can be achieved by considering only those events (x_i, y_i) in $[0, a] \times [b_1, b_2]$, where $0 \le b_1 < b_2 \le b$. Define

$$w_i \equiv \int_0^{x_i}\int_{b_1}^{b_2} \lambda(x, y)\,dy\,dx.$$

Then $\{w_i: b_1 \leq y_i \leq b_2; i = 1, 2, \ldots, N(A)\}$ is a realization of a homogeneous Poisson process in $\mathbb{R}^1$ with unit intensity.

Although inhomogeneous Poisson processes have frequently been fitted to data in the time domain (e.g., Cox and Lewis, 1966; Cox and Isham, 1980; Ogata, 1983; Ogata and Shimazaki, 1984), there are very few examples where inhomogeneous Poisson processes have been fitted to data in $\mathbb{R}^d$, $d \geq 2$: Kooijman (1979) fits a linear intensity function to ecological data, Lawson (1988) models cancer-incidence patterns with an intensity function that depends on distance and direction to a point-source of contamination, and Ogata and Katsura (1988) fit a cubic-spline intensity function to seismological data.

8.5.2* Cox Process

Cox processes (or doubly stochastic point processes) were first considered by Lundberg (1940) and Cox (1955). Informally speaking, a Cox process is an inhomogeneous Poisson process with random mean measure $M(\cdot)$. In what is to follow, a review of its properties is presented; for further details, the reader can consult Grandell (1976).

Let Υ be a collection of Radon measures on X (i.e., measures that are finite on compact sets) and let $\mathscr{M}$ be the smallest σ algebra generated by sets of the form $\{\mu \in \Upsilon: \mu(B) \leq a\}$, for all $B \in \mathscr{X}$ and all $a \in [0, \infty)$. Then, a *random measure* M on X is a measurable mapping from a probability space $(\Omega, \mathscr{A}, P)$ into $(\Upsilon, \mathscr{M})$. A random measure M induces a probability measure $\Pi_M(Y) \equiv P\{M^{-1}(Y)\}$ on $(\Upsilon, \mathscr{M})$, for all $Y \in \mathscr{M}$. Let N be a point process defined on the same probability space; that is, N is a measurable mapping of $(\Omega, \mathscr{A}, P)$ into $(\Phi, \mathscr{N})$.

Definition. The point process N is a *Cox process directed by* M if, conditional on $M = \mu$, N is an inhomogeneous Poisson process with mean measure μ. ■

Suppose N is a Cox process with directing measure M. For each $\mathbf{s} \in X$, define

$$\Lambda(\mathbf{s}) \equiv \lim_{\nu(d\mathbf{s}) \to 0} M(d\mathbf{s})/\nu(d\mathbf{s}),$$

provided the limit exists almost surely. Then Λ is the random intensity function of the Cox process N and, conditional on $\Lambda = \lambda$, N is an inhomogeneous Poisson process with intensity function $\lambda(\cdot)$.

Laplace Functional

Let $L_M(\zeta)$ be the Laplace functional of the random measure M; that is,

$$L_M(\zeta) = \int_{\Upsilon} \exp\left\{-\int_X \zeta(\mathbf{s})\mu(d\mathbf{s})\right\}\Pi_M(d\mu). \qquad (8.5.19)$$

Then, by Eq. (8.5.6), the Laplace functional of a Cox process N directed by M is

$$\begin{aligned} L_N(\zeta) &\equiv E\left[\exp\left\{-\int_X \zeta(\mathbf{s}) N(d\mathbf{s})\right\}\right] \\ &= E\left\{E\left[\exp\left\{-\int_X \zeta(\mathbf{s}) N(d\mathbf{s})\right\}\middle| M\right]\right\} \\ &= L_M(1 - e^{-\zeta}). \end{aligned} \tag{8.5.20}$$

Moment Measures

Let $\mu_M^{(k)}(B_1 \times \cdots \times B_k) = E(M(B_1) \cdots M(B_k))$ be the kth moment measure of the random measure M. From (8.5.20), it is easy to see that $\mu_M(B) = \mu_N(B)$, for all $B \in \mathscr{X}$. From (8.3.15), the second-moment measure of N is

$$\begin{aligned} \mu_N^{(2)}(B_1 \times B_2) &= \lim_{\substack{\alpha_1 \downarrow 0 \\ \alpha_2 \downarrow 0}} \frac{\partial^2}{\partial \alpha_1 \partial \alpha_2} L_M\left(1 - \exp\{-\alpha_1 I_{B_1} - \alpha_2 I_{B_2}\}\right) \\ &= \mu_M^{(2)}(B_1 \times B_2) + \mu_M(B_1 \cap B_2). \end{aligned} \tag{8.5.21}$$

Then, from (8.3.6), the covariance measure is

$$C_N(B_1 \times B_2) = \mu_M^{(2)}(B_1 \times B_2) + \mu_M(B_1 \cap B_2) - \mu_M(B_1)\mu_M(B_2). \tag{8.5.22}$$

Because $\mu_N = \mu_M$ and because of (8.5.22), the first- and second-order intensities of a Cox process are identical to the first- and second-order intensities of the directing measure M: $\lambda_N(\mathbf{s}) = \lambda_M(\mathbf{s})$ and $\lambda_{N,2}(\mathbf{s}, \mathbf{u}) = \lambda_{M,2}(\mathbf{s}, \mathbf{u}) \equiv \lambda_2(\mathbf{s}, \mathbf{u})$. If the random measure M is almost surely stationary, then $\lambda_N(\mathbf{s}) = \lambda$ and $\lambda_2(\mathbf{s}, \mathbf{u}) = \lambda_2^*(\mathbf{s} - \mathbf{u})$. If in addition M is almost surely isotropic, then $\lambda_2(\mathbf{s}, \mathbf{u}) = \lambda_2^o(\|\mathbf{s} - \mathbf{u}\|)$. Thus, from Section 8.3.5, the reduced second moment measure of a stationary isotropic Cox process on $\mathbb{R}^d$ is given by

$$K(h) = \frac{d\pi^{d/2}}{\lambda\Gamma(1 + \frac{d}{2})} \int_0^h u^{d-1} \lambda_2^o(u)\, du, \qquad h \geq 0. \tag{8.5.23}$$

Examples of Cox Processes

The simplest Cox process is the inhomogeneous Poisson process; take $M = \mu$ with probability 1. Next consider the *mixed Poisson process* in which the directing measure $M = W \cdot \mu$, where μ is a fixed Radon measure and W is a nonnegative random variable. A mixed homogeneous Poisson process is a mixed Poisson process where $M = W \cdot \nu$ and ν is Lebesgue measure.

Suppose the bounded region A can be partitioned into disjoint sets $\{A_i: i = 1, 2, \ldots, k\}$, such that the events in each set A_i are a realization of a homogeneous Poisson process with intensity λ_i, where $\lambda_1, \lambda_2, \ldots, \lambda_k$ are independent and identically-distributed random variables. Then the resulting spatial point process on A is a Cox process (Cox and Isham, 1980, pp. 71–72). This can be extended to partitions upon which inhomogeneous Poisson processes are observed, where the intensity functions are independent-and-identically-distributed nonnegative random functions.

Suppose the intensity function of a Poisson process is $\lambda_{\boldsymbol{\theta}}(\mathbf{s})$. Then, a Cox process can be obtained by taking $\boldsymbol{\theta} \in \Theta$ to be a random vector with some joint cumulative distribution function G.

Additional examples of Cox processes can be defined in the time domain. Hawkes (1971) introduced a class of *self-exciting* point processes: Suppose the intensity $\lambda(t)$ depends not only on t, but also on the past realization of the process in the interval $(0, t)$, according to

$$\lambda(t) = p(t) + \int_0^t g(t - s) N(ds), \qquad t > 0, \tag{8.5.24}$$

where the large-scale variation in $\lambda(t)$ is captured by $p(t)$, a function of time t and perhaps other concomitant variables, and the small-scale variation due to interaction with past events is captured by $g(t)$. Because $\lambda(t)$ depends on the past realization of the process, $\lambda(t)$ is a random variable. Furthermore, the probability generating functional of a self-exciting point process is identical to the probability generating functional of a Cox process (Westcott, 1971), so the self-exciting point process is a special case of a Cox process. Ogata (1981) gives a method for simulating self-exciting point processes.

Rudemo (1972) and Freed and Shepp (1982) consider a Cox process N with Markov intensity. That is, let $\{\eta_t\}$ be a Markov chain in continuous time, with a finite state space and stationary transition probabilities. Conditional on $\eta_t = k$, N is defined to be a Poisson process with intensity λ_k, $k = 1, \ldots, K$.

Model Fitting

For Cox processes, two quantities might be inferred. The first is the probability measure generating the directing measure M. However, because only a single realization of the Cox process is usually observed, one often has to settle for state estimation of the intensity. (An exception would be a Cox process thought to be generated by independent-and-identically distributed intensity functions on a partition of A.) State estimation here means prediction of $\Lambda(\mathbf{s})$.

Parametric Model Fitting

For a parameterized intensity function $\lambda_{\boldsymbol{\theta}}(\mathbf{s})$, suppose $\boldsymbol{\theta} \in \Theta$ is a random vector having some parameterized mixing density $g_{\boldsymbol{\alpha}}(\boldsymbol{\theta})$. From Eq. (8.5.16), the "density" of obtaining n events in A located at $\{\mathbf{s}_1, \mathbf{s}_2, \dots, \mathbf{s}_n\}$, conditional on $\boldsymbol{\theta}$, is

$$\left\{\prod_{i=1}^{n} \lambda_{\boldsymbol{\theta}}(\mathbf{s}_i)\right\} \exp\left\{-\int_A \lambda_{\boldsymbol{\theta}}(\mathbf{u}) \nu(d\mathbf{u})\right\}. \tag{8.5.25}$$

So, the (mixture) likelihood of $\boldsymbol{\alpha}$ is given by

$$l(\boldsymbol{\alpha}; A) = \int_{\Theta} \left\{\prod_{i=1}^{n} \lambda_{\boldsymbol{\theta}}(\mathbf{s}_i)\right\} \exp\left\{-\int_A \lambda_{\boldsymbol{\theta}}(\mathbf{u}) \nu(d\mathbf{u})\right\} g_{\boldsymbol{\alpha}}(\boldsymbol{\theta})\, d\boldsymbol{\theta}, \tag{8.5.26}$$

for *one* observation of the Cox process. A maximum likelihood estimator $\hat{\boldsymbol{\alpha}}$ is obtained by maximizing (8.5.26) over $\boldsymbol{\alpha}$. Unfortunately, asymptotic properties such as those discussed by Lindsay (1983) are not appropriate unless the region A can be partitioned into a large number of smaller regions to provide the replication needed.

State estimation for the parameterized Cox process involves prediction of $\boldsymbol{\theta}$. Conditional on $\boldsymbol{\theta}$, the likelihood of $\boldsymbol{\theta}$ is just the likelihood for an inhomogeneous Poisson process. Thus the maximum likelihood estimator $\hat{\boldsymbol{\theta}}$ is obtained by maximizing (8.5.16) over $\boldsymbol{\theta}$. This ignores the mixing mechanism, which can be taken into account using Bayesian decision theory. There, the optimal predictor of $\boldsymbol{\theta}$ minimizes the posterior expected loss. For example, for squared-error loss, $E(\boldsymbol{\theta} | \{\mathbf{s}_1, \mathbf{s}_2, \dots, \mathbf{s}_n\}, n)$ is the optimal (Bayes) predictor, although this does require exact specification of the mixing function (prior) g. Kutoyants (1984, p. 141) gives conditions under which the Bayes predictor for $\boldsymbol{\theta}$ is asymptotically efficient and Gaussian.

Nonparametric State Estimation

Diggle (1985) considers nonparametric state estimation of $\lambda(\cdot)$, the realization of $\Lambda(\cdot)$, assuming a stationary isotropic Cox process with first- and second-order intensities λ and $\lambda_2^o(\cdot)$, respectively. Recall the nonparametric kernel estimator given in Section 8.5.1:

$$\hat{\lambda}_h(\mathbf{s}) = \frac{1}{p_h(\mathbf{s})} \sum_{i=1}^{n} h^{-d} \kappa\left\{\frac{\mathbf{s} - \mathbf{s}_i}{h}\right\}, \qquad \mathbf{s} \in A. \tag{8.5.27}$$

Let the kernel $\kappa(\cdot)$ be a uniform density on the ball $b(\mathbf{0}, 1)$. Diggle shows that the mean-squared error of $\hat{\lambda}_h(\mathbf{s})$ is

$$\mathrm{MSE}\left(\hat{\lambda}_h(s)\right) = \lambda_2^o(0) + \lambda\{1 - 2\lambda K(h)\}/(2h) + \{\lambda/(2h)\}^2 \int_0^{2h} K(u)\, du \tag{8.5.28}$$

for a stationary Cox process on $\mathbb{R}$ and is

$$\mathrm{MSE}(\hat{\lambda}_h(\mathbf{s})) = \lambda_2^o(0) + \lambda\{1 - 2\lambda K(h)\}/(\pi h^2) + (\pi h^2)^{-2} \int\int \lambda_2^o(\|\mathbf{v} - \mathbf{u}\|)\nu(d\mathbf{v})\nu(d\mathbf{u}) \quad (8.5.29)$$

for a stationary Cox process on $\mathbb{R}^2$. Each integral in (8.5.29) is over the disk $b(\mathbf{0}, h)$.

Estimates of $\{\mathrm{MSE}(\hat{\lambda}_h(\mathbf{s})) - \lambda_2^o(0)\}$ can be obtained by substituting in estimates of $K(h)$ and the integral in (8.5.29). The empirical K function $\hat{K}(h)$ can be estimated by methods discussed in Section 8.4, and Berman and Diggle (1989) give an estimator of the integral. By plotting the sum of these estimated terms of (8.5.29) against h, the value of h for which $\widehat{MSE}(\hat{\lambda}_h(\mathbf{s}))$ is smallest can be used to indicate kernel width (Diggle, 1985). Diggle and Marron (1988) show further that minimizing $\widehat{MSE}(\hat{\lambda}_h(\mathbf{s}))$ is equivalent to minimizing a cross-validation score that has been used by others for density estimation (e.g., Rudemo, 1982; Bowman, 1984).

Estimation for Self-Exciting Point Processes

Consider a self-exciting point process whose conditional intensity is of the parametric form

$$\lambda_{\boldsymbol{\theta}}(t) = p_{\boldsymbol{\alpha}}(t) + \int_0^t g_{\boldsymbol{\beta}}(t - s)N(ds), \quad (8.5.30)$$

where the parameter $\boldsymbol{\theta} = (\boldsymbol{\alpha}', \boldsymbol{\beta}')'$. It can be shown that the likelihood function of a self-exciting point process is of the same form as that of the inhomogeneous Poisson process

$$l(\boldsymbol{\theta}; T) = \left\{\prod_{i=1}^{n} \lambda_{\boldsymbol{\theta}}(t_i)\right\} \exp\left\{-\int_0^T \lambda_{\boldsymbol{\theta}}(u)\, du\right\} \quad (8.5.31)$$

(e.g., Ogata, 1983). Then a maximum likelihood estimator $\hat{\boldsymbol{\theta}}$ is obtained by maximizing (8.5.31) over $\boldsymbol{\theta}$. Ogata (1978) gives conditions under which the maximum likelihood estimators are asymptotically consistent and Gaussian. Self-exciting point processes have been fitted to seismological data (Ogata, 1983, 1988; Ogata et al., 1982; Ogata and Katsura, 1986) and telephone-exchange failure data (Helvik and Swensen, 1987).

8.5.3* Poisson Cluster Process

A special case of the Poisson cluster process was introduced by Neyman (1939) to model particle counts in entomology and bacteriology, generalizing Student's (1907) Poisson model for hemocytometer counts. The model was later applied to problems in cosmology by Neyman and Scott (1958); a spatial point process N on $X = \mathbb{R}^d$ is a *Neyman–Scott process* if:

1. Parent events are realized from an inhomogeneous Poisson process with mean measure μ_p.
2. Each parent produces a random number K of offspring, realized independently and identically for each parent according to a discrete probability distribution $\{p_k: k = 0, 1, 2, \dots\}$.
3. The positions of the offspring relative to their parents are independently and identically distributed according to a d-dimensional density function $f(\cdot)$.
4. The final process is composed of the superposition of *offspring* only.

If $f(\cdot)$ is radially symmetric about $\mathbf{0}$ and if the parent process is stationary and isotropic, then the Neyman–Scott process is obviously stationary and isotropic. If $f(\cdot)$ is degenerate at $\mathbf{0}$ and $p_0 = 0$, then a compound Poisson process results (Section 8.7.3).

The more general *Poisson cluster process* is obtained by invoking condition (1), generalizing condition (2) to specify simply that each parent produces a random number of offspring, generalizing condition (3) to allow arbitrary spatial positioning of offspring, and invoking condition (4). In the subsequent text, the properties of the simpler Neyman–Scott process will be investigated.

Probability Generating Functional

The probability generating functional of a Neyman–Scott process on $X = \mathbb{R}^d$ is

$$G_N(\xi) = \exp\left\{-\int_X\left[1 - \eta\left\{\int_X \xi(\mathbf{u} + \mathbf{s})f(\mathbf{u})\nu(d\mathbf{u})\right\}\right]\mu_p(d\mathbf{s})\right\}, \quad (8.5.32)$$

where $\eta(\cdot)$ is the probability generating function of K, the number of events per cluster (Moyal, 1958, 1962; Vere-Jones, 1970). Before proving (8.5.32), first note that a Poisson cluster process N is a superposition of spatial point processes N_i, each formed by clusters around Poisson parent events located at $\mathbf{s}_i$. For a Neyman–Scott process, the $\{N_i\}$ are independent, each N_i is the superposition of a random number K of independent point processes $\{N_{i_k}\}$, and each N_{i_k} contains a single event at a location that is distributed according to the density $f(\mathbf{u}_{i_k} - \mathbf{s}_i)$. Clearly, the probability generating functional of N_{i_k} is

$$\begin{aligned} G_{N_{i_k}}(\xi) &= E\left(\xi(\mathbf{u}_{i_k})\right) \\ &= \int_X \xi(\mathbf{u})f(\mathbf{u} - \mathbf{s}_i)\nu(d\mathbf{u}) \\ &= \int_X \xi(\mathbf{u} + \mathbf{s}_i)f(\mathbf{u})\nu(d\mathbf{u}). \end{aligned} \quad (8.5.33)$$

From (8.5.33) and the independence of the N_{i_k}, the probability generating functional of N_i is

$$G_{N_i}(\xi|\mathbf{s}_i) = \sum_{k=0}^{\infty}\left\{\int_X \xi(\mathbf{u}+\mathbf{s}_i)f(\mathbf{u})\nu(d\mathbf{u})\right\}^k p_k,$$

$$= \eta\left\{\int_X \xi(\mathbf{u}+\mathbf{s}_i)f(\mathbf{u})\nu(d\mathbf{u})\right\},$$

where $\eta(\cdot)$ is the probability generating function of the random variable K, the number of events per cluster. Let $G_p(\cdot)$ be the probability generating functional of the (parent) Poisson process N_p, with mean measure μ_p. Then, the probability generating functional of a Neyman–Scott process N is

$$G_N(\xi) = E\left(\prod_{\mathbf{s}_i \in N_p} G_{N_i}(\xi|\mathbf{s}_i)\right) = G_p(\psi),$$

where $\psi(\mathbf{s}) \equiv G_{N_1}(\xi|\mathbf{s})$, $\mathbf{s} \in \mathbb{R}^d$ (Moyal, 1962). The result (8.5.32) then follows by Eq. (8.5.5).

Equivalence of Neyman–Scott and Cox Processes

Bartlett (1964) shows that certain Neyman–Scott processes and certain Cox processes can be identical. Consider a Cox process N with random intensity

$$\Lambda(\mathbf{u}) = \omega\sum_{i=1}^{\infty} f(\mathbf{u}-\mathbf{s}_i), \qquad \mathbf{u} \in \mathbb{R}^d, \tag{8.5.34}$$

where ω is a positive constant and $\mathbf{s}_1, \mathbf{s}_2, \ldots,$ are points of an inhomogeneous Poisson process with mean measure μ. Then the probability generating functional of N on $X = \mathbb{R}^d$ is

$$G_N(\xi) = E\left(\exp\left\{\int_X \log \xi(\mathbf{u})N(d\mathbf{u})\right\}\right)$$

$$= E\left\{E\left(\exp\left\{\int_X \log \xi(\mathbf{u})N(d\mathbf{u})\right\}\middle|\{\mathbf{s}_i: i=1,2,\ldots\}\right)\right\}. \tag{8.5.35}$$

Note that the conditional expectation in (8.5.35) is the probability generating functional of an inhomogeneous Poisson process with intensity $\Lambda(\cdot)$, given

$\mathbf{s}_1, \mathbf{s}_2, \ldots$. So, by Eq. (8.5.5),

$$E\left(\exp\left\{\int_X \log \xi(\mathbf{u}) N(d\mathbf{u})\right\} \Big| \{\mathbf{s}_i : i = 1, 2, \ldots\}\right)$$
$$= \exp\left\{-\omega \int_X (1 - \xi(\mathbf{u})) \sum_{i=1}^{\infty} f(\mathbf{u} - \mathbf{s}_i) \nu(d\mathbf{u})\right\}$$
$$= \prod_{i=1}^{n} \exp\left\{-\omega\left(1 - \int_X \xi(\mathbf{u} + \mathbf{s}_i) f(\mathbf{u}) \nu(d\mathbf{u})\right)\right\}. \quad (8.5.36)$$

Let

$$\psi(\mathbf{s}) \equiv \exp\left\{-\omega\left(1 - \int_X \xi(\mathbf{u} + \mathbf{s}) f(\mathbf{u}) \nu(d\mathbf{u})\right)\right\}.$$

Then, because $E(\prod_{i=1}^{\infty} \psi(\mathbf{s}_i))$ is a probability generating functional and $\mathbf{s}_1, \mathbf{s}_2, \ldots$ are events of an inhomogeneous Poisson process with mean measure μ, Eq. (8.5.5) yields

$$G_N(\xi) = \exp\left\{-\int_X \left[1 - \exp\left\{-\omega\left(1 - \int_X \xi(\mathbf{u} + \mathbf{s}) f(\mathbf{u}) \nu(d\mathbf{u})\right)\right\}\right] \mu(d\mathbf{s})\right\}. \quad (8.5.37)$$

Equation (8.5.37) is the probability generating functional of a Neyman–Scott process in which the numbers of offspring are Poisson distributed (mean ω) around their parent events according to the density function $f(\cdot)$ and whose parent process is realized from an inhomogeneous Poisson process with mean measure μ; compare (8.5.29). So, by the Equivalence Theorem (Section 8.3.2), the Cox process defined by Eq. (8.5.34) is identical to a particular Neyman–Scott process. Hence, no method of statistical analysis can distinguish between the two processes.

Although a Neyman–Scott process with a Poisson number of offspring is always identical to a Cox process, it is not known in general when a Poisson cluster process is identical to a Cox process. Bartlett (1963) gives an example of a Neyman–Scott process on $\mathbb{R}$ that is *not* a Cox process.

***K* Function and Nearest-Neighbor Distance Function**

In the following paragraphs, assume that the parent process is homogeneous Poisson with intensity ρ and that $f(\cdot)$ is radially symmetric. Hence, the Neyman–Scott process N is stationary and isotropic. The intensity of N is $\lambda = \rho E(K)$, where $E(K)$ is the average number of events per cluster. The joint density of two events $\mathbf{s}$ and $\mathbf{u}$ in the same cluster is given by $f(\mathbf{s}) \cdot f(\mathbf{u})$, and the density of the distance $h = \|\mathbf{v}\| = \|\mathbf{s} - \mathbf{u}\|$ is given by

$$f_2(h) = f_2(\|\mathbf{v}\|) = \int_{\mathbb{R}^d} f(\mathbf{w}) \cdot f(\mathbf{w} - \mathbf{v})\nu(d\mathbf{w}).$$

The expected number of ordered pairs of events within a cluster is $E(K(K - 1))$, so the second-order intensity of the Neyman–Scott process is

$$\lambda_2^o(h) = \lambda^2 + \rho E(K(K-1)) \cdot f_2(h). \tag{8.5.38}$$

By (8.3.32), the K function of N in $\mathbb{R}^d$ is

$$K(h) = \frac{\pi^{d/2}h^d}{\Gamma\left(1 + \frac{d}{2}\right)} + \frac{E(K(K-1)) \cdot F_2(h)}{\left(\rho m_K^2\right)}, \qquad h \geq 0 \tag{8.5.39}$$

(e.g., Diggle, 1983, p. 55), where $F_2(\cdot)$ is the distribution function of the distance between two events in the same cluster and $m_K \equiv E(K)$.

For example, suppose that in $\mathbb{R}^2$ the parent process is homogeneous Poisson with intensity ρ, the mean number of offspring per parent has a Poisson distribution with mean m_K, and f is a bivariate Gaussian density function with mean $\mathbf{0}$ and variance matrix $\sigma^2 I$ (i.e., circular Gaussian distribution). Then, the mean of the squared distance, to an offspring from its parent, is $2\sigma^2$. Because K is a Poisson random variable, $E(K(K - 1)) = m_K^2$. Let $\mathbf{s}_1$ and $\mathbf{s}_2$ be two arbitrary events from the same cluster with coordinates $(x_1, y_1)'$ and $(x_2, y_2)'$, respectively. Then,

$$(x_1 - x_2, y_1 - y_2)'/(2\sigma^2)^{1/2} \sim \text{Gau}(\mathbf{0}, I)$$

and so

$$z \equiv \frac{1}{2\sigma^2}\left\{(x_1 - x_2)^2 + (y_1 - y_2)^2\right\} \sim \chi_2^2.$$

Let $h \equiv (2\sigma^2 z)^{1/2} = \{(x_1 - x_2)^2 + (y_1 - y_2)^2\}^{1/2}$, be the distance between two arbitrary events in the same cluster. Then the density of h is

$$f_2(h) = (h/2\sigma^2)\exp(-h^2/4\sigma^2), \qquad h > 0.$$

Thus, by Eq. (8.5.39), the K function in $\mathbb{R}^2$ is

$$K(h; \sigma^2, \rho) = \pi h^2 + \rho^{-1}\{1 - \exp(-h^2/4\sigma^2)\}, \qquad h > 0 \tag{8.5.40}$$

(e.g., Diggle, 1983, p. 75).

Heinrich (1988) shows that the nearest-neighbor (event-to-event) distribution function (Section 8.2.6) of a stationary Neyman–Scott process in $\mathbb{R}^d$ is

$$G(r) = 1 - \{\eta'(1)\}^{-1} \exp\left\{-\rho \int_{\mathbb{R}^d} [1 - \eta\{H(b(\mathbf{s}, r))\}]\nu(d\mathbf{s})\right\}$$

$$\times \int_{\mathbb{R}^d} \eta'\{H(b(\mathbf{s}, r))\} f(\mathbf{s})\nu(d\mathbf{s}),$$

where η is the probability generating function of the number of offspring per cluster and $H(B) \equiv \int_B f(\mathbf{s})\nu(d\mathbf{s})$; see also Baudin (1981). Heinrich further shows that the empirical nearest-neighbor distribution function $\hat{G}_3(r)$ is consistent for $G(r)$ and asymptotically Gaussian as the sample window A becomes large. Kryscio and Saunders (1983) give similar results under more restrictive conditions.

Simulating a Neyman–Scott Process

The method for simulating a Neyman–Scott process on a bounded region A follows directly from its definition. First, simulate an inhomogeneous Poisson process (Section 8.5.2) with mean measure μ_p to obtain the locations of parent events. To avoid edge effects, this process should be simulated on a region $B \supset A$, so that the contribution of offspring falling in A from parent events outside A is not lost. Second, for each simulated parent event, independently generate a random number K of offspring from the discrete probability distribution $\{p_k\colon k = 0, 1, 2, \ldots\}$. Third, independently locate each offspring around its parent according to the density function $f(\cdot)$. Finally, the realization consists of only those *offspring* that fall in the region A.

Model Fitting

The explicit formula (8.5.39) of the K function under a Neyman–Scott process permits the use of readily available software to obtain least-squares estimates of its parameters. Let $K(h; \boldsymbol{\theta})$ denote a model for the K function and let $\hat{K}(h)$ be a nonparametric estimator obtained from the data. (A number of such estimators are given in Section 8.4.3.) Then, a (modified) least-squares estimator for $\boldsymbol{\theta}$ is obtained by minimizing the *ad hoc* criterion

$$D(\boldsymbol{\theta}) = \int_0^{h_0} \left\{ \left(\hat{K}(h)\right)^c - \left(K(h; \boldsymbol{\theta})\right)^c \right\}^2 dh, \qquad (8.5.41)$$

where c and h_0 are tuning constants (Diggle, 1983, p. 74). The power-transformation parameter c is used to control for heterogeneity of variance of the estimate $\hat{K}(h)$, but no attention is paid to the strong statistical dependence between $\hat{K}(h_1)$ and $\hat{K}(h_2)$. If the sampling variances and covariances of $\hat{K}(\cdot)$ were known, a weighted sum of squares or a generalized sum of squares version of (8.5.41) could be defined (cf. estimation of variogram parameters in Section 2.6.2). Diggle (1983, p. 74) suggests that $c = 1/4$ is appropriate for fitting Neyman–Scott processes. The limit of integration h_0 must be sufficiently small to keep the variance of $(\hat{K}(h))^c$ under control.

Very little is known about the statistical properties of the resulting least-squares estimator $\hat{\boldsymbol{\theta}}$, such as bias, variance, consistency, efficiency, or asymptotic distribution. This lack of distribution theory and of clear guidelines for the choice of tuning constants makes Diggle's least-squares approach difficult to work with. However, the goodness-of-fit of the empirical K function $\hat{K}(h)$ to a fitted K function $K(h; \hat{\boldsymbol{\theta}})$, where $\hat{\boldsymbol{\theta}}$ is any estimator of $\boldsymbol{\theta}$, can be assessed using a modified Cramér–von Mises test statistic

$$k_1 \equiv \int_0^{h_0} \left\{ \left(\hat{K}(h)\right)^c - \left(K(h; \hat{\boldsymbol{\theta}})\right)^c \right\}^2 dh \tag{8.5.42}$$

(Diggle, 1983, p. 77). The goodness-of-fit is determined by comparing values of k_1 to values obtained by Monte Carlo simulation of a Neyman–Scott process with parameter $\hat{\boldsymbol{\theta}}$. Suppose $R - 1$ realizations are simulated. For each realization i, $i = 2, \ldots, R$, calculate the empirical K function $\hat{K}_i(t)$ and hence the Cramér–von Mises statistic k_i. Then rank the k_is from 1 to R; let r be the rank of k_1. The significance level of the goodness-of-fit test is estimated to be $\hat{p} = (R + 1 - r)/R$. If $\hat{p}$ is small, then a poor fit of the model to the data is inferred.

From Baudin (1981) and the probability generating functional of the Neyman–Scott process, a formula for the likelihood function can be obtained. However, its intractability makes it of little use for statistical inference.

The Neyman–Scott process has been fitted to redwood-seedling data by Diggle (1978). In the following paragraphs, I shall attempt to fit such a process to the longleaf-pine data.

Longleaf-Pine Data

Figure 8.19 shows a plot of $\hat{L}(h) \equiv (\hat{K}(h)/\pi)^{1/2} - h$, versus distance h [the estimate $\hat{K}_4(h)$ given by (8.2.21) was used]. This figure suggests that the longleaf pines are clustered; hence, it seems plausible to fit a Neyman–Scott process to the empirical K function. The particular model that resulted in Eq. (8.5.40) was fitted. Recall that for this model, clusters contain a Poisson number of events and the distribution of events around homogeneous Poisson cluster centers is circular Gaussian. From (8.5.41), least-squares estimates of $\boldsymbol{\theta} = (\sigma^2, \rho)'$ were obtained using the Gauss–Newton algorithm with tuning constants $c = 0.25$ and $h_0 = 50$ m. The least-squares estimates of (σ^2, ρ) are $(\hat{\sigma}^2, \hat{\rho}) = (15.3\ \text{m}^2, 0.00252\ \text{m}^{-2})$; however, standard-error expressions are unknown. The value $\hat{\rho} = 0.00252\ \text{m}^{-2}$ corresponds to an average of 101 clusters in 4 ha of forest and an average of 5.8 events per cluster. The mean-squared distance of offspring from their parent events is $2\hat{\sigma}^2 = 30.6$

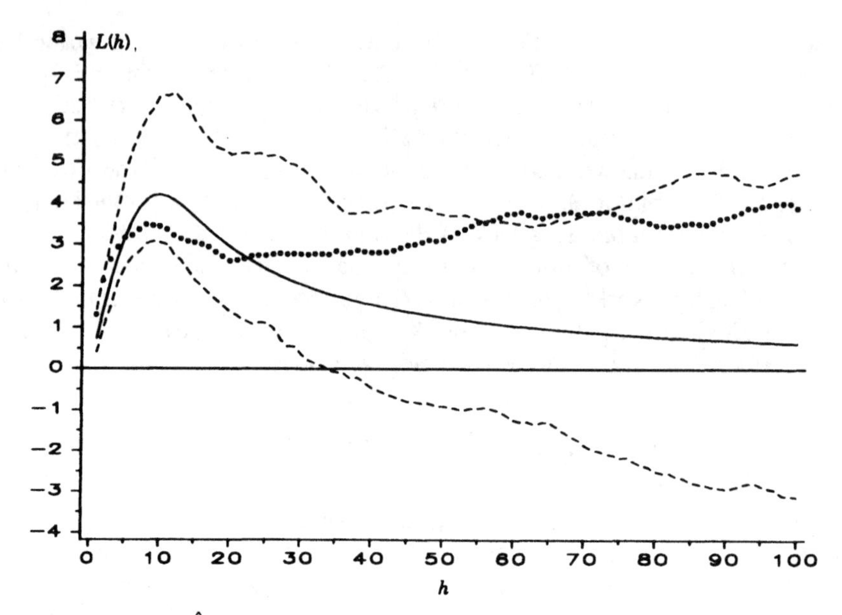

Figure 8.19 Plot of $\hat{L}_4(h)$ versus distance h (circles) and plot of $L(h; \hat{\sigma}^2, \hat{\rho})$ versus h (solid line), for the fitted Neyman–Scott process. Upper and lower envelopes from 99 simulations of the fitted process (dashed lines) are also shown. Units on the horizontal axis and the vertical axis are in meters.

m^2. In Figure 8.19, $\hat{L}_4(h)$ is compared to $L(h; \hat{\sigma}^2, \hat{\rho}) \equiv (K(h; \hat{\sigma}^2, \hat{\rho})/\pi)^{1/2} - h$, where $K(h; \sigma^2, \rho)$ is given by Eq. (8.5.40). In addition, Figure 8.19 gives simulation envelopes for 99 realizations of a Neyman–Scott process with parameters $(\hat{\sigma}^2, \hat{\rho})$. This figure indicates that the Neyman–Scott model provides a poor fit to the data. The empirical K function exceeds the upper simulation envelope for h between 57 and 73 m. For the fitted model, $L(h)$ increases monotonically to a peak at approximately 11 m, then decreases thereafter, asymptotically approaching zero with increasing distance h. In contrast, $\hat{L}_4(h)$ increases to a peak at 8 m, but then fluctuates erratically thereafter and does not appear to approach zero. This lack of fit is also confirmed by the Cramér–von Mises statistic k_1 [Eq. (8.5.42)]. The empirical value $k_1 = 1.72$ exceeds all but two of the simulated values, indicating that departure from the Neyman–Scott process is significant at the 5% level ($\hat{p} = 0.03$).

The failure of the Neyman–Scott process to provide an adequate fit does not result from a poor choice for the distribution of number of offspring per parent or the dispersion of offspring around parents. For a Neyman–Scott process in $\mathbb{R}^2$, (8.5.39) implies

$$L(h) = h \cdot \left\{ 1 + \frac{E(K(K-1)) \cdot F_2(h)}{\pi \rho m_K^2 h^2} \right\}^{1/2} - h. \qquad (8.5.43)$$

Now $F_2(h)$ is bounded between 0 and 1, so the term inside the square root approaches 1 as h increases. Necessarily then, $L(h) \to 0$ as $h \to \infty$, regardless of the choice of parameters.

Extension to Multiple Generations

Felsenstein (1975) extends the Poisson cluster process to multiple discrete generations of events. Events in generation k are assumed to be produced by a Poisson cluster process from parents in generation $k - 1$. Felsenstein assumes that the number of offspring of each parent is a Poisson random variable, but in fact any discrete distribution function may be used.

Consider the following simple example. Let $\mathbf{s}_1, \mathbf{s}_2, \ldots$ be the locations of events in generation $k - 1$. Assume that each event in generation $k - 1$ independently produces a Poisson number of offspring with mean 1, and that these offspring are independently distributed around their parents according to a multivariate density function f. Then, the events of generation k are realized from a Cox process with random intensity

$$\Lambda_k(\mathbf{u}) \equiv \sum_{i=1}^{\infty} f(\mathbf{u} - \mathbf{s}_i). \tag{8.5.44}$$

Felsenstein shows, through Monte Carlo simulation, that as k grows, clusters become larger and more widely separated. Thus, Felsenstein's model is not satisfactory for any biological system in equilibrium.

Kingman (1977) modifies Felsenstein's model to allow the mean number of offspring m_k per parent to depend on the generation k and, say, the spatial location of the parent. (Thus, parents in crowded regions may produce a smaller mean number of offspring than parents in sparse regions.) Then, the events of generation k are realized from a Cox process with random intensity

$$\Lambda_k(\mathbf{u}) \equiv \sum_{i=1}^{\infty} m_k(\mathbf{s}_i) f(\mathbf{u} - \mathbf{s}_i). \tag{8.5.45}$$

8.5.4* Simple Inhibition Point Processes

Apart from the trivial case of regular lattices in $\mathbb{R}^d$, the simplest class of point-process models whose realizations exhibit regularity is the class of hard-core models. For such models, a minimum permissible distance δ is imposed; no two events may be located within distance δ of each other. Matern (1960) was the first to describe formally the hard-core models. For point processes on $\mathbb{R}^d$, he made two specific suggestions:

Model I (Matern, 1960, pp. 47–48)

Let N_0 be a homogeneous Poisson process on $\mathbb{R}^d$ with intensity ρ. Model I is formed by deleting all pairs of events of N_0 that are separated by a distance of less than δ. The remaining events form the (more regular) spatial point process N_{I}. The probability that an arbitrary event of N_0 is retained is $\exp\{-\rho\omega_d\delta^d\}$, where $\omega_d = \pi^{d/2}/\Gamma(1 + \frac{d}{2})$ is the volume of the unit sphere. So, the intensity of N_{I} is

$$\lambda = \rho \exp\{-\rho\omega_d\delta^d\}. \tag{8.5.46}$$

The probability that events located at $\mathbf{s}$ and $\mathbf{u}$ are both retained in N_{I} is

$$k(\|\mathbf{s} - \mathbf{u}\|) \equiv \begin{cases} \exp\{-\rho U_\delta(\|\mathbf{s} - \mathbf{u}\|)\}, & \|\mathbf{s} - \mathbf{u}\| \geq \delta, \\ 0, & \|\mathbf{s} - \mathbf{u}\| < \delta, \end{cases} \tag{8.5.47}$$

where $U_\delta(h)$ is the volume of the union of two solid spheres of radius δ separated by distance h. The second-order intensity of N_{I} is

$$\lambda_2^o(h) = \rho^2 k(h) \tag{8.5.48}$$

(Diggle, 1983, p. 61).

Model II (Matern, 1960, p. 48)

Let N_0 be a homogeneous Poisson process on $\mathbb{R}^d$ with intensity ρ. Independently mark the events $\mathbf{s}$ of N_0 with numbers $Z(\mathbf{s})$ from any absolutely continuous distribution function F. An event $\mathbf{s}$ of N_0 with mark $Z(\mathbf{s})$ is deleted if there exists another event $\mathbf{u}$ with $\|\mathbf{s} - \mathbf{u}\| < \delta$ and $Z(\mathbf{u}) < Z(\mathbf{s})$. The retained events form the (more regular) spatial point process N_{II}. Stoyan and Stoyan (1985) show that the intensity and second-order intensity of N_{II} do not depend on F. The probability that an arbitrary event of N_0 is retained is $\{1 - \exp(-\rho\omega_d\delta^d)\}/(\rho\omega_d\delta^d)$, so that the intensity of N_{II} is

$$\lambda = \frac{\{1 - \exp(-\rho\omega_d\delta^d)\}}{\omega_d\delta^d}. \tag{8.5.49}$$

The probability that events at $\mathbf{s}$ and $\mathbf{u}$ are both retained in N_{II} is

$$k(\|\mathbf{s} - \mathbf{u}\|) = \begin{cases} \dfrac{2U_\delta(\|\mathbf{s} - \mathbf{u}\|)\{1 - \exp(-\rho\omega_d\delta^d)\} \\ \qquad -2\omega_d\delta^d\{1 - \exp(-\rho U_\delta(\|\mathbf{s} - \mathbf{u}\|))\}}{\rho^2\omega_d\delta^d U_\delta(\|\mathbf{s} - \mathbf{u}\|)\{U_\delta(\|\mathbf{s} - \mathbf{u}\|) - \omega_d\delta^d\}}, & \|\mathbf{s} - \mathbf{u}\| \geq \delta, \\ 0, \quad \|\mathbf{s} - \mathbf{u}\| < \delta. \end{cases} \tag{8.5.50}$$

Then the second-order intensity of N_{II} is given by

$$\lambda_2^o(h) = \rho^2 k(h). \tag{8.5.51}$$

Matern–Stoyan Model (Stoyan and Stoyan, 1985)
The following extension of Model II has a variable hard core. Let N_0 be a Poisson process on $\mathbb{R}^d$ with intensity ρ. Independently mark the events $\mathbf{s}$ of N_0 with independent marks $Z(\mathbf{s})$ and $R(\mathbf{s})$, where the marks $Z(\mathbf{s})$ are from an absolutely continuous distribution function F and the *positive* marks $R(\mathbf{s})$ are from a distribution function G. An event $\mathbf{s}$ with marks $Z(\mathbf{s})$ and $R(\mathbf{s})$ is deleted if there exists another event $\mathbf{u}$ such that $\|\mathbf{s} - \mathbf{u}\| < R(\mathbf{s})$ and $Z(\mathbf{u}) < Z(\mathbf{s})$. The retained events form the spatial point process N_{MS}. Stoyan and Stoyan (1985) show that the intensity of N_{MS} is

$$\lambda = \int_0^\infty \frac{\{1 - \exp(-\rho\omega_d r^d)\}}{\omega_d r^d} G(dr) \tag{8.5.52}$$

and the second-order intensity is

$$\lambda_2^o(h) = \rho^2 k(h), \tag{8.5.53}$$

where

$$\begin{aligned} k(h) \equiv &\int_{\{r+t \ge h,\, h>r,\, h>t\}} \{A(r,t,h) + A(t,r,h)\} G(dr)G(dt) \\ &+ \int_{\{r+t\ge h,\, h>r,\, h\le t\}} A(r,t,h)G(dr)G(dt) \\ &+ \int_{\{r+t\ge h,\, h\le r,\, h>t\}} A(t,r,h)G(dr)G(dt) \\ &+ \int_{\{r+t<h\}} \frac{1-\exp(-\rho\omega_d r^d)}{\rho\omega_d r^d} \frac{1-\exp(-\rho\omega_d t^d)}{\rho\omega_d t^d} G(dr)G(dt), \end{aligned}$$

$$\begin{aligned} A(r,t,h) \equiv &\frac{\{\rho\omega_d r^d + b(r,t,h)\}^{-1} - \{b(r,t,h)\exp(\rho\omega_d r^d)\}^{-1}}{\rho\omega_d r^d} \\ &+ \left[b(r,t,h)\{\rho\omega_d r^d + b(r,t,h)\}\exp\{\rho\omega_d r^d + b(r,t,h)\}\right]^{-1}, \end{aligned}$$

$b(r,t.h) \equiv \rho\nu(b(\mathbf{0},t) \setminus b(\mathbf{h},r))$, and $B_1 \setminus B_2 \equiv B_1 \cap B_2^{\mathscr{C}}$; $\mathbf{h}$ is a point in $\mathbb{R}^d$ at distance h from the origin.

Matern–Bartlett Model (Bartlett, 1974; Stoyan, 1988)
Another generalization of Model II can be defined as follows. Let N_0 be a Poisson process on $\mathbb{R}^d$ with intensity ρ. Independently mark the events $\mathbf{s}$ of N_0 with marks $Z(\mathbf{s})$ from a uniform distribution on [0, 1]. The thinning of N_0 is controlled by the marks and a positive nondecreasing continuous function f: $[0, 1] \to [0, \infty)$. An event $\mathbf{s}$ with mark $Z(\mathbf{s})$ is deleted from N_0 if there exists

another event $\mathbf{u}$ with mark $Z(\mathbf{u})$ such that

$$(\mathbf{u}, Z(\mathbf{u})) \in M(\mathbf{s}, Z(\mathbf{s})) \equiv \{(\boldsymbol{\xi}, \zeta): \|\boldsymbol{\xi} - \mathbf{s}\| < f(\zeta), 0 \leq \zeta < Z(\mathbf{s})\}. \tag{8.5.54}$$

The spatial point process N_{MB} consists of the retained events. The intensity of N_{MB} is

$$\lambda = \rho \int_0^1 \exp(-\rho V(r))\, dr \tag{8.5.55}$$

and the second-order intensity is

$$\lambda_2^o(h) = 2\rho^2 \int_0^1 \int_{T(h,r)} \exp(-\rho V(h,r,t))\, dt\, dr. \tag{8.5.56}$$

Here, $V(r)$ is the volume of $M(\mathbf{0}, r)$, $V(h, r, t)$ is the volume of $M(\mathbf{0}, r) \cup M(\mathbf{h}, t)$, $\mathbf{h}$ is a point in $\mathbb{R}^d$ distance h from the origin, and $T(h, r)$ is the set of all real numbers t in $[0, r)$ with $(\mathbf{h}, t) \notin M(\mathbf{0}, r)$ (see Stoyan, 1988).

For Model II and its generalizations, the events of N_0 are ranked by independent marks $Z(\cdot)$. Interaction between two nearby events comes from the deletion of the event with the higher mark.

Other Simple Inhibition Point Processes

Diggle et al. (1976) consider a "simple sequential inhibition" process under which disks of constant radius δ are placed sequentially over a finite region A. At each stage, the next disk center is chosen at random from a uniform distribution over those remaining points in A for which no two disks overlap. The procedure terminates when a prespecified number of disks has been placed, or it is impossible to continue.

In fact, Pielou (1960) had already considered a sequential inhibition point process, but where the disk diameters were allowed to vary. For Pielou's model, events are placed sequentially in the finite region A, as in Diggle et al.'s model, but at each stage the disk is made as large as possible, subject to the restrictions that it not overlap any previously assigned disk and that its radius lies between certain preassigned limits. Pielou found that her model can result in clustering as well as regularity.

These spatial-point-process models reflect the competitive advantage that early-arriving events may have over later arrivals.

Model Fitting

Least-squares estimators of the parameters $\boldsymbol{\theta} = (\rho, \delta)'$ of Matern's models and their extensions can be obtained by fitting the empirical K function

$\hat{K}(h)$ to the K function under the model $K(h; \boldsymbol{\theta})$. Recall

$$K(h) = \frac{d\pi^{d/2}}{\lambda^2\Gamma(1 + \frac{d}{2})} \int_0^h u^{d-1}\lambda_2^o(u)\, du; \qquad (8.5.57)$$

see Section 8.3.5. Because $\lambda_2^o(h)$ is known for each of these models, $K(h; \boldsymbol{\theta})$ can be obtained by applying Eq. (8.5.57). Then, the least-squares estimator for $\boldsymbol{\theta}$ is obtained by finding the $\hat{\boldsymbol{\theta}}$ that minimizes the *ad hoc* criterion

$$D(\boldsymbol{\theta}) \equiv \int_0^{h_0} \left\{ \left(\hat{K}(h)\right)^c - \left(K(h; \boldsymbol{\theta})\right)^c \right\}^2 dh. \qquad (8.5.58)$$

Diggle (1983, p. 74) suggests that $c = 1/2$ is appropriate for these hard-core models. Note that for Matern's hard-core models, the minimum distance between two events d_1 is a sensible estimator for δ (Ripley and Silverman, 1978). As in the case of the Neyman–Scott process (Section 8.5.3), very little is known about the statistical properties (e.g., bias, variance, consistency, etc.) of the least-squares estimator, making it difficult to work with in practice. Monte Carlo methods can be used to assess goodness-of-fit.

8.5.5* Markov Point Process

A more flexible framework for modeling inhibition processes is provided by the Markov point process. The Markov point process, here on a bounded set $A \subset \mathbb{R}^d$, is a natural extension of the Markov notion in time (viz., given the present, the past and future are independent). Extension to countably infinite point processes on $\mathbb{R}^d$ is considered by Preston (1974, 1976) and Daley and Vere-Jones (1988, Chapter 14).

Definition (Ripley and Kelly, 1977). A spatial point process is defined to be *Markov of range* ρ if the conditional intensity at $\mathbf{s}$, given the realization of the process in $A - \{\mathbf{s}\}$, depends only on the events in $b(\mathbf{s}, \rho) - \{\mathbf{s}\}$, where $b(\mathbf{s}, \rho)$ is the closed ball of radius ρ centered at $\mathbf{s}$. ■

A brief presentation of theory, models, inference, and simulation for Markov point processes is given in the subsequent text. The reader who would like to go beyond the treatment given in this section can consult Spitzer (1971), Preston (1974, 1976), Berger (1988), and Daley and Vere-Jones (1988, Chapter 5).

Markov Point Processes: Some Theory

Let $(A, \mathscr{A}, \nu)$ be a measure space, where A is a bounded subset of $\mathbb{R}^d$ and ν is the Lebesgue measure, and let $(A_e, \mathscr{A}_e, \Pi)$ be the corresponding exponential space (described in Section 8.3). Suppose the measure Π is that of a

Poisson process with mean measure ν; that is, a Poisson process with unit intensity. This intensity of one event per unit d-dimensional volume appears (invisibly) in many of the Markov–point-process formulas that follow. In particular, its presence resolves apparent dimensional inconsistencies in formulas like (8.5.69) and (8.5.84). Let each point process on A be defined by a sequence of intensity functions (Radon–Nikodym derivatives with respect to the product measures ν^n, $n = 1, 2, \ldots$), which are $\mathscr{A}_e$-measurable functions $\{\lambda_n(\mathbf{s}_1, \mathbf{s}_2, \ldots, \mathbf{s}_n)\}$ that do not depend on the order of the $\mathbf{s}_i$s. The sequence of intensity functions must satisfy $\sum_{n=0}^{\infty} p_n = 1$, where

$$p_n \equiv \begin{cases} e^{-\nu(A)}, & n = 0, \\ \dfrac{e^{-\nu(A)}}{n!} \displaystyle\int_{A^n} \lambda_n(\mathbf{s}_1, \mathbf{s}_2, \ldots, \mathbf{s}_n)\nu(d\mathbf{s}_1)\nu(d\mathbf{s}_2) \cdots \nu(d\mathbf{s}_n), & \\ & n = 1, 2, \ldots \end{cases} \tag{8.5.59}$$

is $\Pr(N(A) = n)$ (Kelly and Ripley, 1976). Under this model, the point process will contain n events with probability p_n and, conditional on n, the ordered n tuple $(\mathbf{s}_1, \mathbf{s}_2, \ldots, \mathbf{s}_n)$ is distributed on A^n with probability density proportional to $\lambda_n(\mathbf{s}_1, \mathbf{s}_2, \ldots, \mathbf{s}_n)$.

The joint "density" function $f((\mathbf{s}_1, \mathbf{s}_2, \ldots, \mathbf{s}_n), n) = e^{-\nu(A)}\lambda_n(\mathbf{s}_1, \ldots, \mathbf{s}_n)/n!$ can be factored uniquely as

$$\begin{aligned} f((\mathbf{s}_1, \mathbf{s}_2, \ldots, \mathbf{s}_n), n) &= \frac{e^{-\nu(A)}}{\alpha n!} \exp\Bigg\{ \sum_{i=1}^{n} g_1(\mathbf{s}_i) + \sum\sum_{1 \le i < j \le n} g_{1,2}(\mathbf{s}_i, \mathbf{s}_j) \\ &\quad + \cdots + g_{1,2,\ldots,n}(\mathbf{s}_1, \mathbf{s}_2, \ldots, \mathbf{s}_n) \Bigg\}, \qquad n = 1, 2, \ldots \end{aligned} \tag{8.5.60}$$

(e.g., Daley and Vere-Jones, 1988, p. 129; a similar factorization is given in Section 6.4). The normalizing constant α is chosen so that

$$\sum_{n=0}^{\infty} \int_{A^n} f((\mathbf{s}_1, \mathbf{s}_2, \ldots, \mathbf{s}_n), n)\nu(d\mathbf{s}_1)\nu(d\mathbf{s}_2) \cdots \nu(d\mathbf{s}_n) = 1,$$

but it usually does not have a closed-form expression. The Janossy density (e.g., Daley and Vere-Jones, 1988, p. 122), for which the order of the events in A is unimportant (and which is the likelihood), is given by

$$j_n(\{\mathbf{s}_1, \ldots, \mathbf{s}_n\}) \equiv \sum_{\pi} f\big((\mathbf{s}_{\pi(1)}, \ldots, \mathbf{s}_{\pi(n)}), n\big) = n! f((\mathbf{s}_1, \ldots, \mathbf{s}_n), n).$$

Here $\pi(\cdot)$ is a permutation over $\{1, 2, \ldots, n\}$ and $\sum_{\pi}$ denotes the sum over all such permutations.

The point process defined by (8.5.60) is called a *Gibbs process*. Gibbs processes were originally introduced into physics by Gibbs (1902) and have a long history in statistical mechanics (see Rowlinson, 1959; Ruelle, 1969; Preston, 1974, 1976). In its most general form, a Gibbs process is nothing more than a joint probability density written as the exponent of a negative potential energy function and normalized (by a partition function) so that it integrates to 1. Adding a nearest-neighbor condition makes the process Markov. The nearest-neighbor condition refers to the graph structure imposed on the sites that ensures nonclique components of the potential energy function are 0 (see, e.g., Berger, 1988). The Hammersley–Clifford Theorem, proved in Section 6.4, justifies the descriptor "Markov" for point processes defined in this way. (Some authors refer to the Markov point process as a nearest-neighbor Gibbs process or, misleadingly, simply as a Gibbs process.)

Henceforth in this section it will be assumed that the preceding graph structure is defined in terms of Euclidean distance (although more general distances could be chosen). Define $\mathbf{s}_i, \mathbf{s}_j \in A$ to be *neighbors* if $\|\mathbf{s}_i - \mathbf{s}_j\| \le \rho$, for some $\rho > 0$. A *clique* is defined to be a single event or a set of events, all of which are neighbors of each other. By the Hammersley–Clifford Theorem, $g_{1,2,\ldots,k}(\mathbf{s}_{i_1}, \mathbf{s}_{i_2}, \ldots, \mathbf{s}_{i_k}) = 0$, unless the events $\mathbf{s}_{i_1}, \mathbf{s}_{i_2}, \ldots, \mathbf{s}_{i_k}$ form a clique. Further, the point process defined by (8.5.60) is Markov of range ρ.

Baddeley and Moller (1989) extend Markov point processes to allow the neighborhoods to depend on the realization of the process. For example, suppose that the events $\mathbf{s}_1, \ldots, \mathbf{s}_n$ in A are used to form Voronoi polygons (see Delauney triangulation, Section 5.9.2). Then define $\mathbf{s}_i$ and $\mathbf{s}_j$ to be neighbors if their Voronoi polygons adjoin (Ord, 1977).

Strauss Process

Strauss (1975a) considers a Markov point process under which the density $f(\cdot)$, given by (8.5.60), depends only on the number of neighbor pairs, defined by

$$Y_n(\mathbf{s}_1, \mathbf{s}_2, \ldots, \mathbf{s}_n) \equiv \sum_{1 \le i < j \le n}\sum I\big(\|\mathbf{s}_i - \mathbf{s}_j\| \le \rho\big). \qquad (8.5.61)$$

Let $T(\mathbf{s}_n; \mathbf{s}_1, \mathbf{s}_2, \ldots, \mathbf{s}_{n-1})$ be the number of neighbors of $\mathbf{s}_n$ in the set $\{\mathbf{s}_1, \mathbf{s}_2, \ldots, \mathbf{s}_{n-1}\}$. Then

$$T_n(\mathbf{s}_n; \mathbf{s}_1, \mathbf{s}_2, \ldots, \mathbf{s}_{n-1}) = Y_n(\mathbf{s}_1, \mathbf{s}_2, \ldots, \mathbf{s}_n) - Y_{n-1}(\mathbf{s}_1, \mathbf{s}_2, \ldots, \mathbf{s}_{n-1}).$$

Further, suppose that the Papangelou intensity (e.g., Daley and Vere-Jones, 1988, p. 575) of $\mathbf{s}_n$, conditional on $\mathbf{s}_1, \mathbf{s}_2, \ldots, \mathbf{s}_{n-1}$, is a function of T_n alone; that is,

$$j_n(\{\mathbf{s}_1, \ldots, \mathbf{s}_n\}) = j_{n-1}(\{\mathbf{s}_1, \ldots, \mathbf{s}_{n-1}\}) \cdot g(T_n(\mathbf{s}_n; \mathbf{s}_1, \mathbf{s}_2, \ldots, \mathbf{s}_{n-1}))$$

for some function g and Janossy densities $\{j_n\}$. Then, the Janossy density [i.e., the likelihood $l(\boldsymbol{\theta})$, where $\boldsymbol{\theta} = (\beta, \gamma)'$] for the Strauss process is shown by Kelly and Ripley (1976) to take the form

$$l(\boldsymbol{\theta}) \equiv j_n(\{\mathbf{s}_1,\ldots,\mathbf{s}_n\}) = e^{-\nu(A)} \cdot \alpha^{-1} \cdot \beta^n \cdot \gamma^{Y_n} \qquad (8.5.62)$$

for nonnegative parameters β and γ and normalizing constant α. Comparing (8.5.62) with (8.5.60), we see that for a Strauss process, $g_1(\mathbf{s}_i) \equiv \log \beta$,

$$g_{1,2}(\mathbf{s}_i,\mathbf{s}_j) \equiv \begin{cases} \log \gamma, & \text{if } \|\mathbf{s}_i - \mathbf{s}_j\| \le \rho, \\ 0, & \text{if } \|\mathbf{s}_i - \mathbf{s}_j\| > \rho, \end{cases}$$

and higher-order terms are identically equal to 0.

The case $\gamma = 1$ corresponds to a Poisson process with intensity β and $\gamma < 1$ corresponds to regularity of events. If $\gamma = 0$, the result is a simple inhibition process that contains no events at a distance less than or equal to ρ. If $\gamma > 1$, then intuitively the process should result in clustering, but for such values of γ the process is unstable; that is, there is no normalizing constant α in (8.5.62) for which a density can be obtained (Kelly and Ripley, 1976).

One way to form a Strauss cluster process is to condition the process on n, the observed number of events in A (Kelly and Ripley, 1976). Then this conditional process is stable. Another approach is to replace the assumption that the Papangelou intensity of $\mathbf{s}_n$, given the remaining events, depends only on T_n, with the assumption that it depends on n as well as T_n. As in the Strauss process, assume $f(\cdot)$, given by (8.5.60), depends only on Y_n. Then, the Janossy density (i.e., the likelihood) is

$$l(\boldsymbol{\theta}) \equiv j_n(\{\mathbf{s}_1,\ldots,\mathbf{s}_n\}) = e^{-\nu(A)} \cdot \alpha^{-1} \cdot \beta^n \cdot \gamma^{Y_n} \cdot \psi(n), \qquad (8.5.63)$$

where a formula for ψ is given by Kelly and Ripley (1976). As before, $\gamma < 1$ indicates regularity of events, whereas $\gamma > 1$ indicates clustering.

Pair-Potential Markov Point Process

Frequently, the joint density of a Markov point process is expressed in terms of *pair-potential functions* Ψ. Define $g_{1,2}(\mathbf{s}_i,\mathbf{s}_j) \equiv -\Psi(\|\mathbf{s}_i - \mathbf{s}_j\|)$, a function of the distance between events $\mathbf{s}_i$ and $\mathbf{s}_j$, $\mathbf{s}_i \neq \mathbf{s}_j$. Also assume that there are no large-scale effects ($g_1(\mathbf{s}_i) \equiv 0$) and no higher-order interactions. Then, the number of events and the ordered n tuple of their locations are jointly distributed according to the Gibbs grand canonical distribution

$$f((\mathbf{s}_1,\mathbf{s}_2,\ldots,\mathbf{s}_n),n) = \frac{e^{-\nu(A)}}{\alpha n!} \exp\left\{-\sum_{1\le i<j\le n}\sum \Psi(\|\mathbf{s}_i - \mathbf{s}_j\|)\right\}, \qquad (8.5.64)$$

where the normalizing constant α is chosen so that

$$\sum_{n=0}^{\infty} \int_{A^n} f((\mathbf{s}_1,\mathbf{s}_2,\ldots,\mathbf{s}_n),n)\nu(d\mathbf{s}_1)\nu(d\mathbf{s}_2)\cdots\nu(d\mathbf{s}_n) = 1. \qquad (8.5.65)$$

For distances h such that $\Psi(h) > 0$, the model shows inhibition among events, whereas for h such that $\Psi(h) < 0$, the model shows attraction among events. Thus, the pair-potential Markov point process can, in principle, be used to model clustering as well as regularity. The Strauss process defined by (8.5.62) is a special case of (8.5.64), where

$$\Psi(h) = \begin{cases} -\log \gamma, & \text{if } 0 < h \le \rho, \\ 0, & \text{if } h > \rho. \end{cases}$$

Notice that the Gibbs grand canonical distribution has a form similar to that of the autologistic model in Section 6.5. Besag et al. (1982) show that, provided Ψ is continuous except on a set of Lebesgue measure zero, the grand canonical distribution can be obtained as a limit of autologistic processes or auto-Poisson processes (see Section 6.5) on successively finer and finer lattices.

Total Potential Energy and Local Energy

Consider the Markov point process with joint density written in the form of (8.5.60). Then, the *total potential energy* is defined as

$$U_n(\mathbf{s}_1, \ldots, \mathbf{s}_n) \equiv -\left\{\sum_{i=1}^{n} g_1(\mathbf{s}_i) + \sum_{1 \le i < j \le n}\sum g_{1,2}(\mathbf{s}_i, \mathbf{s}_j) + \cdots + g_{1,2,\ldots,n}(\mathbf{s}_1, \ldots, \mathbf{s}_n)\right\}. \quad (8.5.66)$$

Thus, $j_n(\{\cdot\}) \propto \exp\{-U_n(\cdot)\}$. The *local energy* at a point $\mathbf{u} \in \mathbb{R}^d$ is defined as

$$e(\mathbf{u}; \mathbf{s}_1, \ldots, \mathbf{s}_n) \equiv -\left\{g_1(\mathbf{u}) + \sum_{i=1}^{n} g_{1,2}(\mathbf{u}, \mathbf{s}_i) + \sum_{1 \le i < j \le n}\sum g_{1,2,3}(\mathbf{u}, \mathbf{s}_i, \mathbf{s}_j) + \cdots + g_{1,2,\ldots,n+1}(\mathbf{u}, \mathbf{s}_1, \ldots, \mathbf{s}_n)\right\}. \quad (8.5.67)$$

The quantity e can be interpreted as the energy required to add an event $\mathbf{u}$ to the point pattern $\mathbf{s}_1, \ldots, \mathbf{s}_n$ and $\exp(-e)$ is the Radon–Nikodym derivative of the reduced Campbell measure with respect to the product measure $\nu \times \Pi_N$.

The Markov point process is said to be *stable* if a normalizing constant $0 < \alpha < \infty$ in (8.5.60) can be found. A sufficient condition is

$$U_n(\mathbf{s}_1, \mathbf{s}_2, \ldots, \mathbf{s}_n) \ge -cn, \quad (8.5.68)$$

for all $n \ge 0$, all $\mathbf{s}_1, \mathbf{s}_2, \ldots, \mathbf{s}_n \in \mathbb{R}^d$, and some constant $0 \le c < \infty$ (Ruelle,

1969, p. 33). To see this, assume (8.5.68) and compute

$$\sum_{n=0}^{\infty} \frac{e^{-\nu(A)}}{n!} \int_{A^n} \exp\{-U_n(\mathbf{s}_1, \ldots, \mathbf{s}_n)\} \nu(d\mathbf{s}_1) \cdots \nu(d\mathbf{s}_n)$$

$$\leq \sum_{n=0}^{\infty} \frac{e^{-\nu(A)}}{n!} e^{cn} (\nu(A))^n = \exp\{\nu(A)(e^c - 1)\} < \infty.$$

For the special case of pair-potential Markov point processes, $U_n(\mathbf{s}_1, \ldots, \mathbf{s}_n) = \sum\sum_{1 \leq i < j \leq n} \Psi(\|\mathbf{s}_i - \mathbf{s}_j\|)$ and $e_n(\mathbf{u}; \mathbf{s}_1, \ldots, \mathbf{s}_n) = \sum_{i=1}^{n} \Psi(\|\mathbf{u} - \mathbf{s}_i\|)$. Furthermore, a strictly positive pair-potential Ψ (corresponding to regularity of events) always yields a stable process (Kelly and Ripley, 1976), but those with $\Psi(h) < 0$ for some h (i.e., with components of clustering) are frequently unstable (Gates and Westcott, 1986).

One approach to modeling Markov point processes with unstable potentials is to condition the process on n, the observed number of events in A. This approach is justified under the argument that n provides little information about the interactions among events.

Spatial Birth-and-Death Process

The spatial birth-and-death process can be used to justify the modeling of sedentary events (e.g., trees in a forest or towns in a geographical region) via a Markov point process. For a spatial birth-and-death process, assume that the probability of more than one birth or death in a small interval of time $[t, t + \delta)$ is $o(\delta)$ (Preston, 1975). Suppose, at time t, the events of the process are located at $\{\mathbf{s}_1, \mathbf{s}_2, \ldots, \mathbf{s}_n\}$. Define the birth rate to be

$$b(\mathbf{u}; \mathbf{s}_1, \ldots, \mathbf{s}_n) \equiv \lim_{\substack{\nu(d\mathbf{u}) \to 0 \\ \delta \to 0}} \frac{\Pr(\text{birth in } d\mathbf{u} \text{ during } [t, t + \delta) \,|\, \mathbf{s}_1, \ldots, \mathbf{s}_n)}{\delta \nu(d\mathbf{u})}$$

and the death rate to be

$$d(\mathbf{u}; \mathbf{s}_1, \ldots, \mathbf{s}_n) \equiv \lim_{\delta \to 0} \frac{\Pr(\text{death of event } \mathbf{u} \text{ during } [t, t + \delta) \,|\, \mathbf{s}_1, \ldots, \mathbf{s}_n)}{\delta};$$

notice that neither depends on t. Then, the probability that an event is born in a bounded Borel set A during the interval $[t, t + \delta)$ is $\delta \cdot \int_A b(\mathbf{u}; \mathbf{s}_1, \mathbf{s}_2, \ldots, \mathbf{s}_n) \nu(d\mathbf{u}) + o(\delta)$. If events at time t are located at $\mathbf{s}_1, \mathbf{s}_2, \ldots, \mathbf{s}_n, \mathbf{u}$, then the probability that the event $\mathbf{u}$ dies during the interval $[t, t + \delta)$ is $\delta \cdot d(\mathbf{u}; \mathbf{s}_1, \mathbf{s}_2, \ldots, \mathbf{s}_n) + o(\delta)$. Glotzel (1981) shows that a spatial birth-and-death process with birth rate $b(\cdot)$ and death rate $d(\cdot)$ converges in distribution (as $t \to \infty$) to a Markov point process with local energy

$$e(\mathbf{u};\mathbf{s}_1,\mathbf{s}_2,\ldots,\mathbf{s}_n) = -\log\frac{b(\mathbf{u};\mathbf{s}_1,\mathbf{s}_2,\ldots,\mathbf{s}_n)}{d(\mathbf{u};\mathbf{s}_1,\mathbf{s}_2,\ldots,\mathbf{s}_n)}; \qquad (8.5.69)$$

under appropriate conditions, the rate of convergence is geometric (Moller, 1989b).

Simulating a Markov Point Process

The spatial birth-and-death process provides the framework under which Ripley (1977) proposes to simulate a Markov point process on the bounded Borel set $A \subset \mathbb{R}^d$ with $N(A) = n$ fixed. First, select n events from a uniform distribution on A; call this initial point pattern $\phi_n(0)$. At step $(t+1)$, delete one of the n events of $\phi_n(t) \equiv (\mathbf{s}_1,\ldots,\mathbf{s}_n)$ at random, say event $\mathbf{s}_i$, and let $\phi_n(t) - \{\mathbf{s}_i\}$ denote the point pattern formed by removing $\mathbf{s}_i$ from $\phi_n(t)$. Let

$$p(\mathbf{u};\phi_n(t) - \{\mathbf{s}_i\}) \equiv j_n(\phi_n(t) - \{\mathbf{s}_i\},\mathbf{u})/j_{n-1}(\phi_n(t) - \{\mathbf{s}_i\})$$

denote the Papangelou intensity at $\mathbf{u} \in A$ given $\phi_n(t) - \{\mathbf{s}_i\}$, where $j_k(\phi_k(t))$ is the Janossy density of $\phi_k(t)$ and k [e.g., Eq. (8.5.62)]. Define

$$M \equiv \sup_{\mathbf{u}\in A} p(\mathbf{u};\phi_n(t) - \{\mathbf{s}_i\}).$$

Select an event $\mathbf{u}$ from a uniform distribution on A and set $\phi_n(t+1) = \{\phi_n(t) - \{\mathbf{s}_i\},\mathbf{u}\}$, with probability $(1/M)p(\mathbf{u};\phi_n(t) - \{\mathbf{s}_i\})$; otherwise, selection is repeated until a qualifying $\mathbf{u}$ is found. Ultimately, convergence to a Markov point process with Janossy density $j_n(\cdot)$ will occur. Ripley (1979b) provides FORTRAN code to implement the procedure; he also suggests (Ripley, 1987, p. 113) that $10n$ iterations should be adequate for convergence. This algorithm is analogous to the Gibbs sampler on the spatial lattice (Geman and Geman, 1984; Section 7.7.1).

Ogata and Tanemura (1981, 1989) propose a different method of simulation based on the Metropolis algorithm (Metropolis et al., 1953) and the results in Wood (1968). Suppose one starts with n events from, say, a uniform distribution on the torus A; call this initial point pattern $\phi_n(0) \equiv \{\mathbf{s}_i(0): i = 1,2,\ldots,n\}$. At step t, $\phi_n(t) \equiv \{\mathbf{s}_i(t): i = 1,2,\ldots,n\}$ is available. First, select a trial state $\phi'_n(t) \equiv \{\mathbf{s}'_i(t): i = 1,2,\ldots,n\}$, where $\mathbf{s}'_i(t)$ is chosen from a uniform distribution on the d-dimensional square centered at $\mathbf{s}_i(t)$ with sides of length $2\delta > 0$. Then, take $\phi_n(t+1) = \phi'_n(t)$, (1) with probability 1, if $j_n(\phi_n(t)) \le j_n(\phi'_n(t))$ or (2) with probability $p = j_n(\phi'_n(t))/j_n(\phi_n(t))$, if $j_n(\phi_n(t)) > j_n(\phi'_n(t))$. Otherwise, choose a new trial state $\phi'_n(t)$ until a qualifying point pattern $\phi_n(t+1)$ is found. Ogata and Tanemura do not say how many steps are required in practice. Wood (1968) suggests that δ should be chosen so as to reject approximately one-half of the trial realizations.

Gates and Westcott (1986) find that Ogata and Tanemura's simulation method gives inaccurate results for pair-potential Markov point processes that are unstable. Even though $N(A)$ is conditioned to be n, Gates and Westcott show that such potentials should give much more tightly packed clusters than those simulated. However, provided the process is stable, the simulation methods are appropriate.

Maximum Likelihood Estimation for Pair-Potential Markov Point Processes

Suppose the data consist of the locations $\mathbf{s}_1, \mathbf{s}_2, \ldots, \mathbf{s}_n$ of $n = N(A)$ events in a bounded region $A \subset \mathbb{R}^d$. Let $\{\Psi_{\boldsymbol{\theta}}\colon \boldsymbol{\theta} \in \Theta\}$ be a family of parameterized pair-potential functions. The likelihood function is the Janossy density, here given by

$$l(\boldsymbol{\theta}) = \frac{e^{-\nu(A)}}{\alpha(\boldsymbol{\theta})} \exp\left\{ - \sum_{1 \le i < j \le n} \sum \Psi_{\boldsymbol{\theta}}(\|\mathbf{s}_i - \mathbf{s}_j\|) \right\}, \tag{8.5.70}$$

where the normalizing constant is

$$\alpha(\boldsymbol{\theta}) = \sum_{n=0}^{\infty} \frac{e^{-\nu(A)}}{n!} \int_{A^n} \exp\left\{ - \sum_{1 \le i < j \le n} \sum \Psi_{\boldsymbol{\theta}}(\|\mathbf{s}_i - \mathbf{s}_j\|) \right\} \nu(d\mathbf{s}_1) \cdots \nu(d\mathbf{s}_n). \tag{8.5.71}$$

The maximum likelihood estimator of $\boldsymbol{\theta}$ is obtained by finding a $\hat{\boldsymbol{\theta}}$ that maximizes Eq. (8.5.70). This maximization problem requires computation of (8.5.71), which is not usually available in closed form.

An alternative approach is to condition the analysis on $N(A)$, the observed number of events in A, because it usually provides little information about the interactions among events. Conditional on $N(A) = n$, the likelihood function is

$$l_n(\boldsymbol{\theta}) = \frac{n!}{c_n(\boldsymbol{\theta})} \exp\left\{ - \sum_{1 \le i < j \le n} \sum \Psi_{\boldsymbol{\theta}}(\|\mathbf{s}_i - \mathbf{s}_j\|) \right\}, \tag{8.5.72}$$

where the normalizing constant is

$$c_n(\boldsymbol{\theta}) = \int_{A^n} \exp\left\{ - \sum_{1 \le i < j \le n} \sum \Psi_{\boldsymbol{\theta}}(\|\mathbf{s}_i - \mathbf{s}_j\|) \right\} \nu(d\mathbf{s}_1) \cdots \nu(d\mathbf{s}_n). \tag{8.5.73}$$

[More generally, $l_n(\boldsymbol{\theta}) = (n!/c_n(\boldsymbol{\theta}))\exp\{-U_n(\mathbf{s}_1, \cdots, \mathbf{s}_n; \boldsymbol{\theta}\}$.] Maximum likelihood estimation of $\boldsymbol{\theta}$ then requires the evaluation of (8.5.73). Although it is not usually obtainable in closed form, approximate expressions are sometimes available.

Approximate Maximum Likelihood Estimation

One method of approximating the normalizing constant $c_n(\boldsymbol{\theta})$ is through simulation of a pair-potential Markov point process with parameter $\boldsymbol{\theta}$. Let $\phi_n(t)$ be the point pattern of the $(t + \nu_0)$th iterate of Ogata and Tanemura's (1981) simulation method, where approximate equilibrium is reached after the ν_0th iterate. Then the time average,

$$\frac{1}{T}\sum_{t=1}^{T} \exp\{U_n(\phi_n(t);\boldsymbol{\theta})\}, \tag{8.5.74}$$

can be a good approximation of

$$E_{\boldsymbol{\theta}}(\exp\{U_n(\mathbf{s}_1,\ldots,\mathbf{s}_n;\boldsymbol{\theta})\}) = \frac{(\nu(A))^n}{c_n(\boldsymbol{\theta})}, \tag{8.5.75}$$

where the total potential energy $U_n(\mathbf{s}_1,\ldots,\mathbf{s}_n;\boldsymbol{\theta}) \equiv \sum_{1\le i<j\le n}\Psi_{\boldsymbol{\theta}}(\|\mathbf{s}_i - \mathbf{s}_j\|)$ (Ogata and Tanemura, 1981). Maximum likelihood estimation would require simulation of the Markov point process for a range of values of $\boldsymbol{\theta}$, which may be impractical.

Penttinen (1984) suggests a Newton–Raphson-type algorithm for solving the m.l. estimating equation. Assume that $\Psi_{\boldsymbol{\theta}}(\cdot)$ in (8.5.70) is twice differentiable with respect to $\boldsymbol{\theta}$; then so is the total potential energy $U_n(\mathbf{s}_1,\ldots,\mathbf{s}_n;\boldsymbol{\theta}) \equiv \sum\sum_{1\le i<j\le n}\Psi_{\boldsymbol{\theta}}(\|\mathbf{s}_i - \mathbf{s}_j\|)$. Differentiation of both sides of the equation

$$c_n(\boldsymbol{\theta}) = \int_{A^n} \exp\{-U_n(\mathbf{s}_1,\ldots,\mathbf{s}_n;\boldsymbol{\theta})\}\nu(d\mathbf{s}_1)\cdots\nu(d\mathbf{s}_n)$$

yields

$$-\partial c_n(\boldsymbol{\theta})/\partial\boldsymbol{\theta} = c_n(\boldsymbol{\theta})E_{\boldsymbol{\theta}}\{\partial U_n(\mathbf{s}_1,\ldots,\mathbf{s}_n;\boldsymbol{\theta})/\partial\boldsymbol{\theta}\},$$

a result due to Ogata and Tanemura (1984). Then the time average,

$$\frac{1}{T}\sum_{t=1}^{T} \partial U_n(\phi_n(t);\boldsymbol{\theta})/\partial\boldsymbol{\theta}$$

should be a good approximation of $-\partial\log\{c_n(\boldsymbol{\theta})\}/\partial\boldsymbol{\theta}$. Ogata and Tanemura (1984, 1989) give the (approximate) integral equation for $\log c_n(\boldsymbol{\theta})$ and suggest doing many simulations over a range of $\boldsymbol{\theta}$ values to fit an empirically determined functional relationship between $c_n(\boldsymbol{\theta})$ and $\boldsymbol{\theta}$.

Consider now the score function

$$\boldsymbol{\beta}(\boldsymbol{\theta}) \equiv \partial\log\{l_n(\boldsymbol{\theta})\}/\partial\boldsymbol{\theta} = -\partial U_n(\mathbf{s}_1,\ldots,\mathbf{s}_n;\boldsymbol{\theta})/\partial\boldsymbol{\theta} - \partial\log\{c_n(\boldsymbol{\theta})\}/\partial\boldsymbol{\theta}.$$

Then the time average

$$\bar{\boldsymbol{\beta}}_T(\boldsymbol{\theta}) \equiv \frac{1}{T}\sum_{t=1}^{T}\boldsymbol{\beta}_t(\boldsymbol{\theta};\phi_n(t)),$$

where $\boldsymbol{\beta}_t(\boldsymbol{\theta};\phi_n(t)) \equiv -\partial U_n(\mathbf{s}_1,\ldots,\mathbf{s}_n;\boldsymbol{\theta})/\partial\boldsymbol{\theta} + \partial U_n(\phi_n(t);\boldsymbol{\theta})/\partial\boldsymbol{\theta}$, should be a good approximation of the score function.

The maximum likelihood estimator $\hat{\boldsymbol{\theta}}$ solves $\partial \log\{l_n(\boldsymbol{\theta})\}/\partial\boldsymbol{\theta} = \mathbf{0}$. A Newton–Raphson iteration yields the updated value $\hat{\boldsymbol{\theta}}_{k+1}$ based on a previous value $\hat{\boldsymbol{\theta}}_k$:

$$\hat{\boldsymbol{\theta}}_{k+1} = \hat{\boldsymbol{\theta}}_k - \left(\Gamma\left(\hat{\boldsymbol{\theta}}_k\right)\right)^{-1}\boldsymbol{\beta}\left(\hat{\boldsymbol{\theta}}_k\right),$$

where $\Gamma(\boldsymbol{\theta})$ is the square matrix whose (i,j)th element is $\partial^2 \log\{l_n(\boldsymbol{\theta})\}/\partial\theta_i\,\partial\theta_j$.

Approximate the matrix $\Gamma(\boldsymbol{\theta})$ by the time average

$$\bar{\Gamma}_T(\boldsymbol{\theta}) \equiv \frac{1}{T}\sum_{t=1}^{T}\Gamma_t(\boldsymbol{\theta};\phi_n(t)),$$

where $\Gamma_t(\boldsymbol{\theta};\phi_n(t))$ has (i,j)th element

$$\partial^2 U_n(\phi_n(t);\boldsymbol{\theta})/\partial\theta_i\,\partial\theta_j - \partial^2 U_n(\mathbf{s}_1,\ldots,\mathbf{s}_n;\boldsymbol{\theta})/\partial\theta_i\,\partial\theta_j$$
$$-(i,j)\text{th element of }\left\{\boldsymbol{\beta}_t(\boldsymbol{\theta};\phi_n(t)) - \bar{\boldsymbol{\beta}}_T(\boldsymbol{\theta})\right\}\left\{\boldsymbol{\beta}_t(\boldsymbol{\theta};\phi_n(t)) - \bar{\boldsymbol{\beta}}_T(\boldsymbol{\theta})\right\}'.$$

Let $\hat{\boldsymbol{\theta}}_0$ denote an initial guess for $\hat{\boldsymbol{\theta}}$. Then, the Monte Carlo-based Newton–Raphson algorithm is

$$\hat{\boldsymbol{\theta}}_{k+1} = \hat{\boldsymbol{\theta}}_k - \left(\bar{\Gamma}_T\left(\hat{\boldsymbol{\theta}}_k\right)\right)^{-1}\bar{\boldsymbol{\beta}}_T\left(\hat{\boldsymbol{\theta}}_k\right), \qquad k = 1,2,\ldots,$$

where $\phi_n(1),\ldots,\phi_n(T)$ are simulated according to a pair-potential Markov point process with parameter $\hat{\boldsymbol{\theta}}_k$, using either Ripley's (1977) method or Ogata and Tanemura's (1981) method. For a Strauss process [see Eq. (8.5.62)] on a unit square with $n = 60$, $\theta \equiv \log\gamma = 1.6$, and $\rho = 0.08$, Penttinen (1984) achieves convergence to within two significant digits, after four iterations.

Ogata and Tanemura (1981) use the cluster-expansion method of statistical mechanics to obtain an approximation of (8.5.73) (see Mayer and Mayer, 1940; Feynman, 1972). Assume that the events of the point process are sparsely distributed, so that third- and higher-order cluster integrals are negligible. Then, using up to the second-order cluster integral, the normalizing constant is approximately

$$c_n(\boldsymbol{\theta}) \simeq (\nu(A))^n\{1 - b(\boldsymbol{\theta})/\nu(A)\}^{n(n-1)/2}, \tag{8.5.76}$$

where

$$b(\boldsymbol{\theta}) \equiv \frac{d\pi^{d/2}}{\Gamma\left(1+\frac{d}{2}\right)} \int_0^\infty r^{d-1}\{1-\exp(-\Psi_{\boldsymbol{\theta}}(r))\}\,dr. \tag{8.5.77}$$

Ogata and Tanemura (1984) consider approximations to (8.5.73) using higher-order cluster integrals. The approximation given by (8.5.76) and those by Ogata and Tanemura (1984) hold only for stable pair-potentials. For unstable pair-potentials, higher-order interactions become important; in such cases, Gates and Westcott (1986) show that (8.5.76) can differ from the true normalizing constant by several orders of magnitude. Thus, approximations using truncated cluster expansions are not reliable for Markov cluster point processes.

Another sparse-data approximation to $c_n(\boldsymbol{\theta})$ is given by Penttinen (1984), where it is assumed that there exists a $\rho > 0$ such that $\Psi_{\boldsymbol{\theta}}(r) = 0$, for all $r > \rho$. Let $Y_n(\rho)$ denote the number of distinct pairs of events separated by a distance ρ or smaller; that is,

$$Y_n(\rho) \equiv \sum_{1 \le i < j \le n}\sum I(\|\mathbf{s}_i - \mathbf{s}_j\| \le \rho),$$

and let $r_{(1)}, r_{(2)}, \ldots$ denote the $\binom{n}{2}$ ordered interevent distances. Then, under a sparseness assumption that assumes the size of the study region grows like the square of the number of points n (Saunders and Funk, 1977), $Y_n(\rho)$ is approximately Poisson distributed with mean $(1/2)n(n-1)\pi\rho^2$. Furthermore, conditional on $Y_n = m$, $r_{(1)}^2, \ldots, r_{(m)}^2$ are approximately the order statistics from a uniform distribution on $[0, \rho^2]$.

Now, for $U_{1m}, \ldots, U_{mm}$ independent and identically distributed according to a uniform distribution on $[0, \rho^2]$, $m = 1, 2, \ldots,$

$$c_n(\boldsymbol{\theta}) \simeq \Pr(Y_n(\rho) = 0) + \sum_{m=1}^{\infty} \Pr(Y_n(\rho) = m) E\left(\exp\left\{-\sum_{i=1}^{m} \Psi_{\boldsymbol{\theta}}\left(U_{im}^{1/2}\right)\right\}\right).$$

Notice that $E(\exp\{-\Psi_{\boldsymbol{\theta}}(U_{im}^{1/2})\}) = (2/\rho^2)\int_0^\rho r \exp(-\Psi_{\boldsymbol{\theta}}(r))\,dr$ and hence, by independence, the preceding approximation for $c_n(\boldsymbol{\theta})$ can be written as

$$c_n(\boldsymbol{\theta}) \simeq E\left[\left\{(2/\rho^2)\int_0^\rho r \exp(-\Psi_{\boldsymbol{\theta}}(r))\,dr\right\}^{Y_n(\rho)}\right].$$

This takes the form of a probability generating function of the random variable $Y_n(\rho)$, which is approximately Poisson distributed with mean

$(1/2)n(n-1)\pi\rho^2$. Thus,

$$c_n(\boldsymbol{\theta}) \simeq \exp\left[(1/2)n(n-1)\int_0^{\rho} 2\pi r\{\exp(-\Psi_{\boldsymbol{\theta}}(r)) - 1\}\,dr\right].$$

Estimation Based on the K Function

Using an integral characterization of Nguyen and Zessin (1979), Hanisch and Stoyan (1983) show that the K function of a pair-potential Markov point process N on $\mathbb{R}^2$, with pair-potential $\Psi_{\boldsymbol{\theta}}$, is given by

$$K(h;\boldsymbol{\theta}) = 2\pi\lambda^{-2}\int_0^h r\exp(-\Psi_{\boldsymbol{\theta}}(r))\int_{\Phi}\exp(-e_{\boldsymbol{\theta}}(\mathbf{0};\phi)) \times\exp(-e_{\boldsymbol{\theta}}(\mathbf{s}_r;\phi))\Pi_{\boldsymbol{\theta}}(d\phi)\,dr, \quad (8.5.78)$$

where $\mathbf{s}_r$ is any point distance r away from the origin $\mathbf{0}$, $e_{\boldsymbol{\theta}}$ is the local energy given by (8.5.67) with pair-potential $\Psi_{\boldsymbol{\theta}}$, and $\Pi_{\boldsymbol{\theta}}(\cdot)$ is the induced probability measure of the Markov point process N with intensity λ. Equation (8.5.78) is not computable in closed form, but when $\Psi_{\boldsymbol{\theta}}$ is a step function,

$$\Psi_{\boldsymbol{\theta}}(r) = \begin{cases} \infty, & 0 < r \le \rho_0, \\ v_1, & \rho_0 < r \le \rho_1, \\ \vdots & \\ v_k, & \rho_{k-1} < r \le \rho_k, \\ 0, & r > \rho_k, \end{cases} \quad (8.5.79)$$

it can be shown that the K function's first derivative is discontinuous at $\rho_0, \rho_1, \ldots, \rho_k$. Thus, discontinuity points of $\Psi_{\boldsymbol{\theta}}$ may be recognizable from $\hat{K}(\cdot)$, a nonparametric estimator of $K(\cdot)$.

For a Markov point process on $\mathbb{R}^3$, the following relationship between second-order intensity $\lambda_2^o(\cdot;\boldsymbol{\theta})$ and pair-potential $\Psi_{\boldsymbol{\theta}}(\cdot)$ is given in the physics literature by Born and Green (1946):

$$\log\frac{\lambda_2^o(h;\boldsymbol{\theta})}{\lambda^2} + \Psi_{\boldsymbol{\theta}}(h) = \pi\lambda^{-3}\int_0^{\infty}\lambda_2^o(r;\boldsymbol{\theta})\Psi_{\boldsymbol{\theta}}'(r)\int_{-r}^{r}(r^2-v^2) \times\left(\frac{r+v}{v}\right)\left(\lambda_2^o(r+v;\boldsymbol{\theta}) - \lambda^2\right)dv\,dr. \quad (8.5.80)$$

Given $\Psi_{\boldsymbol{\theta}}(h)$, Eq. (8.5.80) can be solved for $\lambda_2^o(h;\boldsymbol{\theta})$ numerically for each h (see, e.g., Kirkwood et al., 1952). Then the K function can be obtained from (8.3.32).

Approximations for the K function have been found for a few Markov point processes. Recall that the pair-potential of a Strauss process is

$$\Psi_{\boldsymbol{\theta}}(h) = \begin{cases} -\log\gamma, & \text{if } 0 < h \leq \rho, \\ 0, & \text{if } h > \rho, \end{cases}$$

where $\boldsymbol{\theta} = (\rho, \gamma)'$. Isham (1984) shows that in $\mathbb{R}^2$ the K function of the Strauss process is approximately

$$K(h;\boldsymbol{\theta}) \simeq \begin{cases} \gamma\pi h^2, & 0 < h \leq \rho, \\ \pi h^2 - (1-\gamma)\pi\rho^2, & h \geq \rho, \end{cases} \tag{8.5.81}$$

with error that is $O((1-\gamma)^2)$. More generally, results of Saunders et al. (1982) and Hanisch and Stoyan (1983) suggest that the K function of pair-potentials of the form (8.5.79) is approximately

$$K(h;\boldsymbol{\theta}) \simeq \begin{cases} 0, & 0 < h \leq \rho_0, \\ \pi \sum_{i=1}^{j-1} (\rho_i^2 - \rho_{i-1}^2) e^{-v_i} + \pi(h^2 - \rho_{j-1}^2) e^{-v_j}, & \rho_{j-1} \leq h \leq \rho_j, \\ & j = 1, \cdots, k, \\ \pi \sum_{i=1}^{k} (\rho_i^2 - \rho_{i-1}^2) e^{-v_i} + \pi(h^2 - \rho_k^2), & h \geq \rho_k, \end{cases} \tag{8.5.82}$$

for a sufficiently large study region A in $\mathbb{R}^2$.

Given an approximate or numerical evaluation of $K(h;\boldsymbol{\theta})$, a least-squares estimator of $\boldsymbol{\theta}$ could be obtained by finding the $\hat{\boldsymbol{\theta}}$ that minimizes the *ad hoc* criterion

$$D(\boldsymbol{\theta}) \equiv \int_0^{h_0} \left\{ \left(\hat{K}(h)\right)^c - \left(K(h;\boldsymbol{\theta})\right)^c \right\}^2 dh \tag{8.5.83}$$

(Section 8.5.3), where $\hat{K}(h)$ is the empirical K function. Diggle (1983) recommends that for repulsive pair-potentials $c = 1/2$ should be chosen, but for attractive pair-potentials $c < 1/2$ should be chosen. In practice, several cs in $(0, 1/2)$ might be tried. The simplex algorithm of Nelder and Mead (1965) can be used to solve the minimization problem [such as in Diggle and Gratton's (1984) SQ algorithm]. When $K(h;\boldsymbol{\theta})$ cannot be evaluated, Diggle and Gratton (1984) suggest that $K(h;\boldsymbol{\theta})$ in (8.5.83) could be replaced by

$$\frac{1}{m} \sum_{i=1}^{m} \hat{K}_i(h;\boldsymbol{\theta}),$$

for various $\boldsymbol{\theta}$, where $\hat{K}_i(h; \boldsymbol{\theta})$ is the empirical K function calculated from the ith simulation of a Markov point process with parameter $\boldsymbol{\theta}$.

Inference for Strauss' Hard-Core Model

Consider a pair-potential Markov point process N on a bounded region $A \subset \mathbb{R}^2$ with

$$\Psi_\rho(h) = \begin{cases} \infty, & 0 < h \leq \rho, \\ 0, & h > \rho, \end{cases}$$

which is the pair-potential of a (hard-core) Strauss process with $\gamma = 0$. If $\rho = 0$, then N is a Poisson process. Let $d_1 \leq d_2 \leq \cdots$ be the ordered interpoint distances in A. Then d_1 is sufficient for ρ and the uniformly most powerful test for $\rho = 0$ against $\rho > 0$ is based on d_1. Furthermore, d_1 is the maximum likelihood estimator for ρ and $n(n-1)(d_1^2 - \rho^2)$ is approximately exponentially distributed with mean $2\nu(A)/\pi$. Thus, confidence intervals for ρ can be computed (Ripley and Silverman, 1978). The smallest interpoint distance d_1 is sensitive to measurement and rounding errors, so inference based on it is not very robust. Distances d_i, $i > 1$, are less sensitive to such errors. For a Poisson process, Silverman and Brown (1978) show that $\{n(n-1)\pi/\nu(A)\}d_i^2$ is approximately chi-squared distributed on $2i$ degrees of freedom (see also Saunders and Funk, 1977).

Takacs–Fiksel Estimation Procedure

Takacs (1986, originally published in 1983 in Russian) and Fiksel (1984a, 1988b) consider a method for estimating $\boldsymbol{\theta}$, of the pair-potential $\Psi_{\boldsymbol{\theta}}$, based on the observation that for the pair-potential Markov point process N with (induced) distribution $\Pi_{\boldsymbol{\theta}}$ and reduced Palm distribution $P_0^!(\cdot; \boldsymbol{\theta})$,

$$\int_\Phi \xi(\phi) P_0^!(d\phi; \boldsymbol{\theta}) = \lambda^{-1} \int_\Phi \xi(\phi) \exp\{-e_{\boldsymbol{\theta}}(\mathbf{0}; \phi)\} \Pi_{\boldsymbol{\theta}}(d\phi), \quad (8.5.84)$$

for all nonnegative measurable functions ξ on Φ, where $e_{\boldsymbol{\theta}}$ is the local energy defined by (8.5.67). Estimation of $\boldsymbol{\theta}$ requires the choice of suitable nonnegative functions $\xi^{(i)}(\phi)$, $i = 1, 2, \ldots, m$, where m is an integer larger than or equal to the dimension of $\boldsymbol{\theta}$. Then, estimators $g_0^{(i)}(\boldsymbol{\theta})$ and $g^{(i)}(\boldsymbol{\theta})$ of the left-hand and right-hand sides, respectively, of (8.5.84) are constructed. The Takacs–Fiksel estimator of $\boldsymbol{\theta}$ is obtained by finding the $\hat{\boldsymbol{\theta}}_{TF}$ that minimizes

$$Q(\boldsymbol{\theta}) \equiv \sum_{i=1}^{m} \left\{ g_0^{(i)}(\boldsymbol{\theta}) - g^{(i)}(\boldsymbol{\theta}) \right\}^2. \quad (8.5.85)$$

Fiksel (1988b) suggests that a suitable choice of $\xi^{(i)}(\phi)$ is

$$\xi^{(i)}(\phi) = \phi(b(\mathbf{0}, r_i)) \exp\{e_{\boldsymbol{\theta}}(\mathbf{0}; \phi)\},$$

where $0 < r_1 < r_2 < \cdots < r_m$ are fixed constant distances and $b(\mathbf{0}, r)$ is the ball of radius r centered at $\mathbf{0}$. Then, for $i = 1, \ldots, m$, the left-hand side of (8.5.84) is

$$g_{\mathbf{0}}^{(i)}(\boldsymbol{\theta}) = \int_{\Phi} \phi(b(\mathbf{0}, r_i)) \exp\{e_{\boldsymbol{\theta}}(\mathbf{0}; \phi)\} P_{\mathbf{0}}^{!}(d\phi; \boldsymbol{\theta})$$

and the right-hand side is

$$g^{(i)}(\boldsymbol{\theta}) = \lambda^{-1} \int_{\Phi} \phi(b(\mathbf{0}, r_i)) \Pi_{\boldsymbol{\theta}}(d\phi) = \nu(b(\mathbf{0}, r_i)) = \frac{\pi^{d/2} r_i^d}{\Gamma(1 + \frac{d}{2})}. \quad (8.5.86)$$

In this case, only $g_{\mathbf{0}}^{(i)}(\boldsymbol{\theta})$ must be estimated. Suppose the pair-potentials $\Psi_{\boldsymbol{\theta}}$ have a known interaction distance R, such that $\Psi_{\boldsymbol{\theta}}(r) = 0$, for all $r \geq R$, and let $R_{\max} \equiv \max(r_m, R)$. Then, for $\phi = \{\mathbf{s}_i: i = 1, 2, \ldots, n\}$, a given realization of the Markov point process in A, an edge-corrected estimator for $g_{\mathbf{0}}^{(i)}(\boldsymbol{\theta})$ is

$$\hat{g}_{\mathbf{0}}^{(i)}(\boldsymbol{\theta}) = \{\nu(A_R)\}^{-1} \sum_{\mathbf{s} \in \phi \cap A_R} \exp\{e_{\boldsymbol{\theta}}(\mathbf{s}; \phi)\} \sum_{\mathbf{u} \in \phi \cap A} I(0 < \|\mathbf{s} - \mathbf{u}\| \leq r_i), \quad (8.5.87)$$

where $A_R \equiv \bigcap\{A \oplus \mathbf{b}: \mathbf{b} \in b(\mathbf{0}, R_{\max})\}$ and $A \oplus \mathbf{b} \equiv \{\mathbf{a} + \mathbf{b}: \mathbf{a} \in A\}$. Equations (8.5.86) and (8.5.87) can then be substituted into (8.5.85) and a least-squares estimator for $\boldsymbol{\theta}$ can be obtained.

Nonparametric Estimation of the Pair-Potential

Diggle et al. (1987) consider a nonparametric estimator of the pair-potential Ψ. Define $g(h) \equiv (\lambda_2^o(h)/\lambda^2) - 1$, where λ and $\lambda_2^o(h)$ are the intensity and second-order intensity, respectively, of the stationary isotropic Markov point process. Define the direct correlation function $c(\cdot)$ to be the solution to the Ornstein–Zernike equation

$$c(h) = g(h) - \lambda(c * g)(h), \quad (8.5.88)$$

where $c * g$ is the two-dimensional convolution of c and g:

$$(c * g)(h) \equiv \int_0^{2\pi} \int_0^{\infty} c(t) g\left((h^2 + t^2 - 2ht\cos\nu)^{1/2}\right) t\, dt\, d\nu.$$

The nonparametric estimator of Diggle et al. is then calculated using the Percus–Yevick approximation

$$\Psi(h) \simeq \frac{g(h) + 1}{g(h) + 1 - c(h)} = \frac{\lambda_2^o(h)}{\lambda_2^o(h) - \lambda^2 c(h)}; \quad (8.5.89)$$

estimation of Ψ involves solving (8.5.88) in terms of a nonparametric estimator $\hat{g}(h)$ and substituting the resulting $\hat{c}(h)$ into Eq. (8.5.89). A nonparametric estimator for $\lambda_2^o(h)$ (and hence for $g(h)$) can be found in Fiksel (1988a); see also Brillinger (1975) and Krickeberg (1982).

Inhomogeneous Markov Point Process

An inhomogeneous pair-potential Markov point process includes terms for both large-scale effects $\xi(\mathbf{s})$ and small-scale effects $\Psi(\|\mathbf{s} - \mathbf{u}\|)$. Conditional on n, the number of events in A, the likelihood function, is given by

$$l_n(\boldsymbol{\alpha}, \boldsymbol{\beta}) = \frac{n!}{c_n(\boldsymbol{\alpha}, \boldsymbol{\beta})} \exp\left\{ -\sum_{i=1}^{n} \xi_{\boldsymbol{\alpha}}(\mathbf{s}_i) - \sum_{1 \le i < j \le n} \Psi_{\boldsymbol{\beta}}(\|\mathbf{s}_i - \mathbf{s}_j\|) \right\}, \quad (8.5.90)$$

where the normalizing constant is

$$c_n(\boldsymbol{\alpha}, \boldsymbol{\beta}) \equiv \int_{A^n} \exp\left\{ -\sum_{i=1}^{n} \xi_{\boldsymbol{\alpha}}(\mathbf{s}_i) - \sum_{1 \le i < j \le n} \Psi_{\boldsymbol{\beta}}(\|\mathbf{s}_i - \mathbf{s}_j\|) \right\} \nu(d\mathbf{s}_1) \cdots \nu(d\mathbf{s}_n). \quad (8.5.91)$$

If $\xi_{\boldsymbol{\alpha}} \equiv 0$, then (8.5.90) is the conditional likelihood of a homogeneous pair-potential Markov point process. Conversely, if $\Psi_{\boldsymbol{\beta}} \equiv 0$, then (8.5.90) is the conditional likelihood of an inhomogeneous Poisson process with intensity $\lambda_{\boldsymbol{\alpha}}(\mathbf{s}) \equiv \exp\{-\xi_{\boldsymbol{\alpha}}(\mathbf{s})\}$.

Maximum likelihood estimation of $(\boldsymbol{\alpha}, \boldsymbol{\beta})$ requires the evaluation of (8.5.91), which is not usually available in closed form. Ogata and Tanemura (1986) use the cluster-expansion method to obtain an approximate value for (8.5.91): Assume that the events of the process are sparsely distributed, so that third- and higher-order cluster integrals are negligible, and that the absolute variation of the spatial inhomogeneity term $\xi_{\boldsymbol{\alpha}}$ is small relative to that of the pair-potential $\Psi_{\boldsymbol{\beta}}$. Let

$$a(\boldsymbol{\alpha}) \equiv \int_A \exp\{-\xi_{\boldsymbol{\alpha}}(\mathbf{s})\} \nu(d\mathbf{s}) \quad (8.5.92)$$

and let

$$b(\boldsymbol{\beta}) \equiv \frac{d\pi^{d/2}}{\Gamma(1 + \frac{d}{2})} \int_0^\infty r^{d-1}\left[1 - \exp\{-\Psi_{\boldsymbol{\beta}}(r)\}\right] dr, \quad (8.5.93)$$

the second-order cluster integral. Then, under the assumptions just specified, the normalizing constant is approximately

$$c_n(\boldsymbol{\alpha}, \boldsymbol{\beta}) \simeq (a(\boldsymbol{\alpha}))^n \left\{1 - \frac{n(n-1)}{2(a(\boldsymbol{\alpha}))^2} b(\boldsymbol{\beta}) \int_A \exp(-2\xi_{\boldsymbol{\alpha}}(\mathbf{s}))\nu(d\mathbf{s})\right\}. \tag{8.5.94}$$

When $\xi_{\boldsymbol{\alpha}} \equiv 0$, then (8.5.94) and (8.5.76) are equivalent up to $O(1/(\nu(A)))$. As in the homogeneous case, the approximation given by (8.5.94) only holds for stable pair-potentials (Gates and Westcott, 1986).

8.5.6* Thinned and Related Point Processes

Consider the result of randomly removing events from a realization of the spatial point process N_0. Then the retained events can be viewed as a realization of some other spatial point process, say N_1. The simplest such process is the thinned point process: Events are retained from N_0 independently of one another with probability $q(\cdot)$, which is a deterministic function of spatial location. Call $q(\cdot)$ the thinning function. If the thinning function is random, N_1 is called an interrupted point process.

Dependent removal of events can be modeled either through extension of the simple inhibition processes of Section 8.5.4 or through a Markov thinning process defined on a bounded subset $A \subset \mathbb{R}^d$.

Thinned Point Process

Let N_0 be any spatial point process on a set X (X is $\mathbb{R}^d$ or a subset of $\mathbb{R}^d$) and let $q(\cdot)$ be a measurable mapping from X onto $[0, 1]$. Independently mark the events $\mathbf{s}$ of N_0 with marks $Z(\mathbf{s})$ that take the value 1 with probability $q(\mathbf{s})$, and 0 otherwise. Define the spatial point process N_1 as follows. Each event $\mathbf{s}$ in the initial process N_0 is retained in the thinned point process N_1 if and only if $Z(\mathbf{s}) = 1$. Then N_1 is called a *q-thinned point process*.

The simplest thinned point process has a thinning function that is independent of spatial location; that is, $q(\mathbf{s}) \equiv q^o$, a constant. Such processes are called *classical* q-thinned point processes by Karr (1986, p. 24). The thinning function $q(\cdot)$ may be a function of concomitant (spatial) variables, of spatial location alone, or of both.

Let $L_{N_0}(\zeta)$ be the Laplace functional of N_0. Then the Laplace functional of the thinned point process N_1 is obtained from the conditional argument

$$\begin{aligned} L_{N_1}(\zeta) &= E_{N_1}\left[\exp\left\{-\sum_i \zeta(\mathbf{s}_i)\right\}\right] \\ &= E_{N_0, Z}\left[\exp\left\{-\sum_i Z(\mathbf{s}_i)\zeta(\mathbf{s}_i)\right\}\right] \\ &= L_{N_0}(-\log(1 - q + qe^{-\zeta})). \end{aligned} \tag{8.5.95}$$

Similar arguments can be used to show that the probability generating functional of N_1 is

$$G_{N_1}(\xi) = G_{N_0}(q\xi). \tag{8.5.96}$$

Suppose N_0 is an inhomogeneous Poisson process with mean measure μ_{N_0}. Then, by (8.5.93) and (8.5.6), the Laplace functional of a q-thinned point process N_1 is

$$L_{N_1}(\zeta) = \exp\left\{-\int_X (1 - e^{-\zeta(\mathbf{s})}) q(\mathbf{s}) \mu_{N_0}(d\mathbf{s})\right\},$$

which is the Laplace functional of an inhomogeneous Poisson process with mean measure $\mu_{N_1}(A) \equiv \int_A q(\mathbf{s}) \mu_{N_0}(d\mathbf{s})$. So, by the Equivalence Theorem of Section 8.3.2, the thinned process is also an inhomogeneous Poisson process. Karr (1985) also shows that if N_0 is a Cox process directed by a random measure M and q is a thinning function, then the q-thinned process N_1 is a Cox process directed by the random measure $q \cdot M$.

Suppose N_0 is a Neyman–Scott process on $X = \mathbb{R}^d$, based on an inhomogeneous Poisson process (mean measure μ) of cluster centers and a Poisson number (mean ω) of offspring, and suppose N_1 is a classical ($q(\mathbf{s}) \equiv q^o$) thinning of N_0. Then, from (8.5.34) and (8.5.96), the probability generating functional of the classical q-thinned process N_1 on $X = \mathbb{R}^d$ is

$$G_{N_1}(\xi) = \exp\left[-\int_X \left[1 - \exp\left\{-q^o\omega\left(1 - \int_X \xi(\mathbf{u} + \mathbf{s}) f(\mathbf{u}) \nu(d\mathbf{u})\right)\right\}\right] \mu(d\mathbf{s})\right],$$

which is the probability generating functional of a Neyman–Scott process with a Poisson number (mean $q^o\omega$) of offspring. So, by the Equivalence Theorem of Section 8.3.2, the classical q-thinned process N_1 is also a Neyman–Scott process.

From (8.5.95), the mean measure of a thinned point process N_1 is

$$\begin{aligned} \mu_{N_1}(A) &= -\lim_{\alpha \downarrow 0} \frac{\partial}{\partial \alpha} L_{N_0}[-\log\{1 - q + q\exp(-\alpha I_A)\}] \\ &= \int_A q(\mathbf{s}) \mu_{N_0}(d\mathbf{s}). \end{aligned} \tag{8.5.97}$$

Similarly, the second-moment measure is

$$\begin{aligned} \mu_{N_1}^{(2)}(A_1 \times A_2) &= \int_{A_1}\int_{A_2} q(\mathbf{s}) q(\mathbf{u}) \mu_{N_0}^{(2)}(d\mathbf{s} \times d\mathbf{u}) \\ &\quad + \int_{A_1 \cap A_2} q(\mathbf{s})(1 - q(\mathbf{s})) \mu_{N_0}(d\mathbf{s}). \end{aligned} \tag{8.5.98}$$

Then the first- and second-order intensities of N_1 are $\lambda_{N_1}(\mathbf{s}) = q(\mathbf{s})\lambda_{N_0}(\mathbf{s})$ and $\lambda_{N_1,2}(\mathbf{s},\mathbf{u}) = q(\mathbf{s})q(\mathbf{u})\lambda_{N_0,2}(\mathbf{s},\mathbf{u})$, respectively. For a classical q-thinned point process where $q(\mathbf{s}) \equiv q^o$, it can be seen that $K_{N_1}(h) = K_{N_0}(h)$, $h \geq 0$. Kenkel (1988) uses the K function to test for departures from classical q-thinning.

Suppose the data consist of the locations $\{\mathbf{s}_i: i = 1,2,\ldots,n\}$ of $n = N(A)$ events in a bounded study region A at time t_0, and of the marks $\{z(\mathbf{s}_i): i = 1,2,\ldots,n\}$, which denote those events that survived to time t_1. Assume that the point pattern at t_1 is a realization of a thinned point process with thinning function $q_{\boldsymbol{\theta}}$, where $q_{\boldsymbol{\theta}}$ belongs to a family of parameterized thinning functions $\{q_{\boldsymbol{\theta}}: \boldsymbol{\theta} \in \Theta\}$. Notice that, under q-thinning, the random marks $Z(\mathbf{s}_1), Z(\mathbf{s}_2),\ldots, Z(\mathbf{s}_n)$ are independent Bernoulli random variables. Consider the problem of estimating $\boldsymbol{\theta}$, conditional on the point pattern at t_0. The conditional likelihood of $\boldsymbol{\theta}$ is given by

$$l(\boldsymbol{\theta}) = \prod_{i=1}^{n} \{q_{\boldsymbol{\theta}}(\mathbf{s}_i)\}^{z(\mathbf{s}_i)}\{1 - q_{\boldsymbol{\theta}}(\mathbf{s}_i)\}^{1-z(\mathbf{s}_i)}. \qquad (8.5.99)$$

Suppose $q_{\boldsymbol{\theta}}(\mathbf{s})$ is a function of $\mathbf{x}(\mathbf{s})$, a vector that may contain functions of the spatial coordinates and explanatory (spatial) variables. Note that the condition $q_{\boldsymbol{\theta}}(\mathbf{s}) \in [0,1]$ must be satisfied for all $\mathbf{s} \in A$ and all $\boldsymbol{\theta} \in \Theta$, which can be achieved by writing

$$q_{\boldsymbol{\theta}}(\mathbf{s}) \equiv \frac{\exp\{\mathbf{x}(\mathbf{s})'\boldsymbol{\theta}\}}{1 + \exp\{\mathbf{x}(\mathbf{s})'\boldsymbol{\theta}\}}, \qquad (8.5.100)$$

a logistic model. Then the conditional likelihood is,

$$l(\boldsymbol{\theta}) = \frac{\prod_{i=1}^{n} \exp\{z(\mathbf{s}_i) \cdot \mathbf{x}(\mathbf{s}_i)'\boldsymbol{\theta}\}}{\prod_{i=1}^{n}[1 + \exp\{\mathbf{x}(\mathbf{s}_i)'\boldsymbol{\theta}\}]}, \qquad (8.5.101)$$

which when maximized yields the maximum likelihood estimator $\hat{\boldsymbol{\theta}}$ of $\boldsymbol{\theta}$.

Karr (1985) considers the state estimation of N_0 given a realization of N_1 and known thinning function q, namely, the problem of minimum mean-squared-error reconstruction of N_0 from a given realization of N_1. Circumstances where this problem arises in practice are less likely, because usually the thinning function is not known.

Interrupted Point Process

Let N_0 be any spatial point process on a set X ($\mathbb{R}^d$ or a subset of $\mathbb{R}^d$). Let $Q(\mathbf{s})$ be a random field on X, which is independent of N_0 and whose realizations $q(\cdot)$ satisfy $0 \leq q(\mathbf{s}) \leq 1$, for all $\mathbf{s} \in X$. For any realization $Q(\mathbf{s}) = q(\mathbf{s})$, independently mark the events of N_0 with marks $Z(\mathbf{s})$ that take the value 1 with probability $q(\mathbf{s})$ and 0 otherwise. Suppose each event $\mathbf{s}$ in the initial point process N_0 is retained in the point process N_1 if and only if $Z(\mathbf{s}) = 1$. The point process N_1 is called an *interrupted point process* generated by the *initial point process* N_0 and the *interrupting field* Q (Stoyan, 1979a).

Suppose N_0 is second-order stationary and isotropic with intensity λ_{N_0} and second-order intensity $\lambda^o_{N_0,2}(h)$. Likewise, let Q be second-order stationary and isotropic with mean $E(Q(\mathbf{s})) = p$ and covariogram $C^o_Q(\|\mathbf{s} - \mathbf{u}\|) \equiv E(Q(\mathbf{s}) \cdot Q(\mathbf{u})) - p^2$. Then, the intensity of the interrupted point process N_1 is $\lambda_{N_1} = p\lambda_{N_0}$. Similarly, the second-order intensity of N_1 is $\lambda^o_{N_1,2}(h) = \lambda^o_{N_0,2}(h)\{C^o_Q(h) + p^2\}$. If K_0 is the K function for N_0, then from Eq. (8.3.32), the K function for N_1 is

$$K_1(h) = K_0(h) + p^{-2}\int_0^h C^o_Q(r)\, dK_0(r), \qquad h \geq 0; \qquad (8.5.102)$$

see Diggle (1983, p. 67).

Equation (8.5.102) provides the basis for estimation of the parameters of an interrupted point process. Let N_0 be a spatial point process with K function $K_0(h; \boldsymbol{\theta})$ and let $C^o_Q(h; \boldsymbol{\psi})$ be the covariogram of the interrupting field; K_0 and C^o_Q are known up to parameters $\boldsymbol{\theta}$ and $\boldsymbol{\psi}$, respectively. For example, Stoyan (1979a) considers the initial process N_0 to be a simple inhibition process (Section 8.5.4), and Diggle (1983, pp. 86–87) considers N_0 to be a Neyman–Scott process (Section 8.5.3). From (8.5.102), the K function $K_1(h; \boldsymbol{\theta}, \boldsymbol{\psi})$ of the interrupted point process N_1 can be found. Then, a least-squares estimator for $(\boldsymbol{\theta}, \boldsymbol{\psi})$ could be obtained by finding the $(\hat{\boldsymbol{\theta}}, \hat{\boldsymbol{\psi}})$ that minimizes the criterion

$$D(\boldsymbol{\theta}, \boldsymbol{\psi}) \equiv \int_0^{h_0} \left\{ \left(\hat{K}_1(h)\right)^c - \left(K_1(h; \boldsymbol{\theta}, \boldsymbol{\psi})\right)^c \right\}^2 dh, \qquad (8.5.103)$$

where $\hat{K}_1(h)$ is the empirical K function of the interrupted point process and the constant c is chosen to obtain approximate homoskedasticity; see Diggle (1983, p. 74).

Simple Dependent-Thinned Point Process

The thinned and interrupted point processes assume that the thinning function or interrupting field is independent of the spatial locations of the events in N_0. The simple inhibition processes of Section 8.5.4 are examples of processes with dependent thinning, where N_0 is a homogeneous Poisson process with given intensity ρ. These models can be extended to allow N_0 to be any spatial point process. For example, Stoyan (1988) assumes that N_0 is a realization of a Matern–Bartlett process (Section 8.5.4), whose dependent thinning is controlled by $f_0(\cdot)$ [see Eq. (8.5.54)]. Then, the realization ϕ_0 of N_0 is thinned further by a Matern–Bartlett process, whose dependent thinning is controlled by $f_1(r) \geq f_0(r)$, $0 \leq r \leq 1$.

Markov Thinned Point Process

A Markov thinned point process also allows dependent thinning. Let $\mathbf{s}_1, \mathbf{s}_2, \ldots, \mathbf{s}_n$ be a realization of N_0 in A and let $Z(\mathbf{s}_i)$ equal 1 if and only if the ith event is retained and equal 0 otherwise. Consider the conditional distribution of $Z(\mathbf{s}_i)$ with respect to all other values $\{Z(\mathbf{s}_j): j \neq i\}$. The thinned point process is defined to be Markov with neighborhood radius ρ if

$$\begin{aligned}\Pr\left(Z(\mathbf{s}_i) = 1 | \{Z(\mathbf{s}_j): j \neq i\}\right) \\ = \Pr\left(Z(\mathbf{s}_i) = 1 | \{Z(\mathbf{s}_j): j \neq i, \|\mathbf{s}_i - \mathbf{s}_j\| \leq \rho\}\right),\end{aligned}$$

for some $\rho > 0$. Notice that, conditional on $\{\mathbf{s}_i: i = 1, 2, \ldots, n\}$, the process $\{Z(\mathbf{s}_i): i = 1, 2, \ldots, n\}$ is a lattice process, so that an autologistic model (Section 6.5) could be used. Then, the joint density of $\{Z(\mathbf{s}_i)\}$ is

$$f(\mathbf{z}) = k^{-1} \exp\left\{ -\sum_{i=1}^{n} z_i \zeta(\mathbf{s}_i) - \sum\sum_{1 \leq i < j \leq n} z_i z_j \Psi\left(\|\mathbf{s}_i - \mathbf{s}_j\|\right) \right\}, \qquad z \in \{0, 1\}^n, \tag{8.5.104}$$

where the normalizing constant is

$$k = \sum_{z_1=0}^{1} \sum_{z_2=0}^{1} \cdots \sum_{z_n=0}^{1} \exp\left\{ -\sum_{i=1}^{n} z_i \zeta(\mathbf{s}_i) - \sum\sum_{1 \leq i < j \leq n} z_i z_j \Psi\left(\|\mathbf{s}_i - \mathbf{s}_j\|\right) \right\}.$$

If $\Psi \equiv 0$, then the model reduces to a thinned point process with thinning function

$$q(\mathbf{s}) = \frac{\exp\{-\xi(\mathbf{s})\}}{\exp\{-\xi(\mathbf{s})\} + 1}, \qquad \mathbf{s} \in A.$$

8.5.7* Other Models

Models for point patterns exhibiting regularity have been proposed by Dacey (1966), Brown and Holgate (1974), and Diggle (1975). The central-place model has been widely applied in geography to model the locations of towns (Dacey, 1966). Brown and Holgate's (1974) thinned-plantation model can be applied to certain forestry problems and Diggle's (1975) model is a superposition of a Poisson process with the vertices of a regular lattice.

Central-Place Process

The central-place model was proposed by geographers to model the locations of towns in a homogeneous region (Christaller, 1933; Losch, 1954) and has been used by Shapiro et al. (1985) to model the locations of blue cones in the retina of Macaque monkeys. It is based on the assumption that the resources of a region are uniformly distributed over the region, so the optimal distribution of towns is a hexagonal grid of points. A stochastic version of this, called the central-place process, was first introduced by Dacey (1966) and is generated as follows:

1. Let H be a hexagonal or square lattice of points.
2. For each vertex $\mathbf{u} \in H$, independently generate one event $\mathbf{s}$ according to a d-dimensional density function $f(\mathbf{s} - \mathbf{u})$, where $f(\cdot)$ is radially symmetric about zero.
3. The central-place process consists of the superposition of events generated from the vertices of H.

Thinned-Plantation Process

Brown and Holgate (1974) consider a spatial point process obtained by thinning a square lattice. Let H be a square lattice of points. Independently mark the vertices $\mathbf{s}$ of H with marks $Z(\mathbf{s})$ that take the value 1 with probability q and 0 otherwise. A point $\mathbf{s} \in H$ is retained if and only if $Z(\mathbf{s}) = 1$. The superposition of retained points defines a spatial point process called a thinned-plantation process.

Superposition of a Poisson Process with a Regular Lattice

Diggle (1975) considers a spatial point process formed by the superposition of a Poisson process with intensity ρ and a point process whose events are the vertices of a square (or triangular, for that matter) lattice with unit spacing. This spatial point process can be used to model a continuous range of spatial point patterns, ranging from extreme regularity when $\rho = 0$ to complete spatial randomness as ρ tends to infinity.

8.5.8* Some Final Remarks

The present review has considered the major stochastic models for univariate spatial point patterns. Although this list of models is extensive, new scientific applications will demand the development of additional models. The models in this section can serve as the building blocks for future model development, for example, by superposition of familiar models. Diggle (1983, pp. 68–69) considers the superposition of a Poisson process with a Neyman–Scott process and the superposition of a Poisson process with a regular lattice has just been considered. Another way to build new models is to carry out dependent or independent thinning of familiar models.

Several of the models considered here are equivalent in the sense that they have identical Laplace functionals. Neyman–Scott processes are often equivalent to Cox processes. A q-thinned Poisson process is equivalent to an inhomogeneous Poisson process. An interrupted Poisson process is equivalent to a Cox process. No statistical analysis based on a single realization of a spatial point pattern can distinguish between two spatial point processes having identical Laplace functionals. This lack of identifiability makes scientific interpretation of model parameters difficult.

If more than one realization of a spatial point process were available, then it might be possible to identify the underlying model. For example, McDonald (1989) gives a rank test for the "Poissonness" of independent-and-identically-distributed realizations of a spatial point process. However, because in the majority of cases only a single realization is available, tests of this sort have limited use.

Because spatial point patterns are often the end result of dynamic processes that occur over time as well as over space, one approach to attacking the nonidentifiability problem would be to model the point pattern over time as well, provided the data were available; see Section 8.8. For some applications, it may be possible to break a point process into several components and analyze each component individually. For example, for a spatial birth-and-death process, individual analyses might be performed on the birth process and the death process (e.g., Rathbun and Cressie, 1990b). Thus, an understanding may be gained of the global process that would not have been possible from analyzing the superposition of the two processes.

The methods for estimating the parameters of point process models can mostly be put in one of two major classes, namely, maximum likelihood or least squares using the K function. Unfortunately, very little is known about the statistical properties of most of these estimators. Are they unbiased, what is their variance, and are they consistent, efficient, and asymptotically Gaussian? Kutoyants (1984) and Ogata (1978) have addressed these questions for maximum likelihood estimators of inhomogeneous Poisson processes and self-exciting process, respectively, in $\mathbb{R}^1$. Krickeberg (1982) and Rathbun and Cressie (1990a) extend Kutoyants' results to d-dimensional space. There are currently no such results for maximum likelihood estimators of Markov point-process parameters, and Ripley (1984) expresses doubts as to their efficiency for such a model. To my knowledge, no results regarding unbiasedness, variance estimation, consistency, efficiency, and asymptotic Gaussianity are available for Diggle's least-squares estimators based on the K function or for any of the other *ad hoc* approaches considered here.

8.6* MULTIVARIATE SPATIAL POINT PROCESSES

A multivariate spatial point pattern consists of the locations of two or more types of events in a bounded study region $A \subset \mathbb{R}^d$. In ecology, for example, event types may be different tree or animal species. Then assume that the locations of types of events are realized from a multivariate point process on a set X ($\mathbb{R}^d$ or a subset of $\mathbb{R}^d$). In this section, I shall review briefly the theory of multivariate spatial point processes, methods of estimating reduced second moment measures, and some multivariate spatial–point-process models. Bivariate spatial point patterns and processes will be used as the paradigm; extensions to multitype events are usually straightforward.

8.6.1* Theoretical Considerations

A bivariate spatial point pattern can be defined either through the spatial locations $\{\mathbf{s}_j^{(i)}: i = 1,2;\ j = 1,2,\dots\}$ of type i ($i = 1,2$) events or through a bivariate counting measure $\boldsymbol{\phi} \equiv (\phi_1, \phi_2)'$ on X ($\mathbb{R}^d$ or a subset of $\mathbb{R}^d$). For Borel sets B_1 and B_2, $(\phi_1(B_1), \phi_2(B_2))$ gives the number of events of type i in B_i ($i = 1,2$); so $(\phi_1(B_1), \phi_2(B_2)) \in \{(n_1, n_2): n_1 = 0,1,2,\dots;\ n_2 = 0,1,2,\dots\}$.

Now think of the bivariate point pattern as a realization of a point process. Let $(\Omega, \mathscr{A}, P)$ be a probability space and let Φ be a collection of locally finite bivariate counting measures on $X \subset \mathbb{R}^d$. On Φ define $\mathscr{N}$, the smallest σ algebra generated by sets of the form $\{\boldsymbol{\phi} \in \Phi: (\phi_1(B_1), \phi_2(B_2)) = (n_1, n_2)\}$, for all Borel sets B_1, B_2 and all $n_1, n_2 \in \{0,1,2,\dots\}$. Then, a *bivariate spatial point process* $\mathbf{N} \equiv (N_1, N_2)'$ on X is a measurable mapping of $(\Omega, \mathscr{A})$ into $(\Phi, \mathscr{N})$. A bivariate spatial point process defined over the probability space $(\Omega, \mathscr{A}, P)$ induces a probability measure $\Pi_{\mathbf{N}}(Y) \equiv P\{\mathbf{N} \in Y\}$, $Y \in \mathscr{N}$, on $(\Phi, \mathscr{N})$.

Multivariate Moment Measures

Moment measures can be defined for bivariate spatial point processes in an obvious way. The *mean measure* is defined as

$$\mu_{N_i}(B) \equiv E(N_i(B)) = \int_{\Phi} \phi_i(B) \Pi_{\mathbf{N}}(d\boldsymbol{\phi}), \qquad i = 1,2, \qquad (8.6.1)$$

which is a measure on $(X, \mathscr{X})$.

Similarly, a *cross second-moment measure* is defined by

$$\begin{aligned}\mu_{N_1N_2}(B_1 \times B_2) &\equiv E(N_1(B_1)N_2(B_2)) \\ &= \int_{\Phi} \phi_1(B_1)\phi_2(B_2)\Pi_{\mathbf{N}}(d\boldsymbol{\phi}), \qquad (8.6.2)\end{aligned}$$

where $B_1, B_2 \in \mathscr{X}$. Note that $\mu_{N_1 N_2}$ is a measure on $(X^2, \mathscr{X}^{(2)})$, where $X^2 \equiv X \times X$ and $\mathscr{X}^{(2)}$ is the smallest σ algebra generated by sets of the form $B_1 \times B_2$ $(B_1, B_2 \in \mathscr{X})$. The usual second-moment measures of N_1 and N_2 are denoted $\mu_{N_1 N_1}$ and $\mu_{N_2 N_2}$, respectively. The *cross-covariance measure* is defined by

$$C_{N_1 N_2}(B_1 \times B_2) \equiv \mu_{N_1 N_2}(B_1 \times B_2) - \mu_{N_1}(B_1) \cdot \mu_{N_2}(B_2), \quad (8.6.3)$$

where $B_1, B_2 \in \mathscr{X}$.

Intensity measures can also be defined for bivariate spatial point processes. Let $d\mathbf{s}$ and $d\mathbf{u}$ be small regions located at $\mathbf{s}$ and $\mathbf{u}$, respectively. Then the *first-order intensity* is defined by

$$\lambda_i(\mathbf{s}) \equiv \lim_{\nu(d\mathbf{s}) \to 0} \mu_{N_i}(d\mathbf{s})/\nu(d\mathbf{s}), \qquad i = 1,2, \quad (8.6.4)$$

provided the limit exists. Similarly, the *cross second-order intensity* is defined by

$$\lambda_{12}(\mathbf{s}, \mathbf{u}) \equiv \lim_{\substack{\nu(d\mathbf{s}) \to 0 \\ \nu(d\mathbf{u}) \to 0}} \frac{\mu_{N_1 N_2}(d\mathbf{s} \times d\mathbf{u})}{\nu(d\mathbf{s})\nu(d\mathbf{u})}, \quad (8.6.5)$$

provided the limit exists. The usual second-order intensities of N_1 and N_2 are denoted $\lambda_{11}(\mathbf{s}, \mathbf{u})$ and $\lambda_{22}(\mathbf{s}, \mathbf{u})$, respectively. If $\mathbf{N}$ is stationary and isotropic, then $\lambda_i(\mathbf{s}) = \lambda_i$ and $\lambda_{ij}(\mathbf{s}, \mathbf{u}) = \lambda_{ij}^o(\|\mathbf{s} - \mathbf{u}\|)$, $i, j = 1, 2$, for all $\mathbf{s}, \mathbf{u} \in X$.

Reduced Palm Distribution

The reduced Palm distribution with respect to spatial point process N_i is defined through the *reduced Campbell measure* with respect to N_i:

$$\mathscr{C}_{N_i}^!(B \times Y) \equiv \int_\Phi \int_B I[(\boldsymbol{\phi} - \delta_\mathbf{s}) \in Y]\phi_i(d\mathbf{s})\Pi_\mathbf{N}(d\boldsymbol{\phi}). \quad (8.6.6)$$

For an event of ϕ_i at $\mathbf{s}$, $\boldsymbol{\phi} - \delta_\mathbf{s}$ represents the bivariate spatial point pattern $\boldsymbol{\phi}$ minus the event of type i at $\mathbf{s}$. Clearly, $0 \le \mathscr{C}_{N_i}^!(B \times Y) \le \mu_{N_i}(B)$, for all $Y \in \mathscr{N}$ and all $B \in \mathscr{X}$, so $\mathscr{C}_{N_i}^!$ is absolutely continuous with respect to μ_{N_i}, $i = 1, 2$, which are assumed to be σ finite. By the Radon–Nikodym Theorem, there exist (a.e.) uniquely determined measures $P_{N_i, \mathbf{s}}^!$ on $(\Phi, \mathscr{N})$ such that

$$\mathscr{C}_{N_i}^!(B \times Y) = \int_B P_{N_i, \mathbf{s}}^!(Y)\mu_{N_i}(d\mathbf{s}), \qquad i = 1, 2. \quad (8.6.7)$$

The measure $P_{N_i, \mathbf{s}}^!$ is called the *reduced Palm distribution* of N_i with respect to s. For a stationary bivariate spatial point process $\mathbf{N}$, the reduced Palm distribution does not depend on s; then write $P^!_{N_i,\mathbf{s}} \equiv P^!_{N_i,\mathbf{0}}$.

Reduced Second Moment Measure

Hanisch and Stoyan (1979) extend the reduced second moment measure to bivariate spatial point processes. Suppose that the bivariate spatial point process $\mathbf{N}$ is stationary and isotropic with intensity $\boldsymbol{\lambda} \equiv (\lambda_1, \lambda_2)'$. Then, the reduced second moment measure of N_i and N_j is defined by

$$\lambda_j \mathscr{K}^{(ij)}(B) \equiv \int_\Phi \phi_j(B) P^!_{N_i,\mathbf{0}}(d\phi), \qquad i,j = 1,2,\ B \in \mathscr{X}, \quad (8.6.8)$$

and $\mathscr{K}^{(ij)}(B) = \mathscr{K}^{(ji)}(-B)$, for all $B \in \mathscr{X}$ (Stoyan and Ohser, 1982). If $B = b(\mathbf{0}, h)$, the closed ball of radius h centered at $\mathbf{0}$, define

$$K^{(ij)}(h) \equiv \mathscr{K}^{(ij)}(b(\mathbf{0}, h)), \qquad h \geq 0. \quad (8.6.9)$$

In words, $\lambda_j K^{(ij)}(h)$ is the expected number of events of N_j within distance h of an event of N_i. The relationship between $K^{(ij)}$ and second-order intensity function λ^o_{ij} is

$$K^{(ij)}(h) = \frac{d\pi^{d/2}}{\lambda_i \lambda_j \Gamma(1 + \frac{d}{2})} \int_0^h u^{d-1} \lambda^o_{ij}(u)\, du, \qquad h \geq 0. \quad (8.6.10)$$

8.6.2* Estimation of the Cross K Function

Corresponding to each of the five estimators for the univariate K function [see Eqs. (8.2.18) through (8.2.21) and (8.4.24)], there are analogous estimators for the cross K function $K^{(12)}$ defined by (8.6.9). In the following, let $\{\mathbf{s}^{(1)}_i: i = 1, 2, \ldots, n_1\}$ and $\{\mathbf{s}^{(2)}_j: j = 1, 2, \ldots, n_2\}$ be the $n_1 = N_1(A)$ events of type 1 and the $n_2 = N_2(A)$ events of type 2, respectively, in a bounded study region A.

Hanisch and Stoyan (1979) extend Ripley's (1976a) estimator $\hat{K}_4$ to obtain the estimator for $K^{(12)}$:

$$\hat{K}^{(12)}_4(h) \equiv \left(\hat{\lambda}_1 \hat{\lambda}_2 \nu(A)\right)^{-1} \sum_{i=1}^{n_1} \sum_{j=1}^{n_2} w\left(\mathbf{s}^{(1)}_i, \mathbf{s}^{(2)}_j\right)^{-1} I\left(\|\mathbf{s}^{(1)}_i - \mathbf{s}^{(2)}_j\| \leq h\right),$$

$$h > 0, \quad (8.6.11)$$

where $\hat{\lambda}_i \equiv n_i/\nu(A)$, $i = 1, 2$. The weight $w(\mathbf{s}^{(1)}_i, \mathbf{s}^{(2)}_j)$ is the proportion of the circumference of a circle centered at $\mathbf{s}^{(1)}_i$, passing through $\mathbf{s}^{(2)}_j$, that is inside the study region A. For a rectangular region A, an explicit formula for $w(\mathbf{s}^{(1)}_i, \mathbf{s}^{(2)}_j)$ is given in Section 8.4.3 and by Diggle (1983, p. 72). The estimator

$\hat{K}_4^{(12)}$ is approximately unbiased provided n_1, n_2, and $\hat{K}_4^{(12)}(h)$ are independent; for small h this is approximately true (Ripley, 1981, p. 159).

Exploiting the fact that $K^{(12)} = K^{(21)}$ when $\mathbf{N}$ is stationary, Lotwick and Silverman (1982) suggest the estimator

$$\tilde{K}_4^{(12)}(h) \equiv \frac{n_1 \hat{K}_4^{(12)}(h) + n_2 \hat{K}_4^{(21)}(h)}{n_1 + n_2}, \qquad h > 0. \qquad (8.6.12)$$

Lotwick and Silverman show that when the component processes N_1 and N_2 are independent Poisson processes, $\tilde{K}^{(12)}$ is more efficient than either $\hat{K}^{(12)}$ or $\hat{K}^{(21)}$.

Stoyan and Ohser (1982) extend Ohser and Stoyan's (1981) estimator $\hat{K}_5$ [see Eq. (8.4.24)] to obtain the following (approximately) unbiased estimator for $K^{(12)}$:

$$\hat{K}_5^{(12)}(h) \equiv \left(\hat{\lambda}_1 \hat{\lambda}_2\right)^{-1} \sum_{i=1}^{n_1} \sum_{j=1}^{n_2} \frac{I\left(\|\mathbf{s}_i^{(1)} - \mathbf{s}_j^{(2)}\| \le h\right)}{\nu\left\{\left(A \oplus \mathbf{s}_i^{(1)}\right) \cap \left(A \oplus \mathbf{s}_j^{(2)}\right)\right\}}, \qquad h > 0, \qquad (8.6.13)$$

where $A \oplus \mathbf{s} \equiv \{\mathbf{a} + \mathbf{s} : \mathbf{a} \in A\}$ and $\hat{\lambda}_i \equiv n_i / \nu(A)$, $i = 1, 2$.

8.6.3* Bivariate Spatial–Point-Process Models

Typically, bivariate spatial–point-process models have been constructed by extending univariate models. The simplest such models are those composed of two independent components, N_1 and N_2. In many respects, the role of independence here is analogous to the role of the homogeneous Poisson process (complete spatial randomness) in Section 8.2.2. One aims to describe departure from independence in terms of either positive or negative dependence between N_1 and N_2. In ecology, for example, positive dependence may result from similar responses of two plant species to spatially varying environmental factors or from the presence of one species that is of benefit to another species. Negative dependence may result from contrary responses of the two species to spatially varying environmental factors or from interspecific competition.

Diggle (1983, p. 92) suggests that there is a "dearth of applicable models for bivariate spatial point patterns... ." By and large, this is still true. Bivariate extensions of univariate models, including the Poisson process, Cox process, Markov process, and interrupted process, are given in the following text.

Independent Spatial Point Processes

Two point processes N_1 and N_2 are independent if

$$\Pi_{\mathbf{N}}(Y_1, Y_2) = \Pi_{N_1}(Y_1)\Pi_{N_2}(Y_2), \qquad (8.6.14)$$

where $Y_i \in \mathcal{N}_i$ and $\mathcal{N}_i$ is the smallest σ algebra generated by sets of the form $\{\phi_i\colon \phi_i(B_i) = n_i\}$, for $n_i = 0, 1, \ldots,$ and all Borel sets $B_i \in \mathcal{X}$, $i = 1, 2$. Suppose N_1 and N_2 are independent point processes. Therefore,

$$\mu_{N_1 N_2}(B_1 \times B_2) = \mu_{N_1}(B_1) \cdot \mu_{N_2}(B_2), \tag{8.6.15}$$

for all $B_1, B_2 \in \mathcal{X}$. Then, the cross-covariance measure is

$$C_{N_1 N_2}(B_1 \times B_2) \equiv 0,$$

for all $B_1, B_2 \in \mathcal{X}$. Equation (8.6.15) implies that, for all $\mathbf{s}, \mathbf{u} \in X$, the second-order intensity is $\lambda_{12}(\mathbf{s}, \mathbf{u}) = \lambda_1(\mathbf{s}) \cdot \lambda_2(\mathbf{u})$. Thus, from (8.6.5), the cross K function is

$$K^{(12)}(h) = \frac{\pi^{d/2} h^d}{\Gamma\left(1 + \frac{d}{2}\right)}, \qquad h \geq 0. \tag{8.6.16}$$

Lotwick (1984) has a counterexample to demonstrate that (8.6.16) does not necessarily imply independence of N_1 and N_2.

Let $G_{ij}(h)$ be the probability that the distance from a randomly chosen event of type i, to the nearest event of type j, is less than or equal to h. Similarly, let G_{0j} be the probability that the distance from a randomly chosen point, to the nearest event of type j, is less than or equal to h. If the point processes N_i and N_j are independent, then

$$G_{ij}(h) = G_{0j}(h), \qquad i \neq j, h \geq 0. \tag{8.6.17}$$

The hypothesis of independence can be tested from quadrat-count data or from distance-based methods. Such methods are appropriate for use in the field, where efficient sparse-sampling methods are desired.

For data based on quadrats, the presence or absence of the two types in each of the quadrats can be summarized in a 2×2 contingency table (Greig-Smith, 1964). Then, independence can be tested in the usual way using, say, Pearson's X^2 statistic. Choice of quadrat size needs some care. To ensure that the four marginal totals of the 2×2 contingency table are as large as possible, quadrats should neither be too large nor too small. To provide a reasonably powerful test, quadrats should be large enough so that both types are present in enough quadrats, but small enough so that both species are absent from enough quadrats. Several quadrat sizes might be tried, but this is usually not practical in the field, where these methods are usually applied.

A distance-based method for testing independence of N_1 and N_2 is given by Goodall (1965), who exploits the relationship (8.6.17). Goodall tests the equality of G_{ij} and G_{0j} by applying a two-sample t test to the square roots of the respective nearest-neighbor distances. Diggle and Cox (1981) show that there is little loss of power using nonparametric tests, such as the Wilcoxon or two-sample Kolmogorov–Smirnov tests. They also note that the randomly chosen event of type i in (8.6.17) can be obtained by choosing it to be the nearest type-i event to a random point.

For mapped data sets, tests based on the cross K function are more powerful. Equation (8.6.10) suggests that estimators of $K^{(12)}$ can be used to investigate independence between N_1 and N_2. Departure from independence can be inferred by comparing an estimator of $K^{(12)}$ to the expression (8.6.16), for which knowledge of the estimator's variance is essential. Under the hypothesis of independence, these variances will depend on the assumed marginal distributions of N_1 and N_2. Lotwick and Silverman (1982) give estimators for the variance of $\tilde{K}_4^{(12)}(h)$ [see Eq. (8.6.12)] when N_1 and N_2 are homogeneous Poisson processes. If the marginal processes are clustered, then the variance of $\tilde{K}_4^{(12)}(h)$ is increased. Conversely, if the marginal processes result in regularity, then the variance of $\tilde{K}_4^{(12)}(h)$ is decreased. Alternatively, an estimate of the variance of any estimator of $K^{(12)}$, assuming independence, can be obtained through independent simulations of the marginal point processes N_1 and N_2.

Lotwick and Silverman (1982) consider a Monte Carlo test for independence, conditional on the observed marginal patterns. Their procedure involves wrapping the sample window A onto a torus and fixing the locations of the type-1 events. Then, each of the type-2 events $\mathbf{s}_j^{(2)}$ is translated to $\mathbf{s}_j^{(2)} + \mathbf{u}$, where $\mathbf{u}$ has a uniform distribution on the torus. For each of k realizations of $\mathbf{u}$, the cross K function $K^{(12)}$ is estimated. Note that, on the torus, no edge corrections are necessary, so $K^{(12)}$ can be estimated from

$$\hat{K}^{(12)}(h) \equiv \left(\hat{\lambda}_1\hat{\lambda}_2\nu(A)\right)^{-1} \sum_{i=1}^{n_1} \sum_{j=1}^{n_2} I\left(\|\mathbf{s}_i^{(1)} - \mathbf{s}_j^{(2)}\| \le h\right), \qquad h > 0,$$

where the distances $\|\mathbf{s}_i^{(1)} - \mathbf{s}_j^{(2)}\|$ are computed toroidally. Simulation envelopes obtained from the k estimates of $K^{(12)}$ can be plotted and compared to the cross K function estimated from the original bivariate point pattern (analogous to the univariate case plotted in Figure 8.14).

Bivariate Poisson Process

A bivariate spatial point process is said to be a bivariate Poisson process if both of its components are homogeneous Poisson processes (Diggle, 1983, p. 93). The component processes need not be independent; models incorporating a dependence structure have been considered by Brown et al. (1981) and Diggle and Cox (1981). Griffiths and Milne (1978) consider a general class of bivariate spatial point processes whose marginal processes N_1 and N_2 are identically distributed Poisson processes.

Brown et al. (1981) construct a bivariate Poisson process that allows either negative or positive dependence between the two types of events. Their model is defined as follows:

1. The events of a primary process are realized from a Poisson process with intensity ρ.
2. Voronoi polygons (see, e.g., the Delauney triangulation in Section 5.9.2) are constructed from the events of the primary process. (Diggle, 1983, p. 6, calls the set of Voronoi polygons a Dirichlet tessellation.)
3. For each Voronoi polygon B, the number of events $N_1(B)$ and $N_2(B)$ of types 1 and 2, respectively, are distributed according to a bivariate Poisson distribution with respective marginal means $\mu_1 = \lambda_1 \nu(B)$ and $\mu_2 = \lambda_2 \nu(B)$.
4. Conditional on $N_1(B)$ and $N_2(B)$, the events of types 1 and 2 are distributed independently and identically according to a uniform distribution on B.
5. The final process is composed of the events of types 1 and 2 only.

The marginal point processes are homogeneous Poisson with intensities λ_1 and λ_2. Positive and negative dependence is incorporated through the bivariate Poisson distribution in step 3.

The linked-pairs model of Diggle and Cox (1981) is a model for positive dependence between two types of events. Their model is defined as follows:

1. The events of type 1 are realized from a Poisson process with intensity λ.
2. For each type-1 event $\mathbf{s}_i^{(1)}$, a type-2 event $\mathbf{s}_i^{(2)}$ is generated independently according to the density function $f(\mathbf{r})$, $\mathbf{r} \in \mathbb{R}^d$, where f is radially symmetric about zero.

The marginal point processes are homogeneous Poisson with identical intensities λ, with the property that event types come in pairs independently and identically displaced from each other.

Bivariate Cox Process

The bivariate Cox process on $\mathbb{R}$ was introduced by Cox and Lewis (1972) and extended to $\mathbb{R}^d$ by Diggle and Milne (1983). Let $\mathbf{M} \equiv (M_1, M_2)'$, where M_1 and M_2 are not necessarily independent random measures on $\mathbb{R}^d$, and let $\boldsymbol{\mu} \equiv (\mu_1, \mu_2)'$ be a realization of $\mathbf{M}$. Then $\mathbf{N} \equiv (N_1, N_2)'$ is a *bivariate Cox process directed by* $\mathbf{M}$ if, conditional on $\mathbf{M} = \boldsymbol{\mu}$, N_1 and N_2 are independent inhomogeneous Poisson processes with respective mean measures μ_1 and μ_2 (Section 8.5.2).

Frequently, bivariate Cox processes are defined in terms of the random intensity $\boldsymbol{\Lambda}(\cdot) \equiv (\Lambda_1(\cdot), \Lambda_2(\cdot))'$, where

$$\Lambda_i(\mathbf{s}) \equiv \lim_{\nu(d\mathbf{s}) \to 0} M_i(d\mathbf{s})/\nu(d\mathbf{s}), \qquad i = 1,2,$$

provided the limits exist almost surely. Then, conditional on $\mathbf{\Lambda}(\cdot) = (\lambda_1(\cdot), \lambda_2(\cdot))'$, N_1 and N_2 are independent inhomogeneous Poisson processes with intensity functions $\lambda_1(\cdot)$ and $\lambda_2(\cdot)$, respectively.

Let $\mu_{M_i}(B) \equiv E(M_i(B))$ and $\mu_{M_iM_j}(B_1 \times B_2) \equiv E(M_i(B_1) \cdot M_j(B_2))$ define the first- and second-moment measures, respectively, of the random measure $\mathbf{M}$. Through a conditioning argument, it is easy to see that $\mu_{N_i}(B_i) = \mu_{M_i}(B_i)$, for all $B_i \in \mathscr{X}$, $i = 1, 2$. Furthermore, conditional independence of N_1 and N_2, given $\mathbf{M} = \boldsymbol{\mu}$, yields,

$$\mu_{N_1N_2}(B_1 \times B_2) = E\{E(N_1(B_1) \cdot N_2(B_2) | \mathbf{M} = \boldsymbol{\mu})\}$$

$$= E\big(\mu_{N_1}(B_1) \cdot \mu_{N_2}(B_2)\big) = \mu_{M_1M_2}(B_1 \times B_2). \quad (8.6.18)$$

Then, the cross-covariance measure is

$$C_{N_1N_2}(B_1 \times B_2) = \mu_{M_1M_2}(B_1 \times B_2) - \mu_{M_1}(B_1) \cdot \mu_{M_2}(B_2). \quad (8.6.19)$$

Therefore, the first- and second-order intensities of a bivariate Cox process $\mathbf{N}$ are identical to the first- and second-order intensities of the directing measure $\mathbf{M}$: $\lambda_{N_i}(\mathbf{s}) = \lambda_{M_i}(\mathbf{s}) \equiv \lambda_i(\mathbf{s})$ and $\lambda_{N_iN_j}(\mathbf{s}, \mathbf{u}) = \lambda_{M_iM_j}(\mathbf{s}, \mathbf{u}) \equiv \lambda_{ij}(\mathbf{s}, \mathbf{u})$, where $\mathbf{s}, \mathbf{u} \in X$. Define $C_{ij}(\mathbf{s}, \mathbf{u}) \equiv \operatorname{cov}(\Lambda_i(\mathbf{s}), \Lambda_j(\mathbf{u}))$, $i, j = 1, 2$. Then $\lambda_{ij}(\mathbf{s}, \mathbf{u}) = C_{ij}(\mathbf{s}, \mathbf{u}) + \lambda_i(\mathbf{s}) \cdot \lambda_j(\mathbf{u})$. If the random intensity $\mathbf{\Lambda}(\cdot)$ is stationary and isotropic, then $\lambda_i(\mathbf{s}) = \lambda_i$, $\lambda_{ij}(\mathbf{s}, \mathbf{u}) = \lambda_{ij}^o(\|\mathbf{s} - \mathbf{u}\|)$, and $C_{ij}(\mathbf{s}, \mathbf{u}) = C_{ij}^o(\|\mathbf{s} - \mathbf{u}\|) = \lambda_{ij}^o(\|\mathbf{s} - \mathbf{u}\|) - \lambda_i\lambda_j$. Therefore, from (8.6.10), the cross K function is

$$K^{(12)}(h) = \frac{\pi^{d/2}h^d}{\Gamma(1 + \frac{d}{2})} + \frac{d\pi^{d/2}}{\lambda_1\lambda_2\Gamma(1 + \frac{d}{2})}\int_0^h u^{d-1}C_{12}^o(u)\,du, \qquad h \geq 0. \quad (8.6.20)$$

Diggle and Milne (1983) consider two special cases of bivariate Cox processes, namely, the linked Cox process that leads to positive dependence, and the balanced Cox process that leads to negative dependence. These are described in the following two paragraphs.

Let the Cox process of type-2 events be stationary and isotropic with random intensity Λ_2. Then, for a linked Cox process, define the process of type-1 events to be a Cox process with random intensity

$$\Lambda_1 = \omega\Lambda_2,$$

for some positive constant ω. It is easily shown that $\omega = \lambda_1/\lambda_2$, the ratio of first-order intensities of N_1 and N_2. Then, $\lambda_{11}^o(h) = \omega\lambda_{12}^o(h) = \omega^2\lambda_{22}^o(h)$ and

$\lambda_{12}^o(h) = \omega(C_{22}^o(h) + \lambda_2^2)$. In $\mathbb{R}^2$, the K functions are

$$K^{(11)}(h) = K^{(12)}(h) = K^{(22)}(h) = \pi h^2 + 2\pi\lambda_2^{-2}\int_0^h uC_{22}^o(u)\,du, \qquad h \geq 0. \tag{8.6.21}$$

For a balanced Cox process, let

$$\Lambda_1(\cdot) + \Lambda_2(\cdot) = \omega,$$

for some fixed constant $\omega > 0$. Then, the superposition of N_1 and N_2 is a homogeneous Poisson process with intensity ω. Furthermore,

$$\lambda_{11}^o(h) = -\lambda_{12}^o(h) + \lambda_1\lambda_2 + \lambda_1^2 = \lambda_{22}^o(h) - \lambda_2^2 + \lambda_1^2$$

and

$$\lambda_{12}^o(h) = -C_{22}^o(h) + \lambda_1\lambda_2.$$

If the covariance function C_{22}^o is nonnegative, then N_1 and N_2 are negatively dependent. In $\mathbb{R}^2$, the cross K function is

$$\begin{aligned} K^{(12)}(h) &= \pi h^2 - 2\pi(\lambda_1\lambda_2)^{-1}\int_0^h uC_{22}^o(u)\,du \\ &= \pi h^2 - (2\lambda_1\lambda_2)^{-1}\sum_{j=1}^{2}\lambda_j^2(K^{(jj)}(h) - \pi h^2), \qquad h \geq 0. \end{aligned} \tag{8.6.22}$$

Mutual Inhibition Point Process

Diggle and Cox (1981) extend the simple sequential inhibition point process for univariate spatial point patterns (Section 8.5.4; Diggle et al., 1976) to form a mutual inhibition process for bivariate spatial point patterns. For a mutual inhibition point process, events of each type are generated in an alternating sequence over the bounded study region A. At each stage, the next event of a given type is realized from a uniform distribution over the region of A that is at least distance δ away from any previously assigned events of the opposite type. The procedure terminates when a prespecified number of events have been placed or when it is impossible to continue. This process can be generalized by taking δ to be a function of event type.

A two-parameter generalization of the mutual inhibition point process is also considered by Diggle and Cox (1981). They define a *survival function*

$$q(h) = \begin{cases} (h/\delta)^\beta, & \text{if } 0 \leq h \leq \delta, \\ 1, & \text{if } h \geq \delta, \end{cases} \tag{8.6.23}$$

a function of the distance h from an event of one type to an event of the opposite type. As before, events are generated in an alternating sequence of types. Suppose that t events of each type have been generated. A trial location for the $(t+1)$th event of a given type is realized from a uniform distribution on A and is retained with probability $\pi_{j=1}^{t} q(h_j)$, where h_j denotes the distance from the trial event to the jth retained event of the opposite type. If the trial event is not retained, new trial events are generated until one is retained. Termination occurs when a prespecified number of events have been placed or it is impossible to continue. This process can be further generalized by taking δ and β to be functions of event type or, more generally, by considering other families of survival functions.

Bivariate Markov Point Process

Extensions of Markov point processes to model bivariate spatial point patterns have been considered by Diggle (1983, pp. 102–103), Isham (1984), and Ogata and Tanemura (1985). Suppose $\{\mathbf{s}_j^{(i)}: j = 1, \ldots, n_i;\ i = 1, 2\}$ are the spatial locations of events of types 1 and 2 in a study region A. Then, conditional on n_1 and n_2, assume the likelihood can be expressed as

$$\frac{n_1!n_2!}{c_{n_1n_2}(\boldsymbol{\tau}_1, \boldsymbol{\tau}_2, \boldsymbol{\eta}_1, \boldsymbol{\eta}_2, \boldsymbol{\gamma})} \times \exp\Bigg\{ -\sum_{i=1}^{n_1} \xi_{\boldsymbol{\tau}_1}^{(1)}(\mathbf{s}_i^{(1)}) - \sum_{j=1}^{n_2} \xi_{\boldsymbol{\tau}_2}^{(2)}(\mathbf{s}_j^{(2)}) - \sum_{1 \le k < l \le n_1} \Psi_{\boldsymbol{\eta}_1}^{(11)}(\|\mathbf{s}_k^{(1)} - \mathbf{s}_l^{(1)}\|) - \sum_{1 \le k < l \le n_2} \Psi_{\boldsymbol{\eta}_2}^{(22)}(\|\mathbf{s}_k^{(2)} - \mathbf{s}_l^{(2)}\|) - \sum_{k=1}^{n_1} \sum_{l=1}^{n_2} \Psi_{\boldsymbol{\gamma}}^{(12)}(\|\mathbf{s}_k^{(1)} - \mathbf{s}_l^{(2)}\|) \Bigg\}, \tag{8.6.24}$$

where $c_{n_1n_2}(\cdot)$ is the normalizing constant analogous to (8.5.73). Large-scale effects are described by $\xi_{\boldsymbol{\tau}_1}^{(1)}$ and $\xi_{\boldsymbol{\tau}_2}^{(2)}$. The pair-potentials $\Psi_{\boldsymbol{\eta}_1}^{(11)}$ and $\Psi_{\boldsymbol{\eta}_2}^{(22)}$ describe the interactions among events of the same type, whereas $\Psi_{\boldsymbol{\gamma}}^{(12)}$ describes interactions between type-1 and type-2 events.

Maximum likelihood estimation of the parameters $\boldsymbol{\tau}_1$, $\boldsymbol{\tau}_2$, $\boldsymbol{\eta}_1$, $\boldsymbol{\eta}_2$, and $\boldsymbol{\gamma}$ of (8.6.24) requires the evaluation of the normalizing constant $c_{n_1n_2}$. An approximation for a stationary process ($\xi_{\boldsymbol{\tau}_1}^{(1)} \equiv 0$ and $\xi_{\boldsymbol{\tau}_2}^{(2)} \equiv 0$), using second-order cluster integrals, can be found in Ogata and Tanemura (1985).

Isham (1984) considers a bivariate version of the Strauss process. Define

$$Y^{(ii)} \equiv \sum\sum_{1 \le k < l \le n_i} I(\|\mathbf{s}_k^{(i)} - \mathbf{s}_l^{(i)}\| \le \rho), \qquad i = 1, 2,$$

and

$$Y^{(12)} \equiv \sum_{k=1}^{n_1} \sum_{l=1}^{n_2} I(\|\mathbf{s}_k^{(1)} - \mathbf{s}_l^{(2)}\| \le \rho).$$

Then the Janossy density [i.e., the likelihood $l(\boldsymbol{\theta})$, where $\boldsymbol{\theta} \equiv (\beta_1, \beta_2, \gamma_{11}, \gamma_{22}, \gamma_{12})'$] of the bivariate Strauss process is defined to be

$$j_{n_1 n_2}\left(\left\{\mathbf{s}_j^{(i)}: j = 1, \ldots, n_i; i = 1, 2\right\}\right)$$
$$\equiv e^{-2\nu(A)} \cdot \alpha^{-1} \cdot \beta_1^{n_1} \cdot \beta_2^{n_2} \cdot \gamma_{11}^{Y^{(11)}} \cdot \gamma_{22}^{Y^{(22)}} \cdot \gamma_{12}^{Y^{(12)}}, \qquad (8.6.25)$$

where β_1, β_2, γ_{11}, γ_{22}, and γ_{12} are positive constants and α is the normalizing constant. The bivariate Strauss process is a bivariate Markov point process with intensities β_1 and β_2, and pair-potentials

$$\Psi^{(ij)}\left(\|\mathbf{s}_k^{(i)} - \mathbf{s}_l^{(j)}\|\right) = \begin{cases} \log \gamma_{ij}, & \text{if } \|\mathbf{s}_k^{(i)} - \mathbf{s}_l^{(j)}\| \le \rho, \\ 0, & \text{if } \|\mathbf{s}_k^{(i)} - \mathbf{s}_l^{(j)}\| > \rho, \ 1 \le i \le j \le 2. \end{cases}$$

Isham (1984) shows that in $\mathbb{R}^2$ the K function of the bivariate Strauss process can be approximated by

$$K^{(ij)}(h) \simeq \begin{cases} \gamma_{ij}\pi h^2, & \text{if } 0 < h \le \rho, \\ \pi h^2 - (1 - \gamma_{ij})\pi\rho^2, & \text{if } h \ge \rho, \end{cases} \qquad (8.6.26)$$

with error that is $O((1 - \gamma_{ij})^2)$, $1 \le i \le j \le 2$.

Bivariate Interrupted Point Process

The bivariate interrupted point process is obtained by randomly removing events from a realization of a bivariate spatial point process $\mathbf{N}_{(0)}$ on a bounded set A (Diggle and Milne, 1983). Let $\mathbf{Q}(\mathbf{s}) \equiv (Q_1(\mathbf{s}), Q_2(\mathbf{s}))'$ be a bivariate random field on X that is independent of $\mathbf{N}_{(0)}$ and whose realizations satisfy $0 \le q_i(\mathbf{s}) \le 1$, for all $\mathbf{s} \in X$, $i = 1, 2$. For any realization $\mathbf{Q}(\mathbf{s}) = \mathbf{q}(\mathbf{s})$, independently mark the events of $\mathbf{N}_{(0)}$ with marks $\mathbf{Z}(\mathbf{s}) \equiv (Z_1(\mathbf{s}), Z_2(\mathbf{s}))'$ such that $Z_i(\mathbf{s})$ takes the value 1 with probability $q_i(\mathbf{s})$ and 0 otherwise. Suppose each event $\mathbf{s}^{(i)}$ in the initial bivariate point process $\mathbf{N}_{(0)}$ is retained in the bivariate point process $\mathbf{N}$ if and only if $Z_i(\mathbf{s}^{(i)}) = 1$, $i = 1, 2$. Then the bivariate point process $\mathbf{N}$ is called a *bivariate interrupted point process* generated by the initial point process $\mathbf{N}_{(0)}$ and the interrupting field $\mathbf{Q}$ (Stoyan, 1979a).

Suppose the bivariate point process $\mathbf{N}_{(0)}$ is second-order stationary and isotropic with intensity $(\lambda_{1(0)}, \lambda_{2(0)})$ and second-order intensity functions $\lambda^o_{i(0), j(0)}(h)$, $i, j = 1, 2$. Likewise, let the interrupting field $\mathbf{Q}$ be second-order stationary and isotropic with mean $E(Q_i(\mathbf{s})) = p_i$ and covariance function $C^o_{Q_iQ_j}(\|\mathbf{s} - \mathbf{u}\|) = E(Q_i(\mathbf{s}) \cdot Q_j(\mathbf{u})) - p_i p_j$. Then, the intensities of the bivariate interrupted point process $\mathbf{N}$ are given by $(\lambda_1, \lambda_2) = (p_1\lambda_{1(0)}, p_2\lambda_{2(0)})$. Similarly, the second-order intensities of $\mathbf{N}$ are given by $\lambda^o_{ij}(h) = \lambda^o_{i(0), j(0)}(h)\{C^o_{Q_iQ_j}(h) + p_i p_j\}$. If $K^{(ij)}_{(0)}$ are the K functions for $\mathbf{N}_{(0)}$, then from

Eq. (8.6.10) the K functions for $\mathbf{N}$ are given by

$$K^{(ij)}(h) = K_{(0)}^{(ij)}(h) + (p_i p_j)^{-1} \int_0^h C_{Q_i Q_j}^{o}(u)\, dK_{(0)}^{(ij)}(u),$$

$$1 \le i \le j \le 2,\ h \ge 0 \quad (8.6.27)$$

(Diggle and Milne, 1983).

8.7* MARKED SPATIAL POINT PROCESSES

This section is concerned with modeling data observed as a marked spatial point pattern, which consists of the locations of events in a bounded study region $A \subset \mathbb{R}^d$ and associated measurements or marks. For example, events may be trees in a forest, towns in a geographic region, or epicenters of earthquakes. The corresponding marks may be, respectively, species types or diameters of trees, population sizes of towns, or magnitudes of earthquakes. Note that a multivariate spatial point pattern (Section 8.6) is a special case of a marked spatial point pattern, where the marks are categorical variables belonging to a finite set. Here, it is assumed that the locations of events $\{\mathbf{s}_i\}$ and their corresponding marks $\{Z(\mathbf{s}_i)\}$ are the realization of some stochastic process of the form

$$\{Z(\mathbf{s}) : \mathbf{s} \in D\},$$

where both $Z(\cdot)$ and D are random (see Section 1.1).

A marked spatial point process will be defined later in terms of random locations of events in a set X ($\mathbb{R}^d$ or a subset of $\mathbb{R}^d$) and corresponding random marks $Z(\cdot)$ in a set $\mathscr{F}$. The most general mark space used in this book will be $\mathscr{F} =$ the set of all closed subsets in $\mathbb{R}^d$ (Section 9.4). However, in this section, $\mathscr{F}$ will be $\mathbb{R}$ or a subset of $\mathbb{R}$ (generalizations to vector marks are obvious). If an event located at $\mathbf{s} \in X$ is marked by a quantity $z \in \mathscr{F}$, then $(\mathbf{s}, z)$ is a point in $X \times \mathscr{F}$. Hence, a marked spatial point process on X can be represented mathematically as a point process on the product space $X \times \mathscr{F}$. This representation is frequently useful.

8.7.1* Theoretical Considerations

A marked spatial point pattern (a realization of a marked spatial point process) can be defined either through the spatial locations of events $\{\mathbf{s}_i: i = 1, 2, \ldots\}$ and their corresponding marks $\{Z(\mathbf{s}_i): i = 1, 2, \ldots\}$ or through a counting measure ϕ on $X \times \mathscr{F}$. Let $\mathscr{X}$ be the Borel sets of X and let $\mathscr{B}$ be the Borel sets of $\mathscr{F}$. Then $\phi(A \times B)$ is the number of events in A whose marks belong to B, where $A \in \mathscr{X}$ and $B \in \mathscr{B}$. Let $(\Omega, \mathscr{A}, P)$ be a probability space and let Φ be a collection of locally finite counting measures on

$X \times \mathscr{F}$ (Section 8.3). On Φ define $\mathscr{N}$, the smallest σ algebra generated by sets of the form $\{\phi \in \Phi: \phi(A \times B) = n\}$, for all $A \in \mathscr{X}$, all $B \in \mathscr{B}$, and all $n \in \{0, 1, 2, \dots\}$.

Definition. A *marked spatial point process* N is a point process on $X \times \mathscr{F}$, that is, a measurable mapping of $(\Omega, \mathscr{A})$ into $(\Phi, \mathscr{N})$. A marked spatial point process defined over $(\Omega, \mathscr{A}, P)$ induces a probability measure $\Pi_N(Y) \equiv P(N \in Y)$ on $(\Phi, \mathscr{N})$, for all $Y \in \mathscr{N}$. ∎

From a model-construction point of view, this definition is not all that helpful because the role of the point process on $X \subset \mathbb{R}^d$ needs to be separated from the role of the random marks on $\mathscr{F}$.

Stationary and Isotropic Marked Spatial Point Processes

Just as for ordinary spatial point processes, assumptions of stationarity and isotropy are often made to reduce the parameter space and to allow parameter estimation from a single realization. For each $\mathbf{s} \in \mathbb{R}^d$, define the translated marked spatial point process $N_{\mathbf{s}}$ in terms of the original marked spatial point process N, as follows. Define $\Pi_{N_{\mathbf{s}}}(A \times B) \equiv \Pi_N(\{\mathbf{a} + \mathbf{s}: \mathbf{a} \in A\} \times B)$, for all $A \in \mathscr{X}$ and $B \in \mathscr{B}$, to be the induced probability distribution of $N_{\mathbf{s}}$. Then, N is said to be *stationary* if $\Pi_N = \Pi_{N_{\mathbf{s}}}$, for all $\mathbf{s} \in \mathbb{R}^d$. An analogous definition of an *isotropic* marked spatial point process follows by replacing the preceding translation operation with the rotation operation (about the origin).

Moment Measures

The moment measures of a marked spatial point process are simple extensions of the moment measures of an ordinary spatial point process. The mean measure is defined by

$$\mu_N(A \times B) \equiv E(N(A \times B)) = \int_{\Phi} \phi(A \times B) \Pi_N(d\phi), \quad (8.7.1)$$

a measure on $(X \times \mathscr{F}, \mathscr{X} \times \mathscr{B})$, where $\mathscr{X} \times \mathscr{B}$ is the smallest σ algebra generated by sets of the form $A \times B$, $A \in \mathscr{X}$ and $B \in \mathscr{B}$. Similarly, the kth moment measure of N is defined by

$$\begin{aligned} \mu_N^{(k)}&((A_1 \times B_1) \times \cdots \times (A_k \times B_k)) \\ &\equiv E(N(A_1 \times B_1) \cdots N(A_k \times B_k)) \\ &= \int_{\Phi} \phi(A_1 \times B_1) \cdots \phi(A_k \times B_k) \Pi_N(d\phi), \end{aligned} \quad (8.7.2)$$

where $A_1, \dots, A_k \in \mathscr{X}$ and $B_1, \dots, B_k \in \mathscr{B}$. Note that $\mu_N^{(k)}$ is a measure on $(X^k \times \mathscr{F}^k, \mathscr{X}^{(k)} \times \mathscr{B}^{(k)})$, where $\mathscr{X}^{(k)} \times \mathscr{B}^{(k)}$ is the smallest σ algebra generated by sets of the form $(A_1 \times B_1) \times \cdots \times (A_k \times B_k)$, $A_i \in \mathscr{X}$, $B_i \in \mathscr{B}$, $i = 1, 2, \ldots, k$.

Factorial moment measures may also be defined for marked spatial point processes. Let $\mathbf{s}_1, \mathbf{s}_2, \ldots, \mathbf{s}_n$ and $Z(\mathbf{s}_1), Z(\mathbf{s}_2), \ldots, Z(\mathbf{s}_n)$ be the events and their associated marks corresponding to the realization ϕ of a marked spatial point process N. Then the kth factorial moment measure of N is defined by

$$\alpha_N^{(k)}((A_1 \times B_1) \times \cdots \times (A_k \times B_k))$$
$$\equiv \int_\Phi \sum_{\substack{\text{distinct} \\ \mathbf{s}_1, \mathbf{s}_2, \ldots, \mathbf{s}_k \in \phi}} I(\mathbf{s}_1 \in A_1, Z(\mathbf{s}_1) \in B_1) \cdots$$
$$I(\mathbf{s}_k \in A_k, Z(\mathbf{s}_k) \in B_k) \Pi_N(d\phi). \quad (8.7.3)$$

Note that $\alpha_N^{(k)}((A \times B)^k) = E[N(A \times B)\{N(A \times B) - 1\} \cdots \{N(A \times B) - k + 1\}]$, the kth factorial moment of the random variable $N(A \times B)$. For $k = 2$,

$$\alpha_N^{(2)}((A_1 \times B_1) \times (A_2 \times B_2)) = \mu_N^{(2)}((A_1 \times B_1) \times (A_2 \times B_2))$$
$$- \mu_N((A_1 \times B_1) \cap (A_2 \times B_2)). \quad (8.7.4)$$

Intensity Functions

Let $d\mathbf{s}$ and $d\mathbf{u}$ be small regions located at $\mathbf{s}$ and $\mathbf{u}$, respectively. Intensity functions are defined relative to mark sets $B \in \mathscr{B}$. The first-order intensity relative to $B \in \mathscr{B}$ is defined by

$$\lambda_B(\mathbf{s}) \equiv \lim_{\nu(d\mathbf{s}) \to 0} \mu_N(d\mathbf{s} \times B)/\nu(d\mathbf{s}), \qquad \mathbf{s} \in X, \qquad (8.7.5)$$

provided the limit exists. Similarly, the second-order intensity, relative to $B_1, B_2 \in \mathscr{B}$, is defined by

$$\lambda_{B_1 B_2}(\mathbf{s}, \mathbf{u}) \equiv \lim_{\substack{\nu(d\mathbf{s}) \to 0 \\ \nu(d\mathbf{u}) \to 0}} \frac{\mu_N^{(2)}((d\mathbf{s} \times B_1) \times (d\mathbf{u} \times B_2))}{\nu(d\mathbf{s})\nu(d\mathbf{u})}, \qquad \mathbf{s}, \mathbf{u} \in X, \quad (8.7.6)$$

provided the limit exists. If N is stationary and isotropic, then $\lambda_B(\mathbf{s}) = \lambda_B$ and $\lambda_{B_1 B_2}(\mathbf{s}, \mathbf{u}) = \lambda^o_{B_1 B_2}(\|\mathbf{s} - \mathbf{u}\|)$, for all $\mathbf{s}, \mathbf{u} \in X$.

Reduced Palm Distribution

The reduced Palm distribution with respect to a mark set $B \in \mathscr{B}$ can be defined through the reduced Campbell measure with respect to the marked spatial point process N:

$$\mathscr{C}_N^!((A \times B) \times Y) \equiv \int_\Phi \int_A I[(\phi - \delta_{\mathbf{s}}) \in Y]\phi(d\mathbf{s} \times B)\Pi_N(d\phi). \quad (8.7.7)$$

Clearly, $0 \le \mathscr{C}_N^!((A \times B) \times Y) \le \mu_N(A \times B)$, for all $Y \in \mathscr{N}$, all $A \in \mathscr{X}$, and all $B \in \mathscr{B}$, so $\mathscr{C}_N^!$ is absolutely continuous with respect to μ_N, which is assumed to be σ finite. By the Radon–Nikodym Theorem, there exist (a.e.) uniquely determined measures $P_{N,\mathbf{s}}^{B!}$ on $(\Phi, \mathscr{N})$ such that

$$\mathscr{C}_N^!((A \times B) \times Y) = \int_A P_{N,\mathbf{s}}^{B!}(Y)\mu_N(d\mathbf{s} \times B). \quad (8.7.8)$$

The measure $P_{N,\mathbf{s}}^{B!}$ is called the reduced Palm distribution of N with respect to the mark set B. For a stationary marked spatial point process N, the reduced Palm distribution does not depend on $\mathbf{s}$; hence, write $P_{N,\mathbf{s}}^{B!} \equiv P_{N,\mathbf{0}}^{B!}$.

Reduced Second Moment Measure

Hanisch and Stoyan (1979) extend the reduced second moment measure to marked spatial point processes. Suppose that the marked spatial point process N is stationary and isotropic. Let B_1 and B_2 be any two mark sets. Then the reduced second moment measure of N with respect to B_1 and B_2 is defined by

$$\lambda_{B_1}\mathscr{K}^{(B_1B_2)}(A) \equiv \int_\Phi \phi(A \times B_2)P_{N,\mathbf{0}}^{B_1!}(d\phi), \qquad A \in \mathscr{X}. \quad (8.7.9)$$

Note that $\mathscr{K}^{(B_1B_2)}(A) = \mathscr{K}^{(B_2B_1)}(-A)$, for all $A \in \mathscr{X}$ and all $B_1, B_2 \in \mathscr{B}$ (Stoyan and Ohser, 1982). If $A = b(\mathbf{0}, h)$, the closed ball of radius h centered at $\mathbf{0}$, define

$$K^{(B_1B_2)}(h) \equiv \mathscr{K}^{(B_1B_2)}(b(\mathbf{0}, h)), \qquad h \ge 0. \quad (8.7.10)$$

Heuristically, $\lambda_{B_2}K^{(B_1B_2)}(h)$ is the expected number of extra events having marks in B_2 within distance h of an event whose mark is in B_1. The relationship between $K^{(B_1B_2)}$ and $\lambda_{B_1B_2}^o$ is

$$K^{(B_1B_2)}(h) = \frac{d\pi^{d/2}}{\lambda_{B_1}\lambda_{B_2}\Gamma(1 + \frac{d}{2})}\int_0^h u^{d-1}\lambda_{B_1B_2}^o(u)\,du. \quad (8.7.11)$$

Mark Distribution

The mark distribution was originally defined by Stoyan (1984a). The definition of the mark distribution is based on Campbell measures and Palm distributions. Assume that the mean measure μ_N is σ finite. Define the Campbell measure of a marked spatial point process N by

$$\mathscr{C}_N((A \times B) \times Y) \equiv \int_Y \phi(A \times B)\Pi_N(d\phi), \quad (8.7.12)$$

a measure on $\mathscr{X} \times \mathscr{B} \times \mathscr{N}$. Note that $\mathscr{C}_N((A \times B) \times \Phi) = \mu_N(A \times B)$, for all $A \times B \in \mathscr{X} \times \mathscr{B}$. Thus, $\mathscr{C}_N$ is absolutely continuous with respect to μ_N. By the Radon–Nikodym Theorem, there exist (a.e.) uniquely defined measures $P_{N,\mathbf{s}}^z$ on $(\Phi, \mathscr{N})$ such that

$$\mathscr{C}_N((A \times B) \times Y) = \int_{A \times B} P_{N,\mathbf{s}}^z(Y)\mu_N(d\mathbf{s} \times dz). \quad (8.7.13)$$

The measure $P_{N,\mathbf{s}}^z$ is called the Palm distribution with respect to $\mathbf{s}$ and z (Stoyan, 1984a). (An analogous definition based on the reduced Campbell measure yields the reduced Palm distribution $P_{N,\mathbf{s}}^{z!}$.) For a stationary marked spatial point process N, the Palm distribution $P_{N,\mathbf{s}}^z$ does not depend on $\mathbf{s}$; hence, write $P_{N,\mathbf{s}}^z \equiv P_{N,\mathbf{0}}^z$.

Stoyan's (1984a) definition of the mark distribution is based on the Palm distribution of a stationary marked spatial point process. For $\phi = \{(\mathbf{s}_i, z_i): i = 1, 2, \ldots\} \in \Phi$ and $\mathbf{s} \in \mathbb{R}^d$, let $\phi_\mathbf{s}$ denote the marked point pattern $\{(\mathbf{s}_i + \mathbf{s}, z_i): i = 1, 2, \ldots\}$ and, for $Y \in \mathscr{N}$, let $Y_{-\mathbf{s}}$ denote the set $\{\phi \in \Phi: \phi_\mathbf{s} \in Y\}$. Then (8.7.13) takes the form

$$\mathscr{C}_N((A \times B) \times Y) = \lambda_0 \int_A \int_B P_{N,\mathbf{0}}^z(Y_{-\mathbf{s}})\mathscr{M}(dz)\nu(d\mathbf{s}), \quad (8.7.14)$$

where λ_0 is the intensity of the associated marginal point process N_0. The distribution $\mathscr{M}$ is called the *mark distribution* or *mark probability measure*. It can be interpreted as yielding the distribution of a mark given there is an event located at a point $\mathbf{s}$. Because the process is stationary, this distribution does not depend upon $\mathbf{s}$. Stoyan (1984a) shows that $P_{N,\mathbf{0}}^B$, the Palm distribution with respect to mark set $B \in \mathscr{B}$, is related to $P_{N,\mathbf{0}}^z$ by

$$P_{N,\mathbf{0}}^B(Y) = \int_B P_{N,\mathbf{0}}^z(Y)\mathscr{M}(dz)/\mathscr{M}(B). \quad (8.7.15)$$

Two-Point Mark Distribution

Definition of the two-point mark distribution is essential to the definition of the mark covariance function that follows. Stoyan (1984a) uses the factorial moment measures to define the two-point mark distribution as follows. Let $\alpha_N^{(2)}$ denote the second factorial moment measure of the marked point process [see (8.7.3)] and let $\alpha_{N_0}^{(2)}$ denote the second factorial moment measure of the associated marginal point process N_0. Note that

$$\alpha_{N_0}^{(2)}(A_1 \times A_2) = \alpha_N^{(2)}((A_1 \times \mathscr{F}) \times (A_2 \times \mathscr{F})), \quad (8.7.16)$$

for all $A_1, A_2 \in \mathscr{X}$. Thus, $\alpha_N^{(2)}$ is absolutely continuous with respect to $\alpha_{N_0}^{(2)}$, assumed σ finite. By the Radon–Nikodym Theorem, there exist uniquely determined measures $\mathscr{M}_{\mathbf{s}_1,\mathbf{s}_2}$ on $(\mathscr{F}^2, \mathscr{B}^{(2)})$ such that

$$\alpha_N^{(2)}((A_1 \times B_1) \times (A_2 \times B_2)) = \int_{A_1 \times A_2} \mathscr{M}_{\mathbf{s}_1,\mathbf{s}_2}(B_1 \times B_2)\alpha_{N_0}^{(2)}(d\mathbf{s}_1 \times d\mathbf{s}_2). \quad (8.7.17)$$

The measure $\mathscr{M}_{\mathbf{s}_1, \mathbf{s}_2}$ is called a *two-point mark distribution*. It can be interpreted as yielding the joint distribution of the marks of two events, given those events are located at $\mathbf{s}_1$ and $\mathbf{s}_2$. If N is a stationary isotropic marked spatial point process, then $\mathscr{M}_{\mathbf{s}_1, \mathbf{s}_2}$ depends only on $h = \|\mathbf{s}_1 - \mathbf{s}_2\|$; in this case, the simpler notation $\mathscr{M}_h$ will be used.

Mark Covariance Function

Stoyan (1984b) uses the two-point mark distribution to define the mark covariance function. For stationary isotropic marked spatial point processes, the mark covariance function is a function only of the distance $\|\mathbf{h}\|$ between events at $\mathbf{s}$ and $\mathbf{s} + \mathbf{h}$:

$$C^o(\|\mathbf{h}\|) \equiv \operatorname{cov}(Z(\mathbf{s}), Z(\mathbf{s} + \mathbf{h})). \tag{8.7.18}$$

Then the mark covariance function can be expressed as

$$C^o(\|\mathbf{h}\|) = k_{zz}(\|\mathbf{h}\|) - \mu_z^2, \tag{8.7.19}$$

where $\mu_z \equiv E(Z(\mathbf{s}))$, the mean of the marks, and $k_{zz}(\|\mathbf{h}\|)$ is the mean of the product of marks of two events distance $\|\mathbf{h}\|$ apart:

$$k_{zz}(\|\mathbf{h}\|) \equiv E(Z(\mathbf{s}) \cdot Z(\mathbf{s} + \mathbf{h})). \tag{8.7.20}$$

The expectation in the definition of μ_z is conditional on the presence of an event at $\mathbf{s}$, for some $\mathbf{s} \in \mathbb{R}^d$. In terms of the mark distribution,

$$\mu_z = \int_{\mathscr{F}} z \mathscr{M}(dz). \tag{8.7.21}$$

Similarly, k_{zz} can be expressed in terms of the two-point mark distribution:

$$k_{zz}(h) = \int_{\mathscr{F}^2} z_1 z_2 \mathscr{M}_h(dz_1 \times dz_2). \tag{8.7.22}$$

In addition to the mark covariance function, the mark variogram could also be defined:

$$2\gamma^o(\|\mathbf{h}\|) \equiv \operatorname{var}(Z(\mathbf{s}) - Z(\mathbf{s} + \mathbf{h})), \qquad \mathbf{h} \in \mathbb{R}^d, \tag{8.7.23}$$

where the variance is conditional on the presence of events at $\mathbf{s}$ and $\mathbf{s} + \mathbf{h}$, for some $\mathbf{s} \in \mathbb{R}^d$. The mark variogram can also be obtained through the two-point mark distribution:

$$2\gamma^o(h) = \int_{\mathscr{F}^2} (z_1 - z_2)^2 \mathscr{M}_h(dz_1 \times dz_2). \tag{8.7.24}$$

Mark-Sum Measure

Assume that the marks are nonnegative. Stoyan (1984a) defines the *mark-sum measure* of ϕ to be

$$\phi_m(A) \equiv \sum_{(\mathbf{s}, Z(\mathbf{s})) \in \phi} Z(\mathbf{s}) I(\mathbf{s} \in A), \qquad A \in \mathscr{X}, \phi \in \Phi. \quad (8.7.25)$$

Moment measures can then be defined for the mark-sum measure. The mean measure is

$$\mu_m(A) \equiv \int_\Phi \sum_{(\mathbf{s}, Z(\mathbf{s})) \in \phi} Z(\mathbf{s}) I(\mathbf{s} \in A) \Pi_N(d\phi) \qquad (8.7.26)$$

and the second-moment measure is

$$\mu_m^{(2)}(A_1 \times A_2) \equiv \int_\Phi \left\{ \sum_{(\mathbf{s}_1, Z(\mathbf{s}_1)) \in \phi} Z(\mathbf{s}_1) I(\mathbf{s}_1 \in A_1) \right\}$$
$$\times \left\{ \sum_{(\mathbf{s}_2, Z(\mathbf{s}_2)) \in \phi} Z(\mathbf{s}_2) I(\mathbf{s}_2 \in A_2) \right\} \Pi_N(d\phi). \quad (8.7.27)$$

Intensity functions can also be defined for the mark-sum measures. Let $d\mathbf{s}$ and $d\mathbf{u}$ be small regions located at $\mathbf{s}$ and $\mathbf{u}$, respectively. The first-order intensity is

$$\lambda_m(\mathbf{s}) \equiv \lim_{\nu(d\mathbf{s}) \to 0} \mu_m(d\mathbf{s})/\nu(d\mathbf{s}), \qquad (8.7.28)$$

provided the limit exists. Similarly, the second-order intensity is

$$\lambda_{m,2}(\mathbf{s}, \mathbf{u}) \equiv \lim_{\substack{\nu(d\mathbf{s}) \to 0 \\ \nu(d\mathbf{u}) \to 0}} \frac{\mu_m^{(2)}(d\mathbf{s} \times d\mathbf{u})}{\nu(d\mathbf{s})\nu(d\mathbf{u})}, \qquad (8.7.29)$$

provided the limit exists. If N is stationary and isotropic, then $\lambda_m(\mathbf{s}) = \lambda_m = \lambda_0 \mu_z$ and $\lambda_{m,2}(\mathbf{s}, \mathbf{u}) = \lambda_{m,2}^o(\|\mathbf{s} - \mathbf{u}\|) = k_{zz}(\|\mathbf{s} - \mathbf{u}\|) \cdot \lambda_{2,N_0}^o(\|\mathbf{s} - \mathbf{u}\|)$, for all $\mathbf{s}, \mathbf{u} \in X$ (Stoyan, 1984a). Here, λ_0 and $\lambda_{2,N_0}^o(\cdot)$ are the first- and second-order intensities of the associated marginal spatial point process N_0.

Stoyan (1984a) defines the mark-sum measure's reduced second moment measure $\mathscr{K}^{(m)}$, for stationary N, through

$$\lambda_m \mu_z \mathscr{K}^{(m)}(A) \equiv \int_0^\infty z \left[\int_\Phi \sum_{\substack{(\mathbf{s}, Z(\mathbf{s})) \in \phi \\ \mathbf{s} \neq \mathbf{0}}} Z(\mathbf{s}) I(\mathbf{s} \in A) P_{N,\mathbf{0}}^z(d\phi) \right] \mathscr{M}(dz),$$
$$(8.7.30)$$

where $P_{N,\mathbf{0}}^{z}$ is the Palm distribution with respect to z and $\mathscr{M}$ is the mark distribution. The reduced second moment measure can also be expressed in terms of $\mathscr{K}_{N_0}$, the reduced second moment measure of the associated marginal spatial point process N_0:

$$\mu_z^2 \mathscr{K}^{(m)}(A) = \int_A \int_{[0,\infty)^2} z_1 z_2 \mathscr{M}_{\mathbf{h}}(dz_1 \times dz_2) \mathscr{K}_{N_0}(d\mathbf{h}). \quad (8.7.31)$$

If $A = b(\mathbf{0}, h)$, the closed ball of radius h centered at $\mathbf{0}$, define

$$K^{(m)}(h) \equiv \mathscr{K}^{(m)}(b(\mathbf{0}, h)), \qquad h \geq 0. \quad (8.7.32)$$

Then, $\lambda_{m,2}^{o} = \{\Gamma(1 + \frac{d}{2})/(d\pi^{d/2}h^{d-1})\}(\lambda_0\mu_z)^2(dK^{(m)}(h)/dh)$.

8.7.2* Estimation of Moment Measures

In what follows, estimators for $K^{(B_1B_2)}(h)$, $K^{(m)}(h)$, $C^o(h)$, and $2\gamma^o(h)$ are considered. Let $\{\mathbf{s}_i: i = 1, \ldots, n\}$ be the locations of $n = N(A)$ events in a bounded study region $A \subset \mathbb{R}^d$ and let $\{Z(\mathbf{s}_i): i = 1, \ldots, n\}$ be the corresponding marks.

Estimating $K^{(B_1B_2)}$

When B_1 and B_2 are disjoint sets in $\mathscr{B}$, estimators of $K^{(B_1B_2)}$ for a marked spatial point process are the same as those for $K^{(12)}$ of a multivariate spatial point process (Section 8.6.2): Take events in B_1 to be type-1 events and events in B_2 to be type-2 events. More generally, however, B_1 and B_2 may not be disjoint.

Hanisch and Stoyan (1979) extend Ripley's (1976a) estimator $\hat{K}_4$, given by (8.4.22), to obtain the following estimator for $K^{(B_1B_2)}(h)$ in the bounded study region A:

$$\hat{K}_4^{(B_1B_2)}(h) \equiv \left(\hat{\lambda}_{B_1}\hat{\lambda}_{B_2}\nu(A)\right)^{-1} \times \sum_{i=1}^{n} \sum_{\substack{j=1 \\ i \neq j}}^{n} w(\mathbf{s}_i, \mathbf{s}_j)^{-1} I\left(\|\mathbf{s}_i - \mathbf{s}_j\| \leq h\right) \times I\left(Z(\mathbf{s}_i) \in B_1\right) I\left(Z(\mathbf{s}_j) \in B_2\right), \qquad h \geq 0, \quad (8.7.33)$$

where the weight $w(\mathbf{s}_i, \mathbf{s}_j)$ is the proportion of the circumference of a circle, centered at $\mathbf{s}_i$ and passing through $\mathbf{s}_j$, that is inside the study region A and

$$\hat{\lambda}_{B_k} \equiv \sum_{i=1}^{n} I\left(Z(\mathbf{s}_i) \in B_k\right)/\nu(A), \qquad k = 1, 2. \quad (8.7.34)$$

For a rectangular region A, an explicit formula for $w(\mathbf{s}_i, \mathbf{s}_j)$ is given in Section 8.4.3 and by Diggle (1983, p. 72).

Stoyan and Ohser (1984) extend Ohser and Stoyan's (1981) estimator $\hat{K}_5$ [see Eq. (8.4.24)] to obtain the (approximately) unbiased estimator for $K^{(B_1B_2)}$:

$$\hat{K}_5^{(B_1B_2)}(h) \equiv \left(\hat{\lambda}_{B_1}\hat{\lambda}_{B_2}\right)^{-1} \times \sum_{i=1}^{n}\sum_{\substack{j=1\\ i\neq j}}^{n} \frac{I(\|\mathbf{s}_i - \mathbf{s}_j\| \le h) I(Z(\mathbf{s}_i) \in B_1) I(Z(\mathbf{s}_j) \in B_2)}{\nu((A \oplus \mathbf{s}_i) \cap (A \oplus \mathbf{s}_j))}, \tag{8.7.35}$$

where $h \ge 0$ and $A \oplus \mathbf{s} \equiv \{\mathbf{a} + \mathbf{s}: \mathbf{a} \in A\}$.

Estimating $K^{(m)}$

Based on the work of Stoyan (1984b), two estimators for $K^{(m)}(h)$ can be defined. One is an extension of Ripley's (1976a) estimator for K [see Eq. (8.4.22)]:

$$\hat{K}_4^{(m)}(h) \equiv \left(\hat{\lambda}_0^2\hat{\mu}_z^2\nu(A)\right)^{-1} \sum_{i=1}^{n}\sum_{\substack{j=1\\ i\neq j}}^{n} w(\mathbf{s}_i, \mathbf{s}_j)^{-1} I(\|\mathbf{s}_i - \mathbf{s}_j\| \le h) Z(\mathbf{s}_i) Z(\mathbf{s}_j); \tag{8.7.36}$$

an explanation of the function w is given following (8.7.33).

The other estimator for $K^{(m)}(h)$ is an extension of Ohser and Stoyan's (1981) estimator $\hat{K}_5(h)$ [see Eq. (8.4.24)]:

$$\hat{K}_5^{(m)}(h) \equiv \left(\hat{\lambda}_0\hat{\mu}_z\right)^{-2} \sum_{i=1}^{n}\sum_{\substack{j=1\\ i\neq j}}^{n} \frac{I(\|\mathbf{s}_i - \mathbf{s}_j\| \le h) Z(\mathbf{s}_i) Z(\mathbf{s}_j)}{\nu((A \oplus \mathbf{s}_i) \cap (A \oplus \mathbf{s}_j))}. \tag{8.7.37}$$

In (8.7.36) and (8.7.37), $\hat{\lambda}_0$ and $\hat{\mu}_z$ are estimators of the intensity of the associated marginal spatial point process and the mean of the marks, respectively. An unbiased estimator for the intensity is

$$\hat{\lambda}_0 = n/\nu(A). \tag{8.7.38}$$

The mean of the marks might be estimated by the simple average

$$\bar{Z} \equiv \frac{1}{n}\sum_{i=1}^{n} Z(\mathbf{s}_i),$$

or a more efficient estimator that incorporates the mark covariances.

Estimating the Mark Covariance Function and the Mark Variogram

Stoyan (1984b) considers two edge-corrected estimators for $k_{zz}(h)$, both derived from its expression as a ratio of second-order intensities. One is based on $\hat{K}_4^{(m)}$ and $\hat{K}_4$:

$$\hat{k}_{zz}(h) \equiv (\hat{\mu}_z)^2 \frac{\hat{K}_4^{(m)}(h+\delta) - \hat{K}_4^{(m)}(h-\delta)}{\hat{K}_4(h+\delta) - \hat{K}_4(h-\delta)}, \tag{8.7.39}$$

where δ is a small positive number.

Stoyan's other estimator for $k_{zz}(h)$ is given by

$$\hat{k}_{zz}(h) \equiv \frac{\displaystyle\sum_{i=1}^{n}\sum_{\substack{j=1\\ i\neq j}}^{n} \frac{I(h-\delta < \|\mathbf{s}_i - \mathbf{s}_j\| \le h+\delta) Z(\mathbf{s}_i) Z(\mathbf{s}_j)}{\nu((A \oplus \mathbf{s}_i) \cap (A \oplus \mathbf{s}_j))}}{\displaystyle\sum_{i=1}^{n}\sum_{\substack{j=1\\ i\neq j}}^{n} \frac{I(h-\delta < \|\mathbf{s}_i - \mathbf{s}_j\| \le h+\delta)}{\nu((A \oplus \mathbf{s}_i) \cap (A \oplus \mathbf{s}_j))}}, \tag{8.7.40}$$

where δ is a small positive number. Notice that (8.7.40) is in the same spirit as Stoyan's second estimator of $K^{(m)}$ [see Eq. (8.7.37)].

Estimators for the covariance function may be obtained using either of Stoyan's estimators for k_{zz}. That is,

$$\hat{C}^o(h) = \hat{k}_{zz}(h) - (\hat{\mu}_z)^2. \tag{8.7.41}$$

Notice that $\hat{C}^o$ requires an estimator for μ_z, one possibility being the simple average $\bar{Z}$.

Estimation of the variogram does not require an estimator for μ_z. An edge-corrected estimator of the same form as (8.7.40) can be obtained for the variogram:

$$2\hat{\gamma}^o(h) \equiv \frac{\displaystyle\sum_{i=1}^{n}\sum_{\substack{j=1\\ i\neq j}}^{n} \frac{I(h-\delta < \|\mathbf{s}_i - \mathbf{s}_j\| \le h+\delta) (Z(\mathbf{s}_i) - Z(\mathbf{s}_j))^2}{\nu((A \oplus \mathbf{s}_i) \cap (A \oplus \mathbf{s}_j))}}{\displaystyle\sum_{i=1}^{n}\sum_{\substack{j=1\\ i\neq j}}^{n} \frac{I(h-\delta < \|\mathbf{s}_i - \mathbf{s}_j\| \le h+\delta)}{\nu((A \oplus \mathbf{s}_i) \cap (A \oplus \mathbf{s}_j))}}. \tag{8.7.42}$$

8.7.3* Marked Spatial-Point-Process Models

The simplest (but least interesting) form of a marked spatial point process has independent-and-identically-distributed (i.i.d.) marks that are independent of the associated marginal spatial point process. For example, the *compound Poisson process* has i.i.d. marks that are positive integers, located at the events of a Poisson process in $\mathbb{R}^d$. The mark distribution is sometimes called a batch-size distribution.

For general marked spatial point processes, the marks may be modeled conditional on the observed spatial point pattern. (The spatial point pattern can be modeled first, using models such as those presented in Section 8.5.) The events of the spatial point pattern can be viewed as locations of an irregularly spaced lattice, so that the marks may be considered as spatial-lattice data and modeled accordingly (Chapter 6).

Stoyan (1984b) and Isham (1987) consider marked spatial point processes in which the marks are completely determined by some local property of the spatial point process. For example, marks may be first-nearest-neighbor distances, which are often used to test for complete spatial randomness of an observed spatial point pattern in a study region A. However, these marks are highly correlated.

Models for marked spatial point processes are even less common than those for multivariate spatial point processes. Extensions of the inhomogeneous Poisson process and the Markov point process have been considered. In addition, analogues to some of the simple inhibition processes and thinned processes presented in Section 8.5 have been defined. Some discussion of these models is now given.

Marked Inhomogeneous Poisson Process

Ogata and Katsura (1988) consider a marked inhomogeneous Poisson process, which they define to be an inhomogeneous Poisson process on $X \times \mathscr{F}$. Let $\{\mu_N(A \times B): A \in \mathscr{X}, B \in \mathscr{B}\}$ be its mean measure (Section 8.7.1) and let μ_{N_0} be the mean measure of the associated marginal spatial point process on X. Recall that $\mathscr{F}$ is $\mathbb{R}$ or a subset of $\mathbb{R}$. Define the first-order intensities

$$\lambda(\mathbf{s}, z) \equiv \lim_{\nu(ds \times dz) \to 0} \mu_N(ds \times dz)/\nu(ds \times dz) \qquad (8.7.43)$$

and

$$\lambda_0(\mathbf{s}) \equiv \lim_{\nu(ds) \to 0} \mu_{N_0}(ds)/\nu(ds), \qquad (8.7.44)$$

provided the limits exist. Then

$$\lambda(\mathbf{s}, z) = f(z|\mathbf{s})\lambda_0(\mathbf{s}), \qquad (8.7.45)$$

where $f(z|\mathbf{s})$ is the conditional density of the marks. Upon parameterizing λ_0 and f with parameters $\boldsymbol{\eta}$ and $\boldsymbol{\tau}$, respectively, the negative loglikelihood for $\boldsymbol{\theta} \equiv (\boldsymbol{\eta}', \boldsymbol{\tau}')'$ becomes

$$\begin{aligned} L(\boldsymbol{\theta}; A) &= -\sum_{i=1}^{n} \log \lambda_{\boldsymbol{\theta}}(\mathbf{s}_i, Z(\mathbf{s}_i)) + \int_{\mathscr{F}} \int_A \lambda_{\boldsymbol{\theta}}(\mathbf{s}, z)\nu(ds \times dz) \\ &= L_1(\boldsymbol{\eta}) + L_2(\boldsymbol{\tau}), \end{aligned} \qquad (8.7.46)$$

where

$$L_1(\boldsymbol{\eta}) \equiv -\sum_{i=1}^{n} \log \lambda_{0,\boldsymbol{\eta}}(\mathbf{s}_i) + \int_A \lambda_{0,\boldsymbol{\eta}}(\mathbf{s})\nu(d\mathbf{s}) \tag{8.7.47}$$

and

$$L_2(\boldsymbol{\tau}) \equiv -\sum_{i=1}^{n} \log f_{\boldsymbol{\tau}}(Z(\mathbf{s}_i)|\mathbf{s}_i). \tag{8.7.48}$$

Ogata and Katsura (1988) note that maximum likelihood estimators for $\boldsymbol{\eta}$ and $\boldsymbol{\tau}$ can be obtained separately by maximizing (8.7.47) and (8.7.48), respectively, provided $\boldsymbol{\eta}$ and $\boldsymbol{\tau}$ have no common components. Specifically, they consider the special case

$$f_{\boldsymbol{\tau}}(z|\mathbf{s}) = \beta_{\boldsymbol{\tau}}(\mathbf{s})\exp\{-\beta_{\boldsymbol{\tau}}(\mathbf{s}) \cdot z\}, \qquad z \geq 0,$$

and

$$\lambda_{0,\boldsymbol{\eta}}(\mathbf{s}) = c \cdot \exp\{g_{\boldsymbol{\eta}}(\mathbf{s})\},$$

where $g_{\boldsymbol{\eta}}$ is a two-dimensional spline function.

Marked Markov Point Process

Extensions of Markov point processes to model marked spatial point patterns have been considered by Fiksel (1984a), Ogata and Tanemura (1985), Mase (1986), and Stoyan (1989).

Ogata and Tanemura (1985) assume that the conditional likelihood based on $\{(\mathbf{s}_i, z(\mathbf{s}_i)): i = 1, 2, \ldots, n\}$, conditional on $N(A) = n$ and $z \equiv (z(\mathbf{s}_1), \ldots, z(\mathbf{s}_n))'$, can be expressed in terms of mark pair-potentials, namely,

$$\frac{n!}{c_n(\boldsymbol{\theta}; z)} \exp\left\{-\sum_{1 \leq i < j \leq n}\sum \Psi_{\boldsymbol{\theta}}(\|\mathbf{s}_i - \mathbf{s}_j\|; z(\mathbf{s}_i), z(\mathbf{s}_j))\right\}, \tag{8.7.49}$$

where $c_n(\boldsymbol{\theta}; z)$ is the normalizing constant analogous to (8.5.73). If the neighborhood structure [implied by nonzero $\Psi_{\boldsymbol{\theta}}$s in (8.7.49)] does not depend on the marks, Baddeley and Moller (1989) show that the mark process, conditional on $\{\mathbf{s}_i: i = 1, \ldots, n\}$, is itself a Markov random field on the spatial lattice formed by the point pattern. Maximum likelihood estimation of $\boldsymbol{\theta}$ in (8.7.49) requires the evaluation of the normalizing constant; an approximation using second-order cluster integrals is given by Ogata and Tanemura (1985).

Fiksel (1984a) extends the Takacs–Fiksel estimator (Section 8.5.5) to the case where marks are present. Let $\Pi_{\boldsymbol{\theta}}$ be the induced distribution of the stationary marked Markov point process N and let $P_{\mathbf{0}}^{z!}(\cdot; \boldsymbol{\theta})$ be the reduced Palm distribution with respect to z. Then

$$\int_{\Phi \times \mathcal{F}} \xi(\phi, z) P_{\mathbf{0}}^{z}(d\phi; \boldsymbol{\theta}) \mathcal{M}(dz)$$

$$= \lambda_0^{-1} \int_{\Phi \times \mathcal{F}} \xi(\phi, z) \exp\{-e_{\boldsymbol{\theta}}(\mathbf{0}, z; \phi)\} \Pi_{\boldsymbol{\theta}}(d\phi) F(dz), \tag{8.7.50}$$

for all nonnegative functions ξ on $\Phi \times \mathscr{F}$. In (8.7.50),

$$e_{\boldsymbol{\theta}}(\mathbf{u}, z; \phi) \equiv \sum_{j=1}^{n} \Psi_{\boldsymbol{\theta}}(\|\mathbf{u} - \mathbf{s}_j\|; z, z(\mathbf{s}_j)),$$

the local energy at $\mathbf{u} \in \mathbb{R}^d$ with mark $z \in \mathscr{F}$, and $\exp\{-e_{\boldsymbol{\theta}}\}$ is the Radon–Nikodym derivative of the reduced Campbell measure with respect to the product measure $\nu \times F \times \Pi_{\boldsymbol{\theta}}$. Estimation of $\boldsymbol{\theta}$ then proceeds by methods similar to those discussed in Section 8.5.5.

Mase (1986) and Stoyan (1989) use a marked Markov point process to model the locations of nonintersecting balls. The marks are the radii of the balls centered on event locations.

Marked Matern Point Process

The simple inhibition point processes of Section 8.5.4 were formulated as marked spatial point processes. Recall, for example, Matern's Model II. Let N_0 be a homogeneous Poisson process on $\mathbb{R}^d$ with intensity ρ. Independently mark the events $\mathbf{s}$ of N_0 with nonnegative numbers $Z(\mathbf{s})$ from any absolutely continuous distribution function F. An event $\mathbf{s}$ of N_0 is deleted if there exists another event $\mathbf{u}$ with $\|\mathbf{s} - \mathbf{u}\| < \delta$ and $Z(\mathbf{u}) < Z(\mathbf{s})$. The retained events form the (more regular) point process N_{II}. Expressions for the first- and second-order intensities of N_{II} can be found in Section 8.5.4 [see Eqs. (8.5.49) and (8.5.51)]. Stoyan and Stoyan (1985) give equations for the mark distribution and mark covariance function of N_{II}. In particular, the mark distribution of N_{II} is given by

$$\mathscr{M}(z) = \frac{\int_0^z \exp\{-\rho\omega_d \delta^d F(t)\} F(dt)}{\int_0^\infty \exp\{-\rho\omega_d \delta^d F(t)\} F(dt)}, \qquad z \geq 0, \tag{8.7.51}$$

where $\omega_d \equiv \pi^{d/2}/\Gamma(1 + \frac{d}{2})$ is the volume of the unit sphere in $\mathbb{R}^d$.

8.8 SPACE–TIME POINT PATTERNS

Marked spatial point patterns are often the result of dynamic processes that occur over time as well as space. Consider the locations and times of events $(\mathbf{s}; t)$ and their corresponding marks $Z(\mathbf{s}; t)$ as a realization of a stochastic process of the form

$$\{Z(\mathbf{s}; t) : \mathbf{s} \in D(t), t \in T\}, \tag{8.8.1}$$

where $Z(\mathbf{s}; t)$ and $D(t)$ are random, and the time index set T may or may not be deterministic. However, most published accounts of modeling marked spatial point patterns have usually only had data available at a single instant of time, thus ignoring the temporal component. Consequently, these models often suffer from nonidentifiability problems; widely divergent models can generate identical realizations (see, e.g., Gurland, 1957; Westcott, 1971; Cliff and Ord, 1981, pp. 90–92; Diggle, 1983, pp. 58–59; Section 8.5). Dynamic space–time models are less susceptible to this type of problem.

Two types of space–time point processes may be distinguished, based on the duration of events over time. For a *space–time shock point process*, events occur instantaneously over both time and space. Conversely, for a *space–time survival point process*, events are born at some random location and time, and then live for a random length of time.

A space–time shock point process might be used to model occurrences of earthquakes over time and space in a given geographic region. Realizations of such processes consist of events with locations $\mathbf{s}_i \in X$ at times $t_i \in U$, $i = 1, 2, \ldots$. Generally, data for such processes consist of the locations of events in a study region A during a time interval $U \equiv [0, t]$. It is mathematically convenient to think of a space–time shock point process as a point process on $X \times U$. Then, moment measures and Palm distributions can be defined as in Section 8.3. Alternatively, the space–time shock point process can be viewed as a marked spatial point process with mark space U. Then, moment measures and Palm distributions can be defined as in Section 8.7. Superposition (integration) of events over $[0, t]$ yields a spatial point pattern, to which the methods of Sections 8.2 through 8.5 could be applied.

Space–time survival point processes have applications in ecology and geography. They can be used to describe the locations and time of births of trees (establishment of towns), their growth, and their thinning due to tree mortality (abandonment of towns). Realizations of such processes often consist of the locations of events present at fixed points of time $t_1, t_2, \ldots, t_n$; for example, see Rathbun and Cressie (1990b), where longleaf-pine data are modeled. (An event is born between t_{i-1} and t_i, say, and is detected from the realization observed at time t_i. The event survives until a time belonging to the interval t_{j-1} and t_j, say; that is, it is no longer present in the realization observed at time t_j.) Then, moment measures and Palm distributions can be thought of as functions of time.

In practice, which of these two types of models one might use will depend on the scale of time over which the process is observed. If earthquake data from a highly unstable region are observed on the scale of microseconds, then a space–time survival process may be appropriate. Conversely, if tree data were observed on a scale of millenia, then a space–time shock process might be considered.

Complete spatiotemporal randomness (cstr) is the simplest space–time shock point process, in that there is an absence of structure in time as well as space. As such, cstr is the natural null hypothesis against which observed space–time point patterns could be tested. Formally, a process with cstr is defined to be a homogeneous Poisson process on $X \times U$, where X is the spatial domain and U is the time domain. Consequently, many of the methods for testing csr (Section 8.4) could be extended to test for cstr.

Space–time clustering is perhaps the most frequently considered alternative to cstr. Space–time clustering is said to exist if, among those events that are close in time, there are events that are closer in space than would be expected due to chance alone (McAuliffe and Afifi, 1984). Methods of testing that compare events $\{(\mathbf{s}_i; t_i)\}$ through distances $\{\|\mathbf{s}_i - \mathbf{s}_j\|: |t_i - t_j| < \delta\}$ are considered by Knox (1964), Ederer et al. (1966), Mantel (1967), Pike and Smith (1974), and McAuliffe and Afifi (1984), and are reviewed by Williams (1984). Chen et al. (1984) consider the power of such tests for cstr against various alternative models for space–time clustering.

Fiksel (1984b) proposes a space–time cluster model for a sequence of earthquakes. Let $\{(\mathbf{s}_i; t_i): i = 1, 2, \ldots\}$ denote the locations and times of events ordered so that $t_1 < t_2 < \cdots$. Assume that the position of the $(n + 1)$th event depends on the earlier events $(\mathbf{s}_1; t_1), \ldots, (\mathbf{s}_n; t_n)$. Let $d\mathbf{s}_{n+1}$ be an infinitesimal d-dimensional region located at $\mathbf{s}_{n+1}$. Define the transition density

$$\psi_n(\mathbf{s}_{n+1} | \mathbf{s}_1, \ldots, \mathbf{s}_n) \equiv \lim_{\nu(d\mathbf{s}_{n+1}) \to 0} \frac{\Pr\{N(d\mathbf{s}_{n+1}) = 1 | \mathbf{s}_1, \ldots, \mathbf{s}_n\}}{\nu(d\mathbf{s}_{n+1})}, \quad (8.8.2)$$

assuming the limit exists. If ψ_n is known, then

$$\Pr\{\mathbf{s}_{n+1} \in A | \mathbf{s}_1, \ldots, \mathbf{s}_n\} = \int_A \psi_n(\mathbf{u} | \mathbf{s}_1, \ldots, \mathbf{s}_n) \nu(d\mathbf{u}). \quad (8.8.3)$$

For example, for a space–time shock point process on $\mathbb{R}^2$, Fiksel considers the parametric form

$$\psi_n(\mathbf{s}_{n+1} | \mathbf{s}_1, \ldots, \mathbf{s}_n) = \frac{\lambda^2}{2\pi n} \sum_{i=1}^{n} \exp\{-\lambda \|\mathbf{s}_{n+1} - \mathbf{s}_i\|\}. \quad (8.8.4)$$

Then, conditional on n, the likelihood is given by

$$l_n(\lambda) = \psi_{n-1}(\mathbf{s}_n | \mathbf{s}_1, \ldots, \mathbf{s}_{n-1}) \cdots \psi_1(\mathbf{s}_2 | \mathbf{s}_1) \quad (8.8.5)$$

and a maximum likelihood estimator can be found by maximizing (8.8.5) with respect to λ.

Space–time models for the locations of rain cells are considered by Smith and Karr (1985), Rodriguez-Iturbe and Eagleson (1987), Jacobs et al. (1988), and Rodriguez-Iturbe et al. (1988). For example, temporal evolution of Smith and Karr's model is governed by a "stochastic climatological process" $\{X(t): t = 1, 2, \ldots\}$, where $X(t)$ is the climatological state on day t. The state space for the climatological process is $\{0, 1\}$, where 0 represents a state in which no rainfall can occur and 1 represents a state in which rainfall can occur. For days on which rainfall can occur, rain cells are realized from a Poisson process with random intensity λ.

A temporal component is implicit in the definition of several spatial point processes in Section 8.5. The Neyman–Scott process consists of point processes at two points in time: The parent process is realized from a Poisson process and the offspring process is clustered around the former parents. This process can then be extended to multiple discrete generations of events at times $t_0, t_1, t_2, \ldots$ (Felsenstein, 1975; Kingman, 1977; and Section 8.5.3).

The thinning processes of Section 8.5.6 also consist of point processes at two points in time. Recall that for such processes events are randomly removed from a realization of an initial point process N_0 at time t_0 to obtain the thinned point process N_1 at time t_1. This process can be extended to multiple intervals of time by applying the thinning process to the extant events at the beginning of each interval, resulting in the surviving events at the end of each interval.

The spatial birth-and-death process of Preston (1975), presented in Section 8.5.5, is a special case of a space–time survival point process, where the probability of more than one birth or death in a small interval of time $[t, t + \delta)$ is $o(\delta)$. The asymptotic distribution of the spatial birth-and-death process is the same as the distribution of a Markov point process with local energy given by Eq. (8.5.69) (Glotzel, 1981).

8.9 SPATIAL POINT PATTERNS, SIMULATED AND REAL

The development of statistical methods can be a sterile exercise without their application to real data. Moreover, statistical analyses of substantive problems often stimulate interesting research questions in statistical theory. Simulation is useful when theoretical properties of statistics are unavailable, because theoretical distributions and moments can be approximated by their empirical versions. Furthermore, simulation can have a role in formal inference, such as the Monte Carlo tests presented in Section 8.2 and throughout Section 8.5.

8.9.1 Simulation of Spatial Point Patterns

An algorithm for simulating a realization from a given spatial point process usually follows directly from its definition. Sections 8.4 and 8.5 present various spatial–point-process models and give descriptions of their associated simulation methods. These are summarized briefly in the following text.

Section 8.4.1 gives a description of how to simulate a homogeneous Poisson process in a bounded study region A. This method is based on the observation that the number of events in A has a Poisson distribution with mean $\lambda\nu(A)$ and that, conditional on $N(A) = n$, the n events in A are independently sampled from a uniform distribution on A. The method given for simulating an inhomogeneous Poisson process in A is based on Lewis and Shedler's (1979) rejection-sampling algorithm, and is described in Section 8.5.1. The simulation of a Cox process requires a method for simulating the (random) intensity function $\Lambda(\cdot)$; Section 3.6.1 considers methods for simulating such random processes. Then, conditional on $\Lambda(\cdot) = \lambda(\cdot)$, the Cox process can be simulated as a realization of an inhomogeneous Poisson process with intensity function $\lambda(\cdot)$. The method for simulating a Neyman–Scott process follows directly from its definition and is given in Section 8.5.3. Simple inhibition point processes are defined by how they are simulated; these are described in Section 8.5.4.

Methods for simulating a Markov point process are covered in Section 8.5.5. Ripley (1977, 1987a) uses a spatial birth-and-death process (Preston, 1975), whereas Ogata and Tanemura (1981) use a method based on the Metropolis algorithm (Metropolis et al., 1953; Wood, 1968). Both methods involve simulating a sequence of realizations $\phi_n(1), \phi_n(2), \ldots$, of point processes, such that $\phi_n(t)$ converges in distribution to a Markov point process, as $t \to \infty$. They differ in how the transition from one state $\phi_n(t)$ to the next state $\phi_n(t + 1)$ is defined.

Methods for simulating thinned, interrupted, and simple dependent-thinned point processes follow directly from their definitions; see Section 8.5.6. The Markov thinned point process, defined in Section 8.5.6, is obtained from a binary Markov random field on the lattice formed by the locations of events in N_0. The binary field can be simulated using the Gibbs sampler described in Section 7.7.1.

8.9.2 Spatial Point Patterns

Very few spatial point patterns have been published in the literature. However, Diggle (1983, Appendix) gives values for six such data sets. These include univariate point patterns formed by Japanese black saplings, redwood seedlings (Strauss, 1975a), and biological cells analyzed by Ripley (1977). Three multivariate data sets are also given: Data from Lansing woods includes the locations of 703 hickories, 514 maples, 346 red oaks, 448 white oaks, 135 black oaks, and 105 miscellaneous trees. The bivariate hamster-tumor data consist of the locations of pyknotic nuclei and nuclei arrested in metaphase. The bramble-cane data includes canes from three age classes.

Upton and Fingleton (1985, pp. 62–63) list the locations and diameters of beadlet anemones, which constitutes a marked point pattern. The longleaf-pine data set presented in Section 8.2.1 is a marked point pattern formed by the locations of the trees and their diameters.

CHAPTER 9

Modeling Objects

The three- or four-dimensional world in which we live is full of objects to be measured and summarized. Very often a parsimonious finite collection of measurements is enough for scientific investigation into an object's genesis and evolution. At the other extreme, images of an object viewed from many different angles are a rich source of data that can overwhelm the scientist or engineer with its complexity.

There is a growing need to describe and model objects through their form as well as their size, yet still retain parsimony in their description and modeling. These considerations are important when deciding the type of data to collect in the first place. Thus, the question of whether data or models come first is not always clear, although new models (modified in the light of data analyses) almost always give rise to new ideas about data collection.

In this chapter, the data will be recordings of objects, often images on a photographic plate or a television screen. Such visual images may be analyzed *per se* or they may be converted into numerical data by defining discrete picture elements (pixels); each pixel is recorded according to its gray level or its (color) frequency and intensity. A further conversion to a two-phase image might be made by recording black or 1 if the gray level of a pixel is above a certain threshold or of a certain color, and white or 0 everywhere else. In the sections that follow, I shall concentrate mainly on 0-1 images and the mathematical sets that are used to model them. (Section 7.4 also addresses the problem of analyzing spatial data defined on pixels. However, it is a low-level type of image analysis where the goal, typically, is not characterization of the form of objects. Rather, one is usually most interested in prediction and classification of various collections of pixels.) Not all of the set models are random, but one that is, the *Boolean model*, provides useful descriptions of objects analyzed in this chapter. Other set models are mentioned briefly in Sections 9.1 and 9.3.

Stereology

How to sample high-dimensional objects so that the loss of information is small, but also so that informative measurements are taken in lower (more manageable) dimensions, is the subject of *stereology*. This very important area deserves the treatment of a whole book; instead, I shall take just a few paragraphs and refer the interested reader to a number of very accessible references.

A set in three dimensions consists of volumes, surfaces, curves, and points. Various quantities could be measured in an attempt to characterize the set; for example, one could measure the specific volume, the specific surface area, or the specific length (the word "specific" here refers to the measurement per unit of volume under consideration). One could also consider the set's connectivity number, curvature, convexity measure, and so forth (e.g., Serra, 1982). Based on data from a one- or two-dimensional probe of the object, it is desired to estimate any or all of these quantities.

Sampling considerations are paramount in stereology. Some procedures are able to take images of the object in a highly regularized manner, such as slices at a fixed spacing (in much the same way a wire egg slicer cuts up a boiled egg). The sampling need not be destructive. The important subject of *tomography* came from an idea to use an x-ray machine to measure the transmission across all lines traversing a solid object, calculate the Fourier transform, and then invert it. This has generated considerable mathematical and statistical discussion; see, for example, Shepp and Kruskal (1979), Herman (1980), and Vardi et al. (1985). A more recent development, which is similar in concept but uses a combination of radio frequency waves and high energy magnetic field strengths, is magnetic resonance imaging (MRI). It is a powerful diagnostic tool that allows noninvasive evaluation of brain tumors, herniated discs in the spine, knee injuries to ligamentous structures, and so forth.

Other sampling designs define random probes; indication and estimation then proceed based on this randomness (e.g, Davy and Miles, 1977). If the object is modeled as a random set, stereological methods can be developed from the probability measure of the set model; see Davy (1978).

Much of the ingenuity and creativity of stereologists has been directed toward reducing or removing estimation bias, with too little attention given to variances of estimators (and estimators of those variances). For more details and recent developments in stereology, the reader should consult Coleman (1979), Weibel (1980), Ambartzumian (1982), Sterio (1984), Jensen et al. (1985), Gundersen (1986), Jensen (1987), and Stoyan et al. (1987, Chapter 11).

Chapter Summary

Section 9.1 presents a number of set models of naturally occurring objects. One of them, the Boolean model, receives particular attention in this chapter. An exploratory analysis of a simulated Boolean model is given in Section 9.2. The mathematical foundations for random set models are laid in Section 9.3, and Section 9.4 discusses the properties of the Boolean model. Sections 9.5 and 9.6 continue this discussion into parameter estimation and inference, respectively. Finally, in Section 9.7, a conditional Boolean model is fitted to successive images of *in vitro* cancerous growth.

9.1 SET MODELS

Consider the data at hand to be a large finite number of vector-valued variables, each one indexed by the spatial location of its pixel (picture element). In this chapter, the vector-valued variable will be assumed real-valued (in fact, usually 0-1). Thus, a random set might be viewed as a 0-1 random function. However, the usual random-function techniques handle the geometry of the random set rather crudely. New (nonlinear) operations are needed to probe and sort complicated sets; mathematical morphology (Matheron, 1975; Serra, 1982) systematizes this approach.

It will be assumed that the resolution of the object's image is so good that it is equivalent to consider the data as being defined in Euclidean space. The problems of loss of information in going from continuous space to discrete space and the reconstruction of continuous images from discrete ones are considered by Switzer (1975) and Serra (1982, Part 2). The sets given in this section are mathematical constructs, defined in $\mathbb{R}^d$. Some knowledge of Euclidean topology will be assumed, to the level of knowing the definitions of open, closed, compact, and convex sets (see, e.g., Royden, 1968). Certain classes of sets are often used to model objects, and some of these are presented in the following text.

9.1.1 Fractal Sets

"Fractal" (from Latin, *fractus*, meaning irregular or fragmented) is a word coined by B. Mandelbrot (see, e.g., Mandelbrot, 1982) to describe sets with abrupt and tortuous edges. Mandelbrot argues that fractals are very good models of many natural phenomena, for example, coastlines with their sharpness and convolutedness, or fine particles observed under an electron microscope. What is needed is some way of characterizing these irregular boundaries. (Although it must be said that, depending on the question being addressed, a set model with smooth boundaries may suffice.)

Most of the applications of mathematics use Euclidean geometry, but suppose instead the boundary of the object is modeled by a curve that has no tangents. Mandelbrot (1982) quantifies these non-Euclidean curves using an irregularity index he calls the *fractal dimension*, which is larger than the usual Euclidean dimension. For example, a six-pointed star in $\mathbb{R}^2$ has a boundary of dimension equal to 1, but by successively adding smaller and smaller triangular spikes to each side of the star a compact set in $\mathbb{R}^2$ is created whose *boundary* has fractal dimension D equal to $(\log 4)/(\log 3) = 1.2618$. Mandelbrot (1982, pp. 36, 37) calls this compact set a triadic Koch island, whose early stages of construction are given in Figure 9.1.

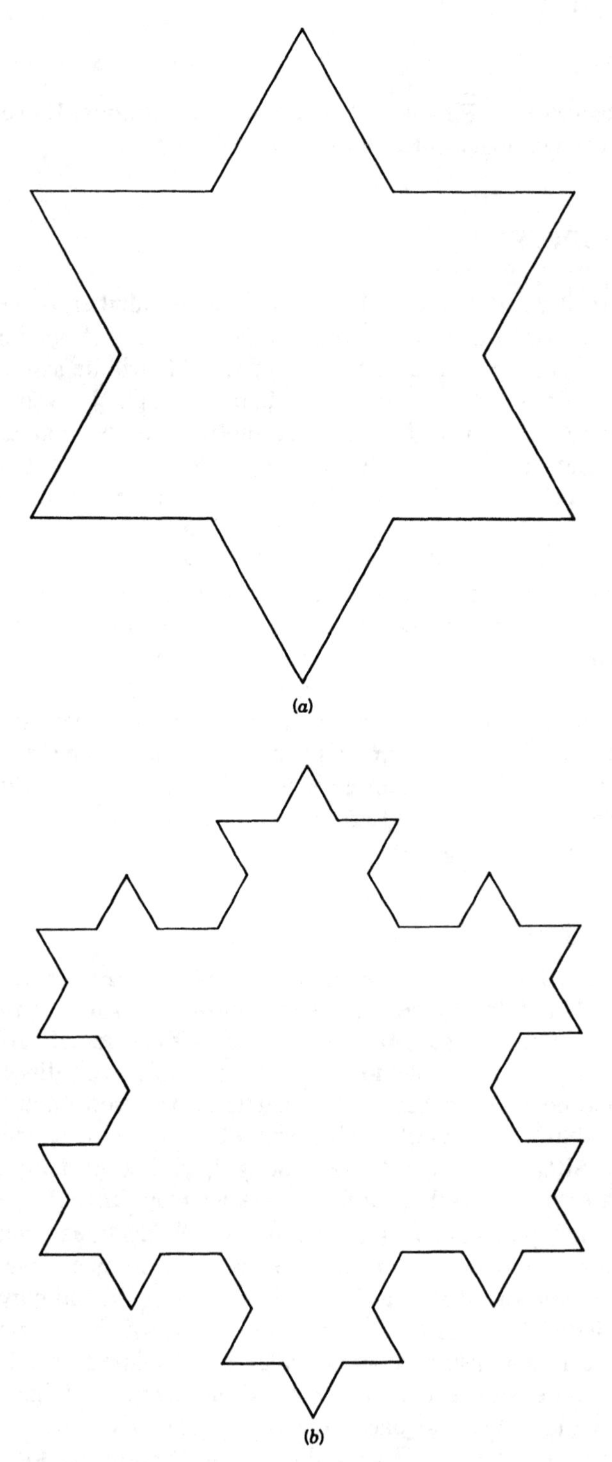

Figure 9.1 Triadich Koch island. (a), (b), and (c) show the initial stages of construction of its boundary.

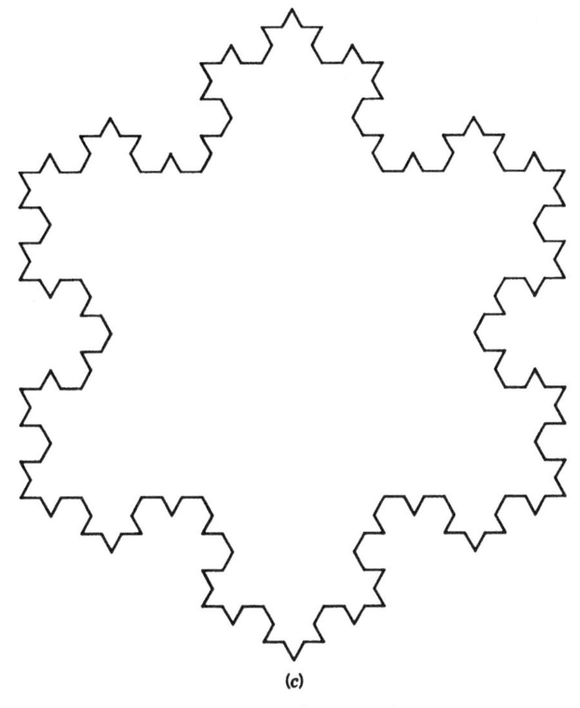

(c)

Figure 9.1 *(continued)*

Fractal Dimension

How is the fractal dimension D obtained? Taylor (1986) has a rigorous definition, but I shall take a more intuitive approach here. Suppose your task is to measure the length of a closed boundary in $\mathbb{R}^2$ and you are given a pair of calipers of *fixed* width ν, say, to do it. Starting at some arbitrary point on the boundary, the caliper allows you to step around the boundary $l(\nu)$ times (disregarding the last incomplete step). Then the length of the boundary using the "yardstick" ν is $\nu \cdot l(\nu)$. Now vary ν, letting it become smaller and smaller. Define the fractal dimension as

$$D \equiv -\lim_{\nu \to 0} \frac{\log l(\nu)}{\log \nu}. \tag{9.1.1}$$

It is clear now why the previously mentioned triadic Koch island has a boundary with fractal dimension $D = (\log 4)/(\log 3)$ (see Mandelbrot, 1982, p. 36).

In $\mathbb{R}^d$, suppose a compact set X has boundary such that it takes $l(\nu)$ $(d-1)$-dimensional cubes of side ν to cover it. Then the fractal dimension D is as defined in (9.1.1). Thus, $l(\nu) \simeq c \cdot \nu^{-D}$, or the measure of the boundary is approximately $c \cdot \nu^{d-1-D}$.

Estimation of the fractal dimension D has often been obtained from a log-log plot of $\{(\log \nu_i, \log l(\nu_i)): i = 1, \ldots, n\}$. By choosing a range of ν values and fitting a straight line through the resulting set of n points and the origin, the estimated fractal dimension is the negative slope of the line [see (9.1.1)]. In $\mathbb{R}^d$, $d \geq 3$, a direct application of this cube-counting method can be very unstable; Dubuc et al. (1989) develop a more reliable variation method in $\mathbb{R}^3$. The fractal dimension might also be estimated from the relationship of volume to surface measure: The volume of X, in units of ν^d, is proportional to $(l(\nu))^{d/D}$ (Mandelbrot, 1977, p. 72). In words, volume $\propto$ (surface measure)$^{d/D}$; for example, in $\mathbb{R}^2$, area $\propto$ (perimeter)$^{2/D}$ or perimeter $\propto$ (area)$^{D/2}$. For further details on the estimation of fractal dimension, see Cutler and Dawson (1990), Ramsay and Yuan (1990), and Theiler (1990).

Self-Similarity

Fractal sets may be self-similar (i.e., a small part of the boundary magnified resembles a larger part of the boundary), but are not necessarily so. For example, for the triadic Koch island it is only necessary to use different shaped triangular spikes at each step to obtain a fractal that is not self-similar. Hutchinson (1981) explores the notions of self-similarity and fractals from a theoretical point of view.

Consider the geostatistical problem treated in Part I of this book. A finite number of observations $\{Z(\mathbf{s}_i): i = 1, \ldots, n\}$ are taken from $\{Z(\mathbf{s}): \mathbf{s} \in D\}$, a potentially uncountable number. Some geophysical questions need to be answered at a scale smaller than the smallest spacing, $\min\{\|\mathbf{s}_i - \mathbf{s}_j\|: 1 \leq i < j \leq n\}$. Assuming a fractal structure for the possible trajectories $\{z(\mathbf{s}): \mathbf{s} \in D\}$, more specifically a *self-similar* fractal structure, knowledge gained from the data at the observed spatial scales can be extrapolated down to the unobserved scales. Because this self-similar fractal structure can only be verified at the scale of observation, it is only a *hope* that similar structure is manifested at the microscale. Until more experience is gained with its applicability, caution should be exercised. Section 5.5 discusses further the fractal properties of random-field trajectories.

Fractal Set Models of Natural Phenomena

Small volumes surrounded by enormous surface areas are frequently the rule in our bodies: To enrich the blood with oxygen, our lungs require a surface area that is approximately the size of two basketball courts. Every cell in our body receives nutrients from an intricate branching network of blood vessels. In these and other examples, the volumes involved are just a small percentage of total body volume, indicating that fractals can be the result of evolution toward efficiency of natural processes.

Mandelbrot (1982) describes how fractal sets might model sponges, tree branching, surfaces of the brain and lung, rivers, watersheds, the vascular system, coastlines, and so on (see also Frontier, 1987; Goodchild and Mark,

1987; Tarboton et al., 1988). His general approach has been one of parsimonious description of a phenomenon rather than explanation of its genesis. However, by summarizing these sets using their fractal dimension, a phenomenon whose genesis is highly complex could be monitored for *changes*.

Random Fractals

Random fractals are usually obtained by considering the trajectories of stochastic processes. For example, a Brownian trail in two dimensions (obtained from the locus of points visited by two-dimensional Brownian motion) has fractal dimension $D = 2$, because eventually almost every trajectory will fill the plane (i.e., will visit every point in $\mathbb{R}^2$). The trajectories of a temporal stochastic process considered as functions of time may have a fractal dimension. Mandelbrot and Van Ness (1968) detail how a fractional Brownian motion can be constructed, where $\mathrm{var}(X(t + \Delta t) - X(t))$ is proportional to $|\Delta t|^{2H}$, $0 < H \leq 1$. The case $H = 1/2$ refers to the usual Brownian motion, whose trajectories have a fractal dimension of 1.5, larger than its topological dimension of 1. In general, the trajectories have a fractal dimension $2 - H$. For $H > 1/2$, fractional Brownian motion no longer has independent increments and displays long-range dependence (or persistence). Mandelbrot (1982) uses higher-dimensional versions of random functions to produce attractive pictures of computer-generated planetscapes; see Section 5.5 for further discussion. Taylor (1986) can be consulted for the appropriate measure theory associated with these types of random fractals.

For random fractals analogous to, but more general than, the deterministic triadic Koch island, the interested reader should consult Falconer (1986). His construction of random fractals is more direct, as he does not use the path properties of stochastic processes.

Finally, it should be pointed out that, for many purposes, modeling an object or process with abrupt and tortuous surfaces is not necessary; for example, Switzer (1976) investigates geometrical measures of the *smoothness* of random functions. If inferences are to be made on only one spatial scale, fractal set models are probably not appropriate. They are very useful, however, when considering phenomena at scales that span orders of magnitude.

9.1.2 Fuzzy Sets

Scientific investigations always have some elements of uncertainty, ambiguity, and vagueness. I think it is fair to say that the more the investigation involves quantification of and by human beings, the more uncertain the data become. For example, the medical assessment of side effects of a drug may be based on patients' responses like "drowsy," "nauseous," and "lousy."

Probability theory is one way of dealing with uncertainty. From the classical axioms that define a probability space $(\Omega, \mathscr{A}, P)$, an elegant theory can be built (e.g., Chung, 1968) that is the underpinning of all the Statistics in this book. Relatively recently, Zadeh (1965) has introduced a different approach, namely, the modeling of uncertainty by fuzzy sets. Its premise is that all data are imprecise, even *after* they have been observed. The uncertainty in Probability and Statistics usually refers to future data; those that have been observed have a precise value.

To quantify data that are vague, Zadeh defined a membership function $\mu_A(\cdot)$, where $0 \le \mu_A(\cdot) \le 1$ and A is a set in $\mathbb{R}^d$. If $\mu_A(\mathbf{s})$ is near zero, then $\mathbf{s}$ is probably not in the set A, whereas if $\mu_A(\mathbf{s})$ is near 1, $\mathbf{s}$ is probably in the set A. For example, suppose a newly manufactured mechanical part is within specification if its diameter s is between 2.9 and 3.0 cm. Let

$$A \equiv \{s: 2.9 \le s \le 3.0\}.$$

A quality-control inspector observes many such parts every day, but has to be selective about the ones that will be taken off the production line to be measured. A perfect inspector is one whose candidate defective parts are all defective and who leaves no defective parts on the production line. In reality, mistakes are made. Therefore, it seems sensible to think of each quality control inspector as having a membership function; the "closer" it is to the ideal

$$\mu_A^*(s) = \begin{cases} 1, & \text{if } s \in A, \\ 0, & \text{elsewhere,} \end{cases}$$

the better the inspector.

The example just given quantifies a subjective reaction to objective data, and there are obvious ways that membership functions could be estimated using techniques developed in bioassay (see, e.g., Finney, 1971). However, a subjective reaction to subjective data, such as might occur in medical diagnosis, brings forth very challenging problems of modeling. Nevertheless, attempts have been made to use fuzzy sets to quantify the diagnostic process *inter alia* by Sanchez (1979), Fordon and Bezdek (1979), and Esogbue and Elder (1980). Shafer's (1976) belief function and its calculus bears a striking resemblance to fuzzy set theory; some discussion of this is given in Shafer (1987).

Definition of a Fuzzy Set

A fuzzy set in $\mathbb{R}^d$ is defined abstractly as $\{(\mathbf{s}, \mu(\mathbf{s})): \mathbf{s} \in \mathbb{R}^d\}$, where $0 \le \mu(\cdot) \le 1$. To distinguish a fuzzy set from a usual set in $\mathbb{R}^d$, denote fuzzy sets by $\tilde{A}$, $\tilde{B}$, and so on. By analogy with the usual set theory, the (fuzzy) union, (fuzzy) intersection, and (fuzzy) complement of fuzzy sets $\tilde{A}_i \equiv \{(\mathbf{s}, \mu_i(\mathbf{s})):$

$\mathbf{s} \in \mathbb{R}^d\}$, $i = 1, 2$, are defined as

$$\tilde{A}_1 \cup \tilde{A}_2 \equiv \{(\mathbf{s}, \max(\mu_1(\mathbf{s}), \mu_2(\mathbf{s}))): \mathbf{s} \in \mathbb{R}^d\},$$

$$\tilde{A}_1 \cap \tilde{A}_2 \equiv \{(\mathbf{s}, \min(\mu_1(\mathbf{s}), \mu_2(\mathbf{s}))): \mathbf{s} \in \mathbb{R}^d\},$$

$$\tilde{A}_1^{\mathscr{C}} \equiv \{(\mathbf{s}, 1 - \mu_1(\mathbf{s})): \mathbf{s} \in \mathbb{R}^d\},$$

respectively. A simple illustration of fuzzy sets in $\mathbb{R}^1$ and the union operation is given in Figure 9.2. However, notice that

$$\tilde{A}_1^{\mathscr{C}} \cup \tilde{A}_1 \neq \{(\mathbf{s}, 1): \mathbf{s} \in \mathbb{R}^d\},$$

although other properties, such as DeMorgan's laws,

$$(\tilde{A}_1 \cup \tilde{A}_2)^{\mathscr{C}} = \tilde{A}_1^{\mathscr{C}} \cap \tilde{A}_2^{\mathscr{C}}$$

$$(\tilde{A}_1 \cap \tilde{A}_2)^{\mathscr{C}} = \tilde{A}_1^{\mathscr{C}} \cup \tilde{A}_2^{\mathscr{C}},$$

are satisfied. There are alternative definitions to these fuzzy set operations, and the interested reader can consult Kaufman (1975) for more details.

Almost any branch of set theory can be "fuzzified," usually in a number of different ways. This has led to a large literature, most of which has paid little real attention to potential applications in psychology, linguistics, speech recognition, artificial intelligence, and so on.

Gray-Tone Functions as Fuzzy Sets

Image data that are received as gray tones are very precise, but could be thought of as fuzzy sets. Suppose the intensity is compressed so that 0 denotes white, 1 denotes black, and various grays correspond to values between 0 and 1. Denote this gray-tone function as $\mu(\cdot)$. Then $\{(\mathbf{s}, \mu(\mathbf{s})): \mathbf{s} \in \mathbb{R}^2\}$ is a surface in $\mathbb{R}^3$ that can be constructed from the image. It can also be thought of as a fuzzy set, and many of the fuzzy operations make sense from an image analysis point of view. Serra (1982, Chapter 12) analyzes image data using several such operations, although his work was developed independently. With the explosion of mathematical results in fuzzy set theory, it would be profitable to catalogue them from this image-analysis viewpoint; Bandemer and Roth (1987) develop a fuzzy exploratory data analysis in this way.

Probability Measures of Fuzzy Sets

The probability of a fuzzy event was defined initially by Zadeh (1968). First, a fuzzy event is a fuzzy set $\tilde{A}$ in $\mathbb{R}^d$, whose membership function

$$\mu: \mathbb{R}^d \to [0, 1]$$

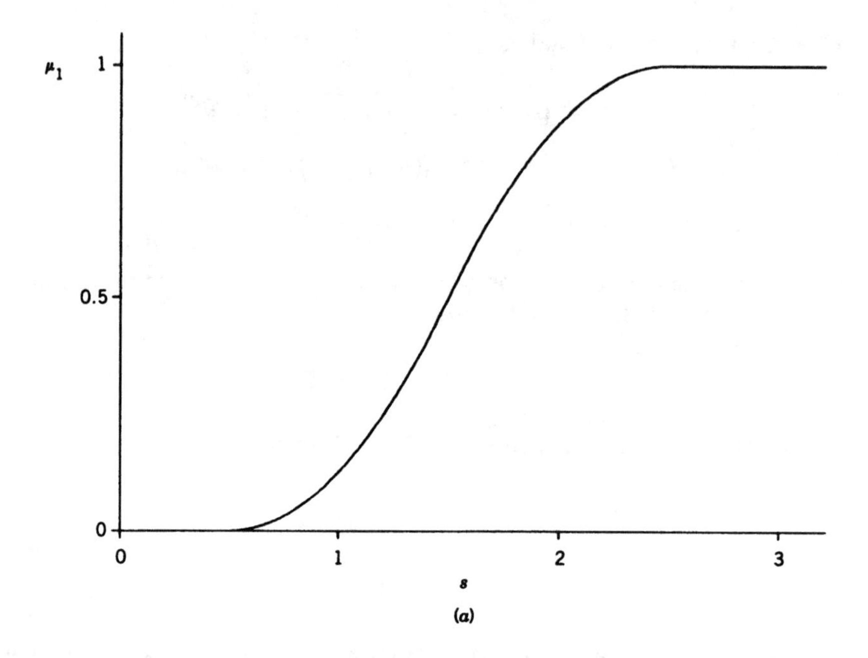

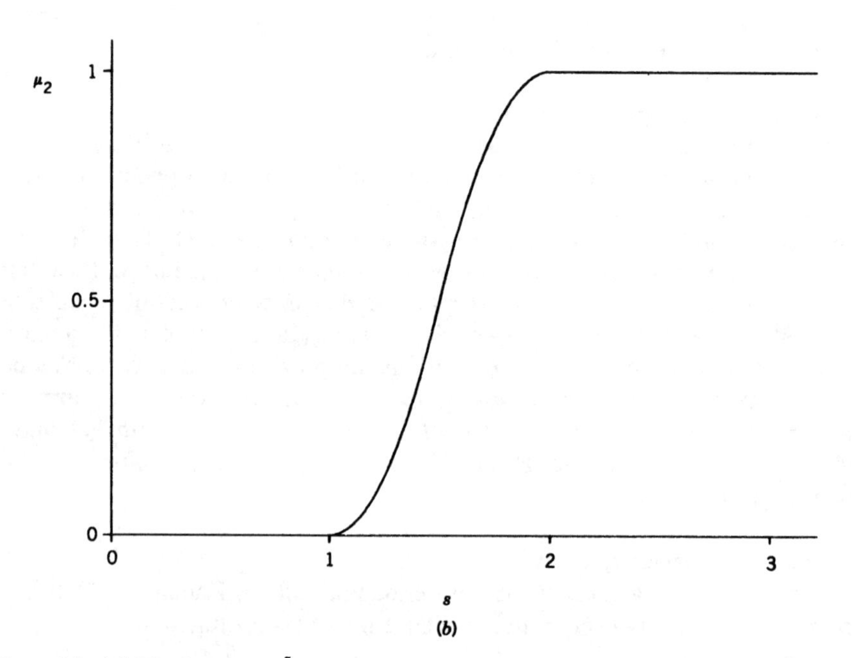

Figure 9.2 (*a*) The fuzzy set $\tilde{A}_1$, with membership function μ_1. (*b*) The fuzzy set $\tilde{A}_2$, with membership function μ_2.

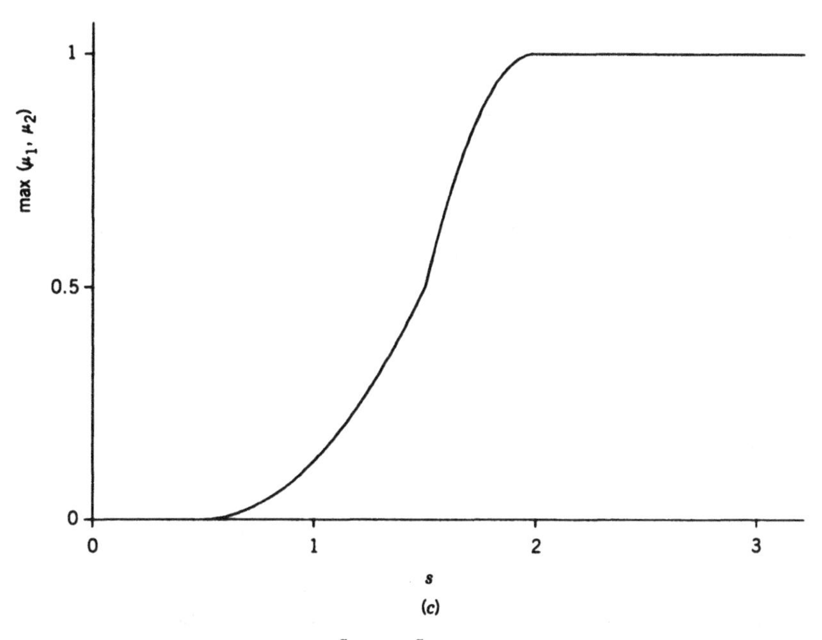

Figure 9.2 (c) The fuzzy union of $\tilde{A}_1$ and $\tilde{A}_2$, with membership function $\max(\mu_1, \mu_2)$.

is measurable (i.e., μ-inverse images of Borel subsets of $[0, 1]$ are Borel subsets of $\mathbb{R}^d$). Second, suppose that P is a probability measure on the Borel subsets of $\mathbb{R}^d$. Finally, the *probability* $\tilde{P}(\tilde{A})$ of a fuzzy event $\tilde{A}$ is defined by

$$\tilde{P}(\tilde{A}) \equiv \int_{\mathbb{R}^d} \mu(\mathbf{s}) P(d\mathbf{s}). \tag{9.1.2}$$

Another way to think of (9.1.2) is as the *expectation* of the membership function. Notice that

$$\tilde{P}(\tilde{A}^{\mathscr{C}}) = 1 - \tilde{P}(\tilde{A}),$$

along with many other familiar probability equalities and inequalities (Zadeh, 1968). However, it is *not* necessarily true that $\tilde{P}(\tilde{A}^{\mathscr{C}} \cap \tilde{B}) = \tilde{P}(\tilde{B}) - \tilde{P}(\tilde{A} \cap \tilde{B})$, which has an effect on the way fuzzy set *independence* should be defined. Buoncristiani (1983) shows that Zadeh's (1968) definition of independence of $\tilde{A}$ and $\tilde{B}$, namely,

$$\tilde{P}(\tilde{A} \cap \tilde{B}) = \tilde{P}(\tilde{A}) \cdot \tilde{P}(\tilde{B}),$$

needs modification and offers instead

$$\tilde{P}(\tilde{A} | \tilde{B}) = \tilde{P}(\tilde{A});$$

the fuzzy conditional probability $\tilde{P}(\tilde{A}|\tilde{B})$ is defined by

$$\tilde{P}(\tilde{A}|\tilde{B}) \equiv \frac{\tilde{P}(\tilde{A} \cap \tilde{B})}{\tilde{P}(\tilde{A} \cap \tilde{B}) + \tilde{P}(\tilde{A}^{c} \cap \tilde{B})}.$$

Random Fuzzy Sets

The most obvious way to generate a well defined random fuzzy set is through a stochastic process $\{Z(\mathbf{s}): \mathbf{s} \in \mathbb{R}^d\}$, where $0 \leq Z(\cdot) \leq 1$. Then, intuitively,

$$\tilde{A}_Z \equiv \{(\mathbf{s}, Z(\mathbf{s})): \mathbf{s} \in \mathbb{R}^d\}$$

is a random fuzzy set [although it takes a little work to obtain the right σ algebra in the image space (Zadeh, 1968, p. 424)]; see also Feron (1976).

It appears that random fuzzy sets are more general, in that one can define a whole sequence of random sets through the level sets

$$A_Z(u) \equiv \{\mathbf{s}: Z(\mathbf{s}) \geq u\}, \qquad 0 \leq u \leq 1.$$

In the rest of this chapter I shall discuss the simpler random sets. Fuzzy sets would provide useful models for the purposes of this chapter if a fuzzy set calculus were developed that has direct application to inference for models of objects.

9.1.3 Random Closed Sets: An Example

Although not as general as a random fuzzy set, the notion of a random closed set (RACS) is sufficiently rich to model highly complex objects, yet often sufficiently simple to allow inferences on model parameters. The general definition of a RACS is given in Section 9.3 as a measurable mapping from a probability space into the set of all closed sets in $\mathbb{R}^d$ equipped with the "hit-or-miss" topology (Matheron, 1971a).

An important and flexible class of random closed sets is the Boolean model. Suppose D is a homogeneous Poisson process in $\mathbb{R}^d$ (see Section 8.3) and Z is a random polygon (random sides and angles) or a random compact set generated by random scaling (e.g, a disk of random radius). Define the displaced random set

$$Z \oplus \mathbf{s} \equiv \{\mathbf{z} + \mathbf{s}: \mathbf{z} \in Z\}, \qquad s \in \mathbb{R}^d$$

and suppose that $Z_1, Z_2, \ldots$ are i.i.d. as Z. Then, the random set

$$X \equiv \bigcup\{Z(\mathbf{s}_i): \mathbf{s}_i \in D\}, \tag{9.1.3}$$

where $Z(\mathbf{s}_i) \equiv Z_i \oplus \mathbf{s}_i$, is known as a Boolean model with "grains" $Z_1, Z_2, \ldots$ (independent RACS, each with probability measure the same as that of Z) and "germs" $\mathbf{s}_1, \mathbf{s}_2, \ldots$ (points of the point process D). A simulated realization of X is given in Figure 9.3, where Z is a disk of random radius.

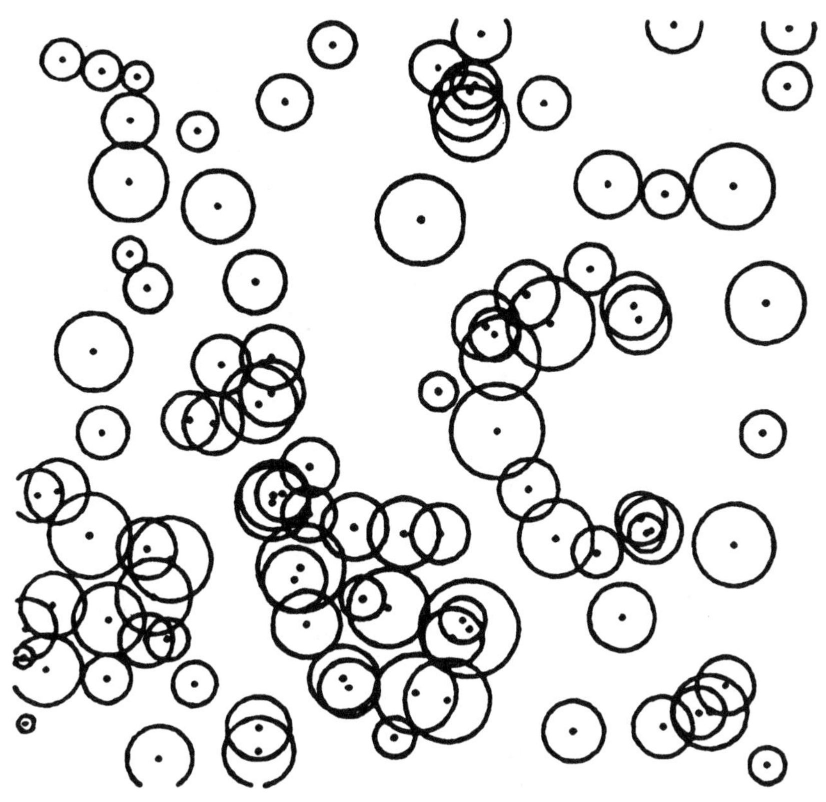

Figure 9.3 The union of the disks shown is a partial realization of a Boolean model in $\mathbb{R}^2$. Germs are points of a homogeneous Poisson process and grains are disks of random radius.

Figure 9.4 is a real data set that could be modeled by the RACS defined in (9.1.3). These data are the result of a cloud-seeding experiment and represent water droplets in a cloud. A glass slide (7.5 × 2.5 cm) was coated with magnesium oxide and mounted on a small airplane that was flown at a known speed through the cloud under study. Thc slide was exposed for a fixed time interval, returned to the laboratory, and examined under a microscope. A droplet causes a crater in the magnesium oxide; the microscope is focused on the surface level of the magnesium oxide so that the bright ring is the edge of the crater and the dark center is its out-of-focus center. Figure 9.4 shows only a very small part of the 7.5 × 2.5-cm slide; the scale is shown on the figure. After adjustment for the spread of the droplets on impact and for occasional oblique impact, it is sensible to think of the data as representing various sized water droplets scattered randomly over the area of interest. Notice that a big crater could obliterate a small one, so that simple computation of the number of observed craters and their size distribution gives a *biased* summary of the droplet intensity and droplet size distribution. However, an analysis of such data, based on the Boolean model (9.1.3), can compensate for these biases.

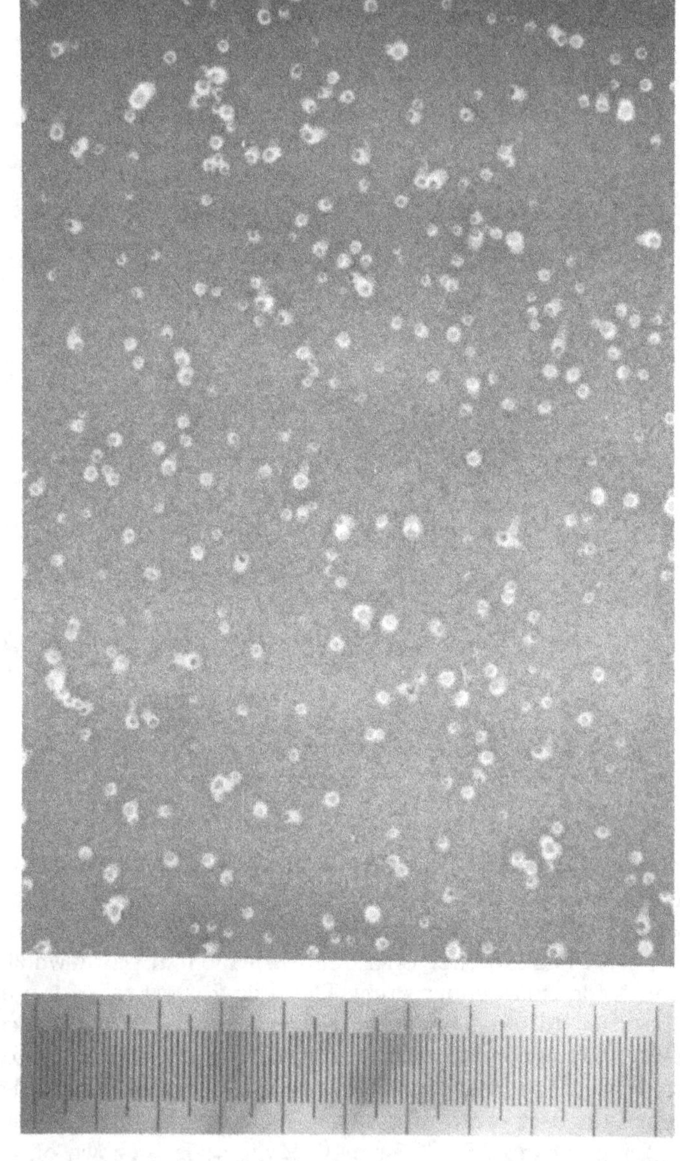

Figure 9.4 Cloud-droplets data. The size and location of a droplet is shown by a bright ring. The overall length of the scale shown is 1 mm. [*Source*: Dr. S. C. Mossop, Division of Atmospheric Research, CSIRO]

The Boolean model and its variants are considered in some detail in this chapter. Section 9.2 presents a simulated example, definitions and properties are given in Section 9.4, inferential aspects are discussed in Sections 9.5 and 9.6, and in Section 9.7 a conditional Boolean model is used to model tumor growth.

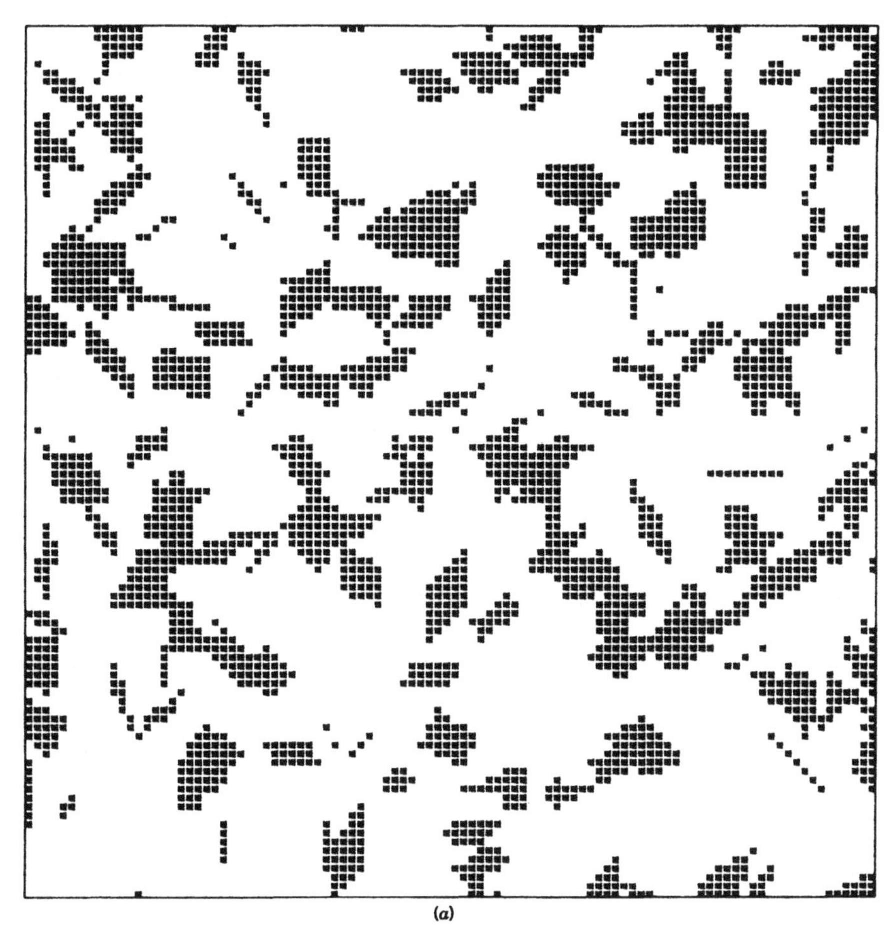

(*a*)

Figure 9.5 (*a*) A realization of a Boolean model (restricted to a square mask) whose grains are random parallelograms. The realization is discretized into 100×100 pixels. [*Source*: Cressie and Laslett, 1987] Reprinted with permission from the *SIAM Review*, volume 29, no. 4, pp. 557–574. Copyright 1987 by the Society for Industrial and Applied Mathematics. All rights reserved.

9.2† RANDOM PARALLELOGRAMS IN $\mathbb{R}^2$

This section gives a brief exploratory data analysis of simulated data, generated according to the Boolean model defined by (9.1.3). The data are analyzed further in Section 9.5.

Figure 9.5*a* and *b* presents two realizations of (9.1.3) restricted to a square mask of 100×100 pixels, where the grains are i.i.d. random compact sets, more specifically, random parallelograms, and the germs are events of a two-dimensional homogeneous Poisson process with intensity $\lambda = 0.02$ per pixel area (see Section 8.3). These data were first presented in Cressie and

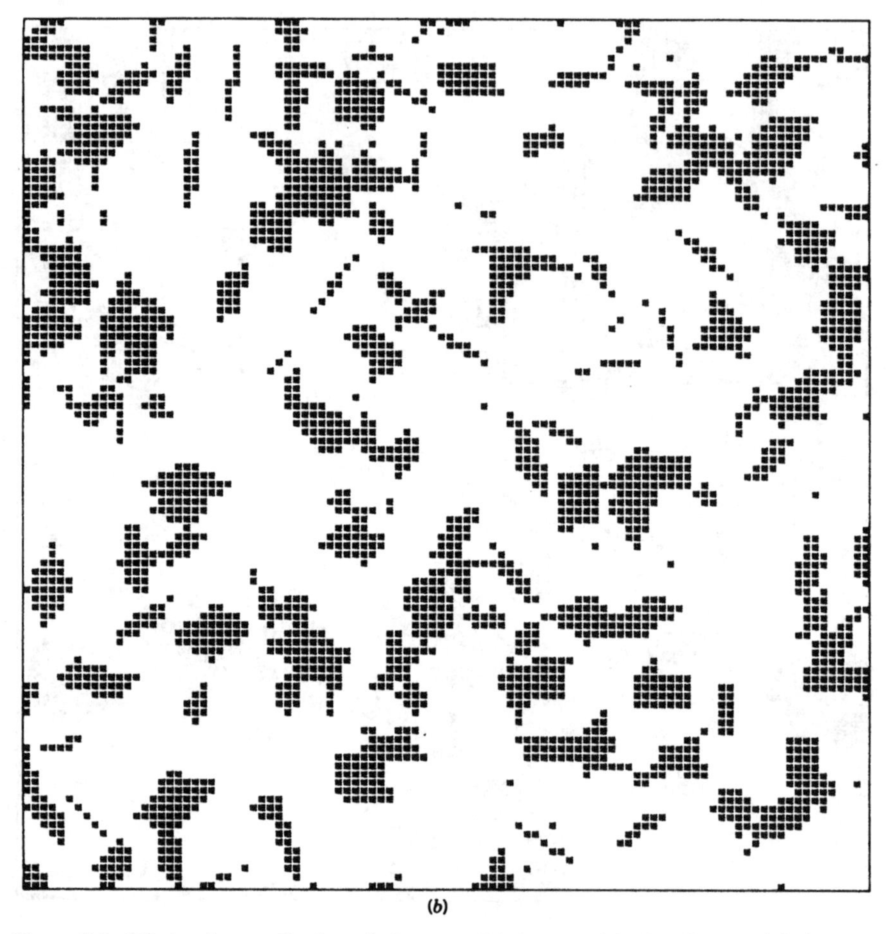

Figure 9.5 (*b*) Another realization of the same Boolean model, described in (*a*). [*Source*: Cressie and Laslett, 1987]

Laslett (1987). The bottom left-hand corner of the random parallelogram was fixed at each point of the realization of the Poisson process. The length of both sides of the parallelogram was uniformly distributed on $[2.5, 7.5]$ and the angle between the adjacent sides was uniform on $[0, \pi)$. Finally, the parallelogram was uniformly oriented on $[0, 2\pi)$. Figure 9.5*a* and *b* shows two realizations *after discretization*.

Faced with data of this kind, the statistician struggles. I shall briefly describe some exploratory techniques that are appropriate for Figures 9.5*a* and *b*. Section 9.5.1 continues the analysis and fits back a Boolean model. If one wants to consider such data as representative of a phenomenon, that is, as carrying information on interpretable "average" quantities (parameters) associated with the phenomenon, then one must turn to a model. This is true for studying any type of variation, but it is particularly difficult here because of the dearth of tractable random-set models available; see Section 9.3 for more discussion.

I shall give some summary statistics for the data presented in Figure 9.5*a*. The most basic one is the specific area, that is, the area of the set per unit of area under consideration. In this case the area under consideration is 10^4 square units. Here

$$\text{specific area} = 0.27050.$$

Notice that 1 − (specific area) = 0.72950 is called the *porosity*. The number of distinct contiguous sets and the boundary length could also be calculated, but to anticipate Section 9.5, it turns out that the specific area of a dilated version of Figure 9.5*a* is of more interest. Define

$$X \oplus tC \equiv \{\mathbf{s} + t\mathbf{c}: \mathbf{s} \in X, \mathbf{c} \in C\},$$

where X is the set in Figure 9.5*a*, C is a solid square of side 2 units with center of gravity centered at the origin, and t is a dilation factor. As t increases, the original set X becomes more and more "bloated". At each value of t, the specific area of $X \oplus tC$ can be calculated; Table 9.1 gives the values for $t = 0, 1, 2, 3, 4$.

Ripley (1988, Chapter 6) gives a host of other summary statistics that are defined in terms of morphological transformations on X with structuring element tB (where B is often chosen to be a disk). Plotting the area of the transformed X against t gives an exploratory way of summarizing the image data, as well as a tool for fitting (or diagnosing the fit of) a random-set model.

From the summary statistics provided by Table 9.1 and assuming Figure 9.5*a* was generated by a Boolean model (which it was), it is possible to estimate various parameters of the model. The theoretical details are given in Section 9.4, followed by an analysis of these data in Section 9.5.

Table 9.1 Specific Areas of the Dilated Set $X \oplus tC$, Where X is Given by Figure 9.5*a*

t	Specific Area
0	0.27050
1	0.55862
2	0.77203
3	0.90052
4	0.96432

9.3* RANDOM CLOSED SETS AND MATHEMATICAL MORPHOLOGY

The statistical and probabilistic theory that ensures the existence of stochastic models and the efficient estimation of their parameters is very well developed for data modeled as independent-and-identically-distributed *random variables*. The interrelationships between any two subcollections are extremely simple, and one only needs to determine the law of any individual to determine the law of the whole. In fact, only the probability of the events "$\{X \leq x\}$, for all $x \in (-\infty, \infty)$" are needed. This theory has been extended for (1) a sample whose variables are vectors, elements of a Banach space, and so forth and (2) a collection of (often real-valued but also vector-valued) random variables whose dependence structure is Markov, (strongly or weakly) stationary, and so forth. Further theoretical and methodological extensions are needed for random sets.

Although the main goal of this chapter is to model (images of) objects, it should be noted that the theory of random sets has much wider potential. For example, Barnett (1986) discusses how economists could benefit by viewing the economy as a random set. Indeed, a sample of n random variables can either be viewed as a set of n points in $\mathbb{R}^1$ or as a vector in $\mathbb{R}^n$. Taking the latter point of view allowed Fisher (1915) and Hotelling (1939) to calculate the distributions of intuitively appealing test statistics. Taking the former point of view, more generally for a random sample of n random vectors in $\mathbb{R}^d$, one can ask questions about the convex hulls of subsets of such points. (Through these, the notion of order statistics can be generalized to higher dimensions or the support of a population distribution can be estimated.) These are addressed *inter alia* by Efron (1965), Fisher (1971), Ripley and Rasson (1977), Eddy and Gale (1981), Moore (1984), Davis et al. (1988), and Brozius (1989). One might also consider each point in $\mathbb{R}^d$ as having a set associated with it and ask questions about the sum of the n sets' indicator functions (Robbins, 1944) or about the coverage of the union of the n sets (see, e.g., Hall, 1988a).

In what is to follow, the main results for modeling and inference for random sets will be reviewed. It will become clear that this subject is still in its infancy, and much remains to be done.

Random Sets and Random Functions

In one sense, a random set is just a special case of a random function (e.g., Adler, 1981) that takes only the values 0 or 1. In fact, if any random function is sliced at some level u and looked at from above, then the boundary of the slice traces out the boundary of a random set. Any analysis of the original random function should be equally possible on these level sets indexed by u, and conversely. Radchenko (1985) considers geometric measures (e.g., length, area, etc.) of level sets of an almost surely continuous random function defined on a metric space X (e.g., $\mathbb{R}^d$) and gives conditions under which these are measurable.

Both random sets and random functions have the concept of covariation; however, beyond this there seems little in common in their analysis [see the discussion later in this section and Adler (1981, p. 71)]. The main reason is that random-function operations, such as convolution and Fourier filtering, are linear, whereas the random-set morphological transformations [i.e., transformations that affect form, such as can be found in Serra (1982, Part 1)] are highly nonlinear.

Recall from Section 2.3 that the variogram of a random function $\{Z(\mathbf{s}): \mathbf{s} \in \mathbb{R}^d\}$ is defined as $\mathrm{var}(Z(\mathbf{s} + \mathbf{h}) - Z(\mathbf{s}))$, which is usually considered to be a function only of the vector $\mathbf{h}$ [known as intrinsic stationarity; see Matheron (1971b) and Section 2.3]. Now suppose that X is a stationary random closed set in $\mathbb{R}^d$ (defined in the following text) and let $Z_X(\cdot)$ denote its indicator function. Then, it is not difficult to prove that $\{Z_X(\mathbf{s}): \mathbf{s} \in \mathbb{R}^d\}$ is a random function satisfying intrinsic stationarity with variogram $2\{p - \Pr(\mathbf{s} \in X, \mathbf{s} + \mathbf{h} \in X)\}$, where $p = \Pr(\mathbf{s} \in X)$. Write $X \oplus \{-\mathbf{h}\}$ for the (random) set X translated to $-\mathbf{h}$. Then the expression for the variogram becomes $2\{p - \Pr(\mathbf{s} \in X \cap X \oplus \{-\mathbf{h}\})\}$. Clearly, $\kappa(\mathbf{h}) \equiv \int \Pr(\mathbf{s} \in X \cap X \oplus \{-\mathbf{h}\})\, d\mathbf{s}$ is the probabilistic analogue of the *geometrical covariance* of a (deterministic) set $A \subset \mathbb{R}^d$, defined as $\nu(A \cap A \oplus \{-\mathbf{h}\})$, where $\nu(\cdot)$ denotes Lebesgue measure in $\mathbb{R}^d$. Notice that $\kappa(\mathbf{h})$ contains information about the surface measure of X as $\mathbf{h} \to \mathbf{0}$: Provided X is almost surely (a.s.) regular (see Serra, 1982, p. 274), then $\{\kappa(\mathbf{h}) - \kappa(\mathbf{0})\}/\|\mathbf{h}\|$ exists; call it $\kappa'_{\mathbf{u}}(\mathbf{0})$, where $\mathbf{u} \equiv \mathbf{h}/\|\mathbf{h}\|$. In $\mathbb{R}^2$,

$$-(1/2)\int_0^{2\pi} \kappa'_{\mathbf{u}}(\mathbf{0})\, d\mathbf{u} = E(\text{perimeter of } X). \tag{9.3.1}$$

Now, knowledge gained from the behavior at the origin of the covariance of a random function usually relates to the behavior of its spectrum at very high frequencies. This is exactly what is happening in the relation (9.3.1), where the random function is the indicator function of a random set: The left-hand side can be interpreted simply as covariance behavior near the origin and the right-hand side pertains to the boundary of X where there is high frequency, that is, where the 0-1 random function undergoes its most drastic changes. However, it is at this stage of an image analysis that a random-function approach fails to capture the full geometric complexities of the image. This is clear when $X \cap X \oplus \{-\mathbf{h}\}$ is written in terms of the *erosion* operation $X \ominus \check{B} \equiv \bigcap\{X \oplus \{-\mathbf{b}\}: \mathbf{b} \in B\}$. The choice of structuring element $B = \{\mathbf{0}, \mathbf{h}\}$, which yields $X \cap X \oplus \{-\mathbf{h}\}$, is one of *many* that could be made in order to "sort" the random set X. The scope of geometric possibilities expands enormously through varying the structuring element B; (linear) spectral analysis in random function theory is just one of a number of possibilities. Even more general ways of probing the image are found in the AFATL Image Algebra (Ritter et al., 1990), which provides a common mathematical environment for the development of image-processing algorithms and an image-processing language.

Random Sets Based on Tessellations

A *tessellation* of $\mathbb{R}^d$ is a partition such that the Euclidean space is divided into mutually exclusive subsets whose union is the whole space. For example, random straight lines in $\mathbb{R}^2$ tessellate the space, as does the Dirichlet tessellation or Delauney triangulation (e.g., Diggle, 1983, p. 6) of some arbitrary point process. Moller (1989a) gives an extensive review of random tessellations in $\mathbb{R}^d$.

Each subset of the tessellation with nonzero Lebesgue measure is called a *tile* of the tessellation; tiles (and their boundaries) are random sets. Further randomness could be introduced by coloring each tile; this might be done independently or according to a Markov random field (e.g., Arak and Surgailis, 1989), defined in Section 6.4. Ambartzumian (1982) can be consulted for constructions of random sets based on tessellations. However, the emphasis in this book is principally on random sets based on point processes and Boolean models.

Size and Shape

In what is to follow in this section, and for that matter in this chapter, the principal goal is the mathematical treatment of objects of varying size and shape. I shall take a random-set-model approach; the calculus associated with it has been called *mathematical morphology* by G. Matheron [a short account of which is given in Watson (1975) and in Section 9.3.1]. A more eclectic viewpoint of size and shape can be found in Lord and Wilson's (1984) book, but it has little development of stochastic modeling. I shall now review briefly some of the ways size and shape have been quantified.

Sometimes the data come as points in $\mathbb{R}^d$, and the collection of points is studied for the size and shape it engenders. These *landmarks* are useful in biology for comparing size and shape of specimens; for example, nine landmarks of primate skulls might be external occipital protuberance, posterior-most point of the foramen magnum, anterior-most point of the foramen magnum, deepest point of pituitary fossa, deepest point of ethmoid fossa, deepest point of curvature of nasal bones, terminus of internasal suture, tip of alveolar bone between upper central incisors, and posterior end of the hard palate. The area of biology that considers such problems is known as *morphometrics*: A single form could be characterized by distances or ratios of distances between landmarks. To compare two forms, one can either superimpose them on each other (Siegel and Benson, 1982; Bookstein, 1986), deform one form into the other (Lew and Lewis, 1977), or use the maximal invariant implied by the underlying group structure (Lele, 1991b). Biologists are often interested in studying the association of size and shape in such situations, a subject known as *allometry*. Mosimann (1970) gives a number of mathematical models that might be used in such studies.

Bookstein (1986) introduces a stochastic spatial component by assuming (unrealistically, it seems to me) that the landmarks (in $\mathbb{R}^2$) are independent bivariate Gaussian variates. Kendall (1984, 1989) has a statistical theory of shape based on k not-totally-coincident points in $\mathbb{R}^d$; for example, $k = 3$,

$d = 2$ yields triangles in the plane. Small (1988) compares Kendall's approach with Bookstein's. Watson (1986) considers the shapes of a random sequence of triangles in the plane.

Algorithms for computing triangles based on planar point sets can be found in, for example, Kirkpatrick and Radke (1985). These descriptive summaries are typically devoid of any stochastic interpretation.

For sets in $\mathbb{R}^d$, the subjective notions of size and shape can be quantified by the experimenter's choice of structuring elements that are used to interact with the image. For example, a purely descriptive (and arbitrary) measure of the size of X might be defined as

$$\Lambda_B(X) \equiv \sup\{\lambda: \{\lambda\mathbf{b} + \mathbf{s}: \mathbf{b} \in B\} \subset A, \text{for some } \mathbf{s}\}.$$

Consider the planar case $d = 2$. When the structuring element B is a unit disk centered at $\mathbf{0}$, $\Lambda_B(X)$ is the maximum inscribed circle within X. When $B = \{\mathbf{s}: \mathbf{s} = \alpha\mathbf{e}; 0 \leq \alpha \leq 1\}$, $\Lambda_B(X)$ is the maximum intercept in the direction $\mathbf{e}$, and when $B = \{\mathbf{0}, \mathbf{e}\}$, $\Lambda_B(X)$ is the maximum distance between two points of X in the direction $\mathbf{e}$. Shape can be investigated using set operations, such as addition, subtraction, opening, closing, and so forth (with various structuring elements), defined in the next section. These have been hard-wired into an opticodigital machine called a texture analyzer (Serra, 1982, p. 24) to avoid tedious programming and to accomplish real-time imaging.

Although summary descriptions of set data can be very insightful, the main goal of this chapter is to introduce various random-set models, along with methods of fitting them to image data. My intention is to progress from description to indication and then to estimation (as described in Section 1.1).

9.3.1* Theory and Methods

A summary of the main definitions and results of random-set theory (Matheron, 1971a, 1975; Kendall, 1974) and mathematical morphology (e.g., Serra, 1982) will be needed. Let E be a locally compact, Hausdorff, and separable space, and define $\mathscr{F}$ to be the set of all closed subsets of E (including the empty set $\varnothing$). For example, let $E = \mathbb{R}^d$. Let $\mathscr{K}$ denote the set of all compact sets and let $\mathscr{K}' \equiv \mathscr{K} \setminus \varnothing$ be the set of all nonempty compact sets. For any set of sets $\mathscr{R}$, $C(\mathscr{R})$ denotes that subset whose sets are convex. For any $A \subset E$, define

$$\mathscr{F}_A \equiv \{F \in \mathscr{F}: F \cap A \neq \varnothing\}, \qquad \mathscr{F}^A \equiv \{F \in \mathscr{F}: F \cap A = \varnothing\}. \quad (9.3.2)$$

For any K compact and any $G_1, \ldots, G_n$ open, generate sets of the form $\mathscr{F}^K \cap \mathscr{F}_{G_1} \cap \cdots \cap \mathscr{F}_{G_n}$. It can be shown that this class of subsets of $\mathscr{F}$ is a *base* for a topology on $\mathscr{F}$ (called the hit-or-miss topology) and that the topological space is compact, Hausdorff, and separable. In fact, it can be shown (Matheron, 1975, p. 28) that all that is needed is hit-or-miss information, either on the set of all compact sets or on the set of all open sets; I shall return to this point later. Equipped with a topology on $\mathscr{F}$, it is now possible to be rigorous about convergence of a sequence of closed sets. Furthermore, by taking countable unions and intersections of the open sets of the topological space $\mathscr{F}$, a σ algebra Σ on $\mathscr{F}$ is generated.

Definition. A *random closed set* or RACS (which is often just called a random set) is defined as a measurable mapping X from a probability space $(\Omega, \mathscr{A}, \mu)$ into the measure space $(\mathscr{F}, \Sigma)$. Let Pr be the law of X; that is, the probability induced on Σ by

$$\Pr(\mathscr{V}) \equiv \mu(X^{-1}(\mathscr{V})), \qquad \mathscr{V} \in \Sigma. \qquad \blacksquare$$

Special cases are random variables, random vectors, simple point processes (e.g., the index set D in Chapter 8), and random subsets of Euclidean space $\mathbb{R}^d$. Matheron's (1975) requirement of a locally compact, Hausdorff, and separable topological space E is weakened by Ross (1986). Zahle (1982, 1988) considers random-set processes in $\mathbb{R}^d$.

Set Operations

Mathematical-morphology preliminaries are now given; more details can be found in Serra (1982) and Giardina and Dougherty (1988). It is important to define the basic set operations in $\mathbb{R}^d$:

Translation.

$$A \oplus \mathbf{h} \equiv \{\mathbf{a} + \mathbf{h} : \mathbf{a} \in A\}. \qquad (9.3.3)$$

Minkowski Addition.

$$A_1 \oplus A_2 \equiv \{\mathbf{a}_1 + \mathbf{a}_2 : \mathbf{a}_1 \in A_1, \mathbf{a}_2 \in A_2\} \qquad (9.3.4)$$

or

$$A_1 \oplus A_2 \equiv \bigcup \{A_1 \oplus \mathbf{a}_2 : \mathbf{a}_2 \in A_2\};$$

see Minkowski (1903) and Hadwiger (1957).

Scalar Multiplication.

$$\lambda A \equiv \{\lambda \mathbf{a} : \mathbf{a} \in A\}. \qquad (9.3.5)$$

Reflection.

$$\check{A} \equiv \{-\mathbf{a} : \mathbf{a} \in A\}. \qquad (9.3.6)$$

Complementation.

$$A^{\mathscr{C}} \equiv \{\mathbf{s} \in \mathbb{R}^d : \mathbf{s} \notin A\}. \qquad (9.3.7)$$

Subtraction.

$$A_1 \ominus A_2 \equiv \left(A_1^{\mathscr{C}} \oplus A_2\right)^{\mathscr{C}} \tag{9.3.8}$$

or

$$A_1 \ominus A_2 \equiv \bigcap \{A_1 \oplus \mathbf{a}_2 : \mathbf{a}_2 \in A_2\};$$

see Hadwiger (1957).
Corrected Subtraction (Erosion).

$$A_1 \circleddash A_2 \equiv A_1 \ominus \check{A}_2. \tag{9.3.9}$$

Opening.

$$A_1 \circ A_2 \equiv (A_1 \circleddash A_2) \oplus A_2. \tag{9.3.10}$$

Closing.

$$A_1 \bullet A_2 \equiv \left(A_1 \oplus \check{A}_2\right) \ominus A_2. \tag{9.3.11}$$

Hausdorff Distance.

$$\rho(A_1, A_2) \equiv \inf\{r: A_1 \subset A_2 \oplus rb(\mathbf{0}, 1) \text{ and } A_2 \subset A_1 \oplus rb(\mathbf{0}, 1)\}, \tag{9.3.12}$$

where $b(\mathbf{0}, 1)$ is the closed ball of radius 1 centered at $\mathbf{0}$.
Volume.

$$\nu(A) \text{ (or } |A|) \equiv \int_A \nu(d\mathbf{s}), \tag{9.3.13}$$

where $\nu(\cdot)$ is Lebesgue measure in $\mathbb{R}^d$.

Hit-or-Miss

The hit-or-miss topology is basic to this theory of random sets. It was chosen because it reflects the way image data in $\mathbb{R}^d$ are analyzed; that is, its roots are in applications. Often there is little to be gleaned from an image or pattern in $\mathbb{R}^d$ just by looking at it (although of course it is the first thing to do). Clearly, some sort of systematic probing is needed, which leads to the use of structuring elements B to check whether "B hits X" ($B \cap X \neq \varnothing$) or "$B$ misses X" ($B \cap X = \varnothing$). Furthermore, suppose $\mathscr{P}(\mathbb{R}^d)$, the set of all subsets of $\mathbb{R}^d$, is equipped with a σ algebra generated by $\mathscr{P}_G = \{P \in \mathscr{P}(\mathbb{R}^d): P \cap G \neq \varnothing\}$, G open. Then the equivalence

$$P \cap G \neq \varnothing \quad \text{if and only if} \quad \bar{P} \cap G \neq \varnothing$$

shows that, in order to study *any* random set with the σ algebra generated by

$\mathcal{P}_G$, it is equivalent to study its closure using the σ algebra Σ. Thus, the hit-or-miss approach virtually *demands* the study of random *closed* sets. This restriction of the type of sets under study is a strength of the approach, because it reflects the reality of the objects being modeled. For example, no experiment can hope to distinguish between X being a disk of the plane versus X being the set of irrational points in that disk.

Choquet's Theorem and the Hitting Function

It can be shown that all the interesting set transformations (dilation, erosion, opening, closing, convexification, etc.) of a RACS X are themselves RACS. Matheron (1975, p. 28) has demonstrated that, provided the set transformation is upper or lower semicontinuous into $\mathcal{F}$, the transform of the RACS X is also a RACS. Therefore, to analyze set data, all one needs is a "bagful" of random-set models,and the rest is, in principle, straightforward. But it is here where the random-set approach has difficulty fulfilling its potential.

How can the models be specified? What are the important events that make two random sets different? For a partial answer, return to the hit-or-miss topology. If $\Pr(X \in \mathcal{F}^K \cap \mathcal{F}_{G_1} \cap \cdots \cap \mathcal{F}_{G_n})$ can be specified for all compact K and all open $G_1, \ldots, G_n$, for all integers n, in a consistent way, then X is well defined. This can be burdensome, but, fortunately, a great reduction of test sets *is* possible.

For any $K \in \mathcal{K}$, define *the hitting function* T_X as

$$T_X(K) \equiv \Pr(X \in \mathcal{F}_K) = \Pr\{X \cap K \neq \varnothing\}. \tag{9.3.14}$$

Then T_X has the following properties (Matheron, 1975, p. 29):

1. $T_X(\varnothing) = 0$ and $0 \leq T_X \leq 1$.
2. T_X is increasing.
3. T_X satisfies the following recurrence relations. For any $n \geq 0$, let $S_n(B_0; B_1, \ldots, B_n)$ denote the probability that X misses B_0 but hits $B_1, \ldots, B_n$. Then

$$\begin{aligned}
S_0(B_0) &= 1 - T_X(B_0) \geq 0, \\
S_1(B_0; B_1) &= T_X(B_0 \cup B_1) - T_X(B_0) \geq 0, \\
&\vdots \\
S_n(B_0; B_1, \ldots, B_n) &= S_{n-1}(B_0; B_1, \ldots, B_{n-1}) \\
&\quad - S_{n-1}(B_0 \cup B_n; B_1, \ldots, B_{n-1}) \geq 0.
\end{aligned}$$

That is, T_X is a Choquet capacity of infinite order. A powerful result, proved independently by Matheron (1971a) and by Kendall (1974), is *Choquet's Theorem* in the context of random-set theory. It says that the converse of the preceding statement is true. In other words, if a given T on $\mathscr{K}$ is a Choquet capacity of infinite order, then there exists a *necessarily unique* P_T on Σ such that

$$P_T(\mathscr{F}_K) = T(K), \quad \text{for all } K \in \mathscr{K}.$$

Ross (1986) generalizes this result to random sets of a not-necessarily-separable space E.

An immediate example of its use is when the RACS X is a simple point process (i.e., with probability 1, no more than one event is at any location) in $\mathbb{R}^d$, almost surely locally finite. Let $N(A)$ denote the number of points of the process in $A \subset \mathbb{R}^d$. Then Choquet's Theorem says the point process is completely specified from

$$T_X(K) = 1 - \Pr(X \cap K = \varnothing) = 1 - \Pr(N(K) = 0), \quad \text{for all } K \in \mathscr{K}.$$

This observation, that the point process is uniquely determined from $\{\Pr(N(K) = 0): K \in \mathscr{K}\}$, was made by Ripley (1976b, p. 989), and is one of the consequences of the Equivalence Theorem for point processes (Section 8.3.2).

It is tempting to bracket Choquet's Theorem with the result that says that a random variable X is well defined once the probabilities of events $\{X \leq x\}$, for all $x \in (-\infty, \infty)$, are specified in a consistent way. But the results, although similar, are not identical. In fact, Choquet's Theorem for a random variable, where the RACS X is a.s. a one point set in $\mathbb{R}^1$, involves test sets $\{[a, b]: -\infty < a \leq b < \infty\}$. More work is needed to modify the necessary test sets down to $\{(-\infty, x]: -\infty < x < \infty\}$. It is in this domain, namely, finding ways to reduce the hitting-function test sets down from the full complement $\mathscr{K}$, where results are scarce. I believe that this has held back considerably the development of random-set *models*.

If something extra is known about the RACS X, say that all of its Minkowski functionals (e.g., volume, surface area, diameter, etc.) almost surely exist and are finite (see Serra, 1982, Chapter V), then in principle this extra knowledge should reduce the number of test sets needed (Molchanov, 1984). For example, Trader and Eddy (1981) considered almost surely compact convex sets and were able to work with events $\{X \subset C\}$, for all $C \in C(\mathscr{K})$. But $\Pr(X \subset C) = \Pr(X \cap C^{\mathscr{C}} = \varnothing) = 1 - \Pr(X \cap C^{\mathscr{C}} \neq \varnothing) = 1 - T_X(C^{\mathscr{C}})$. Not only are the number of test sets reduced from that of Choquet's Theorem, but also $\{C^{\mathscr{C}}: C \in C(\mathscr{K})\}$ is not even contained in $\mathscr{K}$. Trader (1981) has demonstrated the quite general result that just as $\{T_X(K): K \in \mathscr{K}\}$ determines the probability measure of a RACS X, so also does $\Pr\{X \subset K\}$, for all $K \in \mathscr{K}$; that is, so also does $\{T_X(K^{\mathscr{C}}): K \in \mathscr{K}\}$. This is perhaps not so surprising because $K^{\mathscr{C}}$ is an open set, which in turn can be approximated by a sequence of compact sets, and the compact sets themselves are measure determining. Ripley (1981, Section 9.1) also discusses the problem of choice

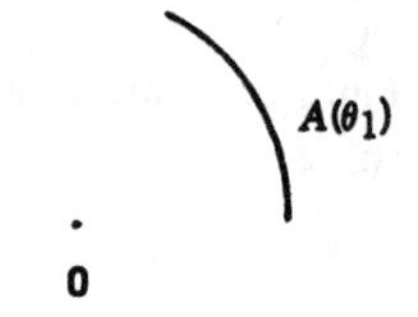

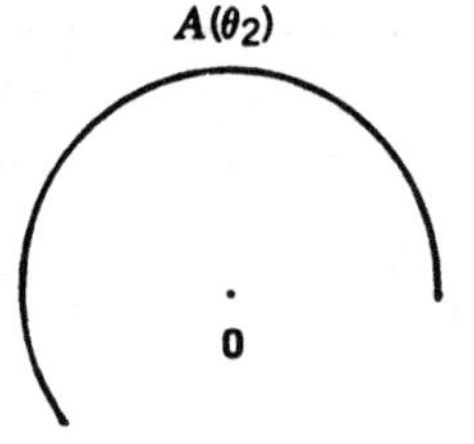

Figure 9.6 Examples of test sets for random rays on the unit disk in $\mathbb{R}^2$. [*Source*: Cressie and Laslett, 1987]. Reprinted with permission from the *SIAM Review*, volume 29, no. 4, pp. 557–574. Copyright 1987 by the Society for Industrial and Applied Mathematics. All rights reserved.

of test sets. A strong result in this regard is owing to Salinetti and Wets (1986), who prove that $\{T_X(U)\colon U \in$ set of all finite unions of closed balls in $E\}$ determines the probability measure of a RACS X.

Those who wish to build models depending on sets more regular than those of $\mathscr{K}$ struggle with the test sets of Choquet's Theorem, and even when a reduction is possible it is not always easy to calculate the hitting function. Suppose that the random set X in $\mathbb{R}^2$ is the random ray obtained by taking a random point on the unit circle according to a distribution function given by $F(\theta) = \Pr\{\text{point} \in \text{arc } [0, \theta]\}$ and joining this point to the origin. The test sets are simply the arcs $\{A(\theta)\colon \theta \in [0, 2\pi]\}$, two of which are pictured in Figure 9.6.

Now generate an independent random *vector* $\mathbf{W}$ in $\mathbb{R}^2$, according to a distribution function $G(\cdot)$. Then the random set $X \oplus \mathbf{W} \equiv \{\mathbf{s} + \mathbf{W}\colon \mathbf{s} \in X\}$ is well defined and has hitting function $T_{X \oplus \mathbf{W}}(K) = \int_{\mathbb{R}^2} T_X(K \oplus \{-\mathbf{w}\}) G(d\mathbf{w})$, $K \in \mathscr{K}$. In fact, the test sets can again be reduced to arcs, but they are no longer necessarily centered at the origin. Hence $\Pr\{X \cap (A(\theta) \oplus \mathbf{s}) \neq \varnothing\}$, for any $\mathbf{s} \in \mathbb{R}^2$ and any $\theta \in [0, 2\pi]$, needs to be calculated. This is not a trivial task once the harmony of both X *and* the test set being centered at the origin is broken.

9.3.2* Inference on Random Closed Sets

Too often, probabilistic research into stochastic modeling is not well counterbalanced with statistical research. Specifically, beyond multivariate analysis and time series, inference techniques for more general stochastic processes are often rudimentary. This is particularly so for random-set models of objects.

In the fortunate circumstance where a number of objects are observed that could be thought of as independent realizations of the same random set, limit theorems are available. An important set parameter to estimate is the expected value $E(X)$ of the random set X. The expected value of a random set is defined to be the set of expected selections, where a selection is any random vector that almost surely belongs to the random set (Aumann, 1965; Artstein and Vitale, 1975). The natural estimator is

$$\bar{X}_n \equiv (X_1 \oplus X_2 \oplus \cdots \oplus X_n)/n,$$

the Minkowski average of n independent copies $X_1, X_2, \ldots, X_n$ of X. Limit theorems for this estimator are necessary for formal inference. Artstein and Vitale (1975), Cressie (1978b), Hess (1979, 1985), Artstein and Hart (1981), and Puri and Ralescu (1983) prove strong laws of large numbers for $\bar{X}_n$ under various conditions. They show that $\bar{X}_n$ converges (with respect to the Hausdroff metric) to the expected *convex hull* of X, with probability 1; Schurger (1983) does likewise for subadditive superstationary X_is. The result is generalized to convergence of weighted sums of random sets by Taylor and Inoue (1985). Notice that the limiting set is necessarily convex. For example, suppose $\mathbf{a}_i$ and $\mathbf{b}_i$ are random vectors in $\mathbb{R}^d$ and let $X_i = [\mathbf{a}_i, \mathbf{b}_i] \equiv \{\mathbf{s}\colon \mathbf{s} = \lambda \mathbf{a}_i + (1-\lambda)\mathbf{b}_i;\ 0 \le \lambda \le 1\}$ be the line segment joining $\mathbf{a}_i$ and $\mathbf{b}_i$. Then $\bar{X}_n$ converges almost surely to a zonoid (by definition, a zonoid is the limit of a sequence of Minkowski sums of line segments), which is convex. A different strong law of large numbers for a possibly nonconvex limiting set is proved by Kovyazin (1986).

Cressie (1979b), Ljasenko (1979), Trader and Eddy (1981), Vitale (1981), Weil (1982), and Artstein (1984) show that, under various conditions, the rate of convergence in the strong law of large numbers is $n^{-1/2}$. Cressie's central limit theorem is geometric in that limiting normalized sets are given, whereas the other authors' theorems are in terms of normalized Hausdorff distances, which lose the geometric subtleties of the limiting process. These latter methods are employed by Giné et al. (1983) to carry over any probability result in a Banach space (law of large numbers, central limit theorem, law of the iterated logarithm, etc.) to Minkowski sums of compact *convex* random sets. The convexity condition is dispensed with in Puri and Ralescu (1985).

To model the limiting random set, one should look for classes of models from among those random sets that are infinitely divisible under Minkowski addition $\oplus$ (Mase, 1979). Giné and Hahn (1985) give a representation for such sets of Lévy–Hincin type.

The limit results referred to in the preceding text are not proved by hitting-function considerations, but rather by direct inspection of the random set $(X_1 \oplus \cdots \oplus X_n)/n$. Norberg's (1984) result, which says that convergence in distribution of RACS requires convergence of hitting functions only on a certain countable class of bounded Borel sets, does not seem helpful in this regard. The difficulty lies in not having a formula for $T_{X_1 \oplus X_2}$ in terms of T_{X_1} and T_{X_2}.

Analogues to extreme-value limit theorems can be obtained by considering limits of $X_1 \cup \cdots \cup X_n$ or $X_1 \cap \cdots \cap X_n$; that is, replace $\oplus$ with $\cup$ or $\cap$ (Eddy, 1982).

Consider the more general problem of convergence of a sequence of random sets $\{Z_n: n \geq 1\}$ to a limiting random set Z. Assuming $\{Z_n\}$ is a set-valued martingale (defined via the conditional expectation of selections), Hiai and Umegaki (1977), Bagchi (1985), and Papageorgiou (1989) give various convergence results.

The importance of the hitting function justifies its estimation. Suppose $X_1, \ldots, X_n$ are independent copies of X, available as data, and define $T_n(K) \equiv \sum_{i=1}^{n} I(X_i \cap K \neq \varnothing)/n$. Molchanov (1987) considers conditions under which $T_n(K)$ converges to $T_X(K) = \Pr(X \cap K \neq \varnothing)$, uniformly in K. This allows one to obtain, for example, a consistent estimator of $\chi(\mathbf{h})$, the covariogram of the pores of a stationary RACS. Now, $\chi(\mathbf{h}) \equiv 1 - T_X(\{\mathbf{0}, \mathbf{h}\})$. Then

$$\hat{\chi}(\mathbf{h}) \equiv \sum_{i=1}^{n} I\big(\mathbf{0} \in X_i^{\mathscr{C}} \cap \big(X_i^{\mathscr{C}} \oplus \{-\mathbf{h}\}\big)\big)\big/n, \qquad \mathbf{h} \in \mathbb{R}^d,$$

is a consistent estimator.

One Observation on X

Usually data consist of just one observation; in this case, inference is impossible unless repeatability in space is assumed. The statistical methods used in Chapters 2 through 8 all have some type of stationarity (actually, ergodicity) assumption to ensure their consistency. Without it, any statistical analysis can only claim to be descriptive.

But stationarity puts its own restrictions on X. Matheron (1975) has shown that a nonempty stationary RACS in $\mathbb{R}^d$ is almost surely *unbounded*. Thus, there are various types of phenomena that are not amenable to stochastic modeling and inference.

The rest of this chapter is concerned with one particular class of unbounded sets, namely the Boolean model and its variants (see Section 9.4 for definitions). There are pragmatic reasons for choosing this class; it will be seen that inference on its parameters is possible from just one observation X (Sections 9.5 and 9.6) and it exhibits a great deal of flexibility for modeling objects. Let X be any RACS in $\mathbb{R}^d$ whose realizations are in the extended convex ring $\mathscr{S}^d$. (That is, for any compact convex set C, $X \cap C$ is a finite union of compact convex sets.) Surprisingly, this rather mild regularity condition is enough to ensure that X is a (generalized) Boolean model of convex grains (Weil and Wieacker, 1984, 1988).

9.4 THE BOOLEAN MODEL

In this section I shall present what is arguably the most important random-set model, namely, the Boolean model. Generalizations to other models will be discussed in Section 9.4.2, but it is clear that a fruitful path to truly broad classes of models has yet to be developed.

The basic ideas behind the construction of this model are found in Avrami (1939), Armitage (1949), Solomon (1953), Matern (1960), and Roach (1968). Marcus (1966, 1967) uses the Boolean model to examine the meteoroidal-impact hypothesis for the origin of lunar craters, Dupac (1980) considers the etching of tracks formed by the fission of randomly located uranium atoms in a fission material, Serra (1980) models ore sintering, and Diggle (1981b) uses it to model the incidence of heather (he calls it a random binary mosaic).

Special cases of the Boolean model were given in Sections 9.1.3 and 9.2. The Boolean model X is obtained by locating independent realizations of a *bounded RACS* Z (of which random polygons, disks, and parallelograms are special cases) at each point of a realization of a *homogeneous Poisson process* D in $\mathbb{R}^d$ (Section 8.3) and then taking the union. That is,

$$X \equiv \bigcup \{Z(\mathbf{s}_i): \mathbf{s}_i \in D\}, \tag{9.4.1}$$

where $Z(\mathbf{s}_i) \equiv Z_i \oplus \mathbf{s}_i$, a realization of the (almost surely) bounded RACS Z (i.e., a *grain*) translated to a point $\mathbf{s}_i$ (i.e., a *germ*) of the Poisson process D. Thus, there are two sources of randomness in the model: (1) the Poisson process characterized by its intensity parameter λ and (2) the probability law of the bounded random grain Z.

The Hitting Function

The hitting function of the Boolean model X, defined by (9.4.1), can be constructed in several stages.

Consider the *homogeneous* Poisson process D in $\mathbb{R}^d$ with intensity λ, which, according to Section 8.3, is characterized as follows. Let $N(A)$ denote the number of points of the process in any subset $A \subset \mathbb{R}^d$. Then, if $A_1 \cap A_2 = \varnothing$, $N(A_1)$ and $N(A_2)$ are independent Poisson random variables with respective means $\lambda|A_1|$ and $\lambda|A_2|$. Also

$$\Pr\{N(d\mathbf{s}) = 0\} = 1 - \lambda \cdot \nu(d\mathbf{s}) + o(\nu(d\mathbf{s})),$$
$$\Pr\{N(d\mathbf{s}) = 1\} = \lambda \cdot \nu(d\mathbf{s}) + o(\nu(d\mathbf{s})),$$

where $\nu(d\mathbf{s})$ is the volume element (Lebesgue measure) of the infinitesimal region $d\mathbf{s}$, located at $\mathbf{s} \in \mathbb{R}^d$.

Now the complement of the hitting function of X is

$$Q_X(K) \equiv \Pr(X \cap K = \varnothing) = \Pr(Z(\mathbf{s}_i) \cap K = \varnothing, \text{ for all } i).$$

In the infinitesimal region ds, two mutually exclusive events may happen:

1. There is no element of $\{\mathbf{s}_1, \mathbf{s}_2, \cdots\}$ in ds. This occurs with approximate probability $1 - \lambda \cdot \nu(ds)$.
2. There is one element of $\{\mathbf{s}_1, \mathbf{s}_2, \cdots\}$ in ds, but $Z(\mathbf{s}_i)$ does not hit K. This occurs with approximate probability $\lambda \cdot \nu(ds) \cdot Q_Z(K \oplus \{-\mathbf{s}\})$.

To the order of magnitude ignored, this yields the probability,

$$1 - \lambda(1 - Q_Z(K \oplus \{-\mathbf{s}\})\nu(ds))$$
$$= \exp\{-\lambda(1 - Q_Z(K \oplus \{-\mathbf{s}\}))\nu(ds)\} + o(\nu(ds)).$$

Finally, $Q_X(K)$ is obtained by taking the product over all infinitesimal regions:

$$Q_X(K) = \exp\left\{-\lambda \int_{\mathbb{R}^d} (1 - Q_Z(K \oplus \{-\mathbf{s}\}))\nu(ds)\right\}. \tag{9.4.2}$$

This formula is fundamental in linking the hitting function of X to the hitting function of Z.

Now define $I_{\check{Z} \oplus K}(\cdot)$ to be the indicator function of $\check{Z} \oplus K \equiv \{-\mathbf{s} + \mathbf{k}: \mathbf{s} \in Z, \mathbf{k} \in K\}$; see (9.3.4) and (9.3.6). But

$$|\check{Z} \oplus K| = \int_{\mathbb{R}^d} I_{\check{Z} \oplus K}(\mathbf{s})\nu(ds),$$

and so

$$E(|\check{Z} \oplus K|) = \int E(I_{\check{Z} \oplus K}(\mathbf{s}))\nu(ds) = \int T_Z(K \oplus \{-\mathbf{s}\})\nu(ds),$$

because $\mathbf{s} \in \check{Z} \oplus K$ if and only if $Z \cap (K \oplus \{-\mathbf{s}\}) \neq \emptyset$. Hence,

$$E(|\check{Z} \oplus K|) = \int (1 - Q_Z(K \oplus \{-\mathbf{s}\}))\nu(ds),$$

where the integrals with respect to $\mathbf{s}$ are taken over $\mathbb{R}^d$. Therefore, from (9.4.2),

$$T_X(K) = 1 - Q_X(K) = 1 - \exp\{-\lambda E(|\check{Z} \oplus K|)\}, \qquad K \in \mathscr{K}. \tag{9.4.3}$$

The flexibility in choice of K makes (9.4.3) a very useful result; for example, the porosity and covariogram of the pores are derived from it in the next section. When K is a straight line segment, $Q_X(K)$ is the probability that one end of the segment is visible to an observer positioned at the other end of the segment. Such visibility problems have been discussed by Yadin and Zacks (1985) and, in a more general context, by Wieacker (1986).

9.4.1* Main Properties

In this section, I shall summarize the main properties of the Boolean model, from which it should be clear why it is central among random set models. More details can be found in Serra (1980).

1. *Porosity q.* Porosity is the probability that a point of the space is in the complement of the RACS X (i.e., in the pores). Let $K = \{\mathbf{s}\}$. Then $q \equiv Q_X(\{\mathbf{s}\}) = \Pr\{\mathbf{s} \in X^{\mathscr{C}}\}$; that is, from (9.4.3),

$$q = e^{-\lambda E(|Z|)}.$$

   This is a coverage property of the Boolean model; Avrami (1939) is the earliest reference I know of where such problems are addressed in $\mathbb{R}^2$. Hall (1988a) contains a complete discussion of coverage, vacancy, counting, and clumping for the Boolean model.
2. *Covariogram of the Pores $\chi(\boldsymbol{h})$.* This is defined to be the probability that two points separated by a vector $\mathbf{h}$ are both in the complement of the RACS X. It measures dependence between pores. Let $K = \{\mathbf{s}, \mathbf{s} + \mathbf{h}\}$ and $\chi(\mathbf{h}) \equiv Q_X(\{\mathbf{s}, \mathbf{s} + \mathbf{h}\})$. Then, from (9.4.3),

$$\begin{aligned}\chi(\mathbf{h}) &= \Pr\{\mathbf{s}, \mathbf{s} + \mathbf{h} \in X^{\mathscr{C}}\}\\ &= \exp\{-\lambda E(|Z \cup Z \oplus \{-\mathbf{h}\}|)\} = q^2 e^{\lambda\kappa(\mathbf{h})},\end{aligned}$$

   where

$$\kappa(\mathbf{h}) \equiv E(|Z \cap Z \oplus \{-\mathbf{h}\}|) = \int \Pr\{\mathbf{s} \in Z \cap Z \oplus \{-\mathbf{h}\}\}\nu(d\mathbf{s}),$$

   the probabilistic analogue of the geometrical covariance.
3. *Stationarity.* Because $E(|\cdot|)$ in (9.4.3) ignores location, X is clearly stationary.
4. *Stability under Dilation.* From (9.4.3), the RACS $X \oplus L \equiv \{\mathbf{s} + \mathbf{l}: \mathbf{s} \in X, \mathbf{l} \in L\}$, where L is any deterministic compact set, is also a Boolean model.
5. *Cross Sections are Boolean Models.* This is true because K in (9.4.3) may belong to a subspace of $\mathbb{R}^d$.
6. *Infinite Divisibility with Respect to Union.* A RACS Y is said to be *infinitely divisible with respect to union* if, for any integer $m > 0$, Y is equivalent to $\bigcup_{i=1}^{m} Y_i$ of m independent stochastically equivalent RACS $\{Y_i: i = 1, \ldots, m\}$. The Boolean model has this property, which shows it to be a candidate model for limits of unions of RACS.

7. *The Boolean Model with Convex Grain Z is Semi-Markov.* A RACS Y is said to be *semi-Markov* if for any K, L, and $M \in \mathscr{K}$, and K and M separated by L (i.e., the segment joining any point of K to any point of M hits L), the RACS $Y \cap K$ and $Y \cap M$ are conditionally independent given $Y \cap L = \varnothing$. The Boolean model with convex grains has this property, which shows that its probability law is really determined by local conditions. Widely separated parts of the set are only weakly dependent. In terms of the functional Q, Matheron (1975, p. 132) has shown that if Y is infinitely divisible with respect to union and if, for any K, M, and $K \cup M \in C(\mathscr{K})$ (the space of compact convex sets),

$$Q_Y(K \cup M)Q_Y(K \cap M) = Q_Y(K)Q_Y(M),$$

then Y is semi-Markov.

8. *The Union of Two Independent Boolean Models with Identically Distributed Grains Z is a Boolean Model.* Furthermore, if the Poisson points of a Boolean model are *thinned* so that they now occur with an intensity $\rho \cdot \lambda (0 \le \rho < 1)$, leaving behind a Boolean model whose union is taken with another independent Boolean model of intensity $(1 - \rho) \cdot \lambda$ and the same probability law for its grains, then the resulting random set is a Boolean model identically distributed to the original one.

A Characterization Theorem

Matheron (1975, p. 148) has provided the following important characterization theorem (presented here in $\mathbb{R}^3$). Any RACS in $\mathbb{R}^3$ that is stationary, infinitely divisible with respect to union, and semi-Markov, is equivalent to

$$X_1 \cup X_2 \cup X_3,$$

where X_1, X_2, and X_3 are stationary and independent, X_1 is a Boolean model with convex grains, X_2 is a union of cylinders with bases that are two-dimensional Boolean models with convex grains (in $\mathbb{R}^2$), and X_3 is a union of cylinders with bases that are one-dimensional Boolean models with convex grains (in $\mathbb{R}^1$). Notice that if the grains in X_2 and X_3 are almost surely points, then the associated cylinders become Poisson lines and Poisson planes, respectively, yielding Poisson-flat processes (studied extensively by Miles, 1969).

9.4.2* Generalizations of the Boolean Model

The choice of mathematical models available to the data analyst is often governed by their tractability rather than their applicability. When the data are sets, this leaning is even more pronounced. Serra (1982, Chapter XIII) has provided users with a menu of models and of examples for which they are appropriate. By far the most important group is that based on the Boolean model, which I shall present and extend here.

One extension already mentioned in Section 9.4.1 is to the Poisson-flat processes of Miles (1969). There, hyperplanes of infinite extent and random orientation are placed at points of a homogeneous Poisson process in $\mathbb{R}^d$. Another extension is due to Cowan (1989), who assumes only that the point process of germs has constant intensity. The point process may be dependent on the grains $\{Z_i\}$, which are assumed to have smooth boundaries. He obtains mean-value formulas for features seen in an observation window of general size and shape, but his assumptions are too weak to allow calculation of hitting functions.

The most fruitful extension of the Boolean model is the so-called *grain-germ model* (Hanisch, 1980). It removes the Poisson assumption, the independence of the Z_is, and allows for nonoverlapping of the grains (e.g., Mase, 1986). Let

$$X = \bigcup \{Z_i \oplus \mathbf{s}_i : (\mathbf{s}_i, Z_i) \in N\}, \tag{9.4.4}$$

where N is a marked point process with mark space $\mathscr{K}$, the set of compact sets. Let Π_N be the induced measure on $(\Phi_{\mathscr{K}}, \mathscr{N}_{\mathscr{K}})$, where $\Phi_{\mathscr{K}}$ is the set of all Radon counting measures ϕ on $\mathbb{R}^d \times \mathscr{K}$ with $\phi(B \times \mathscr{K}) < \infty$ for any bounded Borel set B, and $\mathscr{N}_{\mathscr{K}}$ is the corresponding σ algebra (Section 8.7). Let U be the set of all measurable functions from $\mathbb{R}^d \times \mathscr{K}$ to $[0,1]$ and $V \equiv \{1 - u : u \in U\}$. The functional $G_N: V \to [0,1]$, which is given by

$$G_N(v) = \int_{\Phi_{\mathscr{K}}} \prod_{(\mathbf{s}, Z) \in \phi} v(\mathbf{s}, Z) \Pi_N(d\phi), \qquad v \in V, \tag{9.4.5}$$

is called the probability generating functional of the marked point process N. For X given by (9.4.4), Hanisch is able to relate the complement of the hitting function Q_X to G_N:

For $K \in \mathscr{K}$, let v_K be the mapping given by

$$v_K(\mathbf{s}, Z) = 1 - I_{\check{Z} \oplus K}(\mathbf{s}), \qquad \mathbf{s} \in \mathbb{R}^d, Z \in \mathscr{K}.$$

Then,

$$Q_X(K) = G_N(v_K), \qquad K \in \mathscr{K}. \tag{9.4.6}$$

Special cases of (9.4.4) yield models already studied in the literature:

1. The marked point process N almost surely yields *independent* markings; see Stoyan (1979b) and Mase (1982b).
2. The point process is a Neyman–Scott process where parent events are generated according to a homogeneous Poisson process and offspring

events are generated independently and identically around each initial point; see Neyman and Scott (1958) and Section 8.5.3.

3. The point process is regionally independent, in particular, a (not necessarily homogeneous) Poisson process with intensity function $\{\lambda(\mathbf{s}): \mathbf{s} \in \mathbb{R}^d\}$. If the *weighted* measure of $\check{Z} \oplus K$ is defined as

$$|\check{Z} \oplus K|_\lambda \equiv \int_{\mathbb{R}^d} I_{\check{Z} \oplus K}(\mathbf{s})\lambda(\mathbf{s})\, \nu(d\mathbf{s}), \tag{9.4.7}$$

then the complement of the hitting function is

$$Q_X(K) = \exp\left\{-E\left(|\check{Z} \oplus K|_\lambda\right)\right\}. \tag{9.4.8}$$

Hypothesis testing for constant intensity function in the Boolean model is an area of inference that needs investigation.

4. The point process is a homogeneous Poisson process. Serra (1982, p. 484ff) has a detailed discussion of the model and some of its associates. When the mark process is independent of the point process and consists of independent realizations of an almost surely bounded RACS Z, then the Boolean model (9.4.1) is obtained. One interesting extension is to consider a series of independent Boolean models occurring from the infinite past: At each instant of time the Boolean model fills part of the space. At the next instant, some of the pores (and sets) will be covered by a set from the new Boolean model and some of the sets (and pores) will remain exposed. The procedure is continued until the present, so that finally a tessellation of the space results. This is called the *dead-leaves model*, which has been further extended by Jeulin (1989) to sequential-random-function models.

5. The events are not generated by a point process, but are fixed and finite in number in $\mathbb{R}^d$, namely, $\{\mathbf{a}_1, \ldots, \mathbf{a}_n\}$. The only source of randomness is from the mark process, assumed to be generated by independent realizations of a bounded RACS Z. Then, from (9.4.5),

$$G_N(v_K) = \prod_{i=1}^{n} E(v_K(\mathbf{a}_i, Z_i)).$$

Hence,

$$Q_X(K) = \prod_{i=1}^{n} Q_Z(K \oplus \{-\mathbf{a}_i\}). \tag{9.4.9}$$

6. The point process is homogeneous Poisson, independent from the mark process generated by independent realizations of a bounded RACS Z, with the property that $|Z|$ is *not* random (although in general un-

known). In this case,

$$\log q = -\lambda |Z|,$$

$$\int_{\mathbb{R}^d} \log(\chi(\mathbf{h})/q^2)\, d\mathbf{h} = \lambda |Z|^2,$$

where the porosity q and the covariogram of the pores $\chi(\mathbf{h})$ are given by properties 1 and 2 of Section 9.4.1 and can be estimated from the set data. Hence λ and $|Z|$ can be estimated by the method-of-moments, although little is known about the efficiency of these estimators; Sections 9.5, 9.6, and 9.7 can be consulted for matters of estimation and inference on data of this type.

7. The point process is homogeneous Poisson, independent of the mark process generated by independent realizations of the RACS $b(\mathbf{0}, R)$, the closed ball of random radius R centered at $\mathbf{0}$. The Poisson-intensity parameter λ and the distribution of R can be estimated from the covariance $\chi(\mathbf{h})$ (Serra, 1980), although little is known about the estimators' statistical properties; see Dupac (1980) and Section 9.6.

9.5 METHODS OF BOOLEAN-MODEL PARAMETER ESTIMATION

Several data sets in the form of images have been presented in this chapter so far. The cloud-droplets data of Figure 9.4 are naturally suited to the fitting of a Boolean model. Of the parameters estimated, cloud physicists are very interested in the droplet size distribution; the mean and variance of the droplet radius give some idea of the efficacy of the cloud-seeding treatment. Cloud droplets are three-dimensional but the data are two-dimensional, representing the cumulative projection of droplets over a region of the cloud. Problems of stereology (i.e., estimating d-dimensional quantities from data taken in lower, more manageable dimensions) were discussed generally in remarks made at the beginning of this chapter. Weil (1987) gives an account of the results available for the (generalized) Boolean model.

Cressie and Laslett (1987) generated a Boolean model in $\mathbb{R}^2$ made up of random parallelograms; they are reproduced in Figure 9.5. An analysis of these data is presented in Section 9.5.1; however, some preliminary formulas are needed before I can proceed.

Steiner's Formula

Consider an almost surely compact convex grain Z, whose probability law is invariant under rotations of Z about $\mathbf{0}$. Then the Boolean model X defined by (9.4.1) is stationary and isotropic. Under these circumstances, and for K compact convex, Steiner's formula (Mack, 1954; Serra, 1982, p. 111) in $\mathbb{R}^2$ yields

$$E\{|\check{Z} \oplus K|\} = E(|Z|) + \frac{1}{2\pi} E(P(Z))P(K) + |K|, \tag{9.5.1}$$

where $P(A)$ denotes the perimeter of the set A. When $Z = b(\mathbf{0}, R)$, the disk centered at $\mathbf{0}$ of random radius R, $E(P(Z))/(2\pi) = E(R)$ and $E(|Z|)/\pi = E(R^2)$. Weil and Wieacker (1984) can be consulted for generalizations of (9.5.1) to generalized Boolean models (presented in Section 9.4.2) in higher dimensions.

With one realization of what is believed to be a Boolean model in $\mathbb{R}^2$, the goal is to make inferences about the random set Z. From (9.4.3) and (9.5.1),

$$Q_X(K) = \exp\left[-\lambda\left\{|K| + \frac{1}{2\pi}E(P(Z))P(K) + E(|Z|)\right\}\right], \quad (9.5.2)$$

and hence estimators can be found for λ, $E(P(Z))$, and $E(|Z|)$: For example, let K be $b(\mathbf{0}, t)$, the disk of radius t; then vary t. Theoretically,

$$\begin{aligned} -\log Q_X(b(\mathbf{0}, t)) &= \lambda\{\pi t^2 + E(P(Z))t + E(|Z|)\} & (9.5.3) \\ &\equiv \beta_0 + \beta_1 t + \beta_2 t^2, & (9.5.4) \end{aligned}$$

where $\beta_0 \equiv \lambda E(|Z|)$, $\beta_1 \equiv \lambda E(P(Z))$, and $\beta_2 \equiv \lambda\pi$.

Boolean-Model Parameter Estimators

By scanning the image appropriately, one can obtain an estimate $\hat{Q}_X(b(\mathbf{0}, t))$, for various t, which, when regressed on 1, t, and t^2 yields estimates $\hat{\beta}_0$, $\hat{\beta}_1$, and $\hat{\beta}_2$. Image analyzers (e.g., Serra, 1982, Part II) are specifically built to do this type of task; however, it is not difficult to use conventional computers to analyze random-set data like Figure 9.5. Cressie and Laslett (1987) compute both ordinary-least-squares and generalized-least-squares estimates of quadratic-regression coefficients for Figure 9.5a. Estimates of set parameters are easily obtained from (9.5.4) through the transformations

$$\begin{aligned} \hat{\lambda} &= \hat{\beta}_2/\pi, \\ \hat{E}(P(Z)) &= \hat{\beta}_1/\hat{\lambda}, & (9.5.5) \\ \hat{E}(|Z|) &= \hat{\beta}_0/\hat{\lambda}. \end{aligned}$$

Alternatively, choice of the three different types of test set, K_0 the origin, K_1 the straight line segment of length l, and K_2 the closed square of side l, yields (in $\mathbb{R}^2$)

$$\begin{aligned} Q_X(K_0) &= \exp\{-\lambda E(|Z|)\}, \\ Q_X(K_1) &= \exp\left\{-\lambda\left\{E(|Z|) + \frac{l}{\pi}E(P(Z))\right\}\right\}, & (9.5.6) \\ Q_X(K_2) &= \exp\left\{-\lambda\left\{E(|Z|) + \frac{2l}{\pi}E(P(Z)) + l^2\right\}\right\}, \end{aligned}$$

from which method-of-moments estimators could be obtained. Hall (1985b) and Kellerer (1985) take essentially this approach; however, it leaves no degrees of freedom to assess model adequacy. The same comment applies to the intensity-based method-of-moments estimators given by Stoyan et al. (1987, p. 88).

9.5.1 Analysis of Random-Parallelograms Data

Because the original Boolean model has been discretized onto a square grid, Figure 9.5a was analyzed with $K = tC$, where C is a square of side 2 units centered on $\mathbf{0}$, and $t = 0, 1, 2, 3, 4$ (larger values of t led to $\hat{Q}_X \equiv 0$). Then (9.5.2) gives

$$-\log Q_X(tC) = \lambda E(|Z|) + \{4\lambda E(P(Z))/\pi\}t + 4\lambda t^2. \quad (9.5.7)$$

Consider the estimator

$$\hat{Q}_X(tC) \equiv \sum_{\mathbf{x}_i \in E_t} \{1 - I_{X \oplus tC}(\mathbf{x}_i)\}/N_t, \qquad t = 0, \ldots, 4, \quad (9.5.8)$$

where N_t is the total number of pixels (at locations $\{\mathbf{x}_i: i = 1, \ldots, N_t\}$) in the mask $E_t = E \ominus tC$ [see (9.3.8)] placed over the region of interest E. (In Figure 9.5a and b, E is the bordered region.) Clearly, this is an unbiased estimator of $Q_X(tC)$. Write $\mathbf{Y}' \equiv (-\log \hat{Q}_X(tC): t = 0, \ldots, 4)$ and let T be the 5×3 matrix whose tth row is $(1, t, t^2)$, $t = 0, \ldots, 4$. Then the usual ordinary-least-squares (o.l.s.) estimator is given by $(T'T)^{-1}T'\mathbf{Y}$, which estimates the coefficients of 1, t, and t^2 without taking into account correlations between the elements of $\mathbf{Y}$. Estimates of higher efficiency can be obtained by taking a generalized-least-squares (g.l.s.) approach. Table 9.2 gives the o.l.s. and g.l.s. estimates for the data of Figure 9.5a.

Table 9.2 Summary Statistics and Final Estimates for the Simulated Boolean Model Data of Figure 9.5a[a]

t	$\hat{Q}_X(tC)$	Fitted Quadratic (g.l.s.)	Fitted Quadratic (o.l.s.)
0	0.72950	0.72985	0.72692
1	0.44138	0.44137	0.44430
2	0.22797	0.22678	0.22831
3	0.09948	0.09901	0.09863
4	0.03568	0.03673	0.03582
		Parameter Estimates[a] (g.l.s.)	Parameter Estimates[a] (o.l.s.)
	$\hat{\lambda}$	0.020	0.022
	$\hat{E}(P(Z))$	16.27	14.69
	$\hat{E}(\|Z\|)$	15.47	14.71

Source: Cressie and Laslett (1987). Reprinted with permission from the *SIAM Review*, volume 29, no. 4, pp. 557–574.

[a]The true values of the parameters are $\lambda = 0.02$, $E(P(Z)) = 20.00$, and $E(|Z|) = 15.92$.

Generalized-Least-Squares Estimation

To give the g.l.s. estimator explicitly, some definitions are needed first. From the definition of $\hat{Q}_X(tC)$, given by (9.5.8), it is easily seen that, for the Boolean model,

$$\operatorname{cov}\left(\hat{Q}_X(tC), \hat{Q}_X(uC)\right) = \sum_{\mathbf{x}_i \in E_t} \sum_{\mathbf{x}_j \in E_u} C_{t,u}(\mathbf{x}_i - \mathbf{x}_j)/(N_t N_u), \quad (9.5.9)$$

where

$$C_{t,u}(\mathbf{h}) \equiv \left[\exp\{\lambda E(|(Z \oplus tC) \cap (Z \oplus \{-\mathbf{h}\} \oplus uC)|)\} - 1\right] \times \exp\{-\lambda E(|Z \oplus tC|) - \lambda E(|Z \oplus uC|)\}. \quad (9.5.10)$$

The first term in this expression for $C_{t,u}(\mathbf{h})$ can be simplified for special cases of the structuring element C. Regardless, the expression for the covariance depends on the unknown parameters; recall that the objective is to estimate these parameters. Instead of trying to minimize a complicated objective function with parameters also in the covariance terms, the approach taken here will be to use the *empirical* covariance function in g.l.s. estimation of the parameters.

The pixels of Figure 9.5 are regularly spaced, and so $C_{t,u}(\mathbf{h})$ can be estimated for $\mathbf{h}$ in horizontal, vertical, and diagonal directions. Let

$$\hat{C}_{t,u}(\mathbf{h}) \equiv \left[\sum\sum_{\mathbf{x}_i, \mathbf{x}_j \in M(\mathbf{h})} \{1 - I_{X \oplus tC}(\mathbf{x}_i)\}\{1 - I_{X \oplus uC}(\mathbf{x}_j)\} \Big/ \sum\sum_{\mathbf{x}_i, \mathbf{x}_j \in M(\mathbf{h})} 1\right] - \hat{Q}_X(tC) \cdot \hat{Q}_X(uC), \quad (9.5.11)$$

where $M(\mathbf{h}) \equiv \{(\mathbf{x}_i, \mathbf{x}_j)\colon \mathbf{x}_i \in E_t, \mathbf{x}_j \in E_u, \mathbf{x}_i - \mathbf{x}_j = \mathbf{h}\}$. Substituting (9.5.11) for $C_{t,u}(\mathbf{h})$ in (9.5.9) yields $\widehat{\operatorname{cov}}(\hat{Q}_X(tC), \hat{Q}_X(uC))$. Write

$$\hat{\Sigma} \equiv \left(\widehat{\operatorname{cov}}\left(\hat{Q}_X(tC), \hat{Q}_X(uC)\right)\colon t = 0, \ldots, 4;\ u = 0, \ldots, 4\right), \quad (9.5.12)$$

which is a 5×5 matrix. The error vector is

$$\boldsymbol{\delta}' \equiv \left(\hat{Q}_X(tC) - \exp\{-\beta_0 - \beta_1 t - \beta_2 t^2\}\colon t = 0, \ldots, 4\right). \quad (9.5.13)$$

Cressie and Laslett (1987) proposed a g.l.s. estimator for $\beta_0 = \lambda E(|Z|)$, $\beta_1 = \{4\lambda E(P(Z))/\pi\}$, and $\beta_2 = 4\lambda$ by minimizing $\boldsymbol{\delta}'\hat{\Sigma}^{-1}\boldsymbol{\delta}$ with respect to β_0, β_1, and β_2. This is carried out using a quasi-Newton numerical optimization procedure. (A g.l.s. estimator on the log-transformed scale is discussed in Section 9.6.)

Cressie and Laslett (1987) reported that occasionally $\hat{\Sigma}$ was not positive-definite. Difficulties of this sort are to be expected when empirical covariances are used; for example, a time series $\{z_t\colon t = 1, \ldots, n\}$ has an (approximately) unbiased empirical covariance function $\{\sum_{t=1}^{n-h}(z_{t+h}-\bar{z})\cdot(z_t-\bar{z})/(n-h): h=1,\ldots,n-1\}$, which is not assured of being positive-definite. The usual way around this problem is to fit a positive-definite model to the empirical covariance function. One suggestion made by Diggle (1981b) is to fit a shifted Weibull density to (9.5.11).

9.5.2 Analysis of Heather-Incidence Data

Diggle (1981b) presented an image of the incidence of heather, *Calluna vulgaris*, over a 10 × 20-m rectangle at Jadraas, Sweden. Individual heather plants grow from seedlings into roughly hemispherical bushes. As the individual bushes expand, their branches intermingle, but a map of presence or absence of heather does not record the multiplicity of bushes at any particular spatial location; see Figure 9.7.

Diggle (1981b) fits the Boolean model in $\mathbb{R}^2$,

$$X = \bigcup \{Z(\mathbf{s}_i): \mathbf{s}_i \in D\}, \tag{9.5.14}$$

where $Z(\mathbf{s}_i) \equiv Z_i \oplus \mathbf{s}_i (i = 1, 2, \ldots), Z_1, Z_2, \ldots$ are disks centered at $\mathbf{0}$ whose radii $R_1, R_2, \ldots$ are i.i.d. as R with density function $f(\cdot)$, and D is a homogeneous Poisson process of intensity λ. Define $H(\cdot)$ to be the distribution function of the distance from an arbitrary point to the nearest point occupied by heather. Then $H(u) = 1 - Q_X(b(\mathbf{0}, u))$, where Q_X is given by (9.4.2), and hence,

$$H(u) = 1 - \exp\left\{-\lambda\pi\int_0^\infty (u+r)^2 f(r)\, dr\right\}, \qquad u \geq 0. \tag{9.5.15}$$

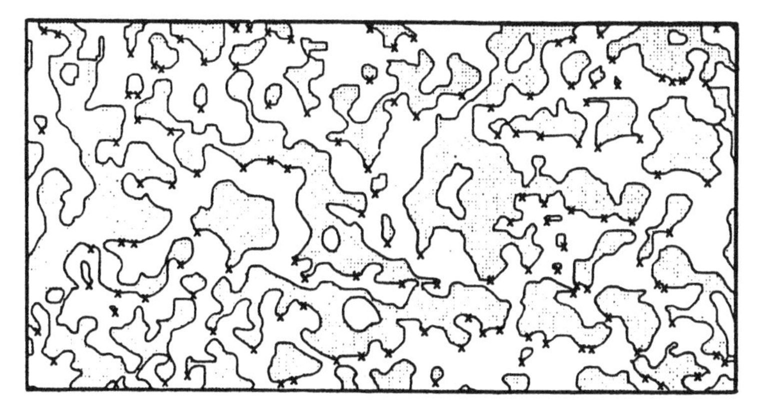

Figure 9.7 Incidence of *Calluna vulgaris* (shaded area) over a 10 × 20-m rectangular area at Jadraas, Sweden. Exposed marker points (see Section 9.5.3) are identified with a cross. [*Source:* Laslett et al., 1985]

Notice that $H(\cdot)$ depends on $f(\cdot)$ only through $E(R)$ and $E(R^2)$. An empirical version $\hat{H}$ of H can be calculated by superimposing a fine grid over the rectangle E in Figure 9.7 and randomly sampling N nodes of the grid. This would yield $U_1, \ldots, U_N$ distances to the nearest heather point (a number of U_is will be zero, should their respective nodes be in the set of heather). Then $\hat{H}(u) = \sum_{i=1}^{N} I(U_i \le u)/N$. Some modification can be made for edge effects by only sampling from $E \ominus b(\mathbf{0}, u_{\max})$, where $u_{\max}$ is the largest value of u considered. Diggle proposed to estimate parameters λ, $E(R)$, and $\{\mathrm{var}(R)\}^{1/2}$ by minimizing

$$\int_0^{u_{\max}} \left(H(u) - \hat{H}(u)\right)^2 du. \tag{9.5.16}$$

He also considers the covariance $\kappa_X(\mathbf{h}) \equiv \Pr(\mathbf{s} \in X,\ \mathbf{s} + \mathbf{h} \in X)$ and calculates that, for the Boolean model (9.5.14),

$$\kappa_X(\mathbf{h}) \equiv \kappa_X^o(h) = \exp\{-2\pi\lambda E(R^2)\}\left[\exp\left\{\lambda \int_{h/2}^{\infty} V_h(r) f(r)\, dr\right\} - 1\right],$$
$$\|\mathbf{h}\| \equiv h \ge 0, \tag{9.5.17}$$

where

$$V_h(r) \equiv 2r^2\left\{\cos^{-1}(h/(2r)) - (h/(2r))\left(1 - h^2/(4r^2)\right)^{1/2}\right\} \tag{9.5.18}$$

is the area of intersection of two disks with common radius r and centers that are a distance h apart. Notice that $\kappa_X(\mathbf{h})$ is related to the porosity q and the covariogram of the pores $\chi(\mathbf{h})$, given in Section 9.4.1, through $\kappa_X(\mathbf{h}) = 1 - 2q + \chi(\mathbf{h})$. The empirical version $\hat{\kappa}_X^o(h)$ can be calculated by measuring the area of $X \cap X \oplus \{-\mathbf{h}\}$ for various choices of $\mathbf{h}$. Using $\hat{\kappa}_X^o(h)$ as a diagnostic, Diggle (1981b) finds that fitting the parameters by minimizing (9.5.16) is unsatisfactory for the heather-incidence data.

His second approach is to minimize

$$\int_0^{h_{\max}} \{\kappa_X^o(h) - \hat{\kappa}_X^o(h)\}^2\, dh, \tag{9.5.19}$$

where he chooses $h_{\max} = 0.1$ units (here, 1 unit $= 10$ m). He fits a parametric form, namely, the shifted Weibull distribution, to $f(\cdot)$:

$$f(r) = \begin{cases} 0, & r < \delta, \\ k\rho(r-\delta)^{k-1}\exp\{-\rho(r-\delta)^k\}, & r \ge \delta, \end{cases} \tag{9.5.20}$$

where δ is a shift parameter that removes disks of small radius and $k > 0$. He obtains a good fit of $\kappa_X^o(h)$ to $\hat{\kappa}_X^o(h)$ and a satisfactory diagnostic comparison of plots of $H(u)$ with $\hat{H}(u)$. For cross-validation purposes, he splits up the area of Figure 9.7 into two unit areas (each 10×10 m); using (9.5.19) and (9.5.20), he obtains

$$\text{plot 1:} \quad (\hat{\lambda}, \hat{\delta}, \hat{k}, \hat{\rho}) = (221, 0.0281, 0.8471, 144.7),$$

$$\text{plot 2:} \quad (\hat{\lambda}, \hat{\delta}, \hat{k}, \hat{\rho}) = (211, 0.0226, 1.011, 128.4).$$

Without some notion of variation of these parameter estimates, it is impossible to tell how well the estimation procedure is doing. Hall (1985b) analyzes the same data set taking essentially the approach described by (9.5.6). Unfortunately, once again no estimate of variation is available. It would be interesting to see how the approach described in Section 9.5.1 (for the random-parallelogram data) compares to those already tried on the heather-incidence data. This has not been carried out because there is evidence that the Boolean model is unsatisfactory. Diggle (1981b) himself provides some of it by simulating the theoretical model and comparing it to the observed image (given by Figure 9.7). Section 9.5.3 provides further evidence by taking a very different approach (but one for which the Boolean model assumption of homogeneous λ is essential) to estimate λ and finding a large discrepancy.

9.5.3* Intensity Estimation in the Boolean Model

This section is concerned with analyzing data in $\mathbb{R}^2$, using the Boolean model

$$X = \bigcup \{Z(\mathbf{s}_i) : \mathbf{s}_i \in D\},$$

where $Z(\mathbf{s}_i) \equiv Z_i \oplus \mathbf{s}_i (i = 1, 2, \ldots), Z_1, Z_2, \ldots$ are independent realizations of an almost surely bounded (not necessarily convex) RACS Z, and D is a homogeneous Poisson process with intensity λ. In particular, it concentrates on estimation of the intensity λ.

Write a generic point $\mathbf{s}$ in $\mathbb{R}^2$ as $\mathbf{s} = (x, y)'$. Assume that each realization of Z has a uniquely identifiable point on its boundary, referred to as its *marker point*. If $y^* \equiv \inf\{y : \mathbf{s} = (x, y)' \in Z\}$, $x^* \equiv \inf\{x : \mathbf{s} = (x, y^*)' \in Z\}$, then $\mathbf{s}^* \equiv (x^*, y^*)'$ is defined to be the marker point of Z and for Z convex corresponds to its most southwesterly point. Marker points have previously been used to construct edge-corrected estimators in image analysis (Miles, 1974; Gundersen, 1978). Notice that $\mathbf{s}^*$ depends on the coordinate system chosen, although final estimates will be little affected by that choice. Further, assume that $|\partial Z| = 0$ almost surely, where ∂Z is the boundary of the random grain Z.

It will be shown that the observed marker points, after suitable transformation, form a planar homogeneous Poisson process with the *same* intensity as the original Boolean model. Although previous methods of analysis described in Sections 9.5.1 and 9.5.2 have given estimates of the intensity, the following method neatly separates out the spatial-location information. The development follows closely that of Laslett et al. (1985).

A Transformation of the Boolean Model

Consider the Boolean model with grain Z and Poisson points (germs) occurring at constant intensity λ per unit area. It follows that the original germ can be replaced with a more convenient point on the boundary of the grain without changing the stochastic properties of the process. In particular, the marker point can be chosen. Define a marker point $\mathbf{s}^*$ to be *exposed* if $\mathbf{s}^*$ belongs to $\overline{X^{\mathscr{C}}}$, the closure of the background.

The idea is to remove X from the two-dimensional image, *except* for the exposed marker points, and close up, along the x direction, the gaps in the background left by the departed set X. Now, instead of points in the background having coordinates (x, y), they will have new coordinates (χ, ψ). The transformation ξ from (x, y) to (χ, ψ) is defined as follows. Assume, without loss of generality, that $x \geq 0$. For $\mathbf{s} = (x, y)' \in \overline{X^{\mathscr{C}}}$, let $L(\mathbf{s}) = (0, x] \times \{y\}$. Then define $\chi \equiv |L(\mathbf{s}) \cap \overline{X^{\mathscr{C}}}|_1$, where $|\cdot|_1$ is one-dimensional Lebesgue measure, $\psi \equiv y$, and $(\chi, \psi) \equiv \xi(x, y)$. For $\mathbf{s} = (x, y)' \notin \overline{X^{\mathscr{C}}}$, let $x_l \equiv \sup\{x'\colon\ x' < x,\ (x', y) \in \overline{X^{\mathscr{C}}}\}$. Then define $\xi(x, y) \equiv \xi(x_l, y)$ and $(\chi, \psi) \equiv \xi(x, y)$.

Let $\Omega \equiv \xi(\overline{X^{\mathscr{C}}} \cap \{(x, y)\colon\ x \geq 0\})$ denote the range of ξ. Although the transformation ξ onto Ω is not invertible, the transformation $\xi\colon \overline{X^{\mathscr{C}}} \to \Omega$ is, and when necessary I shall write $\xi^{-1}(\boldsymbol{\omega})$ for the location, in $\overline{X^{\mathscr{C}}}$, of $\boldsymbol{\omega}$ in Ω. Points and sets in Ω will be denoted by Greek letters and those in $\mathbb{R}^2$ by Latin letters.

The following theorem, owing to G. M. Laslett, can be proved:

Laslett's Theorem If X is a Boolean model with intensity λ and grain Z for which there exists a unique marker point, then the exposed marker points of X (called the *induced point process*) form a homogeneous Poisson process of intensity λ in Ω.

Proof (Laslett et al., 1985). The proof can be accomplished by replacing $\mathbb{R}^2$ with a square grid of points $\{\mathbf{s}(i, j)\colon i = 0, 1, \ldots,\ j = \ldots, -1, 0, 1, \ldots\}$ of density m per unit area. Based on the same density of points, Ω is replaced with $\{\boldsymbol{\omega}(\eta, \kappa)\colon \eta = 0, 1, \ldots,\ \kappa = \ldots, -1, 0, 1, \ldots\}$. As $m \to \infty$, the effect of this discretization is shown to be negligible.

Suppose that at grid point $\mathbf{s}(i, j)$ there are placed independent Bernoulli random variables $Y(i, j)$ such that $\Pr\{Y(i, j) = 1\} = \lambda/m$. As part of the discretization, the Poisson germs in $\mathbb{R}^2$ are replaced with the set $P = \{\mathbf{s}(i, j)\colon Y(i, j) = 1\}$. For each $\mathbf{s}(i, j) \in P$, consider the associated grain $Z_{i,j}(\mathbf{s}(i, j))$.

For the reasons given in the following text, define the subset $A(i_0, j_0)$ of the square grid by

$$A(i_0, j_0) \equiv \{\mathbf{s}(i, j): i \geq 0, j < j_0; \text{or } 0 \leq i < i_0, j = j_0\}.$$

Let $X(i_0, j_0) \equiv \bigcup\{Z_{i,j}(\mathbf{s}(i, j)): \mathbf{s}(i, j) \in P \cap A(i_0, j_0)\}$; that is, take only that part of the discretized model whose marker points are either below the $j = j_0$ line or on the $j = j_0$ line but to the left of i_0. The particular definition used for marker point determines this choice of A.

Replace $\overline{X^{\mathscr{C}}}$ with the set $\overline{X_m^{\mathscr{C}}} \equiv \{\mathbf{s}(i, j): \mathbf{s}(i, j) \notin X(i, j)\}$. Define $\eta_0 \equiv \#\{\mathbf{s}(l, j_0): \mathbf{s}(l, j_0) \in \overline{X_m^{\mathscr{C}}}, \text{and } 0 \leq l \leq i_0\}$ (i.e., η_0 is the number of grid points in the background to the left of i_0 and along the line $j = j_0$); then

$$\mathbf{s}(i_0, j_0) = \xi^{-1}(\boldsymbol{\omega}(\eta_0, \kappa_0)),$$

where $\kappa_0 = j_0$. To conclude the discretization, define Bernoulli random variables $\zeta(\eta_0, \kappa_0) \equiv Y(i_0, j_0)$, where $\mathbf{s}(i_0, j_0) = \xi^{-1}(\boldsymbol{\omega}(\eta_0, \kappa_0))$. The induced point process is then $\{\zeta(\eta, \kappa)\}$.

Let $N(\Theta)$ be the number of marker points in a set $\Theta \subset \Omega$, where Θ contains μ grid points. It is sufficient to derive $\Pr\{N(\Theta) = 0\}$ for arbitrary sets, because probabilities of other events can be obtained combinatorially from these. For the (η_0, κ_0)th point in Ω, let $\Delta(\eta_0, \kappa_0) \equiv \{\boldsymbol{\omega}(\eta, \kappa): \eta \geq 0, \kappa < \kappa_0; \text{ or } 0 \leq \eta < \eta_0, \kappa = \kappa_0\}$, similar to the preceding definition of $A(x_0, y_0)$ (but here in the transformed space). Define $\Theta(\eta, \kappa) \equiv \Theta \cap \Delta(\eta, \kappa)$. Then

$$\Pr\{N(\Theta) = 0\} = \prod_{\eta, \kappa \in \Theta} \Pr\{\zeta(\eta, \kappa) = 0 | N(\Theta(\eta, \kappa)) = 0\}.$$

This is a spatial generalization of the usual sequential conditioning argument found in time-series analysis. (One of the $\Theta(\eta, \kappa)$s will be empty, in which case the conditional probability is interpreted as the unconditional $\Pr\{\zeta(\eta, \kappa) = 0\}$.) Because in the transformed space only $\overline{X_m^{\mathscr{C}}}$ has been retained, a point $\boldsymbol{\omega}(\eta, \kappa)$ in Ω can result from any one of $\{\mathbf{s}(i, j): i \geq \eta, j = \kappa\}$. Therefore,

$$\begin{aligned}
&\Pr\{N(\Theta) = 0\} \\
&\quad = \prod_{\eta, \kappa \in \Theta} \sum_{i \geq \eta} \Pr\{Y(i, \kappa) = 0 | N(\Theta(\eta, \kappa)) = 0, \xi^{-1}(\boldsymbol{\omega}(\eta, \kappa)) = \mathbf{s}(i, \kappa)\} \\
&\qquad\qquad \times \Pr\{\xi^{-1}(\boldsymbol{\omega}(\eta, \kappa)) = \mathbf{s}(i, \kappa) | N(\Theta(\eta, \kappa)) = 0\}.
\end{aligned}$$

The first conditional probability can be simplified because the shapes of the sets A and Δ guarantee that random variables in the conditioning event are independent of the random variable $Y(i, \kappa)$. It is equal to $(1 - \lambda/m)$, not depending on i. Thus,

$$\begin{aligned}\Pr\{N(\Theta) = 0\} \\ &= \prod_{\eta,\kappa\in\Theta} (1-\lambda/m) \sum_{i\geq\eta} \Pr\{\xi^{-1}(\boldsymbol{\omega}(\eta,\kappa)) = \mathbf{s}(i,\kappa) | N(\Theta(\eta,\kappa)) = 0\} \\ &= (1-\lambda/m)^{\mu}. \end{aligned} \qquad (9.5.21)$$

To prove the theorem in $\mathbb{R}^2$, let m tend to infinity. Then, up to integerization, $\mu = m|\Theta|$, so that (9.5.21) converges to $\exp\{-\lambda|\Theta|\}$. Furthermore, the expected number of marker points per unit area in $\mathbb{R}^2$ that coincide with boundaries of other associated primary sets is zero, so that these can be ignored in the limit. Finally, by construction, the limiting point process is simple, and so the result follows from the Equivalence Theorem in Section 8.3.2. ■

The converse of the theorem is not true, and counterexamples are readily derived. Consider a random set process in $\mathbb{R}^2$ produced as follows: Let C_i be the square of side 1 unit having $\mathbf{s}_i = (x_i, y_i)'$, $i = 1, 2, \cdots$, as its top right-hand corner, where $\{\mathbf{s}_1, \mathbf{s}_2, \cdots\}$ are the points of a Poisson process of intensity λ. A new point $\mathbf{s}_i = (x_i + 1/2, y_i + 1/2)'$ is generated if $\{\mathbf{s}_j: j \neq i\} \cap C_i = \varnothing$ (i.e., if the square does not intersect with any other Poisson points). Now let $\{\mathbf{s}_i\} \cup \{\mathbf{s}'_i\}$, $i = 1, 2, \cdots$, be southwesterly marker points for associated grains of unit squares. The union of these grains defines a random-set process. The exposed marker points are a subset of $\{\mathbf{s}_1, \mathbf{s}_2, \cdots\}$ and, as can be proved using the aforementioned techniques, form a Poisson process of intensity λ in the transformed space Ω. However, the random set process is not Boolean nor is it equivalent to a Boolean model, as would be the case if every $\mathbf{s}_i$ gave rise to an $\mathbf{s}'_i$.

Heather-Incidence Example

Figure 9.7 shows a representation of the incidence of heather. These data (discussed briefly in Section 9.5.2) were analyzed by Diggle (1981b), who suggested that they could be treated as a realization of a Boolean model. The exposed marker points are shown on the same figure. The shaded set is removed and the background closed up toward the left-hand vertical axis. The closed-up marker points and the closed-up background are shown in Figure 9.8.

For convenience of arithmetic and in order to reduce possible edge effects, a rectangular subset of the transformed data was used to estimate the intensity. This region, shown with a solid perimeter in Figure 9.8, is exactly half of one 10×10-m unit in area (a quarter of the size of the original sample). The finite extent of the data set means that the full range Ω of the transformation is not available and necessitates masking out some of the points in the right-hand part of Figure 9.8. This does not mean that information is necessarily lost, because one could also close up the back-

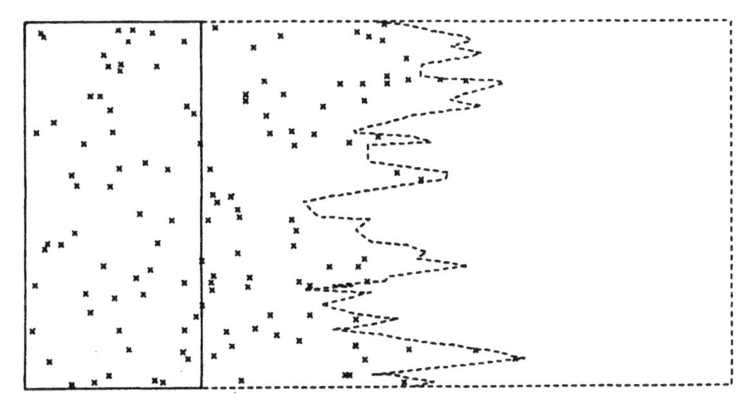

Figure 9.8 Transformed image showing location of marker points, after removal of the heather-incidence set and closing up the background. The rectangle in bold delineates the area used to estimate the intensity. [*Source*: Laslett et al., 1985]

ground of Figure 9.7 toward the right-hand vertical axis and mask out left-hand marker points in the transformed space.

The half unit area of Figure 9.8 contains 58 marker points, so that the heather intensity is estimated to be $\hat{\lambda} = 116$ plants per unit area, substantially less than Diggle's final estimate of $\hat{\lambda} = 211$ plants per unit area. In discussion, Diggle points out that the data contain fewer patches than do simulations of the model, and the lower estimate reflects this, at least in part. Some of the discrepancy must also be due to the quality of the heather-incidence data, which, in fact, was collected on a square grid; the boundary lines in Figure 9.7 are smooth approximations to the actual heather boundary. The fact that this is a discretized map means that some small or hardly exposed particles may fall between grid points and their marker points missed. This will not be a concern when analogue data are taken from actual pictures. Indeed, exposed marker points could be sampled directly as part of the data collection process.

Some Final Remarks

Because the clustering of points from an inhomogeneous Poisson process is an important alternative, one might ask how the previously defined transformation ξ deals with clustered original marker points. Globally speaking, ξ will distort the relative positions of exposed marker points so that clustering (or regularity) may be lost in the transformed space. However, a test for the Boolean model could be constructed at the local level. For example, choose at random a number of origins and axes within a large realization of the process under investigation. After transformation, tests based on first nearest neighbors, in the immediate vicinity of the randomly chosen origins and axes, are recommended. The advantage of distorting the original space via ξ is that the method can ignore poor resolution areas of the picture. In satellite imagery, for example, if cloud cover should block part of an image, the cloud is simply "removed" along with the set, and the preceding results still apply.

In summary, an alternative way of estimating the Boolean-model intensity has been proposed that, when applied to the heather-incidence data, disagrees with previous estimates (Diggle, 1981b; Hall, 1985b). The germs for those data are not well fitted by random disks, which will almost certainly influence the associated intensity estimates; both Diggle and Hall assume a random-disk Boolean model. On the contrary, the approach of this section makes no such strong assumption about grain shape (but it does require constant intensity λ). A measure of the variation of the proposed estimator of λ has yet to be developed.

Good diagnostic methods and distribution theory for estimators are needed for inference on any stochastic model. The Boolean model is no exception; limited results are given in the next section.

9.6 INFERENCE FOR THE BOOLEAN MODEL

The estimation technique of Section 9.5.1 could be liberally described as a method-of-moments. The matching of theoretical moments to sample values in order to estimate parameters is well known in statistics for random variables and random vectors, although it is usually used only when other approaches, such as maximum likelihood, have failed. This is because there is no general theory that yields distributional properties of the method-of-moments estimators nor is there any guarantee that they are asymptotically efficient.

Likelihood

What is the likelihood function for the Boolean model $X = \bigcup\{Z(\mathbf{s}_i): \mathbf{s}_i \in D\}$ defined by (9.4.1)? The random set X can be considered equivalently as a random function

$$\{I_X(\mathbf{s}): \mathbf{s} \in \mathbb{R}^d\},$$

where I_X denotes the indicator function of the set X; see Section 9.3. Thus a particular realization consists of 0-1 values throughout $\mathbb{R}^d$, which in practice is discretized onto a regular grid. In fact, define a new process

$$\xi(\mathbf{s}) = \begin{cases} 1, & \text{with probability } p, \\ 0, & \text{with probability } 1 - p, \quad \mathbf{s} \in \mathbb{Z}^d \cap A, \end{cases} \tag{9.6.1}$$

where $\mathbb{Z}^d$ is the d-dimensional lattice of integers and A is the (rescaled) domain of observation. Rescaling the domain of observation is important because the lattice dimension is scaled to be 1 unit in (9.6.1). Let $|\mathbb{Z}^d \cap A|$ denote the number of lattice points in the domain of observation. In order to equate Boolean-model parameter λ with its discretized version, put $p \cdot |\mathbb{Z}^d \cap A| = \lambda \cdot \nu(A)$ or

$$p = \lambda \cdot \nu(A) / |\mathbb{Z}^d \cap A|, \tag{9.6.2}$$

where $\nu(\cdot)$ is Lebesgue measure. Because in practice the number of lattice points in the domain of observation is chosen to be greater than the expected number of events of the Poisson point process, $p \in (0, 1)$.

Let Z be an almost surely bounded RACS Z and suppose $\{Z_{\mathbf{u}}: \mathbf{u} \in \mathbb{Z}^d \cap A\}$ is a sequence of independent realizations of Z. Define

$$a(\mathbf{s}; \mathbf{u}) \equiv I_{Z_{\mathbf{u}}}(\mathbf{s} - \mathbf{u}) \tag{9.6.3}$$

and

$$\eta(\mathbf{s}) \equiv \sup\{\xi(\mathbf{u}) \cdot a(\mathbf{s}; \mathbf{u}) : \mathbf{u} \in \mathbb{Z}^d \cap A\}, \qquad \mathbf{s} \in \mathbb{Z}^d \cap A. \tag{9.6.4}$$

Then $\eta(\cdot)$ is a discretized version of $I_X(\cdot)$, the indicator function of the Boolean model X, in the domain of observation A. The data are

$$\{\eta(\mathbf{s}) : \mathbf{s} \in \mathbb{Z}^d \cap A\}, \tag{9.6.5}$$

a finite number of 0s and 1s. However, their joint distribution and hence the likelihood cannot be written down in an illuminating way. The sup in (9.6.4) causes the difficulty; see Section 9.4.1, where the marginal distribution of $\eta(\mathbf{s})$ is given by property 1 and the bivariate distribution of $(\eta(\mathbf{s}), \eta(\mathbf{s} + \mathbf{h}))$ is given by properties 1 and 2. Notice that, *in principle*, all the trivariate, quadrivariate, and so on, distributions can be obtained from the recurrence relations that allow the hitting function (9.3.14) to be called a Choquet capacity of infinite order [see property 3 following (9.3.14)], but the calculations are practically prohibitive.

In conclusion, obtaining the likelihood function for (a discretized version of) the Boolean model does not yield tractable results. Moreover, it is not clear that maximum likelihood estimators would be optimal in any sense. In another context, where the parameters of the random set represent set boundaries and the data are i.i.d. uniform vectors over that set in $\mathbb{R}^d$, DeGroot and Eddy (1983) carry out inference based on the likelihood. In the case where the data are exhaustively observed sets (up to the limits of digitization), it seems that more success can be gained from fitting the empirical hitting function to the theoretical hitting function or, more generally, from the method of moments. The properties of the resulting estimators will now be discussed.

Method of Moments

Matching theoretical expressions with empirical expressions and solving for (theoretical) parameters will liberally be called the method-of-moments.

Exact distribution theory for such estimators (as for any other Boolean-model parameter estimators) is unavailable, but some limited results will be presented. Simulations of Boolean models could always be carried out to indicate distributions of statistics under given hypotheses.

Dupac (1980) considers an application of the Boolean model (9.4.1) in $\mathbb{R}^2$ to the analysis of fission tracks. The grains are assumed to be i.i.d. random disks $b(\mathbf{0}, R)$, where R is a random variable with probability density function $f(\cdot)$. Assume a large domain of observation, or equivalently that the domain A is a unit square, λ is large, and $E(\pi R^2)$ is small. The aim is to estimate λ, $E(R)$, and $\text{var}(R)$. Dupac equates three theoretical moments of the Boolean model with their empirical counterparts, namely, the porosity given in Section 9.4.1, and the expectation and the variance of the circular clumps. Approximate variances for $\hat{\lambda}$, $\hat{E}(R)$, and $\widehat{\text{var}}(R)$ are obtained, but some of his assumptions have not been satisfactorily verified. The estimators are certainly biased and covariances between them are nonzero. Unfortunately, the order of his approximations does not allow investigation of these quantities.

Tallis and Davis (1984) consider a Boolean model whose grains are random spheres in $\mathbb{R}^3$. Superimposed upon the problem of parameter determination is the stereological problem of having to infer three-dimensional quantities from two-dimensional probes of the object. Assuming that $\log R$ is $\text{Gau}(\mu, \sigma^2)$, method-of-moment estimators for λ, μ, and σ^2 are obtained. With the exception of the estimator of the intensity parameter λ, biases, variances, and covariances of estimators are not given. Method-of-moments estimation when the Boolean model consists of convex grains is discussed, but no attention is given to the statistical properties of the estimators.

Ohser (1980) considers the Boolean model in $\mathbb{R}^d$, where the grains $Z_1, Z_2, \ldots$ are i.i.d. compact convex sets in $\mathbb{R}^d$. He shows how, by matching the theoretical and empirical hitting function, parameters λ, $E(W_k(Z))$, $k = 1, \ldots, d$, can be estimated in precisely the same way as λ, $E(P(Z))$, and $E(|Z|)$ were estimated in (9.5.5). The quantity $E(W_k(Z))$ is the expectation of the kth Minkowski functional of the random compact convex set Z (Matheron, 1975, p. 78). In $\mathbb{R}^2$, these Minkowski functionals reduce to *perimeter* and *area*, and it is this case that will now be considered in some detail.

Inference in $\mathbb{R}^2$

The data of Figure 9.5 are realizations of a Boolean model in $\mathbb{R}^2$ with random compact convex grain, in particular with random parallelograms. A slightly more general version of (9.5.4) is

$$-\log Q_X(tK_0) = \beta_0 + \beta_1 t + \beta_2 t^2,$$

where $\beta_0 = \lambda E(|Z|)$, $\beta_1 = \lambda E(P(Z))P(K_0)/(2\pi)$, $\beta_2 = \lambda |K_0|$, Q_X is the complement of the hitting function of the Boolean model X, and K_0 is a fixed compact convex set *symmetric* about the origin.

Now, for any compact convex K_0,

$$Q_X(tK_0) = \exp\left(-\lambda E\left(|\check{Z} \oplus tK_0|\right)\right) = Q_{X \oplus t\check{K}_0}(\mathbf{0}), \qquad (9.6.6)$$

because $X \oplus t\check{K}_0$ is a Boolean model with grain $Z \oplus t\check{K}_0$. That is, $Q_X(tK_0)$ is simply the *porosity* of $X \oplus t\check{K}_0$. It is now clear how to estimate $Q_X(tK_0)$:

$$\hat{Q}_X(tK_0) = \text{proportion of pixels not in } X \oplus t\check{K}_0; \qquad (9.6.7)$$

the exact formula (for K_0 the square C) is given by (9.5.8).

Now write

$$\mathbf{Y} = T\boldsymbol{\beta} + \boldsymbol{\delta}, \qquad (9.6.8)$$

where $\mathbf{Y}' \equiv (-\log \hat{Q}_X(tK_0)\colon t = t_1, \ldots, t_n)$, T is an $n \times 3$ matrix whose ith row is $(1\ t_i\ t_i^2)$ and $\boldsymbol{\delta}$ is an $n \times 1$ error vector with $E(\boldsymbol{\delta}) = 0$ and $\text{var}(\boldsymbol{\delta}) = \Sigma$. From the approximation

$$\text{cov}(-\log U_1, -\log U_2) \simeq \{E(U_1)E(U_2)\}^{-1} \cdot \text{cov}(U_1, U_2),$$

the (i, j)th element of Σ is approximately

$$\{Q_X(t_iK_0)Q_X(t_jK_0)\}^{-1} \cdot \text{cov}\left(\hat{Q}_X(t_iK_0), \hat{Q}_X(t_jK_0)\right). \qquad (9.6.9)$$

Equations (9.5.9) and (9.5.10) (for K_0 the square C) allow evaluation of (9.6.9), which is seen to depend on the unknown parameters of X. Let each term in (9.6.9) be estimated by its empirical version, such as (9.5.8), (9.5.9), and (9.5.11), giving

$$\hat{\Sigma} = \left(\{\hat{Q}_X(t_iK_0)\hat{Q}_X(t_jK_0)\}^{-1} \cdot \widehat{\text{cov}}\left(\hat{Q}_X(t_iK_0), \hat{Q}_X(t_jK_0)\right)\right). \qquad (9.6.10)$$

Note that $\hat{\Sigma}$ may not be positive-definite; fitting of a "closest" positive-definite matrix to $\hat{\Sigma}$ is advisable in these cases. How this might be done is worthy of further research [Diggle (1981b) suggests fitting a shifted Weibull to the empirical Boolean-model covariance (9.5.11)].

Parameter Estimates, Bias Estimates, and Variance Estimates

A (estimated) generalized-least-squares estimator of the coefficients $\boldsymbol{\beta}$ in the linear equation (9.6.8) is

$$\hat{\boldsymbol{\beta}} = (T'\hat{\Sigma}^{-1}T)^{-1}T'\hat{\Sigma}^{-1}\mathbf{Y}, \qquad (9.6.11)$$

whose variance matrix can be estimated by

$$\widehat{\text{var}}(\hat{\boldsymbol{\beta}}) = (T'\hat{\Sigma}^{-1}T)^{-1}. \tag{9.6.12}$$

Consider the case where $K_0 = b(\mathbf{0}, 1)$, the unit disk centered at the origin. Then, from (9.5.5),

$$\hat{\theta} \equiv \hat{E}(P(Z)) = \pi\hat{\beta}_1/\hat{\beta}_2. \tag{9.6.13}$$

Using the δ-method (Kendall and Stuart, 1969, pp. 231, 244), it can be shown that

$$\begin{aligned} E(\hat{\theta}) &\simeq \frac{\pi\beta_1}{\beta_2} - \frac{\pi \operatorname{cov}(\hat{\beta}_1, \hat{\beta}_2)}{\beta_2^2} + \frac{\pi\beta_1 \operatorname{var}(\hat{\beta}_2)}{\beta_2^3} \\ &= E(P(Z)) + \text{bias}, \end{aligned} \tag{9.6.14}$$

where [using (9.6.11) and (9.6.12)] the bias can be estimated by

$$\widehat{\text{bias}} = \pi\left\{-\frac{\widehat{\text{cov}}(\hat{\beta}_1, \hat{\beta}_2)}{\hat{\beta}_2^2} + \frac{\hat{\beta}_1 \widehat{\text{var}}(\hat{\beta}_2)}{\hat{\beta}_2^3}\right\}. \tag{9.6.15}$$

Furthermore,

$$\operatorname{var}(\hat{\theta}) \simeq \pi^2\left\{\frac{\operatorname{var}(\hat{\beta}_1)}{\beta_1^2} + \frac{\operatorname{var}(\hat{\beta}_2)}{\beta_2^2} - \frac{2\operatorname{cov}(\hat{\beta}_1, \hat{\beta}_2)}{\beta_1\beta_2}\right\}\frac{\beta_1^2}{\beta_2^2}, \tag{9.6.16}$$

which [using (9.6.11) and (9.6.12)] can be estimated by

$$\widehat{\text{var}}(\hat{\theta}) = \pi^2\left\{\frac{\widehat{\text{var}}(\hat{\beta}_1)}{\hat{\beta}_1^2} + \frac{\widehat{\text{var}}(\hat{\beta}_2)}{\hat{\beta}_2^2} - \frac{2\widehat{\text{cov}}(\hat{\beta}_1, \hat{\beta}_2)}{\hat{\beta}_1\hat{\beta}_2}\right\}\frac{\hat{\beta}_1^2}{\hat{\beta}_2^2}. \tag{9.6.17}$$

Similar expressions can be obtained for $\hat{E}(|Z|) = \pi\hat{\beta}_0/\hat{\beta}_2$ and $\hat{\lambda} = \hat{\beta}_2/\pi$.

The results (9.6.11) and (9.6.12) are general in that no assumption is made about K_0 beyond that it is a symmetric compact convex set in $\mathbb{R}^2$. Slight modifications to this method are made in Section 9.7 to allow inference on tumor-growth parameters.

Design and Diagnostics

A couple of issues arise that warrant further investigation. What are the optimal values of n and $t_1, \ldots, t_n$, given the experimental constraints of pixel dimension? Because of the nontrivial correlation in Σ, the answer is not necessarily to make n as large as possible. The design space is limited to

$[0, t_{max}]$, where t_{max} is the largest t for which $\hat{Q}_X(tK_0) > 0$. Specification of $0 \leq t_1 \leq t_2 \leq \cdots \leq t_n \leq t_{max}$ may be limited by the discreteness of the image data, but if a choice is possible, equally spaced explanatory variables gives a robust design.

Having fit the model by the method-of-moments, are there any diagnostic checks that can be carried out to assess the goodness-of-fit? One suggestion (Diggle, 1981b) is to compare the empirical covariance of $X^{\mathscr{C}}$ [i.e., proportion of pixels in $(X^{\mathscr{C}} \cap X^{\mathscr{C}} \oplus \{-\mathbf{h}\}]$, with the theoretical covariance $\chi(\mathbf{h})$, given in Section 9.4.1. For $Z = b(\mathbf{0}, R)$, where the random radius R has distribution function F, Hall (1988a, Section 5.2) proves that knowing $\chi(\mathbf{h})$ and porosity q, given in Section 9.4, is equivalent to knowing λ and F.

The error term $\boldsymbol{\delta}$ in (9.6.8) is made up of two components, namely, lack of fit of the model and error caused by using an estimate $\hat{Q}_X$ of Q_X. Assuming the model is correct, what asymptotic results are available for Boolean model parameter estimation? A few results are given in the following text.

Asymptotics

Having worked out the biases and variances of estimators, a natural question to ask is whether they tend to zero as more information is accumulated. Sometimes it is not clear what "more information" means; here, the asymptotics are controlled by assuming the measure of the domain of observation $|A|$ tends to ∞ (increasing-domain asymptotics).

Baddeley (1980) proves a limit theorem for $\hat{Q}_X(tK_0)$, assuming $|A| \to \infty$. In his Theorem 2 applied to the Boolean model (9.4.1) (which recall has almost surely bounded grains), it can be seen that

$$\left\{|A|^{1/2}\left(\hat{Q}_X(tK_0) - Q_X(tK_0)\right): 0 \leq t \leq t_0\right\}$$

converges weakly to a zero-mean Gaussian process on $[0, t_0]$, for any positive constant t_0. [To invoke Baddeley's results, notice that $\Pr\{\mathbf{0} \in X \oplus t\check{K}_0\} = 1 - Q_X(tK_0)$.] Because $-\log(\cdot)$ is a continuous function,

$$\left\{|A|^{1/2}\left(-\log \hat{Q}_X(tK_0) + \log Q_X(tK_0)\right): 0 \leq t \leq t_0\right\}$$

also converges weakly to a zero-mean Gaussian process. This justifies the model (9.6.8) and guarantees that the estimates (9.6.11) and (9.6.13) will be consistent and asymptotically Gaussian.

More generally, suppose that the random set X is given by (9.4.1), where D is now an almost surely locally finite random set. (For example, for the Boolean model, D is the Poisson process.) Then Mase's (1982b, Section 3) results can be invoked to prove that $\hat{Q}_X(tK_0)$ is consistent and asymptotically Gaussian, where t is fixed and nonnegative.

Another way to achieve the same asymptotic results for the Boolean model (9.4.1) is to keep $|A|$ fixed, but to let $\lambda \to \infty$ and $E(|Z|) \to 0$ in such a way that $\lambda E(|Z|)$ remains constant. That is, define $Z = (\lambda^{-1/d})Z'$, where the standardized grain Z' satisfies $E(|Z'|) < \infty$, and let $\lambda \to \infty$.

9.7 MODELING GROWTH WITH RANDOM SETS

Growth of an entity can be observed and measured in a number of different ways; it is rarely one-dimensional, leading to ambiguity as to how the process can be summarized and analyzed. For example, cellular organization is known to be important in characterizing tumor growth (Rubin, 1982), but most of the mathematical models have used one-number summaries (e.g., volume, number of cells, leading front, mean radius) of the tumor in (sometimes stochastic) differential equations. These contributions (see, e.g., Laird, 1964; Burton, 1966; Saidel et al., 1976; Sawyer, 1977; Hanson and Tier, 1982; Bartoszynski et al., 1982; Le Cam, 1982; Miller et al., 1982; Adam, 1986; Swan, 1987; Norton, 1988) are typically answering different questions than are being asked in this section and are usually meant for applications where incidence data on many cancer patients are available.

The spatial component of growth models, although less prominent, has not been ignored completely (see, e.g., Shymko and Glass, 1976; Green, 1980; Goodall, 1985; Tautu, 1986; Adam, 1987). Furthermore, interacting-particle models, built at the cellular level (Section 9.7.1), allow simulations of and conjectures about the shape of tumors. Theoretical results have appeared that show the tumor to be asymptotically (i.e., as time tends to infinity) circular. Nevertheless, none of these results is relevant to the inference problems encountered when analyzing data like the sequence of images presented in Figure 9.13.

As has already been mentioned, one application of random-set growth models is to tumor growth. Although the data presented in Section 9.7.2 are of *in vitro* growth, one could imagine monitoring the changes in size and shape of a tumor in an individual via very precise, noninvasive imaging. Necessarily then, the data have to be good quality images sequenced at specified time intervals and taken under identical conditions. *In vivo* data of this type are unknown (to me), partly because tumors have been surgically removed as soon as possible after discovery, partly because the precision of x-ray pictures has been inadequate for detailing tumor *shape*, and partly because the importance of taking pictures under *identical* conditions, sequenced at specified intervals, has only recently been clinically appreciated. Apart from simply looking at the sequence of pictures, how can the clinician analyze such data? This section explores some possibilities in this regard. It is hoped that more theoretical research in image analysis, combined with higher-resolution sensing, will lead to a better description and understanding of tumor growth. Pictures of *in vitro* growth are analyzed in Sections 9.7.2 and 9.7.3 to demonstrate what can be done at present; the fits are remarkably good.

The general growth model is presented in Section 9.7.1 and various simulations illustrate its flexibility. Indeed, the contact processes (Eden, 1961; Williams and Bjerknes, 1972; Harris, 1974) are seen to be special cases of the model.

9.7.1 Random-Set Growth Models

Recall (9.4.1), the definition of the Boolean model. Think of the $\{\mathbf{s}_1, \mathbf{s}_2, \cdots\}$ as *foci* (or germs) of growth; about each $\mathbf{s}_i$ an object (or grain) Z_i grows. With regard to the modeling of tumor growth, these foci are clinically verifiable (Chover and King, 1985). The same principle is now applied to the tumor itself as it grows, namely, within the tumor there are foci about which growth occurs. If this model is sensible, a much higher intensity of foci will occur in malignant tissue than in normal tissue. Let X_1 denote the tumor at time $t = 1$ and $F_1 \subset X_1$ denote the foci, countable in number. Then the tumor at time $t = 2$ is modeled to be

$$X_2 = \bigcup \{Z(\mathbf{s}_i) : \mathbf{s}_i \in F_1\}, \tag{9.7.1}$$

where $Z(\mathbf{s}_i) \equiv Z_i \oplus \mathbf{s}_i$, $i = 1, 2, \ldots$, and $Z_1, Z_2, \ldots$ are independent copies of a random set Z. This model was first proposed by Cressie (1984b), where F_1 consisted of Poisson points, homogeneous on X_1; it was extended by Cressie (1991) to the other cases presented in this section.

Poisson Foci

For Poisson foci in X_1, the model (9.7.1) can be written as

$$X_2 = \bigcup \{Z(\mathbf{s}_i) : \mathbf{s}_i \in D\}, \tag{9.7.2}$$

where D is the *inhomogeneous* Poisson process with intensity function,

$$\lambda(\mathbf{s}) = \begin{cases} \lambda, & \mathbf{s} \in X_1, \\ 0, & \text{otherwise}. \end{cases} \tag{9.7.3}$$

That is, (9.7.2) is a conditional Boolean model. Then (9.4.8) yields the hitting function

$$T_{X_2}(K) = 1 - \exp\left\{-\lambda E\left(\left|(\check{Z} \oplus K) \cap X_1\right|\right)\right\}, \qquad K \in \mathscr{K}, \tag{9.7.4}$$

which is used extensively in analyzing the tumor-growth data of Section 9.7.2.

Simulations of (9.7.2) yield pictures that evoke images of tumor growth. Figure 9.9 shows a progression from an initial disk in *a* through to an irregular set in *d*. Each set is generated from the previous one via the relations (9.7.2) and (9.7.3), with grain Z in each case being a disk centered at the origin of fixed radius. This circular growth is often seen in experiments

(Haji-Karim and Carlsson, 1978; Martinez and Griego, 1980) and has been proved mathematically for certain stochastic models of tumor growth (Richardson, 1973; Schurger, 1979; Bramson and Griffeath, 1980, 1981; Durrett and Liggett, 1981).

It is not expected that simulations such as Figure 9.9 are a realistic picture of *in vivo* tumor growth over long time periods. Initially, the growth is uninhibited, but before long lack of blood supply to parts of the tumor, as well as tissue barriers, result in a more complicated growth pattern. It is not my intention to model such patterns here, but I shall mention other researchers who have; without exception, they were only able to simulate and could not perform statistical inference (cf. Section 9.7.3) with their models. Notable is the work of Duchting (1980) and his co-workers (Duchting and Dehl, 1980; Duchting and Vogelsaenger, 1981, 1983). See also Ransom (1977), who models and then simulates the displacements that occur when interior tumor cells divide.

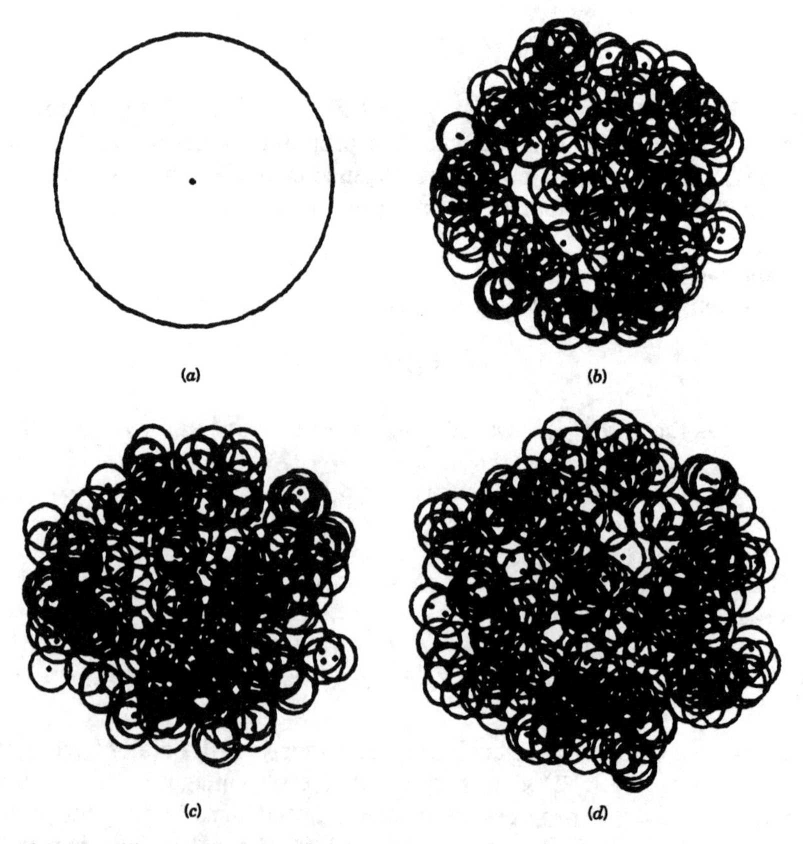

Figure 9.9 A realization of the tumor growth model (9.7.2). (*a*) X_1, a disk centered at the origin. (*b*), (*c*), and (*d*) Realizations at successive times, obtained by iterating (9.7.2). [*Source*: Cressie, 1991]. Reprinted by permission of the Institute of Mathematical Statistics.

It is my hope that the model (9.7.2) and (9.7.3) provides a good description of a variety of growth processes; this will depend on the spatial scale of observations and the type of questions being asked. An important requirement is to be able to make statistical inferences about λ and the probability law of Z from data X_1 and X_2. Section 9.7.3 discusses this important inferential aspect.

Nonrandom Foci

Suppose F_1 in (9.7.1) is an at-most-countable collection of foci, where the foci locations $\{\mathbf{s}_1, \mathbf{s}_2, \dots\}$ are fixed. Define

$$T_Z(K) \equiv \Pr(Z \cap K \neq \varnothing), \qquad K \in \mathscr{K}, \tag{9.7.5}$$

the hitting function of the grain Z. Then, from (9.7.1) and (9.7.5),

$$T_{X_2}(K) = 1 - \prod_{i=1}^{\infty} \{1 - T_Z(K \oplus \{-\mathbf{s}_i\})\}, \qquad K \in \mathscr{K}. \tag{9.7.6}$$

A special case is when F_1 is a finite set of nodes of a regular grid; in $\mathbb{R}^2$,

$$F_1 = \{(i\Delta_1, j\Delta_2): i, j \in \mathbb{Z}\} \cap X_1. \tag{9.7.7}$$

It is easy to simulate

$$X_2 = \bigcup \{Z(\mathbf{s}_i): \mathbf{s}_i \in F_1\}, \tag{9.7.8}$$

where F_1 is given by (9.7.7). Figure 9.10 shows an analogous simulation to Figure 9.9, where now foci are nonrandom and belong to a square grid in X_1. For illustrative purposes, the grains are once again disks of fixed radius.

Grains need not be disks, and if one is modeling tumor growth at the cellular level, it may make more sense to choose them as rectangular blocks (dominoes) of dimension $2\Delta \times \Delta$, where Δ is the square-grid spacing. A growth process can then be generated by (9.7.8), where Z is a north-pointing domino with probability (w.p.) p_1, a south-pointing domino w.p. p_2, an east-pointing domino w.p. p_3, a west-pointing domino w.p. p_4, and a $\Delta \times \Delta$ square w.p. $1 - p_1 - p_2 - p_3 - p_4$. Figure 9.11 shows an analogous simulation to Figure 9.10 (i.e., regular foci), with these dominoes and square as realizations of the random grain.

Contact Processes

A natural question to ask is whether models of the type presented in Figure 9.11 have anything in common with interacting-particle systems (see, e.g., Liggett, 1985; Durrett, 1988). Although the latter are typically continuous-time models, they can be discretized in time by considering small time intervals. That is, they are Markov processes whose state ξ_t at time t is a subset of $\mathbb{Z}^2$

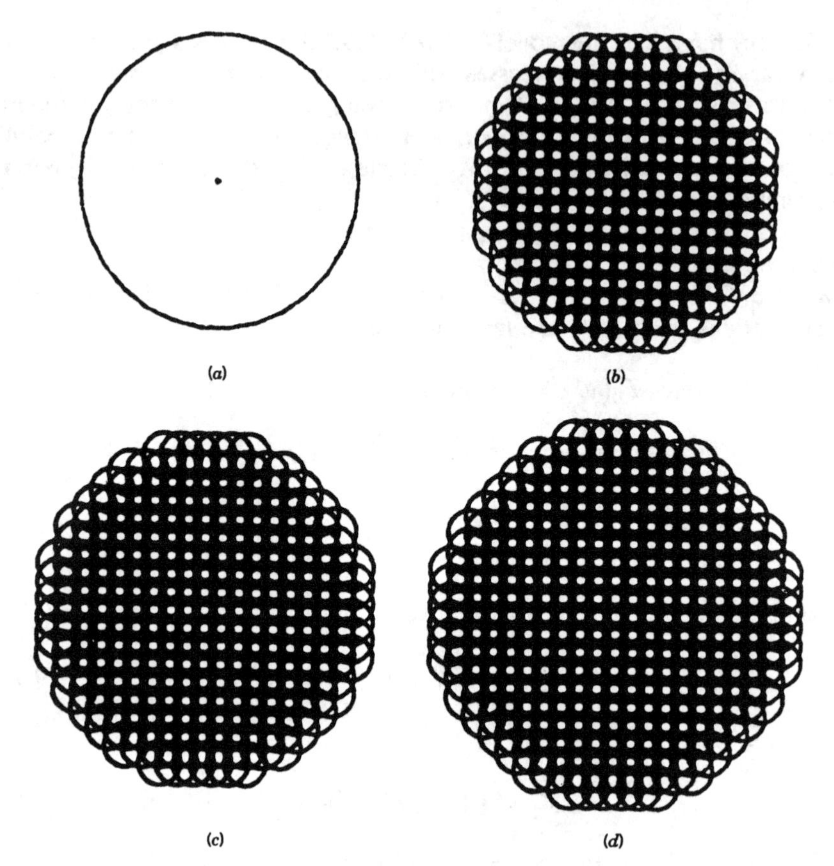

Figure 9.10 Same as in Figure 9.9 except that now growth is simulated by iterating the model (9.7.8). [*Source*: Cressie, 1991]. Reprinted by permission of the Institute of Mathematical Statistics.

(more generally, a subset of $\mathbb{Z}^d$, the set of d-dimensional vectors with integer coordinates). For u small, (1) tumor cells die at rate δ; that is,

$$\Pr(\mathbf{s} \notin \xi_{t+u} | \mathbf{s} \in \xi_t) \simeq \delta \cdot u; \tag{9.7.9}$$

(2) tumor cells are born at rate $r_{\mathbf{s}}(\xi_t)$; that is,

$$\Pr(\mathbf{s} \in \xi_{t+u} | \mathbf{s} \notin \xi_t) \simeq r_{\mathbf{s}}(\xi_t) \cdot u. \tag{9.7.10}$$

Contact processes occur when $r_{\mathbf{s}}(\xi_t) = 0$ if $\mathbf{s}$ is not a nearest neighbor of any point in ξ_t (Harris, 1974).

Suppose ξ_1 is given and ξ_{1+u} is the random set obtained from the Markov process (9.7.9) and (9.7.10); u is a small positive number. Let $F_1 = \xi_1$ in (9.7.8). Does there exist a random grain Z in (9.7.8) that yields the random

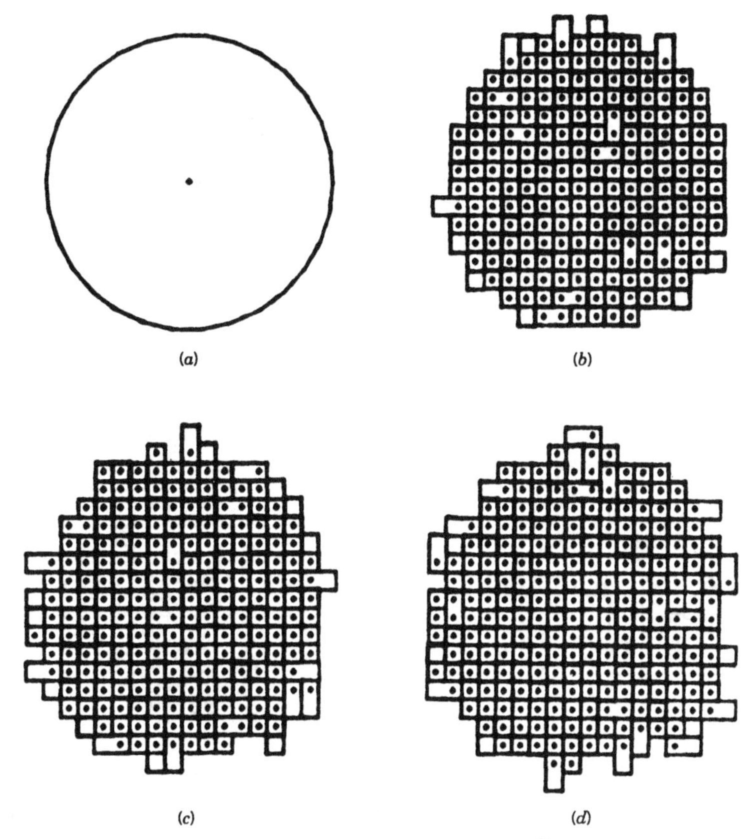

Figure 9.11 Same as in Figure 9.10, except that now the random grain Z is a square w.p. $1/5$ or dominoes in each of the four directions, each w.p. $1/5$. [*Source*: Cressie, 1991]. Reprinted by permission of the Institute of Mathematical Statistics.

set X_2, identical in distribution to ξ_{1+u}? Consider the contact process; here, $r_{\mathbf{s}}(\xi)$ = number of points in $(\xi \cap N_{\mathbf{s}})$, where $N_{\mathbf{s}} \equiv \{\mathbf{v}: \|\mathbf{v} - \mathbf{s}\| = 1\}$. Then clearly such a random grain *does* exist (Fig. 9.11 gives an example where $\delta = 0$; when $\delta > 0$, then $Z = \varnothing$ with positive probability). But the contact process just described is simply the Williams–Bjerknes (1972) tumor growth model (which was built in two dimensions because they restricted attention to the basal layer of an epithelium). Eden's (1961) model is the special case $\delta = 0$. Thus, the process defined by repeated application of (9.7.1) is flexible enough to include various contact processes as special cases.

Much research has been devoted to contact processes, usually with the aim of obtaining asymptotic ($t \to \infty$) properties, such as an asymptotically circular shape in some metric (Richardson, 1973; Schurger, 1979; Bramson and Griffeath, 1980, 1981; Durrett and Liggett, 1981), the critical value of δ above which the tumor eventually dies out (Harris, 1974; Griffeath, 1981; Andjel, 1988), and rates of convergence (Griffeath, 1981).

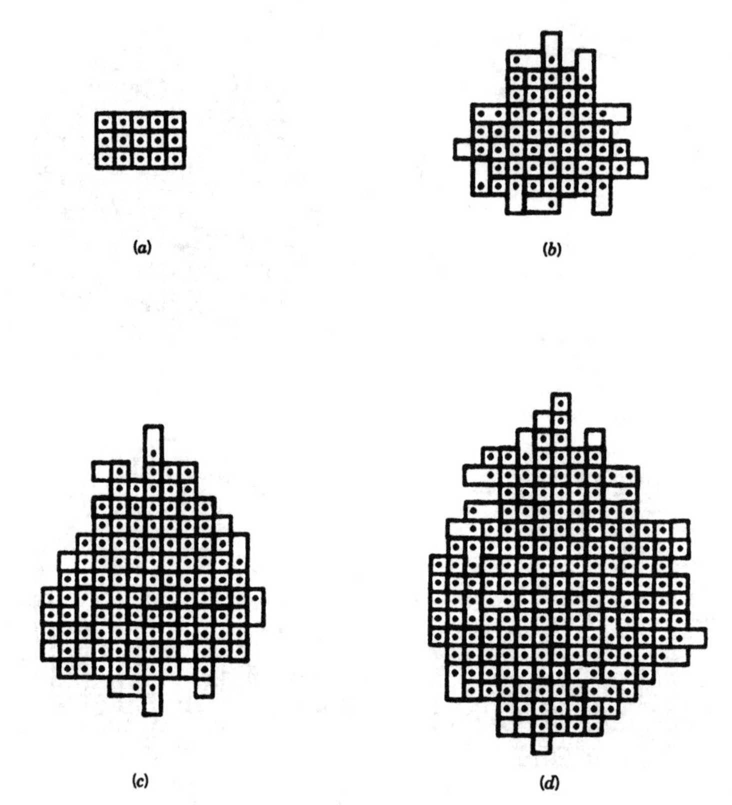

Figure 9.12 Same as in Figure 9.11, except that now (a) X_1 is a rectangle; (b) the set is X_6; (c) the set is X_{11}; (d) the set is X_{16}. [*Source*: Cressie, 1991]. Reprinted by permission of the Institute of Mathematical Statistics.

To demonstrate the asymptotically circular shape previously mentioned, I used the same parameters as in Figure 9.11, but now with a rectangular initial set X_1. Shown in Figure 9.12a–d are generations of the tumor at times $t = 1$, $t = 6$, $t = 11$, and $t = 16$, respectively.

Some Final Remarks

One could now return to the Poisson-foci case and ask the same questions regarding asymptotics that were asked for the contact process. Most notably, is there an asymptotically circular shape in some metric? I conjecture that there is; see Figure 9.9.

One could also combine the two cases and build a random-set model based on *random regular* foci. Suppose F_1 is defined by (9.7.7). Define the thinned set F_1^* as follows. For each $\mathbf{s}_i \in F_1$, $\mathbf{s}_i \in F_1^*$ w.p. p_i, independently of whether $\mathbf{s}_j \in F_1^*$, for any $j \neq i$. Finally, define

$$X_2 = \bigcup \{Z(\mathbf{s}_i): \mathbf{s}_i \in F_1^*\}, \qquad (9.7.11)$$

where $Z_1, Z_2, \ldots$ are independent copies of the random set Z (with hitting function T_Z). The hitting function of X_2 is

$$T_{X_2}(K) = 1 - \prod_{i=1}^{\infty} \left[1 - p_i + p_i\{1 - T_Z(K \oplus \{-\mathbf{s}_i\})\}\right], \qquad K \in \mathscr{K}. \tag{9.7.12}$$

If $\Delta_1 = \Delta_2 = \Delta$ in (9.7.7) and $p_i = \lambda \cdot \Delta^2$, then for Δ small,

$$\begin{aligned} T_{X_2}(K) &\simeq 1 - \exp\left\{-\lambda \int_{X_1} T_Z(K \oplus \{-\mathbf{s}\}) \nu(d\mathbf{s})\right\} \\ &= 1 - \exp\left\{-\lambda E\left(|(\check{Z} \oplus K) \cap X_1|\right)\right\}. \end{aligned}$$

Thus $\lim_{\Delta \to 0} T_{X_2}(K)$ is precisely (9.7.4), the hitting function of the random-set model based on Poisson foci (as it should be).

9.7.2 Tumor-Growth Data

Cancer is a disease involving disordered cell growth and is characterized by the presence of one or more malignant tumors (International Union Against Cancer, 1978). A *tumor* is any abnormal swelling, including those due to passing inflammation. However, the term is usually used to mean *neoplasm* (meaning new growth). Neoplasms are often categorized into two main types. Benign tumors are almost always enclosed in a capsule of fibrous tissue and do not invade surrounding tissue; the cells resemble their tissue of origin and reproduce in a relatively orderly, controlled manner. Malignant tumors directly invade surrounding tissue; their cells usually reproduce in a rapid, disorderly manner and can spread through the body via the lymphatic or blood system to develop secondary tumors (metastases) elsewhere.

A good review of the mathematical theories of carcinogenesis can be found in the papers by Whittemore (1978), Whittemore and Keller (1978), Forbes and Gibberd (1984), and Murdoch et al. (1987). The process of carcinogenesis is generally considered to follow the degeneration of a normal cell to a malignant state through a finite number of intermediate stages; heritable alterations to the cell are accumulated at each stage. What causes this degeneration? It is generally thought to start at the level of DNA, the genetic material of the cell. Normal cells contain DNA segments called protooncogenes that appear to be responsible for regulation of cell growth. A *carcinogen* alters the DNA by transforming protooncogenes to oncogenes (DNA segments that produce cancer when transferred to normal cells). The effects of exposure to a carcinogen are often not seen for many years after the exposure occurred.

Other factors, such as hormones, dietary components, asbestos, and so on (called *promoters*) then modify cells or body defense mechanisms to allow for more rapid growth once the cellular DNA has been altered. Cell organization is also important, but it seems that this spatial component has not been given the attention it deserves (Rubin, 1982, 1985; Takahashi et al., 1986). The implication is that to understand tumor growth in humans, it is necessary to understand the growth process at the supracellular as well as the cellular level. The spatial statistical analyses that follow attempt to do this; they are based largely on Cressie and Hulting (1992).

***In Vitro* Growth of Cell Islands**

Cells from the human breast cancer cell line MCF-7 (Soule et al., 1973) were grown in a flat dish covered with a nutrient medium. This cell line typically grows in compact cell islands, as is seen from the sequence of three images (photographed 72 hours apart) presented in Figure 9.13. These data were

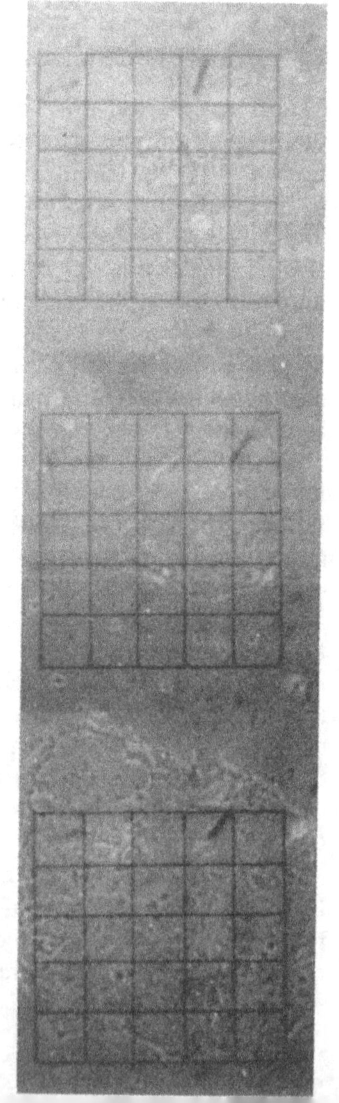

Figure 9.13 *In vitro* growth of human breast cancer cell line MCF-7. The sequence of cell islands X_1, X_2, X_3 were photographed 72 hours apart. Each of the 25 squares on the photographs are 1 mm^2 at 40 × magnification. [*Source*: Dr. G. C. Buehring, University of California, Berkeley]

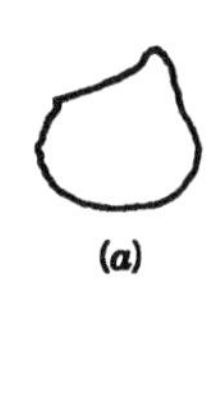

(a)

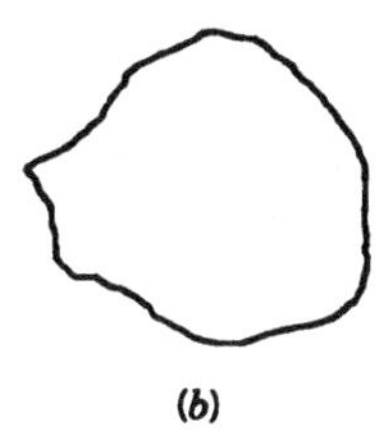

(b)

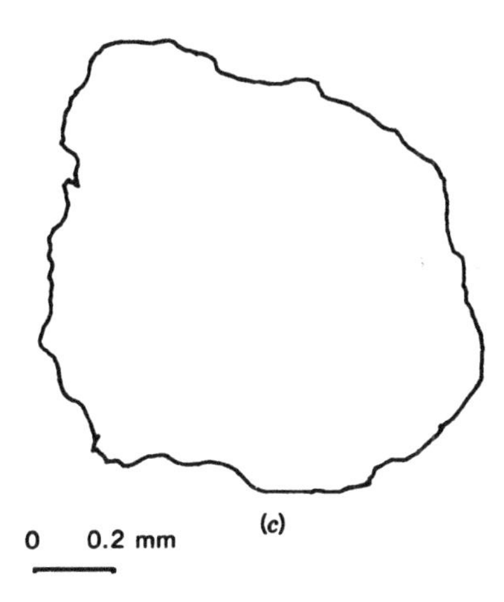

Figure 9.14 Digitized boundary of the cell islands shown in Figure 9.13. (*a*), (*b*), and (*c*) The boundaries of X_1, X_2, and X_3, respectively. [*Source*: Cressie, 1991]. Reprinted by permission of the Institute of Mathematical Statistics.

supplied by Dr. G. C. Buehring (School of Public Health, University of California, Berkeley) and come from experiments similar to those described in Buehring and Williams (1976).

Shown in Figure 9.14 are digitized boundaries of the cell islands of Figure 9.13; details of the digitization are given later in this section. Let $X_i \subset \mathbb{R}^2$ denote the cell-island boundary and its interior, taken at the ith time point, $i = 1, 2, 3$. Recall that time points are 72 hours apart.

The model I shall fit to these data is given by (9.7.2) and (9.7.3), after a transformation to isotropy. My reasons for choosing this model are based on observations made in Section 9.7.1. Because Poisson foci are seen in normal tissue, I assume that the same occurs in malignant tissue, only at very, very much higher intensities λ. Because at the cellular level asymptotic growth is circular, assume that at the supracellular level (after 72 hours of growth) the grains are disks with random radius R.

After transformation to spaces where the model is likely to hold, data of Figure 9.14 are analyzed by matching theoretical hitting functions, given by

(9.7.4), with empirical hitting functions calculated from the images. This yields estimates and standard deviations of parameters λ, $E(R)$, and var(R). Full details can be found in Cressie and Hulting (1992); I shall present a number of their results in succeeding text.

Stochastic Geometric Model of Tumor Growth

The *basic* model describes how X_{t+1}, the time-$(t+1)$ tumor, evolves from X_t, the time-t tumor, and was first suggested by Cressie (1984b). An important modification transforms the plane to ensure isotropy of the growth process, but for the moment consider the untransformed version of the model. Let X_1 denote the initial tumor and D_2 denote a homogeneous Poisson process of intensity λ_2. Then, define the time-2 tumor to be

$$X_2 \equiv \bigcup \{Z_2(\mathbf{s}_i): \mathbf{s}_i \in D_2 \cap X_1\}, \tag{9.7.13}$$

where $Z_2(\mathbf{s}_i) \equiv Z_{i,2} \oplus \mathbf{s}_i$, $i = 1, 2, \dots$, and $\{Z_{i,2}: i = 1, 2, \dots\}$ are independent copies of the RACS Z_2. In general,

$$X_{t+1} \equiv \bigcup \{Z_{t+1}(\mathbf{s}_i): \mathbf{s}_i \in D_{t+1} \cap X_t\}, \qquad t = 1, 2, \dots, \tag{9.7.14}$$

where $Z_{t+1}(\mathbf{s}_i) \equiv Z_{i,t+1} \oplus \mathbf{s}_i$ $(i = 1, 2, \dots)$; $Z_{1,t+1}, Z_{2,t+1}, \dots$ are independent copies of the RACS Z_{t+1}; and D_{t+1} is a homogeneous Poisson process of intensity λ_{t+1}, independent from $Z_{1,t+1}, Z_{2,t+1}, \dots$. Notice that X_{t+1} does not necessarily include all of X_t.

Conditional on observing X_t, the hitting function of X_{t+1} is the complement of

$$Q_{X_{t+1}}(K) = \exp\left\{-\lambda_{t+1} E\left(\left|(\check{Z}_{t+1} \oplus K) \cap X_t\right|\right)\right\}, \qquad K \in \mathscr{K}, \tag{9.7.15}$$

which is a general version of (9.7.4).

Set-Growth Parameters

The parameters of the set-growth process (9.7.14) are those from the probability law of the RACS Z_{t+1} and the probability law of the point process D_{t+1}, having conditioned on the time-t tumor X_t. Suppose that the source of randomness in $\{Z_{i,t+1}: i = 1, 2, \dots\}$ is a scaling random variable R_{t+1}; that is, Z_{t+1} is distributed as $R_{t+1} \cdot B$, where B is a fixed closed set in $\mathbb{R}^d$ and scalar multiplication is defined by (9.3.5). Then the goal is to carry out statistical inference on λ_{t+1}, $E(R_{t+1})$, var(R_{t+1}), and possibly higher-order moments, for each $t = 1, 2, \dots$. This means not only estimating the parameters, but also providing standard deviations for the estimators. It is natural to make B a disk, and indeed there is some clinical evidence (e.g., Haji-Karim and Carlsson, 1978; Martinez and Griego, 1980) that growth of tumors is spheroidal.

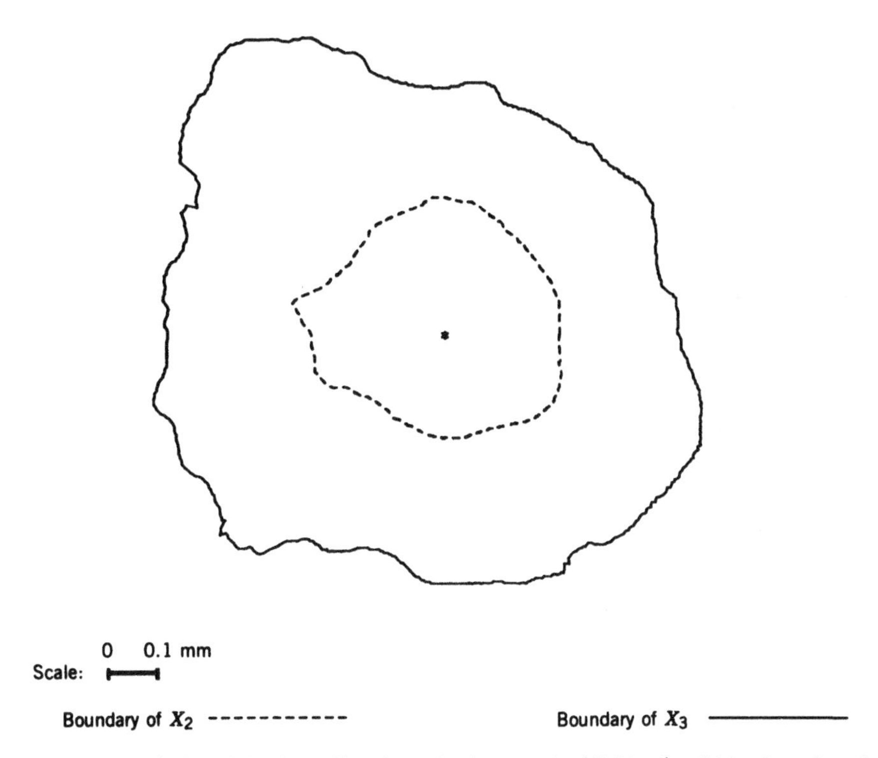

Figure 9.15 Evolution of X_3 from X_2, where the data are the (digitized) cell islands at times 2 and 3.

For the data of Figure 9.14, results will be presented for X_2 growing from X_1 and for X_3 growing from X_2. To illustrate the technique, I shall give full details for the evolution of X_3 from X_2. Figure 9.15 shows X_2 centered at the origin, which is its center of gravity. The outer curve is the boundary of X_3, which evolved from X_2 in a period of 72 hours.

Isotropic Growth on the Transformed Scale

It is natural to analyze these data rotationally rather than translationally, which leads to the question of whether the growth process is isotropic (i.e., no direction of growth is preferred). If the mechanism of growth is something like a contact process (e.g., Harris, 1974), parts of the boundary of X_2 that exhibit higher curvature may grow at a faster rate. However, a transformation of $\mathbb{R}^2$ that transforms X_2 into a set whose boundary has constant curvature (i.e., a disk) would make an isotropy assumption reasonable. This transformation is also important for *comparison* of the two growth periods; in both cases, one starts from the same initial set (i.e., a disk) in the transformed spaces. Finally, but least importantly, the hitting functions are easier to calculate for this isotropic process.

Define $b(\mathbf{s}, r) \equiv \{\mathbf{u}: \|\mathbf{u} - \mathbf{s}\| \leq r\}$ to be the closed ball of radius r whose center is at $\mathbf{s} \in \mathbb{R}^d$. Notice that $\lambda \cdot b(\mathbf{s}, r) = b(\lambda \mathbf{s}, |\lambda| r)$ and $b(\mathbf{s}_1, r_1) \oplus b(\mathbf{s}_2, r_2) = b(\mathbf{s}_1 + \mathbf{s}_2, r_1 + r_2)$.

After transformation, it will be assumed in (9.7.14) that

$$Z_{t+1} = b(\mathbf{0}, R_{t+1}), \tag{9.7.16}$$

a *disk* of random radius R_{t+1}, and that D_{t+1} is a *homogeneous* Poisson process. Then the goal is to estimate λ_{t+1}, $E(R_{t+1})$, $\text{var}(R_{t+1}), \ldots$.

The appropriate transformation that converts X_2 into a disk is not hard to perform. Let the boundary of X_2 be described by the polar coordinates $\{(r_2(\theta), \theta): 0 \leq \theta < 2\pi\}$, where the radius $r_2(\theta)$ is a function of the angle θ; that is, assume X_2 is star-shaped. If it is not and there are several boundary points $r_{2,1}^0, \ldots, r_{2,k}^0$ at $\theta = \theta_0$, then choose $r_2(\theta_0) = \max\{r_{2,1}^0, \ldots, r_{2,k}^0\}$. The *geometric-signature transformation* from $\mathbb{R}^2$ onto $\mathbb{R}^2$ is defined in polar coordinates as

$$S_2(\mathbf{u}) \equiv \left(u \cdot \{r_2/r_2(\theta)\}, \theta\right), \qquad \mathbf{u} = (u, \theta)' \in \mathbb{R}^2, \tag{9.7.17}$$

where

$$r_2 \equiv (1/2)\left[\sup\{r_2(\theta): 0 \leq \theta < 2\pi\} + \inf\{r_2(\theta): 0 \leq \theta < 2\pi\}\right]. \tag{9.7.18}$$

Zahn and Roskies (1972) also take this boundary-description approach to analyzing the size and shape of objects.

The quantity r_2 in (9.7.18) is the average radius of the largest inscribed circle and the smallest circumscribed circle centered at $\mathbf{0}$; $2r_2$ has a clinical interpretation as the mean caliper width of the tumor X_2. Along with the area $|X_2|$ (or its square root), r_2 is a measure of the *size* of the tumor.

In any fixed direction, (9.7.17) preserves ratios of distances from the origin. Figure 9.16 shows a plot of the multiplicative ratio $r_2/r_2(\theta)$, as a function of $\theta \in [0, 2\pi)$. When this transformation is applied to the boundaries of X_2 and X_3, presented in Figure 9.15, the result is Figure 9.17. Notice that X_2 is transformed to the disk of radius r_2. The outer curve shows the boundary of Y_3, the transformation of X_3.

After the transformation, it is reasonable to assume that the stochastic mechanism resulting in Y_3 does not depend on direction. When analyzing spatial data, stationarity is usually an important assumption because it allows subregions of the spatial process to serve as replicates and hence to provide estimates of spatial parameters. Here isotropy is important for the same reason.

Stochastic-Geometric Model on the Transformed Scale

Now assume that the conditional Boolean model given by (9.7.14) and (9.7.16) applies on the *transformed* scale. Specifically, because X_2 is trans-

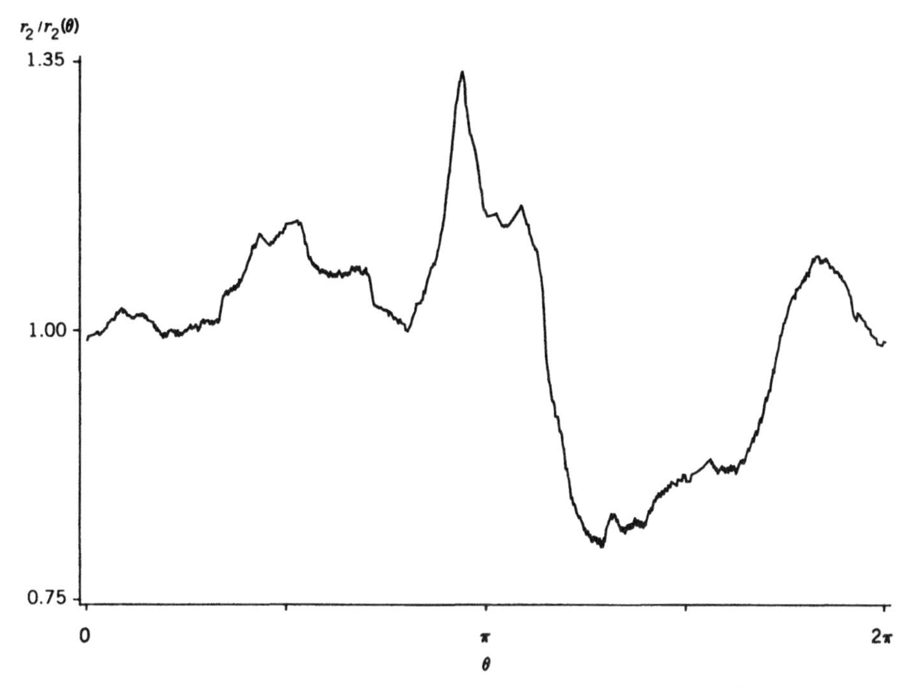

Figure 9.16 Plot shows the geometric signature transformation of the boundary of X_2 (the inner set in Figure 9.15).

formed to $b(\mathbf{0}, r_2)$ (the disk of radius r_2) and because $Z_3 = b(\mathbf{0}, R_3)$, the model for Y_3 is

$$Y_3 = \bigcup \{b(\mathbf{s}_i, R_{3,i}) \colon \mathbf{s}_i \in D_3 \cap b(\mathbf{0}, r_2)\}, \qquad (9.7.19)$$

which is an isotropic RACS. It can be characterized by its hitting function, which from (9.7.15) is the complement of

$$\begin{aligned} &\Pr(Y_3 \cap K = \varnothing \mid \text{time-2 tumor is } b(\mathbf{0}, r_2)) \\ &\quad = \exp\{-\lambda_3 E(|(b(\mathbf{0}, R_3) \oplus K) \cap b(\mathbf{0}, r_2)|)\}, \qquad K \in \mathscr{K}. \end{aligned}$$

Denote this function by Q_{Y_3}. Then for $K = b(\mathbf{s}, l)$, a disk of radius l centered at $\mathbf{s}$,

$$Q_{Y_3}(b(\mathbf{s}, l)) = \exp\{-\lambda_3 E(|b(\mathbf{s}, R_3 + l) \cap b(\mathbf{0}, r_2)|)\}. \qquad (9.7.20)$$

Define

$$q(u, h; \lambda, R) \equiv \exp\{-\lambda E(|b(\mathbf{u}, R + h) \cap b(\mathbf{0}, 1)|)\}, \qquad (9.7.21)$$

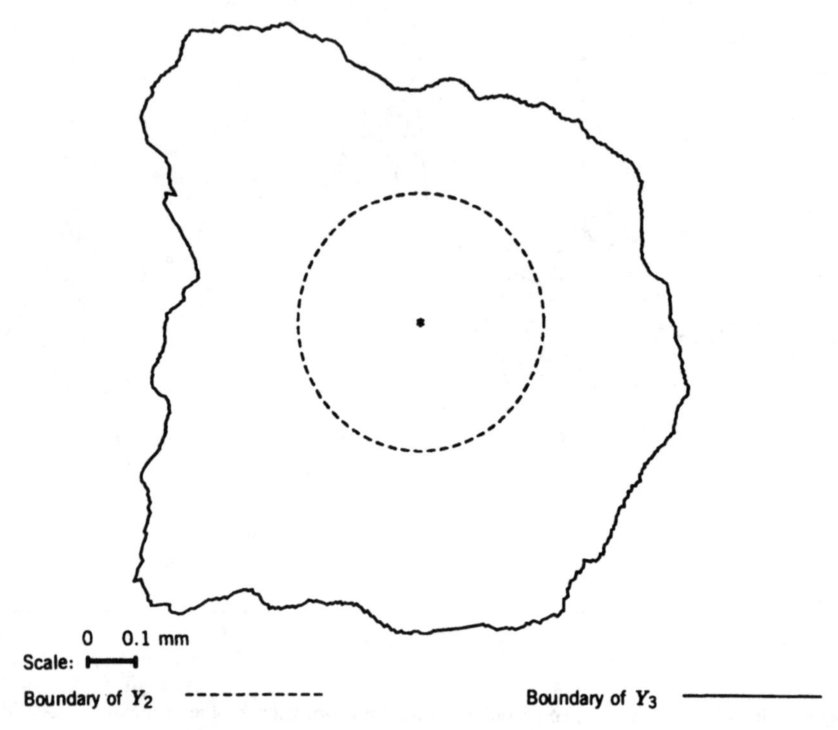

Figure 9.17 Evolution of Y_3 from Y_2, where Y_2 is a unit disk on the transformed scale. These data represent the geometric-signature transformation of the boundaries of X_3 and X_2 presented in Figure 9.15.

where $u \equiv \|\mathbf{u}\|$, $h > 0$, $\lambda > 0$, and $R > 0$. It is easily seen from (9.7.20) that

$$Q_{Y_3}(b(\mathbf{s}, l)) = q\left(s/r_2, l/r_2; \lambda_3 r_2^2, R_3/r_2\right), \tag{9.7.22}$$

where $s \equiv \|\mathbf{s}\|$. Notice that Q_{Y_3} depends on $\mathbf{s}$ only through s because the RACS is isotropic.

By working in units of r_2 (the size of the tumor X_2), one can concentrate on the shape characteristics and transform back to the original units later. Thus, for the moment, assume $r_2 = 1$.

From the exponent of (9.7.20), write

$$A_l(R_3) \equiv |b(\mathbf{s}, R_3 + l) \cap b(\mathbf{0}, 1)|, \tag{9.7.23}$$

the area of intersection of two disks distance s apart, one of radius $(R_3 + l)$ and the other of radius 1. Figure 9.18 shows the geometrical configuration.

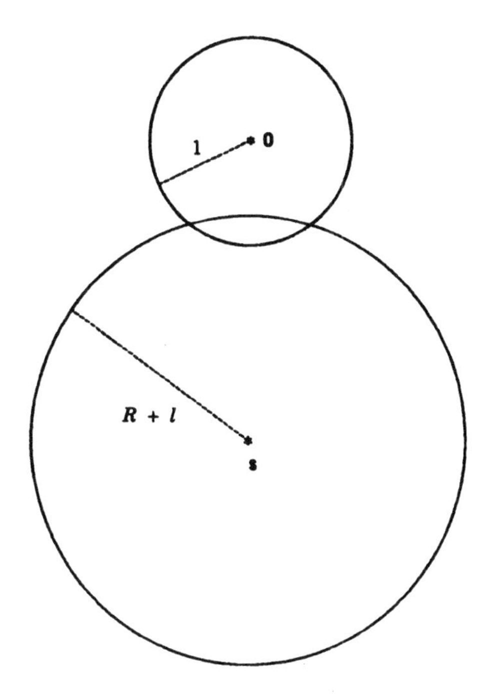

Figure 9.18 Two disks distance s apart, one with radius $R + l$, the other with radius 1. The expected area of intersection is used to define $q(u, h;\ \lambda, R)$ given by (9.7.21).

Theoretical Investigation of the Hitting Function

From (9.7.20), (9.7.22), and (9.7.23),

$$-\log q(s, l;\ \lambda_3, R_3) = \lambda_3 E(A_l(R_3)), \tag{9.7.24}$$

and, provided $R_3 + l \geq 1$ and $|R_3 + l - s| < 1$,

$$\begin{aligned} A_l(R_3) \equiv (\pi/2)\{(R_3 + l)^2 + 1\} &- y(l)\{(R_3 + l)^2 - y(l)^2\}^{1/2} \\ &- (R_3 + l)^2 \sin^{-1}(y(l)/(R_3 + l)) - \sin^{-1}(s - y(l)) \\ &- (s - y(l))\{1 - (s - y(l))^2\}^{1/2}, \end{aligned} \tag{9.7.25}$$

where $y(l) \equiv (s^2 + (R_3 + l)^2 - 1)/(2s)$. Expand (9.7.25) in a Taylor series about some fixed value R_3^0 to obtain

$$\begin{aligned} &-\log q(s, l;\ \lambda_3, R_3) \\ &\quad = \lambda_3\left\{A_l(R_3^0) + E(R_3 - R_3^0)A_l'(R_3^0) + E(R_3 - R_3^0)^2 A_l''(R_3^0)/2! + \cdots\right\} \\ &\quad \equiv \beta_1 x_1(l) + \beta_2 x_2(l) + \beta_3 x_3(l) + \cdots, \end{aligned} \tag{9.7.26}$$

where $x_{j+1}(l) \equiv A_l^{(j)}(R_3^0)/j!$, $j = 0, 1, 2 \ldots$, and

$$\lambda_3 = \beta_1, \tag{9.7.27}$$

$$E(R_3) = (\beta_2/\beta_1) + R_3^0, \tag{9.7.28}$$

$$\mathrm{var}(R_3) = (\beta_3/\beta_1) - (\beta_2/\beta_1)^2. \tag{9.7.29}$$

Notice that (9.7.26) is a linear combination of explanatory variables $\{x_j(\cdot)\}$, whose coefficients are functions of the parameters to be estimated. Formulas for the independent variables are straightforward to evaluate because they are just derivatives of $A_l(\cdot)$ given by (9.7.25).

Empirical Investigation of the Hitting Function

In order to estimate parameters β_1, β_2, and β_3, the theoretical function (9.7.24) will be matched to its empirical version. This is very much in the spirit of the classical method-of-moments technique of parameter estimation.

Isotropy of the RACS (9.7.19) ensures that

$$\begin{aligned} &Q_{Y_3}(b(\mathbf{s}, l)) \\ &\quad \simeq (2\pi)^{-1} |\{\theta: \theta \in [0, 2\pi);\, Y_3 \cap b(\mathbf{s}, l) = \varnothing, \text{where } \mathbf{s} = (s, \theta)'\}|_1 \\ &\quad = (2\pi)^{-1} \left| \left\{\theta: \theta \in [0, 2\pi);\, s > \frac{r_3(\theta)}{r_2(\theta)} + l \right\} \right|_1 \\ &\quad \equiv \hat{q}(s, l;\, \lambda_3, R_3), \end{aligned} \tag{9.7.30}$$

where $|\cdot|_1$ denotes one-dimensional Lebesgue measure. Recall that $r_2 = 1$, and so $\{(r_3(\theta)/r_2(\theta), \theta)\colon \theta \in [0, 2\pi)\}$ in (9.7.30) is the boundary of Y_3. That is, $\hat{q}(s, l;\, \lambda_3, R_3)$ is the proportion of angles θ for which $r_3(\theta)/r_2(\theta) < s - l$. Notice that $\{(r_3(\theta), \theta)\colon \theta \in [0, 2\pi)\}$ is the boundary of X_3, the untransformed time-3 tumor.

The values of the parameters β_1, β_2, and β_3 that bring $\beta_1 x_1(l) + \beta_2 x_2(l) + \beta_3 x_3(l)$ closest to $-\log \hat{q}(s, l;\, \lambda_3, R_3)$ will be used to obtain estimates of λ_3, $E(R_3)$, and $\mathrm{var}(R_3)$. Before I describe how to do this, some of the computational aspects involved in obtaining $\hat{q}(s, l;\, \lambda_3, R_3)$ will be presented briefly.

Computational Aspects

The original data came as photographic images, so they were not initially in a suitable form for computing $\hat{q}(s, l;\, \lambda_3, R_3)$. The information in the photographs had to be converted to numerical data by digitizing the relevant features of each picture. In this case, it is sufficient to quantify the boundaries of the cell islands. The method used consisted of tracing the outline of the cell islands using a pointer on a graphics tablet. The software interface to the device generated a sequence of Cartesian coordinates that define a polygonal approximation to each of the boundaries. (A common reference point on each picture allowed the cell islands to be aligned.) Denote the sampled coordinates as (x_{ik}^0, y_{ik}^0) where $k = 1, 2, 3$ indexes the time and $i = 1, \ldots, n_k$ indexes the coordinate pairs of the boundary at time k. (Digitization of the boundaries represented in Figure 9.14 yielded $n_1 = 213$, $n_2 = 510$, and $n_3 = 704$.) For the computational aspects of this study, X_k refers not to the cell-island image, but to the approximating polygon.

From the boundary description, it is possible to determine the cell island's center of gravity $(\bar{x}_k, \bar{y}_k)$ and its area. Henceforth, assume that the origin of the plane is moved to $(\bar{x}_1, \bar{y}_1)$ when studying the evolution of X_2 from X_1 and to $(\bar{x}_2, \bar{y}_2)$ when studying the evolution of X_3 from X_2. For the data of Figure 9.14, $|X_1| = 0.00465$ mm^2, $|X_2| = 0.01932$ mm^2, and $|X_3| = 0.0969$ mm^2, leading to areal growth factors of 4.155 and 5.016, respectively.

Computation and application of the geometric signature transformation S_2 requires a polar-coordinate representation for the boundaries of X_2 and X_3. Using

$$\theta_{ik}^0 = \begin{cases} \arctan\left(y_{ik}^0 / x_{ik}^0\right), & \text{if } x_{ik}^0 \geq 0, \\ \pi + \arctan\left(y_{ik}^0 / x_{ik}^0\right), & \text{if } x_{ik}^0 < 0, \end{cases}$$

and $r_k(\theta_{ik}^0) = \{(x_{ik}^0)^2 + (y_{ik}^0)^2\}^{1/2}$, each of the sample boundary points (x_{ik}^0, y_{ik}^0) can be converted to polar coordinates $(r_k(\theta_{ik}^0), \theta_{ik}^0)$. For any given $\theta^* \in (0, 2\pi]$, not necessarily a sample boundary point, Cressie and Hulting (1992) give the linear-interpolation formula used to compute $r_k(\theta^*)$. Hence S_2 may be calculated using (9.7.18).

The necessary elements to compute $\hat{q}(s, l; \lambda_3, R_3)$ are now in place. Let $\{\theta_0^*, \ldots, \theta_J^*)$ be the sequence $\theta_j^* = 2\pi(j/J)$, $j = 0, \ldots, J$, of equispaced angles. For fixed s and l, define

$$\delta(s, l, \theta_j^*) = \begin{cases} 1, & \text{if } s > \dfrac{r_3(\theta_j^*)}{r_2(\theta_j^*)} + l, \\ 0, & \text{else}. \end{cases}$$

Then $\hat{q}$, defined by (9.7.30), is evaluated by

$$\hat{q}(s, l; \lambda_3, R_3) = \frac{1}{J} \sum_{j=1}^{J} \delta(s, l, \theta_j^*);$$

a choice of $J = 750$ was made here.

9.7.3 Fitting the Tumor-Growth Parameters

Define

$$W(l) \equiv -\log \hat{q}(s, l; \lambda_3, R_3) \qquad (9.7.31)$$

to be the dependent variable in the linear model

$$W(l) = (x_1(l), x_2(l), x_3(l))(\beta_1, \beta_2, \beta_3)' + \epsilon(l), \qquad (9.7.32)$$

obtained by matching (9.7.26) with (9.7.30). Here, $\epsilon(\cdot)$ is an error term that represents a combination of higher-order Taylor series terms in (9.7.26), lack of model fit, and sampling fluctuations; it is assumed to have mean zero and homoskedastic variance σ^2, assumptions that will be checked by analyses of the residuals.

Choosing the Design Constants

In order to implement estimation through (9.7.32), design constants R_3^0, s, n, and $l_1, \ldots, l_n$ must be chosen. In analyzing the growth from time 2 to time 3, $R_3^0 = 1.2$, $s = 3.2$, $n = 13$, and $l_1, \ldots, l_n = 1.05(0.02)1.29$ were chosen.

The choice of R_3^0 is based on the rate of growth as measured by the areas of the cell islands. Define C_k to be the areal growth factors; that is, $|X_{k+1}| = C_k|X_k|$, $k = 1, 2$. Assuming that X_2 is a disk of radius 1 and X_3 is a disk of radius $1 + R_3$, C_2 is obtained from $\pi(1 + R_3)^2 = C_2\pi$. Substituting in the observed value for C_2 (5.016) and solving for R_3 yields the value $R_3^0 = 1.2$.

It is possible to vary *both* s and l in selecting the remaining design points. I chose to fix s and vary l because numerical evidence demonstrated that $A_l(R_3^0)$ varies approximately as a function of the difference $s - l$. Now if $l \geq 1$, then $R_3 + l$ is always ≥ 1, satisfying one of the conditions of (9.7.25). The other condition can be expressed as $s - R_3 - 1 \leq l \leq s - R_3 + 1$, which suggests taking s so that $s - R_3^0 - 1 \simeq 1$; that is, choose $s = 3.2$. Substituting $s = 3.2$ and $R_3^0 = 1.2$ into the preceding inequality gives an approximate range for l to be $(1.0, 3.0]$. However, for these data, zero estimates for $\hat{q}(s, l; \lambda_3, R_3)$ were obtained for $l > 1.4$, effectively reducing the range to the interval $(1.0, 1.4]$. Because observations are inexpensive, any number of l values may be selected. However, it is wise to avoid taking values of l close to 1.0 and 1.4, as well as values that are too close together. In these two instances, one can expect either imprecise or highly correlated observations, respectively. Moreover, the Taylor series approximation in (9.7.26) gets worse as l gets further from $R_3^0 = 1.2$. Diagnostic plots (see following text) were made for various choices of $l_1, \ldots, l_n$, and the final choice was made on the basis that the linear model (9.7.32) should provide an adequate description of the data.

Table 9.3 Design Constants $l_1, \ldots, l_{13}$, Dependent Variable $W(l)$, and Independent Variables $x_1(l)$, $x_2(l)$, and $x_3(l)$; Time 3 from Time 2

l	$W(l)$	$x_1(l)$	$x_2(l)$	$x_3(l)$
1.05	0.83011	0.030741	0.53380	2.61583
1.07	0.90965	0.042402	0.62957	2.20634
1.09	0.96758	0.055837	0.71218	1.94088
1.11	1.05173	0.070830	0.78582	1.75022
1.13	1.10262	0.087225	0.85279	1.60419
1.15	1.17334	0.104907	0.91454	1.48727
1.17	1.24479	0.123779	0.97204	1.39054
1.19	1.38897	0.143764	1.02598	1.30847
1.21	1.50208	0.164798	1.07686	1.23744
1.23	1.57022	0.186821	1.12509	1.17493
1.25	1.72224	0.209785	1.17095	1.11917
1.27	1.83258	0.233645	1.21469	1.06885
1.29	1.92873	0.258360	1.25652	1.02297

Estimating the Growth Parameters

The data for estimating β_1, β_2, and β_3 in the linear model (9.7.32) are given in Table 9.3. For reasons already given, the error term $\epsilon(\cdot)$ will not necessarily be uncorrelated; however, the choice of the design constants were to make $\epsilon(l_1), \ldots, \epsilon(l_n)$ look as much as possible like white noise. We therefore proceed with ordinary-least-squares estimation of the parameters; more efficient estimation procedures would try to model the serial correlation in the errors, such as in Section 9.5.1.

Estimates of β_1, β_2, and β_3, denoted by b_1, b_2, and b_3, respectively, are given in Table 9.4, along with estimates of the estimators' standard deviations, an estimate of $\sigma^2 \equiv \text{var}(\epsilon(l_i))$, and t-statistics to assess the significance of each coefficient. Writing (9.7.32) as

$$\mathbf{W} = X\boldsymbol{\beta} + \boldsymbol{\epsilon}, \tag{9.7.33}$$

Table 9.4 Estimates of Linear Regression Parameters, their Standard Deviations, and their t-Statistics[a]; Time 3 from Time 2

β_i	b_i	$(\widehat{\text{sd}}(b_i))$	t ($\hat{\sigma}^2 = 5.52 \times 10^{-4}$)
β_1	4.04668	(0.31894)	12.688
β_2	0.57624	(0.06949)	8.293
β_3	0.16165	(0.01660)	9.740

[a] $\hat{\sigma}^2$ denotes the estimated error variance.

where $\text{var}(\boldsymbol{\epsilon}) = \sigma^2 I$, standard results give

$$\mathbf{b} \equiv (b_1, b_2, b_3)' = (X'X)^{-1}X'\mathbf{W}, \tag{9.7.34}$$

$$\widehat{\text{var}}(\mathbf{b}) = (X'X)^{-1}\hat{\sigma}^2, \tag{9.7.35}$$

$$\hat{\sigma}^2 = \mathbf{W}'\big(I - X(X'X)^{-1}X'\big)\mathbf{W}/(n-3). \tag{9.7.36}$$

To obtain estimates of the growth parameters, sample versions of (9.7.27), (9.7.28), and (9.7.29) are invoked:

$$\hat{\lambda}_3 \equiv b_1, \tag{9.7.37}$$

$$\hat{E}(R_3) \equiv (b_2/b_1) + R_3^0, \tag{9.7.38}$$

$$\widehat{\text{var}}(R_3) \equiv (b_3/b_1) - (b_2/b_1)^2. \tag{9.7.39}$$

Because these are obtained from a nonlinear transformation of the original estimators, a δ-method calculation is needed to obtain their approximate biases and variances. Specifically,

$$\widehat{\text{bias}}\big(\hat{\lambda}_3\big) = 0,$$

$$\widehat{\text{var}}\big(\hat{\lambda}_3\big) = \widehat{\text{var}}(b_1);$$

$$\widehat{\text{bias}}\big(\hat{E}(R_3)\big) = -\frac{\widehat{\text{cov}}(b_1, b_2)}{b_1^2} + \frac{b_2\,\widehat{\text{var}}(b_1)}{b_1^3},$$

$$\widehat{\text{var}}\big(\hat{E}(R_3)\big) = \left\{\frac{\widehat{\text{var}}(b_2)}{b_2^2} + \frac{\widehat{\text{var}}(b_1)}{b_1^2} - \frac{2\,\widehat{\text{cov}}(b_1, b_2)}{b_1 b_2}\right\}\frac{b_2^2}{b_1^2};$$

$$\widehat{\text{bias}}\big(\widehat{\text{var}}(R_3)\big) = -\frac{\widehat{\text{cov}}(b_3, b_1)}{b_1^2} + \frac{b_3\,\widehat{\text{var}}(b_1)}{b_1^3} - \frac{3b_2^2\,\widehat{\text{var}}(b_1)}{b_1^4} + \frac{4b_2\,\widehat{\text{cov}}(b_1, b_2)}{b_1^3} - \frac{\widehat{\text{var}}(b_2)}{b_1^2},$$

$$\begin{aligned}\widehat{\text{var}}\big(\widehat{\text{var}}(R_3)\big) = {} & \frac{b_3\,\widehat{\text{var}}(b_1)}{b_1^4} + \frac{\widehat{\text{var}}(b_3)}{b_1^2} + \frac{4b_2^4\,\widehat{\text{var}}(b_1)}{b_1^6} + \frac{4b_2^2\,\widehat{\text{var}}(b_2)}{b_1^4} \\ & - \frac{2b_3\,\widehat{\text{cov}}(b_1, b_3)}{b_1^3} - \frac{4b_3b_2^2\,\widehat{\text{var}}(b_1)}{b_1^5} + \frac{4b_2b_3\,\widehat{\text{cov}}(b_1, b_2)}{b_1^4} \\ & + \frac{4b_2^2\,\widehat{\text{cov}}(b_3, b_1)}{b_1^4} - \frac{4b_2\,\widehat{\text{cov}}(b_2, b_3)}{b_1^3} - \frac{8b_2^3\,\widehat{\text{cov}}(b_1, b_2)}{b_1^5},\end{aligned}$$

Table 9.5 Estimates of Growth Parameters, their Standard Deviations, and their Biases; Time 3 from Time 2

Parameter	Estimator	Estimate	$(\widehat{\text{sd}})$	$[\widehat{\text{bias}}]$
		Standardized		
λ_3	b_1	4.04668	(0.3189)	[0.0]
$E(R_3)$	$(b_2/b_1)+R_3^0$	1.34240	(0.0283)	[0.0022]
$\text{var}(R_3)$	$(b_3/b_1)-(b_2/b_1)^2$	0.01967	(0.0114)	[−0.0015]
		Actual ($r_2 = 0.2604$ mm)		
λ_3	b_1/r_2^2	59.685 mm^{-2}	(4.7035)	[0.0]
$E(R_3)$	$\{(b_2/b_1)+R_3^0\}r_2$	0.3496 mm	(0.00737)	[0.00057]
$\text{var}(R_3)$	$\{(b_3/b_1)-(b_2/b_1)^2\}r_2^2$	0.00133 mm^2	(0.00077)	[−0.00010]

where the values given by (9.7.34) and (9.7.35) are used in the computations. The estimated biases and standard deviations are given in Table 9.5.

Recall it was assumed that $r_2 = 1$. In fact r_2 is not 1, but the simple relationships

$$\lambda_3^{(a)} = \lambda_3^{(s)}/r_2^2, \qquad R_3^{(a)} = R_3^{(s)}r_2 \tag{9.7.40}$$

(where the superscripts (a) and (s) stand for actual and standardized, respectivity), allow reconstruction of *actual* parameter estimates from *standardized* parameter estimates (obtained assuming $r_2 = 1$). The size variable $r_2 = 0.2604$ allows the latter half of Table 9.5 to be constructed.

Regression Diagnostics

The error variance σ^2 is estimated by $\hat{\sigma}^2 = 5.52 \times 10^{-4}$ (on the standardized scale), which is very small when compared to the regression mean-squared error of 8.1303. The adjusted $R^2 = 0.9997$.

The model is a remarkably good fit to the data, as seen by the small estimated standard deviations in Table 9.4. Figure 9.19a shows a plot of the data $W(\cdot)$ versus the fitted relation $b_1x_1(\cdot) + b_2x_2(\cdot) + b_3x_3(\cdot)$, and Figure 9.19$b$ shows the residuals

$$\hat{\epsilon}(l_i) \equiv W(l_i) - b_1x_1(l_i) - b_2x_2(l_i) - b_3x_3(l_i), \qquad i = 1,\ldots,n, \tag{9.7.41}$$

versus the fitted relation. There is no obvious lack of fit.

The simple residual plot of Figure 9.19b can sometimes miss important features unexplained by the fitted regression line. Part of the reason for its failure is that it is only a single two-dimensional summary of a model in four-dimensional space. A series of added variable (AV) and detrended added variable (DAV) plots, designed to assess the role of the explanatory

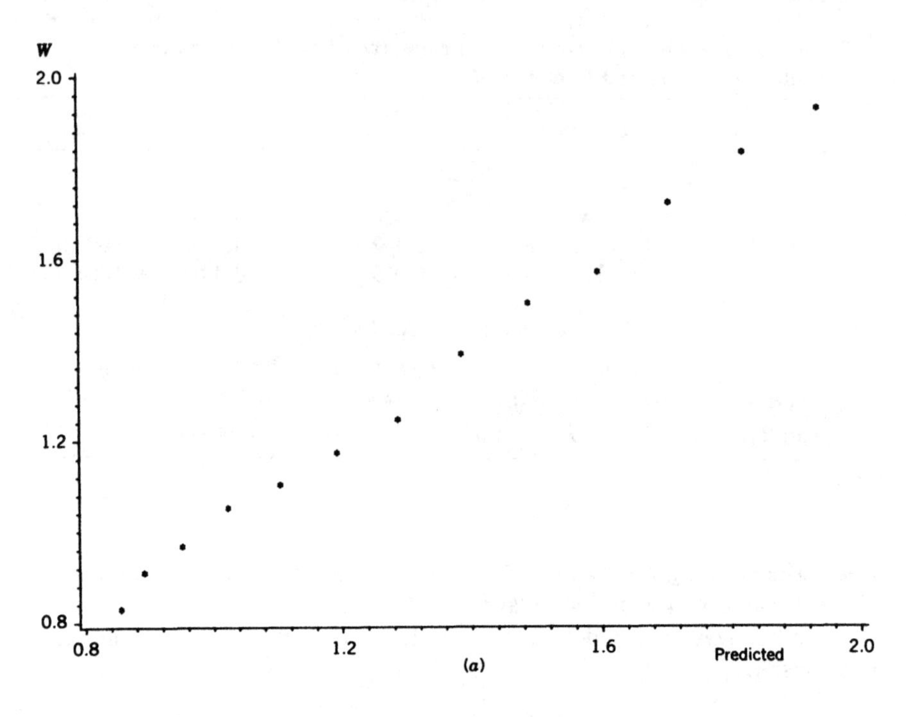

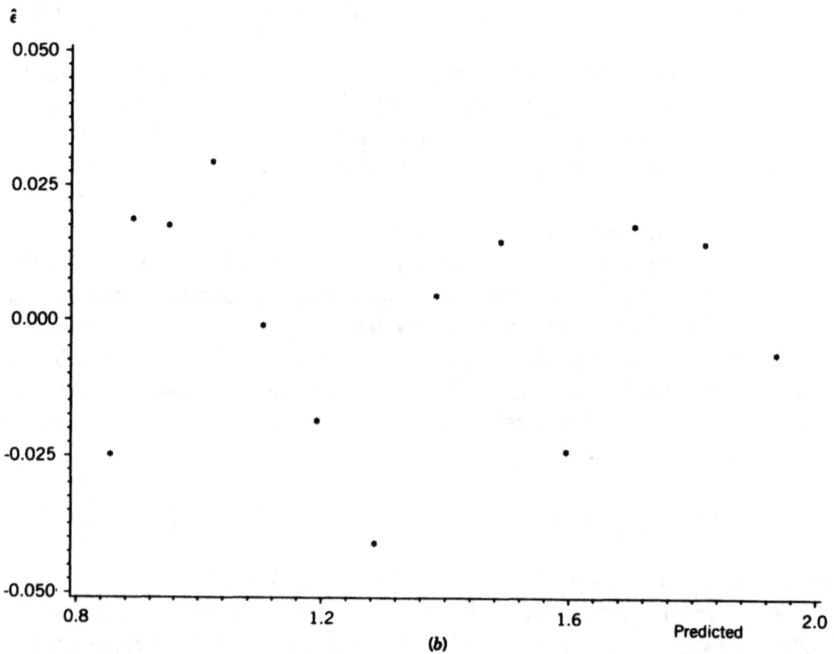

Figure 9.19 (*a*) Plot of data $\{W(l_i)\}$ versus predicted values $\{b_1x_1(l_i) + b_2x_2(l_i) + b_3x_3(l_i)\}$. (*b*) Plot of residuals $\{\hat{\epsilon}(l_i)\}$ versus predicted values $\{b_1x_1(l_i) + b_2x_2(l_i) + b_3x_3(l_i)\}$.

variables x_1, x_2, x_3 in the regression, provides additional visual summaries of the model. The AV plot for x_i shows $\hat{\epsilon}_{W \cdot [i]}$ versus $\hat{\epsilon}_{i \cdot [i]}$, where $\hat{\epsilon}_{W \cdot [i]}$ is the residual from the regression of W on $\{x_j: j \neq i\}$ and $\hat{\epsilon}_{i \cdot [i]}$ is the residual from the regression of x_i on $\{x_j: j \neq i\}$, $i = 1, 2, 3$. The DAV plot shows $\hat{\epsilon}$ versus $\hat{\epsilon}_{i \cdot [i]}$, $i = 1, 2, 3$.

To understand the purpose of these plots, note that $\hat{\epsilon}_{W \cdot [i]} = \hat{\epsilon} + b_i \hat{\epsilon}_{i \cdot [i]}$. Thus, a least-squares line fitted to the AV plot will have slope b_i. Then the DAV plot can be obtained from the AV plot by removing the systematic component. Therefore, the AV plot gives an indication of the consistency and strength of the functional relationship between x_i and W, whereas the associated DAV plot highlights model deficiencies (Cook and Weisberg, 1982; Cook, 1986). Figure 9.20*a* and *b* displays the AV/DAV pair for $i = 1$ (i.e., for the variable x_1). The AV plot suggests a strong linear relationship and the DAV plot of the residuals shows no serious model deficiencies. Similar conclusions may be drawn for x_2 and x_3. The AV/DAV plots were found to be useful in choosing l values during the early stages of modeling. The displays helped to identify nonlinear relationships for extreme l values by clearly depicting the effects of suspected failures in the model assumptions.

Evolution of the Cell Island at Time 2 from the Cell Island at Time 1

Having modeled and fitted the cell island at time 3 from that at time 2, we now consider fitting time 2 from time 1. The sequence of steps was repeated. The multiplicative ratios $\{r_1/r_1(\theta): \theta \in [0, 2\pi)\}$ were determined and $\mathbb{R}^2$ was transformed using the geometric-signature transformation $S_1(\cdot)$ defined analogously to (9.7.17). Again, without loss of generality, assume $r_1 = 1$, and transform X_1 to a unit disk and the boundary of X_2 to $\{(r_2(\theta)/r_1(\theta), \theta): \theta \in [0, 2\pi)\}$, which defines the set Y_2. An analogous stochastic model to (9.7.19) is posited and the function $q(s, l; \lambda_2, R_2)$ now represents the complement of the (conditional) hitting function of Y_2. Matching $-\log q_2(s, l; \lambda_2, R_2)$ with its empirical version $-\log \hat{q}_2(s, l; \lambda_2, R_2)$ yields parameter estimates, standard deviations, and biases for λ_2, $E(R_2)$, and $\text{var}(R_2)$, which are given in standardized form in the first half of Table 9.6. The cell-island size is $r_1 = 0.1097$ mm, from which actual estimates were computed; they are presented in the latter half of Table 9.6. Design constants chosen were $R_2^0 = 1.0$, $s = 3.0$, $n = 22$, and $l_i = 1.05(0.02)1.47$. Diagnostic checking again showed an excellent fit ($\hat{\sigma}^2 = 0.0047$, regression mean-squared error $=$ 26.4464, and adjusted $R^2 = 0.9987$).

Interpretation of the Results

Comparison of Tables 9.5 and 9.6 shows that the character of the tumor has changed significantly (in a statistical sense) from one time interval to the next, in a way that is not obvious from inspection of Figure 9.14. Comparison of standarized estimates for each table avoids a confounding of the size variable. In the first 72 hours (between time 1 and time 2), foci of growth

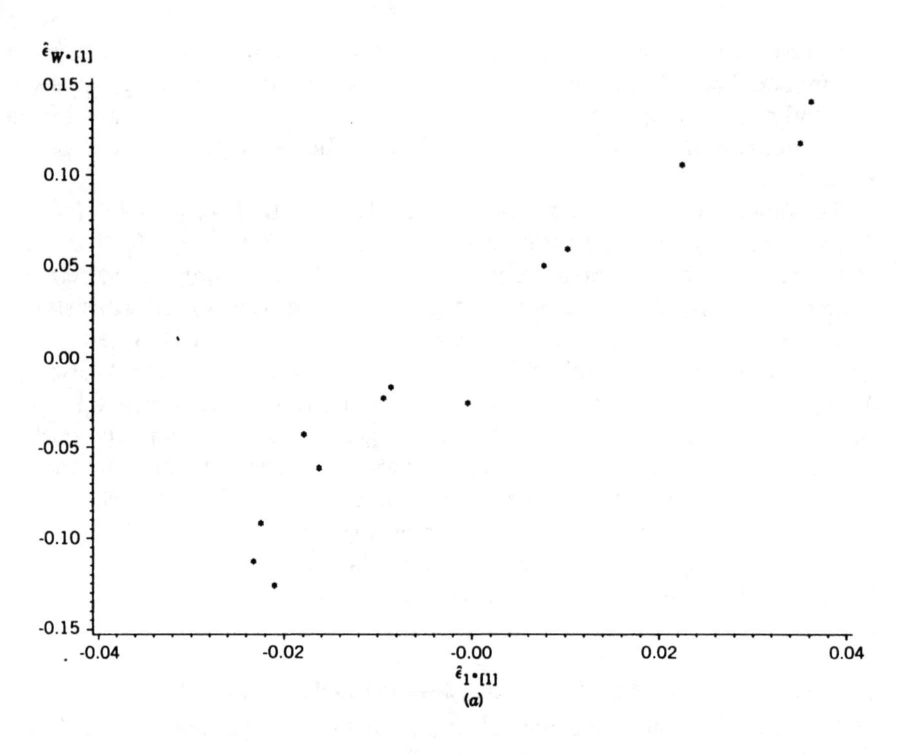

(a)

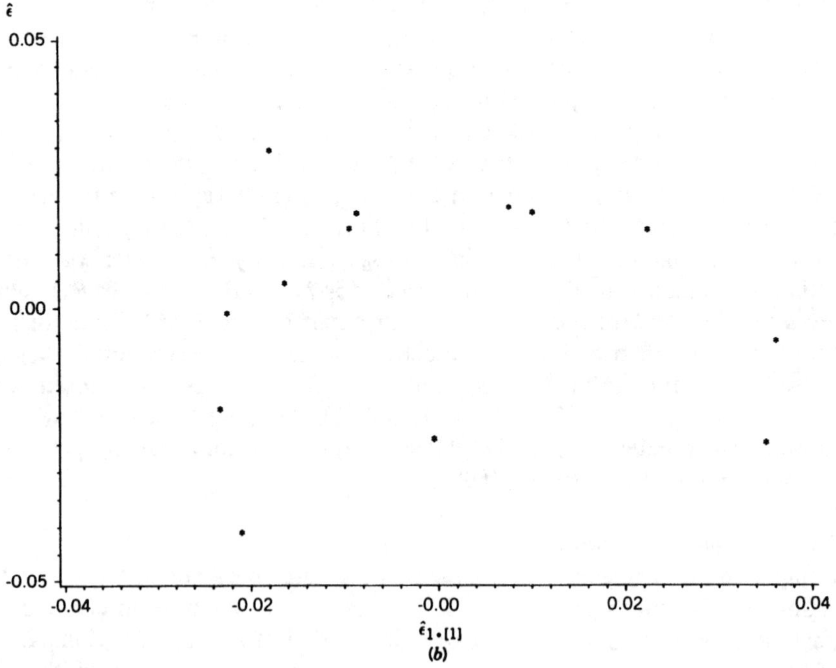

(b)

Figure 9.20 (*a*) Added variable (AV) plot of $\{\hat{\epsilon}_{W\cdot[1]}(l_i)\}$ versus $\{\hat{\epsilon}_{1\cdot[1]}(l_i)\}$. (*b*) Detrended added variable (DAV) plot of residuals $\{\hat{\epsilon}(l_i)\}$ versus $\{\hat{\epsilon}_{1\cdot[1]}(l_i)\}$.

Table 9.6 Estimates of Growth Parameters, their Standard Deviations, and their Biases; Time 2 from Time 1

Parameter	Estimate	$(\widehat{\text{sd}})$	$[\widehat{\text{bias}}]$
	Standardized		
λ_2	5.26194	(0.2968)	[0.0]
$E(R_2)$	1.04758	(0.0205)	[0.0011]
$\text{var}(R_2)$	0.02978	(0.0076)	[−0.00071]
	Actual ($r_l = 0.1097$ *mm*)		
λ_2	438.495 mm^{-2}	(24.733)	[0.0]
$E(R_2)$	0.11492 mm	(0.00225)	[0.00012]
$\text{var}(R_2)$	0.00036 mm^{-2}	(0.00009)	[−0.00001]

occur at a higher intensity, but the average radius of growth is smaller, as compared to the next 72 hours (between time 2 and time 3). Simulation of the model (Section 9.7.1) indicates that greater growth potential occurs when the growth radii are larger, in spite of a drop in the intensity of growth foci. Whereas $|X_2| = 4.15|X_1|$, it is seen that in the later period $|X_3| = 5.01|X_2|$.

The analysis demonstrates that the model (9.7.19) provides an *excellent description* of the image data. Although the growth process does not occur according to a time scale of 72 hours, the dynamic conditional Boolean model may provide conjectures about how the cell island is growing. It appears that foci of growth occur randomly and the growth is isotropic with approximately constant radius of growth (on the transformed scale).

The model allows a tumor to *regress* as well as grow. It would be interesting to see if parameter estimates $\hat{\lambda}$, $\hat{E}(R)$, and $\widehat{\text{var}}(R)$ could be used to monitor the effect of a treatment regime on tumor shape and size.

Some Final Remarks

I see the flexibility of growth or regression within the same model as an advantage. However, if strong growth is expected, it might be seen as a disadvantage of the model that a point in X_t may not necessarily be contained in X_{t+1}. D. Stoyan has suggested the following modification of (9.7.14):

$$X_{t+1}^* = \bigcup \{Z_{t+1}(\mathbf{s}_i): \mathbf{s}_i \in D_{t+1} \cap X_t^*\} \cup X_t^*. \tag{9.7.42}$$

Notice that $X_t^* \subset X_{t+1}^*$. The complement of the hitting function (conditional on X_t^*) is easily seen to be

$$Q_{X_{t+1}^*}(K) = \begin{cases} 0, & \text{if } K \cap X_t^* \neq \emptyset, \\ Q_{X_{t+1}}(K), & \text{if } K \cap X_t^* = \emptyset, \end{cases}$$

where $Q_{X_{t+1}}(K)$ is given by (9.7.15) with X_t^* substituted for X_t. Thus, from a practical point of view, inference on X_{t+1}^* conditional on X_t^* is *identical* with inference on X_{t+1} conditional on X_t. Moreover, the analysis and interpretation of the tumor-growth data remain the same regardless of whether (9.7.14) or (9.7.42) is chosen.

If the mechanism of growth of an object is something like a contact process (Harris, 1974), then growth will be concentrated on or near its boundary. Thus, the model (9.7.19) might be more realistic if Y_2 were an annulus rather than the disk $b(\mathbf{0}, r_2)$. A little thought reveals that, in a certain region of s and l values, (9.7.20) is unaffected by this change.

Suppose $1 - \rho$ is the radius of the interior boundary of the annulus (centered at $\mathbf{0}$) and 1 is the radius of the exterior boundary. Then, for this annulus to replace the unit disk $b(\mathbf{0}, 1)$ in (9.7.23) without change of $A_l(R_3)$, the inequality, $R_3 + l + 1 - s < \rho$ is required. For the choice of design constants given for time 3 from time 2, this implies $\rho > 0.29$; that is, $1 - \rho < 0.71$ (in units or r_2). For the choice of design constants given for time 2 from time 1, this implies $\rho > 0.47$; that is, $1 - \rho < 0.53$ (in units of r_1).

The two preceding remarks could now be combined to yield, on the transformed scale, a modification to the model (9.7.19), namely,

$$Y_3^{**} = \bigcup \{b(\mathbf{s}_i, R_{3,i}): \mathbf{s}_i \in D_3 \cap (b(\mathbf{0}, r_2) \setminus b(\mathbf{0}, \rho r_2))\} \cup b(\mathbf{0}, r_2),$$
$$0 < \rho < 1.$$

Here the foci of growth are restricted to the annulus $b(\mathbf{0}, r_2) \setminus b(\mathbf{0}, \rho r_2)$ *and* $Y_2^{**} = b(\mathbf{0}, r_2)$ is always contained in Y_3^{**}.

In conclusion, the accurate description achieved by these conditional Boolean models is encouraging. Although no claim can be made that an explanation for tumor growth at the supracellular level has been found, the stochastic model has raised interesting questions that warrant translation into biological terms.

References

Abend, K. (1968). Compound decision procedures for unknown distributions and dependent states of nature. In *Pattern Recognition*, L. Kanal, ed. Thompson Book Co., Washington, DC, 207–249.

Abend, K., Harley, T. J., and Kanal, L. N. (1965). Classification of binary random patterns. *IEEE Transactions on Information Theory*, **IT-11**, 538–544.

Abraham, B. (1983). The exact likelihood function for a space–time model. *Metrika*, **30**, 239–243.

Abraham, B. and Box, G. E. P. (1979). Bayesian analysis of some outlier problems in time series. *Biometrika*, **66**, 229–236.

Abramowitz, M. and Stegun, I. A. (1965). *Handbook of Mathematical Functions*. Dover, New York.

Abril, J. C. (1987). The approximate densities of some quadratic forms of stationary random variables. *Journal of Time Series Analysis*, **8**, 249–259.

Adam, J. A. (1986). A simplified mathematical model of tumor growth. *Mathematical Biosciences*, **81**, 229–244.

Adam, J. A. (1987). A mathematical model of tumor growth. II. Effects of geometry and spatial nonuniformity on stability. *Mathematical Biosciences*, **86**, 183–211.

Adler, R. J. (1981). *The Geometry of Random Fields*. Wiley, New York.

Adler, R. J. (1984). The supremum of a particular Gaussian field. *Annals of Probability*, **12**, 436–444.

Agterberg, F. P. (1984). Trend surface analysis. In *Spatial Statistics and Models*, G. L. Gaile and C. J. Willmott, eds. Reidel, Dordrecht, 147–171.

Ahmad, I. A. (1979). Strong consistency of density estimation by orthogonal series methods for dependent variables with applications. *Annals of the Institute of Statistical Mathematics*, **31**, 279–288.

Ahmed, S. and de Marsily, G. (1987). Comparison of geostatistical methods for estimating transmissivity using data on transmissivity and specific capacity. *Water Resources Research*, **23**, 1717–1737.

Aitchison, J. (1986). *The Statistical Analysis of Compositional Data*. Chapman and Hall, London.

Aitchison, J. and Brown, J. (1957). *The Lognormal Distribution*. Cambridge University Press, London.

Aitchison, J. and Ho, C. H. (1989). The multivariate Poisson-log normal distribution. *Biometrika*, **76**, 643–653.

Aitken, A. C. (1935). On least squares and linear combinations of observations. *Proceedings of the Royal Society of Edinburgh A*, **55**, 42–48.

Akaike, H. (1973). Information theory and an extension of the maximum likelihood principle. In *Second International Symposium on Information Theory*, B. N. Petrov and F. Csaki, eds. Akademiai Kiado, Budapest, 267–281.

Albeverio, S., Hoegh-Krohn, R., and Olsen, G. (1981). The global Markov property for lattice systems. *Journal of Multivariate Analysis*, **11**, 599–607.

Ali, M. M. (1979). Analysis of stationary spatial-temporal processes: Estimation and prediction. *Biometrika*, **66**, 513–518.

Ali, M. M. (1984). An approximation to the null distribution and power of the Durbin–Watson statistic. *Biometrika*, **71**, 253–261.

Ali, M. M. (1987). Durbin–Watson and generalized Durbin–Watson tests for autocorrelations and randomness. *Journal of Business and Economic Statistics*, **5**, 195–203.

Allen, D. M. (1971). The prediction sum of squares as a criterion for selecting predictor variables. Technical Report No. 23, Department of Statistics, University of Kentucky, Lexington, KY.

Ambartzumian, R. V. (1982). *Combinatorial Integral Geometry: With Applications to Mathematical Stereology*. Wiley, New York.

Anderson, B. D. O. and Moore, J. B. (1979). *Optimal Filtering*. Prentice-Hall, Englewood Cliffs, NJ.

Andjel, E. D. (1988). The contact process in high dimensions. *Annals of Probability*, **16**, 1174–1183.

Anselin, L. (1988). *Spatial Econometrics: Methods and Models*. Kluwer, Dordrecht.

Anselin, L. and Griffith, D. A. (1988). Do spatial effects really matter in regression analysis? *Papers of the Regional Science Association*, **65**, 11–34.

Ansley, C. F. (1979). An algorithm for the exact likelihood of a mixed autoregressive-moving average process. *Biometrika*, **66**, 59–65.

Arak, T. and Surgailis, D. (1989). Markov fields with polygonal realizations. *Probability Theory and Related Fields*, **80**, 543–579.

Arato, N. M. (1989). Equivalence of Gaussian measures corresponding to generalized random fields. *Applied Mathematics Letters*, **2**, 211–214.

Arato, N. M. (1990). Quadratic functionals of generalized Gaussian random fields. *Computers and Mathematics with Applications*, **19**, 1–7.

Armitage, P. (1949). An overlap problem arising in particle counting. *Biometrika*, **36**, 257–266.

Armstrong, A. C. (1986). On the fractal dimensions of some transient soil properties. *Journal of Soil Science*, **37**, 641–652.

Armstrong, M. and Delfiner, P. (1980). Towards a more robust variogram: A case study on coal. Internal Note N-671, Centre de Geostatistique, Fontainebleau, France.

Armstrong, M. and Diamond, P. (1984). Testing variograms for positive-definiteness. *Journal of the International Association for Mathematical Geology*, **16**, 407–421.

Armstrong, M. and Matheron, G. (1986a). Disjunctive kriging revisited: Part I. *Mathematical Geology*, **18**, 711–728.

Armstrong, M. and Matheron, G. (1986b). Disjunctive kriging revisited: Part II. *Mathematical Geology*, **18**, 729–742.

Armstrong, M. and Wackernagel, H. (1988). The influence of the covariance function on the kriged estimator. *Sciences de la Terre, Serie Informatique Geologique, Nancy*, **No. 27**, 245–262.

Aroian, L. A. (1980). Time series in *m* dimensions: Definitions, problems, and prospects. *Communications in Statistics. Simulation and Computation*, **B9**, 453–465.

Artstein, Z. (1984). Convergence rates for the optimal values of allocation processes. *Mathematics of Operations Research*, **9**, 348–355.

Artstein, Z. and Hart, S. (1981). Law of large numbers for random sets and allocation processes. *Mathematics of Operations Research*, **6**, 485–492.

Artstein, Z. and Vitale, R. (1975). A strong law of large numbers for random compact sets. *Annals of Probability*, **3**, 879–882.

Atkinson, A. (1969). The use of residuals as a concomitant variable. *Biometrika*, **56**, 33–41.

Atkinson, D. (1978). *Epidemiology of Sudden Infant Death in North Carolina: Do Cases Tend to Cluster?* PHSB Studies, No. 16, North Carolina Department of Human Resources, Division of Health Services, Public Health Statistics Branch, Raleigh, NC.

Aumann, R. (1965). Integrals of set-valued functions. *Journal of Mathematical Analysis and Applications*, **12**, 1–12.

Avrami, M. (1939). Kinetics of phase change, I. *Journal of Chemical Physics*, **7**, 1103–1112.

Babu, G. J. (1989). Strong representations for LAD estimators in linear models. *Probability Theory and Related Fields*, **83**, 547–558.

Baczkowski, A. J. and Mardia, K. V. (1987). Approximate lognormality of the sample semi-variogram under a Gaussian process. *Communications in Statistics. Simulation and Computation*, **16**, 571–585.

Baddeley, A. J. (1980). A limit theorem for statistics of spatial data. *Advances in Applied Probability*, **12**, 447–461.

Baddeley, A. J. and Moller, J. (1989). Nearest-neighbor Markov point processes and random sets. *International Statistical Review*, **57**, 89–121.

Baddeley, A. J. and Silverman, B. W. (1984). A cautionary example on the use of second-order methods for analyzing point patterns. *Biometrics*, **40**, 1089–1093.

Bagchi, S. (1985). On a.s. convergence of classes of multivalued asymptotic martingales. *Annales de l'Institut Henri Poincaré*, **21**, 313–321.

Bailey, N. T. J. (1980). Spatial models in the epidemiology of infectious diseases. In *Biological Growth and Spread*, W. Jager, H. Rost, and P. Tautu, eds. *Lecture Notes in Biomathematics*, **No. 38**. Springer, New York, 233–261.

Bailey, R. A. (1985). Restricted randomization for neighbour-balanced designs. *Statistics and Decisions, Supplement Issue No. 2*, **3**, 237–248.

Baker, J. L. and Laflen, J. M. (1983). Water quality consequences of conservation tillage. *Journal of Soil Water Conservation*, **38**, 186–193.

Bandemer, H. and Roth, K. (1987). A method of fuzzy-theory-based computer-aided exploratory data analysis. *Biometrical Journal*, **29**, 497–504.

Bardossy, A. (1988). Notes on the robustness of the kriging system. *Mathematical Geology*, **20**, 189–203.

Bardossy, A. and Bardossy, G. (1984). Comparison of geostatistical calculations with the results of open pit mining at the Iharkut Bauxite District, Hungary: A case study. *Journal of the International Association for Mathematical Geology*, **16**, 173–191.

Bardossy, A., Bogardi, I., and Duckstein, L. (1987). A geostatistical model of reservoir deposition. *Water Resources Research*, **23**, 510–514.

Barker, A. A. (1965). Monte Carlo calculations of the radial distribution functions for a proton–electron plasma. *Australian Journal of Physics*, **18**, 119–133.

Barnard, G. (1963). Comment on "The spectral analysis of point processes" by M. S. Bartlett. *Journal of the Royal Statistical Society B*, **25**, 294.

Barnes, R. J. (1989). Sample design for geologic site characterization. In *Geostatistics, Vol.* **2**, M. Armstrong, ed. Kluwer, Dordrecht, 809–822.

Barnes, R. J. and Johnson, T. B. (1984). Positive kriging. In *Geostatistics for Natural Resources Characterization, Part 1*, G. Verly, M. David, A. Journel, and A. Marechal, eds. Reidel, Dordrecht, 231–244.

Barnett, W. A. (1986). Random set methodology and stochastic geometry in economics and statistics. In *Innovations in Quantitative Economics: Essays in Honor of Robert L. Basmann*, D. J. Slottje, ed. *Advances in Econometrics, Vol.* **5**. JAI Press, Greenwich, CT, 125–141.

Bartels, R. (1977). On the use of limit theorem arguments in economic statistics. *American Statistician*, **31**, 85–87.

Bartko, J. J., Greenhouse, S. W., and Patlak, C. S. (1968). On expectations of some functions of Poisson variates. *Biometrics*, **24**, 97–102.

Bartlett, M. S. (1937). Properties of sufficiency and statistical tests. *Proceedings of the Royal Society A*, **160**, 268–282.

Bartlett, M. S. (1938). The approximate recovery of information from replicated field experiments with large blocks. *Journal of Agricultural Science (Cambridge)*, **28**, 418–427.

Bartlett, M. S. (1955). *An Introduction to Stochastic Processes*. Cambridge University Press, Cambridge.

Bartlett, M. S. (1963). The spectral analysis of point processes. *Journal of the Royal Statistical Society B*, **25**, 264–281.

Bartlett, M. S. (1964). The spectral analysis of two-dimensional point processes. *Biometrika*, **51**, 299–311.

Bartlett, M. S. (1967). Inference and stochastic processes. *Journal of the Royal Statistical Society A*, **130**, 457–474.

Bartlett, M. S. (1968). A further note on nearest neighbour models. *Journal of the Royal Statistical Society A*, **131**, 579–580.

Bartlett, M. S. (1971). Physical nearest-neighbour models and non-linear time-series. *Journal of Applied Probability*, **8**, 222–232.

Bartlett, M. S. (1974). The statistical analysis of spatial pattern. *Advances in Applied Probability*, **6**, 336–358.

Bartlett, M. S. (1975). *The Statistical Analysis of Spatial Pattern*. Chapman and Hall, London.

Bartlett, M. S. (1978). Nearest neighbour models in the analysis of field experiments. *Journal of the Royal Statistical Society B*, **40**, 147–158.

Barton, R. J. and Poor, H. V. (1988). Signal detection in fractional Gaussian noise. *IEEE Transactions on Information Theory*, **IT-34**, 943–959.

Bartoszynski, R., Brown, B., and Thompson, J. (1982). Metastatic and systematic factors in neoplastic progression. In *Probability Models and Cancer*, L. Le Cam and J. Neyman, eds. North-Holland, Amsterdam, 253–264.

Basawa, I. V. and Prakasa Rao, B. L. S. (1980). *Statistical Inference for Stochastic Processes*. Academic Press, New York.

Bastin, G. and Gevers, M. (1985). Identification and optimal estimation of random fields from scattered point-wise data. *Automatica*, **21**, 139–155.

Batchelor, L. D. and Reed, H. S. (1918). Relation of the variability of yields of fruit trees to the accuracy of field trials. *Journal of Agricultural Research*, **12**, 245–283.

Batty, M. (1976). Entropy in spatial aggregation. *Geographical Analysis*, **8**, 1–21.

Baudin, M. (1981). Likelihood and nearest-neighbor distance properties of multidimensional Poisson cluster processes. *Journal of Applied Probability*, **18**, 879–888.

Baumeister, A. A. (1978). Origins and control of stereotyped movements. In *Quality of Life in Severely and Profoundly Mentally Retarded People: Research Foundations for Improvement. Monograph No. 3*, American Association on Mental Deficiency, Washington, DC, 353–384.

Baver, L. D., Gardner, W. H., and Gardner, W. R. (1972). *Soil Physics*, 4th ed. Wiley, New York.

Baxter, M. (1986). Geographical and planning models for data on spatial flows. *The Statistician*, **35**, 191–198.

Beckmann, P. (1973). *Orthogonal Polynomials for Engineers and Physicists*. Golem Press, Boulder, CO.

Beder, J. H. (1988). Estimating a covariance function having an unknown scale parameter. *Communications in Statistics. Theory and Methods*, **17**, 1641–1657.

Bell, W. (1987). A note on overdifferencing and the equivalence of seasonal time series models with monthly means and models with $(0,1,1)_{12}$ seasonal parts when $\theta = 1$. *Journal of Business and Economic Statistics*, **5**, 383–387.

Bellhouse, D. R. (1977). Some optimal designs for sampling in two dimensions. *Biometrika*, **64**, 605–611.

Bennett, R. J. and Haining, R. P. (1985). Spatial structure and spatial interaction: Modelling approaches to the statistical analysis of geographical data. *Journal of the Royal Statistical Society A*, **148**, 1–27.

Beran, J. (1989). A test of location for data with slowly decaying serial correlations. *Biometrika*, **76**, 261–269.

Berger, T. (1988). Information theory in random fields. In *Coding Theory and Applications*, G. Cohen and P. Godlewski, eds. *Lecture Notes in Computer Science*, **No. 311**. Springer, New York, 1–18.

Berman, M. (1983). Comment on "Likelihood analysis of point processes and its applications to seismological data" by Y. Ogata. *Bulletin of the International Statistical Institute*, **50**, Book 3, 412–418.

Berman, M. and Diggle, P. (1989). Estimating weighted integrals of the second-order intensity of a spatial point process. *Journal of the Royal Statistical Society B*, **51**, 81–92.

Bertrand, A. R. (1965). Rate of water intake in the field. In *Methods of Soil Analysis*, C. A. Black et al., eds. *Agronomy Monographs*, **9**, 197–209.

Besag, J. E. (1974). Spatial interaction and the statistical analysis of lattice systems. *Journal of the Royal Statistical Society B*, **36**, 192–225.

Besag, J. E. (1975). Statistical analysis of non-lattice data. *The Statistician*, **24**, 179–195.

Besag, J. E. (1977a). Errors-in-variables estimation for Gaussian lattice schemes. *Journal of the Royal Statistical Society B*, **39**, 73–78.

Besag, J. E. (1977b). Efficiency of pseudolikelihood estimators for simple Gaussian fields. *Biometrika*, **64**, 616–618.

Besag, J. E. (1977c). Comment on "Modelling spatial patterns" by B. D. Ripley. *Journal of the Royal Statistical Society B*, **39**, 193–195.

Besag, J. E. (1981). On a system of two-dimensional recurrence equations. *Journal of the Royal Statistical Society B*, **43**, 302–309.

Besag, J. E. (1986). On the statistical analysis of dirty pictures. *Journal of the Royal Statistical Society B*, **48**, 259–279.

Besag, J. E. (1989a). A candidate's formula: A curious result in Bayesian prediction. *Biometrika*, **76**, 183.

Besag, J. E. (1989b) Towards Bayesian image analysis. *Journal of Applied Statistics*, **16**, 395–407.

Besag, J. E. and Diggle, P. J. (1977). Simple Monte Carlo tests for spatial pattern. *Applied Statistics*, **26**, 327–333.

Besag, J. E. and Gleaves, J. T. (1973). On the detection of spatial pattern in plant communities. *Bulletin of the International Statistical Institute*, **45**, Book 1, 153–158.

Besag, J. E. and Kempton, R. A. (1986). Statistical analysis of field experiments using neighboring plots. *Biometrics*, **42**, 231–251.

Besag, J. E., Milne, R. K., and Zachary, S. (1982). Point process limits of lattice processes. *Journal of Applied Probability*, **19**, 210–216.

Besag, J. E. and Moran, P. A. P. (1975). On the estimation and testing of spatial interaction in Gaussian lattice processes. *Biometrika*, **62**, 555–562.

Bhat, B. R. (1974). On the method of maximum-likelihood for dependent observations. *Journal of the Royal Statistical Society B*, **36**, 48–53.

Bickel, P. J. (1975). One-step Huber estimates in the linear model. *Journal of the American Statistical Association*, **70**, 428–434.

Bickel, P. J. and Doksum, K. A. (1977). *Mathematical Statistics*. Holden-Day, San Francisco, CA.

Bickel, P. J. and Freedman, D. A. (1981). Some asymptotic theory for the bootstrap. *Annals of Statistics*, **9**, 1196–1217.

Billingsley, P. (1968). *Convergence of Probability Measures*. Wiley, New York.

Bilonick, R. A. (1983). Risk qualified maps of hydrogen ion concentration for the New York State area for 1966–1978. *Atmospheric Environment*, **17**, 2513–2524.

Bilonick, R. A. (1985). The space–time distribution of sulfate deposition in the northeastern United States. *Atmospheric Environment*, **19**, 1829–1845.

Bilonick, R. A. (1988). Monthly hydrogen ion deposition maps for the Northeastern U.S. from July 1982 to September 1984. *Atmospheric Environment*, **22**, 1909–1924.

Birkhoff, G. (1967). *Lattice Theory*, 3rd ed. American Mathematical Society, Providence, RI.

Birkhoff, G. D. (1931). Proof of the ergodic theorem. *Proceedings of the National Academy of Sciences, Washington*, **17**, 656–660.

Birnbaum, Z. W. (1975). Testing for intervals of increased mortality. In *Reliability and Fault Tree Analysis*, R. E. Barlow, J. B. Fussell, and N. D. Singpurwalla, eds. SIAM, Philadelphia, PA, 413–426.

Black, T. C. and Freyberg, D. L. (1987). Stochastic modeling of vertically averaged concentration uncertainty in a perfectly stratified aquifer. *Water Resources Research*, **23**, 997–1004.

Blanc-Lapierre, A. and Faure, P. (1965). Stationary and isotropic random functions. In *Bernoulli, Bayes, Laplace Anniversary Volume*, J. Neyman and L. M. Le Cam, eds. Springer, New York, 17–23.

Blum, J. R. (1982). Ergodic theorems. Entry in *Encyclopedia of Statistical Sciences, Vol.* **2**, S. Kotz and N. L. Johnson, eds. Wiley, New York, 541–545.

BMDP Statistical Software, Inc. (1988). *BMDP Statistical Software Manual, Vols.* **1** *and* **2**. University of California Press, Berkeley, CA.

Bochner, S. (1955). *Harmonic Analysis and the Theory of Probability*. University of California Press, Berkeley and Los Angeles, CA.

Bogardi, I., Bardossy, A., and Duckstein, L. (1985). Multicriterion network design using geostatistics. *Water Resources Research*, **21**, 199–208.

Bolthausen, E. (1982). On the central limit theorem for stationary random fields. *Annals of Probability*, **10**, 1047–1050.

Bookstein, F. L. (1986). Size and shape spaces for landmark data in two dimensions. *Statistical Science*, **1**, 181–222.

Borgman, L. E., Taheri, M., and Hagan, R. (1984). Three-dimensional, frequency-domain simulations of geological variables. In *Geostatistics for Natural Resources Characterization, Part 1*, G. Verly, M. David, A. G. Journel, and A. Marechal, eds. Reidel, Dordrecht, 517–541.

Born, M and Green, H. S. (1946). A general kinetic theory of liquids I. The molecular distribution functions. *Proceedings of the Royal Society of London A*, **188**, 10–18.

Bosq, D. (1973). Sur l'estimation de la densite d'un processus stationnaire et melangeant. *Comptes Rendus de l'Academie des Sciences, Paris, A*, **277**, 535–538.

Bosq, D. (1983). Sur la prediction non parametrique de variables aleatoires et de mesures aleatoires. *Zeitschrift für Wahrscheinlichkeitstheorie und verwandte Gebiete*, **64**, 541–553.

Bosq, D. (1989). Estimation et prevision non parametrique d'un processus stationnaire. *Comptes Rendus de l'Academie des Sciences, Paris, I*, **308**, 453–456.

Bowman, A. (1984). An alternative method of cross-validation for the smoothing of density estimates. *Biometrika*, **71**, 353–360.

Box, G. E. P. and Cox, D. R. (1964). An analysis of transformations. *Journal of the Royal Statistical Society B*, **26**, 211–243.

Box, G. E. P. and Jenkins, G. M. (1970). *Time Series Analysis: Forecasting and Control.* Holden-Day, San Francisco, CA.

Box, G. E. P. and Jenkins, G. M. (1976). *Time Series Analysis: Forecasting and Control*, 2nd ed. Holden-Day, San Francisco, CA.

Bramson, M. and Griffeath, D. (1980). The asymptotic behavior of a probabilistic model for tumor growth. In *Biological Growth and Spread*, W. Jager, H. Rost, and P. Tautu, eds. *Lecture Notes in Biomathematics*. **No. 38**. Springer, New York, 165–172.

Bramson, M. and Griffeath, D. (1981). On the Williams–Bjerknes tumor growth model, I. *Annals of Probability*, **9**, 173–185.

Brandsma, A. S. and Ketellapper, R. H. (1979). A biparametric approach to spatial autocorrelation. *Environment and Planning A*, **11**, 51–58.

Bras, R. L. and Rodriguez-Iturbe, I. (1976). Network design for the estimation of areal mean of rainfall events. *Water Resources Research*, **12**, 1185–1195.

Bras, R. L. and Rodriguez-Iturbe, I. (1985). *Random Functions and Hydrology*. Addison-Wesley, Reading, MA.

Bratley, P., Fox, B. L., and Schrage, L. E. (1987). *A Guide to Simulation*, 2nd ed. Springer, New York.

Breiman, L. and Friedman, J. H. (1985). Estimating optimal transformations for multiple regression and correlation. *Journal of the American Statistical Association*, **80**, 580–598.

Brewer, A. C. and Mead, R. (1986). Continuous second order models of spatial variation with application to the efficiency of field crop experiments. *Journal of the Royal Statistical Society A*, **149**, 314–336.

Brillinger, D. R. (1975). Statistical inference for stationary point processes. In *Stochastic Processes and Related Topics*, M. L. Puri, ed. Academic Press, New York, 55–92.

Bronars, S. G. and Jansen, D. W. (1987). The geographic distribution of unemployment rates in the U.S. *Journal of Econometrics*, **36**, 251–279.

Brook, D. (1964). On the distinction between the conditional probability and joint probability approaches in the specification of nearest neighbour systems. *Biometrika*, **51**, 481–483.

Brooker, P. (1985). Two-dimensional simulation by turning bands. *Journal of the International Association for Mathematical Geology*, **17**, 81–90.

Brooker, P. (1986). A parametric study of robustness of kriging variance as a function of range and relative nugget effect for a spherical variogram. *Mathematical Geology*, **18**, 477–488.

Brooks, D. R and Wiley, E. O. (1988). *Evolution as Entropy*, 2nd ed. University of Chicago Press, Chicago, IL.

Brouwer, F., Nijkamp, P., and Scholten, H. (1988). Hybrid log-linear models for spatial interaction and stability analysis. *Metroeconomica*, **39**, 43–65.

Brown, S. and Holgate, P. (1974). The thinned plantation. *Biometrika*, **61**, 253–261.

Brown, T. C., Silverman, B. W., and Milne, R. K. (1981). A class of two-type point processes. *Zeitschrift für Wahrscheinlichkeitstheorie und verwandte Gebiete*, **58**, 299–308.

Brozius, H. (1989). Convergence in mean of some characteristics of the convex hull. *Advances in Applied Probability*, **21**, 526–542.

Bucklew, J. A. and Cambanis, S. (1988). Estimating random integrals from noisy observations: Sampling designs and their performance. *IEEE Transactions on Information Theory*, **IT-34**, 111–127.

Buehring, G. C. and Williams, R. R. (1976). Growth rates of normal and abnormal human mammary epithelia in cell culture. *Cancer Research*, **36**, 3742–3747.

Buoncristiani, J. F. (1983). Probabilities on fuzzy sets and ϵ-fuzzy sets. *Journal of Mathematical Analysis and Applications*, **96**, 24–41.

Burch, S. F., Gull, S. F., and Skilling, J. (1983). Image restoration by a powerful maximum entropy method. *Computer Vision, Graphics, and Image Processing*, **23**, 113–128.

Burgess, T. M. and Webster, R. (1980). Optimal interpolation and isarithmic mapping of soil properties, I: The semi-variogram and punctual kriging. *Journal of Soil Science*, **31**, 315–331.

Burridge, P. (1980). On the Cliff–Ord test for spatial correlation. *Journal of the Royal Statistical Society B*, **42**, 107–108.

Burrough, P. A. (1983). Problems of superimposed effects in the statistical study of the spatial variation of soil. *Agricultural Water Management*, **6**, 123–143.

Burrough, P. A. (1986). *Principles of Geographical Information Systems for Land Resource Assessment*. Clarendon Press, Oxford.

Burton, A. (1966). Rate of growth of solid tumors as a problem of diffusion. *Growth*, **30**, 157–176.

Buxton, B. E. (1982). Coal reserve assessment: A geostatistical study. Masters Thesis, Applied Earth Sciences Department, Stanford University, Stanford, CA, unpublished.

Byth, K. and Ripley, B. D. (1980). On sampling spatial patterns by distance methods. *Biometrics*, **36**, 279–284.

Cambanis, S. (1985). Sampling designs for time series. In *Handbook of Statistics*, **5**, E. J. Hannan, P. R. Krishnaiah, and M. M. Rao, eds. Elsevier, Amsterdam, 337–362.

Campbell, J. B. (1978). Spatial variation of sand content and pH within single continuous delineation of two soil mapping units. *Soil Science Society of America Journal*, **42**, 460–464.

Carlstein, E. (1986). The use of subseries values for estimating the variance of a general statistic from a stationary sequence. *Annals of Statistics*, **14**, 1171–1179.

Carlstein, E. (1988). Law of large numbers for the subseries values of a statistic from a stationary sequence. *Statistics*, **19**, 295–299.

Carr, J. R. and Roberts, K. P. (1989). Application of universal kriging for estimation of earthquake ground motion: Statistical significance of results. *Mathematical Geology*, **21**, 255–265.

Carroll, C. W. (1961). The created response surface technique for optimizing nonlinear, restrained systems. *Operations Research*, **9**, 169–184.

Carroll, R. J. and Ruppert, D. (1982). A comparison between maximum likelihood and generalized least squares in a heteroscedastic linear model. *Journal of the American Statistical Association*, **77**, 878–882.

Carroll, R. J. and Ruppert, D. (1988). *Transformation and Weighting in Regression*. Chapman and Hall, New York.

Carroll, R. J., Wu, C. F. J., and Ruppert, D. (1988). The effect of estimating weights in weighted least squares. *Journal of the American Statistical Association*, **83**, 1045–1054.

Carter, D. S. and Prenter, P. M. (1972). Exponential spaces and counting processes. *Zeitschrift für Wahrscheinlichkeitstheorie und verwandte Gebiete*, **21**, 1–19.

Carter, G. M. and Rolph, J. E. (1974). Empirical Bayes methods applied to estimating fire alarm probabilities. *Journal of the American Statistical Association*, **69**, 880–885.

Castellana, J. V. and Leadbetter, M. R. (1986). On smoothed probability density estimation for stationary processes. *Stochastic Processes and Their Applications*, **21**, 179–193.

Catana, A. J. (1963). The wandering quarter method of estimating population density. *Ecology*, **44**, 349–360.

Chalmond, B. (1986). Regression avec residus spatialement autocorreles et recherche de la tendance spatiale. *Statistique et Analyse des Donnees*, **11**, No. 2, 1–25.

Chan, K. H., Hayya, J. C., and Ord, J. K. (1977). A note on trend removal methods: The case of polynomial regression versus variate differencing. *Econometrica*, **45**, 737–744.

Chan, K. S. and Tong, H. (1987). A note on embedding a discrete parameter ARMA model in a continuous parameter ARMA model. *Journal of Time Series Analysis*, **8**, 277–281.

Chauvet, P. (1982). The variogram cloud. Internal Note N-725, Centre de Geostatistique, Fontainebleau, France.

Chauvet, P. (1989). Quelques aspects de l'analyse structurale des FAI-*k* a 1 dimension. In *Geostatistics, Vol.* **1**, M. Armstrong, ed. Kluwer, Dordrecht, 139–150.

Chen, R., Mantel, N., and Klingberg, M. A. (1984). A study of three techniques for space–time clustering in Hodgkin's disease. *Statistics in Medicine*, **3**, 173–184.

Cheng, C. S. (1983). Construction of optimal balanced incomplete block designs for correlated observations. *Annals of Statistics*, **11**, 240–246.

Chernick, M. R. and Murthy, V. K. (1983). The use of influence functions for outlier detection and data editing. *American Journal of Mathematical and Management Sciences*, **3**, 47–61.

Chover, J. and King, J. H. (1985). The early growth of cancer. *Journal of Mathematical Biology*, **21**, 329–346.

Choynowski, M. (1959). Maps based on probabilities. *Journal of the American Statistical Association*, **54**, 385–388.

Christakos, G. (1984). On the problem of permissible covariance and variogram models. *Water Resources Research*, **20**, 251–265.

Christakos, G. (1987). Stochastic simulation of spatially correlated geoprocesses. *Mathematical Geology*, **19**, 807–831.

Christakos, G. (1989). Optimal estimation of nonlinear state nonlinear observation systems. *Journal of Optimization Theory and Applications*, **62**, 29–48.

Christaller, W. (1933). *Die Zentralen Orte in Suddeutschland*. Fischer, Jena.

Chung, K. L. (1968). *A Course in Probability Theory*. Harcourt, Brace, and World, New York.

Clark, I. (1979). *Practical Geostatistics*. Applied Science Publishers, Essex, England.

Clark, I., Basinger, K. L., and Harper, W. V. (1989). MUCK: A novel approach to co-kriging. In *Proceedings of the Conference on Geostatistical, Sensitivity, and Uncertainty Methods for Ground-Water Flow and Radionuclide Transport Modeling*, B. E. Buxton, ed. Battelle Press, Columbus, OH, 473–493.

Clark, P. J. and Evans, F. C. (1954). Distance to nearest neighbor as a measure of spatial relationships in populations. *Ecology*, **35**, 445–453.

Clayton, D. and Kaldor, J. (1987). Empirical Bayes estimates of age-standardized relative risks for use in disease mapping. *Biometrics*, **43**, 671–681.

Cleveland, W. S. (1971). Projection with the wrong inner product and its application to regression with correlated errors and linear filtering of time series. *Annals of Mathematical Statistics*, **42**, 616–624.

Cleveland, W. S. (1979). Robust locally weighted regression and smoothing scatterplots. *Journal of the American Statistical Association*, **74**, 829–836.

Cleveland, W. S. (1985). *The Elements of Graphing Data*. Wadsworth, Monterey, CA.

Cleveland, W. S. and Devlin, S. J. (1988). Locally weighted regression: An approach to regression analysis by local fitting. *Journal of the American Statistical Association*, **83**, 596–610.

Cleveland, W. S. and McGill, R. (1984). Graphical perception: Theory, experimentation, and application to the development of graphical methods. *Journal of the American Statistical Association*, **79**, 531–554.

Cliff, A. D., Hagget, P., Ord, J. K., Bassett, K. A., and Davies, R. B. (1975). *Elements of Spatial Structure: A Quantitative Approach*. Cambridge University Press, Cambridge.

Cliff, A. D. and Ord, J. K. (1973). *Spatial Autocorrelation*. Pion, London.

Cliff, A. D. and Ord, J. K. (1975). Model building and the analysis of spatial pattern in human geography. *Journal of the Royal Statistical Society B*, **37**, 297–328.

Cliff, A. D. and Ord, J. K. (1981). *Spatial Processes: Models and Applications*. Pion, London.

Clifford, P. and Middleton, R. D. (1989). Reconstruction of polygonal images. *Journal of Applied Statistics*, **16**, 409–422.

Cochran, W. G. (1936). The statistical analysis of the distribution of field counts of diseased plants. *Journal of the Royal Statistical Society, Supplement*, **3**, 49–67.

Cochran, W. G. (1940). The analysis of variance when experimental errors follow the Poisson or binomial laws. *Annals of Mathematical Statistics*, **11**, 335–347.

Cochran, W. G. (1946). Relative accuracy of systematic and stratified random samples for a certain class of populations. *Annals of Mathematical Statistics*, **17**, 164–177.

Cochran, W. G. (1954). Some methods for strengthening the common χ^2 tests. *Biometrics*, **10**, 417–451.

Cohen, F. S. and Cooper, D. B. (1987). Simple parallel hierarchical and relaxation algorithms for segmenting noncausal Markovian random fields. *IEEE Transactions on Pattern Analysis and Machine Intelligence*, **PAMI-9**, 195–219.

Coleman, R. (1979). *An Introduction to Mathematical Stereology*. Memoirs No. 3, Department of Theoretical Statistics, University of Aarhus, Denmark.

Collomb, G. (1983). From non parametric regression to non parametric prediction: Survey of the mean square error and original results on the predictogram. In *Specifying Statistical Models*, J. P. Florens et al., eds. *Lecture Notes in Statistics*, **No. 16**. Springer, New York, 182–204.

Collomb, G. (1985). Non parametric time series analysis and prediction: Uniform almost sure convergence of the window and *k*-NN autoregression estimates. *Statistics*, **16**, 297–307.

Cook, D. G. and Pocock, S. J. (1983). Multiple regression in geographic mortality studies, with allowance for spatially correlated errors. *Biometrics*, **39**, 361–371.

Cook, N. R. (1985). Three-way analyses. In *Exploring Data Tables, Trends, and Shapes*, D. C. Hoaglin, F. Mosteller, and J. W. Tukey, eds. Wiley, New York, 125–188.

Cook, R. D. (1986). Assessment of local influence. *Journal of the Royal Statistical Society B*, **48**, 133–155.

Cook, R. D. and Weisberg, S. (1982). *Residuals and Influence in Regression*. Chapman and Hall, London.

Cormack, R. M. (1979). Spatial aspects of competition between individuals. In *Spatial and Temporal Analysis in Ecology*, R. M. Cormack and J. K. Ord, eds. International Co-operative Publishing House, Fairland, MD, 151–212.

Cornish, E. A. (1936). On the secular variation of the rainfall at Adelaide, South Australia. *Royal Meteorological Society Quarterly Journal*, **62**, 481–498.

Cornish, E. A. (1954). On the secular variation of rainfall at Adelaide. *Australian Journal of Physics*, **7**, 334–346.

Cornish, E. A. (1976). The variability of monthly rainfall in South Australia. *Technical paper No. 41*, CSIRO Division of Mathematical Statistics, Australia.

Cornish, E. A. and Coote, G. G. (1958). The correlation of monthly rainfall with position and altitude of observing stations in South Australia. *Technical paper No. 4*, CSIRO Division of Mathematical Statistics, Australia.

Cornish, E. A., Hill, G. W., and Evans, M. J. (1961). Inter-station correlations of rainfall in Southern Australia. *Technical paper No. 10*, CSIRO Division of Mathematical Statistics, Australia.

Cornish, E. A. and Stenhouse, N. S. (1958). Inter-station correlations of monthly rainfall in South Australia, *Technical paper No.* 5, CSIRO Division of Mathematical Statistics, Australia.

Cowan, R. (1989). Objects arranged randomly in space: An accessible theory. *Advances in Applied Probability*, **21**, 543–569.

Cox, D. R. (1955). Some statistical methods related with series of events. *Journal of the Royal Statistical Society B*, **17**, 129–157.

Cox, D. R. (1970). *The Analysis of Binary Data*. Methuen, London.

Cox, D. R. (1975). Partial likelihood. *Biometrika*, **62**, 269–276.

Cox, D. R. (1981), Statistical analysis of time series: Some recent developments. *Scandinavian Journal of Statistics*, **8**, 93–115.

Cox, D. R. (1985). Theory of statistics: Some current themes. *Bulletin of the International Statistical Institute*, **51**, Book 1, Section 6.3.

Cox, D. R. and Hinkley, D. V. (1974). *Theoretical Statistics*. Chapman and Hall, London.

Cox, D. R. and Isham, V. (1980). *Point Processes*. Chapman and Hall, London.

Cox, D. R. and Lewis, P. A. W. (1966). *The Statistical Analysis of Series of Events*. Methuen, London.

Cox, D. R. and Lewis, P. A. W. (1972). Multivariate point processes. *Proceedings of the Sixth Berkeley Symposium, Vol.* **3**. University of California Press, Berkeley, CA, 401–448.

Cox, T. F. and Lewis, T. (1976). A conditioned distance ratio method for analyzing spatial patterns. *Biometrika*, **63**, 483–491.

Cramér, H. and Leadbetter, M. R. (1967). *Stationary and Related Stochastic Processes*. Wiley, New York.

Craven, P. and Wahba, G. (1979). Smoothing noisy data with spline functions: Estimating the correct degree of smoothing by the method of generalized cross-validation. *Numerische Mathematik*, **31**, 377–403.

Cressie, N. (1977). On some properties of the scan statistic on the circle and the line. *Journal of Applied Probability*, **14**, 272–283.

Cressie, N. (1978a). Estimation of the integral of a stochastic process. *Bulletin of the Australian Mathematical Society*, **18**, 83–93.

Cressie, N. (1978b). A strong limit theorem for random sets. *Advances in Applied Probability, Supplement*, **10**, 36–46.

Cressie, N. (1979a). Straight line fitting and variogram estimation. *Bulletin of the International Statistical Institute*, **48**, Book 3, 573–576.

Cressie, N. (1979b). A central limit theorem for random sets. *Zeitschrift für Wahrscheinlichkeitstheorie und verwandte Gebiete*, **49**, 37–47.

Cressie, N. (1980a). Weighted *M*-estimation in the presence of unequal scale. *Statistica Neerlandica*, **34**, 19–32.

Cressie, N. (1980b). The asymptotic distribution of the scan statistic under uniformity. *Annals of Probability*, **8**, 828–840.

Cressie, N. (1982). Playing safe with misweighted means. *Journal of the American Statistical Association*, **77**, 754–759.

Cressie, N. (1984a). Towards resistant geostatistics. In *Geostatistics for Natural Resources Characterization, Part 1*, G. Verly, M. David, A. G. Journel, and A. Marechal, eds. Reidel, Dordrecht, 21–44.

Cressie, N. (1984b). Modelling sets. In *Multifunctions and Integrands*, G. Salinetti, ed. *Lecture Notes in Mathematics*, **No. 1091**. Springer, New York, 138–149.

Cressie, N. (1985a). Fitting variogram models by weighted least squares. *Journal of the International Association for Mathematical Geology*, **17**, 563–586.

Cressie, N. (1985b). When are relative variograms useful in geostatistics? *Journal of the International Association for Mathematical Geology*, **17**, 693–702.

Cressie, N. (1985c). A geostatistical analysis of the Mercer and Hall wheat data. *Bulletin of the International Statistical Institute*, **51**, *Contributed Papers Volume*, 277–278.

Cressie, N. (1986). Kriging nonstationary data. *Journal of the American Statistical Association*, **81**, 625–634.

Cressie, N. (1987). A nonparametric view of generalized covariances for kriging. *Mathematical Geology*, **19**, 425–449.

Cressie, N. (1988a). Variogram. Entry in *Encyclopedia of Statistical Sciences*, *Vol.* **9**, S. Kotz and N. L. Johnson, eds. Wiley, New York, 489–491.

Cressie, N. (1988b). Spatial prediction and ordinary kriging. *Mathematical Geology*, **20**, 405–421. [Erratum, *Mathematical Geology*, **21**, 493–494 (1989).]

Cressie, N. (1988c). A graphical procedure for determining nonstationarity in time series. *Journal of the American Statistical Association*, **83**, 1108–1116. [Correction, *Journal of the American Statistical Association*, **85**, 272 (1990).]

Cressie, N. (1989a). The many faces of spatial prediction, in *Geostatistics*, *Vol.* **1**, M. Armstrong, ed. Kluwer, Dordrecht, 163–176.

Cressie, N. (1989b). Ergodicity for time series and spatial processes. *Journal of Statistical Computation and Simulation*, **32**, 61–63.

Cressie, N. (1989c). Geostatistics. *American Statistician*, **43**, 197–202.

Cressie, N. (1989d). Erratum: Spatial prediction and ordinary kriging. *Mathematical Geology*, **21**, 493–494.

Cressie, N. (1990a). Reply to Wahba's letter. *American Statistician*, **44**, 256–258.

Cressie, N. (1990b). The origins of kriging. *Mathematical Geology*, **22**, 239–252.

Cressie, N. (1990c). Weighted smoothing of estimated undercount. *Proceedings of Bureau of the Census 1990 Annual Research Conference*. U.S. Bureau of the Census, Washington, DC, 301–325.

Cressie, N. (1991). Modeling growth with random sets. In *Spatial Statistics and Imaging,* A. Possolo, ed. *Lecture Notes-Monograph Series*, Vol. **20**. Institute of Mathematical Statistics, Hayward, CA, 31–45.

Cressie, N. (1992). Smoothing regional maps using empirical Bayes predictors. *Geographical Analysis*, **24**, 75–95.

Cressie, N. and Chan, N. H. (1989). Spatial modeling of regional variables. *Journal of the American Statistical Association*, **84**, 393–401.

Cressie, N. and Glonek, G. (1984). Median based covariogram estimators reduce bias. *Statistics and Probability Letters*, **2**, 299–304.

Cressie, N., Gotway, C. A., and Grondona, M. O. (1990). Spatial prediction from networks. *Chemometrics and Intelligent Laboratory Systems*, **7**, 251–271.

Cressie, N. and Grondona, M. O. (1992). A comparison of variogram estimation with covariogram estimation. In *The Art of Statistical Science*, K. V. Mardia, ed. Wiley, Chichester, 191–208.

Cressie, N. and Guo, R. (1987). Mapping variables. In *Proceedings of the National Computer Graphics Association Conference, Computer Graphics '87, Vol.* **III**. National Computer Graphics Association, McLean, VA, 521–530.

Cressie, N. and Hawkins, D. M. (1980). Robust estimation of the variogram, I. *Journal of the International Association for Mathematical Geology*, **12**, 115–125.

Cressie, N. and Horton, R. (1987). A robust/resistant spatial analysis of soil-water infiltration. *Water Resources Research*, **23**, 911–917.

Cressie, N. and Hulting, F. (1992). A spatial statistical analysis of tumor growth. *Journal of the American Statistical Association*, **87**, 272–283.

Cressie, N. and Keightley, D. D. (1981). Analyzing data from hormone-receptor assays. *Biometrics*, **37**, 235–249.

Cressie, N. and Laslett, G. M. (1986). In search of generalized covariances for kriging. Unpublished manuscript, Department of Statistics, Iowa State University, Ames, IA.

Cressie, N. and Laslett, G. M. (1987). Random set theory and problems of modeling. *SIAM Review*, **29**, 557–574.

Cressie, N. and Lele, S. (1992). New models for Markov random fields. *Journal of Applied Probability*, **29**, 877–884.

Cressie, N. and Read, T. R. C. (1985). Do sudden infant deaths come in clusters? *Statistics and Decisions, Supplement Issue No. 2*, **3**, 333–349.

Cressie, N. and Read, T. R. C. (1989). Spatial data analysis of regional counts. *Biometrical Journal*, **31**, 699–719.

Cressie, N. and Zimmerman, D. L. (1992). On the stability of the geostatistical method. *Mathematical Geology*, **24**, 45–59.

Creutin, J. D. and Obled, C. (1982). Objective analysis and mapping techniques for rainfall fields: An objective comparison. *Water Resources Research*, **18**, 413–431.

Critchley, F., Ford, I., and Rijal, O. (1988). Interval estimation based on the profile likelihood: Strong Lagrangian theory with applications to discrimination. *Biometrika*, **75**, 21–28.

Cross, G. R. and Jain, A. K. (1983). Markov random field texture models. *IEEE Transactions on Pattern Analysis and Machine Intelligence*, **PAMI-5**, 25–39.

Cullis, B. R. and Gleeson, A. C. (1991). Spatial analysis of field experiments—an extension to two dimensions. *Biometrics*, **47**, 1449–1460.

Cushman, J. H. (1986). Multiphase transport in the space of stochastic tempered distributions. *IMA Journal of Applied Mathematics*, **36**, 159–175.

Cutler, D. and Dawson, D. L. (1990). Nearest-neighbor analysis for a family of fractal distributions. *Annals of Probability*, **18**, 256–271.

Dacey, M. F. (1966). A probability model for central place locations. *Annals of the Association of American Geographers*, **56**, 550–568.

Dagbert, M., David, M., Crozel, D., and Desbarats, A. (1984). Computing variograms in folded strata-controlled deposits. In *Geostatistics for Natural Resources Characterization, Part 1,* G. Verly, M. David, A. G. Journel, and A. Marechal, eds. Reidel, Dordrecht, 71–89.

Dahlhaus, R. (1988). Small sample effects in time series analysis: A new asymptotic theory and a new estimate. *Annals of Statistics*, **16**, 808–841.

Dahlhaus, R. and Kunsch, H. R. (1987). Edge effects and efficient parameter estimation for stationary random fields. *Biometrika*, **74**, 877–882.

Dalenius, T., Hajek, J., and Zubrzycki, S. (1960). On plane sampling and related geometrical problems. *Proceedings of the Fourth Berkeley Symposium on Mathematical Statistics and Probability*, *Vol.* **I**. University of California Press, Berkeley, CA, 125–150.

Daley, D. J. (1974). Various concepts of orderliness for point processes. In *Stochastic Geometry*, E. F. Harding and D. G. Kendall, eds. Wiley, New York, 307–321.

Daley, D. J. and Vere-Jones, D. (1972). A summary of the theory of point processes. In *Stochastic Point Processes: Statistical Analysis, Theory, and Applications*, P. A. W. Lewis, ed. Wiley, New York, 299–383.

Daley, D. J. and Vere-Jones, D. (1988). *Introduction to the Theory of Point Processes*. Springer, New York.

David, F. N. and Moore, P. G. (1954). Notes on contagious distributions in plant populations. *Annals of Botany*, **18**, 47–53.

David, M. (1977). *Geostatistical Ore Reserve Estimation*. Elsevier, Amsterdam.

David, M. (1988). *Handbook of Applied Advanced Geostatistical Ore Reserve Estimation*. Elsevier, Amsterdam.

Davis, B. M. (1987). Uses and abuses of cross-validation in geostatistics. *Mathematical Geology*, **19**, 241–248.

Davis, B. M. and Borgman, L. E. (1979). Some exact sampling distributions for variogram estimators. *Journal of the International Association for Mathematical Geology*, **11**, 643–653.

Davis, B. M. and Borgman, L. E. (1982). A note on the asymptotic distribution of the sample variogram. *Journal of the International Association for Mathematical Geology*, **14**, 189–193.

Davis, J. C. (1986). *Statistics and Data Analysis in Geology*, 2nd ed. Wiley, New York.

Davis, M. W. (1987a). Production of conditional simulations via the LU triangular decomposition of the covariance matrix. *Mathematical Geology*, **19**, 91–98.

Davis, M. W. (1987b). Generating large stochastic simulations—the matrix polynomial approximation method. *Mathematical Geology*, **19**, 99–107.

Davis, R. A., Mulrow, E., and Resnick, S. I. (1988). Almost sure limit sets of random samples in $\mathbb{R}^d$. *Advances in Applied Probability*, **20**, 573–599.

Davy, P. J. (1978). Aspects of random set theory. *Advances in Applied Probability, Supplement*, **10**, 28–35.

Davy, P. J. and Miles, R. E. (1977). Sampling theory for opaque spatial specimens. *Journal of the Royal Statistical Society B*, **39**, 56–65.

Dawid, A. P. (1984). Statistical theory. The prequential approach. *Journal of the Royal Statistical Society A*, **147**, 278–290.

DeGroot, M. H. and Eddy, W. F. (1983). Set-valued parameters and set-valued statistics. In *Recent Advances in Statistics*, M. H. Rizvi, J. S. Rustagi, and D. Siegmund, eds. Academic Press, New York, 175–195.

Deguchi, K. (1986). Two-dimensional auto-regressive model for analysis and synthesis of gray-level textures. In *Science on Form: Proceedings of the First International Symposium for Science on Form*, Y. Kato, R. Takaki, and J. Toriwaki, eds. KTK Scientific Publishers, Tokyo, 441–449.

Delfiner, P. (1976). Linear estimation of nonstationary spatial phenomena. In *Advanced Geostatistics in the Mining Industry*, M. Guarascio, M. David, and C. Huijbregts, eds. Riedel, Dordrecht, 49–68.

Delfiner, P., Renard, D., and Chiles, J. P. (1978). *BLUEPACK-3D Manual*. Centre de Geostatistique, Ecole des Mines, Fontainebleau, France.

Delhomme, J. P. (1978). Kriging in the hydrosciences. *Advances in Water Resources*, **1**, 251–266.

Dempster, A. P., Laird, N. M., and Rubin, D. B. (1977). Maximum likelihood for incomplete data via the EM algorithm. *Journal of the Royal Statistical Society B*, **39**, 1–22.

Dent, W. and Min, A. (1978). A Monte Carlo study of autoregressive moving average processes. *Journal of Econometrics*, **7**, 23–55.

Derin, H. and Elliott, H. (1987). Modeling and segmentation of noisy and textured images using Gibbs random fields. *IEEE Transactions on Pattern Analysis and Machine Intelligence*, **PAMI-9**, 39–55.

Deutsch, C. (1989). DECLUS: A Fortran 77 program for determining optimal spatial declustering weights. *Computers and Geosciences*, **15**, 325–332.

Devijver, P. A. (1988). Image segmentation using causal Markov random field models. In *Pattern Recognition*, J. Kittler, ed. *Lecture Notes in Computer Science*, **No. 301**, 131–143.

Devlin, S. J., Gnanadesikan, R., and Kettenring, J. R. (1975). Robust estimation of dispersion matrices and principal components. *Journal of the American Statistical Association*, **76**, 354–362.

Diaconis, P. and Efron, B. (1983). Computer-intensive methods in statistics. *Scientific American*, **248**, 116–130.

Diamond, P. and Armstrong, M. (1984). Robustness of variograms and conditioning of kriging matrices. *Journal of the International Association for Mathematical Geology*, **16**, 809–822.

Dielman, T. E. and Pfaffenberger, R. C. (1989). Small sample properties of estimators in the autocorrelated error model: A review and some additional simulations. *Statistische Hefte*, **30**, 163–183.

Diggle, P. J. (1975). Robust density estimation using distance methods. *Biometrika*, **62**, 39–48.

Diggle, P. J. (1977). The detection of random heterogeneity in plant populations. *Biometrics*, **33**, 390–394.

Diggle, P. J. (1978). On parameter estimation for spatial point processes. *Journal of the Royal Statistical Society B*, **40**, 178–181.

Diggle, P. J. (1979). On parameter estimation and goodness-of-fit testing for spatial point processes. *Biometrics*, **35**, 87–101.

Diggle, P. J. (1981a). Some graphical methods in the analysis of spatial point patterns. In *Interpreting Multivariate Data*, V. Barnett, ed. Wiley, New York, 55–73.

Diggle, P. J. (1981b). Binary mosaics and the spatial pattern of heather. *Biometrics*, **37**, 531–539.

Diggle, P. J. (1983). *Statistical Analysis of Spatial Point Patterns*. Academic Press, New York.

Diggle, P. J. (1985). A kernel method for smoothing point process data. *Applied Statistics*, **34**, 138–147.

Diggle, P. J., Besag, J. E., and Gleaves, J. T. (1976). Statistical analysis of spatial point patterns by means of distance methods. *Biometrics*, **32**, 659–667.

Diggle, P. J. and Cox, T. F. (1981). On sparse sampling methods and tests of independence for multivariate spatial point patterns. *Bulletin of the International Statistical Institute*, **49**, Book 1, 213–229.

Diggle, P. J., Gates, D. J., and Stibbard, A. (1987). A nonparametric estimator for pairwise-interaction point processes. *Biometrika*, **74**, 763–770.

Diggle, P. J. and Gratton, R. J. (1984). Monte Carlo methods of inference for implicit statistical models. *Journal of the Royal Statistical Society B*, **46**, 193–212.

Diggle, P. J. and Hutchinson, M. F. (1989). On spline smoothing with autocorrelated errors. *Australian Journal of Statistics*, **31**, 166–182.

Diggle, P. J. and Marron, J. S. (1988). Equivalence of smoothing parameter selectors in density and intensity estimation. *Journal of the American Statistical Association*, **83**, 793–800.

Diggle, P. J. and Milne, R. K. (1983). Bivariate Cox processes: Some models for bivariate spatial point patterns. *Journal of the Royal Statistical Society B*, **45**, 11–21.

Dixon, E. C. (1989). Modeling under uncertainty: Comparing three acid-rain models. *Journal of the Operational Research Society*, **40**, 29–40.

Dobrushin, R. L. (1968). The description of a random field by means of conditional probabilities and conditions of its regularity. *Theory of Probability and Its Applications*, **13**, 197–224.

Dobrushin, R. L. (1988). A new approach to the analysis of Gibbs perturbations of Gaussian fields. *Selecta Mathematica Sovietica*, **7**, 221–277.

Doguwa, S. I. (1989a). A comparative study of the edge-corrected kernel-based nearest neighbour density estimators for point processes. *Journal of Statistical Computation and Simulation*, **33**, 83–100.

Doguwa, S. I. (1989b). On second order neighbourhood analysis of mapped point patterns. *Biometrical Journal*, **31**, 451–459.

Doguwa, S. I. and Upton, G. J. G. (1988). On edge corrections for the point-event analogue of the Clark–Evans statistic. *Biometrical Journal*, **30**, 957–963.

Donnelly, K. (1978). Simulation to determine the variance and edge-effect of total nearest neighbour distance. In *Simulation Methods in Archeology*, I. R. Hodder, ed. Cambridge University Press, Cambridge, 91–95.

Doob, J. L. (1953). *Stochastic Processes*. Wiley, New York.

Doreian, P. (1980). Linear models with spatially distributed data. *Sociological Methods and Research*, **9**, 29–60.

Douglas, J. B. (1975). Clustering and aggregation. *Sankhyā B*, **37**, 398–417.

Dow, M. M., Burton, M. L., and White, D. R. (1982). Network autocorrelation: A simulation study of a foundational problem in regression and survey research. *Social Networks*, **4**, 169–200.

Dowd, P. A. (1982). Lognormal kriging—The general case. *Journal of the International Association for Mathematical Geology*, **14**, 475–499.

Draper, N. R. and Faraggi, D. (1985). Role of the Papadakis estimator in one- and two-dimensional field trials. *Biometrika*, **72**, 223–226.

Draper, N. and Smith, H. (1981). *Applied Regression Analysis*, 2nd ed. Wiley, New York.

Dubes, R. C. and Jain, A. K. (1976). Clustering techniques: The user's dilemma. *Pattern Recognition*, **8**, 247–260.

Dubin, R. A. (1988). Estimation of regression coefficients in the presence of spatially autocorrelated error terms. *Review of Economics and Statistics*, **70**, 466–474.

Dubrule, O. (1989). A review of stochastic models for petroleum reservoirs. In *Geostatistics, Vol.* **2**, M. Armstrong, ed. Kluwer, Dordrecht, 493–506.

Dubrule, O. and Kostov, C. (1986). An interpolation method taking into account inequality constraints: I. Methodology. *Mathematical Geology*, **18**, 33–51.

Dubuc, B., Zucker, S. W., Tricot, C., Quiniou, J. F., and Wehbi, D. (1989). Evaluating the fractal dimension of surfaces. *Proceedings of the Royal Society of London A*, **425**, 113–127.

Duby, C., Guyon, X., and Prum, B. (1977). The precision of different experimental designs for a random field. *Biometrika*, **64**, 59–66.

Duchon, J. (1977). Splines minimizing rotation-invariant semi-norms in Sobolev spaces. In *Constructive Theory of Functions of Several Variables*, W. Schempp and K. Zeller, eds. *Lecture Notes in Mathematics*, **No. 571**. Springer, Berlin, 85–100.

Duchting, W. (1980). Simulation of cell systems growing in competition. In *Simulation of Systems '79*, L. Dekker, G. Savastano, and G. C. Vansteenkiste, eds. North-Holland, Amsterdam, 149–159.

Duchting, W. and Dehl, G. (1980). Spatial structure of tumor growth: A simulation study. *IEEE Transactions on Systems, Man, and Cybernetics*, **SMC-10**, 292–296.

Duchting, W. and Vogelsaenger, T. (1981). Three-dimensional pattern generation applied to spheroidal tumor growth in a nutrient medium. *International Journal of Bio-Medical Computing*, **12**, 377–392.

Duchting, W. and Vogelsaenger, T. (1983). Aspects of modelling and simulating tumor growth and treatment. *Journal of Cancer Research and Clinical Oncology*, **105**, 1–12.

Dudgeon, D. E. and Mersereau, R. M. (1984). *Multidimensional Digital Signal Processing*. Prentice-Hall, Englewood Cliffs, NJ.

Dunn, R. (1988). Framed rectangle charts or statistical maps with shading. An experiment with graphical perception. *American Statistician*, **42**, 123–129.

Dunstan, S. P. and Mill, A. J. B. (1989). Spatial indexing of geological models using linear octrees. *Computers and Geosciences*, **15**, 1291–1301.

Dupac, V. (1980). Parameter estimation in the Poisson field of discs. *Biometrika*, **67**, 187–190.

Durbin, J. and Watson, G. S. (1950). Testing for serial correlation in least squares regression, I. *Biometrika*, **37**, 409–428.

Durbin, J. and Watson, G. S. (1951). Testing for serial correlation in least squares regression, II. *Biometrika*, **38**, 159–178.

Durlauf, S. N. and Phillips, P. C. B. (1988). Trends versus random walks in time series analysis. *Econometrica*, **56**, 1333–1354.

Durrett, R. (1981). An introduction to infinite particle systems. *Stochastic Processes and Their Applications*, **11**, 109–150.

Durrett, R. (1988). *Lecture Notes on Particle Systems and Percolation*. Wadsworth, Pacific Grove, CA.

Durrett, R. and Liggett, T. M. (1981). The shape of the limit set in Richardson's growth model. *Annals of Probability*, **9**, 186–193.

Dzhaparidze, K. O. (1974). On simplified estimators of unknown parameters with good asymptotic properties. *Theory of Probability and Its Applications*, **19**, 347–358.

Eagleson, P. S. (1984). The distribution of catchment coverage by stationary rainstorms. *Water Resources Research*, **20**, 581–590.

Eaton, M. L. (1985). The Gauss–Markov Theorem in multivariate analysis. In *Multivariate Analysis* **VI**, P. R. Krishnaiah, ed. Elsevier, Amsterdam, 177–201.

Eberhardt, L. L. (1967). Some developments in 'distance sampling'. *Biometrics*, **23**, 207–216.

Eddy, W. F. (1982). Laws of large numbers for intersection and union of random closed sets. Technical Report No. 227, Department of Statistics, Carnegie-Mellon University, Pittsburgh, PA.

Eddy, W. F. and Gale, J. D. (1981). The convex hull of a spherically symmetric sample. *Advances in Applied Probability*, **13**, 751–763.

Eden, M. (1961). A two dimensional growth process. In *Proceedings of the Fourth Berkeley Symposium on Mathematical Statistics and Probability*, **IV**, J. Neyman, ed. University of California Press, Berkeley, CA, 223–239.

Ederer, F., Myers, M. H., and Mantel, N. (1966). A statistical problem in space and time: Do leukemia cases come in clusters? *Biometrics*, **20**, 626–638.

Efron, B. (1965). The convex hull of a random set of points. *Biometrika*, **52**, 331–343.

Efron, B. (1979). Bootstrap methods: Another look at the jackknife. *Annals of Statistics*, **7**, 1–26.

Efron, B. (1982). *The Jackknife, the Bootstrap and Other Resampling Plans*. Society for Industrial and Applied Mathematics, Philadelphia, PA.

Efron, B. and Hinkley, D. V. (1978). Assessing the accuracy of the maximum likelihood estimator: Observed versus expected Fisher information. *Biometrika*, **65**, 457–482.

Efron, B. and Tibshirani, R. (1986). Bootstrap methods for standard errors, confidence intervals, and other measures of statistical accuracy. *Statistical Science*, **1**, 54–75.

Egbert, G. D. and Lettenmaier, D. P. (1986). Stochastic modeling of the space–time structure of atmospheric chemical decomposition. *Water Resources Research*, **22**, 165–179.

Emerson, J. D. and Hoaglin, D. C. (1983). Analysis of two-way tables by medians. In *Understanding Robust and Exploratory Data Analysis*, D. C. Hoaglin, F. Mosteller, and J. W. Tukey, eds. Wiley, New York, 166–210.

Epanechnikov, V. A. (1969). Non-parametric estimation of multivariate probability density. *Theory of Probability and Its Applications*, **14**, 153–158.

Errington, J. C. (1973). The effect of regular and random distributions on the analysis of pattern. *Journal of Ecology*, **61**, 99–105.

Esbensen, K. and Geladi, P. (1989). Strategy of multivariate image analysis (MIA). *Chemometrics and Intelligent Laboratory Systems*, **7**, 67–86.

Esogbue, A. O. and Elder, R. C. (1980). Fuzzy sets and the modelling of physician decision processes, Part II: Fuzzy diagnosis decision models. *Fuzzy Sets and Systems*, **3**, 1–9.

European Inland Fisheries Advisory Commission (EIFAC) (1969). Water quality criteria for European freshwater fish—extreme pH values and inland fisheries. *Water Research*, **3**, 593–611.

Eynon, B. P. and Switzer, P. (1983). The variability of rainfall acidity. *Canadian Journal of Statistics*, **11**, 11–24.

Fahrmeir, L. (1987). Asymptotic likelihood inference for nonhomogeneous observations. *Statistische Hefte*, **28**, 81–116.

Fairfield Smith, H. (1938). An empirical law describing heterogeneity in the yields of agricultural crops. *Journal of Agricultural Science (Cambridge)*, **28**, 1–23.

Falconer, K. J. (1986). Random fractals. *Mathematical Proceedings of the Cambridge Philosophical Society*, **100**, 559–582.

Federer, W. T. and Schlottfeldt, C. S. (1954). The use of covariance to control gradients in experiments. *Biometrics*, **10**, 282–290.

Fedorov, V. V. (1972). *Theory of Optimal Experiments*. Academic Press, New York.

Fellner, W. H. (1986). Robust estimation of variance components. *Technometrics*, **28**, 51–60.

Felsenstein, J. (1975). A pain in the torus: Some difficulties with models of isolation by distance. *American Naturalist*, **109**, 359–368.

Ferguson, T. (1967). *Mathematical Statistics: A Decision Theoretic Approach*. Academic Press, New York.

Feron, R. (1976). Random fuzzy sets. *Comptes Rendus de l'Academie des Sciences A*, **282**, 903–906.

Feynman, R. P. (1972). *Statistical Mechanics: A Set of Lectures*. Benjamin, Reading, MA.

Fienberg, S. E. and Tanur, J. M. (1987). Experimental and sampling structures: Parallels diverging and meeting. *International Statistical Review*, **55**, 75–96.

Fiksel, T. (1984a). Estimation of parameterized pair potentials of marked and non-marked Gibbsian point processes. *Elektronische Informationsverarbeitung und Kybernetik*, **20**, 270–278.

Fiksel, T. (1984b). Simple spatial–temporal models for sequences of geological events. *Elektronische Informationsverarbeitung und Kybernetik*, **20**, 480–487.

Fiksel, T. (1988a). Edge-corrected density estimators for point processes. *Statistics*, **19**, 67–75.

Fiksel, T. (1988b). Estimation of interaction potentials of Gibbsian point processes. *Statistics*, **19**, 77–86.

Fink, A. M. (1988). How to polish off median polish. *SIAM Journal of Scientific and Statistical Computing*, **9**, 932–940.

Finney, D. J. (1971). *Probit Analysis*, 3rd ed. Cambridge University Press, Cambridge.

Fisher, L. (1971). Limiting convex hulls of samples: Theory and function space examples. *Zeitschrift für Wahrscheinlichkeitstheorie und verwandte Gebiete*, **18**, 281–297.

Fisher, R. A. (1915). Frequency distribution of the values of the correlation coefficient in samples from an indefinitely large population. *Biometrika*, **10**, 507–529.

Fisher, R. A. (1925). *Statistical Methods for Research Workers*. Oliver and Boyd, Edinburgh.

Fisher, R. A. (1935). *The Design of Experiments*. Oliver and Boyd, Edinburgh.

Fisher, R. A., Thornton, H. G., and Mackenzie, W. A. (1922). The accuracy of the plating method of estimating the density of bacterial populations. *Annals of Applied Biology*, **9**, 325–359.

Fletcher, R. and Powell, M. J. D. (1963). A rapidly convergent descent method for minimization. *Computer Journal*, **6**, 163–168.

Flinn, P. A. (1974). Monte Carlo calculation of phase separation in a two-dimensional Ising system. *Journal of Statistical Physics*, **10**, 89–97.

Fogerty, A. C., Ford, G. L., Willcox, M. E., and Clancy, S. L. (1984). Liver fatty acids and the sudden infant death syndrome. *American Journal of Clinical Nutrition*, **39**, 201–208.

Follmer, H. (1982). A covariance estimate for Gibbs measures. *Journal of Functional Analysis*, **46**, 387–395.

Forbes, W. F. and Gibberd, R. W. (1984). Mathematical models of carcinogenesis: A review. *Mathematical Scientist*, **9**, 95–110.

Ford, L. R. and Fulkerson, D. R. (1962). *Flows in Networks*. Princeton University Press, Princeton, NJ.

Fordon, W. A. and Bezdek, J. C. (1979). The application of fuzzy set theory to medical diagnosis. In *Advances in Fuzzy Set Theory and Applications*, M. M. Gupta et al., eds. North-Holland, Amsterdam, 445–461.

Forster, F. (1972). Use of a demographic base map for the presentation of areal data in epidemiology. In *Medical Geography*, N. D. McGlashan, ed. Methuen, London, 59–67.

Fox, A. J. (1972). Outliers in time series. *Journal of the Royal Statistical Society B*, **34**, 350–363.

Franke, J. (1984). On the robust prediction and interpolation of time series in the presence of correlated noise. *Journal of Time Series Analysis*, **5**, 227–244.

Franke, R. (1982). Scattered data interpolation: Tests of some methods. *Mathematics of Computation*, **38**, 181–200.

Freed, D. S. and Shepp, L. A. (1982). A Poisson process whose rate is a hidden Markov process. *Advances in Applied Probability*, **14**, 21–36.

Freedman, D. A. and Peters, S. C. (1984). Bootstrapping a regression equation: Some empirical results. *Journal of the American Statistical Association*, **79**, 97–106.

Freeman, M. F. and Tukey, J. W. (1950). Transformations related to the angular and the square root. *Annals of Mathematical Statistics*, **21**, 607–611.

Friedman, J. H. and Stuetzle, W. (1981). Projection pursuit regression. *Journal of the American Statistical Association*, **76**, 817–823.

Frontier, S. (1987). Applications of fractal theory to ecology. In *Developments in Numerical Ecology*, P. Legendre and L. Legendre, eds. Springer, New York, 335–378.

Fuller, W. A. (1976). *Introduction to Statistical Time Series*. Wiley, New York.

Fuller, W. A. and Hasza, D. P. (1981). Properties of predictors for autoregressive time series. *Journal of the American Statistical Association*, **76**, 155–161.

Galiano, E. F. (1983). Detection of multi-species patterns in plant communities. *Vegetatio*, **53**, 129–138.

Gambolati, G. and Galeati, G. (1987). Comment on "Analysis of nonintrinsic spatial variability by residual kriging with application to regional groundwater levels" by Shlomo P. Neuman and Elizabeth A. Jacobson. *Mathematical Geology*, **19**, 249–257.

Gandin, L. S. (1963). *Objective Analysis of Meteorological Fields*. Gidrometeorologicheskoe Izdatel'stvo (GIMIZ), Leningrad (translated by Israel Program for Scientific Translations, Jerusalem, 1965).

Gandolfi, A. (1989). Uniqueness of the infinite cluster for stationary Gibbs states. *Annals of Probability*, **17**, 1403–1415.

Gaposhkin, V. F. (1988). On a relationship between ergodicity of a continuous-time stationary process and a quantized process. *Theory of Probability and Its Applications*, **33**, 377–381.

Gardiner, C. W. (1983). *Handbook of Stochastic Methods*. Springer, Berlin.

Gardner, L. I., Brundage, J. F., Burke, D. S., McNeil, J. G., Visintine, R., and Miller, R. N. (1989). Spatial diffusion of the Human Immunodeficiency Virus infection epidemic in the United States, 1985–1987. *Annals of the Association of American Geographers*, **79**, 25–43.

Garside, M. J. (1971). Some computational procedures for the best subset problem. *Applied Statistics*, **20**, 8–15.

Gastwirth, J. L. and Rubin, H. (1975). The behavior of robust estimators on dependent data. *Annals of Statistics*, **3**, 1070–1100.

Gates, D. J. and Westcott, M. (1981). Negative skewness and negative correlations in spatial competition models. *Journal of Mathematical Biology*, **13**, 159–171.

Gates, D. J. and Westcott, M. (1986). Clustering estimates for spatial point distributions with unstable potentials. *Annals of the Institute of Statistical Mathematics A*, **38**, 123–135.

Gautschi, W. (1957). Some remarks on systematic sampling. *Annals of Mathematical Statistics*, **28**, 385–394.

Geary, R. C. (1954). The contiguity ratio and statistical mapping. *The Incorporated Statistician*, **5**, 115–145.

Geisser, S. (1975). The predictive sample reuse method with applications. *Journal of the American Statistical Association*, **70**, 320–328.

Gel'fand, I. M. and Vilenkin, N. Y. (1964). *Generalized Functions: Applications of Harmonic Analysis*, *Vol.* **4**. Academic Press, New York.

Geman, D. and Geman, S. (1986). Bayesian image analysis. In *Disordered Systems and Biological Organization*, E. Bienenstock, F. F. Soulie, and G. Weisbuch, eds. *NATO ASI Series*, **F20**. Springer, New York, 301–319.

Geman, D., Geman, S., Graffigne, C., and Dong, P. (1990). Boundary detection by constrained optimization. *IEEE Transactions on Pattern Analysis and Machine Intelligence*, **12**, 609–628.

Geman, S. and Geman, D. (1984). Stochastic relaxation, Gibbs distributions and the Bayesian restoration of images. *IEEE Transactions on Pattern Analysis and Machine Intelligence*, **PAMI-6**, 721–741.

Geman, S. and Graffigne, C. (1987). Markov random field image models and their applications to computer vision. In *Proceedings of the International Congress of Mathematicians (1986, Berkeley, CA), Vol.* **II**, A. M. Gleason, ed. American Mathematical Society, Providence, RI, 1496–1517.

Geman, S. and McClure, D. E. (1985). Bayesian image analysis: An application to single photon emission tomography. *Proceedings of the Statistical Computing Section, American Statistical Association*. American Statistical Association, Washington, DC, 12–18.

Geman, S. and McClure, D. E. (1987). Statistical methods for tomographic image reconstruction. *Bulletin of the International Statistical Institute*, **52**, Book 4, 5–21.

Georgiev, A. A. (1984). Speed of convergence in nonparametric kernel estimation of a regression function and its derivatives. *Annals of the Institute of Statistical Mathematics*, **36**, 455–462.

Getis, A. and Boots, B. (1978). *Models of Spatial Processes*. Cambridge University Press, Cambridge.

Getis, A. and Franklin, J. (1987). Second-order neighborhood analysis of mapped point patterns. *Ecology*, **68**, 473–477.

Giardina, C. R. and Dougherty, E. R. (1988). *Morphological Methods in Image and Signal Processing*. Prentice-Hall, Englewood Cliffs, NJ.

Gibbs, J. W. (1902). *Elementary Principles in Statistical Mechanics, Developed with Especial Reference to the Rational Foundation of Thermodynamics*. Charles Scribner's Sons, New York.

Gidas, B. (1985). Nonstationary Markov chains and convergence of the annealing algorithm. *Journal of Statistical Physics*, **39**, 73–131.

Gidas, B. (1988). Consistency of maximum likelihood and pseudo-likelihood estimators for Gibbs distributions. In *Stochastic Differential Systems, Stochastic Control Theory and Applications*, W. Fleming and P. L. Lions, eds. *IMA Volumes in Mathematics and its Applications*, **Vol. 10**. Springer, New York, 129–145.

Gilbert, R. O. (1987). *Statistical Methods for Environmental Pollution Monitoring*. Van Nostrand Reinhold, New York.

Gill, P. S. and Shukla, G. K. (1985a). Efficiency of nearest neighbour balanced block designs for correlated observations. *Biometrika*, **72**, 539–544.

Gill, P. S. and Shukla, G. K. (1985b). Experimental designs and their efficiencies for spatially correlated observations in two dimensions. *Communications in Statistics. Theory and Methods*, **14**, 2181–2197.

Giné, E. and Hahn, M. G. (1985). The Lévy–Hincin representation for random compact convex subsets which are infinitely divisible under Minkowski addition. *Zeitschrift für Wahrscheinlichkeitstheorie und verwandte Gebiete*, **70**, 271–287.

Giné, E., Hahn, M. G., and Zinn, J. (1983). Limit theorem for random sets: An application of probability in Banach space results. In *Probability in Banach Spaces*, **IV**, A. Beck and K. Jacobs, eds. *Lecture Notes in Mathematics*, **No. 990**. Springer, New York, 112–135.

Gish, T. J. and Starr, J. L. (1983). Temporal variability of infiltration under field conditions. In *Advances in Infiltration, Proceedings of the National Conference on Advances in Infiltration*. American Society of Agricultural Engineers, St. Joseph, MI, 122–131.

Giulian, G. G., Gilbert, E. F., and Moss, R. L. (1987). Elevated fetal hemoglobin levels in sudden infant death syndrome. *New England Journal of Medicine*, **316**, 1122–1126.

Glass, L. and Tobler, W. R. (1971). Uniform distribution of objects in a homogeneous field: Cities on a plain. *Nature*, **233**, 67–68.

Gleason, H. A. (1920). Some applications of the quadrat method. *Bulletin of the Torrey Botanical Club*, **47**, 21–33.

Gleeson, A. C. and Cullis, B. R. (1987). Residual maximum likelihood (REML) estimation of a neighbour model for field experiments. *Biometrics*, **43**, 277–288.

Gleeson, A. C. and McGilchrist, C. A. (1980). Bilateral processes on a rectangular lattice. *Australian Journal of Statistics*, **22**, 197–206.

Glotzel, E. (1981). Time reversible and Gibbsian point processes, I: Markovian spatial birth and death processes on a general phase space. *Mathematische Nachrichten*, **102**, 217–222.

Godambe, V. P. (1985). The foundations of finite sample estimation in stochastic processes. *Biometrika*, **72**, 419–428.

Gold, C. M., Charters, T. D., and Ramsden, J. (1977). Automated contour mapping using triangular element data structures and an interpolant over each irregular triangular domain. *Computer Graphics*, **11**, 170–175.

Goldberg, D. E. (1989). *Genetic Algorithms in Search, Optimization, and Machine Learning*. Addison-Wesley, Reading, MA.

Goldberg, J. and Stein, R. (1978). Seasonal variation in SIDS. *Lancet*, **8080**, 107.

Goldberger, A. S. (1962). Best linear unbiased prediction in the generalized linear regression model. *Journal of the American Statistical Association*, **57**, 369–375.

Goldie, C. M. and Morrow, G. J. (1986). Central limit questions for random fields. In *Dependence in Probability and Statistics*, E. Eberlein and M. S. Taqqu, eds. Birkhauser, Boston, 275–289.

Golub, G. H. and Van Loan, C. F. (1983). *Matrix Computations*. The Johns Hopkins University Press, Baltimore, MD.

Gomez, M. and Hazen, K. (1970). Evaluating sulfur and ash distribution in coal seams by statistical response surface regression analysis. U.S. Bureau of Mines Report RI 7377.

Gong, G. and Samaniego, F. J. (1981). Pseudo maximum likelihood estimation: Theory and applications. *Annals of Statistics*, **9**, 861–869.

Goodall, C. R. (1985). Statistical and data digitization techniques applied to an analysis of leaf initiation in plants. In *Computer Science and Statistics: The Interface*, L. Billard, ed. Elsevier, Amsterdam, 253–264.

Goodall, D. W. (1952). Quantitative aspects of plant distribution. *Biological Review*, **27**, 194–245.

Goodall, D. W. (1963). Pattern analysis and minimal area—some further comments. *Journal of Ecology*, **51**, 705–710.

Goodall, D. W. (1965). Plot-less tests of interspecific association. *Journal of Ecology*, **53**, 197–210.

Goodall, D. W. (1970). Statistical plant ecology. *Annual Review of Ecology and Systematics*, **1**, 99–124.

Goodall, D. W. (1974). A new method for the analysis of spatial pattern by random pairing of quadrats. *Vegetatio*, **29**, 135–146.

Goodchild, M. F. and Mark, D. M. (1987). The fractal nature of geographic phenomena. *Annals of the Association of American Geographers*, **77**, 265–278.

Gotway, C. A. (1991). Fitting semivariogram models by weighted least squares. *Computers and Geosciences*, **17**, 171–172.

Gotway, C. A. and Cressie, N. (1990). A spatial analysis of variance applied to soil-water infiltration. *Water Resources Research,* **26,** 2695–2703.

Gotway, C. A. and Cressie, N. (1992). Improved multivariate prediction under a general linear model. *Journal of Multivariate Analysis,* forthcoming.

Gower, J. C. (1988). Statistics and agriculture. *Journal of the Royal Statistical Society A*, **151**, 179–200.

Grandell, J. (1976). Doubly Stochastic Poisson Processes. *Lecture Notes in Mathematics*, **No. 529**. Springer, New York.

Grandell, J. (1977). Point processes and random measures. *Advances in Applied Probability*, **9**, 502–526.

Granger, C. W. J. and Joyeaux, R. (1980). An introduction to long memory time series models and fractional differencing. *Journal of Time Series Analysis*, **1**, 15–30.

Granger, C. W. J. and Newbold, P. (1974). Spurious regressions in econometrics. *Journal of Econometrics*, **2**, 111–120.

Green, P. B. (1980). Organogenesis—a biophysical view. *Annual Review of Plant Physiology*, **31**, 51–82.

Green, P. J. (1985). Linear models for field trials, smoothing and cross-validation. *Biometrika*, **72**, 527–537.

Green, P. J., Jennison, C., and Seheult, A. H. (1985). Analysis of field experiments by least squares smoothing. *Journal of the Royal Statistical Society B*, **47**, 299–315.

Green, P. J. and Titterington, D. M. (1987). Recursive methods in image processing. *Bulletin of the International Statistical Institute*, **52**, Book 4, 51–67.

Greig, D. M., Porteous, B. T., and Seheult, A. H. (1986). Comment on "On the Statistical analysis of dirty pictures" by J. E. Besag. *Journal of the Royal Statistical Society B*, **48**, 282–284.

Greig, D. M., Porteous, B. T., and Seheult, A. H. (1989). Exact maximum *a posteriori* estimation for binary images. *Journal of the Royal Statistical Society B*, **51**, 271–279.

Greig-Smith, P. (1952). The use of random and contiguous quadrats in the study of the structure of plant communities. *Annals of Botany*, **16**, 293–316.

Greig-Smith, P. (1964). *Quantitative Plant Ecology*, 2nd ed. Butterworths, London.

Grenander, U. (1954). On the estimation of regression coefficients in the case of an autocorrelated disturbance. *Annals of Mathematical Statistics*, **25**, 252–272.

Grenander, U. (1981). *Abstract Inference*. Wiley, New York.

Grenander, U. (1989). Advances in pattern theory. *Annals of Statistics*, **17**, 1–30.

Grenander, U. and Rosenblatt, M. (1984). *Statistical Analysis of Stationary Time Series*, 2nd ed. Chelsea Publishing Co., New York.

Griffeath, D. (1981). The basic contact process. *Stochastic Processes and Their Applications*, **11**, 151–185.

Griffith, D. A. (1979). Urban dominance, spatial structure and spatial dynamics: Some theoretical conjectures and empirical implications. *Economic Geography*, **55**, 95–113.

Griffith, D. A. (1983). The boundary value problem in spatial statistical analysis. *Journal of Regional Science*, **23**, 377–387.

Griffith, D. A. and Amrhein, C. G. (1983). An evaluation of correction techniques for boundary effects in spatial statistical analysis: Traditional method. *Geographical Analysis*, **15**, 352–360.

Griffiths, R. C. and Milne, R. K. (1978). A class of bivariate Poisson processes. *Journal of Multivariate Analysis*, **8**, 380–395.

Grimmett, G. (1987). Interacting particle systems and random media: An overview. *International Statistical Review*, **55**, 49–62.

Grimmett, G. (1989). *Percolation*. Springer, New York.

Grondona, M. O. (1989). Estimation and design with correlated observations. Ph.D. Dissertation, Department of Statistics, Iowa State University, Ames, IA, unpublished.

Grondona, M. O. and Cressie, N. (1991). Using spatial considerations in the analysis of experiments. *Technometrics*, **33**, 381–392.

Grondona, M. O. and Cressie, N. (1992). Efficiency of block designs under stationary second-order autoregressive errors. *Sankhyā A*, forthcoming.

Grubb, D. and Magee, L. (1988). A variance comparison of OLS and feasible GLS estimators. *Econometric Theory*, **4**, 329–335.

Gull, S. F. and Daniell, G. J. (1978). Image reconstruction from incomplete and noisy data. *Nature*, **272**, 686–690.

Gull, S. F. and Skilling, J. (1984). The maximum entropy method. In *Indirect Imaging*, J. A. Roberts, ed. Cambridge University Press, Cambridge, 267–280.

Gundersen, H. J. G. (1978). Estimators of the number of objects per area unbiased by edge effect. *Journal of Microscopy*, **111**, 219–223.

Gundersen, H. J. G. (1986). Stereology of arbitrary particles. A review of unbiased number and size estimators and the presentation of some new ones, in memory of William R. Thompson. *Journal of Microscopy*, **143**, 3–45.

Gupta, V. K. and Waymire, E. (1979). A stochastic kinematic study of subsynoptic space–time rainfall. *Water Resources Research*, **15**, 637–644.

Gurland, J. (1957). Some interrelations among compound and generalized distributions. *Biometrika*, **63**, 357–360.

Gustaffson, N. (1981). A review of methods for objective analysis. In *Dynamic Meteorology: Data Assimilation Methods*, L. Bengtsson et al., eds. Springer, New York, 17–76.

Guyon, X. (1982). Parameter estimation for a stationary process on a d-dimensional lattice. *Biometrika*, **69**, 95–105.

Guyon, X. (1987). Estimation d'un champ par pseudo-vraisemblance conditionelle: Etude asymptotique et application au cas Markovien. In *Spatial Processes and Spatial Time Series Analysis (Proceedings of the Sixth Franco-Belgian Meeting of Statisticians, 1985)*, F. Droesbeke, ed. Publications des Facultés Universitaires Saint-Louis, Brussels, Belgium, 15–62.

Guyon, X. and Richardson, S. (1984). Vitesse de convergence du theoreme de la limite centrale pour les champs faiblement dependants. *Zeitschrift für Wahrscheinlichkeitstheorie und verwandte Gebiete*, **66**, 297–314.

Gy, P. M. (1982). *Sampling of Particulate Materials: Theory and Practice*. Elsevier, Amsterdam.

Hadwiger, H. (1957). *Vorlesungen uber Inhalt, Oberflache und Isoperimetrie*. Springer, Berlin.

Haining, R. P. (1978a). Specification and Estimation Problems in Models of Spatial Dependence. *Studies in Geography*, **No. 24**, Northwestern University, Evanston, IL.

Haining, R. P. (1978b). The moving average model for spatial interaction. *Transaction and Papers, Institute of British Geographers, New Series*, **3**, 202–225.

Haining, R. P. (1978c). A spatial model for high plains agriculture. *Annals of the Association of American Geographers*, **68**, 493–504.

Haining, R. P. (1979). Statistical tests and process generators for random field models. *Geographical Analysis*, **11**, 45–64.

Haining, R. P. (1987). Trend-surface models with regional and local scales of variation with an application to aerial survey data. *Technometrics*, **29**, 461–469.

Haining, R. P. (1988). Estimating spatial means with an application to remotely sensed data. *Communications in Statistics. Theory and Methods*, **17**, 573–597.

Haining, R. P., Griffith, D. A., and Bennett, R. J. (1989). Maximum likelihood estimation with missing spatial data and with an application to remotely sensed data. *Communications in Statistics. Theory and Methods*, **18**, 1875–1894.

Hajek, B. and Berger, T. (1987). A decomposition theorem for binary Markov random fields. *Annals of Probability*, **15**, 1112–1125.

Haji-Karim, M. and Carlsson, J. (1978). Proliferation and viability in cellular spheroids of human origin. *Cancer Research*, **38**, 1457–1464.

Hall, P. (1985a). Resampling a coverage pattern. *Stochastic Processes and Their Applications*, **20**, 231–246.

Hall, P. (1985b). Counting methods for inference in binary mosaics. *Biometrics*, **41**, 1049–1052.

Hall, P. (1988a). *Introduction to the Theory of Coverage Processes*. Wiley, New York.

Hall, P. (1988b). On confidence intervals for spatial parameters estimated from nonreplicated data. *Biometrics*, **44**, 271–277.

Hall, P. and Heyde, C. C. (1980). *Martingale Limit Theory and its Applications*. Academic Press, New York.

Hall, P. and Titterington, D. M. (1986). On some smoothing techniques used in image processing. *Journal of the Royal Statistical Society B*, **48**, 330–343.

Halley, E. (1686). An historical account of the trade winds, and monsoons, observable in the seas between and near the tropicks; with an attempt to assign the phisical cause of said winds. *Philosophical Transactions*, **183**, 153–168.

Hamlett, J. M., Horton, R., and Cressie, N. A. C. (1986). Resistant and exploratory techniques for use in semivariogram analyses. *Soil Science Society of America Journal*, **50**, 868–875.

Hammersley, J. M. and Clifford, P. (1971). Markov fields on finite graphs and lattices. Unpublished manuscript, Oxford University.

Hammersley, J. M. and Mazzarino, G. (1983). Markov fields, correlated percolation, and the Ising model. In *The Mathematics and Physics of Disordered Media*, B. D. Hughes and B. W. Ninham, eds. *Lecture Notes in Mathematics*, **No. 1035**. Springer, New York, 201–245.

Hampel, F. R., Ronchetti, E. M., Rousseeuw, P. J., and Stahel, W. A. (1986). *Robust Statistics: The Approach Based on Influence Functions*. Wiley, New York.

Hanisch, K. H. (1980). On classes of random sets and point process models. *Elektronische Informationsverarbeitung und Kybernetik*, **16**, 498–502.

Hanisch, K. H. (1984). Some remarks on estimators of the distribution function of nearest neighbor distance in stationary spatial point processes. *Mathematische Operationsforschung und Statistik, Series Statistics*, **15**, 409–412.

Hanisch, K. H. and Stoyan, D. (1979). Formulas for second-order analysis of marked point processes. *Mathematische Operationsforschung und Statistik, Series Statistics*, **10**, 555–560.

Hanisch, K. H. and Stoyan, D. (1983). Remarks on statistical inference and prediction for a hard-core clustering model. *Mathematische Operationsforschung und Statistik, Series Statistics*, **14**, 559–567.

Hanisch, K. H. and Stoyan, D. (1984). Once more on orientations in point processes. *Elektronische Informationsverarbeitung und Kybernetik*, **20**, 279–284.

Hannan, E. J. (1960). *Time Series Analysis*. Methuen, London.

Hansen, M. H., Madow, W. G., and Tepping, B. J. (1983). An evaluation of model-dependent and probability-sampling inferences in sample surveys. *Journal of the American Statistical Association*, **78**, 776–793.

Hanson, F. and Tier, C. (1982). A stochastic model of tumor growth. *Mathematical Biosciences*, **61**, 73–100.

Hardle, W. and Tuan, P. D. (1986). Some theory on M-smoothing of time series. *Journal of Time Series Analysis*, **7**, 191–204.

Hardy, R. L. (1971). Multiquadric equations of topography and other irregular surfaces. *Journal of Geophysical Research*, **76**, 1905–1915.

Hardy, R. L. and Nelson, S. A. (1986). A multiquadric-biharmonic representation and approximation of disturbing potential. *Geophysical Research Letters*, **13**, 18–21.

Harkness, R. D. and Isham, V. (1983). A bivariate spatial point pattern of ants' nests. *Applied Statistics*, **32**, 293–303.

Harper, W. V. and Furr, J. M. (1986). Geostatistical analysis of potentiometric data in the Wolfcamp Aquifer of the Palo Duro Basin, Texas. Technical Report BMI/ONWI-587, Battelle Memorial Institute, Columbus, OH.

Harris, T. E. (1974). Contact interactions on a lattice. *Annals of Probability*, **2**, 969–988.

Hart, J. F. (1954). Central tendency in areal distributions. *Economic Geography*, **30**, 48–59.

Harville, D. A. (1974). Bayesian inference for variance components using only error contrasts. *Biometrika*, **61**, 383–385.

Harville, D. A. (1977). Maximum likelihood approaches to variance component estimation and to related problems. *Journal of the American Statistical Association*, **72**, 320–340.

Harville, D. A. (1985). Decomposition of prediction error. *Journal of the American Statistical Association*, **80**, 132–138.

Harville, D. A. (1988). Mixed-model methodology: Theoretical justifications and future direction. In *1988 Proceedings of the Statistical Computing Section of the American Statistical Association*, American Statistical Association, Alexandria, VA, 41–49.

Harville, D. A. and Jeske, D. R. (1992). Mean squared error of estimation or prediction under a general linear model. *Journal of the American Statistical Association*, **87**, 724–731.

Haslett, J. (1985). Maximum likelihood discriminant analysis on the plane using a Markovian model of spatial context. *Pattern Recognition*, **18**, 287–296.

Haslett, J. and Horgan, G. (1987). Linear models in spatial discriminant analysis. In *Pattern Recognition Theory and Applications*, P. A. Devijver and J. Kittler, eds. *NATO ASI Series*, **Vol. F30**. Springer, Berlin, 48–55.

Haslett, J. and Raftery, A. E. (1989). Space–time modelling with long-memory dependence: Assessing Ireland's wind power resource. *Applied Statistics*, **38**, 1–21.

Hassner, M. and Sklansky, J. (1980). The use of Markov random fields as models of texture. *Computer Graphics and Image Processing*, **12**, 357–370.

Hastie, T. and Tibshirani, R. (1990). *Generalized Additive Models*. Chapman and Hall, London.

Hastings, W. K. (1970). Monte Carlo sampling methods using Markov chains and their applications. *Biometrika*, **57**, 97–109.

Hawkes, A. G. (1971). Spectra of some self-exciting and mutually exciting point processes. *Biometrika*, **58**, 83–90.

Hawkins, D. M. and Cressie, N. (1984). Robust kriging—a proposal. *Journal of the International Association for Mathematical Geology*, **16**, 3–18.

Hayter, A. J. (1984). A proof of the conjecture that the Tukey–Kramer multiple comparisons procedure is conservative. *Annals of Statistics*, **12**, 61–75.

Hayter, A. J. (1989). Pairwise comparisons of generally correlated means. *Journal of the American Statistical Association*, **84**, 208–213.

Heijmans, R. D. H. and Magnus, J. R. (1986a). Consistent maximum-likelihood estimation with dependent observations. *Journals of Econometrics*, **32**, 253–285.

Heijmans, R. D. H. and Magnus, J. R. (1986b). Asymptotic normality of maximum likelihood estimators obtained from normally distributed but dependent observations. *Econometric Theory*, **2**, 374–412.

Heine, V. (1955). Models for two-dimensional stationary stochastic processes. *Biometrika*, **42**, 170–178.

Heinrich, L. (1984). On a test of randomness of spatial point patterns. *Mathematische Operationsforschung und Statistik, Series Statistics*, **15**, 413–420.

Heinrich, L. (1988). Asymptotic behavior of an empirical nearest-neighbor distance function for stationary Poisson cluster processes. *Mathematische Nachrichten*, **136**, 131–148.

Helson, H. and Lowdenslager, D. (1958). Prediction theory and Fourier series in several variables. *Acta Mathematica*, **99**, 165–202.

Helson, H. and Lowdenslager, D. (1961). Prediction theory and Fourier series in several variables. II. *Acta Mathematica*, **106**, 175–213.

Helvik, B. E. and Swensen, A. R. (1987). Modelling of clustering effects in point processes: An application to failures in SPC—systems. *Scandinavian Journal of Statistics*, **14**, 57–66.

Henderson, R. (1924). A new method of graduation. *Transactions of the Actuarial Society of America*, **25**, 29–40.

Hepple, L. W. (1976). A maximum likelihood model for econometric estimation with spatial data. In *London Papers in Regional Science 6. Theory and Practice in Regional Science*, I. Masser, ed. Pion, London, 90–104.

Herman, G. T. (1980). *Image Reconstructions from Projections*. Academic Press, New York.

Herzfeld, U. C. (1989). A note on programs performing kriging with nonnegative weights. *Mathematical Geology*, **21**, 391–393.

Hess, C. (1979). Theoreme ergodique et loi forte des grands nombres pour des ensembles aleatoires. *Comptes Rendus de l'Academie des Sciences A*, **288**, 519–522.

Hess, C. (1985). Loi forte des grands nombres pour des ensembles aleatoires non bornes a valeurs dans un espace de Banach separable. *Comptes Rendus de l'Academie des Sciences, Paris, Serie I*, 177–180.

Heyl, P. R. and Cook, G. S. (1936). The value of gravity at Washington. *Journal of Research of the U.S. Bureau of Standards*, **17**, 805–839.

Hiai, F. and Umegaki, H. (1977). Integrals, conditional expectations and martingales of multivalued functions. *Journal of Multivariate Analysis*, **7**, 149–182.

Hill, J. R., Hinkley, D. V., Kostal, H., and Morris, C. N. (1984). Spatial estimation from remotely sensed data via empirical Bayes models. In *Proceedings of the NASA Symposium on Mathematical Pattern Recognition and Image Analysis*, L. F. Guseman, Jr., ed. Department of Mathematics, Texas A&M, 115–136.

Hill, M. O. (1973). The intensity of spatial pattern in plant communities. *Journal of Ecology*, **61**, 225–236.

Hines, R. J. O. and Hines, W. G. S. (1989). Repeated sampling of spatial point distributions. *Communications in Statistics. Theory and Methods*, **18**, 2599–2614.

Hines, W. G. S. and Hines, R. J. O. (1979). The Eberhardt statistic and the detection of nonrandomness of spatial point distributions. *Biometrika*, **66**, 73–79.

Hinkley, D. V. (1977). Jackknifing in unbalanced situations. *Technometrics*, **19**, 285–292.

Hjort, N. L. and Mohn, E. (1984). A comparison of some contextual methods in remote sensing classification. In *Proceedings of the Eighteenth International Symposium on Remote Sensing of the Environment*. Centre National d'Etudes Spatiales, Paris, 1693–1702.

Hjorth, U. (1982). Model selection and forward validation. *Scandinavian Journal of Statistics*, **9**, 95–105.

Hodder, I. and Orton, C. (1976). *Spatial Analysis in Archeology*. Cambridge University Press, London.

Hogg, R. V. and Craig, A. T. (1978). *Introduction to Mathematical Statistics*, 4th ed. Macmillan Publishing Co., New York.

Hohn, M. E. (1988). *Geostatistics and Petroleum Geology*. Van Nostrand Reinhold, New York.

Holgate, P. (1965a). Some new tests for randomness. *Journal of Ecology*, **53**, 261–266.

Holgate, P. (1965b). Tests of randomness based on distance methods. *Biometrika*, **52**, 345–353.

Holley, R. (1971). Free energy in a Markovian model of a lattice spin system. *Communications in Mathematical Physics*, **23**, 87–99.

Hope, A. C. A. (1968). A simplified Monte Carlo significance test procedure. *Journal of the Royal Statistical Society B*, **30**, 582–598.

Hopkins, B. (1954). A new method for determining the type of distribution of plant individuals. *Annals of Botany*, **18**, 213–227.

Hotelling, H. (1939). Tubes and spheres in *n*-spaces, and a class of statistical problems. *American Journal of Mathematics*, **61**, 440–460.

Howarth, R. J. and Earle, S. A. M. (1979). Application of a generalized power transformation to geochemical data. *Journal of the International Association for Mathematical Geology*, **11**, 45–62.

Howe, S. and Webb, T. (1983). Calibrating pollen data in climatic terms: Improving the methods. *Quaternary Science Reviews*, **2**, 17–51.

Huang, J. S. (1984). The autoregressive moving average model for spatial analysis. *Australian Journal of Statistics*, **26**, 169–178.

Huang, W. J. and Puri, P. S. (1987). Another look at Poisson processes. *Sankhyā A*, **49**, 133–137.

Huber, P. J. (1964). Robust estimation of a location parameter. *Annals of Mathematical Statistics*, **35**, 73–101.

Huber, P. J. (1972). Robust statistics: A review. *Annals of Mathematical Statistics*, **43**, 1041–1067.

Huber, P. J. (1979). Robust smoothing. In *Robustness in Statistics*, R. L. Launer and G. N. Wilkinson, eds. Academic Press, New York, 33–47.

Huber, P. J. (1981). *Robust Statistics*. Wiley, New York.

Huertas, A. and Medioni, G. (1986). Detection of intensity changes with subpixel accuracy using Laplacian–Gaussian masks. *IEEE Transactions on Pattern Analysis and Machine Intelligence*, **PAMI-8**, 651–664.

Huijbregts, C. J. and Matheron, G. (1971). Universal kriging (An optimal method for estimating and contouring in trend surface analysis). In *Proceedings of Ninth International Symposium on Techniques for Decision-Making in the Mineral Industry*, J. I. McGerrigle, ed. *The Canadian Institute of Mining and Metallurgy, Special Volume* **12**, 159–169.

Hunt, B. R. (1980). Nonstationary statistical image models (and their application to image data compression). *Computer Graphics and Image Processing*, **12**, 173–186.

Hurvich, C. M. and Zeger, S. L. (1987). Frequency domain bootstrap methods for time series. Faculty of Business Administration, Working Paper Series No. 87-115. New York University, New York.

Hutchinson, J. E. (1981). Fractals and self similarity. *Indiana University Mathematics Journal*, **30**, 713–747.

Hutt, S. J. and Hutt, C. (1970). *Direct Observation and Measurement of Behavior*. C. C. Thomas, Springfield, IL.

Iachan, R. (1983). Asymptotic theory of systematic sampling. *Annals of Statistics*, **11**, 959–969.

Iachan, R. (1985). Plane sampling. *Statistics and Probability Letters*, **3**, 151–159.

Ibragimov, I. A. and Linnik, Y. V. (1971). *Independent and Stationary Sequences of Random Variables*. Wolters-Noordhoff, Groningen, Netherlands.

Ibragimov, I. A. and Rozanov, Y. A. (1978). *Gaussian Random Processes*. Springer, New York.

Imhof, J. P. (1961). Computing the distribution of quadratic forms in normal variates. *Biometrika*, **48**, 419–426.

IMSL Inc. (1987). *User's Manual, Math / Library*. IMSL, Houston, TX.

Inoue, K. (1976). Equivalence of measures for some class of Gaussian random fields. *Journal of Multivariate Analysis*, **6**, 295–308.

International Agency for Research on Cancer (1985). Scottish Cancer Atlas. *IARC Scientific Publications*, **No. 72**. Lyon, France.

International Union Against Cancer (1978). *Clinical Oncology*, 2nd ed., Committee on Professional Education of UICC, eds. Springer, New York.

Isaacson, D. L. and Madsen, R. R. (1976). *Markov Chains. Theory and Applications*. Wiley, New York.

Isaaks, E. H. and Srivastava, R. M. (1989). *An Introduction to Applied Geostatistics*. Oxford University Press, Oxford.

Isham, V. (1984). Multitype Markov point processes: Some applications. *Proceedings of the Royal Society of London A*, **391**, 39–53.

Isham, V. (1987). Marked point processes and their correlations, in *Spatial Processes and Spatial Time Series Analysis (Proceedings of the Sixth Franco-Belgian Meeting of Statisticians, 1985)*, F. Droesbeke, ed. Publications des Facultes universitaires Saint-Louis, Brussels, Belgium, 63–75.

Ising, E. (1925). Beitrag zur theorie des ferromagnetismus. *Zeitschrift für Physik*, **31**, 253–258.

Istok, J. D. and Cooper, R. M. (1988). Geostatistics applied to groundwater pollution. III: Global estimates. *Journal of Environmental Engineering*, **114**, 915–928.

Jacobs, B. L., Rodriguez-Iturbe, I., and Eagleson, P. S. (1988). Evaluation of a homogeneous point process description of Arizona rainfall. *Water Resources Research*, **24**, 1174–1186.

Jagers, P. (1973). On Palm probabilities. *Zeitschrift für Wahrscheinlichkeitstheorie und verwandte Gebiete*, **26**, 17–32.

Jaynes, E. T. (1957). Information theory and statistical mechanics. *Physical Review*, **106**, 620–630.

Jennrich, R. I. (1969). Asymptotic properties of nonlinear least squares estimators. *Annals of Mathematical Statistics*, **40**, 633–643.

Jennrich, R. I. and Sampson, P. F. (1976). Newton–Raphson and related algorithms for maximum likelihood variance component estimation. *Technometrics*, **18**, 11–17.

Jensen, E. B. (1987). Design- and model-based stereological analysis of arbitrarily shaped particles. *Scandinavian Journal of Statistics*, **14**, 161–180.

Jensen, E. B., Baddeley, A. J., Gundersen, H. J. G., and Sundberg, R. (1985). Recent trends in stereology. *International Statistical Review*, **53**, 99–108.

Jensen, S. T. (1988). Covariance hypotheses which are linear in both the covariance and the inverse covariance. *Annals of Statistics*, **16**, 302–322.

Jeulin, D. (1989). Sequential random functions models. In *Geostatistics*, *Vol.* **1**, M. Armstrong, ed. Kluwer, Dordrecht, 189–200.

Jewell, W. S. (1988). A heteroscedastic hierarchical model. In *Bayesian Statistics*, **3**, J. M. Bernardo, M. H. DeGroot, D. V. Lindley, and A. F. M. Smith, eds. Oxford University Press, Oxford, 657–663.

Johansen, S. (1966). An application of extreme point methods to the representation of infinitely divisible distributions. *Zeitschrift für Warscheinlichkeitstheorie und verwandte Gebiete*, **5**, 304–316.

Johns, M. V. (1988). Importance sampling for bootstrap confidence intervals. *Journal of the American Statistical Association*, **83**, 709–714.

Johnson, N. L. and Kotz, S. (1972). *Distributions in Statistics: Continuous Multivariate Distributions*. Wiley, New York.

Jones, M. C. and Silverman, B. W. (1989). An orthogonal series density estimation approach to reconstructing positron emission tomography images. *Journal of Applied Statistics*, **16**, 177–191.

Jones, T. A., Hamilton, D. E., and Johnson, C. R. (1986). *Contouring Geologic Surfaces with the Computer*. Van Nostrand Reinhold, New York.

Journel, A. G. (1974). Geostatistics for conditional simulation of ore bodies. *Economic Geology*, **69**, 673–687.

Journel, A. G. (1977). Kriging in terms of projections. *Journal of the International Association for Mathematical Geology*, **9**, 563–586.

Journel, A. G. (1980). The lognormal approach to predicting local distributions of selective mining unit grades. *Journal of the International Association for Mathematical Geology*, **12**, 285–303.

Journel, A. G. (1983). Nonparametric estimation of spatial distributions. *Journal of the International Association for Mathematical Geology*, **15**, 445–468.

Journel, A. G. (1984). mAD and conditional quantile estimators. In *Geostatistics for Natural Resources Characterization, Part 1*, G. Verly, M. David, A. Journel, and A. Marechal, eds. Reidel, Dordrecht, 261–270.

Journel, A. G. (1985). The deterministic side of geostatistics. *Journal of International Association for Mathematical Geology*, **17**, 1–14.

Journel, A. G. (1986). Geostatistics: Models and tools for the earth sciences. *Mathematical Geology*, **18**, 119–140.

Journel, A. G. (1988a). New distance measures: The route towards truly non-Gaussian geostatistics. *Mathematical Geology*, **20**, 459–475.

Journel, A. G. (1988b). Nonparametric geostatistics for risk and additional sampling assessment. In *Principles of Environmental Sampling*, L. H. Keith, ed. American Chemical Society, Washington, DC, 45–72.

Journel, A. G. and Huijbregts, C. J. (1978). *Mining Geostatistics*. Academic Press, London.

Jowett, G. H. (1952). The accuracy of systematic sampling from conveyer belts. *Applied Statistics*, **1**, 50–59.

Justusson, B. I. (1981). Median filtering: Statistical properties. In *Two Dimensional Digital Signal Processing II*, T. S. Huang, ed. *Topics in Applied Physics*, **Vol. 43**. Springer, New York, 161–196.

Kackar, R. N. and Harville, D. A. (1984). Approximations for standard errors of estimators of fixed and random effects in mixed linear models. *Journal of the American Statistical Association*, **79**, 853–862.

Kallenbeg, O. (1986). *Random Measures*, 4th ed. Academic Press, New York.

Karr, A. F. (1985). Inference for thinned point processes, with application to Cox processes. *Journal of Multivariate Analysis*, **16**, 368–392.

Karr, A. F. (1986). *Point Processes and their Statistical Inference*. Marcel Dekker, New York.

Kashyap, R. L. and Chellappa, R. (1983). Estimation and choice of neighbors in spatial-interaction models of images. *IEEE Transactions on Information Theory*, **IT-29**, 60–72 (Corrigendum p. 629).

Kaufman, A. (1975). *Introduction to the Theory of Fuzzy Subsets. Volume* **1**: *Fundamental Theoretical Elements*. Academic Press, New York.

Kawashima, H. (1980). Parameter estimation of autoregressive integrated processes by least squares. *Annals of Statistics*, **8**, 423–435.

Keiding, N., Andersen, P. K., and Frederiksen, K. (1990). Modelling excess mortality of the unemployed: Choice of scale and extra-Poisson variability. *Applied Statistics*, **39**, 63–74.

Kellerer, A. M. (1985). Counting figures in planar random configurations. *Journal of Applied Probability*, **22**, 68–81.

Kelly, F. P. and Ripley, B. D. (1976). A note on Strauss's model for clustering. *Biometrika*, **63**, 357–360.

Kemperman, J. H. B. (1984). Least absolute value and median polish. In *Inequalities in Statistics and Probability*, Y. L. Tong, ed. IMS Monograph Series, Vol. **5**. Institute of Mathematical Statistics, Hayward, CA, 84–103.

Kempthorne, O. (1952). *Design and Analysis of Experiments*. Wiley, New York.

Kempton, R. A. and Howes, C. W. (1981). The use of neighboring plot values in the analysis of variety trials. *Applied Statistics*, **30**, 59–70.

Kendall, D. G. (1974). Foundations of a theory of random sets. In *Stochastic Geometry*, E. F. Harding and D. G. Kendall, eds. Wiley, New York, 322–376.

Kendall, D. G. (1984). Shape manifolds, procrustean metrics, and complex projective spaces. *Bulletin of the London Mathematical Society*, **16**, 81–121.

Kendall, D. G. (1989). A survey of the statistical theory of shape. *Statistical Science*, **4**, 87–99.

Kendall, M. G. and Stuart, A. (1969). *The Advanced Theory of Statistics, Vol.* **1**, 3rd ed. Griffin, London.

Kenkel, N. C. (1988). Pattern of self-thinning in jack pine: Testing the random mortality hypothesis. *Ecology*, **69**, 1017–1024.

Kenkel, N. C., Hoskins, J. A., and Hoskins, W. D. (1989). Edge effects in the use of area polygons to study competition. *Ecology*, **70**, 272–274.

Kent, J. T. (1989). Continuity properties for random fields. *Annals of Probability*, **17**, 1432–1440.

Kernan, W. J., Higby, W. J., Hopper, D. L., Cunningham, W., Lloyd, W. E., and Reiter, L. (1980). Pattern recognition of behavioral events in the nonhuman primate. *Behavior Research Methods and Instrumentation*, **12**, 524–534.

Kernan, W. J., Hopper, D. L., and Lloyd, W. E. (1981). Computer study of the behavioral effects of pharmacologic and toxicologic agents. *Pharmaceutical Technology*, **5**, 60–70.

Kernan, W. J., Mullenix, P. J., Kent, R., Hopper, D. L., and Cressie, N. A. C. (1988). Analysis of the time distribution and time sequence of behavioral acts. *International Journal of Neuroscience*, **43**, 35–51.

Khinchin, A. Y. (1960). *Mathematical Methods in the Theory of Queuing*. Griffin, London.

Kidron, M. and Segal, R. (1984). *The New State of the World Atlas*. Simon and Schuster, New York.

Kiefer, J. (1974). General equivalence theory for optimum designs (approximate theory). *Annals of Statistics*, **2**, 849–879.

Kiefer, J. (1975). Construction and optimality of generalized Youden designs. In *A Survey of Statistical Design and Linear Models*, J. N. Srivastava, ed. North-Holland, Amsterdam, 333–353.

Kiefer, J. (1980). Optimal design theory in relation to combinatorial design. In *Combinatorial Mathematics, Optimal Designs and Their Application,* J. N. Srivastava, ed. *Annals of Discrete Mathematics*, **6**. North-Holland, Amsterdam, 225–241.

Kiefer, J. and Wynn, H. P. (1981). Optimum balanced block and latin square designs for correlated observations. *Annals of Statistics*, **9**, 737–757.

Kiiveri, H. T. and Campbell, N. A. (1989). Covariance models for lattice data. *Australian Journal of Statistics*, **31**, 62–77.

Kimeldorf, G. and Wahba, G. (1970). A correspondence between Bayesian estimation of stochastic processes and smoothing by splines. *Annals of Mathematical Statistics*, **41**, 495–502.

Kindermann, R. and Snell, J. L. (1980). *Markov Random Fields and their Applications*. American Mathematical Society, Providence, RI.

King, M. L. (1981). A small sample property of the Cliff–Ord test for spatial correlation. *Journal of the Royal Statistical Society B*, **43**, 263–264.

Kingman, J. F. C. (1977). Remarks on the spatial distribution of a reproducing population. *Journal of Applied Probability*, **14**, 577–583.

Kirkpatrick, D. G. and Radke, J. D. (1985). A framework for computational morphology. In *Computational Geometry*, G. T. Toussaint, ed. Elsevier, Amsterdam, 217–248.

Kirkwood, J. G., Lewinson, V. A., and Alder, B. J. (1952). Radial distribution functions and the equation of state of fluids composed of molecules interacting according to the Lennard–Jones potential. *Journal of Chemical Physics*, **20**, 929–938.

Kitagawa, G. (1987). Non-Gaussian state-space modeling of nonstationary time series. *Journal of the American Statistical Association*, **82**, 1032–1041.

Kitanidis, P. K. (1983). Statistical estimation of polynomial generalized covariance functions and hydrologic applications. *Water Resources Research*, **19**, 909–921.

Kitanidis, P. K. (1985). Minimum variance unbiased quadratic estimation of covariances of regionalized variables. *Journal of the International Association for Mathematical Geology*, **17**, 195–208.

Kitanidis, P. K. (1986). Parameter uncertainty in estimation of spatial functions: Bayesian analysis. *Water Resources Research*, **22**, 499–507.

Kitanidis, P. K. (1987). Parametric estimation of covariances of regionalized variables. *Water Resources Bulletin*, **23**, 557–567.

Kitanidis, P. K. and Lane, R. W. (1985). Maximum likelihood parameter estimation of hydrologic spatial processes by the Gauss–Newton method. *Journal of Hydrology*, **79**, 53–71.

Kitanidis, P. K. and Vomvoris, E. G. (1983). A geostatistical approach to the inverse problem in groundwater modeling (steady state) and one-dimensional simulations. *Water Resources Research*, **19**, 677–690.

Klein, R. and Press, S. J. (1989). Contextual Bayesian classification of remotely sensed data. *Communications in Statistics. Theory and Methods*, **18**, 3177–3202.

Klimko, L. A. and Nelson, P. I. (1978). On conditional least squares estimation for stochastic processes. *Annals of Statistics*, **6**, 629–642.

Knox, E. G. (1964). The detection of space–time interactions. *Applied Statistics*, **13**, 25–29.

Koenker, R. and Bassett, G. (1978). Regression quantiles. *Econometrica*, **46**, 33–50.

Kolmogorov, A. N. (1941a). The local structure of turbulence in an incompressible fluid at very large Reynolds numbers. *Doklady Akademii Nauk SSSR*, **30**, 301–305. Reprinted (1961), in *Turbulence: Classic Papers on Statistical Theory,* S. K. Friedlander and L. Topping, eds. Interscience Publishers, New York, 151–155.

Kolmogorov, A. N. (1941b). Interpolation and extrapolation of stationary random sequences. *Izvestiia Akademii Nauk SSSR, Seriia Matematicheskiia*, **5**, 3–14. [Translation (1962), Memo RM-3090-PR, Rand Corp. Santa Monica, CA.]

Kooijman, S. A. L. M. (1979). The description of point patterns. In *Spatial and Temporal Analysis in Ecology*, R. M. Cormack and J. K. Ord, eds. International Co-operative Publishing House, Fairland, MD, 305–331.

Koopmans, L. H. (1974). *The Spectral Analysis of Time Series*. Academic Press, New York.

Korezlioglu, H. and Loubaton, P. (1986). Spectral factorization of wide sense stationary processes on $\mathbb{Z}^2$. *Journal of Multivariate Analysis*, **19**, 24–47.

Kovyazin, S. A. (1986). On the limit behavior of a class of empirical means of a random set. *Theory of Probability and Its Applications*, **30**, 814–820.

Krajewski, W. F. (1987). Cokriging radar-rainfall and rain gage data. *Journal of Geophysical Research*, **92**, 9571–9580.

Kramer, W. and Donninger, C. (1987). Spatial autocorrelation among errors and the relative efficiency of OLS in the linear regression model. *Journal of the American Statistical Association*, **82**, 577–579.

Krickeberg, K. (1982). Processus ponctuels en statistique. In *Ecole d'Ete de Probabilites de Saint-Flour X—1980*, P. L. Hennequin, ed. *Lecture Notes in Mathematics*, **No. 929**. Springer, Berlin, 205–313.

Krige, D. G. (1951). A statistical approach to some basic mine valuation problems on the Witwatersrand. *Journal of the Chemical, Metallurgical and Mining Society of South Africa*, **52**, 119–139.

Kruskal, J. B. and Wish, M. (1977). *Multidimensional Scaling*. Sage Publications, Beverley Hills, CA.

Kryscio, R. J. and Saunders, R. (1983). On interpoint distances for planar Poisson cluster processes. *Journal of Applied Probability*, **20**, 513–528.

Kubat, P. (1979). Mean or median? (A note on an old problem.) *Statistica Neerlandica*, **33**, 191–196.

Kulperger, R. J. (1987). Some remarks on regression with autoregressive errors and their residual processes. *Journal of Applied Probability*, **24**, 668–678.

Kunert, J. (1985). Optimal repeated measurements designs for correlated observations and analysis by weighted least squares. *Biometrika*, **72**, 375–389.

Kunert, J. (1987). Neighbor balanced block designs for correlated errors. *Biometrika*, **74**, 717–724.

Kunsch, H. R. (1981). Thermodynamics and statistical analysis of Gaussian random fields. *Zeitschrift für Wahrscheinlichkeitstheorie und verwandte Gebiete*, **58**, 407–421.

Kunsch, H. R. (1983). Asymptotically unbiased inference for Ising models. *Advances in Applied Probability*, **15**, 887–888.

Kunsch, H. R. (1984). Non reversible stationary measures for infinite interacting particle systems. *Zeitschrift für Wahrscheinlichkeitstheorie und verwandte Gebiete*, **66**, 407–424.

Kunsch, H. R. (1985). Statistical analysis of uniformity trials based on parametric models. Unpublished manuscript, Fachgruppe für Statistik, ETH, Zurich.

Kunsch, H. R. (1986). Discrimination between monotonic trends and long-range dependence. *Journal of Applied Probability*, **23**, 1025–1030.

Kunsch, H. R. (1987). Intrinsic autoregressions and related models on the two-dimensional lattice. *Biometrika*, **74**, 517–524.

Kunsch, H. R. (1989). The jackknife and the bootstrap for general stationary observations. *Annals of Statistics*, **17**, 1217–1241.

Kunst, R. M. (1989). The performance of robust filtering: Some Monte Carlo evidence. *Computational Statistics Quarterly*, **5**, 53–76.

Kutoyants, Y. A. (1984). *Parameter Estimation for Stochastic Processes*. Heldermann Verlag, Berlin.

Laird, A. (1964). The dynamics of tumor growth. *British Journal of Cancer*, **28**, 490–502.

Lajaunie, C. (1990). Comparing some approximate methods for building local confidence intervals for predicting regionalized variables. *Mathematical Geology*, **22**, 123–144.

Lamp, J. (1984). Remote sensing in land and soil monitoring. In *Soil Information Systems Technology (Proceedings of the Sixth Meeting of the ISS Working Group on*

Soil Information Systems, Bolkesjo, Norway, 1983), P. A. Burrough and S. W. Bie, eds. Pudoc, Wageningen, 152–158.

Lancaster, H. O. (1958). The structure of bivariate distributions. *Annals of Mathematical Statistics*, **29**, 719–736. [Corrigendum (1964), *Annals of Mathematical Statistics*, **35**, 1388.]

Lantuejoul, C. (1988). On the importance of choosing a change of support model for global reserves estimation. *Mathematical Geology*, **20**, 1001–1019.

Laslett, G. M., Cressie, N., and Liow, S. (1985). Intensity estimation in a spatial model of overlapping particles. Unpublished manuscript, Division of Mathematics and Statistics, CSIRO, Melbourne.

Laslett, G. M. and McBratney, A. B. (1990). Estimation and implications of instrumental drift, random measurement error and nugget variance of soil attributes—a case study for soil pH. *Journal of Soil Science*, **41**, 451–471.

Laslett, G. M., McBratney, A. B., Pahl, P. J., and Hutchinson, M. F. (1987). Comparison of several spatial prediction methods for soil pH. *Journal of Soil Science*, **38**, 325–341.

Laurent, P. J. (1972). *Approximation et Optimisation*. Hermann, Paris.

Lawoko, C. R. O. and McLachlan, G. J. (1983). Some asymptotic results on the effect of autocorrelation on the error rates of the sample linear discriminant function. *Pattern Recognition*, **16**, 119–121.

Lawson, A. (1988). On tests for spatial trend in a nonhomogeneous Poisson process. *Journal of Applied Statistics*, **15**, 225–234.

Lax, D. A. (1985). Robust estimators of scale: Finite-sample performance in long-tailed symmetric distributions. *Journal of the American Statistical Association*, **80**, 736–741.

Le, D. N. and Petkau, A. J. (1988). The variability of rainfall acidity revisited. *Canadian Journal of Statistics*, **16**, 15–38.

Leadbetter, M. R. (1972). On basic results of point process theory. *Proceedings of the Sixth Berkeley Symposium on Mathematical Statistics and Probability, Vol.* **3**. University of California Press, Berkeley, CA, 449–462.

Le Cam, L. (1960). Locally asymptotically normal families of distributions. *University of California Publications in Statistics*, **3**, 37–98.

Le Cam, L. (1982). On some mathematical models of tumor growth and metastasis. In *Probability Models and Cancer*, L. Le Cam and J. Neyman, eds. North-Holland, Amsterdam, 265–286.

Le Cam, L. (1985). Sur l'approximation de familles de mesures par des familles Gaussiennes. *Annales de l'Institut Henri Poincaré*, **21**, 225–287.

Legendre, P. (1987). Constrained clustering. In *Developments in Numerical Ecology*, P. Legendre and L. Legendre, eds. Springer, Berlin, 289–307.

Lele, S. (1989). Non-parametric bootstrap for spatial processes. Technical Report No. 671, Department of Biostatistics, Johns Hopkins University, Baltimore, MD.

Lele, S. (1991a). Jackknifing linear estimating equations: Asymptotic theory and applications in stochastic processes. *Journal of the Royal Statistical Society B*, **53**, 253–267.

Lele, S. (1991b). Some comments on coordinate-free and scale-invariant methods in morphometrics. *American Journal of Physical Anthropology*, **85**, 407–417.

Lemmer, I. C. (1984). Estimating local recoverable reserves via indicator kriging. In *Geostatistics for Natural Resources Characterization, Part 1*, G. Verly, M. David, A. Journel, and A. Marechal, eds. Reidel, Dordrecht, 365–384.

Levene, H. (1960). Robust tests for equality of variances. In *Contributions to Probability and Statistics: Essays in Honor of Harold Hotelling*, I. Olkin et al., eds. Stanford University Press, Stanford, CA, 278–292.

Lévy, P. (1954). Le mouvement Brownien. *Memorial des Sciences Mathematiques*, **126**. Gauthier-Villars, Paris.

Lew, W. and Lewis, J. (1977). An anthropometric scaling method with application to the knee joint. *Journal of Biomechanics*, **10**, 171–184.

Lewis, F. L. (1986). *Optimal Estimation*. Wiley, New York.

Lewis, P. A. W., ed. (1972). *Stochastic Point Processes: Statistical Analysis, Theory, and Applications*. Wiley, New York.

Lewis, P. A. W. and Shedler, G. S. (1979). Simulation of non-homogeneous Poisson processes by thinning. *Naval Research Logistics Quarterly*, **26**, 403–413.

Li, G. and Chen, Z. (1985). Projection-pursuit approach to robust dispersion matrices and principal components: Primary theory and Monte Carlo. *Journal of the American Statistical Association*, **80**, 759–766.

Liao, G. Y., Nodes, T. A., and Gallagher, N. C. (1985). Output distributions of two-dimensional median filters. *IEEE Transactions on Acoustics, Speech, and Signal Processing*, **ASSP-33**, 1280–1295.

Liggett, T. M. (1985). *Interacting Particle Systems*. Springer, New York.

Lill, W. J., Gleeson, A. C., and Cullis, B. R. (1988). Relative accuracy of a neighbour method for field trials. *Journal of Agricultural Science (Cambridge)*, **111**, 339–346.

Lim, J. S. and Malik, N. A. (1981). A new algorithm for 2D maximum entropy power spectrum estimation. *IEEE Transactions on Acoustics, Speech, and Signal Processing*, **ASSP-29**, 401–413.

Lindgren, B. W. (1976). *Statistical Theory*, 3rd ed. MacMillan, New York.

Lindsay, B. G. (1983). The geometry of mixture likelihoods: A general theory. *Annals of Statistics*, **11**, 86–94.

Lin'kov, Y. N. (1985). Asymptotic normality of the logarithm of the likelihood ratio, and hypothesis testing for nonhomogeneous Poisson processes. *Selected Translations in Mathematical Statistics and Probability*, **16**, 113–125.

Ljasenko, N. (1979). Limit theorems for sums of independent random sets. *Theory of Probability and Its Applications*, **24**, 438–440.

Ljung, G. M. and Box, G. E. P. (1979). The likelihood function of stationary autoregressive-moving average models. *Biometrika*, **66**, 265–270.

Lloyd, M. (1967). Mean crowding. *Journal of Animal Ecology*, **36**, 1–30.

Loeve, M. (1963). *Probability Theory*, 3rd ed. Van Nostrand, Princeton, NJ.

Longley, P. A. and Batty, M. (1989). Fractal measurement and line generalization. *Computers and Geosciences*, **15**, 167–183.

Lord, E. A. and Wilson, C. B. (1984). *The Mathematical Description of Shape and Form*. Wiley, New York.

Losch, A. (1954). *The Economics of Location*. Yale University Press, New Haven.

Lotwick, H. W. (1984). Some models for multitype spatial point processes, with remarks on analysing multitype patterns. *Journal of Applied Probability*, **21**, 575–582.

Lotwick, H. W. and Silverman, B. W. (1982). Methods for analysing spatial processes of several types of points. *Journal of the Royal Statistical Society B*, **44**, 406–413.

Ludwig, J. A. and Goodall, D. W. (1978). A comparison of paired- with blocked-quadrat variance methods for the analysis of spatial pattern. *Vegetatio*, **38**, 49–59.

Lundberg, O. (1940). *On Random Processes and their Application to Sickness and Accident Statistics*. Almqvist and Wiksells, Uppsala.

Luxmoore, R., Spalding, B. P., and Munro, I. M. (1981). A real variation and chemical modification of weathered shale infiltration characteristics. *Soil Science Society of America Journal*, **45**, 687–691.

Ma, F., Wei, M. S., and Mills, W. H. (1987). Correlation structuring and the statistical analysis of steady-state groundwater flow. *SIAM Journal on Scientific and Statistical Computing*, **8**, 848–867.

Mack, C. (1954). The expected number of clumps when convex laminae are placed at random and with random orientation on a plane area. *Proceedings of the Cambridge Philosophical Society*, **50**, 581–585.

Magnus, J. R. (1978). Maximum likelihood estimation of the GLS model with unknown parameters in the disturbance covariance matrix. *Journal of Econometrics*, **7**, 281–312.

Magnus, J. R. and Neudecker, H. (1988). *Matrix Differential Calculus with Applications in Statistics and Econometrics*. Wiley, New York.

Major, P. (1981). *Multiple Wiener–Itô Integrals. Lecture Notes in Mathematics*, **No. 849**. Springer, New York.

Malinverno, A. (1989). Testing linear models of sea-floor topography. *Pure and Applied Geophysics*, **131**, 139–155.

Mandelbrot, B. B. (1977). *Fractals: Form, Chance, and Dimension*. Freeman, San Francisco.

Mandelbrot, B. B. (1982). *The Fractal Geometry of Nature*. Freeman, San Francisco.

Mandelbrot, B. B. and Van Ness, J. W. (1968). Fractional Brownian motions, fractional noises and applications. *SIAM Review*, **10**, 422–437.

Mantel, N. (1967). The detection of disease clustering and a generalized regression approach. *Cancer Research*, **27**, 209–220.

Mantoglou, A. (1987). Digital simulation of multivariate two- and three-dimensional stochastic processes with a spectral turning bands method. *Mathematical Geology*, **19**, 129–149.

Mantoglou, A. and Wilson, J. L. (1982). The turning bands method for simulation of random fields using line generation by a spectral method. *Water Resources Research*, **18**, 1379–1384.

Manton, K. G., Woodbury, M. A., Stallard, E., Riggan, W. B., Creason, J. P., and Pellom, A. C. (1989). Empirical Bayes procedures for stabilizing maps of U.S. cancer mortality rates. *Journal of the American Statistical Association*, **84**, 637–650.

Marcus, A. (1966). A stochastic model of the formation and survival of lunar craters. *Icarus*, **5**, 165–200.

Marcus, A. (1967). A multivariate immigration with multiple death process and applications to lunar craters. *Biometrika*, **54**, 251–261.

Mardia, K. V. (1970). *Families of Bivariate Distributions*. Griffin, London.

Mardia, K. V. (1984). Spatial discrimination and classification maps. *Communications in Statistics. Theory and Methods*, **13**, 2181–2197.

Mardia, K. V. (1988). Multi-dimensional multivariate Gaussian Markov random fields with application to image processing. *Journal of Multivariate Analysis*, **24**, 265–284.

Mardia, K. V., Kent, J. T., and Bibby, J. M. (1979). *Multivariate Analysis*, Academic Press, London.

Mardia, K. V. and Marshall, R. J. (1984). Maximum likelihood estimation of models for residual covariance in spatial regression. *Biometrika*, **71**, 135–146.

Mardia, K. V. and Watkins, A. J. (1989). On multimodality of the likelihood in the spatial linear model. *Biometrika*, **76**, 289–295.

Mark, D. M. and Aronson, P. B. (1984). Scale-dependent fractal dimensions of topographic surfaces: An empirical investigation, with applications in geomorphology and computer mapping. *Journal of the International Association for Mathematical Geology*, **16**, 671–683.

Marquiss, M., Newton, I., and Radcliffe, D. A. (1978). The decline of the raven, *Corvus corax*, in relation to afforestation in southern Scotland and northern England. *Journal of Applied Ecology*, **15**, 129–144.

Marr, D. and Hildreth, E. (1980). Theory of edge detection. *Proceedings of the Royal Society of London B*, **207**, 187–217.

Marroquin, J., Mitter, S., and Poggio, T. (1987). Probabilistic solution of ill-posed problems in computational vision. *Journal of the American Statistical Association*, **82**, 76–89.

Marshall, R. J. and Mardia, K. V. (1985). Minimum norm quadratic estimation of components of spatial covariance. *Journal of the International Association for Mathematical Geology*, **17**, 517–525.

Marthaler, H. P., Vogelsanger, W., Richard, F., and Wierenga, P. J. (1983). A pressure transducer for field tensiometers. *Soil Science Society of America Journal*, **47**, 624–627.

Martin, R. J. (1979). A subclass of lattice processes applied to a problem in planar sampling. *Biometrika*, **66**, 209–217.

Martin, R. J. (1982). Some aspects of experimental design and analysis when errors are correlated. *Biometrika*, **69**, 597–612.

Martin, R. J. (1986a). On the design of experiments under spatial correlation. *Biometrika*, **73**, 247–277.

Martin, R. J. (1986b). A note on the asymptotic eigenvalues and eigenvectors of the dispersion matrix of a second-order stationary process on a d-dimensional lattice. *Journal of Applied Probability*, **23**, 529–535.

Martin, R. J. (1990). The use of time-series models and methods in the analysis of agricultural field trials. *Communications in Statistics. Theory and Methods*, **19**, 55–81.

Martin, R. L. and Oeppen, J. E. (1975). The identification of regional forecasting models using space–time correlation functions. *Transactions of the Institute of British Geographers*, **66**, 95–118.

Martinez, A. O. and Griego, R. J. (1980). Growth dynamics of multicell spheroids from three murine tumors. *Growth*, **44**, 112–122.

Mase, S. (1979). Random compact convex sets which are infinitely divisible with respect to Minkowski addition. *Advances in Applied Probability*, **11**, 834–850.

Mase, S. (1982a). Asymptotic properties of stereological estimators of volume fraction for stationary random sets. *Journal of Applied Probability*, **19**, 111–126.

Mase, S. (1982b). Properties of fourth-order strong mixing rates and its application to random set theory. *Journal of Multivariate Analysis*, **12**, 549–561.

Mase, S. (1984). Locally asymptotic normality of Gibbs models on a lattice. *Advances in Applied Probability*, **16**, 585–602.

Mase, S. (1986). On the possible form of size distributions for Gibbsian processes of mutually non-intersecting balls. *Journal of Applied Probability*, **23**, 646–659.

Masreliez, C. J. and Martin, R. D. (1977). Robust Bayesian estimation for the linear model and robustifying the Kalman filter. *IEEE Transactions of Automatic Control*, **AC-22**, 361–371.

Masry, E. (1983). Probability density estimation from sampled data. *IEEE Transactions on Information Theory*, **IT-29**, 696–709.

Matern, B. (1960). Spatial Variation. *Meddelanden fran Statens Skogsforskningsinstitut*, **49**, No. 5. [Second edition (1986), *Lecture Notes in Statistics*, **No. 36**, Springer, New York.]

Matern, B. (1971). Doubly stochastic Poisson processes in the plane. In *Statistical Ecology*, *Vol.* **1**, G. P. Patil, E. C. Pielou, and W. E. Waters, eds. Pennsylvania State University Press, University Park, PA, 195–213.

Matheron, G. (1962). Traite de Geostatistique Appliquee, Tome **I**. *Memoires du Bureau de Recherches Geologiques et Minieres,* **No. 14**. Editions Technip, Paris.

Matheron, G. (1963a). Traite de Geostatistique Appliquee, Tome **II**: Le Krigeage. *Memoires du Bureau de Recherches Geologiques et Minieres,* **No. 24**. Editions Bureau de Recherche Geologiques et Minieres, Paris.

Matheron, G. (1963b). Principles of geostatistics. *Economic Geology*, **58**, 1246–1266.

Matheron, G. (1965). *La Theorie des Variables Regionalisees et ses Applications*. Masson, Paris.

Matheron, G. (1967). Kriging or polynomial interpolation procedures? *Transactions of the Canadian Institute of Mining and Metallurgy*, **70**, 240–244.

Matheron, G. (1969). Le Krigeage Universel. *Cahiers du Centre de Morphologie Mathematique*, **No. 1**. Fontainebleau, France.

Matheron, G. (1971a). Random sets theory and its application to stereology. *Journal of Microscopy*, **95**, 15–23.

Matheron, G. (1971b). The Theory of Regionalized Variables and Its Applications. *Cahiers du Centre de Morphologie Mathematique*, **No. 5**. Fontainebleau, France.

Matheron, G. (1973). The intrinsic random functions and their applications. *Advances in Applied Probability*, **5**, 439–468.

Matheron, G. (1975). *Random Sets and Integral Geometry*. Wiley, New York.

Matheron, G. (1976a). A simple substitute for conditional expectation: The disjunctive kriging. In *Advanced Geostatistics in the Mining Industry*, M. Guarascio, M. David, and C. Huijbregts, eds. Reidel, Dordrecht, 221–236.

Matheron, G. (1976b). Forecasting block grade distributions: The transfer functions. In *Advanced Geostatistics in the Mining Industry*, M. Guarascio, M. David, and C. Huijbregts, eds. Reidel, Dordrecht, 237–251.

Matheron, G. (1981). Splines and kriging: Their formal equivalence. In *Down-to-Earth Statistics: Solutions Looking for Geological Problems*, D. F. Merriam, ed. Syracuse University Geological Contributions, Syracuse, 77–95.

Matheron, G. (1984). Isofactorial models and change of support. In *Geostatistics for Natural Resources Characterization, Part 1*, G. Verly, M. David, A. Journel, and A. Marechal, eds. Reidel, Dordrecht, 449–467.

Matheron, G. (1985). Change of support for diffusion-type random functions. *Journal of the International Association for Mathematical Geology*, **17**, 137–165.

Matheron, G. and Armstrong, M., eds. (1987). *Geostatistical Case Studies*. Reidel, Dordrecht.

Matthes, K., Kerstan, J., and Mecke, J. (1978). *Infinitely Divisible Point Processes*. Wiley, New York.

Mayer, J. E. and Mayer, M. G. (1940). *Statistical Mechanics*. Wiley, New York.

McArthur, R. D. (1987). An evaluation of sample designs for estimating a locally concentrated pollutant. *Communications in Statistics. Simulation and Computation*, **16**, 735–759.

McAuliffe, L. and Afifi, A. A. (1984). Comparison of a nearest-neighbor and other approaches to the detection of space–time clustering. *Computational Statistics and Data Analysis*, **2**, 125–142.

McBratney, A. B. and Webster, R. (1981). Detection of ridge and furrow pattern by spectral analysis of crop yield. *International Statistical Review*, **49**, 45–52.

McBratney, A. B. and Webster, R. (1986). Choosing functions for semivariograms of soil properties and fitting them to sampling estimates. *Journal of Soil Science*, **37**, 617–639.

McBratney, A. B., Webster, R., and Burgess, T. M. (1981). The design of optimal sampling schemes for local estimation and mapping of regionalized variables—I. *Computers and Geosciences*, **7**, 331–334.

McClellan, J. H. (1982). Multidimensional spectral estimation. *Proceedings of the IEEE*, **70**, 1029–1039.

McCullagh, M. J. and Ross, C. G. (1980). Delauney triangulation of a random data set for isarithmic mapping. *Cartographic Journal*, **17**, 93–99.

McCullagh, P. (1983). Quasi-likelihood functions. *Annals of Statistics*, **11**, 59–67.

McCullagh, P. and Nelder, J. A. (1983). *Generalized Linear Models*. Chapman and Hall, London.

McDonald, D. (1989). On nonhomogeneous spatial Poisson processes. *Canadian Journal of Statistics*, **17**, 183–195.

McGilchrist, C. A. (1989). Bias of ML and REML estimators in regression models with ARMA errors. *Journal of Statistical Computation and Simulation*, **32**, 127–136.

McLeod, A. I. (1977). Improved Box–Jenkins estimators. *Biometrika*, **64**, 531–534.

McLeod, A. I., Hipel, K. W., and Comancho, F. (1983). Trend assessment of water quality time series. *Water Resources Bulletin*, **19**, 537–547.

McNeil, D. R. and Tukey, J. W. (1975). Higher order diagnosis of two-way tables, illustrated on two sets of demographic empirical distributions. *Biometrics*, **31**, 487–510.

Mead, R. (1967). A mathematical model for the estimation of interplant competition. *Biometrics*, **23**, 189–205.

Mead, R. (1974). A test for spatial pattern at several scales using data from a grid of contiguous quadrats. *Biometrics*, **30**, 295–307.

Meinhold, R. J. and Singpurwalla, N. D. (1983). Understanding the Kalman filter. *American Statistician*, **37**, 123–127.

Mejia, J. M. and Rodriguez-Iturbe, I. (1974). On the synthesis of random field sampling from the spectrum: An application to the generation of hydrologic spatial processes. *Water Resources Research*, **10**, 705–711.

Mendelsohn, F. (1980). Some aspects of ore reserve estimation. Information Circular No. 147, Economic Geology Research Unit, University of the Witwatersrand, Johannesburg, South Africa.

Mercer, W. B. and Hall, A. D. (1911). The experimental error of field trials. *Journal of Agricultural Science (Cambridge)*, **4**, 107–132.

Metropolis, N., Rosenbluth, A. W., Rosenbluth, M. N., Teller, A. H., and Teller, E. (1953). Equations of state calculations by fast computing machines. *Journal of Chemical Physics*, **21**, 1087–1092.

Miamee, A. G. and Pourahmadi, M. (1988). Wold decomposition, prediction and parameterization of stationary processes with infinite variance. *Probability Theory and Related Fields*, **79**, 145–164.

Micchelli, C. A. (1986). Interpolation of scattered data: Distance matrices and conditionally positive definite functions. *Constructive Approximation*, **2**, 11–22.

Miesch, A. T. (1975). Variograms and variance components in geochemistry and ore evaluation. *Geological Society of America Memoirs*, **142**, 333–340.

Miles, R. E. (1969). Poisson flats in Euclidean spaces. *Advances in Applied Probability*, **1**, 211–237.

Miles, R. E. (1974). On the elimination of edge effects in planar sampling. In *Stochastic Geometry*, E. F. Harding and D. G. Kendall, eds. Wiley, Chichester, 228–247.

Miller, J. J. (1977). Asymptotic properties of maximum likelihood estimates in the mixed model of the analysis of variance. *Annals of Statistics*, **5**, 746–762.

Miller, M. P., Singer, M. J., and Nielsen, D. R. (1988). Spatial variability of wheat yield and soil properties on complex hills. *Soil Science Society of America Journal*, **52**, 1133–1141.

Miller, R. and Kahn, J. (1962). *Statistical Analysis in the Geological Sciences*. Wiley, New York.

Miller, R. G. (1974). The jackknife—a review. *Biometrika*, **61**, 1–15.

Miller, R. G., Halks-Miller, M., Egger, M. J., and Halpern, J. W. (1982). Growth kinetics of glioma cells. *Journal of the American Statistical Association*, **77**, 505–514.

Minkowski, H. (1903). Volumen and oberblache. *Mathematische Annalen*, **57**, 447–495.

Modjeska, J. S. and Rawlings, J. O. (1983). Spatial correlation analysis of uniformity data. *Biometrics*, **39**, 373–384.

Mohammad-Djafari, A. and Demoment, G. (1988). Image restoration and reconstruction using entropy as a regularization functional. In *Maximum-Entropy and Bayesian Methods in Science and Engineering, Vol.* **2**, G. J. Erickson and C. R. Smith, eds. Kluwer, Dordrecht, 341–355.

Molchanov, I. (1984). A generalization of the Choquet theorem for random sets with a given class of realizations. *Theory of Probability and Mathematical Statistics*, **28**, 99–106.

Molchanov, I. (1987). Uniform laws of large numbers for empirical associated functionals of random closed sets. *Theory of Probability and Its Applications*, **32**, 556–559.

Molina, R. and Ripley, B. D. (1989). Using spatial models as priors in astronomical image analysis. *Journal of Applied Statistics*, **16**, 193–206.

Moller, J. (1989a). Random tessellations in $\mathbb{R}^d$. *Advances in Applied Probability*, **21**, 37–73.

Moller, J. (1989b). On the rate of convergence of spatial birth-and-death processes. *Annals of the Institute of Statistical Mathematics*, **41**, 565–581.

Moore, M. (1984). On the estimation of a convex set. *Annals of Statistics*, **12**, 1090–1099.

Moore, M. (1987). Inference statistique dans les processus stochastiques: Apercu historique. *Canadian Journal of Statistics*, **15**, 185–207.

Moore, M. (1988). Spatial linear processes. *Communications in Statistics. Stochastic Models*, **4**, 45–75.

Moran, P. A. P. (1948). The interpretation of statistical maps. *Journal of the Royal Statistical Society B*, **10**, 243–251.

Moran, P. A. P. (1950). Notes on continuous stochastic phenomena. *Biometrika*, **37**, 17–23.

Moran, P. A. P. (1973a). A Gaussian Markovian process on a square lattice. *Journal of Applied Probability*, **10**, 54–62.

Moran, P. A. P. (1973b). Necessary conditions for Markovian processes on a lattice. *Journal of Applied Probability* **10**, 605–612.

Moran, P. A. P. (1976). Another quasi-Poisson plane point process. *Zeitschrift für Wahrscheinlichkeitstheorie und verwandte Gebiete*, **33**, 269–272.

Morgan, J. P. and Chakravarti, I. M. (1988). Block designs for first and second order neighbor correlations. *Annals of Statistics*, **16**, 1206–1224.

Morisita, M. (1959). Measuring of the dispersion and analysis of distribution patterns. *Memoires of the Faculty of Science, Kyushu University, Series E. Biology*, **2**, 215–235.

Morris, C. N. (1983). Parametric empirical Bayes inference: Theory and applications. *Journal of the American Statistical Association*, **78**, 47–55.

Morris, M. D. and Ebey, S. F. (1984). An interesting property of the sample mean under a first-order autoregressive model. *American Statistician*, **38**, 127–129.

Morrison, D. F. (1978). *Multivariate Statistical Methods,* 2nd ed. McGraw-Hill, Tokyo.

Mosimann, J. E. (1970). Size allometry: Size and shape variables with characterizations of the lognormal and generalized gamma distributions. *Journal of the American Statistical Association*, **65**, 930–948.

Mosteller, F. and Tukey, J. W. (1977). *Data Analysis and Regression*. Addison-Wesley, Reading, MA.

Moyal, J. E. (1958). Contribution to the discussion of "Statistical approach to problems of cosmology" by J. Neyman and E. L. Scott. *Journal of the Royal Statistical Society B*, **20**, 36–37.

Moyal, J. E. (1962). The general theory of stochastic population processes. *Acta Mathematica*, **108**, 1–31.

Mueller, P. K., Jansen, J. J., and Allen, M. A. (1984). Utility acid precipitation study program second summary report. UAPSP 109, prepared by Utility Acid Study Program and Electric Power Research Institute, UAPSP Report Center, Damascus, MD.

Muirhead, R. J. (1982). *Aspects of Multivariate Statistical Theory*. Wiley, New York.

Mukherjee, D. and Ratnaparkhi, M. V. (1986). On the functional relationship between entropy and variance with related applications. *Communications in Statistics. Theory and Methods*, **15**, 291–311.

Mukhtar, S., Baker, J. L., Horton, R., and Erbach, D. C. (1985). Soil water infiltration as affected by the use of the paraplow. *American Society of Agricultural Engineers*, **28**, 1811–1816.

Mullenix, P. J. (1981). Structure analysis of spontaneous behavior during the estrus cycle of the rat. *Physiology and Behavior*, **27**, 723–726.

Mullenix, P. J., Moore, P. A., and Tassinari, M. S. (1986). Behavioral toxicity of nitrous oxide in rats following prenatal exposure. *Toxicology and Industrial Health*, **2**, 273–287.

Murdoch, D. J., Krewski, D. R., and Crump, K. S. (1987). Quantitative theories of carcinogenesis. In *Cancer Modeling*, J. R. Thompson and B. W. Brown, eds. Marcel Dekker, New York, 61–89.

Myers, D. E. (1982). Matrix formulation of co-kriging. *Journal of the International Association for Mathematical Geology*, **14**, 249–257.

Myers, D. E. (1984). Cokriging—new developments. In *Geostatistics for Natural Resources Characterization, Part 1*, G. Verly, M. David, A. Journel, and A. Marechal, eds. Reidel, Dordrecht, 295–305.

Myers, D. E. (1989). Vector conditional simulation. In *Geostatistics, Vol.* **1**, M. Armstrong, ed. Kluwer, Dordrecht, 283–293.

Myers, D. E., Begovitch, C. L., Butz, T. R., and Kane, V. E. (1982). Variogram models for regional groundwater geochemical data. *Journal of the International Association for Mathematical Geology*, **14**, 629–644.

Narendra, P. M. (1981). A separable median filter for image noise smoothing. *IEEE Transactions on Pattern Analysis and Machine Intelligence*, **PAMI-3**, 20-29.

Nather, W. (1985). *Effective Observations of Random Fields. Teubner-Texte zur Mathematik*, **Band 72**. Teubner, Leipzig.

National Acid Precipitation Assessment Program (NAPAP) (1988). *Interim Assessment. The Causes and Effects of Acidic Deposition*, *Volumes* **I–IV**. U.S. Government Printing Office, Washington, DC.

Naus, J. I. (1965). Clustering of random points in two dimensions. *Biometrika*, **52**, 263–267.

Nelder, J. A. and Mead, R. (1965). A simplex method for function minimization. *Computer Journal*, **7**, 308–313.

Nelson, C. R. and Kang, H. (1981). Spurious periodicity in inappropriately detrended time series. *Econometrica*, **49**, 741–751.

Nelson, C. R. and Kang, H. (1984). Pitfalls in the use of time as an explanatory variable in regression. *Journal of Business and Economic Statistics*, **2**, 73–82.

Nelson, C. R. and Plosser, C. I. (1982). Trends and random walks in macroeconomic time series. *Journal of Monetary Economics*, **10**, 139–162.

Neuman, S. P. and Jacobson, E. A. (1984). Analysis of nonintrinsic spatial variability by residual kriging with application to regional groundwater levels. *Journal of the International Association for Mathematical Geology*, **16**, 499–521.

Newman, W. I. (1977). A new method for multidimensional power spectral analysis. *Astronomy and Astrophysics*, **54**, 369–380.

Neyman, J. (1939). On a new class of "contagious" distributions, applicable in entomology and bacteriology. *Annals of Mathematical Statistics*, **10**, 35–57.

Neyman, J. and Scott, E. L. (1958). Statistical approach to problems of cosmology. *Journal of the Royal Statistical Society B*, **20**, 1–29.

Nguyen, X. X. (1979). Ergodic theorems for subadditive spatial processes. *Zeitschrift fur Wahrscheinlichkeitstheorie und verwandte Gebiete*, **48**, 133–158.

Nguyen, X. X. and Zessin, H. (1979). Integral and differential characterizations of the Gibbs process. *Mathematische Nachrichten*, **88**, 105–115.

Norberg, T. (1984). Convergence and existence of random set distributions. *Annals of Probability*, **12**, 726–732.

Norton, L. (1988). A Gompertzian model of human breast cancer growth. *Cancer Research*, **48**, 7067–7071.

Norton, S. (1968). On the discontinuous nature of behavior. *Theoretical Biology*, **21**, 229–243.

Norton, S. (1973). Amphetamine as a model for hyperactivity in the rat. *Physiology and Behavior*, **11**, 181–186.

Norton, S. (1977). The structure of behavior of rats during morphine induced hyperactivity. *Communications in Psychopharmacology*, **1**, 333–341.

O'Brien, C. (1987). A test for non-linearity of prediction in time series. *Journal of Time Series Analysis*, **8**, 313–327.

O'Connor, D. P. H. and Leach, B. G. (1979). Geostatistical analysis of 18CC stope block, C.S.A. mine, Cobar, N.S.W. In *Proceedings of a Symposium: Estimation and Statement of Mineral Reserves*. Australian Institute of Mining and Metallurgy, Melbourne, Australia.

Ogata, Y. (1978). The asymptotic behavior of maximum likelihood estimators for stationary point processes. *Annals of the Institute of Statistical Mathematics A*, **30**, 243–261.

Ogata, Y. (1981). On Lewis' simulation method for point processes. *IEEE Transactions on Information Theory*, **IT–27**, 23-31.

Ogata, Y. (1983). Estimation of the parameters in the modified Omori formula for aftershock frequencies by the maximum likelihood procedure. *Journal of Physics of the Earth*, **31**, 115–124.

Ogata, Y. (1988). Statistical models for earthquake occurrences and residual analysis for point processes. *Journal of the American Statistical Association*, **83**, 9–27.

Ogata, Y. (1989). Statistical modeling for standard seismicity and detection of anomalies by residual analysis. *Tectonophysics*, **169**, 159–174.

Ogata, Y., Akaike, H., and Katsura, K. (1982). The application of linear intensity models to the investigation of causal relationships between a point process and another stochastic process. *Annals of the Institute of Statistical Mathematics*, **34**, 373–387.

Ogata, Y. and Katsura, K. (1986). Point-process models with linearly parameterized intensity for application to earthquake data. *Journal of Applied Probability*, **23A**, 291–310.

Ogata, Y. and Katsura, K. (1988). Likelihood analysis of spatial inhomogeneity for marked point patterns. *Annals of the Institute of Statistical Mathematics*, **40**, 29–39.

Ogata, Y. and Shimazaki, K. (1984). Transition from aftershock to normal activity: The 1965 Rat Islands earthquake aftershock sequence. *Bulletin of the Seismological Society of America*, **74**, 1757–1765.

Ogata, Y. and Tanemura, M. (1981). Estimation of interaction potentials of spatial point patterns through the maximum likelihood procedure. *Annals of the Institute of Statistical Mathematics B*, **33**, 315–338.

Ogata, Y. and Tanemura, M. (1984). Likelihood analysis of spatial point patterns. *Journal of the Royal Statistical Society B*, **46**, 496–518.

Ogata, Y. and Tanemura, M. (1985). Estimation of interaction potentials of marked spatial point patterns through the maximum likelihood method. *Biometrics*, **41**, 421–433.

Ogata, Y. and Tanemura, M. (1986). Likelihood estimation of interaction potentials and external fields of inhomogeneous spatial point patterns. In *Pacific Statistical Congress*, I. S. Francis, B. F. J. Manly, and F. C. Lam, eds. Elsevier, Amsterdam, 150–154.

Ogata, Y. and Tanemura, M. (1989). Likelihood estimation of soft-core interaction potentials for Gibbsian point patterns. *Annals of the Institute of Statistical Mathematics*, **41**, 583–600.

Ohser, J. (1980). On statistical analysis of the Boolean model. *Elektronische Informationsverarbeitung und Kybernetik*, **16**, 651–653.

Ohser, J. (1983). On estimators for the reduced second moment measure of point processes. *Mathematische Operationsforschung und Statistik*, *Series Statistics*, **14**, 63–71.

Ohser, J. and Stoyan, D. (1981). On the second-order and orientation analysis of planar stationary point processes. *Biometrical Journal*, **23**, 523–533.

Olea, R. A. (1974). Optimal contour mapping using universal kriging. *Journal of Geophysical Research*, **79**, 695–702.

Olea, R. A. (1984). Sampling design optimization for spatial functions. *Journal of the International Association for Mathematical Geology*, **16**, 369–392.

Oliver, M. A. and Webster, R. (1986). Combining nested and linear sampling for determining the scale and form of spatial variation of regionalized variables. *Geographical Analysis*, **18**, 227–242.

Olshen, R. A., Biden, E. N., Wyatt, M. P., and Sutherland, D. H. (1989). Gait analysis and the bootstrap. *Annals of Statistics*, **17**, 1419–1440.

Omre, H. (1984). The variogram and its estimation. In *Geostatistics for Natural Resources Characterization, Part 1,* G. Verly, M. David, A. Journel, and A. Marechal, eds. Reidel, Dordrecht, 107–125.

Omre, H. (1987). Bayesian kriging—merging observations and qualified guesses in kriging. *Mathematical Geology*, **19**, 25–39.

Omre, H. and Halvorsen, K. B. (1989). The Bayesian bridge between simple and universal kriging. *Mathematical Geology*, **21**, 767–786.

Openshaw, S. and Taylor, P. J. (1979). A million or so correlation coefficients: Three experiments on the modifiable areal unit problem. In *Statistical Applications in the Spatial Sciences*, N. Wrigley, ed. Pion, London, 127–144.

Ord, J. K. (1975). Estimation methods for models of spatial interaction. *Journal of the American Statistical Association*, **70**, 120–126.

Ord, J. K. (1977). Comment on "Modelling spatial patterns" by B. D. Ripley. *Journal of the Royal Statistical Society B*, **39**, 199.

Ord, J. K. and Rees, M. (1979). Spatial processes: Recent developments with applications to hydrology. In *The Mathematics of Hydrology and Water Resources*, E. H. Lloyd, T. O'Donnell, and J. C. Wilkinson, eds. Academic Press, London, 95–118.

Orey, S. (1971), Lecture Notes on Limit Theorems for Markov Chain Transition Probabilities. Van Nostrand, New York.

Owen, A. (1984). A neighborhood-based classifier for Landsat data. *Canadian Journal of Statistics*, **12**, 191–200.

Palm, C. (1943). Intensitatsschwankungen im fernsprechverkehr. *Ericsson Technics*, **44**, 1–189.

Papadakis, J. S. (1937). Methode statistique pour des experiences sur champ. Institut d'Amelioration des Plantes a Thessaloniki (Grece), *Bulletin Scientifique*, **No. 23**.

Papadakis, J. S. (1970). *Agricultural Research: Principles, Methodology, Suggestions.* Edicion Argentina, Buenos Aires.

Papadakis, J. S. (1984). Advances in the analysis of field experiments. *Proceedings of the Academy of Athens*, **59**, 326–342.

Papageorgiou, N. S. (1989). Convergence and representation theorems for set valued random processes. *Stochastic Analysis and Applications*, **7**, 187–210.

Papangelou, F. (1972). Premonitory seismicity of expected increments of point processes and related random change of scale. *Transactions of the American Mathematical Society*, **165**, 483–506.

Papangelou, F. (1974). The conditional intensity of general point processes and an application to line processes. *Zeitschrift für Wahrscheinlichkeitstheorie und verwandte Gebiete,* **28**, 207–226.

Patankar, V. N. (1954). The goodness of fit of frequency distributions obtained from stochastic processes. *Biometrika*, **41**, 450–462.

Patterson, H. D. and Hunter, E. A. (1983). The efficiency of incomplete block designs in National List and Recommended List cereal variety trials. *Journal of Agricultural Science (Cambridge),* **101**, 427–433.

Patterson, H. D. and Thompson, R. (1971). Recovery of interblock information when block sizes are unequal. *Biometrika*, **58**, 545–554.

Patterson, H. D. and Thompson, R. (1974). Maximum likelihood estimation of components of variance. *Proceedings of the 8th International Biometric Conference*, Biometric Society, Washington, DC, 197–207.

Payendeh, B. (1970). Comparison of methods for assessing spatial distribution of trees. *Forest Science*, **16**, 312–317.

Pearce, S. C. (1976). An examination of Fairfield Smith's law of environmental variation. *Journal of Agricultural Science*, **87**, 21–24.

Pentland, A. P. (1984). Fractal-based description of natural scenes. *IEEE Transactions on Pattern Analysis and Machine Intelligence*, **PAMI-6**, 661–674.

Penttinen, A. (1984). Modelling interaction in spatial point patterns: Parameter estimation by the maximum likelihood method. *Jyvaskyla Studies in Computer Science, Economics and Statistics*, **7**, 1–105.

Peskun, P. H. (1973). Optimum Monte-Carlo sampling using Markov chains. *Biometrika*, **60**, 607–612.

Peters, N. E. and Bonelli, J. E. (1982). Chemical composition of bulk precipitation in the north-central and northeastern United States, December 1980 through February 1981. U.S. Geological Survey Circular 874.

Peters, T. J. and Golding, J. (1986). Prediction of sudden infant death syndrome: An independent evaluation of four scoring methods. *Statistics in Medicine*, **5**, 113–126.

Pfeifer, P. E. and Deutsch, S. J. (1980a). Identification and interpretation of first order space–time ARMA models. *Technometrics*, **22**, 397–408.

Pfeifer, P. E. and Deutsch, S. J. (1980b). A three-stage iterative procedure for space–time modelling. *Technometrics*, **22**, 35–47.

Philip, G. M. and Watson, D. F. (1986). Matheronian geostatistics—Quo vadis? *Mathematical Geology*, **18**, 93–117.

Pickard, D. K. (1982). Inference for general Ising models. *Journal of Applied Probability*, **19A**, 345–357.

Pickard, D. K. (1987). Inference for discrete Markov fields: The simplest nontrivial case. *Journal of the American Statistical Association*, **82**, 90–96.

Pielou, E. C. (1959). The use of point-to-plant distances in the pattern of plant populations. *Journal of Ecology*, **47**, 607–613.

Pielou, E. C. (1960). A single mechanism to account for regular, random and aggregated populations. *Journal of Ecology*, **48**, 575–584.

Pielou, E. C. (1977). *Mathematical Ecology*. Wiley, New York.

Pike, M. C. and Smith, P. G. (1974). A case-control approach to examine diseases for evidence of contagion, including diseases with long latent periods. *Biometrics*, **30**, 263–279.

Pitelka, L. F. and Raynal, D. J. (1989). Forest decline and acidic deposition. *Ecology*, **70**, 2–10.

Platt, W. J., Evans, G. W., and Rathbun, S. L. (1988). The population dynamics of a long-lived conifer (*Pinus palustris*). *American Naturalist*, **131**, 491–525.

Pohl, P. (1976). Spontaneous fluctuation in rate of body rocking: A methodological note. *Journal of Mental Deficiency Research*, **20**, 61–65.

Pollard, J. H. (1971). On distance-estimators of density in randomly distributed forests. *Biometrics*, **27**, 991–1002.

Pontius, A. A. (1973). Dysfunction patterns analogous to frontal lobe system and caudate nucleus syndrome in some groups of minimal brain dysfunction. *Journal of American Medical Women's Association*, **28**, 285–291.

Possolo, A. (1986). Estimation of binary Markov random fields. Technical Report No. 77, Department of Statistics, University of Washington, Seattle, WA.

Prasad, N. G. N. and Rao, J. N. K. (1990). On the estimation of mean square error of small area predictors. *Journal of the American Statistical Association*, **85**, 163–171.

Preston, C. J. (1974). *Gibbs States on Countable Sets*. Cambridge University Press, Cambridge.

Preston, C. J. (1975). Spatial birth-and-death processes. *Bulletin of the International Statistical Institute*, **46**, Book 2, 371–391.

Preston, C. J. (1976). Random Fields. *Lecture Notes in Mathematics*, **No. 534**. Springer, New York.

Priestley, M. B. (1981). *Spectral Analysis and Time Series, Volumes* **I**, **II**. Academic Press, London.

Priestley, M. B. (1988). *Non-Linear and Non-Stationary Time Series Analysis.* Academic Press, London.

Puente, C. E. and Bras, R. L. (1986). Disjunctive kriging, universal kriging, or no kriging: Small sample results with simulated fields. *Mathematical Geology*, **18**, 287–305.

Puri, M. L. and Ralescu, D. A. (1983). Strong law of large numbers for Banach space-valued random sets. *Annals of Probability*, **11**, 222–224.

Puri, M. L. and Ralescu, D. A. (1985). Limit theorems for random compact sets in Banach space. *Mathematical Proceedings of the Cambridge Philosophical Society*, **97**, 151–158.

Qian, W. and Titterington, D. M. (1989). On the use of Gibbs Markov chain models in the analysis of images based on second-order pairwise interaction distributions. *Journal of Applied Statistics*, **16**, 267–281.

Quenouille, M. (1949a). Approximate tests of correlation in time series. *Journal of the Royal Statistical Society B*, **11**, 68–84.

Quenouille, M. (1949b). Problems in plane sampling. *Annals of Mathematical Statistics*, **20**, 355–375.

Quenouille, M. (1956). Notes on bias in estimation. *Biometrika*, **43**, 353–360.

Quimby, W. F. (1986). Selected topics in spatial statistical analysis: Nonstationary vector kriging, large scale conditional simulation of three dimensional Gaussian random fields, and hypothesis testing in a correlated random field. Ph.D. dissertation, Department of Statistics, University of Wyoming, Laramie, WY, unpublished.

Radchenko, A. N. (1985). Measurability of a geometric measure of a level set of a random function. *Theory of Probability and Mathematical Statistics*, **31**, 131–140.

Ramsay, J. B. and Yuan, H. J. (1990). The statistical properties of dimension calculations using small data sets. *Nonlinearity*, **3**, 155–176.

Ranneby, B. (1982). Stochastic models of variation in time and space. In *Statistics in Theory and Practice. Essays in Honour of Bertil Matern*, B. Ranneby, ed. Swedish University of Agricultural Sciences, Umea, Sweden, 227–245.

Ransom, R. (1977). A computer model of cell clone growth. *Simulation*, **28**, 189–192.

Rao, C. R. (1961). Combinatorial arrangements analogous to orthogonal arrays. *Sankhyā*, **23**, 283–286.

Rao, C. R. (1972). Estimation of variance and variance components. *Journal of the American Statistical Association*, **67**, 112–115.

Rao, C. R. (1973). *Linear Statistical Inference and Its Applications,* 2nd ed. Wiley, New York.

Rao, C. R. (1979). MINQE theory and its relation to ML and MML estimation of variance components. *Sankhyā B*, **41**, 138–153.

Rao, P. V., Rao, P. S., Davidson, J. M., and Hammond, L. C. (1979). Use of goodness-of-fit tests for characterizing the spatial variability of soil properties. *Soil Science Society of America Journal*, **43**, 274–278.

Rathbun, S. L. and Cressie, N. (1990a). Asymptotic properties of estimators for the parameters of a spatial inhomogeneous Poisson point process. Statistical Laboratory Preprint No. 90-2, Iowa State University, Ames, IA. [Forthcoming in *Advances in Applied Probabilty*.]

Rathbun, S. L. and Cressie, N. (1990b). A space–time survival point process for a longleaf pine forest in southern Georgia. Statistical Laboratory Preprint No. 90-13, Iowa State University, Ames, IA. [Tentatively accepted to appear in *Journal of the American Statistical Association*.]

Record, F. A., Bubenick, D. V., and Kindya, R. J. (1982). *Acid Rain Information Book*. Noyes Data Corporation, Park Ridge, NJ.

Reinsel, G. (1984). Estimation and prediction in a multivariate random effects generalized linear model. *Journal of the American Statistical Association*, **79**, 406–414.

Rendu, J. M. (1979). Normal and lognormal estimation. *Journal of the International Association for Mathematical Geology*, **11**, 407–422.

Renshaw, E. and Ford, E. D. (1983). The interpretation of process from pattern using two-dimensional spectral analysis: Methods and problems of interpretation. *Applied Statistics*, **32**, 51–63.

Renyi, A. (1967). Remarks on the Poisson process. *Studia Scientiarum Mathematicarum Hungarica*, **2**, 119–123.

Rice, S. O. (1954). Mathematical analysis of random noise. In *Selected Papers on Noise and Stochastic Processes*, N. Wax, ed. Dover, New York, 133–294.

Richardson, D. (1973). Random growth in a tessellation. *Mathematical Proceedings of the Cambridge Philosophical Society*, **74**, 515–528.

Ripley, B. D. (1976a). The second-order analysis of stationary point processes. *Journal of Applied Probability*, **13**, 255–266.

Ripley, B. D. (1976b). Locally finite random sets: Foundations for point process theory. *Annals of Probability*, **4**, 983–994.

Ripley, B. D. (1977). Modelling spatial patterns. *Journal of the Royal Statistical Society B*, **39**, 172–192.

Ripley, B. D. (1978). Spectral analysis and the analysis of pattern in plant communities. *Journal of Ecology*, **66**, 965–981.

Ripley, B. D. (1979a). Tests of 'randomness' for spatial point patterns. *Journal of the Royal Statistical Society B*, **41**, 368–374.

Ripley, B. D. (1979b). Simulating spatial patterns: Dependent samples from a multivariate density. *Applied Statistics*, **28**, 109–112.

Ripley, B. D. (1981). *Spatial Statistics*. Wiley, New York.

Ripley, B. D. (1982). Edge effects in spatial stochastic processes. In *Statistics in Theory and Practice. Essays in Honour of Bertil Matern*, B. Ranneby, ed. Swedish University of Agricultural Sciences, Umea, Sweden, 247–262.

Ripley, B. D. (1983). Computer generation of random variables: A tutorial. *International Statistical Review*, **51**, 301–319.

Ripley, B. D. (1984). Spatial statistics: Developments 1980–1983. *International Statistical Review*, **52**, 141–150.

Ripley, B. D. (1987a). *Stochastic Simulation*. Wiley, New York.

Ripley, B. D. (1987b). Spatial point pattern analysis in ecology. In *Developments in Numerical Ecology*, P. Legendre and L. Legendre, eds. Springer, Berlin, 407–429.

Ripley, B. D. (1988). *Statistical Inference for Spatial Processes*. Cambridge University Press, Cambridge, UK.

Ripley, B. D. and Kelly, F. P. (1977). Markov point processes. *Journal of the London Mathematical Society*, **15**, 188–192.

Ripley, B. D. and Rasson, J. P. (1977). Finding the edge of a Poisson forest. *Journal of Applied Probability*, **14**, 483–491.

Ripley, B. D. and Silverman, B. W. (1978). Quick tests for spatial interaction. *Biometrika*, **65**, 641–642.

Rissanen, J. (1984). Universal coding, information, prediction, and estimation. *IEEE Transactions on Information Theory*, **IT-30**, 629–636.

Rissanen, J. (1987). Stochastic complexity. *Journal of the Royal Statistical Society B*, **49**, 223–239.

Ritter, G. X., Wilson, J. N., and Davidson, J. L. (1990). Image algebra: An overview. *Computer Vision, Graphics, and Image Processing*, **49**, 297–331.

Rivoirard, J. (1987). Two key parameters when choosing the kriging neighborhood. *Mathematical Geology*, **19**, 851–856.

Roach, S. A. (1968). *The Theory of Random Clumping*. Methuen, London.

Robbins, H. E. (1944). On the measure of a random set. *Annals of Mathematical Statistics*, **15**, 70–74.

Robinson, P. M. (1983). Nonparametric estimators for time series. *Journal of Time Series Analysis*, **4**, 185–207.

Rodriguez-Iturbe, I., Cox, D. R., and Isham, V. (1988). A point process model for rainfall: Further developments. *Proceedings of the Royal Society of London A*, **417**, 283–298.

Rodriguez-Iturbe, I. and Eagleson, P. S. (1987). Mathematical models of rainstorm events in space and time. *Water Resources Research*, **23**, 181–190.

Rogers, A. (1974). *Statistical Analysis of Spatial Dispersion*. Pion, London.

Rosanov, Y. A. (1967). On the Gaussian homogeneous fields with given conditional distributions. *Theory of Probability and Its Applications*, **12**, 381–391.

Rosenblatt, M. (1985). *Stationary Sequences and Random Fields*. Birkhauser, Boston.

Rosenfeld, A. and Kak, A. C. (1976). *Digital Picture Processing*. Academic Press, New York.

Rosenkrantz, R. D., ed. (1983). *E. T. Jaynes: Papers on Probability, Statistics, and Statistical Physics.* Reidel, Dordrecht.

Ross, D. (1986). Random sets without separability. *Annals of Statistics*, **14**, 1064–1069.

Rothenberg, T. J. (1984). Approximate normality of generalized least squares. *Econometrica*, **52**, 811–825.

Roussas, G. G. (1969). Nonparametric estimation in Markov processes. *Annals of the Institute of Statistical Mathematics*, **21**, 73–87.

Roussas, G. G. (1972). *Contiguity of Probability Measures*. Cambridge University Press, Cambridge.

Rowlinson, J. S. (1959). *Liquids and Liquid Mixtures*. Academic Press, New York.

Royden, H. L. (1968). *Real Analysis,* 2nd ed. MacMillan, New York.

Rubin, H. (1982). Some remarks on cancer as a state of disorganization at the cellular and supracellular levels. In *Probability Models and Cancer*, L. Le Cam and J. Neyman, eds. North-Holland, Amsterdam, 211–219.

Rubin, H. (1985). The significance of cell variation in understanding cancer. In *Proceedings of the Berkeley Conference in Honor of Jerzy Neyman and Jack Kiefer, Vol.* **I**, L. M. Le Cam and R. A. Olshen, eds. Wadsworth, Belmont, CA., 321–330.

Rudemo, M. (1972). Doubly stochastic Poisson processes and process control. *Advances in Applied Probability*, **4**, 318–338.

Rudemo, M. (1982). Empirical choice of histograms and kernel density estimators. *Scandinavian Journal of Statistics*, **9**, 65–78.

Ruelle, D. (1969). *Statistical Mechanics: Rigorous Results.* Benjamin, New York.

Russell, K. G. and Eccleston, J. A. (1987). The construction of optimal incomplete block designs when observations within a block are correlated. *Australian Journal of Statistics*, **29**, 293–302.

Russo, D. and Bresler, E. (1981). Soil hydraulic properties as stochastic processes: I. An analysis of field spatial variability. *Soil Science Society of America Journal*, **45**, 682–687.

Russo, D. and Jury, W. A. (1987a). A theoretical study of the estimation of the correlation scale in spatially variable fields. 1. Stationary fields. *Water Resources Research*, **23**, 1257–1268.

Russo, D. and Jury, W. A. (1987b). A theoretical study of the estimation of the correlation scale in spatially variable fields. 2. Nonstationary fields. *Water Resources Research*, **23**, 1269–1279.

Sabourin, R. (1976). Application of two methods for the interpretation of the underlying variogram. In *Advanced Geostatistics in the Mining Industry*, M. Guarascio, M. David, and C. Huijbregts, eds. Reidel, Dordrecht, 101–109.

Sacks, J. and Schiller, S. (1988). Spatial designs. In *Statistical Decision Theory and Related Topics IV, Vol.* **2**, S. S. Gupta and J. O. Berger, eds. Springer, New York, 385–399.

Sacks, J., Welch, W. J., Mitchell, T. J., and Wynn, H. P. (1989). Design and analysis of computer experiments. *Statistical Science*, **4**, 409–423.

Sacks, J. and Ylvisaker, D. (1966). Designs for regression problems with correlated errors. *Annals of Mathematical Statistics*, **37**, 66–89.

Sacks, J. and Ylvisaker, D. (1968). Designs for regression problems with correlated errors; many parameters. *Annals of Mathematical Statistics*, **39**, 49–69.

Sacks, J. and Ylvisaker, D. (1970). Designs for regression problems with correlated errors. III. *Annals of Mathematical Statistics*, **41**, 2057–2074.

Saidel, G., Liotta, L., and Kleinerman, J. (1976). System dynamics of a metastatic process from implanted tumor. *Journal of Theoretical Biology*, **56**, 417–434.

Salinetti, G. and Wets, R. (1986). On the convergence in distribution of measurable multifunctions (random sets), normal integrands, stochastic processes, and stochastic infima. *Mathematics of Operations Research*, **11**, 385–419.

Salkauskas, K. (1982). Some relationships between surface splines and kriging. In *Multivariate Approximation Theory II*, W. Schempp and K. Zeller, eds. Birkhauser, Basel, 313–325.

Samet, H., Rosenfeld, A., Shaffer, C. A., and Webber, R. E. (1984). A geographic information system using quadtrees. *Pattern Recognition*, **17**, 647–656.

Samper, F. J. and Neuman, S. P. (1989). Estimation of spatial covariance structures by adjoint state maximum likelihood cross validation 1. Theory. *Water Resources Research*, **25**, 351–362.

Sampson, P. D. and Guttorp, P. (1992). Nonparametric estimation of nonstationary spatial covariance structure. *Journal of the American Statistical Association*, **87**, 108–119.

Samra, J. S., Gill, H. S., and Bhatia, V. K. (1989). Spatial stochastic modeling of growth and forest resource evaluation. *Forest Science*, **35**, 663–676.

Sanchez, E. (1979). Inverses of fuzzy relations, application to possibility distributions and medical diagnosis. *Fuzzy Sets and Systems*, **2**, 75–86.

SAS Institute Inc. (1985). *SAS User's Guide: Statistics, Version 5 Edition.* SAS Institute, Cary, NC.

Satyamurthi, K. R. (1979). Density, derived from measured distances, for studying the spatial patterns. *Sankhyā B*, **40**, 197–203.

Saunders, R. and Funk, G. M. (1977). Poisson limits for a clustering model of Strauss. *Journal of Applied Probability*, **14**, 776–784.

Saunders, R., Kryscio, R. J., and Funk, G. M. (1982). Poisson limits for a hard-core clustering model. *Stochastic Processes and Their Applications*, **12**, 97–106.

Sawyer, S. (1977). Rates of consolidation in a selectively neutral migration model. *Annals of Probability*, **5**, 486–493.

Scheffe, H. (1959). *The Analysis of Variance*. Wiley, New York.

Schoenberg, I. J. (1938). Metric spaces and completely monotone functions. *Annals of Mathematics*, **39**, 811–841.

Schoenfelder, C. and Cambanis, S. (1982). Random designs for estimating integrals of stochastic processes. *Annals of Statistics*, **10**, 526–538.

Scholz, F. W. (1978). Weighted median regression estimates. *Annals of Statistics*, **6**, 603–609.

Schurger, K. (1979). On the asymptotic geometrical behavior of a class of contact interaction processes with a monotone infection rate. *Zeitschrift für Wahrscheinlichkeitstheorie und verwandte Gebiete,* **48**, 35–48.

Schurger, K. (1983). Ergodic theorems for subadditive superstationary families of convex compact random sets. *Zeitschrift für Wahrscheinlichkeitstheorie und verwandte Gebiete,* **62**, 125–135.

Schwarz, G. (1978). Estimating the dimension of a model. *Annals of Statistics*, **6**, 461–464.

Searle, S. R. (1971). *Linear Models*. Wiley, New York.

Seheult, A. H. and Tukey, J. W. (1982). Some resistant procedures for analyzing 2^n factorial experiments. *Utilitas Mathematica*, **21B**, 57–98.

Sen, P. K. (1972). On the Bahadur representation of sample quantiles for sequences of ϕ-mixing random variables. *Journal of Multivariate Analysis*, **2**, 77–95.

Serra, J. (1980). The Boolean model and random sets. *Computer Graphics and Image Processing*, **12**, 99–126.

Serra, J. (1982). *Image Analysis and Mathematical Morphology*. Academic Press, London.

Shafer, G. (1976). *A Mathematical Theory of Evidence*. Princeton University Press, Princeton.

Shafer, G. (1987). Belief functions and possibility measures. In *Analysis of Fuzzy Information*, *Vol.* **1**, J. C. Bezdek, ed. CRC Press, Boca Raton, FL, 51–84.

Shannon, C. (1948). A mathematical theory of communication. *Bell System Technical Journal*, **27**, 379–423.

Shapiro, M. B., Schein, S. J., and de Monasterio, F. M. (1985). Regularity and structure of the spatial pattern of blue cones of Macaque retina. *Journal of the American Statistical Association*, **80**, 803–812.

Sharma, M. G., Gander, G. A., and Hunt, G. G. (1980). Spatial variability of infiltration in a watershed. *Journal of Hydrology*, **45**, 101–122.

Sharma, S. C. (1986). The effects of correlation among observations on the consistency property of sample variance. *Communications in Statistics. Theory and Methods*, **15**, 1125–1152.

Sharp, W. E. and Aroian, L. A. (1985). The generation of multidimensional autoregressive series by the herringbone method. *Journal of the International Association for Mathematical Geology*, **17**, 67–79.

Shepp, L. A. and Kruskal, J. (1979). Computerized tomography: The new medical X-ray technology. *American Mathematical Monthly*, **85**, 420–439.

Shimizu, K. and Iwase, K. (1987). Unbiased estimation of the autocovariance function in a stationary generalized lognormal process. *Communications in Statistics. Theory and Methods*, **16**, 2145–2154.

Shinozuka, M. (1971). Simulation of multivariate and multidimensional random processes. *Journal of the Acoustical Society of America*, **49**, 357–367.

Shinozuka, M. and Jan, C. M. (1972). Digital simulation of random processes and its applications. *Journal of Sound and Vibration*, **25**, 111–128.

Shurtz, R. F. (1985). Letter to the Editor. *Journal of the International Association for Mathematical Geology*, **17**, 861–868.

Shymko, R. M. and Glass, L. (1976). Cellular and geometric control of tissue growth and mitotic instability. *Journal of Theoretical Biology*, **63**, 355–374.

Sibson, R. (1981). A brief description of natural neighbor interpolation. In *Interpreting Multivariate Data*, V. Barnett, ed. Wiley, New York, 21–36.

Siegel, A. F. and Benson, R. H. (1982). A robust comparison of biological shapes. *Biometrics*, **38**, 341–350.

Sievers, G. L. (1978). Weighted rank statistics for simple linear regression. *Journal of the American Statistical Association*, **73**, 628–631.

Silverman, B. W. (1978a). Choosing the window width when estimating a density. *Biometrika*, **65**, 1–11.

Silverman, B. W. (1978b). Distances on circles, toruses and spheres. *Journal of Applied Probability*, **15**, 136–143.

Silverman, B. W. (1981). Density estimation for univariate and bivariate data. In *Interpreting Multivariate Data*, V. Barnett, ed. Wiley, New York, 37–53.

Silverman, B. W. (1984). Spline smoothing: The equivalent variable kernel method. *Annals of Statistics*, **12**, 898–916.

Silverman, B. W. (1986). *Density Estimation*. Chapman and Hall, New York.

Silverman, B. W. and Brown, T. C. (1978). Short distances, flat triangles and Poisson limits. *Journal of Applied Probability*, **15**, 815–825.

Silverman, B. W., Jennison, C., Stander, J., and Brown, T. C. (1990). The specification of edge penalties for regular and irregular pixel images. *IEEE Transactions on Pattern Analysis and Machine Intelligence*, **12**, 1017–1024.

Silvey, S. D. (1959). The Lagrangian multiplier test. *Annals of Mathematical Statistics*, **30**, 389–407.

Sinai, Y. G. (1982). *Theory of Phase Transitions: Rigorous Results.* Pergamon, Oxford.

Singh, B. B. and Shukla, G. K. (1983). A test of autoregression in Gaussian spatial processes. *Biometrika*, **70**, 523–527.

Singh, K. (1981). On the asymptotic accuracy of Efron's bootstrap. *Annals of Statistics*, **9**, 1187–1195.

Skellam, J. G. (1952). Studies in statistical ecology, I: Spatial pattern. *Biometrika*, **39**, 346–362.

Skovgaard, I. M. (1985). A second-order investigation of asymptotic ancillarity. *Annals of Statistics*, **13**, 534–551.

Small, C. G. (1988). Techniques of shape analysis on sets of points. *International Statistical Review*, **56**, 243–257.

Smith, J. A. and Karr, A. F. (1985). Parameter estimation for a model of space–time rainfall. *Water Resources Research*, **21**, 1251–1257.

Snedecor, G. W. (1937). *Statistical Methods*. Iowa State College Press, Ames, IA.

Snyder, D. L. (1975). *Random Point Processes*. Wiley, New York.

Solo, V. (1986). Modelling of 2D random fields by parametric cepstrum. *IEEE Transactions on Information Theory*, **IT-32**, 743–750.

Solomon, H. (1953). Distribution of the measure of a random two-dimensional set. *Annals of Mathematical Statistics*, **24**, 650–656.

Solow, A. R. (1985). Bootstrapping correlated data. *Journal of the International Association for Mathematical Geology*, **17**, 769–775.

Solow, A. R. (1986). Mapping by simple indicator kriging. *Mathematical Geology*, **18**, 335–352.

Soltani, A. R. (1984). Extrapolation and moving average representation for stationary random fields and Beurling's Theorem. *Annals of Probability*, **12**, 120–132.

Soule, H. D., Vazquez, J., Long, A., Albert, S., and Brennan, M. (1973). A human cell line from a pleural effusion derived from a breast carcinoma. *Journal of the National Cancer Institute*, **51**, 1409–1413.

Speed, T. P. and Kiiveri, H. T. (1986). Gaussian Markov distributions over finite graphs. *Annals of Statistics*, **14**, 138–150.

Spitzer, F. (1971). Markov random fields and Gibbs ensembles. *American Mathematical Monthly*, **78**, 142–154.

Sposito, V. A. (1987). On median polish and L_1 estimators. *Computational Statistics and Data Analysis*, **5**, 155–162.

Starks, T. H. and Fang, J. H. (1982a). The effect of drift on the experimental semivariogram. *Journal of the International Association for Mathematical Geology*, **14**, 309–320.

Starks, T. H. and Fang, J. H. (1982b). On the estimation of the generalized covariance function. *Journal of the International Association for Mathematical Geology*, **14**, 57–64.

Stein, M. L. (1985). A simple model for spatial–temporal processes with an application to estimation of acid deposition. Technical Report No. 82, Department of Statistics, Stanford University, Stanford, CA.

Stein, M. L. (1986). A simple model for spatial–temporal processes. *Water Resources Research*, **22**, 2107–2110.

Stein, M. L. (1987a). Gaussian approximations to conditional distributions for multi-Gaussian processes. *Mathematical Geology*, **19**, 387–405.

Stein, M. L. (1987b). Minimum norm quadratic estimation of spatial variograms. *Journal of the American Statistical Association*, **82**, 765–772.

Stein, M. L. (1988). Asymptotically efficient prediction of a random field with a misspecified covariance function. *Annals of Statistics*, **16**, 55–63.

Stein, M. L. (1989a). Asymptotic distributions of minimum norm quadratic estimators of the covariance function of a Gaussian random field. *Annals of Statistics*, **17**, 980–1000.

Stein, M. L. (1989b). The loss of efficiency in kriging prediction caused by misspecification of the covariance structure. In *Geostatistics*, *Vol.* **1**, M. Armstrong, ed. Kluwer, Dordrecht, 273–282.

Stein, M. L. and Handcock, M. S. (1989). Some asymptotic properties of kriging when the covariance function is misspecified. *Mathematical Geology*, **21**, 171–190.

Steinschneider, A. (1972). Prolonged apnea and the sudden infant death syndrome: clinical and laboratory observations. *Pediatrics*, **50**, 646–654.

Stephan, F. (1934). Sampling errors and interpretations of social data ordered in time and space. In *Proceedings of the American Statistical Journal, New Series No.* **185A**, F. A. Ross, ed. *Journal of the American Statistical Association*, **29** Suppl., 165–166.

Sterio, D. C. (1984). The unbiased estimation of number and sizes of arbitrary particles using the disector. *Journal of Microscopy*, **134**, 127–136.

Stoffer, D. S. (1985). Maximum likelihood fitting of STARMAX models to incomplete space–time series data. In *Time Series Analysis: Theory and Practice 6*, O. D. Anderson, J. K. Ord, and E. A. Robinson, eds. North-Holland, Amsterdam, 283–296.

Stoffer, D. S. (1986). Estimation and identification of space–time ARMAX models in the presence of missing data. *Journal of the American Statistical Association*, **81**, 762–772.

Stone, C. J. (1977). Consistent nonparametric regression. *Annals of Statistics*, **5**, 595–620.

Stone, M. (1974). Cross-validatory choice and assessment of statistical predictions. *Journal of the Royal Statistical Society B*, **36**, 111–133.

Stoyan, D. (1979a). Interrupted point processes. *Biometrical Journal*, **21**, 607–610.

Stoyan, D. (1979b). Applied stochastic geometry: A survey. *Biometrical Journal*, **21**, 693–715.

Stoyan, D. (1983). Inequalities and bounds for variances of point processes and fibre processes. *Mathematische Operationsforschung und Statistik, Series Statistics*, **14**, 409–419.

Stoyan, D. (1984a). On correlations of marked point processes. *Mathematische Nachrichten*, **116**, 197–207.

Stoyan, D. (1984b). Correlations of the marks of marked point processes—statistical inference and simple models. *Elektronische Informationsverarbeitung und Kybernetik*, **20**, 285–294.

Stoyan, D. (1988). Thinnings of point processes and their use in the statistical analysis of a settlement pattern with deserted villages. *Statistics*, **19**, 45–56.

Stoyan, D. (1989). Statistical inference for a Gibbs point process of mutually non-intersecting discs. *Biometrical Journal*, **31**, 153–161.

Stoyan, D., Kendall, W. S., and Mecke, J. (1987). *Stochastic Geometry and Its Applications*. Wiley, New York and Akademie-Verlag, Berlin.

Stoyan, D. and Ohser, J. (1984). Cross-correlation measures of weighted random measures and their estimation. *Theory of Probability and Its Applications*, **29**, 345–355.

Stoyan, D. and Stoyan, H. (1985). On one of Matern's hard-core point process models. *Mathematische Nachrichten*, **122**, 205–214.

Strauss, D. J. (1975a). A model for clustering. *Biometrika*, **62**, 467–475.

Strauss, D. J. (1975b). Analyzing binary lattice data with the nearest neighbor property. *Journal of Applied Probability*, **12**, 702–712.

Strauss, D. J. (1977). Clustering on coloured lattices. *Journal of Applied Probability*, **14**, 135–143.

Strauss, D. J. (1986). On a general class of models for interaction. *SIAM Review*, **28**, 513–527.

Strauss, D. J. and Ikeda, M. (1990). Pseudolikelihood estimation for social networks. *Journal of the American Statistical Association*, **85**, 204–212.

Student (1907). On the error of counting with a haemacytometer. *Biometrika*, **5**, 351–360.

Student (1914). The elimination of spurious correlation due to position in time or space. *Biometrika*, **10**, 179–180.

Sullivan, J. (1984). Conditional recovery estimation through probability kriging—theory and practice. In *Geostatistics for Natural Resources Characterization, Part 1*, G. Verly, M. David, A. Journel, and A. Marechal, eds. Reidel, Dordrecht, 365–384.

Svedberg, T. (1922). Ett bidrag till de statistika metodernas anvandning inom vaxtbiologien. *Svensk Botanisk Tidskrift*, **16**, 1–8.

Swain, P. H., Vardeman, S. B., and Tilton, J. C. (1981). Contextual classification of multispectral image data. *Pattern Recognition*, **13**, 429–441.

Swallow, W. H. and Monahan, J. F. (1984). Monte Carlo comparison of ANOVA, MIVQUE, REML, and ML estimators of variance components. *Technometrics*, **26**, 47–57.

Swan, G. W. (1987). Tumor growth models and cancer chemotherapy. In *Cancer Modeling*, J. R. Thompson and B. W. Brown, eds. Marcel Dekker, New York, 91–179.

Sweeting, T. J. (1980). Uniform asymptotic normality of the maximum likelihood estimator. *Annals of Statistics*, **8**, 1375–1381.

Swendsen, R. H. and Wang, J. S. (1987). Nonuniversal critical dynamics in Monte Carlo simulations. *Physical Review Letters*, **58**, 86–88.

Switzer, P. (1975). Estimation of the accuracy of qualitative maps. In *Display and Analysis of Spatial Data*, J. C. Davis and M. J. McCullagh, eds. Wiley, New York, 1–13.

Switzer, P. (1976). Geometrical measures of the smoothness of random functions. *Journal of Applied Probability*, **13**, 86–95.

Switzer, P. (1977). Estimation of spatial distributions from point sources with application to air pollution measurement. *Bulletin of the International Statistical Institute*, **47**, Book 2, 123–137.

Switzer, P. (1980). Extension of discriminant analysis for statistical classification of remotely sensed satellite imagery. *Journal of the International Association for Mathematical Geology*, **12**, 367–376.

Switzer, P. (1983). Some spatial statistics for the interpretation of satellite data. *Bulletin of the International Statistical Institute*, **50**, Book 2, 962–972.

Switzer, P. (1986). Statistical image processing. In *Quantitative Analysis of Mineral and Energy Resources*, C. F. Chung, A. G. Fabbri, and R. Sinding-Larsen, eds. Reidel, Dordrecht, 271–282.

Switzer, P. (1989). Non-stationary spatial covariances estimated from monitoring data. In *Geostatistics*, *Vol.* **1**, M. Armstrong, ed. Kluwer, Dordrecht, 127–138.

Switzer, P., Kowalik, W. S., and Lyon, R. J. P. (1982). A prior method for smoothing discriminant analysis classification maps. *Journal of the International Association for Mathematical Geology*, **14**, 433–444.

Symons, M. J., Grimson, R. C., and Yuan, Y. C. (1983). Clustering of rare events. *Biometrics*, **39**, 193–205.

Szidarovsky, F., Baafi, E. Y., and Kim, Y. C. (1987). Kriging without negative weights. *Mathematical Geology*, **19**, 549–559.

Takacs, R. (1986). Estimator for the pair-potential of a Gibbsian point process. *Statistics*, **17**, 429–433.

Takahashi, T., Iwama, N., and Yaegashi, H. (1986). The 3-D microstructure of cancer and its topological properties. In *Science on Form: Proceedings of the First International Symposium for Science on Form*, Y. Kato et al., eds. KTK Scientific Publishers, Tokyo, 543–551.

Takahata, H. (1983). On the rates in the central limit theorem for weakly dependent random fields. *Zeitschrift für Wahrscheinlichkeitstheorie und verwandte Gebiete*, **64**, 445–456.

Takashima, K. (1989). Sample path properties of ergodic self-similar processes. *Osaka Journal of Mathematics*, **26**, 159–189.

Tallis, G. M. and Davis, A. W. (1984). The nucleus problem. *Biometrical Journal*, **26**, 95–100.

Tamura, Y. (1987). An approach to the nonstationary process analysis. *Annals of the Institute of Statistical Mathematics*, **39**, 227–241.

Taqqu, M. S. (1988). Self-similar processes. Entry in *Encyclopedia of Statistical Sciences*, *Vol.* **8**, S. Kotz and N. L. Johnson, eds. Wiley, New York, 352–357.

Tarboton, D. G., Bras, R. L., and Rodriguez-Iturbe, I. (1988). The fractal nature of river networks. *Water Resources Research*, **24**, 1317–1322.

Tautu, P. (1986). Stochastic spatial processes in biology: A concise historical survey. In *Stochastic Spatial Processes*, P. Tautu, ed. *Lecture Notes in Mathematics*, **No. 1212**. Springer, New York, 1–41.

Taylor, C. C. and Burrough, P. A. (1986). Multiscale sources of spatial variation in soil. III. Improved methods for fitting the nested model to one-dimensional semivariograms. *Mathematical Geology*, **18**, 811–821.

Taylor, J. M. G. (1987). Using a generalized mean as a measure of location. *Biometrical Journal*, **29**, 731–738.

Taylor, R. L. and Inoue, H. (1985). Convergence of weighted sums of random sets. *Stochastic Analysis and Applications*, **3**, 379–396.

Taylor, S. J. (1986). The measure theory of random fractals. *Mathematical Proceedings of the Cambridge Philosophical Society*, **100**, 383–406.

Theiler, J. (1990). Statistical precision of dimension estimates. *Physical Reviews*, **A41**, 3038–3051.

Thiebaux, H. J. and Pedder, M. A. (1987). *Spatial Objective Analysis with Applications in Atmospheric Science*. Academic Press. London.

Thomopoulos, S. C. A. (1985). Minimum error entropy estimation and entropic prediction filtering: An optimal predictive coding scheme. *IEEE Transactions on Information Theory*, **IT-31**, 697–703.

Thompson, H. R. (1955). Spatial point processes with applications to ecology. *Biometrika*, **42**, 102–115.

Thompson, H. R. (1956). Distribution of distance to *n*th neighbour in a population of randomly distributed individuals. *Ecology*, **37**, 391–394.

Thompson, J. A. (1956). A note on the balanced incomplete block designs. *Annals of Mathematical Statistics*, **27**, 842–846.

Tiao, G. C., Box, G. E. P., and Hamming, W. J. (1975). A statistical analysis of the Los Angeles ambient carbon monoxide data 1955–1972. *Journal of Air Pollution Control Association*, **25**, 1130–1136.

Tilton, J. C., Vardeman, S. B., and Swain, P. H. (1982). Estimation of context for statistical classification of multispectral image data. *IEEE Transactions on Geoscience and Remote Sensing*, **20**, 445–452.

Titterington, D. M. (1985a). General structure of regularization procedures in image reconstruction. *Astronomy and Astrophysics*, **144**, 381–387.

Titterington, D. M. (1985b). Common structure of smoothing techniques in statistics. *International Statistical Review*, **53**, 141–170.

Tjostheim, D. (1978). Statistical spatial series modelling. *Advances in Applied Probability*, **10**, 130–154.

Tjostheim, D. (1983). Statistical spatial series modelling II: Some further results on unilateral lattice processes. *Advances in Applied Probability*, **15**, 562–584.

Tobler, W. R. and Kennedy, S. (1985). Smooth multidimensional interpolation. *Geographical Analysis*, **17**, 251–257.

Tompson, A. F. B., Ababou, R., and Gelhar, L. W. (1989). Implementation of the three-dimensional turning bands random field generator. *Water Resources Research*, **25**, 2227–2243.

Toutenburg, H. (1982). *Prior Information in Linear Models*. Wiley, New York.

Toyooka, Y. (1982). Prediction error in a linear model with estimated parameters. *Biometrika*, **69**, 453–459.

Trader, D. (1981). Infinitely divisible random sets. Ph.D. Dissertation, Department of Statistics, Carnegie-Mellon University, Pittsburgh, PA, unpublished.

Trader, D. and Eddy, W. F. (1981). A central limit theorem for Minkowski sums of random sets. Technical Report No. 228, Department of Statistics, Carnegie-Mellon University, Pittsburgh, PA.

Trangmar, B. B., Yost, R. S., and Uera, G. (1986). Spatial dependence and interpolation of soil properties in West Sumatra, Indonesia: II. Coregionalization and co-kriging. *Soil Science Society of America Journal*, **50**, 1396–1400.

Trench, W. F. (1964). An algorithm for the inversion of finite Toeplitz matrices. *Journal of the Society for Industrial and Applied Mathematics*, **12**, 515–522.

Tsaknakis, H., Kazakos, D., and Papantoni-Kazakos, P. (1986). Robust prediction and interpolation for vector stationary processes. *Probability Theory and Related Fields*, **72**, 589–602.

Tsay, R. S. and Tiao, G. C. (1984). Consistent estimates of autoregressive parameters and extended sample autocorrelation function for stationary and nonstationary ARMA models. *Journal of the American Statistical Association*, **79**, 84–96.

Tsutakawa, R. K., Shoop, G. L., and Marienfeld, C. J. (1985). Empirical Bayes estimation of cancer mortality rates. *Statistics in Medicine*, **4**, 201–212.

Tubilla, A. (1975). Error convergence rates for estimates of multidimensional integrals of random functions. Technical Report No. 72, Department of Statistics, Stanford University, Stanford, CA.

Tufte, E. (1983). *The Visual Display of Quantitative Information*. Graphics Press, Cheshire, CT.

Tukey, J. W. (1949). One degree of freedom for non-additivity. *Biometrics*, **5**, 232–242.

Tukey, J. W. (1958). Bias and confidence in not-quite large samples (abstract). *Annals of Mathematical Statistics*, **29**, 614.

Tukey, J. W. (1960). A survey of sampling from contaminated distributions. In *Contributions to Probability and Statistics*, I. Olkin, ed. Stanford University Press, Stanford, CA, 448–485.

Tukey, J. W. (1977). *Exploratory Data Analysis*. Addison-Wesley, Reading, MA.

Tunicliffe-Wilson, G. (1989). On the use of marginal likelihood in time series model estimation. *Journal of the Royal Statistical Society B*, **51**, 15–27.

Tyan, S. G. (1981). Median filtering: Deterministic properties. In *Two Dimensional Digital Signal Processing II. Topics in Applied Physics*, **Vol. 43**, T. S. Huang, ed. Springer, New York, 197–217.

Upton, G. J. G. and Fingleton, B. (1985). *Spatial Data Analysis by Example, Volume* **1***: Point Pattern and Quantitative Data.* Wiley, Chichester.

Upton, G. J. G. and Fingleton, B. (1989). *Spatial Data Analysis by Example, Volume* **2***: Categorical and Directional Data.* Wiley, Chichester.

Usher, M. B. (1969). The relation between mean square and block size in the analysis of similar patterns. *Journal of Ecology*, **57**, 505–514.

Usher, M. B. (1975). Analysis of pattern in real and artificial plant populations. *Journal of Ecology*, **63**, 569–586.

Vanmarcke, E. (1983). *Random Fields*. MIT Press, Cambridge, MA.

Vardi, Y., Shepp, L. A., and Kaufman, L. (1985). A statistical model for positron emission tomography. *Journal of the American Statistical Association*, **80**, 8–20.

Varga, R. S. (1962). *Matrix Iterative Analysis*. Prentice-Hall, Englewood Cliffs, NJ.

Vecchia, A. V. (1985). A general class of models for stationary two-dimensional random processes. *Biometrika*, **72**, 281–291.

Vecchia, A. V. (1988). Estimation and model identification for continuous spatial processes. *Journal of the Royal Statistical Society B*, **50**, 297–312.

Velleman, P. F. and Hoaglin, D. C. (1981). *Applications, Basics, and Computing of Exploratory Data Analysis*. Duxbury, Boston, MA.

Vere-Jones, D. (1970). Stochastic models for earthquake occurrence. *Journal of the Royal Statistical Society B*, **32**, 1–45.

Ver Hoef, J. M. and Cressie, N. (1991). Multivariable spatial prediction. Statistical Laboratory Preprint No. 91-13, Iowa State University, Ames, IA. [Forthcoming in *Mathematical Geology*.]

Ver Hoef, J. M., Cressie, N., and Glenn-Lewin, D. C. (1991). Nested ANOVA and variogram analysis for spatial pattern. Statistical Laboratory Preprint No. 91-2, Iowa State University, Ames, IA. [Forthcoming in *Journal of Vegetation Science* under the title "Spatial models for spatial statistics: Some unification."]

Verly, G. (1983). The multigaussian approach and its applications to the estimation of local reserves. *Journal of the International Association for Mathematical Geology*, **15**, 259–286.

Vieira, S. R., Nielsen, D. R., and Biggar, J. W. (1981). Spatial variability of field-measured infiltration rate. *Soil Science Society of America Journal*, **45**, 1040–1048.

Vinod, H. D. (1973). Generalization of the Durbin–Watson statistic for higher order autoregressive processes. *Communications in Statistics*, **2**, 115–144.

Vitale, R. (1981). A central limit theorem for random convex sets. Technical Report, Department of Mathematics, Claremont Graduate School, Claremont, CA.

Vong, R. J., Moseholm, L., Covert, D. S., Sampson, P. D., O'Loughlin, J. F., Stevenson, M. N., Charlson, R. J., Zoller, W. H., and Larson, T. V. (1988). Changes in rainwater acidity associated with closure of a copper smelter. *Journal of Geophysical Research D*, **93**, 7169–7179.

von Neumann, J. (1941). Distribution of the ratio of the mean square successive difference to the variance. *Annals of Mathematical Statistics*, **12**, 367–395.

von Neumann, J., Kent, R. H., Bellinson, H. R., and Hart, B. I. (1941). The mean square successive difference. *Annals of Mathematical Statistics*, **12**, 153–162.

von Neumann, J. and Schoenberg, I. N. (1941). Fourier integrals and metric geometry. *Transactions of the American Mathematical Society*, **50**, 226–251.

Wackernagel, H. (1988). Geostatistical techniques for interpreting multivariate spatial information. In *Quantitative Analysis of Mineral and Energy Resources*, C. F. Chung, A. G. Fabbri, and R. Sinding-Larsen, eds. Reidel, Dordrecht, 393–409.

Wahba, G. (1978). Improper priors, spline smoothing, and the problem of guarding against model errors in regression. *Journal of the Royal Statistical Society B*, **40**, 364–372.

Wahba, G. (1981). Numerical experiments with the thin plate histospline. *Communications in Statistics. Theory and Methods*, **10**, 2475–2514.

Wahba, G. (1983). Bayesian "confidence intervals" for the cross-validated smoothing spline. *Journal of the Royal Statistical Society B*, **45**, 133–150.

Wahba, G. (1985). A comparison of GCV and GML for choosing the smoothing parameter in the generalized spline smoothing problem. *Annals of Statistics*, **13**, 1378–1402.

Wahba, G. (1990). *Spline Models for Observational Data*. SIAM, Philadelphia.

Wahba, G. and Wendelberger, J. (1980). Some new mathematical methods for variational objective analysis using splines and cross validation. *Monthly Weather Review*, **108**, 1122–1143.

Wallin, E. (1984). Isarithmic maps and geographical disaggregation. In *Proceedings of the International Symposium on Spatial Data Handling* (Zurich, Switzerland,

1984). Geographisches Institut, Universitat Zurich-Irchel, Zurich, Switzerland, 209–217.

Warnes, J. J. (1986). A sensitivity analysis for universal kriging. *Mathematical Geology*, **18**, 653–676.

Warnes, J. J. and Ripley, B. D. (1987). Problems with likelihood estimation of covariance functions of spatial Gaussian processes. *Biometrika*, **74**, 640–642.

Watkins, A. J. and Al-Boutiahi, F. H. M. (1990). On maximum likelihood estimation of parameters in incorrectly specified models of covariance for spatial data. *Mathematical Geology*, **22**, 151–173.

Watson, G. S. (1971). Trend-surface analysis. *Journal of the International Association for Mathematical Geology*, **3**, 215–226.

Watson, G. S. (1972). Trend surface analysis and spatial correlation. *Geological Society of America, Special Paper*, **146**, 39–46.

Watson, G. S. (1975). Mathematical morphology. In *A Survey of Statistical Design and Linear Models*, J. N. Srivastava, ed. North-Holland, Amsterdam, 547–553.

Watson, G. S. (1984). Smoothing and interpolation by kriging with splines. *Journal of the International Association for Mathematical Geology*, **16**, 601–615.

Watson, G. S. (1986). The shapes of a random sequence of triangles. *Advances in Applied Probability*, **18**, 156–169.

Watt, A. S. (1947). Pattern and process in the plant community. *Journal of Ecology*, **35**, 1–22.

Wedderburn, R. W. M. (1972). Quasi-likelihood functions, generalized linear models, and the Gauss-Newton method. *Biometrika*, **61**, 439–447.

Weerahandi, S. and Zidek, J. V. (1988). Bayesian nonparametric smoothers for regular processes. *Canadian Journal of Statistics*, **16**, 61–74.

Weibel, E. R. (1980). *Stereological Methods, Volumes* **1**, **2**. Academic Press, London.

Weil, W. (1982). An application of the central limit theorem for Banach-space-valued random variables to the theory of random sets. *Zeitschrift für Wahrscheinlichkeitstheorie und verwandte Gebiete*, **60**, 203–208.

Weil, W. (1987). Point processes of cylinders, particles and flats. *Acta Applicandae Mathematicae*, **9**, 103–136.

Weil, W. and Wieacker, J. A. (1984). Densities for stationary random sets and point processes. *Advances in Applied Probability*, **16**, 324–346.

Weil, W. and Wieacker, J. A. (1988). A representation theorem for random sets. *Probability and Mathematical Statistics*, **9** (1), 147–151.

Wender, P. H. (1971). *Minimal Brain Dysfunction in Children*. Wiley, New York.

Westcott, M. (1971). On existence and mixing results for cluster point processes. *Journal of the Royal Statistical Society B*, **33**, 290–300.

Westcott, M. (1972). The probability generating functional. *Journal of the Australian Mathematical Society*, **14**, 448–466.

Wheatcraft, S. W. and Tyler, S. W. (1988). An explanation of scale-dependent dispersivity in heterogeneous aquifers using concepts of fractal geometry. *Water Resources Research*, **24**, 566–578.

Whittaker, E. T. (1923). On a new method of graduation. *Proceedings of the Edinburgh Mathematical Society*, **41**, 63–75.

Whittemore, A. (1978). Quantitative theories of oncogenesis. *Advances in Cancer Research*, **27**, 55–88.

Whittemore, A. and Keller, J. B. (1978). Quantitative theories of carcinogenesis. *SIAM Review*, **20**, 1–30.

Whitten, E. H. T. (1970). Orthogonal polynomial trend surfaces for irregularly spaced data. *Journal of the International Association for Mathematical Geology*, **2**, 141–152.

Whittle, P. (1954). On stationary processes in the plane. *Biometrika*, **41**, 434–449.

Whittle, P. (1962). Topographic correlation, power-law covariance functions, and diffusion. *Biometrika*, **49**, 305–314.

Whittle, P. (1963). Stochastic processes in several dimensions. *Bulletin of the International Statistical Institute*, **40**, Book 2, 974–994.

Wieacker, J. A. (1986). Intersections of random hypersurfaces and visibility. *Probability Theory and Related Fields*, **71**, 405–433.

Wiener, N. (1949). *Extrapolation, Interpolation, and Smoothing of Stationary Time Series*. MIT Press, Cambridge, MA.

Wiens, J. A., Rotenberry, J. T., and Van Horne, B. (1986). A lesson in the limitations of field experiments: Shrubsteppe birds and habitat alteration. *Ecology*, **67**, 365–376.

Wilkinson, G. N., Eckert, S. R., Hancock, T. W., and Mayo, O. (1983). Nearest neighbor (NN) analysis with field experiments. *Journal of the Royal Statistical Society B*, **45**, 151–178.

Williams, E. R. (1986). A neighbor model for field experiments. *Biometrika*, **73**, 279–287.

Williams, G. W. (1984). Time-space clustering of disease. In *Statistical Methods for Cancer Studies*, R. G. Cornell, ed. Marcel Dekker, New York, 167–227.

Williams, J. S. (1975). Lower bounds on convergence rates of weighted least squares to best linear unbiased estimators. In *A Survey of Statistical Design and Linear Models*, J. N. Srivastava, ed. North-Holland, Amsterdam, 555–570.

Williams, T. and Bjerknes, R. (1972). Stochastic model for abnormal clone spread through epithelial basal layer. *Nature*, **236**, 19–21.

Wilson, E. B. and Hilferty, M. M. (1931). The distribution of chi-square. *Proceedings of the National Academy of Sciences, Washington*, **17**, 684–688.

Wilson, P. D. (1988). Maximum likelihood estimation using differences in an autoregressive-1 process. *Communications in Statistics. Theory and Methods*, **17**, 17–26.

Wold, H. (1938). *A Study in the Analysis of Stationary Time Series*. Almqvist and Wiksells, Uppsala.

Wolff, P. (1968). The serial organization of sucking in the young infant. *Pediatrics*, **42**, 943–956.

Wong, C. S. (1989). Linear models in a general parametric form. *Communications in Statistics. Theory and Methods*, **18**, 3095–3115.

Wong, E. (1968). Two-dimensional random fields and representation of images. *SIAM Journal on Applied Mathematics*, **16**, 756–770.

Wood, W. W. (1968). Monte Carlo studies of simple liquid models. In *Physics of Simple Liquids*, H. N. V. Temperley, J. S. Rowlinson, and G. S. Rushbrooke, eds. North-Holland, Amsterdam, 115–230.

Wu, C. F. J. (1986). Jackknife, bootstrap and other resampling methods in regression analysis. *Annals of Statistics*, **14**, 1261–1295.

Wu, C. F. J. and Wynn, H. P. (1978). The convergence of general step-length algorithms for regular optimum design criteria. *Annals of Statistics*, **6**, 1273–1285.

Yadin, M. and Zacks, S. (1985). The visibility of stationary and moving targets in the plane subject to a Poisson field of shadowing elements. *Journal of Applied Probability*, **22**, 776–786.

Yadrenko, M. I. (1983). *Spectral Theory of Random Fields*. Optimization Software, New York.

Yaglom, A. M. (1955). Correlation theory of processes with random stationary *n*th increments. *Matematicheskii Sbornik*, **37**, 141–196 (in Russian). [Translation in *American Mathematical Society Translations, Series 2,* **Vol. 8** (1958). American Mathematical Society, Providence, RI, 87–141.]

Yaglom, A. M. (1957). Some classes of random fields in *n*-dimensional space, related to stationary random processes. *Theory of Probability and Its Applications*, **2**, 273–320.

Yaglom, A. M. (1962). *An Introduction to the Theory of Stationary Random Functions*. Prentice-Hall, Englewood Cliffs, NJ (as of 1973, published by Dover, New York).

Yajima, Y. (1989). A central limit theorem of Fourier transforms of strongly dependent stationary processes. *Journal of Time Series Analysis*, **10**, 375–383.

Yakowitz, S. J. (1985). Nonparametric density estimation, prediction, and regression for Markov sequences. *Journal of the American Statistical Association*, **80**, 215–221.

Yakowitz, S. J. and Szidarovsky, F. (1985). A comparison of kriging with nonparametric regression methods. *Journal of Multivariate Analysis*, **16**, 21–53.

Yates, F. (1936). A new method of arranging variety trials involving a large number of varieties. *Journal of Agricultural Science (Cambridge),* **26**, 426–455.

Yates, F. (1938). The comparative advantages of systematic and randomized arrangements in the design of agricultural and biological experiments. *Biometrika*, **30**, 444–466.

Yates, S. R. and Warrick, A. W. (1987). Estimating soil water content using cokriging. *Soil Science Society of America Journal*, **51**, 23–30.

Yfantis, E. A., Flatman, G. T., and Behar, J. V. (1987). Efficiency of kriging estimation for square, triangular, and hexagonal grids. *Mathematical Geology*, **19**, 183–205.

Ylvisaker, D. (1975). Designs on random fields. In *A Survey of Statistical Design and Linear Models*, J. N. Srivastava, ed. North-Holland, Amsterdam, 593–607.

Ylvisaker, D. (1987). Prediction and design. *Annals of Statistics*, **15**, 1–19.

Younes, L. (1988). Estimation and annealing for Gibbsian fields. *Annales de l'Institut Henri Poincaré*, **24**, 269–294.

Young, D. S. (1987). Random vectors and spatial analysis by geostatistics for geotechnical applications. *Mathematical Geology*, **19**, 467–479.

Yule, G. U. and Kendall, M. G. (1950). *An Introduction to the Theory of Statistics,* 14th ed. Griffin, London.

Zadeh, L. A. (1965). Fuzzy sets. *Information and Control*, **8**, 338–353.

Zadeh, L. A. (1968). Probability measures of fuzzy events. *Journal of Mathematical Analysis and Applications*, **23**, 421–427.

Zahl, S. (1977). A comparison of three methods for the analysis of spatial pattern. *Biometrics*, **33**, 681–692.

Zahle, M. (1982). Random processes of Hausdorff rectifiable closed sets. *Mathematische Nachrichten*, **108**, 49–72.

Zahle, M. (1988). Random cell complexes and generalised sets. *Annals of Probability*, **16**, 1742–1766.

Zahn, C. T. and Roskies, R. Z. (1972). Fourier descriptors for plane closed curves. *IEEE Transactions on Computers*, **C-21**, 269–281.

Zeger, S. L. (1985). Exploring an ozone spatial time series in the frequency domain. *Journal of the American Statistical Association*, **80**, 323–331.

Zeger, S. L. (1988). A regression model for time series of counts. *Biometrika*, **75**, 621–629.

Zeleny, M. (1982). *Multiple Criteria Decision Making*. McGraw-Hill, New York.

Zellner, A. (1986). Bayesian estimation and prediction using asymmetric loss functions. *Journal of the American Statistical Association*, **81**, 446–451.

Zessin, H. (1983). The method of moments for random measures. *Zeitschrift für Wahrscheinlichkeitstheorie und verwandte Gebiete,* **62**, 395–409.

Zhao, L. C. (1987). Exponential bounds of mean error for the nearest neighbor estimates of regression functions. *Journal of Multivariate Analysis*, **21**, 168–178.

Zimmerman, D. L. (1986). A random field approach to spatial experiments. Ph.D. dissertation, Department of Statistics, Iowa State University, Ames, IA, unpublished.

Zimmerman, D. L. (1989a). Computationally efficient restricted maximum likelihood estimation of generalized covariance functions. *Mathematical Geology*, **21**, 655–672.

Zimmerman, D. L. (1989b). Computationally exploitable structure of covariance matrices and generalized covariance matrices in spatial models. *Journal of Statistical Computation and Simulation*, **32**, 1–15.

Zimmerman, D. L. and Cressie, N. (1992). Mean squared prediction error in the spatial linear model with estimated covariance parameters. *Annals of the Institute of Statistical Mathematics*, **44**, 27–43.

Zimmerman, D. L. and Harville, D. A. (1989). On the unbiasedness of the Papadakis estimator and other nonlinear estimators of treatment contrasts in field-plot experiments. *Biometrika*, **76**, 253–259.

Zimmerman, D. L. and Harville, D. A. (1991). A random field approach to the analysis of field-plot experiments and other spatial experiments. *Biometrics*, **47**, 223–239.

Zimmerman, D. L. and Zimmerman, M. B. (1991). A comparison of spatial semivariogram estimators and corresponding ordinary kriging predictors. *Technometrics,* **33,** 77–91.

Zirschky, J. H. and Harris, D. J. (1986). Geostatistical analysis of hazardous waste site data. *Journal of Environmental Engineering*, **112**, 770–784.

Zyskind, G. (1967). On canonical forms, non-negative covariance matrices and best and simple least squares linear estimators in linear models. *Annals of Mathematical Statistics*, **38**, 1092–1109.

Author Index

Subject Index

CPSIA information can be obtained
at www.ICGtesting.com
Printed in the USA
BVHW042311080719
552903BV00005B/32/P